CW00385633

LTE-Advanced

LTE-Advanced

A Practical Systems Approach to Understanding the 3GPP LTE Releases 10 and 11 Radio Access Technologies

Sassan Ahmadi

AMSTERDAM • BOSTON • HEIDELBERG • LONDON
NEW YORK • OXFORD • PARIS • SAN DIEGO
SAN FRANCISCO • SINGAPORE • SYDNEY • TOKYO
Academic Press is an imprint of Elsevier

Academic Press is an imprint of Elsevier
The Boulevard, Langford Lane, Kidlington, Oxford, OX5 1GB, UK
225 Wyman Street, Waltham, MA 02451, USA
525 B Street, Suite 1800, San Diego, CA 92101-4495, USA

British Library Cataloguing-in-Publication Data
A catalogue record for this book is available from the British Library

Library of Congress Cataloging-in-Publication Data
A catalog record for this book is available from the Library of Congress

ISBN: 978-0-12-405162-1

For information on all Academic Press publications
visit our website at elsevierdirect.com

Typeset by MPS Limited, Chennai, India
www.adi-mps.com

Printed and bound by CPI Group (UK) Ltd, Croydon, CR0 4YY

Contents

Preface

The past two decades have witnessed a phenomenal paradigm shift in the design methodology, and substantial improvement in the performance of mobile broadband wireless access technologies. With the increasing consumer demand for diverse wireless multimedia and data applications, and mobile Internet, improved quality and increasing capacity of the wireless networks have become a priority for the standards development organizations such as 3GPP. 3GPP started the development of the evolved packet system in late 2004, which subsequently resulted in specification and standardization of a new OFDM-based radio access network and an IP-based core network in 3GPP Rel-8, which was further enhanced in the subsequent releases. LTE Rel-10 (i.e., LTE-Advanced) was submitted as a candidate to ITU-R by 3GPP, and was selected as an IMT-Advanced technology in October 2010. The studies and research have continued in order to further advance and improve radio access technologies beyond LTE Rel-10 and HSPA Rel-10 by development of 3GPP Rel-11. The next generation of radio access technologies, informally referred to as systems beyond IMT-Advanced, are currently under investigation. Some of the main considerations in the design and development of new radio interfaces include higher spectral efficiency by using advanced multi-antenna technologies, carrier aggregation, distributed antenna systems, heterogeneous networks and small cells, energy efficiency and green radios, as well as the use of cognitive radios and software-defined radios for more efficient and flexible use of the radio spectrum. The systems beyond IMT-Advanced will encompass the capabilities of previous generations, as well as new communication schemes such as machine-to-machine, machine-to-person, and person-to-machine.

The framework for development of IMT-Advanced and systems beyond IMT-Advanced can be viewed from multiple perspectives including the users, manufacturers, application developers, network operators, and service and content providers. From the user's perspective, there is a demand for a variety of services, content, and applications whose capabilities will increase over time. The users expect services to be ubiquitously available through a variety of delivery mechanisms and service providers using a variety of wireless devices. From a service provision perspective, the domains share some common characteristics. Wireless service provision is characterized by global mobile access (terminal and personal mobility), improved security and reliability, higher service quality, and access to personalized multimedia services, Internet, and location-based services via one or multiple user terminals. Multi-radio operation requires seamless interaction among systems in order for the user to be able to receive/transmit a variety of content via different delivery mechanisms, depending on the device capabilities, location, and mobility, as well as user profile. Different radio access systems can be interconnected via a flexible core network and appropriate interworking functions. In this way, a user can be connected through different radio access systems to the network and utilize the services. The interworking between different radio access systems in terms of horizontal or vertical handover and seamless connectivity with service negotiation, mobility, security, and quality of service management are key requirements of radio-agnostic networks.

Similarity of services and applications across different radio access systems is beneficial not only to users, but also to network operators and content providers, stimulating the current trend toward convergence. Furthermore, similar user experience across different radio access systems leads to large-scale adoption of products and services, common applications, and content. Access to a service or an application may be performed using one system, or may be performed using multiple systems simultaneously. The increasing predominance of IP-based applications has been a key driver for the convergence trend in the core network and access network technologies.

The evolution of legacy IMT-2000 systems and the IMT-Advanced systems has employed several new concepts and functionalities, including adaptive modulation and coding and link adaptation, OFDM-based multiple access schemes, single-/multi-user multi-antenna concepts and techniques, dynamic QoS control, mobility management and handover between heterogeneous radio interfaces (vertical and horizontal), robust packet transmission, error detection and correction, multi-user detection, and interference cancellation. Systems beyond IMT-Advanced may further utilize sophisticated schemes including software-defined radio and reconfigurable RF and baseband processing, adaptive radio interface, mobile ad hoc networks, routing algorithms, and cooperative communication.

This book provides an in-depth description of 3GPP Rel-10/11 standards from a systems engineering point of view. A unique top-down systems approach is used to identify and describe the functional and protocol components of the 3GPP Rel-10/11 radio access technologies, starting from the radio access and core network architecture and then continuing with the theory of operation, interfaces, and practical implementation and deployment considerations. The book contains the background theoretical information about the operation of mobile stations, base stations, and other radio interface nodes. The distinctive features of this book relative to similar publications include an organized and systems engineering approach, self-containment of the technical topics discussed in the book, in-depth description of the theoretical concepts, real-life implementation issues, and practical considerations using clear and unambiguous block diagrams, state diagrams, and signal flow graphs.

Many articles, blogs, white papers, application notes, book chapters, and books have been published on the subject of LTE/LTE-Advanced and fourth generation of cellular networks since 2004, varying from academic theses to network operator analyses and manufacturers' application notes. By their very nature, these publications have viewed these subjects from one particular perspective whether it is academic, operational, or promotional/commercial. A very different and unique approach has been taken in this book, and that is a top-down systems approach to understanding the system operation and the design principles of the underlying functional components of fourth generation radio access networks. This book can be considered as the most up-to-date and complete technical reference for the design of LTE/LTE-Advanced systems. In this book, the protocol layers and functional elements of LTE/LTE-Advanced radio access and core networks are described.

This book is a reflection of the author's several years of participation in the inception, design and development, standardization, and implementation of LTE/LTE-Advanced, as well as numerous books, articles, white papers, public presentations, public blogs and online notes, simulations, etc., that the author has come across and studied since 2004 about LTE/LTE-Advanced, along with the author's 25 + years of experience in communication theory, signal processing, systems engineering, and cellular communication.

Acknowledgments

The author would like to acknowledge and thank his friends in 3GPP, 3GPP2, ITU-R, and IEEE for their encouragement, consultation, and assistance in selection of the content, as well as editing and improving the quality of the chapters.

The author would like to sincerely appreciate the Academic Press (Elsevier) publishing and editorial staff for providing the author with the opportunity to publish this work and for their assistance, cooperation, patience, and understanding throughout this project.

Finally, the author is immeasurably grateful to his family for their unwavering encouragement, cooperation, support, patience, and understanding throughout this project.

Abbreviations and Acronyms

3GPP	3rd generation partnership project
AAA	authorization, authentication, and accounting
ABS	almost blank subframe
AC	access class
ACIR	adjacent channel interference ratio
ACK	acknowledgment
ACLR	adjacent channel leakage ratio
ADC	analog to digital converter
AF	application function
AM	acknowledged mode
AMBR	aggregate maximum bit rate
AMC	adaptive modulation and coding
ANR	automatic neighbor relation
AoA	angle of arrival
AoD	angle of departure
AP	application protocol
APN	access point name
ARFCN	absolute ratio frequency channel number
ARP	allocation and retention priority
ARQ	automatic repeat request
AS	access stratum/angular spread
AWGN	additive white Gaussian noise
BC	broadcast channel (MU-MIMO)
BCCH	broadcast control channel
BCH	broadcast channel
BER	bit error rate
BPSK	binary phase shift keying
BSR	buffer status report
C/I	carrier-to-interference ratio
CA	carrier aggregation
CAZAC	constant amplitude zero autocorrelation
CBC	cell broadcast center
CBE	cell broadcast entity
CC	component carrier
CCE	control channel element

CCO	cell change order
CDD	cyclic delay diversity
CDF	cumulative distribution function
CDM	code division multiplexing
CFI	control format indicator
CFO	carrier frequency offset
CGI	cell global identifier
CIF	carrier indicator field
CM	cubic metric
CMAS	commercial mobile alert service
CMC	connection mobility control
CoMP	coordinated multi-point
CP	cyclic prefix
CQI	channel quality indicator
CRC	cyclic redundancy check
CRE	cell range extension
C-RNTI	cell RNTI
CRS	cell-specific reference signal
CSA	common subframe allocation
CSFB	circuit-switched fallback
CSG	closed subscriber group
CSG ID	closed subscriber group identity
CSI	channel state information
CSI-IM	CSI interference measurement
CSI-RS	CSI reference signal
CSS	CSG subscriber server
DAC	digital to analog converter
DAI	downlink assignment index
DCCH	dedicated control channel
DCI	downlink control information
DeNB	donor eNB
DFT-S-OFDM	DFT spread OFDM
DL	downlink
DL TFT	downlink traffic flow template
DL-SCH	downlink shared channel
DM-RS	demodulation reference signal
DoA	direction of arrival
DPC	dirty paper coding
DRB	data radio bearer
DRX	discontinuous reception
DS-CDMA	direct-sequence code division multiple access
DTCH	dedicated traffic channel
DTX	discontinuous transmission
DwPTS	downlink pilot time slot

EAB	extended access barring
ECCE	enhanced control channel element
ECGI	E-UTRAN cell global identifier
E-CID	enhanced cell-ID (positioning method)
ECM	EPS connection management
ECN	explicit congestion notification
eICIC	enhanced inter-cell interference coordination
EMM	EPS mobility management
eNB	E-UTRAN node B/evolved node B
EPC	evolved packet core
ePDCCH	enhanced physical downlink control channel
ePHR	extended power headroom report
EPRE	energy per resource element
EPS	evolved packet system
E-RAB	E-UTRAN radio access bearer
eREG	enhanced resource-element group
E-SMLC	evolved serving mobile location center
ETWS	earthquake and tsunami warning system
E-UTRA	evolved UTRA
E-UTRAN	evolved UTRAN
EVM	error vector magnitude
FDD	frequency division duplex
FDM	frequency division multiplexing
FFT	fast Fourier transform
FSTD	frequency switched transmit diversity
FTP	file transfer protocol
GBR	guaranteed bit rate
GERAN	GSM edge radio access network
GGSN	gateway GPRS support node
GNSS	global navigation satellite system
GP	guard period
GPS	global positioning system
GSM	global system for mobile communication
GT	guard time
GTP	GPRS tunneling protocol
GUMMEI	globally unique MME identifier
GUTI	globally unique temporary identifier
GW	gateway
HARQ	hybrid ARQ
HeNB	home eNB
HeNB GW	home eNB gateway
HetNet	heterogeneous network
HFN	hyper frame number
HI	HARQ indicator

HO	handover
HRPD	high rate packet data
HSPA	high speed packet access
HSS	home subscriber server
HTTP	hypertext transfer protocol
ICIC	inter-cell interference coordination
ID	identity
IDC	in-device coexistence
IEEE	Institute of Electrical and Electronics Engineers
IETF	Internet Engineering Task Force
IFFT	inverse fast Fourier transform
IMD	intermodulation distortion
IMS	IP multimedia subsystem
IoT	interference over thermal
IP	Internet protocol
IQ	in-phase/quadrature
IRC	interference rejection combining
ISI	inter-symbol interference
ISM	industrial, scientific, and medical
LB	load balancing
LBI	linked EPS bearer identifier
LCG	logical channel group
LCID	logical channel identifier
LCR	low chip rate
LCS	location service
L-GW	local gateway
LIPA	local IP access
LLR	log likelihood ratio
LMU	location measurement unit
LNA	low-noise amplifier
LPPa	LTE positioning protocol annex
LTE	long-term evolution
MAC	medium access control/multiple access channel (MU-MIMO)
MAC-I	message authentication code for integrity
MAP	maximum a posteriori probability
MBMS	multimedia broadcast/multicast service
MBR	maximum bit rate
MBSFN	multimedia broadcast multicast service single frequency network
MCCH	multicast control channel
MCE	multi-cell/multicast coordination entity
MCH	multicast channel
MCS	modulation and coding scheme
MDT	minimization of drive tests

MIB	master information block
MIMO	multiple-input multiple-output
MLD	maximum likelihood detection
MME	mobility management entity
MMEC	MME code
MMSE	minimum mean-square error
MRC	maximal ratio combining
MSA	MCH subframe allocation
MSI	MCH scheduling information
MSP	MCH scheduling period
MTC	machine-type communications
MTCH	multicast traffic channel
M-TMSI	M-temporary mobile subscriber identity
MU-MIMO	multi-user MIMO
NACK	negative acknowledgment
NAS	non-access stratum
NCC	next-hop chaining counter
NDS	network domain security
NH	next hop key
NNSF	NAS node selection function
NR	neighbor cell relation
NRT	neighbor relation table
OCC	orthogonal cover code
OFDM	orthogonal frequency division multiplexing
OFDMA	orthogonal frequency division multiple access
OMC-ID	operation and maintenance center identity
OOB	out-of-band (emission)
OTDOA	observed time difference of arrival (positioning method)
PA	power amplifier
PAPR	peak-to-average power ratio
PBCH	physical broadcast channel
PBR	prioritized bit rate
PCC	primary component carrier
PCCH	paging control channel
PCell	primary cell
PCFICH	physical control format indicator channel
PCH	paging channel
PCI	physical cell identifier
PCRF	policy and charging rules function
PDCCH	physical downlink control channel
PDCP	packet data convergence protocol
PDN	packet data network
PDSCH	physical downlink shared channel
PDU	protocol data unit

P-GW	PDN gateway
PHICH	physical hybrid ARQ indicator channel
PHR	power headroom report
PHY	physical layer
PLMN	public land mobile network
PMCH	physical multicast channel
PMI	precoding matrix indicator
PMI	precoding matrix index
PMIP	proxy mobile IP
PO	paging occasion
PRACH	physical random access channel
PRB	physical resource block
P-RNTI	paging RNTI
PRS	positioning reference signal
PSAP	public safety answering point
PSC	packet scheduling
pTAG	primary timing advance group
PTI	procedure transaction identifier
PUCCH	physical uplink control channel
PUSCH	physical uplink shared channel
PWS	public warning system
QAM	quadrature amplitude modulation
QCI	QoS class identifier
QoS	quality of service
QPP	quadrature permutation polynomial
RAC	radio admission control
RACH	random access channel
RAR	random access response
RA-RNTI	random access RNTI
RAT	radio access technology
RB	radio bearer
RBC	radio bearer control
RBG	resource block group
RE	resource element
REG	resource-element group
RET	remote electrical tilting
RF	radio frequency
RFSP	RAT/frequency selection priority
RI	rank indication
RIM	RAN information management
RLC	radio link control
RLF	radio link failure
RM	rate matching/Reed–Muller (code)
RN	relay node

RNC	radio network controller
RNL	radio network layer
RNTI	radio network temporary identifier
ROHC	robust header compression
RoT	rise over thermal
R-PDCCH	relay physical downlink control channel
RRC	radio resource control
RRM	radio resource management
RS	reference signal
RSRP	reference signal received power
RSRQ	reference signal received quality
RSSI	received signal strength indicator
RTP	real-time transport protocol
RTT	round trip time
RU	resource unit
RV	redundancy version
RX	receiver
S1-MME	S1 for the control-plane
S1-U	S1 for the user-plane
SAE	system architecture evolution
SAP	service access point
SCC	secondary component carrier
SCell	secondary cell
SC-FDMA	single carrier-frequency division multiple access
SCH	synchronization channel
SCM	spatial channel model
SDF	service data flow
SDMA	spatial division multiple access
SDU	service data unit
SeGW	security gateway
SFBC	space-frequency block code
SFN	system frame number
SGSN	serving GPRS support node
S-GW	serving gateway
SI	system information
SIB	system information block
SIC	successive interference cancellation
SIMO	single-input multiple-output
SINR	signal to interference plus noise ratio
SIP	session initiation protocol
SIPTO	selected IP traffic offload
SI-RNTI	system information RNTI
SISO	single-input single-output/soft-input soft-output
SMS	short message service

SN sequence number
SNR signal-to-noise ratio
SON self-organizing network
SPID subscriber profile ID for RAT/frequency priority
SPS C-RNTI semi-persistent scheduling C-RNTI
SR scheduling request
SRB signaling radio bearer
SRS sounding reference signal
SRVCC single radio voice call continuity
sTAG secondary timing advance group
S-TMSI S-temporary mobile subscriber identity
SU scheduling unit
SU-MIMO single-user MIMO
TA tracking area
TAC tracking area code
TAD traffic aggregate description
TAG timing advance group
TAI tracking area identity
TAU tracking area update
TB transport block
TCP transmission control protocol
TDD time division duplex
TDM time division multiplexing
TEID tunnel endpoint identifier
TFT traffic flow template
TI transaction identifier
TIN temporary identity used in next update
TM transparent mode
TMA tower mounted amplifier
TNL transport network layer
TPC transmit power control
TPMI transmitted precoding matrix indicator
TTI transmission time interval
TX transmitter
UCI uplink control information
UDP user datagram protocol
UE user equipment
UL uplink
UL TFT uplink traffic flow template
ULR-Flags update location request flags
UL-SCH uplink shared channel
UM unacknowledged mode
UMTS universal mobile telecommunication system
UpPTS uplink pilot time slot

UTRA	universal terrestrial radio access
UTRAN	universal terrestrial radio access network
VoIP	voice over Internet protocol
VoLTE	voice over LTE
VRB	virtual resource block
X2-C	X2-control plane
X2-U	X2-user plane
ZC	Zadoff–Chu
ZCZ	zero correlation zone
ZF	zero forcing (receiver)

CHAPTER

Introduction to LTE-Advanced

1

CHAPTER CONTENTS

Wireless communication comprises a wide range of technologies, services, and applications that have come into existence to meet the particular needs of users in different deployment scenarios. Wireless systems can be broadly characterized by content and services offered, reliability and performance, operational frequency bands, standards defining those systems, data rates supported, bidirectional and unidirectional delivery mechanisms, degree of mobility, regulatory requirements, complexity, and cost. The number of mobile subscribers has increased dramatically worldwide in the past two decades. The growth in the number of mobile subscribers will be further intensified by adoption of broadband mobile access technologies in largely populated and developing countries. It is envisioned that potentially the entire world population will have access to broadband mobile services depending on economic conditions and favorable cost structures offered by regional network operators. There are already more mobile devices than fixed-line telephones or fixed computing platforms such as desktop computers that can access the Internet. The number of mobile devices is expected to continue to grow more rapidly than nomadic and stationary devices. Mobile terminals will be the most commonly used platforms for accessing and exchanging information. In particular, users will expect a dynamic, continuing stream of new applications, capabilities, and services that are ubiquitous and available across a range of devices using a single subscription and a single identity. Versatile communication systems offering customized and ubiquitous services based on diverse individual needs require flexibility in the technology in order to satisfy multiple demands simultaneously. Wireless multimedia traffic is increasing far more rapidly than voice and will increasingly dominate traffic flows. The paradigm shift from predominantly circuit-switched air-interface design to full IP-based systems has provided users with the ability to more efficiently, more reliably, and more securely utilize multimedia services including e-mail, file transfers, messaging, browsing, gaming, voice over Internet protocol (VoIP), and location-based multicast and broadcast services. These services can be either symmetrical or asymmetrical (in terms of use of radio resources in the downlink or uplink) and real-time or non-real-time with differing quality of service (QoS) requirements. The new applications consume relatively larger bandwidths, resulting in higher data rate requirements [13,14].

International Mobile Telecommunications-Advanced (IMT-Advanced) systems are broadband mobile wireless access systems that include new capabilities and flexibility that go beyond those of legacy IMT-2000 systems [1–7]. IMT-Advanced provided a global framework for development of the new generation of mobile services that offer fast data access and unified messaging, and broadband wireless multimedia in the form of new interactive services anywhere and any-time. Such systems provide access to a wide range of telecommunication services including advanced mobile services and are supported by mobile and fixed net-works, which are entirely based on Internet protocol (IP). The IMT-Advanced systems support low to high mobility applications and a wide range of data rates in accordance with user and service demands in multiple user environments. The IMT-Advanced systems have also capabilities for high-quality multimedia appli-cations within a wide range of services and platforms, providing a significant improvement in performance and QoS.

The capabilities of IMT-2000 systems have been continuously enhanced over the past decade as IMT-2000 technologies are upgraded and deployed. From the radio access perspective, the evolved IMT-2000 systems have been built on the existing systems and further enhanced the radio interface functionalities, and at the same time, new systems have emerged to replace the legacy radio access sys-tems. This evolution has improved the reliability and throughput of the cellular systems and has promoted the development of an increasing number of services and applications. Similarity of services and applications across different IMT technologies and frequency bands is beneficial to users, and a broadly similar user experience leads to a large-scale deployment of products and services. The technologies, applications, and services associated with systems beyond IMT-Advanced could well be radically different from the present, challenging the per-ceptions of what may be considered viable by today's standards and going beyond what has just been achieved by the IMT-Advanced radio systems.

3GPP initiated a project on the long-term evolution (LTE) of the Universal Mobile Telecommunication System (UMTS) radio access network (RAN) in late 2004 to maintain 3GPP's competitive edge over other cellular technologies. The evolved UMTS terrestrial radio access network (E-UTRAN) substantially improved end-user throughputs and system capacity and significantly reduced user-plane and control-plane latencies, bringing in considerably improved user experience with full mobility. With the emergence of IP as the protocol of choice for carrying all types of traffic, the 3rd Generation Partnership Project (3GPP) LTE has provided support for IP-based traffic with end-to-end QoS. Voice traffic is supported mainly as VoIP, enabling better integration with other multimedia services. Unlike its predecessors, which were developed within the framework of a global system for mobile communication (GSM)/UMTS network architecture, 3GPP specified an evolved packet core (EPC) architecture to support the E-UTRAN through reduction in the number of network elements and simplification of functionality, but most importantly allowing for connections and handover to other fixed and wireless access technologies, providing the network operators with the ability to deliver a seamless mobility experience. 3GPP set aggressive performance requirements for LTE that relied on improved physical layer technologies such as orthogonal frequency division multiplexing (OFDM) and single-user and/or multi-user multiple-input multiple-output (MIMO) techniques, and streamlined layer 2/layer 3 protocols and functionalities. The main objectives of LTE were to minimize the system and user equipment (UE) complexities, to allow flexible spectrum allocation in the existing or new frequency bands, and to enable coexistence with other 3GPP and non-3GPP radio access technologies. The LTE Rel-8/9 was used as the baseline for further enhancements under 3GPP Rel-10 to meet the requirements of the IMT-Advanced. A candidate proposal based on the latter enhancements (3GPP LTE-Advanced) was submitted to the International Telecommunication Union-Radio Telecommunications (ITU-R) and was subsequently qualified as an IMT-Advanced technology in 2010 [6,7]. However, concurrent with 3GPP LTE standard development, the operators were rolling out high-speed packet access (HSPA) networks to upgrade their second-generation (2G) and early 3G infrastructure; thus, they were not ready to embrace yet another paradigm shift in the RAN and core network technologies. Therefore, 3GPP continued to improve UMTS technology by adding multi-antenna support at the base station, higher modulation order in the downlink, and multi-carrier support to prolong the life span of 3G systems [15,16].

LTE-Advanced (LTE Rel-10/11) extended the capabilities of LTE Rel-8/9 by introducing new features such as carrier aggregation, which is the most significant and complex improvement provided by LTE-Advanced that can combine up to five 20 MHz radio frequency (RF) carriers to create a virtually wide bandwidth; enhanced multi-antenna techniques in the uplink and downlink (e.g., SU-MIMO and MU-MIMO, reference signals, channel state information measurement/reporting); cell coverage and throughput improvement by means of relay nodes, which connect to relay-enabled eNBs; enhanced inter-cell interference coordination to improve performance of heterogeneous networks; and coordinated multi-point transmission/reception (also known as network MIMO in the literature), which is a set of techniques using different forms of MIMO and beamforming to improve the performance at cell edges using coordinated scheduling, and transmitters/

receivers and antennas that are not collocated to provide greater spatial diversity to improve link reliability and data rate as well as system capacity [17,18]. The framework for the LTE-Advanced air-interface technology was mostly defined by the use of wider bandwidths up to 100 MHz, contiguous/non-contiguous spectrum deployments, also referred to as spectrum aggregation, and a need for flexible spectrum usage. In general, OFDM provides a simple means to increase bandwidth by adding additional subcarriers. Due to the discontinuous nature of the spectrum reserved for IMT-Advanced, the available bandwidth is fragmented. Therefore, the user terminals should be able to filter, process, and decode such a large variable bandwidth. The increased decoding complexity is one of the major challenges of the wider bandwidths [9].

Some early deployments of LTE-Advanced are expected in late 2013 with increasing adoption in 2014 and beyond [13,14]. LTE-Advanced is forward and backward compatible with LTE Rel-8/9, meaning that LTE Rel-8/9 devices will work on LTE-Advanced networks and LTE-Advanced capable terminals will work on the legacy LTE networks. The LTE-Advanced features offer performance gains and may considerably affect the system complexity and cost. For instance, higher order MIMO schemes up to 8×8 in the downlink will improve peak data rates and spectral efficiency; however, this feature also impacts the network side and the complexity of the UE, as the implementation of the antennas differs and additional baseband transmission/reception chains are needed. Moreover, while band aggregation does not affect spectral efficiency (since it is independent of bandwidth), cell-edge user throughput (due to insufficient signal to interference plus noise ratio (SINR) value, secondary cells cannot be utilized), and coverage (since coverage is determined by the link budget of the primary cell), it does increase the cost of network deployment. It is worth mentioning that the peak data rate is improved depending on the number of the aggregated carriers, further affecting the complexity and cost of the UE implementation.

This book begins with an introduction to the LTE-Advanced/IMT-Advanced systems and an overview of the 3GPP standards evolution and standardization process. The unique, top-down systems approach taken in this book necessitates beginning with the review of the overall network architecture and later examination of each and every significant network element in the evolved packet system (EPS) in Chapter 2. Once the access network and core network aspects of the LTE/LTE-Advanced systems are described, we turn our attention to the reference model and protocol structure of LTE/LTE-Advanced and discuss the operation and behavior of each entity (base station, mobile station, femto-cells, and relay stations) as well as functional components and their interactions, logical interfaces, protocol stack, and protocol terminations in Chapter 3. The remaining chapters of this book are organized consistently with the order of the protocol layers on the user-plane and control-plane, starting from the network layer and moving down to the physical layer. Each protocol layer has been described in a separate chapter in order for the readers to clearly understand the functions and operation of each layer. The overall operation of the mobile station and the base station and their corresponding state machines are described in Chapter 4. Perhaps, this chapter is the most important part of the book as far as understanding the general operation of the system is concerned. Chapter 5 describes the radio resource control functions and protocols. The packet data convergence protocol, radio link control, and medium access control sublayers are described in Chapters 6, 7, and 8, respectively. The description of downlink and uplink physical layer algorithms and protocols has been divided into two separate parts (Chapters 9 and 10) for further clarity and structure. There are many commonalities between LTE Rel-8, 9, 10, and 11; however, release-specific features/functionalities, algorithms, or protocols which have been introduced in subsequent releases are clearly distinguished in the content so that they are not confused with the legacy components throughout the book. In this book, the term E-UTRA or LTE refers to both LTE and LTE-Advanced radio access technologies everywhere except where the feature or functionality does not exist or is not supported in the previous releases. The performance evaluation of LTE-Advanced against IMT-Advanced requirements is described in Chapter 11 where the performance metrics are defined and link-level and system-level simulation methodologies and parameters are elaborated. The theory and operation of LTE Rel-11 coordinated multi-point transmission/reception and LTE Rel-10/11 carrier aggregation schemes are described in Chapters 12 and 13. Multi-radio coexistence and enhanced inter-cell interference coordination techniques used in LTE Rel-10/11 are described in Chapter 14. In Chapter 15, we describe the positioning methods and evolved multimedia broadcast/multicast services. Each chapter contains a theoretical description of the features/functions and scrutinizes the real-life and practical implementation and deployment issues and remedies.

1.1 Background on IMT-Advanced standards development

The past two decades have witnessed a rapid growth in the number of subscribers and incredible advancement in the technology of cellular communication from simple, all-circuit-switched, analog first-generation systems with limited voice service capabilities, limited mobility, and small capacity to the fourth-generation (4G) systems with significantly increased capacity, advanced all-digital packet-switched all-IP implementations that offer a variety of multimedia services. With the increasing demand for high-quality wireless multimedia services, the radio access technologies continue to advance with faster pace toward the next generation of systems. The general characteristics envisioned for the fourth generation of the cellular systems included all-IP core networks, support for a wide range of user mobility, significantly improved user throughput and system capacity, reliability and robustness, seamless connectivity, reduced access latencies, etc.

IMT-Advanced, or alternatively 4G, cellular systems are mobile systems that extend and improve upon the capabilities of legacy IMT-2000 standards. The IMT-Advanced systems provide the users with access to a variety of advanced IP-based services and applications. The IMT-Advanced systems can support a wide range of data rates, with different QoS requirements, proportional to user mobility conditions in multi-user environments. The key features of IMT-Advanced systems can be summarized as follows [1]:

- Enhanced cell and peak spectral efficiencies, and cell-edge user throughput to support advanced services and applications
- Lower airlink access and signaling latencies to support delay-sensitive applications
- Support of higher user mobility while maintaining session connectivity
- Efficient utilization of spectrum
- Inter-technology interoperability, allowing worldwide roaming capability
- Enhanced air-interface-agnostic applications and services
- Lower system complexity and implementation cost
- Convergence of fixed and mobile networks
- Capability of interworking with other radio access systems

These features enable IMT-Advanced systems to accommodate emerging applications and services. The capabilities of IMT systems have been continuously enhanced proportional to user demand and technology advancements in the past decade. However, the framework and overall objectives of the systems beyond IMT-Advanced are expected to be a paradigm shift in the design and development of radio interface, network architecture, and topology [1]. Present mobile communication systems have evolved by incremental addition of capabilities and enhancement of features of the baseline (legacy) systems. Examples include the evolution of UMTS family of standards in 3GPP. The systems beyond IMT-Advanced will be realized by functional fusion of the existing IMT systems, nomadic wireless access systems, and other wireless systems with high commonality and seamless interworking, in conjunction with the development of new and enhanced features.

Translating service/application requirements into requirements that are directly related to data transport over a wireless network commonly leads to considering a limited number of QoS attributes, such as user throughput, packet delay and/or delay variations (often referred to as delay jitter), bit/packet error rate, and other similar aspects. This motivates the introduction of service classes that group together services that are similar in terms of their requirements toward a network. Figure 1.1 shows four service classes (conversational, interactive, streaming, and background services) and their characteristics in terms of reliability, bit rate, and latency that were considered for the IMT-Advanced systems. We will further discuss these requirements and characteristics in the next sections. In Figure 1.1, BER denotes bit error rate which is a measure of reliability of communication link and is the ratio of the number of incorrectly received information bits to the total number of information bits sent within a certain time interval [2].

Prominent standards developing organizations such as 3GPP[1], and Institute of Electrical and Electronics Engineers Standards Association (IEEE-SA)[2] have

---

[1]http://www.3gpp.org

[2]http://standards.ieee.org

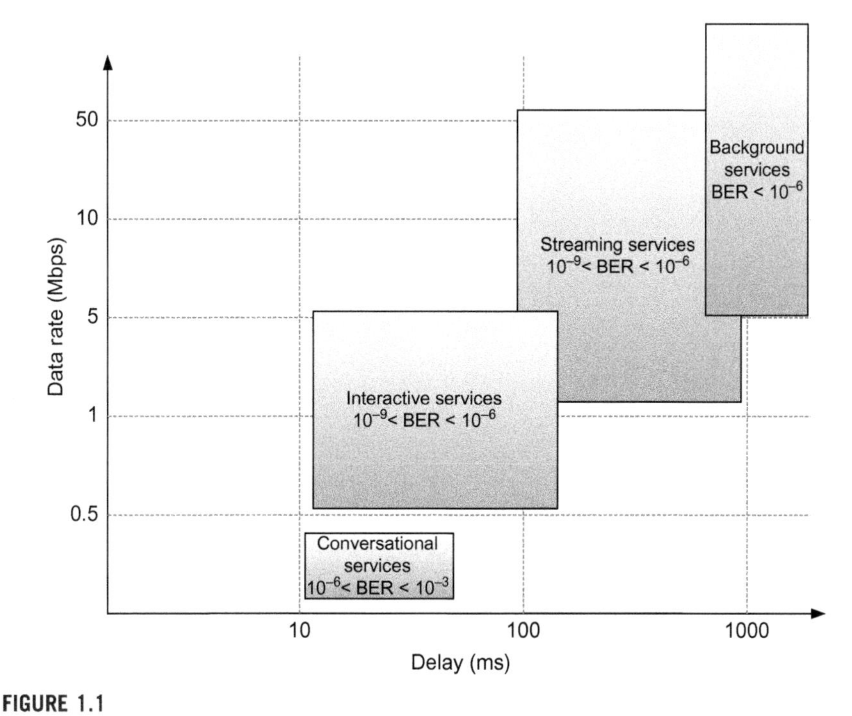

FIGURE 1.1

IMT-Advanced service classes and their characteristics [2].

actively contributed to the design, development, and proliferation of broadband wireless systems in the past decade. A number of broadband wireless access standards for fixed, nomadic, or mobile systems have been developed by these standardization groups and deployed by a large number of operators across the globe [6,7].

As mentioned earlier, the IMT-Advanced framework divided service classes into conversational, interactive, streaming, and background services. There are three service classes in the conversational service category. The basic conversational service class comprises basic services that are dominated by voice communication characteristics. The rich conversational service class consists of services that mainly provide synchronous communication enhanced by additional media such as video and collaborative document viewing. Conversational low-delay class covers real-time services that have very strict delay and delay variation requirements. In the interactive service category, two service classes are distinguished. Interactive services that permit relatively high delay which usually follow a request-response pattern (e.g., web browsing, database query). In such cases, response times in the order of a few seconds are permitted. Interactive services requiring significantly lower delay are remote server access or remote collaboration. In the streaming category, there are two service classes. The differentiating factor between these two classes is the live or non-live nature of

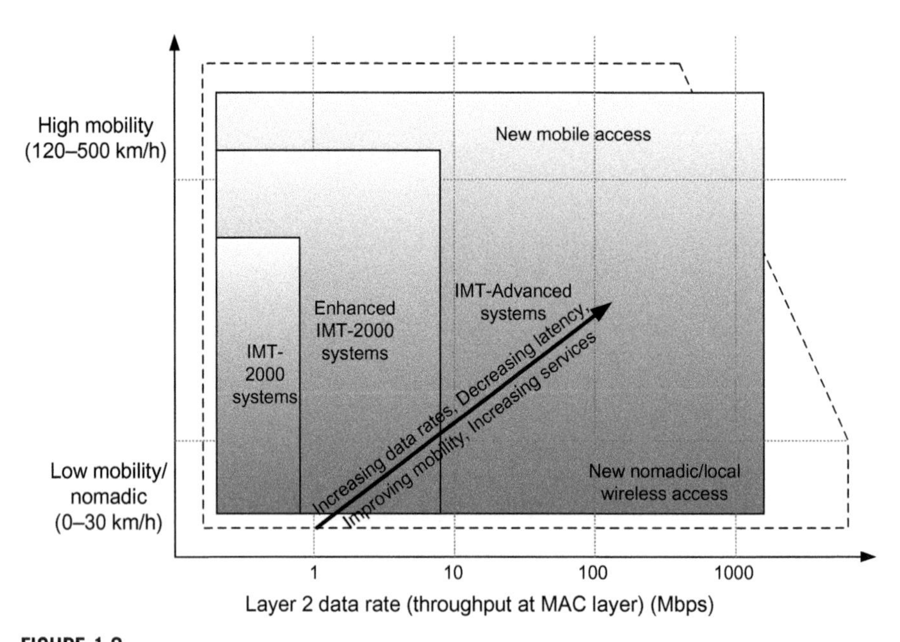

FIGURE 1.2

Illustration of the capabilities and evolution of IMT systems [1].

the content transmitted. In case of live content, buffering possibilities are very limited, which makes the service delay sensitive. In the case of non-live (i.e., pre-recorded) content, playout buffers at the receiver side provide a high robustness against delay and delay jitter. The background class only contains delay-insensitive services so that there is no need for further differentiation.

The advancement of capabilities and evolution of IMT systems are illustrated in Figure 1.2. As it appears from the figure, the services and performance of the systems noticeably increase as the systems evolve from one generation to another.

1.2 Requirements for the IMT-Advanced systems

The service and application requirements must be translated to system design requirements in order to obtain a set of minimum requirements and objective criteria for the design of a new mobile broadband wireless system. The design requirements encompass a wide range of system attributes such as data and signaling transmission latency over the airlink and the core network, system data rates, user cell-edge and average throughputs, and application capacity. The high-level requirements presented in Table 1.1 were defined by ITU-R Working Party 5D, prior to the development of the candidate proposals, for the purpose of consistent definition, specification, and evaluation of the IMT-Advanced systems.

Table 1.1 IMT-Advanced Requirements [3,5]

Cell Spectral Efficiency		
Test Environment	**Downlink (Bits/s/Hz/Cell)**	**Uplink (Bits/s/Hz/Cell)**
Indoor	3	2.25
Microcellular	2.6	1.80
Base coverage urban	2.2	1.4
High speed	1.1	0.7
Peak Spectral Efficiency		
Test Environment	**Downlink (Bits/s/Hz)**	**Uplink (Bits/s/Hz)**
N/A	15	6.75
Cell-Edge User Spectral Efficiency		
Test Environment	**Downlink (Bits/s/Hz)**	**Uplink (Bits/s/Hz)**
Indoor	0.1	0.07
Microcellular	0.075	0.05
Base coverage urban	0.06	0.03
High speed	0.04	0.015
Traffic Channel Link Data Rates		
Test Environment	**Uplink (Bits/s/Hz)**	**UE Speed (km/h)**
Indoor	1.0	10
Microcellular	0.75	30
Base coverage urban	0.55	120
High speed	0.25	350
VoIP Capacity		
Test Environment	**VoIP Capacity (Active Users/Sector/MHz)**	
Indoor	50	
Microcellular	40	
Base coverage urban	40	
High speed	30	

The intent of these requirements was to ensure that IMT-Advanced technologies were able to fulfill the objectives of the fourth generation of cellular systems, as laid out in Recommendation ITU-R M.1645 as the ITU-R vision for IMT-Advanced systems, as well as to set a specific level of minimum performance that each proposed technology needed to achieve [1]. These requirements were not intended to restrict the full range of capabilities or performance that candidate technologies for IMT-Advanced might achieve, nor were they intended to describe how the IMT-Advanced technologies might perform in actual deployments under operating conditions that could be different from those presented in ITU-R recommendations and reports on IMT-Advanced. Satisfaction of these requirements were verified through link and system-level simulations (for VoIP capacity, spectral efficiencies, and mobility), analytical (for user and control-plane latencies as well as handover interruption time), and inspection (for service requirements and bandwidth scalability) [3]. The methodology, guidelines, and common configuration parameters were specified in Report ITU-R M.2135-1 [4].

The requirements for system and user data rates were described in the form of spectral efficiency to make them independent of bandwidth and practical spectrum allocations. Some of the requirements noted in Table 1.1 were evaluated via system-level simulations; however, the mobility requirement was evaluated through link-level simulation. Note that in Table 1.1, a cell or equivalently a sector is defined as a physical partition of the coverage area of a base station. The base station coverage area is typically partitioned into three or more sectors/cells to mitigate interference effects.

In Table 1.1, the cell spectral efficiency is defined as the aggregate throughput, i.e., the number of correctly received bits delivered at the data link layer over a certain period of time, of all users divided by the product of the effective bandwidth, the frequency reuse factor, and the number of cells. The cell spectral efficiency is measured in bits/s/Hz/cell.

The peak spectral efficiency in the above table denotes the highest theoretical data rate normalized by bandwidth (assuming error-free conditions) assignable to a single mobile station when all available radio resources for the corresponding link are utilized, excluding radio resources that are used for physical layer synchronization, reference signals, guard bands, and guard times (collectively known as layer-1 overhead). Bandwidth scalability is the ability to operate with different bandwidth allocations using single or multiple RF carriers. These values were defined assuming an antenna configuration of 4×4 and 2×4 in the downlink and uplink, respectively [3].

The normalized user throughput in Table 1.1 is defined as the number of correctly received bits by a user at the data link layer over a certain period of time, divided by the total spectrum. The cell-edge user spectral efficiency is defined as 5% point of cumulative distribution function (CDF) of the normalized user throughput.

Control-plane latency is defined as the transition time from idle state to connected state. The transition time (assuming downlink paging latency and core network signaling delay are excluded) of less than 100 ms is required for IMT-Advanced systems. The user-plane latency, also known as transport delay, is defined as the one-way transit time between a packet being available at the IP layer of the origin (user terminal in the uplink or base station in the downlink) and the availability of this packet at IP layer of the destination (base station in the uplink or user terminal in the downlink). User-plane packet delay includes delay introduced by associated protocols and signaling assuming the user terminal is in the active mode. The IMT-Advanced systems are required to achieve a transport delay of less than 10 ms in unloaded conditions (i.e., a single user with a single data stream) for small IP packets (e.g., an empty payload + IP header) for both downlink and uplink as hypothetically shown in Figure 1.3. In Figure 1.3, the user-plane latency is measured from point 1 to point 2 or from point 3 to point 4 [3,19].

The following mobility classes were defined in IMT-Advanced systems [3]:

- Stationary: 0 km/h
- Pedestrian: 0–10 km/h
- Vehicular: 10–120 km/h
- High-speed vehicular: 120–350 km/h

A mobility class is supported if the traffic channel link-level data rate, normalized by bandwidth, on the uplink, exceeds the minimum requirements shown in Table 1.1, when the user is moving at the maximum speed in that mobility class in each of the test environments. The requirements were defined assuming an antenna configuration of 4×2 and 2×4 in the downlink and uplink, respectively.

The handover interruption time is defined as the time duration during which a user terminal cannot exchange user-plane packets with any base station. The handover interruption time includes the time required to execute any RAN procedure, radio resource signaling protocol, or other message exchanges between the user terminal and the RAN. For the purposes of determining handover interruption time, interactions with the core network are not considered. It is also assumed that all necessary attributes of the target base station are known at initiation of the handover from the serving base station. The requirements for intra-frequency and inter-frequency handover interruption time were 27.5 and 40 ms (within a spectrum band)/60 ms (between spectrum bands), respectively [3].

In Table 1.1, VoIP capacity is calculated assuming a 12.2 kbps codec with 50% voice activity factor such that the percentage of users in outage is less than 2% where a user is defined to have experienced a voice outage if less than 98% of the VoIP packets have been delivered successfully to the user within a one-way radio access delay bound of 50 ms. The VoIP capacity is the minimum of the calculated capacity for downlink or uplink divided by the effective bandwidth in the respective link direction. These values were defined assuming an antenna configuration of 4×2 in the downlink and 2×4 in the uplink.

1.3 Introduction to 3GPP standards

The 3GPP unites several telecommunications standard development organizations (ARIB, ATIS, CCSA, ETSI, TTA, TTC), also known as organizational partners, and provides their members with a stable technical basis to transpose into their

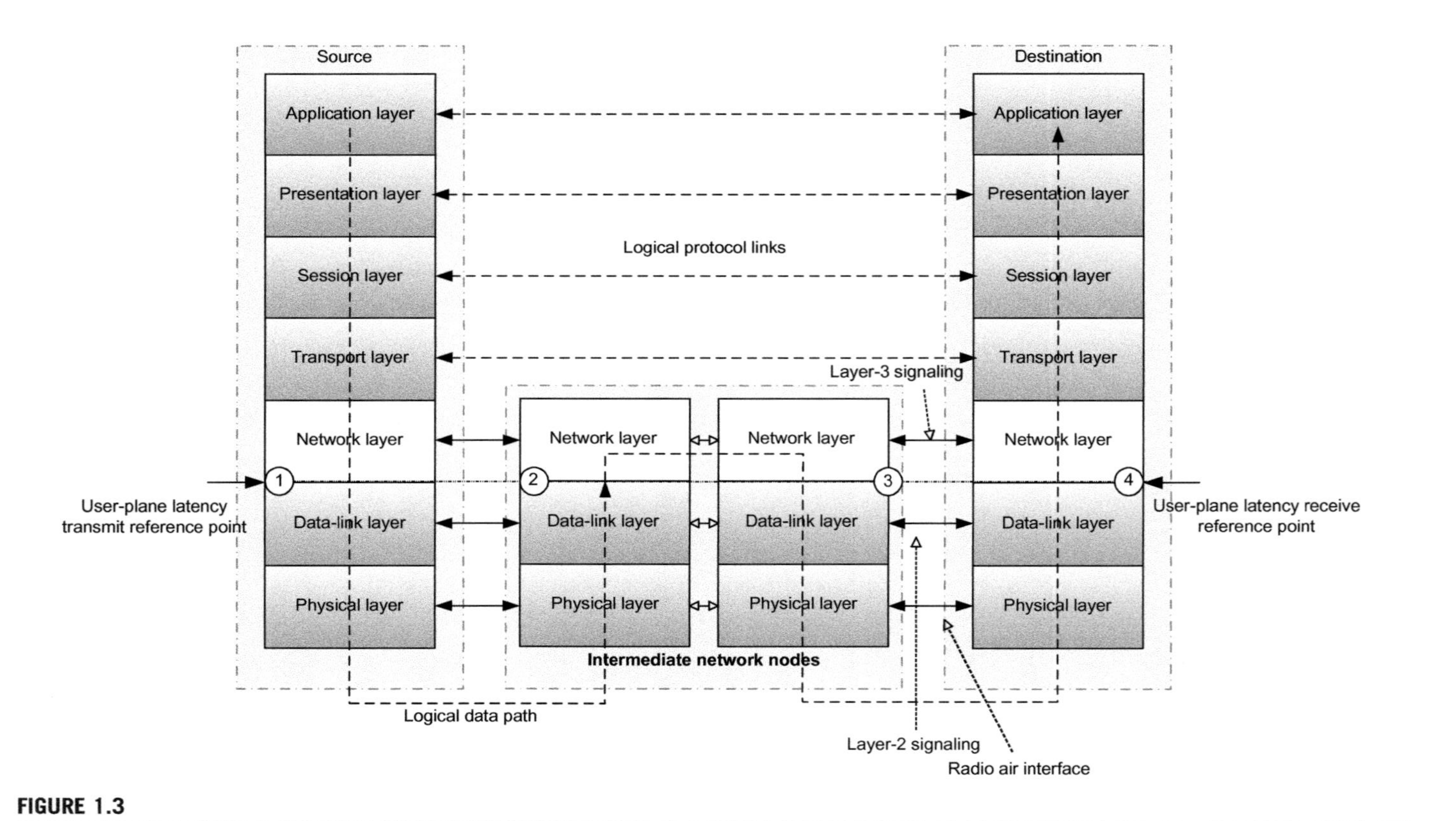

FIGURE 1.3

Graphical interpretation of user-plane latency [19].

own specifications. There are four technical specification groups in 3GPP which include RAN, Service and Systems Aspects (SA), Core Network and Terminals (CT), and GSM EDGE Radio Access Network (GERAN). Each technical group has a set of working groups and has a particular area of responsibility for the reports and specifications under its own terms of reference or group charter.

The original scope of 3GPP was to produce technical specifications and technical reports for 3G mobile systems based on evolved GSM core networks and the radio access technologies that they support, i.e., universal terrestrial radio access (UTRA) with both frequency division duplex (FDD) and time division duplex (TDD) modes. The scope was subsequently amended to include the maintenance and development of GSM technical specifications and technical reports including evolved radio access technologies, e.g., general packet radio service (GPRS) and enhanced data rates for GSM evolution (EDGE). The 3GPP organization was created in December 1998 by approval of the 3rd Generation Partnership Project Agreement. The latest 3GPP scope and objectives document has evolved from this original agreement. Table 1.2 shows the 3GPP organization and the charter of the technical specification groups and their subcommittees.

3GPP uses a system of parallel releases in order to provide device and network developers with a stable platform for implementation and to allow for addition of new features required by the market. Following the introduction of an entirely new air-interface with LTE in 3GPP Rel-8, and with various enhancements thereafter, the upcoming 3GPP Rel-12 and beyond will be as significant as ever in the industry's quest to extend mobile broadband availability, to provide more consistent service quality, and to economically satisfy the demand for fast and reliable mobile Internet, despite the challenges due to spectrum scarcity and other constraints. Successive mobile generations are typically depicted in technological terms. For the world's mobile operators following the development standards path, first laid by ETSI and then 3GPP, first-generation (1G) analog was followed by 2G TDMA-based GSM, then 3G wideband-CDMA-based UMTS, and 4G OFDM-based LTE/LTE-Advanced. However, concurrent with each upgrade in generation have been profound advances in the capabilities, economics, and availability of network, device, and service offerings.

3GPP specified support for several radio access technologies and also the handover between these access networks. The idea was to bring convergence using a unique core network providing various IP-based services over multiple access technologies. In addition to the support of the existing 3GPP radio access networks, 3GPP defined the interworking functions between the E-UTRAN (LTE/LTE-Advanced), GERAN (GSM/GPRS), and UTRAN (UMTS-based technologies WCDMA and HSPA). The EPS also allows the UE (and EPC) to access non-3GPP technologies. Non-3GPP accesses can be divided into two categories: the trusted and the untrusted. The trusted non-3GPP access networks can interact directly with the EPC, whereas the untrusted non-3GPP access network connects to the EPC via an intermediate interworking function and the evolved packet data

Table 1.2 3GPP Organization

Project Coordination Group (PCG) This group is the ultimate decision-making body in 3GPP responsible for final adoption of 3GPP Technical Specification Group work items, to ratify election results, and the resources committed to 3GPP	**GSM/EDGE Radio Access Network (GERAN)** This group is responsible for specifying the radio access part of GSM/EDGE, more specifically RF front-end, layer 1, 2, and 3, internal and external interfaces, conformance test specifications for all aspects of GERAN base stations and terminals, and GERAN specifications for the nodes in the GERAN	**GERAN WG1** This group is responsible for RF aspects of GERAN, internal GERAN interface specifications, specifications for GERAN radio performance and RF system aspects, conformance test specifications for testing of all aspects of GERAN base stations, etc. **GERAN WG2** This group is responsible for protocol aspects of GERAN. It specifies the data link layer protocols and the interfaces between these layers and the physical layer **GERAN WG3** This group is responsible for conformance test specifications for testing of all aspects of GERAN terminals and liaising with other technical groups to ensure overall coordination
	Radio Access Network (RAN) This group is responsible for the definition of the functions, requirements, and interfaces of the UTRA/E-UTRA network in its two modes, FDD and TDD.	**RAN WG1** This group is responsible for the specification of the physical layer of the radio Interface for UE, UTRAN, E-UTRAN, covering both FDD and TDD modes of radio interface. It is also responsible for handling physical layer related UE capabilities and physical layer related parameters used in UE tests developed in the RAN working group **RAN WG2** This group is in charge of the radio interface architecture and protocols, the specification of the radio resource control protocol, radio resource

(Continued)

Table 1.2 (Continued)

		management, and the services provided by the physical layer to the upper layers **RAN WG3** This group is responsible for the overall UTRAN/E-UTRAN architecture and the specification of inter-base station protocols and interface between mobile station and base station **RAN WG4** This group works on the radio aspects of UTRAN/E-UTRAN. It conducts simulations of various practical radio system scenarios and derives the minimum requirements for transmission and reception parameters. Once these requirements are set, the group defines the test procedures that will be used to verify them (only for base stations) **RAN WG5** This group specifies conformance testing at the radio interface for the UE. The test specifications are based on the requirements defined by other groups such as RAN WG4 for the radio test cases, and RAN WG2 and CT WG1 for the signaling and protocols test cases
	Service and System Aspects (SA) The SA working group is responsible for the overall architecture and service capabilities of systems based on 3GPP specifications and has a responsibility for inter-group coordination	**SA WG1 (Services)** This group works on the services and features for 3G/4G and beyond systems. The group sets high-level requirements for the overall system and provides this in a stage 1 description in the form of specifications and reports **SA WG2 (Architecture)** This group is in charge of

(Continued)

Table 1.2 (Continued)

		developing stage 2 of the 3GPP network and identifies the main functions and entities of the network, how these entities are linked to each other, and the information they exchange. The group has a system-wide view, and decides on how new functions integrate with the existing network entities **SA WG3 (Security)** This group is responsible for the security of the 3GPP system, performing analyses of potential security threats to the system, considering the new threats introduced by the IP-based services and systems, and setting the security requirements for the overall 3GPP system **SA WG4 (Codec)** This group develops specifications for speech, audio, video, and multimedia codecs, in both circuit-switched and packet-switched environments **SA WG5 (Telecom Management)** This group specifies the management framework and requirements for management of the 3G system, delivering the architecture descriptions of the telecommunication management network and coordinating across other working groups
	Core Network and Terminals (CT) This group is responsible for specifying terminal interfaces,	**CT WG1** This group is responsible for the 3GPP specifications that define the UE-core network

(Continued)

Table 1.2 (Continued)

	terminal capabilities, and the core network part of 3GPP systems	layer-3 radio protocols and core network side of the reference points **CT WG3** This group specifies the bearer capabilities for circuit and packet-switched data services, and the necessary interworking functions towards both, the UE and the terminal equipment in the external network **CT WG4** This group defines stage 2 and stage 3 aspects within the core network focusing on supplementary services, basic call processing, mobility management within the core network, bearer independent architecture, GPRS between network entities, transcoder free operation, etc. **CT WG6** This group is responsible for development and maintenance of specifications and associated test specifications for the 3GPP smart card applications, and the interface with the mobile terminal

http://www.3gpp.org/Specification-Groups

gateway (ePDG) entity. The main role of the ePDG entity is to provide security mechanisms such as IPsec tunneling of connections with the UE over an untrusted non-3GPP access. 3GPP does not specify which non-3GPP technologies should be considered trusted or untrusted. This decision is made by the operator.

The 3GPP standardization process starts with developing a requirement document where the targets to be achieved are determined (i.e., Stage 1). The system architecture is defined where the underlying building blocks and interfaces are decided (Stage 2). Various working groups are engaged in developing the detailed specifications where every functional component, interface, and signaling protocol is specified (Stage 3). The compliance and interoperability testing and verification follow upon completion of Stage 3 specifications.

1.4 Evolution of 3GPP standards

As mentioned earlier, the growing demand for mobile Internet and wireless multimedia applications has motivated development of broadband wireless access technologies in the recent years. As a result, 3GPP initiated the work on the LTE in late 2004 to maintain 3GPP's competitive edge over other cellular technologies. The 3GPP LTE symbolizes the migration of the UMTS family of standards from systems that supported both circuit-switched and packet-switched voice/data communications to an all-IP, packet-switched system. The development of the LTE air-interface was closely coupled with the 3GPP system architecture evolution (SAE) project to define the overall system architecture and evolved packet core network [10]. Figure 1.4 illustrates the evolution of the 3GPP standards in the past decade, their major features, and release dates. It shows that the 3GPP standards have continuously evolved toward higher performance and data rates, lower access latencies, and increasing capability to support emerging wireless applications. To achieve higher downlink and uplink data rates, 3G operators upgraded their networks with high-speed downlink packet access (HSDPA), which was specified in 3GPP Rel-5, and high-speed uplink packet access (HSUPA), which was specified in 3GPP Rel-6. The HSDPA and HSUPA, collectively known as HSPA, were further upgraded in 3GPP Rel-7 and 8 with additional features such as higher order modulations in the downlink and uplink, downlink open-loop MIMO, improvements of data link layer protocols, and continuous connectivity for packet data users [13,14].

The 3GPP Rel-8 specifications specified LTE (access network) and EPC (core network) which were completed in late 2008 followed by the UE conformance testing specifications. The 3GPP then worked toward development of an advanced standard specification to address the IMT-Advanced requirements and services for the 4th generation of cellular systems [6,7]. The LTE-Advanced was part of 3GPP Rel-10 and was completed in early 2011. The IMT-Advanced compliant technologies may be available for commercial deployment as early as 2014. 3GPP Rel-11 which initially started as an incremental development, turned into a full release with several features such as coordinated multi-point transmission, enhanced carrier aggregation, and further enhanced inter-cell interference coordination. This release was completed in early 2013. 3GPP is currently working on Rel-12 whose main study and work items are described in Section 1.5.

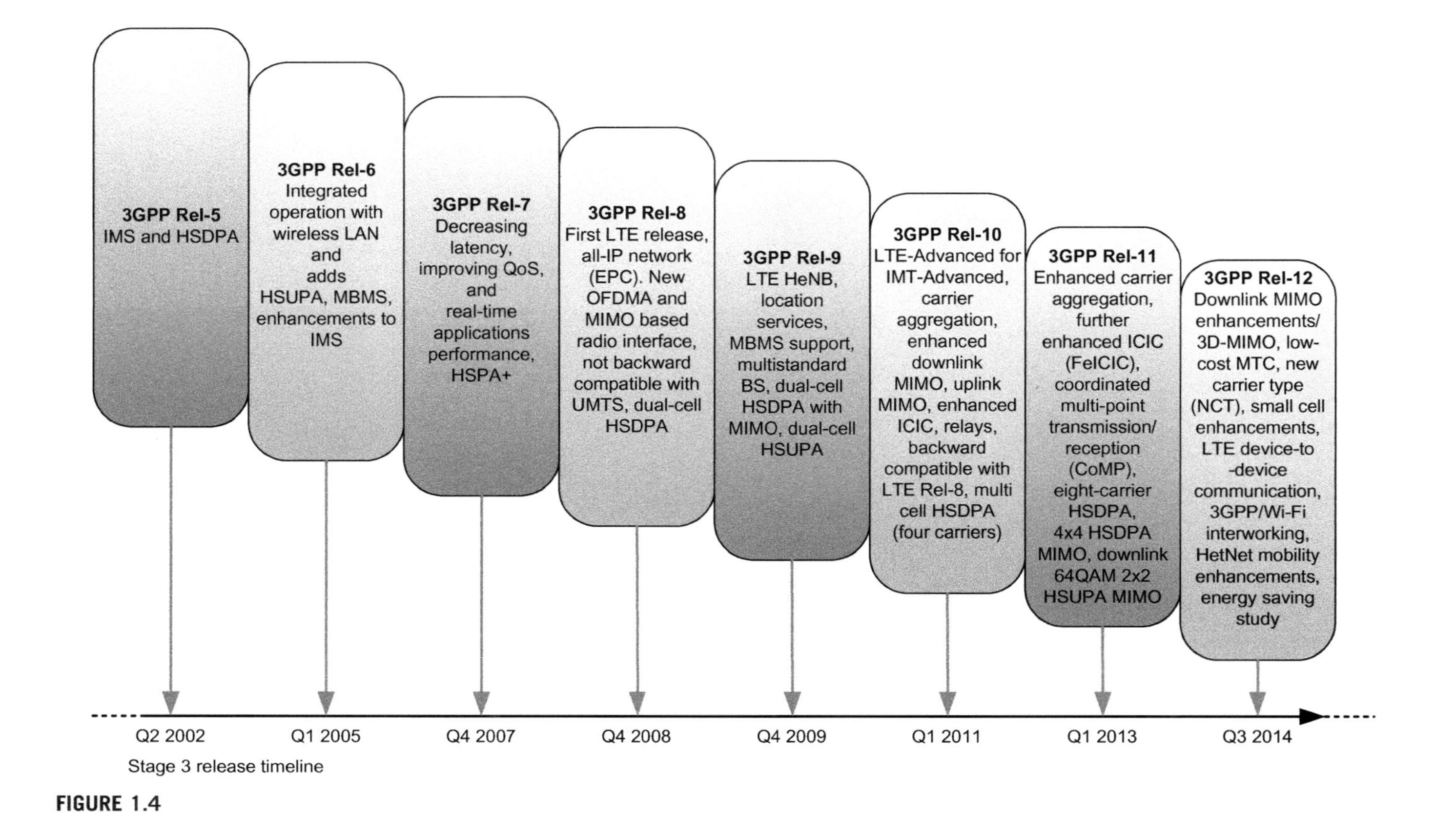

FIGURE 1.4

Evolution of the 3GPP standards in the past decade, their major features and release dates.

Source: *3GPP.*

1.5 A perspective on the future trends in cellular networks

A major challenge facing mobile operators and their technology suppliers is to satisfy the enormous user demands for mobile Internet applications and services. The number of users with mobile broadband wireless devices is exponentially increasing. The use of personal data services is growing with always on, always with you access to an expanding set of applications and services including network-intensive video streaming and cloud services. LTE networks are already providing peak data rates approaching 100 Mbps, but these are only possible under ideal conditions on lightly loaded networks and where UE is close to the base station radio transmitter [11].

While the existing frequency bands are often rearranged for more efficient use, new licensed bands, including higher frequencies for hot-spot zones are introduced, which will be used in conjunction with unlicensed spectrum (if suitable) along with cognitive radio techniques to access and manage the spectrum utilization. A large number of innovative techniques are being studied and researched which can increase spectral efficiency, reduce latency, and improve mobility and connectivity with emphasis on average achievable throughputs across the entire cell including cell-edges. Improving consistency of achievable performance across the network (capacity and throughput), rather than just increasing the peak data rates, is going to be a fundamental improvement demanded by the operators. The coordinated multi-point transmission/reception, and other interference coordination techniques will improve cell-edge performance. The 3D-MIMO and beamforming with large arrays of antenna elements would enable additional frequency reuse within a cell.

While the peak spectral efficiency continues to be important from a technology marketing perspective, the end-user experience is closely associated with the degree of uniformity of service provision; this implies that future enhancements will need to augment cell-edge data rates rather than simply addressing peak data rates. This means that system-level enhancements must generally be targeted considering multi-cell operation and coordination in order to realize useful improvements in spectral efficiency and user experience across the network. As the number of LTE/LTE-Advanced capable devices increases, smaller cell sizes are needed in order to deliver sufficient capacity; such capacity increases resulting from reducing the cell size are often referred to as cell-splitting gains. Since the small cells are embedded in the existing network of macro-cells, the result is a heterogeneous network of macro- and small cells operating at different transmission powers and with different coverage areas. The magnitude of the cell-splitting gains is dependent on suitable cell association/handover decisions to ensure that sufficient users are served by the small cells. In such networks, inter-cell interference coordination becomes significantly more complex than in a homogeneous network [12,16].

The new network topology comprising the small cells is seen as one of the key techniques to increase spectrum efficiency and to improve coverage, bringing the users closer to the network nodes (e.g., pico-cells, femto-cells, or remote radio heads) as a result of the shrinking cell sizes. This would improve the SINR and increase the overall user throughput and capacity. The heterogeneous networks can impact many aspects of the system operation including large traffic and user variations in the cells, larger SINR dynamics, increased handover rates, and so on. One of the methods to deploy small cells overlaid by macro-cells is a frequency-separated local access, where different frequency layers are being used for the small cells. Another deployment scenario for small cells is a frequency-integrated local access, where macro- and pico-cells/femto-cells use the same frequency wherever the small cells are integrated into the network. Depending on which scenario is adopted, different methods are required to assign cell identities and to support the UEs in finding and using the respective small cells. For the small cells, particularly when using higher frequencies, new regulatory requirements might be required and coexistence studies need to be conducted. The use of different TDD schemes can also become beneficial. A more dynamic uplink/downlink subframe allocation in isolated cell clusters can more efficiently match to the instantaneous local traffic conditions. Advanced sensing and resource reservation might be required to avoid severe interference scenarios. The integrated local access is likely to improve performance. The small cells extend the macro-cell coverage with the same physical layer identification or with dynamically assigned virtual cell identities that are reused in different geographical areas. The small cells can provide a fast radio link with a largely improved link budget due to the close proximity of the network node to the user. The macro-cell, on the other hand, can provide cell-wide system information, as well as radio resource control including traffic management and carrier selection and supporting the UE to detect a neighboring small cell. The mobility in heterogeneous networks can be more robust and the operation of small cells may even be transparent to the UE, so the UE may not be aware of the local access. On the other hand, this architecture requires a tight integration of the small cells into the network by low-latency interfaces with a central baseband processing unit [12,16].

Improved backhaul and inter-eNB interfaces can support techniques such as baseband pooling and inter-cell coordination that can more efficiently and effectively manage resources with the large arrays of remote radio heads that will be deployed in high-demand locations. There are initiatives to simplify management across cells with simplified signaling and control maintained at the macrolayer, and introduction of phantom cells in high-density small-cell layers. Local wireless access overlaid by wide area macro-cells providing basic coverage has been identified as one of the key enablers of the future RAN. World Radiocommunication Conference 2015 (WRC-15) is expected to allocate the frequency range beyond 10 GHz to overcome spectrum shortage for broadband wireless communication systems, which may result in the development of a completely new access scheme potentially standardized as part of 3GPP future releases. Access technologies making extensive use of beamforming will be possible candidates since the beamforming gain may compensate the increased path loss at higher frequencies.

Although an increase in available and useful spectrum is expected to be at least threefold until 2020, this is by far not sufficient enough to keep up with the expected growth of traffic [20,21].

Downlink MIMO enhancements and three-dimensional MIMO is one of the 3GPP Rel-12 study items. With continued advancement of multi-antenna technologies, enhancements of the macro-cell eNBs can be realized mostly by exploiting an increased frequency reuse and reduced interference in the spatial domain. Under the new study item, a new three-dimensional channel model will be developed to allow more accurate system-level simulation of multi-antenna technologies. Three-dimensional beamforming through utilization of flexible electronic beam shaping adds the concept of beamforming in the vertical dimension. It has the potential to increase spectrum efficiency of the network through proactive cell shaping and splitting as well as improving the coverage. Other enhancement areas to support this technology may include reference signals, codebooks and feedback, and measurement and reporting for interference coordination in the beam domain as well as RF requirements.

The significant growth of machine-to-machine (M2M) applications (e.g., smart grids, transportation, healthcare services) has compelled the 3GPP and the cellular industry to ensure that the design and development of future releases of LTE can support massive transfer of small, infrequent packets using very low cost, low complexity, and low power devices. The 3GPP version of M2M, called machine type communication (MTC), was already standardized in Rel-11. The latter work covers service requirements, architecture, and security issues. An MTC interworking function and service-capability server, among other functionalities, were defined as part of 3GPP Rel-11 [12,22].

The LTE-Advanced carrier aggregation scheme in Rel-10/11 was limited to backward-compatible carriers only. On the one hand, this enables a smooth transition to new releases; on the other hand, it makes the introduction of new technologies more difficult and cumbersome. One of the consequences of backward compatibility to LTE Rel-8/9 is the necessity to continuously transmit cell-specific reference signals in every downlink subframe across the system bandwidth. This precludes switching off a cell temporarily and represents unnecessary overhead particularly in the case of non-codebook-based beamforming using UE-specific reference signals or multi-user MIMO. A new carrier type allows switching off cells, at least temporarily, and will reduce the overhead and interference from cell-specific reference signals by maximizing the use of UE-specific reference signals that can be precoded with the data subcarriers which are required for advanced multi-antenna technologies. Furthermore, new bandwidth combinations may be introduced to enable more efficient use of fragmented bandwidths. The new carrier type can be operated as an extension carrier along with another LTE/LTE-Advanced carrier, or alternatively as a stand-alone non-backward-compatible carrier in Rel-12 and beyond [16].

Device-to-device (D2D) communication is a new feature in LTE Rel-12 that allows direct communication between user terminals that are in the proximity of each other. In addition to its potential for energy saving, interference reduction, and coverage extension, the key driver for this technology is to ensure that the evolution of LTE meets the growing demands for new applications/services. Once D2D functionality is developed, the market may find new proximity-based applications and services. Focus is mainly given to network-controlled D2D communication. In this case, the attach procedure, e.g., initial access, authentication, connection control, as well as resource reservation, can be handled by the network; therefore, the QoS can be guaranteed and the network operator can remain in control of the transactions (because they own the infrastructure and the spectrum).

Today's Wi-Fi network detection and selection functions are mostly UE-based, and future solutions might be operator controlled. This will empower implementation of more intelligent networks that monitor cell load or transport network load, user QoS requirements, or radio link quality for the different links during the operation. Cellular networks supporting Wi-Fi are seen as a viable solution by operators to off-load network traffic in indoor or dense environments, but until now, interworking has been mainly limited to core network functions such as user authentication/authorization and accounting. Although 3GPP specifications support mobility across 3GPP and non-3GPP access networks, further optimization is needed to improve load balancing, QoS provisioning, mobility and handover across heterogeneous networks, and improved UE power consumption when using Wi-Fi radio in tight cooperation with cellular networks [12,16].

The future evolution path or further enhancements of LTE-Advanced system will be directed toward a balance among determining factors such as technical solutions, practicality, and cost of implementation and deployment. Not every theoretical claim of potential improvement is deemed cost effective and practical to deploy. Factors such as robust conformance testing and real-life performance verification must also be taken into account, and unnecessary market fragmentation must be avoided, in order to ensure that economies of scale are achieved. Such factors would imply that the introduction of multiple alternative options, offering similar performance, should be avoided; careful evaluation must continue to make informed selection of new features [16].

It is evident that the existing mobile broadband radio access systems will continue to evolve and new systems will emerge. The vision, service and system requirements as well as use cases for systems beyond IMT-Advanced (alternatively known as 5G) are being defined. While it is not exactly clear what technologies will be incorporated into the design of such systems and whether the existing radio access technologies will converge into a single universal radio interface, it is envisioned that the future radio interfaces will rely on software-defined radio, cognitive radio system, flexible spectrum sharing, small-cell network deployments, cloud-based services, and cooperative communication concepts as well as reconfigurable RF and baseband circuitry in order to provide higher quality of user experience, higher capacities, and wider range of services with minimal cost and complexity.

Although the LTE family of standards have been designed from the beginning to facilitate energy-efficient implementations of both mobile devices and base stations, energy efficiency is an area with increasing importance due to the rising cost of energy and environmental concerns; thus, the evolution of LTE-Advanced must prioritize such considerations. This will involve minimizing the transmission overheads and processing requirements when network load is light, and in general enhancing the ability of the network to adapt dynamically to the actual load levels. This may include more advanced techniques for load balancing and resource sharing so that parts of the system which are not essential may be switched off whenever possible, while retaining flexibility to bring them online rapidly to handle high traffic demand when needed [16].

A generic framework for coordinated multi-point transmission/reception in the downlink and uplink was developed in LTE Rel-11, which may be deployed as part of LTE-Advanced networks. The gains promised by this method may only be achieved upon availability of low-latency broadband fiber-optic interconnections between coordinated transmission points. Such links are not widely available in contemporary networks; thus, early CoMP deployments should not be overly dependent on their existence. Nevertheless, centralized (alternatively known as cloud) processing may become an important enabler for realizing the full potential of CoMP, and efficient schemes for the transfer of scheduling/precoding information as well as data between transmission points will need to be identified. It is worth noting that centralized processing has the potential to provide some practical benefits in addition to those arising from centralized CoMP scheduling/beamforming and joint system optimization. The centralized processing can potentially reduce the costs of network operation by reducing the number of secure sites needed for network equipment installation, as well as facilitating network reconfiguration and robustness. For the uplink CoMP, each reception point can use interference cancellation techniques to isolate the weaker signals from mobile devices in the neighboring cells and convey them to a central node for combining with the versions of the signal received at other points.

MU-MIMO techniques are the most effective solutions to increase spectral efficiency in a cell. However, some of the promised MU-MIMO gains in practice fade away. The gap between theoretical MU-MIMO performance and that practically achievable in a real-life deployment may be explained by arguing that good spatial separation of users is achieved with correlated arrays of uniformly spaced antenna elements, whereas many antenna configurations deployed in practice are cross-polarized, with low correlation between the polarizations. Furthermore, channel state information at the transmitter is far from perfect, due to the constraints of uplink feedback overhead and latency. The availability of sufficiently spatially separated users with non-empty data queues is limited in practice; nevertheless, as the number of LTE-Advanced capable UEs increases, this will become less of a concern. The studies have shown that for a given channel state information feedback overhead, higher capacity can be achieved with more accurate channel state information feedback from a small number of terminals relative to less accurate feedback from a large number of terminals when using MU-MIMO techniques. Therefore, future improvements in downlink MU-MIMO will depend on significant improvement in the accuracy of the channel state information at the transmitter, given practical channel estimation, realistic antenna configurations, typical mobility scenarios, without incurring unrealistic feedback overhead. In the uplink, MU-MIMO can be implemented without availability of accurate channel state information at the transmitter. The complexity of the uplink MU-MIMO operation is managed at the receiver rather than the transmitters, and sophisticated non-standardized receivers can theoretically attain the maximum multiple access channel capacity [16].

Another promising future trend in cellular networks is software-defined networking (SDN) or network virtualization. The SDN is an emerging network architecture where the control and data planes are decoupled, network intelligence and state are logically centralized, and the underlying network infrastructure is abstracted from the applications. The migration of control layer into accessible computing devices (formerly tightly coupled with individual network devices) enables the underlying infrastructure to be abstracted for applications and network services, which can treat the network as a logical or virtual entity. Figure 1.5 illustrates the concept of a software-defined network. In one embodiment of SDN, the network control-plane hardware can be physically decoupled from the data plane hardware; i.e., a network switch can forward packets and a separate server can run the network control plane. As a result, enterprises and carriers achieve programmability, automation, and network control, enabling them to build highly

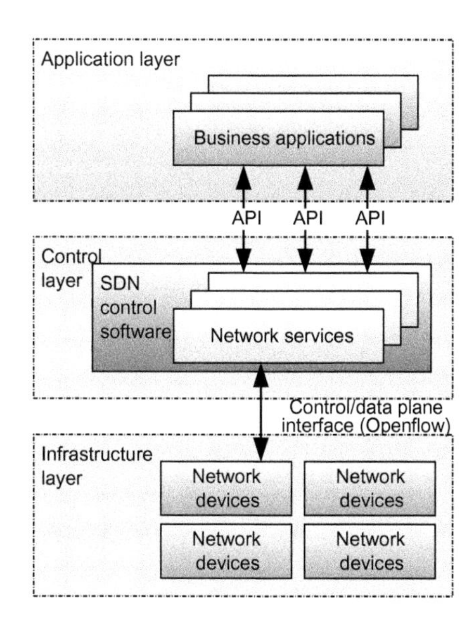

FIGURE 1.5

Illustration of the software-defined network concept [23,24].

scalable, flexible networks that readily adapt to changing business needs. The rationale for this approach is that the decoupling allows for the control plane to be implemented using a different distribution model than the data plane, and the decoupling further allows the control-plane development and run-time environment to be on a different platform than the traditionally low-power management CPUs found on hardware switches and routers [23,24].

The SDN requires some method for the control plane to communicate with the data plane. One example of such a mechanism is open flow, which is a standard interface for controlling computer network switches. The SDN allows network operators to have programmable central control of network traffic without requiring physical access to the network's hardware devices. In addition to abstracting the network, SDN architecture supports a set of application programming interfaces (APIs) that make it possible to implement common network services, including routing, multicast, security, access control, bandwidth management, traffic engineering, QoS, processor and storage optimization, energy usage, and all forms of policy management, which are customized to meet business objectives. The Open Networking Foundation is studying open APIs to promote multivendor management, which opens the door for on-demand resource allocation, self-service provisioning, virtualized networking, and secure cloud services. Using open APIs between the SDN control and the applications layers, the business applications can operate on an abstraction of the network, leveraging network services and capabilities without being tied to the details of their implementation [23,24].

References

ITU Recommendations and Reports

[1] Recommendation ITU-R M.1645, Framework and overall objectives of the future development of IMT-2000 and systems beyond IMT-2000, June 2003.
[2] Recommendation ITU-R M.1822, Framework for services supported by IMT, October 2007.
[3] Report ITU-R M.2134, Requirements related to technical performance for IMT-Advanced radio interface(s), October 2008.
[4] Report ITU-R M.2135-1, Guidelines for evaluation of radio interface technologies for IMT-Advanced, December 2009.
[5] Report ITU-R M.2133, Requirements, evaluation criteria and submission templates for the development of IMT-Advanced, October 2008.
[6] Recommendation ITU-R M.2012, Detailed specifications of the terrestrial radio interfaces of International Mobile Telecommunications Advanced (IMT-Advanced), January 2012.
[7] IMT-Advanced submission and evaluation process <http://www.itu.int/ITU-R/>.
[8] ITU-T Next Generation Networks Global Standards Initiative, <http://www.itu.int/ITU-T/ngn/>.

3GPP Specifications

[9] 3GPP TR 36.913, Requirements for further advancements for Evolved Universal Terrestrial Radio Access (E-UTRA) (LTE-Advanced) (Release 8), March 2009.
[10] 3GPP TR 23.882, 3GPP System Architecture Evolution: Report on Technical Options and Conclusions (Release 8), September 2008.

Articles

[11] K. Mallinson, 2020 Vision for LTE, WiseHarbor. <http://www.wiseharbor.com/forecast.html>, June 2012.
[12] E. Seidel, 3GPP LTE-Standardization in Release 12 and Beyond, Nomor Research GmbH, Munich, Germany. <http://www.nomor.de/lte-standardisation>, January 2013.
[13] 4G Americas, 4G Mobile Broadband Evolution: Release 10, Release 11 and Beyond - HSPA, SAE/LTE and LTE-Advanced. <http://www.4gamericas.org/>, October 2012.
[14] 4G Americas, Mobile Broadband Explosion: The 3GPP Wireless Evolution. <http://www.4gamericas.org/>, August 2012.
[15] P. Bhat, et al., LTE-Advanced: an operator perspective, IEEE Commun. Mag. (February 2012).
[16] M. Baker, From LTE-Advanced to the future, IEEE Commun. Mag. (February 2012).
[17] M. Kottkamp, A. Roessler, J. Schlienz, LTE-Advanced Technology Introduction, Rohde & Schwarz, August 2012.
[18] I.F. Akyildiz, et al., The evolution to 4G cellular systems: LTE-Advanced, Phys. Commun. 3 (4) (December 2010) <http://www.sciencedirect.com/>
[19] White Paper, NGMN Radio Access Performance Evaluation Methodology, NGMN Alliance, January 2008.
[20] White Paper, LTE Backhauling Deployment Scenarios, NGMN Alliance, July 2011.
[21] White Paper, Guidelines for LTE Backhaul Traffic Estimation, NGMN Alliance, July 2011.
[22] White Paper, On Machine-to-Machine Communication, NGMN Alliance, September 2012.
[23] White Paper, Software-Defined Networking: The New Norm for Networks, Open Networking Foundation, April 2012. <https://www.opennetworking.org>.
[24] Software-defined networking, from Wikipedia, <http://en.wikipedia.org/wiki/Software-defined_networking>.

CHAPTER

Network Architecture

2

CHAPTER CONTENTS

LTE-Advanced.

3GPP Rel-8 specifications included enhancements to high-speed packet access (HSPA) core and access networks as well as the introduction of the evolved packet system (EPS) which comprises an Internet protocol (IP)-based, all-packet core network with minimal hierarchy in conjunction with a new orthogonal frequency division multiplexing (OFDM)-based radio access network. The EPS is a simplified architecture (i.e., some intermediate network elements were removed or combined together) with an all-IP network support for higher throughput, lower latency radio access and improved mobility between multiple heterogeneous access networks including evolved universal terrestrial radio access (E-UTRA), 3GPP legacy systems (e.g., GERAN or UTRAN), as well as non-3GPP systems, such as Wi-Fi and CDMA2000.

The network architecture of LTE is based on the functional decomposition principle, where underlying features are divided into logical functional entities without specific implementation assumptions about physical network nodes. That is why 3GPP specified an entirely new packet-switched core network architecture to support the evolved universal terrestrial radio access network (E-UTRAN) through reduction in the number of network elements, simpler functionality, and more importantly allowing connections and handover to other access technologies, giving the service providers the ability to deliver a seamless mobility experience.

The introduction of the evolved packet core (EPC) and all-IP network architecture had significant impacts on mobile services, where voice, data, and video communications are built on the IP-based protocols; interworking of the new architectural entities with 2G/3G legacy systems; scalability required by each of the core elements to address the increasing number of direct connections to user terminals, orders of magnitude of bandwidth increase, and terminal mobility; reliability and availability provided by each network entity to ensure service continuity. To address a fundamentally different set of network and service requirements, the EPC design represented a departure from previous mobile networking paradigms.

This chapter describes the design principles of the EPS, network reference model, logical building blocks of the network architecture as well as theoretical and operational aspects of EPC and LTE/LTE-Advanced radio access network (E-UTRAN) building blocks, functional split, and their interconnections. Under this topic, the network functions, such as mobility management, radio resource management (RRM), quality of service (QoS)/QoE, mobile IP, IP flow mobility, relays and femto-cells, and security, along with practical design/implementation considerations and real-life issues and limitations will be discussed. Furthermore, the operation and management of small cells involving low-power remote radio heads (RRHs) will be covered. The selective IP traffic offload, local IP access (LIPA), and Wi-Fi offloading mechanisms will be discussed, as well. In this chapter, we provide a top-down systems approach to understanding the architecture and operation of the LTE/LTE-Advanced access and core networks. The focus in this chapter is on the network architecture and underlying network functional entities and the network interfaces. The protocols and procedures for network attach/detach, connection establishment/release, mobility management, and other functional details will be discussed in the next chapters.

2.1 Overall network architecture

2.1.1 General concepts

In its most abstract form, the EPS consists of only two nodes in the user-plane, that is, a base station and a core network gateway. The node that performs control-plane functionality is separated from the node that performs user-plane functionality. The basic EPS architecture is illustrated in Figure 2.1. The EPS architecture was designed not only to provide a smooth evolution from the 2G/3G quasi-packet-switched architectures consisting of nodeBs (NBs), radio network controllers (RNCs), SGSNs, and GGSNs, but also to provide support for non-3GPP access (e.g., Wi-Fi), improved policy control and charging, a wider range of QoS capabilities, advanced security/authentication mechanisms, and flexible roaming [70,71].

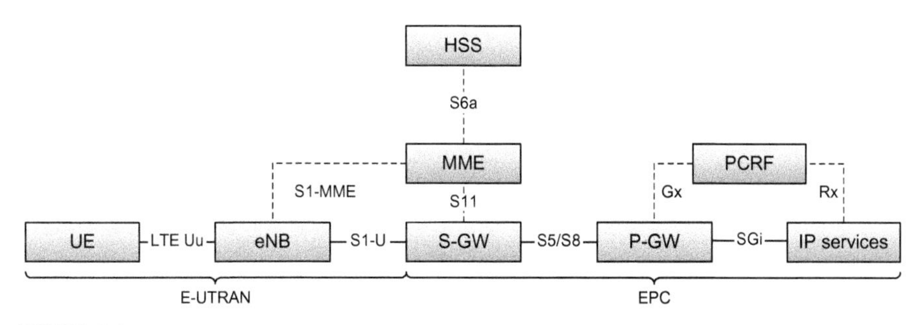

FIGURE 2.1

Generic EPS architecture [70,71].

The introduction of EPC combined with E-UTRAN in many ways represented a radical departure from previous mobile network design philosophies. It heralded the end of the circuit-switched voice-centric design era. 3GPP LTE was meant to use voice over IP (VoIP) for voice services. The EPC treats voice just as other IP-based applications, although an important one that imposes stringent performance and operational requirements [34,37,69].

3GPP LTE was required to match the wired-broadband quality of experience (QoE). This is quite different from providing mainly best-effort low-rate web browsing or short message service (SMS), two data applications for which its predecessor quasi-packet-switched mobile networks were optimized. In LTE, all mobility management functions have been moved to the core network and have become part of mobility management entity (MME). This is a consequence of the split of functions previously performed by the RNC and NB/Base Transceiver Station (BTS). The MME requires a control-plane capacity that is an order of magnitude larger than the SGSN or PDSN, and must ensure interworking with 2G/3G legacy mobile systems (see Figure 2.2).

The long-term evolution of universal mobile telecommunications system (UMTS) systems was expected to provide superior end-to-end QoS management and enforcement in order to deliver multimedia, low-latency, and real-time services, with more granular QoS classes and stricter performance targets. This was going to be achieved while ensuring scalability of users, services, and data sessions. In LTE, service control is provided via the policy and charging rules function (PCRF). This was a major change from previous systems, where service control was realized primarily through user equipment (UE) authentication by the network. The PCRF entity dynamically controls and manages all data sessions and provides appropriate interfaces toward charging and billing systems, as well as enabling new business models. The design of the new air-interface required

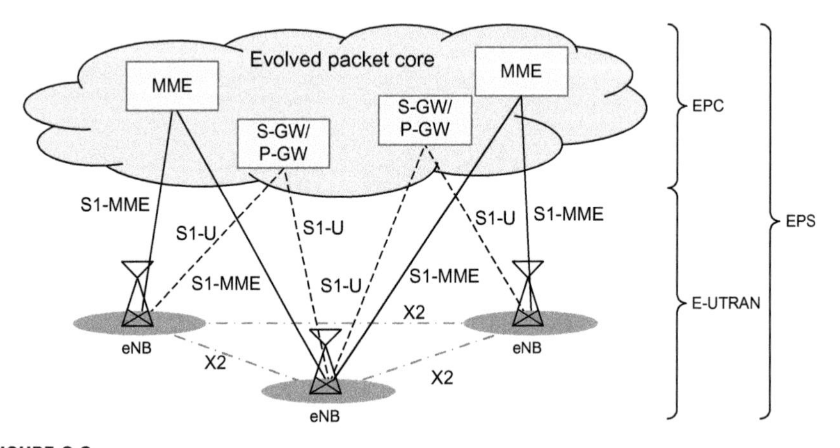

FIGURE 2.2

EPS composition [12].

significantly more capacity in the user-plane and control-plane. The existing 2G/3G mobile core elements cannot satisfactorily address LTE requirements without major upgrades to the platforms. Therefore, the legacy core network implementations must be dramatically upgraded to enable LTE.

While the core network consists of many logical nodes, the access network consists of essentially just one node, the eNB, which provides connections to the UEs. Each of the network elements is interconnected by means of logical interfaces that are standardized in order to allow multi-vendor interoperability. This gives network operators the possibility of sourcing different network elements from different vendors. In fact, network operators may choose in their physical implementations to split or merge these logical network elements depending on commercial considerations. The functional split between the EPC and E-UTRAN is shown in Figure 2.3. The EPC and E-UTRAN network elements are described in the next sections.

The EPC design has radically changed some key networking concepts which were used in the previous generations of cellular core networks and successfully addressed a number of technological challenges which had arisen from significant technological improvements in the radio access side. The LTE radio access system provides more efficient use of the spectrum with wider spectral bands reserved for the broadband systems, resulting in greater system capacity and performance.

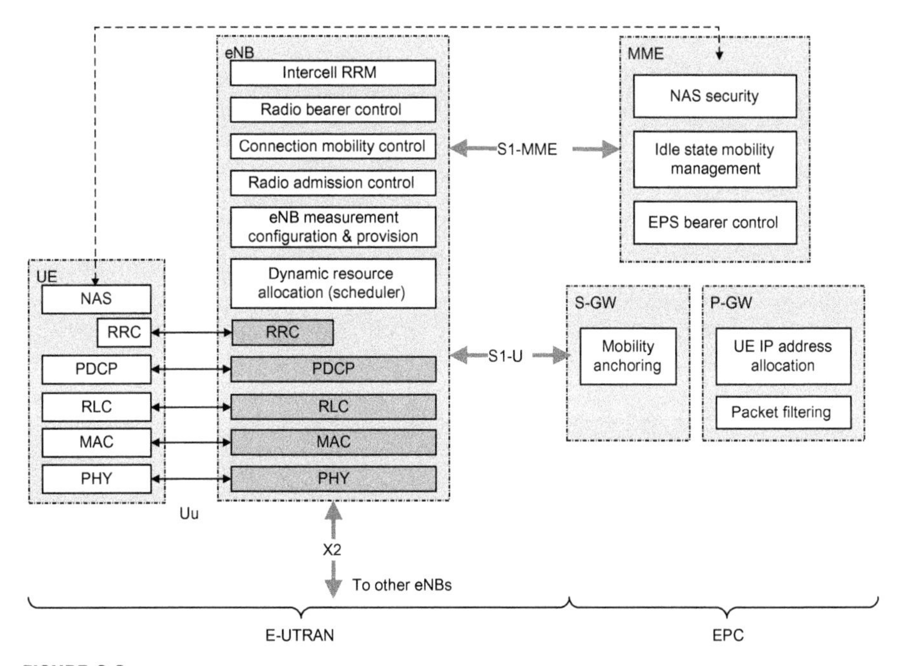

FIGURE 2.3

Functional split between E-UTRAN and EPC entities [12].

At the same time, the core network needs to change in order to provide higher throughput and lower latency; both should come as a result of simplified and improved IP-based network architecture. More specifically, the EPC addressed a number of fundamental IP-related aspects of LTE deployment including the following:

- IP routing and network addressing and real-time management
- Centralized versus distributed network architecture (for MME, serving gateway (S-GW), and packet data network gateway (P-GW) deployment)
- IPv6 (strategy, introduction, and coordination with IPv4)
- End-to-end QoS deployment and coordination with underlying transport
- Layer-2 versus layer-3 connectivity at the transport layer (eNB, S-GW, P-GW, and MME)
- End-to-end security for user-plane and control-plane
- Interconnectivity to external networks and virtual private networks
- Enabling lawful interception of data packets

There are a stringent set of requirements for reliable, scalable, high-performance network elements. Due to the dynamic nature of the user mobility in LTE, coupled with the short duration of multiple data sessions per UE and large-scale deployment targets, the EPC must further address the challenging requirements of dynamic management of mobility, policies, and signaling/data bearers, while ensuring interworking and interoperability with legacy 2G/3G systems.

The EPC is realized through the following four new network elements: (1) S-GW, (2) P-GW, (3) MME, and (4) PCRF. While S-GW, P-GW and MME were introduced in 3GPP Rel-8, PCRF had been introduced in 3GPP Rel-7. The architectures using PCRF have not been widely adopted. The PCRF interoperation with the EPC gateways and the MME is normative in 3GPP Rel-8 and essential for the operation of LTE.

The MME is a node within the EPC that performs signaling and control functions to manage the user access to network connections, assignment of network resources, and management of the mobility states to support tracking, paging, roaming, and handovers. The MME controls all control-plane functions related to subscriber and session management. The MME is the key control-node for the LTE access-network. It is responsible for idle-mode UE tracking and paging procedure. It is involved in the bearer activation/deactivation process and is also responsible for choosing the S-GW for a UE at the initial attach and at the time of intra-LTE handover involving core network node relocation. It is further responsible for user authentication. The non-access stratum (NAS) signaling terminates at the MME and it is also responsible for generation and allocation of temporary identities to UEs. It checks the authorization of the UE to camp on the service provider's public land mobile network (PLMN) and enforces UE roaming restrictions. The MME is the termination point in the network for ciphering/integrity protection for NAS signaling and handles the security key management. Lawful interception of signaling is also supported by the MME. The MME also provides the control-plane function for mobility between LTE and 2G/3G access networks with the S3 interface terminating at the MME from the SGSN. The NAS protocols form the control-plane between the UE and the MME and support the UE mobility and the session management procedures to establish and maintain IP connectivity between the UE and a P-GW. They define the rules for a mapping between parameters during intersystem mobility with 3GPP or non-3GPP access networks. They also provide the NAS security by integrity protection and ciphering of NAS signaling messages.

The home subscriber server (HSS) is a central database that contains user- and subscription-related information. The functions of the HSS include mobility management, call and session establishment support, user authentication, and access authorization. The HSS is based on legacy home location register (HLR) and authentication center (AuC) functions in previous 3GPP releases.

The S-GW is a user-plane element whose primary function is to manage user-plane mobility and to act as a demarcation point between the radio access and core networks. The S-GW maintains data paths between eNBs and the P-GW. From a functional perspective, the S-GW is the termination point of the packet data network interface toward the E-UTRAN. When terminals move in the areas served by eNBs in the E-UTRAN, the S-GW serves as a local mobility anchor (LMA). This means that packets are routed through this point for intra E-UTRAN mobility and mobility with other 3GPP technologies, such as GSM and UMTS. The S-GW routes and forwards user data packets, while acting as the mobility anchor for the user-plane during inter-eNB handovers and as the anchor for mobility between LTE and other 3GPP radio access technologies, that is, terminating S4 interface and relaying the traffic between 2G/3G systems and P-GW. For idle mode UEs, the S-GW terminates the downlink data path and triggers paging when downlink data arrives for a UE. It manages and stores UE contexts, for example, parameters of the IP bearer service and network internal routing information. It also performs replication of the user traffic in case of lawful interception.

The P-GW is the termination point of the packet data interface toward the packet data network. As an anchor point for sessions toward the external packet data networks, the P-GW supports policy enforcement features, such as applying operator-defined rules for resource allocation and usage; packet filtering functions, such as packet inspection for application type detection and charging support. The P-GW provides connectivity from the UE to external packet data networks by being the point of exit and entry of traffic for the UE. A UE may have simultaneous connections with more than one P-GW for accessing multiple packet data networks (PDNs). Another key role of the P-GW is to act as the anchor for mobility between 3GPP and non-3GPP technologies, such as CDMA2000.

In LTE, the data-plane traffic is carried via virtual connections known as service data flows (SDFs). The SDFs are carried over bearers that are virtual containers with unique QoS characteristics. Figure 2.4 illustrates an example scenario

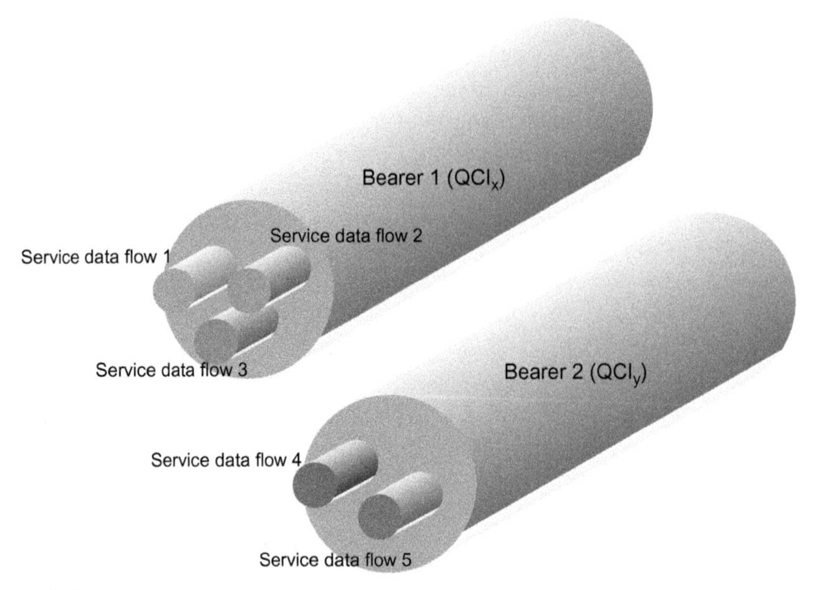

FIGURE 2.4

Relationship between SDFs and bearers [70,71].

where one or more SDFs are aggregated and assigned to one bearer. One bearer, a data path between a UE and a PDN, has three segments as follows: radio bearer between UE and eNB; data bearer between eNB and S-GW (S1 bearer); and data bearer between S-GW and P-GW (S5 bearer). Figure 2.5 illustrates three segments that constitute an end-to-end bearer. The primary role of a P-GW is QoS enforcement for each of these SDFs, while S-GW dynamically manages the bearers. There are two types of EPS bearers: the *default EPS bearer* and the *dedicated EPS bearer*. The default EPS bearer is established during the attach procedure and allocates an IP address to the UE that does not have a specific QoS (only a nominal QoS). A dedicated EPS bearer is typically established during call setup and after transition from idle mode to the connected mode. It does not allocate any additional IP address to the UE and is linked to a specified default EPS bearer but has a specific QoS class. A QoS class identifier (QCI) is a scalar that is used as a reference to access node-specific parameters that control bearer level packet forwarding (e.g., scheduling weights, admission thresholds, queue management thresholds, and link layer protocol configuration), and are preconfigured by the operator owning the access node (eNB).

As shown in Figure 2.5, the EPS bearer transports packets by an S1 bearer between the S-GW and the eNB, and by a radio bearer between a UE and an eNB. There is a one-to-one correspondence between a radio bearer identifier and an S1 bearer. The IP packets mapped to the same EPS bearer receive the same bearer-level packet forwarding treatment, e.g., scheduling policy, queue management policy, rate shaping policy, and radio link control (RLC) configuration. In order to

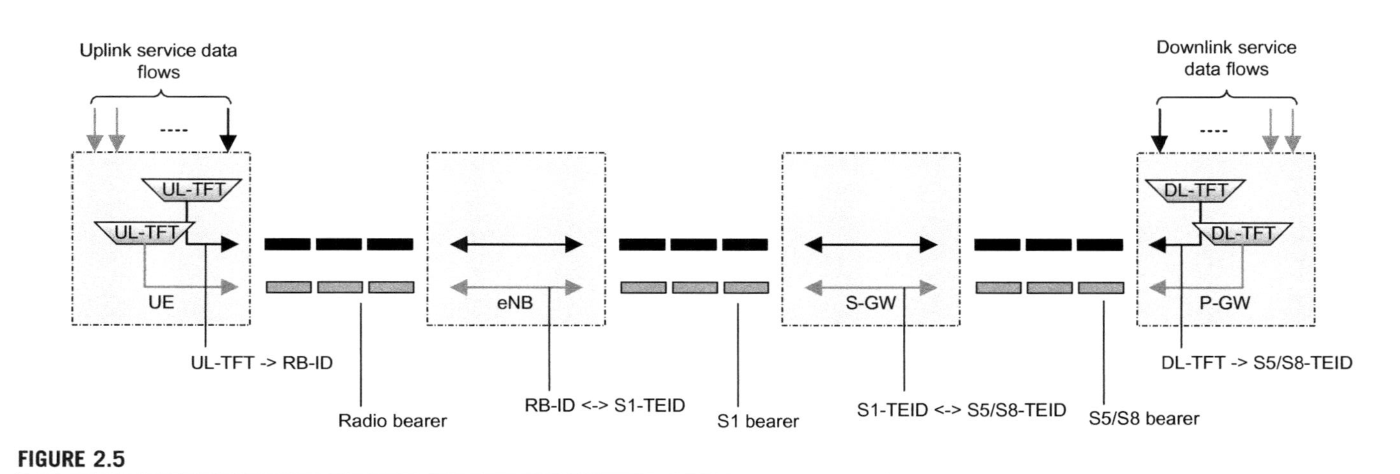

FIGURE 2.5

Bearers across different EPS interfaces [70,71].

provide different bearer-level QoS, a separate EPS bearer should be established. The IP packets must then be filtered and mapped to the appropriate EPS bearers. Packet filtering of different bearers is based on traffic flow templates (TFTs). The TFTs use IP header information, such as source and destination IP addresses and transport control protocol (TCP) port numbers to filter packets, such as VoIP from web-browsing traffic, so that each flow can be mapped to the appropriate EPS bearers with the appropriate QoS settings. An uplink TFT (UL-TFT) associated with each bearer in the UE filters IP packets to EPS bearers in the uplink direction. A downlink TFT (DL-TFT) in the P-GW is a similar set of downlink packet filters. When a UE attaches to the network, the UE is assigned an IP address by the P-GW and a default bearer is established, and it remains in effect during the lifetime of the PDN connection in order to provide the UE with always-on connectivity to that PDN. The initial bearer-level QoS parameter values of the default bearer are assigned by the MME based on subscription data retrieved from the HSS. The policy and charging enforcement function (PCEF) may change these values in interaction with the PCRF or according to local configuration. One or more dedicated bearers can be established at any time during or after completion of the attach procedure. A dedicated bearer can be either a guaranteed bit rate (GBR) or a non-GBR bearer. The default bearer is always a non-GBR bearer. The distinction between default and dedicated bearers should be transparent to the access network. Each bearer has an associated QoS, and if more than one bearer is established for a given UE, then each bearer must also be associated with appropriate TFTs. The dedicated bearers may be established by the network based on a trigger from the IP multimedia subsystem (IMS) domain or a request by the UE. The dedicated bearers for a UE may be provided by one or more P-GWs.

In the E-UTRAN side, the evolved nodeB (eNB) provides the E-UTRA user-plane and control-plane protocol termination toward the UE. The eNBs can optionally be interconnected via X2 logical interfaces. The X2 interface may not be necessarily in the form of a point-to-point physical connection between eNBs. The eNBs are connected by means of the S1 interface to the EPC. More specifically, it is connected to the MME by means of the S1-MME reference point, and to the S-GW by means of the S1-U interface. The S1 interface supports a many-to-many relationship between MMEs/S-GWs and eNBs.

2.1.2 Evolution of the network architecture

The global system for mobile communications (GSM) network architecture was based on a circuit-switched design where a direct connection was established between the call-making and call-receiving entities throughout the telecommunication network, that is, radio access and core networks of the mobile operator and the fixed network. The GSM was a voice-centric system providing circuit-switched telephony, SMS, and some very low-rate data services. The general packet radio service (GPRS) enhancement added packet-switched services to the circuit-switched functions, where data was transported in the form of packets without

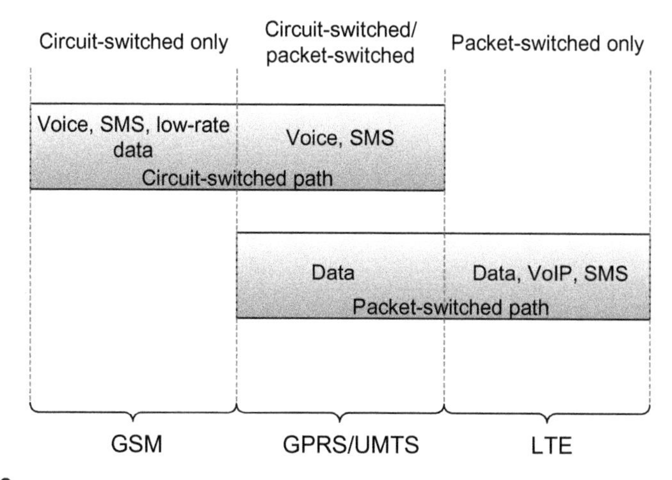

FIGURE 2.6

Evolution of cellular networks and services.

establishing dedicated circuits. This enhancement offered more flexibility and efficiency. In GPRS, voice and SMS are transported via circuit-switched paths; thus, the core network is composed of circuit-switched and packet-switched domains. In UMTS, the dual-domain concept was maintained on the core network side and only a few network elements evolved. In the design of the 3G system long-term evolution, the 3GPP decided to use the IP as the key protocol to transport all services; therefore, the EPC vision did not include a circuit-switched domain and it was an evolution of the packet-switched architecture used in GPRS/UMTS. This paradigm shift in the network design not only had significant impact on the architecture itself but also on the way that the services were provided. Traditional use of circuits to carry voice and short messages needed to be replaced by IP-based solutions in the long term. The evolution of 3GPP network architecture and services are depicted in Figures 2.6–2.8 from different perspectives (services, general architecture, and network elements). As shown in the figures, while the services have migrated toward pure IP-based services, the network architecture has been simplified and redundant entities have been eliminated or integrated into other network elements.

Different options are possible to integrate MME, S-GW, and P-GW logical entities. Each option will result in different handling of control-plane and user-plane packet routing. The choice may depend on operator strategy and/or EPC vendor equipment availability and constraints. Integrating more logical entities together in one physical entity will reduce latency/delay, cost (fewer nodes and interfaces) and signaling load, but will increase the functional complexity per entity/component and will limit the benefit and potential savings achieved by a split architecture, i.e., redundancy, flexibility, and scalability. A possible deployment option may start with some level of integration, to save initial costs, and migrate to more flexible and scalable architecture when the number of users and user traffic grows.

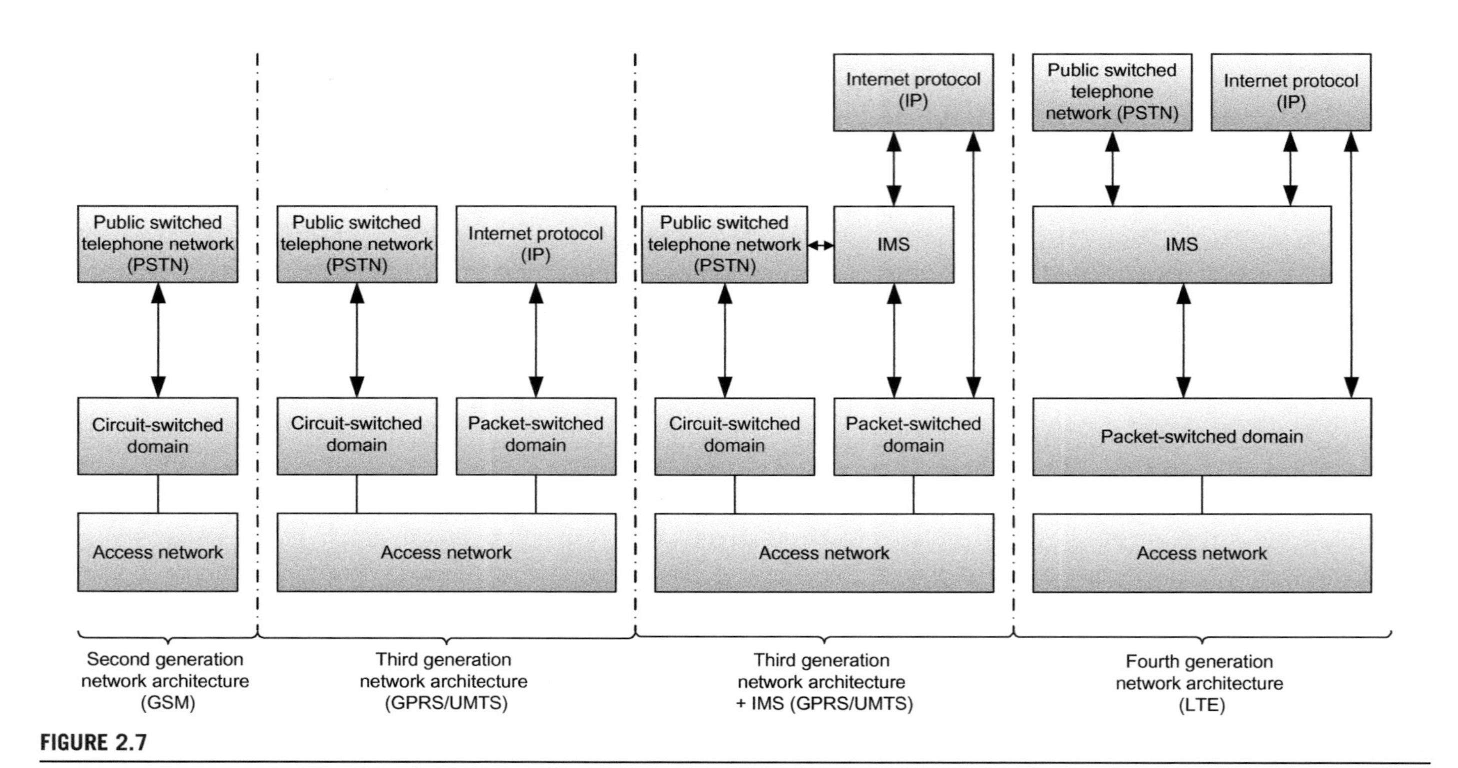

FIGURE 2.7

Evolution of network architecture through cellular generations.

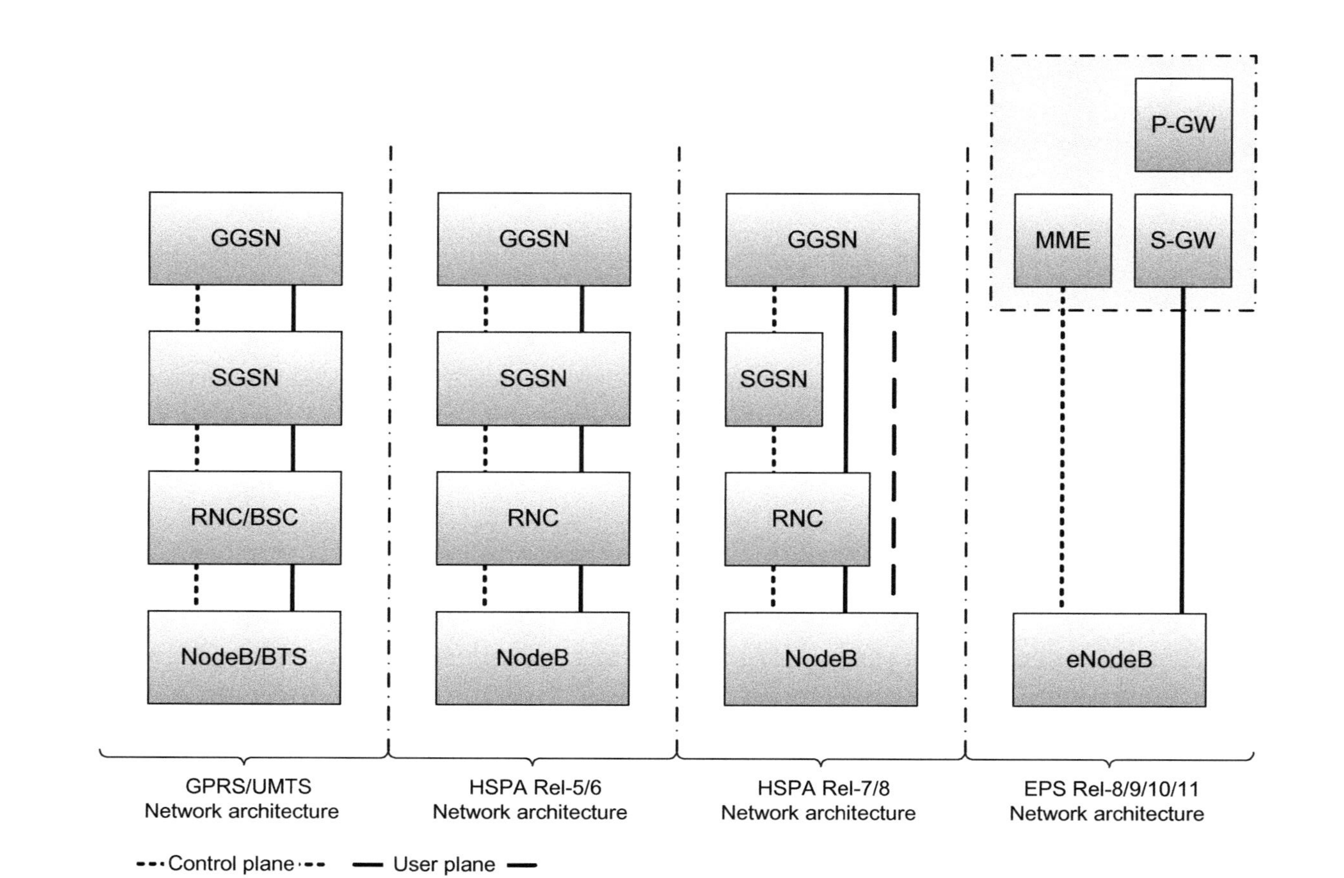

FIGURE 2.8

Evolution of 3GPP network architecture through different releases.

2.2 EPC architecture

2.2.1 Reference model and network interfaces

As mentioned earlier, the EPC is an all-IP mobile core network for the LTE radio access, providing a converged framework for IP-based real-time/non-real-time applications and services. The EPC provides the entire core-network functionality that, in previous mobile network generations, was realized through two separate domains: circuit-switched for voice and packet-switched for data. As shown in Figure 2.7, in LTE, these two distinct mobile core network domains used for separate processing and switching of voice and data are unified in a single IP-based domain. The EPC implementation and deployment is essential for end-to-end IP-based service delivery which complements the LTE radio access network. The EPC promotes the introduction of new services and the enablement of new applications [70,71].

Figure 2.9 shows the EPS reference model, which is a logical representation of the network architecture. The network reference model identifies the functional entities in the architecture and the reference points between the functional elements over which interoperability is achieved. The overall architecture has two distinct components: the access network and the core network. The access network is the E-UTRAN and the core network is known as the EPC. Voice services which were traditionally circuit-switched are handled using IMS infrastructure. Network complexity and latency are reduced as there are fewer nodes in both signaling and data paths. The EPC is further designed to support non-3GPP access for mobile IP. To improve system robustness, security, integrity protection, and ciphering have been added and represented by NAS, which is an additional layer of abstraction to protect important information, such as key and security interworking between 3GPP and non-3GPP networks. Apart from the network entities handling user-plane traffic, the EPC further includes network control entities for storing user subscription information represented by HSS, determining the identity and privileges of a user and tracking activities via authorization, authentication and accounting (AAA) server, and enforcing charging and QoS policies through PCRF. Note that E-UTRAN and EPC together constitute the EPS. The radio access and core networks perform several functions including network access control functions; packet

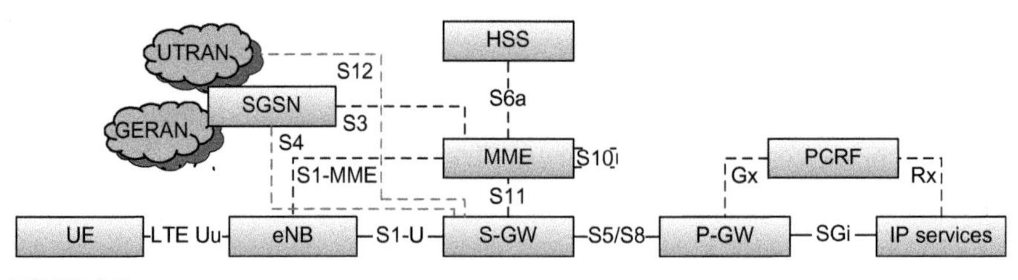

FIGURE 2.9

EPS reference model [6].

routing and transfer functions; mobility management functions; security functions; RRM functions; and network management functions.

Some general principles that were taken into consideration in the design of E-UTRAN architecture as well as the E-UTRAN interfaces are summarized as follows:

- Signaling and data transport networks are logically separated.
- E-UTRAN and EPC functions are fully separated from transport functions. The addressing schemes used in E-UTRAN and EPC are not associated with the addressing schemes of transport functions. The fact that some E-UTRAN or EPC functions may reside in the same physical equipment does not make the transport functions part of the E-UTRAN or the EPC.
- The mobility for RRC connection is fully controlled by the E-UTRAN.
- The interfaces are based on the logical model of the network entities.
- One physical network element can implement multiple logical nodes.

3GPP network architecture defines a reference point as a logical link that connects two groups of functions that reside in different functional entities of the E-UTRAN or EPC. Figure 2.9 illustrates the EPS network reference model and the most commonly used reference points specified by 3GPP as of Rel-11. These reference points are defined in Table 2.1.

2.2.2 Public land mobile network

A PLMN provides land mobile telecommunication services to the public which is established and operated by an administration or private operating agency. A PLMN may be referred to as an extension of networks, e.g., integrated services data digital network (ISDN), corporate, and public PDNs. The PLMN is a collection of mobile switching centers (MSCs) in circuit-switched domain and SGSNs in GPRS core network or SGSNs/MMEs in EPC, in packet-switched domain under a common numbering and routing plan. The MSCs are the functional interfaces between the fixed networks and a PLMN for call setup in circuit-switched domain. The GGSN and the SGSN are the functional interfaces between the fixed telephone networks and a PLMN for packet transmission in a GPRS packet-switched domain. In the case of an EPC packet-switched domain, the P-GW, S-GW, SGSN, and the MME are the functional interfaces between the fixed networks and a PLMN for packet transmission. From a functional point of view, the PLMNs may be considered as independent telecommunication entities even though different PLMNs may be interconnected through an ISDN or public switched telephone network (PSTN) and PDNs for the purpose of forwarding calls or network information. A similar type of interconnection may exist for interaction between the MSCs, SGSNs, or MMEs of a PLMN [1].

The PLMN infrastructure typically consists of a core network and an access network. The core network is logically divided into a circuit-switched domain, a packet-switched domain, and an IMS domain. The access network may include a

Table 2.1 EPS Reference Points [6,33]

Reference Point	Termination Points	Description
S1-MME	eNB and MME	It provides control-plane path between eNB and MME.
S1-U	eNB and S-GW	It enables per-bearer user-plane tunneling and inter-eNB path switching during handover. The S1-U is based on GTP-U protocol.[a]
S3	MME and SGSN	It enables user and bearer information exchange for inter-3GPP access network mobility in idle and/or active mode. This reference point can be used intra-PLMN or inter-PLMN in the case of inter-PLMN handover. The S3 is based on GTP protocol.
S4	S-GW and SGSN	It provides control and mobility support between GPRS core and the 3GPP anchor function of S-GW and is based on GTP protocol. If direct tunneling is not established, it provides the user-plane tunneling capability.
S5	S-GW and P-GW	It provides user-plane tunneling and tunnel management between S-GW and P-GW. It is used for S-GW relocation due to UE mobility and if the S-GW needs to connect to a non-collocated P-GW for PDN connectivity and is based on GTP protocol.
S6a	MME and HSS	It enables transfer of subscription and authentication data for authenticating/authorizing user access to the network services.
Gx	PCRF and P-GW	It provides transfer of QoS policy and charging rules from PCRF to PCEF in the P-GW.
S8	S-GW and P-GW	Inter-PLMN reference point providing user-plane and control-plane between the S-GW in the visited PLMN and the P-GW in the home PLMN. S8 is the inter-PLMN variant of S5 and is based on GTP protocol.
S9	Visited PCRF and home PCRF	It provides transfer of QoS PCC information between the home PCRF and the visited PCRF in order to support local breakout function.
S10	MME and MME	It is used for MME relocation and MME to MME information transfer. This reference point can be used in the form of intra-PLMN or inter-PLMN in the case of inter-PLMN handover.
S11	MME and S-GW	It provides the reference point between MME and S-GW.
S12	UTRAN and S-GW	It provides user-plane tunneling when direct tunnel is established.
S13	MME and EIR	It enables UE identity check procedure between MME and equipment identity register (EIR).[b]

(Continued)

Table 2.1 (Continued)

Reference Point	Termination Points	Description
SGi	P-GW and PDN	It provides the interface between the P-GW and the packet data network. Packet data network may be an operator external public or private packet data network or an intra-operator packet data network as a provision for IMS services.
Rx	PCRF and PDN	This reference point resides between the application function (AF) and the PCRF.

[a]*GPRS tunneling protocol (GTP) is a group of IP-based communications protocols used to carry GPRS within GSM, UMTS, and LTE networks [35]. In 3GPP architecture, GTP and PMIPv6-based interfaces are specified on various reference points. GTP can be decomposed into separate protocols, GTP-C, GTP-U, and GTP'. GTP-C is used within the GPRS core network for signaling between gateway GPRS support node (GGSN) and serving GPRS support node (SGSN). This allows the SGSN to activate a session on a user's behalf (PDP context activation), to deactivate the same session, to adjust QoS parameters, or to update a session for a subscriber joined from another SGSN. GTP-U is used for carrying user data within the GPRS core network and between the radio access network and the core network. The user data transported can be packets in any of IPv4, IPv6, or PPP formats. The primary bearer plane in LTE across the access and the core networks is the evolved GPRS tunneling protocol (eGTP). The user-plane variant of this protocol eGTP-U provides tunneling support for user data between the eNB and the EPC elements. The eGTP-C (GTPv2-C) protocol is responsible for creating, maintaining, and deleting tunnels on multiple Sx interfaces. It is used for control-plane path management, tunnel management, and mobility management. It also controls forwarding relocation messages, serving radio network subsystem (SRNS) context and creating forward tunnels during inter-LTE handovers [9].*

[b]*The EIR is often integrated to the HLR. The EIR maintains a list of mobile phones (identified by their IMEI) which are banned from the network or need to be monitored. This is designed to allow tracking of stolen mobile phones. The EIR is a database that contains information about the identity of the mobile equipment that prevents calls from stolen, unauthorized or defective mobile stations [51].*

single or multiple radio access networks, such as GERAN, UTRAN, or E-UTRAN. The core network, on the other hand, consists of the circuit-switched and packet-switched domains, where the latter includes GPRS core and EPC. These two domains differ in the way that they support user traffic and may include a number of common entities. A PLMN can implement only one domain or both domains. The circuit-switched domain refers to the set of core network entities providing circuit-switched type of connection for user applications and the related signaling. A circuit-switched connection is a connection for which dedicated network resources are allocated at the time of connection establishment and are released at the end of connection.

The packet-switched domain refers to the set of core network entities that provide a packet-switched type of connection for user applications and the related signaling. A packet-switched connection transports user information in the form of bit packets, where each packet can be routed independently from the others.

The network entities specific to the packet-switched domain are the GPRS support entities, i.e., SGSN and GGSN, and EPS-specific entities comprising P-GW, S-GW, and MME. The IMS domain encompasses those core network elements which enable provisioning and delivery of IP-based multimedia services, such as voice, audio, video, text, or chat over the packet-switched domain.

2.2.3 Packet data network gateway

The P-GW is the entity which terminates the SGi interface toward the packet data network. If a UE has multiple PDN connections, then the UE may be assigned to more than one P-GW. The P-GW provides PDN connectivity to GERAN/UTRAN capable UEs as well as E-UTRAN capable UEs using E-UTRAN, GERAN, or UTRAN access networks. The P-GW provides PDN connectivity to E-UTRAN capable UEs using E-UTRAN only over the S5/S8 interface. It may also provide PDN connectivity to UEs using non-3GPP access networks. The P-GW functions include per-user packet filtering; lawful interception; IP address allocation; transport-level packet marking in the uplink and downlink by setting the differentiated services code point based on the QCI of the associated EPS bearer; uplink/downlink service level charging, gating control, rate enforcement; uplink/downlink rate enforcement based on per access point name-aggregate maximum bit rate (APN-AMBR)[1] parameter; downlink rate enforcement based on the accumulated maximum bit rates of a set of SDFs with the same GBR QCI through rate policing/shaping; and dynamic host configuration protocol (DHCP)v4 (server and client) and DHCPv6 (client and server) functions.[2] In addition, the P-GW includes uplink/downlink bearer binding (bearer binding is the generic procedure for associating a bearer in the access network to an SDF) and uplink bearer binding verification functions for the GTP-based network interfaces. The P-GW functions also include user-plane anchor for mobility between 3GPP access and non-3GPP access.

---

[1]The APN-AMBR is a subscription parameter stored per APN in the HSS. It limits the aggregate bit rate that can be provided across all non-GBR bearers and across all PDN connections of the same APN (e.g., excess traffic may be discarded by a rate shaping function). Each of those non-GBR bearers could potentially utilize the entire APN-AMBR, for example, when the other non-GBR bearers do not carry any traffic. The GBR bearers are outside the scope of APN-AMBR. The P-GW enforces the APN-AMBR in downlink. Enforcement of APN-AMBR in uplink is done in the UE and additionally in the P-GW [1].

[2]DHCP is a network protocol that is used to configure devices that are connected to a network so they can communicate on that network using the Internet protocol. It involves clients and a server operating in a client-server model. The clients request configuration settings using the DHCP protocol, such as an IP address, a default route, and one or more domain name system (DNS) server addresses. Once the client implements these settings, the host is able to communicate on the Internet. The DHCP server maintains a database of available IP addresses and configuration information. When the server receives a request from a client, the DHCP server determines the network to which the DHCP client is connected, and then allocates an IP address or prefix that is appropriate for the client, and sends configuration information appropriate for that client. DHCP servers typically grant IP addresses to clients only for a limited interval. DHCP is used for IPv4 and IPv6 [27,74].

2.2.4 Serving gateway

The S-GW is the entity which terminates the interface toward E-UTRAN access network. For each UE associated with the EPS, at any given point of time, there is a single S-GW assigned. The functions of the S-GW include a local mobility anchor point for inter-eNB handover; mobility anchor point for inter-3GPP mobility; EPS connection management (ECM)-IDLE mode downlink packet buffering and initiation of network-triggered service request procedure; lawful interception; packet routing and forwarding; transport-level packet marking in the uplink and downlink; and accounting for interoperator charging; as well as local non-3GPP anchor for the case of roaming when the non-3GPP IP access network is connected to the visited PLMN; event reporting (change of RAT) to the PCRF; uplink and downlink bearer binding toward 3GPP access; uplink bearer binding verification with packet dropping of misbehaving uplink traffic; mobile access gateway (MAG) functions, if proxy mobile IP (PMIP)[3] based S5 or S8 is used; and support of necessary functions for enabling GTP/PMIP chaining functions [6,7].

2.2.5 Mobility management entity

The MME is the main control node that processes the NAS signaling between the UE and the core network which is responsible for idle-mode UE tracking and paging procedure as well as bearer establishment and release [10]. The main functions supported by the MME can be classified into two major categories: (1) functions related to bearer management including the establishment, maintenance, and release of the bearers which are handled by the session management function in the NAS protocol; and (2) functions related to connection management including the establishment of the connection and security association (SA) between the network and UE which is handled by the connection or mobility management function in the NAS protocol [11]. The MME is a control-plane entity within EPS which performs mobility management functions including NAS signaling and security; inter-node signaling for mobility between 3GPP access networks; tracking area list management; P-GW and S-GW selection; SGSN selection for handovers to 2G/3G access networks; roaming; and authentication. The MME's bearer management functions include dedicated bearer establishment. In order to enable 3GPP2 access (in a multi-RAT network), the MME should support high rate packet data (HRPD) access node selection and maintenance for handovers to HRPD and transparent transfer of HRPD signaling messages and transfer of status information between E-UTRAN and HRPD access networks [1].

[3]Network-based mobility management enables IP mobility for a host without requiring its participation in any mobility-related signaling. The network is responsible for managing IP mobility on behalf of the host. The mobility entities in the network are responsible for tracking the movements of the host and initiating the required mobility signaling on its behalf. IETF RFC 5213 specifies a network-based mobility management protocol that is referred to as proxy mobile IPv6 [26].

The MME creates a UE context when a UE is powered on and attaches to the network. It assigns a unique short temporary identity to the UE that identifies the UE context in the MME. This UE context holds user subscription information downloaded from the HSS. The local storage of subscription data in the MME allows faster execution of procedures, such as bearer establishment since it relaxes the need to frequently exchange messages with the HSS. In addition, the UE context also holds dynamic information, such as the list of bearers that are established and the terminal capabilities [70,71].

To reduce the overhead in the E-UTRAN and processing complexity in the UE, all UE-related information in the access network including the radio bearers can be released during long periods of user inactivity, which is known as ECM-IDLE state. The MME retains the UE context and the information about the established bearers during these idle periods. To allow the network to contact an idle-mode UE, the UE provides the network with location update whenever it moves out of its current tracking area using a tracking area update procedure. The MME is responsible for tracking user location while the UE is in the ECM-IDLE state. Upon availability of downlink data for an ECM-IDLE UE, the MME sends a paging message to all eNBs in the tracking area where the UE was recently located and the eNBs page the UE over the radio interface. Upon receipt of the paging message, the UE performs a service request procedure, which results in transitioning to the ECM-CONNECTED state. The UE-related information is thereby created in the E-UTRAN and the radio bearers are reestablished. The MME is responsible for the reestablishment of the radio bearers and updating the UE context in the eNB (see Chapters 4 and 5 for more details).

Security functions are the responsibility of the MME for both signaling and user data. When a UE attaches to the network, a mutual authentication of the UE and the network is performed between the UE and the MME/HSS. This authentication procedure establishes the security keys that are used for encryption of the bearers.

2.2.6 Policy and charging rules function

The PCRF acts as a policy decision point for policy and charging control (PCC) of SDFs/applications and IP bearer resources. It was introduced as an improvement in 3GPP Rel-7 network architecture in terms of policy and charging to allow optimization of interactions between the policy and rules functions. This improvement involved a new network node which integrates policy decision function (PDF) and charging rules function (CRF) in the previous releases. 3GPP Rel-8 further enhanced PCRF functionality by expanding the scope of PCC framework to facilitate non-3GPP access to the network (e.g., Wi-Fi or fixed broadband access). In the 3GPP generic PCC model, policy and charging enforcement function (PCEF), which resides in the P-GW, is the functional entity that supports SDF detection, policy enforcement, and flow-based charging. The PCRF provides the QoS authorization (QCI and bit rates) that decides how a certain data flow will be treated in PCEF and ensures consistency with the user's subscription profile. The application function (AF) represents the network element that supports applications that require dynamic policy and/or charging control.

2.2.7 Home subscriber server

The home subscriber server is the subscription data repository for all permanent user data. The HSS stores the master copy of the subscriber profile, which contains information about the services that are applicable to the user. It also records the location of the user at the level of visited network control node, such as MME. It is the entity containing the subscription-related information to support the network entities that handle calls/sessions. A home network may contain one or several HSSs depending on the number of mobile subscribers, the capacity of the equipment, and the organization of the network. As an example, the HSS provides support to the call control servers in order to complete the routing/roaming procedures by solving authentication, authorization, naming/addressing resolution, location dependencies, etc. The HSS is responsible for holding information related to user identification, numbering and addressing information; user security information; network access control information for authentication and authorization; and user location information at the inter-system level. The HSS supports the user registration and stores inter-system location information as well as the user profile information. The HSS also generates user security information for mutual authentication, communication integrity check, and ciphering. Based on this information, the HSS is also responsible for supporting the call control and session management entities of the different domains and subsystems [1].

2.2.8 Authorization, authentication, and accounting

Trusted and untrusted non-3GPP access networks are IP-based access networks which may use access technologies whose specifications are out of the scope of 3GPP. Whether a non-3GPP access network is trusted or untrusted is not a characteristic of the access network. In non-roaming scenario, it is the home PLMN's operator decision if a non-3GPP IP-based access network is treated as trusted or untrusted. In a roaming scenario, the HSS/3GPP AAA server in home PLMN determines whether a non-3GPP IP access network is viewed as trusted or untrusted. The HSS/3GPP AAA server may consider and utilize the visited PLMN's policy and capability returned from the 3GPP AAA proxy or roaming agreement.

The 3GPP AAA server is located in the home PLMN and provides user support for non-3GPP access with authentication, authorization, and location management in order to allow access to the EPS. It also contains necessary user-related information in order to grant access to a non-3GPP network, and further coordinates the information needed to support mobility between 3GPP and non-3GPP access networks, such as coordination of P-GW information. It interacts with HSS to maintain consistent information for users supporting mobility and service continuity between 3GPP and non-3GPP networks. The 3GPP AAA proxy provides support for roaming non-3GPP access users in the visited PLMN with the necessary authentication, authorization, and location management services in order to get access to the EPS. It may also provide roaming-related information for support of chaining scenarios. If an S-GW is needed for non-3GPP access in the visited network, the 3GPP AAA proxy selects an S-GW for the UE during initial attach or handover process [2].

2.2.9 EPC identifiers

Each entity and bearer in E-UTRAN and EPC is identified with a unique identifier. The identifiers are either permanently provisioned, such as the IMSI and IMEI identities that are assigned by an operator to the UE or they are assigned during the lifetime of UE operation in an operator's network. Figure 2.10 illustrates the identities that are either provisioned or dynamically/semi-statically assigned to the bearers and the entities in EPS. These identities are described in Table 2.2 and the next chapters of this book and further specified by 3GPP in 3GPP TS 23.003 specification [2]. The direction of the arrows in the figure indicates the entity which assigns the identifier and the entity to which the identifier is assigned. The bearers and various bearer identifiers depending on the network interface and the flow direction are also shown in the figure.

Note that the protocols over LTE-Uu and S1 interfaces are divided into two categories: user-plane protocols, which are the protocols implementing the actual E-RAB service carrying user data through the access stratum (AS); and control-plane protocols that are the protocols for controlling the E-RABs and the connection between the UE and the network from different aspects including requesting the service, controlling different transmission resources, handover, etc. Furthermore, a mechanism for transparent transfer of NAS messages is included.

2.3 Roaming architecture and interworking with other networks

In cellular communications, roaming is a general term referring to the extension of service continuity and availability in a location that is different from the home location where the service and the subscriber were registered. Roaming functionality ensures that the wireless device is connected to the network without connection interruption while crossing the PLMN boundaries. The roaming is defined as the ability for a cellular user to automatically make or receive voice calls; send or receive data; or access other services including home data services, when moving outside the geographical coverage area of the home network using a visited network. The roaming feature is technically supported by mobility management,

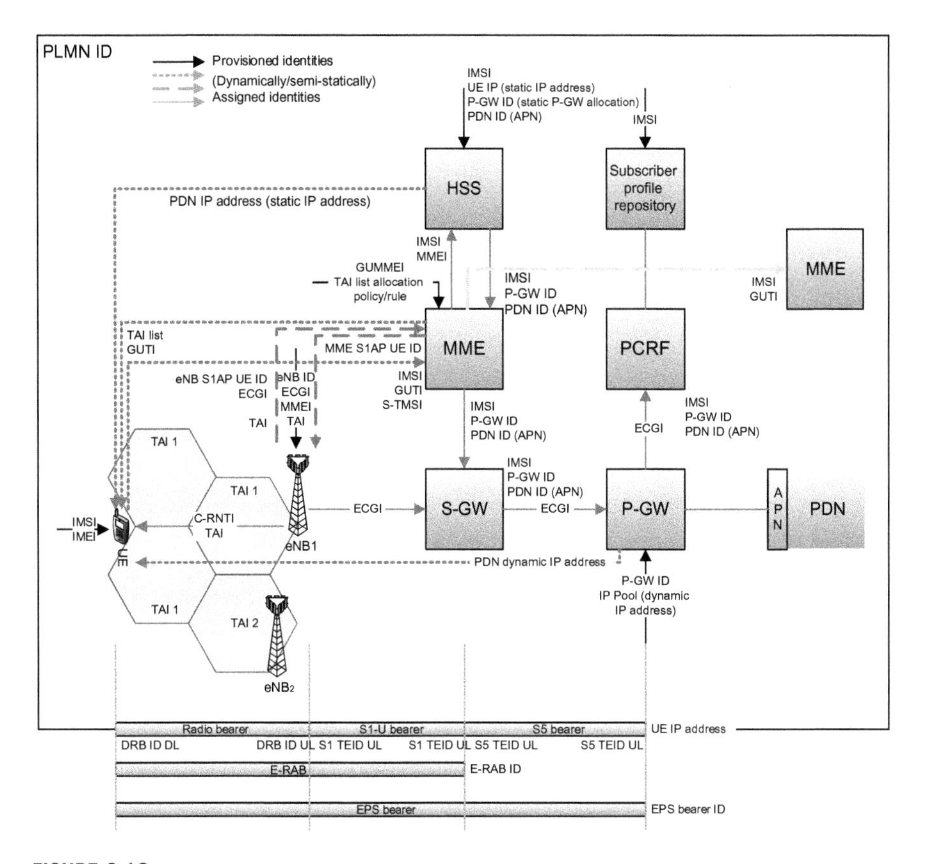

FIGURE 2.10

EPS entities (bearers) and corresponding provisioned/assigned identities [40].

authentication, authorization, and billing functions in the network. While a roaming user is connected to the E-UTRAN, MME, and S-GW of the visited LTE network, the EPS allows the P-GW of either the visited or the home network to be used, as shown in Figure 2.11. Using the home network's P-GW allows the user to access the home operator's services while in a visited network. A P-GW in the visited network allows a local breakout to the Internet in the visited network [70,71].

The EPS also supports interworking with and handover to/from other networks using 3GPP/non-3GPP radio access technologies, such as GSM, UMTS, and CDMA2000. The architecture for interworking with 2G/3G GPRS/UMTS networks is shown in Figure 2.12. The S-GW acts as the mobility anchor for interworking with other 3GPP technologies, such as GSM and UMTS, while the P-GW serves as an anchor allowing seamless mobility to non-3GPP networks, such as CDMA2000 or mobile WiMAX. The P-GW may also support a PMIP-based interface.

Table 2.2 Description of EPC and E-UTRAN Identities [40,41]

Abbreviation	Identifier	Description	Value/Structure	Permanent/ Temporary
IMSI	International mobile subscriber identity	Unique identification of mobile subscriber in the network. The IMSI is composed of three parts: (1) mobile country code (MCC) consisting of three digits. The MCC uniquely identifies the country of residence of the mobile subscriber; (2) mobile network code (MNC) consisting of two or three digits for GSM/UMTS applications. The MNC identifies the home PLMN of the mobile subscriber. The length of the MNC (two or three digits) depends on the value of the MCC. A mixture of two and three digit MNC codes within a single MCC area is not recommended and is outside the scope of this specification; and (3) mobile subscriber identification number (MSIN) identifying the mobile subscriber within a PLMN.	IMSI (not more than 15 digits) = PLMN ID + MSIN = MCC + MNC + MSIN	Permanent
PLMN ID	Public land mobile network identifier	Unique identification of PLMN	PLMN ID (not more than 6 digits) = MCC + MNC	Permanent
MCC	Mobile country code	Assigned by ITU	3 digits	Permanent
MNC	Mobile network code	Assigned by national authority	2 or 3 digits	Permanent
MSIN	Mobile subscriber identification number	Assigned by operator	9 or 10 digits	Permanent

GUTI	Globally unique temporary UE identity	Identifies a UE (between the UE and the MME) on behalf of IMSI for security reasons. The purpose of the GUTI is to provide an unambiguous identification of the UE that does not reveal the UE or the user's permanent identity in the EPS. It also allows the identification of the MME and network. It can be used by the network and the UE to establish the UE's identity during signaling in the EPS.	GUTI (not more than 80 bits) = GUMMEI + M-TMSI	Temporary
TIN	Temporary identity used in next update	GUTI is stored in TIN parameter of UE's context. TIN indicates which temporary ID will be used in the next update	TIN = GUTI	Temporary
S-TMSI	SAE temporary mobile subscriber identity	To locally identify a UE in short-form within an MME group (unique within an MME pool). The S-TMSI is the shortened form of the GUTI to enable more efficient radio signaling procedures (e.g., paging and service request). For paging purposes, the mobile is paged with the S-TMSI.	S-TMSI (40 bits) = MMEC + M-TMSI	Temporary
M-TMSI	MME mobile subscriber identity	Unique within an MME Since the TMSI has only local significance (i.e., within a VLR and the area controlled by a VLR, or within an SGSN and the area controlled by an SGSN, or within an MME and the area controlled by an MME), the structure and coding of it can be chosen by agreement between operator and manufacturer in order to meet local needs.	32 bits	Temporary
GUMMEI	Globally unique MME identity	To identify an MME uniquely in a global sense.	GUMMEI (not more than 48 bits) = PLMN ID + MMEI	Permanent
MMEI	MME identifier	To identify an MME uniquely within a PLMN	MMEI (24 bits) = MMEGI + MMEC	Permanent

(*Continued*)

Table 2.2 (Continued)

Abbreviation	Identifier	Description	Value/Structure	Permanent/ Temporary
MMEGI	MME group identifier	Unique identification of an MME group within a PLMN	16 bits	Permanent
MMEC	MME code	To identify an MME uniquely within an MME group	8 bits	Permanent
IMEI	International mobile equipment identity	To uniquely identify a mobile equipment	IMEI (15 digits) = type approval code (6 digits) + serial number (6 digits) + final assembly code (2 digits) + check digit (1 digit)	Permanent
IMEI/SV	IMEI/software version	Identifies the mobile equipment serial number, hardware model number, and software version number	IMEI/SV (16 digits) = type approval code (6 digits) + serial number (6 digits) + software version (2 digits) + final assembly code (2 digits)	Permanent
ECGI	E-UTRAN cell global identifier	To identify a cell in general (globally unique) EPC can know UE location based on ECGI	ECGI (not more than 52 bits) = PLMN ID + ECI	Permanent
ECI	E-UTRAN cell identifier	To identify a cell within a PLMN	ECI (28 bits) = eNB ID + cell ID	Permanent
Global eNB ID	Global eNB identifier	To globally identify an eNB	PLMN ID + eNB ID	Permanent
eNB ID	eNB identifier	To identify an eNB within a PLMN	20 bits	Permanent
P-GW ID	PDN GW identity	To identify a specific P-GW HSS assigns P-GW for PDN (IP network) connection of each UE	IP address (4 bytes) or fully qualified domain name (variable length)	Permanent
TAI	Tracking area identity	This is the identity used to uniquely identify the tracking areas. The tracking area identity is constructed from the MCC (mobile country code), MNC (mobile network code) and TAC (tracking area code).	TAI (not more than 32 bits) = PLMN ID + TAC P-GW	Permanent

TAC	Tracking area code	To indicate to which tracking area the eNB belongs (per cell) which is unique within a PLMN	16 bits	Permanent
TAI list	Tracking area identity list	UE can move into the cells included in TAI list without location update (TA update) which is globally unique	Variable length	Permanent
PDN ID	Packet data network identity	To identify a PDN (IP network) with which mobile data user communicates PDN identity (APN) is used to determine the P-GW and point of interconnection with a PDN With APN as query parameter to the DNS procedures, the MME will receive a list of candidate P-GWs, and then a P-GW is selected by MME	PDN identity = APN = APN-NI + APN-OI (variable length) APN-NI: APN network identifier APN-OI: APN operator identifier	Permanent
CSG ID	Closed subscriber group ID	A closed subscriber group (CSG) consists of a single cell or a collection of cells within the E-UTRAN and UTRAN that are open to only a certain group of subscribers. A CSG ID is a unique identifier within the scope of PLMN which identifies a CSG in the PLMN associated with a CSG cell or group of CSG cells.	27 bits	Permanent
EPS bearer ID	Evolved packet system bearer identifier	To identify an EPS bearer (default or dedicated) per UE. An EPS bearer identity uniquely identifies an EPS bearer for one UE accessing via E-UTRAN. The EPS bearer identity is allocated by the MME. There is a one-to-one mapping between EPS radio bearer and EPS bearer, and the mapping between EPS radio bearer identity and EPS bearer identity is made by E-UTRAN. The E-RAB ID value, used over S1 and X2 interfaces to identify an E-RAB, is the same as the EPS bearer ID value used to identify the associated EPS bearer.	4 bits	Temporary

(Continued)

Table 2.2 (Continued)

Abbreviation	Identifier	Description	Value/Structure	Permanent/ Temporary
E-RAB ID	E-UTRAN radio access bearer identifier	To identify an E-RAB per UE	4 bits	Temporary
DRB ID	Data radio bearer identifier	To identify a DRB per UE	4 bits	Temporary
LBI	Linked EPS bearer ID	To identify the default bearer associated with a dedicated EPS bearer	4 bits	Temporary
TEID	Tunnel endpoint identifier	To identify the endpoint of a GTP tunnel when the tunnel is established	32 bits	Temporary
eNB S1AP UE ID	eNB S1AP UE identity	This is a temporary identity used to identify a UE over the S1-MME reference point within the eNB and is unique within the eNB.	32 bits	Temporary
MME S1AP UE ID	MME S1AP UE identity	This is a temporary identity used to identify a UE on the S1-MME reference point within the MME and is unique within the MME.	32 bits	Temporary
C-RNTI	Cell-radio network temporary identifier	To identify a UE uniquely in a cell	$0 \times 0001 \sim 0 \times FFF3$ (16 bits)	Temporary

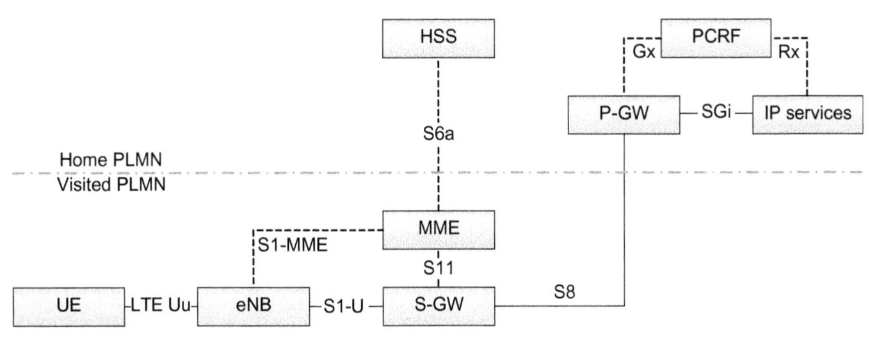

FIGURE 2.11

Example roaming architecture [1,70,71].

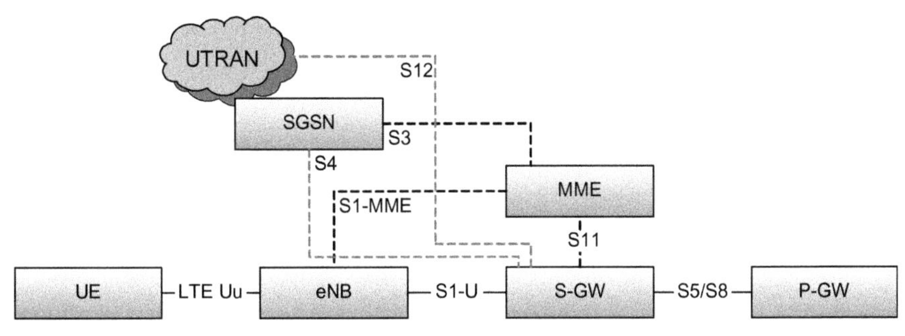

FIGURE 2.12

Interworking of LTE with UMTS network architecture [70,71].

2.4 E-UTRAN architecture

The E-UTRAN access network consists of a network of eNBs, as illustrated in Figure 2.13. For unicast user traffic, there is no centralized controller in the E-UTRAN; thus, the E-UTRAN architecture is considered as a distributed architecture. The eNBs are typically interconnected with each other by means of a logical interface known as X2 and to the EPC by means of the S1 interface. More specifically, the eNB is connected to the MME through S1-MME interface and to the S-GW via S1-U interface. The protocols between the eNBs and the UE are referred to as the AS protocols. The protocols between the EPC and the UE are known as NAS protocols. The E-UTRAN is responsible for all radio-related functions that include RRM, encompassing all functions related to the radio bearers, such as radio bearer control (RBC), radio admission control (RAC), radio mobility control, scheduling and dynamic allocation of radio resources to UEs in both uplink and downlink; header compression, ensuring efficient use of the radio interface by compressing the IP packet headers that could otherwise impose a

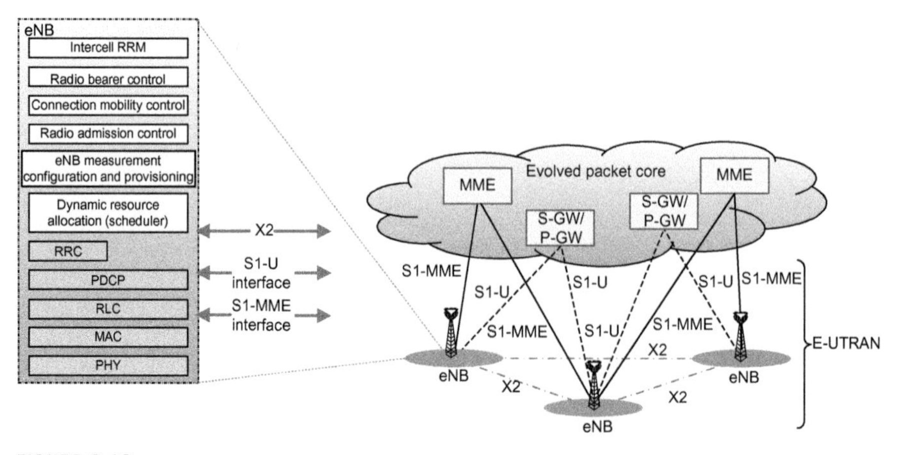

FIGURE 2.13

E-UTRAN architecture and eNB functional decomposition [12].

significant overhead especially for small-packet applications, such as VoIP; security, encrypting all data and control messages sent over the radio interface; and connectivity to the EPC, consisting of the signaling toward MME and the bearer path toward the S-GW [70,71].

As shown in Figure 2.13, on the network side, all of these functions reside in the eNBs, each of which can be responsible for managing multiple cells. Unlike previous 2G/3G technologies, LTE has integrated the radio controller function into the eNB. This allows close interaction between different protocol layers of the radio access network, thus reducing latency and improving efficiency. Such distributed control structure eliminates the need for a processing-intensive central controller, which in turn has the potential to reduce costs, overhead, latency, and complexity. Given that LTE does not support soft handover, there is no need for a centralized data-combining function in the network. One consequence of the absence of a centralized controller node is that, as the UE moves, the network must transfer all information related to a UE (the UE context) together with any buffered data from one eNB to another. Therefore, there are mechanisms in EPS that are provisioned to avoid data loss during handover (e.g., data forwarding).

An important feature of the S1 interface linking the access network to the core network is known as S1-flex [6]. This is a concept whereby multiple core network nodes (MMEs/S-GWs) can serve a common geographical area, being connected by a mesh network to the set of eNBs in that area. An eNB may be served by multiple MMEs/S-GWs. The set of MME/S-GW nodes that serve a common area is called an MME/S-GW pool, and the area covered by such a pool of MMEs/S-GWs is called a pool area. This concept allows the UEs in the cell or cells controlled by one eNB to be shared among multiple core network nodes, thereby providing a possibility for load balancing (LB) and also eliminating a single point of failure for the core network nodes. The UE context typically remains in the same MME as long as the UE is located within the same pool area.

2.4.1 E-UTRAN functions

The functions performed and the services provided by an eNB include RRM (RBC, RAC, connection mobility control (CMC), dynamic allocation and scheduling of resources for the UEs in the uplink and downlink); IP header compression and encryption of user data stream; selection of an MME upon UE attachment to the network when no routing to an MME can be determined from the information provided by the UE; routing of user-plane data toward S-GW; scheduling and transmission of paging messages originated from the MME; scheduling and transmission of broadcast information originated from the MME or network operation and management; measurement and measurement reporting configuration for mobility and scheduling; scheduling and transmission of emergency warning messages originated from the MME; closed-subscriber group (femto-cell) handling; and transport-level packet marking in the uplink. A relay-enabled eNB further performs S1/X2 proxy functionality for supporting the relay nodes (RNs) and S11 termination and S-GW/P-GW functionality for supporting the relay nodes [12]. The functional decomposition of an eNB is shown in Figure 2.13.

2.4.2 E-UTRAN interfaces

This section provides general information about the E-UTRAN interfaces (S1 and X2). The E-UTRAN network interface model consists of two main parts: the transport network layer, corresponding to the transport of the access-network data, and the radio network layer, which encompasses the top-level protocols of the radio interface. In addition, each interface has a user-plane and a control-plane component. The user-plane carries all user data as well as application layer signaling, such as session initiation protocol (SIP) or real-time transport protocol (RTP) control protocol (RTCP) packets [22]. The control-plane handles all messages and procedures related to the radio interface supported features. An example of these messages includes control messages for handover management or bearer establishment. The physical layer is common to both user-plane and control-plane. The user-plane and control-plane use specific protocols which allow definition of a different and independent transport stack and bearers for each of the planes.

The control-plane information is more susceptible to security, reliability, and information loss, whereas the user-plane information can rely on simpler and less secured routing protocols. The E-UTRAN network interfaces are open standards defined by 3GPP. This conformity to open standards allows different vendors to manufacture eNBs, deploy them in a single network, and interconnect them over the X2 interface or with MME or S-GW nodes over the S1 interface.

The S1 interface connects the eNB to the EPC. It can be split into control-plane (S1-MME) and user-plane (S1-U). The S1-MME is a signaling interface which supports a set of functions and procedures between the eNB and the MME. All S1-MME signaling procedures belong to four main groups: (1) bearer-level procedures, corresponding to all procedures related to bearer setup, modification and release, which are typically used during the establishment or the release of a session; (2) handover procedures, which encompass all S1 functions related to user mobility between eNB or with 2G/3G technologies; (3) NAS signaling transport, corresponding to the transport of UE-MME NAS signaling over the S1 interface; and (4) paging procedure, which is used in case of a mobile-terminated call/session. Through the paging procedure, the MME requests the eNB to page a UE in a given set of cells known as a tracking area. The S1-MME interface provides a high level of reliability in order to avoid message retransmission and unnecessary delay in control-plane procedures using stream control transmission protocol (SCTP) [20].

The role of S1-U is to transport user data packets between the eNB and the S-GW. This interface makes use of a very simple GTP over user datagram protocol (UDP)/IP transport protocol which only provides user data encapsulation. There is no flow control or error control, or any mechanism to guarantee data delivery over the S1-U interface. The GTP was inherited from GPRS and UMTS networks. In GPRS networks, GTP is used between the SGSN and the GGSN. In UMTS networks, GTP is also used over the Iu-ps interface between the RNC and the SGSN. S1 application part (S1AP) is the control-plane signaling protocol between the eNB and the MME [13]. It supports the S1-MME interface and utilizes SCTP in the transport layer. The SCTP protocol is for the control-plane, which guarantees delivery of control/signaling messages between the MME and the eNB.

The X2 interface connects multiple eNBs. LTE uses the same protocol structure over both S1 and X2 interfaces, which simplifies the data forwarding operation. The X2 interface has a control-plane (X2-C) and a user-plane (X2-U) component. The X2-U is responsible for the transport of user data packets between eNBs. This interface is only used for a limited period of time, when the terminal moves from the coverage area of one eNB to another and provides buffered packet data forwarding. The X2-U interface utilizes GTP tunneling protocol that is used over the S1-U interface. The X2-C is a signaling interface which supports a set of functions and procedures between eNBs. The X2-C procedures are very limited in number and are all related to user mobility between eNBs, exchange of UE context between nodes including allocated bearers, security information, etc. The need for reliable transport of signaling between nodes results in the use of SCTP protocol for X2-C. X2AP is the control-plane protocol between eNBs over the X2 interface and supports load management and handover coordination between eNBs.

2.4.3 E-UTRAN identifiers

The E-UTRAN (dynamically) assigns unique identities to the UEs at the cell level. These identities temporarily identify the UE resource allocations and assignments as well as UE-specific control signaling during the active state of the UE in the cell. These identities include C-RNTI, which is a unique identification used for identifying RRC connection and scheduling; semi-persistent scheduling C-RNTI that is a unique identification used for semi-persistent scheduling; temporary C-RNTI, TPC-PUSCH-RNTI that is an identification used for the power control of physical uplink shared channel; TPC-PUCCH-RNTI, which is an identification used for the power control of physical uplink control channel; and random access RNTI for random access procedure and contention resolution. During some transient intervals, the UE is temporarily identified with a random value used for contention resolution purposes.

As shown in Figure 2.11, the E-UTRAN further utilizes unique identities for identifying a specific network entity which include GUMMEI that is used to globally identify the MME. The GUMMEI is constructed from the PLMN identity to which the MME belongs, the group identity of the MME group where the MME is part of, and the MMEC of the MME within the MME group. Note that the GUMMEI or S-TMSI containing the MMEC is provided by the UE to the eNB. The E-UTRAN cell global identifier (ECGI) is used to globally identify the cells. The ECGI is constructed from the PLMN identity where the cell belongs to and the cell identity (CI) of the cell. The PLMN information is periodically broadcasted as part of system information. The eNB identifier (eNB ID) is used to identify eNBs within a PLMN and is contained within the CI of its cells. The global eNB ID is used to globally identify the eNBs, and is constructed from the PLMN identity to which the eNB belongs and the eNB ID. The MCC and MNC are the same as that included in the ECGI (see Table 2.2).

In addition to the above identities, there are other identifiers related to the support of home eNBs (HeNBs) or femto-cells and relays. The global eNB ID of a relay node is the same as its serving donor eNB (DeNB), which is a relay-enabled eNB. The tracking area identity is used to identify tracking areas and is constructed from the PLMN identity where the tracking area belongs to, and the TAC of the tracking area. The CSG identity (CSG ID) is an identifier of a CSG femto-cell within a PLMN.

As shown in Figure 2.11, there are unique identifiers that are used to identify the data and signaling radio bearers (SRBs) as well as S1 and S5/S8 bearers on the access network that are assigned to UE services. The value of the E-RAB ID used on S1 and X2 interfaces to identify an E-RAB allocated to the UE is the same as the EPS bearer ID value used at the LTE-Uu interface to identify the associated EPS bearer and the NAS layer. The CI, TAC, CSG ID (if any), and one or more PLMN IDs are broadcast periodically in every E-UTRAN cell as part of system information.

2.4.4 Support of HeNB in E-UTRAN

Small cells are low-power wireless access points that operate in licensed spectrum and are managed by an operator. These cells provide improved cellular coverage and capacity and have applications for homes and enterprises as well as metropolitan public areas. There are various types of small cells including femto-cells, pico-cells, metro-cells and micro-cells, broadly increasing in size from femto-cells (the smallest) to micro-cells (the largest). Small cells also facilitate a new class of mobile services that exploit the technology's ability to detect the presence of, and connect and interact with existing networks [42,56,60].

As shown in Figure 2.14, the E-UTRAN architecture supports deployment of a home eNB gateway (HeNB-GW) to allow an S1 interface between the HeNB and the EPC in order to support a large number of HeNBs in a scalable manner. The HeNB-GW serves as a concentrator for the control-plane, specifically the S1-MME interface. The S1-U interface from the HeNB may be terminated at the HeNB-GW or a direct logical user-plane connection between HeNB and S-GW may be used. Depending on the deployment scenario, the S1 interface in this case is defined as the interface between the HeNB-GW and the core network; between the HeNB and the HeNB-GW; between the HeNB and the core network; or between the eNB and the core network. The HeNB-GW appears to the MME as an eNB, and to the HeNB as an MME. The S1 interface between the HeNB and the EPC is the same, regardless of the HeNB connection to the EPC via a HeNB-GW [12].

The HeNB-GW connects to the EPC such that the inbound and outbound mobility to cells served by the HeNB-GW does not necessarily require inter-MME handovers. One HeNB serves only one cell. In general, the functions supported by the HeNB are the same as those supported by an eNB and the procedures activated between a HeNB and the EPC are similar to those between an eNB and the EPC. Therefore, it can be concluded that HeNB is a (reduced-function and reduced-cost)

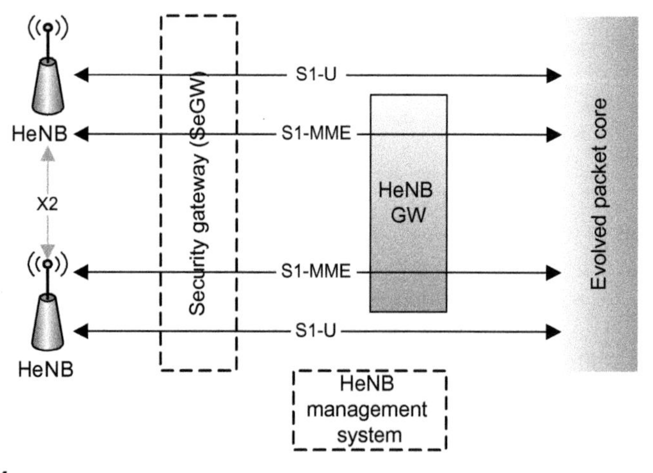

FIGURE 2.14

HeNB logical network architecture [12].

eNB that is installed by the user and not the operator, but is connected to operator's network in the same way that a regular eNB is connected.

3GPP Rel-11 specifications support direct X2-connectivity between HeNBs, independent of whether any of the involved HeNBs are connected to a HeNB-GW [12].

2.4.5 Support of RRHs in E-UTRAN

One recent deployment trend that is foreseen to be widely applied to LTE networks consists of splitting the base station functionalities into a baseband unit, which performs, in particular, the scheduling and the baseband processing functions, and a number of RRHs responsible for all RF transmission/reception aspects. The baseband processing unit is typically located at the center of the cell and is connected via optical fiber to the RRHs. In addition to suppressing the feeder loss (since the power amplifier is immediately close to the antenna), this approach allows a baseband processing unit to manage different radio sites in a central manner. Furthermore, having geographically separated RRHs controlled from the same location enables either centralized baseband processing units jointly managing the operation of several radio sites, or the exchange of very low-latency coordination messages among individual baseband processing units.

Distributed base stations with RRHs can significantly assist mobile operators to overcome cost, performance, and efficiency issues when addressing increasing demand for higher capacities in cellular networks. Multimode radios operating according to GSM, HSPA, LTE, and Wi-Fi standards and advanced software configurability are the key features in the deployment of more flexible and energy-efficient radio networks. Wireless and mobile network operators face the continuing challenge of upgrading their networks to effectively manage high user traffic rates. Mobility and an increased level of multimedia content for end users require end-to-end network adaptations that support both new services and the increased demand for broadband and flat-rate Internet access. In addition, network operators must consider the most cost-effective evolution of the networks toward new generations. Wireless and mobile technology standards are evolving toward higher bandwidth requirements for both peak data rates and cell throughputs. The latest standards HSPA +, mobile WiMAX, and LTE/LTE-Advanced can barely satisfy today's demand for high-performance mobile applications and fast and reliable access to the Internet anywhere and anytime. The network upgrades required to deploy networks based on these standards must balance the limited availability of new spectrum, leverage existing spectrum, and ensure proper operation of the desired wireless standards.

Distributed open base station architecture concepts have evolved in parallel with the evolution of the standards to provide a flexible, low-cost, and scalable modular environment for managing the radio access. For example, the open base station architecture initiative (OBSAI)[4] and the common public radio interface (CPRI)[5] standards introduced standardized interfaces separating the base station server and the RRH part of a base station by an optical fiber.

The RRH concept constitutes a fundamental part of a state-of-the-art base station architecture. The RRH-based system implementation is driven by the need to reduce both capital expenditure (CAPEX) and operational expenditure (OPEX) consistently, which allows an optimized and energy-efficient deployment. Figure 2.15 illustrates an architecture where 2G/3G/4G base stations are connected to RRHs over optical fibers. Either CPRI or OBSAI may be used to carry RF signals to the RRH to cover a three-sector cell. The RRH incorporates a large number of digital interfacing and processing functions. It also includes high-performance, efficient, and frequency-agile analog functions, all packaged into a light device with a small mechanical footprint.

2.4.6 Support of relays in E-UTRAN

In general, RNs can be used to improve urban or indoor user throughput as well as to extend coverage in rural areas. The concept of relaying is not new but the level of sophistication continues to be enhanced. Figure 2.16 shows a typical scenario where an RN is connected wirelessly to the radio access network via a donor cell. The RN can connect to the Donor cell's eNB (DeNB) in one of two ways: in-band (in-channel), in which case the DeNB-to-RN link shares the same carrier frequency with RN-to-UE links, and out-band, in which case the DeNB-to-RN link does not operate on the same carrier frequency as RN-to-UE links. The most basic relay method is the use of a radio repeater, which receives, amplifies, and then retransmits the downlink and uplink signals to overcome areas of poor coverage. The repeater

---

[4]The OBSAI family of specifications provides the architecture, functional description, and minimum requirements for integration of a set of common modules into a base transceiver station. It defines an open, standardized internal modular structure of wireless base stations; a set of standard BTS modules with specified form, fit, and function such that BTS vendors can acquire and integrate modules from multiple vendors in an OEM fashion; defines open, standards-based internal digital interfaces between BTS modules to ensure interoperability and compatibility; and supports different access technologies, such as GSM/EDGE, CDMA2000, WCDMA, LTE/LTE-Advanced, or mobile WIMAX [75].

[5]The CPRI is an initiative to define a publicly available specification that standardizes the protocol interface between the radio equipment control (REC) and the radio equipment (RE) in wireless base stations. This allows interoperability of equipments from different vendors and preserves the software investment made by wireless service providers. Conventional base stations are located adjacent to the antenna in a small compartment at the base of the antenna tower. Finding suitable sites can be a challenge because of the footprint required for the compartment, the need for structural reinforcement of rooftops, and the availability of both primary and backup power sources. The CPRI allows the use of a distributed architecture where base stations, containing the REC, are connected to remote radio heads via low-loss fiber links that carry the CPRI data. This architecture reduces costs for service providers because only the remote radio heads containing the RE need to be situated in environmentally challenging locations. The base stations can be centrally located in less problematic locations where footprint, climate, and availability of power are easily managed [38,44].

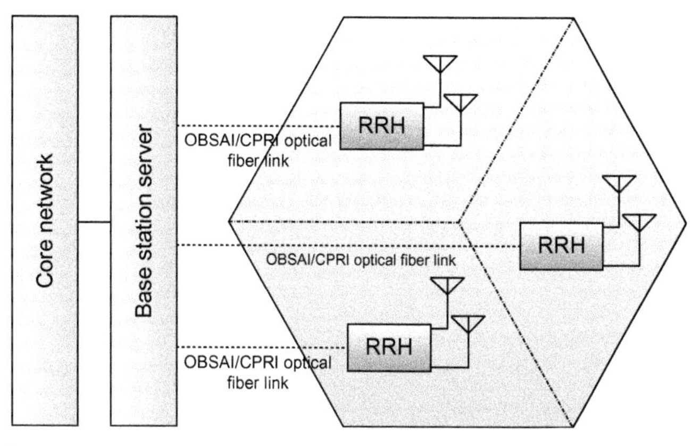

FIGURE 2.15

Example three-sector cell using RRH [38].

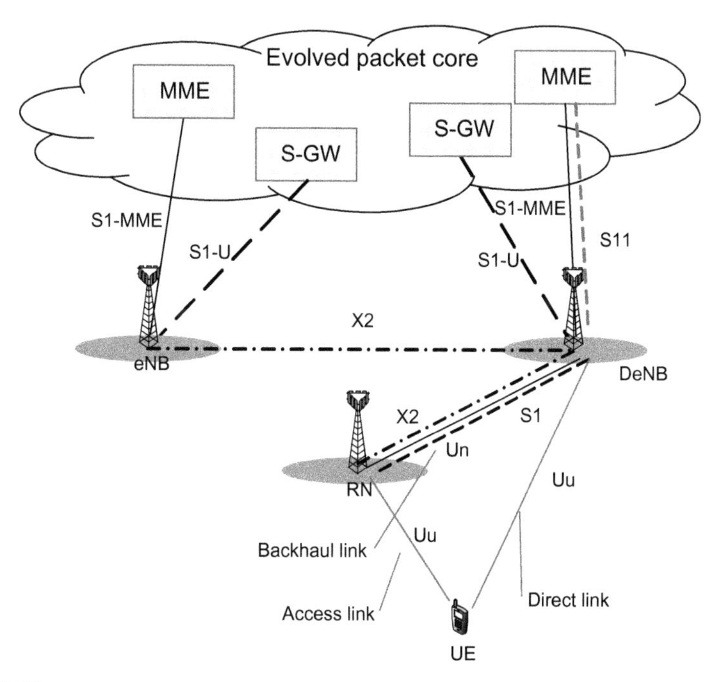

FIGURE 2.16

Support of relay nodes in E-UTRAN architecture [12].

can be located at the cell-edge or in some other areas of poor coverage. Radio repeaters are relatively simple devices operating purely at the RF level. Typically they receive and retransmit an entire frequency band, so they must be sited carefully. In general, repeaters can improve coverage but do not substantially increase capacity. In an in-band relaying scenario and due to self-interference, the RNs cannot simultaneously transmit on the access link and receive on the backhaul link as well as they cannot concurrently receive on the access and transmit on the backhaul link.

E-UTRAN supports relaying functionality by connecting an RN to a serving eNB, known as the donor eNB, via a modified version of the E-UTRA radio interface, where the modified version is denoted as the Un interface. The RN supports most of the eNB functionalities, implying that it terminates the radio protocols of the E-UTRA radio interface on the access link as well as the S1 and X2 interfaces on the backhaul link. In addition to the eNB functionality, the RN also supports a subset of the UE functionality, e.g., physical layer, layer-2, RRC, and NAS functionality, in order to wirelessly connect to the DeNB. An RN may not use another RN as its DeNB and intercell handover of the RN is not supported.

More advanced relays at layer 2 can decode transmissions before retransmitting them. Traffic can then be forwarded selectively to the UE by the RN, thus minimizing the interference created by legacy relays that forwarded all traffic. Depending on the level at which the protocol stack is terminated in the RN, such types of relay may require the development of relay-specific standards. This can be avoided by extending the protocol stack of the RN up to layer 3 to create a wireless router that operates in the same way that a normal eNB operates, using standard air-interface protocols and performing its own resource allocation and scheduling.

The modified E-UTRAN architecture to support RNs is shown in Figure 2.16. The RN terminates the S1, X2, and Un interfaces. The DeNB provides S1 and X2 proxy functionality between the RN and other eNBs, MMEs, and S-GWs. The S1 and X2 proxy functionality includes forwarding UE-specific S1 and X2 signaling messages as well as GTP data packets between the S1 and X2 interfaces associated with the RN and the S1 and X2 interfaces associated with other network nodes. Due to the proxy functionality, the DeNB appears as an MME (for S1-MME), an eNB (for X2) and an S-GW (for S1-U) to the RN. In further development of relay specifications in 3GPP, the DeNB also implements and provides the S-GW/P-GW functions to the associated RNs. This includes creating a session for the RN and managing EPS bearers for the RN, as well as terminating the S11 interface toward the MME serving the RN. The RN and DeNB also map signaling and data packets to the EPS bearers that are set up for the RN. The mapping is based on the existing QoS mechanisms already defined for the UE and the P-GW. The P-GW functions in the DeNB allocate an IP address for the RN for the operation and management functions which may be different than the S1 IP address of the DeNB [12].

The S1 and X2 user-plane packets are mapped to radio bearers over the Un interface. The mapping can be based on the QCI associated with the UE's EPS bearers. The UE's EPS bearers with similar QoS can be mapped to the same Un radio bearer. There is a unique S1 interface relationship between each RN and its DeNB, and between the DeNB and each MME in the MME pool. The DeNB processes and forwards all S1 messages between the RN and the MMEs for all UE-dedicated procedures. The processing of S1AP messages includes modifying S1AP UE IDs, transport layer address and GTP TEIDs, leaving other parts of the message unchanged [12].

All non-UE-dedicated S1AP procedures are terminated at the DeNB, and are handled locally between the RN and the DeNB, and between the DeNB and the MME(s). Upon reception of an S1 non-UE-dedicated message from an MME, the DeNB may trigger corresponding S1 non-UE-dedicated procedure(s) to the RN(s). If more than one RN is involved (in a multi-hop scenario), the DeNB may wait and aggregate the response messages from all involved RNs before responding to the MME. Upon reception of an S1 non-UE-dedicated message from an RN, the DeNB may trigger associated S1 non-UE-dedicated procedure(s) to the MME(s). There is a single X2 interface relation between each RN and its DeNB. In addition, the DeNB may have X2 interface connection to neighboring eNBs. The DeNB processes and forwards all X2 messages between the RN and other eNBs for all UE-dedicated procedures. The processing of X2AP messages includes modifying S1/X2AP UE IDs, transport layer address, and GTP TEIDs, leaving other parts of the message intact [12].

All non-UE-dedicated X2AP procedures are terminated at the DeNB, and are handled locally between the RN and the DeNB, and between the DeNB and other eNBs. Upon reception of an X2 non-cell-related non-UE-associated message from RN or neighbor eNB, the DeNB may trigger associated non-UE-dedicated X2AP procedure(s) to the neighbor eNB or RN(s). Upon reception of an X2 cell-related non-UE-dedicated message from RN or neighbor eNB, the DeNB may pass associated information to the neighbor eNB or RN(s) based on the included cell information. If one or more RN(s) are involved, the DeNB may wait and aggregate the response messages from all participating nodes to respond to the originating node. The S1 and X2 interface signaling packets are mapped to radio bearers over the Un interface. The RN connects to the DeNB via the Un interface using the same radio protocols and procedures as a UE connecting to an eNB [12].

2.5 E-MBMS architecture

Enhanced multimedia broadcast multicast service (E-MBMS) is a point-to-multi-point service in which data is transmitted from a single source to multiple destinations over the radio network. Transmitting the same data to multiple recipients allows network resources to be shared. E-MBMS is realized by addition of new functional entities to 3GPP network architecture. In practice, E-MBMS provides two different services: broadcast and multicast.

The broadcast service can be received by any subscriber located in the area in which the service is offered and multicast services can only be received by users subscribed to the service and joined to the multicast group associated with the service. Both services are unidirectional point-to-multi-point transmissions of multimedia data, text, audio, video, news, etc., from a broadcast multicast service center (BM-SC) to any user located in the service area. For such a service, only the broadcast service providers can be charged, possibly based on the amount of broadcast data size of service area, or broadcast service duration. Multicast is subject to service subscription and requires the end user to explicitly join the group in order to receive the service. Because it is subject to subscription, the multicast service allows the operator to set specific user charging rules for this service.

The E-MBMS content is typically broadcast over a large number of cells that transmit the same data and are required to be synchronized. In that case, the resulting signal will appear to a UE as one transmission over a time-disperse radio channel. This concept is known as multimedia broadcast single frequency network (MBSFN). A group of cells that transmit the same content to multiple users form an MBSFN area. Multiple cells can belong to a single MBSFN area and each cell can be part of up to eight MBSFN areas. There can be up to 256 different MBSFN areas defined in a network, each one with a unique identity. Once defined, MBSFN areas will not change dynamically. It is not required that a terminal simultaneously receive content from multiple MBSFN areas. In other words, an MBSFN area is a geographical area comprising one or several cells that transmit the same content. For example, in Figure 2.17, which illustrates a hypothetical MBSFN network, cells 2 and 3 belong to both MBSFN areas 1 and 2. The MBSFN areas are static and do not vary over time. The usage of MBSFN transmission requires not only time synchronization among the cells participating in an MBSFN area, but also usage of the same set of radio resources in each of the cells for a particular service. This coordination is the responsibility of the multi-cell/multicast coordination entity (MCE), which is a logical node in the radio access network that controls resource allocation and transmission parameters across the cells in the MBSFN area [12,30–33].

Figure 2.18 illustrates the E-MBMS network architecture. As shown in the figure, the MCE can control multiple eNBs, each controlling one or more cells. The BM-SC, located in the core network, is responsible for authorization and authentication of content providers, charging, and overall configuration of the data flow through the core network. The MBMS gateway (MBMS-GW) is a logical node handling multicast of IP packets from BM-SC to all eNBs involved in transmission in an MBSFN area. It also manages session control signaling via the MME. From the BM-SC, the MBMS data is forwarded using IP-based multicast, a method of sending IP packets to multiple receiving network nodes in a single transmission, via the MBMS-GW to the cells. The MBMS is not only efficient from a radio-interface point of view, but also it saves resources in the transport network by avoiding unicast transmission of the same packet to multiple nodes unless necessary. This can lead to significant savings in the transport network.

More specifically, an MBMS-enabled network consists of the following entities and interfaces [12,32]:

- **MCE:** The MCE is a logical entity (this does not preclude the possibility that it may be part of another network element), whose functions are admission

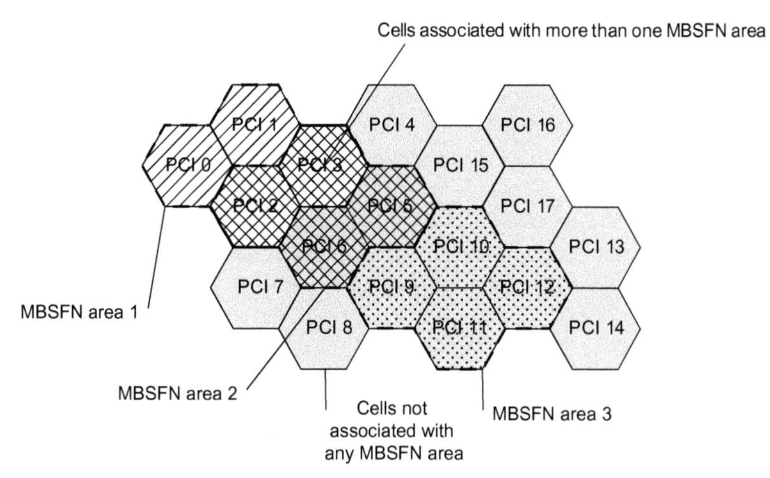

FIGURE 2.17

Illustration of MBSFN areas.

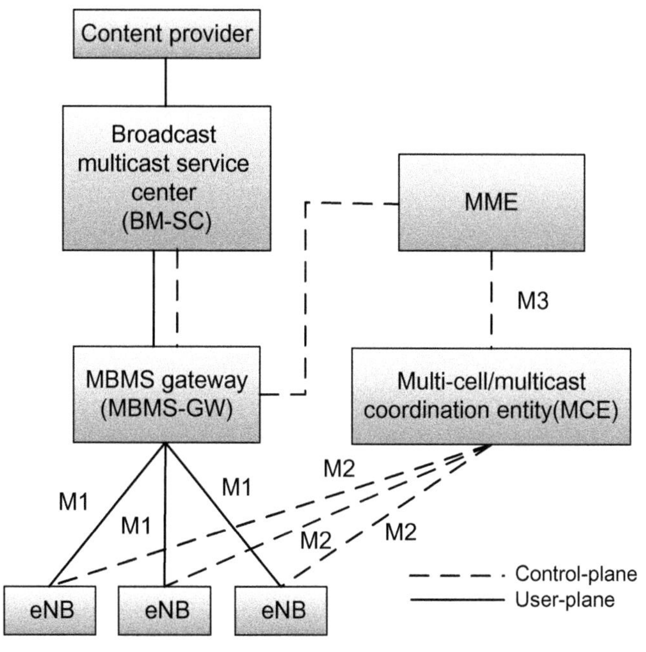

FIGURE 2.18

E-MBMS network architecture [12,32].

control, and allocation of the radio resources used by all eNBs in the MBSFN area for multi-cell MBMS transmissions using MBSFN operation. The MCE decides whether to establish radio bearer(s) for the new MBMS service(s), if the radio resources are not sufficient for the corresponding MBMS service(s), or may allocate radio resources from other radio bearer(s) of ongoing MBMS service(s) according to allocation and retention priority (ARP). In addition to allocation of the time/frequency radio resources, the MCE controls the radio configuration, for example, modulation and coding scheme; counting and acquisition of counting results for MBMS service(s); resumption of MBMS session(s) within MBSFN area(s) based on the ARP and/or the counting results for the corresponding MBMS service(s); suspension of MBMS session(s) within MBSFN area(s) based on the ARP and/or on the counting results for the corresponding MBMS service(s). In the case of distributed MCE architecture, the MCE manages the above functions for a single eNB of an MBSFN.

- **E-MBMS Gateway (MBMS-GW):** The MBMS-GW is a logical entity (this does not preclude the possibility that it may be part of another network element) that is located between the BM-SC and eNBs whose functions include sending/broadcasting of MBMS packets to each eNB participating in the service. The MBMS-GW uses IP multicast as the method of forwarding MBMS user data to the eNBs. The MBMS-GW performs MBMS session control signaling (i.e., session start/update/stop) toward the E-UTRAN via MME.
- **MCE–MME Interface (M3):** The application part of this control-plane interface allows MBMS session control signaling on E-RAB level (i.e., does not convey radio configuration data). The procedures include MBMS session start and stop. The SCTP protocol[6] is used as signaling transport method.
- **MCE–eNB Interface (M2):** The application part of this control-plane interface conveys radio configuration data for the multi-cell transmission mode eNBs and session control signaling. The SCTP protocol is used as signaling transport.
- **MBMS–GW-eNB Interface (M1):** This user-plane interface carries IP packets between the two network entities and uses IP multicast for point-to-multipoint delivery of user packets.

[6]SCTP is a transport layer protocol, performing a similar role as the commonly used protocols TCP and UDP. It provides some of the service features of both protocols, that is, it is message-oriented like UDP and ensures reliable and in-sequence transport of messages with congestion control like TCP. The protocol is defined by IETF RFC 4960 [20]. The SCTP protocol may be characterized as message-oriented, that is, it transports a sequence of messages comprising a group of bytes, rather than transporting an unbroken stream of bytes as does TCP. As in UDP, in SCTP a sender sends a message in one operation, and that exact message is passed to the receiving application process in one operation. In contrast, TCP is a stream-oriented protocol, transporting streams of bytes reliably and in order.

There are two MBMS architectures defined in 3GPP. In the "distributed MCE architecture," the MCE is part of the eNB and the M2 interface is maintained between the MCE and the corresponding eNB, whereas in the "centralized MCE architecture" (shown in Figure 2.18), the MCE is a logical entity and can be deployed as a stand-alone physical entity or collocated in another physical entity, such as eNB. In both cases, the M2 interface is maintained between the MCE and all eNB(s) belonging to the corresponding MBSFN area [12].

2.6 Positioning network architecture

3GPP legacy technologies (GSM and UMTS) have supported location services network architecture for the purpose of positioning of mobile devices since 3GPP Rel-4. With the introduction of EPS in 3GPP Rel-8, control-plane location-service architecture for the EPS was introduced in 3GPP Rel-9. Figure 2.19 illustrates the control-plane location service architecture for the EPS, which introduced new interfaces SLg and SLs; and allowed the EPS control-plane element MME to interconnect with the location service core network elements, enabling location services using the positioning functionality provided by the E-UTRAN.

The location service (LCS) architecture follows a client/server model with the gateway mobile location center (GMLC) acting as the location server providing location information to LCS clients. The GMLC sends location requests to the evolved serving mobile location center (E-SMLC) through the MME to retrieve this location information. The E-SMLC is responsible for interaction with the UE through E-UTRAN to obtain the UE position estimate or to obtain position measurements that help the E-SMLC to estimate the UE position. Note that the GMLC interaction over the interfaces connected to it other than SLg was already available before 3GPP Rel-9 for GSM and UMTS access.

The LCS clients can be part of the core network or external to the core network and can also reside in the UE or attached to the UE. Depending on the location of the LCS client, the location request initiated by the LCS client may be a mobile-originated, mobile-terminated, or network-induced location request. Emergency location service is possible even if the UE does not have a valid service subscription due to regulatory requirements. Support of location service-related functionality in the E-UTRAN, MME, and UE are optional. The LCS is applicable to any target UE no matter if the UE supports LCS. The following is a description of various LCS architectural elements and interfaces shown in Figure 2.19 [30]:

- **GMLC:** This entity receives and processes location requests from LCS clients; obtains routing information from HSS; performs registration authorization; communicates information needed for authorization, location service requests, and location information with other GMLC; checks the target UE privacy profile settings; and depending on roaming support, it may assume the role of requesting GMLC, visited GMLC, or home GMLC.

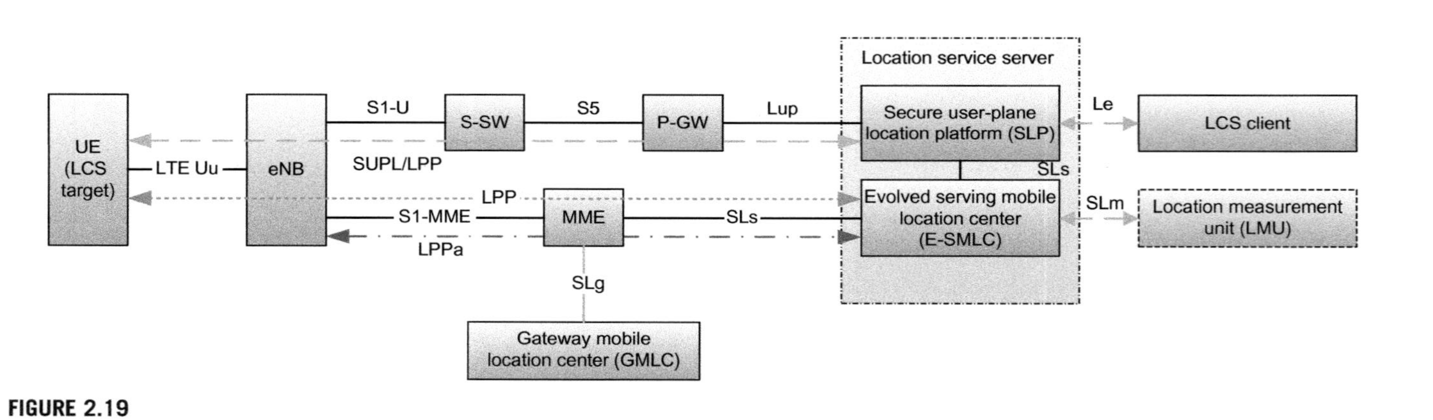

FIGURE 2.19

Positioning network architecture [30].

- **Evolved Serving Mobile Location Center (E-SMLC):** This entity manages the overall coordination and scheduling of resources required for the location determination of a UE that is attached to E-UTRAN. It calculates the final location and estimates velocity as well as the location accuracy. It interacts with the UE to exchange location information applicable to the UE-assisted and the UE-based positioning methods. It interacts with the E-UTRAN to exchange location information applicable to the network-assisted and the network-based positioning methods.
- **Location Measurement Unit (LMU):** The LMU conducts measurements and communicates the measurements to an E-SMLC. All positioning measurements obtained by an LMU are sent to the E-SMLC that requested the measurement report. A UE positioning request may involve measurements by multiple LMUs. The SLm interface between the E-SMLC and the LMU is used for uplink positioning. It is used to transport SLm-AP protocol messages over the E-SMLC-LMU interface. Both stand-alone LMU and LMU integrated into an eNB are supported in LTE Rel-11.

The MME is responsible for UE subscription authorization. It coordinates location services and positioning requests. It handles charging and billing and performs E-SMLC selection (e.g., based on network topology to balance the load on E-SMLC; LCS client type; and requested QoS). It is further responsible for authorization and operation of the LCS services. The HSS is used for storage of location services subscription data and routing information. Two protocols are used for the overall information exchange concerning location services: (1) LTE positioning protocol (LPP); and (2) LTE positioning protocol annex (LPPa). The latter is used for communication between the location server and the eNB. The base station (when using observed time difference of arrival method) is in charge of properly configuring the radio signals that are used by the terminal for positioning measurements, that is, positioning reference signals. It further provides information to the E-SMLC; it enables the device to conduct inter-frequency measurements, if required; and based on the E-SMLC request, it conducts measurements and sends the results to the server. The LPP can be used in user-plane and control-plane methods. The LPP is a point-to-point protocol that allows multiple connections to different devices. The exchanged LPP messages and information can be divided into four categories: (1) UE positioning capability information transfer to the E-SMLC; (2) positioning assisting data delivery from the E-SMLC to the device; (3) location information transfer; and (4) session management.

The SLm-AP protocol, terminated between the E-SMLC and the LMU, is used to support the following functions: (1) delivery of target UE configuration data from the E-SMLC to the LMU; and (2) request positioning measurements from the LMU and delivery of positioning measurements to the E-SMLC. The SLm-AP protocol is directly used between the E-SMLC and the LMU. The MME receives a request for location service associated with a particular target UE from another entity (e.g., GMLC, eNB, or UE), or the MME by itself decides to initiate some location service on behalf of a particular target UE (e.g., for an IMS emergency call from the UE). The MME then sends a location service request to the E-SMLC. The E-SMLC processes the location service request which may include transferring assisting data to the target UE to help UE-based and/or UE-assisted positioning. The E-SMLC then returns the result of the location service to the MME, for example, a position estimate for the UE and/or an indication of any assisting data transferred to the UE. In the case of a location service requested by an entity other than the MME (e.g., UE, eNB, or E-SMLC), the MME returns the location service result to this entity.

2.7 Mobility management

The NAS is a set of protocols in the EPS. The NAS is used to convey non-radio-access-related signaling between the UE and the MME for an E-UTRAN access. The NAS procedures are grouped in two categories: EPS mobility management (EMM) and ECM. The EMM protocol refers to procedures related to mobility over an E-UTRAN access, authentication, and security. The EMM-specific procedures are UE-initiated. These procedures define attach/detach (to/from the EPC) mechanisms. They also introduce the tracking area update mechanism, which updates the location of the UE within the network. In EPS, a UE initiates a tracking area update when it detects that it enters into a new tracking area. The EMM-specific procedures also define periodic tracking area update. The EMM connection management procedures provide several functions to support the connection of the UE to the network.

The states of a UE with respect to mobility and connection establishment are described by the following NAS states: the EMM state (EMM-DEREGISTERED or EMM-REGISTERED) reflects whether the UE is registered in the MME and the ECM state (ECM-IDLE or ECM-CONNECTED) reflects the connectivity of the UE with the EPC. The transition from ECM-IDLE to ECM-CONNECTED not only involves establishment of an RRC connection, but also includes establishment of an S1 connection. The RRC connection establishment is initiated by the NAS and is completed prior to S1 connection establishment, which means that connectivity in RRC_CONNECTED is initially limited to the exchange of control information between the UE and eNB. The UEs are typically moved to ECM-CONNECTED when becoming active. It should be noted that in LTE, the transition from ECM-IDLE to ECM-CONNECTED is performed within 100 ms and the UEs engaged in intermittent data transfer do not need to stay in ECM-CONNECTED, if the ongoing services can tolerate such transfer delays. The RRC connection release is initiated by the eNB following release of the S1connection between the eNB and the core network [12,31].

The EMM states describe the mobility management states that result from the mobility management procedures, for example, attach and tracking area update procedures. The EMM has two states: EMM-REGISTERED and

EMM-DEREGISTERED. In EMM-DEREGISTERED state, the EMM context does not have any valid location or routing information for the UE. Since the location of the UE is not known, the UE is not reachable by the MME. In order to avoid performing an authentication and key agreement (AKA)[7] security procedure in every network attach attempt by the UE, the MME, and the UE may store some UE context in this state. The MME may perform an implicit detach any time after the implicit detach timer expires. The state is changed to EMM-DEREGISTERED in the MME after performing the implicit detach. If all the bearers belonging to a UE are released (e.g., after a handover from E-UTRAN to non-3GPP access), the MME changes the mobility management state of the UE to EMM-DEREGISTERED. This ensures that the UE performs a tracking area update when it reselects E-UTRAN [12,31].

The ECM states describe the signaling connectivity between the UE and the EPC. In EMM-REGISTERED state, there are two EPS connection states: ECM-IDLE and ECM-CONNECTED. In ECM-IDLE, a UE performs cell selection or reselection. There is no NAS signaling connection between the UE and the network. The UE does not have a context in the E-UTRAN and has no S1-MME, and S1-U connection with the network. In EMM-REGISTERED and ECM-IDLE state, the UE performs a tracking area update, if the current TA is not in the list of TAs that the UE has received from the network in order to maintain the registration and enable the MME to page the UE. The UE performs a periodic tracking area update procedure to notify the EPC that it is available [12,31].

2.8 Radio resource management

The purpose of RRM function is to ensure efficient use of the available network resources. In particular, RRM in E-UTRAN provides a means to manage (e.g., assign, reassign, and release) radio resources in single and multi-cell environments. RRM may be treated as a central application at the eNB responsible for interworking between different protocols (RRC, S1AP, and X2AP) so that messages can be properly transferred to different nodes across Uu, S1, and X2 interfaces. RRM may interface with operation and management functions in order to control, monitor, audit, or reset the status due to errors at a protocol stack. The RRM consists of the following main functions [33,66]:

- Radio admission control: The RAC functional module accepts or rejects requests for the establishment of new radio bearers. Admission control is performed according to the required QoS, current system load, and the required service. The RAC function ensures efficient radio resource utilization

[7]AKA is a mechanism which performs authentication and session key distribution in UMTS networks. AKA is a challenge-response-based mechanism that uses symmetric cryptography. AKA is typically run on a UMTS IP multimedia services identity module (ISIM), which resides on a smart card device that also provides tamper resistant storage of shared secrets. AKA is defined in IETF RFC 3310 [21].

by accepting radio bearer requests as long as radio resources are available, and proper QoS enforcement for current sessions by rejecting radio bearer requests, if they cannot be accommodated. It interacts with the RBC module to perform its functions.

- Radio bearer control: The RBC functional module manages the establishment, maintenance, and release of radio bearers. It configures the radio resources depending on the current resource availability and usage, as well as QoS requirements of new and ongoing sessions due to mobility or other reasons. The RBC function releases radio resources associated with radio bearers at session termination, handover, or other similar cases. The establishment, maintenance, or release of radio bearers is tied to the configuration of radio resources associated with them.
- Connection mobility control: The CMC functional module manages radio resources in the idle and connected modes. In the idle mode, this module defines criteria and algorithms for cell selection, reselection, and location registration that assist the UE in selecting or camping on the best cell. In addition, the eNB broadcasts parameters that configure the UE measurement and reporting procedures. In the connected mode, this module manages the mobility of radio connections without disruption of services. Handover decisions may be based on measurement results reported by the UE, by the eNB, and other parameters configured for each cell. Handover decisions may consider additional information, such as neighbor cell load condition, traffic distribution, transport and hardware resources, and operator-defined policies. Inter-RAT RRM can be one of the subsets of this function responsible for managing the resources in inter-RAT mobility and handovers. In the idle mode, the cell reselection algorithms are controlled by setting of parameters (e.g., thresholds and hysteresis values) that define the best cell and/or determine when the UE should select a new cell. The E-UTRAN broadcasts parameters that configure the UE measurement and reporting procedures.
- Dynamic resource allocation (DRA) or packet scheduling (PS): The task of DRA or PS is to allocate and de-allocate resources (including physical resource blocks) to user- and control-plane packets. The scheduling function typically considers the QoS requirements associated with the radio bearers, the channel quality feedback from the UEs, buffer status, inter-cell/intra-cell interference condition, etc. The DRA function may take into account the restrictions or preferences on some of the available resource blocks or resource-block sets due to inter-cell interference coordination (ICIC) considerations.
- Inter-cell interference coordination: ICIC function manages radio resources such that intercell interference effects are minimized. The ICIC is essentially a multi-cell RRM function that considers resource usage status and the traffic load conditions across multiple cells.
- Load balancing: LB function has the task of handling non-uniform distribution of the traffic load across multiple cells. The purpose of LB function is to control the load distribution in such a manner that radio resources are

efficiently utilized; to maintain the QoS of the current sessions as much as possible; and to minimize the probability of call drops. The LB algorithms may result in handover or cell reselection decisions with the purpose of redistributing traffic from overloaded cells to under-utilized cells.

- Inter-RAT radio resource management: Inter-RAT RRM is primarily concerned with the management of radio resources in connection with inter-RAT mobility, notably inter-RAT handover. During inter-RAT handover, the handover decision may take into account the involved RATs resource availability as well as the UE capabilities and the operator policies. Inter-RAT RRM may also include additional functions for inter-RAT load balancing for the idle mode and connected mode UEs.
- Subscriber profile ID (SPID) for RAT/frequency priority: The RRM function maps SPID parameters received via the S1 interface to a locally defined configuration in order to apply specific RRM strategies (e.g., to define RRC_IDLE mode priorities and control inter-RAT/inter-frequency handover in RRC_CONNECTED mode). The SPID is an index referring to user information, such as mobility profile and service usage profile. The SPID information is UE-specific and applies to all of its radio bearers. The RRM strategy in E-UTRAN may be based on user-specific information.

2.9 Quality of service/quality of experience

The end-to-end quality of service experienced by an end user results from a combination of elements throughout the protocol stack and E-UTRAN/EPC components. Thus, the performance evaluation of the service requires a detailed performance analysis of the entire network, that is, from the UE up to the application server (AS) or remote UE. The QoE is a subjective measurement of the quality experienced by a user when he/she uses a communication service. In other words, the QoE is defined as the overall acceptability of an application or a service as perceived subjectively by the end user. When evaluating the quality of a service, the objective is to optimize the operation of the network from a perspective which is purely based on objective parameters, or to determine the quality that the user is actually experiencing and its satisfaction level. The QoE takes into account the satisfaction of a user in terms of both content and application performance. In this sense, the introduction of smartphones has been a quantitative leap in user QoE expectations [43,67,68].

The QoE has been traditionally evaluated through subjective tests carried out with the users in order to assess their satisfaction with a mean opinion score value. This approach can be quite expensive, time-consuming, and cannot be used for real-time decision-making to improve the QoE. New methods have been emerged in recent years to estimate the QoE based on certain performance indicators associated with services. A possible solution to instantaneously evaluate the

QoE is to integrate the QoE analysis tools in the mobile terminal itself. If mobile terminals are able to report the measurements to a central server, the QoE assessment process is significantly simplified. Other solutions are focused on including new network elements (e.g., network analyzers and deep packet inspectors) that are responsible for capturing the traffic from a certain service and analyzing its performance. In some cases, the evaluation of video-streaming quality in the mobile terminals is addressed by monitoring objective parameters, such as packet loss rate or jitter.

As mentioned earlier, the E-UTRAN consists of a network of eNBs which are typically interconnected by means of an X2 interface and to the EPC via an S1 interface. The eNB entity plays a critical role in maintaining the end-to-end QoS. The eNB performs QoS-related functions which include admission control and resource reservation; scheduling and distribution of radio resources among the established bearers; and L1/L2 protocol configuration in accordance with the QoS attributes associated with the bearer. The LTE-enabled UEs may support multiple applications at the same time, each having different QoS requirements. This is achieved by establishing different EPS bearers for each QoS flow. The EPS bearers can be classified into two categories based on the characteristics of the QoS they provide: GBR bearers in which resources are permanently allocated and non-GBR bearers which do not guarantee any particular bit rate. In the access network, it is the eNB's responsibility to ensure that the QoS requirements for a particular bearer over the radio interface are met. Each bearer has an associated QCI, characterized by priority, packet delay budget, and admissible packet loss rate; and an ARP used for call admission control. The IP packets mapped to the same EPS bearer receive the same bearer level packet forwarding treatment (e.g., scheduling policy, queue management policy, and rate shaping policy). Thus, the UE is not only responsible for requesting the establishment of EPS bearers for each QoS flow, but also for performing packet filtering in the uplink into different bearers based on TFTs, as P-GW does for the downlink (see Figures 2.4 and 2.5).

It was mentioned earlier that one EPS bearer/E-RAB is established when the UE connects to a PDN, and that remains established throughout the lifetime of the PDN connection to provide the UE with always-on IP connectivity to that PDN. That bearer is referred to as the default bearer. Any additional EPS bearer/E-RAB that is established to the same PDN is referred to as a dedicated bearer. The default bearer QoS parameters are assigned by the network based upon subscription data. The decision to establish or modify a dedicated bearer can only be taken by the EPC and the bearer level QoS parameter values are always assigned by the EPC.

2.10 Security in E-UTRAN

Security over the E-UTRAN air-interface is provided through cryptographic techniques. The backhaul link from the eNB to the core network uses Internet key exchange (IKE)[8] and the IP security protocol (IPsec)[9] when cryptographic protection is needed. These cryptographic techniques provide end-to-end protection for signaling between the core network and UE. Therefore, the main location where user traffic is exposed to security threats is in the eNB. Moreover, to minimize susceptibility to attacks, the eNB must provide a secure environment that supports the execution of sensitive operations, such as the encryption or decryption of user data and the storage of sensitive data, such as keys for securing UE communication, long-term cryptographic secrets, and vital configuration data. Similarly, the use of sensitive data must be confined to this secure environment. Even with the above security measures in place, one must consider attacks on an eNB, because, if successful, they could give attackers full control of the eNB and its signaling to UEs and other nodes. To limit the effect of a successful attack on one eNB, attackers must not be able to intercept or manipulate user- and control-plane traffic that traverses another eNB, for example, following a handover.

The E-UTRAN and EPS security model is shown in Figure 2.20. The subscriber authentication function in EPS is based on the UMTS AKA protocol. It provides mutual authentication between the UE and core network, ensuring robust charging and guaranteeing that no unauthorized and non-authenticated entities can pose as a valid network node. The EPS AKA provides a root key from which a key hierarchy is derived. The keys in this hierarchy are used to protect signaling and user-plane traffic between the UE and the network. The key hierarchy is derived using cryptographic functions. Furthermore, keys are bound to where, how, and for which purpose they are used. This ensures that keys used for one access network cannot be used in another access network, and that the same key is not used for multiple purposes or with different algorithms. The key hierarchy and bindings also make it possible to routinely and efficiently change the keys used between a UE and eNBs (e.g., during handover) without changing the root key or the keys used to protect signaling between the UE and the core network. For radio-specific signaling, LTE provides integrity, replay protection, and encryption between the UE and the eNB.

The IKE/IPsec protocols can protect the backhaul signaling between the eNB and MME. In addition, LTE-specific protocols provide end-to-end protection of

---

[8]IKE or IKEv2 is the protocol used to set up a security association in the IPsec protocol suite. The IKE protocol is based on a key-agreement protocol and the Internet security association and key management protocol which is a protocol defined by IETF RFC 2408 for establishing security associations and cryptographic keys in an Internet environment. The IKE uses X.509 certificates for authentication which are either pre-shared or distributed to set up a shared session secret from which cryptographic keys are derived. In addition, a security policy for every peer which will connect must be manually maintained [76].

[9]IPsec is a set of protocols for securing IP-based communications by authenticating and encrypting each IP packet of a data stream. IPs also includes protocols for establishing mutual authentication between agents at the beginning of the session and negotiation of cryptographic keys to be used during the session. IPs can be used to protect data flows between a pair of hosts, between a pair of security gateways, or between a security gateway and a host [24,25,52].

signaling between the MME and UE. For user-plane traffic, IKE/IPsec can similarly protect the backhaul from the eNB to the S-GW. Support for integrity, replay protection, and encryption is mandatory in the eNB. The user-plane traffic between the UE and eNB is only protected by encryption, as integrity protection would result in excessive overhead. Nevertheless, it is not possible to intelligently inject traffic on behalf of another user: attackers are essentially blind in the sense that any traffic they try to inject would almost certainly decrypt to invalid information.

2.11 Support of voice and messaging services in LTE/LTE-Advanced

2.11.1 IMS architecture

Voice services in LTE are enabled using the IMS which provides a framework for supporting IP-based services and requires new IMS-specific network elements as part of the dedicated core network architecture. The first version of IMS was introduced in 3GPP Rel-5 specifications; however, many enhancements have been made to the baseline in subsequent releases. During the development of the LTE standards, it was assumed that IMS would be commercially available when the LTE networks were deployed, since voice support in LTE relied on the IMS. However, rollout of IMS infrastructure was slower than expected, making voice support in LTE a real challenge for many network operators. A closely related feature to the support of voice is the support of SMS as another key circuit-switched service which was not supported in LTE. In general, the support of voice and messaging services in LTE using IMS remains one of the major targets for the industry. An industry initiative called voice over LTE (VoLTE) was formed in 2010 by the Global System for Mobile Communications Association (GSMA) which has developed a framework for support of voice and SMS over IMS in LTE, including roaming and interconnection issues [53].

The IMS is an access-independent IP connectivity and service control architecture. It provides the framework for IP-based multimedia services in a mobile network and is the most efficient choice to provide VoIP services in LTE. In this section, we focus on voice and messaging services over IMS. A simplified IMS reference architecture is shown in Figure 2.21, which includes interfaces toward legacy and IP-based networks. This architecture provides entities for session management and routing, service support, databases, and interworking.

The IP connectivity access network (IP-CAN) in the case of LTE would consist of the EPS and the E-UTRAN. The call session control function (CSCF)

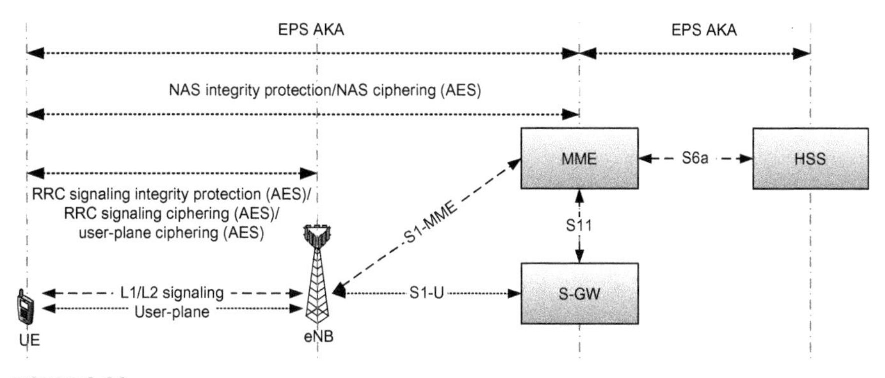

FIGURE 2.20

E-UTRAN and EPS security model [70,71].

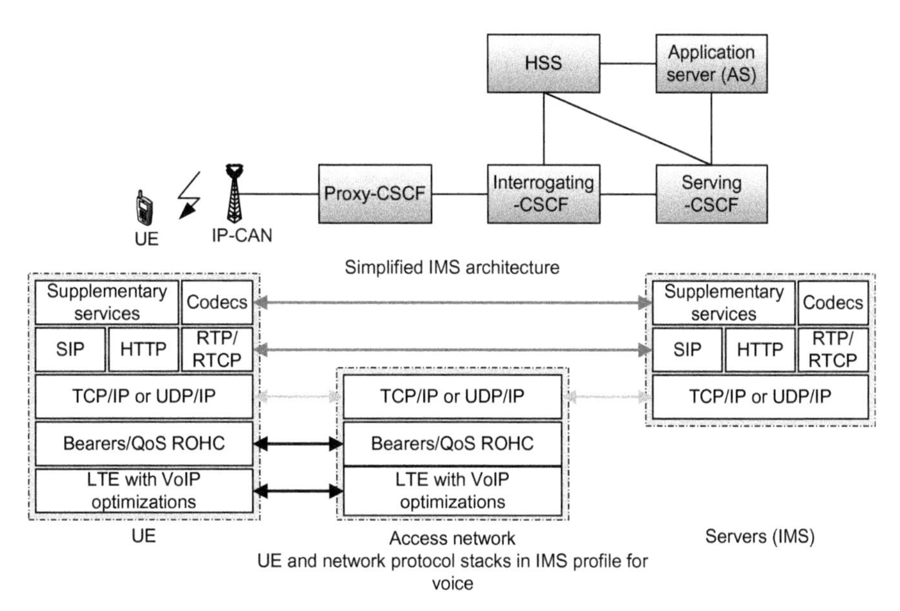

FIGURE 2.21

Simplified IMS architecture and protocol stack [50,65].

entities are the core components of the IMS. There are three CSCF entities defined in IMS architecture:

- Proxy-CSCF (P-CSCF): The P-CSCF is the first point of contact for a user. The P-CSCF behaves as a proxy, that is, it accepts and forwards the requests.
- Interrogating-CSCF (I-CSCF): The I-CSCF is the contact point within an operator's network for all connections destined to a subscriber.

- Serving-CSCF (S-CSCF): The S-CSCF is responsible for handling the registration process, making routing decisions, maintaining sessions, and downloading user information and service profiles from the HSS.

The HSS is the master database for a user. It is comparable to the HLR in a legacy mobile radio network. The HSS contains the subscription-related information required for other network entities that handle calls or sessions. For example, the HSS provides support to the call control servers to complete the routing/roaming procedures by resolving authentication, authorization, naming/addressing resolution, and location dependencies. The application server provides specific IP applications, such as messaging. The purpose of the IMS architecture and the different CSCF entities becomes clearer in the case of roaming. In general, the network operators' internal network structures and user databases are proprietary. The local P-CSCF in the network, which is the point of access for the UE, does not have direct connection to the HSS, rather, it is connected to the HSS via the I-CSCF entity.

The IMS uses a set of IP-based protocols including SIP as a text-based protocol for registration, subscription, notification and initiation of sessions; session description protocol (SDP) as a text-based protocol for negotiating session parameters, such as media type, codec type, bandwidth, IP address and ports, and for media stream setup; RTP and RTCP for transport of real-time applications, such as audio; extensible markup language (XML) configuration access protocol (XCAP), which allows a client to read, write, and modify application configuration data, stored in XML format on a server and maps XML document sub-trees and element attributes to HTTP uniform resource identifiers so that these components can be directly accessed by HTTP; and DHCP for configuring IP addresses for IPv4 and IPv6 [19]. The UDP for messages smaller than 1300 bytes or transport control protocol (TCP) can be used as transport protocols.

As shown in Figure 2.21, the supplementary services requested by the VoLTE profile have been specified in GSMA IR.92 [50,65]. For example, the originating identification presentation (OIP) service provides the terminating user a way to receive identity information to identify the originating user. Another example would be the supplementary service communication hold (HOLD) that enables a user to suspend the reception of media stream(s) from an established IP multimedia session, and resume them at a later time. For supplementary service configuration, the UE and the network must support the XCAP, which is an application layer protocol that allows a client to read, write, and modify application configuration data stored in XML format on a server.

The IMS architecture has been designed to enable operators to provide a wide range of real-time, packet-based services and to track their use in a way that allows both traditional time-based charging as well as packet and service-based charging. It has become increasingly popular both with wire-line and wireless service providers as it is designed to increase carrier revenues, deliver integrated multimedia services, and create an open, standards-based network. The IMS provides a wide range of session border control, including call access control, reachability and security. It also provides a framework for the deployment of both basic calling services and enhanced services, including multimedia messaging; web integration; presence-based services; and push-to-talk. At the same time, it draws on the traditional telecommunications experience of guaranteed QoS; flexible charging mechanisms (time-based, call-collect, and premium rates); and lawful intercept compliance [50,65].

The IMS decomposes the network infrastructure into separate functions with standardized interfaces between them. Each interface is specified as a "reference point," which defines both the protocol over the interface and the functions between which it operates. The standards do not mandate which functions should be collocated, as this depends on the scale of the application, and a single device may contain several functions. The IMS architecture specified by 3GPP can be divided into three main layers, each of which is described by a number of equivalent names: service/application layer; control layer; and transport layer [8].

Application Layer: An application layer consists of application servers that host the IMS services and an HSS. The application layer provides a framework for provisioning and management of services and defines standard interfaces to common functionalities including configuration storage; identity management; user status, such as presence and location, which is held by the HSS billing services and provided by a charging gateway function (CGF) control of voice and video calls and messaging, provided by the control layer.

Control Layer: The control layer is located between the application and transport layers. It routes the call signaling, informs the transport layer of permissible traffic, and generates billing information for the use of the network. At the core of this layer is the CSCF, which comprises the following functions. The P-CSCF is the first point of contact for users with the IMS. The P-CSCF is responsible for security of the messages between the network and the user and allocating resources for the media flows. The I-CSCF is the first point of contact from peered networks. The I-CSCF is responsible for querying the HSS to determine the S-CSCF for a user and may also hide the operator's inner network from peer networks. The S-CSCF is the central entity which is responsible for processing registrations to record the location of each user, user authentication, and call processing including routing of calls to applications. The operation of the S-CSCF is controlled by policy stored in the HSS. This distributed architecture provides an extremely flexible and scalable solution. For example, any of the CSCF functions can generate billing information for each operation. The control layer also controls transport-layer traffic through the resource and admission control subsystem (RACS). This consists of the PDF, which implements local policy on resource usage, for example, to prevent overload of particular access links, and access-RAC function (A-RACF), which controls QoS within the access network.

Transport Layer: The transport layer provides a core QoS-enabled IPv6 network with access from the UE over Wi-Fi and other mobile broadband networks. This infrastructure is designed to provide a wide range of IP multimedia server-based and P2P services. Access into the core network is through border gateways (GGSN/Packet Data Gateway/Broadband Access Server). These entities enforce policy provided by the IMS core network, controlling traffic flows between the access and core networks. Within the transport layer, the interconnect border control function (I-BCF) controls transport level security and informs the RACS of the resources required for a call the interconnect border gateway function (I-BGF), access network inter-connect border gateway function (A-BGF) border gateway functions provide media relay for hiding endpoint addresses.

2.11.2 Single radio voice call continuity

Single radio voice call continuity (SRVCC) is an LTE feature that allows a VoIP/IMS call in the LTE packet-switched domain to be transferred to a legacy circuit-switched domain (GSM/UMTS or CDMA2000). Let us assume that a new LTE network operator is going to migrate its voice services to VoIP over IMS in conjunction with deployment of an LTE radio access network. Without SRVCC features, this operator would need to provide a ubiquitous LTE coverage in order to have a viable VoIP service. However, an SRVCC-enabled LTE network may not require comprehensive LTE coverage in order to provide voice services across the entire network. The SRVCC feature provides the ability to transition a voice call from the VoIP/IMS packet-switched domain to the legacy circuit-switched domain. Variations of SRVCC have been standardized to support both GSM/UMTS and CDMA2000 circuit-switched domains. For an operator with a legacy cellular network, deployment of IMS-based VoIP services in conjunction with the rollout of an LTE network is possible through use of SRVCC, which offers VoIP subscribers coverage over a much larger area than would typically be available during the rollout of a new network [3,48,54,72,76].

The SRVCC feature functions as follows. When an SRVCC-capable UE engaged in a voice call determines that it is moving away from an LTE coverage area (based on neighboring cell measurements), it notifies the serving eNB. The LTE network determines that the voice call has to be moved to the legacy circuit-switched domain, thus, it notifies the MSC server of the need to switch the voice call from the packet-switched to the circuit-switched domain and initiates handover of the LTE voice bearer to the circuit-switched network. The MSC server establishes a bearer path for the mobile in the legacy network and notifies the IMS core network that the UE's call is moving from the packet-switched to the circuit-switched domain. The IMS core network then enacts the necessary interworking functions to facilitate the call transfer from one domain to another. When the mobile arrives in the legacy network, it switches its internal voice processing from VoIP to legacy circuit-switched voice processing and the call is resumed. If the legacy circuit-switched network also has an associated packet-switched capability and is capable of supporting concurrent circuit-switched/packet-switched operations, the subscriber's data sessions can be handed over to the legacy network in conjunction with switching the voice call from the packet-switched to the circuit-switched domain. In this case, when the voice call is terminated and the mobile reenters the LTE coverage, these packet sessions can be handed over to the LTE network again.

The SRVCC feature may not be suitable for the operators that do not plan to migrate their voice service to VoIP. If an operator does plan to migrate to VoIP and also plans to roll out ubiquitous LTE coverage, then the question of whether or not to adopt SRVCC is more complicated. While SRVCC does not require modifications to the existing radio access network, it does require a significant modification of the operator's legacy core network and further requires full deployment of IMS circuit-switched/packet-switched continuity services. Given the cost of these changes, deployment of SRVCC as an interim measure to allow early rollout of VoIP-based services may not make financial sense.

The SRVCC provides voice call continuity between LTE radio access and circuit-switched mode of other radio access technologies (e.g., CDMA2000, GSM, or UMTS) using IMS. Calls remain anchored in IMS, which means that the circuit-switched connections can be treated as standard IMS sessions and the circuit-switched bearer is used as a media for IMS sessions. Using SRVCC, the UE only has to be capable of transmitting/receiving on one access network at a given time. The SRVCC feature provides a solution for LTE networks which do not support circuit-switched voice and messaging services. Circuit-switched legacy technologies complement the LTE coverage with voice using IMS support [72].

The SRVCC functionality is built on the IMS centralized services framework for delivering voice and messaging services to the users regardless of the type of network to which they are attached, and for maintaining service continuity for moving terminals. The call anchoring is done by means of a voice call continuity application server in the home IMS network. During EPS attach, routing information to the voice call continuity application server is provided from the HSS to the MME. During the EPS attach procedure, the network and the UE indicate their support for SRVCC. Some modifications in the core network elements are required to support SRVCC, for example, the CDMA2000 circuit-switched SRVCC interworking function is required [18].

To support GSM and UMTS, some modifications in the MSC server are required. When the E-UTRAN selects a target cell for SRVCC handover, it needs to indicate to the MME that this handover procedure requires SRVCC. Upon receiving the handover request, the MME triggers the SRVCC procedure with the MSC server. The MSC then initiates the session transfer procedure to IMS and coordinates it with the circuit-switched handover procedure to the target cell. Handling of any non-voice packet-switched bearer is by the packet-switched bearer splitting function in the MME. The handover of non-voice packet-switched bearers, if performed, is according to a regular inter-RAT packet-switched handover procedure. The handover procedure from E-UTRAN (as shown in Figure 2.22) is used to initiate the SRVCC procedure. When SRVCC is enacted, the downlink flow of voice packets is switched toward the target circuit-switched network. The call is moved from the packet-switched to the circuit-switched domain, and the UE switches from VoIP to circuit-switched voice. Figure 2.23 shows the simplified procedure for SRVCC for GERAN or UTRAN.

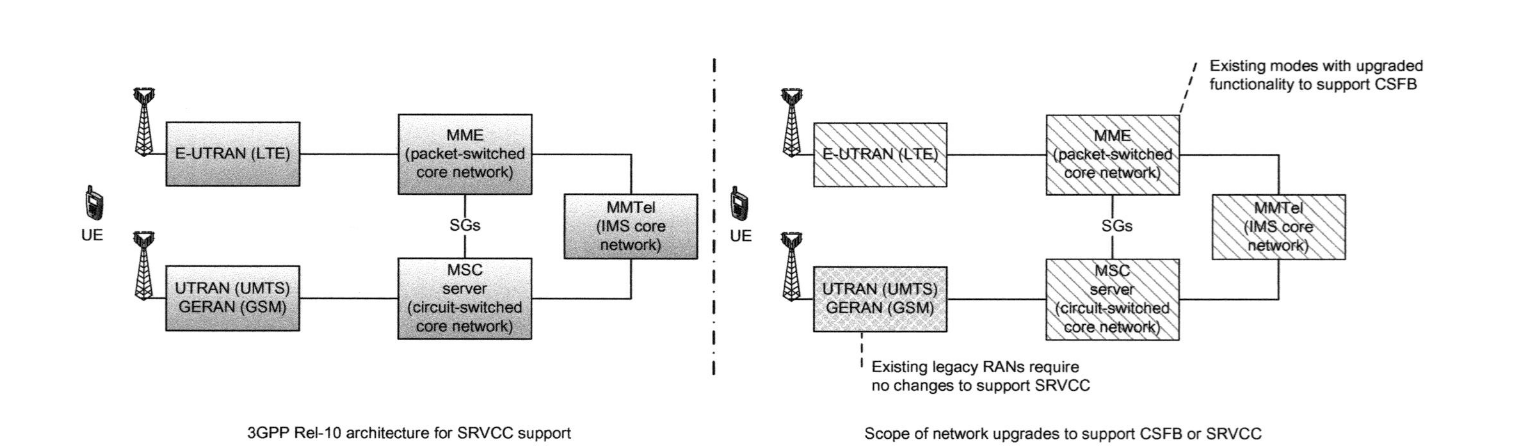

FIGURE 2.22

Network architecture with SRVCC support [72].

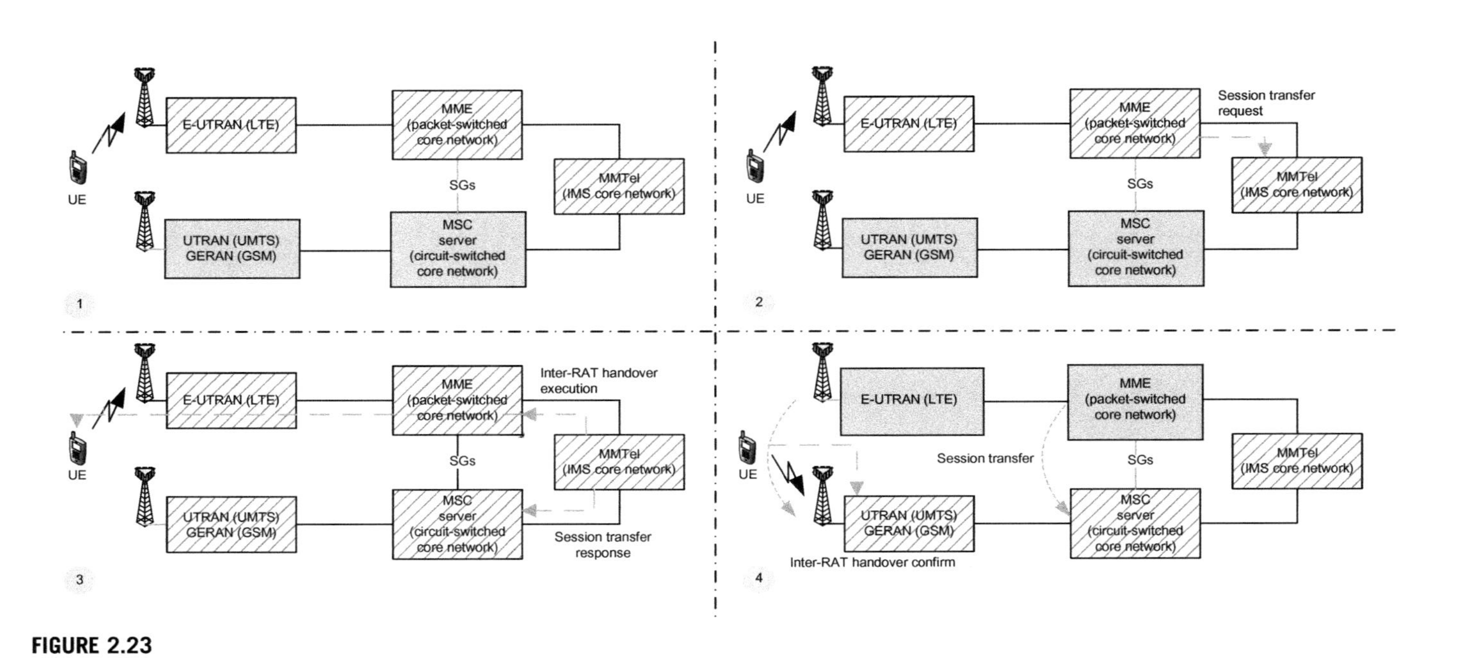

FIGURE 2.23

Call transfer procedure using SRVCC [72].

3GPP Rel-10 architecture has been recommended by GSMA for SRVCC because it reduces both voice interruption time during handover and the dropped call rate compared to earlier configurations. The network controls and moves the UE from E-UTRAN to UTRAN/GERAN as the user moves out of the LTE network coverage area. The SRVCC handover mechanism is entirely network-controlled and calls remain under the control of the IMS core network, which maintains access to subscribed services implemented in the IMS service engine throughout the handover process. 3GPP Rel-10 configuration includes all components needed to manage the time-critical signaling between the user's device and the network, and between network elements within the serving network, including visited networks during roaming. As a result, signaling follows the shortest possible path and is as robust as possible, minimizing voice interruption time caused by switching from the packet-switched core network to the circuit-switched core network, whether the UE is in its home network or roaming. With the industry aligned around the 3GPP standard and GSMA recommendations, SRVCC-enabled user devices and networks will be interoperable, ensuring that solutions work in many scenarios of interest.

As shown in Figure 2.22, the SRVCC functionality can be added to the network by upgrading the mobile soft-switch software as well as the IMS and the LTE/EPC subsystems. The legacy UTRAN/GERAN access networks are not required to be upgraded. When a user with an active voice call using VoLTE moves to an area without LTE coverage, the handover of the call to the network is initiated with inter-RAT handover and is completed with a session transfer. Inter-RAT handover is the conventional method of transferring the UE from E-UTRAN to UTRAN/GERAN. Session transfer is a new mechanism to move access control and voice media anchoring from the EPC to the legacy circuit-switched core network. During the entire voice handover process, the IMS retains control of the user. The handover process is initiated by a session transfer request to the IMS. The IMS/multimedia telephony (MMTel) responds simultaneously with two commands, one to the LTE network and one to the 2G/3G network. The LTE network in which the user's voice call is in progress receives an inter-RAT handover execution command through the MME and the E-UTRAN to instruct the UE to prepare to move to the circuit-switched network. The circuit-switched network, where the user's voice call is being transferred, receives a session transfer response, preparing it to accept the call in progress. With acknowledgments that the commands have been executed, the UE and the IMS, switch to the circuit-switched network to continue the call.

While the implementation of SRVCC does not directly impact the legacy UTRAN/GERAN nodes, an indirect impact arises from methods to help the UE return to the LTE network as soon as the voice call has ended. To help transfer the device back to the LTE network, the legacy RAN can implement functionality to either broadcast LTE system information to the UE, with which the device can perform a cell reselection to an LTE cell after connection release or release connection to the UE and simultaneously redirect the UE to the LTE network.

The inter-RAT handover and session transfer procedures, as well as the number of sub-systems requiring upgrades, have raised concerns about whether SRVCC can meet the voice interruption time of less than 300 ms. It must be noted that the inter-RAT handover in the E-UTRAN and the session transfer in the core network contribute to an interruption time as they break and make the connection. To address this issue, the SRVCC in 3GPP Rel-10 network architecture has minimized the interruption time by initiating the two procedures simultaneously so that they can be executed in parallel. The studies have shown that the session transfer process can be performed in the order of 10 ms; therefore, the voice call interruption time is mainly affected by the inter-RAT handover latency. The inter-RAT handover delay is measured from the time that the UE receives a network handover command to the time that the UE is uplink synchronized with the new radio access network.

Along with the introduction of the LTE radio access network, 3GPP also standardized SRVCC in Rel-8 specifications to provide seamless service continuity when a UE performs a handover from the E-UTRAN to UTRAN/GERAN. With SRVCC, calls are anchored in the IMS network while the UE is capable of transmitting/receiving on only one of those access networks at a given time, where a call anchored in the IMS core can continue in UMTS/GSM networks and outside of the LTE coverage area. Since its introduction in Rel-8, the SRVCC has evolved with each new release, a brief summary of SRVCC capability and enhancements are noted below [48,49].

- **3GPP Rel-8:** Introduces SRVCC for voice calls that are anchored in the IMS core network from E-UTRAN to CDMA2000 and from E-UTRAN/UTRAN (HSPA) to UTRAN/GERAN circuit-switched. To support this functionality, 3GPP introduced new protocol interface and procedures between MME and MSC for SRVCC from E-UTRAN to UTRAN/GERAN, between SGSN and MSC for SRVCC from UTRAN (HSPA) to UTRAN/GERAN, and between the MME and a 3GPP2-defined interworking function for SRVCC from E-UTRAN to CDMA 2000.
- **3GPP Rel-9:** Introduces the SRVCC support for emergency calls that are anchored in the IMS core network. IMS emergency calls, placed via LTE access, need to continue when SRVCC handover occurs from the LTE network to GSM/UMTS/CDMA2000 networks. This evolution resolves a key regulatory exception. This enhancement supports IMS emergency call continuity from E-UTRAN to CDMA2000 and from E-UTRAN/UTRAN (HSPA) to UTRAN/GERAN circuit-switched network. Functional and interface evolution of EPS entities were needed to support IMS emergency calls with SRVCC.
- **3GPP Rel-10:** Introduces procedures of enhanced SRVCC including support of mid-call feature during SRVCC handover (eSRVCC); support of SRVCC packet-switched to circuit-switched transfer of a call in alerting phase

(aSRVCC); MSC server-assisted mid-call feature enables packet-switched/circuit-switched access transfer for the UEs not using IMS centralized service capabilities, while preserving the provision of mid-call services (inactive sessions or sessions using the conference service). The SRVCC in alerting phase feature adds the ability to perform access transfer of media of an instant message session in packet-switched to circuit-switched direction in alerting phase for access transfers.

- **3GPP Rel-11:** Introduces two new capabilities: single radio video call continuity for 3G-circuit-switched network (vSRVCC); and SRVCC from UTRAN/GERAN to E-UTRAN/HSPA (rSRVCC). The vSRVCC feature provides support of video call handover from E-UTRAN to UTRAN-circuit-switched network for service continuity when the video call is anchored in IMS and the UE is capable of transmitting/receiving on only one of those access networks at a given time. Service continuity from UTRAN/GERAN circuit-switched access to E-UTRAN/HSPA was not specified in 3GPP Rel-8/9/10. To overcome this drawback, 3GPP Rel-11 provided support of voice call continuity from UTRAN/GERAN to E-UTRAN/HSPA. To enable video call transfer from E-UTRAN to UTRAN-circuit-switched network, IMS/EPC is evolved to pass relevant information to the EPC side and S5/S11/Sv/Gx/Gxx interfaces are enhanced for video bearer-related information transfer. To support SRVCC from GERAN to E-UTRAN/HSPA, GERAN specifications are evolved to enable a mobile station and base station sub-system to support seamless service continuity when a mobile station hands over from GERAN circuit-switched access to E-UTRAN/HSPA for a voice call. To support SRVCC from UTRAN to E-UTRAN/HSPA, UTRAN specifications are evolved to enable the RNC to perform rSRVCC handover and to provide relative UE capability information to the RNC.

2.11.3 Voice over LTE

Some of the earliest LTE deployments addressed the lack of circuit-switched voice support in LTE by utilizing a dual-radio solution in which the mobile device was required to simultaneously connect to the E-UTRAN and a legacy circuit-switched network, the latter being used to provide voice and messaging services. The advantage of a simultaneous voice and LTE (SVLTE) solution was that circuit-switched voice and packet-switched data services were supported at the same time. However, an SVLTE-based solution would add significant hardware complexity to the mobile device, given that two completely independent baseband and RF chains were required with obvious cost implications. Furthermore, an SVLTE-based solution has proven to be very power-consuming in practice. On the other hand, since the LTE and the legacy cells do not necessarily have identical footprints, the user experience during mobility can be less than desired. There are several options when it comes to solving the LTE voice services problem. Those range from multi-radio solutions, to solutions which tunnel legacy signaling through LTE, to SIP-based

VoIP solutions, to not using LTE for voice at all, instead relying on legacy networks to provide voice services. The VoLTE solution has developed a framework for optimal support of voice and SMS over IMS in E-UTRAN including roaming and inter-operator issues. VoLTE is based on the existing IMS MMTel[10] concepts and has been specified in GSMA profile IR.92 [45–47]. The GSMA IR.92 (IMS profile for voice and SMS) is a profile as opposed to a technical specification. Therefore, rather than specifying new technologies, this document defines how the framework and services already specified by 3GPP should be configured and utilized in a manner that facilitates interoperability among mobile devices and the serving networks, as well as amongst the networks [64,72,76].

The IR.92 profile requires the mobile device to indicate its support for MMTel-based supplementary services during SIP registration, and if the device supports SMS-over-IP it is also required to be indicated. The terminals are further required to provide their IMEI during SIP registration, which allows the operators to implement management policies in the P-CSCF which are device or device class specific. IR.92-compliant devices are required to support network-initiated deregistration, so that the network can force the mobile device to detach in special circumstances.

IR.92 leverages the existing mechanisms for IMS authentication, that is, IMS AKA. VoLTE-enabled UEs are expected to install a universal integrated circuit card (UICC) equipped with the ISIM application. The ISIM application contains the credentials critical to authentication, such as the subscriber's IP multimedia private identity (IMPI), the home carrier's domain name, and a shared secret key, which is used for encryption. If the ISIM application is absent from the UICC, IMS AKA procedures can still be performed using the credentials contained on the UE's USIM.

A mobile-originated call is handled by an MMTel server. This is accomplished by setting the IMS communication service identifier (ICSI) in the communication invitation to point to the MMTel service, resulting in calls being looped by S-CSCF through the MMTel server in each subscriber's home network. In SIP protocol, which IMS relies on, call control and bearer (transport) control are essentially independent. In the context of the IR.92 profile, there are two situations in

[10]IMS MMTel service is a global standard based on the IMS, offering converged, fixed, and mobile real-time multimedia communication using the media capabilities, such as voice, real-time video, text, file transfer and sharing of pictures, audio and video clips. With MMTel, users can add and drop media during a session. One can start with chat, add voice, add another caller, add video, share media and transfer files, and drop any of these without losing, or having to end, the session. MMTel allows a single SIP session to control virtually all MMTel supplementary services and MMTel media. All available media components can easily be accessed or activated within the session. Employing a single session for all media parts means that no additional sessions need to be set up to activate video, to add new users, or to start transferring a file. Although it is possible to manage single-session user scenarios with several sessions, for example, using a circuit-switched voice service that is complemented with a packet-switched video session, a messaging service, or both, there are some benefits in taking MMTel's single-session approach [36].

which it is desirable to have some communication between the two domains. The first is during call session establishment. It is possible for a call-terminating device to accept a call invitation (and to alert the terminating subscriber). This is not normally a problem in legacy circuit-switched systems because call and bearer control are relatively more coupled. In order to minimize the possibility of this occurrence, VoLTE requires the use of SIP pre-conditions framework. This ensures that the terminating subscriber is not notified of the incoming call unless it is certain that there are sufficient resources to support the call (with the required QoS). The second situation in which call and bearer control require some communication is in the context of call drops and call establishment failures. If, for example, communication between the LTE/System Architecture Evolution (SAE) network and the Internet is lost, or if the radio link fails, IR.92 requires the IMS call session to be reestablished or cleared gracefully on the mobile device side and on the network side. VoLTE emergency calls must also be supported by the UE and the network according to regulatory requirements.

The IMS calls are processed by the subscriber's S-CSCF in the home network. The connection to the S-CSCF is, however, indirect and via P-CSCF. Since the P-CSCF varies with time and the mobile device's serving network, an important step in the enablement of voice calling capabilities is the discovery of the P-CSCF.

In legacy systems, telephony features are managed via dedicated signaling messages. Since call control in VoLTE uses completely different signaling, supplementary services and features in VoLTE are managed differently. IR.92 requires the use of XCAP protocol. Using XCAP, the UE connects to the MMTel server via HTTP and utilizes the standard PUT, GET, and DELETE methods to modify XML documents, which describe the supplementary services and their state.

3GPP adaptive multi-rate (AMR) and AMR-wideband (AMR-WB) voice codecs are specified for VoLTE. This choice simplifies interoperability with legacy 3GPP networks, in particular, in the context of transcoder-free operation. In order to ensure interoperability with all legacy telephony networks, in addition to the necessary voice transcoders, support for transmission of dual-tone multi-frequency (DTMF) signaling, which is used for telecommunication signaling over analog telephone lines, is mandatory.

VoLTE requires the use of a particular set of radio bearers, specifically, two signaling radio bearers (SRBs) and two or three data radio bearers (DRBs). The SRBs (SRB1 and SRB2) in E-UTRAN transport the dedicated control channel, which is used for RRC message exchange between the UE and the eNB, as well as NAS message exchange between the UE and the MME. These radio bearers always operate in RLC acknowledged mode. The two (or three) DRBs carry SIP (call control) and XCAP (service management) signaling as well as voice packets. The first DRB corresponds to the default EPS bearer and is used to transport SIP and XCAP signaling messages. The associated QCI index is 5, which means that the bearer operates in non-GBR mode but with high priority, low latency, and very low packet loss. The second DRB corresponds to a dedicated EPS bearer and is used to transport voice packets. The associated QCI index is 1, which means that the bearer operates in GBR mode with high priority and low latency, but with relatively higher packet loss. There may also be an additional DRB operating with QCI index 8/9 (low-priority, high-latency, and very low packet loss) used for the transport of XCAP messages.

All networks and mobile devices are required to utilize a common APN for VoLTE, which in this case is IMS. Unlike the legacy networks, LTE networks employ an always-on concept which means that devices have PDN connectivity virtually from the moment they attempt their initial attach to the core network. During the initial attach procedure, some devices choose to identify the access point through which they prefer to connect. However, mobile devices are not permitted to name the VoLTE APN during initial attach, that is, to utilize the IMS as their main PDN, rather to establish a connection with the IMS application profile separately. Thus, VoLTE-enabled devices must support multiple simultaneous default EPS bearers. Note that because the VoLTE APN is universal, mobile devices will always connect through the visited PLMN's IMS P-GW.

We mentioned earlier that VoLTE sessions utilize two or three DRBs. This implies the use of one default EPS bearer plus one or two dedicated EPS bearers. The default EPS bearer is always used for SIP signaling, and exactly one dedicated EPS bearer is used for voice packets regardless of the number of active voice media streams. The XCAP signaling may be transported over its own dedicated EPS bearer or it may be multiplexed with the SIP signaling on the default EPS bearer, in which case only two EPS bearers are utilized.

The VoLTE-compatible mobile devices are required to operate in dual-stack IPv4/IPv6 mode (i.e., IPv4 and IPv6 networks run in parallel) by default [28]. If the IMS application profile assigns an IPv6 address, the device is required to prefer that address and specifically to utilize it during P-CSCF discovery. Since IP header overhead in VoIP calls can be considerable, VoLTE mandates IP header compression and the use of RoHC protocol for voice packets.

In the EPS network, mobility is implicitly handled for VoLTE by the defined mobility management mechanisms. A challenge arises during mobility to/from UMTS/GSM. UMTS/GSM utilizes a different call model than VoLTE. The call processing and service delivery entities are the MSC/Visited Location Register (VLR) and HLR rather than the CSCFs and MMTel servers. Furthermore, the bearer/transport channels used to exchange voice packets are circuit-switched and not packet-switched. Part of the solution to this problem is known as IMS centralized services. With IMS centralized services, voice calls are always anchored in the IMS, regardless of the domain from which they are originated. This solves the problem of transfer of call control as the mobile device moves between different radio access technologies; nevertheless, the packet-switched bearer challenge still remains. This problem inspired the concept of packet-switched over circuit-switched in UMTS and GSM, where a circuit-switched bearer is utilized to transport voice packets for calls anchored in the IMS. Voice call continuity during radio handover between LTE and the UMTS/GSM circuit-switched domain is managed by an SRVCC mechanism.

In conjunction with support for voice services, VoLTE-compatible devices also must support sending and receiving SMS over IP and instant messaging services. Furthermore, VoLTE-compatible networks must be capable of functioning as an IP short message gateway. Note that SMS over IP represents an alternative mechanism for providing short messaging services to users. Since mobile devices might support both of these SMS over LTE solutions, the operators are required to provision the preferred SMS transport via the IMS management object [53,64].

2.11.4 Circuit-switched fallback

Circuit-switched fallback (CSFB) is an alternative mechanism to move a subscriber from LTE to a legacy radio access technology in order to use circuit-switched voice services (see Figure 2.24). This function is only available, if LTE network coverage is overlapped by GSM, UMTS, or CDMA2000 coverage. The CSFB was already specified in 3GPP Rel-8, with further enhancements defined in later releases [5]. A number of different CSFB mechanisms are available and there are also differences depending on the radio access technology to which the subscriber falls back. The CSFB to GSM, UMTS, and CDMA2000 is defined, but no mechanisms are specified to support CSFB to both UMTS/GSM and CDMA2000 in the same PLMN. In this section, we describe the CSFB mechanisms for UMTS and GSM followed by CSFB procedure to CDMA2000. As mentioned earlier, depending on the target circuit-switched radio access technology, several CSFB solutions exist. As an example, there are three different solutions for CSFB to UMTS, two using the RRC connection release with redirection mechanism, and one using the packet-switched handover mechanism. For CSFB to GERAN, in addition to the RRC connection release with redirection and the packet-switched handover mechanisms, cell change order with or without network-assisted cell change can be used.

The CSFB affects the radio and the core networks. Figure 2.25 illustrates the modified EPS architecture with support for CSFB. It includes the interfaces between the different radio access network types and the core network entities. Note that UTRAN is the UMTS access network and GERAN is the GSM/EDGE access network. To support circuit-switched services, a connection to the MSC server must be established. The MME interfaces with the MSC server via the SGs interface. The CSFB mechanism is implemented using the SGs interface.

The legacy 2G/3G and the EPS can coexist in the form of hybrid networks in the transition period of network upgrades, where the MME serves LTE users, the SGSN provides service to UMTS users when utilizing data services, and the MSC server serves the users when utilizing circuit-switched voice services. The MSC server connects to the operator's telephony network. To support CSFB signaling and SMS transfer for LTE devices, the MME connects to the MSC server. The architecture shown in Figure 2.25 is a simplified schematic of the hybrid E-UTRAN and UTRAN/GERAN networks. The SGs interface between the MSC server and the MME enables the UE to register with both circuit-switched and

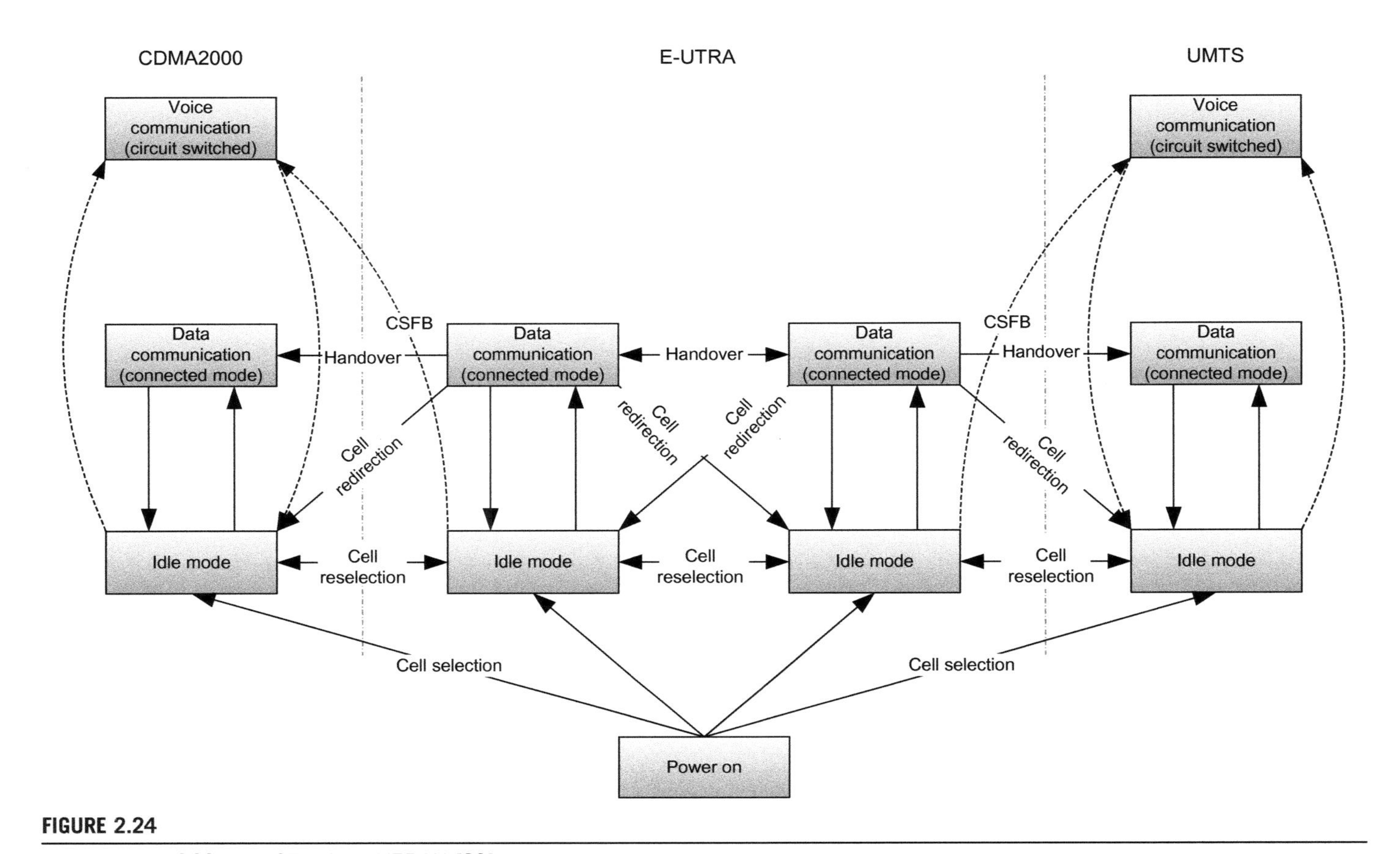

FIGURE 2.24

The concept of CSFB to GERAN or UTRAN [39].

packet-switched networks while on the LTE access network. This interface also enables the delivery of circuit-switched paging messages via the LTE access, as well as SMS, without having the device leave the LTE network. With the default LTE data network connection in operation, a mobile-terminated or incoming circuit-switched voice call triggers a paging message via the LTE network to the UE, as shown in Figure 2.26. The paging message initiates the CSFB operation, as the device sends an extended service request to the network to transition to UMTS/GSM access networks.

Upon transition to circuit-switched mode, the legacy call setup procedures are followed to set up the circuit-switched call. Mobile-originated or outgoing calls follow the same transition from packet-switched LTE to circuit-switched UMTS/GSM, except for the paging step, which is not needed. In UMTS networks, packet-switched data sessions can also move for simultaneous voice and data services. In GSM networks, packet-switched data sessions may be suspended until the voice call ends and the device returns to LTE mode, unless the GSM network supports dual transfer mode, which permits simultaneous voice and data. When the voice call ends, the device returns to LTE mode via idle mode or connected mode mobility procedures, as shown in Figure 2.26.

In EPS network architecture, the CSFB enables the provisioning of voice and traditional circuit-switched domain services. To provide these services, the EPS reuses circuit-switched infrastructure when the UE is served by E-UTRAN. A CSFB-enabled terminal, connected to E-UTRAN, may use GERAN or UTRAN to connect to the circuit-switched domain. This function is only available when the E-UTRAN coverage is overlapped by GERAN or UTRAN coverage. Figure 2.25 illustrates the architecture for CSFB in EPS. The CSFB and IMS-based services can coexist in the same operator's network. Although it is not very straightforward to support CSFB, all participating elements, i.e., UE, MME, MSC, and E-UTRAN must be modified to support additional functionalities.

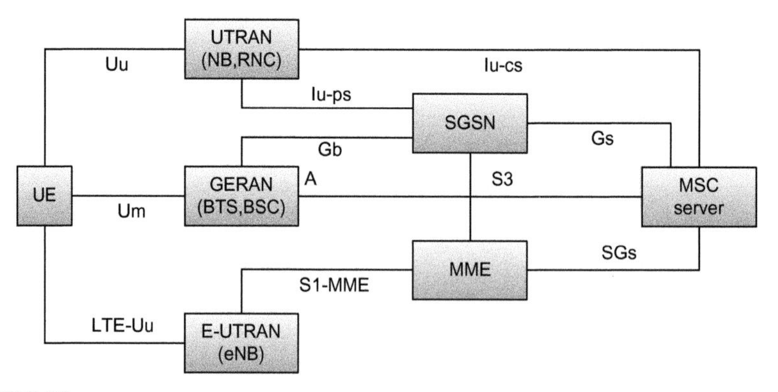

FIGURE 2.25

Modified EPS network architecture to support CSFB [5].

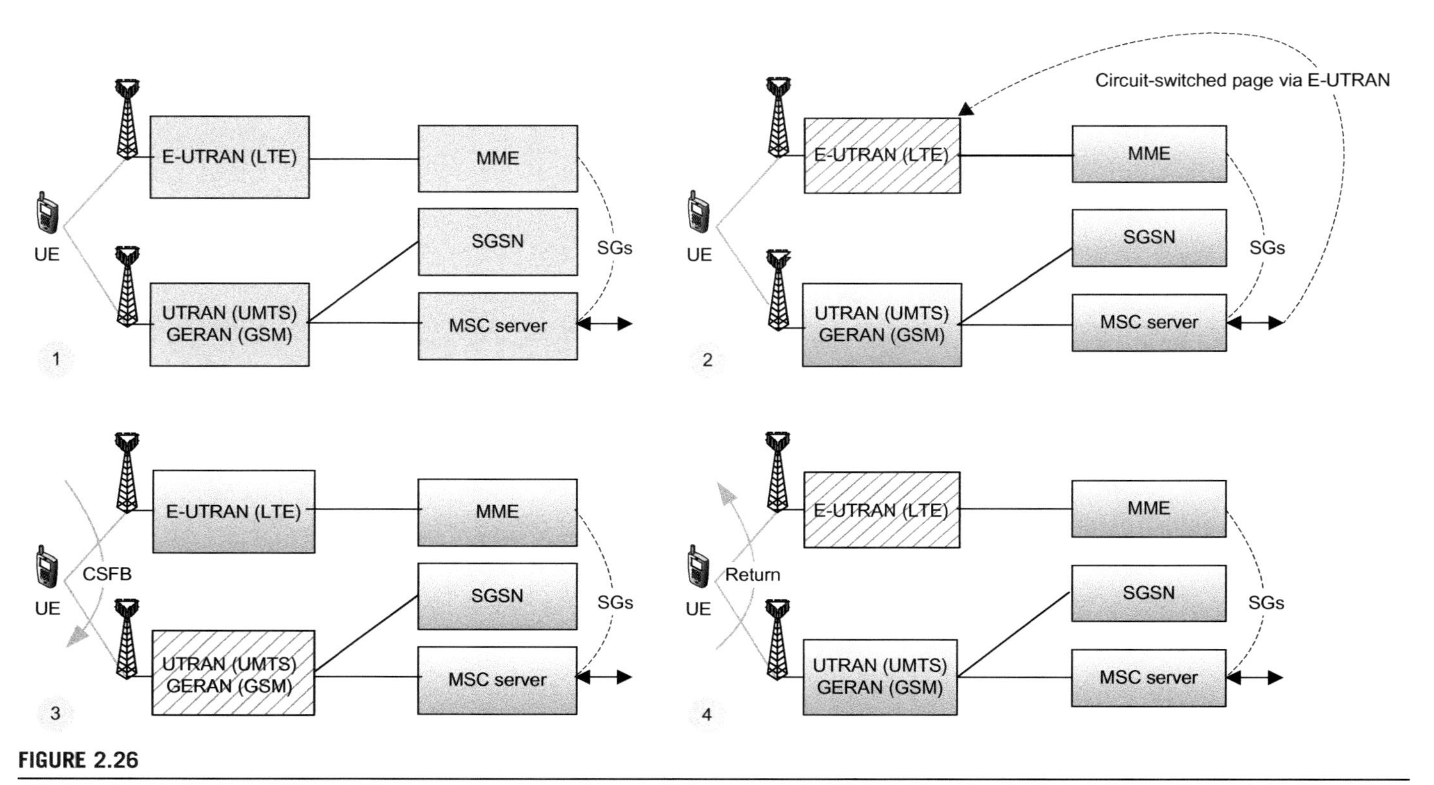

FIGURE 2.26

CSFB call processing in the hybrid network [63].

The CSFB support in EPS includes a new interface SGs, which is the reference point between the MME and MSC server. The SGs interface is used for mobility management and paging procedures between the EPS and the circuit-switched domain. The SGs reference point is also used for delivery of both mobile-originated and mobile-terminated SMS service. The CSFB-enabled network elements are required to support the following additional functions [54]:

- **UE:** Support access to E-UTRAN/EPC as well as access to the circuit-switched domain over GERAN and/or UTRAN; combined procedures for EPS/IMSI attach, update, and detach; and CSFB and SMS procedures for using circuit-switched domain services.
- **MME:** Deriving a VLR number and location area identity (LAI)[11] from the GUTI received from the UE or from a default LAI; maintaining of SGs association toward MSC/VLR for an EPS/IMSI attached UE; initiating IMSI detach at EPS detach; initiating paging procedure toward eNB when MSC pages the UE for circuit-switched services; support of SMS procedures; rejecting CSFB call request; and use of the LAI and a hash value from the IMSI to determine the VLR number when multiple MSC/VLRs serve the same LAI.
- **MSC:** Maintaining SGs association toward MME for an EPS/IMSI attached UE and support of SMS procedures specified by 3GPP.
- **E-UTRAN:** Forwarding paging request and SMS to the UE and directing the UE to the target circuit-switched-capable cell.

A CSFB and IMS-capable UE would follow the procedures for domain selection for UE originating session/calls. If a UE is configured to use SMS over IP services and it is registered to IMS, then it would send SMS over IMS, even if it is EPS/IMSI attached. The home operator has the option to activate/deactivate the UE configuration to use SMS over IP by means of device management in order to allow alignment with home PLMN support of SMS over IP. When UE performs CSFB procedure for a mobile-originated call for the purpose of an emergency call, it needs to indicate to the MME that this CSFB request is for an emergency purpose.

When the UE is paged via LTE with an incoming call, or when the user initiates an outgoing call, the device switches from LTE to UMTS/GSM mode. The acquisition of the UMTS/GSM network and setup of the call can be done in one of following two ways: handover or redirection. In the handover procedure, the target cell is prepared in advance and the device can enter that cell directly in the connected mode. Inter-RAT measurements of signal strength may be conducted by the UE while on the LTE system prior to executing the handover. In the redirection procedure, only the target frequency is indicated to the device.

[11]Each location area of a public land mobile network has its own unique identifier which is known as its LAI. This internationally unique identifier is used for location update of mobile subscribers. It is composed of a three decimal digit mobile country code, a two to three digit mobile network code that identifies the PLMN in that country, and a location area code which is a 16 bit number, thereby allowing 65,536 location areas within one PLMN.

The device is then allowed to select any cell on the indicated frequency, or may even try other frequencies/RATs, if no cell can be found on the target frequency. Once a cell is found, the device initiates normal call setup procedures. Inter-RAT measurements of signal strength are not needed prior to redirection. Consequently, CSFB with redirection may require less time to identify the best cell compared to the handover procedure [49,63,76].

Redirection-based CSFB has variations with differing call setup latencies. In 3GPP Rel-8 basic release and redirection approach, the UE follows LTE Rel-8 procedures and reads the system information which is broadcast through system information block (SIB) messages prior to accessing the target cell. The device can alternatively follow LTE Rel-8 procedures but may read only the mandatory SIBs 1, 3, 5, and 7, and skip other SIBs prior to access to the target cell. In the latter case, the neighbor cell information in SIB11 is delivered to the UE via measurement control messaging once the UE is in the connected mode on the target cell. This approach can be either implicitly supported by the UE and the network, or explicitly through deferred measurement report control signaling.

In 3GPP Rel-9 enhanced connection release with redirection and system information tunneling approach, the device follows LTE Rel-9 procedures, where SIB information can be tunneled from the target access network via the core network to the source access network, and can be included in the redirection message sent to the device. This can avoid reading any SIBs on the target cell. The solutions deployed today are predominantly based on 3GPP Rel-8 connection release with redirection with SIB skipping, in order to achieve acceptable call setup times, reliability, and to simplify deployments. While it is anticipated that 3GPP Rel-9 enhanced connection release with redirection will be deployed soon, a handover-based CSFB approach has not found any traction, since both release with redirection with SIB skipping and enhanced connection release with redirection with system information tunneling can address any call setup latency issues that may arise as a result of the basic release with redirection solution.

Another key issue for the voice call user experience is call setup reliability, which is the ability to successfully establish an incoming or outgoing call on the first attempt, or within a time frame that does not indicate call setup failure. The target is to at least match legacy performance, which is in the 98% range. With CSFB switching between LTE and UMTS networks, there are two primary challenges to call setup success: changes in inter-RAT conditions between measurement and acquisition for handover-based CSFB, and mismatches between LTE and 3G geographic signal coverage areas that require server-side updates. With handover-based CSFB, inter-RAT measurements can change between the time the measurement is conducted using LTE and the time UMTS/GSM/CDMA2000 voice network acquisition is attempted. During this time, the cell identified and prepared for handover may become unavailable, resulting in connection failure [63].

In handover-based CSFB, the measurement is conducted before the inter-RAT transition. If the inter-RAT conditions adversely change, there is a higher probability of handover failure, especially in high mobility scenarios. In practice, it is shown the failure rate of inter-frequency handover or inter-RAT handover are relatively higher, suggesting that delays between inter-RAT measurement and network acquisition can increase setup failures. In contrast, redirection-based CSFB can result in higher call setup success rates relative to handover-based CSFB, because redirection-based CSFB solutions conduct inter-RAT measurements immediately before attempting access to the identified cell. Since the 3GPP Rel-9 system information tunneling and 3GPP Rel-8 redirection-based CSFB methods have shown satisfactory call setup times and higher reliability, the industry has taken the redirection-based CSFB approach [63].

CSFB to CDMA 1xRTT is supported for UEs with both single and dual receiver chain configurations. UEs with single receiver configuration are not able to camp in CDMA 1xRTT when they are active in E-UTRAN. The EPC provides mechanisms for the UE to perform registration procedure with CDMA 1xRTT, receive paging message, and send/receive SMS, while the UE is camped in E-UTRAN. UEs with dual receiver configuration can camp in CDMA 1xRTT while they are active in E-UTRAN, they may not be able to stay in E-UTRAN when they handle a circuit-switched call and/or perform registration signaling, and/or are sending or receiving SMS in CDMA 1xRTT.

The CSFB for CDMA 1xRTT in EPS enables the delivery of circuit-switched domain services (e.g., circuit-switched voice) by reusing CDMA 1xRTT circuit-switched infrastructure when the UE is served by the E-UTRAN. A CSFB-enabled terminal, while connected to E-UTRAN, may register in the CDMA 1xRTT circuit-switched domain in order to be able to use CDMA 1xRTT access to establish one or more circuit-switched services in the circuit-switched domain. The CSFB function is only available where E-UTRAN coverage overlaps with CDMA 1xRTT coverage. 3GPP specifications also specify the architecture for SMS over S102 interface in EPS. The mobile-originated or terminated SMS is tunneled in EPS over S102 and does not cause any CSFB to CDMA 1xRTT and consequently does not require any overlapped CDMA 1xRTT coverage. The CSFB to CDMA 1xRTT and IMS-based services can coexist in the same operator's network [5].

The CSFB to CDMA 1xRTT and SMS over S102 in EPS function are realized by reusing the S102 reference point between the MME and the 1xCS interworking solution (IWS) [16–18]. The reference architecture described in Figure 2.27 is similar to the SRVCC architecture for E-UTRAN to 3GPP2 1xCS, with the additional feature that the S102 session is extended similar to preregistration for S101. The S102 is the reference point between the MME and the 1xCS IWS which provides a tunnel between MME and 3GPP2 1xCS IWS to relay 3GPP2 1xCS signaling messages [14–18].

2.12 LIPA, selected IP traffic offload, and IP flow mobility

Femto-cells have been recently introduced to provide enhanced indoor coverage and to offload macro-cells. 3GPP Rel-10 network architecture introduced three

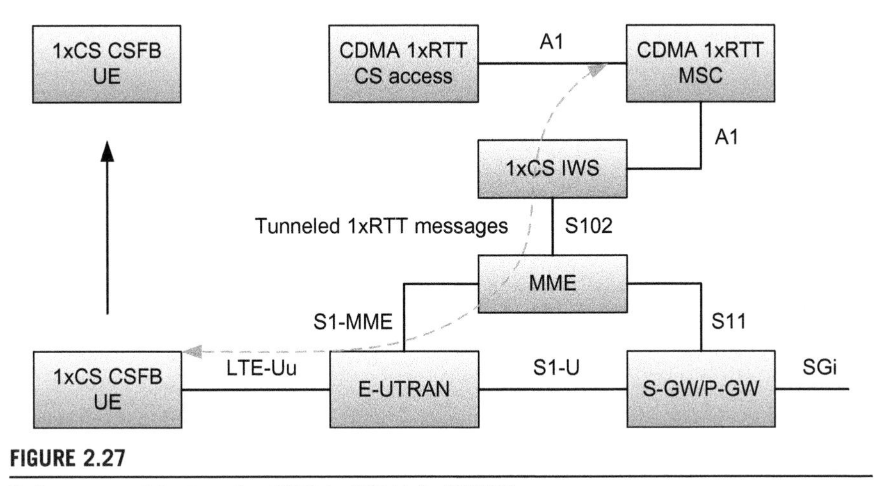

FIGURE 2.27

Reference architecture for CSFB to CDMA 1xRTT [5].

important concepts, namely, LIPA, selected IP traffic offload (SIPTO), and IP flow mobility [57,59].

2.12.1 Local IP access

LIPA is the ability for an IP-enabled UE to directly access a consumer's home local area network as well as the Internet (through the operator's IP core network) using the radio-link of a femto-cell or a home eNB (HeNB). The use of LIPA enables improved performance, more innovative services that interconnect mobile stations and home local area networks, and offloading of traffic from the operator's IP core network which is ultimately through the Internet. Enabling LIPA benefits mobile operators, subscribers, and Internet service providers. For mobile operators, the ability to offer LIPA via an HeNB would mean higher revenues by way of value-added services without having to invest significantly on upgrading network infrastructure to cope with the higher throughput and speed that such applications would demand. For subscribers, LIPA would mean faster and more secure data transfer, since traffic within the home network would not travel outside the subnet. Users could add customized high-speed applications such as file transfer, video streaming, and device sharing without involving the operator's core network, thus avoiding associated bottlenecks and possibly yielding lower data costs.

The requirements for LIPA can be generally classified into two main categories: LIPA to the home network and IP access to the Internet. For LIPA to the home network, the UE should be able to exchange data via the operator's core network and LIPA to the home-based network simultaneously, depending on the destination. The HeNB should decide, based on certain criteria, whether to route the data via the core network or to use LIPA. Furthermore, the UE should be able to access other devices in the home network via the HeNB using the HeNB only as an access medium. Access to local IP through the HeNB E-UTRAN interface should only be granted to UEs with a valid subscription. This means that no specific modifications need to be done to the UE to support LIPA. In addition, it should be possible for a device in the home-based network to contact a UE via LIPA. Therefore, there is not any impact to the UE in accessing packet data services through the cellular core network, resulting in minimal changes to the protocol stacks in the core network.

IP access to the Internet should be possible without going through the operator's network. Simultaneous access from a UE to both the operator's core network and LIPA to the Internet are also possible. The operator or the HeNB owner, within the limits set by the operator, should be able to enable/disable LIPA to the Internet per HeNB. Furthermore, LIPA to the Internet should support a scenario where the HeNB and backhaul are provided by the same operator, as well as a scenario where the HeNB and backhaul are provided by different operators.

Basic functionality for LIPA was specified in 3GPP Rel-10. LIPA signifies the capability of a UE to obtain access to a local residential/enterprise IP network, subsequently called a local network, that is connected to one or more HeNBs. The LIPA functionality is being extended in 3GPP Rel-12 to allow access to the local network when a UE is in the coverage area of a macro-eNB and to provide mobility support (see Figure 2.28). LIPA allows a UE to work with devices in the local network, e.g., printers, video cameras, or a local web-server. LIPA allows the UE to discover supporting devices and to be discovered, and does not require the local network to be connected to the Internet but achieves IP connectivity with the UE through one or more HeNBs of the mobile operator. 3GPP Rel-10 only specified the support of LIPA when the UE accesses the local network via HeNB.

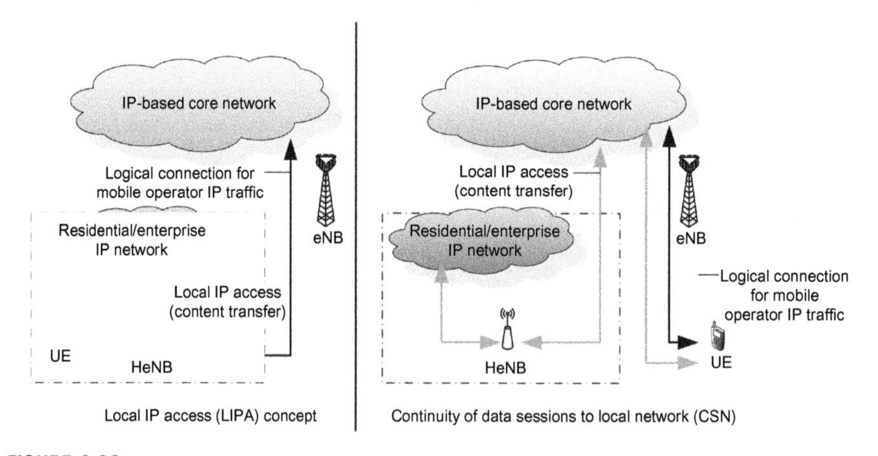

FIGURE 2.28

LIPA and continuity of data sessions concepts [73].

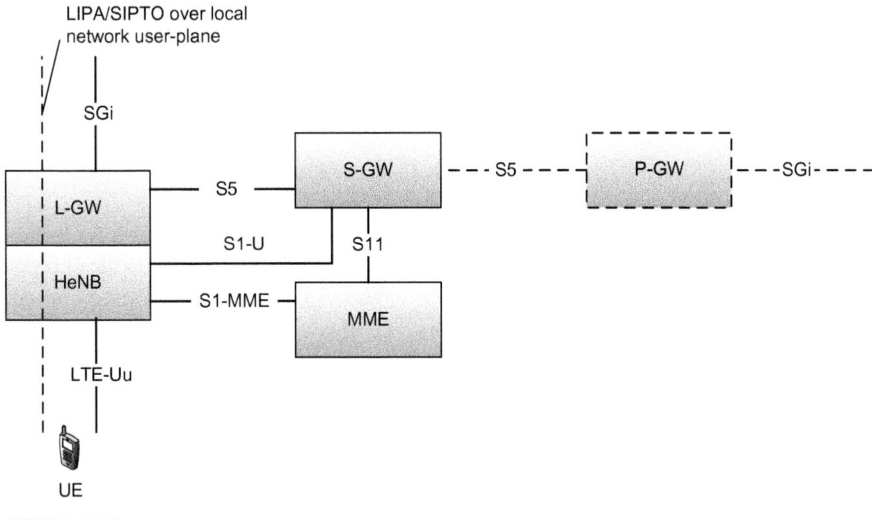

FIGURE 2.29

Network architecture to support LIPA or SIPTO over local network with L-GW collocated with the HeNB [73].

An operator may want to further provide access to the local network to a UE that is in the coverage area of a macro network. In 3GPP Rel-10, the UE was required to be able to maintain IP connectivity to the local network when moving between HeNBs within the same local network. However, access to the local network may be lost as a UE moves out of HeNB coverage into the macro network, even if other services (e.g., telephony, data services, and SIPTO) may continue after a handover to the macro network, resulting in an unsatisfactory user experience. A 3GPP Rel-12 study item will allow continuation of data sessions to the local network, when the UE moves between HeNB and the macro network (see Figure 2.28).

From an architectural perspective, a HeNB subsystem comprises a HeNB entity and it may optionally include a HeNB gateway (HeNB-GW) and a local gateway (L-GW). The LIPA and SIPTO functionalities over the local network are achieved using an L-GW that may be collocated with the HeNB. Figure 2.29 illustrates the network architecture to support LIPA and/or SIPTO over the local network with L-GW function collocated with the HeNB. It is shown that the HeNB subsystem is connected by means of an S1 interface to the EPC, and more specifically to the MME by means of an S1-MME interface and to the S-GW by means of an S1-U interface. When LIPA or SIPTO functions over the local network are activated (when the L-GW is collocated with the HeNB), the L-GW has an S5 interface with the S-GW. The local GW is the gateway toward the IP networks (e.g., residential/enterprise networks and Internet) associated with the HeNB. The L-GW performs some P-GW type functions including UE IP address allocation; DHCPv4 (server and client), and DHCPv6 (client and server) functions; packet screening, as well as ECM-IDLE mode downlink packet buffering; and ECM-CONNECTED mode direct tunneling toward the HeNB [6].

The architecture for SIPTO over the local network with L-GW function collocated with a HeNB is depicted in Figure 2.29. For LIPA traffic, the addition of an L-GW function in EPS that can be either collocated with the HeNB or as a standalone node can provide partial P-GW and S-GW downlink data buffering functions. The S-GW in the core network serves the core network traffic. The L-S11 interface between the L-GW and the MME is used to manage the session for LIPA traffic. For the LIPA PDN connection, the L-GW needs to be located close to the HeNB, and to establish the connection through an L-S11 interface. When the UE is in the idle mode, the LIPA downlink packets are buffered in the L-GW. Upon UE transition to the connected mode, the packets buffered in L-GW are forwarded over the path between the L-GW and the HeNB. It is assumed that the user-plane does not go through the HeNB-GW for the direct tunnel [6].

The LIPA is realized using the L-GW entity that may be collocated with the HeNB. LIPA is established when the UE requests a new PDN connection to an APN for which LIPA functionality is permitted. In this case, the network selects the L-GW associated with the HeNB and enables a direct user-plane path between the L-GW and the HeNB. A HeNB supporting the LIPA function includes the L-GW address in the UE-related control messaging with the MME. The protocol option (i.e., GTP or PMIP) supported on the S5 interface between L-GW and S-GW is configured by the MME. As of 3GPP Rel-11, no interface between the L-GW and the PCRF is specified, and there is no support for dedicated bearers on the PDN connection used for LIPA. The L-GW rejects any UE requests for bearer resource modification.

The direct user-plane path between the HeNB and the collocated L-GW is enabled with a correlation ID parameter that is associated with the default EPS bearer on a PDN connection used for LIPA. Upon establishment of the default EPS bearer, the MME sets a correlation ID equal to the P-GW TEID (GTP-based S5) or the P-GW generic routing encapsulation (GRE) key (PMIP-based S5). The correlation ID is then signaled by the MME to the HeNB as part of E-RAB establishment and is stored in the E-RAB context in the HeNB. The correlation ID is used in the HeNB for matching the radio bearers with the direct user-plane path connections from the collocated L-GW. If the UE is roaming and if the HSS allows LIPA roaming for this UE in the visited PLMN, then the MME may provide LIPA for this UE.

LIPA is supported for APNs that are valid only when the UE is connected to a specific CSG. The MME releases a LIPA PDN connection to an APN, if it detects that the UE's LIPA CSG authorization data for this APN has changed and the LIPA PDN connection is no longer allowed in the current cell. As mobility of the LIPA PDN connection is not supported in 3GPP Rel-11, the LIPA PDN connection is released when the UE moves away from the HeNB coverage area. Prior to handover procedure, the HeNB requests through an intra-node signaling with the collocated L-GW to release the LIPA PDN connection. The HeNB determines that the UE has a LIPA PDN connection from the presence of the correlation ID in the UE E-RAB context. The L-GW then initiates and completes the release of the LIPA PDN connection using the P-GW-initiated bearer deactivation procedure. During the handover, the source MME checks whether the LIPA PDN connection has been released. If it has not been released and the handover is an S1-based or an inter-RAT handover, the source MME rejects the handover. The direct signaling from the HeNB to the L-GW is only possible since mobility of the LIPA PDN connection is not supported. During idle state mobility events, the MME/SGSN deactivates the LIPA PDN connection, when it detects that the UE has moved away from the HeNB coverage area [6].

2.12.2 Selected IP traffic offload

The SIPTO refers to the ability to selectively forward different types of traffic via alternative routes to/from the terminal. Home network flavor of SIPTO depends on a HeNB. In this mode, specific traffic, determined by the operator's policy and/or subscription, is transferred to/from eNB directly to Internet/intranet, bypassing the mobile operator access/core network. Macro network flavor of SIPTO covers the ability to offload traffic next to an RNC or S-GW as opposed to traversing the operator's core network. SIPTO extends what LIPA aims to achieve by providing an alternate path for selected traffic. The primary objective is to add capacity to the primary RAN-core network path and reduce the unit cost of transport.

SIPTO for HeNB is similar to LIPA in terms of necessary changes to network capabilities and additional functionality in the form of an L-GW collocated with HeNB (Figure 2.30). Considering substantial complexity and limited benefit of SIPTO for HeNB as opposed to LIPA, 3GPP decided to mark that as a future study item beyond Rel-10. The SIPTO function enables an operator to offload certain types of traffic at a network node close to the UE's point of attachment to the access network. SIPTO beyond E-UTRAN can be achieved by selecting a set of network gateways (S-GW and P-GW) that is geographically/topologically close to a UE's point of attachment. SIPTO beyond E-UTRAN corresponds to a traffic offload through a P-GW located in the mobile operator's core network. SIPTO applies to the non-roaming, provided that appropriate roaming agreements are in place between the operators, and the roaming cases. Offload of traffic for a UE is available for UTRAN and E-UTRAN accesses only. When the UE enters to UTRAN/E-UTRAN from another type of access network (e.g., from GERAN), it is the responsibility of the new SGSN/MME to decide whether to perform deactivation with reactivation request for a given PDN connection, depending on whether SIPTO is permitted for that APN. Realization of SIPTO beyond E-UTRAN relies on the same architecture models and principles that govern the local breakout.

In order to select a set of appropriate network gateways (S-GW and P-GW) based on geographical/topological proximity to the UE, the gateway selection function uses the UE's current location information. In order to allow or prohibit SIPTO on per user and per APN basis by an operator, subscription data in the

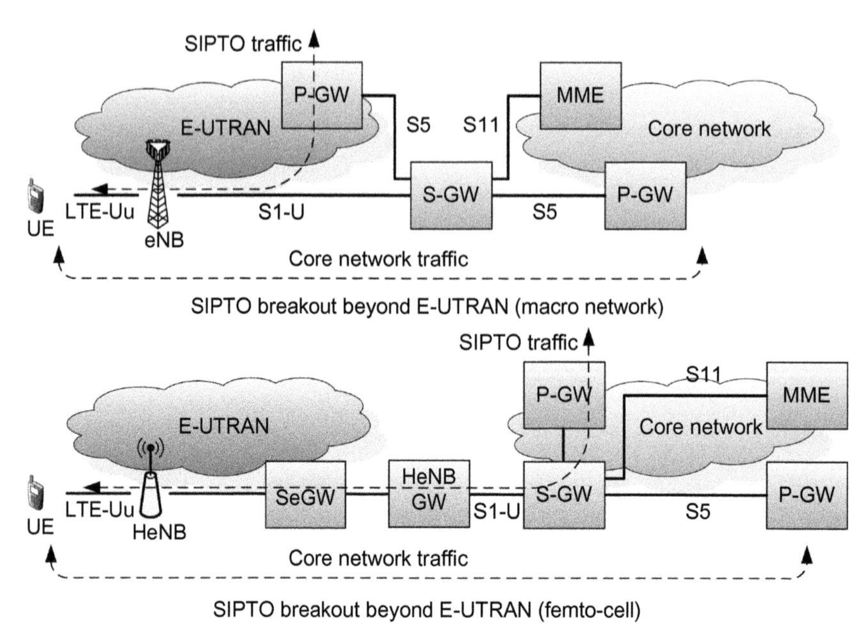

FIGURE 2.30

Illustration of SIPTO variants [62].

HSS is configured to indicate to the MME if offload is allowed or prohibited. If the SIPTO permission information from the HSS conflicts with MME's configuration for that UE, then SIPTO is not used. If HSS indicates the visited PLMN address is not allowed, then the MME does not provide SIPTO. The MME may be configured on an APN basis in order to use SIPTO (e.g., to handle the case where the HSS is not configured with SIPTO information for the UE). For SIPTO beyond E-UTRAN, as a result of UE mobility (e.g., detected by the MME at tracking area update procedure), the target MME may redirect a PDN connection toward a different gateway that is more appropriate for the UE's current location.

The SIPTO at the local network function enables an IP-capable UE connected via a HeNB to access a specific IP network or the Internet without the user-plane going through the mobile operator's network. The subscription data in the HSS is configured per user and per APN to indicate to the MME if offload at the local network is allowed. SIPTO at the local network can be achieved by selecting an L-GW function collocated with the HeNB or selecting stand-alone network gateways (with S-GW and L-GW collocated) residing in the local network. In both cases, the selected IP traffic is offloaded via the local network. Specific to the HeNB sub-system, the applicability of SIPTO at the local network does not depend on CSG membership and the feature can be applied to any UE, as long as the HeNB is configured for open or hybrid access mode. There is no interface between the L-GW and the PCRF, and there is no support for dedicated bearers on the PDN connection used for SIPTO at the local network. The L-GW rejects any UE request for bearer resource modification. In 3GPP Rel-11, SIPTO at the local network is intended for offloading Internet traffic only, thus the L-GW does not provide APN-specific connectivity. Therefore, if the subscription data in the HSS indicates that offload at the local network is allowed, this implies that the related APN is typically used for providing Internet connectivity.

2.12.3 IP flow mobility

An IP flow consists of a set of packets that are exchanged between two nodes and that match a given flow description. In 3GPP Rel-8 mobility architecture, the granularity of access system connectivity was per PDN basis, that is, the UE could not access the same PDN via multiple interfaces, which meant that the UE could not treat individual IP flows within a PDN connection separately; therefore, while performing handover from one access network to another, it had to move all the IP flows within a PDN connection together [57]. In other words, in Rel-8 EPS, the UE could not communicate using multiple radio access networks simultaneously. While the UE could establish a single PDN connection or multiple simultaneous PDN connections, all traffic exchanged by the UE, regardless of the PDN connection it belonged to, was routed through the same radio access network.

IP flow mobility introduced the concept of treating IP flows individually within a PDN connection. It specified that the mobility of a PDN connection is handled per IP flow, which meant that within a PDN connection, the UE could establish IP flows over multiple access networks; selectively remove IP flows from an access system; selectively transfer IP flows between access systems; and transfer all IP flows from a certain access system. The following assumptions are valid while defining the functionality of multi-access PDN connectivity and IP flow mobility: (1) the number of non-3GPP interfaces simultaneously active with a 3GPP interface is limited to one; and (2) all IP flows are toward one, and only one PDN. However, in this framework a single IP flow cannot be split and routed via different access networks, and IP flows cannot be established toward multiple PDNs. The following use cases have been identified for IP flow mobility in 3GPP:

- Adding IP flow to an existing one via a different access network: The UE is reachable through both 3GPP and non-3GPP access networks. It has an active IP flow through the 3GPP (or non-3GPP) access networks. At some point in time, a new IP flow is initiated through non-3GPP (or 3GPP) access toward the same PDN. The non-3GPP (or 3GPP) access of the UE may or may not have any ongoing session through it, as depicted in Figure 2.31.
- Removing IP flow established via a different access network: The UE is reachable through both 3GPP and non-3GPP access networks and has simultaneously active sessions through both RANs. Each of the RANs can have one or more ongoing sessions. At some point in time, one IP flow via non-3GPP (or 3GPP) access is removed.

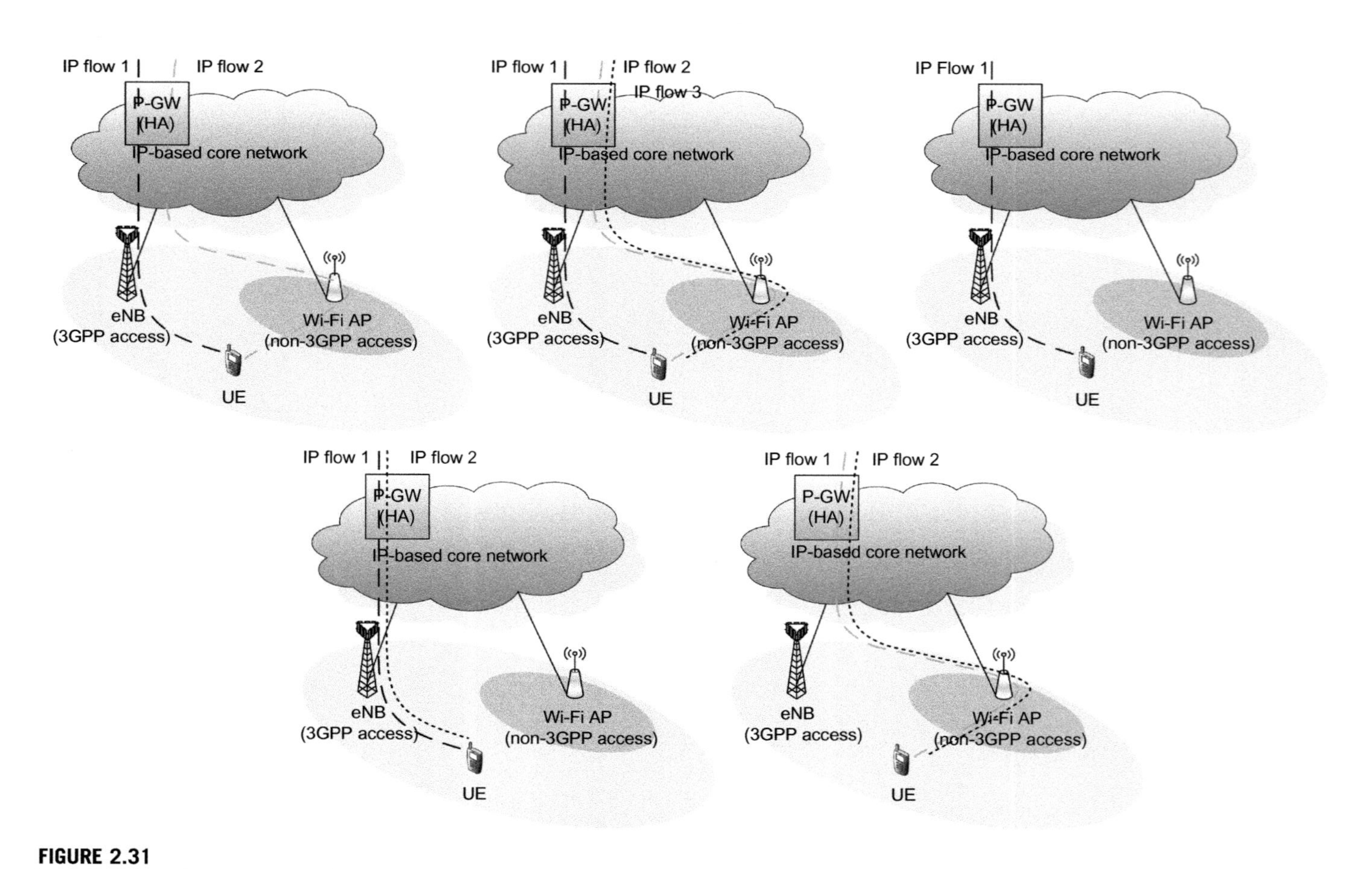

FIGURE 2.31

Adding, removing, and transferring of IP flows in various IP flow mobility scenarios using 3GPP and non-3GPP access networks [57].

- IP flow mobility between access networks: this use case involves selective transfer of IP flows between RANs. The state of the UE's interfaces can be one of the followings: (1) both radio interfaces are active simultaneously; or (2) only one radio interface is active. In the case of both radio interfaces being active simultaneously, the UE is reachable through both 3GPP and non-3GPP RANs and has simultaneously active sessions through both RANs. It has two IP flows through the 3GPP (or non-3GPP) access and one IP flow through the non-3GPP (or 3GPP) RAN. At some point in time, one IP flow is moved from the 3GPP (or non-3GPP) RAN to the non-3GPP (or 3GPP) RAN. The use case is depicted in Figure 2.31. In the case of only one active radio interface, the UE is camping on a 3GPP (or non-3GPP) RAN with at least two active IP flows. The UE is moving toward the coverage of a non-3GPP RAN. At some point in time, the UE detects a non-3GPP (or 3GPP) RAN and moves a subset of active IP flows (i.e., one IP flow) to the newly detected RAN.
- Moving all IP flows from one access network to another: The UE is reachable through both 3GPP and non-3GPP RANs. It has one or more active IP flows through the non-3GPP access. The 3GPP access of the UE may or may not have any ongoing session. The UE is moving away from the coverage of the non-3GPP RAN. At some point in time, due to loss of coverage of the non-3GPP RAN, the UE moves all IP flows going through the non-3GPP access to the 3GPP access.

As mentioned earlier, the granularity of access network connectivity and inter-system mobility is per PDN connection basis. This implies that when a handover occurs, all IP flows belonging to the same PDN connection are moved from the source access system to the target access system. With IP flow mobility, it is possible to have a finer granularity in access network connectivity and inter-system mobility. The handover procedures can be applied to a single or multiple IP flows belonging to the same PDN connection. This implies that some IP flows of one PDN connection can be routed via one access network, while simultaneously some IP flows of the same PDN connection can be routed via another access system. To achieve IP flow mobility, the inter-system mobility signaling is enhanced in order to carry routing filters. The extensions to DSMIPv6 mobility signaling are needed to carry routing filters when the UE is connected to multiple access networks simultaneously [4].

There are two possible approaches to IP flow mobility, client-based and network-based. The former relies on an IP host-centric solution, introducing a mobility client in the host and a mobility agent in the core network. The latter relocates the IP mobility client functionality from the host to the network, thus making the mobile device agnostic to any IP mobility signaling. Client-based IP mobility solutions require the user terminal to be involved in the management of the mobility, running a specialized stack that is able to detect, signal, and react to changes of point of attachment. Dual stack for mobile IPv6 is standardized by the IETF to provide client-based IP mobility support [29]. Network-based IP mobility solutions locate the mobility management control of the terminal in the network.

In this way, the terminal is not required to perform any kind of signaling (e.g., binding updates) to react to changes of its point of attachment to the network, as these changes are transparent to the mobile terminal IP stack. Proxy MIPv6 (PMIPv6) is the protocol standardized by the IETF to provide network-based IP mobility support [59].

2.12.4 Wi-Fi offloading

Wi-Fi offload offers a new way to extend the capacity and coverage of an LTE network. Wi-Fi networks are integrated to complement LTE networks and allow the use of each technology's advantages according to the actual demand. For example, the Wi-Fi is optimized for indoor environments, whereas the LTE network is designed for complete coverage in all areas. Depending on the trust relationship, the Wi-Fi may be integrated as a trusted or a non-trusted access technology. In the newest releases of 3GPP network architecture, a paradigm change has occurred. Instead of a handover for the complete connection, single traffic flows are routed over one access technology while the remaining ones are routed over the other. Therefore, the elaborated QoS feature in LTE may be used for delay or jitter sensitive IP flows, such as voice or video conference, whereas less time critical services may be routed over the cost-effective Wi-Fi when available. Corresponding handover procedures for the partial offload of certain flows are controlled by the network operator to ensure best user experience.

The user authentication is done using USIM credentials and according to LTE authentication procedures. Therefore, in the ideal case, the user does not perceive the offload and only recognizes an enhanced data rate using the mobile services. The Wi-Fi offloading refers to a scenario where part of the IP traffic to or from a device is routed via Wi-Fi access and another part via 3GPP access. Wi-Fi offloading covers both scenarios where the IP traffic via Wi-Fi radio is anchored in the EPC (i.e., seamless offloading) and the scenario where it is not anchored (i.e., non-seamless offloading). There is an access network discovery and selection function (ANDSF) entity to provide the UE with the access network discovery information and the policy on how to use the available access networks, available access networks, and preferred routing of the IP traffic per APN and per IP flow [58,61].

3GPP Rel-8 introduced a client-server-based protocol, dual stack mobile IP (DSMIP), to enable seamless handover between 3G and Wi-Fi. DSMIP is a mobility protocol specified in IETF that provides IP address preservation for IPv4 and IPv6 sessions, allowing the user to roam independently in IPv4 and IPv6 accesses. The solution does not require any support from Wi-Fi access network. If the cellular radio access network supports a home agent (HA), the only new requirement is that the UE (i.e., client) and the HA (i.e., server) are dual-stack capable. Applications and accesses can be either IPv4 or IPv6 compliant depending on the deployment models of the operators. Therefore, the changes required are only made to the UE and GGSN/P-GW. The HA can be implemented as a functionality in the GGSN/P-GW or a stand-alone entity. The HA is the anchor that binds the permanent identifier of the node (i.e., home address or HoA) with the local address based on the node's location (i.e., care-of-address or CoA). The exposed IP address accessible by the application remains the same.

The DSMIP can be used over trusted and/or untrusted non-3GPP access (e.g., Wi-Fi) and/or 3GPP access (e.g., HSPA and LTE). Wi-Fi mobility in 3GPP Rel-8 can be implemented either using DSMIP over an H1 or S2c interface. The H1 interface is compatible with pre-Rel-8 GGSN and PDG, and the S2c interface is part of the EPS/LTE architecture. The DSMIP provides a better user experience compared to application-based switching, since it preserves the IP address whenever the network is changed. Furthermore, the DSMIP is easily portable to multiple devices since it does not depend on the application.

3G/Wi-Fi seamless offload is a further enhancement to the solution based on DSMIP that enables seamless handover between 3G and Wi-Fi. It also provides the possibility to move selected IP traffic (i.e., HTTP, video, and VoIP) while supporting simultaneous 3G and Wi-Fi access networks. The main components of the solution are DSMIP with extensions for seamless IP flow mobility, ANDSF extensions for providing operator control, and PCC enhancements for QoS [61]. The 3G/Wi-Fi seamless offload solution does not require any PCC support. When IP flow mobility is used and PCC function is deployed, the PCC architecture is enhanced to handle multiple simultaneous connections for a single IP session.

In dual radio scenarios, such as Wi-Fi and cellular, the two access networks are not aware of each other and one access network cannot control the device protocol state in the other access network. Also, Wi-Fi is not a controlled access network and works opportunistically in unlicensed spectrum; therefore, the only element in the network that is known is when a Wi-Fi is available, the radio quality and the performance that can be achieved through that Wi-Fi is available only to the device. As a result, the device is best positioned to decide the connectivity options. In this context, network discovery and selection procedures can be performed by a connection manager in the device. This kind of connectivity management is out of the scope of 3GPP and it is up to the vendors to provide a solution. More intelligent connection managers can search and prioritize the best available links based on predefined requirements, such as QoS, bandwidth, latency, jitter, power consumption, and operator policies. [61].

2.13 Mobile IP

Mobility support within EPS is based on a mobile IP framework. Mobile IP is an IETF protocol that allows mobile users to move from one network to another while maintaining their IP addresses [55]. The two versions of mobile IPs, i.e., Mobile IPv4 and Mobile IPv6, are described in IETF RFC 3344 [23] and IETF RFC 3775 [29], respectively. The mobile IP allows transparent routing of IP datagrams over the Internet. Each mobile node is identified by its home address

(HoA), irrespective of its current location in the Internet. When away from home, a mobile node is associated with a CoA, which provides information about its current location. The mobile IP specifies how a mobile node registers with its HA and how the HA routes datagrams to the mobile node through a tunnel. Using mobile IP, nodes may change their point-of-attachment to the Internet without changing their IP address, allowing the application and transport-layer protocols to seamlessly maintain connection while moving. The general characteristics of mobile IP can be summarized as follows:

- Transparency of user mobility to the transport and application layer protocols.
- Interoperability with stationary hosts running conventional IPs.
- Scalability across the Internet.
- Security by preventing an attacker from impersonating a mobile host.
- Macro mobility by ensuring long-term connection while away from HA.

Node mobility is realized without propagating host-specific routes throughout the Internet. Using mobile IP, a mobile device will have two addresses, that is, a primary or permanent HoA and a secondary or temporary CoA, which is associated with the network that the mobile node is visiting. There are two types of entities in mobile IP:

- An HA that stores information about mobile nodes whose permanent HoA is in the HA's network.
- A foreign agent (FA) that stores information about mobile nodes visiting its network. The FAs also advertise CoAs.

A node that wishes to communicate with the mobile node uses the permanent HoA of the mobile node as the destination address for outbound packets. Since the HoA logically corresponds to the network associated with the HA, conventional IP routing mechanisms forward these packets to the HA. Instead of forwarding these packets to a destination that is physically in the same network as the HA, the HA redirects these packets toward the FA. The HA looks for the CoA in the mobility binding table [55] and then tunnels the packets to the mobile node's CoA by appending a new IP header to the original IP packet, which preserves the original IP header. The packets are detected at the end of the tunnel by removing the IP header added by the HA and are delivered to the mobile node.

The mobile node directly sends packets to the other communicating node through the FA without involvement of the HA using its permanent HoA as the source address for the IP packets, that is, triangular routing. The FA can utilize reverse tunneling by sending the mobile node's packets to the HA, which forwards them to the communicating node. This mechanism is needed in networks whose gateway routers have ingress filtering enabled, and hence the source IP address of the mobile host needs to belong to the subnet of the foreign network; otherwise, the packets would be discarded by the router.

The mobile IP defines an authenticated registration procedure through which a mobile node informs its HA of its CoA, router discovery (which allows mobile nodes to discover prospective HA and FAs), and the rules for routing packets to and from mobile nodes, including the specification of one mandatory and several optional tunneling mechanisms.

IP mobility enables routing data packets to the destination when moving to a foreign network. A central point is to maintain the IP address of a UE fixed so that there is no need for upper layers (above the IP layer) to adapt to the network change. Consequently, upper layer connections can continue without notification about the receiver's mobility. In general, IP mobility support can be realized with two different approaches: (1) client-based IP mobility; and (2) network-based IP mobility. This distinction does not depend on the underlying radio access technology; rather it is completely handled in the IP stack.

2.13.1 Client-based IP mobility

In the client-based mobility of IPv6, the UE carries the mobility extensions in its own IP stack. Central to this approach is to split the IP address into the HoA, which is the permanent IP address obtained from the home network, and the CoA, which corresponds to a temporary IP address and is obtained from the visited network. The administration of these addresses is done in an HA. As long as the UE is in the home network, the HA routes the data packets directly to the UE using the HoA. When the UE changes the network, it informs the HA with a proxy binding update (PBU) message about its new IP address, which is stored by the HA in the binding cache. This is essentially a lookup table which is searched for each incoming packet. If there is a CoA entry for this UE in the binding cache, the packet is instead forwarded to its CoA. As an optimization, it is also possible to route IP traffic between a correspondent node (CN) and the UE directly; in this case, a binding cache has to be additionally established in the CN itself.

In an extension, mixed networks with both IPv4 and IPv6 are supported with the dual stack mobile IPv6 protocol (DSMIPv6). This is essential to integrate most existing networks built on IPv4 into the new networks, which will be more and more equipped with IPv6 protocols. For a client-based mobility EPC access with Wi-Fi, DSMIPv6 is mandatory. With this approach, it is possible to offload data traffic from an LTE network to a Wi-Fi connection. However, this works only for a complete offload, that is, it is either possible to communicate over the LTE connection or over the Wi-Fi connection, but not over both. The reason is that in this architecture the Wi-Fi network is considered as a foreign network, to which all data packets are forwarded when there is a corresponding entry in the binding cache of the HA.

2.13.2 Network-based IP mobility

A completely different approach to support user mobility is the network-based IP mobility. Contrary to the client-based IP mobility, the network takes all necessary steps to route the data packets to the destination; thus, there is no need for the client to do any signaling upon network change. There are two approaches for network-based IP mobility: PMIP version 6 (PMIPv6); and the GTP.

The PMIP (PMIPv6) is specified in IETF RFC 5213 [26]. Routing is based on two additional entities, the LMA, which works in a similar way as the HA in the client-based mobility approach, and the MAG, which implements the necessary mobility functions in the visited network. When the UE changes the network, the MAG is contacted by the new base station and informs the LMA about the change of location. Similar to the client-based mobility, this architecture has to be extended in order to implement the IP flow mobility capabilities, that is, moving selected IP flows from one access technology to another, and consequently, installing the required filters for flow routing. In order to send and receive data packets from and to any of its interfaces, the IETF has decided to adapt the logical interface (LIF). This is a software entity which hides the physical interface to the IP layer. This means that the mobile IP stack binds its sessions to the LIF, thus for the UE there is only one single interface to the IP network. The LIF controls the flow mobility in the UE. It is part of the connection manager in the operating system and has no impact on the IP stack. It represents a virtual interface which hides all flow mobility movements to the higher layers. A second aspect to introduce flow mobility to PMIPv6 is to provide signaling extensions to the MAG. This is necessary because the MAG will only forward traffic from and to a UE if the prefix has been delegated to the UE by this MAG. However, in IP flow mobility, this delegation might have been done by a different MAG before the flow handover. Signaling between the LMA and the target MAG resolves this issue.

3GPP has developed and specified the network-based mobility protocol called GTP. The PMIPv6 and client-based mobility were introduced as alternatives in LTE. The HA or LMA in the 3GPP EPC are located in the P-GW for both trusted and untrusted access. This is in contrast to the location of the MAG, where in the trusted access it is located in the Wi-Fi network and in the untrusted access is in the evolved packet data gateway (ePDG) network entity. Consequently, the connection between the LMA and the MAG can always be regarded as trusted, because the ePDG is under control of the EPC network operator. Up to 3GPP Rel-9 network architecture, the Wi-Fi offload was the only way to offload data from the EPC to Wi-Fi. The IP flow mobility was introduced with 3GPP Rel-10 network architecture for client-based mobility. Network-based mobility is not yet supported as of 3GPP Rel-11 network architecture and is being studied in 3GPP Rel-12.

2.14 Network architecture evolution and small cells

While macro-cellular networks will continue to provide essential wide-area coverage and support for high-mobility users, operators have started to look at other solutions to increase capacity in high-traffic areas. Wi-Fi offload is the most widely and successfully adopted solution in areas where subscriber density and usage is high, such as urban areas and locations such as airports or stadiums. Operators are also exploring compatible solutions such as femto-cells to improve cellular coverage and capacity in residential areas, as well as small cell underlays to address the increasing demand for high capacity per area, and complement and strengthen their Wi-Fi and macro-cellular deployments. Mobile operators can further improve network utilization by actively managing the traffic beyond the radio access network within the core, using content caching, tiered pricing, and policy enforcement. These solutions do not increase the capacity per se, but make data transmission more efficient, allowing operators to process more content within the same network infrastructure. The integration of Wi-Fi and LTE small cells within the cellular core helps operators optimize network utilization across different radio access technologies (RATs), thus providing further improvement in performance, and creating a seamless multi-RAT experience for their subscribers [56,58,60,61]. While no single product or technology by itself can accommodate the current and future increase in data traffic, many viable solutions, working in concert, can be considered. The challenge for mobile operators is not the decision of which solution to select, rather it is how to best integrate multiple technologies within their networks, how to find the right balance to maximize their cumulative benefits, and how to leverage existing assets to facilitate the evolution of their networks.

Macro-cellular networks today typically provide coverage throughout a mobile operator's coverage area. Their capacity is adequate where the number of subscribers per cell site is relatively small. While high density areas account only for a small part of operators' footprint, they are of crucial importance. This is where subscribers are densely concentrated and where they use data services more extensively. The main challenge for mobile operators is to identify and correct capacity shortfalls in these high-traffic areas.

In order to increase network capacity and to meet the increasing demand for mobile wireless applications, a new approach is taken which relies on lower-power, shorter-range equipment, such as Wi-Fi access points or LTE/LTE-Advanced small cells installed closer to subscribers, in dense deployment areas. By limiting the range, the impact of interference is reduced and the capacity per area is increased. A single sector in a macro cell may have a comparable capacity to a Wi-Fi access point or a small cell, but it spreads it over a larger area, leading inevitably to lower capacity per area. Small cell deployments are one of the most viable strategies of mobile operators worldwide as they provide the capacity increase in the near term, along with the flexibility and compact form factor needed for highly localized deployments in high-traffic environments. Yet, there are several challenges that need to be addressed when planning for a small cell underlay network to enable the operator to maximize the benefits of the new network and to limit the costs. These include securing mounting assets where a high number of nodes are required and typically small cells are not mounted on telecommunication assets like cell towers that operators are accustomed to managing or leasing; backhaul for the small cells network for dense urban environments; as well as integration with the cellular core network. The small cell underlay network must be fully integrated with the EPC and support self-organizing networks to give the operator the ability to manage user data traffic, support mobility, mitigate interference, and implement policy consistently across different radio access networks.

Most mobile operators no longer see Wi-Fi offload and small cells as competing solutions. They realize both are needed, with each playing an important role within a multi-RAT, multi-layer cellular network, spanning from macro base stations at one end, to residential femto cells at the other. Enhanced small cells can be deployed with macro coverage and stand-alone, both indoor and outdoor, and support both ideal and non-ideal backhauls. Enhanced small cells can also be deployed sparsely or densely. An enhanced small cell may benefit from the presence of overlaid macro cells, and it should also work without macro coverage, for example, in indoor scenarios. Cooperative mechanisms, between macro cell and small cell, as well as among small cells, may be beneficial for the above-mentioned use cases.

References

3GPP Specifications[12]

[1] 3GPP TS 23.002, Technical Specification Group Services and System Aspects; Network architecture (Release 11), December 2012.

[2] 3GPP TS 23.003, Numbering, Addressing and Identification (Release 11), December 2012.

[3] 3GPP TS 23.216, Single Radio Voice Call Continuity (SRVCC); Stage 2 (Release 11), December 2012.

[4] 3GPP TS 23.261, IP Flow Mobility and Seamless Wireless Local Area Network (WLAN) Offload; Stage 2 (Release 11), September 2012.

[5] 3GPP TS 23.272, Circuit-switched Fallback in Evolved Packet System; Stage 2 (Release 11), December 2012.

[6] 3GPP TS 23.401, General Packet Radio Service (GPRS) enhancements for Evolved Universal Terrestrial Radio Access Network (E-UTRAN) access (Release 11), December 2012.

[7] 3GPP TS 23.402, Architecture Enhancements for non-3GPP Accesses (Release 11), December 2012.

[8] 3GPP TR 23.882, 3GPP System Architecture Evolution: Report on Technical Options and Conclusions (Release 8), September 2008.

[9] 3GPP TS 29.274, Evolved General Packet Radio Service (GPRS) Tunneling Control Protocol for Control Plane (GTPv2-C) Stage 3 (Release 11), December 2012.

[10] 3GPP TS 24.301, Non-Access-Stratum (NAS) protocol for Evolved Packet System (EPS); Stage 3 (Release 11), December 2012.

[11] 3GPP TS 33.401, 3GPP System Architecture Evolution: Security Architecture (Release 11), December 2012.

[12]3GPP specifications can be found at the following URL: http://www.3gpp.org/ftp/Specs/archive/

[12] 3GPP TS 36.300, Evolved Universal Terrestrial Radio Access (E-UTRA) and Evolved Universal Terrestrial Radio Access Network (E-UTRAN); Overall description; Stage 2 (Release 11), December 2012.

[13] 3GPP TS 36.413, Evolved Universal Terrestrial Radio Access Network (E-UTRAN); S1 Application Protocol (S1AP), (Release 11), December 2012.

3GPP2 Specifications[13]

[14] 3GPP2 C.S0024-A v3.0, cdma2000 High Rate Packet Data Air Interface Specification, September 2006.

[15] 3GPP2 C.S0005-A v6.0, Upper Layer (Layer 3) Signaling Standard for CDMA2000 Spread Spectrum Systems, February 2002.

[16] 3GPP2 A.S0009-C v4.0, Interoperability Specification (IOS) for High Rate Packet Data (HRPD) Radio Access Network Interfaces with Session Control in the Packet Control Function, April 2011.

[17] 3GPP2 A.S0013-C v4.0, Interoperability Specification (IOS) for CDMA2000 Access Network Interfaces—part 3 Features, August 2012.

[18] 3GPP2 X.S0042-A v1.0, Voice Call Continuity between IMS and Circuit Switched System, August 2008.

IETF Specifications[14]

[19] IETF RFC 2460, Internet Protocol, Version 6 (IPv6) Specification, S. Deering, December 1998.

[20] IETF RFC 4960, Stream Control Transmission Protocol, R. Stewart, September 2007.

[21] IETF RFC 3310, Hypertext Transfer Protocol (HTTP) Digest Authentication Using Authentication and Key Agreement (AKA), A. Niemi, September 2002.

[22] IETF RFC 3261, SIP: Session Initiation Protocol, J. Rosenberg et al., June 2002.

[23] IETF RFC 3344, IP Mobility Support for IPv4, C. Perkins, August 2002.

[24] IETF RFC 2138, Remote Authentication Dial in User Service (RADIUS), C. Rigney et al., April 1997.

[25] IETF RFC 3588, Diameter Base Protocol, P. Calhoun et al., September 2003.

[26] IETF RFC 5213, Proxy Mobile IPv6, S. Gundavelli et al., August 2008.

[27] IETF RFC 2131, Dynamic Host Configuration Protocol, R. Droms, March 1997.

[28] IETF RFC 5555, Mobile IPv6 Support for Dual Stack Hosts and Routers, H. Soliman, June 2009.

[29] IETF RFC 3775, Mobility Support in IPv6, D. Johnson et al., June 2004.

Books

[30] S. Sesia, I. Toufik, M. Baker, LTE—The UMTS Long Term Evolution*: From Theory to Practice*, second ed., John Wiley & Sons, 2011.

[31] C. Johnson, Long Term Evolution in Bullets, second ed., Create Space Independent Publishing Platform, 2012.

[32] E. Dahlman, S. Parkvall, J. Skold, *4G, LTE/LTE-Advanced for Mobile Broadband*, Academic Press, 2011.

[13]3GPP2 specifications can be found at the following URL: http://www.3gpp2.org/Public_html/specs/index.cfm

[14]IETF specifications can be found at the following URL: http://www.ietf.org/rfc.html

[33] T. Ali-Yahiya, Understanding LTE and Its Performance, Springer, 2011.

Articles

[34] LTE and Beyond: LTE, 4G, MME, P-GW, S-GW, Interfaces and beyond. <http://www.lteandbeyond.com/>.
[35] GPRS Tunneling Protocol, <http://en.wikipedia.org/wiki/GPRS_Tunnelling_Protocol>.
[36] (IMS) Multimedia Telephony (MMTel), from Wikipedia, <http://en.wikipedia.org/wiki/Multimedia_telephony>.
[37] K. Suzuki, et al., *Core network (EPC) for LTE*, NTT DoCoMo Tech. J. 13 (3) (2011).
[38] C.F. Lanzani, et al., Remote Radio Heads and the Evolution Towards 4G Networks, Altera, 2009<http://www.altera.com>.
[39] LTE System Description, <http://www.sharetechnote.com/>.
[40] LTE Identifiers, NMC Consulting Group. <http://www.nmcgroups.com>, 2011.
[41] LTE Identifiers, EventHelix.com Inc. <http://www.eventhelix.com/>, 2012.
[42] Small Cell Forum, <http://www.smallcellforum.org/resources-technical-papers>.
[43] G. Gómez, et al., *Towards aQoE-Driven resource control in LTE and LTE-A networks*, J. Comput. Netw. Commun, (2013).
[44] Common Public Radio Interface (CPRI), <http://www.altera.com/technology/high_-speed/protocols/cpri/pro-cpri.html>.
[45] GSMA, 2010, IR.92 IMS Profile for Voice and SMS V3.0. Available at: <http://www.gsma.com/newsroom/wp-content/uploads/2012/06/IR9230.pdf>.
[46] GSMA, 2011, IR.94 IMS Profile for Conversational Video Service V1.0. Available at: <http://www.gsma.com/newsroom/wp-content/uploads/2012/03/ir9410.pdf>.
[47] GSMA, 2011, IR.64 IMS Service Centralization and Continuity Guidelines V2.0. Available at: <http://www.gsma.com/newsroom/wp-content/uploads/2012/03/ir6420.pdf>.
[48] LteWorld, Evolution of Single Radio Voice Call Continuity. <http://lteworld.org/blog/evolution-single-radio-voice-call-continuity-srvcc>, December 2012.
[49] LteWorld, Understanding CS Fallback in LTE, September 2009.
[50] IP Multimedia Subsystem, from Wikipedia. <http://en.wikipedia.org/wiki/IP_Multimedia_Subsystem>.
[51] Network Switching Subsystem, from Wikipedia. <http://en.wikipedia.org/wiki/Network_Switching_Subsystem>.
[52] Internet Protocol Security (IPsec), <http://en.wikipedia.org/wiki/Ipsec>.
[53] C. Gessner, O. Gerlach, Voice and SMS in LTE, Rohde & Schwarz, 2011.
[54] Motorola White Paper, LTE Inter-technology Mobility: Enabling Mobility between LTE and other Access Technologies, Motorola Inc., November 2008.
[55] Mobile IP, from Wikipedia, <http://en.wikipedia.org/wiki/Mobile_ip>.
[56] T. Nakamura, et al., *Trends in small cell enhancements in LTE advanced*, IEEE Commun. Mag. (2013).
[57] T. Ahmed, et al., Multi access data network connectivity and IP flow mobility in evolved packet system (EPS), Proceedings of WCNC 2010, April 2010.
[58] A. Schumacher, J. Schlienz, WLAN Traffic Offload in LTE, Rohde & Schwarz, 2012.
[59] A. De La Oliva, et al., *IP flow mobility: smart traffic offload for future wireless networks*, IEEE Commun. Mag, 2011.

[60] Dealing with Density: The Move to Small-Cell Architectures, Ruckus Wireless White Paper, 2011.
[61] A 3G/LTE Wi-Fi Offload Framework: Connectivity Engine (CnE) to Manage Inter-System Radio Connections and Applications, Qualcomm White Paper, June 2011.
[62] R. Gupta, N. Rastogi, LTE-Advanced—LIPA and SIPTO, Aricent Group, 2012.
[63] Circuit-switched fallback, the first phase of voice evolution for mobile LTE devices, Qualcomm White Paper, 2012.
[64] Anritsu Whitepaper, Voice over LTE (VoLTE), Anritsu Corporation, September 2012.
[65] G. Bertran, The IP Multimedia Subsystem in Next Generation Networks, May 2007.
[66] N. Kottapalli, Diameter and LTE Evolved Packet System, Radisys White Paper, September 2011.
[67] D. Soldani, Bridging QoE and QoS for Mobile Broadband Networks, Huawei Technologies, 2010.
[68] M. Grega, et al., Quality of Experience Evaluation for Multimedia, Overview of Telecommunications, Telecommunication News, No. 4/2008, April 2008.
[69] S. Tripathi, et al., LTE E-UTRAN and its Access Side Protocols, Radisys White Paper, September 2011.
[70] Introduction to Evolved Packet Core, Alcatel-Lucent White Paper, 2009.
[71] The LTE Network Architecture: A comprehensive tutorial, Alcatel-Lucent White Paper, 2009.
[72] VoLTE with SRVCC: the second phase of voice evolution for mobile LTE devices, Qualcomm White Paper, October 2012.
[73] S. Kundalkar, N. Madhur Raj, LIPA: Local IP Access via Home Node B, Radisys White Paper, September 2011.
[74] Dynamic Host Configuration Protocol, from Wikipedia, <http://en.wikipedia.org/wiki/Dynamic_Host_Configuration_Protocol>.
[75] Open Base Station Architecture Initiative, from Wikipedia, <http://en.wikipedia.org/wiki/OBSAI>.
[76] Internet Key Exchange, from Wikipedia, <http://en.wikipedia.org/wiki/Internet_Key_Exchange>.

CHAPTER

E-UTRAN and EPC Protocol Structure

3

CHAPTER CONTENTS

This chapter presents an overview of the LTE/LTE-Advanced protocol structure with the intent of describing where important functions reside in the E-UTRAN and EPC nodes, protocol layer terminations at various access and core network nodes, as well as the functional split between the E-UTRAN and the EPC. The E-UTRAN radio protocols can be separated into control-plane and user-plane entities, where the user-plane protocols are typically responsible for carrying user data, and control-plane protocols are used to transfer signaling and control information.

In general, a communications protocol is a system of rules for message exchange within or between network nodes. Individual protocols within a protocol suite are often designed with a single purpose in mind. This modularity facilitates the design and evaluation of protocols. Because each protocol module usually (virtually or physically) communicates with its peer entity across a communication link, they are commonly seen as layers in a stack of protocols, where the lowest protocol layer always deals with physical interaction of the hardware

LTE-Advanced.

across the communication link. Every higher layer protocol adds more features. User applications usually deal only with the top-most layers. In a practical implementation, protocol stacks are often divided into three major groups: media, transport, and applications.

In the context of protocol structure, we will frequently use the terms "service" and "protocol." It must be noted that services and protocols are distinct concepts. A service is a set of primitives or operations that a layer provides to the layer(s) with which it is interfaced. The service defines what operations a layer performs without specifying how the operations are implemented. It is further related to the interface between two adjacent layers. A protocol, in contrast, is a set of rules presiding over the format and interpretation of the information/messages that are exchanged by peer entities within a layer. The entities use protocols to implement their service definitions. Thus, a protocol is related to the implementation of a service. The protocols and functional elements defined by 3GPP standards correspond to all layers of the OSI[1] seven-layer network reference model. Figure 3.1 illustrates the mapping of the lower layer OSI protocols to those of the LTE/LTE-Advanced systems. As shown in the figure, what 3GPP considers as layer 2 and layer 3 protocols is mapped to the OSI data link layer. The higher layer protocols in the 3GPP stack are the application and transport layers. The presentation and session layers are often abstracted in practice.

According to ITU-T Recommendation X.200, there are seven layers, labeled 1 to 7, with layer 1 at the bottom of the protocol stack [18]. Each layer is generically known as an N layer. An N + 1 entity (at layer N + 1) requests services from an N entity at layer N. At each level, two entities (N-entity peers) interact by means of the N protocol by transmitting protocol data units (PDU). A service data unit (SDU) is a specific unit of data that has been passed down from an OSI layer to a lower layer, and which the lower layer has not yet encapsulated into a PDU. An SDU is a set of data that is sent by a user of the services of a given layer, and is transmitted semantically unchanged to a peer service user.

---

[1]Open systems interconnection (OSI) is a standard description or a "reference model" for the computer networks which describes how messages should be transmitted between any two nodes in the network. Its original purpose was to guide product implementations to ensure consistency and interoperability between products from different vendors. This reference model defines seven layers of functions that take place at each end of a communication link. Although OSI is not always strictly adhered to in terms of grouping the related functions together in a well-defined layer, many if not most products involved in telecommunication make an attempt to describe themselves in relation to the OSI model. OSI was officially adopted as an international standard by the International Organization of Standards (ISO) and it is presently known as Recommendation X.200 from ITU-T. The layers of OSI model are classified into two groups. The upper four layers are used whenever a message passes from or to a user. The lower three layers (up to the network layer) are used when any message passes through the host computer [17–19].

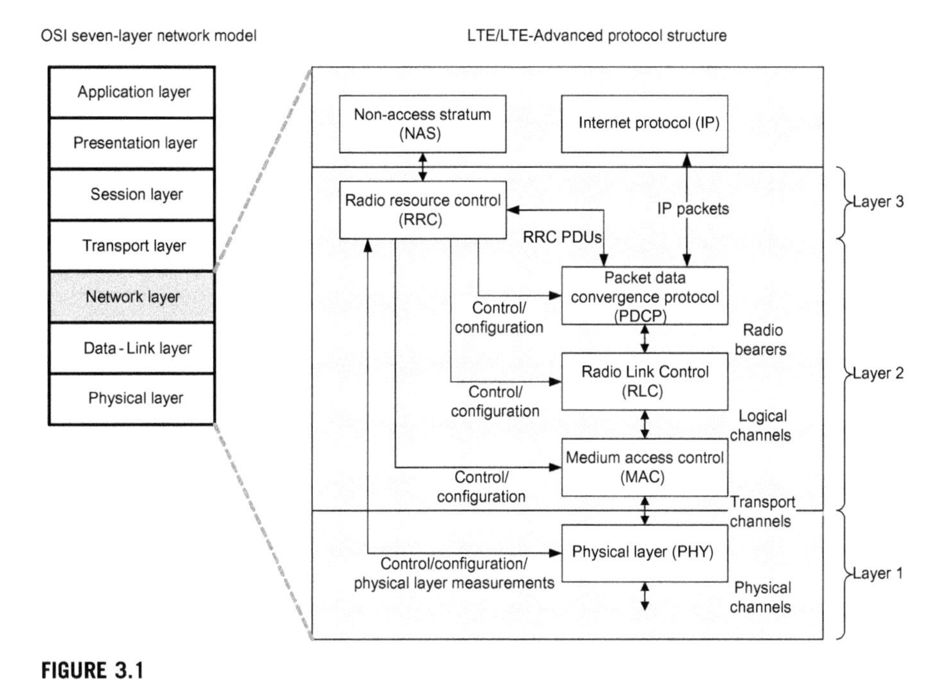

FIGURE 3.1

Mapping of lower E-UTRA protocol layers to OSI seven-layer network model.

The PDU at a layer N is the SDU of layer N − 1. In fact, the SDU is the payload of a given PDU. That is, the process of changing an SDU to a PDU consists of an encapsulation process, performed by the lower layer. All data contained in the SDU becomes encapsulated within the PDU. The layer N − 1 adds headers/sub-headers and padding bits (if necessary to adjust the size) to the SDU, transforming it into the PDU of layer N. The added headers/sub-headers and padding bits are part of the process used to make it possible to get data from a source to a destination.

As shown in Figure 3.2, a PDU is an information exchange between peer entities of the same protocol layer located at the source and destination. On the downward direction, the PDU is the data unit generated for the next lower layer. On the upward direction, it is the data unit received from the previous lower layer. An SDU, on the other hand, is a data unit exchanged between two adjacent protocol layers. On the downward direction, the SDU is the data unit received from the previous higher layer. On the upward direction, it is the data unit sent to the next higher layer.

This chapter provides a top-down systematic description of 3GPP EPC and E-UTRAN reference models, protocol structures, and functional decomposition, starting at the most general level and working toward details or specifics of the protocol layers, their functional constituents and interconnections. The main focus in this chapter is on the protocol aspects of LTE/LTE-Advanced. The functional details and implementation of the protocols will be described in Chapters 4–10.

3.1 EPS reference model and protocol structure

The evolved UMTS terrestrial radio access network consists of eNBs or equivalently E-UTRA base stations, providing the E-UTRA user-plane and control-plane protocol termination toward the UE, or alternatively, mobile station. As part of the evolution from and enhancement of the legacy UMTS systems, some radio network controller (RNC) functions have been included in the eNB to reduce the architectural complexity and further reduce the latency across the network. Figure 3.3 illustrates the EPS reference model comprising the network entities and the standardized interfaces between network elements. At a high level, the EPS encompasses EPC which is the core network and E-UTRAN, which is the radio access network. While the core network consists of many logical nodes, the access network is only made up a collection of eNBs, which are connected to the UEs via the LTE-Uu air-interface. Each of these network elements is

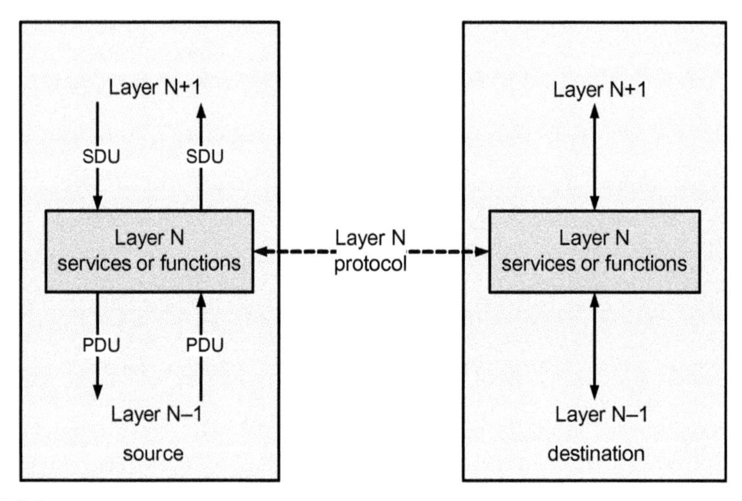

FIGURE 3.2

Illustration of service, protocol, PDU, and SDU concepts.

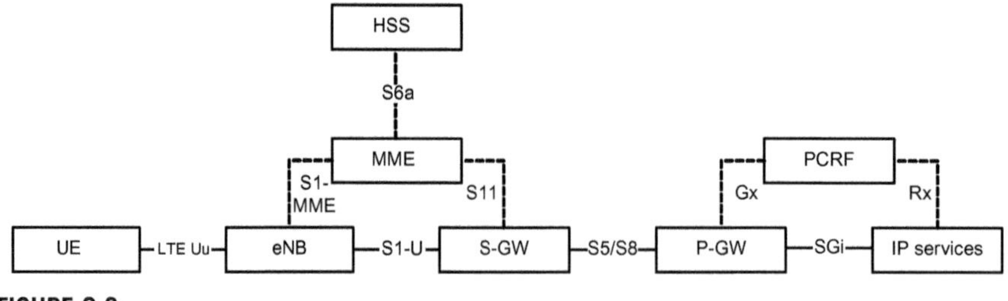

FIGURE 3.3

Reference model for EPS network elements and interfaces [9,18].

interconnected by means of interfaces that are standardized in order to facilitate interoperability among network equipment manufactured by different vendors.

In practice, a network operator, in its actual implementation/deployment, may split or merge some of the logical network elements depending on technical/economical/commercial considerations, but the overall functions remain the same. There are different options to split/combine MME, S-GW, and P-GW functionalities. Each option will result in different handling of control and user-plane packet routing. The choice may depend on operator's strategy and/or EPC vendor equipment availability and constraints. Integrating more logical entities into one physical entity will reduce latency, cost (fewer nodes and interfaces), and signaling overhead, but will increase the functional complexity per entity/component and will limit the benefit and potential savings achieved by a modular architecture, i.e., more redundancy, flexibility, and scalability. Hypothetically, a network implementation may begin with some level of integration of functional entities for simplicity, and then may be upgraded to more flexible and scalable architecture when EPS traffic grows.

In the reference model shown in Figure 3.3, the eNBs are typically responsible for radio resource management (RRM) that includes radio bearer control, radio admission control, connection management, dynamic allocation of resources to UEs in both uplink and downlink (i.e., scheduling); header compression and encryption of user payloads; selection of an MME at UE attachment when no routing to an MME can be determined from the information provided by the UE; routing of user-plane data toward S-GW; scheduling and transmission of paging messages (originated from the MME); scheduling and transmission of broadcast information (originated from the MME); and measurement and reporting for support of mobility and scheduling.

The MME is the key control-node for the 3GPP LTE access-network. It is responsible for idle-mode UE tracking and paging procedure including retransmissions. The MME functions include non-access stratum (NAS) signaling and security, inter-core-network signaling for mobility between 3GPP access networks, idle mode UE accessibility (including control and execution of paging retransmission), tracking area list management (for UE in idle and active mode), P-GW and S-GW selection, MME selection for handovers with MME change, roaming, authentication of the users, and bearer management functions including dedicated bearer establishment [1]. The NAS signaling terminates at the MME and it is also responsible for generation and allocation of temporary identities to the UEs. It verifies the authorization of the UE to camp on the service provider's PLMN and enforces UE roaming restrictions.

The S-GW routes and forwards user data packets, while also acting as the mobility anchor for the user-plane during inter-eNB handovers and as the anchor for mobility between LTE and other 3GPP technologies. The S-GW functions include local mobility anchor point for inter-eNB handover, mobility anchoring for inter-3GPP mobility, E-UTRAN idle mode downlink packet buffering and initiation of network triggered service request procedure, lawful interception, packet routing and forwarding, and transport-level packet marking in the uplink and the downlink [1].

The P-GW provides connectivity of the UE to external packet data networks by acting as the point of exit and entry of traffic for the UE. A UE may have simultaneous connectivity with more than one P-GW for accessing multiple packet data networks. The P-GW functions include per-user packet filtering, lawful interception, IP address allocation to the UE, and transport-level packet marking in the downlink [1].

The policy control and charging rules function (PCRF) is responsible for enforcing network policy and flow-based charging functions in conjunction with the P-GW. The PCRF provides the QoS rules and provisions in accordance with the user's subscription profile to the policy and charging enforcement function (PCEF), which resides in the P-GW. The P-GW uses this QoS policy to assign the bearer-level QoS parameters. Note that in a typical scenario, multiple applications may be running in a UE at any time, each one having different QoS requirements. For example, a UE can be engaged in a VoIP call while at the same time browsing a web page or downloading an FTP file. VoIP has more stringent requirements for QoS in terms of delay and jitter compared to web browsing and FTP, while the latter requires much lower packet-loss rate. In order to support multiple QoS requirements, different bearers are set up within the EPS, each associated with a specific set of QoS attributes [8].

The home subscriber server (HSS) contains users' subscription data such as the EPS-subscriber QoS profile and any access restrictions for roaming. It also holds information about the PDNs to which the user can connect.

The reference model shown in Figure 3.3 also includes a number of standard reference points as follows [10]:

- S1-MME is a reference point for the control-plane protocol between E-UTRAN and MME.
- S1-U is a reference point between E-UTRAN and S-GW which is used for per-bearer user-plane tunneling and inter-eNB path switching during handover.
- S5 interface provides user-plane tunneling and tunnel management between the S-GW and P-GW. It is used for the S-GW relocation due to UE mobility and when the S-GW needs to connect to a non-collocated P-GW for PDN connectivity.
- S6a reference point enables transfer of subscription and authentication data for authenticating/authorizing user access between MME and HSS.
- Gx interface provides transfer of QoS policy and charging rules from PCRF to PCEF in the P-GW.
- S8 is an inter-PLMN reference point providing user- and control-plane between the S-GW in the visited PLMN and the P-GW in the home PLMN. The S8 interface is the inter-PLMN variant of S5.
- S11 is the reference point between MME and S-GW.
- SGi is the reference point between the P-GW and the packet data network.
- Rx is the reference point that resides between the application function and the PCRF.

The EPC and E-UTRAN network elements and their associated protocol layers are described in more detail in the following sections.

3.1.1 Functional split between E-UTRAN and EPC

The functional split between the EPC and E-UTRAN is shown in Figure 3.4. The main functions in each entity and their termination points in the network have been shown in the figure. An IP packet for a UE is encapsulated in an EPC-specific protocol and tunneled between the P-GW and the eNB for transmission to the UE. Different tunneling protocols are used across different interfaces. A 3GPP-specific tunneling protocol called GPRS tunneling protocol (GTP) is used over S1 and S5/S8 interfaces [7]. The protocol stack for the control plane between the UE and MME is shown in Figure 3.5. This includes both access stratum (AS) and NAS protocols. The lower layers perform the same functions as for the user-plane with the exception that there is no header compression function for the control-plane. The radio resource control (RRC) protocol is known as layer 3 in the AS protocol stack and in 3GPP RAN terminology [2]. It is the main controlling function in the AS responsible for establishing the radio bearers and

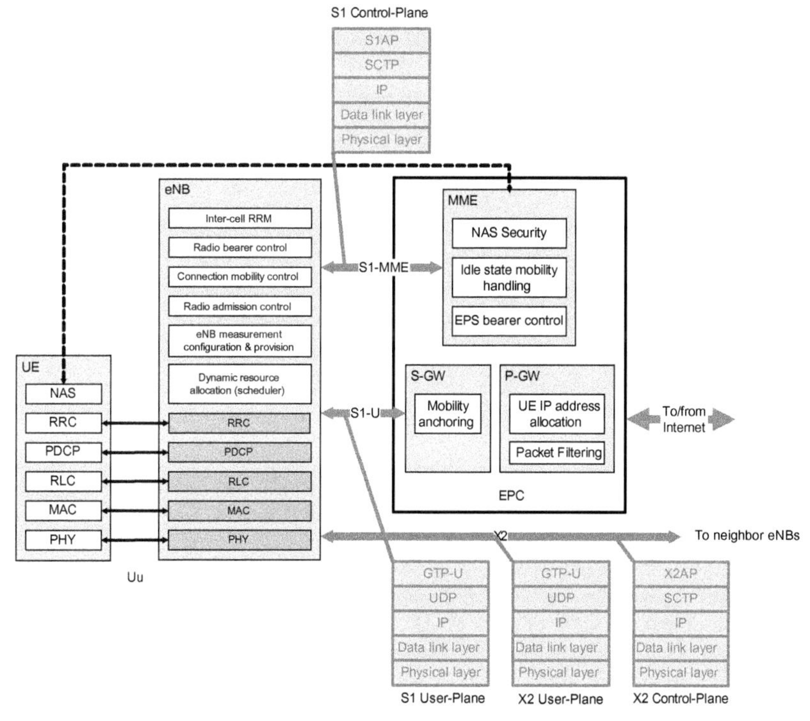

FIGURE 3.4

Protocol layers and functional split in UE, eNB, MME, S-GW, and P-GW [1].

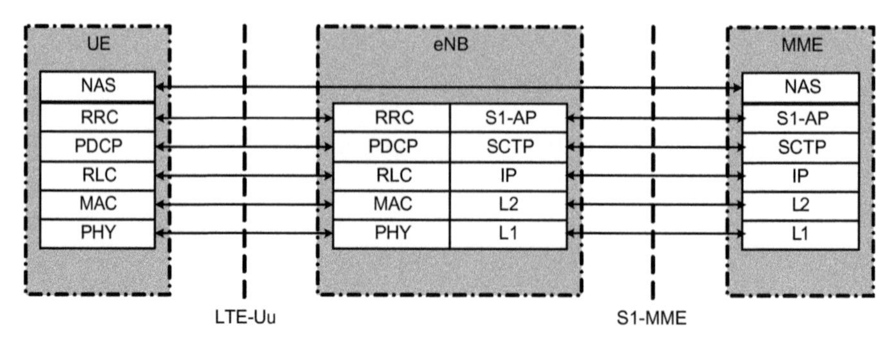

FIGURE 3.5

E-UTRAN control-plane protocol stack [10].

configuring all lower layers using RRC signaling between the eNB and the UE [16]. The E-UTRAN user-plane protocol stack is shown in Figure 3.6, consisting of the packet data convergence protocol (PDCP) [3], radio link control (RLC) [5], and medium access control (MAC) [4] sublayers that are terminated in the eNB on the network side. In the absence of any centralized control node, data buffering during handover due to user mobility in the E-UTRAN must be performed in the eNB itself. Data protection during handover is the responsibility of the PDCP sublayer. The RLC and MAC sublayers both restart in a new cell after the handover.

The S1 interface supports a multi-point connection among MMEs/S-GWs and eNBs. The S1-MME carries S1 application protocol (S1-AP) messages, using SCTP[2] over IP to provide guaranteed data delivery. Each SCTP association between an eNB and an MME can support multiple UEs. The S1-MME is responsible for EPC bearer setup and release procedures, handover signaling, paging, and NAS signaling transport. The S1-U consists of GTP-U protocol running on top of user datagram protocol (UDP), which provides best-effort data delivery. One GTP tunnel is established for each radio bearer in order to carry user traffic between the eNB and the selected S-GW.

[2]Stream control transmission protocol (SCTP) is a protocol for transmitting multiple streams of data at the same time between two end points that have established a connection in a network. SCTP is designed to facilitate a telephone connection over the Internet and specifically to support the telephone system's Signaling System 7—SS7—on an Internet connection. A telephone connection requires that signaling information, which controls the connection, be sent along with voice and other data at the same time. SCTP is also intended to make it easier to manage connections over a wireless network and to manage the transmission of multimedia data. SCTP is a standard protocol (IETF RFC 4960) developed by the IETF. SCTP manages reliable transport, ensuring complete arrival of data units that are sent over the network [13]. Unlike TCP, SCTP ensures complete and concurrent transmission of several streams of data (in units called messages) between connected end points. SCTP also supports multi-homing, which means that a connected end point can have alternate IP addresses [10,14].

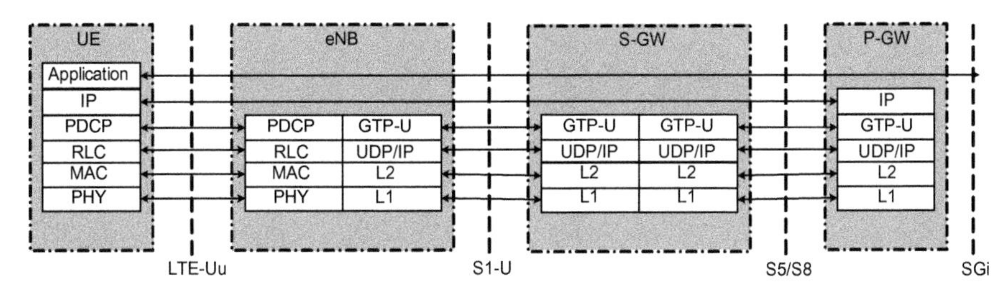

FIGURE 3.6

E-UTRAN user-plane protocol stack [10].

The GTP is a group of IP-based transport protocols that are used to carry IP packets within GSM, UMTS, and LTE networks [7]. As we discussed in Chapter 2, in 3GPP architecture, GTP and Proxy Mobile IPv6-based interfaces are specified on various interface points [12]. GTP can be decomposed into separate protocols: GTP-C and GTP-U. GTP-C is used within the GPRS core network for signaling between gateway GPRS support nodes (GGSN) and serving GPRS support nodes (SGSN). This allows the SGSN to activate a session on a user's behalf (PDP context activation), to deactivate the same session, to adjust QoS parameters, or to update a session for a subscriber who has just arrived from another SGSN. GTP-U is used for carrying user data within EPC and the GPRS core network, and between the radio access network and the core network. The user data that is transported can be packets in any of IPv4, IPv6, or PPP formats. Different GTP variants are implemented by eNBs, S-GW, P-GWs, RNCs, SGSNs, and GGSNs within 3GPP networks. The mobile stations are connected to the network without being aware of GTP. GTP can be used with UDP or transport control protocol (TCP) [11].

3.1.2 Protocol stack for S1 user-plane

The S1 user-plane interface (S1-U) is a standard reference point between the eNB and the S-GW. The S1-U interface provides non-guaranteed delivery of user-plane PDUs between the eNB and the S-GW. The user-plane protocol stack on the S1 interface is shown in Figure 3.7. The transport network layer is built on IP transport and GTP-U is used on top of UDP/IP to carry the user-plane PDUs between the eNB and the S-GW.

3.1.3 Protocol stack for S1 control-plane

The S1 control-plane interface (S1-MME) is a standard reference point between the eNB and the MME. The control-plane protocol stack of the S1 interface is shown in Figure 3.8. The transport network layer is built on IP transport, similar to the user-plane, but for the reliable transport of signaling messages, SCTP is

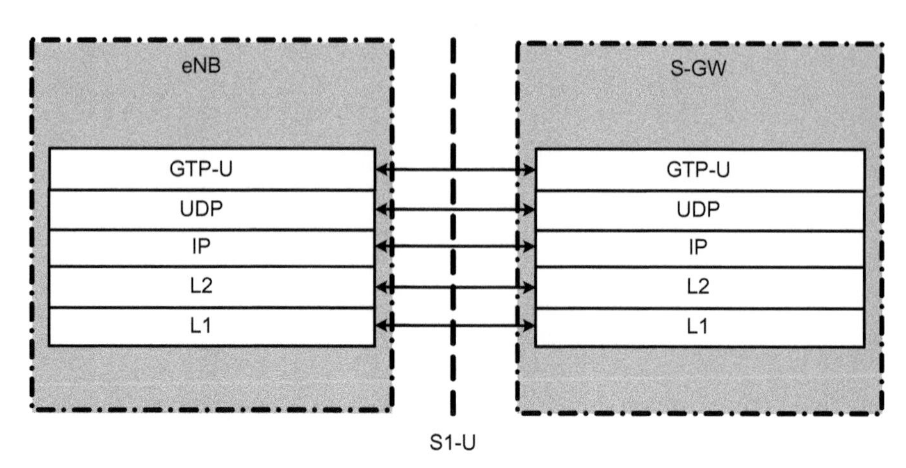

FIGURE 3.7

S1-U protocol architecture [1].

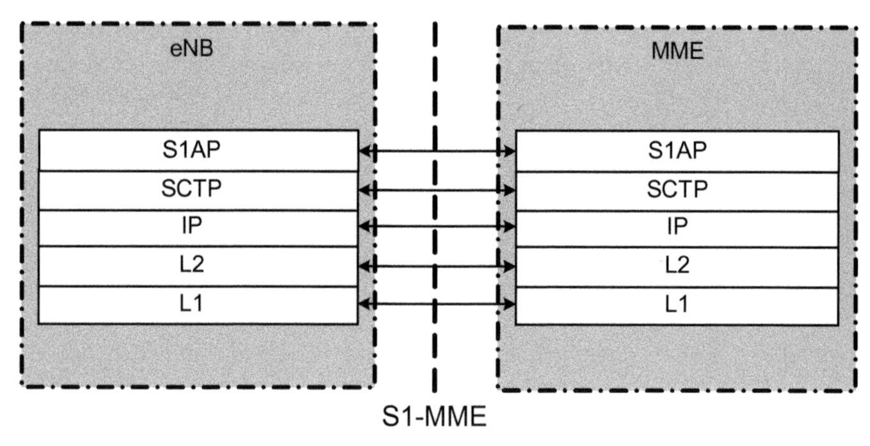

FIGURE 3.8

S1-MME protocol structure [1].

added on top of the IP layer. The application layer signaling protocol is referred to as S1-AP.

The SCTP layer provides the guaranteed delivery of application layer messages. The IP layer point-to-point transmission is used to deliver the signaling PDUs. A single SCTP association per S1-MME interface instance is used with one pair of stream identifiers for S1-MME common procedures. Only a few pairs of stream identifiers should be used for S1-MME-dedicated procedures. MME communication context identifiers that are assigned by the MME for S1-MME-dedicated procedures and eNB communication context identifiers that are assigned by the eNB for S1-MME-dedicated procedures are used to distinguish

UE-specific S1-MME signaling transport bearers. The communication context identifiers are conveyed in the respective S1AP messages. If the S1 signaling transport layer notifies the S1AP layer that the signaling connection is broken, the MME changes the state of the UEs which used this signaling connection to the ECM-IDLE state and the eNB releases the RRC connection with those UEs. In the case of relay nodes (RNs), there is one S1 interface between the RN and the relay-enabled eNB, and one S1 interface between the relay-enabled eNB and each of the MMEs in the MME pool. The S1 interface carries non-UE-associated S1AP signaling between the RN and the relay-enabled eNB as well as UE-associated S1AP signaling for UEs connected to the RN. The S1 interface between the relay-enabled eNB and an MME carries non-UE-associated S1AP signaling and UE-associated S1AP signaling for UEs connected to the RN and for UEs connected to the relay-enabled eNB [1].

3.1.4 Protocol stack for S5/S8 interface

The S5/S8 interface is defined between the S-GW and the P-GW. The S5 interface is used in non-roaming scenarios where the S-GW is located in the home network, or in roaming scenarios with both S-GW and P-GW located in the visited network. The latter scenario is also referred to as local breakout. The S8 interface is the roaming variant of S5 and is used in roaming scenarios with the S-GW in the visited network and the P-GW in the home network. When the GTP variant of the protocol is used, the S5/S8 interface provides the functionality associated with creation/deletion/modification/change of bearers for an individual user connected to the EPS. These functions are performed on a per PDN connection for each terminal. The S-GW provides the local anchor for all bearers for a single terminal and manages them toward the P-GW. Figure 3.9 shows the protocol stack over S5/S8 interface. Chapter 2 provides more information on the PMIP variant of S5/S8 [15].

3.1.5 Protocol stack for S11 interface

The S11 reference point defines the interface between the MME and the S-GW. This interface exclusively uses GTP-C and for LTE access. Due to the separation of the control- and user-plane functions between MME and S-GW, the S11 interface is used to create a new session (i.e., to establish the necessary resources for the session), and then to manage those sessions (i.e., modify, delete, and change any sessions for a terminal for each PDN connection) that has established connection within EPS. The S11 interface is always triggered by some events either directly from the NAS-level signaling from the terminal, such as a device attaching to the EPS network, adding new bearers to an existing session or creating a connection toward a new PDN, handover scenarios, or it may be triggered during network-initiated procedures such as P-GW-initiated bearer modification procedures. As such, the S11 interface coordinates the control- and user-plane procedures for a terminal during the period that the terminal is considered

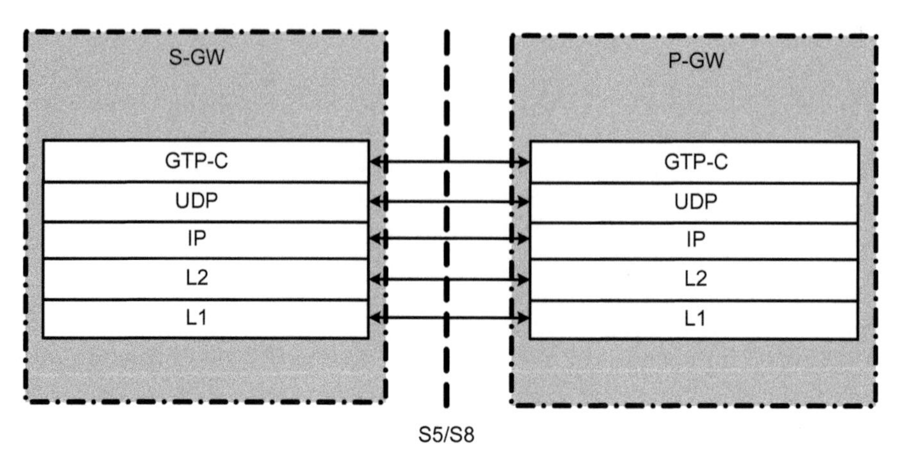

FIGURE 3.9

S5/S8 protocol structure [10].

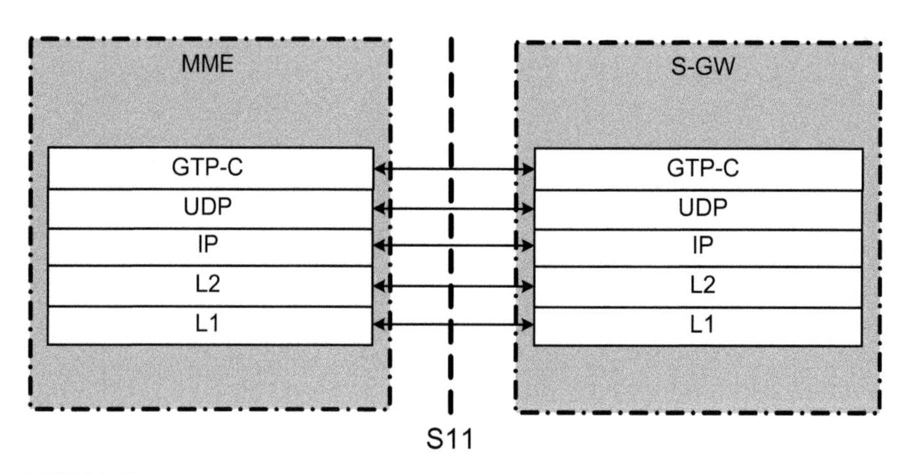

FIGURE 3.10

S11 protocol structure [15].

active/attached in the EPS. In the case of handover, the S11 interface is used to relocate the S-GW when appropriate, establish direct or indirect forwarding tunnel for user-plane traffic, and manage the user data traffic flow. Figure 3.10 shows the protocol stack across the S11 interface.

3.1.6 Protocol stack for X2 user-plane

The X2 user-plane interface defines data/control transport between eNBs. The X2-U interface provides non-guaranteed delivery of user-plane PDUs (due to use

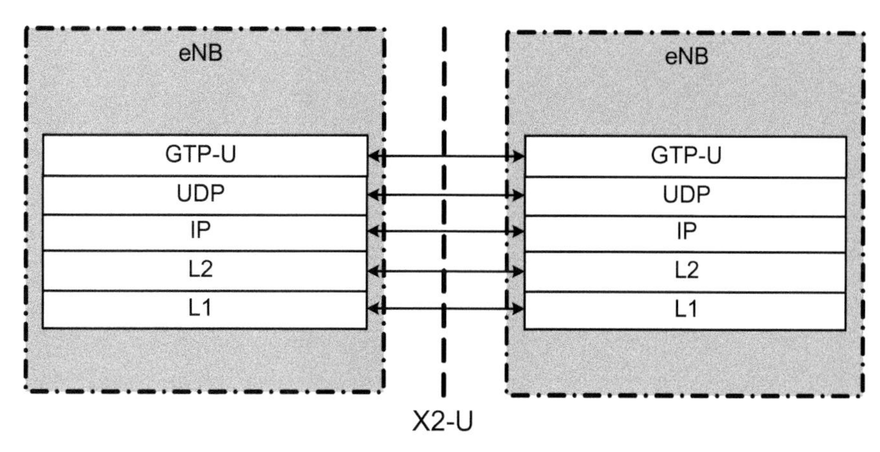

FIGURE 3.11

X2-U protocol structure [1].

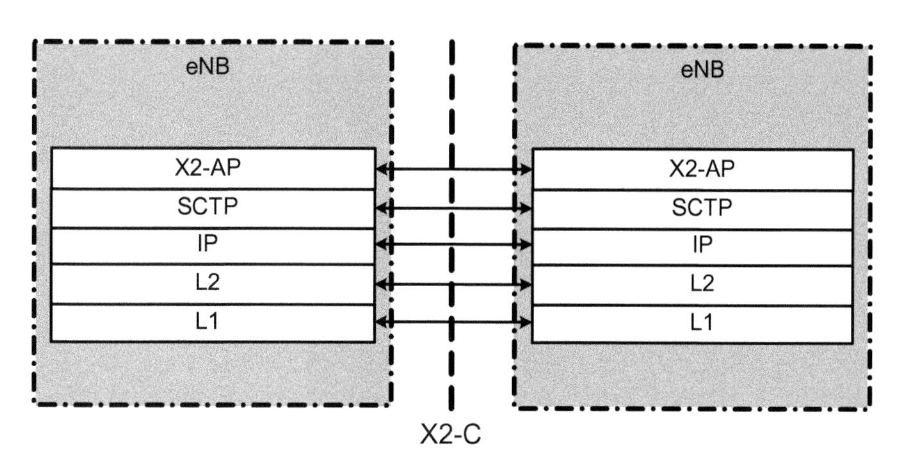

FIGURE 3.12

X2-C protocol structure [1].

of UDP transport protocol). The user-plane protocol stack on the X2 interface is shown in Figure 3.11. The transport layer is built on top of the IP layer and GTP-U is used on top of UDP/IP to carry the user-plane PDUs. The X2-U interface protocol stack is identical to the S1-U protocol stack [1].

The X2 reference point is a logical interface and even though it is normally shown as a direct connection between eNBs, it is usually routed via the same transport connection as the S1 interface to the site. The X2 interface control and user-plane protocol stacks are shown in Figures 3.11 and 3.12. X2 is an open interface and is often used for control-plane information exchange between neighboring eNBs; however, in the case of seamless handover, it can be temporarily used for user data forwarding. The key difference between the user-plane and control-plane X2 interface protocol stacks is the use of SCTP for control-plane information transmission between eNBs. The use of SCTP enables reliable delivery of control-plane information between eNBs, while for data forwarding, the UDP is considered sufficient. The X2AP covers the radio-related signaling while the user-plane GTP is used for user data transport. The X2-AP functionalities are as follows [1,10]:

- Intra-LTE mobility management where the handover messages between eNBs are transmitted on the X2 interface.
- Load management to enable inter-cell interference coordination by providing information on the resource status, overload, and traffic condition between different eNBs.
- Setting up and resetting of the X2 interface.
- Error handling for covering specific or general error cases.

The X2 interface has a key role in the intra-LTE handover operation. The source-eNB uses the X2 interface to send the handover request message to the target-eNB. If the X2 interface does not exist between the serving and the target-eNBs, then procedures need to be initiated to set up one before handover can be initiated. The handover request message informs the target-eNB to reserve resources and to send the handover request acknowledgment message assuming resources are available.

3.1.7 Protocol stack for X2 control-plane

The X2 control-plane interface defines control and signaling transport between two neighbor eNBs. The control-plane protocol stack of the X2 interface is shown in Figure 3.12. The SCTP is used for reliable transport of control and signaling information. The application layer signaling protocol is referred to as X2-AP. A single SCTP association per X2-C interface instance is used with one pair of stream identifiers for X2-C common procedures. Only a few pairs of stream identifiers may be used for X2-C dedicated procedures. Source-eNB communication context identifiers that are assigned by the source-eNB for X2-C dedicated procedures, and target-eNB communication context identifiers that are assigned by the target-eNB for X2-C dedicated procedures, are used to distinguish UE-specific X2-C signaling transport bearers. The communication context identifiers are conveyed in the respective X2-AP messages.

Inter-cell interference coordination in E-UTRAN is performed through the X2 interface. When the interference conditions change, the eNB signals the new condition to its neighbor eNBs, e.g., the neighbor eNBs for which an X2 interface is configured due to mobility requirements. When a time-domain inter-cell interference coordination scheme (e.g., eICIC) is used to mitigate inter-cell interference, the reference eNB signals its almost blank subframe (ABS) patterns to the neighboring eNBs, so that the receiving eNBs can exploit the ABS pattern of the reference eNB and adapt their activities accordingly in order to minimize the inter-cell interference. The load indication procedure is used to transfer interference coordination information between neighboring eNBs managing intra-frequency cells [1].

3.2 E-UTRAN and NAS protocols

The E-UTRAN overall architecture is described in 3GPP TS 36.300 [1,10]. Some general principles taken into consideration in the design of E-UTRAN architecture, as well as the E-UTRAN interfaces, are as follows:

- Signaling and data transport networks are logically separated.
- E-UTRAN and EPC functions are separated from transport functions. The addressing schemes used in E-UTRAN and EPC are not associated with the addressing schemes of transport functions. Some E-UTRAN or EPC functions reside in the same equipment.
- The mobility for RRC connection is controlled by the E-UTRAN.
- The interfaces are based on the logical model of the entity which is controlled through this interface.
- One physical network element can implement multiple logical nodes.

The E-UTRA radio access network consists of a network of eNBs, as illustrated in Figure 3.13. For unicast type of user traffic, there is no centralized controller in E-UTRAN. The eNBs are interconnected through X2 interface and to the EPC by means of S1 interface. The protocols operating between the eNBs and

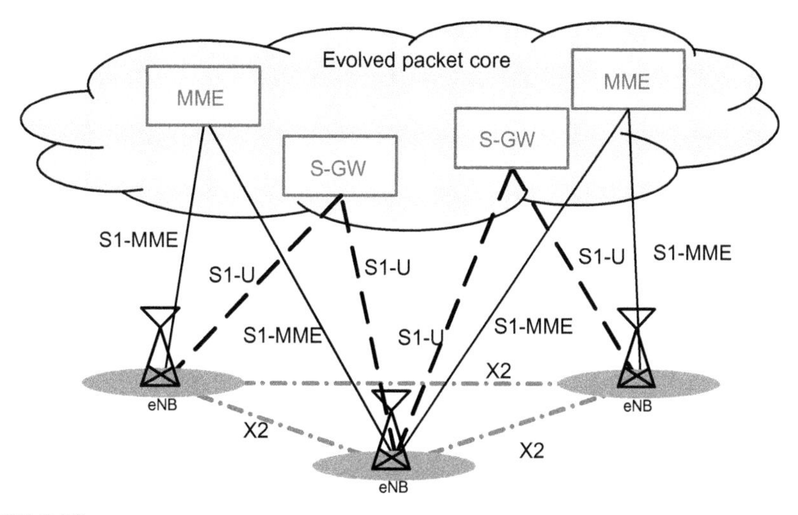

FIGURE 3.13

Radio access network entities and interfaces [1].

the UE are known as the AS protocols. The E-UTRAN is responsible for all radio-related functions, which can be summarized briefly as follows:

- Radio resource management which encompasses all functions related to the radio bearers, such as radio bearer control, admission control, mobility control, scheduling and dynamic allocation of resources to UEs in both uplink and downlink.
- Header compression which helps to ensure efficient use of the radio interface by compressing the IP packet headers that could otherwise cause a significant overhead, especially for small-packet applications such as VoIP.
- Security which ensures all data sent over the radio interface is encrypted.
- Connectivity to the EPC which consists of the signaling toward MME and the bearer path toward the S-GW.

On the network side, all of these functions reside in the eNBs, each of which can be responsible for managing multiple cells. Unlike the previous 3GPP releases, LTE integrates the radio controller function into the eNB. This allows tight interaction between different protocol layers of the radio access network, thus reducing latency and improving efficiency. Such distributed control eliminates the need for a complex central controller with high processing power, which in turn has the potential of reducing costs. Furthermore, since LTE does not support soft handovers similar to UMTS, there is no need for a centralized data-combining function in the network. One drawback for the lack of a centralized controller node is that, as the UE moves, the network must transfer the UE context, together with any buffered data, from one eNB to another. There are mechanisms and provisions in E-UTRA to avoid data loss during handover.

The general protocol model for E-UTRAN interfaces is depicted in Figure 3.14. The structure is based on the principle that the layers and planes are logically independent of each other. Therefore, as needed and when required, 3GPP can easily alter protocol stacks and planes to satisfy future requirements [1,10].

The protocol structure consists of two main layers, the radio network layer and transport network layer. E-UTRAN functions are realized in the radio network layer, and the transport network layer represents standard transport technology that is selected to be used for E-UTRAN. The control-plane includes the application protocol, i.e., S1-AP and X2-AP, and the signaling bearer for transporting the application protocol messages. The application protocol is used for setting up bearers (i.e., E-RAB[3]) in the radio network layer. The bearer parameters in the application protocol are not directly tied to the user-plane technology; rather, they are generic bearer parameters. The user-plane includes the data

[3]E-UTRAN radio access bearer (E-RAB) uniquely identifies the concatenation of an S1 bearer and the corresponding data radio bearer. When an E-RAB exists, there is a one-to-one mapping between the E-RAB and an EPS bearer of the non-access stratum as defined in 3GPP TS 24.301 [6].

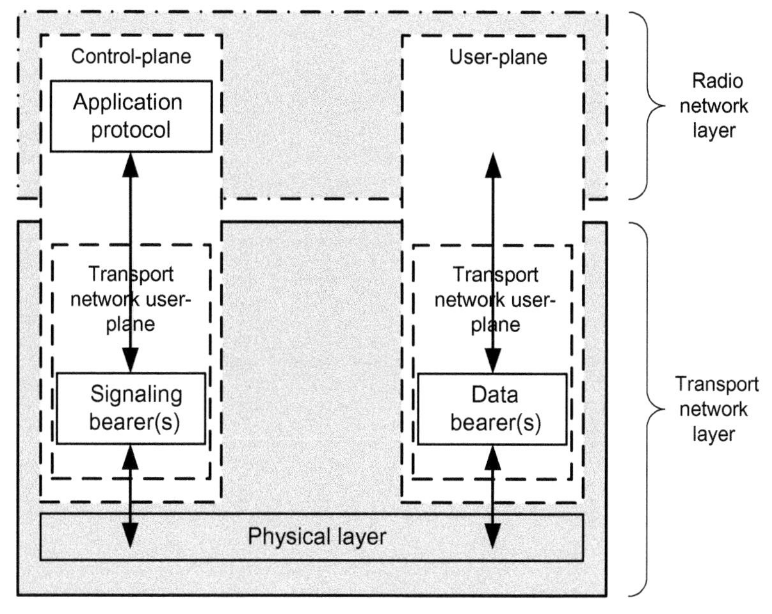

FIGURE 3.14

General protocol model for E-UTRAN interfaces [1].

bearer(s) for the data stream(s). The data stream is characterized by a tunneling protocol in the transport network layer.

As mentioned earlier, the eNB typically performs the following functions [1]:

- Radio resource management functions, including radio bearer control, radio admission control, connection management, dynamic allocation of resources to UEs in both uplink and downlink (i.e., scheduling).
- Header compression and encryption of user payloads.
- Selection of an MME at UE attachment when no routing to an MME can be determined from the information provided by the UE.
- Routing of user-plane data toward S-GW.
- Scheduling and transmission of paging messages (originated from the MME).
- Scheduling and transmission of broadcast information (originated from the MME).
- Measurement and reporting for support of mobility and scheduling.

As illustrated in Figure 3.13, the eNBs are interconnected with each other through an X2 interface. The X2 interface allows eNBs to communicate directly with each other and coordinate their activities. The X2 interface is split into separate control- and user-planes. The X2 control-plane carries X2AP messages between eNBs and uses SCTP for reliable delivery of messages. The X2AP is used to manage inter-eNB mobility and handovers, UE context transfers, inter-cell interference management, and various error handling functions. The X2 user-plane uses GTP-U to tunnel user traffic between eNBs.

The main services and functions of the NAS sublayer include EPS bearer control, ECM-IDLE mobility handling, paging origination, and configuration and control of security [6]. The NAS procedures, and in particular the connection management procedures, are in essence similar to those of UMTS. The main change from UMTS is that EPS facilitates combination of some procedures to enable faster establishment of connections and bearers. The MME creates a UE context when a UE is turned on and attaches to the network. It assigns a unique short temporary identity to the UE that identifies the UE context in the MME. The UE context contains user subscription information obtained from the HSS. The local storage of subscription data in the MME allows faster execution of procedures such as bearer establishment. In addition, the UE context also holds dynamic information such as the list of bearers that are established and the terminal capabilities.

3.3 LTE-Uu interface protocols

The protocols over LTE-Uu and S1 interfaces are divided into two groups: user-plane protocols that are the protocols implementing the actual E-RAB service, i.e., carrying user data through the AS, and control-plane protocols, which are responsible for controlling the E-RABs and the connection between the UE and the network from different aspects including requesting the service, controlling different transmission resources, and handover. Figures 3.15 and 3.16 illustrate the LTE-Uu user-plane and control-plane protocol stacks, respectively. In the control-plane, the NAS functional block is used for network attachment, authentication, setting up bearers, and mobility management. All NAS messages are ciphered and integrity protected by the MME and the UE [1,5]. There is also a mechanism for transparent transfer of NAS messages that is included in the latter group. The E-RAB service is offered from service access point (SAP) to SAP by the AS.

The layer 2 functions in LTE are classified into MAC, RLC, and PDCP functions [1–5]. Figures 3.17 and 3.18 illustrate the structure of layer 2 protocols in LTE downlink and uplink. The SAP for between two adjacent protocol layers is marked with a circle at the interface between the sublayers. The SAP between the physical layer and the MAC sublayer provides the transport channels. The SAP between the MAC sublayer and the RLC sublayer provides the logical channels. The multiplexing of several logical channels on the same transport channel is performed by the MAC sublayer. Each logical channel is defined by the type of information that is transferred. The logical channels are generally classified into two groups: (1) control channels (for the transfer of control-plane information); and (2) traffic channels (for the transfer of user-plane information).

The RRC sublayer in the eNB, makes handover decisions based on neighbor cell measurements reported by the UE, performs paging of the users over the

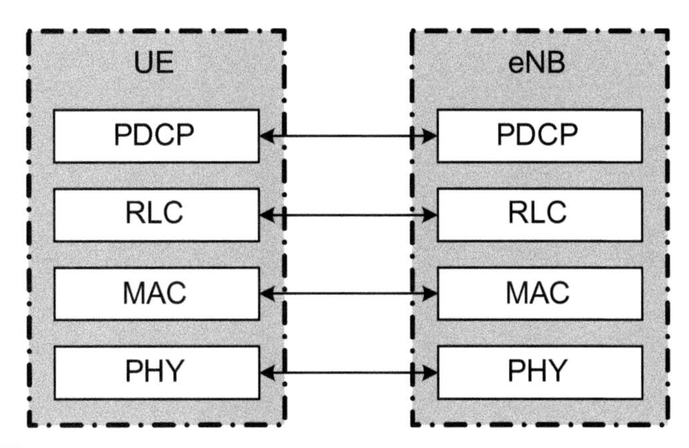

FIGURE 3.15

LTE-Uu interface user-plane protocol stack [1].

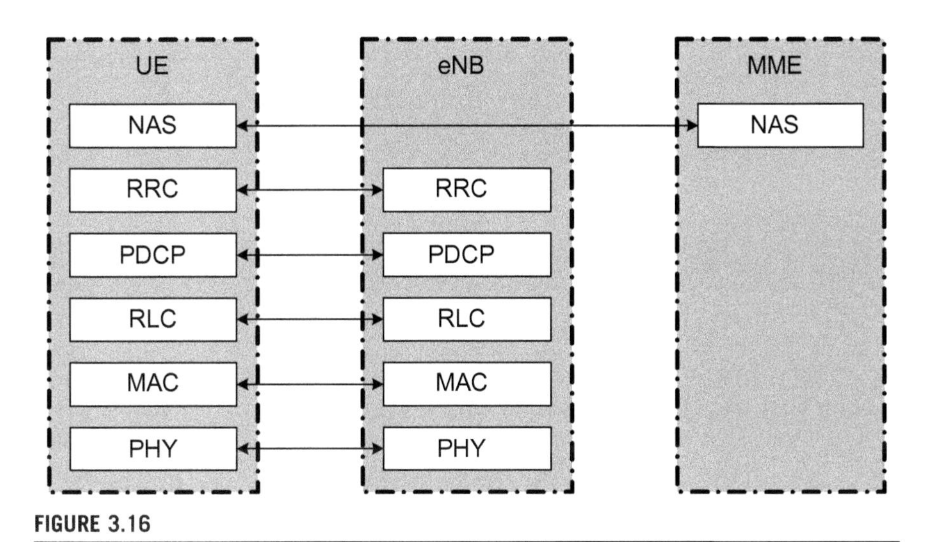

FIGURE 3.16

LTE-Uu interface control-plane protocol stack [1].

air-interface, broadcasts system information, controls UE measurement and reporting functions such as the periodicity of channel quality indicator reports, and further allocates cell-level temporary identifiers to active users. It also executes transfer of UE context from the serving eNB to the target-eNB during handover and performs integrity protection of RRC messages. The RRC sublayer is responsible for setting up and maintenance of radio bearers. Note that the RRC sublayer in 3GPP protocol hierarchy is considered as layer 3.

The services and functions provided by the PDCP sublayer in the user-plane include header compression and decompression of IP packets, transfer of user

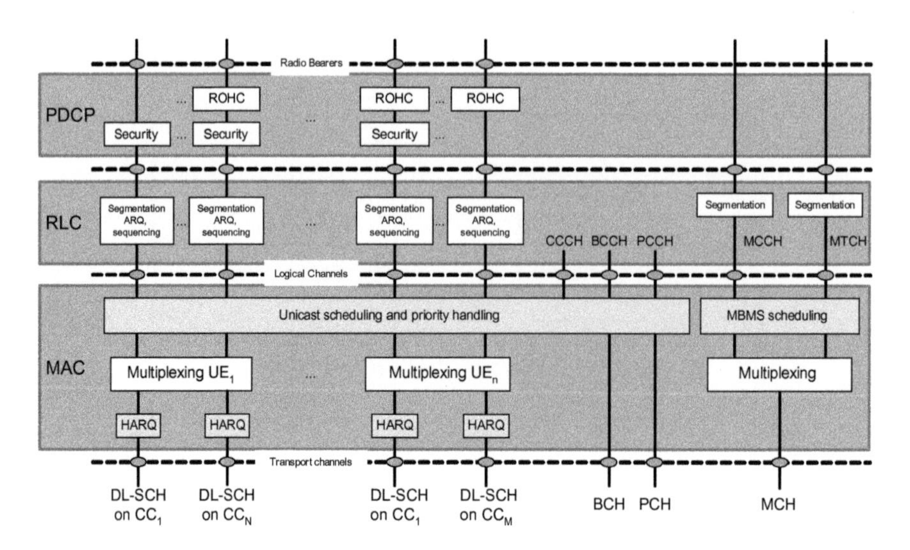

FIGURE 3.17

LTE layer 2 structure in the downlink [1].

data between NAS and RLC sublayer, sequential delivery of upper layer PDUs and duplicate detection of lower layer SDUs following a handover for RLC acknowledged mode, retransmission of PDCP SDUs following a handover for RLC acknowledged mode, and ciphering. The services and functions provided by the PDCP for the control-plane include ciphering and integrity protection and transfer of control-plane data where PDCP receives PDCP SDUs from RRC and forwards it to the RLC sublayer and vice versa [1,3,13].

The RLC sublayer is used to format and transport traffic between the UE and the eNB [1,5]. The RLC sublayer provides three different reliability modes for data transport, i.e., acknowledged mode (AM), unacknowledged mode (UM), and transparent mode (TM). The UM is suitable for transport of real-time services since such services are delay sensitive and cannot tolerate delay due to ARQ retransmissions. The acknowledged mode is appropriate for non-real-time services such as file transfers. The transparent mode is used when the size of SDUs are known in advance such as for broadcasting system information. The RLC sublayer also provides sequential delivery of SDUs to the upper layers and eliminates duplicate packets from being delivered to the upper layers. It may also segment the SDUs.

The MAC sublayer is responsible for data transfer and logical channel multiplexing, HARQ retransmissions and uplink/downlink scheduling, random access procedure, and maintenance of uplink timing. It is also responsible for multiplexing/de-multiplexing of data across multiple component carriers when carrier aggregation is used. From a physical layer perspective, the MAC sublayer provides and receives services in the form of transport channels. The data in a transport channel is organized into transport blocks. By varying the transport format of

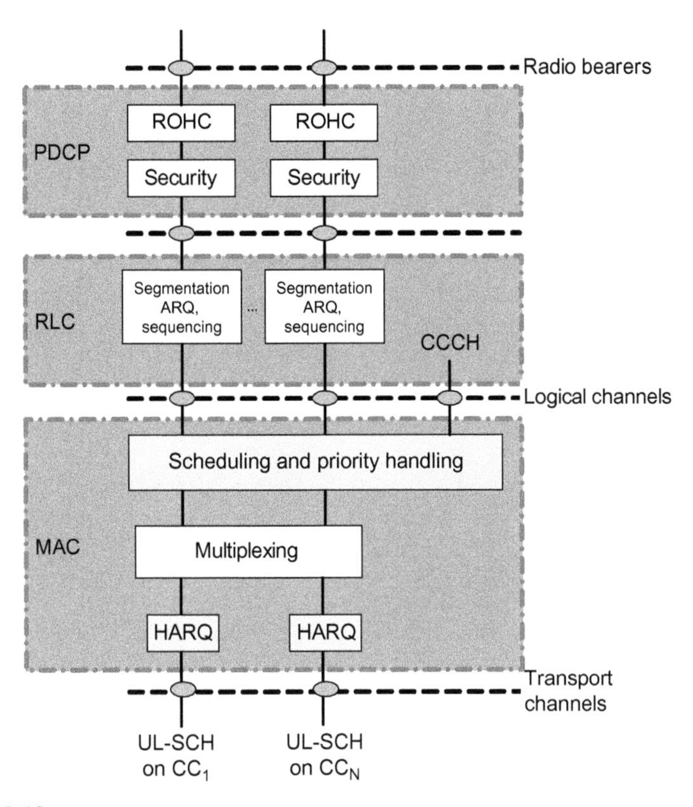

FIGURE 3.18

LTE layer 2 structure in the uplink [1].

the transport blocks, the MAC sublayer can realize different data rates and reliability levels. The MAC sublayer receives the RLC SDUs mapped to various logical channels in the downlink, and generates the MAC PDUs that further become the transport blocks in the physical layer.

The main difference between MAC and RRC control lies in the signaling reliability. The signaling corresponding to state transitions and radio bearer configurations should be performed by the RRC sublayer due to higher signaling reliability.

LTE-Advanced extended the capabilities of LTE Rel-8 with support of carrier aggregation, where two or more component carriers are aggregated in order to support wider transmission bandwidths up to 100 MHz. A user terminal may simultaneously receive or transmit one or multiple component carriers depending on its capabilities. From the UE perspective, the layer 2 aspects of HARQ are similar to those of LTE Rel-8. There is one transport block (in the absence of spatial multiplexing, up to two transport blocks in the case of spatial multiplexing) and one independent HARQ entity per scheduled component carrier. Each transport block is mapped to a single component carrier on which all HARQ retransmissions may take place. A UE may be scheduled over multiple component carriers simultaneously, but at most one random access procedure is performed at any time. Whenever a UE is configured with only one component carrier, LTE Rel-9 DRX is the baseline. In other cases, the same DRX operation will be applied to all configured component carriers. Therefore, the layer 2 structure of the LTE-Advanced is similar to that of LTE Rel-8, except the addition of the carrier aggregation functionality; however, the multi-carrier nature of the physical layer is only exposed to the MAC sublayer through transport channels, where one HARQ entity is required per component carrier [1,4].

The logical and transport channel classification in E-UTRA are illustrated in Figures 3.19 and 3.20, respectively. Each logical channel is defined by the type of information that is transferred. The logical channels are generally classified into two groups: (1) control channels (for the transfer of control-plane information); and (2) traffic channels (for the transfer of user-plane information) as shown in Figure 3.19 [1]. The control channels are exclusively used for transfer of control-plane information. The control channels supported by MAC can be classified as follows:

- Broadcast control channel (BCCH): A downlink channel for broadcasting system information.
- Paging control channel (PCCH): A downlink channel that transfers paging information and system information change notifications. This channel is used for paging when the network does not know the location of the UE.

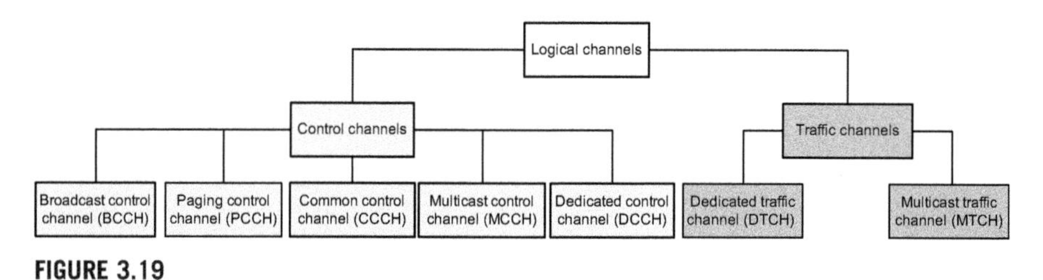

FIGURE 3.19

E-UTRA logical channel structure [1].

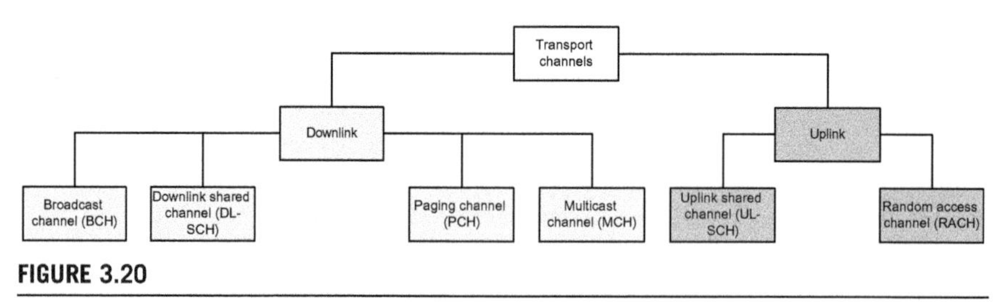

FIGURE 3.20

E-UTRA transport channel structure [1].

- Common control channel (CCCH): A channel for transmitting control information between UEs and eNBs. This channel is used for UEs that have no RRC connection with the network.
- Multicast control channel (MCCH): A point-to-multipoint downlink channel used for transmitting MBMS control information from the network to the UE for one or several MTCHs. This channel is only used by UEs that are subscribed to MBMS.
- Dedicated control channel (DCCH): A point-to-point bidirectional channel that transmits dedicated control information between a UE and the network. It is used by UEs that have RRC connection.

The traffic channels are exclusively used for the transfer of user-plane information. The traffic channels supported by MAC can be classified as follows (as shown in Figure 3.19):

- Dedicated traffic channel (DTCH): A point-to-point bi-directional channel dedicated to a single UE for the transfer of user information.
- Multicast traffic channel (MTCH): A point-to-multipoint downlink channel for transmitting broadcast/multicast data from the network to the UE. This channel is only used by UEs that receive MBMS.

The physical layer provides information transfer services to the MAC and higher layers. The physical layer transport services are described by how and with what characteristics data is transferred over the radio interface. This should be clearly distinguished from the classification of what is transported which relates to the concept of logical channels at the MAC sublayer. As shown in Figure 3.20, downlink transport channels can be classified as follows:

- Broadcast channel (BCH) is characterized by fixed, predefined transport format, and is required to be broadcast in the entire coverage area of the cell.
- Downlink shared channel (DL-SCH) is characterized by support for HARQ protocol, dynamic link adaptation by varying modulation, coding, and transmit power, possibility for broadcast in the entire cell, possibility to use beamforming, dynamic and semi-static resource allocation, UE DRX to enable power saving, and MBMS transmission.
- Paging channel (PCH) is characterized by support for UE DRX in order to enable power saving, requirement for broadcast in the entire coverage area of the cell and is mapped to physical resources which can also be used dynamically for traffic or other control channels.
- Multicast channel (MCH) is characterized by requirement to be broadcast in the entire coverage area of the cell, support for macro-diversity combining of MBMS transmission on multiple cells, and support for semi-static resource allocation.

The uplink transport channels are classified as follows (see Figure 3.20):

- Uplink shared channel (UL-SCH) is characterized by possibility to use beamforming, support for dynamic link adaptation by varying the transmit power

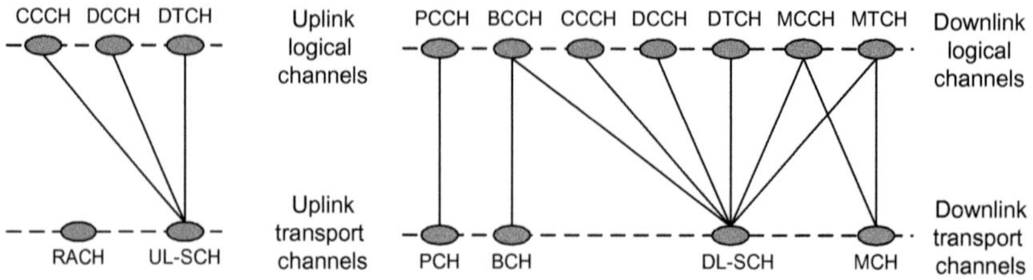

FIGURE 3.21

Logical to transport channel mapping [1].

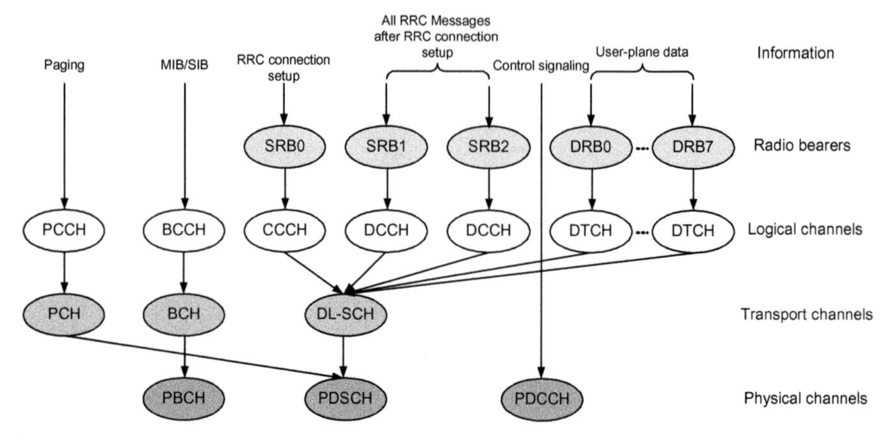

FIGURE 3.22

Mapping of downlink information to radio and signaling bearers as well as various logical, transport, and physical channels [1–5].

and modulation and coding schemes, support for HARQ, support for both dynamic and semi-static resource allocation.

- Random access channel (RACH) is characterized by limited control information and collision risk.

The mapping of the logical channels to the transport channels in the downlink and uplink is shown in Figure 3.21.

The mapping of downlink and uplink information to radio and signaling bearers as well as various logical, transport, and physical channels at different layers of protocol stack are illustrated in Figures 3.22 and 3.23.

3.4 Support for relay nodes

The E-UTRAN architecture was modified in LTE Rel-10 to include support for relaying function. The intermediate relay nodes are connected to a serving eNB

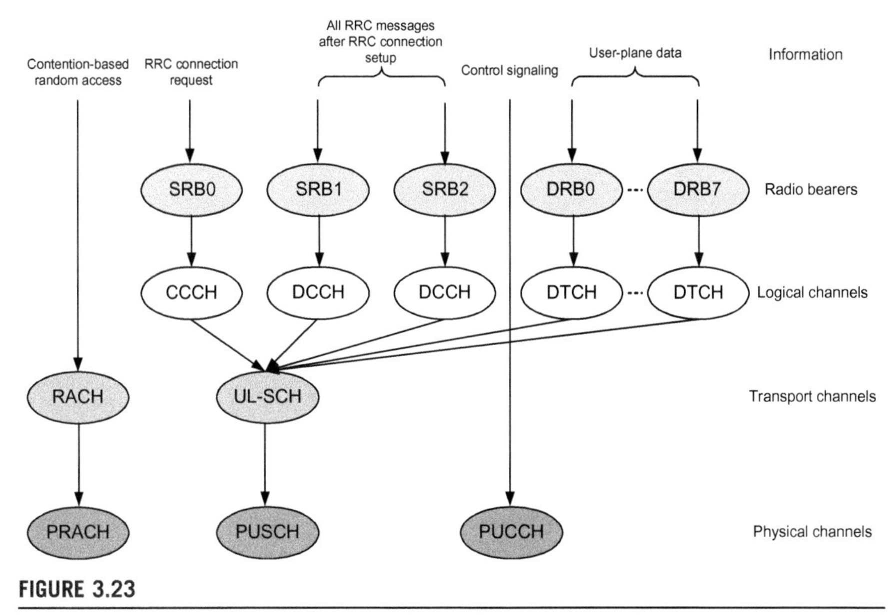

FIGURE 3.23

Mapping of uplink information to radio and signaling bearers as well as various logical, transport, and physical channels [15].

called the donor eNB (DeNB), via a modified version of the E-UTRA radio interface known as Un interface. The relay entity supports the eNB functionality, i.e., it terminates the radio protocols of the E-UTRA radio interface and the S1 and X2 interfaces. In addition to the eNB functionality, the relay node (RN) also supports a subset of the UE functionality, e.g., physical layer, layer 2, RRC, and NAS functionality, in order to wirelessly connect to the DeNB.

The modified architecture for supporting the relay nodes is shown in Figure 3.24. The RN terminates S1, X2, and Un interfaces. The DeNB provides S1 and X2 proxy functionality between the RN and other eNBs, MMEs, and S-GWs. The S1 and X2 proxy functionality includes forwarding UE-dedicated S1 and X2 signaling messages as well as GTP data packets between the S1 and X2 interfaces associated with the RN and the S1 and X2 interfaces associated with other network nodes. Due to the proxy functionality, the DeNB appears as MME (for S1-MME), eNB (for X2), and S-GW (for S1-U) to the RN. The DeNB can provide the S-GW/P-GW type functions needed for the RN operation. This includes creating a session for the RN and managing EPS bearers for the RN, as well as terminating the S11 interface toward the MME serving the RN. The RN and DeNB map signaling and data packets to EPS bearers that are set up for the RN. The mapping is based on the existing QoS mechanisms defined for the UE and the P-GW. The P-GW functions in the DeNB can allocate an IP address for the RN for the purpose of operation and management, which may be different from the S1 IP address of the DeNB.

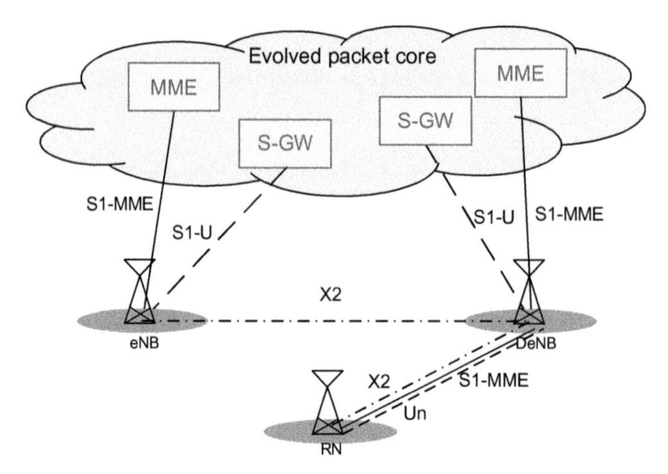

FIGURE 3.24

Modified E-UTRAN architecture supporting relay nodes [1,9].

The S1 user-plane protocol stack for supporting RNs is shown in Figure 3.25. There is a GTP tunnel associated with each UE EPS bearer, connecting the S-GW associated with the UE to the DeNB, which is switched to another GTP tunnel in the DeNB, going from the DeNB to the RN (one-to-one mapping). The X2 user-plane protocol stack for supporting RNs is also shown in Figure 3.25. There is a GTP forwarding tunnel associated with each UE EPS bearer subject to forwarding, connecting the other eNBs to the DeNB, which is switched to another GTP tunnel in the DeNB, going from the DeNB to the RN (one-to-one mapping). The S1 and X2 user-plane packets are mapped to radio bearers over the Un interface. The mapping can be based on the QoS class identifier associated with the UE EPS bearers. The UE EPS bearer with similar QoS can be mapped to the same Un radio bearer [1,9].

The S1 control-plane protocol stack for supporting RNs is shown in Figure 3.26. There is a single S1 interface between each RN and the DeNB, and there is a single S1 interface between the DeNB and the MME. The DeNB processes and forwards all S1 messages between the RN and the MMEs for all UE-dedicated procedures. The processing of S1-AP messages includes modifying S1-AP UE identifiers, transport-layer address, and GTP tunnel endpoint identifiers (TEIDs) without changing other parts of the messages. All non-UE-dedicated S1-AP procedures are terminated at the DeNB and are handled locally between the RN and the DeNB, and between the DeNB and the MME(s). Upon reception of an S1 non-UE-dedicated message from an MME, the DeNB may trigger corresponding S1 non-UE-dedicated procedure(s) to the RN(s). If more than one RN is involved, the DeNB may wait and aggregate the response messages from all participating RNs before responding to the MME. Upon reception of an S1 non-UE-dedicated message from an RN, the DeNB may trigger associated S1 non-UE-dedicated procedure(s) to the MME(s).

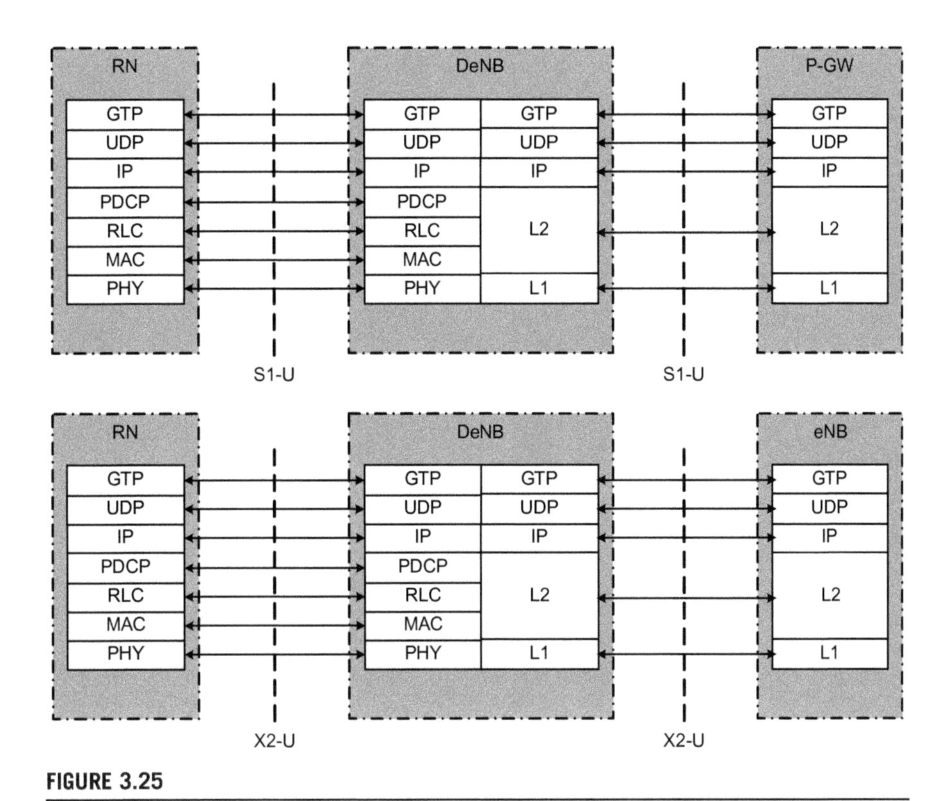

FIGURE 3.25

Modified S1-U and X2-U protocol structure for support of relay nodes [1].

The X2 control-plane protocol stack for supporting RNs is shown in Figure 3.26. There is a single X2 interface between each RN and the corresponding DeNB. The DeNB may have additional X2 interfaces to the neighboring eNBs. The DeNB processes and forwards all X2 messages between the RN and other eNBs for all UE-dedicated procedures. The processing of X2AP messages includes modifying S1AP or X2AP UE identifiers transport-layer address, and GTP TEIDs without changing other parts of the message. All non-UE-dedicated X2AP procedures are terminated at the DeNB and are handled locally between the RN and the corresponding DeNB, or between the DeNB and other eNBs. If one or more RNs are involved, the DeNB may wait and aggregate the response messages from all participating RNs to respond to the originating node [1,9].

The RN connects to the DeNB via a Un interface using the same radio protocols and procedures used by a UE to connect to an eNB. In addition to UE-type protocol functionalities, the RRC sublayer is able to configure and reconfigure an RN subframe configuration through the RN reconfiguration procedure over the Un interface, e.g., downlink subframe configuration and an RN-specific control channel for transmissions between the RN and the DeNB. The RN may request such a configuration from the DeNB during the RRC

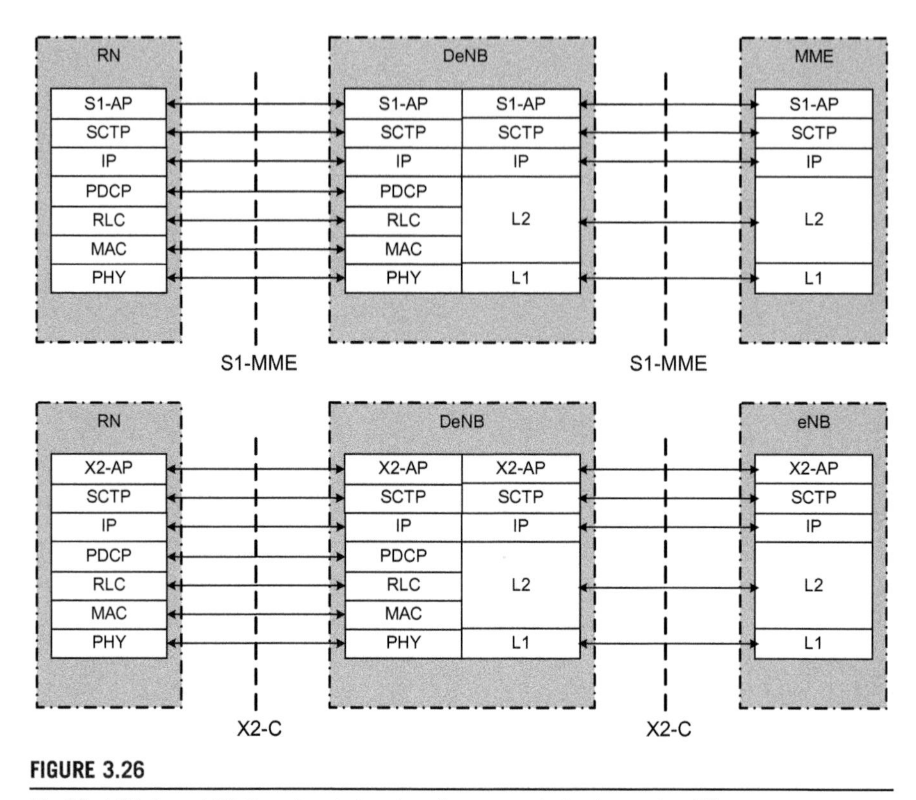

FIGURE 3.26

Modified S1-C and X2-C protocol structure for support of relay nodes [1].

connection establishment, and the DeNB may initiate the RRC signaling for such a configuration. The RN subframe configuration over the Un interface may be temporarily misaligned with the MBSFN subframes configured in the relay cell because a new subframe configuration might have been applied earlier by the RN on the Un reference point. The RRC sublayer of the Un interface can further send updated system information in a dedicated message to an RN with an RN subframe configuration. The PDCP sublayer of the Un interface can provide integrity protection for the user-plane. The integrity protection is configured per data radio bearer [1,9].

3.5 Support for HeNB nodes

Small cells are low-power wireless access points that operate in licensed spectrum and are operator-managed. Small cells can provide improved cellular coverage, and capacity, which have applications for homes and enterprises as well as metropolitan and rural public spaces. Various types of small cells include femto-cells, pico-cells, metro-cells, and micro-cells—broadly increasing in size from femto-cells (the

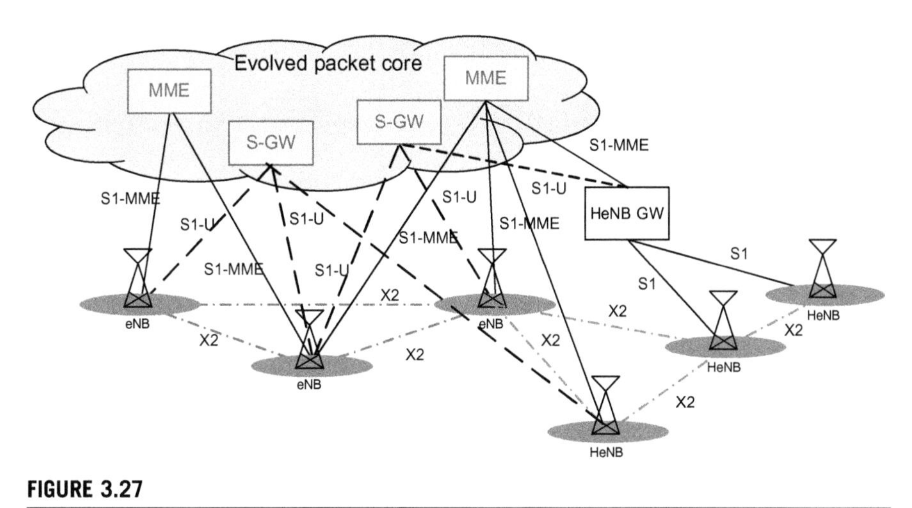

FIGURE 3.27

Modified E-UTRAN architecture with support of HeNB-GW and HeNB [1].

smallest) to micro-cells (the largest). Any or all of these small cells can be based on femto-cell technology, referred to as home eNB (HeNB) in 3GPP literature. Small cells also facilitate a new breed of mobile service that exploits the technology's ability to detect presence and connect and interact with the existing networks.

As shown in Figure 3.27, the E-UTRAN architecture may be extended to include a home eNB gateway (HeNB-GW) to provide an S1 interface between the HeNB entities (also known as femto-cell in the literature) and the EPC [1]. The HeNB-GW serves as a concentrator for the control-plane and the S1-MME interface. The S1-U interface from the HeNB may be terminated at the HeNB-GW, or a direct logical user-plane connection between the HeNB and S-GW may be used. In the modified architecture, the S1 interface is defined between the HeNB-GW and the core network, between the HeNB and the HeNB-GW, between the HeNB and the core network, or between the eNB and the core network. The HeNB-GW would appear to the MME as an eNB and to the HeNB as an MME. The S1 interface between the HeNB and the EPC is the same, regardless of the way HeNB is connected to the EPC.

The HeNB-GW is connected to the EPC in such a way that inbound and outbound mobility to cells served by the HeNB-GW does not require inter-MME handovers. Each HeNB serves only one cell. The functions and protocols supported by the HeNB are the same as those supported by an eNB, with the exception that a new reference point S5 may be used between the HeNB and the S-GW, if local IP access is supported. LTE Rel-11 supports direct X2-connectivity between HeNBs, irrespective of the HeNB-GW existence [1].

If the HeNB supports the local IP access feature, then it is required to support the S5 interface with the S-GW and the SGi interface with the IP network. As mentioned earlier, a HeNB performs the same functions as an eNB with some additional functions and restrictions in the case of connection to the HeNB-GW,

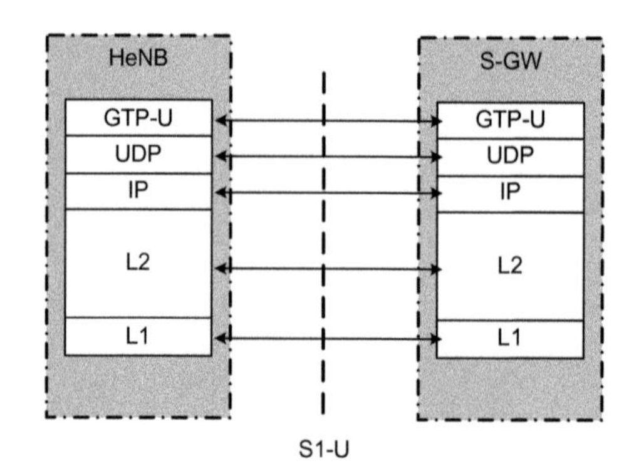

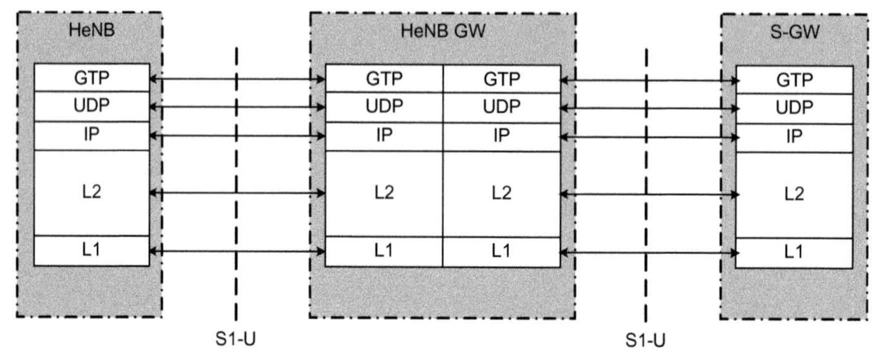

FIGURE 3.28

S1-U interface between HeNB and S-GW with and without HeNB-GW [1].

such as discovery of a suitable serving HeNB-GW, connection to a single HeNB-GW at any time, using the same tracking area code and PLMN ID supported by the HeNB-GW, etc. HeNBs may be deployed without network planning. A HeNB may be moved from one geographical area to another, and therefore it may need to connect to different HeNB-GWs depending on its location. The HeNB may support fixed broadband access network interworking function to signal tunnel information to the MME. The tunnel information includes the HeNB IP address.

The HeNB-GW performs the following functions [1]:

- Relaying UE-associated S1 application messages between the MME serving the UE and the HeNB serving the UE, except the UE context release request message received from the HeNB.
- Terminating non-UE-associated S1 application procedures in the direction of the HeNB or the MME.
- Terminating the S1-U interface with the HeNB and with the S-GW.
- Supporting tracking area code and PLMN ID used by the HeNB.
- Routing the S1 path-switch request message toward the MME based on the globally unique MME identifier of the source MME received from the HeNB.

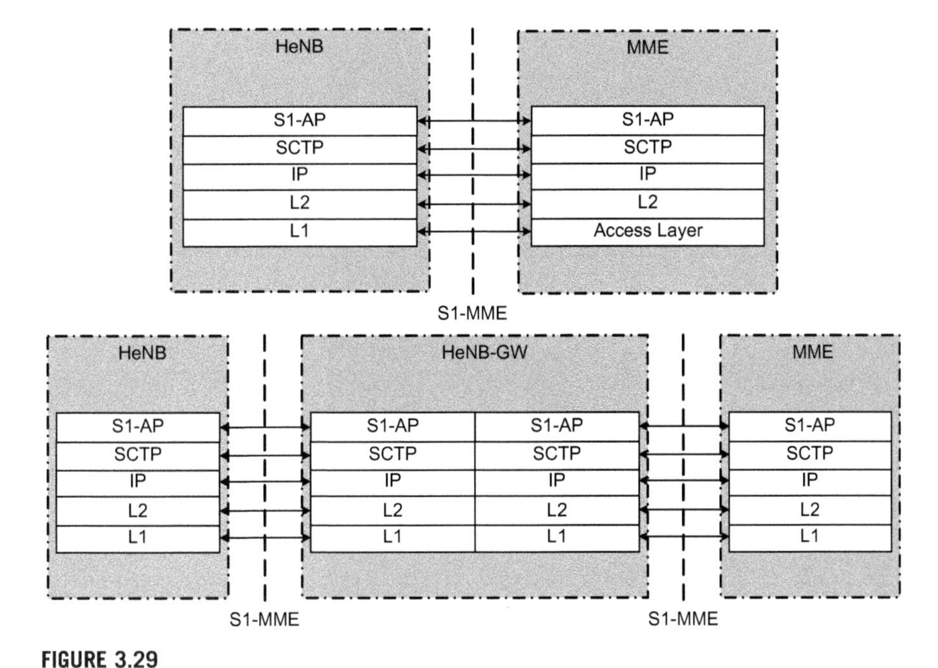

FIGURE 3.29

S1-MME interface between HeNB and MME with and without HeNB-GW [1].

As shown in Figure 3.28, the user-plane interface between the HeNB, HeNB-GW, and the S-GW can be defined with and without the HeNB-GW.

The HeNB-GW may optionally terminate the user-plane toward the HeNB or the S-GW, and forward the user-plane data between the HeNB and the S-GW. As shown in Figure 3.29, the control-plane interface between the HeNB and MME can be defined irrespective of the HeNB-GW presence. When the HeNB GW is not present, all S1AP procedures are terminated at the HeNB and the MME. The HeNB-GW forwards control-plane information between the HeNB and the MME.

References

3GPP Specifications

[1] 3GPP TS 36.300, Evolved Universal Terrestrial Radio Access (E-UTRA) and Evolved Universal Terrestrial Radio Access Network (E-UTRAN); Overall description; Stage 2 (Release 11), December 2012.

[2] 3GPP TS 36.331, Evolved Universal Terrestrial Radio Access (E-UTRA); Radio Resource Control (RRC); Protocol specification (Release 11), December 2012.

[3] 3GPP TS 36.323, Evolved Universal Terrestrial Radio Access (E-UTRA); Packet Data Convergence Protocol (PDCP) specification (Release 11), December 2012.

[4] 3GPP TS 36.321, Evolved Universal Terrestrial Radio Access (E-UTRA); Medium Access Control (MAC) protocol specification (Release 11), December 2012.
[5] 3GPP TS 36.322, Evolved Universal Terrestrial Radio Access (E-UTRA); Radio Link Control (RLC) protocol specification (Release 11), December 2012.
[6] 3GPP TS 24.301, Non-Access-Stratum (NAS) protocol for Evolved Packet System (EPS); Stage 3 (Release 11), December 2012.
[7] 3GPP TS 29.060, General Packet Radio Service (GPRS); GPRS Tunneling Protocol (GTP) across the Gn and Gp interface (Release 11), December 2012.
[8] 3GPP TS 23.203, Policy and charging control architecture (Release 11), December 2012.
[9] 3GPP TR 36.806, Evolved Universal Terrestrial Radio Access (E-UTRA); Relay architectures for E-UTRA (LTE-Advanced) (Release 9), April 2010.
[10] 3GPP TS 36.401, Evolved Universal Terrestrial Radio Access Network (E-UTRAN); Architecture description (Release 11), December 2012.

IETF Specifications

[11] IETF RFC 768, J. Postel, User Datagram Protocol, August 1980.
[12] IETF RFC 2460, S. Deering, Internet Protocol, Version 6 (IPv6) Specification, December 1998.
[13] IETF RFC 3095, C. Bormann, RObust Header Compression (ROHC): Framework and four profiles: RTP, UDP, ESP, and uncompressed, July 2001.
[14] IETF RFC 4960, R. Stewart, Stream Control Transmission Protocol, September 2007.

Books

[15] Magnus Olsson, et al., SAE and the Evolved Packet Core: Driving the Mobile Broadband Revolution, Academic Press, 2009.
[16] A.S. Tanenbaum, Computer Networks, fifth ed., Prentice Hall, 2010.

Articles

[17] Open System Interconnection Reference Model, Wikipedia, <http://en.wikipedia.org/wiki/OSI_model>.
[18] White paper, The LTE network architecture: A comprehensive tutorial, Alcatel Lucent, 2009.
[19] ITU-T Recommendation X.200: Open Systems Interconnection, Basic Reference Model: The basic model, July 1994.

CHAPTER

System Operation and UE States

4

CHAPTER CONTENTS

LTE-Advanced.

This chapter provides an in-depth systems view of the fundamental procedures, signaling, and protocols utilized by a mobile terminal for network discovery; cell selection/cell reselection; network attach/detach; authentication and security establishment; capability negotiation; signaling related to control-plane and user-plane connection establishment (access stratum (AS) and non-access stratum (NAS)); and IP connection establishment using state diagrams and signal flow graphs. We will also discuss radio resource management, paging, UE tracking, mobility, and power management functions. We will further explain practical considerations and real-life issues of network and terminal operation followed by references for additional reading. In this chapter, we will only focus on the procedures and will leave the description of the details of functional processing at different stages to the next chapters. An attempt has been made to characterize the behavior of the mobile terminal and the E-UTRAN in various operating conditions and states including idle mode and connected mode procedures, EPS connection and mobility management states, service establishment and QoS control, as well as new topics including mobility in heterogeneous networks.

State diagrams are used to describe the behavior of a system. They can describe possible states of a system and transitions between them as certain events occur or based on the system inputs. The system described by a state diagram must be composed of a finite number of states. However, in some cases, the state diagram may represent a reasonable abstraction of the system. There are many forms of state diagrams which differ slightly and have different semantics. State diagrams can be used to graphically represent finite state machines (i.e., a model of behavior composed of a finite number of states, transitions between those states, and actions). A state is defined as a finite set of procedures or functions that are executed in a unique order. In the state diagram, each state may have some inputs and outputs, where deterministic transitions to other states or the same state happen based on certain conditions. In this chapter, the notion of mode is used to describe a substate or a collection of procedures/protocols that are associated with a certain state. The unique definition of states and their corresponding modes and protocols, and internal and external transitions, is imperative to the unambiguous behavior of the system. Also, it is important to show the reaction of the system to unsuccessful execution of certain procedures.

4.1 General UE behavior

When a UE is powered on or upon loss of coverage, it needs to select an appropriate PLMN and radio access technology. The main input to PLMN and RAT selection procedures is provided by the user in the form of preferred RAT, or is obtained from the universal subscriber identity module (USIM) card in the form of the most recently used PLMN. In general, a UE at the power-on stage has some information and settings, and requires some other information in order to select the network and attach to it. The information that the UE obtains from its USIM includes the network operator's PLMN list and subscription information. In addition to the information on the USIM, the UE has some stored information in its memory (additional knowledge), which includes most recently used frequency band and PLMN; tracking area code (TAC); cell identifier; SAE temporary mobile subscriber identity (S-TMSI) [1]; and inter-RAT frequency band. The UE capability determines what the UE is capable of in terms of frequency bands, duplexing schemes, multi-antenna schemes, data buffering capacities, etc. This information includes the UE category, frequency band, synchronization signal sequences, general radio resource information, multiantenna mode parameters (e.g., codebooks), duplex mode, and preamble sequence generation algorithm.

In order to perform cell selection, the UE needs to obtain the frequency and timing synchronization information, system bandwidth, number of eNB transmit antennas, cell IDs (e.g., cell radio network temporary identifier (C-RNTI), physical cell ID, TAC), PLMN ID, signaling/data radio resource information, RACH_ROOT_SEQUENCE, and physical random access channel configuration to perform a random access procedure.

During cell selection, the UE searches for a suitable cell in the selected PLMN and RAT, and chooses that cell in order to initially gain access to the available services by tuning to its control channels, i.e., the UE camps on that cell. The UE then registers in the tracking area of the selected cell and RAT by performing the attach procedure. The outcome of this process is the successful location registration where the selected PLMN becomes the registered PLMN. The UE now continuously searches for the best cell on which to camp. The particular cell reselection process depends on the cell reselection scenario, UE capabilities (e.g., multi-RAT support, multi-band support), subscriber priority class, etc.

While the baseline for general UE behavior in LTE is inherited from GSM and UMTS, an LTE-specific aspect of the UE behavior is handling of priorities between available RATs and frequency layers, where a set of broadcast system information parameters configure a default RAT and frequency layer priority list. The RRC signaling is used to configure a UE-specific priority list [9]. The priority list provides the network with the means to direct the UEs to camp on particular RATs and frequency layers, facilitating load balancing functions. In the above priority scheme, inter-frequency and inter-RAT cell reselection relies on two sets of (cell-specific and UE-specific) parameters. The cell-specific set of parameters are common to all UEs camping on a specific RAT and frequency, whereas the UE-specific set of parameters allow the operators to control the UE behavior depending on UE class and operator-defined policies.

Figure 4.1 illustrates the state transitions of a UE during normal operation. The figure shows the transition from power-on to RRC connection setup and establishment of user-plane and control-plane with the serving cell; RRC connection release and transition to RRC_IDLE state; cell reselection and transition to another cell; handover to another cell while in RRC_CONNECTED state; and cell redirection from one cell to another upon RRC connection release. We will discuss cell selection, cell reselection, handover, and cell redirection procedures as well as other aspects of LTE system operation in the idle and connected modes in the next sections.

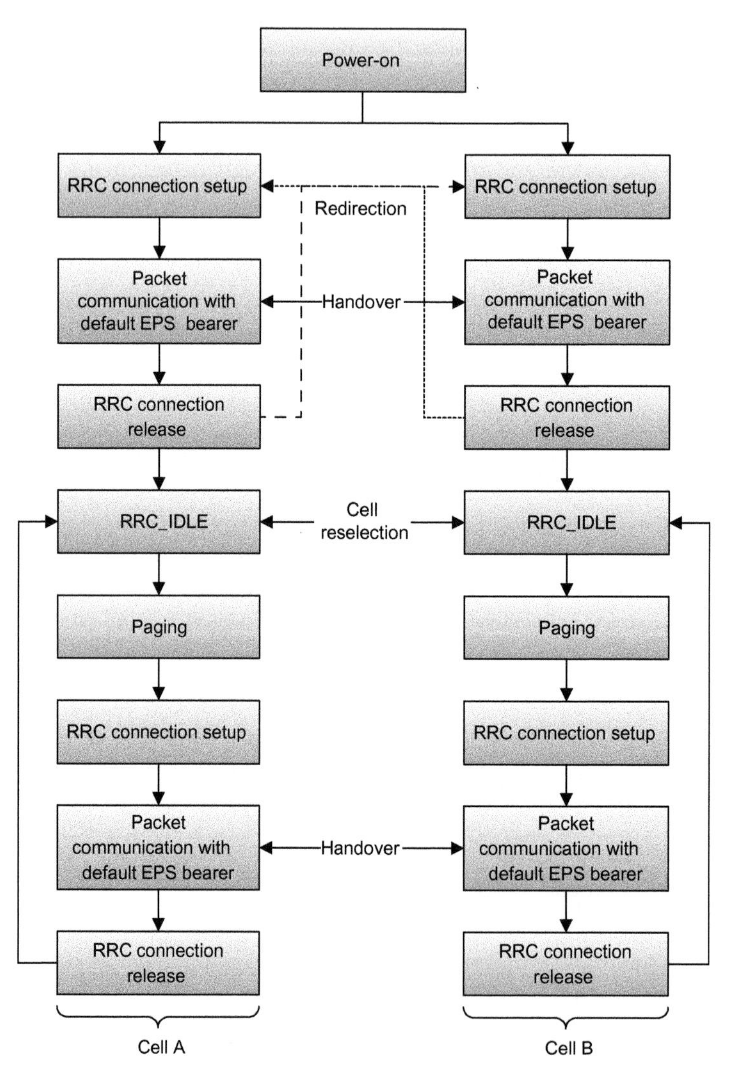

FIGURE 4.1

General UE state machine [32].

4.2 UE initialization procedure (cell search and cell selection)

The UE's USIM card stores the network operator's PLMN ID list and subscription information. The subscription information defines the policy for system selection and helps a UE to acquire a particular network. This feature enables, for example, an existing 3G network operator to deploy LTE by configuring the policy in order to initially instruct the UE to select the LTE network. In addition to the above information, the UE also knows the list of the frequency bands supported by LTE and other RATs such as GSM/UMTS, the most recently used LTE frequency band(s), PLMN IDs, TACs, cell IDs, S-TMSI, and inter-RAT frequency bands, as well as the UE category, synchronization signal sequences, general radio resource information, general MIMO parameters such as codebooks, duplex modes, and preamble sequence generation algorithm. However, the UE needs to acquire network frequency and timing synchronization information, system bandwidth, number of eNB transmit antennas, C-RNTI, cell ID, TAC, RACH_ROOT_SEQUENCE, specific radio resources used for control signaling, and PRACH configuration in order to be able to connect to a particular LTE network. In certain cases, the UE might have experienced link failure and has to make an attempt to reacquire the network using the most recent information that it had used earlier.

As shown in Figure 4.2, in order to select a PLMN, the UE must perform a cell search procedure in the supported frequency bands. The procedure requires

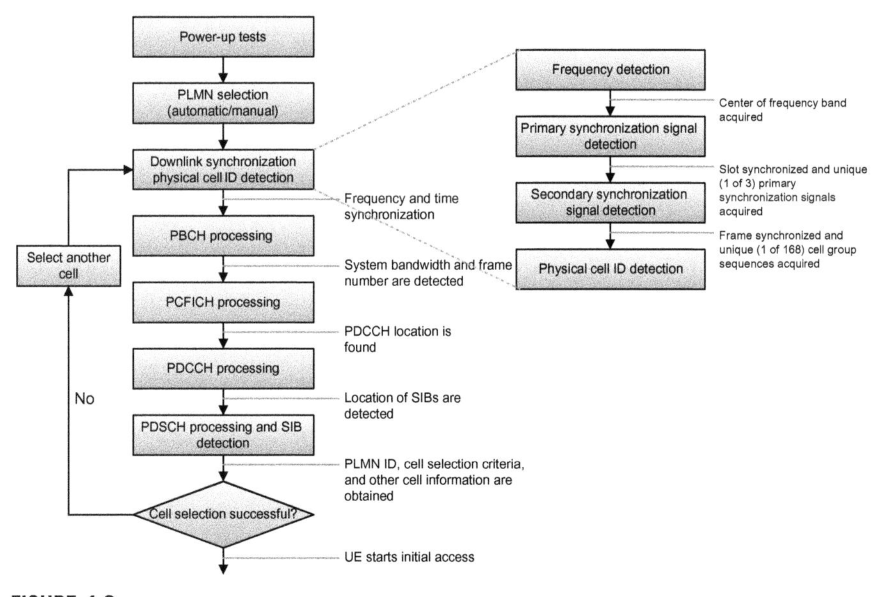

FIGURE 4.2

UE initial cell search and selection [20–22,33].

that the UE attain time and frequency synchronization with the specific cell. This enables decoding of the PBCH which carries the master information block (MIB) that contains the critical system information necessary to decode transmissions on a PDSCH. System information blocks (SIBs) are scheduled for transmission on PDSCH via SI messages. System information block type 1 (SIB1) contains a list of PLMN IDs along with a flag to show if a specific PLMN is reserved for operator use.

The selection of a PLMN can be manual or automatic. In both cases, the UE scans all possible RF channels that it supports based on its capability. During the search, and specifically for LTE, all detected PLMNs with RSRP ≥ -110 dBm are reported to the UE's NAS as high-quality PLMNs without an associated RSRP value.[1] PLMNs with an RSRP < -110 dBm are reported along with an associated RSRP value. The order in which the UE searches for PLMNs (either intra- or inter-band) is not defined but can be optimized based on information that the UE already has such as the last frequency on which it successfully camped. In both automatic and manual modes, the search for PLMNs can be superseded by the NAS at any time.

When the UE is in automatic mode and multiple PLMNs are detected, the order in which selection is performed is governed by a set of rules such that the last registered PLMN is tried first, then the home PLMN, followed by a user-defined and operator-defined list of PLMNs in priority order, followed by high-

[1]Reference signal received power (RSRP) is an important UE physical layer measurement and is the linear average (in watts) of the downlink cell-specific reference signals across the channel bandwidth. Since the cell-specific reference signals exist only for one symbol at a time, the measurement is made only on those resource elements that contain cell-specific reference signals. It is not mandated for the UE to measure every reference signal on the relevant sub-carriers. Instead, accuracy requirements have to be met. There are requirements for both absolute and relative RSRP. The absolute requirements range from ± 6 to ± 11 dB depending on the noise level and environmental conditions. Measuring the difference in RSRP between two cells on the same frequency (intra-frequency measurement) is a more accurate operation for which the requirements vary from ± 2 to ± 3 dB. The requirements are relaxed to ± 6 dB when the cells are on different frequencies (inter-frequency measurement).

Knowledge of absolute RSRP provides the UE with essential information about the strength of cells from which path loss can be calculated and used in the algorithms for determining the optimum power settings for operating the network. Reference signal received power is used both in idle and connected modes. The relative RSRP is used as a parameter in multi-cell scenarios [34].

Although RSRP is an important measure, on its own it gives no indication of signal quality. The reference signal received quality (RSRQ) provides this measure and is defined as the ratio of RSRP to the E-UTRA carrier received signal strength indicator (RSSI). The RSSI parameter represents the entire received power including the desired signal power from the serving cell as well as all co-channel power and other sources of noise. Measuring RSRQ becomes particularly important near the cell-edge when decisions need to be made, regardless of absolute RSRP, to perform a handover to another cell. Reference signal received quality is used only during connected states. Intra- and inter-frequency absolute RSRQ accuracy varies from ± 2.5 to ± 4 dB, which is similar to the inter-frequency relative RSRQ accuracy of ± 3 to ± 4 dB [34].

quality PLMNs in random order, followed by other PLMNs in order of decreasing RSRP are tried. If a UE is camped on a PLMN that is not the highest priority, it will periodically search for higher priority PLMNs and report the results to the UE's NAS. The interval between searches is stored on the USIM and set by the service provider to be between 6 min and 8 hour, in increments of 6 min [33].

Cell search is a process through which the UE acquires the RF carrier frequency, cyclic prefix length (or OFDM symbol timing), subframe and frame timing, and the physical cell identifier. The E-UTRA uses a hierarchical cell search mechanism which is conducted by acquiring the primary synchronization signals followed by acquiring the secondary synchronization signals. In order to support cell search over scalable transmission bandwidth, E-UTRA requires the UEs to support a minimum bandwidth of 20 MHz in the downlink. The cells in E-UTRA are classified as follows [5]:

- Serving cell: The cell on which the UE is currently camped.
- Suitable cell: A cell that the UE may camp on to obtain normal service, which is a cell whose measured cell attributes satisfy the cell selection criteria; the cell is either part of the selected PLMN, registered or an equivalent PLMN; the cell is not barred or reserved, and the cell is not part of a tracking area that is in the list of forbidden tracking areas for roaming. If more than one PLMN ID is broadcast in the cell, the cell is considered to be part of all tracking areas with tracking area identities (TAIs) constructed from the PLMN IDs and the TAC broadcast in the cell.
- Acceptable cell: A cell whose measured cell attributes satisfy the cell selection criteria and is not barred. The UE may camp on this cell to obtain limited service to originate emergency calls and receive earthquake and tsunami warning system (ETWS) notifications.
- Barred cell: A cell that the UE is not allowed to camp on.
- Restricted cell: A cell on which camping is allowed, but access attempts are disallowed for UEs whose access classes are indicated as barred.
- Reserved cell: A cell on which camping is not allowed, except for particular UEs, if that is indicated in the system information.
- Closed-subscriber group (CSG) cell: A CSG cell is accessible by the members of the CSG for that CSG ID. A cell in which only members of the CSG can obtain normal service. Depending on local regulation, the CSG cell can provide emergency bearer services also to subscribers who are not members of the CSG [3].

Following successful completion of the cell search, the UE acquires PBCH and obtains the MIB, which carries the overall downlink transmission bandwidth, PHICH configuration, and the system frame number (SFN). The UE then decodes PDCCH in order to locate the PDSCH resources carrying SIBs which are common radio resource and mobility configurations for all UEs in the cell (see Figure 4.2).

Synchronization signals are transmitted in every radio frame. The synchronization signals are used for time and frequency synchronization during initialization. As part of the system acquisition process, the UE synchronizes to OFDM symbol, slot, subframe, half-frame, and radio frame (in that order) using synchronization signals. Two synchronization signals are defined for system acquisition in E-UTRA: (1) Primary synchronization signal is used to obtain slot, subframe, and half-frame boundary. It also provides cell ID within the cell ID group. (2) Secondary synchronization signal is used to obtain the radio frame boundary. It also enables the UE to determine the cell ID group, which may range from 0 to 167. The synchronization signals are transmitted in the 0th and 5th subframes of every radio frame. In the frequency domain, the synchronization signals occupy the center 1.08 MHz (equivalent to six resource blocks) of the radio channel.

In the time domain, primary and secondary synchronization signals are transmitted in the 0th and 10th slot (1st slot within 0th and 5th subframe) of every radio frame. Within these slots, the primary and secondary synchronization signals are transmitted in the last two symbols. The primary synchronization signal is transmitted in the last symbol of 0th and 10th slot. The secondary synchronization signal is transmitted in one before the last symbol of 0th and 10th slot. The PCBH is mapped to the first four symbols in the 1st slot of the frame (just after primary and secondary synchronization signals). In the frequency domain, the center 1.08 MHz of the radio channel is reserved for primary and secondary synchronization signals. This is equivalent to 6 resource blocks or 72 subcarriers. However, the primary and secondary synchronization signals only require 62 subcarriers around the DC subcarrier. The remaining subcarriers (10 subcarriers) in these resource blocks are not used for transmission. They provide a guard band between primary and secondary synchronization signal transmission and other data transmissions. There are 504 unique physical-layer cell IDs. The physical-layer cell IDs are divided into 168 unique physical-layer cell ID groups, each group contains three unique IDs. The grouping is done such that each physical-layer cell ID is part of one and only one physical-layer cell ID group. A physical-layer cell ID $N_{ID}^{cell} = 3N_{ID}^{(1)} + N_{ID}^{(2)}$ is thus uniquely defined by a number $N_{ID}^{(1)}$ in the range of 0−167, representing the physical-layer cell ID group, and a number $N_{ID}^{(2)}$ in the range of 0−2, denoting the physical-layer ID within the physical-layer cell ID group [20−22,33].

The essential system information which allows the other channels in the cell to be configured and operated is carried by PBCH. Therefore, the reliability and coverage of PBCH is crucial to the successful operation of E-UTRA systems. The SI is divided into two categories: (1) the MIB, which consists of a limited number of the most frequently transmitted parameters essential for initial access to the cell which is carried in PBCH; and (2) the other SIBs which are multiplexed with unicast data and are transmitted in PDSCH.

The detectability of the PCBH without UE's prior knowledge of the system bandwidth is achieved by mapping PBCH to the center of the band using 72 subcarriers, which corresponds to the minimum system bandwidth of six resource blocks, regardless of the actual system bandwidth. The UE will have to identify the system center frequency from the synchronization signals. Minimizing the impact on total overhead is achieved by deliberately keeping the amount of information carried by PBCH to a minimum, since achieving stringent coverage requirements for a large quantity of data would result in a high system overhead. The size of MIB is 24 bits and is transmitted every 40 ms. In order to achieve reliable reception, time diversity, forward error correction, and antenna diversity mechanisms are used to transmit PBCH. Time diversity is introduced by spreading the transmission of PBCH over a period of 40 ms with three retransmissions at 10 ms intervals. This significantly reduces the likelihood of missing/misdetection of MIB due to multipath fading, interference, and noise in the radio channel, even when the mobile terminal is moving at high speeds.

The forward error correction coding for PBCH utilizes a tail-biting convolutional coder, considering that the number of information bits to be encoded is small and turbo codes are typically not suitable for small payloads. The basic code rate is 1/3, after which a high repetition factor is applied to the systematic and parity bits such that each MIB is coded at a very low code rate (1/48 over a 40 ms period) to provide strong error protection. Antenna diversity may be utilized at both the eNB and the UE. The UE performance requirements specified for E-UTRA assume that all UEs can achieve a level of decoding performance proportional to dual-antenna receive diversity although it is understood that in low-frequency deployments, e.g., below 1 GHz, the advantage obtained from receive diversity is reduced due to the correspondingly higher correlation between the antennas. This enables wider or more reliable cell coverage to be achieved with fewer cell sites.

Transmit antenna diversity may also be employed at the eNB to further improve coverage. Depending on eNB implementation, it may transmit PBCH with two or four transmit antennas using space-frequency block codes (SFBC). The set of resource elements designated to PBCH transmission is independent of the number of transmit antennas at the eNB, thereby any resource elements which may be used for cell-specific reference signals transmission are avoided by PBCH, irrespective of the actual number of transmit antenna ports used by the eNB. The number of transmit-antennas used by the eNB is blindly detected by the UE by decoding each SFBC scheme corresponding to different number of transmit antennas (i.e., one, two, or four). The detection of the number of transmit antennas is further facilitated by masking the cyclic redundancy check (CRC) bits of each MIB with a unique masking sequence representing the number of transmit antenna ports.

Achieving low latency and minimal impact on UE power consumption are assisted through use of a low code rate with repetition, wherein the set of coded bits are divided into four subsets, each of which is self-decodable. Each subset of the coded bits is transmitted in a different radio frame during the 40 ms transmission period. This means that if the SINR is sufficiently high to allow the UE to successfully decode the MIB in less than four radio frames, then the UE does not need to receive the other parts of the PBCH transmission in the remainder of the 40 ms period. On the other hand, if the SINR is low, the UE can take advantage of soft-combing of other parts of the MIB transmission until the MIB is successfully decoded. The UE does not know in advance the timing of the 40 ms transmission interval for each MIB on PBCH; therefore, this information is determined implicitly from the scrambling and bit positions, which are reinitialized every 40 ms. The UE can initially determine the 40 ms timing by performing four separate decodings of the PBCH using each of the four possible phases of the PBCH scrambling code and checking the CRC for each decoding.

When a UE initially attempts to select a cell by acquiring PBCH, different approaches may be taken to perform the necessary blind decoding. A simple approach is to perform the decoding using a soft combing of the PBCH instances over four radio frames, advancing a 40 ms sliding window one radio frame at a time until the sliding window aligns with the 40 ms period of the PBCH and the decoding succeeds. However, this would result in a 40–70 ms delay before the PBCH can be decoded. A faster approach would be to attempt to decode the PBCH from the first radio frame, which should be possible provided that the SINR is sufficiently high. If the decoding fails for all four possible scrambling sequences, the PBCH from the first frame could be soft-combined with the next PBCH instances. It is evident that the latter approach may be much faster at the expense of slightly more complexity.

We learned so far that initial frequency and timing acquisition are based on the synchronization signals (Figure 4.3). A minimum bandwidth of 1.08 MHz (six resource blocks) is used for the synchronization and broadcast channels irrespective of the actual system bandwidth. The UE acquires a coarse frequency synchronization; OFDM symbol, slot, and subframe timing; and the index to a cell in the cell ID group through successful detection and processing of the primary synchronization signals. The UE further acquires the half-frame timing; radio frame timing; cell ID group; and FDD/TDD mode information via the secondary synchronization signals. Note that there is no eNB antenna configuration information on the secondary synchronization signals. The eNB antenna configuration is conveyed in PBCH by applying different CRC masks. The cell search procedure is complete when PBCH CRC check passes.

In E-UTRA, the MIB, SIB1, and SIB2 broadcast messages are transmitted for any cell in the network in such a way that the location (subframe) where the SIB is transmitted should not be the same subframe where another SIB is transmitted. As shown in Figure 4.4, the SIB scheduling follows these principles: (1) MIB is transmitted at fixed cycles every four frames starting from SFN 0; (2) SIB1 is also transmitted at fixed cycles every eight frames starting from SFN 0; (3) all other SIBs are transmitted at the cycles between 80 and 5120 ms that are specified by SIB scheduling information elements in SIB1. The SIB1 uses a fixed schedule with a periodicity of 80 ms with retransmissions within 80 ms intervals.

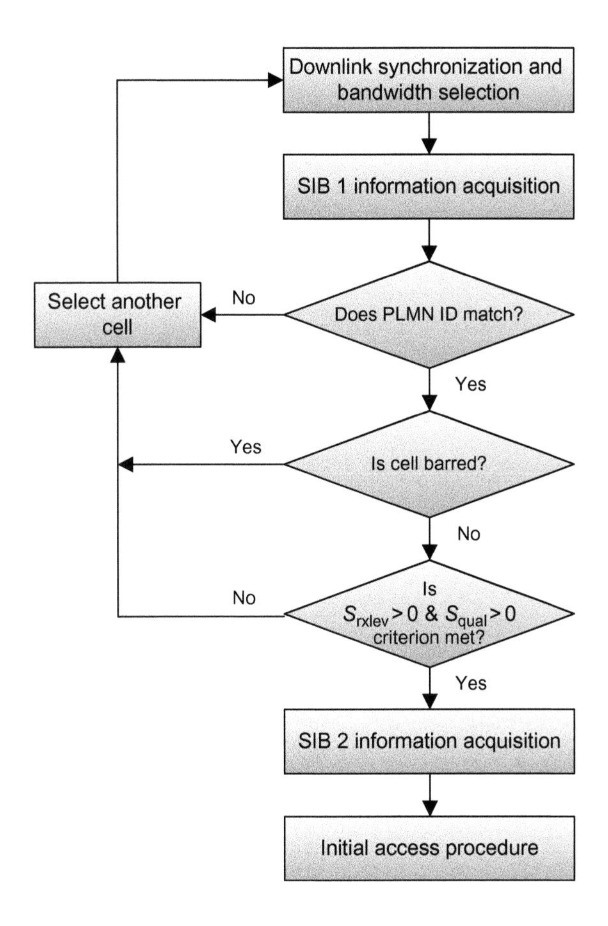

FIGURE 4.3

Cell selection summary [20–22,33].

The first transmission of SIB1 is scheduled in the 5th subframe of radio frames for which SFN mod8 = 0 and retransmissions are scheduled in the 5th subframe of all other radio frames for which SFN mod2 = 0. This means that even though SIB1 periodicity is 80 ms, different redundancy versions of the SIB1 are transmitted every 20 ms. It means that at layer-3 level, the SIB1 is transmitted every 80 ms, but at layer-1 level, it is seen every 20 ms [9].

The transmission periods for other SIBs are determined by the *schedulingInfoList* information element in SIB1 as shown in the following example. In this example, SIB2 and SIB3 are transmitted every 160 and 320 ms, respectively:

```
+-schedulingInfoList::= SEQUENCE OF SIZE(1..maxSI-Message[32]) [2]
|+-SchedulingInfo::= SEQUENCE
||+-si-Periodicity::= ENUMERATED [rf16]
||+-sib-MappingInfo::= SEQUENCE OF SIZE(0..maxSIB-1[31]) [0]
|+-SchedulingInfo::= SEQUENCE
```

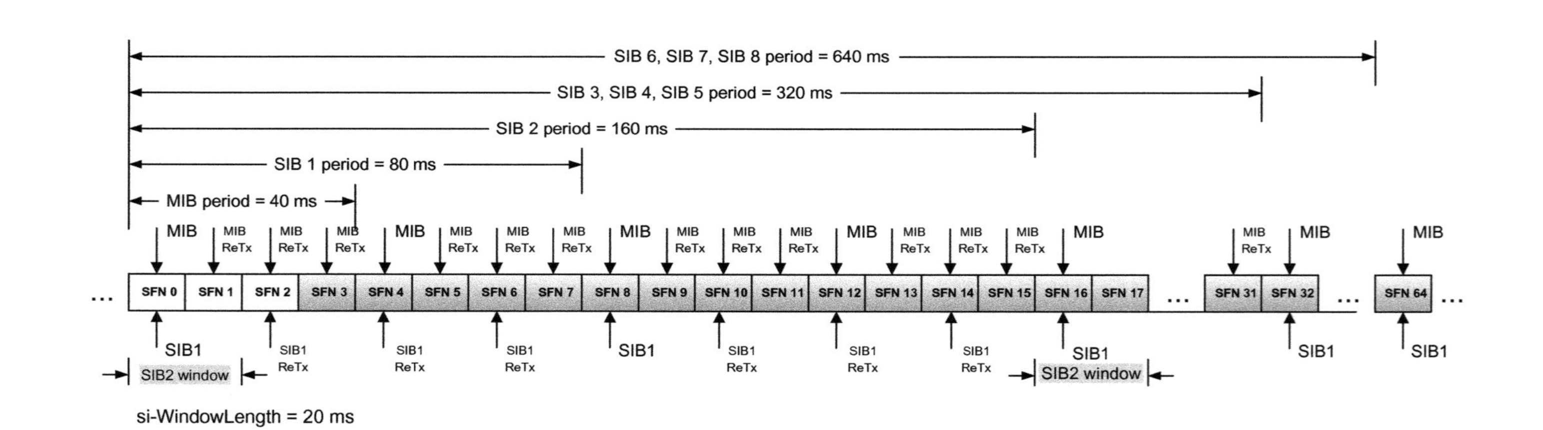

FIGURE 4.4

System information transmission timing (example).

```
|+-si-Periodicity::= ENUMERATED [rf32]
|+-sib-MappingInfo::= SEQUENCE OF SIZE(0..maxSIB-1[31]) [1]
|+-SIB-Type::= ENUMERATED [sibType3]
+-tdd-Config::= SEQUENCE OPTIONAL:Omit
+-si-WindowLength::= ENUMERATED [Ms20]
```

You would also note that the *sib-MappingInfo* information element in the first node is not specified, but the first entity of *schedulingInfoList* should always be for SIB2 as specified in 3GPP TS 36.331 [9]. There is no mapping information of SIB2 since it is always present in the first SI message listed in *schedulingInfoList* list.

Understanding the overall transmission cycle and location of SIBs is related to interpretation of the *si-WindowLength* parameter, which specifies that the corresponding SIB should be transmitted somewhere within the window length starting at the SFN specified by *si-Periodicity*. However, this parameter does not specify the exact subframe number for the SIB transmission. The subframe for a specific SIB transmission is determined by an algorithm defined in 3GPP TS 36.331 [9], and described in the following. When acquiring an SI message, the UE is required to determine the start of the *SI-window* for the corresponding SI message by calculating the entry number n which corresponds to the order of entry in the list of SI messages configured by the *schedulingInfoList* in SIB1. It then calculates the integer-valued number $x = (n-1)w$, where w is given by *si-WindowLength*. The SI-window starts at subframe number $x \bmod 10$ in the radio frame for which SFN mod $T_{\text{SIB}} = \lfloor x/10 \rfloor$, where T_{SIB} is given by *si-Periodicity* of the corresponding SI message. Note that E-UTRAN should configure an SI-window of 1 ms only if all SIs are scheduled before the 5th subframe in radio frames for which SFN mod2 = 0. Therefore, the UE must decode the downlink shared channel using the SI-RNTI from the start of the SI-window and continue until the end of the SI-window whose absolute length in time is given by *si-WindowLength*, or until the SI message was correctly received, with the exception of the following subframes: (1) the 5th subframe in radio frames for which SFN mod2 = 0; (2) any MBSFN subframes; and (3) any uplink subframes in TDD. If the SI message was not received by the end of the SI-window, the UE repeats this process at the next SI-window occasion for the corresponding SI message [9,21,32].

The UE uses one of the following two cell selection procedures [5]:

1. Initial cell selection: This procedure requires no prior knowledge of E-UTRA RF carriers. The UE scans all RF channels in the E-UTRA bands according to its capabilities to find a suitable cell. On each carrier frequency, the UE needs only to search for the strongest cell. Once a suitable cell is found, that cell is selected.
2. Stored information cell selection: This procedure requires stored information of carrier frequencies and also information on cell parameters from previously received measurement control information elements or from previously detected cells. Once a suitable cell is found, the UE selects that cell. If no suitable cell is found, the initial cell selection procedure is enacted.

The SIB1 specifies the criteria for determining whether a cell is suitable for selection [9]. A cell is considered suitable when $S_{rxlev} > 0$ & $S_{qual} > 0$ where for LTE, S_{rxlev} denotes the cell selection receive level value in dB and S_{qual} is the cell selection quality value in dB. The detailed cell selection and reselection criteria are described in Chapter 5. In practice, a UE receives signals from several cells belonging to different network operators. The UE only knows, after reading the SIB1 content (PLMN ID), if this cell belongs to the operator's network to which it is subscribed. Therefore, the UE will initially look for the strongest cell per carrier, then for the PLMN ID by decoding the SIB1 content to decide whether that is a suitable network. It will then compute the $S_{rxlev} > 0$ & $S_{qual} > 0$ criterion and decide whether the cell is suitable.

As an example, assume that the UE belongs to network operator A and there are two other operators which are operating LTE networks but at different frequencies. The UE receives signals of all base stations but at different power levels. Based on the above definition, the UE will select the strongest cell that may belong to network operator C; however, it will reject the cell after decoding the SIB1 content and observing that the PLMN ID saved on the USIM card does not match to that of the cell. The UE will now proceed to the next strongest signal, which may belong to network operator B, and when the PLMN ID does not correspond to the UE's stored information, it will continue with the cell belonging to network operator A. Once the PLMN ID matches and PLMN is selected, the UE will continue to use the information in SIB1 and SIB2 to evaluate the cell selection criteria based on signals received from eNBs belonging to network A. If the parameters transferred and belonging to eNB_1 do not fulfill $S_{rxlev} > 0$ & $S_{qual} > 0$ criterion, the UE will continue with demodulating and decoding the information provided by eNB_2 and if the $S_{rxlev} > 0$ & $S_{qual} > 0$ criterion is fulfilled, the UE will camp on this cell (see Figure 4.3).

4.3 EPS connection and mobility management

The states of a UE with respect to mobility and connection establishment are described by the following NAS states: the EPS mobility management (EMM) state (EMM-DEREGISTERED or EMM-REGISTERED) reflects whether the UE is registered at the MME and the EPS connection management (ECM) state (ECM-IDLE or ECM-CONNECTED) reflects the connectivity of the UE with the EPC. The AS and NAS states and state transitions are illustrated in Figure 4.5. The transition from ECM-IDLE to ECM-CONNECTED not only involves establishment of the RRC connection, but also includes establishment of S1 connection. The RRC connection establishment is initiated by the NAS and is completed prior to S1 connection establishment, which means that connectivity in RRC_CONNECTED is initially limited to the exchange of control information between the UE and eNB. The UEs are typically moved to ECM-CONNECTED

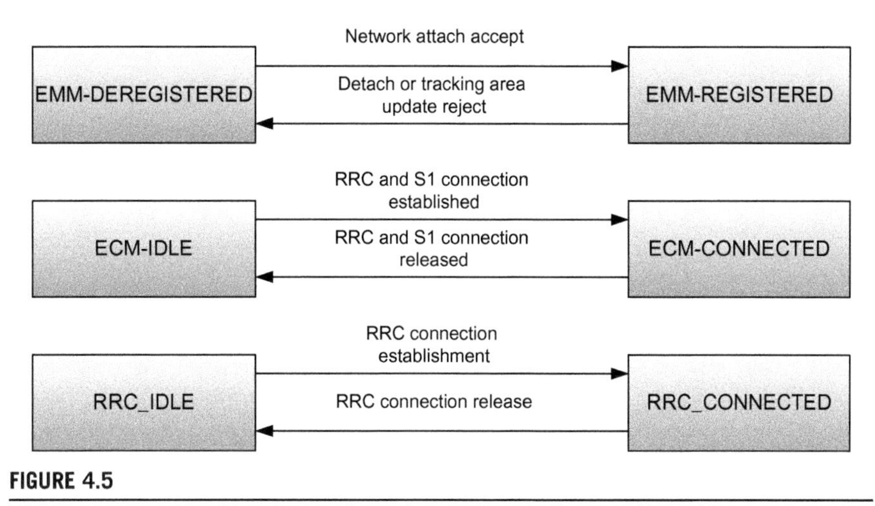

FIGURE 4.5

AS and NAS states and state transition [21].

when becoming active. It should be noted that in LTE, the transition from ECM-IDLE to ECM-CONNECTED is performed within 100 ms and the UEs engaged in intermittent data transfer do not need to stay in ECM-CONNECTED, if the ongoing services can tolerate such transfer delays. RRC connection release is initiated by the eNB following release of the S1 connection between the eNB and the core network [20].

4.3.1 EPS mobility management

The EMM states describe the mobility management states that result from the mobility management procedures, e.g., attach and tracking area update procedures. The EMM has two states: EMM-REGISTERED and EMM-DEREGISTERED as shown in Figure 4.5. In EMM-DEREGISTERED state, the EMM context does not have any valid location or routing information for the UE. Since the location of the UE is not known, the UE is not reachable by the MME. In order to avoid performing an authentication and key agreement (AKA)[2] security procedure in every network attach attempt by the UE, the MME and the UE may store some UE context in this state. The MME may perform an implicit detach any time after the implicit detach timer expires. The state is changed to EMM-DEREGISTERED in the MME after performing the implicit detach. If all bearers belonging to a UE are released (e.g., after a handover from E-UTRAN to

[2]AKA is a mechanism which performs authentication and session key distribution in UMTS networks. AKA is a challenge-response based mechanism that uses symmetric cryptography. AKA is typically run in a UMTS IP multimedia services identity module (ISIM), which resides on a smart card device that also provides tamper resistant storage of shared secrets. AKA is defined in IETF RFC 3310.

non-3GPP access [11]), the MME changes the mobility management state of the UE to EMM-DEREGISTERED. This ensures that the UE performs a tracking area update when it reselects E-UTRAN [4].

In EMM-REGISTERED state, the UE has registered with the MME following an attach procedure. When the UE transitions to this state, a UE context and a default EPS bearer have been established and the UE can receive services that require registration with the EPC. The location of the UE is known by the MME at the tracking area level. In the EMM-REGISTERED state, an idle UE is required to perform a tracking area update with the network. The detach procedure or a tracking area update reject changes the EMM state from EMM-REGISTERED to EMM-DEREGISTERED. In the EMM-REGISTERED state, the UE always has at least one active PDN connection (IP connection) and EPS security context [4].

In general, the ECM and EMM states are independent of each other. Transition from EMM-REGISTERED to EMM-DEREGISTERED can occur regardless of the ECM state, e.g., by explicit detach signaling in ECM-CONNECTED or by implicit detach locally in the MME during ECM-IDLE. However, there are some dependencies between the two states, e.g., to transition from EMM-DEREGISTERED to EMM-REGISTERED the UE has to be in the ECM-CONNECTED state [20].

4.3.2 EPS connection management

The ECM states describe the signaling connectivity between the UE and the EPC. In EMM-REGISTERED state, there are two EPS connection states as follows: ECM-IDLE and ECM-CONNECTED as shown in Figure 4.5. In ECM-IDLE, a UE performs cell selection or reselection. There is no NAS signaling connection between the UE and the network. The UE does not have a context in the E-UTRAN and has no S1-MME and S1-U connection with the network. In EMM-REGISTERED and ECM-IDLE state, the UE performs a tracking area update, if the current TA is not in the list of TAs that the UE has received from the network in order to maintain the registration and enable the MME to page the UE. It performs a periodic tracking area update procedure to notify the EPC that the UE is available. The UE performs a tracking area update, if the RRC connection was released with release cause "*load balancing TAU required*" or performs a tracking area update if there are any changes in the UE's core network capability information or the UE-specific DRX parameters. The UE performs a tracking area update when it manually selects a CSG femto-cell, and the CSG ID[3] of that cell is absent from both the UE's allowed CSG list and the UE's operator CSG list. The UE answers to paging from the MME by performing a service request procedure and performs the service request procedure in order to establish the radio bearers when uplink user data is available for transmission.

---

[3]A CSG ID is a unique identifier within the scope of one PLMN defined in 3GPP TS 23.003 [1] which identifies a CSG in the PLMN associated with a cell or group of cells to which access is restricted to members of the CSG.

The UE and the MME enter the ECM-CONNECTED state when the signaling connection is established between the UE and the MME. Initial NAS messages that initiate a transition from ECM-IDLE to ECM-CONNECTED state are *Attach Request*, *Tracking Area Update Request*, *Service Request*, or *Detach Request*. When the UE is in ECM IDLE state, the UE and the network may be unsynchronized, i.e., the UE and the network may have different sets of established EPS bearers. When the UE and the MME transition to the ECM-CONNECTED state, the set of EPS bearers are synchronized between the UE and the network.

When signaling connection is established between the UE and the network, the UE and the MME make transition to the ECM-CONNECTED state. In the ECM-CONNECTED state, the UE location is known in MME at cell level. The UE can perform handover and tracking area update procedures. The UE has a signaling connection with the MME. A service request upon availability of uplink data or paging message upon availability of downlink data change the UE state from ECM-IDLE to ECM-CONNECTED. The UE-assisted measurements are required for handover, cell selection, and reselection. The signaling connection is made up of two parts: an RRC connection and an S1-MME connection. The UE transitions to the ECM-IDLE state when its signaling connection to the MME has been released or broken. This release or failure is explicitly indicated by the eNB to the UE or detected by the UE. The S1 release procedure changes the state at both UE and MME from ECM-CONNECTED to ECM-IDLE. The UE may not receive the indication for the S1 release due to radio link error or out of coverage causes. In this case, there can be temporary mismatch between the ECM state in the UE and the ECM state in the MME [4,5].

4.4 UE states

During normal operation, an LTE UE can be in one of the two (steady-state) RRC states: RRC_IDLE and RRC_CONNECTED, as shown in Figure 4.6. The emphasis on the steady-state descriptor is due to the fact that when a UE is powered on or upon radio link failure, it goes to a transient state with no attachment to a PLMN or a cell, thus with no connection to the network. The UE starts scanning and selecting a PLMN and then proceeds to camp on a cell (cell selection/reselection) through downlink synchronization and system information acquisition. In this case, the UE has virtually moved to the RRC_IDLE state since it still has no RRC connection, and thus performs uplink synchronization and RRC connection setup procedures. In RRC_IDLE, the UE monitors a paging channel and/or SIB1 content to detect system information change and other system notifications. The UE further monitors control channels associated with the shared data channel to determine whether data is scheduled for it. The RRC_CONNECTED is a state where the UE transmits or receives unicast control/data in the uplink or downlink and may be configured with a customized DRX pattern. The UE monitors control channels associated with the shared data channel to determine if data is scheduled for it, provides channel quality and feedback information, performs

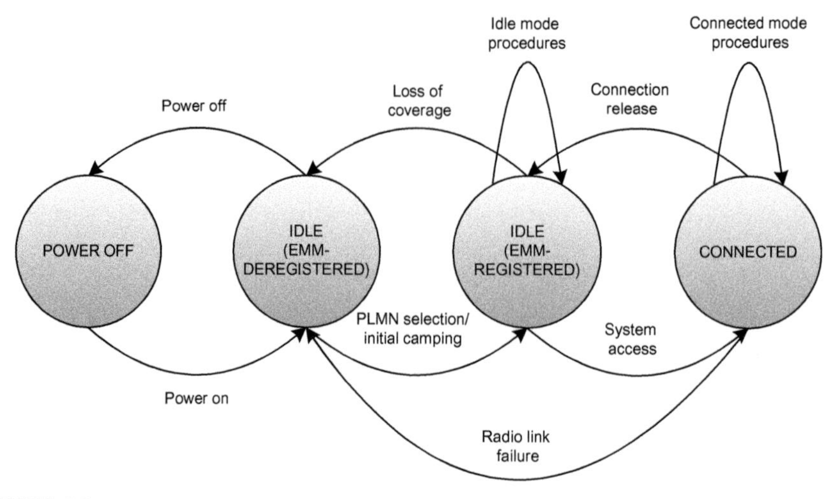

FIGURE 4.6

E-UTRA UE states and state transitions [20–22,32].

neighboring cell measurements and measurement reporting, and acquires the system information. One or multiple IP addresses are assigned to the UE as well as a temporary identifier called C-RNTI that is used for signaling purposes between the UE and the network.

Prior to transition to RRC steady-states, the UE uses the information stored on its USIM and the local memory for system selection and camping. Upon initial power-up, the UE uses the above static information to initiate system selection. In a practical scenario, the IMSI on the USIM may be used to determine the home PLMN code and may serve as the basis for PLMN selection. The USIM may have optionally an equivalent home PLMN list and priorities among them to determine the PLMN selection. The USIM may also provide the allowed RAT for each PLMN where there can be multiple RATs for a given PLMN. The UE considers the supported RAT and frequency band to trigger system selection. The system selection procedures are divided between NAS and AS. Over the air, signaling support provided by the AS assists NAS in camping on a PLMN. The selected PLMN is known as the registered PLMN and is stored on the USIM for future system selections.

The idle mode procedures depend on the UE's camping status. As shown in Figure 4.6, there are two idle states that a UE may transition to: (1) Idle with EMM-DEREGISTERED; and (2) Idle with EMM-REGISTERED. When the UE is in the Idle EMM-DEREGISTERED state, no suitable cell on any PLMN or RAT has been selected; therefore, this state triggers the cell search procedure in order to detect a cell that is suitable to camp on. When the UE is in the Idle EMM-REGISTERED state, it has acquired a suitable cell belonging to a non-barred/unrestricted PLMN. While in this state, the UE performs cell reselection on the selected PLMN; searches for high-priority PLMNs and a suitable cell, if another

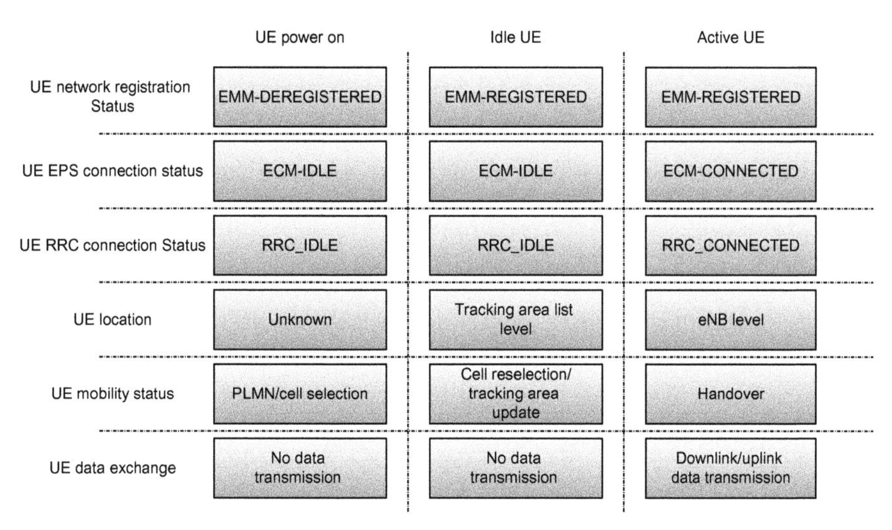

FIGURE 4.7

Summary of the UE and EPS states and the UE status in each state.

PLMN is selected by NAS; monitors paging messages; updates SI; initiates access procedures to transition to connected mode; and performs location update.

In the state diagram shown in Figure 4.6, the connected state refers to the EMM-REGISTERED + ECM-CONNECTED + RRC_CONNECTED state, the idle/EMM-REGISTERED refers to the EMM-REGISTERED + ECM-IDLE + RRC_IDLE state, and the idle/EMM-DEREGISTERED refers to the EMM-DEREGISTERED + ECM-IDLE + RRC_IDLE state. It must be noted that the Idle/EMM-DEREGISTERED state is a transient state, whereas the Idle/EMM-REGISTERED and connected states are steady-state states. Figure 4.7 provides a more thorough perspective into UE status in each of the UE and EPS states. As shown in the figure, when the UE is powered on, it must find a suitable PLMN, select a cell and register with the network. An idle UE might have just terminated its data exchange with the network, releasing RRC and ECM signaling connections. An active UE has attached to the EPS core network and established a default EPS bearer. In this state, the UE is EMM-REGISTERED + ECM-CONNECTED + RRC_CONNECTED, and a UE context is created and stored in the UE, eNB, and MME. In addition, the figure also shows the mobility and location status and information in each state of the UE and EPS mobility and connection management.

4.4.1 Idle mode procedures

The idle mode procedures can be divided into four main tasks, namely PLMN selection; cell selection and reselection; location registration; and support for manual femto-CSG selection. The relationship between these processes is illustrated in Figure 4.8. A UE supporting CSG selection selects a CSG femto-cell either automatically based on the list of allowed CSG IDs, or manually based on user selection of CSG within the list of available CSGs.

When a UE is switched on, a PLMN is selected by NAS, where associated RAT(s) may be set for the selected PLMN. The NAS provides a list of equivalent PLMNs (if any) that the AS may use for cell selection and cell reselection. During cell selection, the UE searches for a suitable cell belonging to the selected PLMN and selects a suitable cell to gain network access. This process is known as camping. The UE must then proceed with registration with the network by means of a NAS registration procedure in the tracking area of the chosen cell, and as the outcome of a successful location registration, the selected PLMN becomes the registered PLMN. If the UE finds a more suitable cell, according to the cell reselection criteria, it reselects that cell and camps on it. If the new cell does not belong to at least one tracking area to which the UE is registered, location registration is repeated. The UE searches for higher priority PLMNs at regular time intervals and searches for a suitable cell if another PLMN has been selected by NAS. Camping on a cell is necessary in order to gain access to network services. Three levels of services may be provisioned for a UE: (1) limited service (emergency calls, ETWS, and commercial mobile alert service (CMAS) on an acceptable cell); (2) normal service (for public use on a suitable cell); and (3) operator service (for operators only on a reserved cell) [5].

The search for available femto-CSGs may be triggered by the UE's NAS to support manual femto-CSG selection. If the UE experiences loss of coverage in

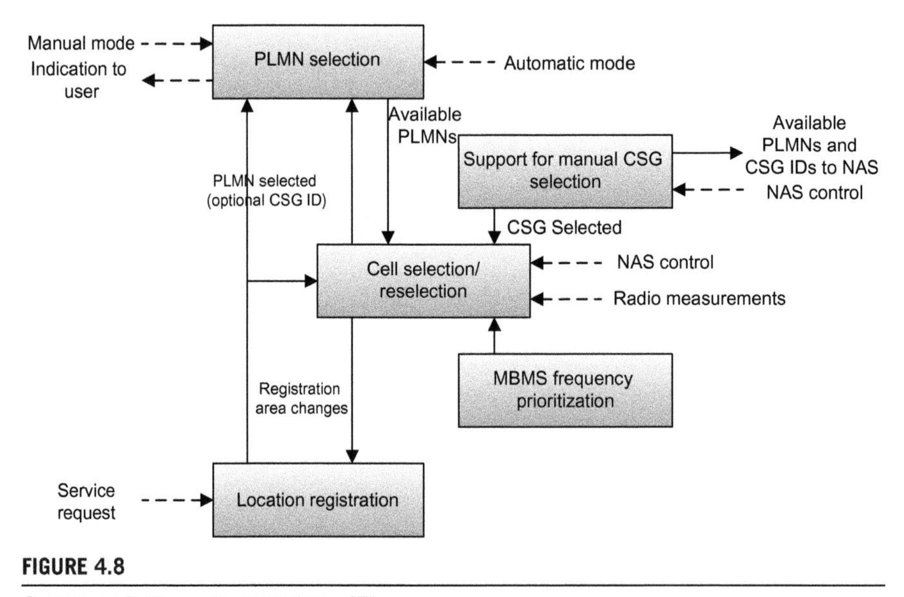

FIGURE 4.8

Overview of idle mode procedures [5].

the registered PLMN, either a new PLMN is selected automatically in automatic mode or an indication of which PLMNs are available is given to the user, so that a manual selection can be made. The purpose of camping on a cell in the idle mode is to enable the UE to receive SI from the network. Furthermore, if the network receives a call or downlink data for a registered UE, it knows the set of tracking areas in which the UE is camped, thus it can then send a paging message for the UE on the control channels of all cells in the set of tracking areas to which the UE is associated. The UE will then receive the paging message because it is tuned to the control channel of a cell in one of the registered tracking areas and the UE can respond on that control channel. It further enables the UE to receive MBMS services while in the idle mode. The tracking area update procedure and signaling is depicted in Figure 4.9. The process starts with RRC connection establishment and is followed by the UE's request for tracking area update,

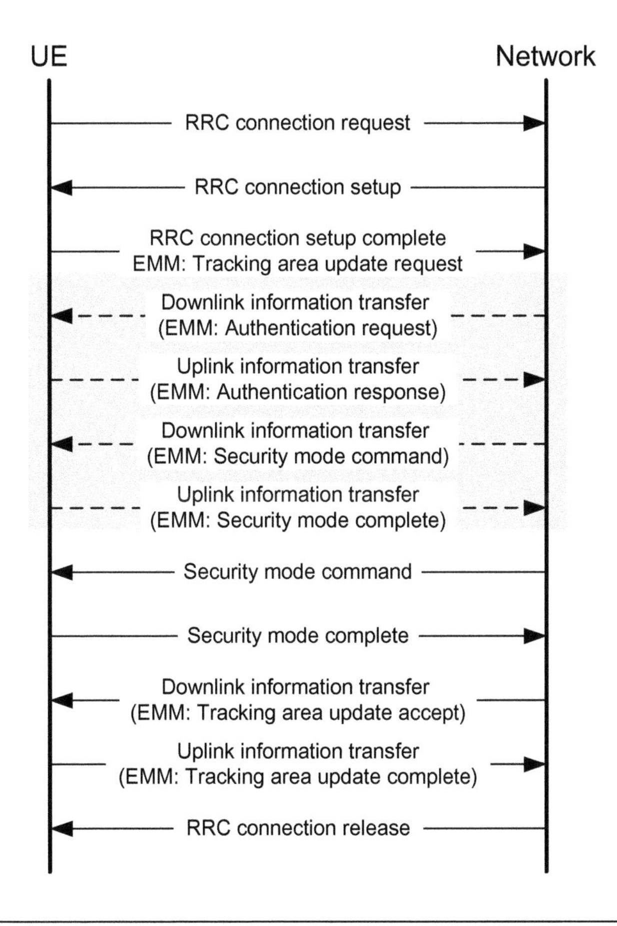

FIGURE 4.9

Tracking area update procedure and signaling [5,25].

authentication and security commands, and finally release of the RRC connection upon successful update of tracking area information. If the UE is unable to find a suitable cell to camp on, or if the location registration fails, it attempts to camp on any cell irrespective of the PLMN ID to receive limited services.

The procedures performed by the UE while in the idle mode, as shown in Figure 4.8, are divided into AS and NAS functions. Table 4.1 provides the functional division between the UE's NAS and AS functions in the idle mode.

4.4.1.1 Tracking area update

After the UE successfully attaches to the network, the UE can move around in the assigned tracking area. If it detects a different tracking area, it needs to inform the network of the new tracking area. Similarly, the network can request the UE to update its tracking area periodically, which helps the network to locate the UE faster whenever a mobile-terminated call is received by the network for the UE since it can just page the UE in the last reported tracking area. During the tracking area update procedure, the MME may initiate an authentication procedure and set up a security context. The tracking area update procedure and signaling is illustrated in Figure 4.9.

4.4.1.2 Paging

The paging procedure is used by the network to request establishment of NAS signaling connection to the UE. If there is an IP packet that arrives for a UE from an external network to the P-GW and there is no existing dedicated bearer for the UE, it will forward the IP packet to the S-GW on the default bearer. Once the packet has reached the S-GW on the default bearer, the S-GW detects the need to create a dedicated bearer and sends the *Downlink Data Notification* message to the MME in order to page the UE and to create the dedicated bearers. At this time, the MME has to ensure that the UE establishes an RRC connection, so the MME sends a *Paging Request* message to all eNBs associated with the last known tracking area in which the UE was located [25].

In other words, when there is a need to deliver downlink data to an ECM-IDLE UE, the MME sends a paging message to all eNBs in its current tracking area, and the eNBs page the UE over the radio interface. Upon receipt of a paging message, the UE performs a service request procedure, which results in transition to ECM-CONNECTED state. The UE-related information is thereby created in the E-UTRAN, and the radio bearers are reestablished. The MME is responsible for reestablishment of the radio bearers and updating the UE context in the eNB. To accelerate the idle-to-active transition and bearer establishment, EPS supports concatenation of the NAS and AS procedures for bearer activation. Some relationship between the NAS and AS protocols is employed to allow procedures to run simultaneously rather than sequentially. For example, the bearer establishment procedure can be executed by the network without waiting for the completion of the security procedure. The security functions are the responsibility of the MME for both signaling and user data. When a UE attaches to the network, a mutual authentication of the UE and the network is performed between the UE and the

Table 4.1 AS and NAS Functions in Idle Mode [5]

Idle Mode Task	UE's NAS Functions	UE's AS Functions
PLMN selection	Maintain a list of PLMNs in priority order Select a PLMN using automatic/manual mode and request AS to select a cell belonging to this PLMN Evaluate reports of available PLMNs from AS for PLMN selection Maintain a list of equivalent PLMN IDs	Search for available PLMNs Search in this set of RAT(s) and other RAT(s) under the selected PLMN Perform measurements for PLMN selection Synchronize to a broadcast channel to identify scanned PLMNs Report available PLMNs with associated RATs to UE's NAS upon request from NAS or autonomously
Cell selection	Control cell selection by indicating RATs associated with the selected PLMN to be used initially in the search of a cell in the cell selection process NAS maintains lists of forbidden registration areas and a list of CSG IDs and their associated PLMN ID on which the UE is allowed (CSG whitelist) and provides these lists to AS	Perform measurements for cell selection Detect and synchronize to a broadcast channel Receive and process broadcast information Forward NAS system information to UE's NAS Search for a suitable cell. The cells broadcast one or more PLMN IDs in their system information If associated RATs are set for the selected PLMN, perform the search among these RATs and other RATs for that PLMN The cell is camped on when it is found
Cell reselection	Control cell reselection by, for example, maintaining lists of forbidden registration areas Maintain a list of equivalent PLMN IDs and provide the list to AS Maintain a list of forbidden registration areas and provide the list to AS Maintain a list of CSG IDs and their associated PLMN ID on which the UE is allowed (CSG whitelist) to camp and provide the list to AS	Perform measurements to support cell reselection Detect and synchronize to a broadcast channel Receive and process broadcast information and forward NAS system information to UE's NAS Change cell when a more suitable cell is found

(*Continued*)

Table 4.1 (Continued)

Idle Mode Task	UE's NAS Functions	UE's AS Functions
Location registration	Register the UE as active after power on Register the UE's presence in a registration area, for instance regularly or when entering a new tracking area Maintain lists of forbidden registration areas Deregister UE when shutting down	Report registration area information to NAS
Support for manual femto-CSG selection	Provide request to search for available femto-CSGs Evaluate reports of available femto-CSGs from AS for CSG selection Select a femto-CSG and request AS to select a cell belonging to this CSG	Search for cells with the specified CSG ID Read the HNB name from BCCH on SIB9 if a cell with a CSG ID is found Report CSG ID of the found cell broadcasting a CSG ID together with the HeNB name and PLMNs to NAS Upon selection of a femto-CSG by NAS, select any cell belonging to the selected CSG fulfilling the cell selection criteria and not barred or reserved for operator use

MME/HSS. This authentication function also establishes the security keys that are used for encryption of the bearers [28]. The UE is normally paged using its S-TMSI. The *Paging Request* message also contains a UE ID index value in order for the eNB to calculate the paging occasions at which the UE will listen to paging messages (see Chapter 5 for more details).

4.4.1.3 Cell reselection

The primary purpose of cell reselection, regardless of intra-frequency, inter-frequency, or inter-RAT nature, is to ensure that the UE camps on or connects to the best cell in terms of radio condition, e.g., in terms of path loss or RSRP metric. The UE should conduct some measurements to support this procedure. A UE in RRC_IDLE state performs cell reselection. The UE measures some attributes of the serving and neighbor cells to enable the reselection process. There is no need to inform the UEs of neighboring cells in the serving cell system information to enable the UEs to search and conduct measurements on neighbor cells, i.e., E-UTRAN relies on the UE to detect the neighbor cells. For the search and measurement of inter-frequency neighbor cells, only the carrier frequencies need to be indicated. Measurements may be avoided, if the serving cell attributes fulfill particular search or measurement criteria.

Cell reselection identifies the cell that the UE may camp on. The cell reselection procedure is based on certain criteria which involve measurements of the serving and neighbor cells. Intra-frequency cell reselection is based on ranking of the cells, whereas inter-frequency cell reselection is based on absolute priorities where a UE attempts to camp on the highest priority frequency available. Absolute priorities for reselection are provided only by the registered PLMN and are valid within the registered PLMN. These priorities are provided by the system information and are valid for all UEs in a cell. Specific priorities per UE can be signaled in the *RRC Connection Release* message. A validity time can be associated with UE-specific priorities [5].

For inter-frequency neighbor cells, it is possible to indicate layer-specific cell reselection parameters (e.g., layer-specific offset). These parameters are common to all neighbor cells on a frequency. A neighbor cell list can be provided by the serving cell to assist specific cases for intra- and inter-frequency neighbor cells. This neighbor cell list contains cell-specific cell reselection parameters (e.g., cell-specific offset) for specific neighbor cells. Black lists can also be provided to prevent the UE from reselecting specific intra- and inter-frequency neighbor cells. Cell reselection can be speed-dependent (speed detection based on UTRAN solution). Cell reselection parameters are applicable for all UEs in a cell, but it is possible to configure specific reselection parameters per UE group or per UE. Cell access restrictions apply similar to UTRAN, which consist of access class barring and cell reservation (e.g., for cells "reserved for operator use") which are applicable for the UEs in RRC_IDLE mode [5,20].

4.4.1.4 Idle mode DRX

E-UTRA supports DRX in the downlink in order to reduce UE power consumption during periods of inactivity. In the DRX mode, the UE monitors only PDCCH during a predefined on-duration interval within one of two possible DRX cycles. The UE may turn off its receiver when it does not expect to receive any transmission in the downlink over PDCCH. The UE can transition to DRX mode in one of the following ways: (1) implicitly based on the expiration of certain configured timers; or (2) explicitly based on the transmission of a DRX command in the form of a MAC control element. A UE can enter the DRX mode of operation during the active mode (DRX in RRC_CONNECTED state) as well as in the idle mode (DRX in RRC_IDLE state). The DRX in the connected state saves UE battery power, especially when the traffic is characterized by long periods of inactivity such as web-browsing traffic. This is achieved by interleaving short periods of active monitoring of the PDCCH with longer periods of inactivity. While in the idle mode utilizing DRX, the UE may wake up periodically and check the paging channel to determine whether it has any pending downlink traffic. Generally, an active mode DRX interval is shorter than an idle mode DRX interval.

It is important to note that E-UTRA supports DRX operation in both idle and connected modes. The DRX in RRC_IDLE state is used when monitoring the paging messages during paging occasions, whereas the DRX in RRC_CONNECTED state is used to conserve power and determines the subframes during which the UE monitors PDCCH using C-RNTI, TPC-PUCCH-RNTI, TPC-PUSCH-RNTI, SI-RNTI, or SPS C-RNTI for a possible data/control message. When idle mode DRX is used, the UE needs to monitor only one P-RNTI per DRX cycle.

There are a number of timers associated with the DRX mode. The DRX inactivity timer controls the number of subframes for which a UE continues to monitor the downlink before the absence of activity triggers the UE to enter the DRX mode. Upon entering the DRX mode, the UE monitors the downlink for a number of subframes defined by the on-duration timer in every short DRX cycle or long DRX cycle. The short DRX cycle is optional. If it is configured, the UE will follow the short DRX cycle for a specific number of cycles controlled by the short DRX cycle timer before switching to the long DRX cycle, if there is no activity. Other timers prevent the UE from entering the DRX mode. These are associated with HARQ downlink retransmissions. If the UE has received data that it was unable to decode and this was not associated with a broadcast HARQ process, then following the expiration of the HARQ round-trip-time timer, which specifies the earliest time at which a UE can expect a retransmission, the DRX retransmission timer is started. The UE must wait for this timer to expire and continue to monitor the downlink before it can enter the DRX mode. In addition to the above timers, the network can order the UE to enter the DRX mode with the DRX command MAC control element. All timers and parameters described in this context are configured by RRC signaling.

During the DRX procedure, the UE maintains a DRX cycle that is defined as a number of subframes. The UE monitors PDCCH for a specific number of subframes in a DRX cycle. This time interval is called on-duration and can range from 1 to 200 subframes. The UE may turn off its receiver for the rest of the DRX cycle. The UE maintains two DRX cycles referred to as short DRX cycle and long DRX cycle, which have different durations. The short DRX cycle is optional and targeted for applications such as VoIP which typically require relatively small transmissions of data at short but regular intervals. If configured, the UE starts with a short DRX cycle (2–640 subframes) when it enters the DRX mode. When the configurable short-DRX timer expires, the UE transitions into the long DRX cycle (10–2560 subframes). The UE can transition to the DRX mode either upon expiration of configured timers or following a DRX command in the form of a MAC control element.

4.4.2 Connected mode procedures

The RRC_CONNECTED is a state where the UE transfers unicast control/data in the uplink or downlink. The UE may be configured with an individualized DRX or discontinuous transmission pattern, or both. The UE monitors control channels associated with the shared data channel to determine if data is scheduled for it, provides channel quality and feedback information, performs neighbor cell measurements and measurement reporting, and acquires the system information. One or multiple IP addresses are assigned to the UE as well as a temporary identifier called C-RNTI that is used for signaling purposes between the UE and the network. In brief, the RRC_CONNECTED state is characterized as follows:

- UE is known in EPC and E-UTRAN/eNB, i.e., UE context is stored in eNB.
- UE location is known at cell level.
- Unicast data transfer with UE is possible.
- Short/long DRX cycles are supported for power saving.
- Mobility is controlled by the network.

In RRC_CONNECTED state, the network controls UE mobility, i.e., the network decides when the UE must move to another cell, which may be on another frequency or RAT. For network-controlled mobility in RRC_CONNECTED state, handover is the only procedure that is defined. The network triggers the handover procedure based on radio conditions or network load. The network may configure the UE to perform measurement reporting, including the configuration of measurement gaps. The network may also initiate handover without receiving measurement reports from the UE. Before sending the handover message to the UE, the source eNB prepares one or more target cells. The target eNB generates the message used to perform the handover including the AS configuration to be used in the target cell. The source eNB transparently forwards the handover message/information received from the target to the UE. The source eNB may initiate data forwarding for some of the user data radio bearers (DRBs). After receiving the handover message, the UE attempts to access the target cell at the first available random access channel opportunity according to the random access resource selection procedure for asynchronous handover. When allocating a dedicated preamble for the random access in the target cell, the E-UTRA ensures it is available from the first random access opportunity the UE may utilize. Upon successful completion of the handover, the UE sends a message used to confirm the handover. Following successful completion of handover, the PDCP SDUs may be retransmitted in the target cell. This only applies to user DRBs using RLC acknowledged mode. Furthermore, the sequence number and the hyper-frame number (HFN)[4] are reset except for the user DRBs using RLC acknowledged mode, for which both sequence number and HFN continue. For a limited period of time, the source eNB maintains the UE context to allow the UE to return in case of handover failure. If handover failure is detected, the UE will attempt to resume the RRC connection either with the source eNB or with another cell using

[4]An HFN is an overflow counter mechanism that is used in the eNB and UE in order to limit the actual number of sequence number bits that are needed to be sent over the radio air-interface. The HFN needs to be synchronized between the UE and eNB.

the RRC connection reestablishment procedure. The connection reestablishment would succeed, if the accessed cell is prepared [3].

The paging procedure is used to transmit paging information to a UE in RRC_IDLE and/or to inform mobile terminals in RRC_IDLE or RRC_CONNECTED states about an system information change and/or to inform about an ETWS primary and/or secondary notification. The paging information is provided to upper layers, which in response may establish an RRC connection to receive an incoming call. The E-UTRAN initiates the paging procedure by transmitting the paging message at the UE's paging opportunities. The E-UTRAN may address multiple mobile terminals within a paging message by including one paging record for each UE. The E-UTRAN may also indicate a change of system information and/or provide an ETWS notification.

The RRC connection establishment involves SRB1 bearer establishment. The procedure is also used to transfer the initial NAS dedicated information/message from the UE to the E-UTRAN. The UE initiates the RRC connection establishment procedure when the upper layers request establishment of an RRC connection while the UE is in the RRC_IDLE state.

The RRC connection reestablishment is used to reestablish the RRC connection including resumption of SRB1 operation and the reactivation of security. A UE in RRC_CONNECTED state, for which security has been activated, may initiate the procedure in order to continue the RRC connection. The connection reestablishment succeeds if the serving/target node is prepared, i.e., it has a valid UE context. If the E-UTRAN accepts the connection reestablishment, SRB1 operation resumes while the operation of other radio bearers are suspended. If the AS security has not been established, the UE cannot initiate the procedure and directly transitions to the RRC_IDLE state [3].

4.5 Network attach procedures

After establishing an RRC connection, the UE can send an *Attach Request* message to the MME in order to register with the network. The UE is required to further request PDN connectivity along with the network attach request. After all necessary signaling connections are established, the EPC may trigger security functions. When a UE connects to the LTE radio access network, it is automatically connected to the PDN and this connected state is continuously maintained. In other words, as the terminal is registered in the network or alternatively attached to the core network through the LTE radio access network, a communication path to the PDN (providing IP connectivity) is established. The PDN to which a connection is established can be configured on a per-subscriber basis, or the terminal can select it during the attach procedure. This PDN is called the default PDN. With the always-on connection function, only the radio link part of the connection is released after a certain time interval elapses with no terminal activity; however, the IP connectivity between the terminal and the network is sustained. By doing so, only the radio link needs to be reconfigured when the terminal resumes data communication in order to reduce the connection setup time. Furthermore, the IP address obtained when the terminal attaches can be used until it detaches, so it is always possible to receive packets using that IP address [29,35].

The procedure for a UE when initially attaching to the core network until connection to the PDN is established is shown in Figure 4.10. The details of attach procedure can be described as follows [25]:

1. The UE establishes an RRC connection with the eNB.
2. The UE sends the *Attach Request* message together with a *PDN Connectivity Request* message for PDN (IP) connectivity through the established RRC connection. As part of this process, the eNB establishes an S1 logical connection with the MME for the UE. In other words, when the UE establishes an RRC connection with the eNB for sending and receiving RRC and NAS messages, it sends an *Attach Request* message to the MME. The UE and the MME perform the required security procedures, including authentication, encryption, and integrity protection. The purpose of the attach request is to register with the EPS network, whereas the purpose of the PDN connectivity request is to establish connectivity with the PDN for transporting user data. In order to transport NAS messages, the UE requests AS to establish NAS signaling connection.
3. If the UE has valid security parameters, the *Attach Request* message must be integrity protected in order to allow validation of the UE by the MME, KSI_{ASME}, NAS sequence number, and NAS message authentication code are included if the UE has valid EPS security parameters. NAS sequence number indicates the sequential number of the NAS message. If the UE does not have a valid EPS security association, then the *Attach Request* message is not integrity protected. The UE network capabilities also indicate the supported NAS and AS security algorithms. The PDN type indicates the requested IP version (IPv4 or IPv6 [17]). If the network is not able to identify the UE with the ID given in the *Attach Request* message, it initiates the identification followed by authentication and security activation procedures.
4. A single eNB can be associated with multiple MMEs within a single PLMN or across multiple PLMNs. Furthermore, the eNB does not maintain any information about the UE in the RRC_IDLE state. During initial attach, the UE is not registered with any MME. As a result, the eNB selects an MME in the selected PLMN based on the load status of MMEs. The load information is provided by the MME(s) on a periodic basis. If an eNB only connects to a single MME, then the MME selection will not be required. An MME pool area is defined as an area consisting of a number of tracking areas. This is an area within which a UE may be served without changing the serving MME. In this pool of MMEs, the eNB can be connected to multiple MMEs and has a choice to select any MMEs. The load balancing among MMEs enables the

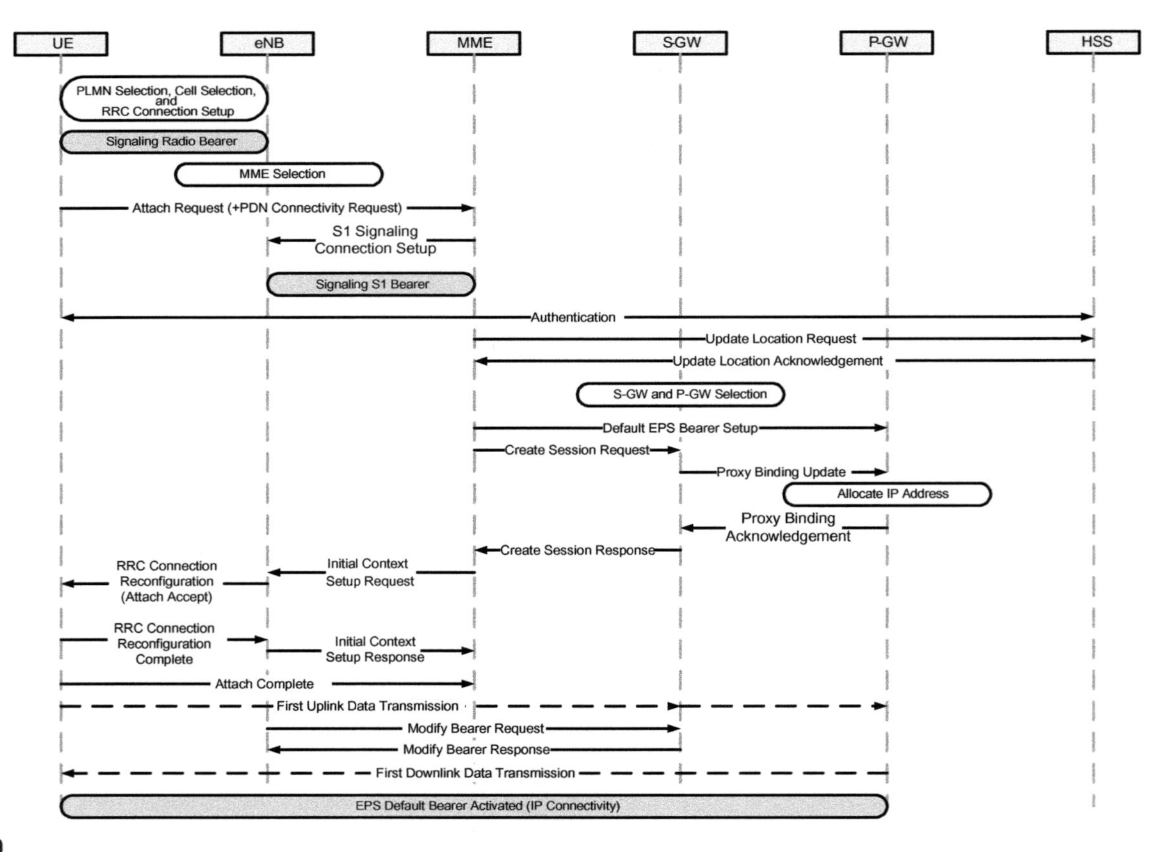

FIGURE 4.10

Attach procedure [10,28,29].

network to ensure equally loaded MMEs within a pool area. This is achieved by setting a weight factor for each MME such that the probability of selecting an MME is proportional to its weight factor. The weight factor is typically set according to the capacity of an MME relative to other MMEs.

5. The eNB and MME are connected via an S1-MME interface and use S1 application protocol [12]. An S1-MME interface is established when the eNB and MME connect with each other during initial provisioning.
6. The MME updates the home subscriber server with the location of the UE using the *Update Location Request* message based on the Diameter protocol.[5] It also requests the subscriber profile from the HSS using this message. The HSS updates its database with the current location of the UE and sends the subscriber profile information to the MME in the Diameter *Update Location Acknowledgement* message.
7. There can be multiple P-GWs and S-GWs for a given access point name (APN) within a single PLMN; only one S-GW and one P-GW serve a UE with an APN. After subscriber data is downloaded, the MME initiates EPS-bearer setup with P-GW. To accomplish this, the MME selects a P-GW in the subscribed default APN and then selects an S-GW. Note that a wireless network connects with multiple external networks, including the Internet and specific enterprise networks. An APN identifies these external networks during a data session. The APN specifies the external networks that a mobile device can access. It also defines which type of IP address to use, which security mechanisms to invoke, and which fixed-end connections to use (fixed-end connections refer to the connections between an operator's wireless network and the customer network). The APNs available to a device are specified as part of the subscriber account. A mobile station can have access to more than one APN, where each packet data protocol context (data session) specifies which individual APN to use.
8. To establish a transmission path to the default PDN, the MME sends a *Create Session Request* message to the S-GW.
9. When the S-GW receives the create session request from the MME, it requests proxy binding update from the P-GW, i.e., a request message sent by a mobile access gateway to a mobile node's local mobility anchor for establishing a binding between the mobile node's home network prefixes assigned to a given interface of a mobile node and its current care-of address. The P-GW allocates an IP address to the terminal and notifies the

[5]The Diameter protocol is designed to provide an authentication, authorization, and accounting (AAA) framework for applications such as network access or IP mobility. Diameter is also intended to work in local AAA as well as roaming scenarios. Diameter protocol was designed as an improved version of the RADIUS protocol; however, it does not address flaws within the RADIUS model. The basic concept behind Diameter is to provide a base protocol that can be extended in order to provide AAA services to new access technologies. Diameter does not use the same RADIUS protocol data unit but is backward compatible with RADIUS to facilitate network migration [19].

S-GW of this information in a *Proxy Binding Acknowledgement* message, i.e., a reply message sent by a local mobility anchor in response to a *Proxy Binding Update* message that it received from a mobile access gateway. This process establishes a continuous core network communication path between the P-GW and the S-GW for the allocated IP address [30].

10. A radio access bearer is established from S-GW to the eNB, where a create session response message is sent to the MME. The *Create Session Response* message contains information required to configure the radio access bearer from the eNB to the S-GW, including information elements issued by the S-GW and the IP address allocated to the terminal.
11. The MME sends the information in the *Create Session Response* message to the eNB in an initial context setup request message. Note that this procedure also contains other notifications such as *Attach Accept*, which is the response to the *Attach Request*. When the terminal receives *Attach Accept*, it sends an *Attach Complete* response to the MME, notifying that the process is complete.
12. The eNB establishes the radio data link and sends the *Attach Accept* message to the terminal. It also configures the radio access bearer from the eNB to the S-GW and sends an initial context setup response to the MME. The initial context setup response contains information elements issued by the eNB required to establish the radio access bearer from the S-GW to the eNB.
13. The MME sends the information in the initial context setup response to the S-GW in a modify bearer request message. The S-GW completes configuration of the previously prepared radio bearer from the S-GW to the eNB and sends a modify bearer response to the MME.

Through the above steps, a communication path from the terminal to the P-GW is established, enabling communication with the default PDN. If the terminal has no activity for a certain period of time, the always-on connection function described above releases the radio control link and the LTE radio access bearer, while maintaining the core network communication path.

After the terminal has established a connection to the default PDN, it is possible to initiate another connection to a different PDN. In this way, it is possible to manage PDNs according to service agreements. For example, the IMS PDN, which provides voice services through a packet network, may be used as the default PDN, and a different PDN may be used for Internet access. To establish a connection to a PDN other than the default PDN, the procedure is the same as the attach procedure shown in Figure 4.10, except no location update is required.

Once the attach procedure succeeds, a context is established for the UE in the MME, a default bearer is set up between the UE and the P-GW, and an IP address is assigned to the UE. At this time, the UE has IP connectivity and it can start using IP-based Internet or IMS services.

If the UE does not require services from the network, e.g., when the UE is switched off, it needs to deregister itself from the network by initiating the detach procedure. The detach procedure is initiated by the UE by sending a *Detach*

Request message. The detach type information element included in the message indicates whether detach is due to a switch off. The network and the UE deactivate the EPS-bearer context(s) for this UE locally without additional signaling between the UE and MME. If the detach type is "switch off," then the MME does not send the *Detach Accept* message; otherwise, the message is sent to the UE. While processing the detach message from the UE, the MME initiates the release of the EPS bearers in the network and it also flushes the UE context held at the eNB.

4.5.1 Random access procedure

A UE may access network services once it is synchronized in downlink and uplink directions. After the PLMN and cell selection phase, the UE is synchronized with the network in the downlink direction. Subsequently, it needs to synchronize with the network in the uplink direction. The random access procedure over PRACH is performed by the UE for this purpose. The random access procedure is characterized as one procedure independent of cell size and is the same for FDD and TDD systems. There are two types of random access procedures:

- *Contention-based random access procedure*: In this case, multiple UEs may simultaneously attempt to access the network, resulting in potential collisions. The contention is resolved among the contending UEs using the following procedure [3]:
  1. Transmission of random access preamble on RACH in the uplink. The RACH preambles are grouped based on the size of the L3 message that the UE would like to transmit; this provides an indication of resource requirements for the UE to the network.
  2. Transmission of random access response generated by eNB MAC on DL-SCH. This is an indication to the UE that the eNB has received its preamble and conveys the resources reserved for this UE.
  3. Transmission of the first scheduled uplink transmission on UL-SCH. The UE sends the *RRC Connection Request* message using the resources allocated by the eNB in the previous step. In the *RRC Connection Request* message, the UE sends its identifier to the network which is used by the eNB to resolve the contention in the next step.
  4. Contention resolution in the downlink. The eNB sends back the received UE identifier to resolve the contention, and at this point the UE which has received its identifier continues with the transmission while others will back off and try again.
- *Non-contention-based random access procedure*: The network initiates this procedure in the case of a handover of a UE from one eNB to another in order to reduce the handover latency. The network usually reserves a set of RACH preambles for this purpose, where one is assigned and sent to the UE. There are no collisions with other UEs' preamble transmissions in the uplink

because the eNB controls the process. Once the UE receives the assigned random access preamble, it sends it to the eNB. The eNB then generates and sends a *Random Access Response* message.

As shown in Figure 4.11, during the first step of the contention-based procedure, the UE transmits a preamble randomly selected from 64 possible preamble sequences which are divided into two groups. The group from which the UE selects the preamble depends on the size of the message it would like to transmit and on the calculated path loss to the eNB. Once the preamble is transmitted, the UE monitors the downlink beginning three subframes after the end of the preamble transmission for a duration controlled by a configurable window. If no downlink response is detected, the preamble is retransmitted with power increased by a configurable parameter. This process continues until the maximum number of random access preamble transmissions is reached. If the preamble is successfully detected by the eNB, then a *Random Access Response* message is sent in the downlink, allocating resources for the UE to transmit its message or enforcing a backoff. In the final step, contention resolution must be performed for a contention-based procedure.

The UE sets its initial power level to the following value *preambleInitialReceivedTargetPower − pathloss + DELTA_PREAMBLE*, where *DELTA_PREAMBLE* is a fixed value that depends on the type of preamble format. The UE selects the first available RACH resource for the transmission. If there is no response within three subframes + *ra-ResponseWindowSize* (defined in number of subframes), the UE assumes that the transmission was unsuccessful. It selects another RACH preamble, performs the backoff procedure, and transmits the new preamble. The power level is increased by *powerRampingStep*. This process can be repeated until *preambleTransMax* preambles are transmitted, in which case a failure is declared to the upper layers.

Once the random access preamble is detected by the eNB, it responds with a *Random Access Response* message. This message provides information that the UE needs to transmit in the uplink including the timing advance and the grant specifying which resources to utilize. To resolve a possible contention, the UE sends an *RRC Connection Request* message using a random number (or the S-TMSI if available) as its identity. A contention resolution timer is started and is monitored in the downlink for the *Contention Resolution* message. If this message is received and contains the UE ID transmitted in step 3, the contention is resolved. The temporary C-RNTI contained in the *Random Access Response* message becomes the C-RNTI. If the *Contention Resolution* message containing the correct UE identity is not received before the contention resolution timer[6] expires, the UE assumes the access failed and the random access procedure needs to be repeated.

[6]The contention resolution timer specifies the number of consecutive subframe(s) during which the UE monitors the physical downlink control channel after Msg3 is transmitted [6].

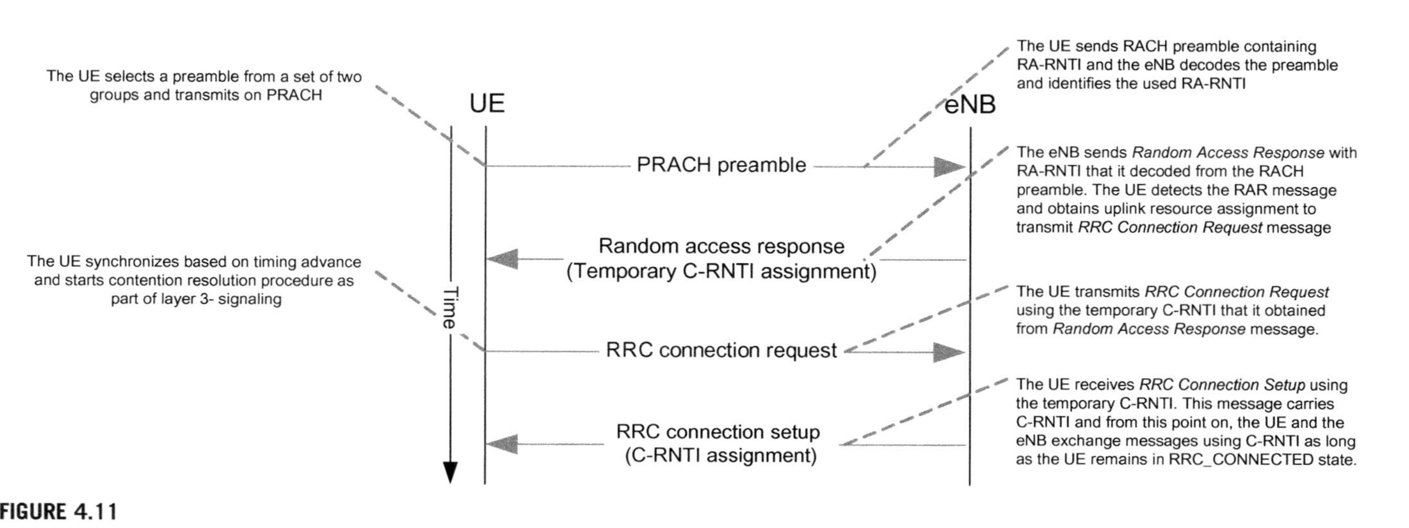

FIGURE 4.11

Contention-based random access procedure [3,25,32].

4.5.2 RRC connection establishment, modification, and release

One of the main functions of the RRC sublayer is to establish, maintain, modify, and release connections with E-UTRAN for transport of user data and control messages. The RRC connection establishment starts with the establishment of signaling radio bearer 1 (SRB1) and the transfer of the initial NAS message. This NAS message triggers the setup of an S1 connection, which normally initiates a subsequent step during which E-UTRAN activates AS security and further establishes signaling radio bearer 2 (SRB2) along with one or more DRBs associated with the default and optionally dedicated EPS bearers. Figure 4.12 illustrates the RRC connection establishment, NAS signaling, and RRC control messages. As shown in the figure, the RRC connection establishment procedure consists of three steps: the *RRC Connection Request* message from the UE; the *RRC Connection Setup* message from the eNB; and the *RRC Connection Setup Complete* message from the UE. Upon successful completion of the RRC connection establishment procedure, the UE is in the RRC_CONNECTED state and is provided with periodic timing alignment commands by the eNB. The UE provides channel quality feedback to the eNB and conducts neighboring cell measurements for handover. Note that NAS messages (ECM/EMM) are encapsulated in RRC messages; thus RRC connection establishment needs to happen prior to NAS signaling transfer. Furthermore, it must be noted that signaling radio bearers (SRBs) are established before DRBs. Upon receiving the UE context from the EPC, the E-UTRAN activates security that includes ciphering and integrity protection using the initial security activation procedure.

It must be noted that an SRB is defined as a radio bearer that is only used for transmission of RRC and NAS messages. More specifically, the following three SRB types are defined in E-UTRA [3,9]:

1. SRB0 is used for transport of RRC messages associated with the common control logical channel.
2. SRB1 is used for transport of RRC messages including piggybacked NAS messages as well as NAS messages prior to the establishment of SRB2 where all are associated with the dedicated control logical channel.
3. SRB2 is used for transport of NAS messages using the dedicated control logical channel, has a lower priority than SRB1, and is always configured by E-UTRAN after security activation.

The UE in the RRC_IDLE state does not have any allocated SRB or DRB. The UE requires (dedicated) SRBs and DRBs to exchange data/messages with the network. In particular, for transfer of NAS messages, SRBs are established. They carry registration and signaling for user-plane setup. The user-plane application data is transported on DRBs. An E-UTRAN radio access bearer (E-RAB)

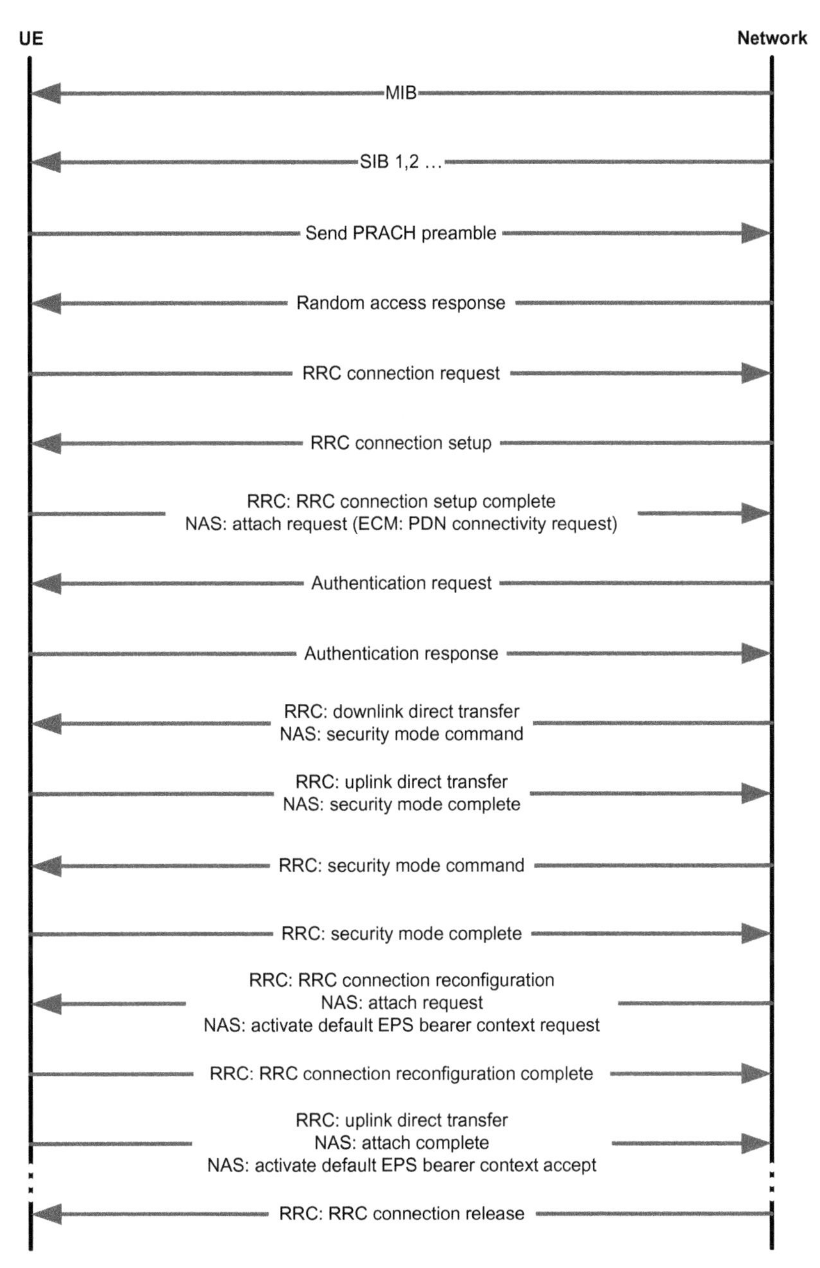

FIGURE 4.12

Sequence of signaling for RRC connection setup, reconfiguration, and release [32].

uniquely identifies the concatenation of an S1 bearer and the corresponding DRB. When an E-RAB exists, there is a one-to-one mapping between the E-RAB and an EPS bearer. To establish, modify, or release DRBs, the E-UTRAN applies the RRC connection reconfiguration procedure.

To allow a wide range of UE implementations, different UE capabilities are specified in the E-UTRA. The UE capabilities are divided into a number of parameters, which are sent from the UE at the establishment of a connection and if/when the UE capabilities are changed during an ongoing connection. The UE capabilities may then be used by the network to select a configuration that is supported by the UE.

In E-UTRAN, the UE radio network-related capability information are transferred using RRC signaling from the UE to the eNB. In order to avoid sending the UE capabilities over the radio interface between the UE and the eNB each time that the UE performs a transition to the connected mode (i.e., when the UE-specific context is created in the eNB), the eNB stores the UE capabilities in the MME so that they are safeguarded when the UE is in the idle mode. When the UE returns to the connected mode, the UE capabilities should be downloaded to the eNB. Before the UE can fully use the services of the network, it has to declare its capabilities (e.g., supported bit rates, antenna configurations, bandwidths, and supported radio access types) to the network. When the network has knowledge of the UE capabilities, it can optimize transmissions in conjunction with services to the UE according to its capabilities.

In general, the UE capabilities can be classified into mainly two categories depending on which layer of the protocol hierarchy the given capability is related to. The access stratum-related capabilities are the access technology-dependent parts of the capability information, such as terminal power class and supported frequency band. The AS capabilities are required by the eNB. The non-access stratum-related capabilities are a set of capability information containing the non-access parts of the UE capability such as supported security algorithms. The NAS capabilities are used by the EPC.

The AS part of the UE capabilities needs to be present in the eNB, while the UE is in the connected mode. Moreover, when a handover is made from the source eNB to the target eNB, the UE capability information needs to be transferred from the source eNB to the target eNB. However, for a UE in the idle mode, there is no need to maintain any UE information, including the UE capabilities, in the eNBs, but only in the MME. Therefore, when a UE transitions to the active state again, the information of the UE, including the UE capabilities, must be recreated in the eNB [20,21].

The *UE Capability Enquiry* message is used to request the transfer of UE radio access capabilities for E-UTRA as well as for other RATs (UTRA, GERAN-CS, GERAN-PS, and CDMA2000). As the UE's AS capability may contain several RATs capabilities, the information can be very large. To reduce the overhead on the air-interface during transition from RRC_IDLE state to RRC_CONNECTED state, the MME stores the UE's AS capability and provides it to the eNB during initial UE context setup over the S1 interface. If the MME does not have a valid UE's AS capability, it does not provide UE's AS capability to the eNB. The eNB has to acquire the UE's AS capability from the UE using a *UE Capability Enquiry* message. The eNB sends a *UE Capability Enquiry* message to transfer the UE's radio access capability information to the E-UTRAN. The eNB identifies the RAT for which it is requesting the capabilities. This message supports requests for E-UTRAN, UTRAN, GERAN, and CDMA2000 1xRTT. The UE responds with a *UE Capability Information* message, which includes the requested RAT container(s), provided that the requested RAT by the eNB is supported. The information obtained is used to configure MAC and physical-layer attributes (receive and transmit capabilities such as single/dual radio, and dual receiver) of the connection. It also enables efficient measurement control, preventing unnecessary UE wake-ups to conduct measurements [9].

4.5.3 Service establishment and QoS

There are two types of EPS bearers in E-UTRA. One is called "default EPS bearer" and the other is known as "dedicated EPS bearer," which can be described as follows. The default EPS bearer is established during the initial attach procedure to allocate an IP address to the UE, and does not have a specific QoS (only nominal QoS is applied). This is similar to the primary PDP context in UMTS [24,32,36]. The dedicated EPS bearer is typically established during a call setup following transition from the idle mode, which does not allocate any additional IP address to the UE and is linked to a specified default EPS bearer. However, it has a specific and usually guaranteed QoS. This is similar to the secondary PDP context in UMTS [24,28].

An EPS bearer is the level of granularity for bearer-level QoS control in the EPC/E-UTRAN. It means that the service data flows (SDFs), which are mapped to the same EPS bearer, receive the same bearer-level packet forwarding treatment (e.g., scheduling policy, queue management policy, rate shaping policy, and RLC configuration). One EPS bearer is established when the UE connects to a PDN and is sustained throughout the lifetime of the PDN connection to provide the UE with always-on IP connectivity to that PDN. This bearer is referred to as the *default* bearer. Any additional EPS bearer that is established to the same PDN is referred to as a *dedicated* bearer. The initial bearer-level QoS parameter values of the default bearer are assigned by the network, based on subscription data. The decision to establish or modify a dedicated bearer can only be taken by the EPC, and the bearer-level QoS parameter values are always assigned by the EPC [20].

An EPS bearer is referred to as a guaranteed bit rate (GBR) bearer, if dedicated network resources related to a GBR value that is associated with the EPS bearer are permanently allocated (e.g., by an admission control function in the eNB) during bearer establishment or modification. Otherwise, an EPS bearer is referred to as a non-GBR bearer. A dedicated bearer can either be a GBR or a non-GBR bearer, while a default bearer is a non-GBR bearer.

The GBR bearers are mainly used to carry voice, video, and real-time gaming services through dedicated bearers or static scheduling. Each GBR bearer is associated with the bit rate that is expected to be provided by the bearer, while the maximum bit rate (MBR) indicates the upper limit of the bit rate that can be provided by that bearer. Non-GBR bearers are mainly used to carry various data services. To improve bandwidth utilization, EPS introduces the concept of aggregated MBR (AMBR). AMBR is an IP connectivity access network (IP-CAN), which refers to an access network that provides IP connectivity, session level QoS parameter of every PDN connection. Multiple EPS bearers for the same PDN connection share the same AMBR value. Each non-GBR bearer can potentially make use of the whole AMBR if other EPS bearers do not transfer any services. Therefore, an AMBR actually restricts the total bit rate of all the bearers sharing this AMBR [24].

The EPS-bearer service layout is depicted in Figure 4.13. An uplink traffic flow template (TFT) in the UE binds an SDF to an EPS bearer in the uplink direction. Multiple SDFs can be multiplexed into the same EPS bearer by including multiple uplink packet filters in the uplink TFT. A downlink TFT in the P-GW binds an SDF to an EPS bearer in the downlink direction. Similar to the uplink, multiple SDFs can be multiplexed into the same EPS bearer by including multiple downlink packet filters in the downlink TFT. An E-RAB transports the packets of an EPS bearer between the UE and the EPC. When an E-RAB is established, there is a one-to-one mapping between the E-RAB and an EPS bearer. A DRB transports the packets of an EPS bearer between a UE and an eNB. When a DRB exists, there is a one-to-one mapping between this DRB and the EPS bearer/E-RAB [10].

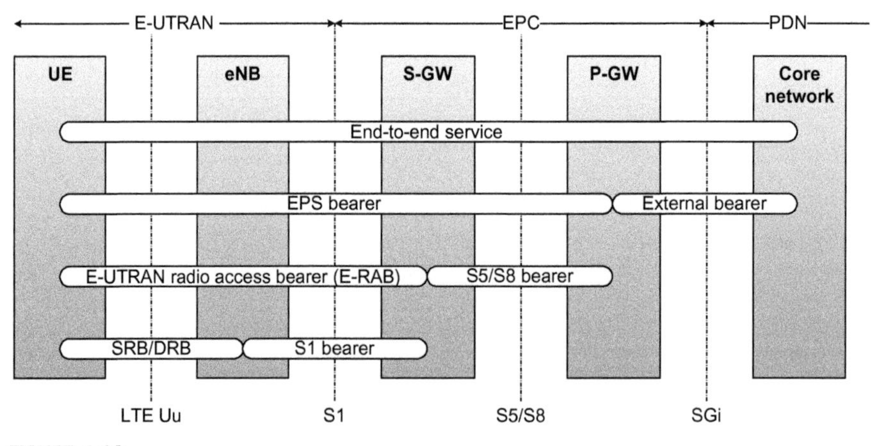

FIGURE 4.13

EPS-bearer service structure [3].

The bearer level (i.e., per bearer or per aggregate) QoS parameters are QCI, ARP, GBR, and AMBR [24]. Each EPS bearer/E-RAB (GBR and non-GBR) is associated with the following bearer level QoS parameters [3]:

- QoS class identifier (QCI): A scalar that is used as a reference to access node-specific parameters that control bearer-level packet forwarding behavior (e.g., scheduling weights, admission thresholds, queue management thresholds, and link layer protocol configuration), which have been preconfigured by the operator. A one-to-one mapping of QCI values to standardized characteristics is captured in 3GPP TS 23.401 [10].
- Allocation and retention priority (ARP): The primary purpose of ARP is to decide whether a bearer establishment/modification request can be accepted or needs to be rejected in case of resource limitation. In addition, the ARP can be used by the eNB to decide which bearer(s) to drop during resource limitation conditions such as handover. The ARP priority level ranges from 1 to 15, where 1 is the highest level of priority.

Each GBR bearer is additionally associated with the following bearer level QoS parameters [3]:

- Guaranteed bit rate (GBR): The bit rate that is expected to be provided by a GBR bearer.
- Maximum bit rate (MBR): The maximum bit rate that is expected to be provided by a GBR bearer. The MBR can be greater than or equal to the GBR.

Each APN access by a UE is associated with per APN aggregate maximum bit rate (APN-AMBR) QoS parameter. The APN-AMBR limits the aggregate bit rate across all non-GBR bearers and across all PDN connections of the same APN (excess traffic may be discarded). Each UE in the EMM-REGISTERED state is associated with a per UE aggregate maximum bit rate (UE-AMBR) QoS parameter. The GBR and MBR denote the bit rate of traffic per bearer, while UE-AMBR/APN-AMBR denote the bit rate of traffic per group of bearers. The UE-AMBR limits the aggregate bit rate across all non-GBR bearers of a UE (excess traffic may be discarded by a rate shaping function). Each of these QoS parameters has an uplink and a downlink component.

UE-AMBR and ARP are network parameters that are not known at the UE. The standardized QCI characteristics are not signaled on any interface. They are guidelines for the preconfiguration of node-specific parameters for each QCI. Standardizing a QCI with corresponding characteristics ensures that applications/services mapped to that QCI receive the same minimum level of QoS in multi-vendor network deployments and in case of roaming. Each QCI (GBR or non-GBR) is associated with a priority level, with priority level one being the highest priority level. Scheduling between different SDF aggregates are primarily based on the packet delay budget (PDB), which refers to the permissible latency of data packets transported between UE and P-GW. For E-UTRAN, the priority level of a QCI may be used as the basis for assigning the uplink priority per radio bearer. The purpose of the PDB is to support the configuration of scheduling and link layer functions (e.g., setting scheduling priority weights, and HARQ target operating points). For a specific QCI, the PDB is the same for both uplink and downlink. Packet loss rate (PLR) is defined as the rate of SDUs not successfully transported to the upper layer to the total number of SDUs processed by the link layer of the transmitting end. Therefore, PLR reflects the upper bound of tolerable packet loss rate in non-congested cases and is identical in both uplink and downlink for the same QCI [36].

4.6 Data call establishment procedures

4.6.1 UE-originated data call procedure

After successfully attaching to the network, the UE can request services from the network using a service request procedure. When the UE requests resources from the network to initiate a data call, it can utilize a NAS service request procedure. Another example of this procedure is during initiation of mobile-originated or mobile-terminated fallback procedures, if they are supported by the network. The call flow for this procedure is shown in Figure 4.14. This procedure can be described as follows [25]:

1. The UE establishes an RRC connection with the eNB.
2. The UE sends a service request message to the MME to request (dedicated) bearer resources by including the bearer resource allocation request. The eNB establishes an S1 logical connection with the MME for this UE. Note that the UE may also send the bearer resource allocation request to the MME as a stand-alone message. At this point, the network can initiate an optional identification followed by authentication and security mode procedures.
3. After the completion of the authentication and security mode procedure, the MME initiates activation of the default bearer with the S-GW/P-GW by sending the eGTP-C[7] modify bearer request message to the S-GW. Upon

[7]GPRS tunneling protocol (GTP) is a group of IP-based communications protocols used to carry general packet radio service (GPRS) within GSM, UMTS, and LTE networks [2]. In 3GPP architecture, GTP and Proxy Mobile IPv6-based interfaces are specified on various reference points. GTP can be decomposed into separate protocols: GTP-C; GTP-U; and GTP'. GTP-C is used within the GPRS core network for signaling between gateway GPRS support nodes (GGSN) and serving GPRS support nodes (SGSN). This allows the SGSN to activate a session on a user's behalf (PDP context activation), to deactivate the same session, to adjust QoS parameters, or to update a session for a subscriber joined from another SGSN. GTP-U is used for carrying user data within the GPRS core network and between the radio access network and the core network. The user data transported can be packets in any IPv4, IPv6, or PPP formats. The primary bearer-plane in LTE across the access and the core networks is the evolved GPRS tunneling protocol (eGTP). The user-plane variant of this protocol, eGTP-U, provides tunneling support for user data between the eNB and the EPC elements. The eGTP-C (GTPv2-C) protocol is responsible for creating, maintaining, and deleting tunnels on multiple Sx interfaces. It is used for control-plane path management, tunnel management, and mobility management. It also controls forwarding relocation messages, serving radio network subsystem (SRNS) context, and creating forward tunnels during inter-LTE handovers [2,27].

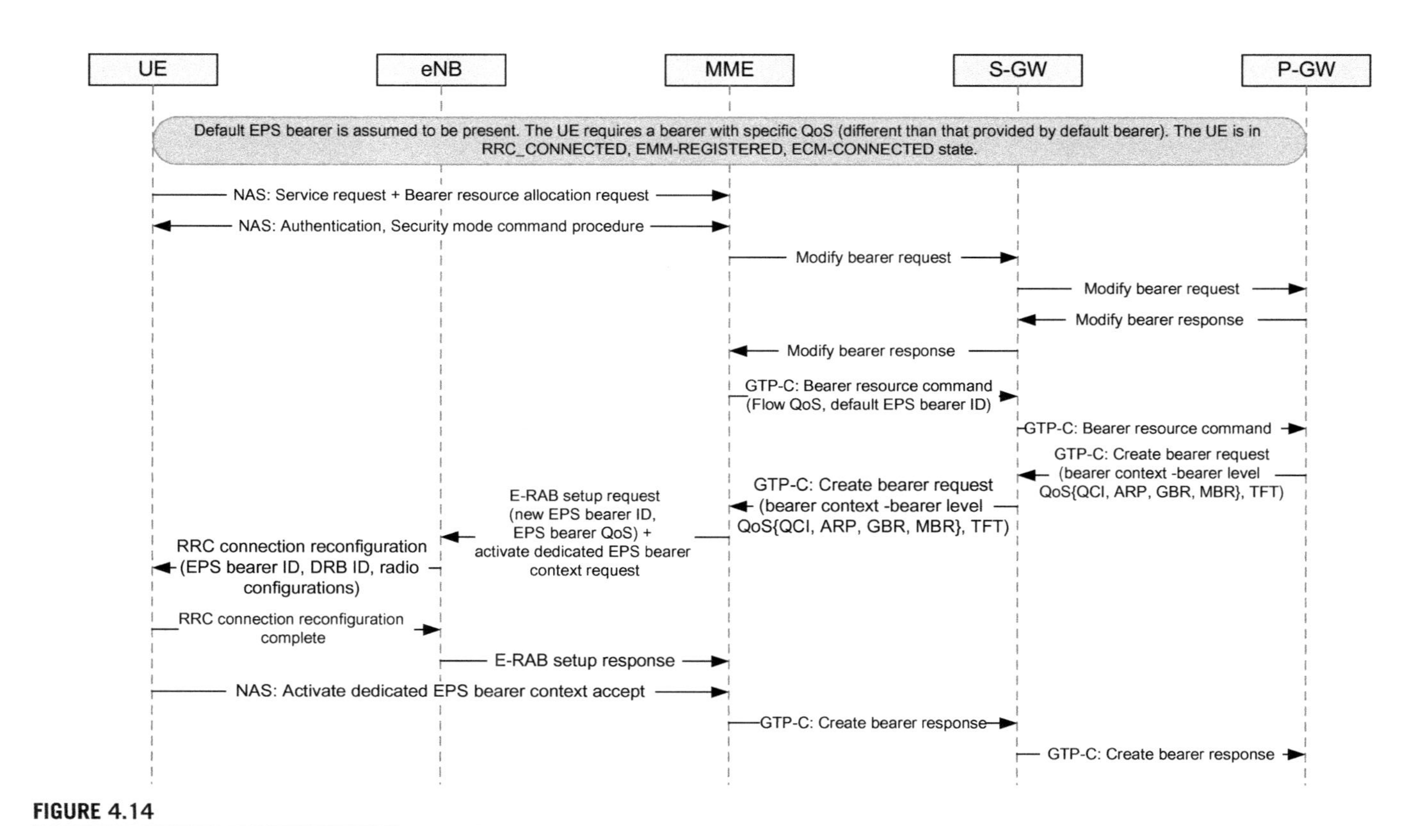

FIGURE 4.14

Call flow for mobile-originated data call [25].

receiving the modify bearer request, the S-GW activates the required resources and forwards the modify bearer request to the P-GW.

4. The P-GW processes the modify bearer request message and activates the required resources. Note that the IP address is allocated during the attach procedure. It responds to S-GW with the modify bearer response message and the S-GW forwards this message to the MME.
5. The MME now initiates the dedicated bearer establishment procedure by sending the eGTP-C bearer resource command to the S-GW. The S-GW processes the bearer resource command and forwards it to the P-GW.
6. The P-GW responds to the S-GW with a create bearer request message after allocating the dedicated bearer resource (TFT initialization). The S-GW processes the create bearer request and forwards it to the MME for further processing.
7. The MME now sends the E-RAB setup request to the eNB to set up the bearer between the eNB and the S-GW. It also piggybacks the NAS-activate-dedicated EPS-bearer context request to the UE.
8. The eNB allocates the resources for the radio bearers using an RRC connection reconfiguration request message to the UE and includes the received NAS message. The UE establishes the radio bearers and responds with an RRC connection reconfiguration complete message to the eNB.
9. At this time, the radio bearers are established between the eNB and the UE; thus the eNB sends the E-RAB setup response to the MME. The UE sends the activate dedicated EPS-bearer context accept NAS message to the MME via the eNB.
10. The MME sends a create bearer response to the S-GW to complete the dedicated bearer activation. The S-GW then forwards the message to the P-GW.

Once the UE concludes a data call, it can trigger the release of the dedicated bearers by sending the *Bearer Resource Modification Request* message to the MME, which can then take care of releasing the dedicated bearer with the S-GW and P-GW.

4.6.2 UE-terminated data call procedure

For a UE-terminated call, the network sends the paging request to all the eNBs associated with the last known tracking area where the UE was located. When receiving the *Paging Request* message from the MME, the eNB sends the paging message over the radio interface in the cells which are contained within one of the tracking areas provided in that message. The UE is normally paged using its S-TMSI or IMSI. The paging message also contains a UE ID index value in order for the eNB to calculate the paging occasions at which the UE listens to the paging message. The procedure is depicted in Figure 4.15, and described as follows [25]:

1. The P-GW/S-GW receives the incoming IP packet addressed to a UE.
2. The S-GW sends a downlink data notification to the MME requesting creation of a dedicated UE bearer.

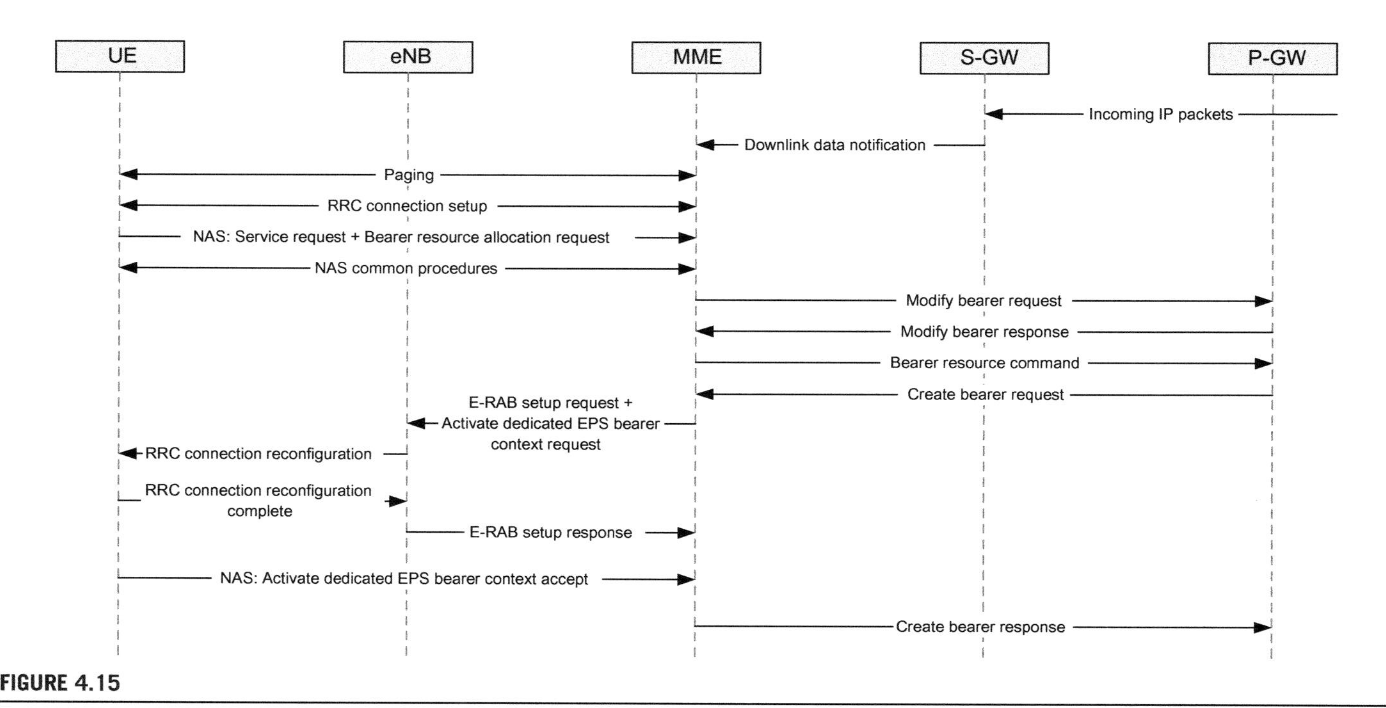

FIGURE 4.15

Call flow for mobile-terminated data call [25].

3. The MME now sends a paging message to notify the UE about the incoming IP packet.
4. Once the UE receives the paging message on the radio interface, it establishes an RRC connection with the eNB.
5. The UE sends the service request message to the MME and includes the bearer resource allocation request to request the dedicated bearer establishment. From this point onward, the message sequence is the same as that outlined in the mobile-originated data call.

4.7 Mobility procedures

E-UTRA supports X2-based and S1-based handover types. The X2-based handover is the least complex form of handover from an operation point of view, and typically would be used when there is an X2 interface between the source eNB and the target eNB. The handover is negotiated directly between the two eNBs; when the UE context is established on the target eNB, the MME is notified in order to switch the path. When no X2 interface exists between the network nodes, the S1-based handover will be utilized with signaling between the target and source eNBs going through the MME.

All handovers in E-UTRA are hard handovers, i.e., the air-interface connection with the source eNB should be disconnected before reestablishing a new connection with the target eNB [20]. The procedure for reestablishing the connection with the target node is identical for both X2-based and S1-based handovers, and requires the use of the random access procedure to synchronize and begin uplink transmissions. The UE is typically notified of the specific random access signature to use so that a non-contention-based procedure can be performed. From the core network perspective, handover can involve changes to the serving MME and S-GW, and this can be due to mobility and the geographic location of the network nodes, or due to loading and congestion control mechanisms. Figure 4.16 illustrates the overall concept of handovers and different handover types in E-UTRA. The X2-based handover procedure is performed without EPC involvement, i.e., preparation messages are directly exchanged between the eNBs. The release of the resources at the serving eNB (or source eNB) during the handover completion phase is triggered by the target eNB.

In order to support mobility, a UE may conduct and report to eNB a series of measurements at the physical layer which include RSRP and E-UTRA-carrier received signal strength indicator. These measurements in E-UTRA are classified as follows: intra-frequency E-UTRAN measurements; inter-frequency E-UTRAN measurements; and inter-RAT measurements. Measurement commands are used by E-UTRAN to instruct the UE to start measurements, modify measurements, or stop measurements. The reporting criteria that are used include event triggered reporting, periodic reporting, and event triggered

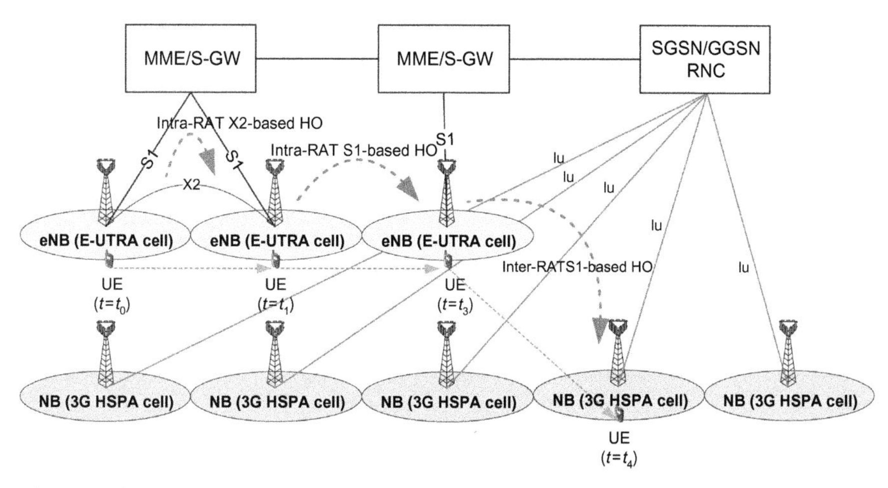

FIGURE 4.16

Overview of LTE handover scenarios [30].

periodic reporting. In the RRC_CONNECTED state, network-controlled UE-assisted handovers are performed and various DRX cycles are supported. In the RRC_IDLE state, cell reselections are performed and DRX is supported. The E-UTRAN can configure the UE to perform radio quality measurements for handover purposes. Different measurement objects are defined in the specification [9], and one object can be configured for each intra-frequency, inter-frequency, or inter-RAT measurement.

Depending on whether the UE needs transmission/reception gaps to perform the relevant measurements, measurements are classified as gap assisted or non-gap assisted. A non-gap-assisted measurement is a measurement on a cell that does not require transmission/reception gaps to allow the measurement to be performed. A gap-assisted measurement is a measurement on a cell that does require transmission/reception gaps to allow the measurement to be performed. Gap patterns are configured and activated by the RRC sublayer [3].

The measurements to be performed by a UE for intra/inter-frequency mobility can be controlled by E-UTRAN, using broadcast or dedicated control signaling. In the RRC_IDLE state, a UE follows the measurement parameters defined for cell reselection. In the RRC_CONNECTED state, a UE follows the measurement configurations specified by RRC sublayer and managed by the E-UTRAN. Intra-frequency and inter-frequency neighbor cell measurements are defined as follows [3]:

- Neighbor-cell measurements performed by the UE are intra-frequency measurements when the source and target cell operate on the same carrier frequency.

- Neighbor-cell measurements performed by the UE are inter-frequency measurements when the neighbor cell operates on a different carrier frequency compared to the source cell.

The mobility procedures in E-UTRA can be divided into idle mode and connected mode mobility for an "attached" UE. The idle mode mobility is based on UE autonomous cell reselections according to the parameters provided by the network. The UE transition between the idle and the connected mode is controlled by the network according to the UE activity and mobility.

Figure 4.17 illustrates the difference between the idle mode and connected mode mobility schemes. As shown in the figure, mobility in the RRC_IDLE state is UE-controlled using cell reselection, while in the RRC_CONNECTED state it is controlled by the E-UTRAN using handover. Furthermore, the network knowledge of the location of the UE in the RRC_IDLE state has a granularity of a tracking area, whereas the network knowledge of the UE location in the RRC_CONNECTED state has a granularity of a cell. A UE in the RRC_IDLE

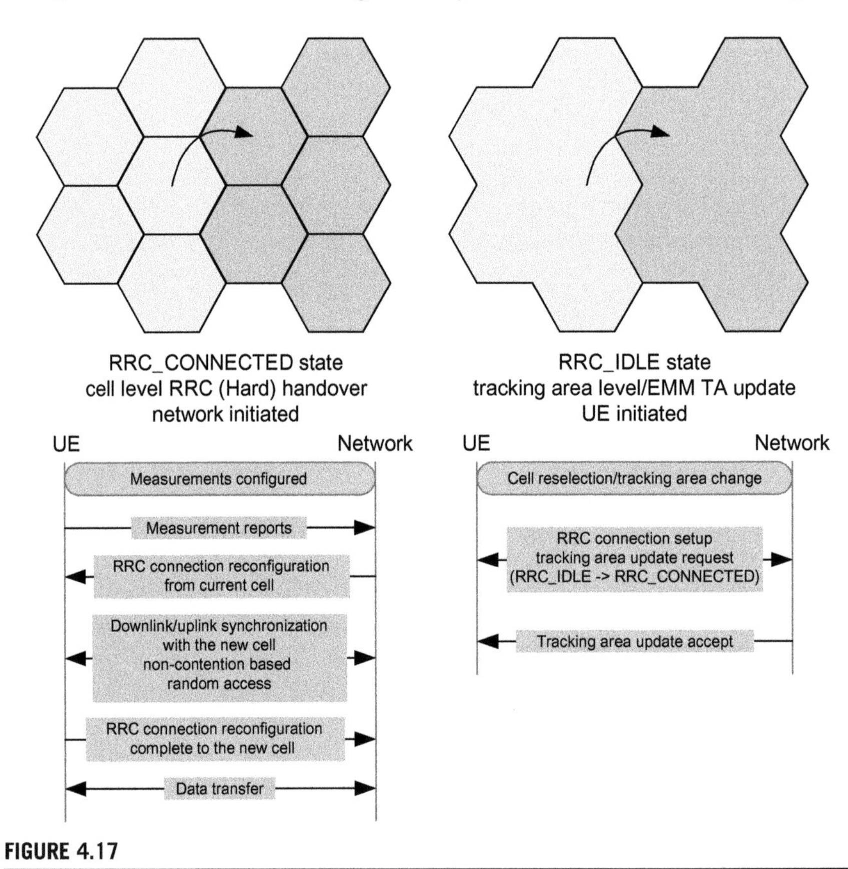

FIGURE 4.17

Overview of idle mode and connected mode mobility procedures [37].

state performs cell reselection and a tracking area update while crossing the tracking area boundaries. A UE in the RRC_CONNECTED state is configured to conduct radio link measurements and send them to the serving eNB. It can further receive/transmit data in the downlink/uplink and perform a non-contention-based random access procedure to stay uplink-synchronized with the network.

In the RRC_IDLE state, if equal priorities are assigned to multiple cells, the cells are ranked based on radio link quality. Equal priorities are not applicable between frequencies of different RATs. The UE does not consider frequencies for which it does not have an associated priority [20].

In RRC_CONNECTED, the E-UTRAN decides to which cell a UE should be transferred in order to maintain a reliable and satisfactory radio link quality. The E-UTRAN may take into account not only the radio link quality, but also other criteria such as UE capability, subscriber type, and access restrictions. Although E-UTRAN may trigger handover without measurement information from the UE, which is referred to as blind handover, typically it configures the UE to report measurements of the candidate target cells. Besides the handover procedure, the E-UTRA also allows a UE to be redirected to another frequency or RAT upon RRC connection release. Redirection during connection establishment is not supported, since at that time the E-UTRAN may not have all relevant information such as the capabilities of the UE and the type of subscriber [20].

4.7.1 Idle mode mobility procedures

In the RRC_IDLE state, cell reselection between frequencies is based on absolute priorities, where each frequency has an associated priority. Cell-specific default values of the priorities are broadcast via the system information. In addition, E-UTRAN may assign UE-specific values upon connection release, considering certain criteria such as UE capability or subscriber type. In case equal priorities are assigned to multiple cells, the cells are ranked based on radio link quality. Equal priorities are not applicable between frequencies of different RATs. The UE does not consider frequencies for which it does not have an associated priority. Besides cell reselection priority of a frequency, there are no idle mode mobility-related parameters that may be assigned via dedicated signaling.

In general, the transition from RRC_IDLE to RRC_CONNECTED is under the assumption that the RRC protection keys and user-plane protection keys are generated, while keys for NAS protection as well as higher-layer keys are assumed to be already available at the MME. On the other hand, in transitions from RRC_CONNECTED to RRC_IDLE, the eNBs delete the security keys that they store; thus the states of idle mode UEs are only stored in the MME.

In LTE-Advanced, the requirement for UE transition latency from the idle mode, with allocated IP address, to connected mode is less than 50 ms including the establishment of the user-plane and excluding the S1 transfer delay. The transition latency requirement from dormant mode (inactivity periods in short/long

DRX mode) to active mode (continuous transmission/reception periods) in connected mode is less than 10 ms. These requirements are stricter than those for earlier releases of LTE. In the dormant mode, the UE has an established RRC connection and radio bearers. The UE has already been identified at cell level, but it may be in DRX mode to save power during inactivity periods.

4.7.2 Connected mode mobility procedures

In the RRC_CONNECTED state, the network controls the UE mobility, i.e., the network decides when and where the UE should initiate a handover process. For network-controlled mobility in the RRC_CONNECTED state, the network triggers the handover process based on radio conditions or network load condition. To facilitate this process, the network may configure the UE to perform and report measurements including the configuration of measurement gaps. The network may also initiate a blind handover, i.e., without waiting for the measurement reports from the UE. Before sending the handover message to the UE, the source eNB prepares one or more target cells. The target eNB generates the message that is used to perform the handover, i.e., the message includes the AS configuration to be used in the target cell. The source eNB transparently (i.e., does not modify values or content) forwards the handover information received from the target cell to the UE. When appropriate, the source eNB may initiate data forwarding for (a subset of) DRBs in order to reduce the handover interruption time.

After receiving the handover message, the UE attempts to access the target cell at the first available random access channel opportunity according to random access resource selection, i.e., the handover is asynchronous. Consequently, when allocating a dedicated preamble for the random access in the target cell, E-UTRA ensures it is available from the first random access opportunity that the UE may use. Upon successful completion of the handover, the UE sends a message to confirm the handover. After successful completion of the handover, PDCP SDUs may be retransmitted in the target cell. This only applies to DRBs using RLC acknowledged mode [3,7]. The source eNB should, for some time, maintain a context to enable the UE to return in case of handover failure. In the case of handover failure, the UE attempts to resume the RRC connection either in the source or another cell using the RRC reestablishment procedure. This connection resumption succeeds only if the accessed cell is prepared, i.e., it concerns a cell of the source eNB or of another eNB for which handover preparation has been performed.

In LTE, the UE always connects to a single cell; thus the switching of a UE's connection from a source cell to a target cell is a hard handover. In a hard handover process, the eNB which controls the source cell requests the target eNB to prepare for the handover. The target eNB subsequently generates the RRC message in order to instruct the UE to perform the handover, and the message is transparently forwarded by the source eNB to the UE. In another type of handover supported by LTE, the UE spontaneously decides to connect to the target cell, where it requests that the connection be continued. The UE applies this connection reestablishment procedure only after terminating the connection to the source cell. The latter procedure only succeeds if the target cell has been prepared in advance for the handover. In addition, LTE also allows a UE to be redirected to another frequency or RAT upon connection release (see Figure 4.1). Redirection during connection establishment is not supported, since at that time the E-UTRAN may not have all relevant information such as UE capability and the type of subscriber. However, the redirection may be performed while AS security has not been activated. When redirecting the UE to UTRAN or GERAN, the E-UTRAN may provide the system information for one or more cells on the relevant frequency. If the UE selects one of the cells for which the system information is provided, then it does not need to acquire it [3,20,21].

4.7.3 HetNet mobility

In general, the mobility issues are the same in heterogeneous networks and (homogeneous) macro-cell networks, i.e., managing inter-cell handovers in the connected mode and idle mode mobility (Figure 4.18). The presence of a small-cell layer and small-cell sizes creates a new opportunity and gives rise to additional challenges. The first issue is how to manage idle mode UE mobility between a small cell and a macro-cell. Camping strategy is an essential part of

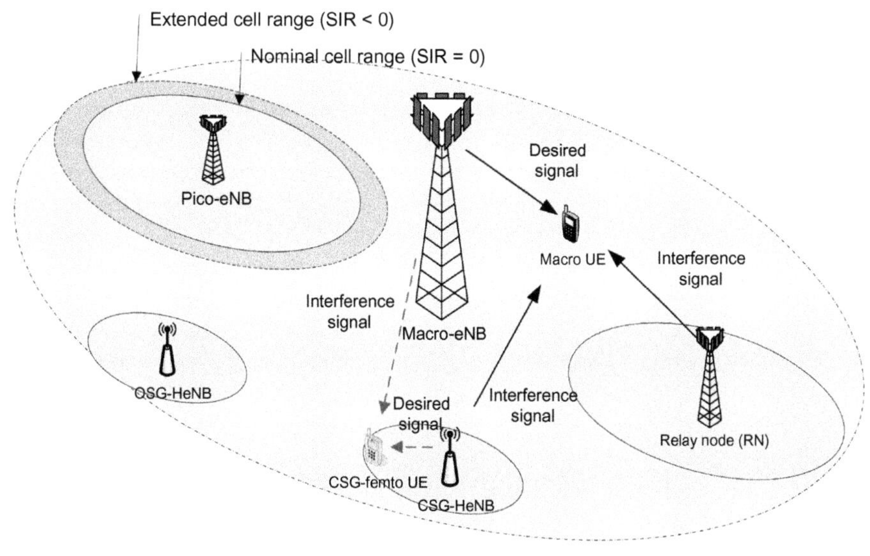

FIGURE 4.18

Illustration of a heterogeneous network and UE mobility.

idle mode cell reselection parameters as it affects the way UEs reselect and camp on macro-cells versus small cells. This fundamentally determines the idle mode UE distribution (cell loading) between macro-cells and small-cells, and affects relative loading between the two layers. The camping strategy also has a potential impact on UE power consumption as well as frequency of handovers. Increased number of handovers due to small-cell sizes creates a potential source of increased dropped call rate, even when handover success rate remains unchanged. A higher failure rate has been observed in the studies for outbound mobility from a small cell for high-speed UEs. This can be mitigated by having the higher-speed UEs confined to the macro-cell layer, but achieving this requires speed-based traffic management which is a challenge for LTE networks. In a limited number of scenarios, depending on camping strategy, operator preference, etc., connected mode UEs in the macro-layer may need to be handed over to a small cell. In such cases, if the small-cell operates on a different RF carrier, efficient small-cell discovery that minimizes impact on UE battery life is also necessary [38]. The number of radio link failures and handover events per UE per second for macro-only and HetNet scenarios is shown in Figure 4.19.

One of the observations which was made during heterogeneous networks mobility studies in 3GPP (results shown in Figures 4.19 and 4.20) suggests that among various possibilities for mobility in a heterogeneous network, the outbound mobility from a small cell has a relatively higher failure rate than other types of heterogeneous network mobility. In particular, the failure rate is higher for fast moving UEs [13]. While decisions regarding the need for enhancements and possible solutions are still under investigation, this is a practical case that a fast moving UE needs to be kept away from a small-cell layer in order to improve the robustness of heterogeneous network mobility.

Mobility management is performed to support UE mobility, which involves informing the network of UEs' present locations at the cell ID (or alternatively HeNB) level, providing the network with UE IDs, and maintaining physical channels. As discussed earlier, LTE mobility management is classified into two states depending on the UE's corresponding RRC connection state, i.e., idle mode and connected mode mobility management. For each of the states, intra-frequency, inter-frequency, or inter-RAT mobility procedures can be performed based on the UE measurements of available frequencies/RATs.

In connected mode mobility management, the mobile network ensures continuity of transmission of physical signals/channels and provides uninterrupted communication services to UEs through handover when the UEs move in the network. Handover is a procedure where the serving cell of a UE in the connected mode is changed. In LTE, the handover decisions are performed by the eNBs without involving the MME or serving gateway, which is used as an anchor node. That is, the eNB delivers the associated configuration through signaling on the control-plane, UEs perform measurements accordingly and complete the handover procedures under the control of the eNB. A UE that is powered on but does not have an RRC connection to the access network is in

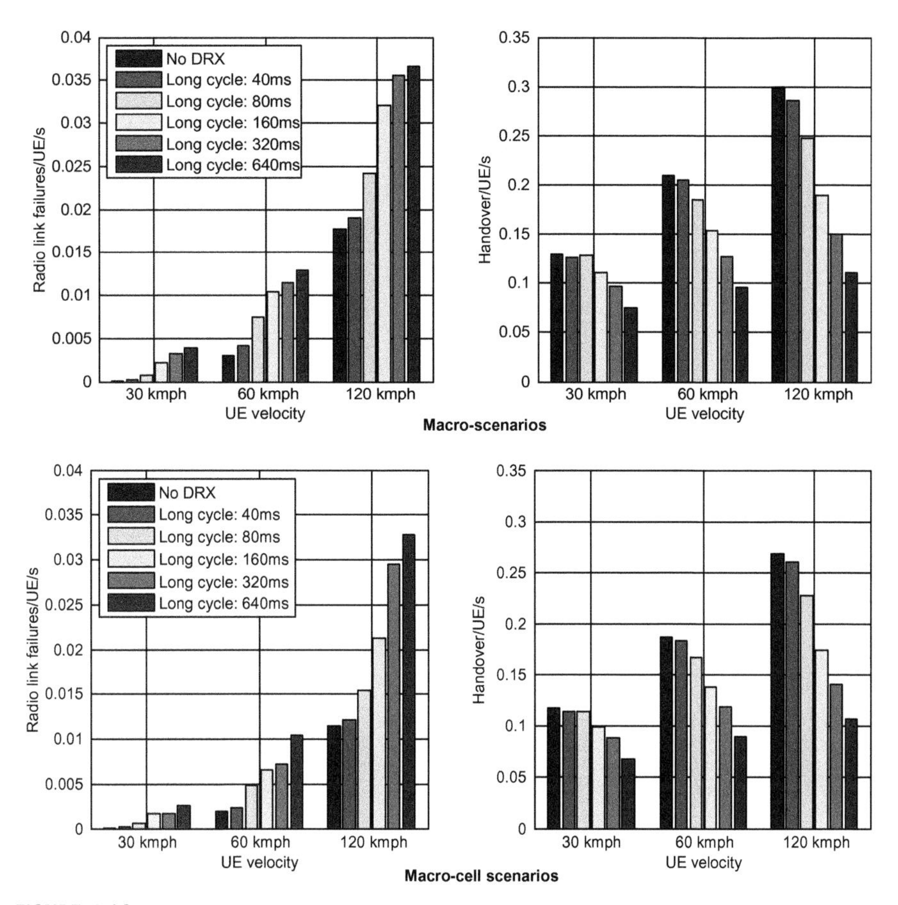

FIGURE 4.19

Comparison of radio link failure and handover events in macro- and HetNet scenarios with various DRX configurations [13].

the idle mode. In this mode, UEs report their locations to the network and the eNB broadcasts system information, which can be used by the UEs to select suitable cells to camp on.

Small cells can be introduced to overlay the existing LTE macro-cell networks either by sharing the same RF carrier or using a dedicated RF carrier. If the existing macro-node RF carrier is shared by small cells, interference management techniques must be used to avoid performance degradation due to inter-cell interference. A notable exception to this rule is when the same-carrier small cell is deployed to overlay an existing macro-cell coverage hole. In this case, the two cells will be sufficiently isolated from the RF perspective. If small cells are deployed on a dedicated LTE carrier, neighbor relationship(s) must be established

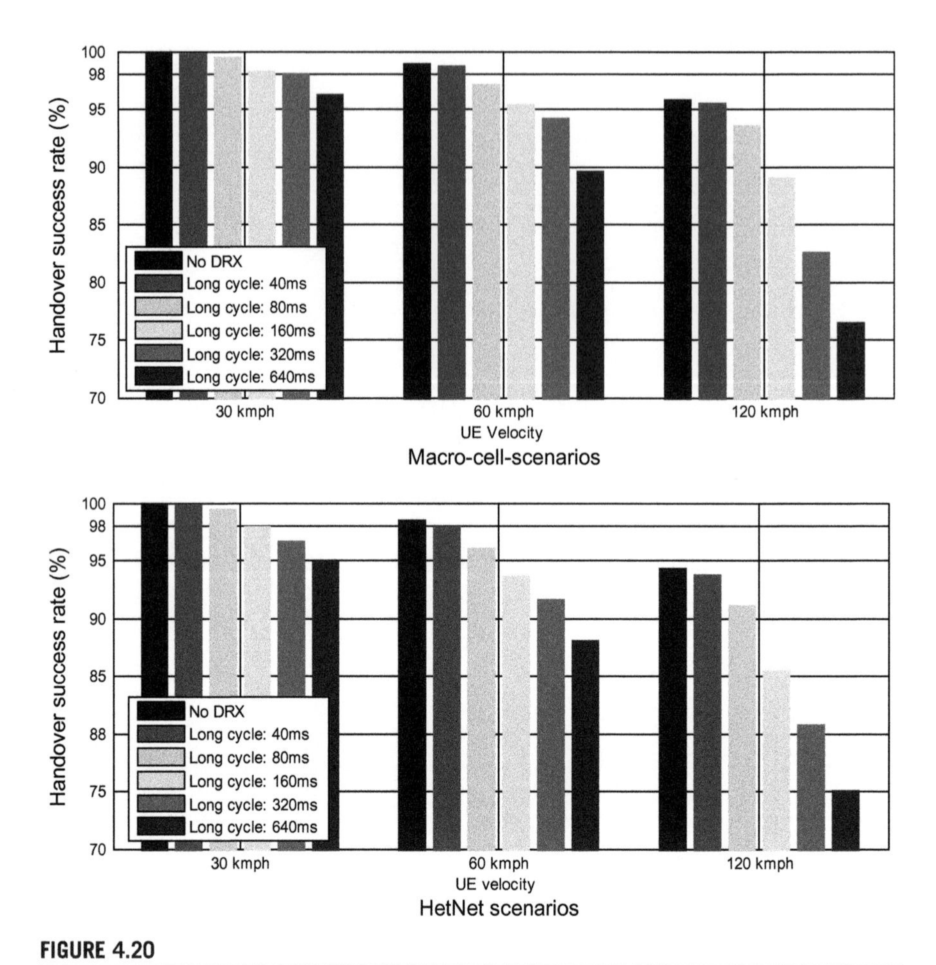

FIGURE 4.20

Comparison of handover success rate in macro-cell and HetNet scenarios with various DRX configurations [13].

with the existing LTE macro-cells (by advertising small-cell frequency in SIB5) to enable UE's in the macro-cell network to search for the small cells.

There are a number of options for establishing neighbor relationship with the macro-cell RF carriers. One option is that all macro-cell carriers list small-cell carrier as neighbor in SIB5. Another option is to allow only a fraction of macro-cell carriers to list small-cell carriers as neighbor in SIB5. For cells on the remaining RF carriers, inter-macro-carrier load-balancing techniques can be used to redistribute the load.

Depending on the option chosen, the camping and load balancing between the small cell and macro-carriers can be impacted. The first option provides immediate offloading to all macro-carriers in a small-cell coverage area, but the risk of small cells becoming overloaded increases. The second option only provides immediate capacity relief to neighbor macro-cells.

During the cell selection procedure, the UE determines the best cell that can be used to establish an RRC connection. The UE will typically select a cell based on the RSRP that is calculated from measurements performed on the cell-specific reference signals broadcast by the cell with a nominal transmit power, i.e., $\text{Cell ID}_{\text{serving-cell}} = \arg\max_k\{\text{RSRP}(k)\}$. Because of the substantially higher power emitted by the macro-cell, the downlink coverage of the small cell is reduced. This issue can further create asymmetry in the uplink coverage that is unaffected by the transmit power differences of the cells. LTE Rel-8 allows addition of a bias to the cell selection with the UE determining the serving cell based on the following criterion: $\text{Cell ID}_{\text{serving-cell}} = \arg\max_k\{\text{RSRP}(k) + \Delta_{\text{bias}}(k)\}$[38]. The cell selection bias can range from 0 to 20 dB and can be communicated to the UE via higher layer signaling or can be cell specific, which in this case is broadcast as part of system information. The effect of the bias is to extend the downlink small-cell coverage as the UE artificially increases the measured RSRP from the small cell by the bias value (see Figure 4.18). At the same time, the downlink signal-to-interference plus noise ratio, when an RRC connection establishment is attempted is degraded. In cases where the small cell is sufficiently far from the macro-cell, the range of the small cell would be adequate without the bias, especially if the transmit power of the small cell is increased. In the latter case, increase of the transmit power is to some extent another method to increase the range of pico-cells.

For UEs in the connected state that are moving out of cell coverage, handovers are required to move their radio link from one cell to another. It is preferable to keep a UE in the same layer (small cell or macro-cell) as it performs handovers, unless there is a necessity to move it from one layer to the other. Same layer handovers use the normal coverage-based intra-frequency handover framework used in a macro-cell network.

In summary, the UE initiates a handover procedure, based on signal and time thresholds defined and transmitted to it by the serving cell. These are the handover hysteresis, time to trigger, and the cell individual offset. The latter is a vector quantity sent by the serving cell that contains the offsets of the serving and neighbor cells that all UEs in this cell must use in determining whether the measured RSRP satisfies the conditions of handover event A3 that results in the UE sending a measurement report that triggers the handover [9]. The individual offset values can be positive or negative. The measured RSRPs are filtered and individual offset values are added according to the following criterion: $\text{Cell ID}_{\text{serving-cell}} = \arg\max_k\{\text{RSRP}(k) + \Delta_{\text{individual-offset}}(k)\}$[38]. The challenge is to optimally configure the cells with these parameters such that the handover success rate and the dropped call rate meet the target values. Load distribution depends, among other factors, on the UE average mobility condition, potentially estimated from MME layer events and usually applied to disable ping-pong effects [38]. The time that a UE stays connected with a cell following a handover is used as the metric to evaluate the ping-pong behavior. The time-of-stay in cell A is the time interval from the moment when the UE successfully sends a handover complete message, i.e., *RRC Connection Reconfiguration Complete* message, to cell A, until the UE successfully sends a handover complete message to cell B. The minimum time of stay connected with a cell models the time needed to allow a UE to establish a reliable connection with the cell, plus the time required for efficient data transmission. If a UE performs a handover from cell B to cell A, and then makes a handover back from cell A to cell B and the time connected to cell A was less than the minimum time-of-stay, it is considered as a ping-pong effect. In general, if the UE's time-of-stay in a cell is less than the minimum time-of-stay, the handover may be considered as unnecessary, resulting in unnecessary use of network resources [13].

4.7.4 S1-based handover procedure

The E-UTRAN supports mobility within the LTE system, as well as mobility to/from other systems using both 3GPP and non-3GPP radio access technologies [11]. The mobility procedures also involve the network interfaces. This section discusses the procedures over the S1 interface to support mobility. In this case, the UE will assist in the handover decision by measuring the neighbor cells and reporting the measurements to the network, which in turn decides upon the handover timing and the target cell/node. The parameters to measure and the thresholds for reporting are determined by the network.

The E-UTRA supports two types of handover for UEs in the connected mode: S1-based and X2-based handover (Figure 4.21). The X2-based handover

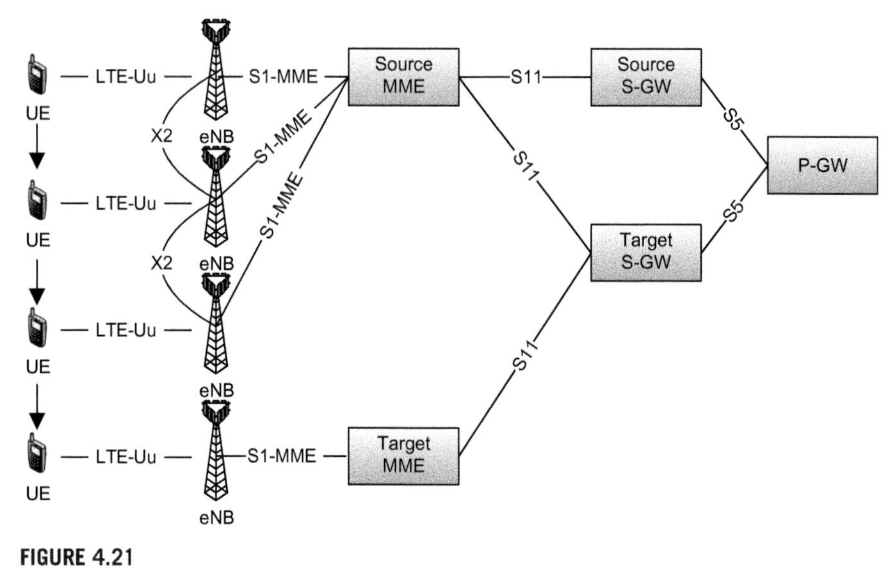

FIGURE 4.21

Illustration of an S1-based mobility scenario [23].

procedure is typically used for inter-eNB handover. However, when there is no X2 interface between the two eNBs, or if the source eNB has been configured to initiate handover to a particular target eNB through the S1 interface, then an S1-based handover will be triggered. The S1-based handover procedure, shown in Figure 4.22, consists of a preparation phase, where the core network resources are prepared at the target node, followed by an execution phase, and a completion phase. The *Status Transfer* message sent from the source eNB carries some PDCP status information that is needed at the target eNB in scenarios where PDCP status preservation applies for the S1-based handover [8]. This is consistent with the information that is sent within the X2 status transfer message used for the X2-based handover. As a result, handling of the handover by the target eNB as seen from the UE is exactly the same, regardless of the type of handover (S1 or X2). The status transfer procedure is assumed to be triggered concurrently with the start of data forwarding after the source eNB has received the *Handover Command* message from the source MME.

In an inter-MME handover, two MMEs (the source and target MME) are involved, as shown in Figure 4.21. The source MME controls the source eNB and the target MME manages the target eNB. In this case, both MMEs are connected to the same S-GW. This handover is triggered when the UE moves from one MME area to another MME area. The data forwarding can be either direct or indirect, depending on the availability of a direct path for the user-plane data between the source eNB and the target eNB. The *Handover Notify* message, which is sent later by the target eNB upon confirmation of the arrival of the UE at the target node, is forwarded by the MME to trigger path update in the S-GW to the target eNB. In contrast to the X2-based handover, the message is not acknowledged and the resources at the source eNB side are released later upon reception of a *Release Resource* message triggered by the source MME [23,28] (Figures 4.23 and 4.24).

Due to lack of direct connection between the source and target eNBs, the MME has to be involved in the process. The S1-based handover procedure is used when the X2-based handover cannot be used. The source eNB initiates a handover by sending a *Handover Required* message over the S1-MME reference point. This procedure may relocate the MME and/or the S-GW. The source MME selects the target MME. The MME should not be relocated[8] during inter-eNB handover unless the UE leaves the MME pool area where the UE is served. The MME (target MME for MME relocation) determines if the S-GW needs to be relocated. If the S-GW needs to be relocated, the MME selects the target S-GW. The source eNB decides which data packets corresponding to the downlink/uplink EPS bearers are going to be forwarded to the target eNB. Packet forwarding can take place either directly from the source eNB to the target eNB, or indirectly from the source eNB to the target eNB via the source and target S-GWs, or if the S-GW is not relocated, only through a single S-GW. The availability of a direct

[8]Relocation is defined as switching of the communication equipment such as area switches during communication.

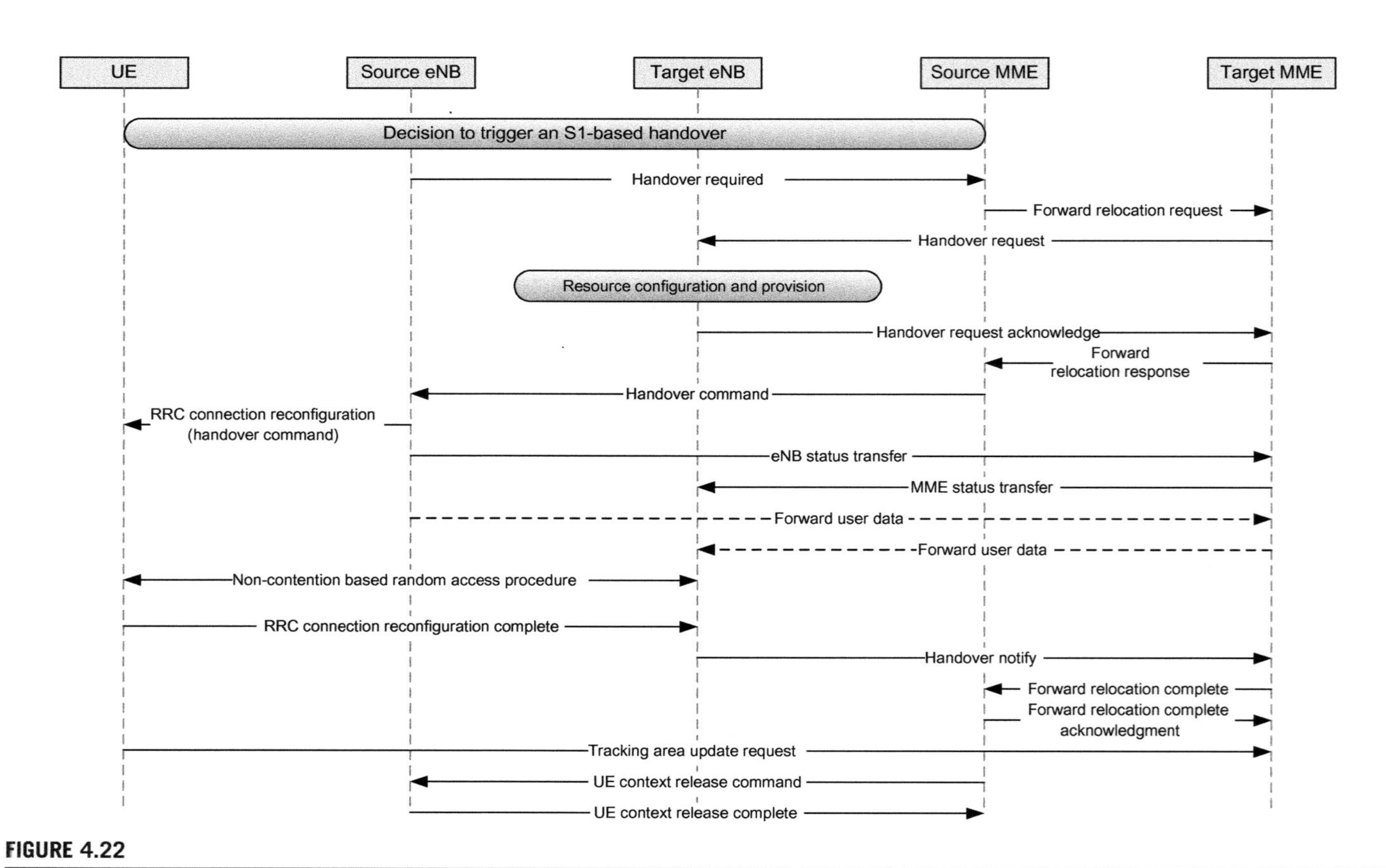

FIGURE 4.22

S1-based handover procedure [26,28].

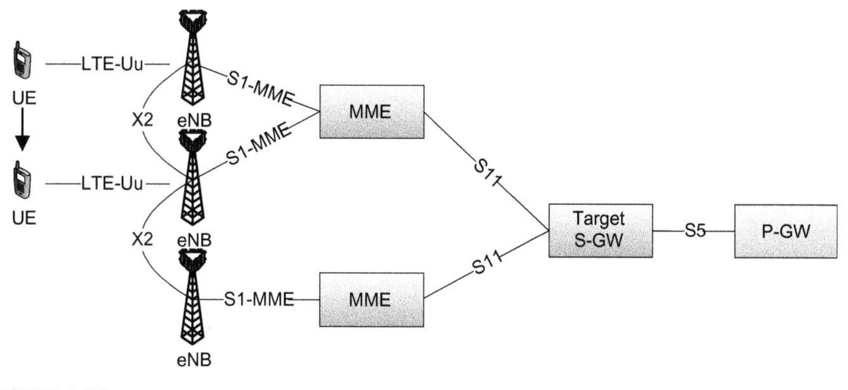

FIGURE 4.23

Illustration of X2-based handover with no S-GW change [23].

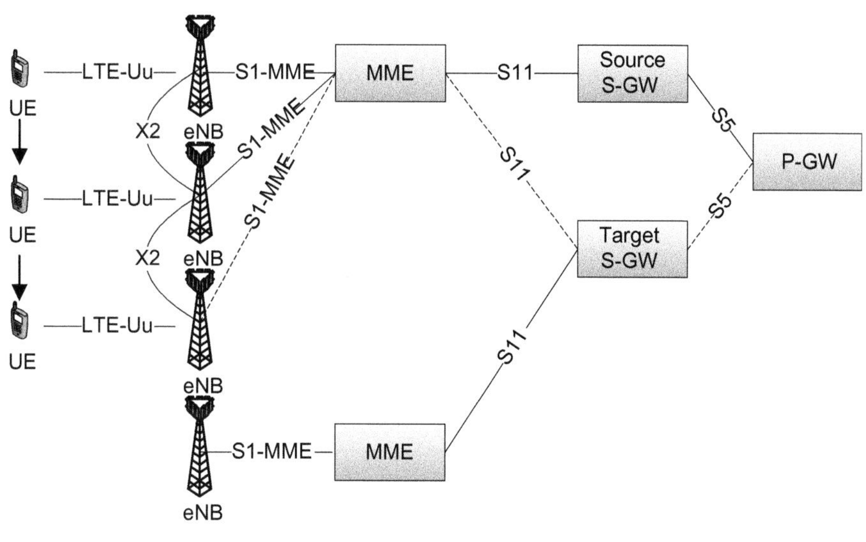

FIGURE 4.24

Illustration of an X2-based handover with change of S-GW [23].

forwarding path is determined in the source eNB and is indicated to the source MME. If X2 connectivity is available between the source and target eNBs, a direct forwarding path is typically available. If a direct forwarding path is not available, indirect forwarding may be used. The source MME uses the information from the source eNB to determine whether to apply indirect forwarding. The source MME informs the target MME whether indirect forwarding should be utilized.

4.7.5 X2-based handover without S-GW change

This procedure is used to hand over a UE from a source eNB to a target eNB using the X2 interface when the MME and S-GW are unchanged, which would be possible only if there is direct connectivity between the source and target eNBs with the X2 interface. The X2 handover procedure is performed without MME involvement, which implies that preparation messages are directly exchanged between the source and the target eNBs. The release of the resources at the source side during the handover completion phase is triggered by the target eNB. The call flow for this type of handover is shown in Figure 4.25. This procedure can be summarized as follows [26,28]:

1. Based on the measurement reports from the UE, the source eNB determines that the UE should be handed over to a target eNB using the handover procedures. Note that each network operator's criteria for handover might be different from other operators. The source eNB may send a *Resource Status Request* message to determine the loading condition at the target eNB. Based on the received *Resource Status Response* message, the source eNB can make a decision to proceed with the handover procedure using the X2 interface.
2. The source eNB sends a *Handover Request* message to the target eNB, transferring the necessary information such as UE context, which includes the security context and radio bearer context to prepare the handover at the target side. The target eNB checks the resource availability and reserves the resources and sends back the *Handover Request Acknowledgement* message including a transparent container that is sent to the UE as an RRC message to perform the handover. The container may include a new C-RNTI, target eNB

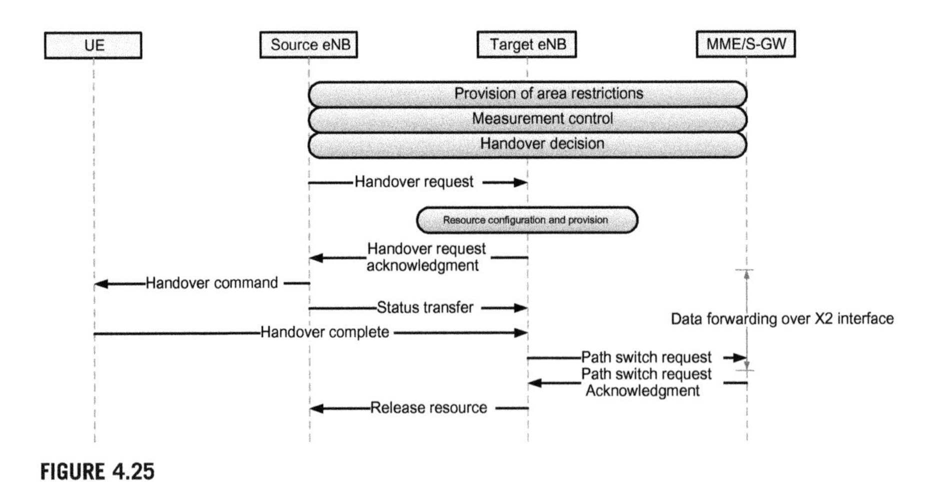

FIGURE 4.25

X2-based handover procedure [28].

security algorithm identifiers for the selected security algorithms, a dedicated RACH preamble, and access parameters.

3. The source eNB generates an *RRC Connection Reconfiguration* message to perform the handover including the mobility control information. The source eNB performs the necessary integrity protection and ciphering of the message and sends it to the UE. The source eNB further sends the *Status Transfer* message to the target eNB to convey the PDCP and HFN status of the E-RABs. For those bearers requiring in-sequence delivery of packets, the *Status Transfer* message provides the sequence number and the HFN that the target eNB should assign to the first packet with no sequence number. The first packet can either be the one received over the target S1 path or the packet received over X2, if data forwarding over X2 is used.
4. The source eNB starts forwarding the downlink data packets to the target eNB for all data bearers (which are being established in the target eNB during the *Handover Request* message processing). In the meantime, the UE tries to access the target eNB using the non-contention-based random access procedure. If it succeeds in accessing the target cell, it sends an *RRC Connection Reconfiguration Complete* to the target eNB.
5. The target eNB sends a *Path Switch Request* message to the MME to inform that the UE has changed cells, including the TAI + E-UTRAN cell global identifier of the target. The MME determines that the S-GW can continue to serve the UE [31].
6. The MME sends a *Modify Bearer Request* message to the S-GW, including the eNB address and TEIDs for downlink user-plane for the accepted EPS bearers. If the P-GW requested the UE's location information, the MME also includes the *User Location Information* in this message.
7. The S-GW sends the downlink packets to the target eNB using the newly received addresses and TEIDs (path switched in the downlink to the target eNB), and the *Modify Bearer Response* to the MME. The S-GW sends one or more end marker packets on the old path to the source eNB and then can release any user-plane/transport network layer resources toward the source eNB.
8. The MME responds to the target eNB with a *Path Switch Request Acknowledge* message to notify the completion of the handover. The target eNB now requests the source eNB to release the resources using the *UE Context Release* message, completing the handover procedure. The target eNB requests the source eNB to release the resources using the *UE Context Release* message.

These procedures are used to hand over a UE from a source eNB to a target eNB using the X2 interface. In these procedures, the MME stays unchanged. Two procedures are defined depending on whether the S-GW is changed. In addition to the X2 interface between the source and target eNB, the procedures rely on the presence of an S1-MME interface between the MME and the source eNB, as well as between the MME and the target eNB.

If during an X2-based handover the serving PLMN changes, the source eNB informs the target eNB of the PLMN selected to be the new PLMN. When the UE receives the handover command, it will remove any EPS bearers for which it did not receive the corresponding EPS radio bearers in the target cell. As part of handover execution, downlink/uplink packets are forwarded from the source eNB to the target eNB. When the UE has arrived in the target eNB area, downlink data forwarded from the source eNB can be sent to it. Uplink data from the UE can be delivered via the (source) S-GW to the P-GW, or optionally forwarded from the source eNB to the target eNB. Only the handover completion phase is affected by a potential change of the S-GW, the handover preparation and execution phases are identical.

4.7.6 X2-based handover with S-GW change

Handover through the X2 interface is triggered by default for intra-LTE mobility unless there is no X2 interface established or the source eNB is configured to use S1-based handover. The X2-based handover procedure is illustrated in Figure 4.25. Similar to an S1-based handover, the procedure is composed of a preparation phase, an execution phase, and a completion phase. The key features of an X2-based handover include direct transfer between the source and target eNBs (facilitating preparation phase); data forwarding per bearer in order to minimize packet loss, where the MME is only informed at the end of the handover procedure when the handover is successfully completed in order to trigger the path switch; and direct release of resources at the source side triggered by the target eNB. For those bearers requiring in-sequence delivery of packets, the *Status Transfer* message provides the sequence number and the HFN that the target eNB should assign to the first packet with no sequence number that it must deliver. The first packet can either be the one received over the target S1 path or received over X2, if data forwarding over X2 is used. When the source eNB sends the *Status Transfer* message, it pauses its packet transmit/receive status update, i.e., it stops assigning PDCP sequence numbers to downlink packets and stops delivering uplink packets to the network. The mobility over X2 can be categorized according to its resilience to packet loss. A handover is seamless, if it minimizes the interruption time during the inter-node transfer of a UE, or the handover can be lossless, if no packets are lost. These handover types rely on data forwarding of user-plane downlink packets. The source eNB may decide to operate in one of these modes on a per EPS-bearer basis, based on the QoS received over S1 for this bearer and the service characteristics.

If the source eNB selects the seamless handover mode for one bearer, it informs the target eNB in the *Handover Request* message to establish a

GTP tunnel[9] to enable downlink data forwarding. The target eNB may send an acceptance in the *Handover Request Acknowledgement* message at the tunnel endpoint where the forwarded data is expected to be received. The tunnel endpoint may be different from the one that has been set up as the termination point of the new bearer which is established over the target S1 [28].

Upon receipt of the *Handover Request Acknowledgement* message, the source eNB can start forwarding the arriving data over the source S1 path toward the indicated tunnel endpoint, concurrent with the transmission of the handover trigger to the UE over the radio interface. The forwarded data is going to be available at the target eNB for early delivery to the UE. When forwarding is in use and in-sequence delivery of packets is required, the target eNB is assumed to first deliver the packets forwarded over X2 before delivering the packets received over the target S1 path, once the S1 path switch has been completed. The end of the forwarding is signaled over X2 to the target eNB prior to switching the S1 path [28].

If the source eNB selects the lossless handover mode for one bearer, it will additionally forward (over X2) those user-plane downlink packets that are processed through the PDCP sublayer, but they are still locally buffered because they have not been delivered and acknowledged by the UE. Those packets are forwarded together with their assigned PDCP sequence number included in a GTP extension header field. They are sent over X2 prior to the arriving packets from the previous S1 path. The same mechanisms described above for seamless handover are used for the GTP tunnel establishment. The end of forwarding is also managed in the same way, since in-sequence packet delivery applies to lossless handovers, as well. The target eNB must ensure that all packets, including the ones received with a sequence number over X2, are delivered in sequence to the target node.

When lossless handover is used, the target eNB may not deliver some of the forwarded downlink packets received via X2 over the radio interface, if it is informed by the UE that those packets have already been received via the source eNB. This new feature is known as downlink selective retransmission in E-UTRA literature [20,28]. In the uplink, the target eNB may instruct the UE not to retransmit packets already received from the source eNB. In order to use uplink selective retransmission, the source eNB forwards the user-plane uplink packets that are received out of sequence to the target eNB via a new GTP tunnel. At the request of the target eNB, the source eNB establishes a new forwarding tunnel. In this

[9]GTP-U protocol is used over S1-U, X2, S4, S5, and S8 interfaces of the evolved packet system. GTP-U tunnels are used to carry encapsulated PDUs and signaling messages between a given pair of GTP-U tunnel endpoints. The tunnel endpoint identifier (TEID) which is present in the GTP header indicates which tunnel a particular PDU belongs to. The transport bearer is identified by the GTP-U TEID and the IP address (source TEID, destination TEID, source IP address, destination IP address) [2,31].

case, the target eNB includes a GTP tunnel endpoint where it expects the forwarded uplink packets to be received in the *Handover Request Acknowledgement* message. The source eNB then indicates in the *Status Transfer* message for this bearer, the list of sequence numbers corresponding to the packets to be forwarded. Based on this information, the target eNB informs the UE of the packets that are not going to be retransmitted, making the overall retransmission scheme more efficient [20,28].

4.7.7 Inter-RAT mobility procedures

In principle, the inter-RAT handover procedures when an LTE system is the source or target are the same as the intra-LTE handover process per se. One of the differences is that upon handover to an LTE, the entire security association configuration needs to be signaled, whereas within an LTE it is possible to signal incremental information, where only the changes to the configuration are signaled. If ciphering was not activated in the previous RAT, the E-UTRAN activates ciphering as part of the handover procedure. The E-UTRAN also establishes SRB1, SRB2, and one or more DRBs, i.e., at least one DRB is associated with the default EPS bearer. Figure 4.26 illustrates the UE states in E-UTRA, UMTS, and GSM, as well as the mobility support between E-UTRAN, UTRAN, and

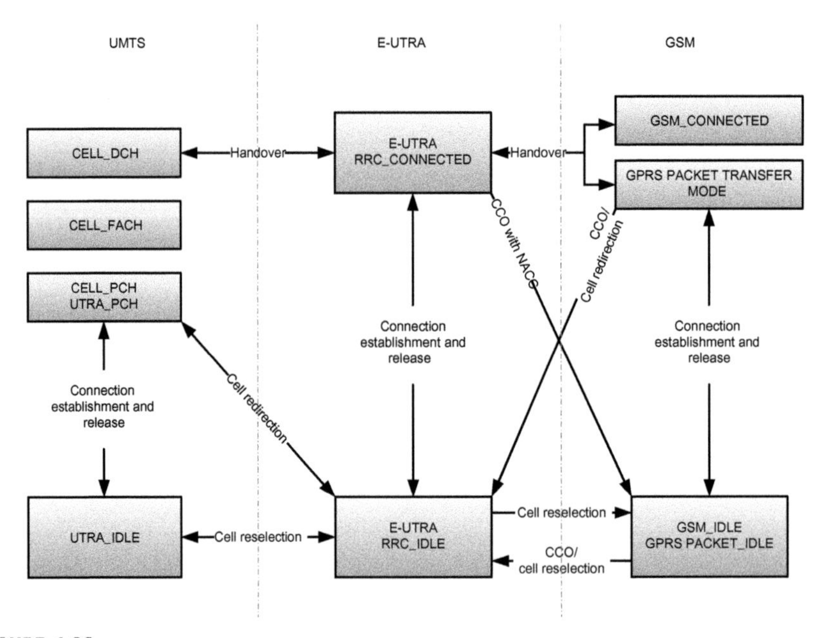

FIGURE 4.26

E-UTRA states and inter-RAT mobility procedures [3].

GERAN [10]. The cell change order (CCO) message is used to command the UE to change to another RAT [3].

In the context of multi-RAT operation, the terms cell selection, cell reselection, cell redirection, handover, and circuit-switched fallback (CSFB) are frequently used. In order to avoid confusion when using these terms, and to better understand the use cases, we use the illustration shown in Figure 4.27 to underline the significance and implication of each term. While this figure may oversimplify the process, it can give the reader some intuitive understanding about multi-cell, multi-RAT operation of E-UTRA and interaction with other radio access technologies. As shown in Figure 4.27, when a UE is powered on, it performs a cell search to find and select a suitable cell to camp on. The cell reselection, on the other hand, is an idle mode procedure and a mechanism to change cell after a UE is camped on a cell. Handover is a connected mode procedure and is performed between connected states of the same or different radio access technologies, e.g., E-UTRA RRC_CONNECTED state to UMTS CELL_DCH state. Redirection is a process which changes the UE state from connected to idle and in many cases for CSFB purposes, e.g., the UE is redirected from the E-UTRA RRC_CONNECTED state to the UMTS/GSM idle mode. In the case of CSFB, a handover is not required and possibly not supported by the network and/or the UE. In such a case, redirection is the default mechanism to utilize. Furthermore, in some cases, the redirection can be faster than handover due to reasons such as inter-RAT

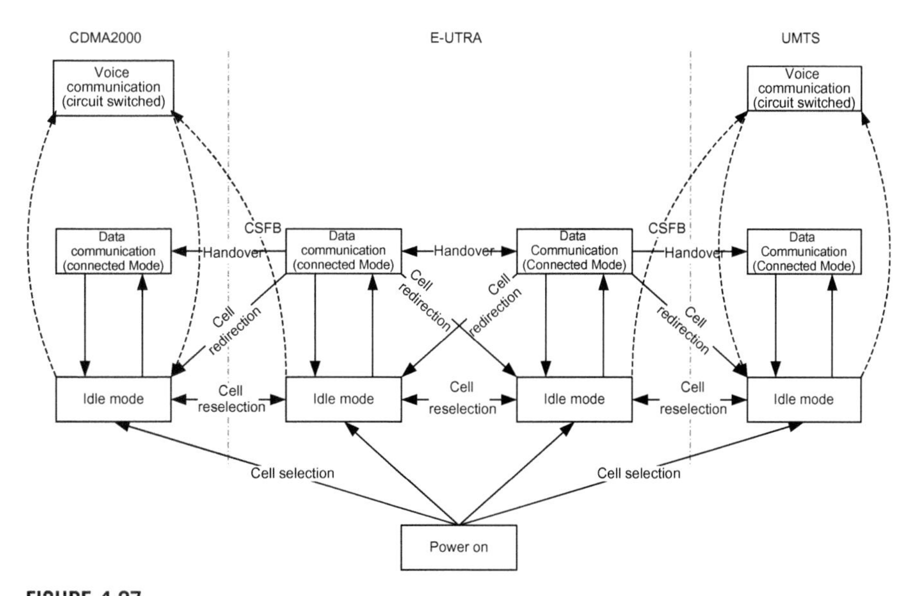

FIGURE 4.27

Illustration of cell selection, cell reselection, cell redirection, circuit-switched fallback, and handover in E-UTRA [32].

measurement delay/cell detection, layer-3 filtering,[10] time-to-trigger, handover preparation in the network, especially in cases where mobile originated or terminated the call, which requires the UE to transition to the RRC_CONNECTED state and to perform CSFB is triggered from the E-UTRA idle mode.

The CSFB functionality enables the support of voice service without deployment of IMS. This is a possible scenario for many operators when LTE is initially deployed because those operators may not support IMS in their networks. The CSFB functionality and IMS can be deployed simultaneously, meaning that an operator can gradually roll out an IMS system while still supporting a fallback mechanism whenever necessary. The UE informs the network that it wishes to perform a "Combined Attached" during initial registration with the MME. In practice, this means that the device requests the network to register its presence in UMTS/GSM circuit-switched networks, as well. When a mobile-terminated call for a subscriber (with multi-RAT support) arrives at the mobile switching center, the incoming call is signaled to the MME. The UE is paged in the LTE system, if it is in the idle mode, or is notified of the call, if it is in the connected mode. It responds by requesting CSFB for the call to proceed. For mobile-originated calls, the UE establishes an RRC connection, if in the idle mode and then notifies the MME that CSFB support for the call is required.

The mobility between E-UTRAN, CDMA2000 1xRTT [15,16], and CDMA2000 high rate packet data (HRPD) or CDMA 1xEV-DO [14] radio access technologies is shown in Figure 4.28. In general, the procedure for handover from LTE to another RAT supports both handover and CCO with network assisted cell change (NACC). The CCO or NACC procedure is applicable only for handover to GERAN. Handover from LTE is performed only after security has been activated. In the case of CSFB to CDMA2000, the procedure includes support for parallel handover (i.e., to 1xRTT and HRPD), for handover to 1xRTT combined with redirection to HRPD, and for redirection to HRPD (see Figure 4.28). The UE may assist by sending a *Measurement Report* message. In the case of handover (as opposed to CCO), the source eNB requests the target node to prepare for the handover in advance. As part of the handover preparation request, the source eNB provides information about the applicable inter-RAT UE capabilities, as well as information about the currently established bearers to the target node. In response, the target node generates the handover command and returns this to the source eNB. The source eNB sends mobility from the E-UTRA command to the

[10]The UE is required to apply a layer 3 filtering to each measurement quantity that it performs, which does not include quantities configured exclusively for UE receive–transmit time difference measurements. The UE is required to filter the measured result, before using for evaluation of reporting criteria or for measurement reporting, by the following formula $F_n = (1 - a)F_{n-1} + aM_n$, where M_n is the latest received measurement result from the physical layer, F_n is the updated filtered measurement result, that is used for evaluation of reporting criteria or for measurement reporting, and F_{n-1} is the previous filtered measurement result. The value of F_0 is set to M_1 when the first measurement result from the physical layer is received. The parameter a is defined as $a = 1/2^{k/4}$, in which k is the *filter coefficent* for the corresponding measurement quantity [9].

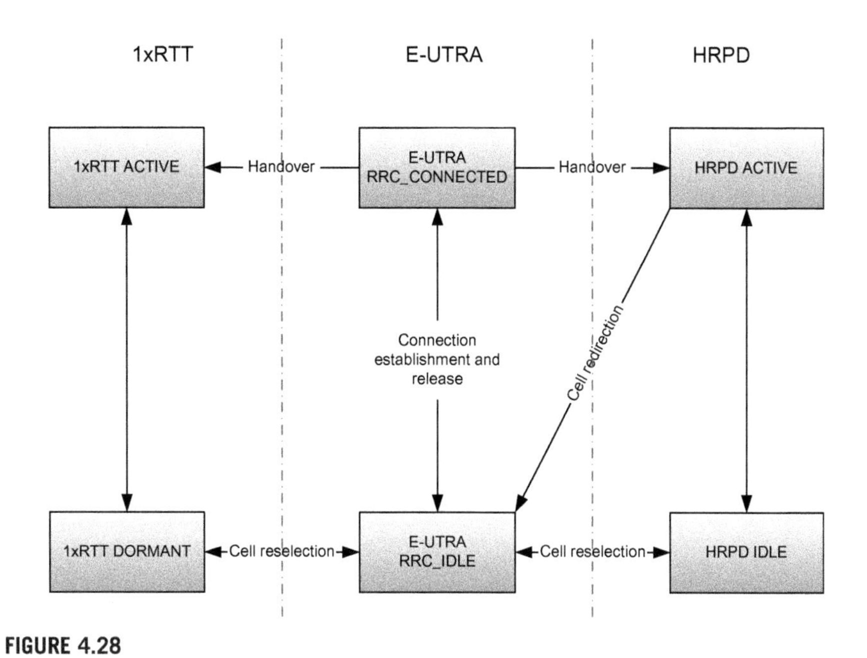

FIGURE 4.28

Mobility procedures between E-UTRA and CDMA2000 1xRTT/HRPD [3].

UE, which includes the inter-RAT message received from the target cell and some inter-RAT parameters in the case of CCO. Upon receiving the message, the UE starts the timer T304 and connects to the target node, either using the received radio configuration (handover) or by initiating connection establishment (CCO) according to the applicable specifications of the target RAT. The upper layers in the UE are informed by the AS of the bearers established by the target RAT base station. As a result, the UE is informed whether some of the previously established bearers were not admitted by the target RAT base station [3,20].

Having discussed inter-RAT operation, let us take LTE-to-UMTS inter-RAT handover as an example and describe the procedure involved in this operation. In this scenario, the source eNB connects to the source MME and source S-GW while the target UMTS radio network controller (RNC) connects to the target SGSN and target S-GW; however, both source and target S-GWs connect to the same P-GW. As shown in Figure 4.29, the source eNB configures the UMTS measurements in the UE using the *RRC Connection Reconfiguration* message. This message sets the thresholds (serving cell and inter-RAT neighbor cell) for activation, and further sets the events and provides UMTS center frequency and the scrambling code to be used. The UE measures the neighboring UMTS cells and sends RRC measurement reports back to the source eNB. The source eNB then makes a decision to perform a handover to the UMTS. In order to prepare the target UMTS cell, the source eNB sends a *Handover Required* message to the source MME. This message contains the target UMTS cell identifier. The source MME then contacts the target SGSN using the UMTS cell-identifier and allocates the resources using a forward relocation request. Once the source MME receives confirmation (*Relocation Complete* message from target SGSN), the source MME sends a *Handover Command* message to the source eNB. The source eNB then sends an RRC *Mobility from E-UTRAN Command* to the UE, indicating the move to UTRAN. The UE then performs UTRAN access procedures to synchronize and attach to the target RNC and upon successful completion of the process, the UE sends a *Handover to UTRAN Complete* message to the target RNC [20,28,30]. Note that when the source eNB receives the handover command from the source MME, it begins forwarding data packets received from the S-GW. Following this process, downlink packets for the terminal that arrive at the S-GW are forwarded to the terminal through S-GW→source eNB→S-GW→target SGSN→target RNC (Figure 4.29).

Following the above procedure, the terminal switches to 3G, and when the radio link configuration is completed, notification that it has connected to the 3G radio access system is sent over each of the links to the source MME. The source MME sends a forward relocation complete acknowledgment to the target SGSN. A given period of time after receiving this message, the target SGSN releases the resources related to data forwarding. The target SGSN then sends a modify bearer request to the S-GW to switch the path between the S-GW and eNB to that between the S-GW and the target SGSN. This message contains information elements required to configure the path from the S-GW to the target SGSN, including those issued by the target SGSN. When the S-GW receives this signal, it configures a communications path from the S-GW to the target SGSN. In this way, the new communication path becomes S-GW→target SGSN→target RNC→terminal, and data transmission to the target 3G radio access system begins. Note that after this point, data forwarding is no longer needed, thus the S-GW sends a packet to the eNB with an "end marker" attached, and when the eNB receives this packet, it releases its resources related to data forwarding. The S-GW sends a modify bearer response to the target SGSN, indicating that the handover procedure has completed. The MME also releases eNB resources that are no longer needed [29].

In the above inter-RAT handover procedure, we described the handover procedure between 3GPP radio access systems in which the S-GW does not change; however, handovers with S-GW relocation are also possible. In those cases, the P-GW provides the anchor function[11] for path switching similar to switching to non-3GPP access systems.

In a UMTS-to-LTE inter-RAT handover, the source RNC connects to the source SGSN and source S-GW, the target eNB connects to the target MME and target S-GW, and both source and target S-GWs connect to the same P-GW.

[11]Anchor function is a function which switches the communication path according to the area where the terminal is located and forwards packets for the terminal to that area.

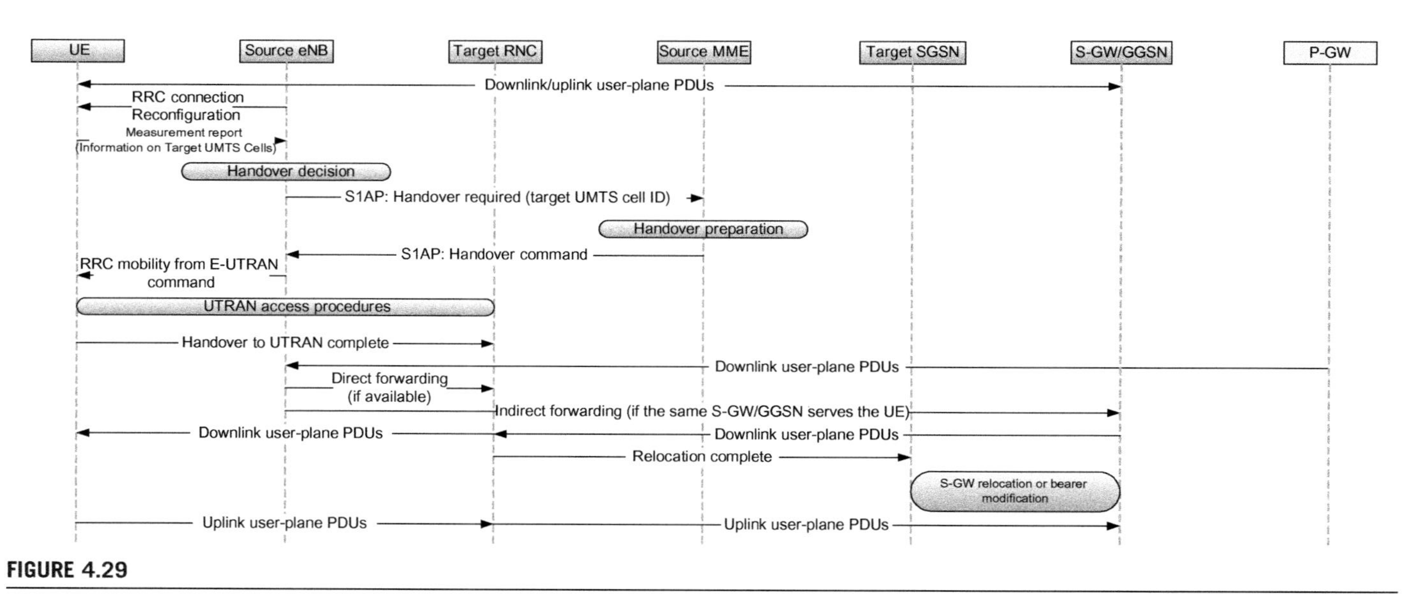

FIGURE 4.29

E-UTRA to UMTS handover procedure [29].

4.8 Overview of IP packet processing procedures

The IP packets are mapped to one or more radio bearers that are provided by E-UTRAN according to the QoS requirements of the corresponding services. An overview of IP packet processing stages in the E-UTRA downlink is illustrated in Figure 4.30. It must be noted that depending on the content

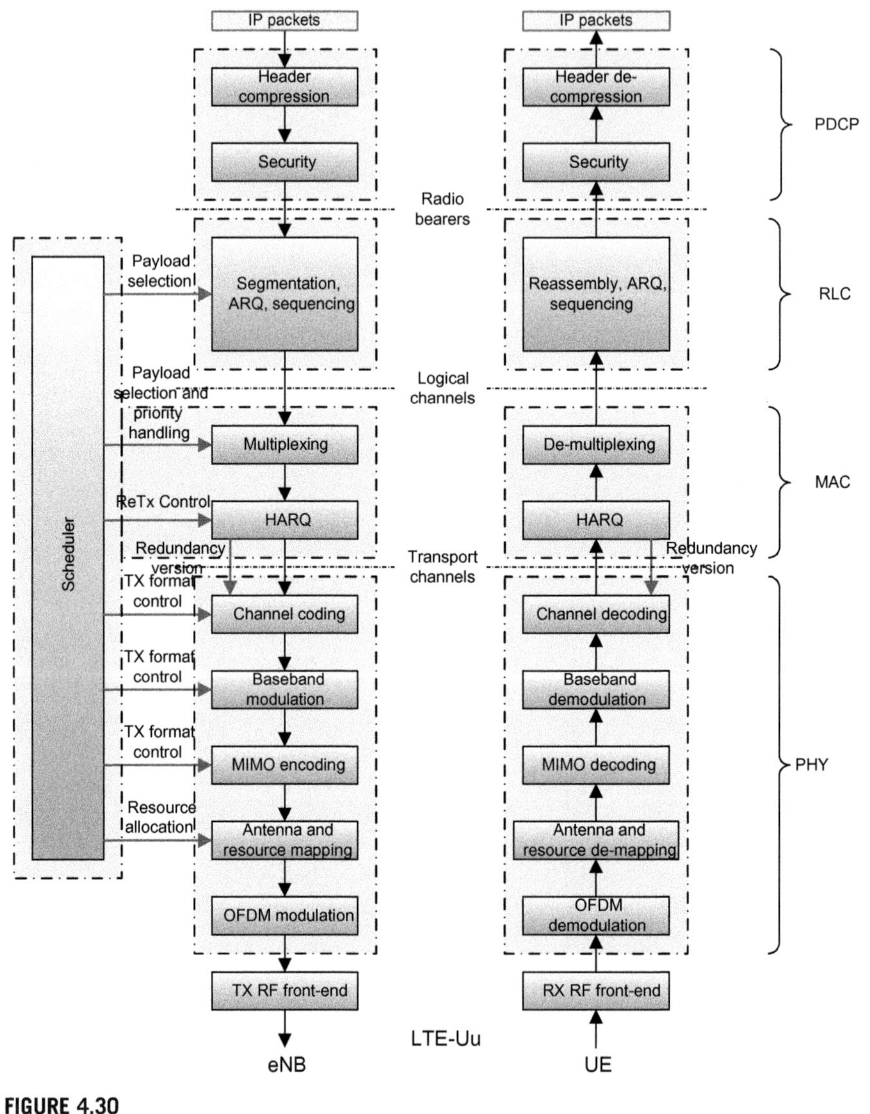

FIGURE 4.30

Overall downlink IP packet processing procedures [22].

of the packets, some of the functions shown in this figure may not be applicable. For example, MAC scheduling and HARQ soft combining functions are not relevant to broadcast of the system information or other control-plane packets. The uplink data packet processing is similar to the downlink with the exception of some differences related to transmission format selection (modulation and coding scheme, transmission modes, multi-antenna processing, etc.) due to dissimilarities in downlink/uplink multiple access schemes and other physical-layer functions.

As shown in Figure 4.30, the IP packets arriving at the PDCP sublayer are processed through the header-compression module in order to make efficient use of radio resources by reducing the number of redundant bits transmitted over the radio interface. The header-compression mechanism is based on the robust header compression (ROHC) scheme [18]. The PDCP sublayer is also responsible for ciphering and, for the control-plane packets, integrity protection of the transmitted data, as well as in-sequence delivery and duplicate removal for handover. A PDCP header is added to PDCP SDUs, carrying information required for deciphering of data in the terminal. At the receiver side, the PDCP performs the corresponding deciphering and decompression functions. There is one PDCP entity per radio bearer configured for a terminal [8].

The PDCP PDUs then go through the RLC sublayer which is responsible for segmentation or concatenation, retransmission, duplicate detection, and in-sequence delivery of the packets to the higher layers. The RLC provides services to the PDCP in the form of radio bearers. There is one RLC entity per radio bearer configured for a terminal. The addition of an RLC header facilitates in-sequence delivery (per logical channel) of the PDCP SDUs in the terminal and identification of the RLC PDUs in the case of retransmissions. The RLC PDUs are forwarded to the MAC sublayer, which multiplexes a number of RLC PDUs and attaches a MAC header to form a transport block. The transport-block size depends on the instantaneous rates selected by the link adaptation mechanism, which affects both MAC and RLC processing [7].

The MAC sublayer performs multiplexing of logical channels, HARQ retransmissions, and uplink/downlink scheduling services. Note that the scheduling function is only present in the eNB for both uplink and downlink packet scheduling services. The HARQ protocol is present in both transmitting and receiving ends of the MAC sublayer. The MAC sublayer provides services to the RLC sublayer in the form of logical channels [6].

The physical layer performs encoding/decoding, modulation/demodulation, scrambling/de-scrambling, interleaving/de-interleaving, multi-antenna processing, resource mapping/de-mapping, and OFDM/SC-FDMA modulation/de-modulation functions. The physical layer provides services to the MAC sublayer in the form of transport channels. The physical layer attaches a CRC to the transport block for error detection purposes, performs coding and modulation, and transmits the resulting signal using multiple transmit antennas [22].

References

3GPP Specifications[12]

[1] 3GPP TS 23.003, Numbering, Addressing and Identification (Release 11), December 2012.
[2] 3GPP TS 29.274, Evolved General Packet Radio Service (GPRS) Tunneling Control Protocol for Control Plane (GTPv2-C) Stage 3 (Release 11), December 2012.
[3] 3GPP TS 36.300, Evolved Universal Terrestrial Radio Access (E-UTRA) and Evolved Universal Terrestrial Radio Access Network (E-UTRAN); Overall description; Stage 2 (Release 11), December 2012.
[4] 3GPP TS 24.301, Non-Access-Stratum (NAS) protocol for Evolved Packet System (EPS); Stage 3 (Release 11), December 2012.
[5] 3GPP TS 36.304, Evolved Universal Terrestrial Radio Access (E-UTRA); UE Procedures in Idle Mode (Release 11), December 2012.
[6] 3GPP TS 36.321, Evolved Universal Terrestrial Radio Access (E-UTRA); Medium Access Control (MAC) protocol specification (Release 11), December 2012.
[7] 3GPP TS 36.322, Evolved Universal Terrestrial Radio Access (E-UTRA); Radio Link Control (RLC) protocol specification (Release 11), December 2012.
[8] 3GPP TS 36.323, Evolved Universal Terrestrial Radio Access (E-UTRA); Packet Data Convergence Protocol (PDCP) specification (Release 11), December 2012.
[9] 3GPP TS 36.331, Evolved Universal Terrestrial Radio Access (E-UTRA); Radio Resource Control (RRC); Protocol specification (Release 11), December 2012.
[10] 3GPP TS 23.401, General Packet Radio Service (GPRS) enhancements for Evolved Universal Terrestrial Radio Access Network (E-UTRAN) access (Release 11), March 2012.
[11] 3GPP TS 23.402, Architecture Enhancements for non-3GPP Accesses (Release 11), December 2012.
[12] 3GPP TS 36.413, Evolved Universal Terrestrial Radio Access Network (E-UTRAN); S1 Application Protocol (S1AP) (Release 11), December 2012.
[13] 3GPP TR 36.839, Evolved Universal Terrestrial Radio Access (E-UTRA); Mobility Enhancements in Heterogeneous Networks (Release 11), December 2012.

3GPP2 Specifications[13]

[14] 3GPP2 C.S0024-A v3.0, cdma2000 High Rate Packet Data Air Interface Specification, September 2006.
[15] 3GPP2 C.S0005-A v6.0, Upper Layer (Layer 3) Signaling Standard for cdma2000 Spread Spectrum Systems, February 2002.
[16] 3GPP2 C.S0004-A v6.0, Signaling Link Access Control (LAC) Standard for cdma2000 Spread Spectrum Systems, February 2002.

[12]3GPP specifications can be found at the following URL: http://www.3gpp.org/ftp/Specs/archive/.

[13]3GPP2 specifications can be found at the following URL: http://www.3gpp2.org/Public_html/specs/index.cfm.

IETF Specifications[14]

[17] IETF RFC 2460, S. Deering, Internet Protocol, Version 6 (IPv6) Specification, December 1998.

[18] IETF RFC 3095, C. Bormann, RObust Header Compression (ROHC): Framework and four profiles: RTP, UDP, ESP, and uncompressed, July 2001.

[19] IETF RFC 6733, V. Fajardo, Diameter Base Protocol, October 2012.

Books

[20] S. Sesia, I. Toufik, M. Baker, *LTE—The UMTS Long Term Evolution: From Theory to Practice*, second ed., John Wiley & Sons, 2011.

[21] C. Johnson, Long Term Evolution in Bullets, second ed., Create Space Independent Publishing Platform, 2012.

[22] E. Dahlman, S. Parkvall, J. Skold, 4G, LTE/LTE-Advanced for Mobile Broadband, Academic Press, 2011.

Articles

[23] LTE and Beyond: LTE, 4G, MME, PGW, SGW, interfaces and beyond. <http://www.lteandbeyond.com/>.

[24] H. Tao, Z. Zhijiang, L. Yunjie, *QoSmechanism in EPS*, ZTE Commun. (2009).

[25] V. Srinivasa Rao, R. Gajula, Protocol signaling procedures in LTE, Radisys Corp. (2011).

[26] V. Srinivasa Rao, R. Gajula, Interoperable UE handovers in LTE, Radisys Corp. (2011).

[27] GPRS tunneling protocol. <http://en.wikipedia.org/wiki/GPRS_Tunnelling_Protocol>.

[28] The LTE Network Architecture: A comprehensive tutorial, Alcatel-Lucent, 2009.

[29] J. Laganier, et al., Mobility management for All-IP core network, NTT DoCoMo Tech. J. 11 (3) (2009).

[30] K. Suzuki, et al., Core network (EPC) for LTE, NTT DoCoMo Tech. J. 13 (3) (2011).

[31] LTE Identifiers, NMC Consulting Group. <http://www.nmcgroups.com>, 2011.

[32] LTE system description. <http://www.sharetechnote.com/>.

[33] A. Roessler, Cell Search and Cell Selection in UMTS LTE, Rohde & Schwarz, 2009.

[34] Agilent Technologies, 3GPP Long Term Evolution: System Overview, Product Development, and Test Challenges, September 2009.

[35] 3G and 4G wireless blog <http://3g4g.blogspot.com/>.

[36] W. Brouwer, QoS in LTE, Alcatel Lucent, December 2010.

[37] S. Amberkar, Telecom Tutorials, <http://samiramberkar-tutorials.blogspot.com/>.

[38] 4G Americas White Paper, developing and integrating a high performance hetnet. <http://www.4gamericas.org/>, October 2012.

---

[14] IETF specifications can be found at the following URL: http://www.ietf.org/rfc.html.

CHAPTER

Radio Resource Control Functions

5

CHAPTER CONTENTS

The radio resource control (RRC) sublayer in E-UTRAN protocol stack performs control-plane functions such as broadcast of system information related to access stratum (AS) and non-access stratum (NAS); paging; establishment, maintenance and release of an RRC connection between the UE and E-UTRAN; signaling radio bearer (SRB) management; security handling; mobility management including UE measurement reporting and configuration; connected mode handover; idle-mode mobility control; multimedia broadcast multicast service (MBMS) notification services and radio bearer management for MBMS; quality of service (QoS) management and NAS direct message transfer from NAS to the UE and vice versa [1]. Figure 5.1 illustrates the RRC sublayer and its interactions with other protocol layers in the E-UTRA protocol stack in terms of configuration of lower-layer functions and transfer of upper-layer control/signaling data via lower layers. The RRC sublayer provides the upper layers with the following services: (1) broadcast of common control information; (2) notification of idle-mode UEs and (3) transfer of dedicated control information. At the same time, the RRC sublayer receives the following services from lower layers: (1) integrity protection and ciphering from packet data convergence protocol (PDCP) sublayer; and (2) reliable and in-sequence transfer of information,

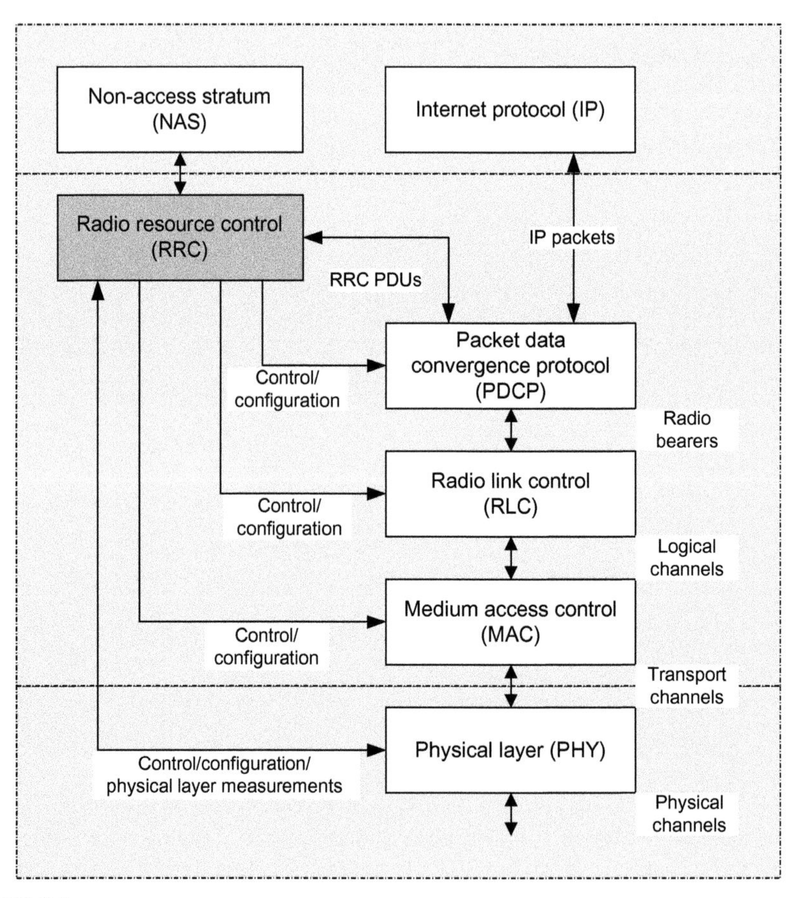

FIGURE 5.1

The RRC sublayer in the LTE protocol stack.

without introducing duplicates and support for segmentation and concatenation from the radio link control (RLC) sublayer.

More specifically, the RRC sublayer performs the following functions [1]:

- Broadcast of system information including NAS common information and the information pertinent to UEs in RRC_IDLE state such as cell selection/reselection parameters, neighbor cell information, and information related to UEs in RRC_CONNECTED state such as common channel configuration information.
- RRC connection control that includes the following tasks:
  - Paging of the UEs in the idle mode.
  - Establishment, modification, or release of the RRC connection which includes assignment and/or modification of the UE identity; establishment, modification, or release of SRBs; and access class barring.
  - Initial security activation which includes initial configuration of AS integrity protection and ciphering [3].
  - Maintaining RRC connection while the UEs are moving which includes intra-frequency/inter-frequency handover-associated security handling, i.e., key/algorithm change, specification of RRC context information transferred between network nodes.
  - Establishment, modification, or release of data radio bearers (DRBs).
  - Radio configuration and control in the lower layers including automatic repeat request (ARQ), hybrid automatic repeat request (HARQ), and discontinuous reception (DRX) configuration.
  - Cell management including change of primary cell; addition, modification, or release of secondary cells; and addition, modification, or release of secondary timing advanced groups when carrier aggregation is supported.
  - QoS control including assignment and modification of downlink/uplink semi-persistent scheduling configuration; assignment and modification of parameters for uplink rate control in the UE, i.e., allocation of a priority and a prioritized bit rate for each radio bearer.
  - Recovery from radio link failure (RLF).
- Inter-radio access technology (inter-RAT) mobility including security activation and transfer of RRC context information.
- Measurement configuration and reporting, which includes establishment, modification, or release of intra-frequency/inter-frequency/inter-RAT measurements; setup and release of measurement gaps; and measurement reporting.
- Support of self-configuration and self-optimization as well as support of measurement logging and reporting for network performance optimization.

In order to simplify the operation and to reduce state transition latency, LTE simplified the legacy systems and defined two operational states for the UE: RRC_CONNECTED and RRC_IDLE. A UE is in the RRC_CONNECTED state when an RRC connection or control-plane has been established; otherwise, the UE is said to be in the RRC_IDLE state. The UE states only correspond to the steady-state behavior and the transient procedures for cell search and selection, as well as network attach/detach, are not designated as states.

In this chapter, we will take a top-down systematic approach to analyzing and understanding the services and functions of the RRC sublayer in LTE/LTE-Advanced and will provide practical and implementation perspectives on LTE RRC sublayer protocols.

5.1 RRC states and state transitions

In LTE, a user terminal at any given time during normal operation can be in one of the two (steady-state) RRC states as shown in Figure 5.2. The emphasis on the steady-state descriptor is due to the fact that the state diagram shown in Figure 5.2 does not include the transient states of the mobile terminal. In fact, when a UE is powered on or upon RLF, it goes to a transient state with no attachment to a public land mobile network (PLMN) or a cell, and thus has no connection to the network. The UE starts scanning and selecting a PLMN and then proceeds to camp on a cell (cell selection/reselection) through downlink synchronization and system information acquisition. In this case, the UE has virtually moved to the RRC_IDLE state since it still has no RRC connection and thus performs uplink synchronization and RRC connection setup procedures. The RRC states can further be characterized as follows [1]:

- RRC_IDLE
  - UE-specific DRX is configured by NAS.
  - UE-controlled mobility is performed.
  - UE monitors a paging channel to detect incoming calls.
  - System information change is tracked and earthquake and tsunami warning system (ETWS) notification is monitored.
  - UE performs neighbor cell measurements and cell selection or reselection and acquires the system information.

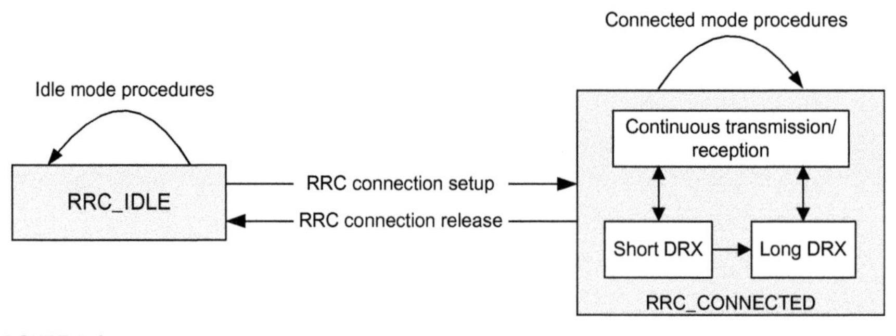

FIGURE 5.2

UE (steady-state) RRC states in LTE/LTE-Advanced.

- RRC_CONNECTED
  - Bi-direction transfer of unicast data in the uplink and downlink.
  - At lower layers, the UE may be configured with a UE-specific DRX.
  - Network-controlled mobility, i.e., handover and cell change order (CCO) with optional network-assisted cell change to GERAN.
  - Neighbor cell measurements.

In RRC_IDLE, the UE monitors a paging channel and/or system information block type 1 (SIB1) content to detect system information change, and for ETWS capable UEs, ETWS notifications. The UE further monitors control channels associated with the shared data channel to determine whether data is scheduled for it.

Prior to transition to RRC steady-states, the UE uses the information stored on its universal subscriber identity module (USIM) and the mobile equipment (ME) for system selection and camping. Upon initial power-up, the UE uses the static information on the USIM and ME to initiate system selection. In a practical scenario, the international mobile subscriber identity (IMSI) on the USIM may be used to determine the home PLMN code, and may serve as the basis for PLMN selection. The USIM may have optionally an equivalent home PLMN list and priorities among them to determine the PLMN selection. The USIM may also provide the allowed radio access technology (RAT) for each PLMN where there can be multiple RATs for a given PLMN. The UE considers the supported RAT and frequency band to trigger system selection. The system selection procedures are divided between NAS and AS. Over the air, signaling support provided by the AS assists NAS in camping on a PLMN. The selected PLMN is known as the registered PLMN and is stored on the USIM for future system selections.

If a PLMN is selected, but the UE cannot register with it because the registration is rejected, the PLMN is added to the prohibited PLMN list, which is retained even when the UE is switched off or the USIM is removed. A prohibited PLMN, which is typically a visited PLMN, can usually be selected only with user manual intervention. The ME uses the registered PLMN or a list of equivalent PLMNs (obtained during registration, all considered equal priority PLMNs) for PLMN selection (not for initial power-up), cell selection/reselection, and handover. If registration cannot be completed on any PLMN, the UE indicates a *No Service* to the user and waits until a new PLMN is detected, or tracking areas (TAs) of an allowed PLMN that are not in the prohibited TA list are found. The UE selects the registered PLMN or equivalent PLMN to initiate system selection during recovery from RLF or at power-up.

The selection of a PLMN can be manual or automatic. In both cases, the UE scans all possible RF channels that it supports based on its capability [5]. During the search, all detected PLMNs with reference signal received power (RSRP) ≥ -110 dBm are reported to the NAS as high-quality PLMNs without an associated RSRP value. The PLMNs with RSRP < -110 dBm are reported along with an associated RSRP value [10]. The order in which the UE searches for PLMNs based on either intra- or inter-band is not defined but can be optimized based on some information that is available to the UE including the last frequency on which it successfully camped. In both automatic and manual modes, the search for PLMNs can be discontinued by the NAS at any time. When the UE is in automatic mode and multiple PLMNs are detected, the order in which the selection is made is controlled by a set of rules such as last-registered PLMN (or equivalent) which is initially attempted, then home (or equivalent) PLMN is tried, followed by a user-defined and/or operator-defined list of PLMNs in priority order, followed by high-quality PLMNs in random order, followed by other PLMNs in the order of decreasing RSRP. If a UE is camped on a PLMN that is not the highest priority, it will periodically search for higher priority PLMNs and report the results to the NAS. The time interval between searches is stored on the USIM and set by the service provider to be between 6 minutes and 8 hours, in increments of 6 minutes.

5.1.1 RRC_CONNECTED state

The RRC_CONNECTED state is a state where the UE transfers unicast control/data in the uplink or downlink. The UE may be configured with a customized DRX or discontinuous transmission (DTX) pattern, or both. The UE monitors control channels associated with the shared data channel to determine if data is scheduled for it, provides channel quality and feedback information, performs neighbor cell measurements and measurement reporting, and acquires the system information. One or multiple IP addresses are assigned to the UE as well as a temporary identifier called cell radio network temporary identifier (C-RNTI) that is used for signaling purposes between the UE and the network. In brief, the RRC_CONNECTED state is characterized as follows:

- UE is known in EPC and E-UTRAN/eNB, i.e., UE context is stored in eNB.
- UE location is known at cell level.
- Unicast data transfer with UE is possible
- Short/long DRX is supported for power saving.
- Mobility is controlled by the network.

In RRC_CONNECTED, the network controls UE mobility, i.e., the network decides when the UE must move to another cell, which may be on another frequency or RAT. For network-controlled mobility in the RRC_CONNECTED state, handover is the only procedure that is defined. The network triggers the handover procedure based on radio conditions or network load. The network may configure the UE to perform measurement reporting including the configuration of measurement gaps. The network may also initiate handover without receiving measurement reports from the UE. Before sending the handover message to the UE, the source eNB prepares one or more target cells. The target eNB generates the message used to perform the handover, including the AS configuration to be used in the target cell. The source eNB transparently forwards the handover message/information received from the target to the UE. The source eNB may initiate data forwarding for some of the user DRBs. After receiving the handover message, the UE attempts to access the target cell at the first available random access channel (RACH) opportunity according to the random access resource selection procedure for asynchronous handover. When allocating a dedicated preamble for the random access in the target cell, the E-UTRA ensures it is available from the first RACH opportunity that the UE may utilize. Upon successful completion of the handover, the UE sends a message used to confirm the handover. Following successful completion of the handover, the PDCP service data units (SDUs) may be retransmitted in the target cell. This only applies for user DRBs using radio link control acknowledged mode (RLCAM) mode. Furthermore, the sequence number (SN) and the hyper-frame number (HFN)[1] are reset, except for the user DRBs using RLCAM mode for which both SN and HFN continue. For a limited period of time, the source eNB maintains the UE context to allow the UE to return in case of handover failure. If handover failure is detected, the UE will attempt to resume the RRC connection with either the source eNB or another cell using the RRC connection reestablishment procedure. The connection reestablishment would succeed, if the accessed cell is prepared.

The paging procedure is used to transmit paging information to a UE in RRC_IDLE and/or to inform mobile terminals in the RRC_IDLE and RRC_CONNECTED states about a system information change and/or to inform about an ETWS primary and/or secondary notification. The paging information is provided to upper layers, which in response may establish an RRC connection to receive an incoming call. The E-UTRAN initiates the paging procedure by transmitting the paging message at the UE's paging opportunities. The E-UTRAN may address multiple mobile terminals within a paging message by including one paging record for each UE. The E-UTRAN may also indicate a change of system information and/or provide an ETWS notification.

The RRC connection establishment involves SRB1 establishment. The procedure is also used to transfer the initial NAS dedicated information/message from the UE to the E-UTRAN. The UE initiates the RRC connection establishment procedure when the upper layers request establishment of an RRC connection while the UE is in the RRC_IDLE state.

The RRC connection reestablishment is used to reestablish the RRC connection including resumption of SRB1 operation and the reactivation of security. A UE in RRC_CONNECTED, for which security has been activated, may initiate the procedure in order to continue the RRC connection. The connection reestablishment succeeds, if the serving/target is prepared, i.e., it has a valid UE context. If the E-UTRAN accepts the connection reestablishment, SRB1 operation resumes while the operation of other radio bearers is suspended. If the AS security has not

[1]An HFN is an overflow counter mechanism that is used in the eNB and UE in order to limit the actual number of sequence number bits that are needed to be sent over the radio air-interface. The HFN needs to be synchronized between the UE and eNB.

Table 5.1 Summary of the Differences Between MAC and RRC Control [1]

	MAC Control		RRC Control
Control Entity	**MAC**		**RRC**
Signaling type	PDCCH	MAC control PDU	RRC message
Signaling reliability	$\sim 10^{-2}$ (no retransmission)	$\sim 10^{-3}$ (with HARQ)	$\sim 10^{-6}$ (with ARQ)
Control latency	Very short	Short	Long
Extensibility	None	Limited	High
Security	No integrity protection; no ciphering	No integrity protection; no ciphering	Integrity-protected ciphering

been established, the UE cannot initiate the procedure and directly transitions to the RRC_IDLE state.

The main difference between medium access control (MAC) and RRC control lies in the signaling reliability. The signaling corresponding to state transitions and radio bearer configurations should be performed by the RRC sublayer due to signaling reliability. The different characteristics of MAC and RRC control are summarized in Table 5.1.

5.1.2 RRC_IDLE state

The RRC_IDLE is a state where a UE-specific DRX may be configured by upper layers. In the idle mode, the UE saves power and does not inform the network of each cell change. The network knows the location of the UE to the granularity of a few cells, known as the tracking area. The UE monitors a paging channel to detect incoming traffic, performs neighbor cell measurements and cell selection/reselection, and acquires system information. In brief, the RRC_IDLE state is characterized as follows [1,4]:

- UE is known in EPC and has an assigned IP address.
- UE is not known in E-UTRAN/eNB.
- User terminal's location is known at tracking-area level.
- Unicast data transfer with UE is not possible.
- UE can be reached through paging in TAs controlled by EPC.
- UE-based cell selection and TA update are provided to EPC.

5.2 RRC connection management

One of the main functions of the RRC sublayer is to establish, maintain, modify, and release connections with E-UTRAN for transport of user data and control messages. There are two NAS states that reflect the state of a UE with respect to connection establishment: (1) evolved packet system (EPS) mobility management (EMM) state (EMM-DEREGISTERED or EMM-REGISTERED) which reflects whether the UE is registered in the mobility management entity (MME); and (2) EPS connection management (ECM) state (ECM-IDLE or ECM-CONNECTED) which signifies the connectivity of the UE with the core network.

The transition from ECM-IDLE to ECM-CONNECTED not only involves establishment of the RRC connection, but also includes establishment of S1 connection. The RRC connection establishment is initiated by the NAS and is completed prior to S1 connection establishment, which means that connectivity in RRC_CONNECTED is initially limited to the exchange of control information between UE and E-UTRAN. The UEs are typically moved to ECM-CONNECTED when becoming active. It should be noted that in LTE, the transition from ECM-IDLE to ECM-CONNECTED is performed within 100ms. The UEs engaged in intermittent data transfer do not need to stay in ECM-CONNECTED, if the ongoing services can tolerate such transfer delays. The LTE design objective was to support similar battery power consumption levels for UEs in RRC_CONNECTED as UEs in RRC_IDLE. The RRC connection release is initiated by the eNB following release of the S1 connection between the eNB and the core network [19].

The RRC connection establishment starts with the establishment of SRB1 and the transfer of the initial NAS message. This NAS message triggers the setup of an S1 connection, which normally initiates a subsequent step during which E-UTRAN activates AS security and further establishes SRB2, along with one or more DRBs associated with the default and optionally dedicated EPS bearers. Figure 5.3 illustrates the RRC connection establishment and the transport paths of user application data, NAS signaling, and RRC control messages. Note that NAS messages (ECM/EMM) are encapsulated in RRC messages (Figure 5.4), thus the RRC connection establishment needs to happen prior to NAS signaling transfer. Furthermore, it must be noted that SRBs are established before DRBs. The integrity protection algorithm in LTE/LTE-Advanced is common for signaling radio bearers SRB1 and SRB2. The ciphering algorithm is common for all radio bearers (i.e., SRB1, SRB2, and DRBs). However, neither integrity protection nor ciphering applies to SRB0 [1].

More specifically, the RRC connection establishment involves the establishment of SRB1. E-UTRAN completes the RRC connection establishment prior to establishing an S1 connection, i.e., prior to receiving the UE context from EPC. As a result, AS security is not activated during the initial phase of the RRC connection.

Upon receiving the UE context from the EPC, the E-UTRAN activates security that includes ciphering and integrity protection using the initial security activation procedure. The RRC messages to activate security are integrity-protected, while ciphering is started only after completion of the procedure. That is, the response to the message used to activate security is not ciphered, while the

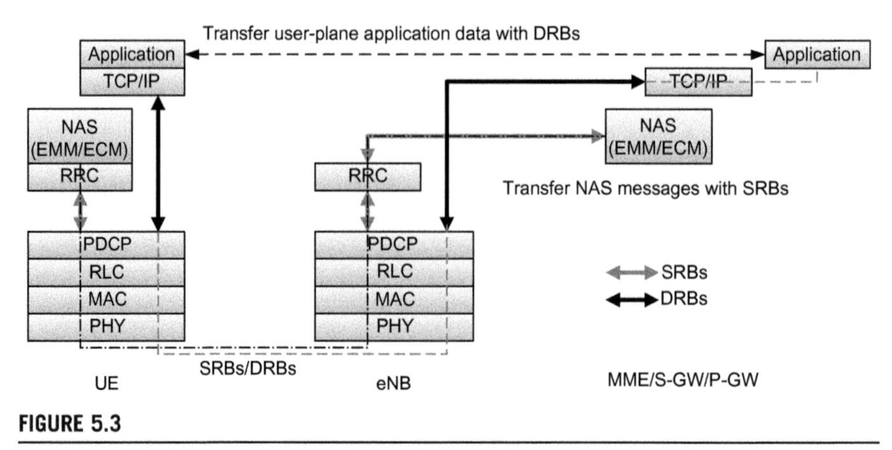

FIGURE 5.3

Illustration of different radio bearers in LTE RRC_CONNECTED state.

subsequent messages used to establish SRB2 and DRBs are both integrity-protected and ciphered. Following initiation of the security activation procedure, E-UTRAN proceeds to establish SRB2 and DRBs. The E-UTRAN will apply ciphering and integrity protection for the RRC connection reconfiguration messages used to establish SRB2 and DRBs. The E-UTRAN should release the RRC connection, if the initial security activation and/or the radio bearer establishment fail. For SRB2 and DRBs, security is always activated from the beginning, i.e., the E-UTRAN does not establish these bearers prior to activating security.

Once the security activation procedure is performed, the E-UTRAN may configure a carrier-aggregation-enabled UE with one or more secondary cells in addition to the primary cell that was initially configured during RRC connection establishment. The primary cell is used to provide the security inputs and upper-layer system information. The secondary cells are used to provide additional downlink and optionally uplink radio resources for user data transmission [1].

In order to better understand the RRC connection configuration/reconfiguration/release process, we need to briefly review the EPS mobility and connection management states. The EMM states describe the mobility management states that result from the mobility management procedures, e.g., attach and TA update procedures. There are two EMM states defined in LTE: EMM-DEREGISTERED and EMM-REGISTERED. The ECM states describe the signaling connectivity between the UE and the EPC. LTE defines two ECM states: ECM-IDLE and ECM-CONNECTED.

In general, the ECM and EMM states are independent of each other. Transition from EMM-REGISTERED to EMM-DEREGISTERED can occur regardless of the ECM state, e.g., by explicit detach signaling in ECM-CONNECTED or by

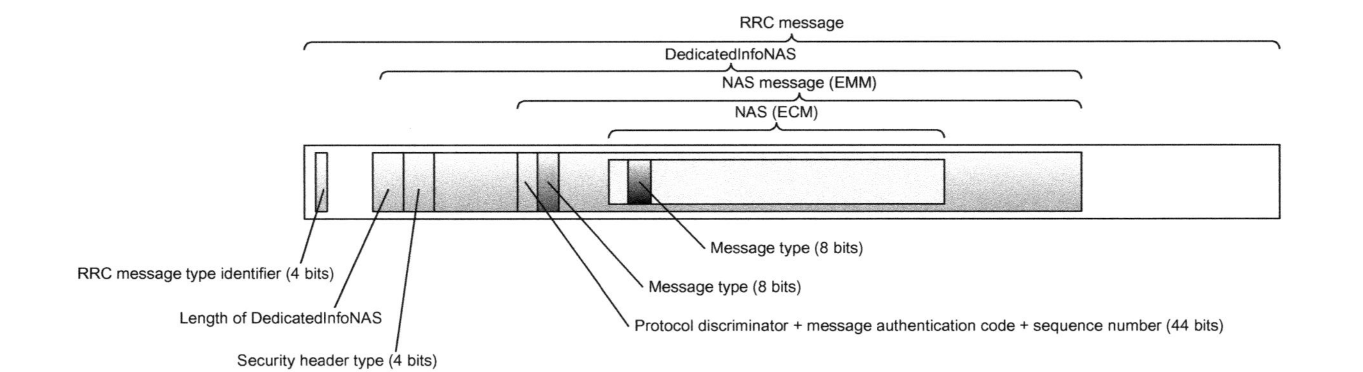

FIGURE 5.4

Encapsulation of NAS signaling/messages in RRC messages [27].

implicit detach locally in the MME during ECM-IDLE. However, in order to transition from EMM-DEREGISTERED to EMM-REGISTERED, the UE has to be in the ECM-CONNECTED state. In the EMM DEREGISTERED state, the EMM context in MME holds no valid location or routing information for the UE. The UE is not reachable by an MME, as the UE location is not known. In the EMM-DEREGISTERED state, some UE context can still be stored in the UE and MME, e.g., to avoid running a key agreement (AKA) procedure during every attach procedure.

A UE is in ECM-IDLE state when no NAS signaling connection between the UE and the network exists. In the ECM-IDLE state, a UE performs cell selection/reselection and PLMN selection. There is no UE context in E-UTRAN for the UE in the ECM-IDLE state. There is no S1-MME and no S1-U connection for the UE in the ECM-IDLE state. The UE location is known in the MME with an accuracy of a tracking area. The UE performs the TA update procedure when the tracking area identity (TAI) in the EMM system information is not in the list of TAs that the UE registered with the network, or when the UE performs a handover to an E-UTRAN cell and the UE's temporary identity indicates P-TMSI. For a UE in the ECM-CONNECTED state, there is a signaling connection between the UE and the MME. The signaling connection consists of two parts: an RRC connection and an S1-MME connection [7].

The RRC connection establishment process has been illustrated in Figure 5.5. The UE attempts to synchronize with the downlink and acquire the system information. The NAS layer in the UE (see Figure 5.3) triggers connection establishment, which may be in response to a paging message. If the access is not barred, the lower layers in the UE perform a contention-based random access procedure. If the contention-based random access procedure is successful, the UE sends the *RRC Connection Request* message. This message includes an initial identity, which is typically the UE's system architecture evolution temporary mobile subscriber identity (S-TMSI) or a random number. If E-UTRAN accepts the connection request, it returns the *RRC Connection Setup* message that includes the initial radio resource configuration including SRB1. Instead of signaling each individual parameter, the E-UTRAN may order the UE to apply a default configuration, i.e., a configuration for which the parameter values are specified in the RRC specification [1]. The UE responds with the *RRC Connection Setup Complete* message with an encapsulated NAS message, an identifier of the selected PLMN, and an identifier of the registered MME. Based on the last two parameters, the eNB decides with which MME it should establish the S1 connection.

The E-UTRAN sends the *Security Mode Command* message to activate integrity protection and ciphering. This message, which is integrity-protected but not ciphered, indicates which algorithms must be used. The UE verifies the integrity protection of the *Security Mode Control* message and configures lower layers to apply integrity protection and ciphering to all subsequent messages with the exception that ciphering is not applied to the response message, i.e., the *Security Mode*

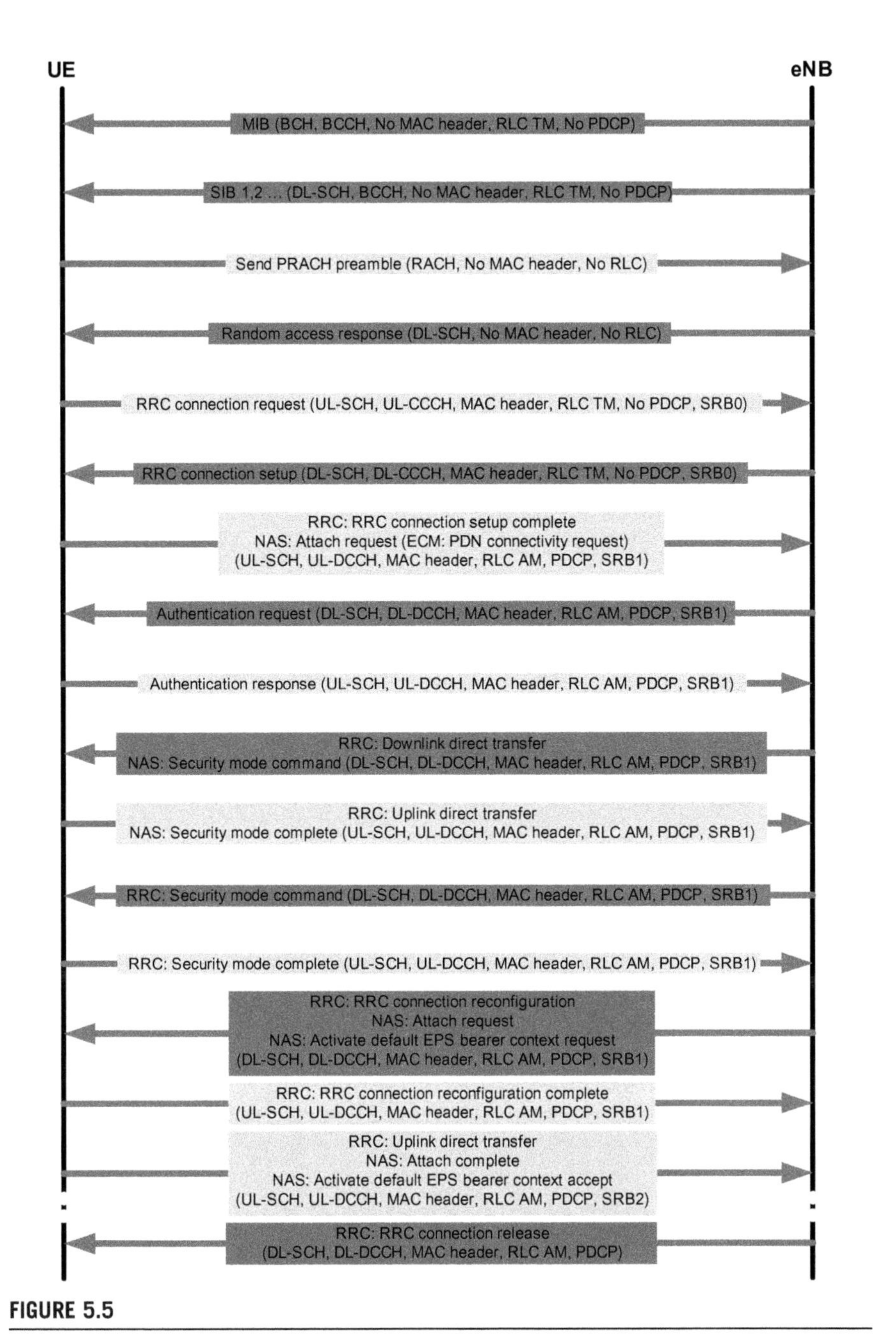

FIGURE 5.5

Sequence of signaling for RRC connection setup, reconfiguration, and release [27].

Complete or *Security Mode Failure* messages are not ciphered. The E-UTRAN sends the message including a radio resource configuration used to establish SRB2 and one or more DRBs. This message may also include other information such as a piggybacked (encapsulated) NAS message or a measurement configuration. The E-UTRAN may send the *RRC Connection Reconfiguration* message prior to receiving the *Security Mode Complete* message. In this case, the E-UTRAN should release the connection when one or both procedures fail. The UE ultimately returns the *RRC Connection Reconfiguration Complete* message [19,27].

5.2.1 Signaling and data radio bearers

A signaling radio bearer is defined as a radio bearer that is only used for transmission of RRC and NAS messages. More specifically, the following three SRB types are defined in E-UTRA [1]:

- SRB0 is used for transport of RRC messages associated with the common control logical channel.
- SRB1 is used for transport of RRC messages including piggybacked NAS messages as well as NAS messages prior to the establishment of SRB2 where all are associated with a dedicated control logical channel.
- SRB2 is used for transport of NAS messages using a dedicated control logical channel, has a lower priority than SRB1, and is always configured by E-UTRAN after security activation.

The SRB types and their usage are summarized in Table 5.2. The UE in the RRC_IDLE state does not have any allocated SRB or DRB. The UE requires (dedicated) SRBs and DRBs to exchange data/messages with the network. In particular, for transfer of NAS messages, SRBs are established. They carry registration and signaling for user-plane setup. The user-plane application data is transported on DRBs. An E-UTRAN radio access bearer (E-RAB) uniquely identifies the concatenation of an S1 bearer and the corresponding DRB. When an E-RAB exists, there is a one-to-one mapping between this E-RAB and an EPS

Table 5.2 E-UTRA SRBs and Their Usage [1]

SRB	Message Type	Associated Logical Channel
SRB0	RRC messages (*it is used when the RRC connection is established*)	CCCH
SRB1	RRC messages, RRC messages with piggybacked NAS messages, NAS messages prior to SRB2 setup	DCCH
SRB2	NAS messages, RRC messages with piggybacked NAS messages (*lower priority than SRB1*)	DCCH

bearer of the NAS [4]. To establish, modify, or release DRBs, the E-UTRAN applies the RRC connection reconfiguration procedure.

DRBs are established by the eNB in the RRC_CONNECTED state based on the relevant information from the MME. This information includes the desired QoS for the radio bearer, which enables the eNB to appropriately configure the PDCP, RLC, and MAC parameters. In the RRC_CONNECTED state, the DRBs can be modified and released based on the signaling information received from the MME. Some restrictions may apply with regard to the modification of existing DRBs. All DRBs are implicitly released when the RRC connection is released. When establishing a DRB, the E-UTRAN decides how to transfer the packets of an EPS bearer across the radio interface. An EPS bearer is uniquely mapped to a DRB, a DRB is then mapped to a dedicated traffic logical channel, all logical channels are mapped to a downlink or uplink shared transport channel, which are mapped to the corresponding physical downlink or uplink shared channel. A UE in the RRC_CONNECTED state, for which security has been activated, may initiate the procedure in order to continue the RRC connection. The connection reestablishment succeeds only if the serving cell is prepared, i.e., it has a valid UE context. In the case where the E-UTRAN accepts the connection reestablishment, SRB1 operation resumes while the operation of other radio bearers remains suspended. If the AS security has not been activated, the UE does not initiate the procedure, but rather directly transitions to the RRC_IDLE state.

Before concluding this section, it would be useful if we summarize the EPS bearer architecture and understand the termination points of each bearer in the LTE system. The EPS bearer service architecture is illustrated in Figure 5.6, in which the following bearers are defined [1,4]:

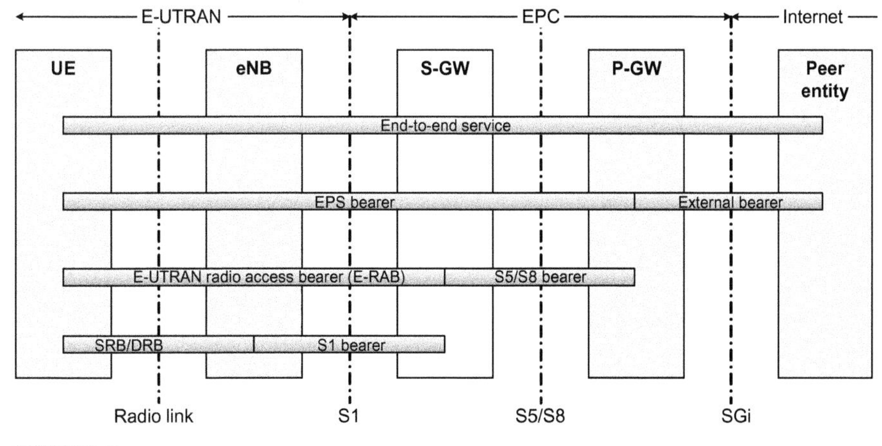

FIGURE 5.6

EPS bearer service architecture [4].

- An uplink traffic flow template (TFT) binds a service data flow to an EPS bearer in the uplink direction, where multiple service data flows can be multiplexed into the same EPS bearer by including multiple uplink packet filters[2] in the uplink TFT.
- A downlink TFT in the P-GW binds a service data flow to an EPS bearer in the downlink direction, where multiple service data flows can be multiplexed into the same EPS bearer by including multiple downlink packet filters in the downlink TFT.
- An E-RAB transports the packets of an EPS bearer between the UE and the EPC. When an E-RAB exists, there is a one-to-one mapping between the E-RAB and an EPS bearer.
- A DRB transports the packets of an EPS bearer between a UE and the eNB. When a DRB exists, there is a one-to-one mapping between this DRB and the EPS bearer/E-RAB.
- An S1 bearer transports the packets of an E-RAB between an eNB and an S-GW.
- An S5/S8 bearer transports the packets of an EPS bearer between an S-GW and a P-GW.
- A UE stores the mapping between an uplink packet filter and a DRB to create the binding between the service data flow and the DRB in the uplink.
- The P-GW stores the mapping between the downlink packet filter and S5/S8a bearer to create the binding between a service data flow and the S5/S8a bearer in the downlink.
- The eNB stores a one-to-one mapping between the DRB and the S1 bearer to create the binding between the DRB and the S1 bearer in both the uplink and downlink.
- The S-GW stores a one-to-one mapping between the S1 bearer and the S5/S8a bearer to create the binding between the S1 bearer and the S5/S8a bearer in both the uplink and downlink.

---

[2]Service data flow (SDF) is a fundamental concept in the 3GPP's definition of QoS and policy management. SDFs represent the IP packets related to a user service (web browsing, e-mail, etc.). SDFs are bound to specific bearers based on policies defined by the network operator. This binding occurs at the P-GW and the UE using traffic flow templates. TFTs contain packet filtering information to identify and map packets to specific bearers. The filters are configurable by the network operator, but at a minimum will contain five parameters, as follows:

- The source IP address.
- The destination IP address.
- The source port number.
- The destination port number.
- The protocol identification (i.e., TCP or UDP).

The policy and charging enforcement function in the P-GW filters packets.

In the downlink, the piggybacked (or encapsulated) NAS messages are used for bearer establishment, modification, or release. In the uplink, the piggybacked NAS messages are used for transferring the initial NAS messages during connection setup. The NAS messages transferred via SRB2 are also contained in RRC messages, which do not include any RRC protocol control information. Once security is established, all RRC messages on SRB1 and SRB2, including those containing NAS or non-3GPP messages, are integrity-protected and ciphered by PDCP sublayer. The NAS independently applies integrity protection and ciphering to the NAS messages.

Random access is specified entirely in the MAC sublayer including initial transmission power estimation [6]. The RRC connection establishment includes establishment of SRB1. The E-UTRAN completes the RRC connection establishment prior to completing the establishment of the S1 connection, i.e., prior to receiving the UE context from the EPC. Therefore, the AS security is not activated during the initial phase of the RRC connection. During the initial phase of the RRC connection, the E-UTRAN may configure the UE to perform measurement reporting. However, the UE only accepts handover messages when security has been activated. Upon receiving the UE context from the EPC, the E-UTRAN activates ciphering and integrity protection using the initial security activation procedure. The RRC messages to activate security are integrity-protected, while ciphering starts only after completion of the procedure, i.e., the response to the message used to activate security is not ciphered, while the subsequent messages used to establish SRB2 and DRBs are both integrity-protected and ciphered.

Following the activation of the initial security, the E-UTRAN establishes SRB2 and user DRBs, i.e., the E-UTRAN may perform this function prior to receiving the confirmation of the initial security activation from the UE. The E-UTRAN applies both ciphering and integrity protection for the RRC connection reconfiguration messages used to establish SRB2 and user DRBs. The E-UTRAN should release the RRC connection, if the initial security activation and/or the radio bearer establishment fails (i.e., security activation and user DRB establishment are triggered by a joint S1 procedure, which does not support partial success). For SRB2 and user DRBs, the security is always activated from the beginning, i.e., the E-UTRAN does not establish these bearers prior to activating security. The release of the RRC connections is initiated by the E-UTRAN. The procedure may be used to redirect the UE to another frequency or radio access technology (RAT). In certain cases, the UE may discontinue the RRC connection, i.e., it may transition to RRC_IDLE without notifying the E-UTRAN.

In the cases where the UE fails to establish a connection to the network (e.g., RLF, handover failure, RLC unrecoverable error, or reconfiguration compliance failure), the UE initiates the RRC connection reestablishment procedure, provided that security is active. If security is not active when one of the above failures occurs, the UE transitions to the RRC_IDLE state. In order to reestablish the RRC connection, the UE starts a timer known as T311 (see Section 5.6) and performs cell selection. In this case, the UE should prioritize searching of LTE frequencies.

However, no requirements are specified to prevent the UE from searching for other RATs. Upon finding a suitable cell on an LTE frequency, the UE stops timer T311, starts timer T301 and initiates a contention-based random access procedure, and sends an *RRC Connection Reestablishment Request* message. In the *RRC Connection Reestablishment Request* message, the UE includes the UE identity used in the previous cell, the identity of that cell, a short message authentication code, and a cause. The E-UTRAN uses the reestablishment procedure to continue SRB1 and to reactivate security without changing algorithms. A subsequent RRC connection reconfiguration procedure is used to resume operation on radio bearers other than SRB1 and to reactivate measurements. If the cell in which the UE initiates the reestablishment is not prepared (i.e., no UE context is available), the E-UTRAN will reject the procedure, causing the UE to transition to the RRC_IDLE state [19].

5.2.2 Quality of service control

An EPS bearer/E-RAB is the level of granularity for bearer-level QoS control in the EPC/E-UTRAN. As a result, service data flows mapped to the same EPS bearer receive the same bearer-level packet forwarding treatment, e.g., scheduling policy, queue management policy, rate shaping policy, and RLC configuration. One EPS bearer/E-RAB is established when the UE connects to a PDN and remains throughout the lifetime of the PDN connection to provide the UE with always-on IP connectivity to the PDN. This bearer is referred to as the *default* bearer. Any additional EPS bearer/E-RAB that is established to the same PDN is referred to as a *dedicated* bearer. The initial bearer-level QoS parameter values of the default bearer are assigned by the network based on subscription information. The decision to establish or modify a dedicated bearer can only be taken by the EPC, and the bearer-level QoS parameter values are always assigned by the EPC.

An EPS bearer/E-RAB is referred to as a guaranteed bit rate (GBR) bearer, if dedicated network resources related to a GBR value that is associated with the EPS bearer/E-RAB are permanently allocated by the admission control function in the eNB upon bearer establishment/modification. Otherwise, an EPS bearer/E-RAB is referred to as a non-GBR bearer. A dedicated bearer can either be a GBR or a non-GBR bearer, while a default bearer is always a non-GBR bearer. Each EPS bearer/E-RAB (GBR and non-GBR) is associated with the following bearer-level QoS parameters:

- QoS class identifier (QCI): A metric that is used as a reference to access node-specific parameters that control bearer-level packet forwarding function, e.g., scheduling weights, admission thresholds, queue management thresholds, link layer protocol configuration. These parameters are preconfigured by the network operator.
- Allocation and retention priority (ARP): The primary purpose of ARP is to decide whether a bearer establishment/modification request can be accepted or needs to be rejected in the case of resource limitation. In addition, the ARP

can be used by the eNB to decide which bearers to drop under resource shortage circumstances.

- Each GBR bearer is additionally associated with the bit rate that can be expected to be provided by that bearer.
- Each access point name (APN) access by a UE is associated with a per APN aggregate maximum bit rate (APN-AMBR).
- Each UE in the EMM-REGISTERED state is associated with a per UE-AMBR.

The GBR denotes bit rate of traffic per bearer while UE-AMBR/APN-AMBR denotes bit rate of traffic per group of bearers. Each of those QoS parameters has an uplink and a downlink component.

In Table 5.3, the delay between a policy control enforcement function (PCEF) and a radio base station (about 20 ms) should be subtracted from a given packet delay budget to derive the packet delay budget that applies to the radio interface. This delay is the average between the case where the PCEF is located close to the radio base station (approximately 10 ms) and the case where the PCEF is located far from the radio base station, e.g., in the case of roaming with home routed traffic (the one-way packet delay between Europe and the US west coast is roughly 50 ms). When calculating the average delay, it is assumed that roaming is a less typical scenario. It is expected that subtracting this average delay of 20 ms from a given packet delay budget will lead to desired end-to-end performance in most

Table 5.3 QoS Classes and Their Attributes and Usage in E-UTRA [11]

QCI	Resource Type	Priority	Packet Delay Budget (ms)	Packet Error Rate	Usage
1	GBR	2	100	10^{-2}	Conversational voice
2		4	150	10^{-3}	Conversational video (live streaming)
3		3	50	10^{-3}	Real-time gaming
4		5	300	10^{-6}	Non-conversational video (buffered streaming)
5	Non-GBR	1	100	10^{-6}	IMS signaling
6		6	300	10^{-6}	Video (buffered streaming), TCP-based (e.g., web-browsing, e-mail, chat, FTP, peer-to-peer file sharing, progressive video)
7		7	100	10^{-3}	Voice, video (live streaming), interactive gaming
8		8	300	10^{-6}	Video (buffered streaming), TCP-based (e.g., web-browsing, e-mail, chat, FTP, peer-to-peer file sharing, progressive video)
9		9			

typical cases. Note that the packet delay budget defines an upper bound. Actual packet delays for GBR traffic should typically be lower than the packet delay budget specified for a QCI as long as the UE has sufficient radio channel quality.

The rate of non-congestion related packet losses that may occur between a radio base station and a PCEF should be considered negligible. A packet error rate value specified for a standardized QCI therefore applies completely to the radio interface between a UE and radio base station.

The QCIs 1–5 and 7 are typically associated with an operator-controlled service, i.e., a service where the service data flow aggregate's uplink/downlink packet filters are known when the service data flow aggregate is authorized. In the case of E-UTRAN, this refers to the time at which a corresponding dedicated EPS bearer is established or modified.

In the case of QCI 6, if the network supports multimedia priority services, then this QCI may be used for the prioritization of non-real-time data (i.e., transport control protocol (TCP)-based services/applications) of subscribers. This QCI can be used for a dedicated premium bearer (e.g., associated with premium content) for any subscriber or a subscriber group. In this case, the service data flow aggregate's uplink/downlink packet filters are known when the service data flow aggregate is authorized. Alternatively, this QCI can be used for the default bearer of a UE/PDN for premium subscribers.

The QCI 9 is typically used for the default bearer of a UE/PDN for non-privileged subscribers. Note that AMBR can be used as a tool to provide differentiation between subscriber groups connected to the same PDN with the same QCI on the default bearer.

5.3 RRC mobility management

In order to support mobility, a UE may conduct and report to the serving eNB a series of measurements at the physical layer which include an RSRP and E-UTRA carrier received signal strength indicator (RSSI). These measurements in E-UTRA are classified as follows [4]:

- Intra-frequency E-UTRAN measurements.
- Inter-frequency E-UTRAN measurements.
- Inter-RAT measurements.

Measurement commands are used by E-UTRAN to order the UE to start measurements, modify measurements, or stop measurements. The reporting criteria that are used include event-triggered reporting, periodic reporting, and event-triggered periodic reporting. In the RRC_CONNECTED state, network-controlled UE-assisted handovers are performed and various DRX cycles are supported. In the RRC_IDLE state, cell reselections are performed and DRX is supported.

The E-UTRAN can configure the UE to perform radio signal quality measurements for handover purposes. Different measurement quantities are defined in the specification [10], and one quantity can be configured for each intra-frequency, inter-frequency, or inter-RAT measurement. The quantity configuration also defines applicable filtering, used for all reporting events associated with that quantity. The typical metrics for handover decision are as follows [8,10]:

- *Reference signal received power* (RSRP) is the linear average over the power contributions of the resource elements that carry cell-specific reference signals within the considered measurement frequency bandwidth.
- *Reference signal received quality* (RSRQ) is the ratio $N \times \text{RSRP}/(E\text{-UTRA}$ carrier RSSI) where N is the number of resource blocks equivalent to the E-UTRA carrier RSSI measurement bandwidth. The measurements in the numerator and denominator are made over the same set of resource blocks.
- *UTRA FDD Common Pilot Channel (CPICH) received signal code power* (RSCP) is the received power on one code measured on the primary CPICH. The reference point for the RSCP is the antenna connector of the UE.
- *UTRA FDD carrier RSSI* is the received wideband power, including thermal noise and noise generated in the receiver, within the bandwidth defined by the receiver pulse shaping filter. The reference point for the measurement is the antenna connector of the UE.
- E_c/N_0*UTRA FDD CPICH* is the received energy per chip divided by the power density in the band. If receiver diversity is not used by the UE, the CPICHE_c/N_0 is identical to the CPICH RSCP/UTRA carrier RSSI. The measurements are performed on the primary CPICH. The reference point for the CPICHE_c/N_0 measurement is the antenna connector of the UE.
- *GSM carrier RSSI* is the wideband received power within the relevant channel bandwidth. The measurements are conducted on a GSM BCCH carrier. The reference point for the RSSI measurement is the antenna connector of the UE.

The UE reports measurement information consistent with the measurement configuration provided by the E-UTRAN. The E-UTRAN provides the measurement configuration applicable for a UE in RRC_CONNECTED by means of dedicated signaling. The UE may be requested to perform the following types of measurements [4]:

- Intra-frequency measurements: Measurements at the downlink carrier frequency of the serving cell.
- Inter-frequency measurements: Measurements at frequencies that differ from the downlink carrier frequency of the serving cell.
- Inter-RAT measurements of UTRA frequencies.
- Inter-RAT measurements of GERAN frequencies.
- Inter-RAT measurements of CDMA2000 high rate packet data (HRPD) or 1xRTT frequencies.

The measurement configuration includes the following parameters [4]:

- *Measurement objects:* The objects on which the UE performs the measurements. For intra-frequency and inter-frequency measurements, a measurement object is a single E-UTRA carrier frequency. Associated with this carrier frequency, E-UTRAN can configure a list of cell-specific offsets and a list of blacklisted cells. The blacklisted cells are not considered in event evaluation or measurement reporting. For inter-RAT UTRA measurements, a measurement object is a set of cells on a single UTRA carrier frequency. For inter-RAT GERAN measurements, a measurement object is a set of GERAN carrier frequencies. For inter-RAT CDMA2000 measurements, a measurement object is a set of cells on a single (HRPD or 1xRTT) carrier frequency.
- *Reporting configurations*: A list of reporting configurations where each reporting configuration consists of the following [4]:
  - *Reporting criterion*: The criterion that triggers the UE to send a measurement report. This can be either periodical or a single event description.
  - *Reporting format*: The quantities that the UE includes in the measurement report and associated information (e.g., number of cells to report).
- *Measurement identities*: A list of measurement identities where each measurement identity links one measurement object with one reporting configuration. By configuring multiple measurement identities, it is possible to link more than one measurement object to the same reporting configuration, as well as to link more than one reporting configuration to the same measurement object. The measurement identity is used as a reference number in the measurement report.
- *Quantity configurations:* One quantity configuration is configured per RAT type. The quantity configuration defines the measurement quantities and associated filtering used for all event evaluation and related reporting of that measurement type. One filter can be configured per measurement quantity.
- *Measurement gaps:* Periods that the UE may use to perform measurements, i.e., no uplink or downlink transmissions are scheduled.

The E-UTRAN only configures a single measurement object for a given frequency, i.e., it is not possible to configure two or more measurement objects for the same frequency with different associated parameters, e.g., different offsets and/or blacklists. The E-UTRAN may configure multiple instances of the same event by configuring two reporting configurations with different thresholds. The UE maintains a single measurement object list, a single reporting configuration list, and a single measurement identities list. The measurement object list includes measurement objects that are specified per RAT type, including an intra-frequency object (i.e., the object corresponding to the serving frequency), inter-frequency object(s), and inter-RAT objects. Similarly, the reporting configuration list includes E-UTRA and inter-RAT reporting configurations. Any measurement object can be linked to any reporting configuration of the same RAT type. Some reporting configurations may not be linked to a measurement object.

Some measurement objects may not be linked to a reporting configuration. The measurement procedures distinguish the type of cells, i.e., the serving cell, listed cells (cells listed within the measurement objects), and detected cells (cells that are not listed within the measurement objects but are detected by the UE on the carrier frequencies indicated by the measurement objects).

Figure 5.7 illustrates the state transitions of a UE during normal operation. The figure shows the transition from power-on to RRC connection setup and establishment of user-plane and control-plane with the serving cell (cell selection); RRC connection release and transition to RRC_IDLE; cell reselection and transition to another cell, handover to another cell while in the RRC_CONNECTED state; and redirection from one cell to another upon RRC

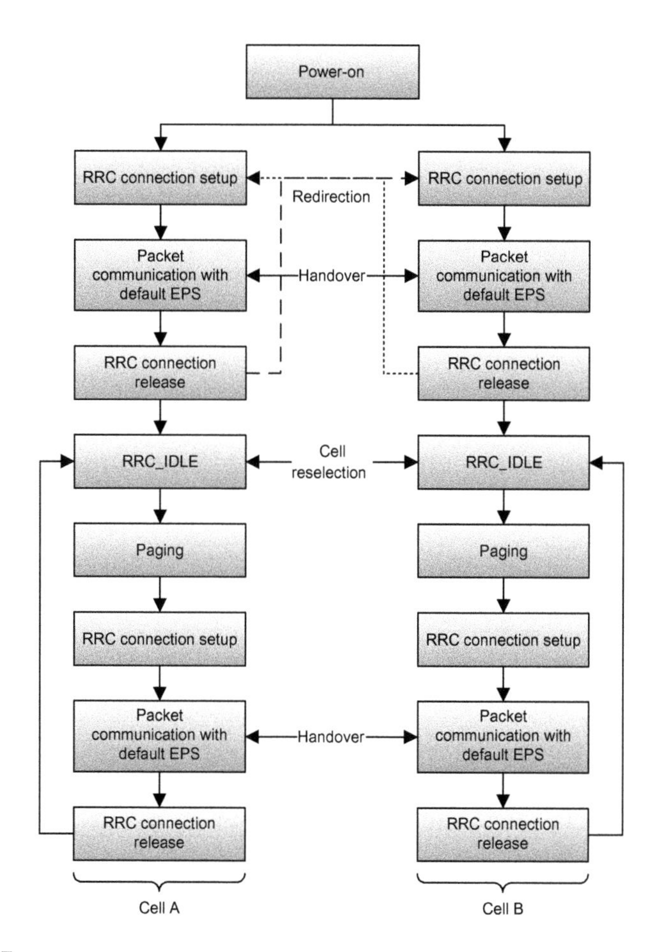

FIGURE 5.7

Overview of handover/cell selection/redirection process in E-UTRA [27].

connection release. We will explain later that cell selection, cell reselection, handover, and redirection are different processes that should not be confused.

For E-UTRA, the UE measures and reports on the serving cell, listed cells, and detected cells. For inter-RAT UTRA, the UE measures and reports on listed cells. For inter-RAT GERAN, the UE measures and reports on detected cells. For inter-RAT CDMA2000, the UE measures and reports on listed cells. For inter-RAT UTRA and CDMA2000, the UE measures and reports detected cells for the purpose of self-organizing networks. The closed subscriber group (CSG) femto-cells are not included within the neighbor list. Furthermore, the assumption is that for non-femto-cell deployments, the physical cell identity is unique within the area of a large macro cell.

The intra-E-UTRAN mobility support for mobile terminals in ECM-CONNECTED handles all necessary steps for relocation or handover procedures, such as processes that precede the final handover decision on the serving network, preparation of resources on the target network, commanding the UE to the new radio resources, and finally releasing resources on the serving network side. It contains mechanisms to transfer the EU context between eNBs and to update node relations on control-plane and user-plane. The UE conducts measurements of attributes of the serving and neighbor cells to enable the process [4].

Depending on whether the UE needs transmission/reception gaps to perform the relevant measurements, measurements are classified as gap-assisted or non-gap-assisted. A non-gap-assisted measurement is a measurement on a cell that does not require transmission/reception gaps to allow the measurement to be performed. A gap-assisted measurement is a measurement on a cell that does require transmission/reception gaps to allow the measurement to be performed. Gap patterns are configured and activated by the RRC sublayer [4]. The handover procedure is performed without EPC involvement, i.e., preparation messages are directly exchanged between the eNBs. The release of the resources at the serving eNB (or source eNB) during the handover completion phase is triggered by the eNB.

The measurements to be performed by a UE for intra/inter-frequency mobility can be controlled by E-UTRAN, using broadcast or dedicated control signaling. In the RRC_IDLE state, a UE follows the measurement parameters defined for cell reselection. In the RRC_CONNECTED state, a UE follows the measurement configurations specified by RRC sublayer and are managed by the E-UTRAN. Intra-frequency and inter-frequency neighbor cell measurements are defined as follows [4]:

- Neighbor cell measurements performed by the UE are intra-frequency measurements when the current and target cell operates on the same carrier frequency.
- Neighbor cell measurements performed by the UE are inter-frequency measurements when the neighbor cell operates on a different carrier frequency compared to the current cell.

5.3.1 Mobility in RRC_CONNECTED state

In the RRC_CONNECTED state, the network controls the UE mobility, i.e., the network decides when and where the UE should initiate a handover process. For network-controlled mobility in the RRC_CONNECTED state, the network triggers the handover process based on radio conditions or network loading conditions. To facilitate this process, the network may configure the UE to perform and report measurements including the configuration of measurement gaps. The network may also initiate a blind handover, i.e., without waiting for the measurement reports from the UE. Before sending the handover message to the UE, the source eNB prepares one or more target cells. The target eNB generates the message that is used to perform the handover, i.e., the message includes the AS configuration to be used in the target cell. The source eNB transparently (i.e., does not modify values or content) forwards the handover information received from the target cell to the UE. When appropriate, the source eNB may initiate data forwarding for (a subset of) the DRBs, in order to reduce the handover interruption time.

After receiving the handover message, the UE attempts to access the target cell at the first available RACH opportunity according to random access resource selection, i.e., the handover is asynchronous. Consequently, when allocating a dedicated preamble for the random access in the target cell, E-UTRA ensures that the preamble is available from the first random access opportunity that the UE may use. Upon successful completion of the handover, the UE sends a message to confirm the handover. After the successful completion of the handover, the PDCP SDUs may be retransmitted in the target cell. This only applies to DRBs using RLC AM mode [1,4]. The source eNB should, for some time, maintain the UE context to enable the UE to return in case of handover failure. In the case of handover failure, the UE attempts to resume the RRC connection either in the source or another cell using the RRC reestablishment procedure. The connection resumption attempt succeeds only if the accessed cell is prepared, i.e., a cell belonging to the source eNB or another eNB has been prepared for the handover.

In RRC_CONNECTED state, the E-UTRAN decides to which cell a UE should hand over in order to maintain the radio link. Similar to RRC_IDLE state, the E-UTRAN may consider not only the radio link quality, but also other aspects such as UE capability, subscriber type, and access restrictions. Although the E-UTRAN may trigger handover without measurement information (blind handover), typically it configures the UE to report measurements of the candidate target cells. In LTE, the UE always connects to a single cell, thus the switching of a UE's connection from a source cell to a target cell is a hard handover. In a hard handover process, the eNB which controls the source cell requests the target eNB to prepare for the handover. The target eNB subsequently generates the RRC message to order the UE to perform the handover, and the message is transparently forwarded by the source eNB to the UE. In another type of handover supported by LTE, the UE spontaneously decides to connect to the target cell, where it then requests that the connection be continued. The UE applies this connection reestablishment procedure only after terminating the connection to the source cell. The latter procedure only succeeds if the target cell has been prepared in advance for the handover. In addition, LTE also allows a UE to be redirected to another frequency or RAT upon connection release (see Figure 5.7). Redirection during connection establishment is not supported, since at that time the E-UTRAN may not have all the relevant information such as UE capability and the type of subscriber. However, the redirection may be performed while AS security has not been activated. When redirecting the UE to UTRAN or GERAN, the E-UTRAN may provide the system information for one or more cells on the relevant frequency. If the UE selects one of the cells for which the system information is provided, it does not need to acquire it [19].

5.3.2 Mobility in RRC_IDLE state

In the RRC_IDLE state, cell reselection between frequencies is based on absolute priorities, where each frequency has an associated priority. Cell-specific default values of the priorities are provided via the system information. In addition, E-UTRAN may assign UE-specific values upon connection release, considering factors such as UE capability or subscriber type. In case equal priorities are assigned to multiple cells, the cells are ranked based on radio link quality. Equal priorities are not applicable between frequencies of different RATs. The UE does not consider frequencies for which it does not have an associated priority. This is useful in situations where a neighboring frequency is applicable only for UEs of one of the sharing networks. Besides the cell reselection priority of a frequency, there are no idle-mode mobility-related parameters that may be assigned via dedicated signaling.

In general, the transition from RRC_IDLE to RRC_CONNECTED state is under the assumption that the RRC protection keys and user-plane protection keys are generated, while keys for NAS protection as well as higher-layer keys are assumed to be already available at the MME. On the other hand, in transitions from RRC_CONNECTED to RRC_IDLE state, the eNBs delete the stored keys and the state for the idle-mode UEs is only maintained in the MME. It is also assumed that eNB no longer stores state information about the corresponding UE and deletes the current keys from its memory.

In LTE-Advanced, the requirement for UE transition latency from the idle mode, with allocated IP address, to connected mode is less than 50 ms including the establishment of the user-plane and excluding the S1 transfer delay. The transition latency requirement from dormant sub-state (inactivity periods in short/long DRX mode) to active sub-state (continuous transmission/reception periods) in connected mode is less than 10 ms. These requirements are stricter than those for earlier releases of LTE. In the dormant sub-state, the UE has an established RRC connection and radio bearers. The UE has already been identified at cell level, but it may be in DRX mode to save power during inactivity periods. The UE may be either synchronized or unsynchronized (see Figure 5.2).

Although LTE already fulfills the latency requirements of IMT-Advanced systems, several new techniques are utilized in LTE-Advanced to further reduce the user-plane and control-plane latencies as follows [12,21]:

- Combined *RRC Connection Request* and *NAS Service Request* allow those two messages to be processed in parallel at the eNB and MME, respectively, reducing overall transition latency from idle mode to connected mode by approximately 20 ms.
- Reduced processing delays in different nodes form the major part of the delay (about 75% for the transition from idle mode to connected mode assuming a combined request); thereby any improvement has a large impact on the overall latency.
- Reduced RACH scheduling period from 10 to 5 ms results in decreasing of the average waiting time for the UE to initiate the procedure to transit from idle mode to connected mode by 2.5 ms.

Furthermore, a shorter cycle of physical uplink control channel helps reduce the average waiting time for a synchronized UE to request radio resources in the connected mode, expediting the transition from the dormant sub-state in the connected mode.

5.3.3 Intra-RAT mobility and handover

LTE supports two types of handover: X2-based handover and S1-based handover. The X2-based procedure is the least complex handover and typically would be used when an X2 interface exists between the source eNB and the target eNB. The handover is negotiated directly between the two eNBs; when the UE context is established on the target eNB, the MME is notified in order to switch the path. When no X2 interface exists between nodes, the S1-based procedure will be utilized with signaling between the target and source eNBs passing through the MME.

All handovers in LTE are hard handovers, i.e., the air-interface connection with the source eNB is disconnected before reestablishing a new connection with the target eNB. The procedure for reestablishing the connection with the target node is identical for both X2-based and S1-based handovers and requires the use of the random access procedure to synchronize and begin uplink transmissions. The UE is typically notified of the specific random access signature to use so that a contention-free procedure can be performed. From the core network perspective, handover can involve changes to the serving MME and S-GW and this can be due to mobility and the geographic location of the network nodes, or due to loading and congestion control mechanisms.

Figure 5.8 illustrates the handover procedure between eNBs within the same MME/S-GW. More information on thc handover procedure can be found in Chapter 4. After the downlink path is switched at the S-GW, downlink packets on the forwarding path and on the new direct path may arrive interchanged at the

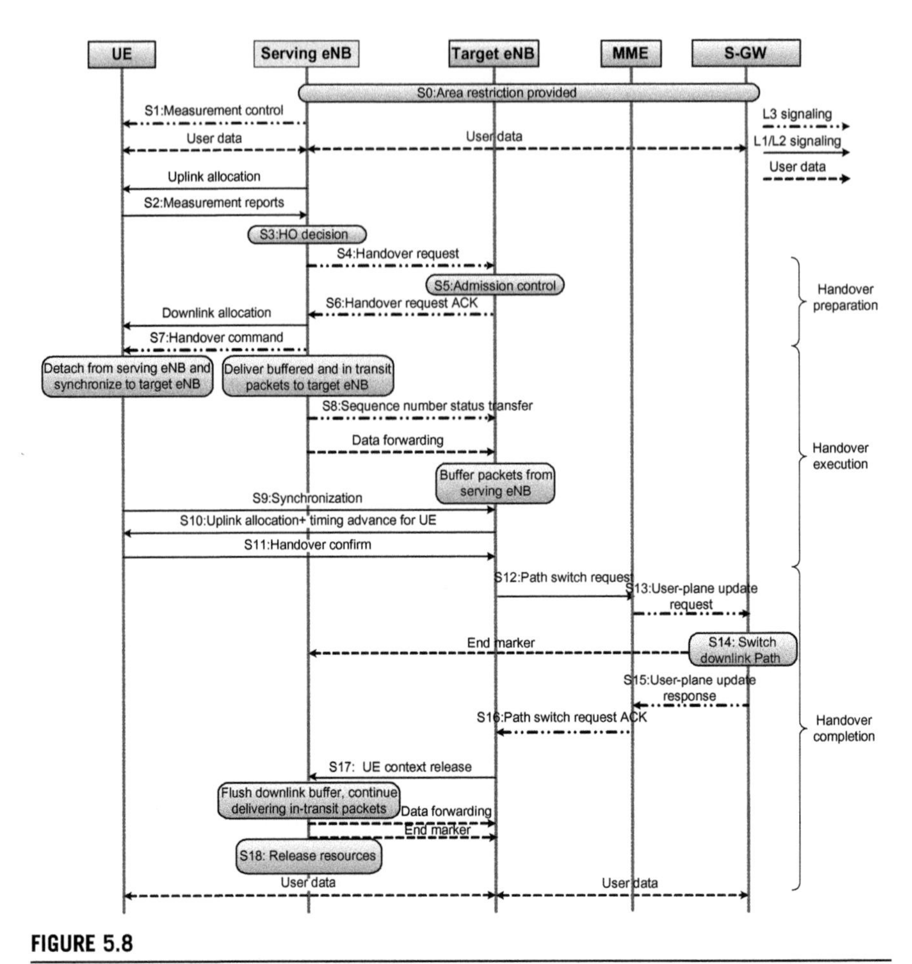

FIGURE 5.8

Intra-MME/S-GW handover procedures [4].

target eNB (see Figure 5.8). The target eNB should first deliver all forwarded packets to the UE before delivering any of the packets received on the new direct path. Upon handover, the serving eNB forwards in an orderly manner all downlink PDCP SDUs that have not been acknowledged by the UE to the target eNB. The serving eNB discards any remaining downlink RLC PDUs. Similarly, the source eNB does not forward the downlink RLC context to the target eNB.

The X2-based handover procedure can be divided into four stages (see Figure 5.8):

1. *Decision*: The E-UTRAN decides to execute a handover based on measurement reports received from the UE.
2. *Preparation*: The serving eNB notifies the target eNB that a handover is requested. It includes a list of bearers that will be transferred and whether

downlink data forwarding is necessary. The target eNB acknowledges the handover request and responds with a list of bearers that are admitted along with downlink and uplink GTP tunnel endpoints to enable data forwarding.

3. *Execution*: Once the resources have been set up on the target side, the UE is notified and detaches from the source eNB. Downlink packets that are received by the source from the S-GW are forwarded to the target eNB. PDCP status and hyperframe number information are exchanged to enable lossless handover, if required. At this point, the MME is notified of the changes in order to define direct uplink and downlink paths for the user-plane with the S-GW.
4. *Completion*: Resources in the source eNB are released following successful completion of the handover.

The message flow for an S1-based handover consists of the same four phases used for an X2-based handover. For an S1-based handover, however, there is no direct path between the eNBs; therefore, the signaling must go through the MMEs. The main stages of an S1-based handover are as follows:

1. *Decision*: The E-UTRAN decides to initiate the handover based on a measurement reports and notifies the serving MME.
2. *Preparation*: The serving MME identifies and then notifies the target MME, which begins preparation of the target resources. Because a change of S-GW is also required, the target MME also initiates preparation of resources in the target S-GW. The final preparation step is the target MME notifying the source MME of the new addresses and tunnel endpoints to enable data forwarding.
3. *Execution*: The source MME begins the execution phase of the handover by notifying the source eNB which subsequently notifies the UE. Data forwarding of downlink data begins. The UE detaches from the source eNB and synchronizes with the target eNB using random access procedures. Once established, the target MME begins the process of defining direct uplink and downlink paths for the P-GW to the target S-GW.
4. *Completion*: After successful completion of the handover, resources in the source eNB and source MME are released.

5.3.4 Inter-RAT mobility and handover

In principle, the inter-RAT handover procedures when an LTE system is the source or target are the same as the handover process in LTE. One of the differences is that upon handover to LTE, the entire AS configuration needs to be signaled, whereas within LTE it is possible to signal incremental information, where only the changes to the configuration are signaled. If ciphering was not activated in the previous RAT, the E-UTRAN activates ciphering as part of handover procedure. The E-UTRAN also establishes SRB1, SRB2, and one or more DRBs, i.e., at least one DRB is associated with the default EPS bearer. Figure 5.9 illustrates the RRC states in E-UTRA and shows the mobility support between E-UTRAN, UTRAN, and GERAN where CCO message is used to command the UE to change to another RAT [4].

The mobility support between E-UTRAN, CDMA2000 1xRTT [17], and CDMA2000 HRPD or CDMA 1xEV-DO [13] radio access technologies are shown in Figure 5.10. In general, the procedure for handover from LTE to another RAT supports both handover and CCO with network-assisted cell change (NACC). The CCO/NACC procedure is applicable only for handover to GERAN. Handover from LTE is performed only after security has been activated. In the case of circuit-switched fallback to CDMA2000, the procedure includes support for parallel handover (i.e., to 1xRTT and HRPD), for handover to 1xRTT combined with redirection to HRPD, and for redirection to HRPD (see Figure 5.10). The UE may send a *Measurement Report* message. In the case of handover (as opposed to CCO), the source eNB requests the target node to prepare for the handover in advance. As part of the handover preparation request, the source eNB provides information about the applicable inter-RAT UE capabilities, as well as information about the currently established bearers to the target node. In response, the target node generates the handover command and returns this to the source eNB. The source eNB sends mobility from E-UTRA command to the UE, which includes the inter-RAT message received from the target cell and some inter-RAT

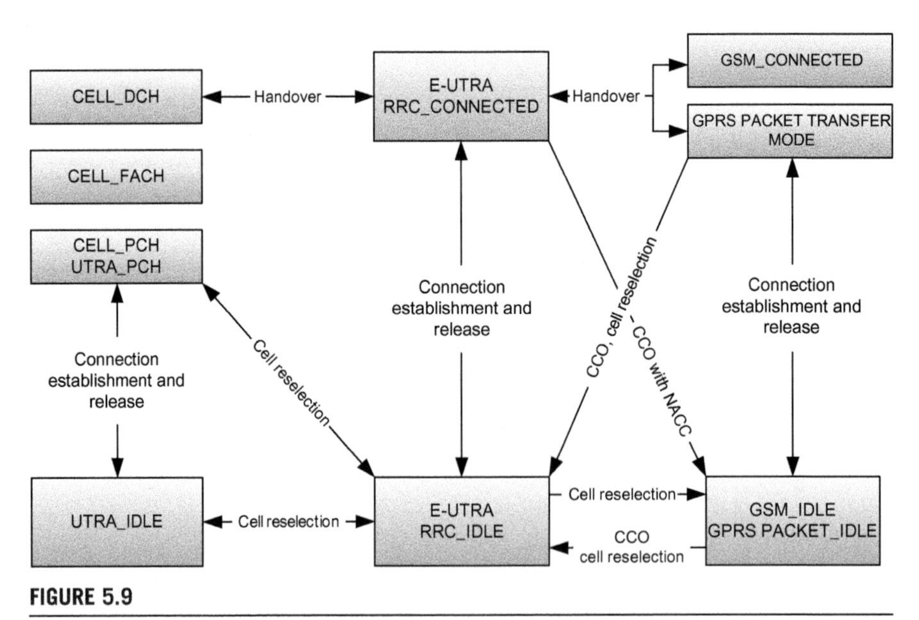

FIGURE 5.9

E-UTRA states and inter-RAT mobility procedures [1,4].

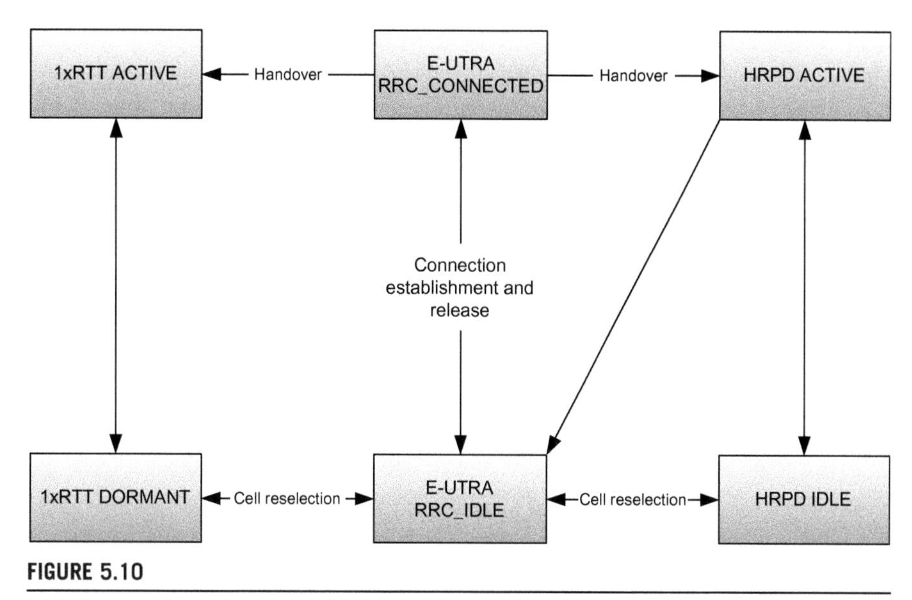

FIGURE 5.10

Mobility procedures between E-UTRA and CDMA2000 [1,4].

parameters in case of CCO. Upon receiving the message, the UE starts the timer T304 (see Section 5.6) and connects to the target node, either using the received radio configuration (handover) or by initiating connection establishment (CCO) according to the applicable specifications of the target RAT. The upper layers in the UE are informed by the AS of the bearers established by the target RAT base station. As a result, the UE is informed whether some of the previously established bearers were not admitted by the target RAT base station [19].

The inter-RAT handover procedures support signaling, conversational, and non-conversational services. The mobility between E-UTRA and non-3GPP systems other than CDMA2000 has also been studied in the literature [13–17]. In addition to the state transitions shown in Figures 5.9 and 5.10, there is support for connection release with redirection information from E-UTRA RRC_CONNECTED state to GERAN, UTRAN, and CDMA2000 (HRPD Idle/1xRTT Dormant modes).

5.4 Measurements and measurement reporting

Measurements to be performed by a UE for intra/inter-frequency mobility are controlled by the E-UTRAN, using broadcast or dedicated control signaling. In the RRC_IDLE state, a UE follows the measurement parameters defined for cell reselection by thc E-UTRAN via broadcast. The use of dedicated measurement control for the RRC_IDLE state is possible through the provision of UE-specific priorities. In the RRC_CONNECTED state, a UE follows the RRC measurement configurations defined by the E-UTRAN. Intra/inter-frequency neighbor cell measurements are defined as follows [4]:

- Intra-frequency neighbor cell measurements are performed by the UE when the current and target cell operate on the same carrier frequency. The UE conducts these measurements without measurement gaps.
- Inter-frequency neighbor cell measurements are performed by the UE when the neighbor cell operates on a different carrier frequency relative to the current cell. The UE would need measurement gaps to conduct these measurements.

The classification of the measurements to non-gap-assisted or gap-assisted depends on the UE capability and the current operating frequency. The UE determines whether a particular cell measurement needs to be performed in a transmission/reception gap, and the scheduler needs to know whether gaps are needed. The relationship between the source and target base stations' carrier frequency and bandwidth may fall into one of the following scenarios [4]:

- *Scenario A*: The current and the target cells have the same carrier frequency and cell bandwidth. This is an intra-frequency scenario without measurement gap.
- *Scenario B*: The current and the target cells have the same carrier frequency but the bandwidth of the target cell is smaller than the bandwidth of the current cell. This is an intra-frequency scenario without measurement gap.
- *Scenario C*: The current and the target cells have the same carrier frequency but the bandwidth of the target cell is larger than the bandwidth of the current cell. This represents an intra-frequency scenario without measurement gap.
- *Scenario D*: The current and the target cells have different carrier frequencies and the bandwidth of the target cell is smaller than the bandwidth of the current cell and the bandwidth of the target cell is within the bandwidth of the current cell. This is an inter-frequency scenario where measurement gap is needed.
- *Scenario E*: The current and the target cells have different carrier frequencies and the bandwidth of the target cell is larger than the bandwidth of the current cell and the bandwidth of the current cell within bandwidth of the target cell is an inter-frequency scenario where a measurement gap is needed.
- *Scenario F*: The current and the target cells have different carrier frequencies and non-overlapping bandwidths. This represents an inter-frequency scenario where a measurement gap is needed.

Measurement gap patterns are configured and activated by the RRC sublayer. When carrier aggregation is configured, the term "current cell" may refer to any serving cell of the configured set of serving cells. More specifically, the impact of carrier aggregation on the definition of intra- and inter-frequency measurements are as follows [4]:

- Intra-frequency neighbor cell measurements: Neighbor cell measurements performed by the UE are considered intra-frequency measurements when one of the serving cells of the configured set and the target cell operate on the same carrier frequency. The UE conducts such measurements without measurement gaps.

- Inter-frequency neighbor cell measurements: Neighbor cell measurements performed by the UE are called inter-frequency measurements when the neighbor cell operates on a different carrier frequency than any serving cell of the configured set. The UE may not be able to conduct measurements without measurement gaps.

In a system with frequency reuse one, the general assumption is that mobility is mainly within the same frequency layer (i.e., between cells with the same carrier frequency). The neighbor cell measurements are needed for cells that have the same carrier frequency as the serving cell in order to ensure sustainable mobility support. The search for neighbor cells with the same carrier frequency as the serving cell, and measurements of the relevant attributes for identified cells are needed. To support mobility between different frequency layers (i.e., between cells with a different carrier frequency), the UE may need to conduct neighbor cell measurements during idle periods that are provided by DRX or packet scheduling (i.e., gap-assisted measurements).

Measurement reporting is configured as either event-triggered or periodic. Different events are defined to monitor the quality of the serving cell and neighbor cells for intra-LTE mobility (events A1 to A6) and for inter-RAT mobility (events B1and B2). For some measurements (e.g., inter-frequency), measurement gaps may be required to monitor radio quality, depending on the capabilities of the terminal. The E-UTRAN only configures a single measurement object for a given frequency, i.e., it is not possible to configure two or more measurement objects for the same frequency with different associated parameters such as different offsets and/or blacklists. Measurement gaps contain gaps every N LTE frames (i.e., the gap periodicity is an integer multiple of 10 ms) with 6 ms duration (Table 5.4; Figure 5.11). The triggers used by the eNB to schedule a measurement gap include the following:

- Inter-LTE mobility (event-triggered reporting criteria):
  - Event A1: Serving cell becomes better than absolute threshold.
  - Event A2: Serving cell becomes worse than absolute threshold.
  - Event A3: Neighbor cell becomes better than an offset relative to serving cell.
  - Event A4: Neighbor cell becomes better than absolute threshold.
  - Event A5: Serving cell becomes worse than one absolute threshold and neighbor cell becomes better than another absolute threshold.
  - Event A6: Neighbor cell becomes better than an offset relative to the secondary cell (SCell).
- Inter-RAT mobility (event-triggered reporting criteria):
  - Event B1: Neighbor cell becomes better than absolute threshold.
  - Event B2: Serving cell becomes worse than one absolute threshold and neighbor cell becomes better than another absolute threshold.

Table 5.4 Measurement Gap Pattern in E-UTRA [4]

Gap Pattern ID	Measurement Gap Length (ms)	Measurement Gap Repetition Period (ms)	Minimum Available Time for Inter-Frequency and Inter-RAT Measurements During 480 ms Period (ms)	Measurement Purpose
0	6	40	60	Inter-frequency E-UTRAN FDD and TDD, UTRAN FDD, GERAN, LCR TDD, HRPD, CDMA2000 1x
1	6	80	30	Inter-frequency E-UTRAN FDD and TDD, UTRAN FDD, GERAN, LCR TDD, HRPD, CDMA2000 1x

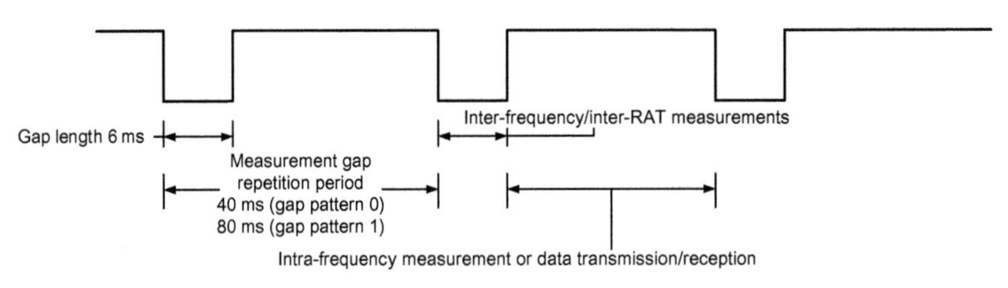

FIGURE 5.11

Illustration of measurement gap in LTE.

The UE may be configured to provide a number of periodic reports after being triggered by an event (event-triggered periodic reporting). If the UE requires measurement gaps to identify and measure inter-frequency and/or inter-RAT cells, the E-UTRAN provides a single measurement gap pattern with constant gap duration for concurrent monitoring of all frequency layers and RATs. During the measurement gaps, the UE does not transmit any data and is not expected to tune its receiver on the E-UTRAN serving carrier frequency. Inter-frequency and inter-RAT measurement requirements within this section rely on the UE being configured with one measurement gap pattern. As an example, the mobility measurement reporting for event A3 is illustrated in Figure 5.12. The figure depicts the process by which the UE reports an event A3. The UE measurements start when the RSRP of the serving cell falls below a certain level as defined by the measurement configuration parameter *s-Measure*. Event A3 is detected when the

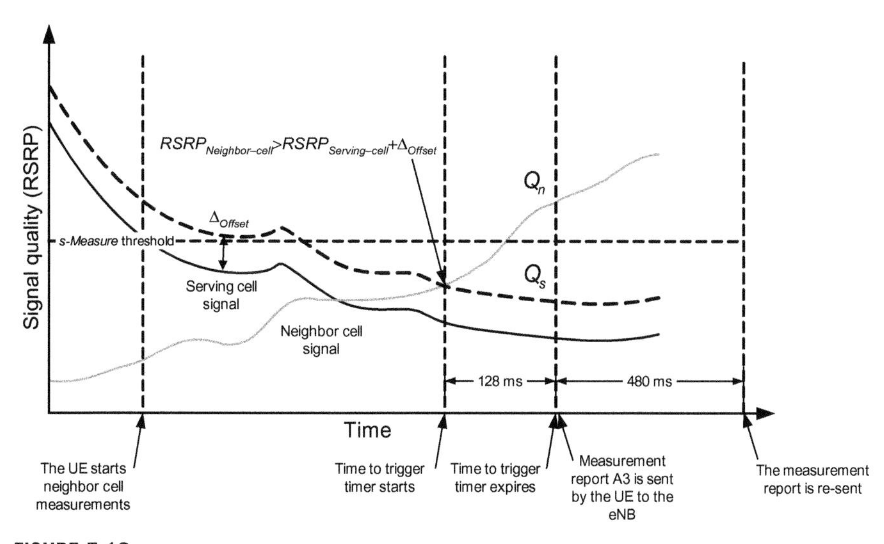

FIGURE 5.12

Illustration of mobility measurement reporting event A3 (example).

RSRP of the neighbor cell is greater than that of the serving cell plus the offset defined in the *Measurement Report Configuration* message. When this condition is met, the *Time-to-Trigger* timer is started and it expires after (for example) 128 ms. If the conditions to satisfy the event persist throughout the period that the timer is running, a measurement report is sent upon expiration of the timer. If the eNB does not respond to the initial measurement report, additional reports are sent every (for example) 480 ms, as configured [1]. Note that no hysteresis is defined for this event. The UE sends a *Measurement Report* message to inform the eNB that an event A3 has been triggered. The report includes the *Measurement Identity* which identifies the *Measurement Object*. The measured RSRPs of both the serving cell and the neighbor cell are specified in this report and correspond to the measured values at the time the report was generated, not when they were first detected. In this example, the measurement report will be re-sent every 480 ms unless the eNB responds. For the serving cell the RSRP and RSRQ is always included regardless of the triggering quantity.

To avoid UE activity outside the DRX cycle, the reporting criteria for neighbor cell measurements should match the current DRX cycle. The UE is required to apply a layer 3 filtering to each measurement quantity that it performs, which does not include quantities configured exclusively for UE receive–transmit time difference measurements. The UE is required to filter the measured result, before using for evaluation of reporting criteria or for measurement reporting, by the following formula $F_n = (1 - a)F_{n-1} + aM_n$, where M_n is the latest received measurement result from the physical layer, F_n is the updated filtered measurement result that is used for evaluation of reporting criteria or for measurement reporting, and

F_{n-1} is the previous filtered measurement result. The value of F_0 is set to M_1 when the first measurement results from the physical layer are received. The parameter a is defined as $a = 1/2^{k/4}$, in which k is the *filterCoefficent* for the corresponding measurement quantity received by the *quantityConfig*. The UE is further required to adapt the filter such that the time characteristics of the filter are preserved at different input rates, observing that the *filterCoefficentk* assumes a sample rate equal to 200 ms. If k is set to zero, no layer 3 filtering is performed. The filtering is performed in the same domain as that used for evaluation of reporting criteria or for measurement reporting, i.e., logarithmic filtering for logarithmic measurements. The filter input rate is implementation dependent in order to fulfill the performance requirements [1].

5.5 RRC procedures

The following sections describe the main RRC procedures as defined in E-UTRA.

5.5.1 System information

System information in E-UTRA is grouped into the *master information block* (MIB) and a number of *system information blocks* (SIBs) [1,4]:

- *Master information block* defines the most essential physical layer information of the cell that is required to receive other system information parts. The MIB provides the system frame number, downlink system bandwidth, and PHICH configuration.
- *System information block type 1* (SIB1) contains information for evaluating whether a UE is allowed to access a cell and further defines the scheduling of other SIBs. More specifically, SIB1 contains cell access-related information (PLMN identity list, PLMN identity, TA code, cell identity, cell status), cell selection information (minimum receiver level), and scheduling information (system information (SI) message type and periodicity, SIB mapping, SI window length).
- *System information block type 2* (SIB2) includes common and shared channel information such as access barring information (access probability factor, access class barring list, access class barring time), semi-static common channel configuration (random access parameter, PRACH configuration), and uplink frequency information (uplink EARFCN, uplink bandwidth, additional emission).
- *System information block type 3* (SIB3) comprises cell reselection information, mainly related to the serving cell. More specifically, it contains cell reselection common parameters for intra-frequency, inter-frequency and/or inter-RAT cell reselection, as well as intra-frequency cell reselection information not related to the neighbor cells.

- *System information block type 4* (SIB4) consists of information about the serving frequency and intra-frequency neighbor cells relevant to cell reselection including common and cell-specific cell reselection parameters. The information element includes cells with specific reselection parameters as well as blacklisted cells.
- *System information block type 5* (SIB5) contains information about other E-UTRA frequencies and inter-frequency neighbor cells relevant for cell reselection (including cell reselection parameters common for a frequency as well as cell-specific reselection parameters).
- *System information block type 6* (SIB6) contains information about UTRA frequencies and UTRA neighbor cells used in cell reselection including common and cell-specific cell reselection parameters.
- *System information block type 7* (SIB7) contains information about GERAN frequencies pertinent to cell reselection including cell reselection parameters for each frequency.
- *System information block type 8* (SIB8) contains information about CDMA2000 frequencies and CDMA2000 neighbor cells relevant for cell reselection including common and cell-specific cell reselection parameters.
- *System information block type 9* (SIB9) contains a home eNB name.
- *System information block type 10* (SIB10) contains an earthquake and tsunami warning system (ETWS) primary notification.
- *System information block type 11* (SIB11) contains an ETWS secondary notification.
- *System information block type 12* (SIB12) contains a commercial mobile alert service (CMAS) warning notification.
- *System information block type 13* (SIB13) contains MBMS-related information.
- *System information block type 14* (SIB14) contains information about *Extended Access Barring* for access control.
- *System information block type 15* (SIB15) contains information related to mobility procedures for MBMS reception. This information element contains the MBMS service area identities (SAI) of the current and/or neighboring carrier frequencies.
- *System information block type 16* (SIB16) contains information related to GPS time and coordinated universal time (UTC). The UE may use the parameters provided in this system information block to obtain UTC, GPS, or to obtain the local time.

The MIB is mapped to BCCH and is carried on BCH, while all other system information messages are mapped to BCCH and are dynamically carried on DL-SCH where they can be identified through system information RNTI (SI-RNTI). The MIB and SIB1 use a fixed schedule with a periodicity of 40 and 80 ms, respectively, while the scheduling of other system information messages is flexible and signaled in SIB1. The eNB may schedule DL-SCH transmissions associated with logical channels other than BCCH in the same subframe as that used for BCCH. The paging message is used to inform the UEs in RRC_IDLE and the UEs in RRC_CONNECTED about a system information change. The system information may also be provided to the UE by means of dedicated signaling and upon handover.

Depending on the deployment, some of the SIBs may not be transmitted. For example, SIB9 is not relevant to an operator-deployed node, and SIB13 is not necessary if an MBMS service is not provided in the cell. Similar to the MIB, the SIBs are periodically broadcast. The period of an SIB transmission depends on how fast the UEs need to acquire the corresponding system information when accessing a cell. In general, lower-order SIBs are more time critical and are transmitted more often relative to higher-order SIBs. As an example, SIB1 is transmitted every 80 ms, whereas the transmission period of higher-order SIBs is flexible and can be different in different networks. The different SIBs are then mapped to different *System Information* messages, which correspond to the actual transport blocks that are transmitted on DL-SCH. SIB1 is always mapped, by itself, on to the first *System Information* message, whereas the remaining SIBs may be group-wise multiplexed on to the same SI subject to the following constraints: (1) the SIBs are mapped to the same *System Information* message, if they have the same transmission period, e.g., two SIBs with a transmission period of 320 ms can be mapped to the same *System Information* message, whereas an SIB with a transmission period of 160 ms must be mapped to a different *System Information* message; (2) the total number of information bits that is mapped to a single *System Information* message must not exceed the permissible size of the transport block. It should be noted that the transmission period of a given SIB may be different in different networks. For example, different operators may have different requirements regarding the period when different types of neighbor cell information is transmitted. Furthermore, the amount of information that can fit into a single transport block depends on the deployment parameters such as cell bandwidth and cell size [20].

The change of system information (other than ETWS) only occurs at specific radio frames. The system information may be transmitted a number of times with the same content within a modification period as defined by its scheduling. When the network changes some of the system information, it notifies the UEs about this change. In the next modification period, the network transmits the updated system information, as illustrated in Figure 5.13 where different colors indicate

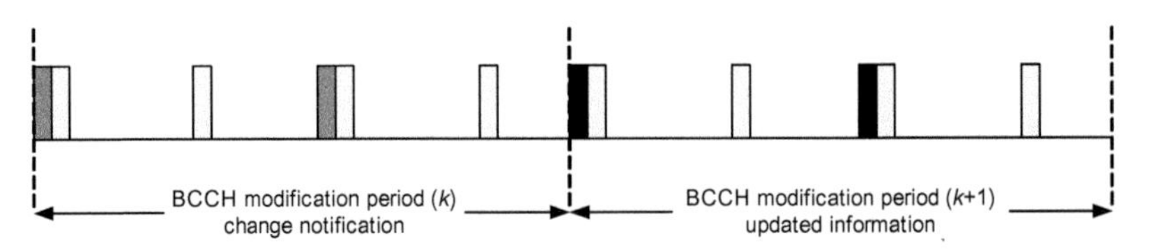

FIGURE 5.13

Change of system information timing [4].

different system information. Upon receiving a change notification, the UE acquires the new system information immediately from the start of the next modification period. The UE applies the previously acquired system information until the UE acquires the new system information.

The paging message is used to inform UEs in RRC_IDLE and UEs in RRC_CONNECTED about a system-information change. If the UE receives a paging message indicating a system information modification, the UE by default expects that the system information will change at the next modification period boundary (see Figure 5.13). Although the UE may be informed about changes in system information, no further details such as which system information is changing are provided. The *SIB1* includes a value tag that indicates if a change has occurred in the *System Information* messages. The UEs may use the tag upon return from out-of-coverage to verify whether the previously stored *System Information* messages are still valid. Furthermore, the UE considers stored system information to be invalid after 3 hours from the moment it was successfully confirmed as valid, unless otherwise specified.

The E-UTRAN may not update the tags upon change of some system information such as ETWS information, or regularly changing parameters such as CDMA2000 system time. Similarly, the E-UTRAN may not include the system information modification within the paging message upon change of some system information. The UE verifies that stored system information remains valid by either checking the tags in *SIB1* after the modification period boundary or by attempting to find the system information modification indication during the modification period in case no paging is received in every modification period. If no paging message is received by the UE during a modification period, the UE may assume that no change of system information will occur at the next modification period boundary. If a UE in RRC_CONNECTED state, during a modification period, receives one paging message, it may assume from the presence/absence of system information modification whether a change of system information other than ETWS information will occur in the next modification period.

The UE performs the system information acquisition procedure to acquire the AS and NAS system information that is broadcast by the E-UTRAN. The procedure applies to UEs in RRC_IDLE and UEs in RRC_CONNECTED. The UE applies the system information acquisition procedure upon cell selection and reselection, after handover completion, after entering E-UTRA from another RAT, upon return from out-of-coverage, upon receiving a notification that the system information has changed, upon receiving an indication about the presence of an ETWS notification, upon receiving a request from CDMA2000 upper layers, and upon exceeding the maximum validity duration. Unless explicitly stated in the procedural specification, the system information acquisition procedure overwrites any stored system information [1].

If the UE is in the RRC_IDLE state, it must ensure that it has a valid version of the *MIB* and *SIB1*, as well as *SIB2* through *SIB8*, depending on the support of other radio access technologies. If the UE is in the RRC_CONNECTED state, it must ensure that it has the *MIB*, *SIB1*, and *SIB2* as well as *SIB8*, depending on the UE's support of CDMA2000 [18].

In LTE, the system information is classified into five categories as follows: (1) information valid across multiple cells; (2) information needed at cell search; (3) information needed prior to cell camping; (4) information needed before cell access; and (5) information needed while camping on a cell. From the UE perspective, the information that is needed at cell selection and prior to camping are very similar. Before a UE can camp on a cell, it needs to know if the access is allowed to that cell. Thus, it would be very beneficial to know all access restrictions already at cell search phase. In order to support full mobility within the serving frequency layer, the UEs need to perform a cell search periodically, and thus it is important that the information needed for a cell search is readily available to reduce cell search latency and minimize UE power consumption.

More specifically, before a UE can camp on a cell, it needs to know any access-related parameters in order to avoid camping on cells where access is restricted, i.e., any cell access restriction parameters such as TA identity, cell barring status, and cell reservation status. Thus, the UE needs to know whether the cell is barred or reserved in order to avoid camping on a barred cell. Also barring time might be needed in order to ensure that the UE does not have to poll barring time frequently from the system information. Another option is that barring status is indicated also in the neighbor cell list and radio access limitation parameters, i.e., any radio condition parameters that limit the access to the cell [1].

When a UE has camped on a cell, it needs to continue measuring the neighbor cells in order to stay camped. In order for the UE to start mobility procedures, it needs to receive parameters such as reporting periods, reporting event parameters, and time to trigger. The UEs in the RRC_IDLE state need cell reselection parameters. The UEs in the RRC_CONNECTED state need parameters of the neighbor cells for handover and for error recovery cases. Neighbor cell lists are needed to start neighbor cell measurements. UEs in different states may use different sets of neighbor cell lists. The LTE system information can be classified into two distinctive groups in terms of timing: static and flexible. The static part is sent more frequently, e.g., once per frame, in the cell and has a limited capacity for information transfer. The flexible part has a flexible amount of scheduled resources available and thus encompasses most of the system information. The flexible part has different types of information elements which require independent scheduling in order to allow sufficiently fast reception and efficient resource utilization.

5.5.2 Paging

The DRX reduces UE power consumption in the idle mode. The UE monitors only one paging occasion (PO) every DRX cycle. This is the shortest of the UE-specific DRX cycle and the default DRX cycle broadcast in SIB2. The default

Table 5.5 UE and Network Identifiers in LTE/LTE-Advanced [27]

Abbreviation	Identifier	Description	Value/Structure
IMSI	International mobile subscriber identity	Unique identification of mobile subscriber in the network	IMSI (not more than 15 digits) = PLMN ID + MSIN = MCC + MNC + MSIN
PLMN ID	PLMN identifier	Unique identification of PLMN	PLMN ID (not more than 6 digits) = MCC + MNC
MCC	Mobile country code	Assigned by ITU	3 digits
MNC	Mobile network code	Assigned by national authority	2 or 3 digits
MSIN	Mobile subscriber identification number	Assigned by operator	9 or 10 digits
GUTI	Globally unique temporary UE identity	Identifies a UE (between the UE and the MME) on behalf of IMSI for security reasons	GUTI (not more than 80 bits) = GUMMEI + M-TMSI
TIN	Temporary identity used in next update	GUTI is stored in TIN parameter of UE's context. TIN indicates which Temporary ID will be used in the next update	TIN = GUTI
S-TMSI	SAE temporary mobile subscriber identity	To locally identify a UE in short form within an MME group (unique within an MME Pool)	S-TMSI (40 bits) = MMEC + M-TMSI
M-TMSI	MME mobile subscriber identity	Unique within an MME	32 bits
GUMMEI	Globally unique MME identity	To identify an MME uniquely in a global sense	GUMMEI (not more than 48 bits) = PLMN ID + MMEI
MMEI	MME identifier	To identify an MME uniquely within a PLMN	MMEI (24 bits) = MMEGI + MMEC

(Continued)

Table 5.5 (Continued)

Abbreviation	Identifier	Description	Value/Structure
MMEGI	MME group identifier	Unique identification of an MME group within a PLMN	16 bits
MMEC	MME code	To identify an MME uniquely within an MME Group	8 bits
C-RNTI	Cell-radio network temporary identifier	To identify a UE uniquely in a cell	0 × 0001 ~ 0xFFF3 (16 bits)
IMEI	International mobile equipment identity	To identify an ME (Mobile Equipment) uniquely	IMEI (15 digits) = TAC + SNR + CD
IMEI/SV	IMEI/ software version	To identify an ME (Mobile Equipment) uniquely	IMEI/SV (16 digits) = TAC + SNR + SVN
ECGI	E-UTRAN cell global identifier	To identify a cell in global (Globally Unique) EPC can know UE location based on ECGI	ECGI (not more than 52 bits) = PLMN I D + ECI
ECI	E-UTRAN cell identifier	To identify a cell within a PLMN	ECI (28 bits) = eNB ID + cell ID
PGW ID	PDN GW identity	To identify a specific PDN GW (P-GW) HSS assigns P-GW for PDN (IP network) connection of each UE	IP address (4 bytes) or FQDN (variable length)
TAI	Tracking area identity	To identify TA which is globally unique	TAI (not more than 32 bits) = PLMN ID + TAC P-GW
TAC	Tracking area code	To indicate to which TA the eNB belongs (per cell) which is unique within a PLMN	16 bits

(Continued)

Table 5.5 (Continued)

Abbreviation	Identifier	Description	Value/Structure
TAI list	Tracking area identity list	UE can move into the cells included in TAI list without location update (TA update) which is globally unique	Variable length
PDN ID	Packet data network identity	To identify a PDN (IP network) with which mobile data user communicates PDN Identity (APN) is used to determine the P-GW and point of interconnection with a PDN With APN as query parameter to the DNS procedures, the MME will receive a list of candidate P-GWs, and then a P-GW is selected by MME	PDN Identity = APN = APN−NI + APN−OI (variable length) APN−NI: APN Network Identifier APN−OI: APN Operator Identifier
EPS bearer ID	Evolved packet system bearer Identifier	To identify an EPS bearer (Default or Dedicated) per UE	4 bits
E-RAB ID	E-UTRAN radio access bearer identifier	To identify an E-RAB per UE	4 bits
DRB ID	Data radio bearer identifier	To identify a DRB per UE	4 bits
LBI	Linked EPS bearer ID	To identify the default bearer associated with a dedicated EPS bearer	4 bits

(Continued)

Table 5.5 (Continued)

Abbreviation	Identifier	Description	Value/Structure
TEID	Tunnel endpoint Identifier	To identify the endpoint of a GTP tunnel when the tunnel is established	32 bits

Table 5.6 E-UTRA Paging Parameters [2]

	FDD				TDD			
N_s	PO when $i\_s=0$	PO when $i\_s=1$	PO when $i\_s=2$	PO when $i\_s=3$	PO when $i\_s=0$	PO when $i\_s=1$	PO when $i\_s=2$	PO when $i\_s=3$
1	9	N/A	N/A	N/A	0	N/A	N/A	N/A
2	4	9	N/A	N/A	0	5	N/A	N/A
4	0	4	5	9	0	1	5	6

paging cycle can be 320, 640, 1280, or 2560 ms. The PO is defined in terms of paging frame (PF) and subframe index. Every UE uses its own IMSI (Table 5.5) to determine the PF, and within the PF, the PO. The network uses the S-TMSI, which is part of the globally unique temporary identifier (GUTI), to send the paging message to the UE. The GUTI is assigned to the UE during the attach procedure. If the network uses the IMSI for paging, the UE locally detaches and initiates the attach procedure.

One PO is a subframe containing a PDCCH, whose CRC is scrambled with P-RNTI, for the purpose of addressing the paging message. One PF is one radio frame, which may contain one or multiple PO(s). When idle-mode DRX is enabled, the UE needs only to monitor one PO per DRX cycle [2]. The parameters PF and PO are determined by following equations using the DRX parameters provided in the *System Information* message:

- PF is given by following equation: SFN mod $T = (T/N)(\text{UE\_ID} \bmod N)$.
- i_s is an index which points to a PO (Table 5.6) and is derived from the following equation: $i\_s = \lfloor \text{UE\_ID}/N \rfloor \bmod N_s$.
- $T = \min(T_{\text{UE-specific-paging-cycle}}, T_{\text{Cell-specific-paging-cycle}})$.

The DRX parameters stored in the UE must be updated whenever the DRX parameter values are changed in the system information. If the UE has no IMSI, for instance, when making an emergency call without USIM, the UE uses, by default, UE_ID $=0$ in the above equations. In the above equations, T denotes the DRX cycle of the UE. The value of T is determined by the smallest of the

UE-specific DRX values, if allocated by upper layers, and a default cell-specific DRX value is broadcast in the system information. If UE-specific DRX is not configured by upper layers, the default value is applied; $nB = \{4T, 2T, T, T/2, T/4, T/8, T/16, T/32\}$ is used as one of parameters to derive the PF and PO and is broadcast in SIB2; $N = \min(T, nB)$indicates the number of PFs within the paging cycle of a UE; $N_s = \max(1, nB/T)$(i.e., number of POs (subframes) per PF); and UE_ID = IMSI mod 1024 is the same as the UE Identity Index provided to the eNB by the MME through the S1-AP *Paging* message. Note that IMSI is given as sequence of digits of type integer and in the above equation is interpreted as a decimal integer number, where the first digit given in the sequence represents the highest order digit [2]. Table 5.6 provides the relationship between the paging parameters in FDD and TDD (for all downlink/uplink ratios) systems.

As we just learned, in the idle mode, the UE does not have to continuously monitor the paging channel. Based on the paging control channel configuration information that is broadcast in SIB2, the UE can go to the DRX mode between POs. Let us try to understand this process through an illustrative example. Figure 5.14 illustrates an example where the paging cycle defined by the *defaultPagingCycle* parameter is set to $T = T_{\text{UE-specific-paging-cycle}} = T_{\text{Cell-specific-paging-cycle}} = 128$ frames. The total number of POs per paging cycle is defined by parameter nB, which is broadcast in SIB2. In this example (assuming an FDD system), we assume that:

$$nB = T; \quad N = 128, N_s = 1$$
$$\text{IMSI of UE (in decimal)} = 262022008880715$$
$$\text{UE\_id} = 262022008880715 \bmod 1024 = 587$$
$$i\_s = 0 \rightarrow \text{PO} = 9$$

Thus, one PO is available in every paging cycle. The specific SFN number that each UE should monitor during each paging cycle is derived from the UE's IMSI. The maximum number of POs that can be defined per radio frame is 4 ($nB = 4T$), and these will only be transmitted in subframes 0, 4, 5, and 9 (see Table 5.6). Therefore, in this example, whenever SFN mod 128 = 75, the UE will wake up and look for a possible paging message addressed to it in the 9th subframe of PF (i.e., SFN = 75, 203, 331, ...) The value of N_s is cell-specific and is set according to the cell load and paging capacity. The choice of 4th and 9th subframes for lower paging capacity configurations is due to the fact that these subframes are adjacent to the 5th and 0th subframes which carry the primary and secondary synchronization signals. This allows the UE to detect the synchronization signals at the same time that it wakes up to check for a paging message. The 0th and the 5th subframes are not selected for narrowband configurations because the 0th subframe carries the MIB and the 5th subframe includes SIB1. For wideband systems, there are usually sufficient resources to accommodate both MIB/SIB1 and the paging messages.

We now change of our focus to tracking the UEs in the RRC_IDLE mode and how paging messages are sent to these UEs in the network. The MME keeps track

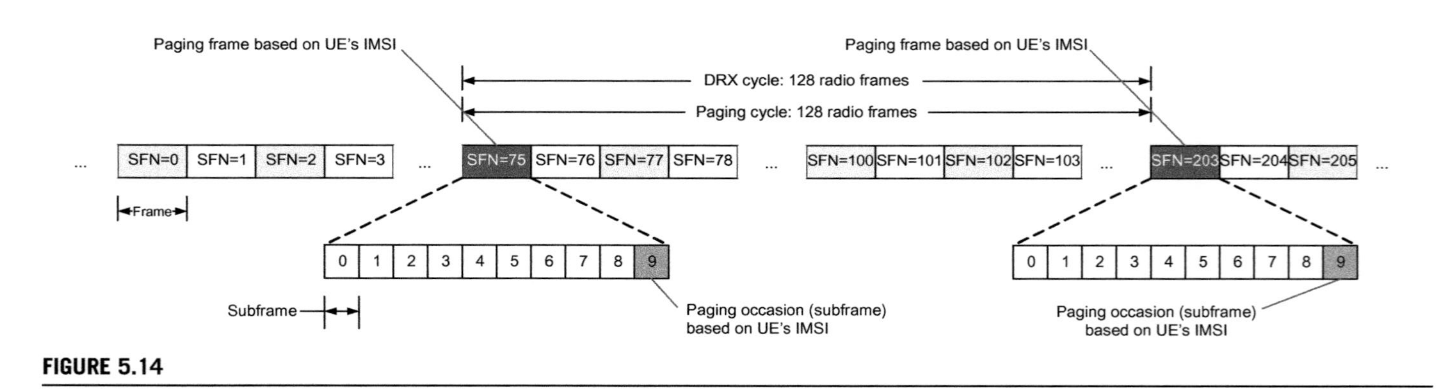

FIGURE 5.14

Illustration of an example paging procedure.

of the location of all UEs in its service area. When a UE registers in the network, the MME creates an entry for the UE and signals the location to the home subscriber server (HSS) in the UE's home network. The MME requests the appropriate resources to be set up in the eNB, as well as in the S-GW which it selects for the UE. The MME then continues tracking the UE's location either at the level of eNB, if the UE remains connected or at the level of a tracking area, which is a group of eNBs which send the paging messages to the UEs in the idle mode (Figure 5.15). The MME controls the setup and release of resources based on the UE's activity. The MME also participates in control signaling for handover of an active mode UE between eNBs, S-GWs, or other MMEs. The MME is involved in every eNB change. An idle-mode UE reports its location either periodically or when it moves to another TA. If data is received from the external networks for an idle-mode UE, the MME will be notified, and it subsequently requests the eNBs in the UE's TA to page the UE [22].

The UE's location in the RRC_IDLE state is known by the MME with the granularity of a TA. The size of the TA can be optimized in network planning

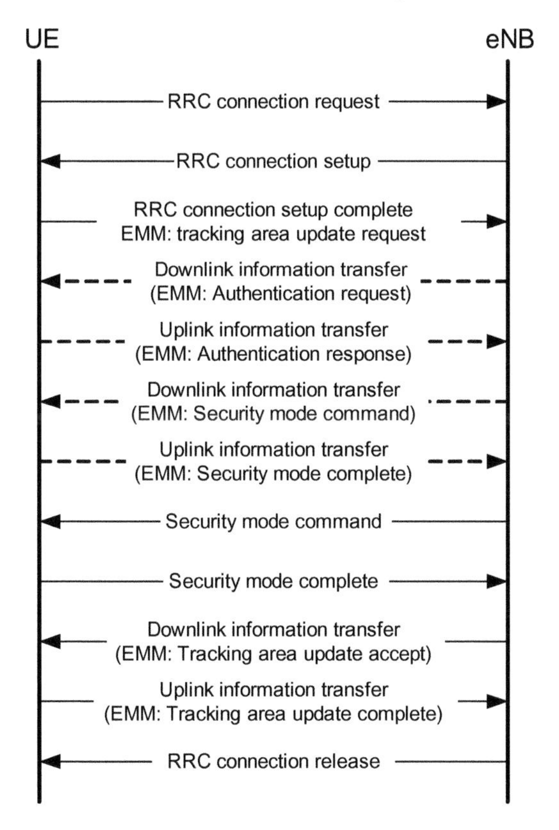

FIGURE 5.15

Tracking Area Update procedure in E-UTRA [1].

phase. A larger TA is beneficial to avoid frequent TA update signaling. On the other hand, a smaller TA is beneficial to reduce the paging signaling load for the incoming packet calls. The UE can be assigned multiple TAs to avoid unnecessary TA updates at the TA boundaries, e.g., when the UE moves back and forth between cells of two different TAs. The UE can also be assigned both an LTE TA and a UTRAN routing area to avoid signaling when changing between the two systems [22].

When an idle-mode UE moves across a TA boundary, this event will trigger the TA update procedure, which is initiated by the UE (Figure 5.16). In this case, the UE sends to the (new) target resources a *Tracking Area Update Request* message. The target MME requests and receives the bearer context information from the (old) source MME. Authentication may be performed at this point, if the source MME does not authenticate the UE. After authentication, the target MME sends a *Create Session Request* message to the target S-GW, which triggers a *Modify Bearer Request* message to be sent to the P-GW. The P-GW responds to the S-GW with a *Modify Bearer Response* message which, in turn, responds to the MME with a *Create Session Response* message. Then, the location of the UE needs to be updated in the HSS. This is accomplished by the target MME sending an *Update Location Request* to the HSS, which cancels the location of the UE in the source MME and confirms the location update to the target MME. Finally, the *Tracking Area Update* is confirmed with a *Tracking Area Update Complete* NAS PDU sent as part of a

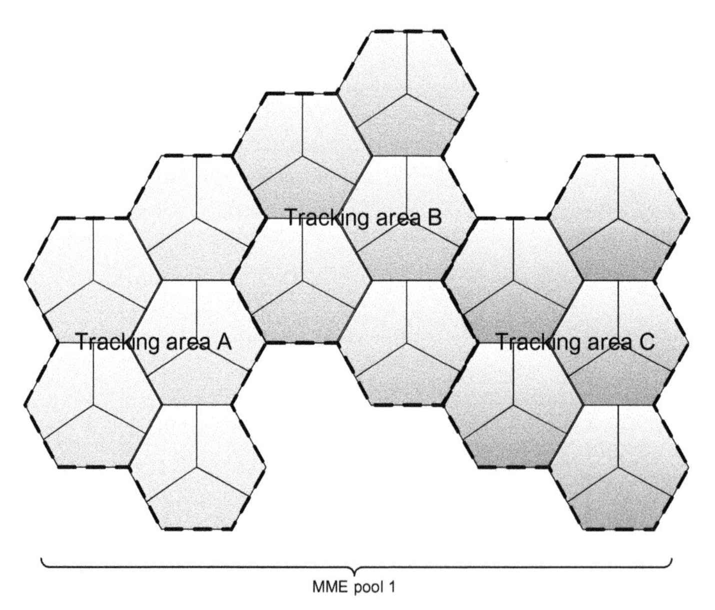

FIGURE 5.16

Illustration of TAs and MME pool [18,22].

downlink information transfer message from the target eNB to the UE. The UE initiates the *Tracking Area Update* procedure in two scenarios:

- When the TA has changed (the UE reselects a cell that does not belong to the TAI list provided by the MME in the *Attach Accept*). The TAI is used to identify TAs. The TAI is constructed from the PLMN identity which the TA belongs to and the tracking area code (TAC) of the TA.
- When timer T3412 expires (also provided in the *Attach Accept* or the last *Tracking Area Update Accept*).

In EPS and during normal operation, the UE registers with an MME group, which may have multiple MMEs sharing the traffic load (see Figure 5.16). This group may have one or more TAs. This implies that the UE can roam in the coverage areas of the eNBs broadcasting these TAs without having to re-register with the network. Note that a UE in the ECM-IDLE (or RRC_IDLE) state is paged in all cells of the TAs in the TAI list. When the UE moves into an eNB coverage area that broadcasts a different TAI, registration with the network is required. In the context of EPS, this is known as the TA update procedure. An EMM-REGISTERED UE also performs periodic TA updates with the network after the periodic TAU timer expires. The UE is still covered by the eNB that is broadcasting the TAC assigned by the registering MME. If GUTI is allocated or changed, the *Tracking Area Update Complete* message must be sent back to the eNB.

At this point, since we have referred to various UE and network identifiers in this chapter, it is worthwhile to review the definition, construction, and application of those identifiers to help better understanding of the procedures described in this chapter. The classification and structure of those identifiers are depicted in Figure 5.17 and summarized in Table 5.5.

5.5.3 PLMN and cell selection

Before the UE can camp on a cell, the cell must meet a number of requirements. Depending on the UE's subscription or access class, the UE may or may not have access to certain cells due to access barring, closed-subscriber-group femto-cell, or reserved cell restrictions. The UE measures the RSRP based on cell-specific reference signals to determine whether the candidate cell satisfies a certain criterion *S*. The criterion *S* is also considered during cell reselection. The cell selection criterion *S* is satisfied when the following conditions are met [2]:

$$S_{\text{rxlev}} > 0 \quad S_{\text{qual}} > 0$$

where

$$S_{\text{rxlev}} = Q_{\text{rxlevmeas}} - (Q_{\text{rxlevmin}} + Q_{\text{rxlevminoffset}}) - P_{\text{compensation}}$$

$$S_{\text{qual}} = Q_{\text{qualmeas}} - (Q_{\text{qualmin}} + Q_{\text{qualminoffset}})$$

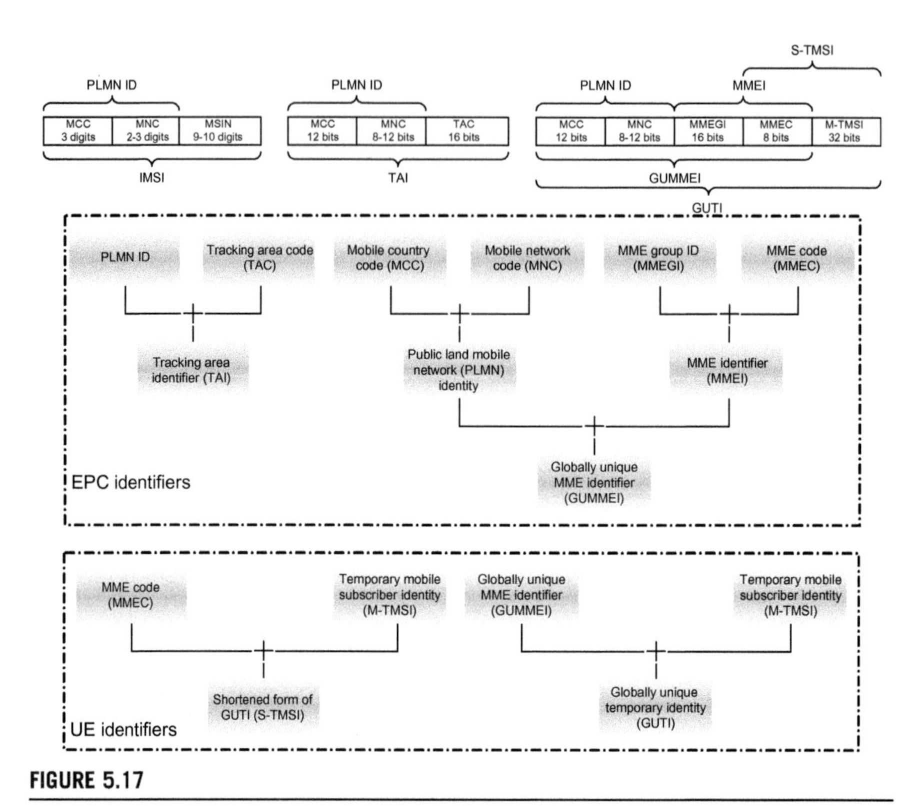

FIGURE 5.17

Classification of UE and network identifiers in LTE/LTE-Advanced [26,27].

In the above equations

- S_{rxlev}: Cell selection receive level value (dB).
- S_{qual}: Cell selection quality value (dB).
- $Q_{rxlevmeas}$: Measured cell receive level value (RSRP).
- $Q_{qualmeas}$: Measured cell quality value (RSRQ).
- $Q_{rxlevmin}$: Minimum required receive level in the cell (dBm).
- $Q_{qualmin}$: Minimum required quality level in the cell (dB).
- $Q_{rxlevminoffset}$: Offset to the signaled $Q_{rxlevmin}$ taken into account in the S_{rxlev} evaluation as a result of a periodic search for a higher priority PLMN while camped normally in a visited PLMN.
- $Q_{qualminoffset}$: Offset to the signaled $Q_{qualmin}$ taken into account in the S_{qual} evaluation as a result of a periodic search for a higher priority PLMN while camped normally in a visited PLMN.
- $P_{compensation} = \max(P_{Emax} - P_{PowerClass}, 0)$ (dB).
- P_{Emax}: Maximum transmit power level that the UE may use when transmitting in the uplink in the cell (dBm) as defined in 3GPP TS 36.101 [8].

- $P_{\text{PowerClass}}$: Maximum RF output power of the UE (dBm) according to the UE power class as defined in 3GPP TS 36.101 [8].

The signaled values $Q_{\text{rxlevminoffset}}$ and $Q_{\text{qualminoffset}}$ are only applied when a cell is evaluated for cell selection as a result of a periodic search for a higher priority PLMN while camped normally in a visited PLMN. During this periodic search for a higher priority PLMN, the UE may check the criterion S of a cell using parameter values stored from a different cell of this higher priority PLMN.

After the UE is camped for normal service, it performs attach and PDN connectivity procedures to register its location with the selected PLMN and obtain TA information. This enables the PLMN to reach the UE. In addition, the UE is required to measure the RSRP level of the serving cell and evaluate the cell selection criterion S defined for the serving cell at least every DRX cycle. The UE is further required to identify new intra-frequency cells and perform RSRP measurements on the identified intra-frequency cells without an explicit intra-frequency neighbor list containing physical cell identities. The LTE specifies the time duration for a UE to detect a cell, as well as the frequency at which the UE measures RSRP. This is true for cells on the serving frequency or other frequencies. The standard also specifies the RSRP levels that are used for the detection of the cells. The UE is required to apply filtering on RSRP during measurements. The NAS is informed, if cell selection and reselection results in changes in the received system information relevant to NAS.

SIB1, SIB3, and SIB4 provide parameters that are used during cell reselection. SIB3 contains measurement rule parameters. The E-UTRAN requires the UEs to detect and measure neighbor cells when the serving cell S_{rxlev} is lower than a specific threshold. The UE may choose not to perform the measurements, if the serving cell's metric is greater than the threshold. After conducting the measurements, the UE applies the ranking criterion and the highest ranked cell is selected during cell reselection. Note that hysteresis is applied through $Q_{\text{Hysteresis}}$ and Q_{offset} before the new cell is selected. During the DRX cycle, if the UE has determined that the serving cell does not fulfill the cell selection criterion S, it initiates measurements of all neighbor cells, irrespective of the measurement rules currently limiting UE measurement activities. If a UE in the RRC_IDLE state searches for 10 seconds without finding a new suitable cell based on searches and measurements using the intra-frequency, inter-frequency, and inter-RAT information indicated in the *System Information*, the UE is considered to be out of service area.

Figure 5.18 demonstrates an intra-frequency cell reselection procedure. In this figure, the cell-ranking criterion R_s for a serving cell and R_n for neighbor cells is defined by [2]:

$$
\begin{aligned}
R_{\text{s}} &= Q_{\text{meas}(s)} + Q_{\text{Hysteresis}} \\
R_{\text{n}} &= Q_{\text{meas}(n)} - Q_{\text{offset}}
\end{aligned}
$$

- Q_{meas}: RSRP measurement quantity used in cell reselections.

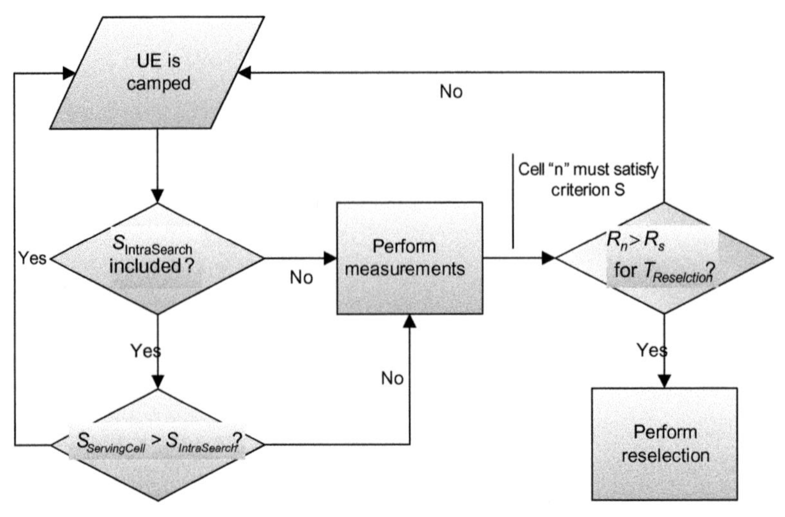

FIGURE 5.18

Illustration of intra-frequency cell reselection [1].

- Q_{offset}: For intra-frequency: $Q_{\text{offset}} = Q_{\text{offset}(s,n)}$ if $Q_{\text{offset}(s,n)}$ (the offset between the two cells) is valid; otherwise $Q_{\text{offset}} = 0$. For inter-frequency: $Q_{\text{offset}} = Q_{\text{offset}(s,n)} + Q_{\text{offset}_{\text{frequency}}}$ if $Q_{\text{offset}(s,n)}$ is valid; otherwise, $Q_{\text{offset}} = Q_{\text{offset}_{\text{frequency}}}$ (frequency-specific offset for equal priority E-UTRAN frequencies).
- $Q_{\text{Hysteresis}}$: Hysteresis value for ranking criteria.

After a UE is powered-up and a PLMN is selected, the cell selection process is initiated. This process allows the UE to select a suitable cell to camp on in order to access network services. During this process, the UE can use initial cell selection or stored information for cell selection. Initial cell selection requires no prior knowledge of E-UTRAN RF channels and the UE is required to scan all RF channels in the E-UTRA bands according to its capabilities to find a suitable cell [8,9]. On each carrier frequency, the UE only searches for the strongest cell. Once a suitable cell is found, the cell is selected. Cell selection based on stored information utilizes the UE stored information of carrier frequencies and cell parameters from previously received measurement control information elements or from previously detected cells. If the UE cannot find any suitable cell based on the latter approach, the initial cell selection procedure is performed. The UE normally operates in its home PLMN. A visited PLMN is selected when the coverage is lost. The visited PLMN may also be manually selected by the user.

5.5.4 Cell reselection

Inter-frequency cell reselection is based on absolute priorities that are provided in SIB5 and are valid for all UEs in the serving cell. In addition, UE-specific priorities may be signaled as part of the *RRC Connection Release* message. The dedicated signaling priorities are deleted when the UE enters the RRC_CONNECTED state, timer T320 expires, or when PLMN selection is performed. Only the frequencies listed in SIB5 are considered for inter-frequency reselection. The list can contain a maximum of eight inter-frequencies that the UE may be allowed to monitor in E-UTRAN. The parameters provided in SIB3 are also considered for ranking evaluations.

Once a cell is selected or reselected, the UE is not allowed to reselect a new cell for at least one second. Furthermore, the latter constraint ensures that reselection decisions are not made in response to instantaneous RF signal variations. For inter-frequency neighbor cells, it is possible to indicate a cell-specific offset to be considered during reselection. These parameters are common to all cells on a different frequency. Blacklists can be provided to prevent the UE from reselecting the specific intra/inter-frequencies. Cell reselection can also be mobility dependent (based on UE speed). SIB6, SIB7, and SIB8 provide the related information for UTRAN/GERAN/CDMA2000, respectively.

If the $S_{\mathrm{ServingCell}}$ of the E-UTRA serving cell (or other cells on the same frequency layer) is greater than $S_{\mathrm{nonIntraSearch}}$, the UE may not search or measure inter-RAT frequency layers of equal or lower priority. If the UE measures the UTRA FDD cells, it is required to measure CPICH E_c/I_0 and CPICH RSCP of detected UTRA FDD cells in the neighbor cell list. Hysteresis, thresholds, suitability, and priorities are considered during the reselection decision to a UTRA cell. For GSM, the signal level of the GSM BCCH carrier of each GSM neighbor cell indicated in the measurement control system information of the serving cell is measured. For HRPD (1xRTT), the UE is required to measure CDMA2000 HRPD pilot strength of HRPD cells (CDMA2000 1x RTT pilot strength of CDMA2000 1X cells) in the neighbor cell list. In addition, the UE may be requested to pre-register with the HRPD network, where the network identifies the pre-registration zone for the UE. Pre-registration reduces the time required for cell reselection and handover by enabling HRPD and point-to-point protocol (PPP) session establishment in advance. The E-UTRA node acts as a relay agent to transfer 1xEV-DO *Session Setup* messaging.

As mentioned earlier, after camping on a suitable E-UTRAN cell, the UE uses a NAS registration procedure to register its presence in the TA of the selected cell. This allows the PLMN to page the user in the registered TA(s). If the location registration is successful, the selected PLMN becomes the registered PLMN. During registration, if the network declares that the TA is forbidden, the TA is added to a list of forbidden TAs for roaming and is stored in the UE. The UE must then search for a suitable cell in the same PLMN and in a TA that is not in the forbidden lists. If a PLMN is not allowed, the PLMN is added to the forbidden PLMN list that is stored on the USIM and is not deleted, even when the UE is switched off. If the UE cannot find another PLMN, it enters a limited-service state and can only originate emergency calls or receive an ETWS message.

Figure 5.19 illustrates an inter-frequency (intra-LTE/inter-RAT) cell reselection procedure. The parameters used in the figure are defined as follows [2]:

- $T_{\text{ReselectionRAT}}$: This specifies the cell reselection timer value. For each target E-UTRA frequency and for each RAT (other than E-UTRA), a specific value for the cell reselection timer is defined, which is applicable when evaluating reselection within E-UTRAN or other RAT. This parameter is not sent in *System Information*, rather it is used in reselection rules by the UE for each RAT.
- $\text{Threshold}_{(x)\text{high}}$: This parameter specifies the S_{rxlev} threshold (in dB) used by the UE when reselecting a higher priority RAT/frequency than the current serving frequency.
- $\text{Threshold}_{(x)\text{low}}$: This parameter specifies the S_{rxlev} threshold (in dB) used by the UE when reselecting a lower priority RAT/frequency than the current serving frequency.
- $\text{Threshold}_{(\text{Serving})\text{low}}$: This parameter specifies the S_{qual} threshold (in dB) used by the UE on the serving cell when reselecting a lower priority RAT/ frequency.

C1 : $cellReselectionPriority_{new\text{-}frequency} > cellReselectionPriority_{serving\text{-}frequency}$

C2 : $S_{nonServingCell(x)} > Threshold_{(x)high}$ for $T_{ReselectionRAT}$

C3: $(S_{ServingCell(x)} < Threshold_{(Serving)low})$ & $(S_{nonServingCell(x)} > Threshold_{(x)low})$ for $T_{ReselectionRAT}$

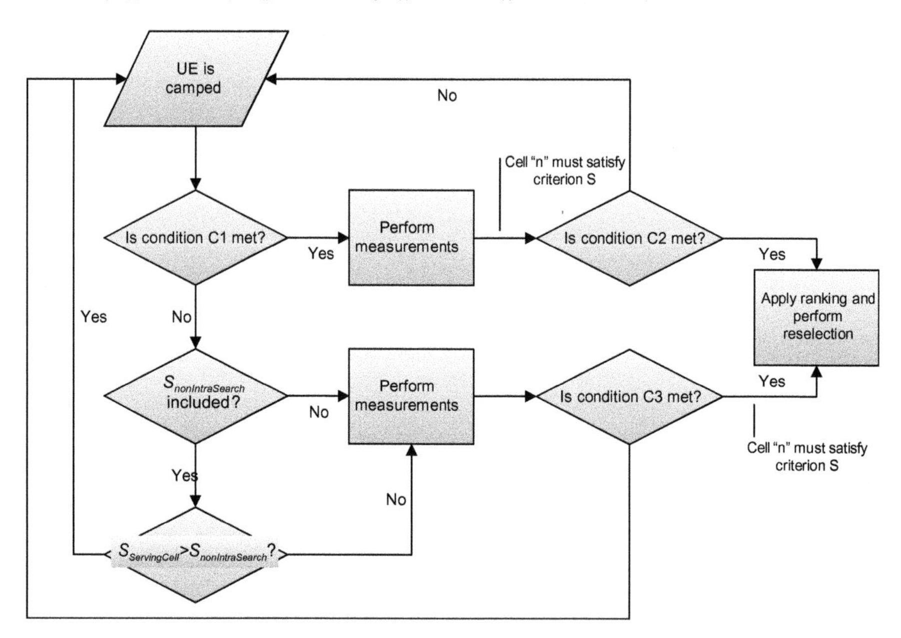

FIGURE 5.19

Illustration of inter-frequency (intra-LTE/inter-RAT) cell reselection procedure (*Note that ranking is only applied to cells with the same reselection priorities*) [1].

– $S_{\mathrm{IntraSearch}}$: This parameter specifies the S_{rxlev} threshold (in dB) for intra-frequency measurements.
– $S_{\mathrm{nonIntraSearch}}$: This parameter specifies the S_{rxlev} threshold (in dB) for E-UTRAN inter-frequency and inter-RAT measurements.

5.5.5 Radio link failure

A UE may declare an RLF in a number of scenarios including the following [1]:

- Downlink physical layer failure when the BLER of PDCCH is greater than 10%: Physical layer problems are detected by monitoring the cell-specific reference signals and estimating the theoretical PDCCH BLER for specific configurations. In-sync and out-of-sync indicators are generated based on the DRX cycle (if configured) and the thresholds Q_{out} (10%) and Q_{in} (2%). If the UE has not been configured with a DRX mode, and the downlink radio link quality estimated during the past 200 ms time interval becomes worse than threshold Q_{out}, the physical layer of the UE sends an out-of-sync indication to the higher layers within 200 ms. When the downlink radio link quality estimated over the past 100 ms time interval becomes better than the threshold Q_{in}, the physical layer of the UE sends an in-sync indication to the higher layers within 100 ms. Two successive indications from the physical layer will be separated by at least 10 ms. The transmitter is turned off within 40 ms after timer T310 expires. If the UE is configured with DRX, the evaluation period for Q_{out} and Q_{in} depends on the DRX configuration. In this case, two successive in-sync or out-of-sync indications are separated by at least the maximum of 10 ms or the DRX cycle.
- Random access problems: If the uplink timing alignment timer expires, then the UE must perform a random access procedure in order to adjust its timing and to obtain an uplink grant before any new data can be transmitted in the uplink. An RLF will be declared, if the UE transmits a maximum number of preamble retransmissions without receiving any response from the eNB.
- Maximum RLC retransmissions: An RLF is declared following the completion of the maximum number of ARQ retransmissions by the RLC sublayer. Following an unsuccessful initial transmission of an RLC PDU, the UE's RLC entity retransmits the PDU a number of times and then declares an RLF upon failure of the receiving entity to confirm reception.
- Handover failure (T304 expiration): Timer T304 is started by the UE after receiving the *RRC Connection Reconfiguration* command ordering a handover. The UE detaches from the source cell and, following successful acquisition of the target cell, uses a random access procedure to gain access to the target cell. If the UE is unable to successfully access the target cell before the expiration of timer T304, an RLF is declared.

Once an RLF has been declared, timer T311 is started. This is only possible, if the AS security was activated prior to the RLF. In this case, SRB1 is suspended and SRB2 and all DRBs are released. If the AS security was not activated, the

UE transitions directly to the idle mode. Upon start of timer T311, the UE attempts cell reselection. If it is unsuccessful before the expiration of timer T311, the UE transitions to the idle mode. If it is successful, the UE starts timer T301 and begins the connection reestablishment procedure with the selected cell. If timer T301 expires without successful reestablishment of the connection, the UE transitions to the idle mode. If successful, SRB1 resumes and the recovery mechanism is complete. The SRB2 and any DRB can then be established.

5.6 RRC timers and constants

A number of LTE/LTE-Advanced RRC functions rely on specific timers to initiate and terminate certain procedures as a result of success or failure of certain tasks, and to measure the run time of certain processes during the operation of the system at the UE and the eNB. The use of timers would prevent a process taking longer than normally expected to perform a task or a series of tasks. The timer would sometimes allow repetition of the associated process/task a permissible number of times before terminating the task. Once a timer associated with a process/task expires when the process has not been successfully concluded, the expiration of the timer terminates the process/task and triggers another process/task. Table 5.7 summarizes the RRC timers specified in LTE Rel-11 specification [1], the conditions to trigger their start and stop, and the actions that the entity would take upon expiration of the timer. Some processes may have several timers associated with them to control different aspects of the process.

5.7 RRC message structure

The RRC messages generally consist of three parts (Figure 5.20):

1. Message type: This information helps the receiver to identify the type of message and other information elements that might be included in the message.
2. RRC transaction identifier: This numerical information is used to separate and identify several ongoing signaling procedures.
3. Other information elements: The RRC message may also include other information elements such as establishment cause, radio resource configuration for SRB1, UE identity, selected PLMN identity, and registered MME information to identify the MME to which the UE is registered.

As an example, the following is an RRC information element that configures the number of transmit antennas and the transmission mode [1,27]:

```
||+-antennaInfo::= CHOICE [explicitValue] OPTIONAL:Exist
||| +-explicitValue::= SEQUENCE [1]
|||    +-transmissionMode::= ENUMERATED [tm4]
|||    +-codebookSubsetRestriction::= CHOICE [n2TxAntenna-tm4] OPTIONAL:Exist
|||    | +-n2TxAntenna-tm4::= BIT STRING SIZE(6) [111111]
|||    +-ue-TransmitAntennaSelection::= CHOICE [setup]
|||      +-setup::= ENUMERATED [closedLoop]
```

Table 5.7 RRC Timers [1]

Timer	Start Condition	Stop Condition	Actions upon Expiration
T300	Transmission of *RRC Connection Request*	Reception of *RRC Connection Setup* or *RRC Connection Reject* message, cell reselection and upon termination of connection establishment by upper layers	Reset MAC, release MAC configuration, and reestablish RLC for all radio bearers that are established and inform upper layers about the failure to establish the RRC connection
T301	Transmission of *RRC Connection Reestablishment Request*	Reception of *RRC Connection Reestablishment* or *RRC Connection Reestablishment Reject* message as well as when the selected cell becomes unsuitable	Go to RRC_IDLE
T302	Reception of *RRC Connection Reject* while performing RRC connection establishment	Upon entering RRC_CONNECTED and upon cell reselection	Inform upper layers about barring mitigation
T303	Access barred while performing RRC connection establishment for mobile originating calls	Upon entering RRC_CONNECTED and upon cell reselection	Inform upper layers about barring mitigation
T304	Reception of *RRC Connection Reconfiguration* message including the *Mobility Control Info* or reception of *Mobility from E-UTRA Command* message including *Cell Change Order*	Criterion for successful completion of handover to E-UTRA or CCO is met (*this criterion is specified in the target RAT in the case of inter-RAT*)	In the case of CCO from E-UTRA or intra E-UTRA handover, initiate the *RRC Connection Reestablishment* procedure
T305	Access barred while performing RRC connection establishment for mobile originating signaling	Upon entering RRC_CONNECTED and upon cell reselection	Inform upper layers about barring mitigation
T310	Upon detecting physical layer problems, i.e., upon	Upon receiving N311 (i.e., *the maximum number of consecutive*	If security is not activated go to RRC_IDLE, else initiate the

(Continued)

Table 5.7 (Continued)

Timer	Start Condition	Stop Condition	Actions upon Expiration
	receiving N310 (i.e., *the maximum number of consecutive "out-of-sync" indications received from lower layers*)	*"in-sync" indications received from lower layers*) consecutive in-sync indications from lower layers; upon triggering the handover procedure, and upon initiating the connection reestablishment procedure	connection reestablishment procedure
T311	Upon initiating the *RRC Connection Reestablishment* procedure	Selection of a suitable E-UTRA cell or a cell using another RAT	Enter RRC_IDLE
T320	Upon receiving t320 or upon cell selection/reselection to E-UTRA from another RAT with validity time configured for dedicated priorities	Upon entering RRC_CONNECTED, when PLMN selection is performed upon request by NAS, or upon cell selection/reselection to another RAT	Discard the cell reselection priority information provided by dedicated signaling
T321	Upon receiving *measConfig* including a *reportConfig* with the purpose set to *reportCGI*. CGI: Cell Global Identification	Upon acquiring the information needed to set all fields of *cellGlobalId* for the requested cell, upon receiving *measConfig* that includes removal of the *reportConfig* with the purpose set to *reportCGI*	Initiate the measurement reporting procedure; stop performing the related measurements and remove the corresponding *measId*
T325	Timer started/restarted upon receiving *RRC Connection Reject* message with *Reprioritization Timer*	–	Stop reprioritization of all frequencies or E-UTRA signaled by *RRC Connection Reject*
T330	Upon receiving *Logged Measurement Configuration* message (*Related to Minimization of Drive Tests feature*)	Upon log volume exceeding the suitable UE memory, upon initiating the release of *Logged Measurement Configuration* procedure	Perform the actions specified in 3GPP TS 36.331
T340	Upon transmitting *UE Assistance Information* message with *powerPrefIndication* set to *normal*	–	–

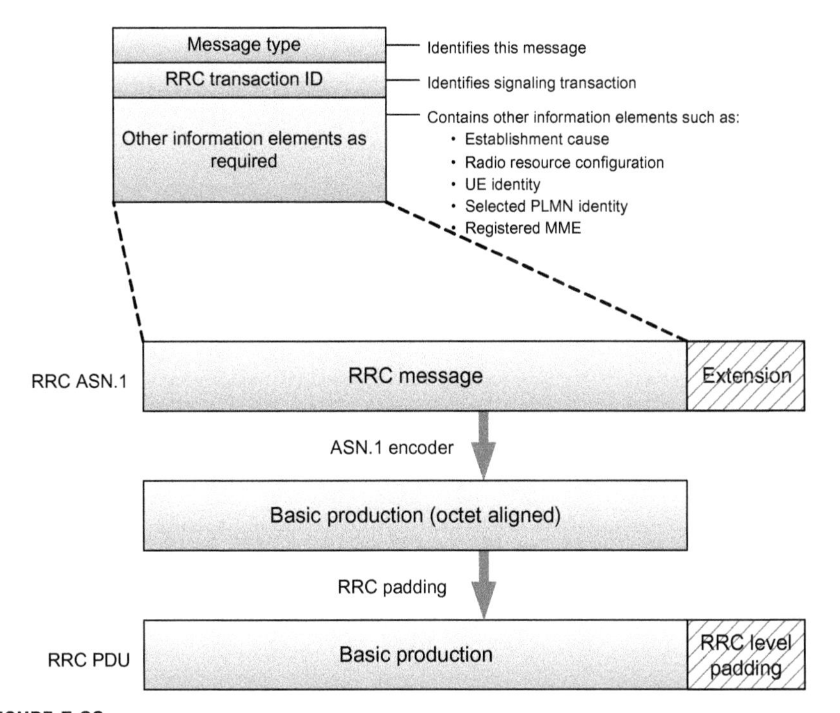

FIGURE 5.20

RRC message structure and processing of RRC messages [1].

The LTE RRC PDU contents are encoded using Abstract Syntax Notation One[3] (ASN.1) as specified in ITU-T Rec. X.680 [23] and X.681 [24]. Transfer syntax for RRC PDUs is derived from their ASN.1 definitions using

[3]ASN.1 is a standard for describing a message that can be sent or received in a network. ASN.1 is divided into two parts: (1) the rules of syntax for describing the contents of a message in terms of data type and content sequence or structure; and (2) how each data item is encoded in a message. ASN.1 is defined in two ISO standards for applications intended for the open systems interconnection framework [23,24]. The following is an example of a message definition specified with ASN.1 notation:

Report:: = SEQUENCE {author OCTET STRING, title OCTET STRING, body OCTET STRING, biblio Bibliography}

In this example, Report is the name of this type of message. SEQUENCE indicates that the message is a sequence of data items. The first four data items have the data type of OCTET STRING, meaning each is a string of bytes. The bibliography data item is another definition named Bibliography that is used.

Bibliography:: = SEQUENCE {author OCTET STRING title OCTET STRING publisher OCTET STRING year OCTET STRING}

Other data types that can be specified include INTEGER, BOOLEAN, REAL, and BIT STRING. An ENUMERATED data type is one that takes one of several possible values. Other data items can be specified as optional.

packed encoding rules as specified in ITU-T Rec. X.691 [25] as shown in Figure 5.20. The "basic production" is obtained by applying unaligned packed encoding rules to the abstract syntax value (the ASN.1 description) as specified in X.691. It always contains a multiple of 8 bits. A transmitter compliant with LTE Rel-11 sets the extension part empty, unless explicitly indicated on a PDU type basis. Transmitters compliant with a later version may send non-empty extensions. If the encoded RRC message does not fill a transport block, the RRC sublayer adds padding bits. This applies to PCCH and BCCH. Padding bits are set to zero and the number of padding bits is a multiple of 8 [1].

References

3GPP Specifications[4]

[1] 3GPP TS 36.331 Evolved Universal Terrestrial Radio Access (E-UTRA); Radio Resource Control (RRC); Protocol specification (Release 11), December 2012.

[2] 3GPP TS 36.304, Evolved Universal Terrestrial Radio Access (E-UTRA); UE Procedures in Idle Mode, (Release 11), December 2012.

[3] 3GPP TS 33.401, 3GPP System Architecture Evolution (SAE): Security architecture (Release 12), December 2012.

[4] 3GPP TS 36.300, Evolved Universal Terrestrial Radio Access (E-UTRA) and Evolved Universal Terrestrial Radio Access Network (E-UTRAN) Overall description, Stage 2 (Release 11), December 2012.

[5] 3GPP TS 36.306, Evolved Universal Terrestrial Radio Access (E-UTRA); User Equipment (UE) Radio Access Capabilities (Release 11), December 2012.

[6] 3GPP TS 36.321, Evolved Universal Terrestrial Radio Access (E-UTRA); Medium Access Control (MAC) Protocol specification (Release 11), December 2012.

[7] 3GPP TS 23.401, General Packet Radio Service (GPRS) enhancements for Evolved Universal Terrestrial Radio Access Network (E-UTRAN) access (Release 11), December 2012.

[8] 3GPP TS 36.101, Evolved Universal Terrestrial Radio Access (E-UTRA); User Equipment (UE) radio transmission and reception (Release 11), December 2012.

[9] 3GPP TS 36.104, Evolved Universal Terrestrial Radio Access (E-UTRA); Base Station (BS) radio transmission and reception (Release 11), December 2012.

[10] 3GPP TS 36.214, Evolved Universal Terrestrial Radio Access (E-UTRA); Physical layer – Measurements (Release 11), December 2012.

[11] 3GPP TS 23.203, Policy and charging control architecture (Release 11), December 2012.

[12] 3GPP TR 36.912, Further Advancements for E-UTRA (LTE-Advanced) (Release 11), September 2012.

[4]3GPP specifications can be found at the following URL: http://www.3gpp.org/ftp/Specs/archive/.

3GPP2 Specifications[5]

[13] 3GPP2 C.S0024-A v3.0, cdma2000 High Rate Packet Data Air Interface Specification, September 2006.
[14] 3GPP2 C.S0057-E v1.0, Band Class Specification for cdma2000 Spread Spectrum Systems, October 2010.
[15] 3GPP2 C.S0005-A v6.0, Upper Layer (Layer 3) Signaling Standard for cdma2000 Spread Spectrum Systems, February 2002.
[16] 3GPP2 A.S0008-C v4.0, Interoperability Specification (IOS) for High Rate Packet Data (HRPD) Radio Access Network Interfaces with Session Control in the Access Network, April 2011.
[17] 3GPP2 C.S0004-A v6.0, Signaling Link Access Control (LAC) Standard for cdma2000 Spread Spectrum Systems, February 2002.

Books

[18] C. Johnson, Long Term Evolution in Bullets, second ed., Create Space Independent Publishing Platform, 2012.
[19] S. Sesia, I. Toufik, M. Baker, LTE, the UMTS Long Term Evolution, From Theory to Practice, second ed., John Wiley & Sons, 2011.
[20] E. Dahlman, S. Parkvall, J. Skold, 4G, LTE/LTE-Advanced for Mobile Broadband, Academic Press, 2011.
[21] H. Holma, A. Toskala, LTE Advanced, 3GPP Solution for IMT-Advanced, John Wiley & Sons, 2012.
[22] H. Holma, A. Toskala, LTE for UMTS, Evolution to LTE-Advanced, John Wiley & Sons, 2011.

Articles

[23] ISO 8824/ITU X.680 specifies the message syntax. <http://www.itu.int/ITU-T/studygroups/com17/languages/X.680-0207.pdf>.
[24] ISO 8825/ITU X.690specifies the basic encoding rules for ASN.1. <http://www.itu.int/ITU-T/studygroups/com17/languages/X.690-0207.pdf>.
[25] ITU-T Recommendation X.691, ASN.1 Encoding Rules: Specification of Packed Encoding Rules (PER), 1999.
[26] LTE Identifiers, NMC Consulting Group, 2011. <http://www.nmcgroups.com>.
[27] LTE System Description. <http://www.sharetechnote.com/>.

[5]3GPP2 specifications can be found at the following URL: http://www.3gpp2.org/Public_html/specs/index.cfm.

CHAPTER

Packet Data Convergence Protocol Functions

CHAPTER CONTENTS

The packet data convergence protocol (PDCP) sublayer is part of the LTE layer 2 protocols (see Figure 6.1), which is responsible for the IP header compression of user-plane data packets in order to reduce the number of information bits transmitted over the air-interface. The header compression mechanism is based on the Internet Engineering Task Force (IETF) standard robust header compression (ROHC) [8–11]. The PDCP sublayer is also responsible for ciphering and integrity protection of control-plane RRC messages, as well as in-sequence delivery and duplicate removal. At the receiver side, the PDCP performs the corresponding deciphering and decompression operations. There is one PDCP entity per radio bearer (RB) configured for a terminal. More specifically, the PDCP sublayer provides the following services to other protocol layers on the user-plane: header compression and decompression using ROHC protocol; transfer of user data; in-sequence delivery of upper-layer PDUs at PDCP reestablishment procedure for RLC acknowledged mode (AM); duplicate detection of lower-layer service data units (SDUs) at PDCP reestablishment procedure for RLC AM; retransmission of PDCP SDUs during handover for RLC AM; ciphering and deciphering; and timer-based SDU discarding in the uplink. The main services and functions of the PDCP on the control-plane include ciphering and integrity protection, as well as transfer of control-plane data. Figure 6.1 illustrates the location of PDCP sublayer in the LTE/LTE-Advanced protocol stack.

The LTE radio-access network provides one or more radio bearers to which IP packets are mapped according to their QoS requirements. Note that there are two

LTE-Advanced.

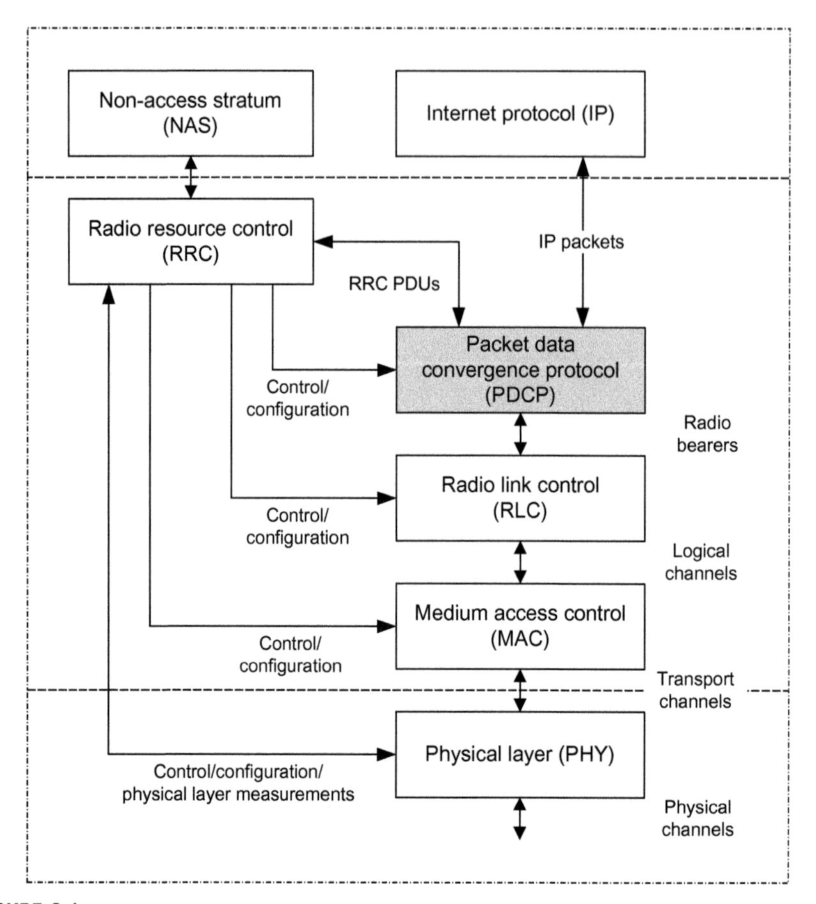

FIGURE 6.1

PDCP sublayer in LTE protocol stack.

types of radio bearers: data radio bearers (DRBs) carrying user-plane data; and signaling radio bearers (SRBs) carrying control-plane information. Each radio bearer is associated with one PDCP entity. Each PDCP entity is associated with one or two (one for each direction) RLC entities depending on the radio bearer characteristic (i.e., unidirectional or bi-directional) and the RLC mode [1–3]. The PDCP entities are located in the PDCP sublayer and are configured by upper layers [2]. A control service access point (SAP) provides the PDCP interface with the RRC sublayer.

As shown in Figure 6.2, the PDCP entities are located in the PDCP sublayer. Several PDCP entities may be defined for a UE. Each PDCP entity carries user-plane data and may be configured to use header compression. Each PDCP entity transports the data of one radio bearer. The LTE standard supports ROHC protocol for header compression as specified by IETF standards [9–11]. Each PDCP entity uses one ROHC instance. A PDCP entity is associated with either the control-plane

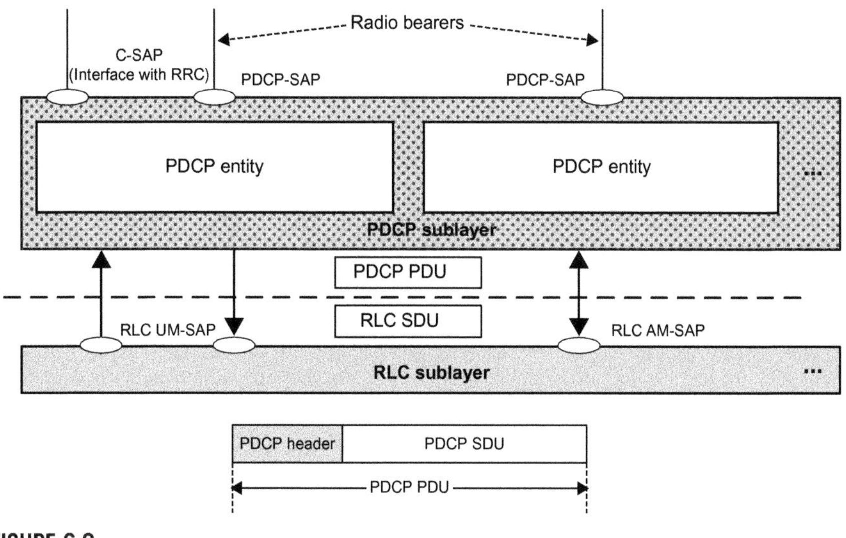

FIGURE 6.2

PDCP sublayer and structure of PDCP PDU [2].

or the user-plane depending on which radio bearer is carrying the data. Figure 6.3 shows the functional decomposition of the PDCP entity in the PDCP sublayer. The figure is based on the radio interface protocol architecture defined in [1]. The PDCP provides services to the RRC sublayer and upper layers in the user-plane at the UE or eNB. As mentioned earlier, PDCP services provided to upper layers include transfer of user-plane and control-plane data, payload header compression, ciphering, and integrity protection.

To provide a better understanding of the PDCP services concerning RRC messages and user data processing, let us further decompose the PDCP functions on the control-plane and user-plane. Figure 6.4 illustrates the PDCP functions of each PDCP entity separately for control-plane and user-plane information. Note that integrity protection and verification functions are only applied to the RRC messages in the downlink and uplink directions, respectively. The header compression and decompression functions are only applied to user-plane data in the downlink and uplink directions, respectively.

The maximum size of a PDCP SDU is 8188 octets [2]. The PDCP uses the services provided by the RLC sublayer. The lower protocol layers provide the following services to the PDCP sublayer:

- Acknowledged data transfer service including indication of successful delivery of PDCP PDUs.
- Unacknowledged data transfer service.
- In-sequence delivery, except at reestablishment of lower layers.
- Duplicate discarding, except at reestablishment of lower layers.

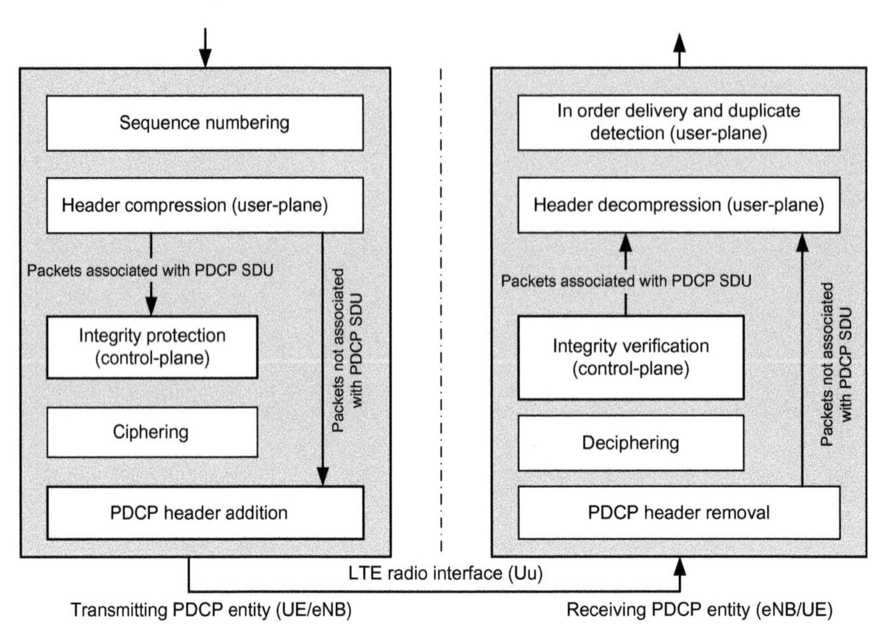

FIGURE 6.3

Functional decomposition of PDCP entities [2].

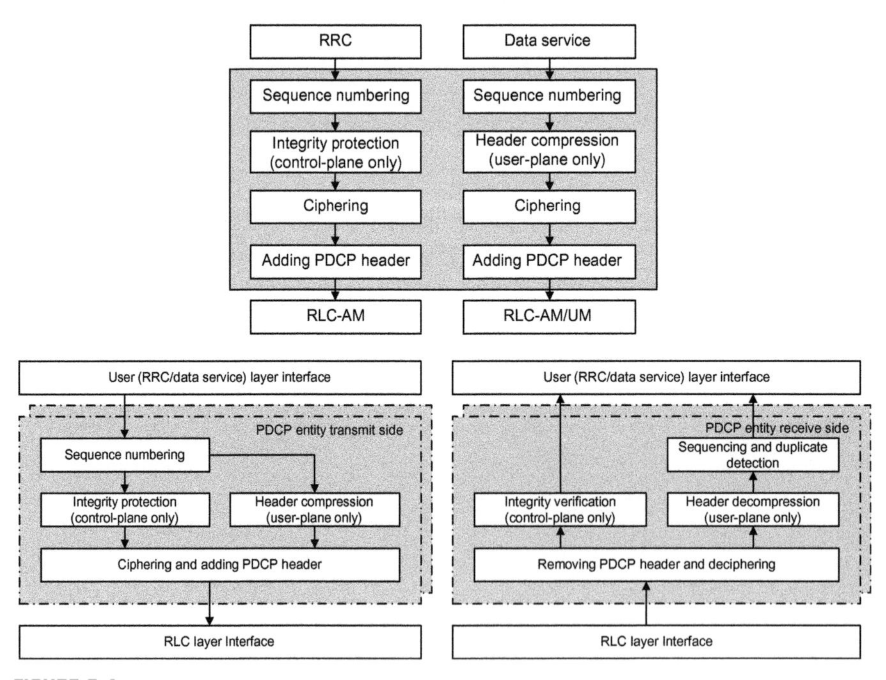

FIGURE 6.4

Illustration of PDCP functions on the control-plane and user-plane.

The PDCP is used for transmission of SRBs (signaling radio bearers carrying control-plane data) and DRBs (data radio bearers carrying user-plane data) that are mapped to dedicated control channel (DCCH) and dedicated traffic channel (DTCH) logical channels. The PDCP is not used for other types of logical channels.

The IETF ROHC protocol is utilized for header compression in LTE [9–11]. There are multiple header compression algorithms, called profiles, defined for the ROHC protocol. Each profile is specific to the particular network layer, transport layer, or upper layer protocol combination, e.g., transmission control protocol/IP (TCP/IP) or real-time transport protocol/user datagram protocol/IP (RTP/UDP/IP) [13,14]. The multiplexing of different flows (with or without header compression) over the ROHC channels, as well as association of a specific IP flow with a specific context state during initialization of the compression algorithm for that flow, are described in [9–11]. The implementation of ROHC functionality and the supported header compression profiles are not specified in 3GPP LTE specifications. Table 6.1 provides the description of ROHC profiles that are used in LTE.

PDCP entities associated with DRBs can be configured by upper layers to use header compression. IETF RFC 4995 provides configuration parameters that are mandatory and must be configured by upper layers between compressor and decompressor entities at the transmitter and receiver sides [11]. Those parameters define the ROHC channel. The ROHC channel is a unidirectional channel, i.e., there is one

Table 6.1 Supported Header Compression Protocols and Profiles [2]

Profile Identifier	Usage	Reference
0 × 0000	No compression	RFC 4995
0 × 0001	RTP/UDP/IP	RFC 3095, RFC 4815
0 × 0002	UDP/IP	RFC 3095, RFC 4815
0 × 0003	ESP[a]/IP	RFC 3095, RFC 4815
0 × 0004	IP	RFC 3843, RFC 4815
0 × 0006	TCP/IP	RFC 4996
0 × 0101	RTP/UDP/IP	RFC 5225
0 × 0102	UDP/IP	RFC 5225
0 × 0103	ESP/IP	RFC 5225
0 × 0104	IP	RFC 5225

[a]*Encapsulating security payload (ESP) is a key component of the IPsec protocol suite. The ESP protocol provides confidentiality and integrity by encrypting data to be protected and places the encrypted data in the data portion of the IP ESP payload. Depending on the user's security requirements, this mechanism may be used to encrypt either a transport-layer segment (e.g., TCP and UDP), or the entire IP datagram. Encapsulating the protected data is necessary to provide confidentiality for the entire original datagram. ESP also supports encryption-only and authentication-only configurations, but using encryption without authentication is not recommended because of insecurity. Unlike the authentication header (AH), ESP does not protect the IP packet header. However, in the tunnel mode, where the entire original IP packet is encapsulated with a new packet header, ESP protection is applied to the entire inner IP packet (including the inner header), while the outer header remains unprotected. ESP operates directly on top of IP using IP number 50 [15].*

channel for the downlink and one for the uplink. Therefore, there is one set of parameters for each channel and the same values are used for both channels belonging to the same PDCP entity.

6.1 Ciphering and integrity protection functions

As shown in Figure 6.5, the ciphering function includes both ciphering and deciphering and is performed in the PDCP sublayer. In LTE, the RLC SDUs are ciphered. For the control-plane, the data unit, that is ciphered, is the data part of the PDCP PDU and the message authentication code for integrity (MAC-I), i.e., a 32-bit field that carries the message authentication code. For the user-plane, the data unit that is ciphered is the data part of the PDCP PDU; ciphering is not applicable to the PDCP control PDUs. The ciphering algorithm and key to be used by the PDCP entity are configured by upper layers, and the ciphering method is applied according to the security architecture of 3GPP system architecture evolution [4]. The ciphering function is activated by upper layers. After security activation, the ciphering function is applied to all PDCP PDUs indicated by upper layers for the downlink and the uplink transmissions. The parameters that are required by the PDCP for ciphering are defined in [2], and are input to the ciphering algorithm. The required inputs to the ciphering function include the *COUNT* (a combination of SN and HFN which is 32 bits) and *DIRECTION* (one-bit direction of the transmission). The parameters required by the PDCP which are provided by upper layers are *BEARER* (defined as the RB identifier which is 5 bits) and *KEY* (the 128-bit ciphering keys for the control-plane and for the user-plane) [2].

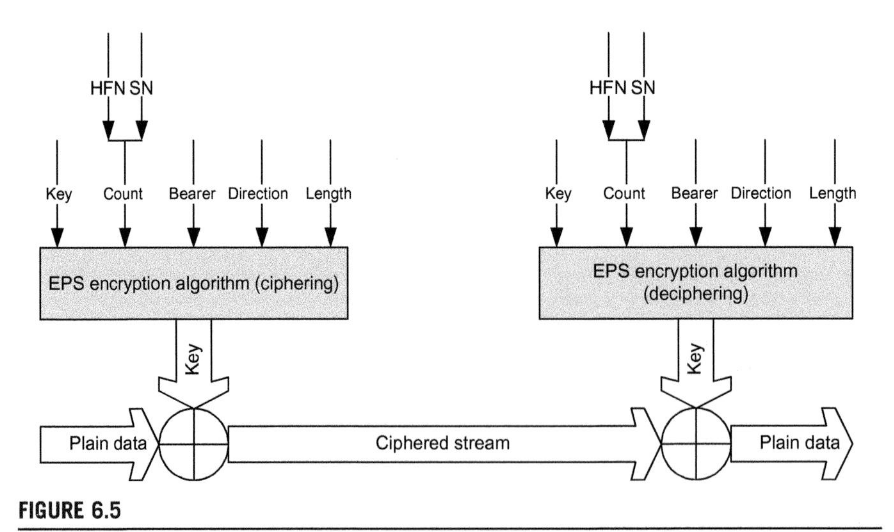

FIGURE 6.5

Illustration of the ciphering and deciphering procedures.

As shown in Figure 6.6, the integrity protection function includes both integrity protection and integrity verification and is performed in the PDCP sublayer for PDCP entities associated with SRBs. The data unit that is integrity protected is the PDU header and the data part of the PDU prior to ciphering. The integrity protection algorithm and key to be used by the PDCP entity are configured by upper layers and the integrity protection method is applied according to the security architecture of 3GPP system architecture evolution [4]. The integrity protection function is activated by upper layers. Following the security activation, the integrity protection function is applied to all PDUs including and subsequent to the PDU indicated by upper layers for the downlink and the uplink transmissions. As the RRC message which activates the integrity protection function is itself integrity protected with the configuration included in that RRC message, the message must be decoded by the RRC sublayer before the integrity protection verification can be performed for the PDU in which the message was received. The parameters that are required by the PDCP for integrity protection are defined in [2], and are input to the integrity protection algorithm. The required inputs to the integrity protection function include the *COUNT* value and *DIRECTION* (direction of the transmission). The parameters required by the PDCP which are provided by upper layers are *BEARER* (defined as the RB identifier) and the *KEY*.

During transmission, the UE computes the value of the MAC-I field, and at reception, it verifies the integrity of the PDCP PDU by calculating the X-MAC, i.e., computed MAC-I, based on the input parameters. If the calculated X-MAC is the same as the received MAC-I, integrity protection is verified successfully. When a PDCP entity receives a PDCP PDU that contains reserved or invalid values, it discards the received PDU. The PDCP data PDU is used to transport the PDCP SDU SN and user-plane data containing an uncompressed PDCP SDU,

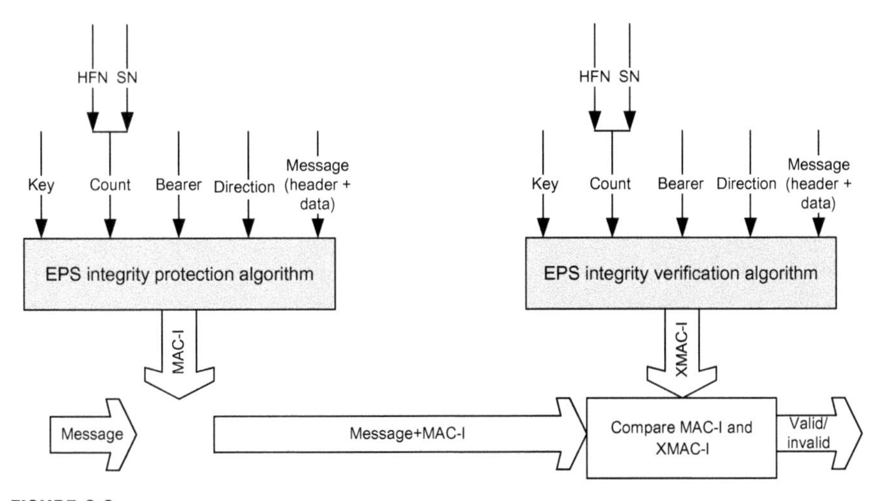

FIGURE 6.6

Integrity protection and verification procedures.

user-plane data containing a compressed PDCP SDU, or control-plane data as well as MAC-I field for SRBs. The PDCP control PDU is used to convey the PDCP status report identifying missing PDCP SDUs following PDCP reestablishment and header compression control information, e.g., ROHC feedback [2].

6.2 Header compression

In some services and applications such as voice over IP (VoIP), interactive gaming, multimedia messaging, etc., the data payload of the IP packet is almost the same size or even smaller than the header itself. Over the end-to-end connection comprising multiple hops, these protocol headers are extremely important but over a single point-to-point link, these headers serve no useful purpose. It is possible to compress these headers, and thus save the bandwidth and use expensive radio resources more efficiently. The header compression also provides other important benefits, such as reduction in packet loss and improved interactive response time [5,8–11]. Payload header compression is the process of suppressing the repetitive portion of payload headers at the sender and restoring them at the receiver side of a low-bandwidth/capacity-limited link. The use of header compression has a well-established history in transport of IP-based payloads over capacity-limited wireless links where more bandwidth efficient transport methods are required. The IETF has developed several header compression protocols that are widely used in telecommunication systems [5,8–11,19–22].

Let us consider some examples of the compression ratios that can be achieved using header compression schemes. The IP version 4 protocol header consists of 20 bytes [6], when combined with the UDP[1] header of 8 bytes and RTP[2] header

[1]The UDP is part of the IP suite, i.e., the set of network protocols used for the Internet. The use of UDP would allow computer applications to send datagrams to other hosts on an IP network without requiring prior signaling to set up data paths. The UDP is defined in IETF RFC 768. The UDP uses a simple transmission model without implicit handshaking dialogues for guaranteeing reliability, ordering, or data integrity. Thus, the UDP provides an unreliable service, and datagrams may be lost or arrive out of order or duplicated. The UDP assumes that error checking and correction is either not necessary or is performed at the application layer, avoiding the overhead of such processing at the network interface level. Delay-sensitive applications such as VoIP and interactive gaming often use UDP because dropping packets would avoid potentially variable and long delays. If error correction schemes are needed at the transport layer, an application may use the TCP which is designed for this purpose. Unlike the TCP, UDP is compatible with packet broadcasting and multicasting. Common applications include media streaming, IPTV, VoIP, and interactive gaming [14,16].

[2]The RTP defines a standardized packet format for delivering audio and video over the Internet. It is defined by IETF RFC 3550 [12]. The RTP is used in communication and entertainment systems that involve media streaming such as telephony, video teleconference applications, and web-based push-to-talk features. Relying on signaling protocols such as session initiation protocol (SIP), the RTP is one of the key ingredients of voice and media transport over IP networks. The RTP is usually used in conjunction with the RTP control protocol (RTCP). While RTP carries the media streams (e.g., audio and video), the RTCP is used to monitor transmission statistics and QoS information. When both protocols are used in conjunction, RTP is usually originated and received on even port numbers, whereas RTCP uses the next highest odd port number [14,17].

of 12 bytes results in an IPv4/UDP/RTP header size of 40 bytes. A header compression scheme typically compresses such headers to 2–4 bytes in the steady-state. Note that the RTP payload size of some commonly used voice codecs is approximately 20–40 bytes for active speech; therefore, the 40 byte IPv4/UDP/RTP protocol header size would be a relatively large overhead. Using header compression in such cases would result in major bandwidth savings. The amount of overhead for small packets when using IP version 6 (IPv6), with increased header size of 40 bytes due to increased IP address space [7], would be even larger. In low-bandwidth or congested networks, the use of header compression may yield better response times due to smaller packet sizes. A small packet may further reduce the probability of packet loss [21]. It has been observed that in applications such as video transmission over wireless links, the use of header compression does not improve the video quality in spite of lower bandwidth usage. For voice transmission, the voice quality may improve while utilizing a lower transmission bandwidth. In summary, header compression helps improve network transmission efficiency, quality, and speed by decreasing protocol header overhead, reducing the packet loss, decreasing interactive response time, and increasing network core users per channel bandwidth which means less infrastructure deployment costs. Figure 6.7 illustrates the structure of IPv6, UDP, and RTP headers.

The IP together with transport protocols such as TCP or UDP and application-layer protocols (e.g., RTP) are described in the form of payload headers. The information carried in the header helps the applications to communicate over large distances connected by multiple links or hops in the network. This information consists of source and destination addresses, ports, protocol identifiers, SNs, error checksums, etc. Under nominal conditions, most of the information carried in packet headers remains the same or changes in specific patterns. By observing the fields that remain constant or change in specific patterns, it is possible either not to send them in each packet, or to represent them in a smaller number of bits than would have been originally required. This process is referred to as compression.

The process of header compression uses the concept of flow context, which is a collection of information about field values and change patterns of field values in the packet header. This context is formed on the compressor and the decompressor side for each packet flow. The first few packets of a newly identified flow are used to build the context on both sides. These packets are sent without compression. The number of these first few packets, which are initially sent uncompressed, is closely related to link characteristics like bit error rate and round trip time. Once the context is established on both sides, the compressor compresses the payload headers as much as possible. By taking into account the link conditions and feedback from the decompressor, the compressed packet header sizes may vary. At certain intervals and in the case of error recovery, uncompressed packet headers are sent to reconstruct the context and revert back to normal operational mode, which is sending compressed packet headers. The header

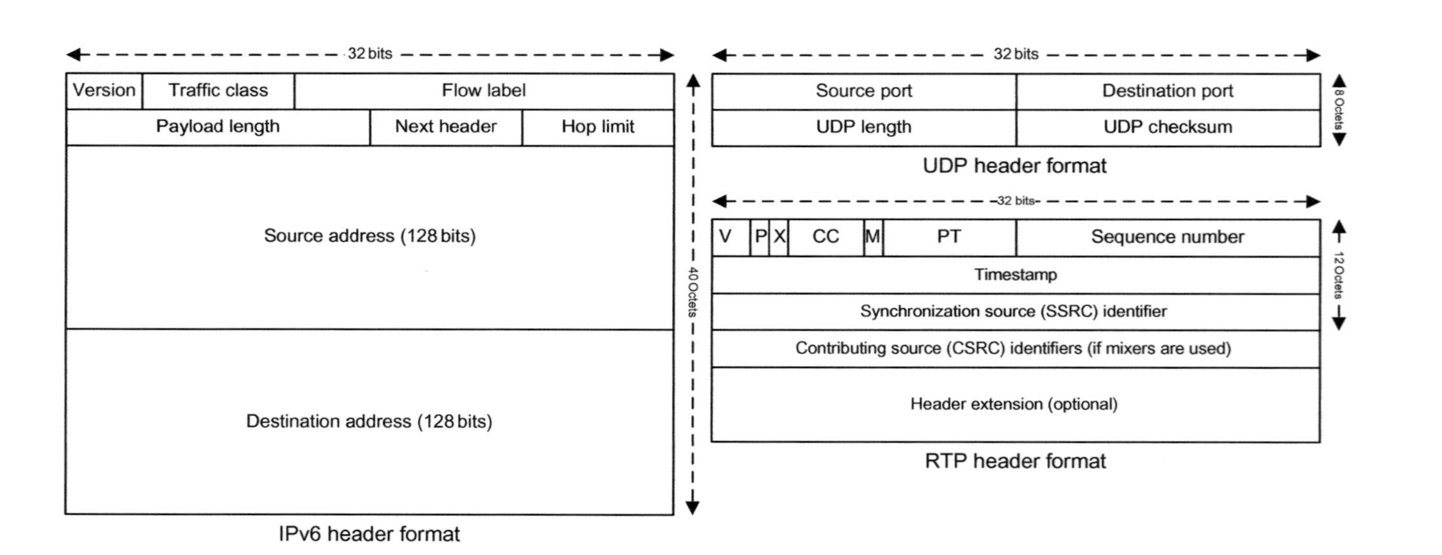

FIGURE 6.7

IPv6, UDP, and RTP header formats [13].

compression module is a part of the protocol stack on the devices. It is a feature which must be negotiated before it can be used on a link. Both end points must agree if they support header compression and on the related parameters to be negotiated.

Considering the end-to-end connectivity over IP, the header compression does not introduce any changes in the data payload when it compresses and decompresses the header. The header compression is a hop-to-hop process and is not applied in an end-to-end connection. At each hop in the IP network, it becomes necessary to decompress the packet to be able to perform operations such as routing, QoS negotiation, and parameter adjustment, etc. Header compression is best suited for specific links in the network characterized by relatively low bandwidth, high bit error rates, and long round trip times. The performance of a header compression scheme can be described with three parameters [18,19]: compression efficiency; robustness; and compression transparency. The compression efficiency is determined by how much the header sizes are reduced by the compression scheme. The compression transparency is a measure of the extent to which the scheme ensures that the decompressed headers are semantically identical to the original headers. If all decompressed headers are semantically identical to the corresponding original headers, the transparency is considered to be 100%. Compression transparency is high when damage propagation is low. When the context of the decompressor is not consistent with the context of the compressor, decompression may fail to reproduce the original header. This condition can occur when the context of the decompressor has not been initialized properly, or when packets have been lost or damaged between compressor and decompressor.

6.3 Robust header compression

The ROHC scheme reduces the size of the transmitted IP/UDP/RTP header by removing redundancies. This mechanism starts by classifying header fields into different classes according to their variation pattern. The fields that are classified as *inferred* are not sent. The *static* fields are sent initially and then are not sent again, and the fields with varying information are always sent. The ROHC mechanism is based on a context[3] which is maintained by both ends, i.e., the compressor and the decompressor (see Figure 6.8). The context encompasses the entire header and ROHC information. Each context has a context identifier, which

[3]The context of the compressor is the state it uses to compress a header. The context of the decompressor is the state it uses to decompress a header. Either of these or the combinations of these two are usually referred to as "context." The context contains relevant information from previous headers in the packet stream, such as static fields and possible reference values for compression and decompression. Moreover, additional information describing the packet stream is also part of the context, for example information about how the IP Identifier field changes and the typical inter-packet increase in sequence numbers or timestamps [14].

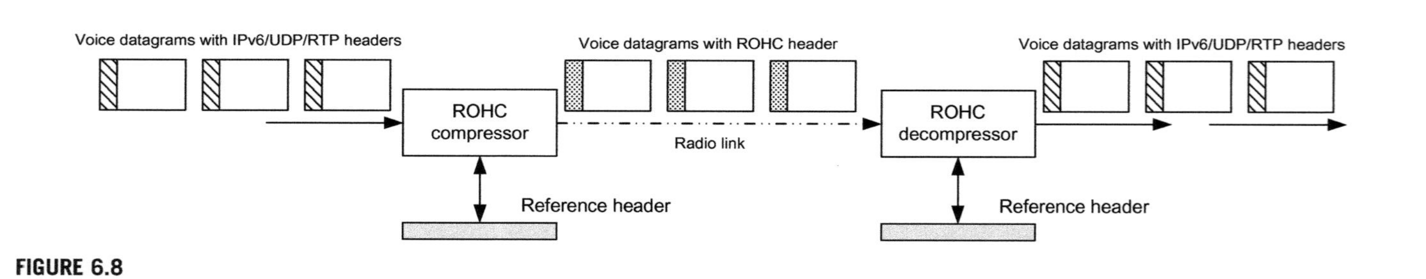

FIGURE 6.8

Example ROHC compression/decompression of IP/UDP/RTP headers for communication over a radio link.

identifies the flows. The ROHC scheme operates in one of the following three operational modes [8–11]:

1. Unidirectional mode (U): In the unidirectional mode of operation, packets are only sent in one direction, from compressor to decompressor. This mode therefore makes ROHC usable over links where a return path from decompressor to compressor is unavailable or undesirable.
2. Optimistic mode (O): The bi-directional optimistic mode is similar to the unidirectional mode, except that a feedback channel is used to send error recovery requests and (optionally) acknowledgments of significant context updates from decompressor to compressor. The O-mode aims to maximize compression efficiency and sparse usage of the feedback channel.
3. Reliable mode (R): The bi-directional reliable mode differs in many ways from the previous two. The most important differences are a more intensive usage of the feedback channel and a stricter logic at both compressor and decompressor that prevents loss of context synchronization between compressor and decompressor except for very high residual bit error rates.

The U-mode is used when the link is unidirectional or when feedback is not possible. For bi-directional links, the O-mode uses positive feedback packets (acknowledgment (ACK)) and R-mode uses positive and negative feedback packets (ACK and negative acknowledgment (NACK)). ROHC always starts header compression using the U-mode, even if it is used in a bi-directional link. ROHC does not make retransmission when an error occurs and the erroneous packet is dropped. The ROHC feedback is used only to indicate to the compressor side that there was an error and probably the context was damaged. After receiving a negative feedback, the compressor always reduces its compression level.

The ROHC compressor has three compression states as follows [8–11]:

1. Initialization and refresh (IR) where the compressor has just been created or reset and full packet headers are sent.
2. First order (FO) where the compressor has detected and stored the static fields such as IP addresses and port numbers on both sides of the connection.
3. Second order (SO) where the compressor is suppressing all dynamic fields such as RTP SNs, and sending only a logical SN and partial checksum to cause the other side to generate the headers based on prediction and to verify the headers of the next expected packet.

Each compression state uses a different header format in order to send the header information. The IR compression state establishes the context, which contains static and dynamic header information. The FO compression state provides the change pattern of dynamic fields. The SO compression state sends encoded values of SN and time stamp (TS), forming the minimal size packets (see Figure 6.9). Using this header format, all header fields can be generated at the other end of the radio link using the previously established change pattern. When some updates or errors occur, the compressor returns to upper compression states.

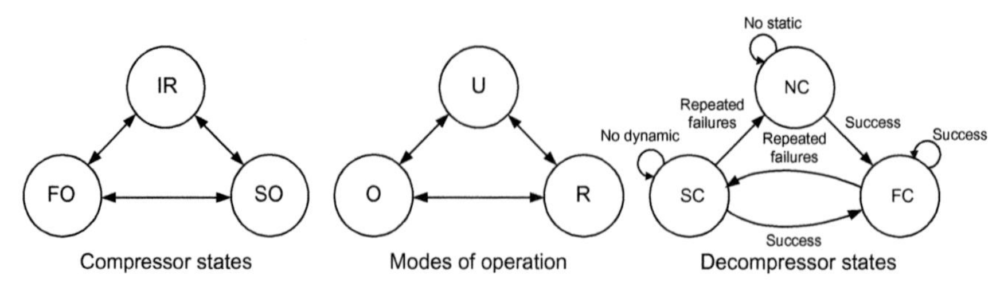

FIGURE 6.9

ROHC state machines [8].

It only transitions to the SO compression state after retransmitting the updated information and reestablishing the change pattern in the decompressor.

In the U-mode, the feedback channel is not used. To increase the compression level, an optimistic approach is used for the compressor to ensure that the context has been correctly established at the decompressor side. This means that the compressor uses the same header format for a number of packets. Since the compressor does not know whether the context is lost, it also uses two timers, to be able to return to the FO and IR compression states. The decompressor works at the receiving end of the link and decompresses the headers based on the header fields' information of the context. Both compressor and decompressor use a context to store all the information about the header fields. To ensure correct decompression, the context should be always synchronized. The decompressor has three states as follows: (1) no context (NC) where there is no context synchronization; (2) static context (SC) where the dynamic information of the context has been lost; and (3) full context (FC) when the decompressor has all the information about header fields. In the FC state, the decompressor transitions to the initial states as soon as it detects corruption of the context. The decompressor uses the "*k out of n*" rule by looking at the last *n* packets with cyclic redundancy check (CRC) failures. If *k* CRC failures have occurred, it assumes the context has been corrupted and transitions to an initial state (SC or NC). The decompressor also sends feedback according to the operation mode (see Figure 6.9).

The values of the ROHC compression parameters that determine the efficiency and robustness are not defined in ROHC specification and are not negotiated initially but are stated as implementation dependent. The values of these parameters stay fixed during the compression process. Those compression parameters are as follows [8–11]:

- *L*: In U-mode and O-mode the ROHC compressor uses a confidence variable *L* in order to ensure the correct transmission of header information.
- *Timer_1 (IR_TIMEOUT)*: In U-mode, the compressor uses this timer to return to the IR compression level and periodically resends static information.
- *Timer_2 (FO_TIMEOUT)*: The compressor also uses another timer in U-mode and this timer is used to go downward to the FO compression level if the compressor is working in the SO compression level.

- *Sliding Window Width (SWW)*: The compressor while compressing header fields like *Sequence Number* and *Time Stamp* uses Window-based least significant bit *(W_LSB)* encoding that uses a sliding window of width equal to *SWW*.
- *W_LSB* encoding is used to compress those header fields whose change pattern is known. When using this encoding, the compressor sends only the least significant bits. The decompressor uses these bits to construct the original value of the encoding fields.
- *k* and *n*: The ROHC decompressor uses a "*k out of n*" failure rule, where *k* is the number of packets received with an error in the last *n* transmitted packets. This rule is used in the state machine of the decompressor to determine the damage of the context and to move to an initial state after sending a negative acknowledgment to the compressor, if a bi-directional link is used. The decompressor does not declare a context corruption and remains in its current state until *k* packets arrive with error in the last *n* packets.

6.4 PDCP PDU packet formats

The PDCP PDU packets are used to transport user-plane data or control-plane information corresponding to the RRC sublayer or NAS. Depending on the type of information, a PDCP PDU packet may include a PDCP SDU SN and user-plane data containing an uncompressed PDCP SDU; or user-plane data containing a compressed PDCP SDU; or control-plane data and a MAC-I field for SRBs. The PDCP control PDU conveys a PDCP status report indicating which PDCP SDUs are missing following PDCP reestablishment or header compression control information, e.g., interspersed ROHC feedback. Figure 6.10 illustrates various forms of PDCP PDU packets [2].

As shown in Figure 6.10, the PDCP PDU packet is a bit string that is octet-aligned. The bit string is supposed to be read from left to right, and then in the reading order of the lines. The bit order of each parameter field within a PDCP PDU is represented with the first and most significant bit in the leftmost bit, and the last and least significant bit in the rightmost bit. The PDCP SDUs are bit strings that are octet-aligned. A compressed or uncompressed SDU is included in a PDCP PDU from the first bit onward.

The PDCP PDU for carrying control-plane information may contain the signaling messages from the RRC and NAS layers where the signaling messages are carried on either SRB 1 or SRB 2 and there is a 5-bit SN field and 32-bit MAC-I checksum field for integrity protection attached to the end of the PDCP control-plane PDU.

The PDCP PDU for carrying user-plane information may contain DRB. A one-bit D/C field (D: user-plane data; C: control information generated at the PDCP layer) is included in the PDCP PDUs carrying user-plane information.

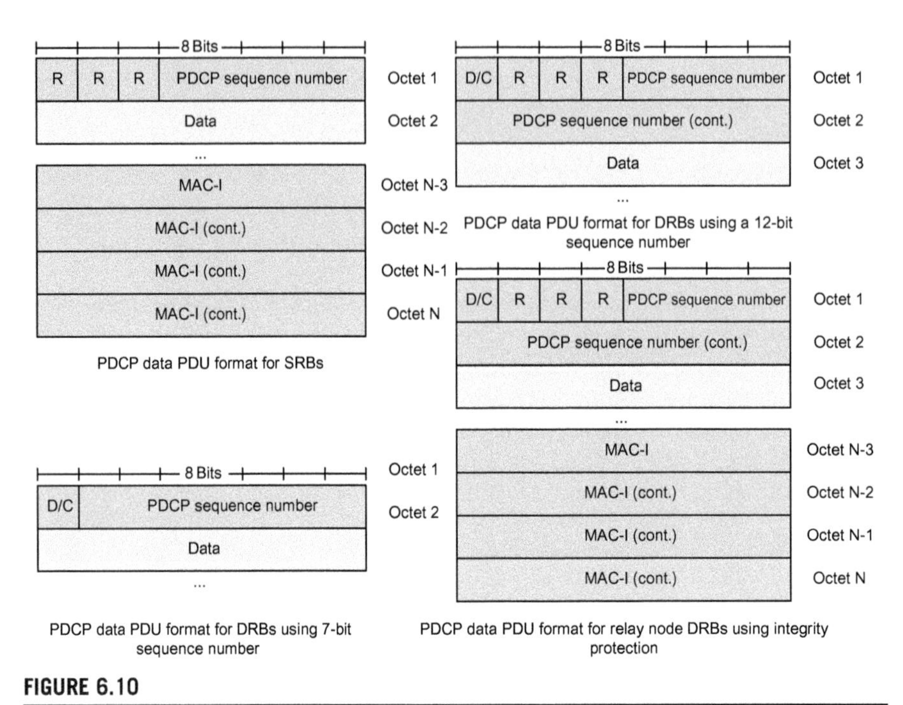

FIGURE 6.10

PDCP PDU packet formats [2].

There are two formats corresponding to a 7-bit or a 12-bit sequence number, and the length of the sequence number is configured on a DRB basis. The unused fields are marked as reserved.

The PDCP data PDU with a 12-bit SN is used for carrying data from DRBs mapped to RLC AM or RLC UM SDUs. The PDCP data PDU with a 7-bit SN is used for transporting data from DRBs mapped on RLC UM. The PDCP control PDU is used to carry one interspersed ROHC feedback packet and is applicable to DRBs mapped to RLC AM or RLC UM SDUs. The PDCP control PDU is used to transport one PDCP status report when a 12-bit SN is utilized. Another form of the PDCP control PDU may be utilized to carry one PDCP status report when a 15-bit SN is utilized, which is applicable to DRBs mapped to RLC AM SDUs. The PDCP data PDU may be used to carry DRBs for relay nodes when integrity protection is configured. The latter format shown in Figure 6.10 is applicable to DRBs mapped to RLC AM or RLC UM SDUs [2].

6.5 LTE security aspects

The E-UTRAN security has been designed based on the following principles. The keys used for NAS and AS protection are dependent on the algorithm with which they are used [1,4]. The eNB keys are cryptographically separated from the EPC keys that are used for NAS protection, making it impossible to use the eNB key to drive an EPC key. The AS and NAS keys are derived in the EPC/UE from key material that was generated by a NAS (EPC/UE) level authentication and key agreement (AKA) procedure[4] (K_{ASME}), and is identified with a key identifier (KSI_{ASME}). The eNB key (K_{eNB}) is sent from the EPC to the eNB when the UE is entering the ECM-CONNECTED state, i.e., during RRC connection or S1 context setup. Separate AS and NAS level security mode procedures are used. The AS level security mode procedure configures AS security (i.e., the RRC and user-plane), and NAS level security mode procedure configures NAS security. Both integrity protection and ciphering for the RRC are activated within the same AS security mode command procedure, but not necessarily within the same message [1].

The user-plane ciphering is activated at the same time as the RRC ciphering. The keys that are stored in the eNBs never leave a secure environment, and user-plane data ciphering/deciphering is conducted within the secure environment where the related keys are stored. The information related to the eNB keys is sent between the eNBs during ECM-CONNECTED state in intra-E-UTRAN mobility. An SN denoted as *COUNT* is used as input to the ciphering and integrity protection. A given SN must only be used once for a given eNB key, except for identical retransmission on the same radio bearer in the same direction. The same SN can be used for both ciphering and integrity protection. An HFN, i.e., an overflow counter mechanism, is used in the eNB and UE in order to limit the actual number of SN bits that are needed to be sent over the radio air-interface. The HFN is synchronized between the UE and eNB. If corruption of keys is detected, the UE has to restart the radio level attach procedure, e.g., similar radio level procedure to the RRC_IDLE to RRC_CONNECTED mode transition or initial attach. Since subscriber identity module (SIM) access is not supported in E-UTRAN, an idle-mode UE not equipped with USIM cannot attempt to reselect the E-UTRAN. To prevent handover to the E-UTRAN, the UE which is not equipped with USIM indicates E-UTRA support in UE capability signaling in other radio access

[4]The AKA mechanism is based on challenge-response mechanisms and symmetric cryptography. The AKA scheme is typically used in a UMTS universal subscriber identity module (USIM), or a cdma2000 removable user identity module (RUIM). The third generation AKA provides substantially longer key lengths and mutual authentication compared to the second generation mechanisms such as GSM AKA. The introduction of AKA inside EAP further enables several new applications, including the use of the AKA as a secure PPP authentication method in devices that already contain an identity module, the use of the third generation mobile network authentication infrastructure in the context of wireless LANs as well as relying on AKA and the existing infrastructure in a seamless way with any other technology that can use EAP. In AKA, the identity module and the home environment have agreed on a secret key in advance. The "home environment" refers to the home operator's authentication network infrastructure. Furthermore, the actual authentication process starts by having the home environment produce an authentication vector, based on the secret key and a sequence number. The authentication vector contains a random part, an authenticator part used for authenticating the network to the identity module, an expected result part, a 128-bit session key for integrity check, and a 128-bit session key for encryption [1,4].

technologies (RATs). A simplified key derivation algorithm is illustrated in Figure 6.11, where [1]:

- K_{NASint} is a key, which is only used for the protection of NAS traffic with a particular integrity algorithm. This key is derived by the UE and MME from K_{ASME}, as well as an identifier for the integrity algorithm.
- K_{NASenc} is a key, which is only used for the protection of NAS traffic with a particular encryption algorithm. This key is derived by the UE and MME from K_{ASME}, as well as an identifier for the encryption algorithm.
- K_{eNB} is a key derived by the UE and MME from K_{ASME}. K_{eNB} may also be derived by the target eNB from the next hop (NH) during handover. K_{eNB} is used for the derivation of K_{RRCint}, K_{RRCenc}, and K_{UPenc}, for the derivation of K_{eNB^*} upon handover.
- K_{eNB^*} is a key derived by the UE and source eNB from either K_{eNB}, or from a fresh NH. K_{eNB^*} is used by the UE and target eNB as a new K_{eNB} for RRC and user-plane traffic.
- K_{UPenc} is a key which is only used for the protection of user-plane traffic with a particular encryption algorithm. This key is derived by the UE and eNB from K_{eNB}, as well as an identifier for the encryption algorithm.
- K_{RRCint} is a key which is only used for the protection of RRC traffic with a particular integrity algorithm. K_{RRCint} is derived by the UE and eNB from K_{eNB}, as well as an identifier for the integrity protection algorithm.

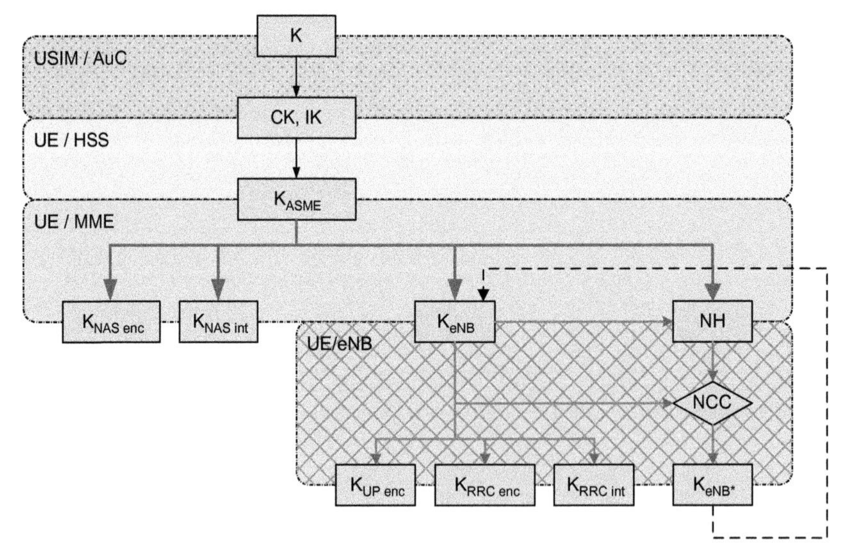

FIGURE 6.11

Key derivation procedure in LTE [1].

- K_{RRCenc} is a key which is only used for the protection of RRC traffic with a particular encryption algorithm. K_{RRCenc} is derived by the UE and eNB from K_{eNB} as well as an identifier for the encryption algorithm.
- The NH is used by the UE and eNB in the derivation of K_{eNB^*} for the provision of *Forward Security*. The NH is derived by the UE and MME from K_{ASME} and K_{eNB} when the security context is established, or from K_{ASME} and the previous NH.
- Next hop chaining count (NCC) is a counter related to NH, i.e., the amount of key chaining that has been performed, which allows the UE to be synchronized with the eNB and to determine whether the next K_{eNB^*} needs to be based on the current K_{eNB} or a fresh NH.

The AuC, HSS, and MME entities shown in Figure 6.11 are authentication center (i.e., the network entity that can authenticate subscribers in a mobile network), home subscriber service (i.e., the main IP multimedia sub-system (IMS) database which also acts as a database in EPC that combines legacy home location register and AuC functions for circuit-switched and packet-switched domains), and mobility management entity, respectively [1].

In Figure 6.11, CK and IK denote keys that have been agreed during the AKA procedure. The MME invokes the AKA procedures by requesting authentication vectors from the home environment (HE), if no unused evolved packet system (EPS) authentication vectors have been stored. The HE sends an authentication response back to the MME that contains a fresh authentication vector including a base-key named K_{ASME}. Consequently, the EPC and the UE share K_{ASME}. From K_{ASME}, the NAS keys, and indirectly K_{eNB} keys and NH are derived. The K_{ASME} is never transported to an entity outside of the EPC, but K_{eNB} and NH are transported to the eNB from the EPC when the UE transitions to ECM-CONNECTED state. From the K_{eNB}, the eNB and UE can derive the user-plane and RRC keys.

The RRC and user-plane keys are refreshed during handover. K_{eNB^*} is derived by the UE and source eNB from target *physical cell identifier*, target frequency and K_{eNB}, or alternatively from target *physical cell identifier*, target frequency and NH. K_{eNB^*} is then used as a new K_{eNB} for RRC and user-plane traffic at the target. If the UE transitions to the ECM-IDLE state, all keys are deleted in the eNB. The reuse of COUNT is avoided for the same radio bearer identity in the RRC_CONNECTED mode without K_{eNB} change, and this is left to eNB implementation. If HFN becomes unsynchronized in the RRC_CONNECTED mode between the UE and eNB, the UE is forced to transition to the idle mode. Table 6.2 shows the security termination points in LTE.

Integrity protection for user-plane is not required and thus it is not supported between the UE and serving gateway, or for the transport of user-plane data between the eNB and the serving gateway on the S1 interface (i.e., the interface between eNB and MME). In general, upon RRC_IDLE to RRC_CONNECTED state transition, RRC protection and user-plane protection keys are generated, while keys for NAS protection as well as higher layer keys are assumed to be already available in the MME. These higher layer keys may have been established in the MME as a

Table 6.2 Security Termination Points in LTE [1]

	Ciphering	Integrity Protection
NAS signaling	Required and terminated in MME	Required and terminated in MME
User-plane data	Required and terminated in eNB	Not required
RRC signaling (AS)	Required and terminated in eNB	Required and terminated in eNB
MAC signaling (AS)	Not required	Not required

result of an AKA run, or transferred from another MME during handover or idle mode mobility. Upon RRC_CONNECTED to RRC_IDLE state transition, the eNB deletes the keys that it stores such that the context for idle-mode UE only resides in the MME. The state and current keys of the UE are assumed to have been deleted at the eNB upon such transition. In particular, the eNB and UE delete NH, K_{eNB}, K_{RRCenc}, K_{RRCint}, and K_{UPenc} and related NCC, while MME and UE keep K_{ASME}, K_{NASint}, and K_{NASenc}.

The RRC and user-plane keys are derived based on the algorithm identifiers and K_{eNB} which results in new RRC and user-plane keys in each handover. Source eNB and UE independently create K_{eNB^*}. The K_{eNB^*} is provided to target eNB during the handover preparation time. Both target eNB and UE consider the new K_{eNB} to be equal to the received K_{eNB^*}. If AS keys (K_{UPenc}, K_{RRCint}, and K_{RRCenc}) need to be changed in RRC_CONNECTED state, an intra-cell handover is used. Inter-RAT handover from UTRAN to E-UTRAN is only supported after activation of integrity protection in UTRAN.

References

3GPP Specifications[5]

[1] 3GPP TS 36.300, Evolved Universal Terrestrial Radio Access (E-UTRA) and Evolved Universal Terrestrial Radio Access Network (E-UTRAN); Overall description; Stage 2 (Release 11), December 2012.

[2] 3GPP TS 36.323, Evolved Universal Terrestrial Radio Access (E-UTRA); Packet Data Convergence Protocol (PDCP) specification (Release 11), December 2012.

[3] 3GPP TS 36.322, Evolved Universal Terrestrial Radio Access (E-UTRA); Radio Link Control (RLC) protocol specification (Release 11), December 2012.

[4] 3GPP TS 33.401, 3GPP System Architecture Evolution (SAE): Security architecture (Release 12), December 2012.

[5]3GPP specifications can be downloaded from http://www.3gpp.org/ftp/Specs/archive/.

IETF Specifications[6]

[5] IETF RFC 2507, M. Degermark et al., IP Header Compression. <http://www.rfc-editor.org/rfc/rfc2507.txt>, February 1999.

[6] IETF RFC 791, Internet Protocol. <http://www.ietf.org/rfc/rfc0791.txt>, September 1981.

[7] IETF RFC 2460, S. Deering, R. Hinden, Internet Protocol, Version 6 (IPv6) specification, December 1998.

[8] IETF RFC 3095, C. Bormann, et al., RObust Header Compression (ROHC): Framework and four profiles: RTP, UDP, ESP, and uncompressed. <http://www.ietf.org/rfc/rfc3095.txt>, July 2001.

[9] IETF RFC 3759, L-E. Jonsson, RObust Header Compression (ROHC): Terminology and Channel Mapping Examples. <http://www.ietf.org/rfc/rfc3759.txt>, April 2004.

[10] IETF RFC 4815, L.-E. Jonsson et al., RObust Header Compression (ROHC): Corrections and Clarifications to RFC 3095, February 2007.

[11] IETF RFC 4995, L.-E. Jonsson et al., The RObust Header Compression (ROHC) Framework, July 2007.

[12] IETF RFC 3550, H. Schulzrinne et al., RTP: A Transport Protocol for Real-Time Applications, July 2003.

Books

[13] D.E. Comer, Internetworking with TCP/IP Vol. 1: Principles, Protocols, and Architecture, fifth ed., Prentice Hall, July 2005.

[14] C. Perkins, RTP: Audio and Video for the Internet, Addison-Wesley Professional, June 2003.

Articles

[15] Encapsulating security payload (ESP). <http://en.wikipedia.org/wiki/IPsec#Encapsulating_Security_Payload>.

[16] User datagram protocol (UDP), Wikipedia. <http://en.wikipedia.org/wiki/User_Datagram_Protocol>.

[17] Real-time transport protocol (RTP), Wikipedia. <http://en.wikipedia.org/wiki/Real-time_Transport_Protocol>.

[18] A. Minaburo, et al., Proposed behavior for robust header compression over a radio link, IEEE Int. Conf. Commun. 7 (2004).

[19] D.E. Taylor, et al., Robust header compression (ROHC) in next-generation network processors, IEEE/ACM Trans. Network. 13 (4) (2005).

[20] S. Ayed, et al., Enhancing robust header compression over IEEE 802 networks, IEEE International Conference on Wireless and Mobile Computing, Networking and Communications, 2006. (WiMob'2006), June 2006.

[21] A. Minaburo, L. Nuaymi, et al., Configuration and analysis of robust header compression in UMTS, 14th IEEE Proceedings on Personal, Indoor and Mobile Radio Communications (PIMRC-2003), vol. 3, September 2003.

[22] P. Fortuna, M. Ricardo, Header compressed VoIP in IEEE 802.11, IEEE Wireless Commun. 16 (2009).

---

[6]IETF specifications can be downloaded from http://www.ietf.org/rfc.html.

CHAPTER

Radio Link Control Functions 7

CHAPTER CONTENTS

The radio link control (RLC) sublayer is located between packet data convergence protocol (PDCP) and medium access control (MAC) sublayers as shown in Figure 7.1. The RLC sublayer is responsible for transferring upper layer protocol data units (PDUs) generated by radio resource control (RRC) or PDCP sublayers [3,4]. The RLC sublayer can operate in one of the three modes of operation defined as transparent mode (TM), unacknowledged mode (UM), and acknowledged mode (AM). Depending on the mode of operation, the RLC entity controls the usage of error correction, concatenation, segmentation, resegmentation, duplicate detection, and in-sequence delivery of service data units. Other functions of the RLC sublayer include protocol error detection, recovery, and service data unit (SDU) discarding. The RLC entity can be reestablished by the RRC such that a reset will cause the transmit/receive buffers to discard any non-transmitted or

LTE-Advanced.

incompletely transmitted SDUs. Furthermore, RLC can also be instructed to discard any unmapped RLC SDU by the upper layers such as PDCP [10].

The RLC sublayer communicates with the PDCP sublayer through a service access point (SAP), and with the MAC sublayer via logical channels [1]. The RLC sublayer reformats PDCP PDUs in order to fit them into the payload size permissible by the MAC sublayer, i.e., the RLC transmitter segments and/or concatenates the PDCP PDUs, and the RLC receiver reassembles the RLC PDUs to reconstruct the PDCP PDUs. In addition, the RLC reorders the RLC PDUs if they are received out of sequence due to a hybrid automatic repeat request (HARQ) operation performed in the MAC sublayer. The functions of the RLC sublayer are performed by RLC entities. An RLC entity is configured in one of three data transmission modes: transparent mode, unacknowledged mode, or acknowledged mode.

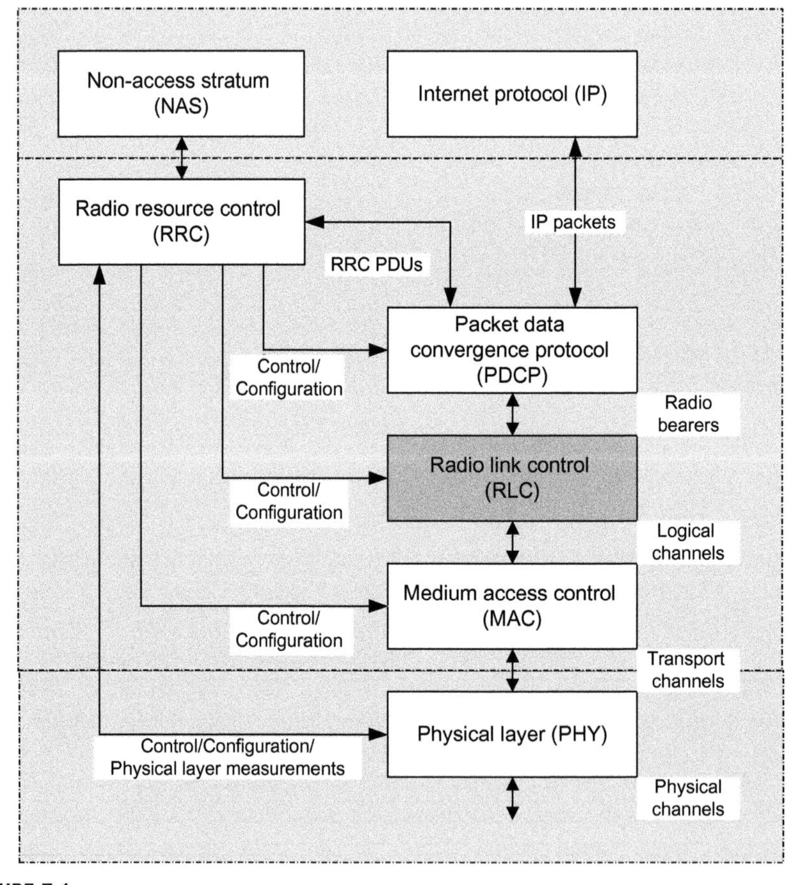

FIGURE 7.1

The RLC sublayer in the LTE protocol stack.

The main services and functions of the RLC sublayer include transfer of upper layer PDUs, error correction through automatic repeat request (ARQ) (AM data transfer), concatenation, segmentation and reassembly of RLC SDUs (UM and AM data transfer), resegmentation of RLC data PDUs (AM data transfer), reordering of RLC data PDUs (UM and AM data transfer), duplicate detection (UM and AM data transfer), protocol error detection (AM data transfer), RLC SDU discard (UM and AM data transfer), and RLC reestablishment [8].

The segmentation and concatenation functions are meant to generate RLC PDUs of appropriate size from the incoming RLC SDUs. One option would be to define a fixed PDU size; however, if the chosen size is too large, it will not be possible to efficiently support small-packet applications and low data rates. Moreover, excessive padding would be required in some scenarios. On the other hand, a single small PDU size would result in an excessive overhead due to the the protocol header included in each PDU. To avoid these disadvantages, which are especially important given the large variety of data applications and services supported by LTE, the RLC PDU size is designed to dynamically vary. The RLC header includes information fields such as sequence number (SN), which is used by the reordering and retransmission mechanisms to sort the PDUs. The reassembly function at the receiver side performs the reverse operation to reassemble the SDUs from the received PDUs [11].

Retransmission of missing PDUs is one of the main functions of the RLC sublayer. Although most of the errors can be detected and corrected by the HARQ protocol at the physical layer and MAC sublayer [1,5,7], there are benefits in having a complementary retransmission mechanism such as ARQ. By inspecting the SNs of the received PDUs, missing PDUs can be detected and a retransmission can be requested from the transmitting entity. Different services have different requirements, i.e., for some services (e.g., FTP), error-free delivery of data is essential, whereas for other applications (e.g., streaming services and VoIP), a small amount of missing packets will not cause a significant degradation.

In-sequence delivery implies that data blocks are delivered by the receiver in the same order as they were transmitted. This is an essential task of the RLC sublayer. The HARQ processes operate independently and the transport blocks may be delivered out of sequence to the RLC sublayer. In-sequence delivery implies that SDU n should be delivered prior to SDU $n + 1$. This is an important feature, as many applications require the data to be received in the same order as it was transmitted. As an example, TCP transport protocol can limitedly handle IP packets arriving out of sequence with some performance degradation; however, for some streaming applications in-sequence delivery is critical. The main idea behind in-sequence delivery is to store the received PDUs in a buffer until all PDUs with lower SN have been delivered. In practice, the RLC retransmission module, when operating in the acknowledged mode, operates on the same buffer as the in-sequence delivery mechanism [11].

The RLC PDUs are exchanged with the MAC sublayer, and the data transfer between the two protocol layers in the uplink and downlink is through logical channels. The RLC PDUs are of variable size and can be formatted based on the transport block size available to the MAC sublayer considering the physical channel conditions. The MAC sublayer notifies the RLC sublayer when a transmission opportunity becomes available, including the size of the RLC PDUs that can be transmitted in the current transmission opportunity.

In this chapter, we will take a top-down systematic approach to describe various functions and services of the LTE/LTE-Advanced RLC sublayer, starting with the RLC architecture and then analyzing each function and service separately to provide an in-depth understanding of this E-UTRA protocol layer.

7.1 RLC architecture

The functions of the RLC sublayer are controlled by the RRC sublayer and are performed by RLC entities. For each RLC entity configured at the eNB, there is a corresponding RLC entity configured at the UE. As shown in Figure 7.2, an RLC entity receives RLC SDUs from the upper layer and sends RLC PDUs to its peer RLC entity via lower layers. An RLC PDU can either be an RLC data PDU or an RLC control PDU. In the transmitter side, the RLC entity receives RLC SDUs from the upper layer through a single SAP between the RLC and the upper layer, and after forming the RLC data PDUs from the received RLC SDUs, the RLC entity delivers the RLC data PDUs to the lower layer through a single logical channel. Similarly, in the receiver side the RLC entity receives RLC data PDUs from a lower layer, through a single logical channel, and after constructing the RLC SDUs from the received RLC data PDUs, the RLC entity delivers the RLC SDUs to the upper layer through a single SAP between the RLC and the upper layer. Depending on the link direction, an RLC entity delivers or receives RLC control PDUs to or from lower layers through the same logical channel that it delivers or receives the RLC data PDUs.

The RLC sublayer receives data from upper layer radio bearers (signaling and data) called SDUs. The transmission entities in the RLC sublayer convert them to RLC PDUs after performing functions such as segmentation, concatenation, and adding RLC headers. In the other direction, the receiving entities receive RLC PDUs from the MAC sublayer. After performing reordering, the RLC PDUs are reassembled into RLC SDUs and are delivered to the upper layer. For each mode, there is a transmitting entity and a receiving entity. The transmitting and receiving entities operate independently for TM and UM. For AM, the transmitting and receiving entities operate together since retransmissions have to be performed. Therefore, feedback received by the receiving entity is given to the transmitting entity to make retransmission decisions.

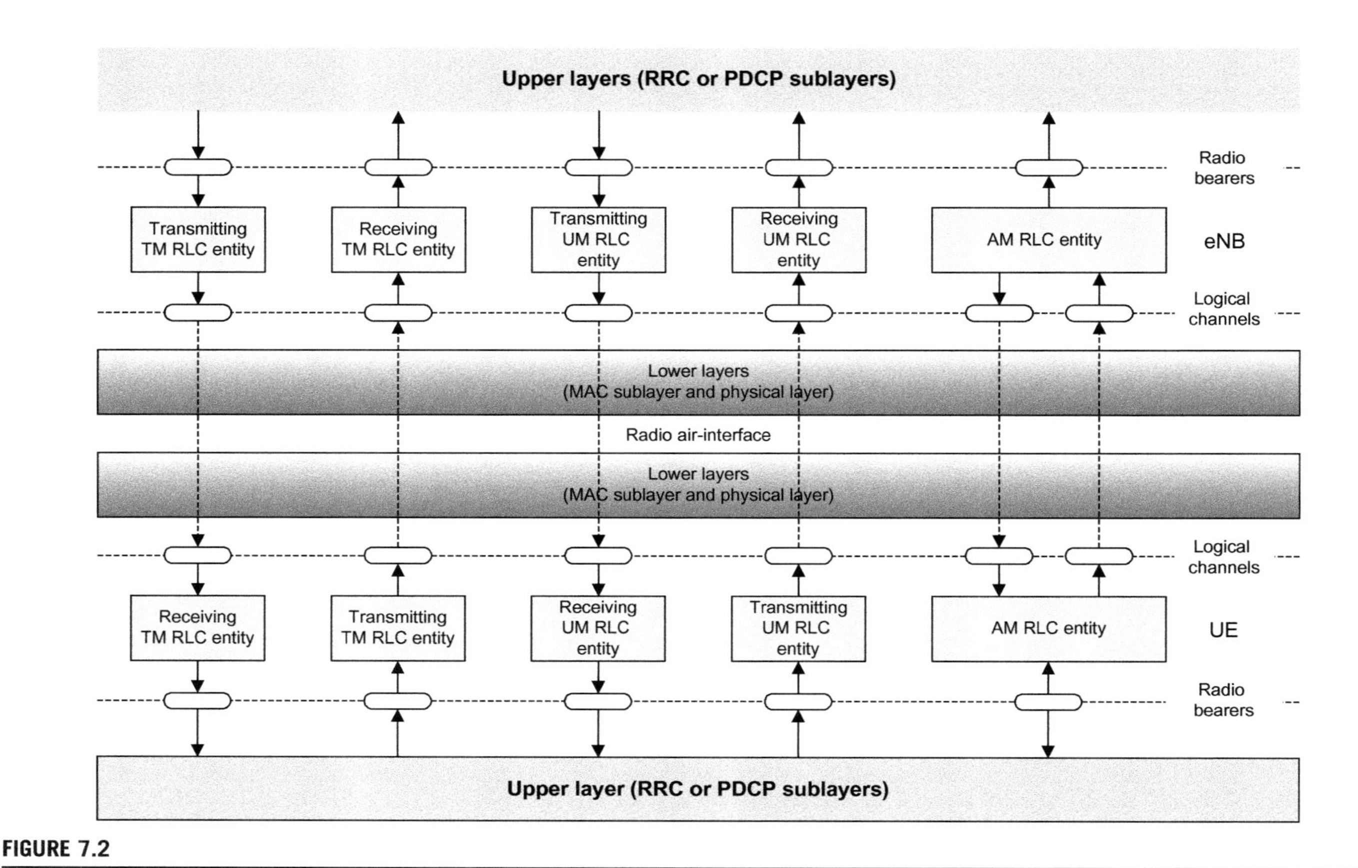

FIGURE 7.2

RLC sublayer architecture [2].

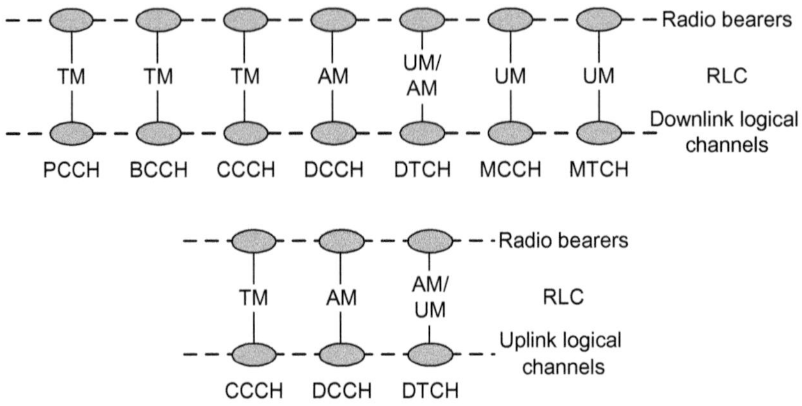

FIGURE 7.3

Downlink and uplink RLC channel mapping.

As mentioned earlier, an RLC entity can be configured to perform data transfer in TM, UM, or AM modes. As a result, an RLC entity is categorized as a TM RLC entity, a UM RLC entity, or an AM RLC entity depending on the mode of data transfer that the RLC entity is configured to operate. As shown in Figure 7.2, a TM/AM/UM RLC entity is configured either as a transmitting TM/AM/UM RLC entity or a receiving TM/AM/UM RLC entity (note that the operation modes must match on both sides). The transmitting RLC entity receives RLC SDUs from the upper layer and sends RLC PDUs to its peer receiving RLC entity via lower layers. The receiving RLC entity delivers RLC SDUs to the upper layer and receives RLC PDUs from its peer transmitting RLC entity via lower layers [2].

Figure 7.3 shows the RLC mode association with different logical channels in the downlink. The TM usage is limited to common control signaling channels of BCCH, PCCH, and CCCH. In LTE, the UM is applied only to traffic channels. The AM is applicable to traffic as well as dedicated control channels. Figure 7.3 also shows RLC mode association with different logical channels in the uplink where only AM and UM are applicable. The UM is utilized for common control signaling in the form of CCCH and can be additionally applied to traffic channels. The AM is applicable to traffic as well as dedicated control channels.

7.2 RLC transfer modes

As we discussed earlier in this chapter, each radio bearer in the LTE is processed in one of the following three transfer modes:

1. **Transparent Mode:** This mode can be regarded as null RLC since it is simply a pass through. None of the major RLC functions are applicable to this

mode. The RLC sublayer does not add any header or other overhead. The use of TM is limited to the common signaling channels such as BCCH, PCCH, and CCCH.

2. **Unacknowledged Mode:** This mode provides all RLC functions except retransmissions and resegmentation; therefore providing unreliable service. It would typically be employed for real-time applications such as VoIP, which are error tolerant and delay sensitive, but still require other functions such as in-sequence delivery and duplicate detection.
3. **Acknowledged Mode:** This mode provides a reliable transmission by offering error correction and retransmissions. It provides the entire RLC functions and is suited for applications that require reliability such as TCP/IP packet data services (e.g., browsing or e-mail).

The octet aligned RLC SDUs of variable sizes are supported for all RLC entity types (i.e., TM, UM, and AM RLC entities). The RLC PDUs are formed upon notification of a transmission opportunity and are delivered to the MAC sublayer. In the following sections, we discuss RLC functions associated with each of the above transfer modes in more detail. Table 7.1 summarizes the RLC functions that are supported by each RLC transfer mode [2].

7.2.1 Transparent mode

The TM entity consists of simply a transmission buffer to hold the RLC SDUs until a transmission opportunity becomes available at the lower layers. No other processing is done by the transmitting TM RLC entity. The receiving TM RLC entity passes the received PDUs to the higher layers. The TM entity does not segment or concatenate RLC SDUs. Therefore, each RLC SDU is an RLC PDU. The framing and identification of the SDU boundary must be performed by the upper layers. As shown in Figure 7.4, a TM RLC entity can be configured to deliver/receive RLC PDUs through BCCH, downlink/uplink CCCH, and PCCH logical channels. When a transmitting TM RLC entity forms transparent mode data

Table 7.1 Support of RLC Functions in Each Mode [2]

RLC Functions	TM	UM	AM
Transfer of upper layer PDUs	Yes	Yes	Yes
Error correction through ARQ	No	No	Yes
Concatenation, segmentation, and reassembly of RLC SDUs	No	Yes	Yes
Resegmentation of RLC data PDUs	No	No	Yes
Reordering of RLC data PDUs	No	Yes	Yes
Duplicate detection	No	Yes	Yes
RLC SDU discard	No	Yes	Yes
RLC reestablishment	Yes	Yes	Yes
Protocol error detection	No	No	Yes

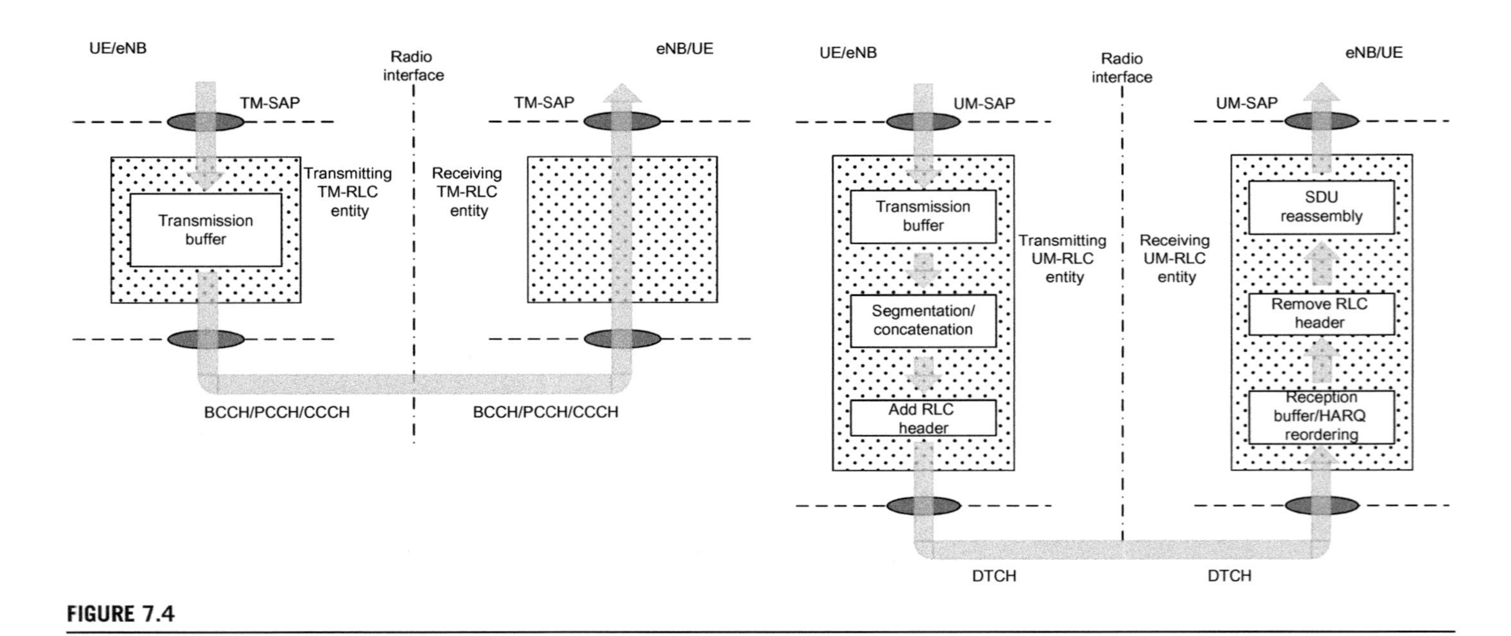

FIGURE 7.4

Illustration of LTE RLC TM and UM entities [2].

(TM data) PDUs from RLC SDUs, it does not segment or concatenate the RLC SDUs, and does not include any RLC headers in the TM data PDUs [2,8].

7.2.2 Unacknowledged mode

A UM RLC entity can be configured to deliver/receive RLC PDUs through downlink/uplink DTCH logical channels. When a transmitting UM RLC entity constructs unacknowledged mode data (UM data) PDUs from RLC SDUs, it segments and/or concatenates the RLC SDUs (so that the UM data PDUs can match the total size of RLC PDU specified by the lower layer protocol at a particular transmission opportunity), and includes relevant RLC headers in the UM data PDU. When a receiving UM RLC entity receives UM data PDUs, it detects whether duplicate UM data PDUs have been received and discards duplicate PDUs, it reorders the UM data PDUs, if they are received out of sequence, it detects the loss of UM data PDUs at lower layers and avoids excessive reordering delays, it reassembles RLC SDUs from the reordered UM data PDUs and delivers the RLC SDUs to the upper layer protocol in ascending order of the RLC SN, and it discards received UM data PDUs that cannot be reassembled into an RLC SDU due to loss of a UM data PDU which belonged to a particular RLC SDU at lower layers. At the time of RLC reestablishment, the receiving UM RLC entity reassembles RLC SDUs from the UM data PDUs that are received out of sequence and delivers them to the upper layer protocol, discards any remaining UM data PDUs that could not be reassembled into RLC SDUs, initializes relevant state variables, and stops relevant timers [2].

As shown in Figure 7.4, the UM transmitting entity places the received RLC SDUs in the transmit buffer. When a transmission opportunity becomes available, it may perform segmentation or concatenation of RLC SDUs depending on SDU size as well as the size of transmission opportunity. After segmentation/concatenation, an RLC header is added. The RLC header includes information fields such as SN and length. The resulting RLC PDU is passed onto the MAC sublayer for transmission. The UM receiving entity holds received UM RLC PDUs in the receive buffer. The PDUs may be out of order due to HARQ retransmissions. Therefore, they are reordered based on their SNs. After removing RLC headers, the data fields of RLC PDUs are assembled back into SDUs, undoing any segmentation and concatenation, and are delivered to the upper layers.

Figure 7.5 shows an example segmentation/concatenation operation on RLC SDUs. In this example, the nth RLC SDU is segmented with the first part transmitted as part of the PDU with SN = 1, and the second part transmitted in the PDU with SN = 2. Two additional complete SDUs ($n+1$, $n+2$) are concatenated in this PDU. The last SDU $n+3$ is segmented with the first segment included as the last data field in the PDU with SN = 2 and the other segment is included in the next PDU with SN = 3. For the PDU with SN = 2, the header is set accordingly with the framing info (FI) set to "11" to indicate that the RLC PDU includes segments of SDUs at both ends. Multiple length indicators (LI) are

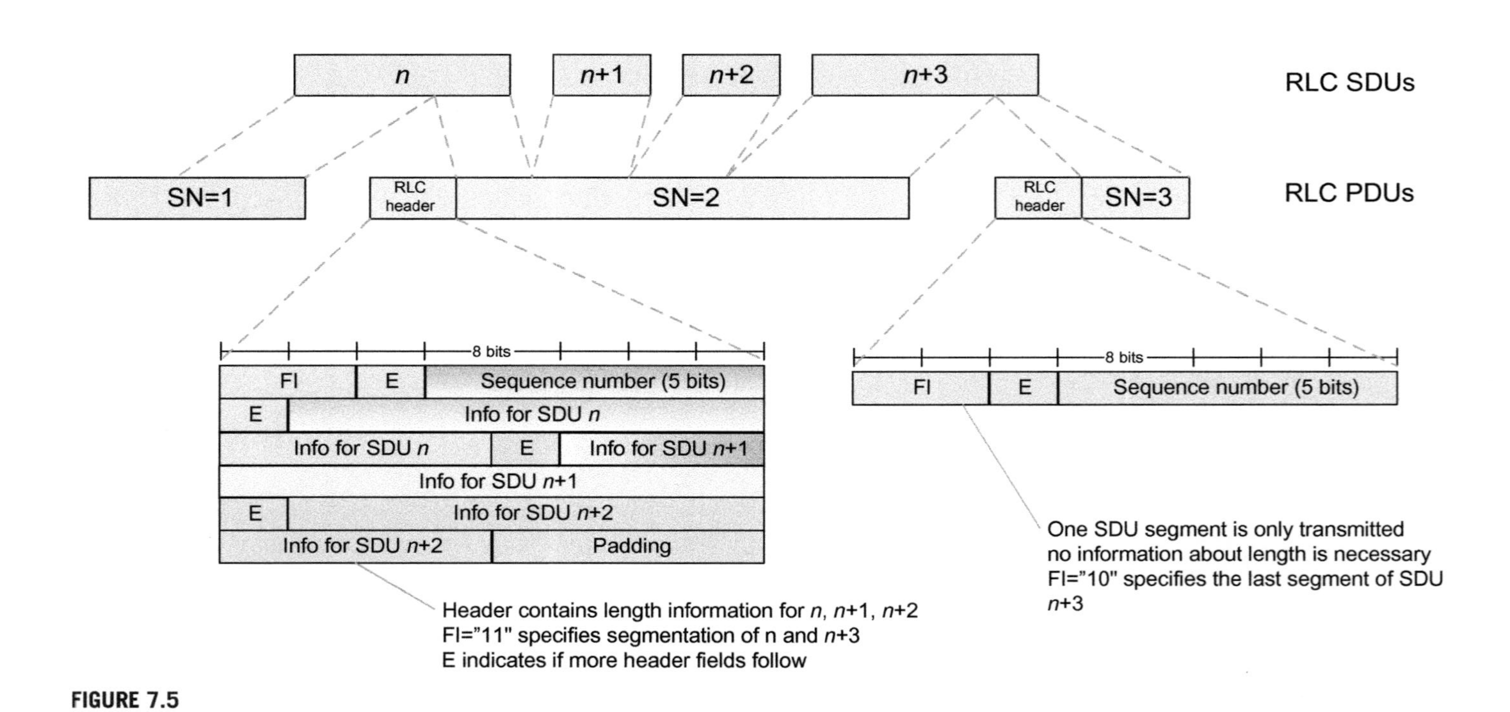

FIGURE 7.5

Illustration of UM segmentation and concatenation functions.

included. Each LI identifies the length of one complete or partial SDU included in the PDU except the last one. For the PDU with SN = 3, no extension fields are necessary as it contains only data from one SDU. In this case, the FI field is set to "10" to indicate that the last byte of the PDU data corresponds to the last byte of the SDU.

Figure 7.6 illustrates an example UM reordering process, which is controlled by parameters *UM_Window_Size* and *T_reordering*. The reordering window specifies the number of PDUs that can be received without advancing the window. If the window advances, any PDUs outside the window, regardless of their status, are delivered to the upper layers. The reordering timer determines how long the receiving entity waits for a missing PDU to be transmitted. It is started as soon as a sequence gap is detected. Upon expiration of the timer, the receiving entity will deliver any PDUs with contiguous packets before or after the sequence gap to the upper layers. In this example, the transmitting entity sends PDUs 0 through 3. PDUs 0 and 3 are lost, while PDUs 1 and 2 are successfully received. The receiving entity starts the *T_reordering* timer once the gap due to PDU 0 is detected. However, before the expiration of *T_reordering*, PDU 0 is delivered (out of sequence) following a HARQ retransmission, thus contiguous PDUs 0, 1, and 2 are delivered to the upper layers. The reordering window is then advanced. Next, the transmitter sends three more PDUs (4, 5, and 6) which are successfully received. The *T_reordering* timer is now started due to the sequence gap caused by PDU 3. Finally, *T_reordering* expires, causing the receiver to stop waiting for PDU 3, and deliver contiguous PDUs beyond PDU 3 (PDUs 4, 5, and 6) to the upper layers.

7.2.3 Acknowledged mode

As shown in Figure 7.7, an AM RLC entity can be configured to deliver or receive RLC PDUs through downlink or uplink DCCH/DTCH logical channels. When the transmitting side of an AM RLC entity constructs acknowledged mode data (AM data) PDUs from RLC SDUs, it segments and/or concatenates the RLC SDUs so that the AM data PDUs fit within the total size of the RLC PDU at a particular transmission opportunity. The transmitting side of an AM RLC entity supports retransmission of RLC data PDUs (i.e., ARQ). If the RLC data PDU that needs to be retransmitted does not fit within the total size of the RLC PDU at a particular transmission opportunity, the AM RLC entity can resegment the RLC data PDU into AM data PDU segments where the number of resegmentations is not limited [2].

In general, when the transmitting side of an AM RLC entity forms AM data PDUs from RLC SDUs received from an upper layer or AM data PDU segments from RLC data PDUs to be retransmitted, it includes relevant RLC headers in the RLC data PDU. The receiving side of an AM RLC entity receives RLC data PDUs and detects and discards the duplicate RLC data PDUs. It reorders the RLC data PDUs, if they are received out of sequence, and detects the missing RLC data

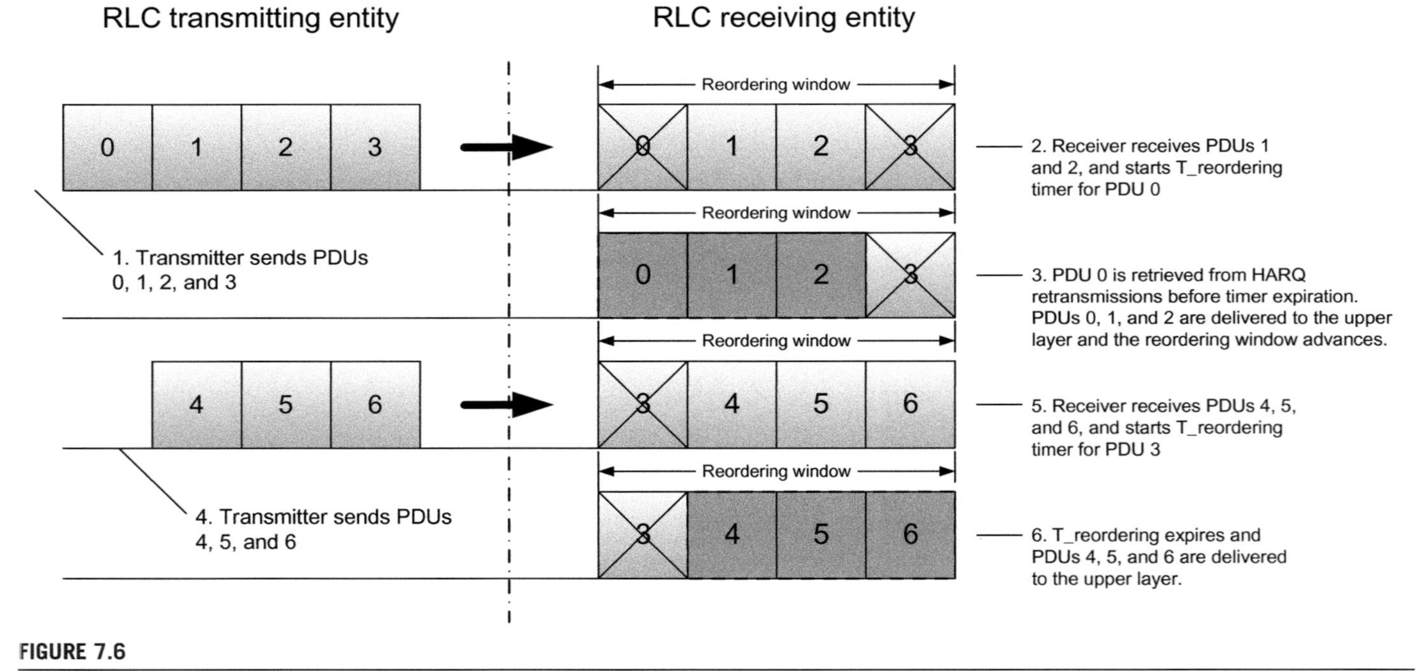

FIGURE 7.6

Illustration of the UM reordering process.

PDUs at the lower layers and sends retransmission requests to its peer AM RLC entity. It further reassembles RLC SDUs from the reordered RLC data PDUs and delivers the RLC SDUs to the upper layer in sequence. At the time of RLC reestablishment, the receiving side of an AM RLC entity reassembles RLC SDUs from the RLC data PDUs that are received out of sequence and delivers them to the upper layer, discarding the remaining RLC data PDUs that could not be reassembled into RLC SDUs, initializing relevant state variables, and stopping relevant timers.

The transmitting AM RLC entity places the received RLC SDUs in the transmission buffer. When a transmission opportunity becomes available, the SDUs in the transmission buffer are segmented or concatenated depending on the permissible size of the packet. An RLC header is added to each PDU prior to passing them to the MAC sublayer for transmission. The RLC PDUs are also placed in the retransmission buffer in case retransmission is necessary. When the receiver

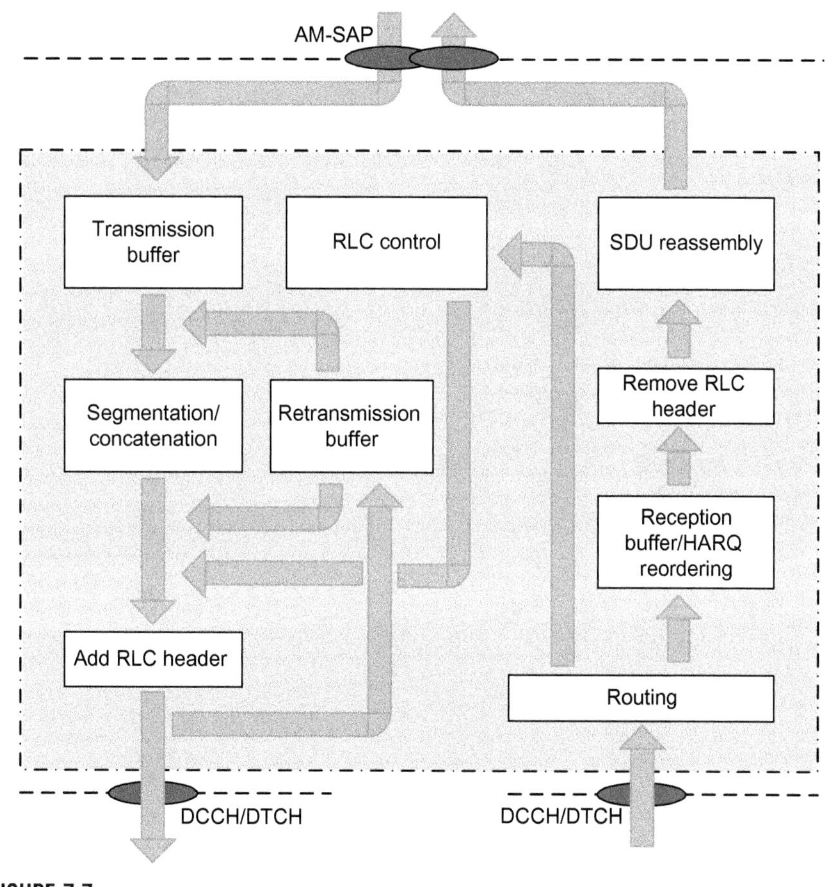

FIGURE 7.7

Illustration of LTE RLC AM entity [2].

sends an ACK or NACK, the transmitting entity attempts a retransmission. If an ACK is received for the entire RLC PDU, then that PDU is flushed from the retransmission buffer. If an NACK is received for part of a PDU or the entire PDU, the transmitting entity schedules a retransmission. If the size of transmission opportunity does not allow the entire RLC PDU to be re-sent, then resegmentation is possible whereby a single PDU is divided into multiple segments. Each segment can then be transmitted as a separate PDU with the RLC header indicating how the segmentation was carried out. Otherwise, the entire RLC PDU is scheduled for retransmission [2].

The receiving RLC entity accumulates the received RLC PDUs in the receive buffer, and then it performs the reordering before passing the complete SDU to the higher layers. Status PDUs are sent by the receiving entity to acknowledge the received PDUs and to signify missing PDUs or parts of missing PDU segments. Figure 7.8 shows an example illustration of the formation of an RLC AM status PDU. The status report shows that the receiving entity has received PDUs up to (but not including) SN = 8. Moreover, *NACK_SN* fields are included to indicate that PDUs with SN = 0, 3, and 6 are entirely missing. A *NACK_SN* is included for the PDU with SN = 5, but this is accompanied by *SOstart* and *SOend* to indicate that only a portion of the PDU between bytes 50 and 99 is missing.

The status PDU is sent from the receiving AM RLC entity to indicate successfully received as well as missing PDUs or portions of missing PDUs assuming resegmentation was used. The status PDU is sent when the poll bit is set in a PDU sent from the transmitting RLC entity while timer *T_status_prohibit* has expired, or *T_reordering* expires (indicating a missing PDU) and *T_status_prohibit* has already expired. *T_status_prohibit* is a timer that ensures duplicate status PDUs are not sent unnecessarily, and is set based on the RLC round trip time. It is restarted whenever a status PDU is sent. The transmitting AM entity sets the poll bit to request the receiving AM entity to send status PDUs. The frequency of the poll bit inclusion is controlled by the following parameters or events: (1) if the number of

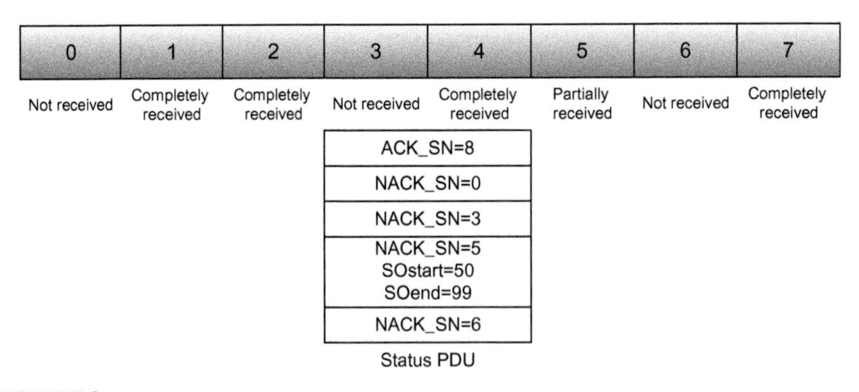

FIGURE 7.8

Illustration of an RLC AM status PDU.

PDUs sent since the last status report is received exceeds *Poll_PDU*; (2) if the total number of bytes of the PDUs sent since receiving the last status report exceeds *Poll_Byte*; (3) if the *T_poll_retransmit* timer expires and there are no PDUs to be sent in the transmission or retransmission buffer, the transmitting AM RLC entity resends the last PDU which was sent with poll bit set; (4) if no PDUs can be sent, the poll bit is set in the last PDU. This trigger is important in LTE since variable size RLC PDUs are supported [2].

In RLC AM, both transmit and receive entities maintain a window for flow control and reordering purposes. Figure 7.9 illustrates the mechanism for advancing the transmit/receive windows, as well as the behavior of the receiving RLC entity when encountering sequence gaps (missing packets) in the data stream. The sequence gaps (missing packets) would be filled using HARQ or RLC retransmissions that are triggered by status reporting. For simplicity, the length of transmit

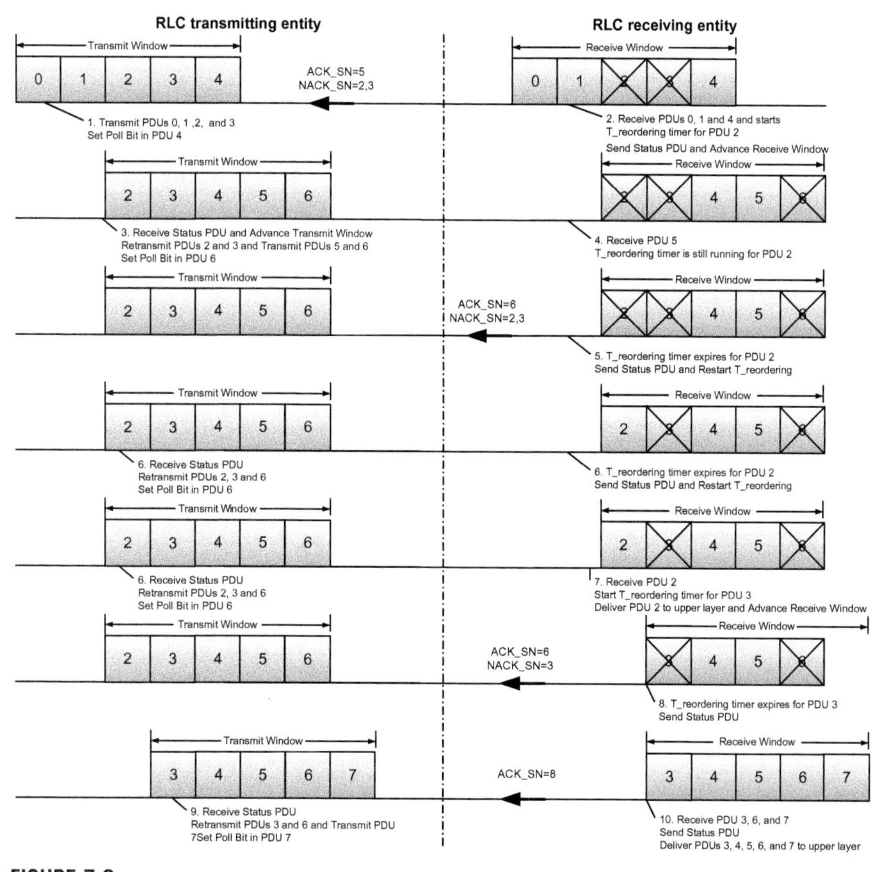

FIGURE 7.9

Illustration of an RLC AM reordering process.

and receive windows in this example is equal to 5 packets, whereas in reality the actual size of the RLC AM window is 512.

Based on the example shown in Figure 7.9, the process of RLC AM reordering can be described using the following steps:

1. Five PDUs are put in the transmit window for transmission.
2. The receiver receives PDUs 0, 1, and 4, but does not receive PDUs 2 and 3. PDUs 0 and 1 are delivered to the higher layers. A status PDU is sent indicating the next not-received PDU is SN = 5 and the missing PDUs are 2 and 3. The *T_reordering* timer is started, since a gap as result of the missing PDU is created within the receive window.
3. The transmitter retransmits the missing PDUs 2 and 3, and transmits new PDUs 5 and 6 as the transmit window can be advanced by two PDUs.
4. The receiver does not receive PDUs 2, 3, and 6, but receives PDU 5. The *T_reordering* timer associated with PDU 2 is still running.
5. The *T_reordering* timer associated with PDU 2 expires. This triggers the transmission of another status PDU, informing the transmitter that the next not-received PDU is 6, and that gaps exist due to PDUs 2 and 3.
6. After receiving the status PDU, the transmitter retransmits PDUs 2, 3, and 6.
7. The receiver successfully receives PDU 2, delivers it to the upper layers and advances the receive window. The *T_reordering* timer is then started for PDU 3.
8. The *T_reordering* timer expires as PDU 3 was never received. This causes the receiver to transmit a status PDU informing the transmitter that the next not-received PDU is 6, and that a gap exists due to PDU 3.
9. The transmitter receives the status PDU and retransmits PDUs 3 and 6, and transmits PDU 7 as the transmit widow can now be advanced.
10. The receiver successfully receives PDUs 3, 6, and 7. A status PDU is transmitted notifying the transmitter that all PDUs up to and including PDU 7 have been received.

The RLC sublayer in LTE has the ability to resegment previously transmitted PDUs. This enables the RLC sublayer to create segments whose size is suited to the transmission opportunity reported by the MAC sublayer. The example illustrated in Figure 7.10 shows the process of resegmentation with the following steps:

1. The transmitter sends an RLC PDU with SN = 101 with a length of 100 bytes.
2. The receiver sends a status PDU containing *NACK_SN* = 101.
3. As the indicated transmission opportunity now will only permit a transmission of 25 bytes, the RLC decides to resegment the PDU. The transmitted SN is still 101, but now the segment field is included (SO = 0) indicating that this is a resegmented PDU, and this is the first part of the original PDU. The LSF field is set to 0 to indicate that this is not the last segment.
4. The transmitter sends the second segment (25 bytes) of the RLC PDU with SO = 25. The LSF is still set to 0.

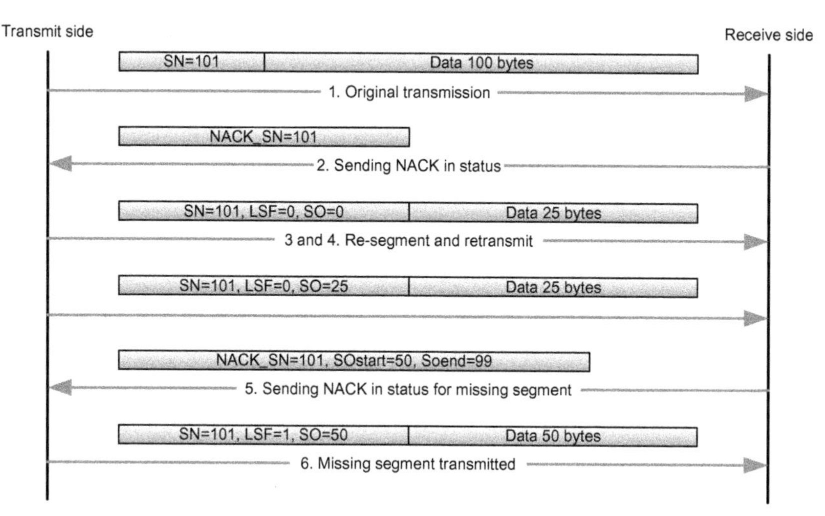

FIGURE 7.10

Illustration of an RLC AM resegmentation process.

5. The receiver sends a status PDU with *NACK_SN* = 101. It also includes *SOstart* = 50 and *SOend* = 99 to indicate the portion of the RLC PDU that has not been received yet.
6. The transmitter sends the third segment (50 bytes) of the RLC PDU with SO = 50. The LSF is set to 1 to indicate that this is the last segment of the RLC PDU. The receiver may send a status PDU, acknowledging the RLC PDU, if it has received all segments correctly.

7.3 RLC ARQ procedure

The ARQ is an error detection and correction mechanism that is incorporated in the AM RLC entity to improve the reliability and robustness of packet transmission. In the following sections, we briefly describe the theory of ARQ and later explain the ARQ procedures and relevant state variables and timers [12,13].

7.3.1 ARQ principles

ARQ is an error-control mechanism for data transmission which uses acknowledgments (ACK) (or negative acknowledgments (NACK)) and timeouts to achieve reliable data transmission over an unreliable communication link. In an ARQ scheme, the receiver uses an error detection code, typically a cyclic redundancy check (CRC), to detect if the received packet is in error [6,12,13]. If no error is detected in the received data, the transmitter is notified by sending a positive ACK. If an error is detected, the receiver discards the packet and sends a

NACK to the transmitter and requests a retransmission. An ACK or a NACK is a short message sent by the receiver to the transmitter to indicate whether it has correctly or incorrectly received a data packet, respectively. Timeout is a predetermined time interval after the sender sends the packet; if the sender does not receive an ACK before the timeout, it usually retransmits the packet until it receives an ACK or exceeds a predefined number of retransmissions. There are three types of ARQ protocol as follows [12,13]:

1. **Stop-and-Wait ARQ** is the basic form of ARQ protocol where the sender sends one packet at a time and then waits for an ACK or NACK signal from the receiver before sending the same or a new packet. The receiver sends an ACK signal following receipt of a good packet. If the ACK does not reach the sender before the timeout, the sender resends the same packet.
2. **Go-Back-N ARQ** is a form of ARQ protocol in which the sender continuously sends a number of packets (determined by the duration of transmission window) without receiving an ACK signal from the receiver. The receiver process keeps track of the sequence number of the next packet it expects to receive and sends the sequence number with every ACK it sends. The receiver will ignore any packet that does not have the exact sequence number it expects, whether that packet is a duplicate of a packet it has already acknowledged or a packet with a sequence number higher than the one expected. Once the sender has sequentially sent all the packets in its transmission window, it will check whether all of the packets are acknowledged and will resume sequential transmission of the packets starting with the next sequence number to the one that was last acknowledged.
3. **Selective Repeat ARQ** is a form of ARQ protocol for the transmission and acknowledgment of packets or fragments of a packet where the sending process continues to send a number of packets specified by a window size even after a packet is lost. Unlike Go-Back-N ARQ, the receiving process will continue to accept and acknowledge packets sent after an initial error. The receiver process keeps track of the sequence number of the earliest packet it has not received, and sends that number with every acknowledgment it sends. If a packet from the sender does not reach the receiver, the sender continues to send subsequent packets until it has exhausted its window. The receiver continues to fill its receiving window with the subsequent packets, replying each time with an acknowledgment containing the sequence number of the earliest missing packet. Once the sender has sent all the packets in its window, it resends the packet number given by the acknowledgments, and then continues where it left off.

Figure 7.11 provides an example illustration of the ARQ variants. It is shown how efficiently different ARQ schemes utilize the communication channel and use the ARQ buffers in the transmitter and receiver. The major advantage of ARQ over forward error correction (FEC) schemes is that error detection requires much simpler decoding mechanisms and much less redundancy than error correcting codes. Furthermore, ARQ is adaptive in the sense that information is

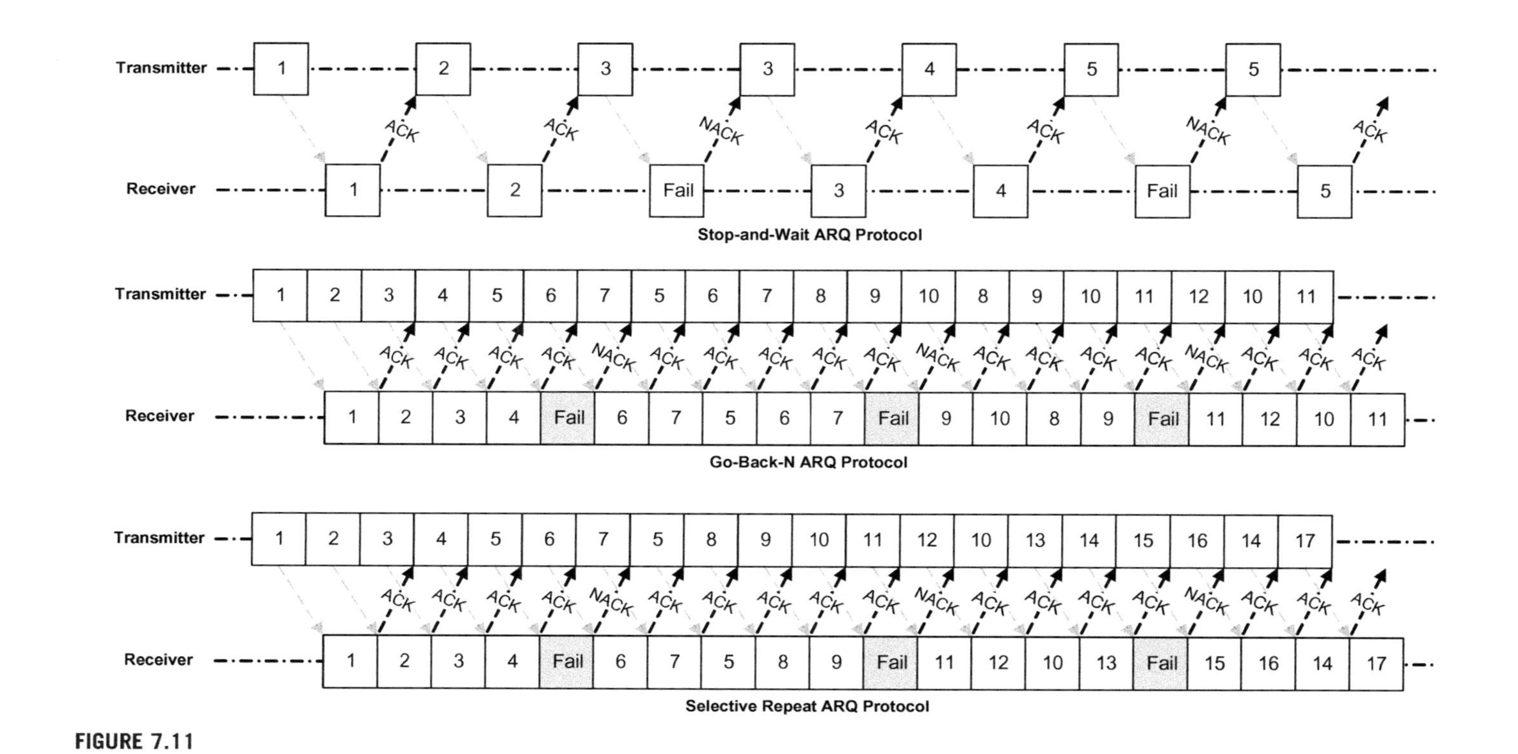

FIGURE 7.11

Illustration of different ARQ schemes [12,13].

retransmitted only when error occurs. On the other hand, FEC may be desirable instead of, or in addition to, error detection, for any of the following reasons: (1) a feedback channel is not available or ARQ delay is not tolerable; (2) the retransmission scheme is not conveniently implemented; and (3) the expected number of errors without correction would require excessive retransmissions [12]. The ARQ can be used for each connection between the UE and the eNB. Since the use of ARQ may increase the latency due to more reliable transmission, the ARQ mechanism is usually disabled for delay-sensitive applications such as VoIP or interactive gaming.

7.3.2 ARQ operation

The ARQ processes in the transmitter and receiver entities maintain separate state machines to track the transmission, retransmission, and discarding of the packets exchanged between the two entities. The state machines on each side employ several state variables and counters that are described in this section. The transmitting side of each AM RLC entity maintains the following state variables [2]:

1. Acknowledgment state variable *VT(A)*: This state variable holds the value of the sequence number of the next AM data PDU for which a positive acknowledgment will be received in-sequence, and it serves as the lower edge of the transmitting window. It is initially set to zero and is updated whenever the AM RLC entity receives a positive acknowledgment for an AM data PDU with $SN = VT(A)$.
2. Maximum send state variable *VT(MS)*: This state variable equals to $VT(A) + AM\_Window\_Size$, and it serves as the higher edge of the transmitting window.
3. Send state variable *VT(S)*: This state variable holds the value of the *SN* to be assigned for the next newly generated AM data PDU. It is initially set to zero and is updated whenever the AM RLC entity delivers an AM data PDU with $SN = VT(S)$.
4. Poll send state variable *POLL_SN*: This state variable holds the value of *VT(S)*-1 upon the most recent transmission of an RLC data PDU with the poll bit set to one. It is initially set to zero.

The transmitting side of each AM RLC entity maintains the following counters [2]:

1. *PDU_WITHOUT_POLL* counter: This counter is initially set to zero. It counts the number of AM data PDUs sent since the most recent poll bit was transmitted.
2. *BYTE_WITHOUT_POLL* counter: This counter is initially set to zero. It counts the number of data bytes sent since the most recent poll bit was transmitted.
3. *RETX_COUNT* counter: This counter counts the number of retransmissions of an AM data PDU. There is one *RETX_COUNT* counter per PDU that needs to be retransmitted.

The receiving side of each AM RLC entity maintains the following state variables [2]:

1. Receive state variable *VR(R)*: This state variable holds the value of the *SN* following the last in-sequence completely received AM data PDU, and it serves as the lower edge of the receiving window. It is initially set to zero and is updated whenever the AM RLC entity receives an AM data PDU with $SN = VR(R)$.
2. Maximum acceptable receive state variable *VR(MR)*: This state variable equals to $VR(R) + AM\_Window\_Size$, it holds the value of the *SN* of the first AM data PDU that is beyond the receiving window, and serves as the higher edge of the receiving window.
3. *t-Reordering* state variable *VR(X)*: This state variable holds the value of the *SN* following the *SN* of the RLC data PDU which triggered *t-Reordering*.
4. Maximum status transmit state variable *VR(MS)*: This state variable holds the highest possible value of the *SN* which can be indicated by *ACK_SN* when a status PDU needs to be constructed. It is initially set to zero.
5. Highest received state variable *VR(H)*: This state variable holds the value of the *SN* following the *SN* of the RLC data PDU with the highest *SN* among received RLC data PDUs. It is initially set to zero.

Figure 7.10 illustrates the application of some of the ARQ state variables and timers in RLC AM. In the process of transmission of packets between peer RLC entities, the transmitting side of an AM RLC entity may receive an NACK, which is a notification of reception failure by its peer AM RLC entity, for an AM data PDU or part of an AM data PDU through a status PDU received from its peer AM RLC entity. In this case, if the sequence number of the corresponding AM data PDU falls within the range $VT(A) \leq SN < VT(S)$, the AM data PDU or part of the AM data PDU for which a NACK was received is retransmitted.

When retransmitting a portion of an AM data PDU, the transmitting side of an AM RLC entity divides the AM data PDU into a number of segments, where each segment forms a new AM data PDU segment which can fit within the size of RLC PDU permissible by the lower layer at a particular transmission opportunity and delivers the new AM data PDU segment to the lower layer.

An AM RLC entity can poll its peer AM RLC entity in order to trigger status reporting at the peer AM RLC entity. Upon assembly of an AM data PDU or AM data PDU segment, the transmitting side of an AM RLC entity includes a poll in the RLC data PDU, when the transmission buffer and the retransmission buffer become empty after transmission of the RLC data PDU, or when no new RLC data PDU can be transmitted after the transmission of the last RLC data PDU.

An AM RLC entity sends status PDUs to its peer AM RLC entity in order to provide positive and/or negative acknowledgments for RLC PDUs. The RRC signaling determines whether the status-prohibit function can be used for an AM RLC entity. The triggers to initiate status reporting include polling from a peer AM RLC entity (this ensures that the RLC status report is transmitted after

HARQ reordering), as well as detection of reception failure of an RLC data PDU (the receiving side of an AM RLC entity triggers a status report when *t-Reordering* expires). A status PDU can be triggered when polling from a peer entity is received or reception failure of RLC data PDU is detected (*t-Reordering* timer expiration). As long as the *t-StatusProhibit* timer is running, a status PDU cannot be sent to the lower layer. When a status PDU is sent to the lower layer, *t-StatusProhibit* is restarted. Note that the *t-StatusProhibit* timer is configured by the upper layer [2].

When constructing a status PDU, the AM RLC entity must include an *NACK_SN* which is set to the *SN* of the AM data PDU for the AM data PDUs with sequence numbers satisfying $VR(R) \leq SN < VR(MS)$ that have not been completely received, in increasing *SN* order of PDUs and increasing byte segment order within PDUs, starting with $SN = VR(R)$ up to the point where the resulting status PDU would still fit the total size of the RLC PDU(s) permissible by the lower layer protocol, or for an AM data PDU for which no byte segments have been received. For an uninterrupted sequence of byte segments of a partly received AM data PDU that have not been received, a set of *NACK_SN*, *SOstart*, and *SOend* are included in the status PDU. The *ACK_SN* is set to the *SN* of the next not-received RLC data PDU, which is not indicated as missing in the resulting status PDU [2].

7.4 RLC PDU packet formats

The RLC sublayer provides TM, UM, and AM data transfer services including indication of successful delivery of PDUs to the RRC or PDCP sublayers. The MAC sublayer provides data transfer services to the RLC sublayer, as well as notification of a transmission opportunity (along with the total size of the RLC PDUs). The RLC PDUs can be categorized into RLC data and control PDUs. RLC data PDUs are used by TM, UM, and AM RLC entities to transfer upper layer PDUs. RLC control PDUs are used by AM RLC entity to perform ARQ procedures. The status PDU is used by the receiving side of an AM RLC entity to inform the peer AM RLC entity about correctly received, as well as lost, RLC data PDUs. Figure 7.12 provides the high-level aspects of packet processing in LTE layer 2 protocols, by illustrating IP packet processing stages in various sublayers.

The RLC SDUs are bit strings that are octet aligned. An RLC SDU is included into an RLC PDU starting from its first bit onward. In the following sections, we discuss the formats of the RLC PDUs when operating in different data transfer modes.

7.4.1 Transparent mode PDU format

The TM data PDU only consists of a data field and does not include any RLC headers. The structure of RLC TM data PDUs is shown in Figure 7.13.

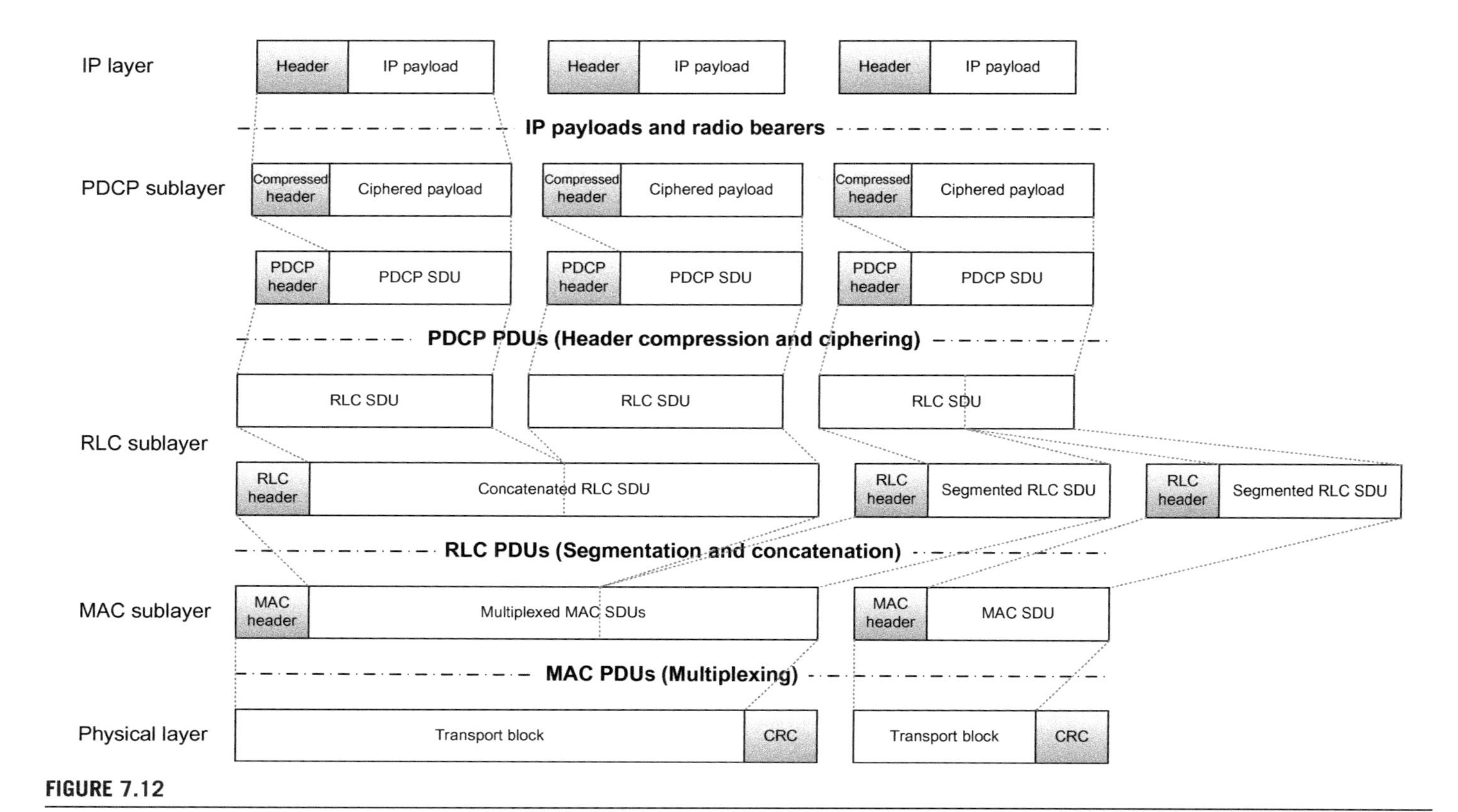

FIGURE 7.12

IP packet processing in different LTE layer 2 sublayers [8,9].

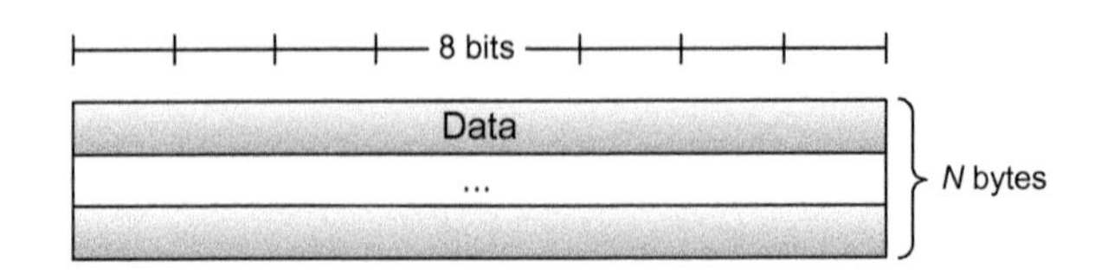

FIGURE 7.13

TM data PDU format [2,9].

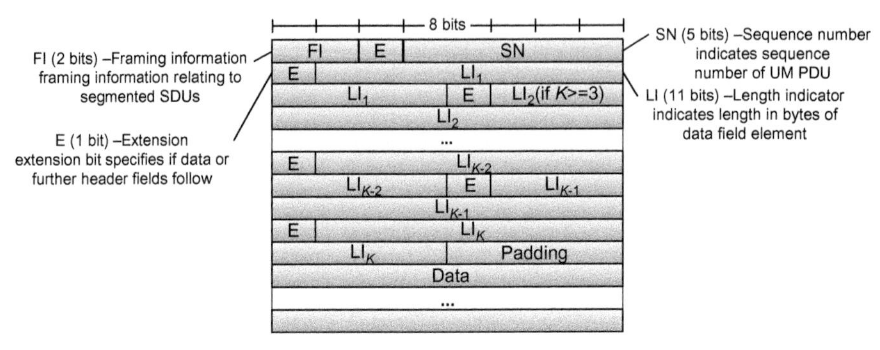

FIGURE 7.14

Unacknowledged mode data PDU format [2,9].

7.4.2 Unacknowledged mode PDU formats

The UM data PDU formats for a single and partial SDU are shown in Figure 7.14. There are different formats based on the length of SN, which may be 5 or 10 bits. The framing info (FI) field indicates whether a PDU contains a segment of RLC SDU either at the beginning or at the end. When multiple SDUs are concatenated into a single PDU, header extension fields are used to specify the size of each SDU with an 11-bit length indicator (LI) specified for each SDU except the last one. A single-bit extension (E) indicator specifies whether more header fields follow each LI. The LI for the last data field is not necessary as the size of the PDU will be signaled by the MAC layer. The UM data PDU header consists of a fixed part, i.e., information fields that are present for every UM data PDU, and an extension part, i.e., information fields that are present for a UM data PDU when necessary. The fixed part of the UM data PDU header is octet aligned and consists of FI, E, and SN fields. The extension part of the UM data PDU header is octet aligned and consists of E(s) and LI(s) fields. A UM data PDU header includes an extension part only when more than one data field elements are present. In that case, E and LI fields are present for every data field element except the last one. Furthermore, when a UM data PDU header consists of an odd number of LI(s), four padding bits will follow after the last LI field [2].

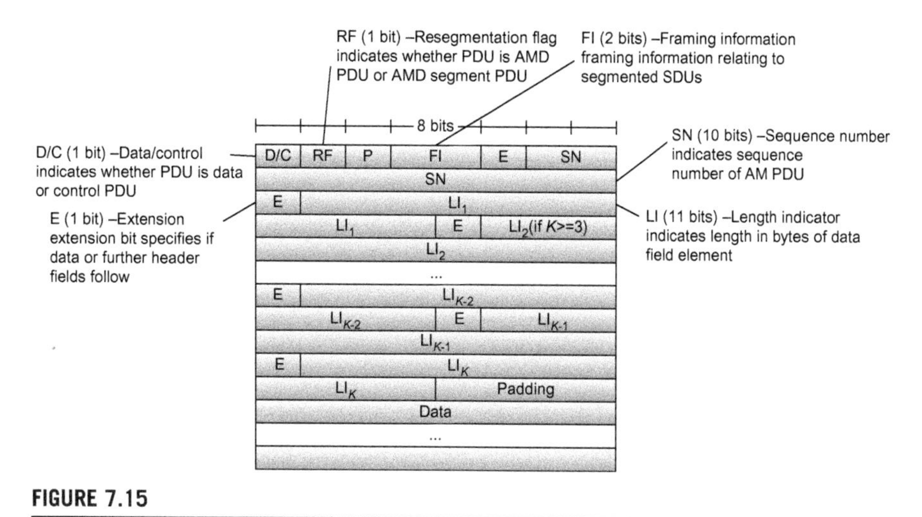

FIGURE 7.15

Acknowledged mode data PDU format [2,9].

7.4.3 Acknowledged mode PDU formats

Figure 7.15 shows the PDU format for transmitting AM data PDUs for the first time, or retransmitting AM data PDUs without resegmentation. The RLC AM transfer mode always operates with 10-bit SN. There is a one-bit D/C field that indicates whether the PDU carries data bits from upper layers or is a control PDU. The resegmentation flag (RF) is a one-bit field that indicates whether this PDU contains retransmitted PDUs that have been segmented. For first time transmissions, this flag is set to zero. The polling bit (P) is set whenever the transmitting AM entity requests that the receiving AM entity sends a status PDU. The FI field indicates whether parts of the RLC SDU are included in the PDU either at the beginning or at the end, and the interpretation is identical to the UM case. Similarly, the extension bit (E) indicates whether LI fields are included in the PDU. If multiple RLC SDUs are concatenated in the RLC PDU, the E bit and the LI are included for each SDU. The E bit indicates whether this is the last LI field before a data field [2].

The AM data PDU resegmentation format is used when an RLC PDU is resegmented for retransmission. The RLC header is identical to the normal RLC data PDU with the exception of the following differences (Figure 7.16):

- Re-segmentation flag (RF): Set to one to indicate resegmentation has been done.
- Last segment flag (LSF): The LSF field indicates whether this AM PDU segment contains the last segment of the retransmitted RLC PDU.
- Segment offset (SO): This 15-bit field indicates the position of the segment within the original RLC PDU. Specifically, the SO field indicates the position within the data field of the original AM data PDU to which the first byte of the data field of the AM data PDU segment belongs.

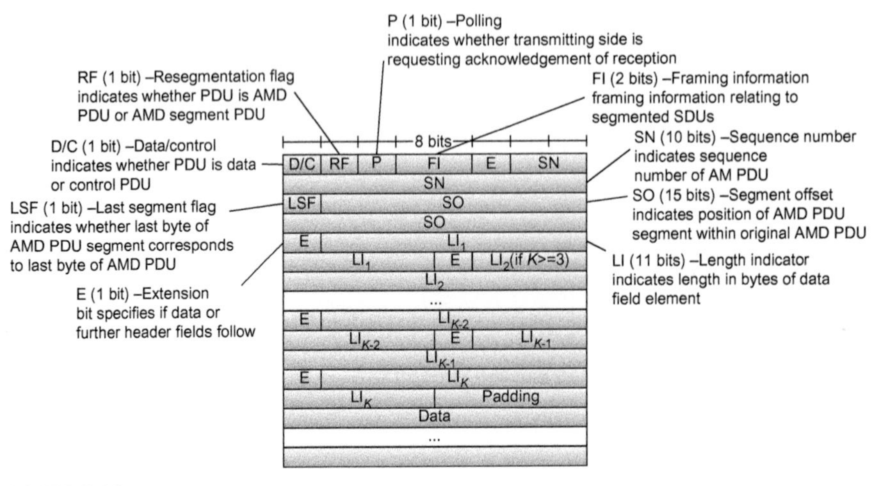

FIGURE 7.16

Acknowledged mode data PDU segment format [2,9].

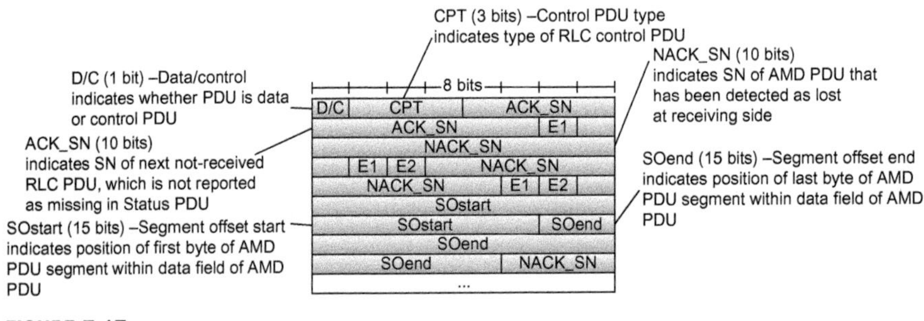

FIGURE 7.17

Status PDU format [2,9].

7.4.4 Status PDU format

The status PDU is sent from the receiving AM entity to the transmitting AM entity to provide information about received and missing RLC PDUs. The structure of status PDU is shown in Figure 7.17. The status PDU can also indicate missing segments of a PDU. The status PDU contains the following fields [2]:

- **D/C:** Data or control flag is set to one to indicate a control PDU.
- **ACK_SN:** The SN of the PDU up to which the receiver has received all the PDUs excluding those indicated by NACK_SN field(s).
- **NACK_SN:** Indicates the SN of a PDU which is not completely or only partially received. NACK_SN indicates gaps in the sequence of PDUs acknowledged by the ACK_SN.

Table 7.2 RLC Constants and Timers [2]

Timers and Constants	Description and Usage	Value
T_poll_retransmit	Used by transmitting side of AM RLC entity to retransmit a poll	0 to 200 ms
T_reordering	Used by receiving side of AM RLC entity and UM RLC entity to detect loss of RLC PDUs at the lower layer	0 to 500 ms
T_status_prohibit	Used by receiving side of AM RLC entity to prohibit transmission of a status PDU	0 to 500 ms
AM_Window_Size	Used by both transmitting side and receiving side of each AM RLC entity to calculate higher edge of window from lower edge	512
UM_Window_Size	Used by receiving UM RLC entity to define SNs of those UM data PDUs that can be received without causing an advancement of the receiving window	16 for 5-bit SN 512 for 10-bit SN

- **E1:** Extension field which indicates whether a set of NACK_SN, E1, and E2 fields follow.
- **E2:** Extension field which indicates whether SOstart and SOend fields are included following NACK_SN.
- **SOstart and SOend:** Indicate the start and stop positions within a segmented PDU which has not been successfully received. The SN is specified by the preceding NACK_SN.

7.5 RLC variables, constants, and timers

You may have noticed that the overall data flow of the RLC sublayer seems relatively easy to understand. But understanding the RLC sublayer mechanism in such detail that would help with troubleshooting and performance optimization in a real implementation would not be a simple task. One of the challenges is to intimately comprehend the definition and usage of the state variables, timers, and counters.

The state variables related to the RLC AM data transfer mode can take values from 0 to 1023. All arithmetic operations on state variables related to AM data transfer are adjusted by the AM modulus defined as *final_value* = [value from arithmetic operation]mod1024. The state variables related to the RLC UM data transfer mode can take values from 0 to $2^{(\text{sn-FieldLength})} - 1$. All arithmetic operations on state variables related to UM data transfer are adjusted by the UM modulus defined as *final_value* = [value from arithmetic operation]mod$2^{(\text{sn-FieldLength})}$. The AM and UM

Table 7.3 Description of RLC Transfer Mode Parameters [2]

Parameter	Category	Description
VT(A)	State variable	ACK state variable This state variable holds the value of the *SN* of the next AM data PDU for which a positive ACK is going to be received in-sequence, and it serves as the lower edge of the transmitting window It is initially set to zero and is updated whenever the AM RLC entity receives a positive ACK for an AM data PDU with $SN = VT(A)$
VT(MS)	State variable	Maximum send state variable $VT(MS) = VT(A) + AM\_Window\_Size$ It serves as the higher edge of the transmitting window
VT(S)	State variable	Send state variable This state variable holds the value of the *SN* to be assigned for the next newly generated AM data PDU. It is initially set to zero and is updated whenever the AM RLC entity delivers an AM data PDU with $SN = VT(S)$
VR(R)	State variable	Receive state variable This state variable holds the value of the *SN* following the last in-sequence completely received AM data PDU. It serves as the lower edge of the receiving window. It is initially set to zero and is updated whenever the AM RLC entity receives an AM data PDU with $SN = VR(R)$
VR(MR)	State variable	Maximum acceptable receive state variable $VR(MR) = VR(R) + AM\_Window\_Size$ It holds the value of the *SN* of the first AM data PDU that is beyond the receiving window and serves as the higher edge of the receiving window
VR(X)	State variable	*t-Reordering* state variable This state variable holds the value of the *SN* following the *SN* of the RLC data PDU which triggered *t-Reordering*
VR(MS)	State variable	Maximum status transmit state variable This state variable holds the highest possible value of the *SN* which can be indicated by *ACK_SN* when a status PDU needs to be constructed. It is initially set to zero

(*Continued*)

Table 7.3 (Continued)

Parameter	Category	Description
VR(H)	State variable	Highest received state variable This state variable holds the value of the *SN* following the *SN* of the RLC data PDU with the highest *SN* among received RLC data PDUs. It is initially set to zero
VT(US)	State variable	This state variable holds the value of the *SN* to be assigned for the next newly generated UM data PDU. It is initially set to zero and is updated whenever the UM RLC entity delivers a UM data PDU with *SN* = *VT(US)*
VR(UR)	State variable	UM receive state variable This state variable holds the value of the SN of the earliest UM data PDU that is still considered for reordering. It is initially set to zero
VR(UX)	State variable	UM *t-Reordering* state variable This state variable holds the value of the *SN* following the *SN* of the UM data PDU which triggered t-Reordering
VR(UH)	State variable	UM highest received state variable This state variable holds the value of the *SN* following the *SN* of the UM data PDU with the highest *SN* among received UM data PDUs, and it serves as the higher edge of the reordering window. It is initially set to zero
POLL_SN	State variable	Poll send state variable This state variable holds the value of *VT(S)*-1 upon the most recent transmission of an RLC data PDU with the poll bit set to one. It is initially set to zero
PDU_WITHOUT_POLL	Counter	This counter is initially set to zero. It counts the number of AM data PDUs sent since the most recent poll bit was transmitted
BYTE_WITHOUT_POLL	Counter	This counter is initially set to zero. It counts the number of data bytes sent since the most recent poll bit was transmitted
RETX_COUNT	Counter	This counter counts the number of retransmissions of an AM data PDU. There is one *RETX_COUNT* counter per PDU that needs to be retransmitted
AM_Window_Size	Constant	*AM_Window_Size* = 512
UM_Window_Size	Constant	*UM_Window_Size* = 16 when *SN* = 5 bits *UM_Window_Size* = 512 when *SN* = 10 bits *UM_Window_Size* = 0 when UM RLC is for MTCH or MCCH

(*Continued*)

Table 7.3 (Continued)

Parameter	Category	Description
t-PollRetransmit	Timer	Used by the transmitting side of an AM RLC entity in order to retransmit a poll
t-Reordering	Timer	This timer is used by the receiving side of an AM RLC entity and receiving UM RLC entity in order to detect loss of RLC PDUs at the lower layer. Only one t-Reordering per RLC entity is running at a given time
t-StatusProhibit	Timer	This timer is used by the receiving side of an AM RLC entity in order to prohibit transmission of a status PDU
maxRetxThreshold	Configurable parameters	This parameter is used by the transmitting side of each AM RLC entity to limit the number of retransmissions of an AM data PDU
pollPDU	Configurable parameters	This parameter is used by the transmitting side of each AM RLC entity to trigger a poll for every *poll PDU* PDUs
pollByte	Configurable parameters	This parameter is used by the transmitting side of each AM RLC entity to trigger a poll for every poll bytes
sn-FieldLength	Configurable parameters	This parameter gives the UM *SN* field size in bits

data PDUs are numbered with integer-valued SNs in the range of 0 to 1023 for AM data PDU and 0 to $2^{(\text{sn-FieldLength})} - 1$ for UM data PDU. A summary of RLC constants and timers as well as state parameters are provided in Tables 7.2 and 7.3.

References

3GPP Specifications

[1] 3GPP TS 36.321, Evolved Universal Terrestrial Radio Access (E-UTRA); Medium Access Control (MAC) Protocol Specification (Release 11), December 2012.

[2] 3GPP TS 36.322, Evolved Universal Terrestrial Radio Access (E-UTRA); Radio Link Control (RLC) Protocol Specification (Release 11), December 2012.

[3] 3GPP TS 36.323, Evolved Universal Terrestrial Radio Access (E-UTRA); Packet Data Convergence Protocol (PDCP) Specification (Release 11), December 2012.

[4] 3GPP TS 36.331, Evolved Universal Terrestrial Radio Access (E-UTRA); Radio Resource Control (RRC) Protocol Specification (Release 11), December 2012.

[5] 3GPP TS 36.211, Evolved Universal Terrestrial Radio Access (E-UTRA), Physical Channels and Modulation (Release 11), December 2012.

[6] 3GPP TS 36.212, Evolved Universal Terrestrial Radio Access (E-UTRA) Multiplexing and Channel Coding (Release 11), December 2012.

[7] 3GPP TS 36.213, Evolved Universal Terrestrial Radio Access (E-UTRA); Physical Layer Procedures (Release 11), December 2012.

[8] 3GPP TS 36.300, Evolved Universal Terrestrial Radio Access (E-UTRA) and Evolved Universal Terrestrial Radio Access Network (E-UTRAN) Overall description, Stage 2 (Release 11), December 2012.

Books

[9] C. Johnson, Long Term Evolution in Bullets, second ed., Create Space Independent Publishing Platform, July 2012.

[10] S. Sesia, I. Toufik, M. Baker, LTE, the UMTS Long Term Evolution: From Theory to Practice, second ed., John Wiley & Sons, August 2011.

[11] E. Dahlman, S. Parkvall, J. Skold, 4G: LTE/LTE-Advanced for Mobile Broadband, Academic Press, May 2011.

[12] S. Lin, D. Costello, Error Control Coding, second ed., Prentice Hall, June 2004.

Articles

[13] R. Comroe, D. Costello, ARQ schemes for data transmission in mobile radio systems, IEEE J. Sel. Areas Commun. 2 (4) (July 1984).

CHAPTER

Medium Access Control Functions

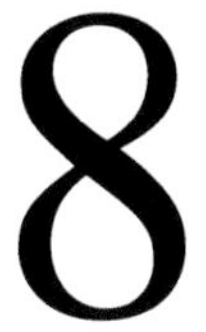

CHAPTER CONTENTS

The layer 2 in LTE are classified into medium access control (MAC), radio link control (RLC), and packet data convergence protocol (PDCP) functions [2,3]. The MAC sublayer is responsible for data transfer and logical channel multiplexing, hybrid automatic repeat request (HARQ) retransmissions and uplink/downlink scheduling, random access procedure, and maintenance of uplink timing. It is also responsible for multiplexing/demultiplexing data across multiple component carriers when carrier aggregation is used. From a physical layer (PHY) perspective, the MAC sublayer provides and receives services in the form of transport channels. A transport channel is defined by how and with what characteristics the information is transmitted over the radio interface. The data in a transport channel is organized into transport blocks. In each transmission time interval (TTI), one or two transport blocks of dynamic size are transmitted over the radio interface to a terminal depending on whether spatial multiplexing is supported. Each transport block is associated with a transport format, specifying how the transport block is physically processed and transmitted over the radio interface. The transport format includes information about the transport block size, the modulation and coding scheme (MCS), and the multi-antenna mapping [5,6]. By varying the transport format, the MAC sublayer can realize different data rates and reliability. Figure 8.1 illustrates the MAC sublayer in the LTE protocol stack. The interaction of the MAC sublayer with other layers in the protocol stack, as well as the service access points with the physical layer and RLC sublayers, are shown in the figure.

The functions of the MAC sublayer in LTE can be more systematically classified as follows:

- Channel mapping: MAC sublayer maps logical channels carrying RLC protocol data units (PDUs) to transport channels. For transmission, multiple service data units (SDUs) from logical channels are mapped to the transport block carried over transport channels. In the receiving end, transport blocks from transport channels are demultiplexed and assigned to the corresponding logical channels.
- Scheduling: MAC sublayer performs all scheduling-related functions in the uplink and downlink, and thus it is responsible for transport format selection associated with all transport channels. This includes HARQ functionality. Since the scheduling services are implemented at the eNB, the MAC sublayer is responsible for reporting scheduling-related information, such as UE buffer status and power headroom. It also handles prioritization from both an inter-UE and intra-UE logical channel perspective.
- Random access procedures: MAC sublayer is responsible for random access procedures in the uplink that can be performed following a contention or non-contention-based process.
- Uplink timing maintenance: UE needs to maintain timing synchronization with the cell at all times. The MAC sublayer performs required procedures for periodic synchronization.

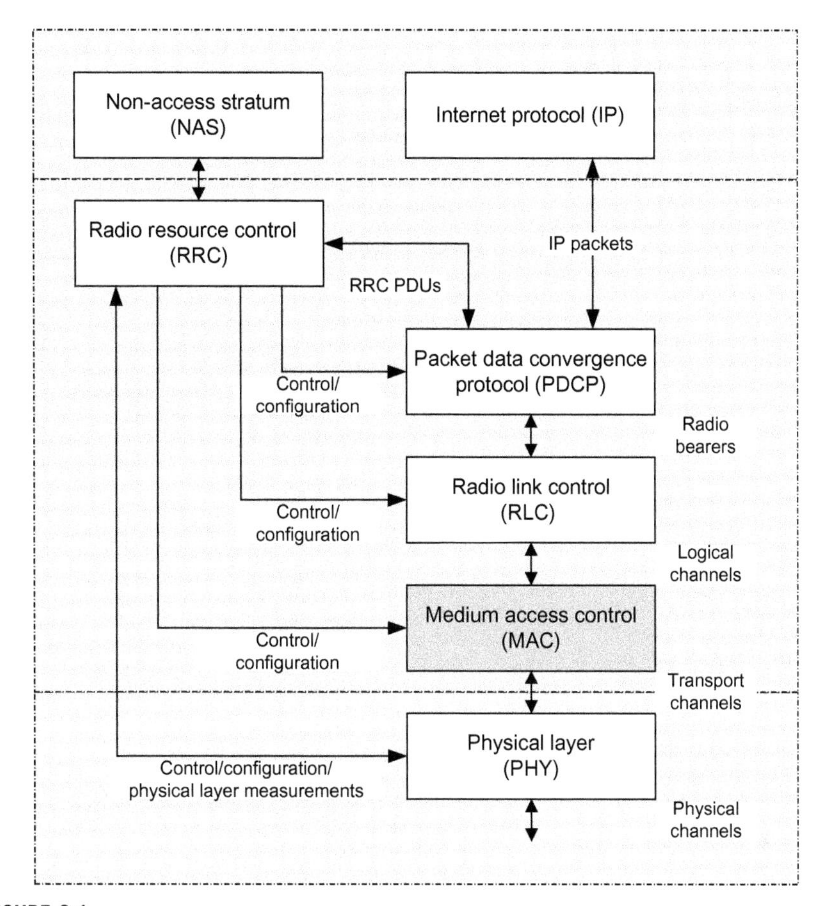

FIGURE 8.1

The MAC sublayer in the LTE protocol stack.

Figure 8.2 illustrates the IP packet processing in different parts of LTE layer 2. As shown in the figure, the MAC sublayer receives the RLC SDUs mapped to various logical channels in the downlink and generates the MAC PDUs that further become the transport blocks in the physical layer. In the following sections, we take a top-down systematic approach to study and understand various MAC sublayer services in LTE/LTE-Advanced and how those functions interact and interoperate with other protocol layers.

8.1 MAC architecture and services

The architecture of the LTE MAC sublayer is shown in Figure 8.3. It comprises a HARQ entity, a multiplexing/demultiplexing entity, a logical channel

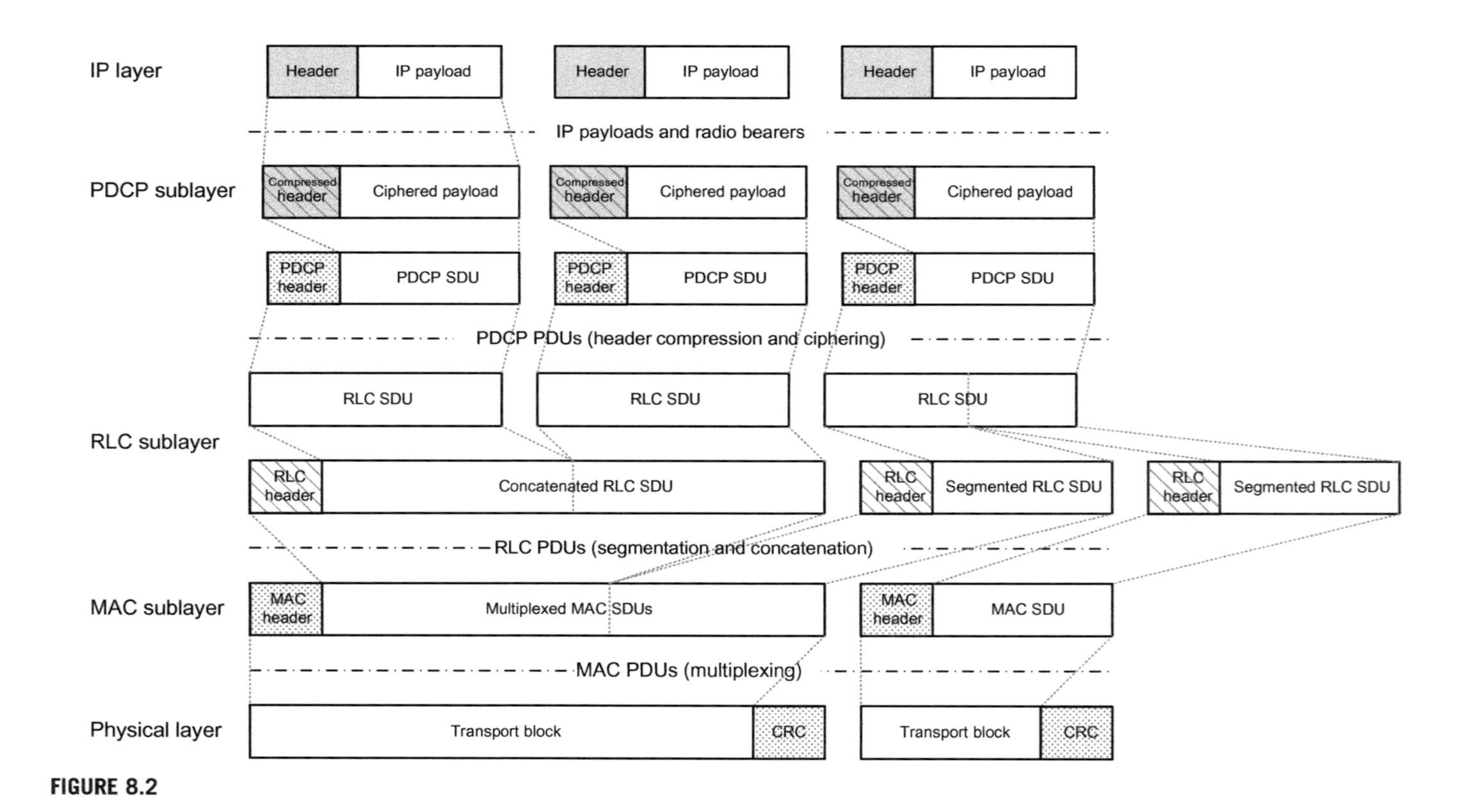

FIGURE 8.2

IP packet processing in different sublayers of LTE Layer 2 [8].

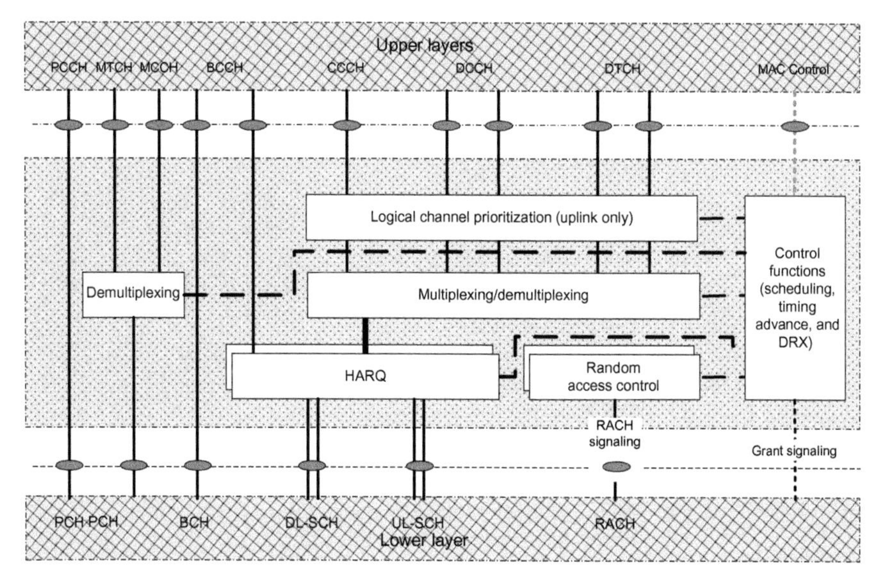

FIGURE 8.3

LTE UE MAC sublayer functions and services [1].

prioritization entity, a random access control entity, and a controller which performs various control functions, such as scheduling, timing advance maintenance, and discontinuous reception (DRX). The HARQ entity is responsible for downlink/uplink HARQ operations. The transmit HARQ operation includes transmission and retransmission of transport blocks, and receiving and processing of acknowledgment/negative-acknowledgment (ACK/NACK) signaling. The receive HARQ operation includes reception of transport blocks, combining of the received data and generation of ACK/NACK signaling. In order to enable continuous transmission while previous transport blocks are being decoded, up to eight HARQ processes in parallel are used to support a multi-process stop-and-wait HARQ protocol, which implies that in each HARQ process, upon transmission of a transport block, the transmitter stops further transmissions and waits for feedback from the receiver. When a NACK is received, or when a certain time period elapses without receiving any feedback, the transmitter retransmits the transport block. Each HARQ process is responsible for a separate stop-and-wait operation and manages a separate buffer.

In general, HARQ protocols can be categorized as synchronous or asynchronous, and the retransmissions in each case can be adaptive or non-adaptive. In a synchronous HARQ scheme, the retransmission(s) for each process occur at predefined time intervals relative to the initial transmission time. In this method, there is no need to signal information, such as the HARQ process number, as this can be inferred from the transmission timing. In contrast, in an asynchronous HARQ scheme, the retransmissions can occur at any time relative to the initial transmission, thus additional signaling is required so that the receiver can correctly associate each retransmission with the corresponding initial transmission. The synchronous HARQ schemes have the advantage of reducing the signaling overhead, whereas asynchronous HARQ schemes allow more flexibility in scheduling. In an adaptive HARQ scheme, transmission attributes, such as the modulation, coding, and resource allocation can be changed at each retransmission in response to variations in the channel conditions. In a non-adaptive HARQ scheme, the retransmissions are performed without explicit signaling of new transmission attributes, i.e., either the same transmission format is reused or the attributes are deterministically changed. The adaptive HARQ schemes offer higher scheduling gains at the expense of increased signaling overhead. In LTE/LTE-Advanced, asynchronous adaptive HARQ is used for the downlink and synchronous HARQ for the uplink. In the uplink, the retransmissions may be either adaptive or nonadaptive depending on the signaling of new transmission attributes.

In the multiplexing and demultiplexing entity, data from several logical channels is multiplexed or demultiplexed to/from one transport channel. The multiplexing entity generates MAC PDUs from MAC SDUs when radio resources are available for a new transmission based on the decisions of the logical channel prioritization entity. The demultiplexing entity reassembles the MAC SDUs from MAC PDUs and distributes them to the appropriate RLC entities.

The logical channel prioritization entity prioritizes the data from logical channels and decides the amount and the source of data that should be included in each MAC PDU. The decisions are sent to the multiplexing/demultiplexing entity. The random access control entity is responsible for controlling the random access procedure. The controller entity is also responsible for a number of functions including DRX, scheduling procedure, and maintaining the uplink timing alignment.

To support priority handling, multiple logical channels, where each logical channel has its own RLC entity, are multiplexed into one transport channel by the MAC sublayer. At the receiver, the MAC sublayer performs the corresponding demultiplexing and forwards the RLC SDUs to their respective RLC entity for in-sequence delivery and other RLC functions. The demultiplexing function at the receiver is enabled through addition of a MAC header, shown in Figure 8.2. There is a subheader in the MAC header associated with each RLC PDU. The subheader further contains the logical channel ID (LCID) from which the RLC PDU was originated, and the length of the PDU in octets.

In addition to multiplexing functions, the MAC sublayer can also include MAC control elements (CEs) in the transport blocks transmitted via the transport channels. A MAC control element is used for in-band control signaling, such as timing advance commands and random access response (RAR) message. The control elements are identified with reserved values in the LCID field, where the LCID value indicates the type of control information. The MAC multiplexing functions were extended to handle multiple component carriers in the case of carrier aggregation. The basic principle of carrier aggregation is independent processing of the component carriers in the physical layer, including control signaling,

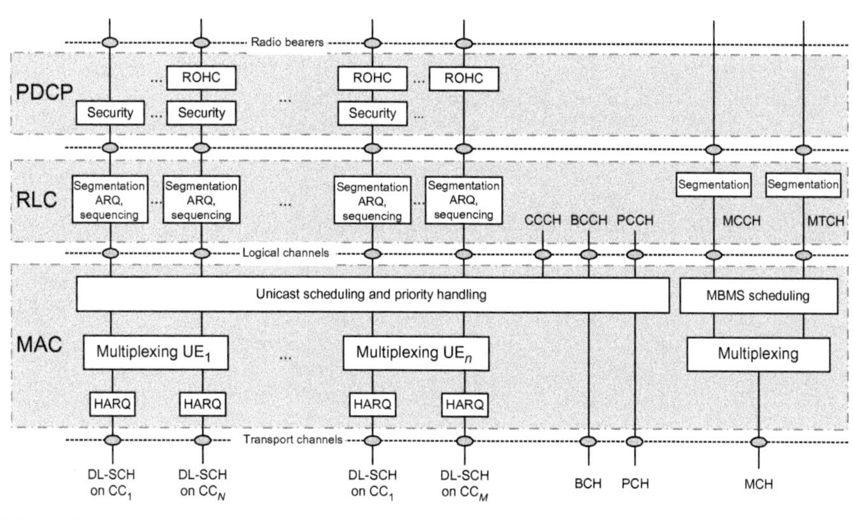

FIGURE 8.4

LTE Layer 2 structure with carrier aggregation support in the downlink [8].

scheduling, and HARQ retransmissions, while carrier aggregation is invisible to the RLC and PDCP. Carrier aggregation is therefore mainly seen in the MAC sublayer, where the logical channels, including any MAC control elements, are multiplexed to one or two transport blocks per component carrier with each component carrier having its own HARQ entity (see Figures 8.4 and 8.5).

The E-UTRA defines one MAC entity in the UE and one MAC entity in the eNB. These MAC entities control the following transport channels: BCH; DL-SCH; PCH; UL-SCH; and RACH. The exact functions performed by the MAC entities are different in the UE and the eNB. As shown in Figures 8.4 and 8.5, the MAC sublayer provides services, such as data transfer and radio resource allocation, to the upper layers. The physical layer provides the MAC sublayer with data transfer services, signaling of HARQ feedback, signaling of scheduling request, and measurements. The access to the data transfer services is through the use of transport channels. The characteristics of a transport channel are defined by its transport format which specifies the physical layer processing to be applied to the transport channel, such as channel coding and interleaving, and any other service-specific rate matching.

8.2 Logical and transport channel mapping

The LTE MAC sublayer provides data transfer services to the RLC sublayer through logical channels. Figures 8.6 and 8.7 depict classification of logical and transport channels in LTE, respectively. Each logical channel type is defined by the type of information which is transferred. The logical channels are generally classified into two groups: (1) control channels (for the transfer of control-plane

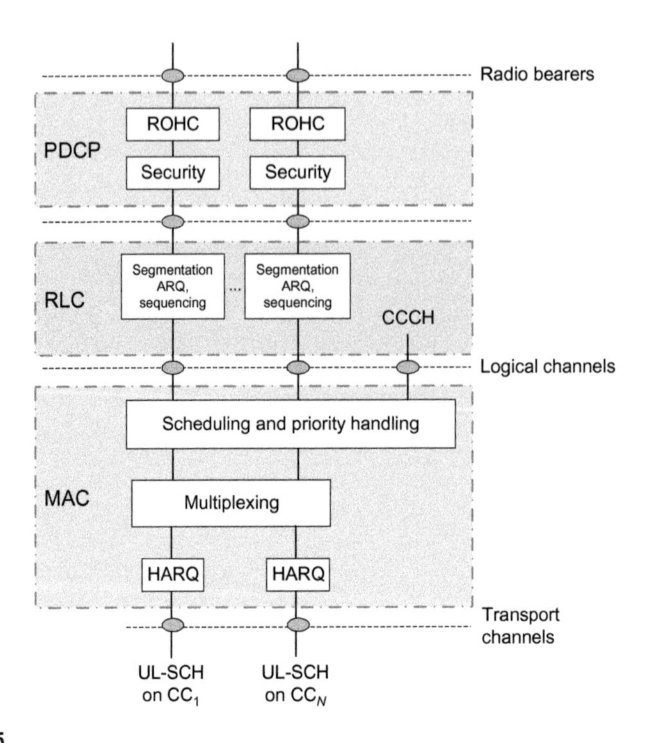

FIGURE 8.5

LTE Layer 2 structure with carrier aggregation support in the uplink [8].

information); and (2) traffic channels (for the transfer of user-plane information), as shown in Figure 8.6.

The control channels are exclusively used for transfer of control-plane information. The control channels supported by MAC can be classified as follows (Figure 8.6):

- Broadcast Control Channel (BCCH): A downlink channel for broadcasting system control information.
- Paging Control Channel (PCCH): A downlink channel that transfers paging information and system information (SI) change notifications. This channel is used for paging when the network does not know the location of the UE.
- Common Control Channel (CCCH): A logical channel for transmitting control information between UEs and eNBs. This channel is used for UEs having no RRC connection with the network.
- Multicast Control Channel (MCCH): A point-to-multipoint downlink channel used for transmitting multimedia broadcast multicast service (MBMS) control information from the network to the UE, for one or several multicast traffic channels. This channel is only used by UEs that are subscribed to MBMS.

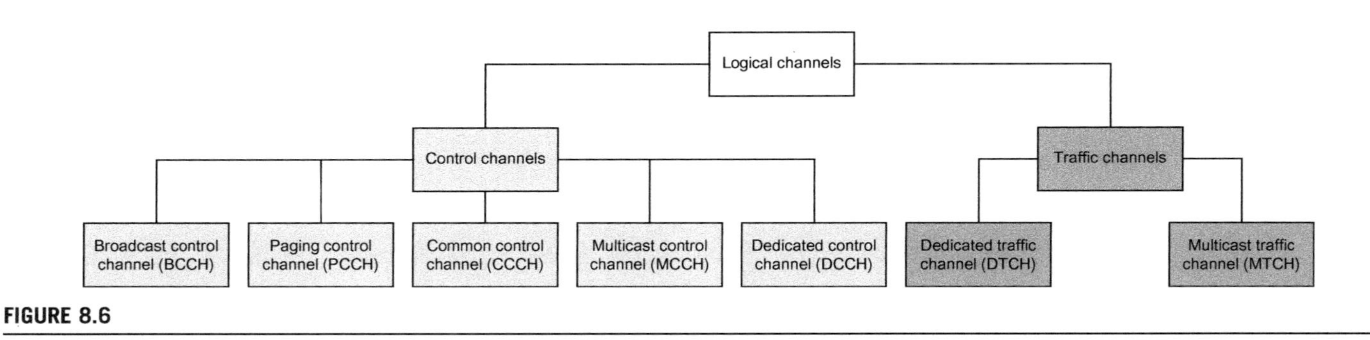

FIGURE 8.6

Classification of LTE logical channels [8].

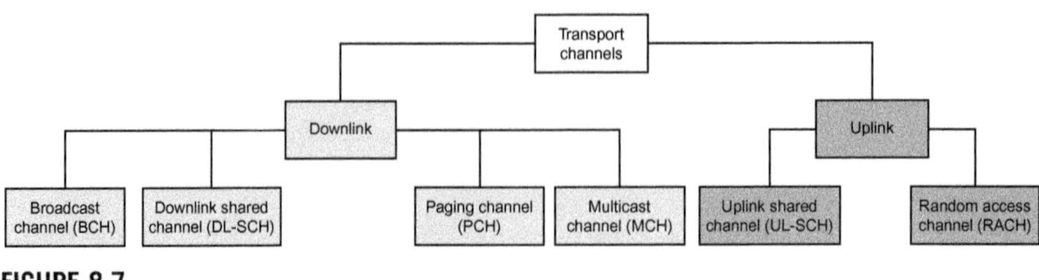

FIGURE 8.7

Classification of LTE transport channels [8].

- Dedicated Control Channel (DCCH): A point-to-point bi-directional channel that transmits dedicated control information between a UE and the network. It is used by UEs that have already established RRC connection.

The traffic channels are exclusively used for the transfer of user-plane information. The traffic channels supported by MAC can be classified as follows (as shown in Figure 8.6):

- Dedicated Traffic Channel (DTCH): A point-to-point bi-directional channel dedicated to a single UE for the transfer of user information.
- Multicast Traffic Channel (MTCH): A point-to-multipoint downlink channel for transmitting traffic data from the network to the UE. This channel is only used by UEs that receive MBMS.

As shown in Figure 8.7, downlink transport channels can be classified as follows:

- Broadcast Channel (BCH) that is characterized by a fixed, predefined transport format and is required to be broadcast in the entire coverage area of the cell.
- Downlink Shared Channel (DL-SCH) that is characterized by support for HARQ, support for dynamic link adaptation by varying the modulation, coding, and transmit power, possibility for broadcast in the entire cell, possibility to use beamforming, support for both dynamic and semi-static resource allocation, support for UE DRX to enable power saving, and support for MBMS transmission.
- Paging Channel (PCH) that is characterized by support for UE DRX in order to enable power saving, requirement for broadcast in the entire coverage area of the cell, mapped to physical resources which can also be used dynamically for traffic or other control channels.
- Multicast Channel (MCH), which is required to be broadcast in the entire coverage area of the cell, and is characterized by support for macro-diversity combining of MBMS transmission on multiple cells, and support for semi-static resource allocation.

The uplink transport channels are classified as follows (Figure 8.7):

- Uplink Shared Channel (UL-SCH), which is characterized by the possibility to use beamforming, support for dynamic link adaptation by varying the transmit

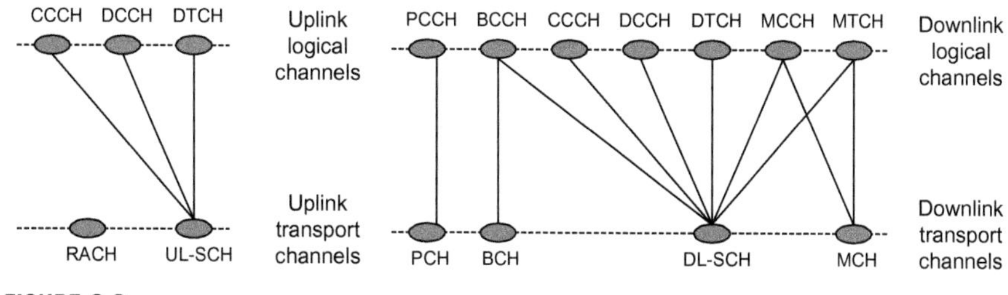

FIGURE 8.8

Mapping of logical to transport channels in the downlink and uplink [8].

power and MCS, support for HARQ, and support for both dynamic and semi-static resource allocation.

- Random Access Channel (RACH) that is characterized by limited control information and collision risk.

The mapping of logical channels to transport channels depends on the multiplexing that is configured by the RRC sublayer (see Figure 8.8) [4].

8.3 Downlink/uplink data transfer

The resource allocation on DL-SCH and UL-SCH is performed by the eNB scheduler. In the downlink, the allocation information pertaining to the physical downlink shared channel (PDSCH) within the same subframe is carried on the physical downlink control channel (PDCCH). The UE may be addressed using different radio network temporary identifiers (RNTIs). During normal operation, either cell radio network temporary identifier (C-RNTI) or semi-persistent scheduling (SPS) C-RNTI may be used depending on whether the transmission is dynamically scheduled, or is periodic with a predefined scheduling pattern. During the RACH procedure, DL-SCH transmissions may also be scheduled using the random access RNTI (RA-RNTI) or the temporary C-RNTI. Furthermore, if the system information is being transmitted via DL-SCH, then the SI-RNTI is utilized. The allocation information that is sent on the PDCCH using various downlink control information (DCI) formats includes resource block allocation, MCS, and HARQ information. Only one HARQ process may be scheduled in any subframe corresponding to either one or two transport blocks [10–12].

The allocation of the uplink grant is managed by the eNB. The allocation information is carried in the PDCCH. Similar to the downlink, the uplink transmissions can be identified using different RNTIs. For normal operation, either C-RNTI or SPS-C-RNTI may be used depending on the scheduling type. During the contention resolution (part of RACH procedure), temporary C-RNTI may be used to address a UE. The allocation information for the uplink is sent using a single DCI format that includes the allocated resource blocks, MCS, HARQ information, and whether frequency hopping is enabled.

The BCCH is a downlink logical channel that carries system information. The MAC sublayer maps the BCCH channel to BCH and DL-SCH transport channels. When mapped to BCH, BCCH broadcasts a master information block (MIB). The MIB carries essential parts of the system information that are necessary for initial system acquisition, such as system bandwidth, system frame number (SFN), and physical hybrid-ARQ indicator channel (PHICH) configuration information. The periodicity of MIB transmission is 40 ms. The remaining parts of the system information are transmitted in the form of system information blocks (SIBs), which are sent via BCCH and are mapped to DL-SCH. The SIBs are sent via system information messages, which can contain multiple SIBs. The system information messages are recognized on the DL-SCH by SI-RNTI. The system information message transmission is scheduled at regular intervals as specified in SIB1, which is transmitted on DL-SCH every 80 ms [10–12].

All SIBs are carried on the BCCH and are mapped to DL-SCH. The BCCH carries SIB1 that is mapped to the SI message 1 (SI-1), whereas other SIBs are mapped to other system information messages. The SI messages are identified by SI-RNTI, which has a predefined value of *0xFFFF*. The SIB1 is always transmitted with a period of 80 ms and contains the scheduling and multiplexing information of other SIBs. A broadcast HARQ process is utilized for all system information transmissions such that ACK/NACK feedback is not required. The redundancy version of a received downlink assignment in a specific TTI is determined by $RV_k = \lceil (3k/2) \rceil \mathrm{mod}4$ where k depends on the type of SI message. For the SIB1 message $k = (\mathrm{SFN}/2)\mathrm{mod}4$. For all other SI messages $k = l\mathrm{mod}4|_{l=0,1,\ldots,N_{\mathrm{SW}}-1}$ where l denotes the subframe number within the SI window, which has a duration defined by N_{SW}. The SI window is defined in SIB1 and specifies the starting point and time duration during which the SIs will be transmitted [11].

The paging control channel is a logical channel that is mapped by the MAC sublayer to the paging transport channel, which is transmitted on DL-SCH. For PCH, a unique radio network paging identifier with a value of *0xFFFE* is signaled in the PDCCH allocation message. Multiple UEs may be paged simultaneously using a single paging allocation with the UE identity indicated within the paging message.

8.3.1 HARQ principles

While the ARQ error control mechanism is simple and provides high transmission reliability, the throughput of ARQ schemes drops rapidly with increasing channel error rates and the ARQ latency, due to retransmissions, could be excessively high and intolerable for some delay-sensitive applications. Systems using forward error correction (FEC), on the other hand, can maintain constant throughput regardless of channel error rate. However, FEC schemes have some drawbacks. High reliability is hard to achieve with FEC and requires the use of long and powerful error correction codes that increase the complexity of implementation [10–12]. The drawbacks of ARQ and FEC can be overcome, if the two error control schemes are properly combined.

In order to achieve increased throughput and lower latency in packet transmission, the HARQ scheme was designed to combine the ARQ error control mechanism and FEC coding. A HARQ system consists of an FEC subsystem contained in an ARQ system. In this approach, the average number of retransmissions is reduced by using FEC through correction of the error patterns that occur more frequently; however, when the less frequent error patterns are detected, the receiver requests a retransmission where each retransmission carries the same or some redundant information to help the packet detection. The HARQ uses FEC to correct a subset of errors at the receiver and relies on error detection to detect the remaining errors. Most practical HARQ schemes utilize CRC codes for error detection and convolutional or turbo codes for error correction. The HARQ schemes are typically classified into three groups depending on the content of subsequent retransmissions, as follows [10–12]:

1. Type I HARQ: In type I HARQ schemes, the same data packet is transmitted in all the retransmissions. Soft combining may be used to improve the reliability. This type of HARQ is also referred to as chase combining (CC) where the blocks of data along with the CRC code are encoded using a FEC encoder before transmission. If the receiver is unable to correctly decode the data block, a retransmission is requested. When a retransmitted coded block is received, it is combined with the previously received block corresponding to the same information bits (using for example maximal ratio combining method) and fed to the decoder. Since each retransmission is an identical replica of the original transmission, the received E_b/N_0, i.e., the energy per information bit divided by the noise spectral power density, increases per each retransmission, improving the likelihood of correct decoding. In chase combining HARQ, the redundancy version of the encoded bits is not changed from one transmission to the next; therefore, the puncturing pattern remains the same. The receiver uses the current and all previous HARQ transmissions of the data block in order to decode the information bits. The process continues until either the information bits are correctly decoded and pass the CRC test, or the maximum number of HARQ retransmissions is reached. When the maximum number of retransmissions is reached, the MAC sublayer resets the process and continues with fresh transmission of the same data block.

   A number of parallel channels for HARQ can help improve the throughput as one process is awaiting an acknowlegment; another process can utilize the channel and transmit other subpackets. Figure 8.9 illustrates the operation of the chase combining HARQ scheme and how the retransmission of the same coded bits changes the combined energy per bit E_b while maintaining the effective code rate intact.
2. Type II HARQ: In the original type II HARQ scheme, an alternate parity/user data retransmission scheme is used. With proper choice of the channel encoder, the user information may also be recovered by parity bits. However, most type II algorithms adopt an incremental parity retransmission scheme, in which only additional parity bits are sent in subsequent retransmissions.

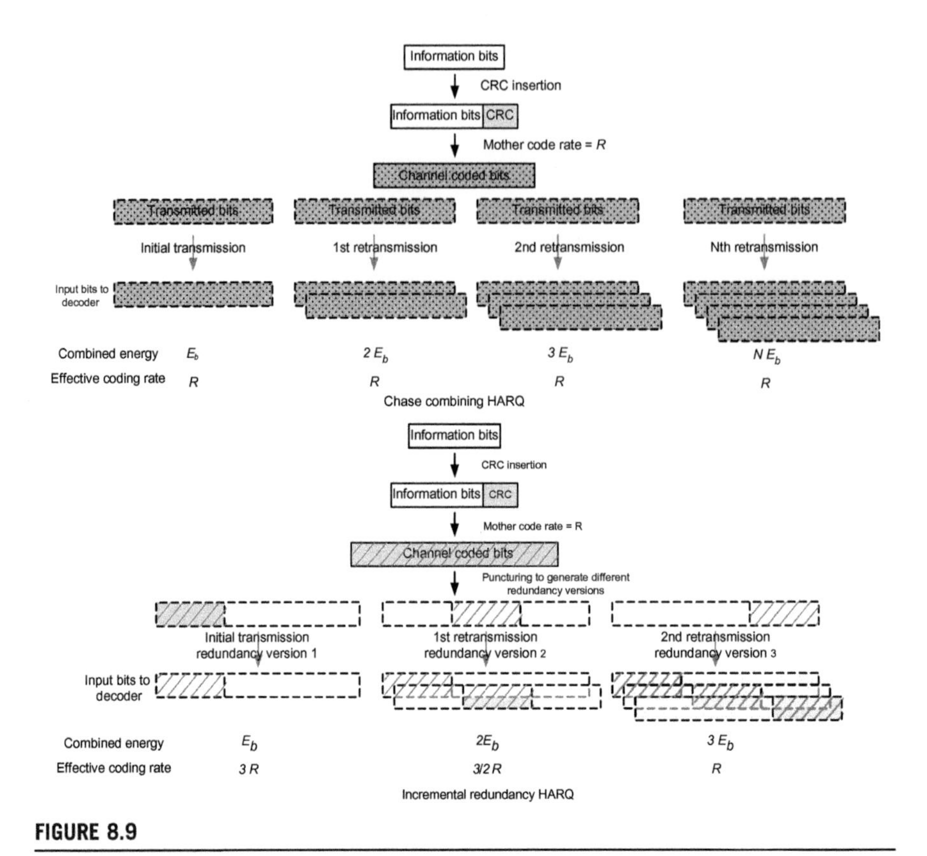

FIGURE 8.9

Illustration of the chase combining (CC) and incremental redundancy (IR) HARQ schemes [12].

Therefore, after each retransmission, a richer set of parity bits is available at the receiver, improving the probability of reliable decoding. In incremental schemes, however, information cannot be recovered from parity bits alone.

In an incremental redundancy HARQ scheme, a number of coded bits with increasing redundancy, where each represents the same set of input bits, are generated and transmitted to the receiver, when a retransmission, is requested to assist the receiver with the decoding of the information bits. The receiver combines each retransmission with the previously received bits belonging to the same packet. Since each retransmission carries additional parity bits, the effective code rate is lowered by each retransmission, as shown in Figure 8.9. The incremental redundancy is based on a low-rate code and the different redundancy versions are generated by puncturing the channel coder output. In the example shown in Figure 8.9, the basic code rate is R and 1/3 of the coded bits are transmitted in each retransmission. Aside from increasing the received signal to noise ratio E_b/N_0 by each retransmission due to combining, there is a coding gain[1] attained as result of each retransmission. It must be noted that chase combining is a special case of incremental redundancy HARQ where the retransmissions are identical copies of the original coded bits.

3. Type III HARQ: In a type III HARQ scheme, both user data and complementary parity bits are included in every retransmission. The type III HARQ schemes are less efficient than incremental redundancy schemes due to repeated transmission of user data bits. Also in this case, after every retransmission, a richer set of parity bits is available at the receiver, improving the probability of reliable decoding. Moreover, by adopting combining techniques, reliability may be further improved. In a type III incremental scheme, information can be recovered from every transmission. In a self-decodable HARQ retransmission scheme, the user data may be recovered from every single transmission.

8.3.2 Downlink HARQ timing and signaling protocols

LTE uses an adaptive asynchronous HARQ scheme in the downlink. Therefore, the retransmissions are scheduled in the same way as new transmissions, i.e., the transmission/retransmissions may occur at any time and at an arbitrary frequency location over the downlink bandwidth. The use of asynchronous and adaptive HARQ in the LTE downlink was partially motivated by the desire to avoid scheduling conflicts with transmission of SI, multicast, and broadcast content over single frequency network (MBSFN) subframes. Instead of dropping a retransmission that otherwise would have collided with MBSFN subframes or SI, the eNB can schedule the retransmission in time and/or frequency to avoid the overlap in resources. The HARQ scheme applies to the data transmissions on PDSCH, where the ACKs are received via the physical uplink control channel (PUCCH) or physical uplink shared channel (PUSCH). There is one HARQ entity at the UE for each serving cell. There are multiple serving cells and multiple HARQ entities when the UE supports carrier aggregation [12].

Soft combining is supported through transmission of an explicit new data indicator (NDI) flag, toggled for each new transport block. In addition to the NDI, HARQ-related downlink control signaling consists of the HARQ process number (3 bits for FDD, 4 bits for TDD) and the redundancy version (2 bits), both explicitly signaled in the scheduling assignment for each downlink transmission. Downlink spatial multiplexing implies transmission of two transport blocks in parallel on a component carrier. In this case, it is possible to retransmit only one of the transport blocks, since error occurrences in transmission of the transport blocks are typically uncorrelated; thus each transport block has separate new data and redundancy version indicators. However, there is no need to signal the process number separately, given that each HARQ process consists of two sub-processes in

[1]The coding gain is defined as the difference between E_b/N_0 required to achieve a given bit error rate (BER) in a coded system and the E_b/N_0 required to achieve the same BER in an uncoded system.

the case of spatial multiplexing, once the process number for the first transport block is known, the process number for the second transport block is derived implicitly. Transmissions on downlink component carriers are acknowledged independently. On each component carrier, transport blocks are acknowledged by transmitting one or two bits on the uplink. In the absence of spatial multiplexing, there is only a single transport block within a TTI and consequently only a single acknowledgment bit is required. However, in the case of spatial multiplexing, there are two transport blocks per TTI, each requiring its own HARQ feedback. The total number of bits required for HARQ acknowledgments depends on the number of component carriers and the transmission mode of each component carrier. Since downlink component carriers are scheduled separately using the PDCCH, the HARQ process numbers are signaled independently for each component carrier. It is obvious that the terminal is not expected to transmit a HARQ acknowledgement in response to transmission of system information, paging messages, and other broadcast information. Therefore, HARQ feedback is only sent in the uplink for unicast transmission [12].

LTE utilizes incremental-redundancy HARQ with a 1/3 turbo encoder used for FEC. The transport block CRC is used to detect errors. The receiver only receives different punctured versions of the same turbo-encoded data; each of these retransmissions is self-decodable. Thus, it falls into the category of a type III HARQ described earlier. The maximum number of retransmissions is limited to four, with an initial coding rate of 1/2 or 3/4. The maximum number of simultaneous downlink HARQ processes (number of PDSCH transmissions) is limited to eight for FDD mode, as specified in [7].

8.3.3 Uplink HARQ timing and signaling protocols

The operation of the LTE uplink HARQ protocol is different from the downlink counterpart. The choice of uplink HARQ protocol was made by taking into consideration the lower overhead of a synchronous non-adaptive scheme compared to an asynchronous adaptive scheme. As a result, uplink retransmissions always occur at fixed intervals. In the case of FDD mode, the uplink retransmissions occur eight subframes after the previous transmission attempt for the same HARQ process. The set of resource blocks used for the retransmission on each component carrier are identical to the initial transmission. The control signaling required in the downlink for an uplink retransmission is a HARQ feedback, which is transmitted on PHICH. In the case of a NACK on PHICH, the data is retransmitted. Spatial multiplexing is handled in the same way as in the downlink where two transport blocks are concurrently transmitted in the uplink, each with its own MCS and NDI, but sharing the same HARQ process number. The two transport blocks are individually acknowledged, thus two bits of information are needed in the downlink to acknowledge an uplink transmission when spatial multiplexing is utilized. The uplink carrier aggregation is another example where multiple ACKs/NACKS are required, i.e., one or two ACK/NACK bits per uplink component carrier. In the TDD mode, the uplink transmissions in different subframes may need to be acknowledged in the same downlink subframe. Thus, multiple PHICHs may be needed as each PHICH is capable of transmitting one bit of HARQ feedback. Each of the PHICHs is transmitted on the same downlink component carrier that was used to schedule the initial uplink transmission. Despite the fact that the default mode of operation for the uplink is synchronous, non-adaptive HARQ, it is also possible to operate the uplink HARQ in a synchronous, adaptive mode, where the resource block set and MCS for the retransmissions are changed according to the transmission conditions. Although non-adaptive retransmissions are typically used, due to their very low downlink control signaling overhead, adaptive retransmissions are sometimes useful in order to avoid the uplink frequency resource fragmentation or to avoid collisions with random access allocated resources. A UE is scheduled for an initial transmission in subframe n and if the transmission is not correctly received, a retransmission is required in subframe $n+8$ in FDD mode (the TDD retransmission timing depends on the subframe configuration). With non-adaptive HARQ, the retransmissions occupy the same part of the uplink spectrum as the initial transmission. In this case, the spectrum may be fragmented, which limits the bandwidth available for other terminals, unless the other terminal is capable of multi-cluster transmission (LTE Rel-10 and onward UEs). Example HARQ operation and timing in the uplink for an FDD and TDD system are shown in Figures 8.10 and 8.11, respectively.

Figure 8.12 depicts an example uplink HARQ process using adaptive and non-adaptive schemes to better understand this operation. In the adaptive case, the

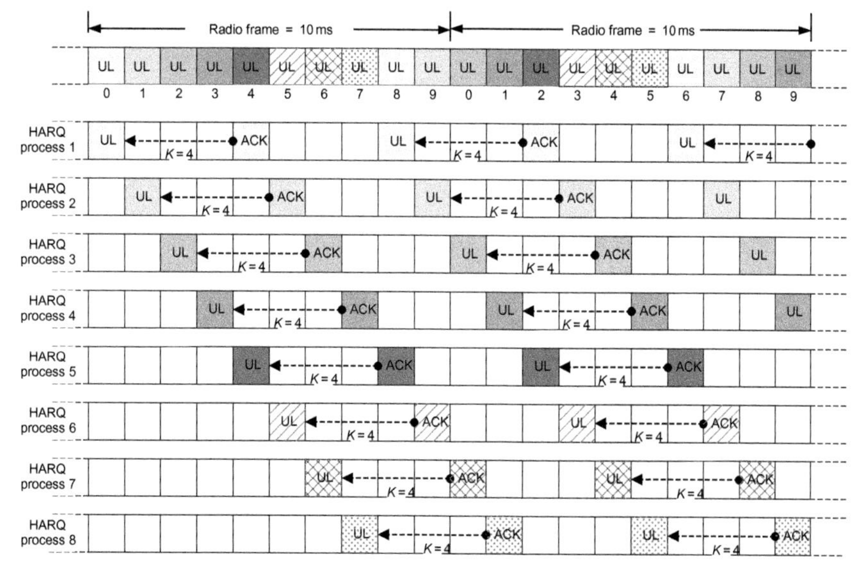

FIGURE 8.10

Example uplink HARQ timing for an FDD system [10].

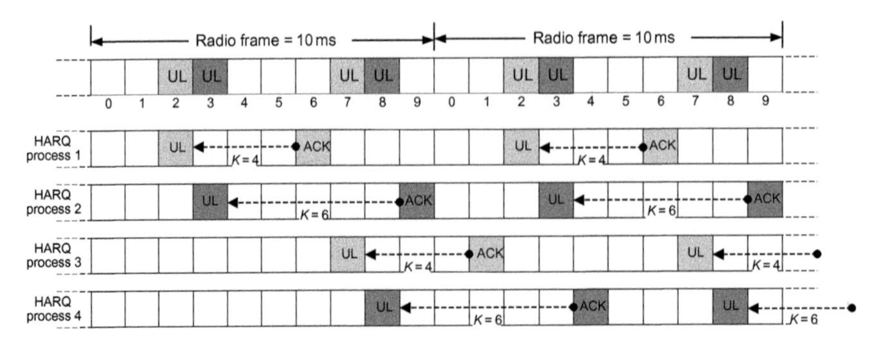

FIGURE 8.11

Example uplink HARQ timing for a TDD system with subframe configuration 1 [10].

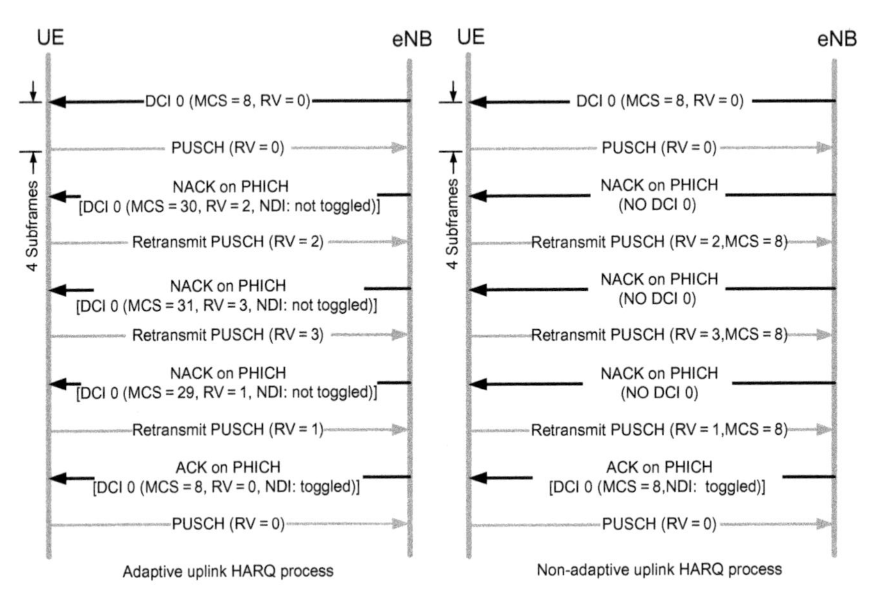

FIGURE 8.12

Illustration of example adaptive and non-adaptive uplink HARQ processes [21].

main idea is that each uplink retransmission uses a different redundancy version (RV) and the RV is determined by DCI format 0 (i.e., uplink grant). In the case of non-adaptive uplink HARQ, each uplink retransmission uses a different RV and the RV is determined by a predefined sequence specified in [1]. An important question is how the UE determines whether it is supposed to use adaptive retransmission or non-adaptive retransmission. The UE uses adaptive retransmission if it detects DCI format 0 and NDI is not toggled. In this case, the UE ignores HARQ feedback in PHICH and it retransmits based on DCI format 0 information. The

UE determines that it is supposed to use non-adaptive retransmission when it receives negative HARQ feedback in PHICH and does not receive DCI format 0.

The adaptive and non-adaptive HARQ schemes can be supported by having the terminal save the content of its transmission buffer until it receives an NDI on PDCCH. The decision on whether the data in the transmission buffer should be retransmitted is made based on the NDI included in the uplink scheduling grant which is sent in the PDCCH. The NDI is toggled for each new transmission. If the NDI is toggled, the terminal flushes the transmission buffer and transmits a new data packet. However, if the NDI does not indicate transmission of a new transport block, the previous transport block is retransmitted. The negative HARQ ACK on PHICH could instead be seen as a single-bit scheduling grant for retransmissions, where the set of bits to transmit and all the resource information are known from the previous transmission attempt [12].

The NDI is explicitly transmitted in the uplink grant. However, unlike the downlink case, the redundancy version is not explicitly signaled for each retransmission and cannot be inferred from a single-bit HARQ feedback on PHICH. Instead, given that the uplink HARQ protocol is synchronous, the redundancy version follows a predefined pattern, starting with zero when the initial transmission is scheduled by PDCCH. Whenever a retransmission is requested by sending a NACK on PHICH, the next redundancy version in the sequence is used. However, if a retransmission is explicitly scheduled by PDCCH overriding PHICH, there is a possibility that a different redundancy version may be used. Uplink grants for retransmissions use the same format as normal grants (for initial transmissions). The transport block size is already known from the initial transmission and cannot change between retransmission attempts.

The uplink HARQ retransmissions are associated with the HARQ feedback as seen by the UE on PHICH and the uplink grant sent in PDCCH. The information in the uplink grant takes precedence over retransmissions. The UE adheres to the following rules:

1. If the HARQ feedback is an ACK and no grant is detected for the HARQ process on PDCCH, then the UE maintains the data in its soft buffer until the PDCCH grant indicates whether the previous transmission was successful.
2. If the HARQ feedback is a NACK and no PDCCH grant is detected for the HARQ process, then the UE retransmits using the same PUSCH resources as the last transmission. The redundancy version used follows a known sequence.
3. If the PDCCH indicates a new transmission on PDCCH for a specific HARQ process, then regardless of HARQ feedback on PHICH, it is treated as a new transmission. Any data in the soft buffer is flushed.
4. If the PDCCH indicates a retransmission for a specific HARQ process, then irrespective of HARQ feedback, the UE performs a retransmission according to the parameters on the PDCCH.

In the case of carrier aggregation, if cross-carrier scheduling is not used and each uplink component carrier is scheduled on its corresponding downlink component carrier, the uplink component carriers are not associated with the same

PHICH resource. However, when cross-carrier scheduling is utilized, transmissions on multiple uplink component carriers may need to be acknowledged on a single downlink component carrier. In order to avoid PHICH collisions, the scheduler ensures that different reference signal phase rotations (cyclic shifts) or different resource block starting positions are used for different uplink component carriers. For semi-persistent scheduling, the reference signal phase rotation is always set to zero, since the semi-persistent scheduling is only supported on the primary component carrier. If multiple component carriers are scheduled for uplink data transmission in a subframe, there will be one or two (with spatial multiplexing) PHICHs transmitted per uplink component carrier. In general, LTE transmits the PHICH on the same component carrier that was used for the uplink grant, scheduling the corresponding uplink data transmission. Thus, a terminal is only required to monitor the component carriers that include uplink scheduling grants (especially as the PDCCH may override PHICH to support adaptive retransmissions in some cases), which helps to reduce the terminal's power consumption [10–12].

The transmission of HARQ feedback in the uplink in response to PDSCH transmissions is illustrated in Figure 8.13 for an FDD system. The data on PDSCH is transmitted to the terminal in subframe n and received by the terminal after the propagation delay in subframe n. The terminal attempts to decode the received signal, using soft combining of previous retransmissions, and then it transmits a HARQ ACK/NACK in uplink subframe $n+4$. It must be noted that the start of an uplink subframe at the terminal is offset by the timing advance relative to the start of the corresponding downlink subframe at the terminal. Upon reception of the HARQ feedback, the eNB can transmit new data or retransmit the previous data in subframe $n+8$. Thus, the HARQ round-trip time (RTT) is 8 ms for an FDD system.

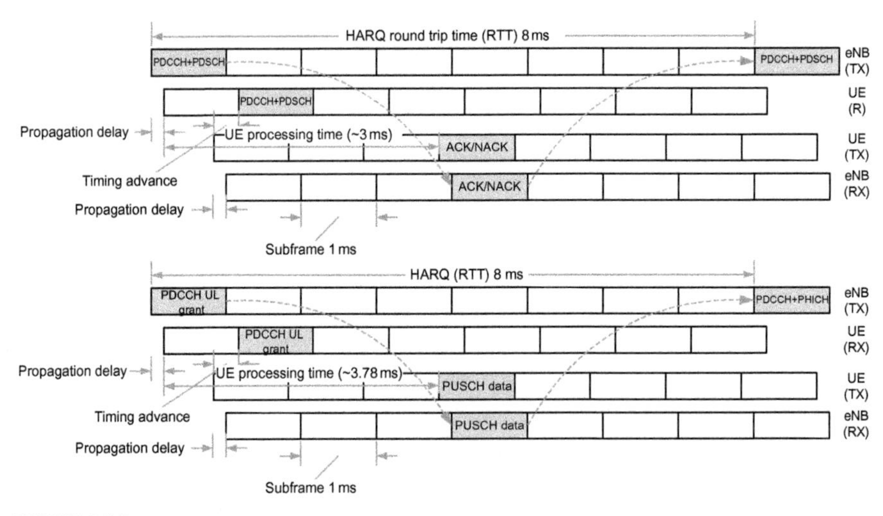

FIGURE 8.13

Example timing diagram for downlink and uplink HARQ schemes (FDD mode) [11].

In practice, the UE processing time depends on the value of the timing advance (to compensate variable propagation delay), or alternatively on its distance to the base station. Since the UE is required to operate at any distance up to the maximum cell radius of 100 km (corresponding to 0.67 ms) supported by the standard, the UE is designed for the worst-case scenario. That means UE baseband processing time is approximately 2.3 ms. For the eNB, the baseband processing time is 3 ms [11].

In LTE TDD, the time association between the data transmission and the ACK/NACK cannot be maintained due to the variable numbers of downlink and uplink subframes being present in a frame. The uplink and downlink delay between data and ACK/NACK is dependent on the TDD subframe configuration. Therefore, a fixed delay between a transmission and the HARQ ACK/NACK is not possible in LTE TDD. In TDD mode, the delay between the transmission and the HARQ ACK/NACK depends on both TDD subframe configuration and the subframe in which the data was transmitted. A fixed delay cannot be assured as subframes are allocated to downlink and uplink depending on the subframe configuration. As an example, in subframe configuration 1 (see Table 8.1), there are a number of downlink subframes whose nearest uplink subframe (four or more subframes apart) is seven subframes apart.

Table 8.1 shows the HARQ ACK/NACK timing delays for the downlink and uplink of a TDD system assuming the initial data transmission occurred in subframe n, i.e., the transmission of ACK/NACK occurs in subframe $n + k$. It can be noticed that the delays are dependent on the configuration and the subframe in which the original data transmission has occurred. For example, in configuration 1, the delay from the eNB side can range from four to six subframes, while the delays from the UE side can be from four to seven subframes. As a result, the delay for transmission of ACK/NACK can range from four to seven subframes for the downlink HARQ and from four to thirteen subframes for the uplink HARQ. The higher number in the uplink case is due to the larger downlink asymmetry in the configuration. Therefore, in TDD systems, the acknowledgment of a transport block in subframe n is transmitted in subframe $n + k$, where $k \geq 4$ and is selected such that $n + k$ is an uplink subframe when the acknowledgment is transmitted from the terminal on PUCCH or PUSCH, and a downlink subframe when the acknowledgment is transmitted from the eNB on PHICH. The value of k depends on the downlink–uplink subframe configuration, as shown in Table 8.1 for downlink and uplink transmissions.

In FDD systems, multiple HARQ processes can be activated in both downlink and uplink directions. Due to the fixed timing associations between the original and the retransmission on the uplink, the maximum number of HARQ processes has been set to eight. However, in TDD systems, due to the variable delays experienced with respect to ACK/NACK arrivals as well as the retransmissions, the number of HARQ processes is dependent on the configuration. The maximum number of HARQ processes allowed per configuration for downlink and uplink is given in Table 8.1. Typically, the retransmissions can be scheduled four subframes (or more) after the reception of a NACK message. In LTE TDD, the numbers of downlink and uplink subframes in a frame are not equal. For instance, configurations 0–5 have more downlink subframes compared to uplink subframes. Consequently, data

Table 8.1 Number of HARQ Processes and Timing for Different TDD Configurations [12]

Subframe Configuration (DL:UL)	Downlink Subframe Number										Uplink Subframe Number										Number of Processes	
	0	1	2	3	4	5	6	7	8	9	0	1	2	3	4	5	6	7	8	9	Uplink	Downlink
0 (2:3)	4	6	–	–	–	4	6	–	–	–	–	–	4	7	6	–	–	4	7	6	7	4
1 (3:2)	7	6	–	–	4	7	6	–	–	4	–	–	4	6	–	–	–	4	6	–	4	7
2 (4:1)	7	6	–	4	8	7	6	–	4	8	–	–	6	–	–	–	–	6	–	–	2	10
3 (7:3)	4	11	–	–	–	7	6	6	5	5	–	–	6	6	6	–	–	–	–	–	3	9
4 (8:2)	12	11	–	–	8	7	7	6	5	4	–	–	6	6	–	–	–	–	–	–	2	12
5 (9:1)	12	11	–	9	8	7	6	5	4	13	–	–	6	–	–	–	–	–	–	–	1	15
6 (5:5)	7	7	–	–	–	7	7	–	–	5	–	–	4	6	6	–	–	4	7	–	6	6

transmission from multiple downlink subframes needs to be acknowledged in a certain uplink subframe and vice versa for certain cases for the uplink data transmission.

The transmission of multiple ACK/NACK messages in uplink or downlink subframes is a unique feature of LTE TDD compared with FDD due to the above-mentioned limitation. Two methods have been specified in the standard to deal with this issue: ACK/NACK bundling and ACK/NACK multiplexing. For the eNB, transmission of separate ACK/NACKs is not a major problem as the eNB is not power constrained and can send multiple ACK/NACKs in a subframe. This method, where separate ACK/NACKs are sent in the same subframe for transmissions in different subframes, is called ACK/NACK multiplexing. Thus, in the downlink ACK/NACK transmission, it is by default ACK/NACK multiplexing. However, in the uplink, the UEs might be power constrained depending on their location within the cell. For UEs which are not constrained by power and coverage, multiple PUCCH messages relating to each HARQ process can be generated, i.e., separate ACK/NACK messages can be sent. For coverage-limited UEs, ACK/NACK bundling is used since fewer bits need to be sent in this case. Multiplexing implies that independent ACKs/NACKs for each of the received transport blocks are sent to the eNB, which allows independent retransmission of erroneous transport blocks. However, it also suggests that multiple bits need to be transmitted from the terminal, which may limit the uplink coverage. This is the motivation for the bundling mechanism.

Bundling of ACKs/NACKs implies that the outcome of the decoding of downlink transport blocks from multiple downlink subframes can be combined into a single HARQ feedback transmitted in the uplink [12]. A logical AND operation on all ACK/NACK messages is typically performed and the result is sent. For example, in TDD subframe configuration 1, the transmissions in subframes 0 and 1 are acknowledged in subframe 7. A logical AND of these messages is performed and sent. This implies that both transport blocks will be retransmitted by the eNB as it does not have the individual acknowledgment status of the downlink transmissions. This is a disadvantage of ACK/NACK bundling as more retransmissions than necessary might be sent. The main advantage of the ACK/NACK bundling method is the fact that the limited power resources of a UE will now be used only in a fewer resource blocks as compared to a larger number of resource blocks. This can improve the uplink coverage of the UE since there will be more power per resource block that will be available to the UE in the case of ACK/NACK bundling [12].

In order to transmit on the UL-SCH the UE must have a valid uplink grant (except for non-adaptive HARQ retransmissions) which it may receive dynamically on the PDCCH or in an RAR message. To perform requested transmissions, the MAC sublayer receives HARQ information from lower layers. When the physical layer is configured for uplink spatial multiplexing, the MAC sublayer can receive up to two grants (one per HARQ process) for the same TTI from lower layers [1].

At a given TTI, if an uplink grant is indicated for the TTI, the HARQ entity identifies the HARQ processes for which a transmission should occur. It also routes the received HARQ feedback (ACK/NACK information), modulation and coding, and resource allocation information (conveyed by the physical layer) to the appropriate HARQ processes. When TTI bundling is configured, the parameter *TTI_BUNDLE_SIZE* provides the number of TTIs of a TTI bundle. The TTI bundling operation relies on the HARQ entity for invoking the same HARQ process for each transmission that is part of the same bundle. Within a bundle, HARQ retransmissions are non-adaptive and triggered without waiting for feedback from previous transmissions according to the *TTI_BUNDLE_SIZE*. The HARQ feedback of a bundle is only received for the last TTI of the bundle (i.e., the TTI corresponding to *TTI_BUNDLE_SIZE*), regardless of a transmission in that TTI (e.g., when a measurement gap occurs). A retransmission of a TTI bundle is also a TTI bundle. TTI bundling is not supported when the UE is configured with one or more secondary cells (SCells) with configured uplink [1].

Each HARQ process in the uplink maintains a state variable *CURRENT_TX_NB*, which indicates the number of transmissions that have taken place for the MAC PDU currently in the buffer, and a state variable *HARQ_FEEDBACK*, which indicates the HARQ feedback for the MAC PDU currently in the buffer. When the HARQ process is established, *CURRENT_TX_NB* set to zero. The sequence of redundancy versions is 0, 2, 3, and 1 (Figure 8.12). The variable *CURRENT_IRV* is an index to the sequence of redundancy versions. This variable is updated in modulo 4. New transmissions are performed on the resource and with the MCS indicated on PDCCH or RAR message. Adaptive retransmissions are performed on the resource and, if provided, with the MCS indicated on PDCCH. Non-adaptive retransmission is performed on the same resource and with the same MCS as was used for the last transmission attempt [1].

8.4 DRX operation

LTE supports DRX in the downlink, to save the UE battery power during periods of inactivity. In the DRX mode, the UE only monitors PDCCH during a predefined on-duration interval within one of two possible DRX cycles. The UE may turn off its receiver when it does not expect to receive any transmission in the downlink over PDCCH. The UE can transition to DRX mode in one of the following two ways: (1) implicitly based on the expiration of certain configured timers; or (2) explicitly based on the transmission of a DRX command in the form of a MAC control element.

As shown in Figure 8.14, there are a number of timers associated with the DRX mode. The DRX inactivity timer controls the number of subframes for which a UE continues to monitor the downlink before an absence of activity triggers the UE to enter the DRX mode. Upon entering the DRX mode, the UE monitors the downlink for a number of subframes defined by the on-duration timer in every short DRX cycle or long DRX cycle. The short DRX cycle is optional. If it is configured, the UE will follow the short DRX cycle for a specific number of cycles controlled by the short DRX cycle timer before switching to the long DRX

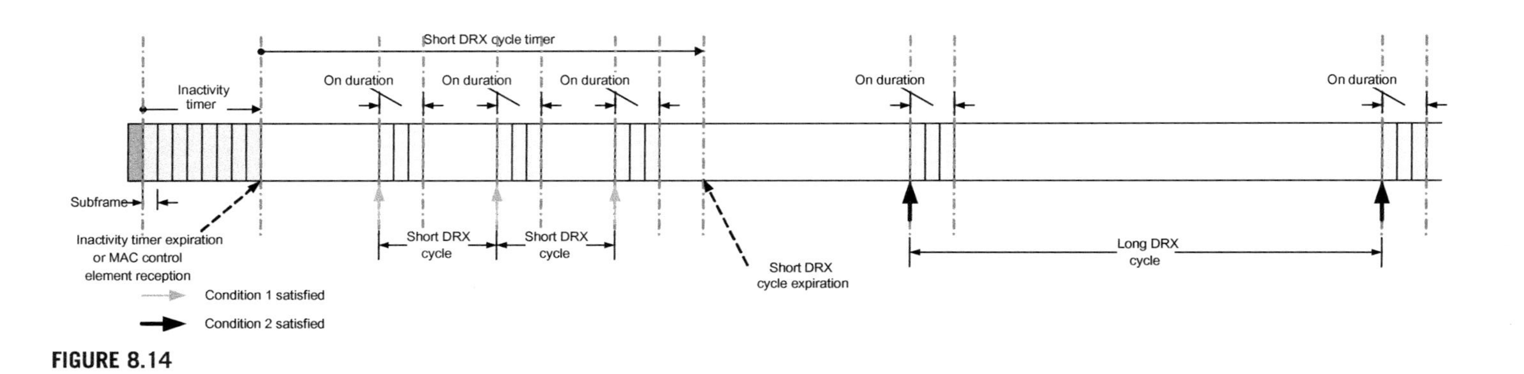

FIGURE 8.14

Illustration of short and long DRX cycles and timers [10].

cycle, if there is no activity. Other timers prevent the UE from entering the DRX mode. These are associated with HARQ downlink retransmissions. If the UE has received data that it was unable to decode and this was not associated with a broadcast HARQ process, then following the expiration of the HARQ RTT timer, which specifies the earliest time at which a UE can expect a retransmission, the DRX retransmission timer is started. The UE must wait for this timer to expire and continue to monitor the downlink before it can enter the DRX mode. Moreover, the UE cannot enter the DRX mode under the following conditions:

- If downlink transmissions are in progress or if the UE has sent a scheduling request.
- If the UE is expecting an uplink grant for HARQ retransmission.
- If the RACH procedure is in progress and the contention resolution timer is running.

In addition to the above timers, the network can order the UE to enter the DRX mode with the DRX command MAC control element. All timers and parameters described in this context are configured by RRC signaling.

A UE may be configured by the RRC sublayer for DRX, which controls the way the UE monitors PDCCH and looks for its C-RNTI, TPC-PUCCH-RNTI, TPC-PUSCH-RNTI, and SPS-C-RNTI (if configured). When the UE is in the RRC_CONNECTED state and if DRX is configured, the UE is allowed to monitor PDCCH intermittently; otherwise, the UE monitors PDCCH continuously. The RRC signaling controls DRX operation by configuring the following timers: *onDurationTimer*; *drx-InactivityTimer*; *drx-RetransmissionTimer* (for each downlink HARQ process except for the broadcast process); *longDRX-Cycle*; *drxStartOffset*; and optionally *drxShortCycleTimer* and *shortDRX-Cycle*. Note that the same activity time applies to all activated serving cells in carrier aggregation scenarios [1].

In Figure 8.14, Conditions 1 and 2 are defined as follows:

Condition 1: $(10\text{SFN} + n_{\text{subframe}})\text{mod}(\textit{shortDRX-Cycle}) = (\textit{drxStartOffset})\text{mod}(\textit{shortDRX-Cycle})$.

Condition 2: $(10\text{SFN} + n_{\text{subframe}})\text{mod}(\textit{longDRX-Cycle}) = \textit{drxStartOffset}$.

The DRX parameters and timers in the above equations are defined as follows [1]:

- *Active time*: The time interval during which the UE monitors the PDCCH in subframes containing a PDCCH.
- *DRX cycle*: The periodic repetition of the on-duration followed by a potential period of inactivity.
- *drx-InactivityTimer*: The number of consecutive subframes containing a PDCCH, after the subframe in which a PDCCH indicates an initial uplink/downlink user data transmission for the UE.
- *drx-RetransmissionTimer*: The maximum number of consecutive subframes containing a PDCCH immediately after a downlink retransmission is expected by the UE.

- *drxShortCycleTimer*: The number of consecutive subframes that the UE is required to follow the *short DRX cycle*.
- *drxStartOffset:* Specifies the subframe where the DRX cycle starts.
- *onDurationTimer*: The number of consecutive subframes containing a PDCCH at the beginning of a DRX cycle. This is the time duration after waking up from DRX that the UE monitors the PDCCH region in the subframe. If the UE successfully decodes a PDCCH, the UE stays awake and starts the inactivity timer.

From the above parameters, the on-duration and inactivity timer have fixed lengths, while the active time has variable length based on scheduling decision and the UE decoding success. Only on-duration and inactivity timer duration are signaled to the UE by the eNB. There is only one DRX configuration applied in the UE at any time and the UE is required to apply an on-duration upon wake-up from the DRX cycle. This is also applicable for the case where the UE has only one real-time service that is being handled through the allocation of predefined resources. This would allow other signaling, mechanisms, such as RRC signaling, to be sent during the remaining portion of the active time. The new transmissions can only occur during the active time, so that when the UE is waiting for one retransmission, it does not have to be awake during the HARQ round-trip time. If the PDCCH has not been successfully decoded during the on-duration, the UE must follow the DRX configuration, i.e., the UE can enter the DRX cycle, if allowed by the DRX configuration. This applies also for the subframes where the UE has been allocated predefined resources.

If the UE successfully decodes a PDCCH for a first transmission, it stays awake and starts the inactivity timer, even if a PDCCH is successfully decoded in the subframes where the UE has also been allocated predefined resources, until a MAC control message informs the UE to reenter the DRX cycle, or until the inactivity timer expires. In both cases, if a short DRX cycle is configured, the UE first follows the short DRX cycle and after a longer period of inactivity the UE follows the long DRX cycle; otherwise, the UE directly enters the long DRX cycle. When the DRX is configured, the network should detect whether the UE remains in the E-UTRAN coverage by requesting the UE to send periodic signals to the network. In the case of carrier aggregation, whenever a UE is configured with only one serving cell, i.e., primary cell, LTE Rel-8/9 DRX scheme applies. In other cases, the same DRX operation applies to all configured and activated serving cells (i.e., identical active time for PDCCH monitoring) [8].

During the DRX procedure, the UE maintains a DRX cycle that is defined as a number of subframes. The UE monitors the PDCCH for a specific number of subframes in a DRX cycle. This time interval is called on-duration and can range from 1 to 200 subframes. The UE may turn off its receiver for rest of the DRX cycle. The UE maintains two DRX cycles, referred to as short DRX cycle and long DRX cycle, which have different durations. The short DRX cycle is optional and targeted for applications, such as voice over IP (VoIP) which typically require transmission of relatively small amounts of data at short but regular intervals. If configured, the UE starts with a short DRX cycle (2–640 subframes) when it enters the DRX mode.

When the configurable short DRX timer expires, the UE transitions to the long DRX cycle (10–2560 subframes). The UE can transition to the DRX mode either upon expiration of configured timers or following a DRX command in the form of a MAC control element. Figure 8.15 illustrates different scenarios in DRX operation.

The downlink radio link quality of the primary cell must be monitored by the UE for the purpose of indicating out-of-sync/in-sync status to higher layers. In non-DRX operation mode, the physical layer in the UE in every radio frame assesses the radio link quality over the previous time period against thresholds (Q_{out} and Q_{in}). In DRX operation mode, the physical layer in the UE, at least once every DRX period assesses the radio link quality over the previous time period, against thresholds (Q_{out} and Q_{in}). The threshold Q_{out} is defined as the level at which the downlink radio link cannot be reliably received, corresponding to a 10% block error rate of a hypothetical PDCCH transmission taking into account the

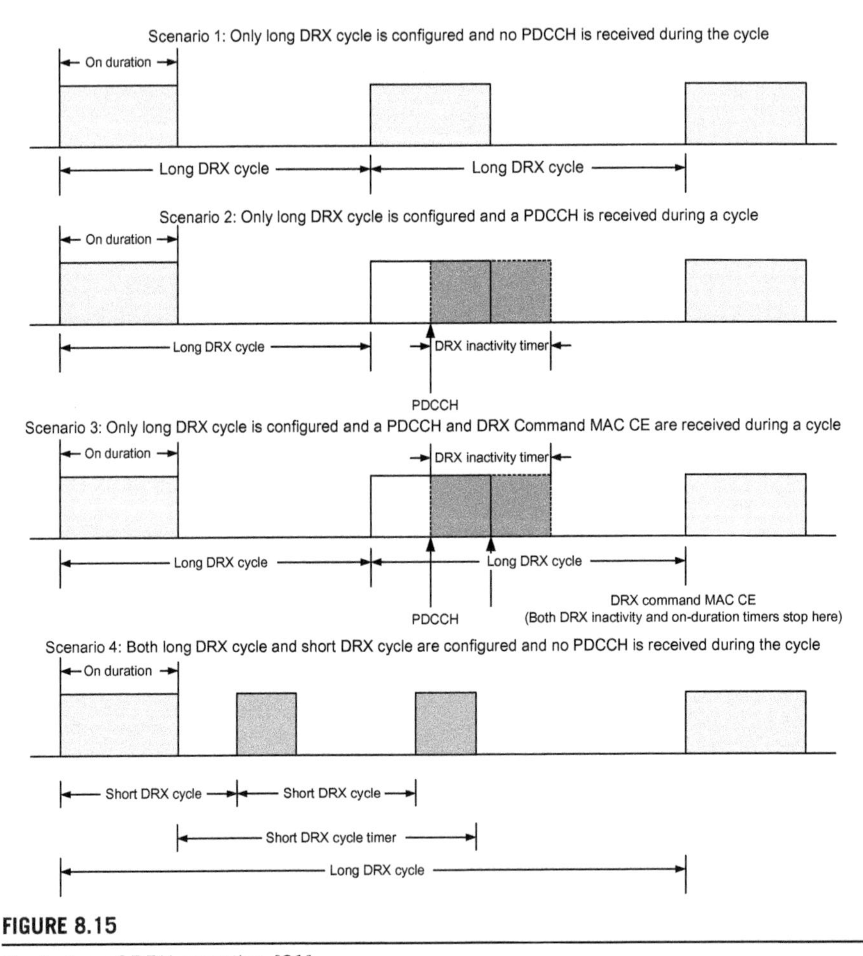

FIGURE 8.15

Illustration of DRX operation [21].

physical control format indicator channel (PCFICH) errors with transmission parameters specified in [9]. The threshold Q_{in} is defined as the level at which the downlink radio link quality can be significantly more reliably received than at Q_{out} and corresponds to a 2% block error rate of a hypothetical PDCCH transmission taking into account PCFICH errors with transmission parameters specified in [9].

To investigate the effect of DRX on packet delay, let's assume that the inter-packet arrival times T_{arrival} follow an exponential distribution with a mean of $1/\lambda$ ms, the extra delay T_{d} (in milliseconds), caused by active mode of DRX, can be computed as $T_{\text{d}} = T_{\text{arrival}} \bmod T_{\text{DRX-cycle}}$. The probability distribution of the extra delay T_{d} can be expressed as follows [13]:

$$p_{\text{d}}(k) = \frac{1}{T_{\text{DRX-cycle}}} \sum_{i=0}^{T_{\text{DRX-cycle}}-1} \frac{\lambda}{a_i^k (1 - a_i \mathrm{e}^{-\lambda})} \tag{8.1}$$

where a_i is the ith order-n root of unity[2] and $p_{\text{d}}(k)$ is the probability that the extra delay is equal to k subframes.

The energy saving of the UE due to use of the DRX mechanism can be expressed as follows [13]:

$$Energy\_Saving\% = 100 \frac{N_{\text{DRX}} E_{\text{DRX}} + N_{\text{Active}} E_{\text{Active}}}{(N_{\text{DRX}} + N_{\text{Active}}) E_{\text{Active}}} \tag{8.2}$$

It is assumed that there are N_{DRX} frames during which the UE is in the DRX mode and N_{Active} frames during which the UE is in the normal operation mode. Furthermore, we assume that the energy consumed per frame is E_{DRX} and E_{Active} during the DRX and active modes, respectively. The ratio of E_{DRX} and E_{Active} is directly related to the number of hardware sub-systems (or modules), that are powered down during the DRX mode [13,14,18].

8.5 MAC control elements

MAC control element are control commands and reports that enable MAC operation. They are transmitted as part of the DL-SCH or UL-SCH and can be piggy-backed to the data payloads. The following CEs have been defined in MAC specification as part of LTE Rel-11:

Downlink

- Timing advance: This CE is used to provide initial and periodic timing adjustment commands to the UE in the uplink.

[2]In mathematics, a root of unity, often called a de Moivre number, is a complex number whose nth power is equal to one. Roots of unity are used in many branches of mathematics, and are especially important in number theory, field theory, and discrete Fourier transform. The notion of root of unity also applies to any algebraic ring with a multiplicative identity element, namely a root of unity is any element of finite multiplicative order. An nth root of unity, where $n = 1, 2, 3, \ldots$ is a positive integer, is a complex number z satisfying the equation $z^n = 1$. An nth root of unity is primitive, if it is not a kth root of unity for some smaller integer k, i.e., $z^k \neq 1 (k = 1, 2, 3, \ldots, n - 1)$ [20].

- DRX command: This CE initiates DRX mode at the UE and conveys DRX commands to the UE.
- UE contention resolution identity: This CE is used during the RACH procedure to resolve possible contention between multiple UEs that are trying to simultaneously access the network.
- MBMS dynamic scheduling information: This CE is transmitted for each MCH to inform the MBMS-capable UEs of scheduled data transmission on MTCH.
- Activation/deactivation: This CE is used to activate or deactivate secondary cells for the carrier-aggregation-enabled UEs.

Uplink

- Buffer status report (BSR): This CE is used to report the UE's soft buffer status and to enable uplink scheduling.
- Power headroom: This CE is used to report the UE's transmit power relative to maximum permissible value, or whether the UE is currently power limited.
- C-RNTI report: This is used to identify a UE when sending information over CCCH and for the UE to transmit a dedicated C-RNTI during the random access procedure for the purpose of contention resolution.
- Extended power headroom: This CE was specified to support carrier aggregation in LTE Rel-10. In the case of carrier aggregation, the UE reports power headroom of each component carrier independently.

As shown in Figure 8.16, each MAC CE included in the MAC PDU has an associated one-byte subheader in the MAC header with the format R/R/E/LCID, where the LCID indicates the type of MAC CE. The values of LCID for DL-SCH and UL-SCH are listed in Table 8.2.

8.5.1 Uplink timing advance control element

After initial synchronization on RACH, the UE must maintain a reasonably accurate uplink time alignment. This is achieved using the timing advance control element which is sent as part of a MAC PDU on DL-SCH, enabling the eNB to periodically send timing adjustment commands to the UE. The network may also send timing adjustment commands as part of an RAR message. When the

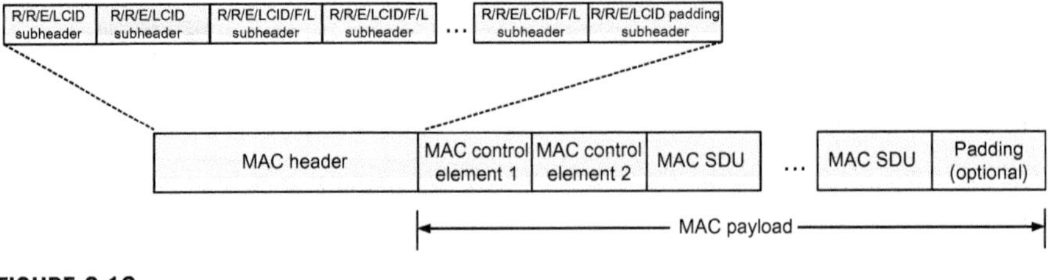

FIGURE 8.16

An example MAC PDU containing MAC header, MAC CEs, MAC SDUs, and padding [1].

Table 8.2 Values of the LCID for DL-SCH and UL-SCH [1]

Values of LCID for DL-SCH		Values of LCID for UL-SCH	
Index	**LCID Values**	**Index**	**LCID Values**
00000	CCCH	00000	CCCH
00001–01010	Identity of the logical channel	00001–01010	Identity of the logical channel
01011–11010	Reserved	01011–11000	Reserved
11011	Activation/deactivation	11001	Extended PHR
11100	UE contention resolution identity	11010	PHR
11101	Timing advance command	11011	C-RNTI
11110	DRX command	11100	Truncated BSR
11111	Padding	11101	Short BSR

command is sent as a control element, the timing advance is a 6-bit number that adjusts timing (advances or delays) relative to the current UE radio frame timing. If the command is sent via an RAR message, the timing advance is an 11-bit number that specifies absolute timing relative to the downlink radio frame. The eNB calculates the preferred timing advance based on uplink transmissions from the UE. The eNB also configures the UE's timing alignment timer. If the UE does not receive a timing advance MAC control element before the timing alignment timer expires, it is no longer considered to be uplink-synchronized with the cell.

The UE must apply the timing advance command when it receives it from the eNB. The command received in the nth frame is applied in the $(n + 6)$th frame. The command provides incremental value which adjusts the current timing alignment value. The incremental value may be negative or positive. Positive value advances and negative value delays the uplink transmission timing. If the UE does not receive a timing advance MAC control element before the timing alignment timer expires, it is no longer considered to be synchronized with the cell. When the timer expires, the UE must flush HARQ buffers and reinitialize the HARQ processes, release all PUCCH and sounding reference signal resources. If any uplink transmission is required, the UE must perform contention-based RACH synchronization procedures to obtain timing alignment and an uplink grant before it can transmit in the uplink.

The timing advance command MAC control element in LTE Rel-11 has a fixed size and consists of a single octet as shown in Figure 8.17. The information fields in this control element consist of a timing advance group[3] (TAG) identity (TAG ID), which indicates the TAG identity of the addressed TAG and the timing advance command. The TAG containing the PCell has the TAG identity value of zero. The

[3]A timing advance group is a group of serving cells that is configured by RRC signaling and uses the same timing reference cell and the same timing advance value. Primary timing advance group (pTAG) is the timing advance group containing the PCell. Secondary timing advance group (sTAG) is the timing advance group not containing the PCell [8].

length of this field is 2 bits. The timing advance command indicates the index value of timing advance $T_A = (0, 1, 2, \ldots, 63)$ that provides the amount of timing adjustment that the UE has to apply. The length of this field is 6 bits. Note that in the earlier releases of LTE, the leftmost 2 bits were reserved.

8.5.2 BSR control element

The BSR MAC control elements are used to report soft buffer size of either a single logical channel group (LCG) or a set of LCGs. An LCG is a group of logical channels identified by a unique 2-bit LCG ID. The mapping of logical channels to an LCG is set up during RRC configuration. There are up to four LCGs that can be set up per UE. The BSR reports aggregate bits across all logical channels of an LCG. As shown in Figure 8.18, there are two formats for the BSR: short/truncated BSRs indicate a buffer size of one LCG, and long BSR indicates buffer sizes of all four LCGs. The buffer size is encoded in 6 bits and includes all data in a logical channel buffer excluding the RLC/MAC headers. The structure of BSR MAC control elements are shown in Figure 8.18.

The UE generates three different types of BSRs as follows:

1. *Regular BSRs* are reported when data becomes available for a logical channel with higher priority than the logical channels for which data exists in the transmission buffer and has already been reported, when a serving cell changes, requiring the UE to report its buffer sizes to the new cell so that uplink resources can be scheduled, or when the timer *RETX_BSR_TIMER* expires and the UE has data available for transmission. This timer is restarted whenever there is an uplink grant and is used for retransmission of a BSR, if there is no allocation for the previous BSR sent by the UE. The regular BSRs can be short or long, depending on what data is available for transmission, and result in generation of a scheduling request at the physical layer.

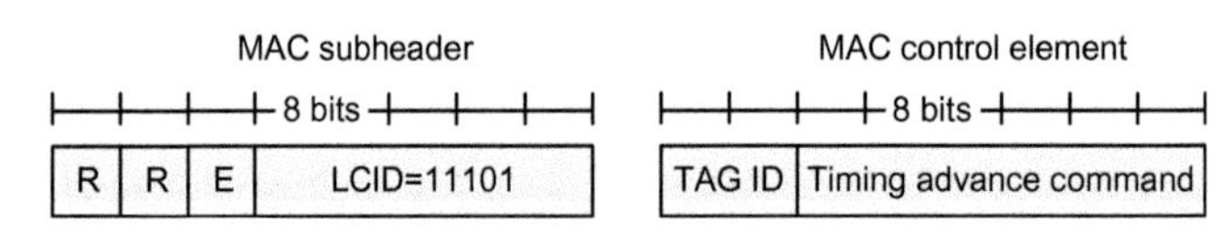

FIGURE 8.17

Uplink timing advance control element [1].

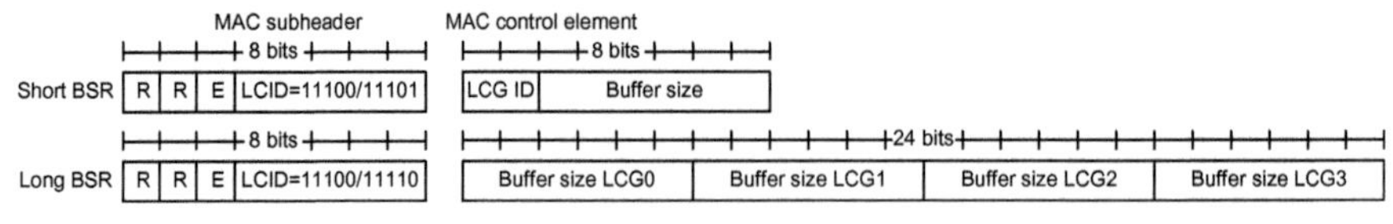

FIGURE 8.18

BSR control elements [1].

2. *Padding BSRs* enable the utilization of bits that would otherwise be used for padding. If the number of padding bits to be sent is equal to or greater than BSR MAC CEs, then BSRs can be sent in lieu of padding. Short, long, or truncated types of BSRs can be sent depending on the size of padding. The truncated BSRs are sent when data is available in more than one LCG, but the size of padding precludes sending long BSRs. In that case, the buffer size of the highest priority LCG is reported.
3. *Periodic BSRs* are sent when the configurable timer *PERIODIC_BSR_TIMER* expires. In this case, short or long BSRs can be sent.

8.5.3 Power headroom control element

The UE power headroom report (PHR) control element is used to report the power headroom available in the UE. The PHR is encoded in 6 bits with a reporting range from −23 dB to +40 dB in steps of 1 dB. Positive values indicate the difference between the maximum UE transmit power and current UE transmit power. Negative values indicate the difference between the maximum UE transmit power and the calculated UE transmit power. The UE transmit power is calculated assuming the UE were to transmit according to the current grant with allocated HARQ and RV configuration. The PHR can be configured either for periodic report or when the downlink path loss changes by a specific amount. For periodic reporting, a report is triggered by the expiration of the *PERIODIC_PHR_TIMER*, which can be configured with values between 10 ms and infinity. For threshold reporting, a PHR is triggered when the path loss changes by *DL_PathlossChange* (1, 3, 6, or infinite dB), provided that the configurable timer *PROHIBIT_PHR_TIMER* has

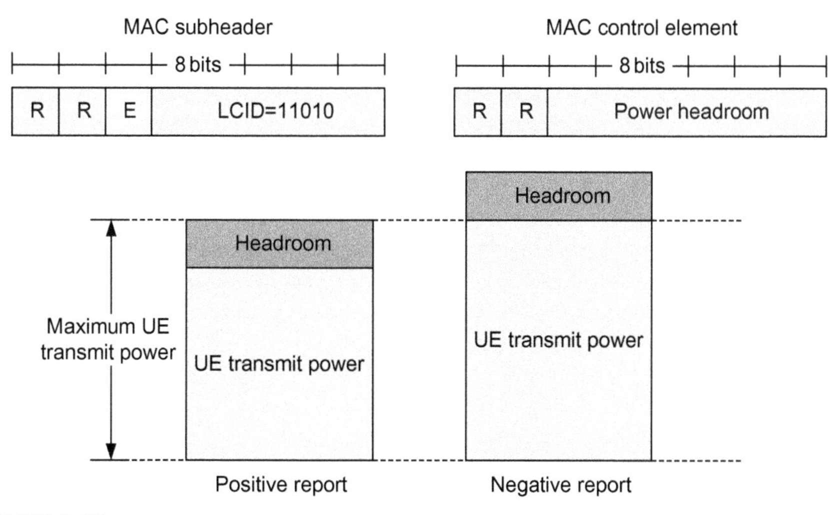

FIGURE 8.19

Power headroom control element [1].

expired. The *PROHIBIT_PHR_TIMER* timer starts when a PHR has been transmitted and can have values between 0 and 1000 ms. The structure of a PHR MAC CE is shown in Figure 8.19, as well as the graphical interpretation of the positive and negative PHRs.

8.5.4 Extended power headroom control element

The extended power headroom MAC control element was specified to support carrier aggregation in LTE Rel-10. In the case of carrier aggregation, the UE reports the power headroom of each component carrier independently. The PHRs provide the serving eNB with information about the difference between the UE maximum power and the estimated transmit power on PUSCH, or the transmit power on PUSCH and PUCCH on the primary carrier. There are two types of PHRs to support simultaneous transmission of PUSCH and PUCCH. Type-1 power headroom is used to report PUSCH transmit power and Type-2 power headroom is used when reporting PUSCH and PUCCH transmit powers. Figure 8.20 shows the order at which the power headroom corresponding to the primary cell and secondary cells are reported in the extended power headroom MAC control element. Note that P_{CMAX_k} in the figure denotes the maximum configured UE transmit power on the kth component carrier. The information fields within the extended power headroom MAC control element are as follows [1]:

- C_i: This bit indicates the presence of a power headroom field for the ith secondary component carrier (secondary cell). The C_i bit is set to one when the power headroom for the ith secondary cell is reported, i.e., the ith active secondary component carrier.
- The reserved bit R is set to zero.
- The information field V indicates whether the reported value of the power headroom is based on a real transmission or a reference. For Type-1 power headroom, V = "0" indicates real transmission on PUSCH and $V = 1$ is indicative of a PUSCH reference format. For Type-2 power headroom, V = "0" indicates real transmission on PUCCH and V = "1" indicates that a PUCCH reference format is used. Furthermore, for both Type-1 and Type-2 power headroom, V = "0" indicates the presence of an octet containing the associated P_{CMAX_k} field, and V = "1" indicates that the octet containing the associated P_{CMAX_k} field is omitted.
- Power Headroom: This 6-bit field indicates the power headroom value.
- P: This field indicates whether the UE applies power backoff due to power management. The UE sets P to one if the corresponding P_{CMAX_k} field might have had a different value if no power backoff due to power management had been applied.
- P_{CMAX_k}: This field indicates the maximum configured UE transmit power on the kth component carrier that is used for calculation of the preceding power headroom value.

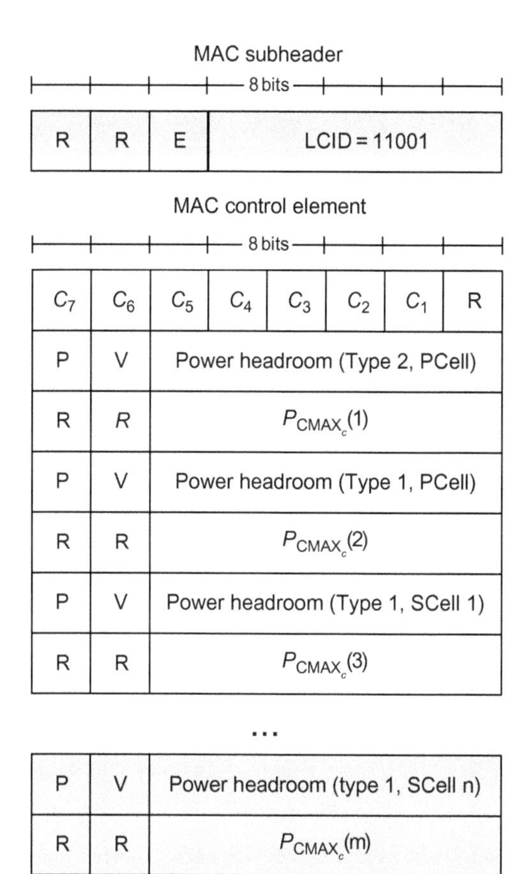

FIGURE 8.20

Extended power headroom control element [1].

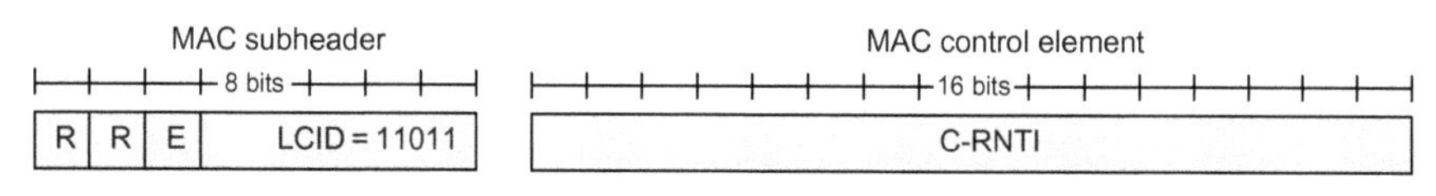

FIGURE 8.21

C-RNTI control element [1].

8.5.5 C-RNTI control element

The C-RNTI MAC control element is used to identify a UE when sending information over CCCH, and for the UE to transmit a dedicated C-RNTI during the random access procedure for the purpose of contention resolution. This CE has a fixed 16-bit size and consists of a single field, as shown in Figure 8.21.

8.5.6 Activation/deactivation control element

LTE Rel-10 specified an activation/deactivation mechanism for secondary cells, i.e., activation/deactivation does not apply to the primary cell, in order to enable reasonable UE battery consumption when carrier aggregation is configured for a UE. When a secondary cell is deactivated, the UE does not need to receive PDCCH or PDSCH on that secondary component carrier and cannot transmit in the corresponding uplink of that cell. However, the UE is required to perform channel quality indicator (CQI) measurements on the deactivated secondary cell. When a secondary cell is activated, the UE is required to monitor PDCCH in order to receive PDSCH and to perform CQI measurements on that cell.

If the UE is configured with one or more secondary cells, the network may activate and deactivate the configured secondary cells. The primary cell is always activated. The network activates and deactivates the secondary cells by sending the activation/deactivation MAC control element. Furthermore, the UE maintains an *sCellDeactivationTimer* per configured secondary cell and deactivates the associated secondary cell upon expiration of its timer. The same initial timer value applies to each instance of the *sCellDeactivationTimer* and it is configured by RRC sublayer. The configured secondary cells are initially deactivated upon addition and after a handover.

The activation/deactivation MAC control element is identified by a MAC PDU subheader with a unique LCID, as shown in Figure 8.22. It has a fixed size and consists of a single octet containing a 7-bit carrier activation/deactivation bitmap and one reserved bit which is set to zero. If the *i*th secondary cell is configured, then the C_i bit in this MAC control element indicates activation of the *i*th secondary cell when set to one.

8.5.7 MCH scheduling information control element

The structure of MCH scheduling information MAC control element is illustrated in Figure 8.23. This CE is transmitted for each MCH to inform the MBMS-capable UEs of scheduled data transmission on MTCH and has a variable size. For each MTCH the information fields include a 5-bit LCID which indicates the LCID of the MTCH and an 11-bit stop MTCH indicating the ordinal number of the subframe within the MCH scheduling period, counting only the subframes allocated to the MCH, where the corresponding MTCH stops, with the value of zero corresponding to the first subframe. The special stop MTCH value 2047 indicates that the corresponding MTCH is not scheduled. The range of values between 2043–2046 is reserved.

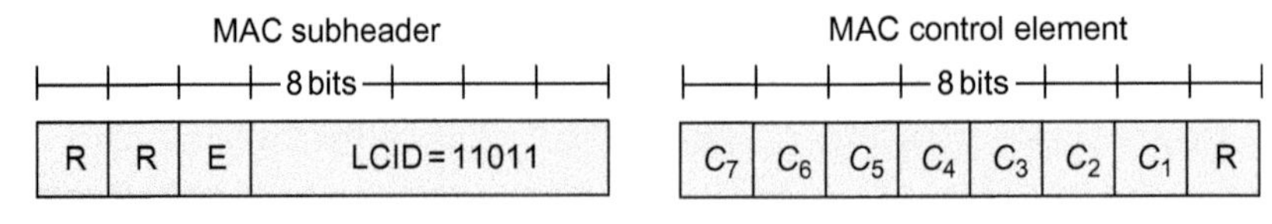

FIGURE 8.22

Activation/deactivation control element [1].

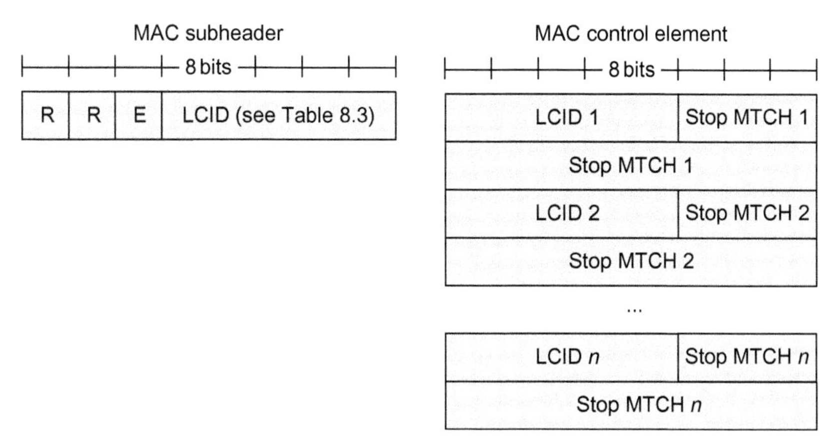

FIGURE 8.23

MCH scheduling information control element [1].

Table 8.3 Values of LCID for MBMS scheduling control element [1]

Index	LCID Values
00000	MCCH (assigned to MTCH, if there is no MCCH on MCH)
00001–11100	MTCH
11101	Reserved
11110	MCH scheduling information
11111	Padding

The MCH may be transmitted in subframes that are configured by an upper layer for MCCH or MTCH transmission. The transmission of an MCH occurs in a set of subframes defined by a *PMCH-Config* parameter. An MCH scheduling information MAC control element is included in the first subframe allocated to the MCH within the MCH scheduling period to indicate the position of each MTCH and unused subframes on the MCH. The UE assumes that the first scheduled MTCH starts immediately after the MCCH or the MCH scheduling information MAC control element, if the MCCH is not present, and the other scheduled MTCH(s) start immediately after the previous MTCH, at the earliest in the subframe where the previous MTCH stops [1] (Table 8.3).

8.5.8 UE contention resolution identity control element

The UE contention resolution identity MAC control element is used during the RACH procedure to resolve possible contention among multiple UEs that are trying to simultaneously access the network. This control element has a fixed 48-bit size and consists of a single field, as shown in Figure 8.24, containing the uplink CCCH SDU.

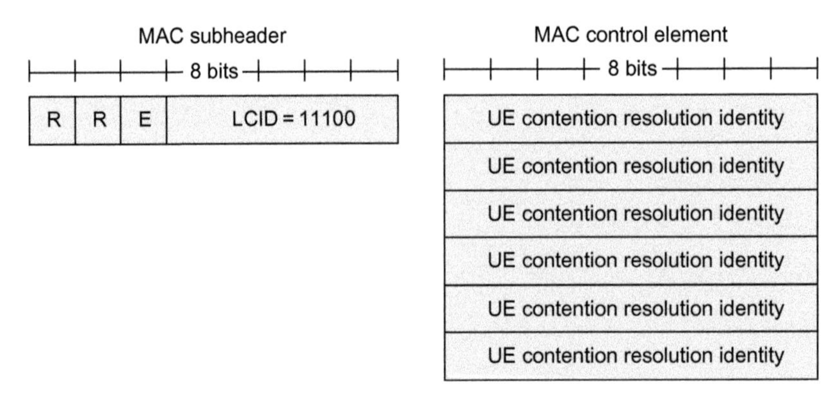

FIGURE 8.24

UE contention resolution identity control element [1].

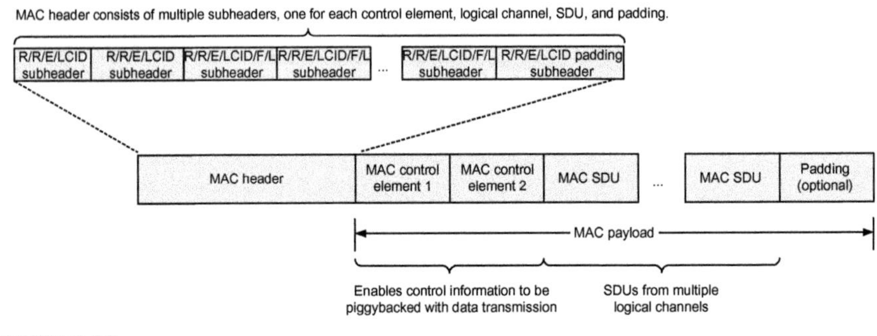

FIGURE 8.25

MAC PDU formation [1].

8.6 MAC PDU packet formats

A MAC PDU consists of a MAC header, zero or more MAC SDUs, zero or more MAC control elements, and optional padding. The MAC header and MAC SDUs may have variable sizes. As shown in Figure 8.25, a MAC PDU header consists of one or more MAC PDU subheaders where each subheader corresponds to a MAC SDU, a MAC control element, or padding. A MAC PDU subheader consists of the six header fields R/R/E/LCID/F/Length except for the last subheader in the MAC PDU and for fixed-size MAC control elements. The last subheader in the MAC PDU and subheaders for fixed-size MAC control elements as well as the subheader for padding consist of the four header fields R/R/E/LCID. Note that a MAC PDU is a bit string that is byte-aligned.

The structure of the MAC PDU used on UL-SCH and DL-SCH is shown in Figure 8.25. Each MAC PDU corresponds to one transport block and consists of a header, control elements, logical channel SDUs (corresponding to RLC PDUs), and

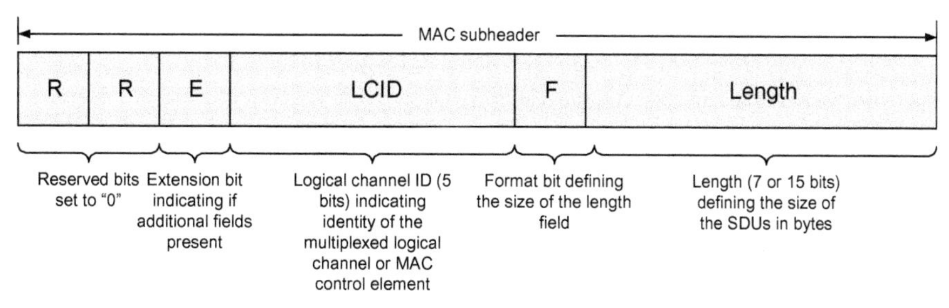

FIGURE 8.26

Structure of a MAC header [1].

optional padding. MAC control elements are used to piggyback control information, such as buffer status or power headroom reports. The MAC header consists of several subheaders. There is one subheader for each MAC control element and each MAC SDU in the PDU. In addition, there can be a subheader to specify padding. For the control element, the subheader identifies the specific control element being sent. For the MAC SDUs, the logical channel number and the size of the SDU are specified. All subheaders are grouped at the start of the MAC PDU in the MAC header. These are followed by the control elements, which are then followed by the MAC SDUs. Any padding occurs at the end of the PDU unless only one or two bytes of padding are required. Then one or two bytes of fixed-size padding subheaders can be included before the subheader corresponding to the first MAC SDU.

The subheader formats for MAC PDUs are shown in Figure 8.26. There are two formats: [R/R/E/LCID] which is used for fixed-length MAC SDUs and MAC control elements; and [R/R/E/LCID/F/Length] that is used for variable length MAC SDUs. The size of the first format is one byte. The size of the second format is either two or three bytes. In each case, the two R bits are reserved bits and are set to "0." The E bit is used to indicate if there are more subheader fields in the MAC header. A value of "1" denotes that another subheader would follow. A value of "0" indicates that a control element or MAC SDU starts at the next byte. The LCID field indicates either the logical channel number associated with the MAC SDU or, in the case of a control element, the specific type of MAC control element. The length field can be either 7 or 15 bits long and specifies the length of the MAC SDU in bytes. The size of the Length field is determined by the F field, which is "0" for 7 bits (128 bytes or less), or "1" for 15 bits.

8.7 MAC scheduling services

The MAC scheduler is responsible for scheduling the transfer of user data and control signaling via transport blocks in downlink and uplink subframes over the LTE air-interface. The MAC scheduler assigns the downlink and uplink physical layer resources used by the eNB and UEs within the cell. The scheduling decisions are based on information from various sources within the eNB radio protocol stack and the UE. In the downlink, the MAC scheduler allocates resources which are used by the eNB MAC sublayer to send transport blocks to specific UEs via DL-SCH. These downlink transport blocks contain:

- Upper-layer data: All upper-layer data, whether this is user data or signaling messages, are fed into the MAC sublayer through logical channels. Whenever one of these logical channels has any data to send over the LTE air-interface, a buffer status indication is provided to the MAC scheduler.
- MAC sublayer data: This can either take the form of a HARQ retransmission or a MAC control element.

In the above cases, the MAC sublayer will provide an indication to the scheduler that it has data ready to send over the LTE air-interface. In the uplink, the MAC scheduler allocates resources to specific UEs which enable them to construct and send transport blocks to the eNB via UL-SCH. Each UE must request the uplink resources (except when semi-persistent scheduling is used) needed to send a transport block. This can be done in a number of ways. When a UE does not have an active connection to a cell, it performs the random access procedure to attach (or reattach) to the cell. When the UE has uplink control resources on the PUCCH physical channel, it sends a scheduling request indication to request uplink resources. When the UE already has resources to send data in the uplink on the UL-SCH, it may indicate to the scheduler that it has more data to send, and therefore requires more uplink resources. The UE does this by sending a BSR MAC control element. Based on HARQ feedback at the eNB, the scheduler might need to allocate uplink resources for HARQ retransmissions by the UE [12].

The output of the scheduling algorithm is a scheduling assignment per uplink and downlink subframe, which is signaled to UEs on the PDCCH. The downlink assignment information will be used by the eNB MAC sublayer to generate the transport blocks which are passed to the eNB physical layer and subsequently sent over the air. The uplink assignment information indicates which UEs have been allocated uplink resources within the subframe. It is also used by the eNB MAC sublayer to identify uplink transmissions in the corresponding uplink subframe. Before any user data can be transferred over the air, the scheduler needs to be configured with a cell-specific configuration, a UE-specific configuration, and finally a bearer-specific configuration for each active UE bearer. Reconfiguration of each UE and/or bearer is also possible. In order to make full use of the information available within the MAC scheduler, certain event-triggered or periodic feedback information is provided to higher layers. Based on this information, the higher layers are able to reconfigure the operation of the cell to provide a more efficient use of the air-interface. For monitoring purposes, the MAC scheduler may also output detailed information about subframe utilization. This may include but is not limited to the following: subframe map showing utilized physical resource blocks (PRBs); a percentage of utilized versus unutilized PRBs; and the number of logical channels/MAC control elements which have not been scheduled.

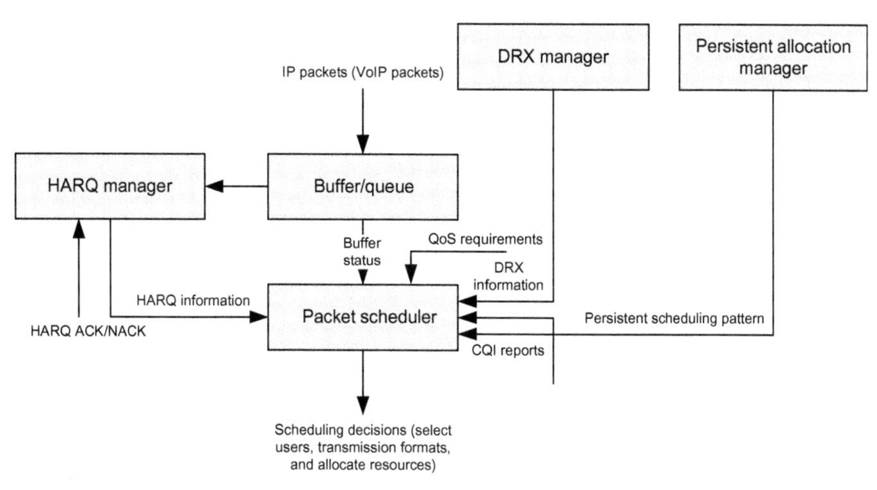

FIGURE 8.27

Main entities involved in packet scheduling in the eNB [15–17].

The scheduler in the eNB distributes the available radio resources in one cell among the UEs, and among the radio bearers of each UE. The details of the scheduling algorithm are left to the eNB implementation, but the signaling to support the scheduling is standardized. In principle, the eNB allocates downlink or uplink radio resources to each UE based on the downlink data buffered in the eNB and the BSRs received from the UE. In this process, the eNB considers the QoS requirements of each configured radio bearer and selects the size of the MAC PDU. The typical mode of scheduling is dynamic scheduling by means of downlink assignment messages for the allocation of downlink transmission resources and uplink grant messages for the allocation of uplink transmission resources; these are valid for specific subframes [12].

Figure 8.27 illustrates the interaction between the main entities involved in eNB scheduling services. These entities include packet scheduler, persistent allocation manager, HARQ manager, and DRX manager. Each newly arrived VoIP packet in the eNB buffer is associated with a discard timer. The discard timer specifies the maximum time a packet is allowed to be buffered in the eNB before it should be dropped. The HARQ manager deals with the new transmissions and pending retransmissions. The DRX manager is responsible for informing the packet scheduler when a specific UE is awake and can receive the L1/L2 control signaling, and thus can be scheduled dynamically. Besides informing each UE of the persistent resource allocation pattern by RRC signaling, the persistent allocation manager also forwards the persistent allocation patterns to the packet scheduler. The packet scheduler is the central controlling entity and makes scheduling decisions based on the buffer status, HARQ information, DRX information, QoS requirements, and persistent resource allocation patterns [13].

These messages also indicate whether the scheduled data is a fresh transmission of a new transport block or a retransmission, using a single-bit NDI. If the value of the NDI is changed relative to its previous value for the same HARQ process, the transmission is the start of a new transport block. These messages are transmitted on PDCCH using a C-RNTI to identify the addressee (UE). This type of scheduling is efficient for service types, such as transport control protocol (TCP) or the signaling radio bearers (SRBs), in which the traffic is bursty and dynamic. The dynamic uplink transmission resource grants are valid for specific single subframes for initial transmissions, although they may also imply a resource allocation in later subframes for HARQ retransmissions [12].

8.7.1 Dynamic and semi-persistent scheduling

The UE may be dynamically scheduled on a per subframe basis, with the control information signaled on the PDCCH. In this case, the UE will be addressed using the C-RNTI. However, the UE may also receive a semi-persistent grant or allocation where the UE is addressed using the SPS-C-RNTI. In this case, the scheduling control information is signaled once via the PDCCH. The UE can then transmit/receive periodically following a pattern defined by the downlink semi-persistent scheduling interval parameter using the same configuration until modified or released. When a dynamic allocation or grant is received that corresponds to a subframe where an semi-persistent scheduling scheduled transmission would occur, the dynamic grant takes precedence. Additionally, any retransmissions would be scheduled dynamically via the PDCCH.

VoIP is characterized by small packet size and very stringent delay and jitter requirements, and a large number of simultaneous users. In LTE, the scheduling of the data traffic is based on the shared-channel transmission and the channel-dependent resource allocation. The eNB allocates radio resources in units of resource blocks to UEs, and sends the scheduling information to each UE on PDCCH. Such dynamic channel-dependent scheduling can exploit the selectivity in both the time and frequency domain, and significantly improve the system throughput. However, applying dynamic scheduling directly for VoIP packets is not desirable. This is mainly due to the small packet size and the constant inter-arrival time of VoIP packets, which results in large control overhead as control information should be transmitted per TTI. Many UEs may be scheduled simultaneously, so the limited resources available on the PDCCH for control signaling become a bottleneck for the VoIP capacity.

The semi-persistent scheduling is a feature that significantly reduces control channel overhead for applications that require persistent radio resource allocations, such as VoIP. In LTE, both downlink and uplink radio resources are fully scheduled since downlink and uplink traffic channels are shared channels. This means that the PDCCH must provide access grant information to indicate which users should decode PDSCH in each subframe, and which users are allowed to transmit on PUSCH in each subframe. Without semi-persistent scheduling, every downlink or uplink resource block allocation must be granted via an access grant message on the PDCCH. This is sufficient for most bursty best effort types of applications, which generally have large packet sizes and thus typically only a few users must be scheduled in each subframe. However, for applications that require persistent allocations of small packets, such as VoIP, the access grant control channel overhead can be greatly reduced with semi-persistent scheduling.

Semi-persistent scheduling introduces a persistent resource allocation that a user should expect on the downlink or can transmit on the uplink. There are many different ways in which semi-persistent scheduling can set up persistent allocations. Note that speech codecs typically generate a speech packet every 20 ms. In LTE, HARQ RTT is 8 ms, which means retransmissions of resource blocks that have failed to be decoded can occur every 8 ms. Once speech packets stop arriving (i.e., silence period), these radio resources can be reassigned to other users. When the user begins talking again, a new semi-persistent scheduling set of radio resources would be assigned for the duration of the new talk spurt. Note that dynamic scheduling of best effort data can occur on top of semi-persistent scheduling, but the semi-persistent scheduling allocations would take precedence over any scheduling conflicts.

In wireless networks, persistent scheduling takes advantage of the traffic characteristics of the application (e.g., VoIP) to increase the number of users running that application in the system. The periodic nature of the packet arrivals from a VoIP source allows the needed resource to be allocated persistently for the period of an active talk spurt. Consequently, base stations can send the allocation information once at the beginning of a talk spurt and avoid sending different allocation information for each subsequent packet. Significant amounts of radio resources that would otherwise be occupied by unnecessary overhead can be used to accommodate VoIP packets from other users. Because VoIP packets are usually small in size, the savings from reduced overhead can significantly increase the overall system capacity. The capacity gain from persistent scheduling, however, can be compromised significantly due to dynamic link adaptation, which is a very common technique in wireless systems to adapt to dynamic variation in wireless channel condition. The underlying concept of persistent and dynamic scheduling is illustrated in Figure 8.28.

When semi-persistent scheduling is enabled by the RRC sublayer, the following information is provided: semi-persistent scheduling C-RNTI; uplink semi-persistent

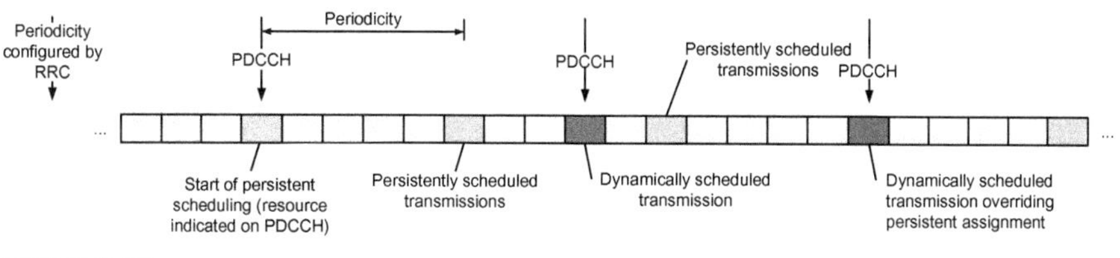

FIGURE 8.28

Illustration of an example of semi-persistent and dynamic scheduling [10].

scheduling interval; the number of empty transmissions before implicit release; downlink semi-persistent scheduling interval; and the number of configured HARQ processes for semi-persistent scheduling. When semi-persistent scheduling for uplink or downlink is disabled by the RRC sublayer, the corresponding configured grant or configured assignment is discarded. Semi-persistent scheduling is supported on the PCell only, and is not supported for relay node communication with the E-UTRAN in combination with a relay node subframe configuration.

After a semi-persistent downlink assignment is configured for a UE, the UE's downlink assignments occur every *N*th subframe in a subframe whose number n_{subframe} satisfies the following equation $(10\text{SFN} + n_{\text{subframe}}) = [(10\text{SFN}_{\text{start-time}} + n_{\text{subframe}_{\text{start-time}}}) + NT_{\text{semiPersistSchedIntervalDL}}]\text{mod}10240$, where $\text{SFN}_{\text{start-time}}$ and $n_{\text{subframe}_{\text{start-time}}}$ are the SFN and subframe numbers at which the configured downlink assignment was initiated/reinitiated. Following configuration of a semi-persistent scheduled uplink grant for the UE, the UE will sequentially have uplink grants every *N*th subframe for which $(10\text{SFN} + n_{\text{subframe}}) = [(10\text{SFN}_{\text{start-time}} + n_{\text{subframe}_{\text{start-time}}}) + NT_{\text{semiPersistSchedIntervalUL}} + Subframe - Offset(N\ \text{mod}2)]\text{mod}10240$, where $\text{SFN}_{\text{start-time}}$ and $n_{\text{subframe}_{\text{start-time}}}$ are the SFN and subframe number at which the configured uplink grants were initialized/reinitialized. The UE is required to release the configured uplink grant immediately after a certain number of consecutive new MAC PDUs, each containing zero MAC SDUs, have been provided by the multiplexing and assembly entity on the semi-persistent scheduled resource. The retransmissions for semi-persistent scheduling can continue after releasing the configured uplink grant [1].

8.7.2 Proportional fair scheduling algorithm

Proportional fair is a commonly-used scheduling algorithm, which is based on maintaining a balance between two competing interests: maximizing the overall system throughput while at the same time providing all users with a minimum level of service. This is done by assigning each data flow a data rate or a scheduling priority (depending on the implementation) that is inversely proportional to its anticipated resource consumption. The proportional fair (PF) scheduling algorithm calculates a metric for all active users in a given scheduling interval. The user with the highest metric is allocated the resource available in the given interval, while the metrics for all users are updated before the next scheduling interval, and the process repeats. To adapt this simple algorithm to orthogonal frequency division multiple access systems, the definition of scheduling interval and scheduling resource are extended to encompass a two-dimensional OFDMA frame structure. For OFDMA systems, the scheduling interval is typically a radio frame and multiple users may be allocated in the same radio frame. Therefore, in the simplest extension to OFDMA systems, two modifications are made to the PF scheduling algorithm: (1) frames are equally partitioned into regular, fixed scheduling resources that are scheduled sequentially until all available resources are assigned; (2) the metric is updated after scheduling each partition. Note that the number of resources eventually allocated to a user depends on the metric update process and does not preclude a single user from obtaining multiple or all the resources in a frame. For system simulations with the assumption of a fixed overhead allowing up to $N_{\text{Partitions}}$ resource partitions, each partition assignment should be considered as a separate packet transmission. The number of partitions $N_{\text{Partitions}}$, the time constant of the filter used in the metric computation, and the number of active users are simulation parameters. At any scheduling instant t, the scheduling metric $M_i(t)$ for subscriber i used by the PF scheduler is given by the following expression:

$$M_i(t) = \frac{R_{\text{instantaneous}_i}(t)}{[R_{\text{average}_i}(t)]^{\alpha}} \tag{8.3}$$

where $R_{\text{instantaneous}_i}(t)$ is the data rate that can be supported at scheduling instant t for the ith user. $R_{\text{instantaneous}_i}(t)$ is a function of the CQI feedback, and consequently a function of the MCS that can meet the packet error rate requirement. Parameter $R_{\text{average}_i}(t)$ denotes the throughput function smoothed by a low-pass filter at the scheduling instant t for user i. The parameter α is a fairness exponent factor with a default value of one. For a scheduled subscriber $R_{\text{average}_i}(t)$ is computed as follows:

$$R_{\text{average}_i}(t) = \frac{1}{N_{\text{PF}}} R_{\text{instantaneous}_i}(t) + \left(1 - \frac{1}{N_{\text{PF}}}\right) R_{\text{average}_i}(t-1) \tag{8.4}$$

For an unscheduled subscriber, the throughput function is given by:

$$R_{\text{average}_i}(t) = \left(1 - \frac{1}{N_{\text{PF}}}\right) R_{\text{average}_i}(t-1) \tag{8.5}$$

The latency scale of the PF scheduler N_{PF} is given by:

$$N_{\text{PF}} = \frac{T_{\text{PF}} N_{\text{Partitions}}}{T_{\text{Frame}}} \tag{8.6}$$

where T_{PF} represents the latency time scale in units of seconds and T_{Frame} is the frame duration of the system. In some implementations, the scheduler may give priority to HARQ retransmissions.

8.8 RNTI mapping

The users in LTE radio access networks are assigned temporary identifiers for identification while protecting their privacy and confidentiality. Depending on the state of the UE, different types of temporary identifiers are used. Table 8.4 summarizes the RNTIs and their usage in a LTE radio access network. Each RNTI is a 16-bit string that is assigned to a UE depending on the state of the UE in the LTE radio access network. For example, the RA-RNTI is used on PDCCH when RAR messages are transmitted. It unambiguously identifies which time-frequency resource was utilized by the UE to transmit the random access preamble. The DCI formats 1A and 1C with CRCs scrambled with RA-RNTI allocate DL-SCH resources to RAR messages. There is a one-to-one mapping between the RA-RNTI and the time-frequency resource used by the UE when transmitting the random access preamble. The one-to-one association allows the eNB to address RAR messages to a specific UE. The Msg3 acronym in Table 8.4 denotes the *RRC Connection Request* message transmitted on UL-SCH containing a C-RNTI MAC control element or CCCH SDU, submitted from the upper layer and associated with the UE contention resolution identity, as part of a random access procedure [10–12].

The P-RNTI is used during the paging procedure, where the UE searches for DCI formats 1A and 1C whose CRC bits are scrambled with P-RNTI. These DCI

Table 8.4 Various RNTIs and Their Usage in LTE [1]

RNTI	Usage	Transport Channel	Logical Channel
Paging RNTI (P-RNTI)	Paging and system information change notification	PCH	PCCH
SI-RNTI	Broadcast of SI	DL-SCH	BCCH
MBMS RNTI (M-RNTI)	MCCH information change notification	N/A	N/A
RA-RNTI	Random Access Response	DL-SCH	N/A
Temporary C-RNTI	Contention resolution (when no valid C-RNTI is available)	DL-SCH	CCCH
Temporary C-RNTI	Msg3 transmission (RRC connection request)	UL-SCH	CCCH, DCCH, DTCH
Cell RNTI (C-RNTI)	Dynamically scheduled unicast transmission	DL-SCH, UL-SCH	DCCH, DTCH
C-RNTI	Triggering of PDCCH ordered random access	N/A	N/A
SPS C-RNTI	Semi-persistently scheduled unicast transmission (activation, reactivation, and retransmission)	DL-SCH, UL-SCH	DCCH, DTCH
SPS C-RNTI	Semi-persistently scheduled unicast transmission (deactivation)	N/A	N/A
TPC-PUCCH-RNTI	Physical layer uplink power control	N/A	N/A
TPC-PUSCH-RNTI	Physical layer uplink power control	N/A	N/A

formats allocate DL-SCH resources to the PCCH and PCH, i.e., the UE proceeds to decode a paging message following detection of a DCI format whose CRC bits are scrambled with P-RNTI. The paging message may include paging information as well as system information change notification, earthquake and tsunami warning system notification, and commercial mobile alert system.

The SI-RNTI is used during the transmission of SIBs on DL-SCH. The UE searches for DCI formats 1A and 1C whose CRC bits are scrambled with SI-RNTI. DCI formats 1A and 1C allocate DL-SCH resources to BCCH. Upon detection of these DCI formats, the UE proceeds to decode the system information message.

The M-RNTI is used during transmission of an MCCH change notification. The UE searches for a DCI format 1C whose CRC is scrambled with M-RNTI. This DCI format includes a bit string of 8 bits, where each bit corresponds to one of the MBSFN areas listed in SIB13. A change in MCCH is signaled by setting the corresponding bit to one.

The temporary C-RNTI is allocated to the UE within the RAR message. The temporary C-RNTI is used for contention resolution when a C-RNTI has not yet been assigned and specifically in the transmission of the RRC connection request on CCCH during initial access. After transmitting Msg3 in the uplink, the UE searches for DCI formats 1A or 1C whose CRC bits are scrambled with the UE's temporary C-RNTI. The successful detection of the temporary C-RNTI is considered success of the contention resolution and assignment of a C-RNTI to the UE. The temporary C-RNTI is also used for scrambling of PUSCH during initial Msg3 transmission on CCCH.

The C-RNTI is assigned to the UE following successful completion of the random access procedure. The C-RNTI is reassigned during the handover procedure by specifying a new value within the *RRC Connection Reconfiguration* message. The C-RNTI is used to identify the UE in all downlink and uplink control and data transmissions when the UE is in the RRC_CONNECTED state. The UE uses the same C-RNTI in all serving cells when supporting carrier aggregation.

The SPS-C-RNTI is allocated to a UE using an *RRC Connection Setup* or an *RRC Connection Reconfiguration* message, when the resources are going to be allocated for more than a single subframe. The transmit power control (TPC) PUSCH/PUCCH RNTI is used for power control of PUSCH/PUCCH. The CRC bits of DCI formats 3 and 3A are scrambled with TPC-PUSCH-RNTI/TPC-PUCCH-RNTI to convey the power control commands to the UE. TPC-PUSCH-RNTI/TPC-PUCCH-RNTI can be allocated using either the *RRC Connection Setup* or *RRC Connection Reconfiguration* messages [10]. The mapping of different types of RNTI to logical channels is depicted in Figure 8.29.

8.9 Random access procedure

The UE must be synchronized in the uplink before it can send signaling or data. For the uplink, the UEs within a cell's coverage area are physically distributed at different distances from the eNB location. As a result, each UE will experience a

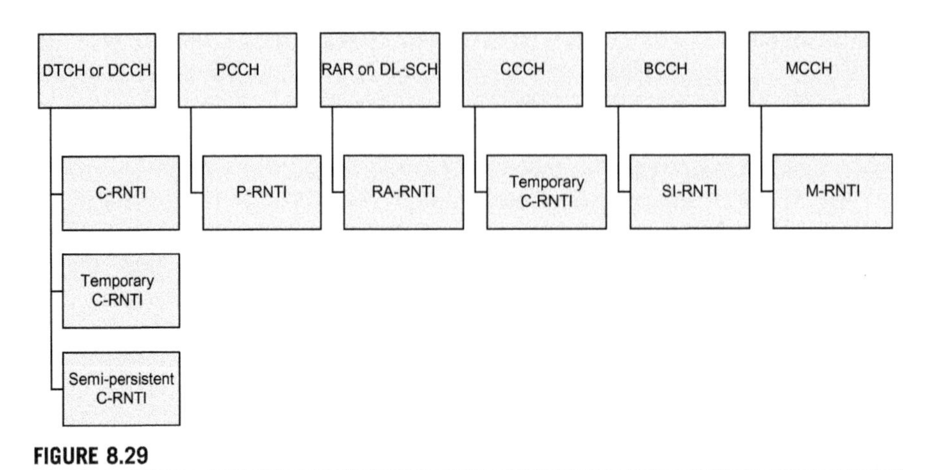

FIGURE 8.29

RNTI mapping to control or traffic logical channels.

different propagation delay when accessing the network. Without synchronization, subframes transmitted from different UEs would likely overlap and interfere with each other. To overcome this problem, the UE performs uplink synchronization procedures on RACH during initial access to the network. In LTE, random access is used for several purposes including initial access when establishing a radio link, i.e., transitioning from RRC_IDLE to RRC_CONNECTED state, reestablishing the radio link after a radio link failure, performing handover when uplink synchronization, needs to be established with the target cell, establishing uplink synchronization if uplink or downlink data arrives when the terminal is in the RRC_CONNECTED state and the uplink is not synchronized, positioning during RRC_CONNECTED state which requires a random access procedure, e.g., when timing advance is needed for UE positioning, and scheduling request if no dedicated scheduling-request resources have been configured on PUCCH.

The process is initiated by sending RACH preambles using physical random access channel (PRACH). The procedure may be contention-based or contention-free. In the contention-based procedure, the UE selects a preamble from a set containing two groups of preamble sequences. If two UEs select the same preamble in the same RACH resource, a collision can result that must be resolved before either UE can access the system. The contention-based RACH is generally used during initial access or if the UE has lost uplink synchronization. In contention-free synchronization, the UE uses a dedicated preamble and it is likely to be already connected to the network and must resynchronize following a handover to a new cell.

Figure 8.30 shows the RACH process at a high level. During the first step the UE transmits a preamble randomly selected from 64 possible preamble sequences divided into two groups. The group from which the UE selects the preamble depends on the size of the message that it would like to transmit and on the

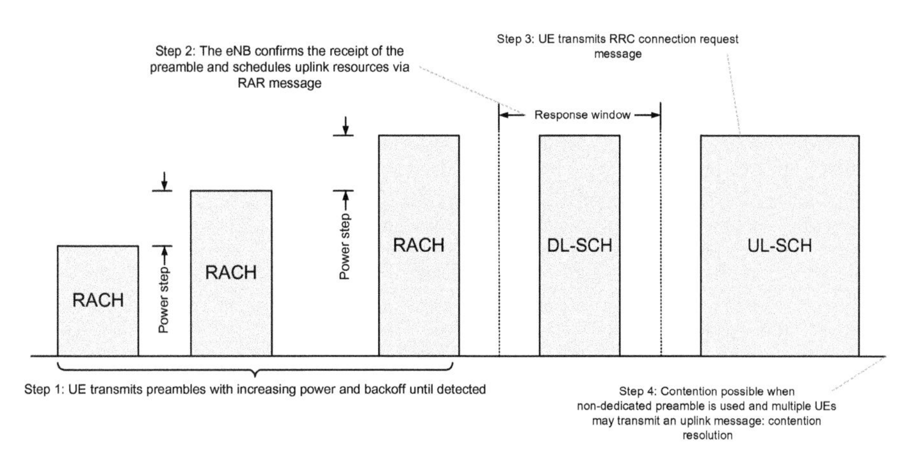
Step 2: The eNB confirms the receipt of the preamble and schedules uplink resources via RAR message
Step 3: UE transmits RRC connection request message
Response window
Power step
Power step
RACH
RACH
RACH
DL-SCH
UL-SCH
Step 1: UE transmits preambles with increasing power and backoff until detected
Step 4: Contention possible when non-dedicated preamble is used and multiple UEs may transmit an uplink message: contention resolution

FIGURE 8.30

Illustration of a contention-based random access procedure [21].

calculated path loss to the eNB. Once the preamble is transmitted, the UE monitors the downlink beginning three subframes after the end of the preamble for a duration controlled by a configurable window. If no downlink response is detected, then the preamble is retransmitted with a power that is increased by a configurable parameter. This process continues until the defined maximum number of preamble transmissions is reached. If the preamble is successfully detected by the eNB, then the downlink RAR message is sent by the eNB allocating resources for the UE to transmit its message or enforcing a backoff. In the final step, contention resolution must be performed for a contention-based access attempt. Therefore, as soon as the random access preamble is transmitted and regardless of the possible occurrence of a measurement gap, the UE is required to monitor PDCCH on the primary cell for RAR(s) identified by RA-RNTI during the RAR window (in time) which starts at the subframe that contains the end of the preamble transmission plus three subframes and has the length *ra-ResponseWindowSize* subframes (Figure 8.30) [1].

The UE sets its initial power level to the following value *PREAMBLE_INITIAL_RECEIVED_TARGET_POWER* + *DELTA_PREAMBLE* where *DELTA_PREAMBLE* depends on the type of preamble format. The UE selects the first available RACH resource for the transmission. If there is no response within three subframes + *ra_Response_Window* (defined in number of subframes), the UE assumes the previous transmission was unsuccessful. It selects another RACH preamble, performs the backoff procedure, and transmits again. The power level is increased by *POWER_RAMP_STEP*. This process can be repeated until the *PREAMBLE_TRANSMISSION_COUNTER* limit is reached. In that case, an access failure is declared to the upper layers.

In LTE Rel-10 and when carrier aggregation is supported, the random access procedure is initiated by a PDCCH order or by the MAC sublayer. The random access procedure on a secondary cell is only initiated by a PDCCH order (i.e., non-contention-based random access procedure). If the UE receives a PDCCH order masked with the UE's C-RNTI, and for a specific serving cell, the UE initiates a random access procedure on that serving cell. For random access on the primary cell, a PDCCH order or RRC message optionally indicates the *ra-PreambleIndex* and the *ra-PRACH-MaskIndex* parameters; whereas for random access on a secondary cell, a PDCCH order indicates the *ra-PreambleIndex* (with a value different from 000000) and the *ra-PRACH-MaskIndex* parameters. For the primary TAG, preamble transmission on PRACH and reception of a PDCCH order are only supported for the primary cell [1]. In LTE Rel-11, the random access procedure is also performed on a secondary cell to establish timing alignment for the corresponding secondary TAG. Furthermore, a non-contention-based random access procedure can be performed for obtaining timing advance alignment for a secondary TAG [8].

8.9.1 Contention-based RACH procedure

The UE selects a preamble from the set of preambles. Two groups of sequences are available for a contention-based RACH procedure defined as Group A and Group B. The size of the Group A preamble set is defined by the *sizeofRA-PreamblesGroupA* parameter. The remaining preambles, up to *numberofRA-Preambles*-1, belong to Group B. Group B preambles are utilized if the required message size exceeds *MESSAGE_SIZE_GROUP_A* and the path loss is less than $P_{\max} - PREAMBLE\_INITIAL\_RECEIVED\_TARGET\_POWER - DELTA\_PREAMBLE\_MSG3 - messagePowerOffsetGroupB$ where $P_{\max}$ is the maximum UE transmit power. This enables the eNB to estimate the size of the message that the UE wants to transmit. If the eNB detects a preamble, it sends a RAR message in the DL-SCH. The MAC PDU containing the RAR message is addressed using the RA-RNTI which is calculated as $RA-RNTI = 1 + t\_id + 10f\_id$ where $0 \leq t\_id < 10$ is the index of the first subframe of RACH resource within a frame, and $0 \leq f\_id < 6$ is the specified RACH resource within that subframe. This forms a one-to-one mapping between the selected RACH resource and the RA-RNTI. The RAR message contains an identifier to enable identification of the preamble sent by the UE, the absolute timing advance, a temporary C-RNTI, and an uplink grant. If the preamble identifier matches the preamble sent by the UE, the UE applies the timing advance and starts the contention resolution procedure (see Figure 8.31).

During contention resolution, the UE transmits an uplink message on CCCH that includes a terminal identifier. If this matches the identifier used in the response from the eNB, then the UE has successfully completed the contention resolution; otherwise, the UE must restart the random access procedure. Upon successful contention resolution, the temporary C-RNTI assigned as part of the RAR message will become the C-RNTI.

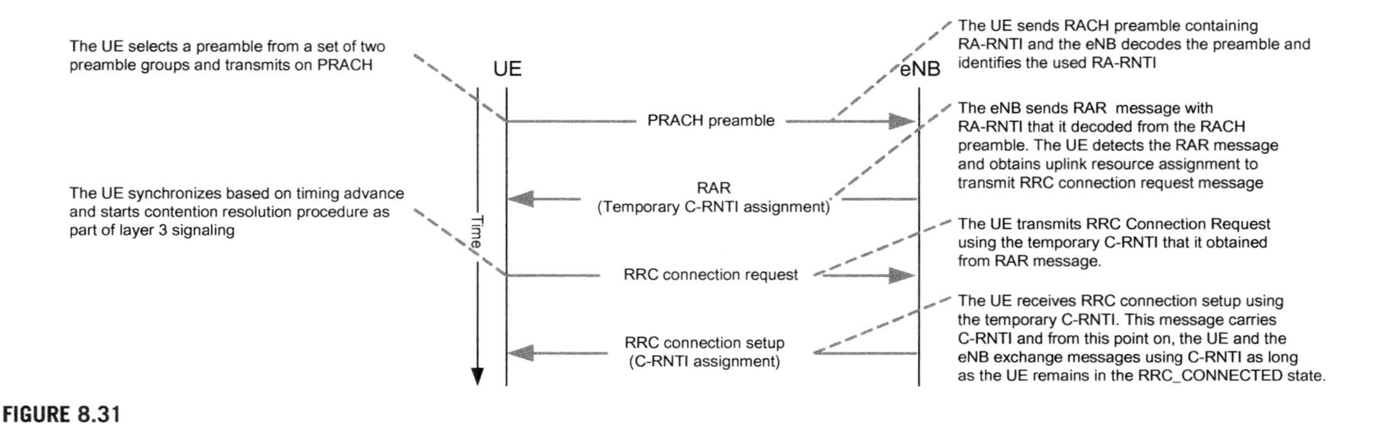

FIGURE 8.31

Breakdown of the contention-based random access procedure [21].

8.9.2 Contention-free RACH procedure

During handovers, the UE must synchronize to the target cell using RACH procedures. However, contention-based random access may result in an unacceptable delay. Therefore, a contention-free procedure is defined. In contention-free RACH, the RRC sublayer defines which RACH preamble sequence and which PRACH resources to utilize. The preamble is dedicated to the UE and is different from those used for Groups A and B in the contention-based process, thus avoiding any possible collision. The PRACH resource to be used for the dedicated transmission is indicated as PRACH Mask, which specifies the subframe number in which the UE can transmit the preamble.

Once the UE transmits the dedicated preamble in the first available RACH resource after receiving the command, it waits for a response in the form of an RAR message from the eNB. The same timing and window specification is enforced in this procedure as those associated with the contention-based procedure. Once the RAR message is received on PDCCH, the UE checks whether the random access preamble identifier (RAPID) contained in the message matches the preamble that was sent and, if so, it applies the timing advance and synchronizes with the eNB.

As mentioned earlier, the MAC PDU that carries the RAR message on PDCCH contains a number of RARs, since multiple UEs may transmit on the uplink RACH resource at the same time. As shown in Figure 8.32, the MAC PDU starts with the MAC header, which contains a number of subheaders. For each RAR message, there is a subheader that identifies the RACH Preamble Identifier that was used for the transmission in uplink. The corresponding RAR payload contains the detailed information on the timing advance, uplink grant, and temporary C-RNTI. The MAC header may also include a backoff indicator (BI) subheader which indexes a backoff ranging from 0 to 960 ms. It is applicable to all the UEs that use a specific RACH resource.

The MAC subheader for the RAR message is 8 bits and can have one of the two formats shown in Figure 8.33. Both RAR formats begin with an E field, which is a single bit that indicates if another subheader is present after the current subheader, and a T field, which is a single bit that defines if this is a subheader that assigns uplink resources or informs all UEs that utilize the RACH resource to

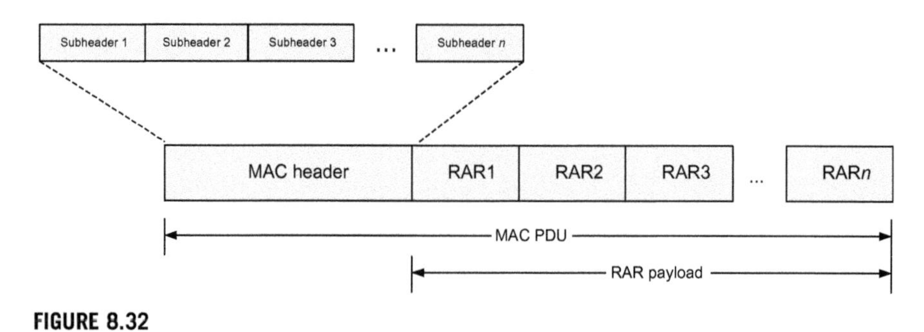

FIGURE 8.32

RAR message structure [1].

backoff. If resources are assigned in a RAR payload, then T is set to "1" and the following field is the 5-bit RAPID value. In this case, there will be an RAR payload to specify the specific grant. If the eNB is enforcing a backoff for all UEs utilizing the RACH resource, then T is set to "0" and the following field consists of two single-bit R (reserved) fields followed by a 4-bit BI that indexes a table of 13 possible backoff periods ranging from 0 to 960 ms.

As shown in Figure 8.34, the RAR payload consists of 48 bits described as follows:

- R: A single reserved bit set to "0."
- TA: The 11-bit timing advance specifies the absolute timing advance that should be applied to the uplink transmissions in units of $16T_s$ (sampling time).
- Uplink grant: 20 bits that specify attributes of the uplink grant. This includes hopping flag (1 bit), defines if frequency hopping is used in the uplink, Fixed-size resource block assignment (10 bits), defines the uplink resource blocks, Truncated MCS (4 bits), TPC command for scheduled PUSCH (3 bits), uplink delay (1 bit), controls which uplink subframe is used for the initial transmission, and CQI request (1 bit), specifies if a CQI report is sent in PUSCH transmission for non-contention-based RACH.
- Temporary C-RNTI: Temporary identifier utilized by the UE until contention resolution is complete.

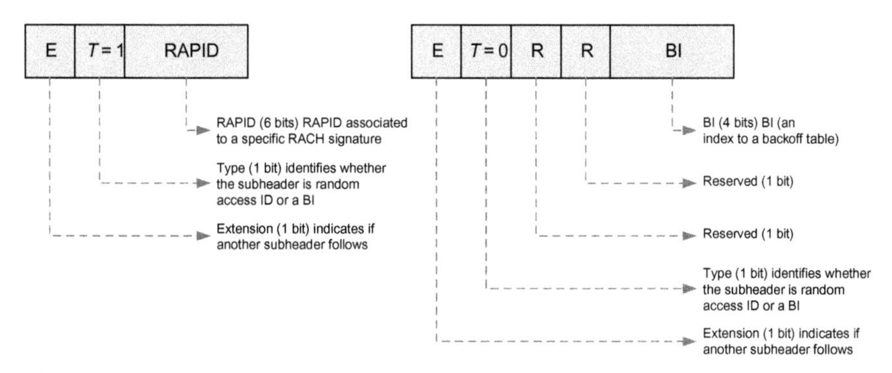

FIGURE 8.33

Structure of RAR subheaders [1].

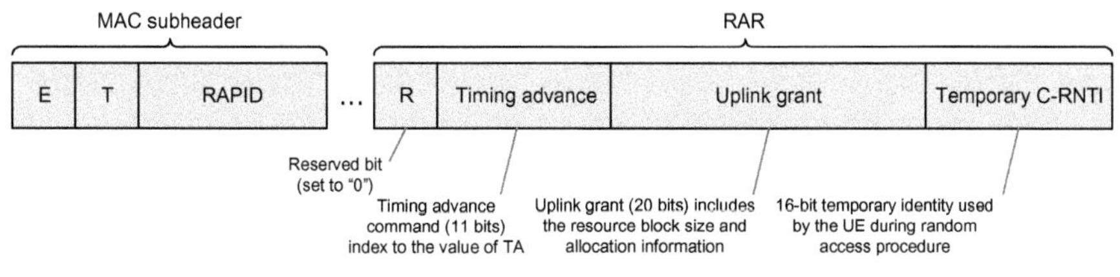

FIGURE 8.34

Structure of RAR payload [1].

References

3GPP Specifications

[1] 3GPP TS 36.321, Evolved Universal Terrestrial Radio Access (E-UTRA); Medium Access Control (MAC) Protocol Specification (Release 11), December 2012.

[2] 3GPP TS 36.322, Evolved Universal Terrestrial Radio Access (E-UTRA); Radio Link Control (RLC) Protocol Specification (Release 11), December 2012.

[3] 3GPP TS 36.323, Evolved Universal Terrestrial Radio Access (E-UTRA); Packet Data Convergence Protocol (PDCP) Specification (Release 11), December 2012.

[4] 3GPP TS 36.331, Evolved Universal Terrestrial Radio Access (E-UTRA); Radio Resource Control (RRC) Protocol Specification (Release 11), December 2012.

[5] 3GPP TS 36.211, Evolved Universal Terrestrial Radio Access (E-UTRA), Physical Channels and Modulation (Release 11), December 2012.

[6] 3GPP TS 36.212, Evolved Universal Terrestrial Radio Access (E-UTRA) Multiplexing and Channel Coding (Release 11), December 2012.

[7] 3GPP TS 36.213, Evolved Universal Terrestrial Radio Access (E-UTRA); Physical Layer Procedures (Release 11), December 2012.

[8] 3GPP TS 36.300, Evolved Universal Terrestrial Radio Access (E-UTRA) and Evolved Universal Terrestrial Radio Access Network (E-UTRAN) Overall Description, Stage 2 (Release 11), December 2012.

[9] 3GPP TS 36.133, Evolved Universal Terrestrial Radio Access (E-UTRA); Requirements for Support of Radio Resource Management (Release 11), December 2012.

Books

[10] C. Johnson, Long Term Evolution in 10, second ed., Create Space Independent Publishing Platform, July 2012.

[11] S. Sesia, I. Toufik, M. Baker, LTE, the UMTS Long Term Evolution: From Theory to Practice, second ed., John Wiley & Sons, August 2011.

[12] E. Dahlman, S. Parkvall, J. Skold, 4G: LTE/LTE-Advanced for Mobile Broadband, Academic Press, May 2011.

Articles

[13] Y. Fan, P. Lundén, M. Kuusela, M. Valkama, Efficient semi-persistent scheduling for VoIP on E-UTRA downlink, in: Proceedings of the IEEE VTC-2008, Calgary, Canada, September 2008.

[14] Y. Fan, M. Kuusela, P. Lundén, M. Valkama, Downlink VoIP support for Evolved UTRA, in: Proceedings of the IEEE WCNC-2008, Las Vegas, NV, April 2008.

[15] A. Pokhariyal, T.E. Kolding, Performance of downlink frequency domain packet scheduling for the UTRAN long term evolution, in: Proceedings of the IEEE PIMRC-2006, Helsinki, Finland, September 2006.

[16] K.I. Pedersen, et al., Frequency domain scheduling for OFDMA with limited and noisy channel feedback, in: Proceedings of the IEEE VTC-2007, Baltimore, MD, September 2007.

[17] Y. Sun, et al., Multi-user scheduling for OFDMA downlink with limited feedback for Evolved UTRA, in: Proceedings of the IEEE VTC-2006, Montreal, Canada, September 2006.

[18] C.S. Bontu, E. Illidge, DRX mechanism for power saving in LTE, IEEE Commun. Mag. (June 2009).
[19] C. Zhong, et al., A new DRX scheme for LTE-advanced carrier aggregation systems with multiple services, IEEE Vehicular Technology Conference, 2011.
[20] Root of unity, from Wikipedia. <http://en.wikipedia.org/wiki/Root_of_unity>.
[21] LTE system description. <http://www.sharetechnote.com>.

CHAPTER

Downlink Physical Layer Functions

CHAPTER CONTENTS

LTE-Advanced.

This chapter describes the physical layer protocols and functional processing in LTE/LTE-Advanced downlink direction. As shown in Figure 9.1, the physical layer is the lowest protocol layer in baseband signal processing that interfaces with the physical media through which the signal is transmitted and received. The physical layer receives MAC PDUs (transport blocks) and processes them through channel coding, interleaving, baseband modulation, layer mapping for multi-antenna operation, eigen-precoding, resource element, and antenna mapping functions. The choice of appropriate modulation and coding scheme, and multi-antenna transmission mode, are critical to achieve the desired reliability/robustness and system/webuser throughput in mobile wireless data communications. Typical mobile radio channels tend to be dispersive and time-variant and exhibit severe Doppler effects, multipath delay variation, intra-cell and inter-cell interference, and fading.

A good and robust design of the physical layer ensures that the system can normally operate and overcome the above deleterious effects, and can provide the maximum throughput and lowest latency under various operating conditions. The chapters on physical layer in this book are dedicated to systematic design of physical layer protocols and functional blocks of fourth generation cellular systems, the theoretical background on physical layer procedures, and performance evaluation of physical layer components. The theoretical background is provided to make the book self-contained, and to ensure that the reader understands the underlying theory governing the operation of various functional blocks and procedures. Additional references are provided for further study. While the focus is mainly on the techniques that were incorporated in the design of LTE/LTE-Advanced physical layer, the author has attempted to take a more generic and systematic a IMT-Advanced and beyond cellular systems, so that the reader can understand and apply the learning to the design and implementation of any OFDM-based physical layer irrespective of the radio access technology. Physical layer processes both control- and user-plane signals; however,

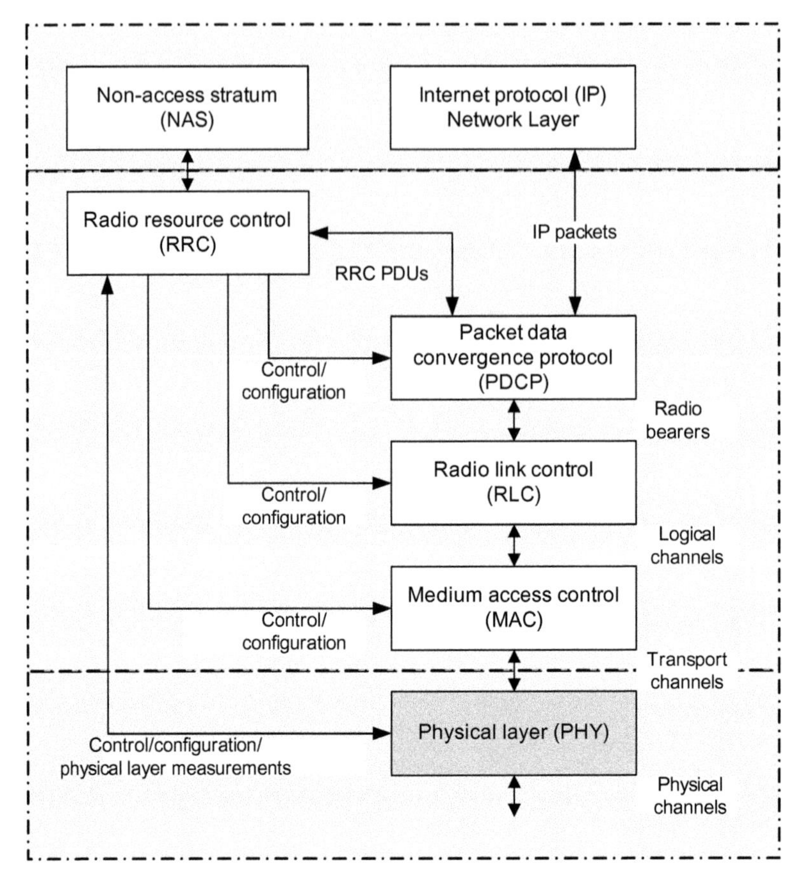

FIGURE 9.1

Physical Layer in LTE Protocol Stack

due to different design requirements and reliability and performance criteria, the procedures tend to be different. We will underline these differences and will describe the methods for selecting the best transmission format for each physical channel.

9.1 Overview of downlink physical layer processing

This section provides an overview of physical layer processing in LTE/LTE-Advanced downlink. The processing steps and the associated control signaling are depicted in Figure 9.2. The main criteria in the design of the LTE/LTE-Advanced physical layer were to increase the application throughput and capacity, reduce access latency, support higher user mobility, minimize intra-cell and inter-cell interference, improve reliability of control and data channel coverage especially

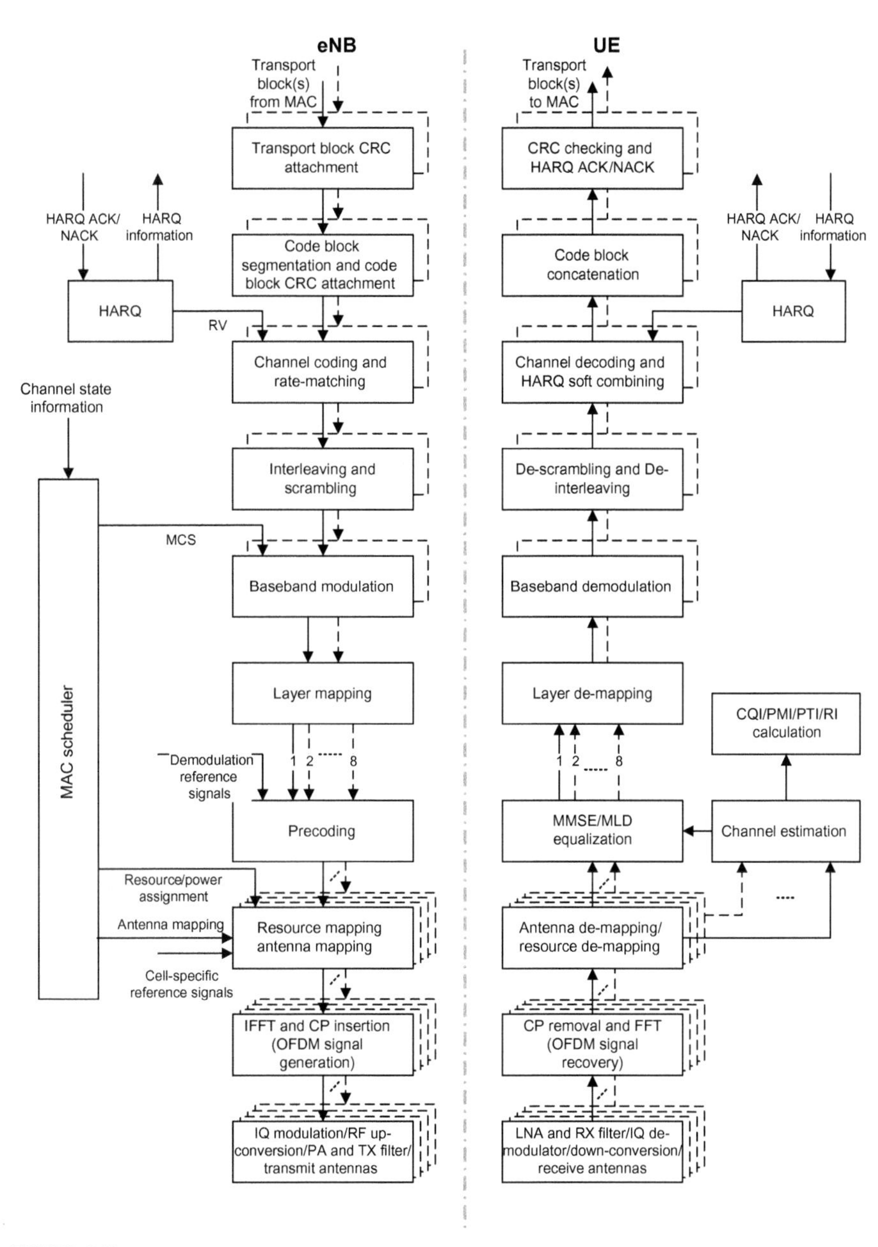

FIGURE 9.2

Downlink physical layer processing in LTE-Advanced [3,156].

at the cell-edge, and reduce the complexity and signaling overhead. The overall performance evaluations provided in Chapter 11 suggest that the desired goals have been successfully achieved. In the next sections of this chapter, we describe in detail the functional blocks and protocols of the downlink physical layer and their interactions based on the order that the information is processed as shown in Figure 9.2. It must be noted that although the overall procedure shown in Figure 9.2 is similar for LTE Rel-8/9 and LTE-Advanced Rel-10/11, the number of transmission layers, transmission modes, and multi-antenna processing are different. These differences will be identified in description of each function in the upcoming sections.

9.2 Characteristics of wireless channels

In a wireless communication system, a signal can travel from the transmitter to the receiver over multiple paths. This phenomenon is referred to as multipath propagation where signal attenuation varies on different paths. This effect, also known as multipath fading, can cause stochastic fluctuations in the received signal's magnitude, phase, and angle of arrival. The propagation over different paths is caused by scattering, reflection, diffraction, and refraction of the radio waves by stationary and moving objects, as well as the medium. It is obvious that different propagation mechanisms result in different channel and path loss models. As a result of wave propagation over multipath fading channels, the radio signal is attenuated due to mean path loss, as well as macroscopic and microscopic fading.

In an ideal free-space propagation model, the attenuation of RF signal energy between the transmitter and receiver follows the inverse-square law. The received power expressed in terms of transmitted power is attenuated proportional to the inverse of $L_s(d)$, which is called free space loss. When the receiving antenna is isotropic, the received signal power can be expressed as follows [19,48,49]:

$$P_{RX} = P_{TX}G_{TX}G_{RX}\left(\frac{\lambda}{4\pi d}\right)^2 = \frac{P_{TX}G_{TX}G_{RX}}{L_s(d)} \quad (9.1)$$

where P_{TX} and P_{RX} denote the transmitted and the received signal power, G_{TX} and G_{RX} denote the transmitting and receiving antenna gains, d is the distance between the transmitter and the receiver, and λ is the wavelength of the RF signal.

Macroscopic fading is caused by shadowing effects of buildings and natural obstructions, and is modeled by the local mean of a fast fading signal. The mean path loss $\overline{L}_p(d)$ as a function of distance d between the transmitter and receiver is proportional to the nth power of d relative to the reference distance d_0. In logarithmic scale, it can be expressed as follows:

$$\overline{L}_p(d) = L_s(d_0) + 10n\log(d/d_0) \quad \text{(dB)} \quad (9.2)$$

The reference distance d_0 corresponds to a point located in the far-field of the antenna typically 1 km for large cells, 100 m for micro-cells, and 1 m for indoor channels. In Eq. (9.2), $\overline{L}_p(d)$ is the mean path loss, which is typically $10n$ dB per decade attenuation for $d \gg d_0$. The value of n depends on the frequency, antenna heights, and propagation environment, and is equal to 2 in free space. Studies show that the path loss $L_p(d)$ is a random variable with log-normal distribution about the mean path loss $\overline{L}_p(d)$. Let $X \sim \mathbb{N}(0, \sigma^2)$ denote a zero-mean Gaussian random variable with standard deviation σ when measured in decibels, then:

$$L_p(d) = L_s(d_0) + 10n \log(d/d_0) + X \quad \text{(dB)} \tag{9.3}$$

The value of X is often derived empirically based on measurements. A typical value for σ is 8 dB. The parameters that statistically describe path loss due to large-scale fading (macroscopic fading) for an arbitrary location with a specific transmitting–receiving antenna separation include the reference distance d_0, the path loss exponent n, and the standard deviation σ of X [19,48,49].

Microscopic fading refers to the rapid fluctuations of the received signal in time and frequency, and is caused by scattering objects between the transmitting and receiving antennas. When the received RF signal is a superposition of independent scattered components plus a line-of-sight (LoS) component, the envelope of the received signal $r(t)$ has a Rician probability distribution function (PDF), and is referred to as Rician fading. As the magnitude of the LoS component approaches zero, the Rician PDF approaches a Rayleigh PDF. Thus:

$$f(r) = \frac{r(K+1)}{\sigma^2} \exp\left[-K - \frac{(K+1)r^2}{2\sigma^2}\right] I_0\left(\frac{2r}{\sigma}\sqrt{K(K+1)/2}\right) \quad r \geq 0 \tag{9.4}$$

where K and $I_0(r)$ denote the Rician factor and zero-order modified Bessel function of the first kind. In the absence of LoS path ($K = 0$), the Rician PDF reduces to Rayleigh distribution.

Time-varying fading due to scattering objects or transmitter/receiver motion results in Doppler spread. The time spreading effect of small-scale or microscopic fading is manifested in the time-domain as multipath delay spread, and in the frequency-domain as channel coherence bandwidth. Similarly, the time variation of the channel is characterized in the time-domain as channel coherence time, and in the frequency-domain as Doppler spread. In a fading channel, the relationship between maximum excess delay time τ_m and symbol time τ_s can be viewed in terms of two different degradation effects, i.e., frequency-selective fading and frequency non-selective or flat fading. A channel is said to exhibit frequency-selective fading if $\tau_m > \tau_s$. This condition occurs whenever the received multipath components of a symbol extend beyond the symbol's time duration. Such multipath dispersion of the signal results in inter-symbol interference (ISI) distortion. In the case of frequency-selective fading, mitigating the distortion is possible because many of the multipath components are separable by the receiver. A channel is said to exhibit frequency non-selective or flat fading if $\tau_m < \tau_s$. In this case, all received multipath components of a symbol arrive within the symbol time duration; therefore, the components are not resolvable. In this case, there is no channel-induced ISI distortion, since the signal time spreading does not result in significant overlap among adjacent received symbols. There is still performance degradation, because the irresolvable phasor components can add up destructively to yield a substantial reduction in signal-to-noise ratio (SNR). Also, signals that are classified as exhibiting flat fading can sometimes experience frequency-selective distortion.

Figure 9.3 illustrates a multipath-intensity profile $\Lambda(\tau)$ versus delay τ where the term delay refers to the excess delay. It represents the signal's propagation delay that exceeds the delay of the first signal arriving at the receiver. For a typical wireless communication channel, the received signal usually consists of several discrete multipath components. The received signals are composed of a

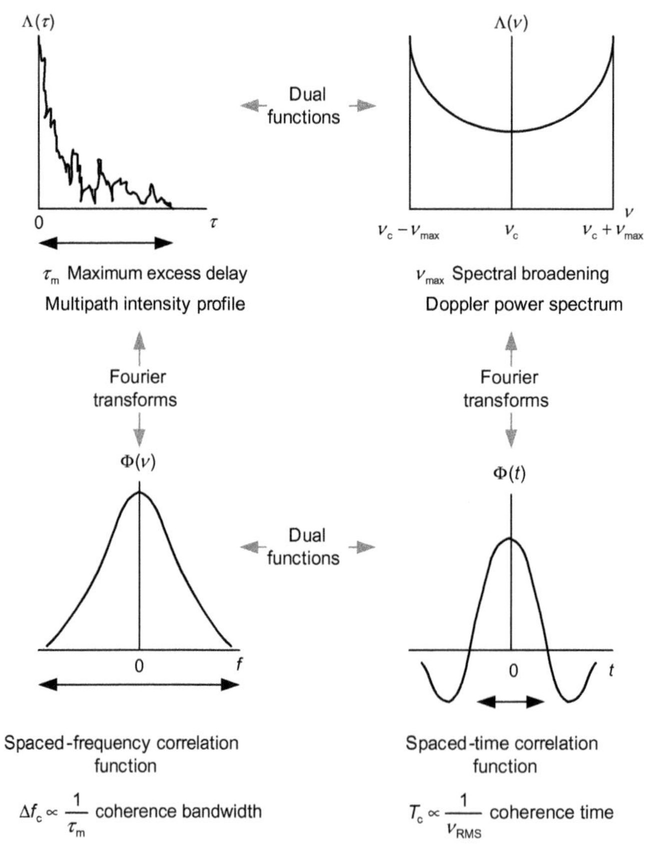

FIGURE 9.3

Illustration of the duality principle in time- and frequency-domains [19,48,49].

continuum of multipath components in some channels such as the tropospheric channel. In order to perform measurements of the multipath intensity profile, a wideband signal, i.e., a unit impulse or Dirac delta function, is used. For a single transmitted impulse, the time τ_m between the first and last received component is defined as the maximum excess delay during which the multipath signal power typically falls to some level 10–20 dB below that of the strongest component. Note that for an ideal system with zero excess delay, the function $\Lambda(\tau)$ would consist of an ideal impulse with weight equal to the total average received signal power. In the literature, the Fourier transform of $\Lambda(\tau)$ is referred to as the spaced-frequency correlation function $\Phi(\upsilon)$ [19]. The spaced-frequency correlation function $\Phi(\upsilon)$ is the channel's response to a pair of sinusoidal signals separated in frequency by υ. The coherence bandwidth Δf_c is a measure of the frequency range over which spectral components have a strong likelihood of amplitude correlation. In other words, a signal's spectral components over this range are affected by the channel in a similar manner. Note that Δf_c and τ_m are inversely proportional ($\Delta f_c \propto 1/\tau_m$). The maximum excess delay τ_m is not the best indicator of how any given wireless system will perform over a communication channel, because different channels with the same value of τ_m can exhibit different variations of signal intensity over the delay span. A useful measure of delay spread is most often characterized in terms of the root mean square (RMS) delay spread τ_{RMS} where [20]:

$$\tau_{RMS} = \sqrt{\frac{\int_0^{\tau_m} (\tau - \overline{\tau})^2 \Lambda(\tau) d\tau}{\int_0^{\tau_m} \Lambda(\tau) d\tau}} \tag{9.5}$$

in which the average multipath delay is calculated as follows:

$$\overline{\tau} = \frac{\int_0^{\tau_m} \tau \Lambda(\tau) d\tau}{\int_0^{\tau_m} \Lambda(\tau) d\tau} \tag{9.6}$$

An exact relationship between coherence bandwidth and delay spread does not exist and must be derived from signal analysis of actual signal dispersion measurements in specific channels. If coherence bandwidth is defined as the frequency interval over which the channel's complex-valued frequency transfer function has a correlation of at least 0.9, the coherence bandwidth is approximately $\Delta f_c \approx 1/(50\tau_{RMS})$. A common approximation of Δf_c corresponding to a frequency range over which the channel transfer function has a correlation of at least 0.5 is $\Delta f_c \approx 1/(5\tau_{RMS})$.

A channel is said to exhibit frequency-selective effects if $\Delta f_c < 1/\tau_s$, where the inverse symbol rate is approximately equal to the signal bandwidth W. In practice, W may differ from $1/\tau_s$ due to filtering or data modulation. Frequency-selective fading effects arise whenever a signal's spectral components are not affected equally by the channel. This occurs whenever $\Delta f_c < W$. Frequency non-selective or flat fading degradation occurs whenever $\Delta f_c > W$. Hence, all of the signal's spectral components will be affected by the channel in a similar manner. Flat-fading does not introduce channel-induced ISI distortion, but performance degradation can still be expected due to loss in SNR whenever the signal experiences fading. In order to avoid channel-induced ISI distortion, the channel is required to exhibit flat-fading by ensuring that $\Delta f_c > W$. Therefore, the channel coherence bandwidth Δf_c sets an upper limit on the transmission rate that can be used without incorporating an equalizer in the receiver.

Figure 9.3 shows function $\Phi(t)$ which is known as the spaced-time correlation function, which is the autocorrelation function of the channel's response to a sinusoid. This function specifies the extent to which there is correlation between the channel's response to a sinusoid sent at time t_1 and the response to a similar sinusoid sent at time t_2, where $\Delta t = t_2 - t_1$. The coherence time is a measure of the expected time duration over which the channel's response is essentially invariant. To estimate $\Phi(t)$, a sinusoidal signal is transmitted through the channel and the autocorrelation function of the channel output is calculated. The function $\Phi(t)$ and the parameter T_c provide information about the speed of fading-channel variation. Note that for an ideal time-invariant channel, the channel's response would be highly correlated for all values of Δt and $\Phi(t)$ would be a constant function. If one ideally assumes uniformly distributed scattering around a mobile station with linearly polarized antennas, then the Doppler power spectrum (i.e., the inverse Fourier transform of spaced-time correlation function $\Lambda(\nu)$) has a U-shaped distribution, as shown in Figure 9.3. In a time-varying fading channel, the channel response to a pure sinusoidal tone spreads over a finite frequency range $\nu_c - \nu_{max} < \nu < \nu_c + \nu_{max}$, where ν_c and ν_{max} denote the frequency of the sinusoidal tone and the maximum Doppler spread, respectively. The RMS bandwidth of $\Lambda(\nu)$ is referred to as Doppler spread and is denoted by ν_{RMS} that can be calculated as follows:

$$\nu_{RMS} = \sqrt{\frac{\int_{\nu_c-\nu_{max}}^{\nu_c+\nu_{max}} (\nu-\overline{\nu})^2 \Lambda(\nu) d\nu}{\int_{\nu_c-\nu_{max}}^{\nu_c+\nu_{max}} \Lambda(\nu) d\nu}} \tag{9.7}$$

where:

$$\overline{\nu} = \frac{\int_{\nu_c-\nu_{max}}^{\nu_c+\nu_{max}} \nu \Lambda(\nu) d\nu}{\int_{\nu_c-\nu_{max}}^{\nu_c+\nu_{max}} \Lambda(\nu) d\nu} \tag{9.8}$$

The coherence time is typically defined as the time lag for which the signal autocorrelation coefficient reduces to 0.7. The coherence time is inversely proportional to Doppler spread $T_c \approx 1/\nu_{RMS}$. A common approximation for the value of coherence time as a function of Doppler spread is $T_c = 0.423/\nu_{RMS}$. It can be observed that the functions on the right-hand side of Figure 9.3 are dual of the functions on the left-hand side (duality principle).

The angle of spread refers to the spread in angle of arrival (AoA) of the multipath components at the receiver antenna array. At the transmitter, on the other hand, the angle of spread refers to the spread in the angle of departure (AoD) of the multipath components that leave the transmit antennas. If the angle spectrum function $\Theta(\theta)$ denotes the average power as a function of AoA, then the RMS angle spread can be calculated as follows:

$$\theta_{RMS} = \sqrt{\frac{\int_{-\pi}^{+\pi} (\theta - \bar{\theta})^2 \Theta(\theta) d\theta}{\int_{-\pi}^{+\pi} \Theta(\theta) d\theta}} \tag{9.9}$$

where:

$$\bar{\theta} = \frac{\int_{-\pi}^{+\pi} \theta \Theta(\theta) d\theta}{\int_{-\pi}^{+\pi} \Theta(\theta) d\theta} \tag{9.10}$$

The angle of spread causes space-selective fading, which manifests itself as variation of signal amplitude according to the location of antennas. The space-selective fading is characterized by the coherence distance D_c which is the spatial separation for which the autocorrelation coefficient of the spatial fading reduces to 0.7. The coherence distance is inversely proportional to the angle of spread $D_c \propto 1/\theta_{RMS}$.

In Figure 9.3, a duality between multipath intensity function $\Lambda(\tau)$ and Doppler power spectrum $\Lambda(\nu)$ is identified. It means that the two functions exhibit similar behavior across the time-domain and frequency-domain. As the $\Lambda(\tau)$ function identifies expected power of the received signal as a function of delay, $\Lambda(\nu)$ identifies expected power of the received signal as a function of frequency. Similarly, the spaced-frequency correlation function $\Phi(f)$ and spaced-time correlation function $\Phi(t)$ are dual functions. It implies that $\Phi(f)$ represents channel correlation in frequency-domain and $\Phi(t)$ corresponds to channel correlation function in time-domain in a similar manner.

Having reviewed some underlying concepts and characteristics of the wireless channels, let us now focus on some practical aspects of communication (data transmission) over wireless channels, and in particular the use of the above properties in the formation of a channel vector/matrix. As mentioned earlier, a wireless channel exhibits time, frequency, and space-selective variations known as fading. This fading arises due to Doppler, delay, and angle spreads in the scattering environment. The channel spreading can be observed by sending a single impulse in frequency or continuous-wave signal in time or angle (point source) through the channel and receiving a signal spread along the spectral, temporal, or spatial dimension, respectively. A frequency-flat solution, however, can be applied to a frequency-selective channel by decomposing the transmission band into multiple narrow, frequency-flat sub-bands. Specifically, we can apply the solution per sub-carrier in systems employing OFDM. In a rich scattering environment, a frequency-flat MIMO wireless channel can be modeled as a complex Gaussian random process, represented as a time-varying matrix. The channel at each time instance is a Gaussian random variable, specified by the mean and its covariance. A non-zero channel mean signifies the presence of a direct LoS, or a cluster of strong paths, and the channel envelop has the Rician statistics, while zero mean corresponds to the Rayleigh statistics. The channel covariance, on the other hand, encompasses the correlation among the antennas at the transmitter and the receiver. Assuming the channel is stationary, the channel temporal variation can be captured by the channel auto-covariance, measuring the correlation between two channel instances separated by a delay. At zero delay, the channel auto-covariance coincides with the channel covariance [129].

Considering the channel state information at the transmitter (CSIT) at the transmit time in the form of a channel estimate and the estimation error covariance derived from a channel measurement at an initial time and the channel statistics, the main source of irreducible error in channel estimation is the random time variation of the channel between the initial measurement and its use by the transmitter. Therefore, we assume that the initial channel measurement is error free. The error in the channel estimate depends only on the time delay and the channel time selectivity, or the Doppler spread.

Let $\mathbf{H}$ denote the $N_{RX} \times N_{TX}$ channel matrix in a system with N_{TX} transmit and N_{RX} receive antennas. The channel has statistical mean $\overline{\mathbf{H}} = \mathscr{E}[\mathbf{H}]$ and covariance $\mathbf{R_H} = \mathscr{E}[\mathbf{hh}^H] - \overline{\mathbf{h}}\,\overline{\mathbf{h}}^H$, where the lower-case letter denotes the vectorized version of the upper-case matrix. Let us further assume that we have an initial, accurate channel measurement $\mathbf{H}(0)$. The channel auto-covariance $\mathbf{R}(\tau)$ at time delay τ then indicates the correlation between the initial measurement $\mathbf{H}(0)$ and the current channel $\mathbf{H}(\tau)$, defined as $\mathbf{R}(\tau) = \mathscr{E}[\mathbf{h}(\tau)\mathbf{h}^H(0)] - \overline{\mathbf{h}}(\tau)\overline{\mathbf{h}}^H(0)$. Intuitively, when this correlation is strong (i.e., $\mathbf{R}(\tau)$ is large when measured in a suitable norm) then $\mathbf{H}(0)$ is useful for estimating $\mathbf{H}(\tau)$. The strongest correlation is when the delay is zero (i.e., if $\tau \to 0$, then $\mathbf{R}(\tau) \to \mathbf{R}(0)$). In a scalar system, $\mathbf{R}(\tau)$ and $\mathbf{R}(0)$ reduce to scalars $r(\tau)$ and $r(0)$, respectively, and they are related through $r(\tau) = \rho(\tau)r(0)$, where $|\rho(\tau)| \leq 1$ is the temporal correlation coefficient [129].

An important assumption about channel temporal homogeneity is that the temporal correlation-coefficient $\rho(\tau)$ between any pair of transmit-antenna and receive-antenna is identical. This assumption is based on the premise that the channel temporal statistics can be expected to be the same for all antenna pairs. It is now possible to separate the temporal correlation from the spatial correlation in the channel auto-covariance as $\mathbf{R}(\tau) = \rho(\tau)\mathbf{R}(0)$. The temporal correlation $\rho(\tau)$ is a function of the time delay τ and the channel Doppler spread. As an example, in

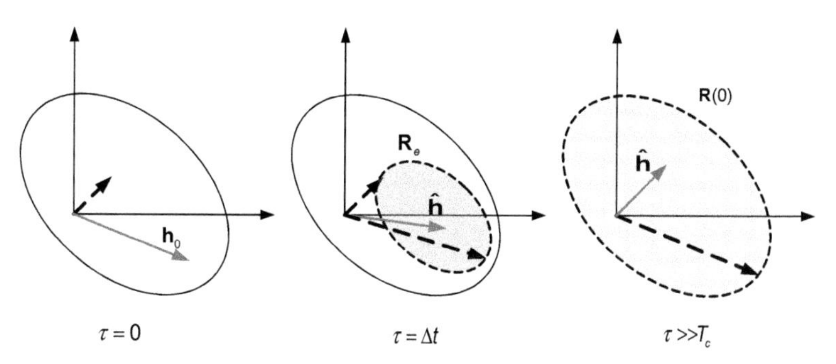

FIGURE 9.4

Illustration of a dynamic CSIT model [129].

Jake's model [21], $\rho(\tau) = J_0(2\pi\tau v)$, where v is the channel Doppler spread and $J_0(\cdot)$ is the zero-order Bessel function of the first kind [35–37].

It can be shown that an estimate of the channel at time τ together with the estimation error covariance using minimum mean-squared error (MMSE) criterion results in the following $\hat{\mathbf{H}}(\tau) = \rho\mathbf{H}(0) - (1-\rho)\overline{\mathbf{H}}$ and $\mathbf{R}_e(\tau) = (1-\rho^2)\mathbf{R}(0)$. The two quantities $\hat{\mathbf{H}}(\tau)$ and $\mathbf{R}_e(\tau)$ function effectively as a new channel mean and a new channel covariance; thus they are referred to as the effective mean and the effective covariance, respectively, and constitute the CSIT. This CSIT ranges from perfect channel knowledge when $\rho = 1$, to pure stochastic when $\rho = 0$. Since the CSIT depends on the value of ρ which captures the channel time variation, it is called dynamic CSIT. The correlation factor ρ functions as a measure of CSIT quality. When $\rho = 1$, the channel estimate coincides with $\mathbf{H}(0)$ and is error free. As ρ approaches to zero, the effect of the initial channel measurement diminishes and the estimate asymptotically approaches the channel mean $\overline{\mathbf{H}}$. In parallel, the estimation error covariance $\mathbf{R}_e$ is zero when $\rho = 1$, and approaches zero as the correlation factor approaches zero. Figure 9.4 illustrates the CSIT variation as a function of the time delay τ. Several special cases of dynamic CSIT are of interest as follows: (1) perfect CSIT, in which the effective covariance is zero and the effective mean is the instantaneous channel matrix; (2) mean CSIT, in which the effective mean is non-zero and arbitrary, but the effective covariance is the identity matrix, corresponding to uncorrelated antennas; (3) covariance CSIT, in which the effective covariance matrix is non-identity and arbitrary, but the effective mean is zero, corresponding to Rayleigh fading. The general case in which both the mean and covariance matrices are arbitrary is referred to as statistical CSIT [129].

9.3 LTE/LTE-Advanced operating frequencies and band classes

The LTE/LTE-Advanced were designed to operate in the frequency bands defined in Table 9.1. The requirements were defined for 1.4, 3, 5, 10, 15, and 20 MHz

Table 9.1 E-UTRA Operating Frequencies, Band Classes, and Supported Bandwidths [1,2]

E-UTRA Operating Band (Band Class)	Uplink Operating Band eNB Receive/UE Transmit (MHz) $f_{UL\text{-}low} - f_{UL\text{-}high}$	$N_{Offset\text{-}DL}$	N_{DL}	Downlink Operating Band eNB Transmit/UE Receive (MHz) $f_{DL\text{-}low} - f_{DL\text{-}high}$	$N_{Offset\text{-}UL}$	N_{UL}	Duplex Scheme	Supported Bandwidths (MHz)
1	1920–1980	0	0–599	2110–2170	18000	18000–18599	FDD	5, 10, 15, 20
2	1850–1910	600	600–1199	1930–1990	18600	18600–19199	FDD	1.4, 3, 5, 10, 15, 20
3	1710–1785	1200	1200–1949	1805–1880	19200	19200–19949	FDD	1.4, 3, 5, 10, 15, 20
4	1710–1755	1950	1950–2399	2110–2155	19950	19950–20399	FDD	1.4, 3, 5, 10, 15, 20
5	824–849	2400	2400–2649	869–894	20400	20400–20649	FDD	1.4, 3, 5, 10
6	830–840	2650	2650–2749	875–885	20650	20650–20749	FDD	5, 10
7	2500–2570	2750	2750–3449	2620–2690	20750	20750–21449	FDD	5, 10
8	880–915	3450	3450–3799	925–960	21450	21450–21799	FDD	1.4, 3, 5, 10
9	1749.9–1784.9	3800	3800–4149	1844.9–1879.9	21800	21800–22149	FDD	5, 10, 15, 20
10	1710–1770	4150	4150–4749	2110–2170	22150	22150–22749	FDD	5, 10, 15, 20
11	1427.9–1447.9	4750	4750–4949	1475.9–1495.9	22750	22750–22949	FDD	5, 10
12	699–716	5010	5010–5179	729–746	23010	23010–23179	FDD	1.4, 3, 5, 10
13	777–787	5180	5180–5279	746–756	23180	23180–23279	FDD	5, 10
14	788–798	5280	5280–5379	758–768	23280	23280–23379	FDD	5, 10
15	Reserved			Reserved			FDD	–
16	Reserved			Reserved			FDD	–
17	704–716	5730	5730–5849	734–746	23730	23730–23849	FDD	5, 10
18	815–830	5850	5850–5999	860–875	23850	23850–23999	FDD	5, 10, 15
19	830–845	6000	6000–6149	875–890	24000	24000–24149	FDD	5, 10, 15
20	832–862	6150	6150–6449	791–821	24150	24150–24449	FDD	5, 10, 15, 20

21	1447.9–1462.9	6450	6450–6599	1495.9–1510.9	24450	24450–24599	FDD	5, 10, 15
22	3410–3490	6600	6600–7399	3510–3590	24600	24600–25399	FDD	5, 10, 15, 20
23	2000–2020	7500	7500–7699	2180–2200	25500	25500–25699	FDD	1.4, 3, 5, 10
24	1626.5–1660.5	7700	7700–8039	1525–1559	25700	25700–26039	FDD	5, 10
25	1850–1915	8040	8040–8689	1930–1995	26040	26040–26689	FDD	1.4, 3, 5, 10, 15, 20
...				Not Specified				
33	1900–1920	36000	36000–36199	1900–1920	36000	36000–36199	TDD	5, 10, 15, 20
34	2010–2025	36200	36200–36349	2010–2025	36200	36200–36349	TDD	5, 10, 15
35	1850–1910	36350	36350–36949	1850–1910	36350	36350–36949	TDD	1.4, 3, 5, 10, 15, 20
36	1930–1990	36950	36950–37549	1930–1990	36950	36950–37549	TDD	1.4, 3, 5, 10, 15, 20
37	1910–1930	37550	37550–37749	1910–1930	37550	37550–37749	TDD	5, 10, 15, 20
38	2570–2620	37750	37750–38249	2570–2620	37750	37750–38249	TDD	5, 10, 15, 20
39	1880–1920	38250	38250–38649	1880–1920	38250	38250–38649	TDD	5, 10, 15, 20
40	2300–2400	38650	38650–39649	2300–2400	38650	38650–39649	TDD	5, 10, 15, 20
41	2496–2690	39650	39650 –41589	2496–2690	39650	39650 –41589	TDD	5, 10, 15, 20
42	3400–3600	41590	41590–43589	3400–3600	41590	41590–43589	TDD	5, 10, 15, 20
43	3600–3800	43590	43590–45589	3600–3800	43590	43590–45589	TDD	5, 10, 15, 20

bandwidth with a specific configuration in terms of the number of physical resource blocks (PRBs). Using carrier aggregation, a number of contiguous and/or non-contiguous frequency bands can be aggregated to create a virtually larger bandwidth. The channel raster is 100 kHz, which means the center frequency must be a multiple of 100 kHz [1]. To support transmission in paired and unpaired spectra, two duplexing schemes are supported, i.e., frequency division duplex (FDD), allowing both full and half-duplex terminal operation, as well as time division duplex (TDD). Table 9.2 gives the band classes where the LTE/LTE-Advanced systems can be deployed.

Figure 9.5 illustrates the relation between the channel bandwidth (BW_{Channel}) and the transmission bandwidth, i.e., the number of permissible PRBs (N_{RB}). The channel edges are defined as the lowest and highest frequencies of the carrier separated by the channel bandwidth; thus $f_C \pm (1/2)BW_{\text{Channel}}$. In the case of carrier aggregation, the aggregated channel bandwidth $BW_{\text{Channel-CA}}$ is defined as $BW_{\text{Channel-CA}} = f_{\text{edge-high}} - f_{\text{edge-low}}$. The lower band edge $f_{\text{edge-low}}$ and the upper band edge $f_{\text{edge-high}}$ of the aggregated channel bandwidth are used as the reference points in frequency for transmitter and receiver requirements, and are subsequently defined as follows: $f_{\text{edge-low}} = f_{C\text{-low}} - f_{\text{offset-low}}$ and $f_{\text{edge-high}} = f_{C\text{-high}} + f_{\text{offset-high}}$. The lower and upper frequency offsets depend on the transmission bandwidth configurations of the lowest and highest assigned edge component carrier, and are defined as $f_{\text{offset-low}} = 0.18N_{\text{RB-low}}/2 + BW_{\text{Guard-band}}$ and $f_{\text{offset-high}} = 0.18N_{\text{RB-high}}/2 + BW_{\text{Guard-band}}$, where $N_{\text{RB-low}}$ and $N_{\text{RB-high}}$ denote the transmission bandwidth configurations for the lowest and highest assigned component carrier, respectively, and $BW_{\text{Guard-band}}$ denotes the nominal guard band. Note that the 0.18 multiplier in the equations is the bandwidth of a PRB in MHz. The aggregated transmission bandwidth is defined as the number of the aggregated resource blocks within the entire assigned aggregated channel bandwidth, and is defined per carrier aggregation band class.

The frequency separation between carriers depends on the deployment scenario and the size of the frequency block available to the network operator, as well as the channel bandwidth. The nominal channel spacing between two adjacent E-UTRA carriers is defined as $\Delta f_{\text{spacing}} = (BW_{\text{Channel(1)}} + BW_{\text{Channel(2)}})/2$, where $BW_{\text{Channel(1)}}$ and $BW_{\text{Channel(2)}}$ parameters are the channel bandwidths of the two respective E-UTRA carriers. The channel spacing can be adjusted to optimize performance in a particular deployment scenario. For intra-band contiguous carrier aggregation bandwidth, the nominal channel spacing between two adjacent E-UTRA component carriers is specified as follows:

$$\Delta f_{\text{spacing-CA}} = 0.3 \left\lfloor \frac{BW_{\text{Channel(1)}} + BW_{\text{Channel(2)}} - 0.1\left|BW_{\text{Channel(1)}} - BW_{\text{Channel(2)}}\right|}{0.6} \right\rfloor \ (\text{MHz}) \tag{9.11}$$

The channel spacing for intra-band contiguous carrier aggregation can be adjusted to any integer multiple of 300 kHz (less than the nominal channel spacing)

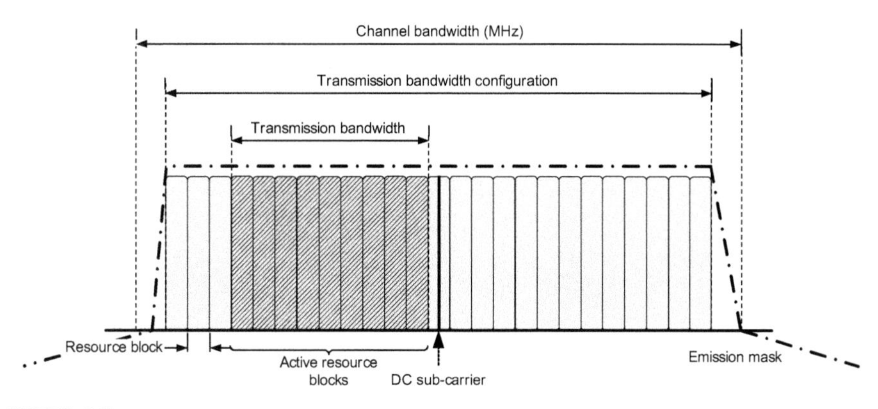

FIGURE 9.5

Relationship between channel bandwidth and transmission bandwidth in E-UTRA [1].

to optimize performance in a particular deployment scenario. The E-UTRA channel raster (including the carrier aggregation scenarios) is 100 kHz for all bands, which means that the carrier center frequency must be an integer multiple of 100 kHz.

The carrier frequency in the downlink/uplink directions is designated by the E-UTRA absolute radio frequency channel number (EARFCN) in the range of 0–65,535. The relation between EARFCN and the downlink carrier frequency (in MHz) is given by $f_{DL} = f_{DL-low} + 0.1(N_{DL} - N_{Offset-DL})$, where f_{DL-low}, N_{DL}, and $N_{Offset-DL}$ parameters are given in Table 9.1. Similarly, the relationship between EARFCN and the uplink carrier frequency (in MHz) for the uplink is defined as $f_{UL} = f_{UL-low} + 0.1(N_{UL} - N_{Offset-UL})$, where f_{UL-low}, N_{UL}, and $N_{Offset-UL}$ parameters are provided in Table 9.1. Note that N_{DL} and N_{UL} in the latter equations are the downlink and uplink EARFCN, respectively.

The channel numbers that designate carrier frequencies which are extremely close to the operating band edges are not used. This implies that the first 7, 15, 25, 50, 75, and 100 channel numbers at the lower operating band edge and the last 6, 14, 24, 49, 74, and 99 channel numbers at the upper operating band edge are not utilized for channel bandwidths of 1.4, 3, 5, 10, 15, and 20 MHz, respectively.

The fragmentation of the suitable and permissible frequency bands for deployment of LTE systems across the globe and the disparity of the operators' network deployments and legacy systems have made significant challenges against worldwide proliferation and accessibility of 4G networks (see Figure 9.6). Figure 9.6 depicts the distribution of the useable frequency bands for commercial LTE/LTE-Advanced deployments across the globe as of August 2012 [161]. In general, typical user equipment has to support multiple communication RF bands and standards (e.g., LTE/LTE-Advanced, UMTS, and 1xEV-DO) in order to allow operation across different regions and operator networks. Enabling the mobile terminals to cover all desired RF

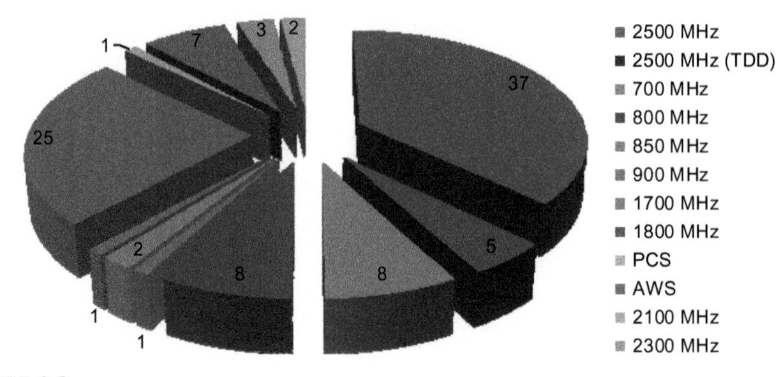

FIGURE 9.6

Distribution of frequency bands for E-UTRA commercial deployments across the world [161].

bands does not only result in an increased chip area due to more complex RF integrated circuit implementation, but also a larger phone form factor to accommodate external filters (see Figure 9.7 as an example of triple-band LTE transceiver). In the case of FDD systems, common handhelds use bulk acoustic wave duplex filters[1] to separate the receive signal from the transmit signal with a nominal frequency separation. These filters are not tunable, and thus require one duplexer per frequency constellation. The selectivity provided by the duplexer must reach a certain level such that third order inter-modulation frequency components are suppressed. These signals are generated by the transmit signal leaking into the receiver and an external interferer in a certain frequency band, thus degrading sensitivity of the entire receiver. Therefore, it is necessary to use either very expensive duplexers that could provide the desired selectivity, or to place an external (i.e., external to RFIC) low-noise amplifier followed by an additional surface acoustic wave filter[2] which is a more common solution. In general, it is not trivial to design extremely broadband RF filters, power or low-noise amplifiers, mixers, etc., to cover a number of contiguous and/or non-contiguous RF bands with optimal performance across the entire operating bands. Figure 9.8 shows the measured and simulated performance of a triple-fed

[1]Bulk acoustic wave filters are electro-mechanical devices typically operating in the frequency range of 2–16 GHz which can be implemented by ladder or lattice filters. Two main variants of bulk acoustic wave filter are thin film bulk acoustic resonator, or FBAR, and solid mounted bulk acoustic resonators [167].

[2]Surface acoustic wave filters are electro-mechanical devices commonly used in RF systems. The electrical signal is converted to a mechanical wave in a device constructed of a piezoelectric crystal or ceramic. This wave is delayed as it propagates across the device, before being converted back to an electrical signal by further electrodes. The delayed outputs are recombined to produce a direct analog implementation of a finite impulse response filter. This hybrid filtering technique has found applications in analog sampled filters. The surface acoustic wave filters are limited to frequencies up to 3 GHz [167].

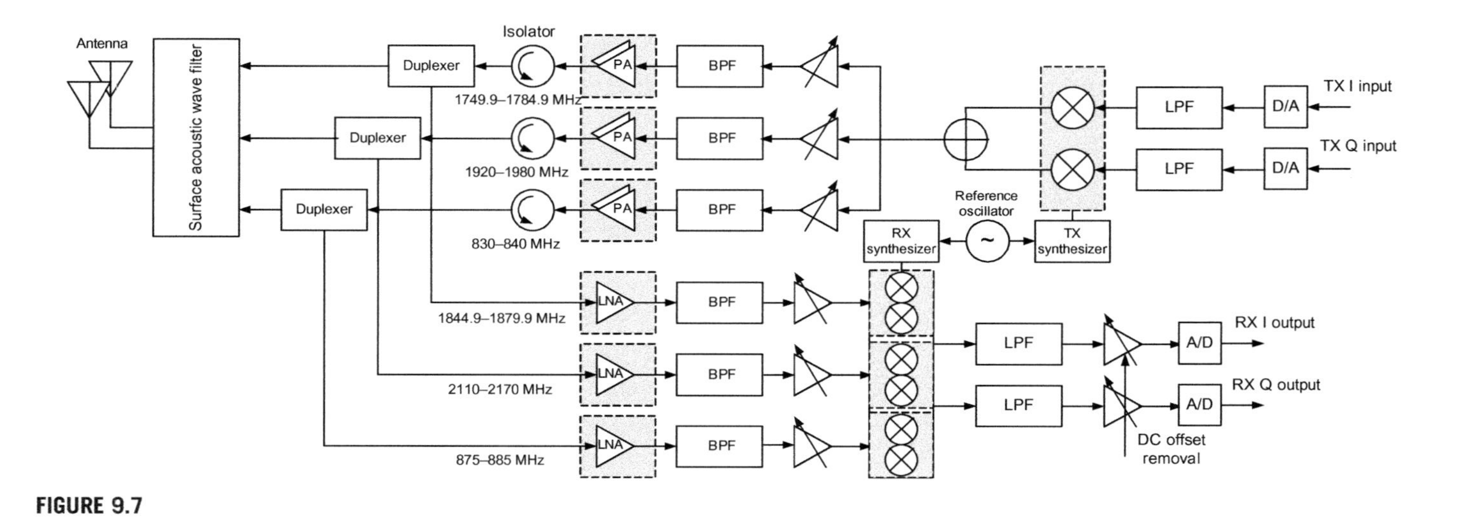

FIGURE 9.7

Example of a Triple-band LTE Transceiver [170].

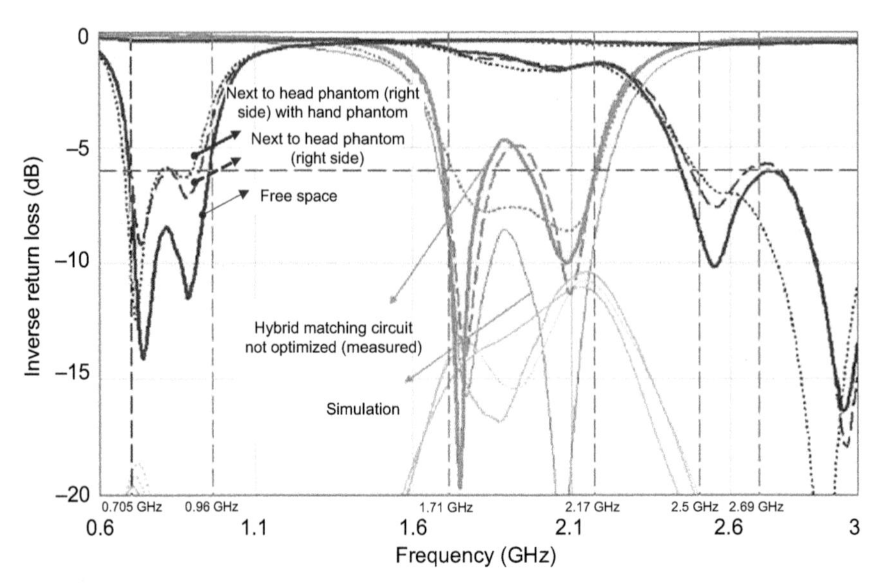

FIGURE 9.8

Simulated and measured inverse return loss of an example triple-fed antenna at 0.8, 1.9, 2.6 GHz [170].

antenna in terms of input return loss[3] (i.e., an indication of the quality of input impedance matching) under three different practical scenarios and in different frequency bands. It can be seen that satisfactory impedance matching is only achieved in a certain frequency region and not across the entire frequency bands, especially if the antennas are covered by hand or head phantoms. This example underlines the practical challenge of designing extremely wideband RF systems to fit into very small user-equipment's form factor in order to enable multi-band multi-mode LTE operation [170].

[3]Return loss is the loss of signal power resulting from the reflection caused at a discontinuity in a transmission media. This discontinuity can be a mismatch with the terminating load or with a device inserted in the line. It is usually expressed as the ratio of the incident power P_i and the reflected power P_r, i.e., $\mathrm{RL\ (dB)} = 10\log(P_i/P_r)$ in dB. The return loss is related to both the standing wave ratio (SWR) and reflection coefficient (Γ). Increasing the value of return loss corresponds to lower SWR. Return loss is a measure of how well the input/output impedances of devices or transmission lines are matched. A match is good, if the return loss value is high. A high return loss is desirable and results in a lower insertion loss. The ratio of the amplitude of the reflected wave V_r to the amplitude of the incident wave V_i is known as the reflection coefficient Γ, which is defined as $\Gamma = V_r/V_i$. When the source and load impedances are known values, the reflection coefficient is given by $\Gamma = (Z_L - Z_0)/(Z_L + Z_0)$ where Z_0 is the impedance of the transmission line toward the source and Z_L is the impedance toward the load. The return loss is the negative of the magnitude of the reflection coefficient in dB. Since power is proportional to the square of the voltage, the return loss is given by $\mathrm{RL(dB)} = -20\log(|\Gamma|)$ [169].

The performance of mobile device antennas is an important factor in the design and implementation of the device (in terms of practical form factor) as well as the ultimate communication quality that a user may expect from that device. This requires adopting an antenna design that yields high performance under realistic conditions in order to ensure satisfaction of LTE/LTE-Advanced system requirements. This challenge will be further intensified in the case of multi-band antenna array design which is required to fit into the small form factor of a mobile device and to provide satisfactory performance (near-field/far-field radiation patterns, gain, return loss, parasitic/spurious effects, etc.) over a large bandwidth.

9.4 Principles of OFDM

The orthogonal frequency division multiplexing (OFDM) is a form of multi-carrier modulation technique that was first introduced over 50 years ago. In a conventional serial data transmission system, the information bearing symbols are transmitted sequentially, with the frequency spectrum of each symbol occupying the entire available bandwidth. Figure 9.9 illustrates an unfiltered quadrature amplitude modulation (QAM) signal spectrum. It is in the form of $\sin(\pi f T_u)/\pi f T_u$ with zero crossing points at integer multiples of $1/T_\mathrm{u}$, where T_u is the QAM symbol period. The concept of OFDM is to transmit the data bits in parallel QAM modulated sub-carriers using frequency division multiplexing. The carrier spacing is carefully selected so that each sub-carrier is located on other sub-carriers' zero-crossing points in the frequency-domain. Although there are spectral overlaps among sub-carriers, they do not interfere with each other if they are sampled at the sub-carrier frequencies. In other words, they maintain spectral orthogonality. As shown in Figure 9.9, the OFDM signal in the frequency-domain is generated through aggregation of N_FFT parallel QAM-modulated sub-carriers, where adjacent sub-carriers, are separated by sub-carrier spacing $1/T_\mathrm{u}$. An example of OFDM signal spectrum as seen on a vector signal analyzer is shown in Figure 9.10.

Since an OFDM signal consists of many parallel QAM sub-carriers, the mathematical expression of the signal in the time-domain can be expressed as follows:

$$s(t) = \mathrm{Re}\left\{ e^{j\omega_c t} \sum_{k=-(N_\mathrm{FFT}-1)/2}^{(N_\mathrm{FFT}-1)/2} \alpha_k \, e^{j2\pi k(t-t_\mathrm{g})/T_\mathrm{u}} \right\} \quad mT_\mathrm{u} \leq t \leq (m+1)T_\mathrm{u} \tag{9.12}$$

where $s(t)$ denotes the OFDM signal in the time-domain, α_k is the complex-valued data that is QAM-modulated and transmitted over sub-carrier k, N_FFT is the number of sub-carriers in the frequency-domain, ω_c is the RF carrier frequency, and T_g is the guard interval or the cyclic prefix (CP) length. For a large number of sub-carriers, direct generation and demodulation of the OFDM signal would require arrays of coherent sinusoidal generators which can become excessively complex and expensive. However, one can notice that the OFDM signal is actually the real part of the inverse discrete Fourier transform (IDFT) of the original complex-valued data

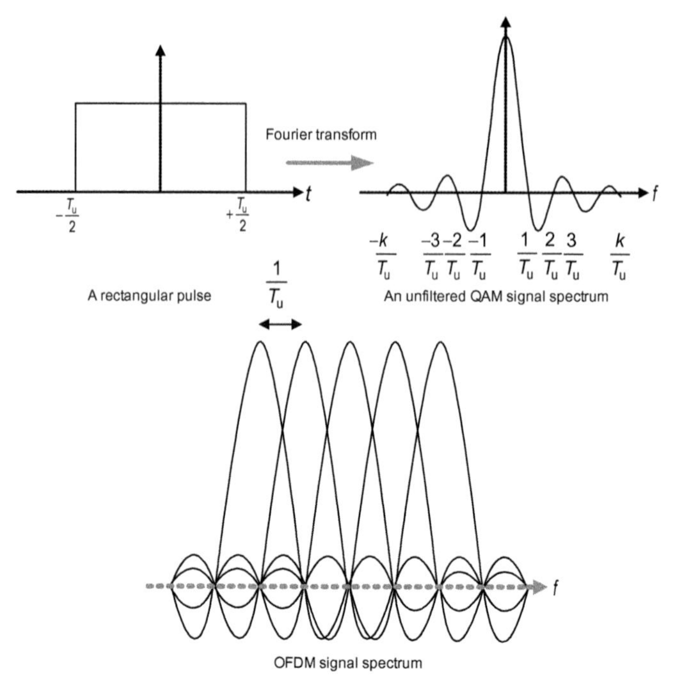

FIGURE 9.9

Illustration of OFDM concept [52].

symbols $\{\alpha_k\}|_{k=-(N_{\mathrm{FFT}}-1)/2,\ldots,(N_{FFT}-1)/2}$. It can be seen that there are $N < N_{\mathrm{FFT}}$ sub-carriers, each carrying the corresponding data α_k. The inverse of the sub-carrier spacing $\Delta f = 1/T_{\mathrm{u}}$ is defined as the OFDM useful symbol duration T_{u}, which is N_{FFT} times longer than that of the original input data symbol duration.

Since IDFT is used in the OFDM modulator, the original data is defined in the frequency-domain, while the OFDM signal $s(t)$ is defined in the time-domain. The IDFT can be implemented via a computationally efficient fast Fourier transform (FFT) algorithm. The orthogonality of sub-carriers in OFDM can be maintained and individual sub-carriers can be completely separated and demodulated by an FFT at the receiver when there is no ISI introduced by the communication channel. In practice, linear distortions such as multipath delay cause ISI between OFDM symbols, resulting in loss of orthogonality and an effect that is similar to co-channel interference. However, when delay spread is small, i.e., within a fraction of the OFDM useful symbol length, the impact of ISI is insignificant, although it depends on the order of modulation implemented by the sub-carriers (see Figure 9.11). A simple solution to mitigate multipath delay is to increase the OFDM effective symbol duration such that it is much larger than the delay spread; however, when the delay spread is large, it requires a large number of sub-carriers and a large FFT size. Meanwhile, the system might become sensitive to Doppler shift and carrier frequency offset (CFO). An alternative approach to mitigate

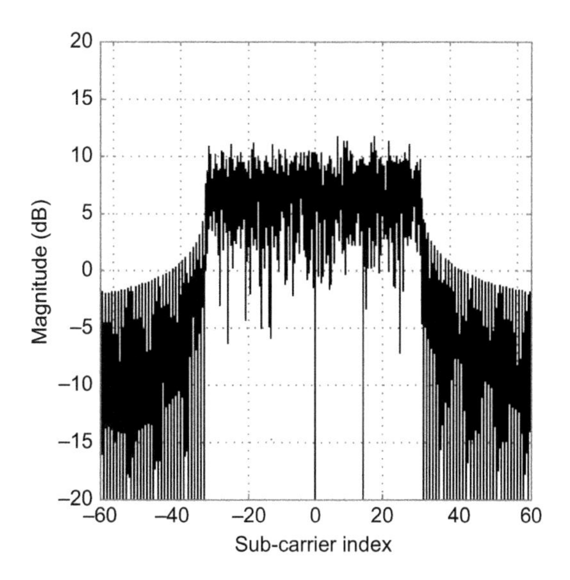

FIGURE 9.10

A snapshot of OFDM signal in the frequency-domain [158].

multipath distortion is to generate a cyclically extended guard interval, where each OFDM symbol is prefixed with a periodic extension of the signal itself, as shown in Figure 9.12 where the tail of the symbol is copied to the beginning of the symbol. The OFDM symbol duration is then defined as $T_s = T_u + T_g$, where T_g is the guard interval or cyclic prefix. When the guard interval is longer than the channel impulse response or the multipath delay, the ISI can be effectively eliminated. The ratio of the guard interval to useful OFDM symbol duration depends on the deployment scenario and the frequency band. Since the insertion of the guard intervals will reduce data throughput, T_g is usually selected less than $T_u/4$. The cyclic prefix should absorb most of the signal energy dispersed by the multipath channel. The entire ISI energy is contained within the cyclic prefix if its length is greater than that of the channel RMS delay spread ($T_g > \tau_{RMS}$). In general, it is sufficient to have most of the delay spread energy absorbed by the guard interval, considering the inherent robustness of large OFDM symbols to time dispersion.

The mapping of the modulated data symbol into multiple sub-carriers also allows an increase in the symbol duration. The symbol duration obtained through an OFDM scheme is much larger than that of a single carrier modulation technique with a similar transmission bandwidth. In general, when the channel delay spread exceeds the guard time, the energy contained in the ISI will be much smaller with respect to the useful OFDM symbol energy, as long as the symbol duration is much larger than the channel delay spread, i.e., $T_s \gg \tau_{RMS}$. Although large OFDM symbol duration is desirable to mitigate the ISI effects caused by time dispersion, large OFDM symbol duration can further reduce the ability to alleviate the effects of fast fading, particularly if the symbol period is large compared to the channel coherence

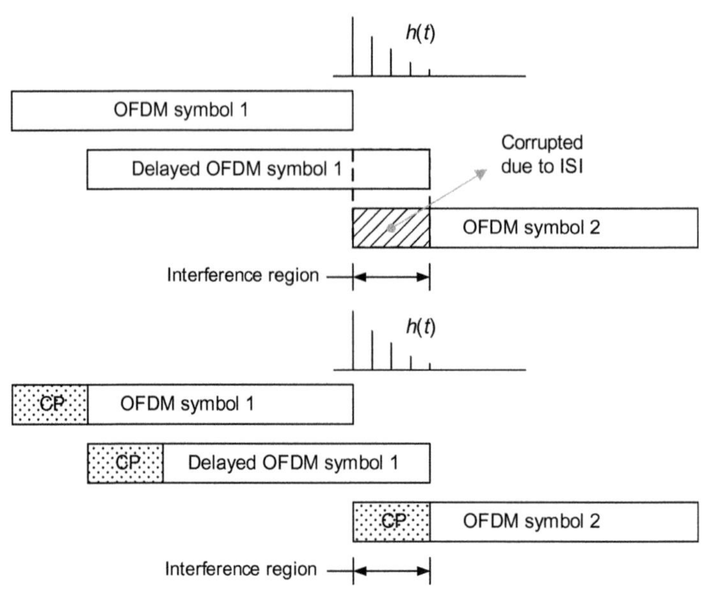

FIGURE 9.11

Illustration of the effect of guard interval for eliminating ISI ($h(t)$ is the hypothetical channel impulse response).

time, then the channel can no longer be considered as time-invariant over the OFDM symbol duration; therefore, this will introduce the inter-sub-carrier orthogonality loss. This can affect the performance in fast fading conditions. Hence, the symbol duration should be kept smaller than the minimum channel coherence time. Since the channel coherence time is inversely proportional to the maximum Doppler spread, the symbol duration T_s must, in general, be chosen such that $T_s \ll 1/\nu_{RMS}$. The large number of OFDM sub-carriers makes the bandwidth of the individual sub-carriers small relative to the overall signal bandwidth. With an adequate number of sub-carriers, the inter-carrier spacing is much smaller than the channel coherence bandwidth. Since the channel coherence bandwidth is inversely proportional to the channel delay spread τ_{RMS}, the sub-carrier separation is generally designed such that $\Delta f \ll 1/\tau_{RMS}$. In this case, the fading on each sub-carrier is flat and can be modeled as a complex-valued constant channel gain. The individual reception of the QAM symbols transmitted on each sub-carrier is therefore simplified to the case of a flat-fading channel. This enables a straightforward introduction of advanced MIMO schemes. Furthermore, in order to mitigate Doppler spread effects, the inter-carrier spacing should be much larger than the RMS Doppler spread $\Delta f \gg \nu_{RMS}$. Since the OFDM sampling frequency is typically larger than the actual signal bandwidth, only a subset of sub-carriers are used to carry QAM symbols. The remaining sub-carriers are left inactive prior to the IFFT, and are referred to as guard sub-carriers. The split between the active and the inactive

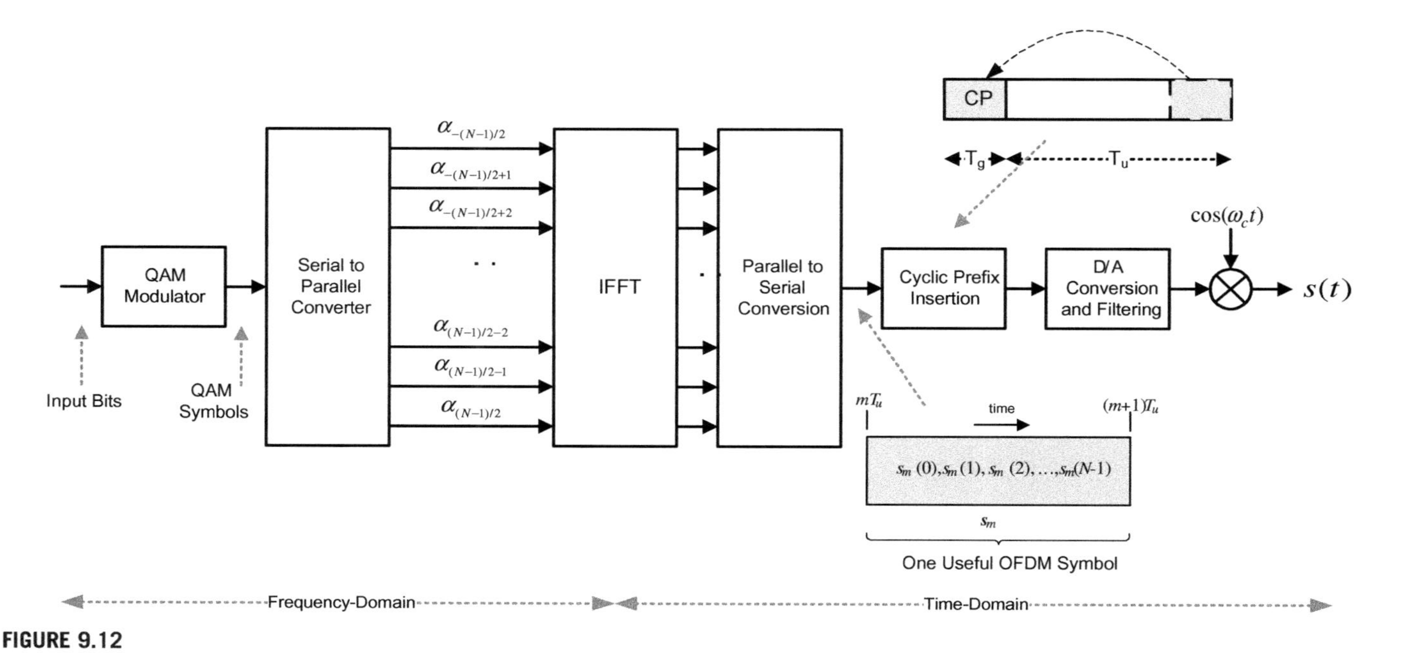

FIGURE 9.12

OFDM signal generation.

sub-carriers is determined based on the spectral sharing and regulatory constraints, such as the bandwidth allocation and the spectral mask.

An OFDM transmitter diagram is shown in Figure 9.12. The incoming bit stream is first QAM modulated to form the complex-valued QAM symbols. The QAM symbols are converted from serial to parallel with $N < N_{\mathrm{FFT}}$ complex-valued numbers per block, where N_{FFT} is the size of FFT/IFFT operation. Each block is processed by an IFFT and the output of the IFFT forms an OFDM symbol, which is converted back to serial data for transmission. A guard interval or cyclic prefix is inserted between symbols to eliminate ISI effects caused by multipath distortion. The discrete symbols are filtered and converted to analog for RF up-conversion.

The reverse process is performed at the receiver. A one-tap equalizer is usually used for each sub-carrier to correct channel distortion. The tap coefficients are calculated based on channel information. As shown in Figures 9.9 and 9.10, one can understand that the OFDM signal spectrum is approximately rectangular. Since the OFDM modulator is an IFFT processor, its role is to convert data from the frequency-domain to the time-domain, and then transmit them in the time-domain over the channel. If a spectrum analyzer, which converts the input from the time-domain to the frequency-domain, is used to monitor an OFDM signal, what will be displayed is the original data.

When there is multipath distortion, a conventional single carrier wideband transmission system suffers from frequency selective fading. A complex adaptive equalizer must be used to equalize the in-band fading. The number of taps required for the equalizer is proportional to the symbol rate and the multipath delay. For an OFDM system, if the guard interval is larger than the multipath delay, the ISI can be eliminated and orthogonality can be maintained among sub-carriers. Since each OFDM sub-carrier occupies a very narrow spectrum, in the order of a few kHz, even under severe multipath distortion, they are only subject to flat fading. In other words, the OFDM converts a wideband frequency-selective fading channel to a series of narrowband frequency non-selective fading sub-channels by using the parallel multi-carrier transmission scheme (Figure 9.13). Since OFDM data sub-carriers are statistically independent and identically distributed, based on the central limit theorem, when the number of sub-carriers N_{FFT} is large, the OFDM signal distribution tends to be Gaussian.

In an OFDM system, the SNR is a measure of channel quality and is a key factor of link-level error assessment. There are different methods for calculation of SNR in single-antenna and multi-antenna transmission systems. For single-input/single-output systems, the SNR can be viewed as the received SNR, i.e., the received SNR before the detector. The post-processing SNR is often used for MIMO links, and represents the SNR after combining in the receiver and measures the likelihood that a coded message is decoded successfully. In link-level simulations, the SNR γ is typically calculated using the following method. Let vector $\mathbf{x} = (x_1, x_2, \ldots, x_{N_{\mathrm{TX}}})^H \in \mathbb{C}^{N_{\mathrm{TX}} \times 1}$ denote the transmit signal, where $x_k \in \mathbb{C}\,\forall k = 1, 2, \ldots, N_{\mathrm{TX}}$ is the complex-valued transmitted symbols from the kth transmit antenna. Note that N_{TX} is the number of transmit antennas. It can be shown that the total transmit signal power can be obtained as $\sigma_{\mathbf{x}}^2 = \mathrm{trace}(\mathbf{R}_{\mathbf{x}})$,

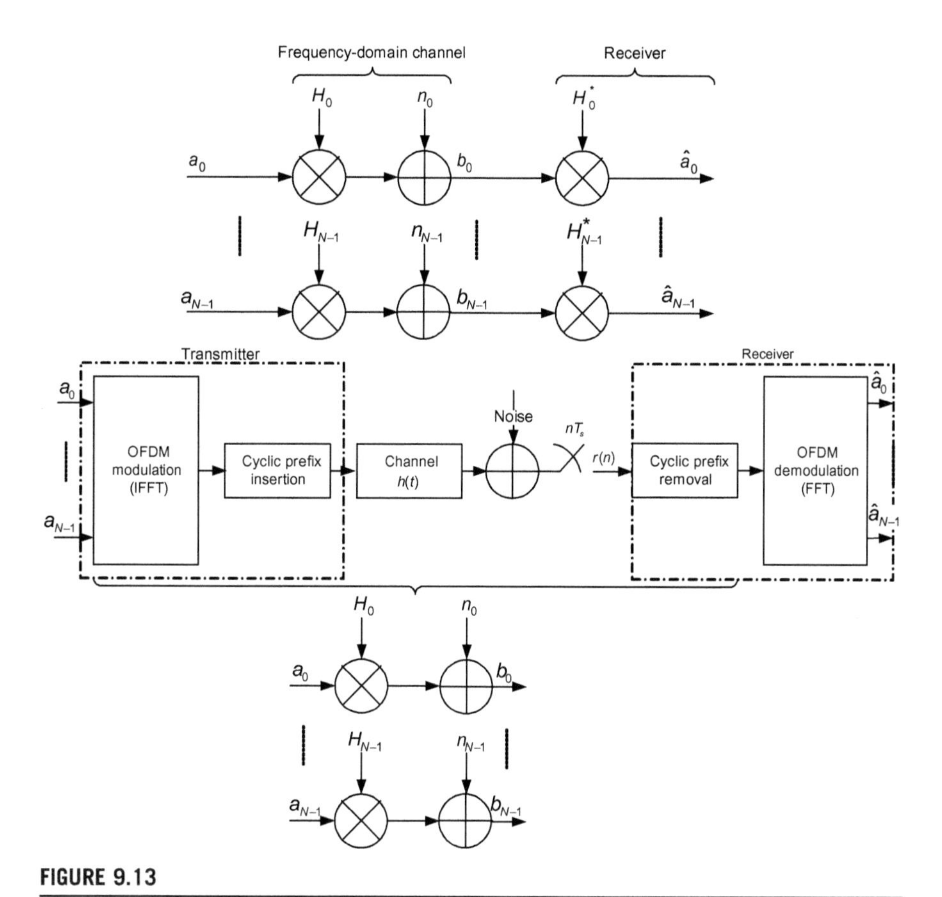

FIGURE 9.13

Single-tap equalization of the OFDM signal [14].

where $\mathbf{R_x}$ denotes the autocorrelation matrix of the transmitted signal. The transmit power from the kth antenna is given as $\sigma_{x_k}^2 = \mathscr{E}\{|x_k|^2\} = 1/N_{TX}$ (uniformly distributed power). If $\mathbf{H}$ represents the channel matrix with $||\mathbf{H}||_F^2 = N_{TX}N_{RX}$[4] in which N_{TX} and N_{RX} denote the number of transmit and receive antennas, respectively, the received signal vector $\mathbf{y}$ can be calculated as $\mathbf{y} = \mathbf{Hx} + \mathbf{v}$ where the complex-valued Gaussian-distributed noise vector $\mathbf{v} \sim \mathbb{Z}(\mathbf{0}, \sigma_\mathbf{v}^2\mathbf{I})$ denotes the noise vector with respect

[4]The p-norm of matrix $\mathbf{H}$ is defined as $\|\mathbf{H}\|_p = \left(\sum_{i=0}^{M-1}\sum_{j=0}^{N-1}|h_{ij}|^p\right)^{1/p}$. In the special case when $p = 2$, the norm is called the Frobenius norm and for $p = \infty$ is called the maximum norm. The Frobenius norm or Hilbert–Schmidt norm of matrix $\mathbf{H}$ are similar, although the latter term is often reserved for the operators on a Hilbert space. In general, this norm can be defined in the following forms: $\|\mathbf{H}\|_p = \sqrt{\sum_{i=0}^{M-1}\sum_{j=0}^{N-1}|h_{ij}|^2} = \sqrt{\text{trace}(\mathbf{H}^H\mathbf{H})} = \sqrt{\sum_{i=0}^{\min(M,N)-1}\varsigma_i^2}$ where $\mathbf{H}^H$ denotes the conjugate transpose of matrix $\mathbf{H}$ and ς_i are the singular values of matrix $\mathbf{H}$. The Frobenius norm is further similar to the Euclidean norm on $\mathbb{R}^N$ and is obtained from an inner product on the space of all matrices. The Frobenius norm is sub-multiplicative and is very useful for numerical linear algebra [31,32,34].

to the size of the FFT N_{FFT} and the number of used sub-carriers before the detector N_{Used}. We define the complex-valued Gaussian-distributed noise vector $\mathbf{n} \sim \mathbb{Z}(\mathbf{0}, \sigma_{\mathbf{n}}^2 \mathbf{I})$ to be the noise after the FFT operation. The receive SNR before the detector is given as $\gamma_{\text{pre-FFT}} = \frac{||\mathbf{Hx}||_F^2}{N_{\text{RX}} \sigma_{\mathbf{v}}^2} = \sigma_{\mathbf{v}}^{-2}$ whereas the SNR after the FFT operation is given as $\gamma_{\text{post-FFT}} = \frac{||\mathbf{Hx}||_F^2}{N_{\text{RX}} \sigma_{\mathbf{n}}^2} = \sigma_{\mathbf{n}}^{-2}$. It can be seen that the ratio of pre-FFT and post-FFT SNRs $\frac{\gamma_{\text{pre-FFT}}}{\gamma_{\text{post-FFT}}} = \frac{\sigma_{\mathbf{v}}^2}{\sigma_{\mathbf{n}}^2} = \frac{N_{\text{FFT}}}{N_{\text{Used}}}; N_{\text{Used}} \leq N_{\text{FFT}}$ is always positive, implying that the FFT operation suppresses the noise and enhances the SNR [142].

9.4.1 Peak-to-average power ratio of OFDM signals

The peak-to-average power ratio (PAPR) for a single-carrier modulation signal depends on its constellation and the pulse shaping filter roll-off factor. For a Gaussian distributed OFDM signal, the cumulative distribution function (CDF)[5] of PAPR for 99.0%, 99.9%, and 99.99% are approximately 8.3, 10.3, and 11.8 dB, respectively. Since the OFDM signal has a high PAPR, it could be clipped in the transmitter power amplifier (PA), because of its limited dynamic range or non-linearity. Higher output back-off is required to prevent performance degradation and inter-modulation products leaking into adjacent channels. Therefore, RF PAs should be operated in a very large linear region. Otherwise, the signal peaks leak into the non-linear region of the PA causing signal distortion. This signal distortion introduces inter-modulation among the sub-carriers and out-of-band emission. Thus, the PAs should be operated with large power back-offs. On the other hand, this leads to very inefficient amplification and expensive transmitters. Thus, it is highly desirable to reduce the PAPR. In addition to inefficient operation of the PA, a high PAPR requires a larger dynamic range for the receiver analog-to-digital (A/D) converter. To reduce the PAPR, several techniques have been proposed and used such as clipping,[6] channel coding, temporal windowing, tone reservation,[7] and tone injection. However,

---

[5]The CDF of the real-valued random variable X is defined as $x \rightarrow F_X(x) = P(X \leq x), \forall x \in \mathbb{R}$, where the right-hand side represents the probability that random variable X takes on a real value less than or equal to x. The CDF of X can be defined in terms of the probability density function $f(x)$ as $F(x) = \int_{-\infty}^{x} f(x)dx$. The complementary CDF (CCDF), on the other hand, is defined as $P(X > x) = 1 - F_X(x)$ [30].

[6]Since the OFDM signal has a high PAPR, it may be clipped in the transmitter power amplifier, because of its limited dynamic range or non-linearity. Higher output back-off is required to prevent BER degradation and intermodulation products spilling into adjacent channels. However, clipping of an OFDM signal has a similar effect to impulse interference, against which an OFDM system is inherently robust. Computer simulations show that for a coded OFDM system, clipping of 0.5% of the time results in a BER degradation of 0.2 dB. At 0.1% clipping, the degradation is less than 0.1 dB [52].

[7]In the tone reservation method, the transmitter and the receiver reserve a subset of tones or sub-carriers for generating PAPR reduction signals. Those reserved tones are not used for data transmission [53].

most of these methods are unable to achieve simultaneously a large reduction in PAPR with low complexity, low coding overhead, without performance degradation, and without transmitter-receiver symbol handshake. The PAPR ξ of the OFDM signal, defined in Eq. (9.12), is calculated as follows:

$$\xi = \left. \frac{\max|s(t)|^2}{\mathscr{E}\{|s(t)|^2\}} \right|_{mT_u \le t \le (m+1)T_u} \tag{9.13}$$

In the above equation, $\mathscr{E}\{.\}$ denotes the expectation operator, and m is an integer. From the central-limit theorem, for large values of N_{FFT}, the real and imaginary values of $s(t)$ (from Eq. (9.12) and without the $\text{Re}\{\cdot\}$ operator) would have Gaussian distribution. Consequently, the amplitude of the OFDM signal has a Rayleigh distribution with zero mean and a variance of N_{FFT} times the variance of one complex sinusoid. Assuming the samples to be mutually uncorrelated, the CDF for the peak power per OFDM symbol is given by:

$$P(\xi > \gamma) = [1 - (1 - e^{-\gamma})^{N_{FFT}}] \tag{9.14}$$

From the above equation, it can be seen that a large PAPR occurs only infrequently due to relatively large values of N_{FFT} used in practice.

Figure 9.14 shows the CDF of OFDM PAPR for BPSK and QPSK modulation assuming a 40 MHz channel bandwidth and $N_{FFT} = 128$. We further applied a 90° phase rotation to the sub-carriers in the upper 20 MHz of the

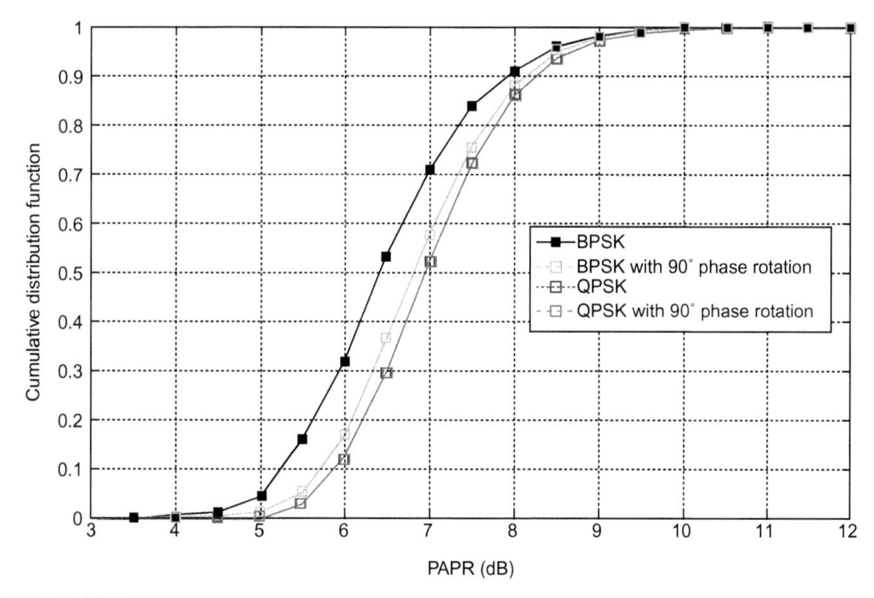

FIGURE 9.14

CDF of OFDM PAPR with BPSK/QPSK modulation and $N_{FFT} = 128$ [159].

channel, and investigated the effect on the PAPR reduction [159]. It is shown that large PAPR values are less likely to occur with large FFT sizes, as suggested by Eq. (9.14).

9.4.2 Effect of non-linearity on OFDM signals

Power amplifiers generally have a non-linear amplitude response, where the output power is saturated for large input powers. Most applications require operation in the linear region of the PA response where the output power is a linear function of the input. The larger the linear operation region, or alternatively the higher the saturation point, the more expensive the PA. Therefore, it is imperative to reduce the PAPR of the OFDM signal before processing through the PA. Figures 9.15 and 9.16 illustrate examples of PA AM/AM response and QAM signal constellation fuzziness when the OFDM signal with large PAPR is processed through a PA with a low saturation point (Figure 9.16). In this example, 64QAM modulation, 256-point IFFT with $T_{cp} = (1/4)T_u$, and PA saturation point of 6 dB are assumed.

The modulation accuracy or the permissible signal constellation fuzziness is often measured in terms of an error vector magnitude (EVM) metric. The EVM may be defined as the square root of the ratio of the mean error vector power to

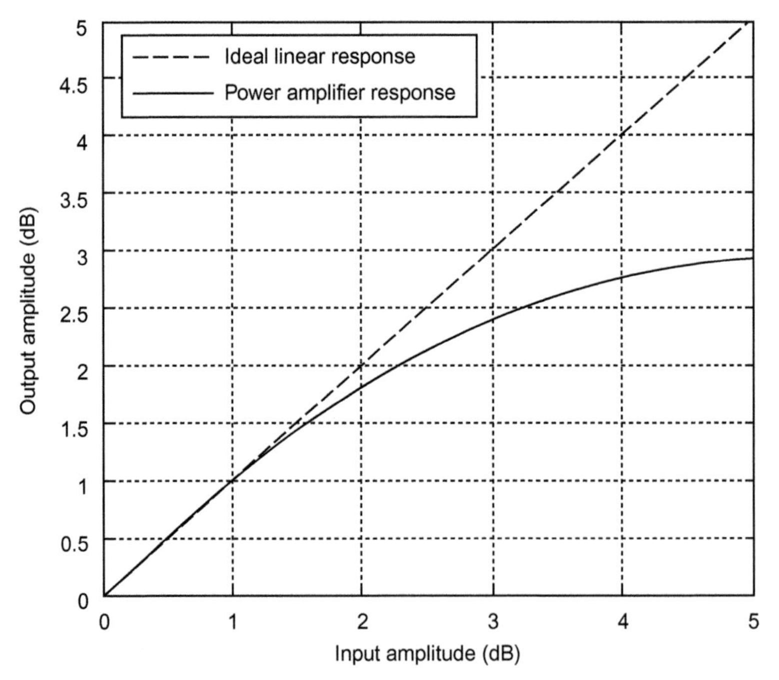

FIGURE 9.15

Example AM/AM response of a power amplifier (saturation point = 6 dB) [141].

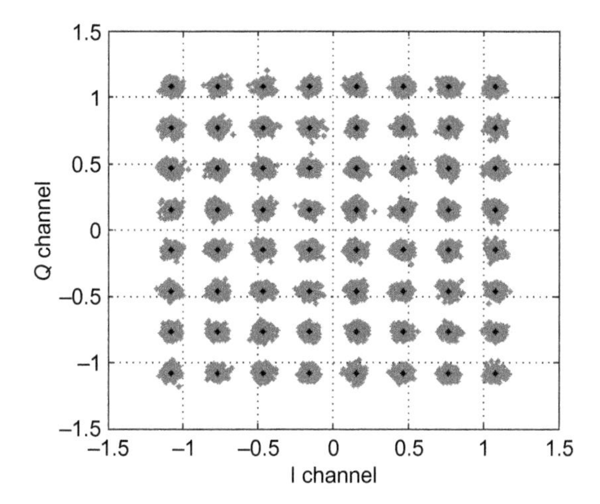

FIGURE 9.16

64QAM signal constellation before and after power amplification [141].

the mean reference signal power expressed as a percentage, i.e., the EVM can be expressed as $\mathrm{EVM} = \sqrt{\mathscr{E}|\mathbf{e}|^2/\mathscr{E}|\mathbf{s}|^2}$ in mathematical form. In other words, the EVM defines the average constellation error with respect to the farthest constellation point (i.e., the distance between the reference signal and measured signal points in the I–Q plane), and is calculated as follows [1]:

$$\mathrm{EVM} = \sqrt{\frac{\sum_{i=0}^{N_{\text{samples}}-1} (\Delta I_i^2 + \Delta Q_i^2)}{N_{\text{samples}} r_{\max}^2}} \tag{9.15}$$

where N_{samples} is the number of symbols used in the measurement period, ΔI_i and ΔQ_i refer to the in-phase (I) and quadrature (Q) components of the ith error vector, and $r_{\max}$ denotes the maximum constellation amplitude. The EVM is measured over the continuous portion of a burst occupying at least one-fourth of the transmission frame at maximum power settings. The permissible EVM can be estimated from the transmitter implementation margin, if the error vector is considered noise, which is added to the channel noise. The implementation margin is the excess power needed to maintain the carrier-to-noise ratio (C/N) intact, when going from the ideal to the realistic transmitter. The EVM cannot be measured at the antenna connector, but should be measured by an ideal receiver with a certain carrier recovery loop bandwidth specified as a percentage of the symbol rate. The measured EVM includes the effects of the transmitter filter accuracy, D/A converter, modulator imbalances, untracked phase noise, and PA non-linearity.

As mentioned earlier, the EVM is a measure of the difference between the reference waveform and the measured waveform. In practice, before calculating the EVM, the measured waveform is corrected by the sample timing offset and RF frequency offset. Then the IQ origin offset is removed from the measured waveform before calculating the EVM. The measured waveform is further modified by selecting the absolute phase and absolute amplitude of the transmit chain. The EVM value is defined after the front-end IDFT as the square root of the ratio of the mean error vector power to the mean reference signal power expressed as a percentage value.

The basic EVM measurement interval in the time-domain depends on the type of physical channel. The measurement interval is equal to the length of one preamble sequence for the physical random access channel (PRACH), and is equal to the length of one slot for the physical uplink control channel (PUCCH) and physical uplink shared channel (PUSCH) in the time-domain. When the PUSCH or PUCCH transmission slot is shortened due to multiplexing with Sounding Reference Signal (SRS), the EVM measurement interval is accordingly reduced by one symbol. The PUSCH or PUCCH EVM measurement interval is also reduced when the mean power, modulation, or allocation between slots is expected to change. In the case of PUSCH transmission, the measurement interval is reduced by a time interval equal to the sum of 5 μs and the applicable exclusion period adjacent to the point where the power change is expected to occur. The PUSCH exclusion period is applied to the signal obtained after the front-end IDFT. In the case of PUCCH transmission with power change, the PUCCH EVM measurement interval is reduced by one symbol adjacent to the point where the power change is expected to occur [1].

9.4.3 Effect of carrier frequency offset on OFDM signals

As mentioned earlier, an OFDM system transmits information as a series of OFDM symbols. The time-domain samples $x_m(n)$ of the mth OFDM symbol are generated by performing IDFT on the information symbols $X_m(k)|_{k=0,1,\ldots,N_{FFT}-1}$, as follows [132]:

$$x_m(n) = \frac{1}{N_{FFT}} \sum_{k=0}^{N_{FFT}-1} X_m(k) e^{j2\pi k(n-N_{CP})/N_{FFT}}; \quad \forall 0 \le n \le N_{FFT} + N_{CP} - 1 \qquad (9.16)$$

where N_{FFT} and N_{CP} are the number of data samples and cyclic prefix samples, respectively. The OFDM symbol $x_m(n)$ is transmitted through a channel $h_m(n)$ and is perturbed by a Gaussian noise $z_m(n)$. The channel $h_m(n)$ is assumed to be block stationary, i.e., it is time invariant over each OFDM symbol. With this assumption, the output $y_m(n)$ of the channel can be represented as $y_m(n) = h_m(n)^* x_m(n) + z_m(n)$, where $h_m(n)^* x_m(n) = \sum_{k=-\infty}^{+\infty} h_m(k) x_m(k-n)$ and $z_m(n)$ is a zero-mean additive white Gaussian noise (AWGN) with variance σ_z^2.

Since the channel impulse response $h_m(n)$ is assumed to be block stationary, the channel response does not change over each OFDM symbol; however, the channel response $h_m(n)$ may vary across different OFDM symbols, thus it is a function of the OFDM symbol index m.

When the receiver oscillator is not perfectly synchronized to the transmitter oscillator, there can be a CFO $\Delta f_{CFO} = f_{TX} - f_{RX}$ between the transmitter carrier frequency f_{TX} and the receiver carrier frequency f_{RX}. Furthermore, there may be a phase offset θ_0 between the transmitter carrier and the receiver carrier. The mth received symbol $y_m(n)$ can be represented as $y_m(n) = e^{j\{2\pi\Delta f_{CFO}[n+m(N_{FFT}+N_{CP})]T_s+\theta_0\}}[h_m(n)^* x_m(n)] + z_m(n)$, where T_s is the sampling period. The CFO Δf_{CFO} can be represented relative to the sub-carrier bandwidth $(1/N_{FFT}T_s)$ by defining the relative frequency offset $\delta_{CFO} = \Delta f_{CFO} N_{FFT} T_s$. The CFO attenuates the desired signal and introduces the ICI, thus decreasing the SNR. The SNR of the kth sub-carrier can be expressed as $\mathrm{SNR}_k(\delta_{CFO}) = (\phi^2_{N_{FFT}}(\delta_{CFO}) P_h \sigma_x^2 / (1 - \phi^2_{N_{FFT}}(\delta_{CFO})) P_h \sigma_x^2 + \sigma_z^2)$, where $\phi_{N_{FFT}}(\delta_{CFO}) = (\sin(\pi\delta_{CFO}) / N_{FFT}\sin(\pi\delta_{CFO}/N_{FFT}))$ in order to demonstrate the dependence of the SNR on the frequency offset [132]. In the latter equation, P_h, σ_x^2, and σ_z^2 denote the total average power of channel impulse response, variance of the signal, and variance of the additive noise, respectively. The sub-carrier index k is dropped since the SNR is the same for all sub-carriers. From this SNR expression, it is clearly seen that the effect of the frequency offset is to decrease the signal power by $\phi^2_{N_{FFT}}(\delta_{CFO})$ and to convert the decreased power to interference power. The SNR depends not only on the frequency offset δ_{CFO}, but also on the number of sub-carriers; however, as N_{FFT} increases $\phi^2_{N_{FFT}}(\delta_{CFO})$ converges to $\mathrm{sinc}^2(\delta_{CFO})$. Therefore, the SNR converges to $\mathrm{SNR}(\delta_{CFO}) = (\mathrm{sinc}^2(\delta_{CFO}) P_h \sigma_x^2 / (1 - \mathrm{sinc}^2(\delta_{CFO})) P_h \sigma_x^2 + \sigma_z^2)$ as N_{FFT} becomes increasingly large [132]. The effect of CFO on OFDM signal in the frequency-domain, which appears as inter-carrier interference, for different values of δ_{CFO} is shown in Figure 9.17. Figure 9.18 illustrates the variation of $\mathrm{SNR}(\delta_{CFO})$ as a function of δ_{CFO} with $\gamma = P_h \sigma_x^2 / \sigma_z^2$ as a parameter. In the above equations, the power of inter-carrier interference as a function of relative CFO is defined as $P_{ICI}(\delta_{CFO}) = (1 - \mathrm{sinc}^2(\delta_{CFO})) P_h \sigma_x^2$. In practice, the sub-carrier spacing is not the same among different tones due to mismatched oscillators (i.e., frequency offset), Doppler shift, and timing synchronization errors, resulting in inter-carrier interference and loss of orthogonality. It can be seen that the ICI increases with the increase of the OFDM symbol duration (or alternatively decrease of sub-carrier spacing) and the frequency offset. The effects of timing offset are typically less than that of the frequency offset, provided that the cyclic prefix is sufficiently large.

It can be shown that the ICI power can be calculated as a function of the generic Doppler power spectrum $\Lambda(\nu)$ as follows [14]:

$$P_{ICI} = \int_{-\nu_{max}}^{\nu_{max}} \Lambda(\nu)(1 - \mathrm{sinc}^2(T_s \nu)) d\nu \tag{9.17}$$

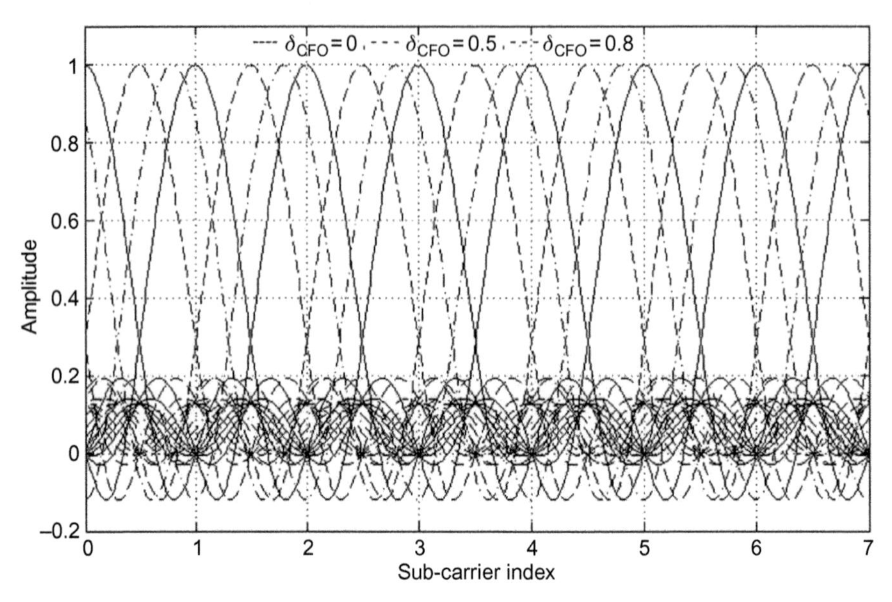

FIGURE 9.17

The effect of carrier frequency offset on OFDM signal for various δ_{CFO} values [159].

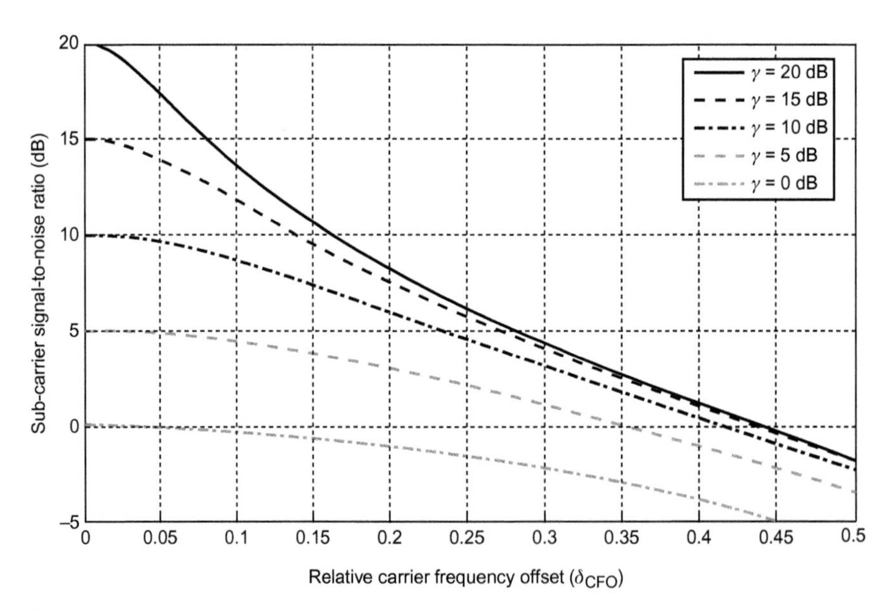

FIGURE 9.18

Variation of the SNR as a function of relative carrier frequency offset δ_{CFO} [159].

where ν_{max} denotes the maximum Doppler frequency. We further assume that the transmitted signal power is normalized. It can be noted that the ICI generated as a result of CFO is a special case of the above equation when $\Lambda(f) = \delta(f - f_{CFO})$, in which $\delta(f)$ represents the Dirac delta function. Using the classic Jakes' model of Doppler spread where the spaced-time correlation function is defined as $\Phi(t) = J_0(2\pi\nu t)$ in which $J_0(x)$ denotes the zero-order Bessel function of the first kind [131], the ICI power can be written as follows:

$$P_{\mathrm{ICI}_{\mathrm{Jakes}}} = 1 - 2\int_0^1 (1 - f)J_0(2\pi\nu_{max}T_s f)\mathrm{d}f \tag{9.18}$$

which approximately gives an upper bound on the ICI power due to Doppler spread. Comparison of the power of ICI generated by CFO and Doppler spread suggests that the ICI impairment due to the former is higher than the latter.

The effect of CFO on EVM for an OFDM system with $E_b/N_0 =$ 30 dB, $N_{FFT} = 64$ has also been investigated and is shown in Figure 9.19. It is shown that the EVM increases with increasing values of relative CFO even at such relatively high SNR.

Figure 9.20 shows the bit error rate sensitivity of an OFDM system under AWGN channel to CFO. The increasing values of relative CFO degrade the BER performance of the system.

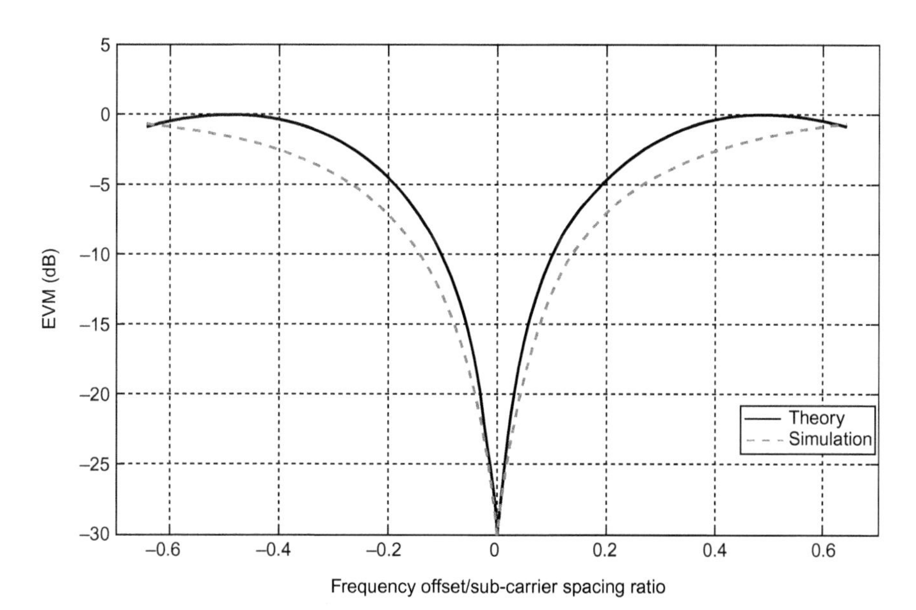

FIGURE 9.19

Error vector magnitude versus frequency offset in OFDM ($E_b/N_0 = 30$ dB, $N_{FFT} = 64$) [159].

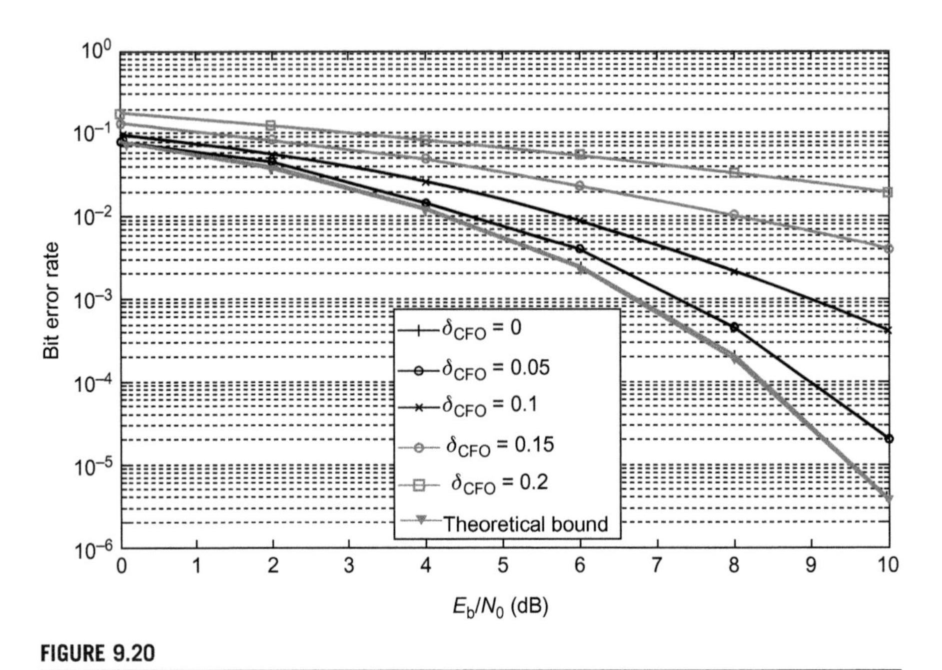

FIGURE 9.20

BER sensitivity to relative CFO δ_{CFO} in an OFDM system in AWGN channel [141].

9.4.4 Effect of oscillator phase noise on OFDM signals

Oscillators are used in typical radio circuits to drive the RF mixers which are utilized for the up-conversion/down-conversion of the band-pass signals. Ideally, the spectrum of the oscillator is expected to have an impulse at the frequency of oscillation with no frequency components elsewhere. However, the spectrum of a practical oscillator's output does exhibit spectrum variation around the oscillation frequency due to phase noise. The impact of local oscillator phase noise on the performance of an OFDM system has been extensively studied in the literature [159]. It has been shown that phase noise may have significant effects on OFDM signals with small sub-carrier spacing (i.e., large OFDM symbol duration in time). Long symbol duration is required for implementing a long guard interval that can mitigate long multipath delay in single frequency operation without excessive reduction of data throughput. The studies suggest that phase noise in OFDM systems can result in two effects: a common sub-carrier phase rotation on all the sub-carriers; and a thermal-noise-like sub-carrier de-orthogonality.

The common phase error, i.e., constellation rotation, on the demodulated sub-carriers is caused by the phase noise spectrum from DC up to the frequency of sub-carrier spacing. This low-pass effect is due to the long integration time of the OFDM symbol duration. This phase error can in principle be corrected by using pilots within the same symbol (in-band pilots). The phase error causes sub-carrier constellation blurring rather than rotation. It results from the phase noise spectrum contained within the system bandwidth. This part of the phase noise is more crucial, since it cannot be easily corrected. The SNR degradation caused by the common phase error can be quantified as $S/N_{\text{phase-rotation}} = 1/(I(\alpha)\beta\,\Delta f)$, where Δf is the sub-carrier spacing, β denotes the upper bound of the phase noise spectral mask, α is the ratio of the equivalent spectrum mask noise bandwidth and sub-carrier spacing, and $I(\alpha) = \int_{-\alpha}^{+\alpha} \text{sinc}(\pi x)\text{d}x$ with $I(0.5) = 0.774$, $I(1) = 0.903$, and $I(\infty) = 0.774$. It can be seen that, when $\alpha > 1$, the common phase error decreases as the sub-carrier spacing decreases [52].

To mathematically model the effect of the phase noise, let us consider the noisy output of an oscillator which contains phase noise $\phi(t)$ as follows $v(t) = A_c \cos(\omega_c t + \phi(t))$. Let us further assume that the stochastic variation of the phase can be modeled as the output of a system with a step function impulse response and input perturbation $n(t)$ as follows: $\phi(t) = \int_{-\infty}^{t} n(\tau)\text{d}\tau$. Based on the above assumption, the single-sided power spectral density of the phase can be written as follows: $S_\phi(f) = S_n(f)/(2\pi f)^2$, in which $S_n(f)$ denotes the noise power spectral density function. As an example, if $S_n(f)$ is modeled as white noise, then $S_\phi(f) \approx f^{-2}$, and if $S_n(f)$ is modeled as flicker noise, then $S_\phi(f) \approx f^{-3}$. Considering that the power spectral density of the phase is difficult to observe, one may alternatively look at the power spectral density of the oscillator's noisy output $v(t)$. It can be shown that the power spectral density of $v(t)$ can be calculated as follows [159]:

$$S_v(f) = \sum_{k=-\infty}^{k=+\infty} \left(\frac{a_k^2 + b_k^2}{2}\right) \frac{\beta k^2 f_c^2}{\beta^2 \pi^2 k^4 f_c^4 + (f - k f_c)^2} \tag{9.19}$$

where $\{a_k\}$ and $\{b_k\}$ denote the Fourier series coefficients of $v(t)$ and β is a constant. Given that we are only interested in evaluating $S_v(f)$ at f_c, the above equation can be simplified as follows:

$$S_v(f) = \left(\frac{a_1^2 + b_1^2}{2}\right) \frac{\beta f_c^2}{\beta^2 \pi^2 f_c^4 + (f - f_c)^2} \tag{9.20}$$

The above function is a Lorentz distribution.[8] We now define function $\Omega(f)$ as the ratio of noise power in the 1 Hz bandwidth at offset f from center frequency to carrier power which is expressed in dBc/Hz. As theoretically expected, having a higher phase noise in the signal does not increase the total power. A signal with higher phase noise will have smaller power near f_c and will have a broader spectrum around the center frequency. Conversely, a signal with lower phase noise

[8]The Cauchy distribution is a continuous probability distribution which is also known as the Lorentz distribution, Cauchy–Lorentz distribution, or Lorentzian function. It describes the distribution of a random variable that is the ratio of two independent standard normal random variables and has the probability density function defined as $f(x; 0, 1) = 1/[\pi(1 + x^2)]$ [30].

has a sharper peak at the center frequency with less deviation. Therefore, $\Omega(f)$ can be expressed as follows:

$$\Omega(f) = \frac{1}{\pi}\frac{\gamma}{\gamma^2 + (f - f_c)^2} \qquad \gamma = \beta\pi f_c^2 \tag{9.21}$$

It can be shown that $\int_{-\infty}^{+\infty}\Omega(f)df = 1$ [159]. Figure 9.21 illustrates the ratio of the noise power to carrier power at $f_c = 500$ Hz for different values of β. It can be seen from the figure, for higher values of β, the spectrum becomes wider with smaller magnitude of the main lobe of the spectrum. Note that a wider main lobe does not increase the total power of the carrier.

In summary, an OFDM system has the following advantages [52]:

- It is sufficiently flexible in meeting various design requirements such as low complexity, bandwidth efficiency and scalability, spectrum shaping, and low sensitivity to various impairments.
- OFDM is well suited to multi-antenna transmission schemes.
- It requires no adaptation to instantaneous channel response for low-order modulations. It is robust to impulse interference and channel variations. Channel estimation is required for high-order QAM modulations.
- As a multi-tone transmission technique, OFDM is less sensitive to shift in sampling time in comparison to serial transmission techniques.

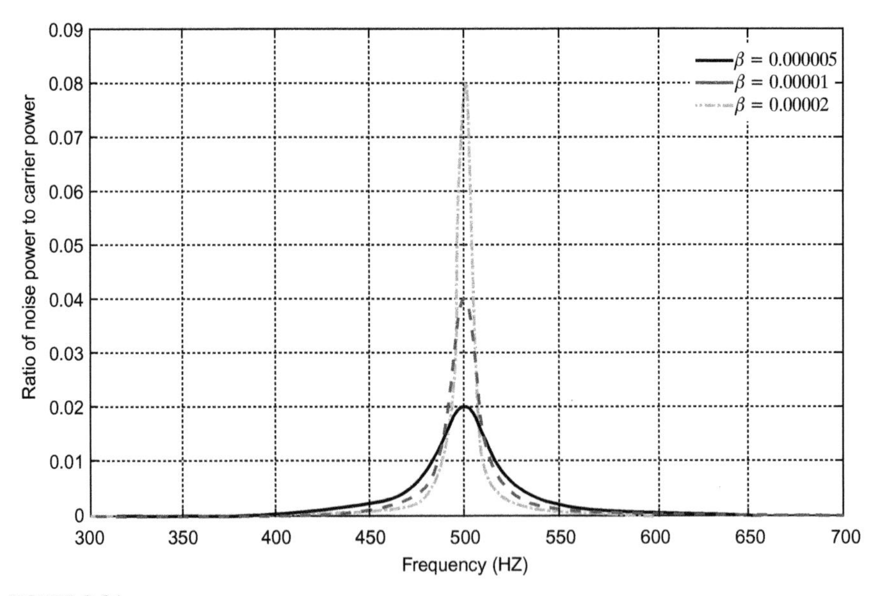

FIGURE 9.21

Plot of the noise power to carrier power $\Omega(f)$ at $f_c = 500$ Hz for different values of β [159].

- The bandwidth efficiency of an OFDM system approaches the Nyquist rate as the FFT size increases (this, however, increases system complexity and is subject to an upper limit due to channel variations and CFO).
- For fixed bandwidth efficiency, the complexity (multiplications per symbol) of an FFT-based OFDM system grows logarithmically with the increase of the channel multipath delay spread, in comparison with a linear increase of complexity for equalizers for a single carrier system.
- A properly coded and interleaved OFDM system can exceed the performance of many other practical systems. Higher transmitter output back-off is required in comparison to a single carrier system, because of the high PAPR of an OFDM system.
- As a parallel transmission technique, OFDM is more sensitive to CFO and tone interference than that of a single carrier system.

9.5 Multiple access schemes

Orthogonal frequency division multiple access (OFDMA) is the multi-user variant of the OFDM scheme where multiple access is achieved by assigning subsets of sub-carriers to different users, allowing simultaneous data transmission from several users. In OFDMA, the radio resources are two-dimensional (2D) regions over time (an integer number of OFDM symbols) and frequency (a number of contiguous or non-contiguous sub-carriers). The difference between OFDM and OFDMA is illustrated in Figure 9.22. Similar to OFDM, OFDMA employs multiple closely spaced sub-carriers, but the sub-carriers are divided into groups of sub-carriers where each group is called a resource block. The grouping of sub-carriers into groups of resource blocks is referred to as sub-channelization. The sub-carriers that form a resource block do not need to be physically adjacent. In the downlink, a resource block may be allocated to different users. In the uplink, a user may be assigned to one or more resource blocks.

Sub-channelization defines sub-channels that can be allocated to mobile stations depending on their channel conditions and service requirements. Using sub-channelization, within the same time slot (i.e., an integer number of OFDM symbols), an OFDMA system can allocate more transmit power to user devices with lower SNR, and less power to user devices with higher SNR. Sub-channelization also enables the base station to allocate higher power to sub-channels assigned to indoor mobile terminals, resulting in better indoor coverage.

There are various multiple access schemes that have been used in cellular systems in the past few years which allow the network to share the available radio resources (i.e., time, frequency, code, and space) among a number of users in the cell in the downlink and uplink directions. Figure 9.23 illustrates the concept of resource sharing in various multiple access schemes. As mentioned earlier, OFDMA has been a promising multiple-access scheme that has recently been

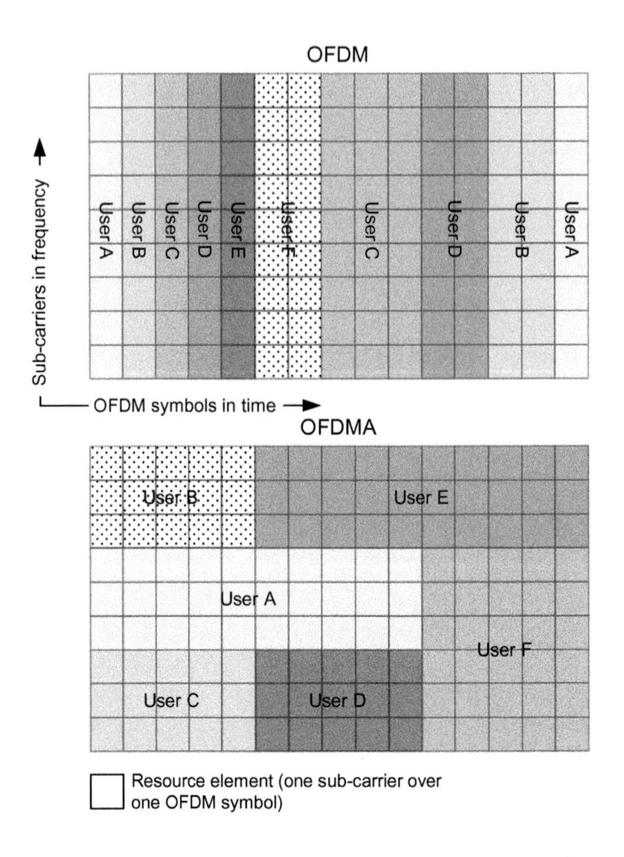

FIGURE 9.22

Illustration of OFDM and OFDMA concepts.

used in mobile broadband radio access technologies. The E-UTRA uses OFDMA and SC-FDMA as the multiple access schemes in the downlink and uplink directions, respectively. The OFDMA/SC-FDMA parameters for the E-UTRA have been chosen to facilitate coexistence with the UTRA on the same platform and sharing the same temporal sampling frequency of 30.72 MHz. The E-UTRA OFDMA/SC-FDMA parameters are given in Table 9.2. The following parameters are compatible with UMTS sampling frequency/chip rate of 3.84 Mcps and frame timing.

The OFDMA/SC-FDMA parameters in E-UTRA satisfy general design criteria for OFDM systems as follows:

$$\begin{aligned} &T_{CP} \geq \tau_{RMS} && \text{(to prevent ISI due to delay spread)} \\ &\Delta f \gg \nu_{max} && \text{(to mitigate ICI due to Doppler spread)} \\ &T_{CP}\,\Delta f \ll 1 && \text{(for spectral efficiency improvement)} \end{aligned} \quad (9.22)$$

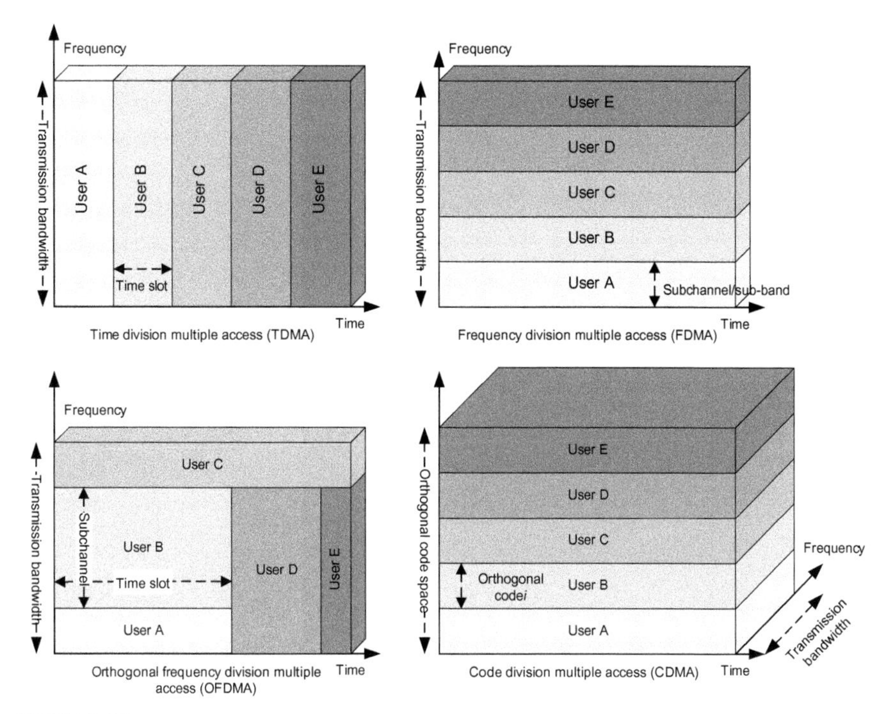

FIGURE 9.23

An illustration of the concept of different multiple access schemes.

In OFDMA, an OFDM symbol is constructed of sub-carriers, the number of which are determined by the FFT size. There are several sub-carrier types: (1) data sub-carriers are used for data transmission; (2) pilot or reference-signal sub-carriers are utilized for channel estimation and coherent detection; and (3) null sub-carriers are not used for pilot/data transmission. The null sub-carriers, including the DC sub-carrier, are used for guard bands. The number of used (or occupied) sub-carriers is always less than the FFT/IFFT size. The guard bands are used to allow spectrum sharing and to reduce the adjacent channel interference and out-of-band emissions. The sampling frequency is selected to be greater than or equal to the channel bandwidth. The number of time samples in a radio frame is always an integer, and to further simplify the design of analog transmit filter the sampling frequency is scaled by a factor greater than one (e.g., the sampling frequency for 20 MHz bandwidth is 30.72 MHz). As given in Table 9.2, the value of the over-sampling factor depends on the channel bandwidth. The E-UTRA supports a large number of channel bandwidths. The channel bandwidths that are integer multiples of 5 MHz are more important. The bandwidths that are not given

Table 9.2 E-UTRA OFDMA/SC-FDMA Parameters [3]

Parameter		Value					
Channel bandwidth (MHz)		1.4	3	5	10	15	20
Number of resource blocks $N_{RB}^{DL/UL}$		6	15	25	50	75	100
Number of occupied sub-carriers		72	180	300	600	900	1200
IFFT/FFT size		128	256	512	1024	1536	2048
Sub-carrier spacing Δf (kHz)		15 (7.5)					
Sampling frequency (MHz)		1.92	3.84	7.68	15.36	23.04	30.72
Samples/slot		960	1920	3840	7680	11520	15360
CP size T_{CP}	Normal CP ($\Delta f = 15$ kHz)	5.21 μs (first symbol of the slot) 4.69 μs (other symbols of the slot) $N_s = 7$ symbols/slot $N_{CP}(l) = \begin{cases} 160 & l = 0 \\ 144 & l = 1, 2, \ldots, 6 \end{cases}$					
	Extended CP ($\Delta f = 15$ kHz)	16.67 μs $N_s = 6$ symbols/slot $N_{CP}(l) = 512 \quad l = 0, 1, \ldots, 5$					
	Extended CP ($\Delta f = 7.5$ kHz)	33.33 μs $N_s = 3$ symbols/slot $N_{CP}(l) = 1024 \quad l = 0, 1, 2$					

in Table 9.2 are supported via tone dropping, i.e., dropping the sub-carriers at the edges of the frequency band based on 10 and 20 MHz systems (e.g., for 6 MHz channel bandwidth, the parameter of 10 MHz is used and the edge sub-carriers are dropped to fit the spectrum block).

9.6 Duplex modes

One of the key elements of any radio communication system is the way in which radio communications are maintained in both directions. For cellular systems, it is necessary to enable transmission of data in both directions simultaneously, and this places a number of constraints on the schemes that may be used to control the transmission flow. As a result, the duplex scheme forms a very basic part of the overall specification for a cellular or any radio communication system. The term duplex refers to the bidirectional communications between two devices. When unpaired spectrum, or alternatively the same RF carrier is used for downlink and uplink communications, the transmit/receive functions are time-multiplexed.

When paired spectrum, or alternatively two RF carriers are used for downlink and uplink communications, the transmit/receive functions are frequency-multiplexed. The FDD is a duplex scheme in which uplink and downlink transmissions occur simultaneously using different frequencies. The downlink and uplink frequencies are separated by a sufficiently large frequency offset (see Table 9.1 for various E-UTRA bands). For the FDD scheme to properly operate, it is necessary that the frequency separation, i.e., channel separation between the transmission and reception frequencies, be sufficient to prevent the receiver being excessively affected by the transmitter signal. Receiver blocking is an important issue with FDD schemes, and often highly selective filters may be required. For cellular systems using FDD, filters are required within the base station and also the user terminal to ensure sufficient isolation of the transmitter signal without desensitizing the receiver. While implementation cost is not a significant constraint for the base stations, placing a filter into the user terminals is more of an issue. The use of an FDD system does enable simultaneous transmission and reception of signals. However, two channels are required and this may not always use the available spectrum efficiently. The spectrum used for FDD systems is allocated by the regulatory authorities. Since there is a frequency separation between the uplink and downlink frequencies, it is not normally possible to reallocate spectrum to change the balance between the capacity of the uplink and downlink directions, if there are changing capacity requirements for each direction.

The TDD is a duplex scheme where uplink and downlink transmissions occur at different times, but may share the same frequency. In other words, the downlink and uplink transmissions are multiplexed in time and are not concurrent. While FDD transmissions require a large frequency separation between the transmit and receive frequencies, TDD schemes require a guard time or guard interval between transmission and reception. This must be sufficient to allow the signals traveling from the remote transmitter to arrive before a transmission is started and the receiver is shut down. Although this delay is relatively short, when switching between transmission and reception many times a second, even a small guard time can reduce the efficiency of the system, because a percentage of the time must be used for the guard interval. For small-sized cells, e.g., up to a mile or so, the guard interval is normally small and acceptable. For large cell sizes, it may become an issue.

The guard interval required for TDD will comprise two main elements: (1) a time allowance for the propagation delay for any transmission from a remote transmitter to arrive at the receiver. This will depend upon the distances involved (i.e., cell radius). (2) A time allowance for the transceiver to switch from receive to transmit mode. The switching times can vary considerably depending on the implementation, but can take a few microseconds. As a result, TDD is not normally suitable for use over very large cell sizes as the guard time increases and the spectral efficiency decreases.

It is often found that traffic in both directions is not balanced. Typically there is more data transmitted in the downlink direction of a cellular system. This means that, ideally, the capacity should be greater in the downlink direction. Using a TDD system, it is possible to change the capacity in either direction relatively easily by changing the number of time slots allocated to each direction. Often this is dynamically configurable so it can be altered to meet the demand. A further aspect to be noted with TDD transmission is that of latency. Since the data may not be immediately transmitted, a result of the time multiplexing between transmit and receive circuitry, there may be a small delay between the time that the data is generated until it is actually transmitted in a TDD system. Typically this may be a few milliseconds depending upon the frame timing. Both TDD and FDD have their advantages, and each can be used in different deployment scenarios. Before deciding on a particular type of duplex scheme, it is necessary to analyze the advantages and disadvantages of each duplex mode. Table 9.3 summarizes the relative advantages/disadvantages of TDD and FDD systems.

The E-UTRA supports TDD and FDD schemes with a great extent of commonality in the baseband processing. In order to reduce the implementation complexity and cost of FDD terminals, and further to increase the reuse of baseband functional elements, a half-duplex FDD (H-FDD) operation is supported where the downlink and uplink transmissions are not simultaneous but occur in two different frequencies. As shown in Figure 9.24, classic H-FDD operation does not efficiently utilize the radio resources on the downlink and uplink RF carriers. The complementary grouping and scheduling of users would allow efficient use of downlink and uplink resources in an H-FDD operation. The various duplex schemes are illustrated in Figure 9.24. For H-FDD operation, a guard period (GP) is virtually created by the UE by not receiving the last OFDM symbol(s) of a downlink subframe immediately preceding an uplink subframe in which the UE is active. The length of the GP is the sum of the maximum round-trip propagation delay in the cell, transmit-to-receive, and receive-to-transmit switching delay at the UE.

In E-UTRA, the differences between TDD and FDD schemes are mainly in the physical layer, and therefore are transparent to the higher layers. As a result, there are no significant operational differences between the two modes in the system architecture. At the physical layer, the fundamental design goal was to achieve as much commonality between the two modes as possible. The key design differences between the two stem from the need to support various TDD UL/DL allocations and provide coexistence with other TDD systems. In this regard, several additional features are exclusively developed for E-UTRA TDD. For instance, the FDD and TDD modes differ in the time placement of the downlink synchronization signals. In FDD, the primary and secondary synchronization signals are contiguously placed within one subframe, while in TDD, the two signals are placed in different subframes, separated by two OFDM symbols. However, this separation of the primary and secondary synchronization signals does not affect the performance of synchronization channels relative to FDD systems.

The system-level simulations also show that the sector throughput and cell-edge throughput of E-UTRA TDD and FDD mode are quite similar for best effort traffic when comparing a 10 + 10 MHz FDD system to a 20 MHz TDD system. The spectral efficiency of the TDD mode is slightly degraded due to the

Table 9.3 Comparison of TDD and FDD Attributes

Attribute	TDD	FDD
Paired spectrum	Does not require paired spectrum as both transmit and receive occur on the same channel.	Requires paired spectrum with sufficient frequency separation to allow simultaneous transmission and reception.
Hardware cost	Lower cost as no duplexer[a] is needed to isolate the transmitter and receiver.	Duplexer is needed and the implementation cost is higher
Channel reciprocity	Channel propagation is the same in both directions which enables estimation of the downlink channel from the uplink channel.	Channel characteristics are different in both directions due to use of different frequencies.
UL/DL asymmetry	It is possible to dynamically change the UL and DL ratio based on the traffic volume in each direction.	UL/DL bandwidths are determined by frequency allocation designated by the regulatory authorities. It is therefore not possible to dynamically adapt to the traffic volume.
Guard period/ frequency separation	Guard period required to ensure uplink and downlink transmissions do not interfere. Large guard period will limit the capacity. Larger guard period is normally required if distances are increased to accommodate larger propagation delays. Note that a guard band in frequency-domain is required to suppress interference to adjacent bands.	Frequency separation is required to provide sufficient isolation between uplink and downlink. However, large frequency separation does not impact the capacity. Note that a guard band in frequency-domain is required to suppress interference to adjacent bands.
Discontinuous transmission	Discontinuous transmission is required to allow both uplink and downlink transmissions. This can degrade the performance of the RF power amplifier in the transmitter.	Continuous transmission is possible.
Switching point synchronization	Base stations are required to be synchronized with respect to the uplink and downlink transmission times. If neighboring base stations use different uplink and downlink assignments and share the same channel, then interference may occur between cells.	Not applicable.

[a]*A diplexer is a passive device that implements frequency domain multiplexing. Two ports are multiplexed onto a third port. The signals on each input port occupy non-overlapping frequency bands. Consequently, the signals on the input ports can coexist on the output port without interfering with each other. On the other hand, a duplexer is a device that allows bidirectional communication over a single channel. In radar and radio communications systems, the duplexer isolates the receiver from the transmitter while permitting them to share a common antenna. Most radio repeater systems include a duplexer. Duplexers are designed for operation in the frequency band used by the receiver and transmitter, and must be capable of handling the output power of the transmitter. They must provide sufficient isolation between transmitter and receiver to prevent receiver desensitization [34].*

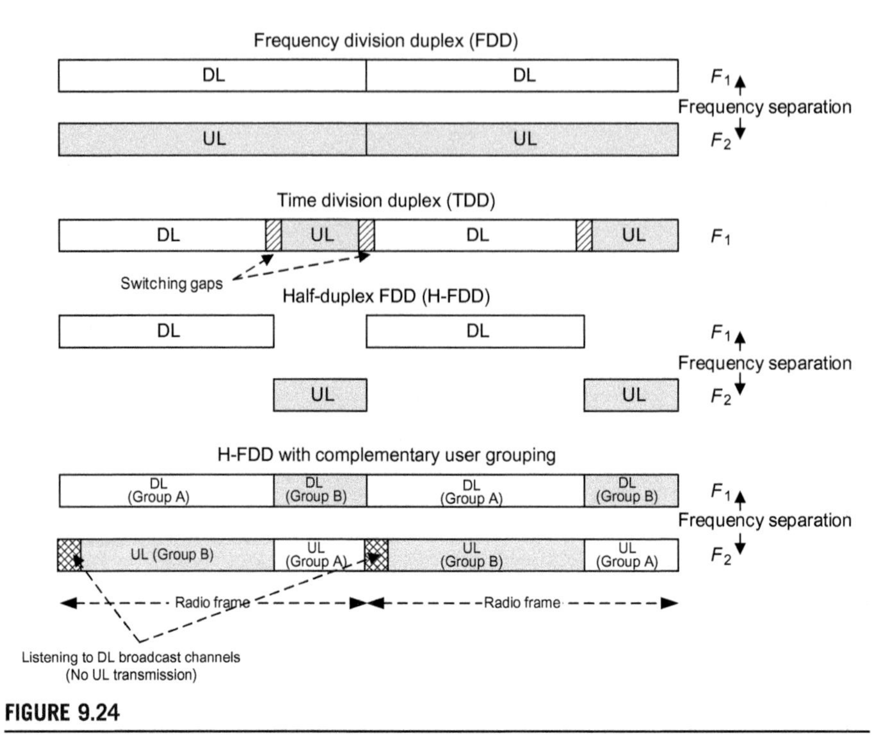

FIGURE 9.24

Illustration of the concept of different duplexing schemes.

additional overhead corresponding to the GP. Another advantage of TDD systems is the flexibility they may offer to network operators in terms of adjusting the DL/UL ratio. This feature allows operators to configure the DL/UL ratio to suit the user traffic in their network.

A common misperception regarding TDD systems is that the uplink coverage of the TDD systems is inferior relative to their FDD counterpart. The link-budget analysis for both systems in the outdoor and indoor environments with compatible system configuration and transmitter and receiver settings shows that the TDD system matches the uplink coverage of the FDD counterpart [10]. In fact, the E-UTRA TDD and FDD modes utilize the same slot/subframe structure and resource block size, thus the same amount of power is transmitted during the same time interval, implying that the coverage will be the same. What has been fueling this misconception is the single-user uplink data rate in an unloaded network at the cell-edge, where an FDD system will show a cell-edge performance that is twice that of the TDD system. It must be noted that the link budget analysis focuses only on the cell-edge single user performance when that user terminal is at the maximum limit of its uplink power, hence it can only be given access to five resource blocks (one resource block every 1 ms for 5 ms, or half of the length of the 10 ms E-UTRA radio frame with DL:UL = 1:1 ratio) to meet the 200 mW maximum power requirement. In comparison, in an FDD system at the cell-edge, one single device will be able to send 200 mW continuously for 10 ms, assuming that the scheduler can give the device access to 10 resource blocks of 1 ms with no waiting time. However, in real-life scenarios, the cells have more than one user and the performance of FDD and TDD systems will be very similar, as the FDD device will not be able to access the full-frame 10 uplink resource blocks similar to an unloaded network. In that case, because the TDD channel is twice as wide as the FDD channel, the user is likely to be given access to as many resource blocks as an FDD system despite the half-uplink time in the frame compared to an FDD frame. For network operators, an E-UTRA TDD will provide the same coverage and very similar sector throughput relative to an E-UTRA FDD system. The TDD with its adaptable DL:UL ratio would allow an optimized and efficient use of the uplink and downlink bandwidths as the operators are able to match the asymmetry of data traffic and to maximize spectrum utilization and efficiency in almost all scenarios [163].

9.7 Frame structure

Downlink and uplink transmissions are structured into radio frames with 10 ms duration. The E-UTRA supports two frame structures, namely, type 1 and type 2 for FDD and TDD duplex schemes, respectively. The radio frames are numbered using a system frame number (SFN). The SFN is defined using a string of 10 bits which provide an integer value from 0 to 1023. Therefore, the SFN repeats once every 10.24 s, i.e., 1024 frames. The eight most significant bits of the SFN are broadcast within the master information block (MIB) of the system information on the broadcast channel (BCH). The remaining bits of SFN are deduced from the four frame cycles used to transmit the MIB (i.e., the first radio frame corresponds to "00" and the fourth radio frame corresponds to "11"). As given in Table 9.2, using the normal cyclic prefix, there are seven OFDM symbols within each slot. In the case of multi-cell aggregation, the same frame structure is used across all serving cells. Transmissions in multiple cells can be aggregated up to four secondary cells (in this context cell refers to RF carriers) in addition to the primary cell.

To provide enhanced multicast and broadcast services, E-UTRA has the capability to transmit multicast and broadcast content over a single frequency network (MBSFN), where a time-synchronized common waveform is transmitted from multiple eNBs, allowing macro-diversity combining of multi-cell transmissions at the UE. The cyclic prefix is utilized to cover the difference in the propagation delays, which makes the MBSFN transmission appear to the UE as a transmission from a single large cell. In LTE, transmission on a dedicated carrier for MBSFN with the possibility to use a longer CP with a sub-carrier bandwidth of 7.5 kHz is supported, as well as transmission of MBSFN on a carrier with both MBMS and unicast transmissions using time division multiplexing [7].

9.7.1 Frame structure type 1

Frame structure type 1 is applicable to both full-duplex FDD and H-FDD schemes. In frame structure type 1, as illustrated in Figure 9.25, each 10 ms radio frame is divided into 10 equally-sized subframes ($T_{frame} = 307,200T_s = 10T_{subframe}$). Each subframe consists of two equally sized and consecutive 0.5 ms slots ($T_{slot} = 15360T_s$). Each slot in turn consists of a number of OFDM symbols which can be either seven (normal cyclic prefix) or six (extended cyclic prefix). As shown in Figure 9.25, the useful symbol time is $T_u = 2048T_s = 66.7\ \mu s$. For a normal cyclic prefix, the first symbol has a cyclic prefix lengths of $T_{CP} = 160T_s = 5.2\ \mu s$. The remaining six OFDM symbols have a cyclic prefix of length $T_{CP} = 144T_s = 4.7\ \mu s$. The reason for different cyclic prefix lengths of the first symbol is to make the overall slot length in terms of time units divisible by 15,360. For the extended mode, the cyclic prefix is $T_{CP} = 512T_s = 16.67\ \mu s$. The length of the cyclic prefix is longer than the delay spread of a few microseconds typically encountered in practice. The normal cyclic prefix is used in urban cells and high data rate applications, while the extended cyclic prefix is used in special cases like multi-cell broadcast and in very large cells (e.g., rural areas, low data rate

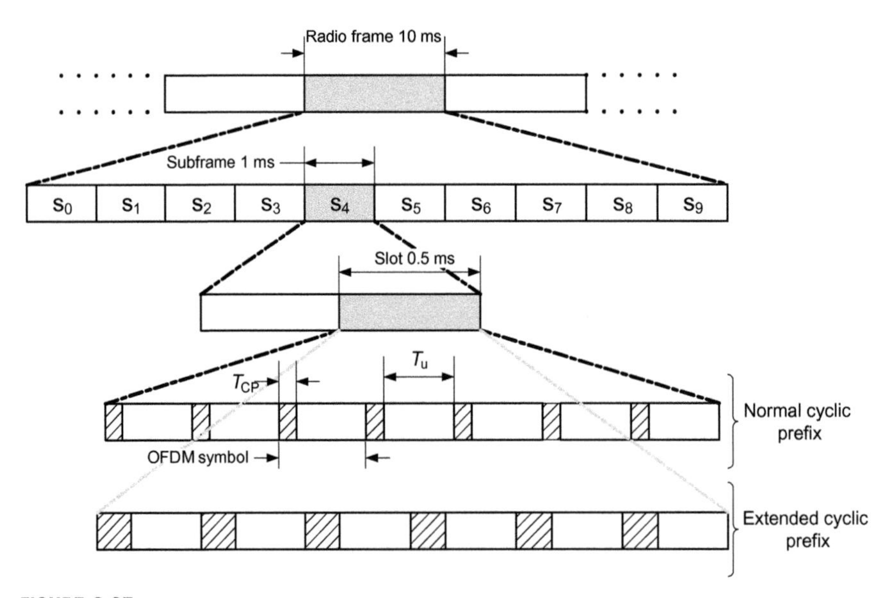

FIGURE 9.25

Frame structure type 1 with different cyclic prefix sizes [3,10].

applications). The cyclic prefix reduces the physical layer capacity by approximately 7.5% in the case of a normal cyclic prefix.

One way to reduce the relative overhead due to cyclic prefix insertion is to reduce the sub-carrier spacing with a corresponding increase in the OFDM symbol duration. However, this will increase the sensitivity of the OFDM transmission to frequency deviations resulting from fast channel variations and large Doppler spread, as well as different types of frequency errors due to RF circuitry. For FDD systems, 10 subframes are separately available for downlink and uplink transmissions in each 10 ms interval. Uplink and downlink transmissions are separated in the frequency-domain. In H-FDD operation, the UE cannot transmit and receive at the same time, while there are no such restrictions in full-duplex FDD.

The transmission time interval (TTI) of the E-UTRA downlink/uplink is 1 ms. The TTI is defined as the duration of the transmission of the physical layer encoded packets over the radio air-interface. In other words, the TTI refers to the length of an independently decodable transmission on the air-link. The data on a transport channel is organized into transport blocks. In each TTI, at most one transport block of variable size is transmitted over the radio air-interface to or from a terminal in the absence of spatial multiplexing. In the case of spatial multiplexing, up to two transport blocks per TTI are transmitted [3].

In the frequency-domain, the number of sub-carriers N_{FFT} ranges from 128 to 2048, depending on the channel bandwidth, with 512 and 1024 for 5 and 10 MHz, respectively, being most commonly used in practice. The sub-carrier spacing is 15 kHz. The sampling rate is chosen as $f_s = 15,000 N_{\mathrm{FFT}}$, resulting in a sampling rate that is an integer multiple or sub-multiple of the WCDMA chip rate of 3.84 Mcps. The OFDM parameters have been chosen such that FFT sizes and sampling rates are easily scalable for all operation bandwidths, while at the same time facilitating the implementation of multi-mode terminals with a common reference clock. Not all the sub-carriers are modulated or used. The DC sub-carrier is not used, as well as a number of sub-carriers on either side of the frequency band; thereby, approximately 10% of sub-carriers are used as guard sub-carriers.

9.7.2 Frame structure type 2

Frame structure type 2 is applied to a TDD duplex scheme where the downlink and uplink transmissions are time-multiplexed over the extent of a radio frame (Figure 9.26). Each radio frame of length 10 ms ($T_{\mathrm{frame}} = 307,200 T_s$) comprises two half-frames, where each half-frame consists of five subframes of length 1 ms. The supported downlink/uplink ratios (i.e., the ratio of downlink and uplink subframes in a frame) are listed in Table 9.4 and shown in Figure 9.27, where for each subframe in a radio frame, "DL" denotes the subframe reserved for downlink transmissions, "UL" denotes the subframe designated to uplink transmissions, and "S" denotes a special subframe comprising three fields downlink pilot time slot (DwPTS), GP, and uplink pilot time slot (UpPTS). The length of DwPTS and

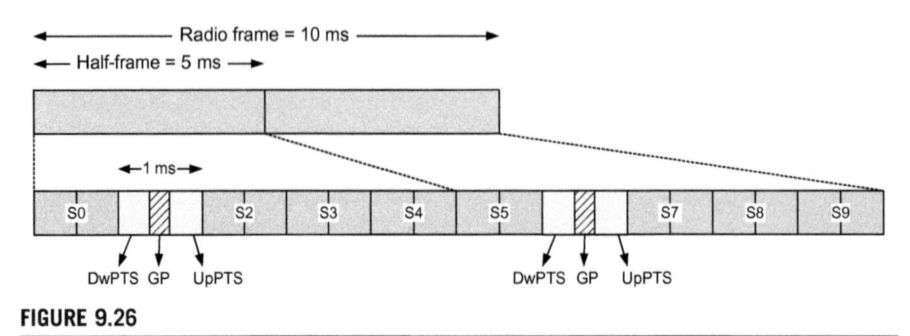

FIGURE 9.26

Frame structure type 2 [3].

Table 9.4 E-UTRA TDD Permissible DL/UL Ratios [3]

E-UTRA TDD Frame Configuration	Switching-Point Periodicity (ms)	Subframe Number									
		0	1	2	3	4	5	6	7	8	9
0	5	DL	S	UL	UL	UL	DL	S	UL	UL	UL
1	5	DL	S	UL	UL	DL	DL	S	UL	UL	DL
2	5	DL	S	UL	DL	DL	DL	S	UL	DL	DL
3	10	DL	S	UL	UL	UL	DL	DL	DL	DL	DL
4	10	DL	S	UL	UL	DL	DL	DL	DL	DL	DL
5	10	DL	S	UL	DL	DL	DL	DL	DL	DL	DL
6	5	DL	S	UL	UL	UL	DL	S	UL	UL	DL

UpPTS are given in Table 9.5, assuming that the total length of DwPTS, GP, and UpPTS is always 1 ms. Each subframe i is defined as two slots $2i$ and $2i+1$ of length 0.5 ms.

The E-UTRA TDD supports downlink/uplink ratios with both 5 and 10 ms downlink/uplink switching-point periodicity. In the case of 5 ms switching-point periodicity, there are two special subframes within one radio frame. Subframes 0 and 5 and DwPTS are always reserved for downlink transmission due to essential downlink physical channel mapping. The UpPTS interval and the subframe immediately following the special subframe are always reserved for uplink transmission. In case multiple cells are aggregated, and in order to minimize inter-cell interference, the UE may assume the same downlink/uplink ratio across all the cells and that the GPs of the special subframes in different cells have a minimum overlap of $1456T_s$. For the switching from uplink to downlink transmission, no special subframe is provisioned, but the GP includes the sum of switching times from DL to UL and UL to DL. The switching from UL to DL is achieved by provisioning an appropriate timing advance at the UE.

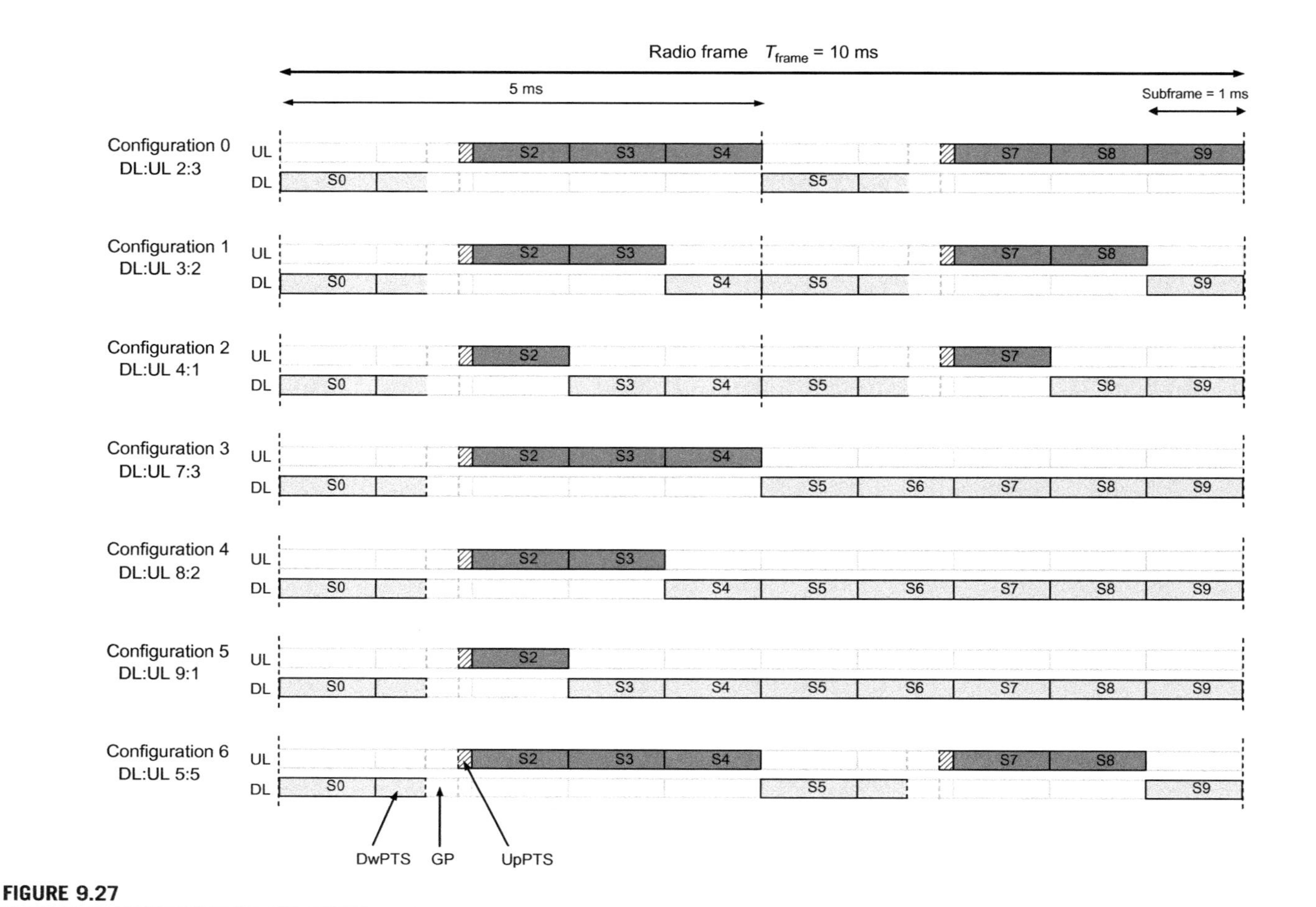

FIGURE 9.27

Illustration of various frame structure type-2 configurations [10].

Table 9.5 Various Special Subframe Configurations (DwPTS/GP/UpPTS) [3]

Special Subframe Configuration	Normal Cyclic Prefix in Downlink				Extended Cyclic Prefix in Downlink			
	DwPTS (Duration/ Number of OFDM Symbols)	GP (Number of OFDM Symbols)	UpPTS: Normal Cyclic Prefix in Uplink (Duration/ Number of OFDM Symbols)	UpPTS: Extended Cyclic Prefix in Uplink (Duration/ Number of OFDM Symbols)	DwPTS (Duration/ Number of OFDM Symbols)	GP (Number of OFDM Symbols)	UpPTS: Normal Cyclic Prefix in Uplink (Duration/ Number of OFDM Symbols)	UpPTS: Extended Cyclic Prefix in Uplink (Duration/ Number of OFDM Symbols)
0	$6592T_s$ 3	10			$7680T_s$ 3	8		
1	$19760T_s$ 9	4			$20480T_s$ 8	3		
2	$21952T_s$ 10	3	$2192T_s$ 1	$2560T_s$ 1	$23040T_s$ 9	2	$2192T_s$ 1	$2560T_s$ 1
3	$24144T_s$ 11	2			$25600T_s$ 10	1		
4	$26336T_s$ 12	1			$7680T_s$ 3	7		
5	$6592T_s$ 3	9			$20480T_s$ 8	2		
6	$19760T_s$ 9	3			$23040T_s$ 9	1	$4384T_s$ 2	$5120T_s$ 2
7	$21952T_s$ 10	2	$4384T_s$ 2	$5120T_s$ 2	$12800T_s$			
8	$24144T_s$ 11	1			— —	—	— —	— —
9	$13168T_s$ 6	6			— —	—	— —	— —

Within a radio frame, an E-UTRA TDD system switches multiple times between downlink and uplink transmissions. In this process, different signal propagation times between eNB and various UEs must be taken into consideration in order to prevent interference with other subframes. The timing advance feature prevents conflicts as a result of switching from uplink to downlink transmission. The UEs are informed by the eNB as to when they must start their uplink transmission. The greater the distance between the eNB and the UE, the earlier the UE starts transmitting. This helps to ensure that all signals reach the eNB with proper timing. When switching from the downlink to the uplink, a GP is inserted between the DwPTS and the UpPTS fields. The duration of the GP depends on the signal propagation round-trip time from the eNB to the UE, as well as on the time the UE requires to switch from receiving to transmitting states. The duration of the GP is configured by the network based on the cell size. Because the overall length of the special subframe remains constant, and the GP length varies based on cell size, the lengths of the DwPTS and UpPTS have to be adjusted. There are ten different special subframe configurations specified for E-UTRA TDD, as given in Table 9.5. The length of the GP determines the maximum supportable cell size. The E-UTRA should support cell sizes of up to 100 km, requiring a GP of approximately 666.7 μs. This is possible by choosing configuration 0 in Table 9.5 for the special subframe.

While the GP separates uplink and downlink transmissions, the other special fields are used for data/control signaling transmission. The DwPTS field carries synchronization and user data, as well as the downlink control channel for transmitting scheduling and control information. The UpPTS field is used for transmitting PRACH and SRS.

An example of coexistence with a legacy TD-SCDMA system [171] is shown in Figure 9.28, which illustrates switching point alignment between the two systems. It must be noted that in order to reduce the number of special subframe patterns in LTE TDD, only few legacy TD-SCDMA frame configurations are supported; therefore, the support is limited to the most common deployment scenarios, e.g., TD-SCDMA DL:UL = 5:2 and DL:UL = 4:3 configurations [163]. Despite the unequal size of the slots in TD-SCDMA and LTE TDD, the use of an

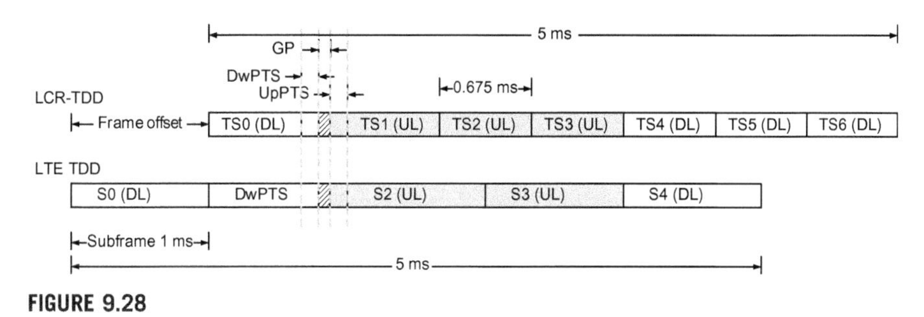

FIGURE 9.28

Example of LTE TDD coexistence with a TD-SCDMA system [163].

appropriate frame offset (as shown in Figure 9.28) allows alignment of the special subframes to the possible extent such that uplink and downlink transmissions of the two systems are synchronized [163].

9.7.3 MBSFN subframe structure

In E-UTRA frame structure, each downlink subframe and the DwPTS fields in the case of TDD is divided into a control region, consisting of the first OFDM symbols of the subframe, and a data region, comprising the remaining OFDM symbols of the subframe. A subset of the downlink subframes in a radio frame on a carrier supporting physical downlink shared channel (PDSCH) transmission can be configured as MBSFN subframes by higher layers. Each MBSFN subframe is divided into a non-MBSFN region and an MBSFN region. The non-MBSFN region spans the first one or two OFDM symbols in an MBSFN subframe. The MBSFN region in an MBSFN subframe is defined as the OFDM symbols not used for the non-MBSFN region.

The MBSFN subframes were originally designed to support MBSFN transmissions. However, they have also been found useful in other areas such as E-UTRA relaying functionality or the new carrier types. The MBSFN subframes are viewed as a generic tool, not necessarily related to MBSFN transmissions, which can provide time–frequency resources for enabling non-backward compatible functionalities relative to LTE Rel-8. An MBSFN subframe, as illustrated in Figure 9.29, consists of a control region of length one or two OFDM symbols, which is identical to normal subframes, followed by an MBSFN data region whose content will depend on the usage of the MBSFN subframe. The reason for maintaining the same control region in the MBSFN subframes is in consideration for normal

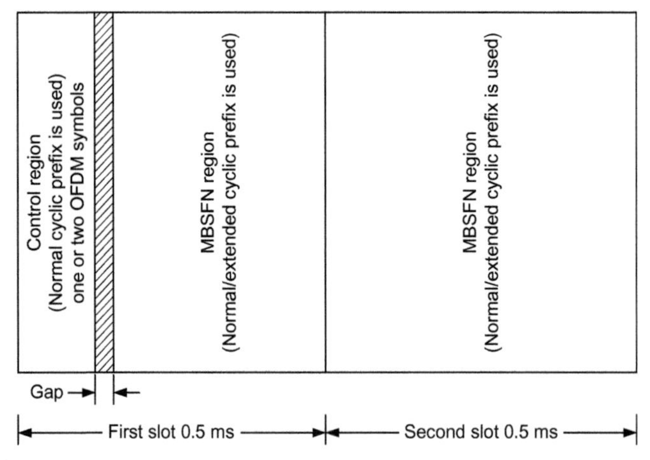

FIGURE 9.29

MBSFN subframe structure [15].

operation of LTE Rel-8 UEs, and in order to transmit control signaling necessary for uplink transmissions. All user terminals will be capable of receiving the control region of an MBSFN subframe. This is the main reason for which MBSFN subframes have been found useful for introducing new and non-backward compatible types of signaling and transmission modes in the new releases of LTE, without compromising the backward compatibility with earlier releases. The user terminals that are not capable of receiving transmissions in the MBSFN data region will simply ignore those transmissions.

Information about the set of subframes that are configured as MBSFN subframes in a cell is provided as part of the system information. In principle, an arbitrary pattern of MBSFN subframes can be configured with the pattern repeating after 40 ms. However, since physical signals and other essential information for system operation need to be transmitted in an orderly fashion to the UEs to find and attach to a cell, subframes where such information is mapped cannot be configured as MBSFN subframes. Therefore, subframes 0, 4, 5, and 9 for FDD and subframes 0, 1, 5, and 6 for TDD cannot be configured as MBSFN subframes, leaving the remaining six subframes as candidates for MBSFN subframes.

9.8 Resource structure, allocation, mapping

The minimum time–frequency resource unit used for downlink transmission is a resource element, which is defined as one sub-carrier over one OFDM symbol. For both TDD and FDD duplex schemes, a group of 12 sub-carriers contiguous in frequency over one slot in time form a PRB, i.e., a two-dimensional region corresponding to one slot in the time-domain and 180 kHz (i.e.,12×15 kHz) in the frequency-domain. The transmissions are allocated in units of PRB. More specifically, the data or control signaling in each slot is transmitted via one or several resource elements of $12 \times N_{RB}^{DL}$ sub-carriers over N_s^{DL} OFDM symbols. The value of the parameter N_{RB}^{DL} depends on the downlink transmission bandwidth configured for the cell, and is given as $\min(N_{RB}^{DL}) \leq N_{RB}^{DL} \leq \max(N_{RB}^{DL})$, where $\min(N_{RB}^{DL}) = 6$ for 1.4 MHz RF channels, and $\max(N_{RB}^{DL}) = 110$ for 20 MHz channels. The set of permissible values for N_{RB}^{DL} are given in Table 9.2. The number of OFDM symbols in a slot N_s^{DL} depends on the cyclic prefix length and the sub-carrier spacing configuration as given in Table 9.2.

There are a number of physical channels and physical signals that are mapped to predetermined (or configurable) locations in time and frequency over a radio frame that are used to assist proper operation of the system and the mobile terminals. The downlink physical channels correspond to a set of resource elements carrying data/control signaling information originating from higher protocol layers. The following downlink physical channels are defined in LTE/LTE-Advanced:

- Physical broadcast channel (PBCH): It carries the MIB portion of the system information. It is mapped to six resource blocks which are centered around

DC sub-carrier in subframe 0 and utilizes the resource elements which are not reserved for transmission of reference signals or other downlink control signaling. The BCH transport block is mapped to four subframes within a 40 ms interval.

- Physical control format indicator channel (PCFICH): It carries the size of PDCCH and is mapped to the first OFDM symbol in each downlink or special subframe. It contains the information on the number of OFDM symbols used for the downlink control region. The exact location of PCFICH is determined by the cell identity and system bandwidth.
- Physical hybrid-ARQ indicator channel (PHICH): It carries HARQ feedback corresponding to uplink data transmissions.
- Physical downlink control channel (PDCCH): It informs the UE(s) about transport format, resource allocation, HARQ feedback, and uplink grants. It is mapped to the first $L = 1, 2, \text{or } 3$ OFDM symbols in each downlink subframe. The number of the OFDM symbols for PDCCH is obtained via PCFICH. Multiple PDCCHs may be monitored by a UE. It also carries uplink power control commands.
- Physical downlink shared channel (PDSCH): It carriers downlink data transmission for different users as well as upper layer signaling.
- Physical multicast channel (PMCH): It is used to broadcast MBMS services.

In addition to the above physical channels, a set of downlink physical signals corresponding to a predetermined set of resource elements are specified and used by the physical layer which do not carry information originating from higher layers but are extremely important for proper LTE/LTE-Advanced system operation. The downlink physical signals include a variety of reference signals (also known as pilots which allow the UE to estimate the communication channel and coherently detect/decode data and control information that can be either cell-specific or UE-specific) and synchronization signals that provide timing and frequency synchronization, as well as physical identity of the cell to UEs attempting to select/reselect the cell. Figure 9.30 depicts the location of the downlink physical channels and signals in time and frequency within a type 1 frame structure.

9.8.1 Physical and virtual resource blocks

Each resource element is uniquely identified by the index pair (k, l) in a slot where $k = 0, \ldots, 12 \times N_{\mathrm{RB}}^{\mathrm{DL}} - 1$ and $l = 0, \ldots, N_{\mathrm{s}}^{\mathrm{DL}} - 1$ are the indices corresponding to frequency and time dimensions, respectively. Throughout this book, we assume that LTE/LTE-Advanced sub-carrier spacing is 15 kHz and there are 12 sub-carriers in each physical and virtual resource block (VRB) per slot. The structure of resource blocks in the downlink is shown in Figure 9.31. LTE/LTE-Advanced defines two types of resource blocks: physical and VRBs where in the latter, the resource blocks over a slot are permuted across frequency dimension to

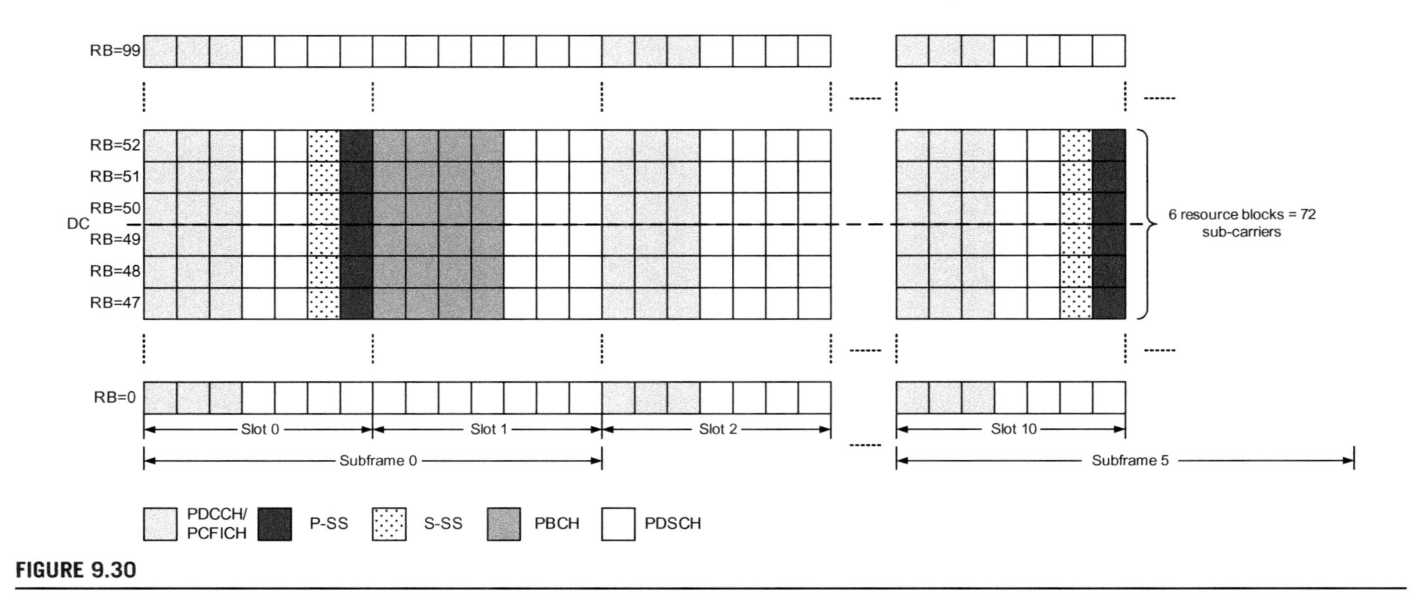

FIGURE 9.30

Physical resource structure and channel mapping (FDD, $N_s^{DL} = 7$, $N_{RB}^{DL} = 100$) [158].

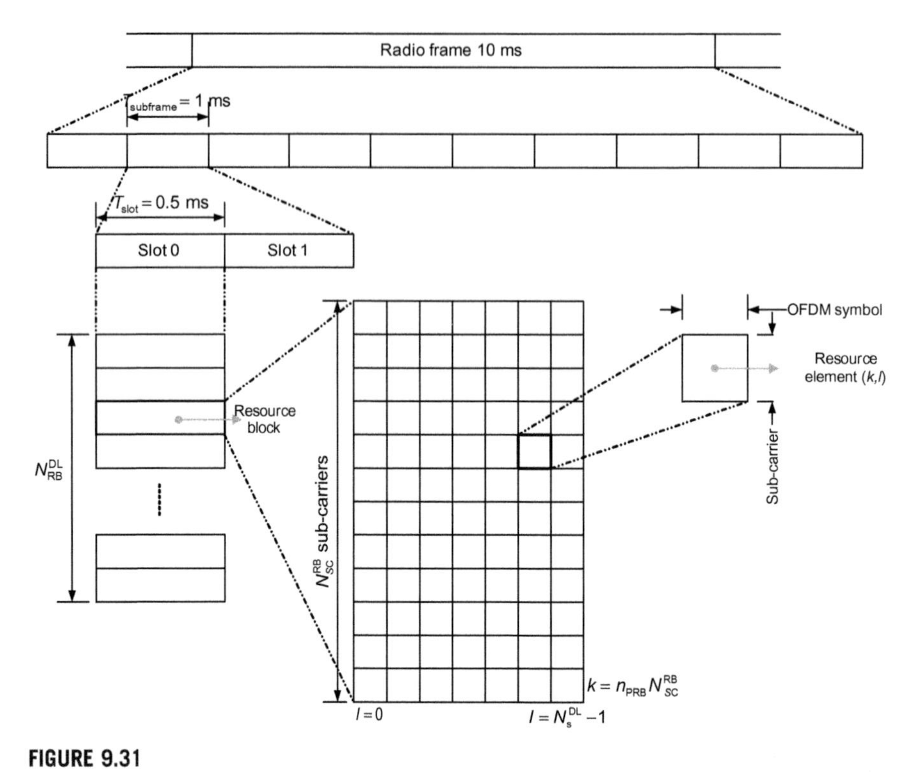

FIGURE 9.31

Resource block structure in time and frequency dimensions [3,162].

take advantage of frequency diversity. Another unit of resource allocation in the downlink is resource block group (RBG), which is a group of resource blocks. Allocating resources in RBG quanta reduces signaling overhead by reducing the number of bits required to uniquely address and identify each RBG relative to individual resource block identification/addressing. The size and number of RBGs depend on the channel bandwidth.

A PRB is defined as N_s^{DL} consecutive OFDM symbols in the time-domain and 12 (or 24 in the case of 7.5 kHz sub-carrier spacing) consecutive sub-carriers in the frequency-domain. A PRB consists of $12 \times N_s^{DL}$ resource elements corresponding to one slot in the time-domain and 180 kHz in the frequency-domain. The PRBs are numbered from 0 to $N_{RB}^{DL} - 1$ in the frequency-domain (see Figure 9.31). The relation between the PRB number n_{PRB} in the frequency-domain and resource elements (k, l) in a slot is given by $n_{PRB} = \lfloor k/12 \rfloor$.

A VRB has the same size as a PRB. There are two types of VRBs: VRBs of localized type and VRBs of distributed type. For each type of VRB, a pair of VRBs over two slots in a subframe is assigned together by a single VRB number n_{VRB}. VRBs of localized type are mapped directly to PRBs such that VRB index

n_{VRB} is identical to PRB index n_{PRB}. The VRBs are numbered from 0 to $N_{\mathrm{VRB}}^{\mathrm{DL}} - 1$, where $N_{\mathrm{VRB}}^{\mathrm{DL}} = N_{\mathrm{RB}}^{\mathrm{DL}}$. In the localized allocations, there is a one-to-one mapping between virtual and PRBs, whereas in the distributed case of resource allocation, the VRB numbers are mapped to PRB numbers according to a certain rule and inter-slot hopping is further utilized. In the latter, the first part of a VRB pair is mapped to one PRB and the other part of the VRB pair is mapped to another PRB which is at a predefined gap distance (which causes the inter-slot hopping) in order to achieve frequency diversity gain. This mechanism is especially important for small resource block allocations due to inherently less frequency diversity.

The distributed VRBs are mapped to PRBs such that consecutive VRBs are not mapped to sequential PRBs in the frequency-domain. A single VRB pair may be further split in the frequency-domain. Allocation of distributed resources includes a mapping from logically consecutive VRB pairs to physically non-consecutive PRB pairs to provide frequency diversity among VRB pairs. This is performed by a virtual block-interleaver or permutation scheme which operates on resource block pairs. The allocation of distributed resources further includes splitting of each resource-block pair such that the two resource blocks of the resource-block pair are transmitted with a certain frequency gap which provides additional frequency diversity for a single VRB pair. The latter can be interpreted as the introduction of frequency hopping on a slot basis. We will see later in this section that the associated PDCCH downlink control information (DCI) indicates whether allocated VRBs are of a localized or distributed type in the case of type 2 resource allocations. Thus, it is possible to dynamically switch between distributed and localized allocation, and also mix distributed and localized transmission for different terminals within the same subframe. The exact size of the frequency gap in Figure 9.32 depends on the overall downlink system bandwidth as given in Table 9.6. The gaps have been chosen based on the following criteria: (1) the gap size should be in the order of half of the total bandwidth (in terms of the number of resource blocks) in order to provide sufficient frequency diversity gain; and (2) the gap size should be an integer multiple of P^2, where parameter P is the size of a RBG for resource allocation types 0 and 1. The reason for this constraint is to ensure that distributed transmission as described above and transmissions based on downlink allocation types 0 and 1 can coexist on the same subframe. Due to the constraint that the gap size should be an integer multiple of P^2, the gap size N_{gap} may deviate from precisely half of the system bandwidth when the number of resource blocks is an odd number. In those cases, only a fraction of available resource blocks within the system bandwidth can be used for distributed transmission. An example of distributed resource allocation is shown in Figure 9.32.

The permissible values of gap size N_{gap} are given in Table 9.6. Note that for $6 \le N_{\mathrm{RB}}^{\mathrm{DL}} \le 49$, only one gap value $N_{\mathrm{gap}} = N_{\mathrm{gap}}^{(1)}$ is defined, whereas for $50 \le N_{\mathrm{RB}}^{\mathrm{DL}} \le 110$, two gap values $N_{\mathrm{gap}}^{(1)}$ and $N_{\mathrm{gap}}^{(2)}$ are defined where the choice of the gap value is signaled as part of the downlink scheduling assignment. VRBs of

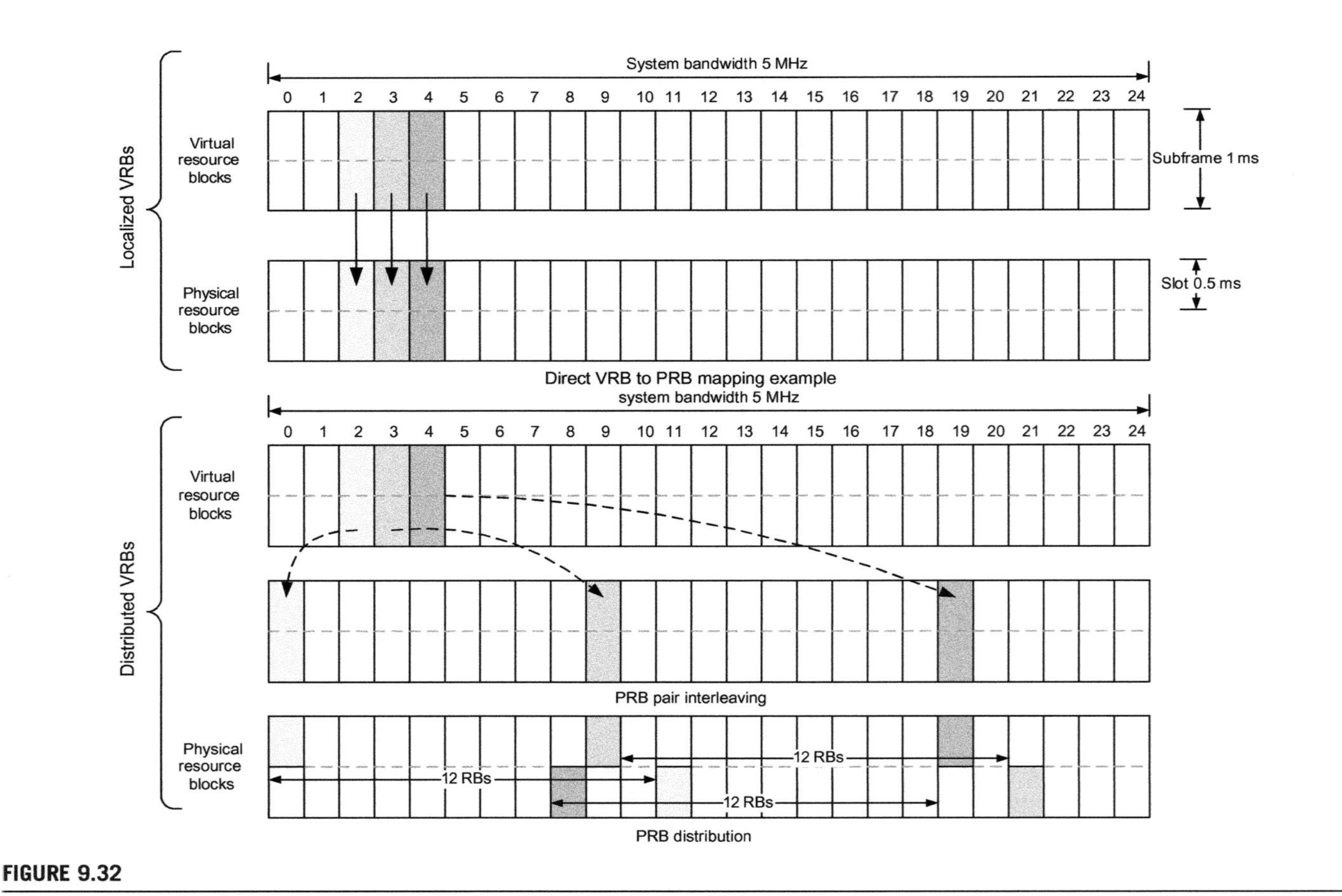

FIGURE 9.32

Illustration of localized and distributed VRB to PRB mapping ($N_{VRB}^{DL} = N_{RB}^{DL} = 25$) [15].

Table 9.6 Distributed VRB Gap Size as a Function of System Bandwidth [3,5]

Total Number of Resource Blocks N_{RB}^{DL}	Gap Size N_{gap}		Resource Block Group Size P
	$N_{gap}^{(1)}$	$N_{gap}^{(2)}$	
6–10	$\lceil N_{RB}^{DL}/2 \rceil$	–	1
11	4	–	2
12–19	8	–	2
20–26	12	–	2
27–44	18	–	3
45–49	27	–	3
50–63	27	9	3
64–79	32	16	4
80–110	48	16	4

distributed type are numbered from 0 to $N_{VRB}^{DL} - 1$, where $N_{VRB}^{DL} = 2\min(N_{gap}, N_{RB}^{DL} - N_{gap}^{(1)})$ or $N_{VRB}^{DL} = 2N_{gap}^{(2)} \left\lfloor N_{RB}^{DL}/2N_{gap}^{(2)} \right\rfloor$ depending on the value of N_{gap}.

Figure 9.33 illustrates the interleaving table for a 10 MHz channel bandwidth when using $N_{gap}^{(1)}$. The VRB indices are written to the table row-by-row and are sequentially read off the table column-by-column. The nulls in the table are ignored and do not generate outputs. When using $N_{gap}^{(1)}$, the VRBs associated with the first half of the output indices from the interleaving table are mapped to the lower set of PRBs starting from the first PRB. The VRBs corresponding to the second half of the output indices from the interleaving table are then mapped to the upper set of PRBs ending at the last PRB. This mapping is applied to the first slot of the subframe. The mapping is cyclically shifted by the gap value $N_{gap}^{(1)}$ for the second slot of the subframe [13]. As shown in Figure 9.34, when using $N_{gap}^{(2)}$ in a 10 MHz bandwidth, two interleaving tables each with six rows are used to define the mapping from VRBs to PRBs. There are some null entries that are inserted in each table [3].

To better understand the difference between the localized and distributed allocations, consider the example shown in Figure 9.35. The frequency response of a multipath fading channel varies with time and frequency due to path loss, shadowing, and user mobility effects. Let us assume that user A and user B are two users with different channel conditions. The base station receives channel quality reports (typically in the form of CINR or SINR measurements) from the mobile stations in the cell. The base station scheduler may adopt either of the following allocation schemes depending on the channel condition reports and other considerations that will be discussed later in this chapter. One allocation strategy is to allocate user A and user B in the sub-channels where the corresponding SINR is

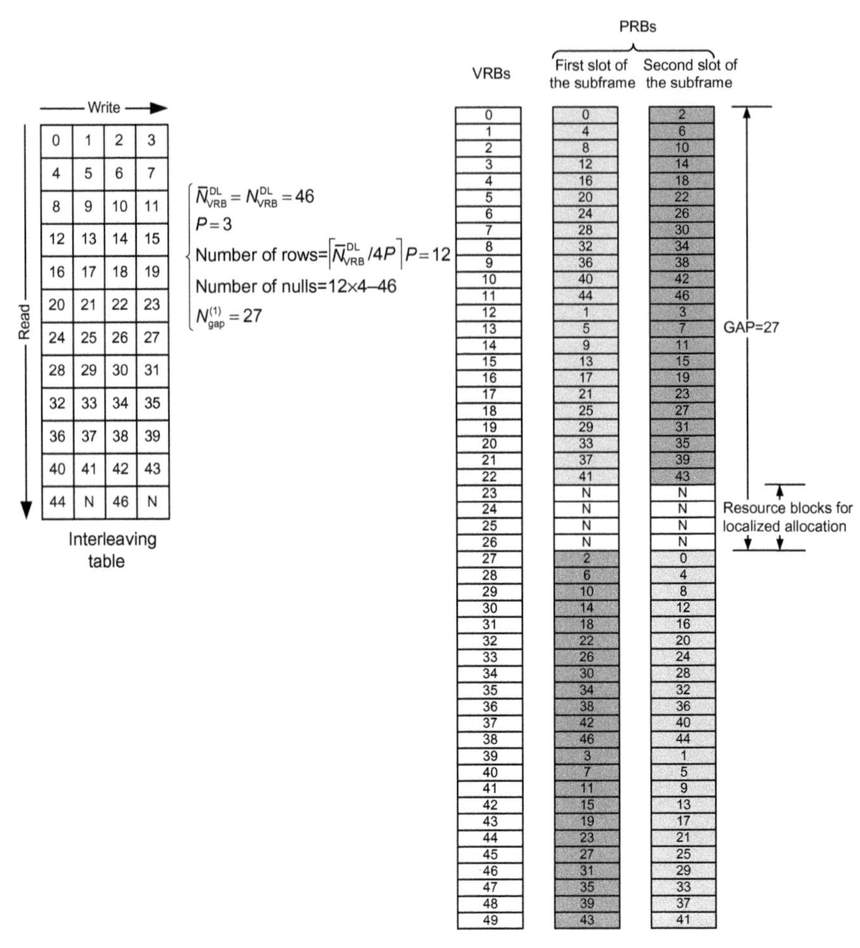

FIGURE 9.33

Illustration of distributed VRB to PRB mapping procedure in 10 MHz using $N_{gap}^{(1)}$ [13].

the best for that user. This is when a frequency-selective scheduling gain can be achieved through allocation of a group of physically adjacent sub-carriers in the sub-channels with relatively strong SINR for that user. Another strategy is to permute the sub-carriers over the entire channel, form a logical group of distributed sub-carriers and assign them to a user. In this case, frequency diversity gain can be achieved through use of distributed resource units for this user.

9.8.2 Resource allocation

One of the main design objectives for signaling of resource allocations, in terms of a set of resource blocks in each subframe, to the active UEs in the cell is to find a

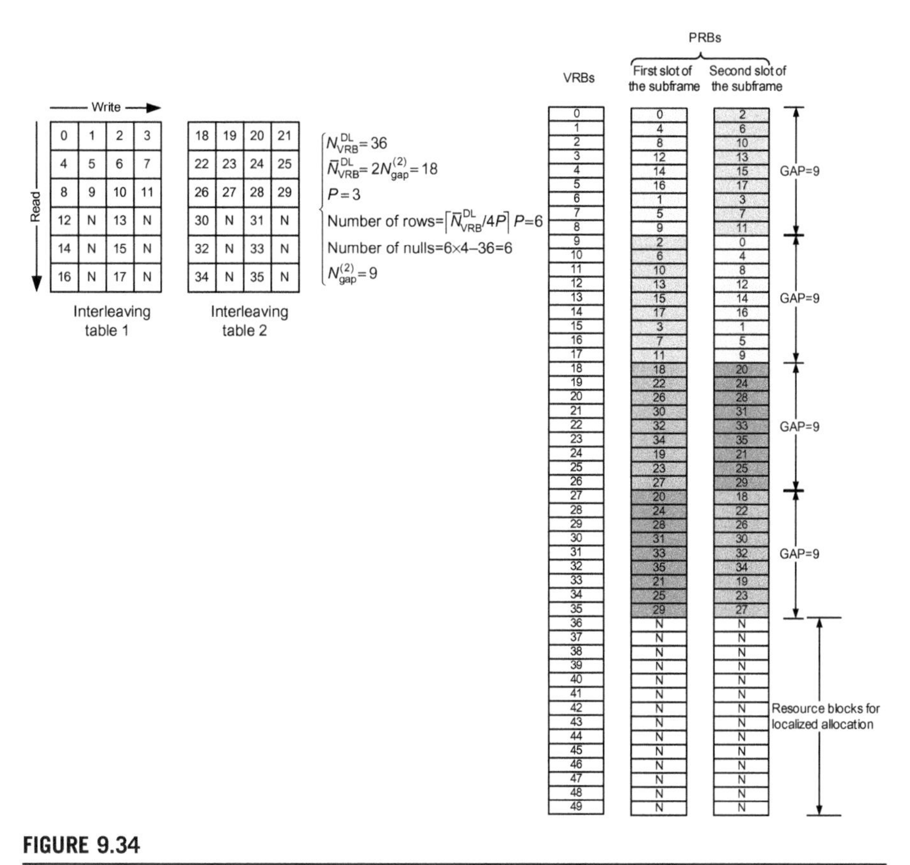

FIGURE 9.34

Illustration of distributed VRB to PRB mapping procedure in 10 MHz using $N_{gap}^{(2)}$ [13].

good trade-off between flexibility and signaling overhead. Indications of localized/distributed resource allocations to different UEs are transmitted via PDCCH. While the exact use of PDCCH depends on the scheduling algorithms implemented in the eNB, it is nevertheless possible to outline some general principles of this operation. The resource allocation field in downlink control information (carried via PDCCH or ePDCCH) is interpreted by the UE depending on the DCI format. A resource allocation field in each DCI includes two parts, a resource allocation header field and information pertaining to the actual resource block assignment. The DCI formats 1, 2, 2A, 2B, 2C and 2D with type 0, and DCI formats 1, 2, 2A, 2B, 2C and 2D with type 1 resource allocation (carried via PDCCH) have the same format and are distinguished from each other via the single bit resource allocation header field which exists depending on the downlink system bandwidth. The PDCCH with DCI format 1A, 1B, 1C, and 1D support type 2 resource allocation, while PDCCH with DCI format 1, 2, 2A, 2B, 2C and 2D support type 0 or type 1

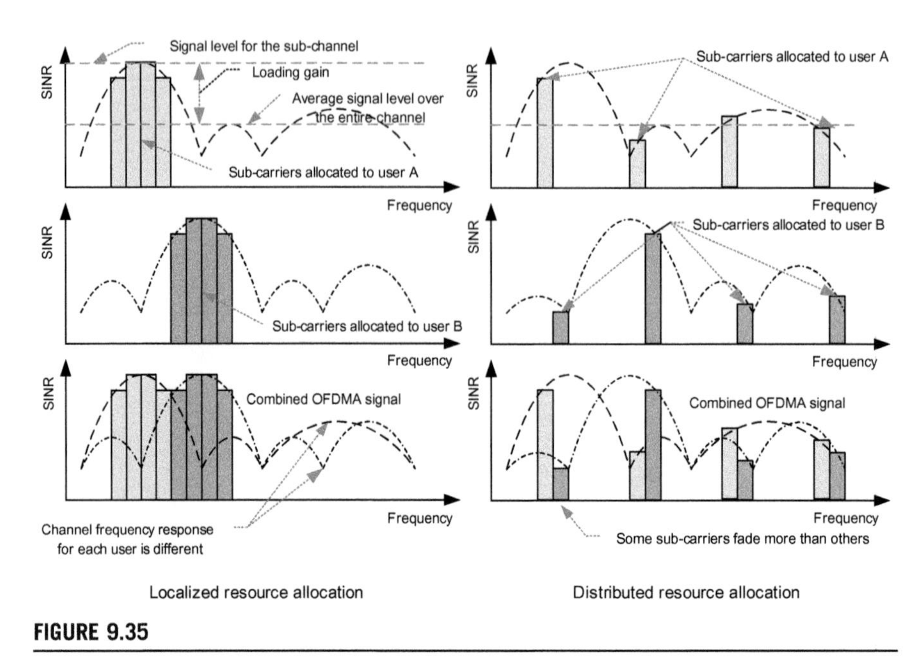

FIGURE 9.35

Comparison of localized and distributed allocation schemes (example) [18].

resource allocation. The DCI formats with a type 2 resource allocation do not have a resource allocation header field [5]. LTE/LTE-Advanced specify three resource allocation schemes in the downlink, where their usage depends on the number resources assigned to the UE, and whether a localized or distributed allocation must be used for the UE considering the channel conditions that the UE is experiencing. In LTE Rel-11, the DCI formats 1, 2, 2A, 2B, 2C and 2D with type 0 and DCI formats 1, 2, 2A, 2B, 2C and 2D with type 1 resource allocation (carried via ePDCCH) have the same format and are distinguished from each other via the single-bit resource allocation header field which exists depending on the downlink system bandwidth. The ePDCCH with DCI format 1A, 1B, and 1D have a type 2 resource allocation.

9.8.2.1 Resource allocation type 0

In resource allocation type 0, the resource block assignment information includes a bitmap indicating the RBGs that are allocated to the scheduled UE where an RBG is a set of consecutive VRBs of localized type. Note that in this case, the granularity of resource assignment is an RBG. The RBG size P is a function of the system bandwidth as given in Table 9.6. The total number of RBGs is given by $N_{RBG} = \lceil N_{RB}^{DL}/P \rceil$. The bitmap is of size N_{RBG} bits and each bit corresponds to a unique RBG in the set. The RBGs are indexed in the order of increasing frequency and non-increasing RBG sizes starting at the lowest frequency (i.e., RBG_0 to $RBG_{N_{RBG}-1}$ are mapped to the most-significant-bit and least-significant-bit of the bitmap, respectively). The RBG is allocated to a UE, if the corresponding bit value in the bitmap is set to "1." Therefore, each bit in the bitmap will select a small contiguous group of resource blocks whose size depends on the bandwidth. The maximum number of resource blocks assigned using type 0 allocation is the entire bandwidth. As an example, consider a total of 50 resource blocks (i.e., 10 MHz system bandwidth), the number of bits in the bitmap is 17. Each bit in the 17-bit bitmap selects a group of three consecutive resource blocks, with the exception of the last group which will only contain two resource blocks, as shown in Figure 9.36.

9.8.2.2 Resource allocation type 1

In resource allocation type 1, a bitmap indicates PRBs inside a selected RBG subset $0 \leq p < P$. The maximum number of resource blocks assignable through type 1 resource allocation is a subset of the entire bandwidth, i.e., a type 1 allocation, even with all bits in the bitmap set to "1" does not span the entire bandwidth. Each bit in the bitmap will select a single resource block from a set of contiguous group of resources whose size and separation depend on the total bandwidth. In resource allocation type 1, the resource block assignment signaling consists of three parts: (1) select a RBG subset; (2) whether to apply an offset when interpreting the bitmap using a shift flag; and (3) the bitmap that indicates to the UE the specific PRBs inside the selected RBG subset.

The size of the bitmap used to address VRBs in a selected RBG subset is given by $N_{RB}^{(1)} = \lceil N_{RB}^{DL}/P \rceil - \lceil \log_2(P) \rceil - 1$. The addressable VRBs within a selected RBG subset start from an offset $n_{shift}(p)$ to the smallest VRB number within the selected RBG subset. The offset is interpreted in terms of the number of VRBs, and is given by $n_{shift}(p) = N_{RB}^{RBG_{subset}}(p) - N_{RB}^{(1)}$ (if the value of the shift bit is non-zero) where the parameter $N_{RB}^{RBG_{subset}}(p)$ denotes the number of VRBs in RBG subset p and is given as follows [5]:

$$N_{RB}^{RBG_{subset}}(p) = \begin{cases} P\left(\left\lfloor \dfrac{N_{RB}^{DL}-1}{P^2} \right\rfloor + 1\right) & p < \left\lfloor \dfrac{N_{RB}^{DL}-1}{P} \right\rfloor \bmod P \\ P\left\lfloor \dfrac{N_{RB}^{DL}-1}{P^2} \right\rfloor + (N_{RB}^{DL}-1) \bmod P + 1 & p = \left\lfloor \dfrac{N_{RB}^{DL}-1}{P} \right\rfloor \bmod P \\ P\left\lfloor \dfrac{N_{RB}^{DL}-1}{P^2} \right\rfloor & p > \left\lfloor \dfrac{N_{RB}^{DL}-1}{P} \right\rfloor \bmod P \end{cases} \tag{9.23}$$

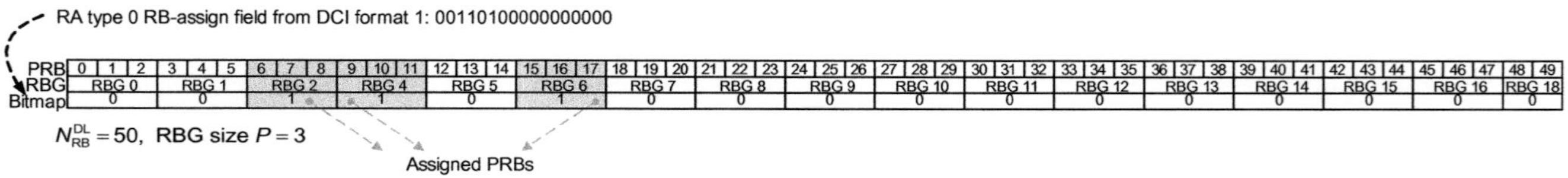

RA type 0 RB-assign field from DCI format 1: 00110100000000000

PRB 0 1 2 3 4 5 6 7 8 9 10 11 12 13 14 15 16 17 18 19 20 21 22 23 24 25 26 27 28 29 30 31 32 33 34 35 36 37 38 39 40 41 42 43 44 45 46 47 48 49

RBG: RBG 0, RBG 1, RBG 2, RBG 4, RBG 5, RBG 6, RBG 7, RBG 8, RBG 9, RBG 10, RBG 11, RBG 12, RBG 13, RBG 14, RBG 15, RBG 16, RBG 18

Bitmap: 0 0 1 1 0 1 0 0 0 0 0 0 0 0 0 0 0

$N_{RB}^{DL} = 50$, RBG size $P = 3$

Assigned PRBs

FIGURE 9.36

Example resource allocation type 0 [3,5].

When RBG subset p is addressed, the ith bit ($i = 0, 1, \ldots, N_{RB}^{(1)} - 1$) in the bitmap field corresponds to VRB number $n_{VRB}^{RBG_{subset}}(p) = \lfloor [i + n_{shift}(p)]/P \rfloor P^2 + pP + (i + n_{shift}(p)) \bmod P$. In comparison to resource allocation type 0, the bitmap size for type 1 is always shorter by $\lceil \log_2 P \rceil + 1$ number of bits. As an example, consider a total of 50 resource blocks, the number of bits in the bitmap is 14 (i.e., three bits shorter relative to type 0). Each bit in the 14-bit bitmap will select an individual resource block inside the selected subset, as demonstrated in Figure 9.37.

9.8.2.3 Resource allocation type 2

In resource allocation type 2, the resource block assignment information comprises a set of contiguously allocated localized VRBs or distributed VRBs. When the resource allocation is signaled in PDCCH/ePDCCH with DCI format 1A, 1B, or 1D, a 1-bit flag indicates whether localized VRBs or distributed VRBs are assigned (value "0" indicates localized and value "1" indicates distributed VRB assignment), while distributed VRBs are always assigned in the case of resource allocation signaled with DCI format 1C. A localized VRB assigned to a UE can vary from a single VRB up to a maximum number of VRBs over the system bandwidth. For DCI format 1A, the distributed VRBs allocated to a UE may vary from a single VRB up to N_{VRB}^{DL}, if the DCI CRC bits are scrambled with P-RNTI, RA-RNTI, or SI-RNTI. With DCI format 1B, 1D, and the CRC scrambled with C-RNTI, or with DCI format 1A and the CRC scrambled with C-RNTI, SPS C-RNTI or temporary C-RNTI, distributed VRB allocations allocated to a UE vary from a single VRB up to N_{VRB}^{DL} VRBs, if $6 \leq N_{RB}^{DL} < 50$, and vary from a single VRB up to 16 if $50 \leq N_{RB}^{DL} < 110$. Using DCI format 1C, the number of distributed VRB allocations for a UE may range from N_{RB}^{step} up to $\lfloor N_{VRB}^{DL}/N_{RB}^{step} \rfloor N_{RB}^{step}$ with an increment of N_{RB}^{step}, where $N_{RB}^{step} = 2$ if $6 \leq N_{RB}^{DL} < 50$, and $N_{RB}^{step} = 4$ if $50 \leq N_{RB}^{DL} < 110$ [5].

For PDCCH/ePDCCH with DCI format 1A, 1B, or 1D, the type 2 resource allocation field consists of a parameter referred to as resource indication value (RIV) which corresponds to the starting resource block index RB_{start} and the length of the allocation in terms of the number of virtually contiguous resource blocks L_{RB}. The RIV is defined by $RIV = N_{RB}^{DL}(L_{RB} - 1) + RB_{start}$ if $(L_{RB} - 1) \leq \lfloor N_{RB}^{DL}/2 \rfloor$; otherwise $RIV = N_{RB}^{DL}(N_{RB}^{DL} - L_{RB} + 1) + (N_{RB}^{DL} - 1 - RB_{start})$, where $1 \leq L_{RB} < N_{VRB}^{DL} - RB_{start}$. For DCI format 1C; however, the RIV in type 2 resource allocation field corresponds to starting resource block index $RB_{start} = \{0, N_{RB}^{step}, 2N_{RB}^{step}, \ldots, (\lfloor N_{VRB}^{DL}/N_{RB}^{step} \rfloor - 1)N_{RB}^{step}\}$ and the length in terms of virtually contiguous resource blocks $L_{RB} = \{N_{RB}^{step}, 2N_{RB}^{step}, \ldots, \lfloor N_{VRB}^{DL}/N_{RB}^{step} \rfloor N_{RB}^{step}\}$. In other words, in type 2 resource allocation, PRBs are not directly allocated. The type 2 allocation supports both localized and distributed VRB allocation which are distinguished by a flag. The information regarding the starting point of VRB and the length in terms of contiguously allocated VRB can be derived from RIV signaled within the DCI.

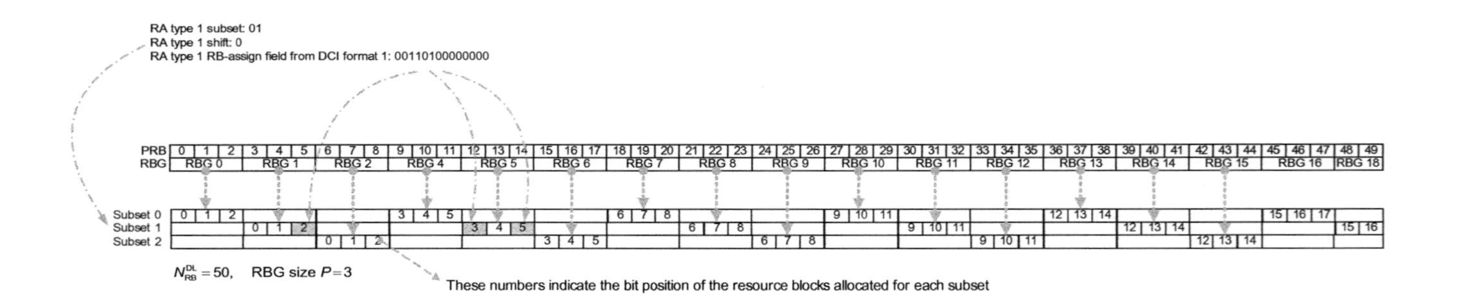

FIGURE 9.37

Example resource allocation type 1 [158].

As an example, consider a total of 50 resource blocks. Let us assume that the eNB is going to assign six localized VRBs starting from resource block 9 in the frequency-domain. Using the above equations for DCI format 1A, the RIV value is calculated as RIV = 259. This RIV is signaled in DCI format 1A, and the UE can unambiguously derive the starting resource block and the number of allocated resource blocks from RIV, as shown in Figure 9.38.

9.8.2.4 PRB bundling

When a UE is configured for operation in transmission mode 9, it may assume that the precoding granularity is multiple resource blocks in the frequency-domain when configured for PMI/RI reporting. In a given serving cell, if a UE is configured in transmission mode 10 and if PMI/RI is reported for all configured CSI processes, then the UE may assume that precoding granularity is multiple resource blocks; otherwise, the precoding granularity is going to be one resource block in the frequency-domain. Fixed bandwidth-dependent precoding resource block groups (PRGs) P' partition the available resource blocks and each PRG consists of a number of consecutive PRBs. If $N_{\mathrm{RB}}^{\mathrm{DL}} \bmod P' > 0$, then the size of one of the PRGs is adjusted to $N_{\mathrm{RB}}^{\mathrm{DL}} - P'\lfloor N_{\mathrm{RB}}^{\mathrm{DL}}/P' \rfloor$. The PRG size is in descending order starting at the lowest frequency index. The UE may assume that the same precoder is applied to all scheduled PRGs within the bundle. The size of the PRG depends on the system bandwidth, and is given in Table 9.7.

9.8.2.5 Resource element group

Resource element groups (REGs) are the basic unit of resource allocation for PCFICH, PHICH, and PDCCH that are represented by an index pair (k', l'), where k' is the sub-carrier index of the resource element within the REG with the lowest sub-carrier index, and l' is the OFDM symbol index of the REG. In other words, resource-element groups are used for defining the mapping of control channels to resource elements. The mapping of a symbol-quadruplet $\{a(i), a(i+1), a(i+2), a(i+3)\}$ to a resource-element group is defined such that elements $a(i)$ are mapped to resource elements (k, l) of the resource-element group not used for cell-specific reference signals (CRSs) in increasing order of indices k and l, as shown in Figure 9.39.

A REG always contains four consecutive resource elements which are not occupied by a CRS on any antenna port in use. All resource elements within a resource block in the first three OFDM symbols are allocated to REGs. Therefore, the number of resource elements within each REG and the number of REGs within an OFDM symbol are affected by the number of CRSs present on all antenna ports. The number and location of cell-specific reference signals are dependent on the number of antenna ports and the type of cyclic prefix used. Each antenna port has a unique CRS associated with it. Considering the dependency of REG arrangement on CRS configuration, the REG structure for one, two, or four antenna port

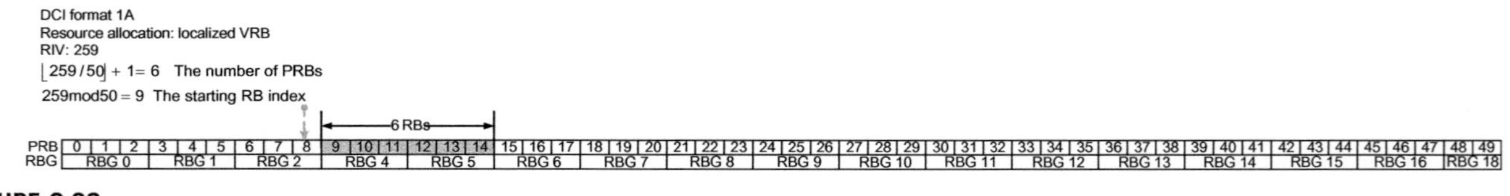

FIGURE 9.38

Example resource allocation type 2 [158].

Table 9.7 Precoding Resource Block Group Size [5]

Number of Resource Blocks N_{RB}^{DL}	Precoding Resource Block Group Size P'
≤10	1
11–26	2
27–63	3
64–110	2

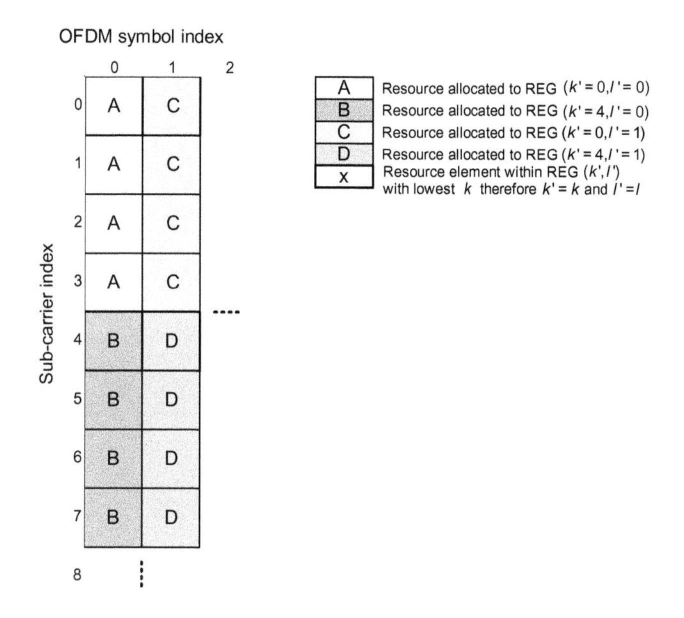

FIGURE 9.39

Resource element group structure [3,160].

configuration is different. However, the REG arrangement for each resource block within a subframe and for each antenna port is identical.

When antenna port 0 or antenna ports 0 and 1 are used, it is assumed that the CRSs are present on both antenna ports 0 and 1. This leads to an REG arrangement for each resource block on the first slot of each subframe, as shown in Figure 9.40. The CRSs are present within the first OFDM symbol. Since four resource elements not including CRSs are required to form a REG, the 12 resource elements over the first symbol in a resource block are divided into two REGs, each containing six resource elements (including two CRSs). On the

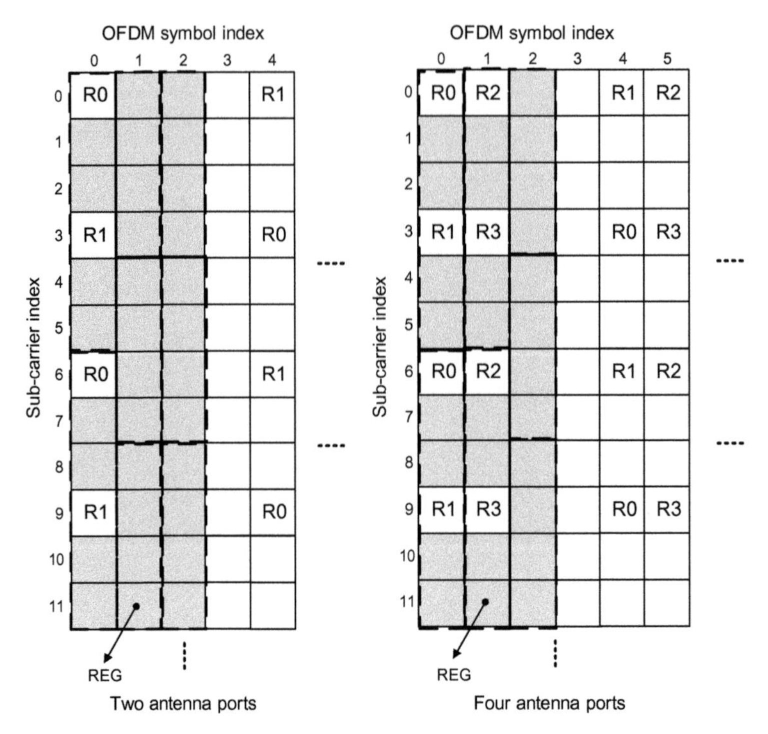

FIGURE 9.40

Resource element group structure with two and four antenna ports [3,160].

second and third OFDM symbols, no CRS is present; therefore; the 12 resource elements over each OFDM symbol are divided into three REGs, each containing four resource elements.

The REG organization in each resource block for four antenna port configuration is illustrated in Figure 9.40. The REG allocation within the first and second OFDM symbols is the same as one or two antenna port configuration. Four CRSs are present on the second OFDM symbol; therefore, eight resource elements are available for the mapping of control information. The 12 reference elements are divided into two REGs, each containing six resource elements. The third OFDM symbol contains no reference signals, so three REGs are available. An extended cyclic prefix subframe contains 12 OFDM symbols as opposed to 14 for the normal cyclic prefix. Since the number of CRSs in a normal or extended cyclic prefix subframe is the same, the limited number of OFDM symbols in an extended cyclic prefix subframe requires the OFDM symbol spacing of the CRSs to be reduced compared to a normal cyclic prefix. This reduction in spacing causes CRSs to be present within the fourth OFDM symbol of an extended cyclic prefix subframe, whereas in a normal cyclic prefix subframe no

CRSs are present on the fourth OFDM symbol of the subframe. The number of cell-specific reference symbols within the first three OFDM symbols is identical for normal or extended cyclic prefix, thus the REG configurations are identical.

9.9 Time and frequency synchronization

Synchronization signals are transmitted in every radio frame. The synchronization signals are used for time and frequency synchronization during initialization. As part of the system acquisition process, the UE synchronizes to the OFDM symbol, slot, subframe, half-frame, and radio frame (in that order) using a synchronization signal. Two synchronization signals are defined for system acquisition in E-UTRA: (1) a primary synchronization signal is used to obtain slot, subframe, and half-frame boundary. It also provides cell identity within the cell identity group. (2) A secondary synchronization signal is used to obtain the radio frame boundary. It also enables the UE to determine the cell identity group, which may range from 0 to 167. The synchronization signals are transmitted in the 0th and 5th subframes of every radio frame. In the frequency-domain, the synchronization signals occupy the center 1.08 MHz (equivalent to six resource blocks) of the radio channel (Figure 9.41).

Figure 9.41 shows the location of the primary and secondary synchronization signals in a radio frame. In the time-domain, primary and secondary synchronization signals are transmitted in the 0th and 10th slot (1st slot within 0th and 5th subframe) of every radio frame. Within these slots, the primary and secondary synchronization signals are transmitted in the last two symbols. The primary synchronization signal is transmitted in the last symbol of the 0th and 10th slot. The secondary synchronization signal is transmitted in one before the last symbol of the 0th and 10th slot. The PBCH is mapped to the first four symbols in slot 1 of the frame (just after the primary and secondary synchronization signals). In the frequency-domain, the center 1.08 MHz of the radio channel is reserved for primary and secondary synchronization signals. This is equivalent to six resource blocks or 72 sub-carriers. However, the primary and secondary synchronization signals only require 62 sub-carriers around the DC sub-carrier. The remaining sub-carriers (10 sub-carriers) in these resource blocks are not used for transmission. They provide a guard band between primary and secondary synchronization signal transmissions and other data transmissions.

There are 504 unique physical layer cell identities. The physical layer cell identities are grouped into 168 unique physical layer cell-identity groups, each group contains three unique identities. The grouping is done such that each physical layer cell identity is part of one and only one physical layer cell-identity group. A physical layer cell identity $N_{\mathrm{ID}}^{\mathrm{cell}} = 3N_{\mathrm{ID}}^{(1)} + N_{\mathrm{ID}}^{(2)}$ is thus uniquely defined by a number $N_{\mathrm{ID}}^{(1)}$ in the range of 0–167, representing the physical layer cell-identity group, and a number $N_{\mathrm{ID}}^{(2)}$ in the range of 0–2, denoting the physical layer identity within the physical layer cell-identity group.

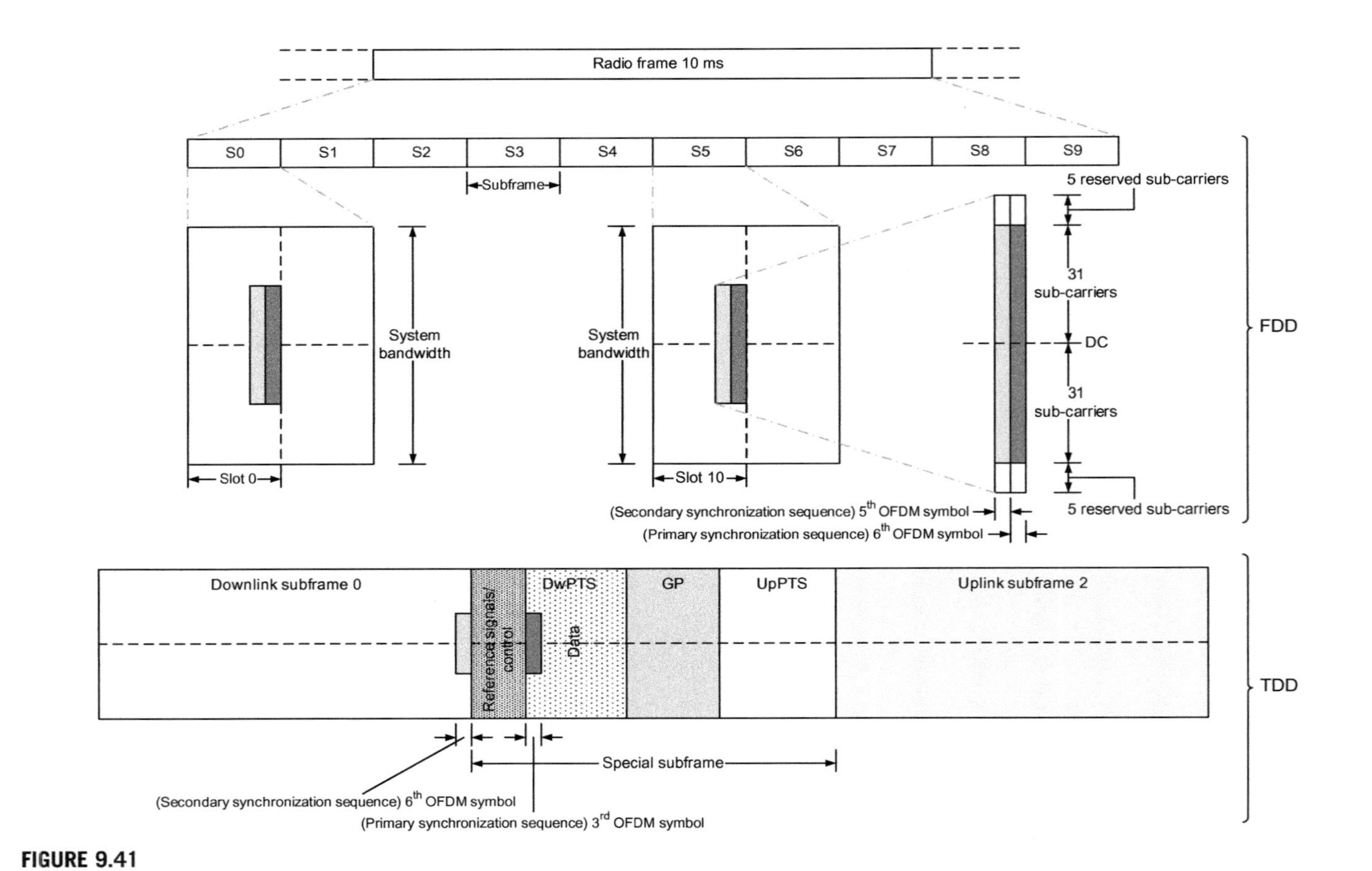

FIGURE 9.41

Location and structure of primary and secondary synchronization signals [3,174].

9.9.1 Primary synchronization signals

The primary synchronization signal is generated from a frequency-domain Zadoff–Chu sequence[9] according to the following equation [3]:

$$d_{\mathrm{u}}(n) = \begin{cases} \mathrm{e}^{-j\frac{\pi un(n+1)}{63}} & n = 0, 1, \ldots, 30 \\ \mathrm{e}^{-j\frac{\pi u(n+1)(n+2)}{63}} & n = 31, 32, \ldots, 61 \end{cases} \tag{9.24}$$

where the Zadoff–Chu root sequence index $u = 25, 29, 34$ if $N_{\mathrm{ID}}^{(2)} = 0, 1, 2$, respectively. The mapping to resource elements depends on the frame structure. The sequence $d(n)$ is mapped to the resource elements according to $a_{\mathrm{kl}} = d(n), n = 0, \ldots, 61$ and $k = n - 31 + 6N_{\mathrm{RB}}^{\mathrm{DL}}$. As shown in Figure 9.41, for frame structure type 1, the primary synchronization signal is mapped to the last OFDM symbol in slots 0 and 10. For frame structure type 2, the primary synchronization signal is mapped to the third OFDM symbol in subframes 1 and 6. The

---

[9]A Zadoff–Chu sequence is a complex-valued sequence with constant amplitude property whose cyclically shifted versions exhibit low cross-correlation. Thus, under certain conditions, the cyclically shifted versions of each sequence remain orthogonal to one another. A Zadoff–Chu sequence that has not been shifted is referred to as a root sequence. The uth root Zadoff–Chu sequence of prime length N is defined as follows [172]:

$$x_u(n) \triangleq \begin{cases} \mathrm{e}^{-\mathrm{j}[\pi un(n+1)]/N} & 0 \le n < N-1 \quad (N \text{ is an odd integer}) \\ \mathrm{e}^{-\mathrm{j}\pi un^2/N} & 0 \le n < N-1 \quad (N \text{ is an even integer}) \end{cases}$$

where N is an integer, which denotes the length of the Zadoff–Chu sequence. It is easily verified that $x_u(n)$ is periodic with period N, i.e., $x_u(n) = x_u(n+N), \forall n$. In other words, the sequence index u is prime relative to N. For a fixed value of u, the Zadoff–Chu sequence has an ideal periodic autocorrelation property (i.e., the periodic autocorrelation is zero for all time shifts other than zero). For different values of index u, the Zadoff–Chu sequences are not orthogonal, but rather exhibit low cross-correlation. If the sequence length N is selected as a prime number, there are different sequences with periodic cross-correlation of $1/\sqrt{N}$ between any two sequences regardless of time shift. The Zadoff–Chu sequences are a subset of constant amplitude zero autocorrelation (CAZAC) sequences. The properties of Zadoff–Chu sequences can be summarized as follows [172]:

- They are periodic with period N, if N is a prime number, i.e., $x_u(n+N) = x_u(n)$.
- Given N is a prime number, the DFT of a Zadoff–Chu sequence is another Zadoff–Chu sequence conjugated and time-scaled multiplied by a constant factor, i.e., $X_u[k] = x_u^*(vk)X_u[0]$ where v is the multiplicative inverse of u modulo N. It can be shown that $x_u^*(vk) = x_v^*(k)\mathrm{e}^{\mathrm{j}\pi(1-v)k/N}$.
- The autocorrelation of a prime-length Zadoff–Chu sequence with a cyclically shifted version of itself also yields zero autocorrelation sequence, i.e., it is non-zero only at one instant which corresponds to the cyclic shift zero.
- The cross-correlation between two prime-length Zadoff–Chu sequences, i.e., different u, is constant and equal to $1/\sqrt{N}$.
- The Zadoff–Chu sequences have low PAPR.

resource elements a_{kl} over OFDM symbols used for transmission of the primary synchronization signal where $k = n - 31 + 6N_{RB}^{DL}, n = \{-5, -4, \ldots, -1, 62, 63, \ldots, 66\}$ are reserved as guard sub-carriers.

9.9.2 Secondary synchronization signals

The secondary synchronization signal is generated from interlacing of two length-31 M-sequences[10] $s_0^{(m_0)}(n)$ and $s_1^{(m_1)}(n)$. There are 168 sequences that are generated from cyclic shifts of sequences $s_0^{(m_0)}(n)$ and $s_1^{(m_1)}(n)$. The cyclic shifts are generated based on cell-identity group number, which ranges from 0 to 167. The two sequences are scrambled, with the scrambling sequences $c_0(n)$ and $c_1(n)$ that are obtained based on the primary synchronization signal identity $N_{ID}^{(2)} \in \{0, 1, 2\}$. The second sequence is further scrambled with additional sequence $z_1^{(m_0)}(n)$. The two sequences are interleaved and mapped to 62 sub-carriers around the DC sub-carrier using BPSK modulation. As shown in Figure 9.42, five sub-carriers on either side are not used, providing a guard band between the primary and secondary synchronization sequences and other data-bearing sub-carriers. The secondary synchronization signals are also used to detect the radio frame boundary. To achieve this, a different variation of the sequence is generated by switching $s_0^{(m_0)}(n)$ and $s_1^{(m_1)}(n)$ in slot 10. The switch produces a different sequence so the UE can distinguish between slot 0 and slot 10.

More specifically, the secondary synchronization signal $d(0), d(1), \ldots, d(61)$ is an interleaved concatenation of two length-31 binary sequences as shown in Figure 9.42. The concatenated sequence is scrambled with a scrambling sequence given by the primary synchronization signal. The combination of two length-31 sequences defining the secondary synchronization signal differs between subframe 0 and subframe 5 according to the following rule:

$$d(2n) = \begin{cases} s_0^{(m_0)}(n)c_0(n) & \text{in subframe } 0 \\ s_1^{(m_1)}(n)c_0(n) & \text{in subframe } 5 \end{cases} \quad (9.25)$$

[10]Maximum length sequences are pseudo-random binary sequences that are generated using maximal linear feedback shift registers. The *M*-sequences are periodic and reproduce every binary sequence that can be reproduced by the shift registers (i.e., for length-*m* registers, they produce a sequence of length $2^m - 1$). An *M*-sequence is spectrally flat with the exception of a near-zero DC term. Since *M*-sequences are periodic and shift registers cycle through every possible binary value with the exception of the zero vectors, the registers can be initialized to any state with the exception of the zero vectors. A binary polynomial over *GF*(2) can be associated with the linear feedback shift register. The degree of polynomial is equal to the length of the shift register and the coefficients that are either 0 or 1 correspond to the taps of the register [173].

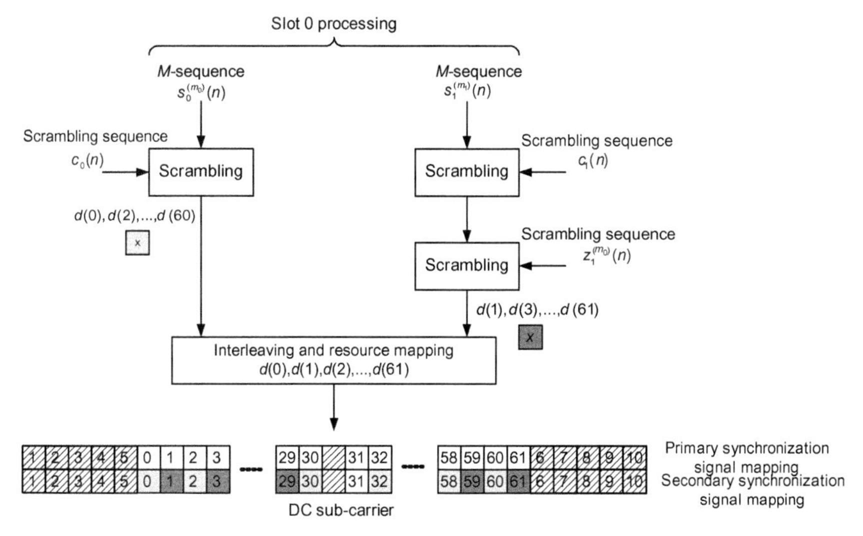

FIGURE 9.42

Secondary synchronization signal generation and mapping.

and:

$$d(2n+1) = \begin{cases} s_1^{(m_1)}(n)c_1(n)z_1^{(m_0)}(n) & \text{in subframe } 0 \\ s_0^{(m_0)}(n)c_1(n)z_1^{(m_1)}(n) & \text{in subframe } 5 \end{cases} \tag{9.26}$$

where $0 \le n \le 30$ and indices m_0 and m_1 are derived from the physical layer cell-identity group $N_{\text{ID}}^{(1)}$. The two sequences $s_0^{(m_0)}(n)$ and $s_1^{(m_1)}(n)$ are derived from two different cyclic shifts of an M-sequence $\tilde{s}(n)$. The two scrambling sequences $c_0(n)$ and $c_1(n)$ depend on the primary synchronization signal and are characterized by two different cyclic shifts of the M-sequence $\tilde{c}(n)$. The scrambling sequences $z_1^{(m_0)}(n)$ and $z_1^{(m_1)}(n)$ are defined by a cyclic shift of the M-sequence $\tilde{z}(n)$. The mapping of the sequence to resource elements depends on the frame structure. In a subframe for frame structure type 1 and in a half-frame for frame structure type 2, the same antenna port used for transmission of the primary synchronization signal is used for transmission of the secondary synchronization signal. The sequence $d(n)$ is mapped to resource elements according to $a_{kl} = d(n)$, $\forall n = 0, \ldots, 61$ and $k = n - 31 + 6N_{\text{RB}}^{\text{DL}}$, where $l = N_{\text{s}}^{\text{DL}} - 2$ in slots 0 and 10 for frame structure type 1 and $l = N_{\text{s}}^{\text{DL}} - 1$ in slots 1 and 11 for frame structure type 2 [3].

The primary synchronization signals are detected using non-coherent detection, whereas the secondary synchronization signals can be detected using either coherent or non-coherent detection schemes. Figure 9.43 shows an example algorithm for coherent detection of the secondary synchronization signals. The channel compensation block is used for coherent detection. The received secondary synchronization signals in the frequency-domain can be expressed as

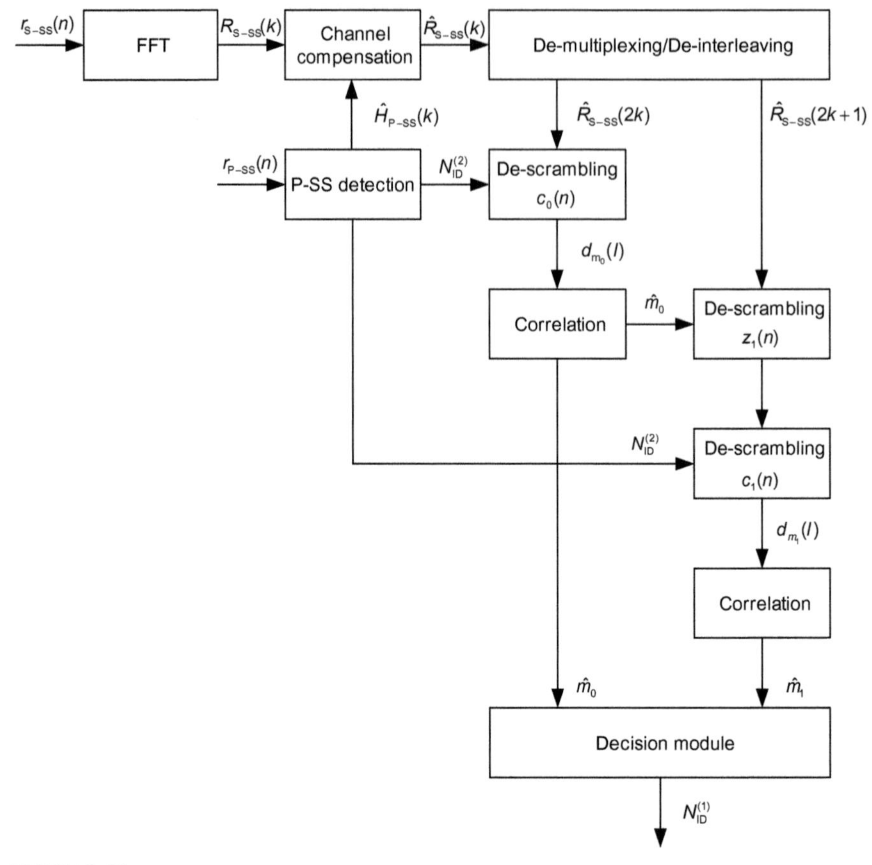

FIGURE 9.43

Example algorithm for coherent detection of secondary synchronization signal [174].

$r_{S-SS}(k) = H(k)d_{S-SS}(k) + w(k)$, where $d_{S-SS}(k)$ is the secondary synchronization signal in the frequency-domain, $H(k)$ is channel frequency response during secondary synchronization signal transmission, and $w(k)$ is the additive white Gaussian noise. For coherent detection, the UE estimates the channel frequency response $\hat{H}(k)$ by using the received primary synchronization sequence. From the UE perspective, the secondary synchronization signal detection is performed after the primary synchronization signal detection; thus the channel can be assumed to be known based on the primary synchronization signal sequence. In the frequency-domain, channel compensated secondary synchronization signal is given by $\hat{R}_{S-SS}(k) = \hat{H}(k)R_{S-SS}(k)$. After channel compensation, the de-interleaved and de-scrambled signal can be expressed as $d_{m_0}(l) = \hat{R}_{S-SS}(2k)c_0(k), d_{m_1}(k) = \hat{R}_{S-SS}(2k+1)c_1(k)z_1^{(\hat{m}_0)}(k)$. The output of coherent detection is the cross-correlation between the de-scrambled signal and the cyclic shift of a secondary synchronization sequence that can be represented as follows:

$$
\begin{aligned}
\hat{m}_0 &= \arg \max_i \left| \sum_{l=0}^{30} d_{m_0}(l) s_i(l) \right|^2 \\
\hat{m}_1 &= \arg \max_i \left| \sum_{l=0}^{30} d_{m_1}(l) s_i(l) \right|^2
\end{aligned}
\tag{9.27}
$$

where $\hat{m}_0$ and $\hat{m}_1$ are the ith secondary synchronization signal sequence. Once $\hat{m}_0$ and $\hat{m}_1$ are identified, the frame timing and cell identity group $N_{ID}^{(1)}$ can be detected (Figure 9.44).

The E-UTRA supports normal and extended cyclic prefix types for various deployment scenarios. The timing of the secondary synchronization signals will change depending on the cyclic prefix type, as shown in Figure 9.44. Prior to successful detection of the secondary synchronization signal, the cyclic prefix type is unknown to the UE and it must be blindly detected by examining the secondary synchronization signals at two possible locations in time. A simple algorithm for determining the cyclic prefix type is introduced in this section as an example. Note that E-UTRA specifications do not specify the receiver architecture and detection methods. This algorithm is based on the principle developed by Van de Beek, where the correlation between the cyclic prefix and its repetition part of the data symbol is calculated for two choices of cyclic prefix type [175]. For each cyclic prefix type, the correlation function is normalized by the associated received secondary synchronization signal power, because the range of values over which the summation is calculated varies with the cyclic prefix type. Let $r(n)$ denote the secondary synchronization signal at the receiver in the time-domain, we define [174]:

$$
\begin{aligned}
\rho_i &= \sum_{n=0}^{L_i-1} r(n+\tau_i) r^*(n+\tau_i+N) \\
P_i &= \sum_{n=0}^{L_i-1} |r(n+\tau_i+N)|^2
\end{aligned}
\tag{9.28}
$$

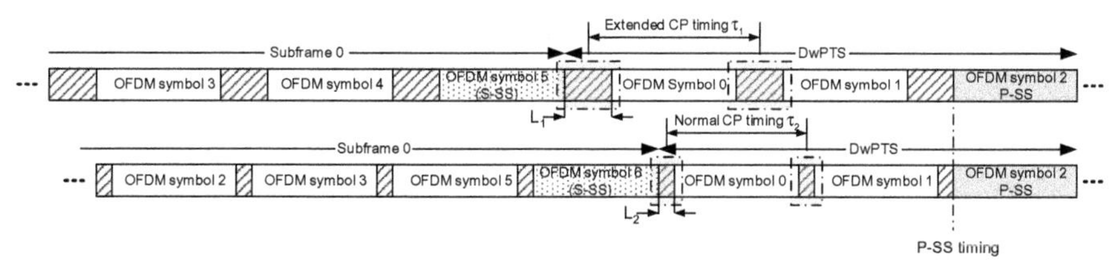

FIGURE 9.44

Illustration of the cyclic prefix detection algorithm [174].

where τ_1 and τ_2 represent the normal and extended cyclic prefix timing, respectively, and L_1 and L_2 denote the corresponding cyclic-prefix lengths. The normalized correlation function $U_i = \text{Re}\{\rho_i\}/P_i$ with the larger value indicates the type of cyclic prefix. Therefore, a normal cyclic prefix is considered, if $U_1 > U_2$ and an extended cyclic prefix is assumed, if $U_2 > U_1$. The main advantage of this algorithm is that a correlation value can be accumulated by using an OFDM symbol within one radio frame in order to improve the detection performance at low SNR. In addition, this algorithm can reduce the computational complexity compared to blind detection using secondary synchronization signal detection at the two possible instances in time.

9.10 Downlink reference signals and channel estimation

To facilitate estimation of the multipath communication channel and reliable, coherent detection of data/control signaling, an OFDM system uses reference signals (or pilot symbols). The pilot symbols provide an estimate of the channel frequency response at the pilot locations over the time–frequency grid. It is possible to estimate the channel at other time–frequency locations using interpolation techniques. In E-UTRA, the CRSs are assigned to specific positions within a subframe, depending on the physical cell identity and the specific antenna port from which they are transmitted. The unique positioning of the pilots ensures that they do not interfere with each other and can be used to provide a reliable estimate of the complex-valued channel gains scaling each resource element within the transmitted grid. Using known pilot symbols to estimate the channel matrix, it is possible to equalize the effects of the channel and to reduce noise and interference effects of the received resource blocks. Downlink spatial multiplexing in LTE Rel-10 was enhanced to support up to eight transmission layers, together with an enhanced reference-signal structure. Relying on cell-specific reference signals for higher order spatial multiplexing is less attractive, as the reference signal overhead is not proportional to the instantaneous transmission rank, but rather to the maximum supported transmission rank. LTE Rel-10, therefore, introduced extensive support of UE-specific reference signals for demodulation of up to eight layers. In addition, feedback of channel-state information (CSI) is based on a separate set of reference signals, known as CSI reference signals (CSI-RSs). The CSI-RSs are relatively sparse in frequency but regularly transmitted from all antennas at the base station, while the UE-specific reference signals are denser in frequency but only transmitted when data is transmitted on the corresponding layer. Separating the reference-signal structures supporting demodulation and channel estimation helps to reduce reference-signal overhead, especially for high degrees of spatial multiplexing, and allows for implementation of various beam-forming schemes. In LTE Rel-8, the reference signal is added to the signal after precoding, one CRS per antenna port. From the received CRS, the UE can estimate the channel matrix and determine how the radio channel affected the signal. Using the estimated channel matrix together with the knowledge about the code-book used for precoding, the UE can demodulate the received signal and regenerate the original information. In LTE Rel-10, the demodulation reference signals (DM-RSs) are added to the different data streams before precoding. The knowledge about the reference signal will then provide information about the combined effect of radio channel and precoding, thus prior knowledge about the precoder is not required by the receiver, this case is referred to as non-codebook-based precoding.

The LTE/LTE-Advanced systems make use of reference signals for downlink and uplink channel estimation. The downlink reference signals are used to measure CSI and channel estimation for demodulation. The design of downlink reference signals in LTE Rel-8 primarily relied on a set of cell-specific reference signals that could be used for both purposes. The CRS are wideband reference signals transmitted across the entire downlink system bandwidth in every subframe, and are defined for up to four antenna ports. The presence of cell-specific reference signals in every subframe, while necessary for decoding essential physical channels, results in unreasonably high overhead. To support eight-layer spatial multiplexing in LTE-Advanced, new low duty-cycle wideband reference signals, known as CSI-RSs, were designed for CSI measurements. Furthermore, a new set of UE-specific or DM-RSs were designed to enable channel estimation for demodulation. The LTE Rel-10 DM-RS design efficiently supports high-rank transmissions because it is present only in those PRBs in which higher rank is used [121]. The enhanced PDCCH (ePDCCH) DM-RSs with an identical structure to Rel-10 DM-RSs were added in Rel-11 to support various use cases involving the new FDM-based control channel structure.

LTE/LTE-Advanced defines an antenna port as a logical entity which is distinct from a physical antenna and is associated with a specific set of reference signals such that the channel over which a symbol on the antenna port is conveyed can be distinguished from the channel over which another symbol on the same antenna port is conveyed. There is one resource grid per antenna port. The set of downlink antenna ports supported depends on the reference signal configuration in the cell as given in Table 9.8 [3].

The set of downlink antenna ports supported depends on the reference signal configuration in the cell as follows [3]:

- Cell-specific reference signals support one, two, or four antenna port configurations and are transmitted on antenna ports $p=0$, $p\in\{0,1\}$, and $p\in\{0,1,2,3\}$, respectively.
- MBSFN reference signals are exclusively transmitted on antenna port $p=4$.
- UE-specific or DM-RSs are transmitted on antenna port $p=5, 7, \text{or } 8$, or one or several of $p\in\{7,8,9,10,11,12,13,14\}$.
- Positioning reference signals are exclusively transmitted on antenna port $p=6$.
- CSI reference signals support configuration of one, two, four, or eight antenna ports, and are transmitted on antenna ports $p=15$, $p=\{15,16\}$, $p=\{15,16,17,18\}$, and $p=\{15,16,17,18,19,20,21,22\}$, respectively.

Table 9.8 Downlink Antenna Ports and Their Use Cases [3,13]

3GPP Release	Downlink Antenna Port	Associated Reference Signals	Use Cases
8	0–3	Cell-specific reference signals	Single-layer transmission, transmit diversity, SU-MIMO/MU-MIMO
8	4	MBSFN reference signal	Multimedia broadcast and multicast services
8	5	UE-specific reference signal/demodulation reference signal	Single-layer beamforming
9	6	Positioning reference signal	Location-based services
9	7,8	UE-specific reference signals/demodulation reference signals	Dual-layer beamforming, MU-MIMO
10	9–14	UE-specific reference signals/demodulation reference signals	Multi-layer beamforming, MU-MIMO
10	15–22	CSI reference signals	CSI measurement and reporting
11	107–110	ePDCCH demodulation reference signals	ePDCCH-based transmission and demodulation

- ePDCCH DM-RSs are transmitted on antenna ports $p \in \{107, 108, 109, 110\}$, and are used to support ePDCCH transmission and demodulation.

Based on the above discussion, six types of downlink reference signals are defined as follows, which will be discussed in detail in the next sections:

- Cell-specific reference signals (CRS)
- MBSFN reference signals
- UE-specific or DM-RSs associated with PDSCH
- Positioning reference signals
- CSI-RSs
- DM-RSs associated with ePDCCH

Note that there is one reference signal transmitted per downlink antenna port.

9.10.1 Reference signal design criteria

In this section, we discuss the theoretical criteria for efficient pilot pattern design for OFDM systems. Pilot-based channel estimation is defined as the use of the channel samples estimated at the pilot tones to reconstruct channel samples at the remaining data tones. As a result, the pilot pattern design is essentially a conventional sampling rate selection problem in the two-dimensional signal processing space. In order to avoid aliasing during reconstruction of the channel time–frequency function, the pilot tone selection should follow the two-dimensional sampling theorem. When multiple antennas are used, the receiver must estimate the channel impulse response (or the transfer function) from each of the transmit antennas in order to correctly detect the signal. This is achieved through distributing reference signals (or pilot tones) among the transmit antennas.

Let $\Delta f = 1/T_u$ and $T_{symbol} = T_g + T_u$ denote the sub-carrier spacing and the OFDM symbol duration (inclusive of the guard interval), respectively. Let us further assume that the pilot sub-carriers are transmitted at integer multiples of sub-carrier spacing and OFDM symbol duration in frequency and time directions, respectively (i.e., $f_p = m/T_u$ and $T_p = nT_{symbol}$, where m and n are integers). The (m, n) pair represents the pilots' separation in terms of sub-carrier spacing and OFDM symbol duration. From the sampling theorem point of view, the channel's two-dimensional delay-Doppler response $h(\tau, \nu)$ can be fully reconstructed, if the two-dimensional transform function $H(t,f)$ is sampled greater than or equal to the Nyquist rate across time and frequency dimensions. Hence, the time-domain sampling rate must be greater than or equal to the channel's maximum Doppler spread, i.e., $T_p \leq 1/\nu_{max}$ (the sampling rate in time must be less than the coherence time) and the frequency-domain sampling rate must be greater than or equal to the channel's maximum delay spread, i.e., $f_p \leq 1/\tau_{max}$ (the sampling rate in frequency must be less than coherence bandwidth). Assuming a wide-sense stationary uncorrelated scattering channel model, and assuming the channel to be constant over one OFDM symbol, the frequency response $H(t,f)$ of the L-path channel is given as [23,140]:

$$H(t,f) = \frac{1}{\sqrt{L}} \sum_{l=1}^{L} e^{j(\psi_l + 2\pi\nu_l t - 2\pi f \tau_l)} \tag{9.29}$$

where ψ_l, ν_l, and τ_l denote the phase, Doppler frequency, and delay of the lth path, respectively. All these parameters are independent random variables. In general, the pilot signals are over-sampled in order to ensure a good trade-off between performance and overhead. Therefore, the choice of (m, n) depends on the channel's maximum delay spread and maximum Doppler spread, and must satisfy the following equation according to the two-dimensional sampling theorem [23]:

$$\begin{aligned} n &\leq \frac{1}{2T_{symbol}\nu_{max}} \\ m &\leq \frac{T_u}{2\tau_{max}} \end{aligned} \tag{9.30}$$

It should be noted that the pilot density for a regular pattern can be calculated using Eq. (9.30). From Eq. (9.30), it can be seen that for large values of ν_{max}

(i.e., large Doppler spread, or alternatively small channel coherence time, means that the channel is time-varying), n should be small in order to appropriately track channel time variations. On the other hand, for large values of τ_{max} (i.e., large delay spread, or alternatively small coherence bandwidth, means that the channel is frequency selective), m should be small in order to closely follow channel frequency variation. In a regularly spaced pilot pattern, the pilot symbols are evenly spaced in frequency and in time.

Using the default OFDM symbol duration and the sub-carrier spacing given in Table 9.2 (OFDM parameters $T_{symbol} = 71.36$ μs and $\Delta f = 1/T_u = 15$ kHz), and assuming that the channel maximum delay spread $\tau_{max} = 5$ μs for the radio channels under consideration in LTE/LTE-Advanced evaluation methodology, the most appropriate range for the values of (m, n) can be determined. Since an E-UTRA system is required to support vehicle speeds up to 350 km/h, the maximum Doppler spread is estimated as $\nu_{max} \approx 810$ Hz at 2.5 GHz carrier frequency. From Eq. (9.30) and the above design parameters, the following pilot spacing constraints are obtained $n \leq 8, m \leq 6$. Figure 9.45 illustrates a regularly spaced

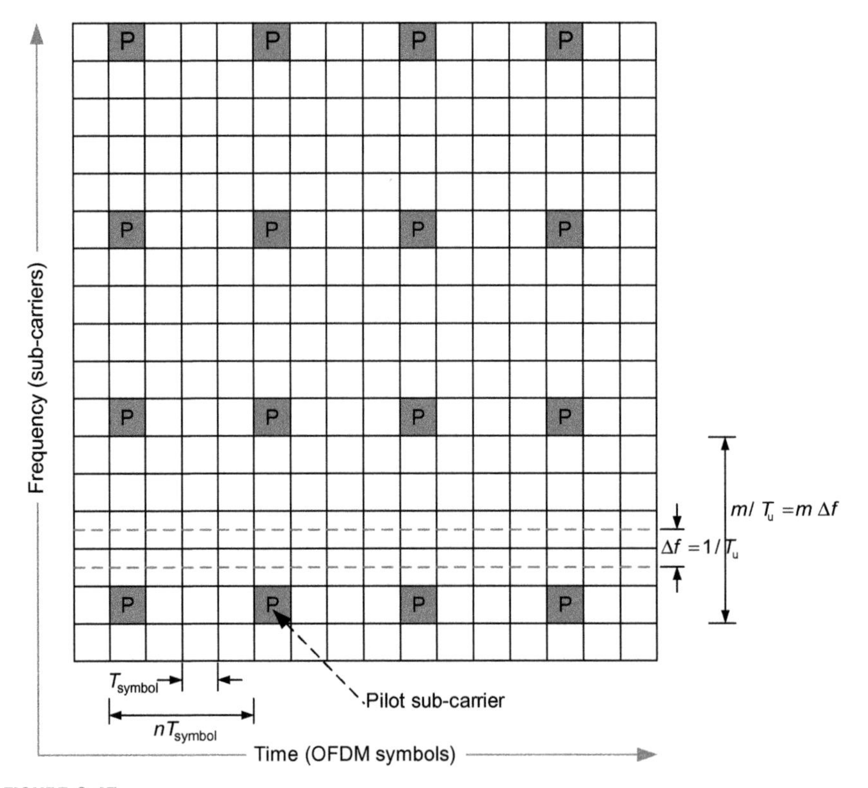

FIGURE 9.45

An example of regular pilot pattern for one spatial layer with $n = 4$ and $m = 5$.

two-dimensional pilot pattern for one spatial stream in the time–frequency domain where $n = 4$ and $m = 5$. The above criteria can be generalized to the case of multi-antenna transmission where each transmit antenna is associated with a group of pilot sub-carriers.

In the case of irregular pilot patterns, the pilot symbols can be irregularly placed in time, in frequency, or in both. The irregular patterns can be designed according to a certain criteria, yielding a specific pattern. Other important considerations in the design of pilot structures for support of multi-antenna OFDM systems include the following:

- Identical pilot pattern for each PRB: Pilot spacing in time and frequency must be in proportion with the resource block size. If the location of the pilot tones within a resource block is not maintained the same across all resource blocks within the system bandwidth, or alternatively the pilot sub-carriers have different positions within each resource block, the filtering and interpolation operations during channel estimation become excessively complex.
- Pilot density: Proper pilot overhead must be considered as a trade-off between accurate channel estimation under various mobility conditions and higher throughput.
- Types of pilots: Pilots are typically classified as common (cell-specific) and dedicated (UE-specific). Each UE in the cell can estimate the channel over the entire bandwidth using the common pilots, while dedicated pilots in the UE assigned resource block can be used for channel estimation. In conventional MIMO schemes, the dedicated pilots are typically adaptively precoded. The dedicated pilots can be used in conjunction with common pilots, or in place of common pilots.
- Pilot power boosting: In general, the power of pilot sub-carriers is boosted relative to data sub-carriers in order to enhance the channel estimation with pilot hopping or shifting. To avoid symbol to symbol power fluctuation when using pilot boosting, it is desirable to place the pilot sub-carriers regularly on every symbol, or to employ power adjustment for data sub-carriers over a symbol.
- Per antenna power balance: When using multiple transmit antennas, it is important to maintain balanced power distribution across antennas. For this purpose, if there are two transmit antennas, each OFDM symbol should contain the same number of pilot tones associated with different antenna ports.

As an example and by taking the above additional criteria into consideration, the structure and link-level performance (i.e., bit error rate versus SNR curves) of several hypothetical common pilot patterns for two transmit antennas and two different multi-antenna transmission schemes are shown in Figure 9.46. In this example, the pilot overhead of type A and type B patterns is 11.11% (i.e., 5.5% per antenna port), and the transmit power per antenna is balanced. The type C

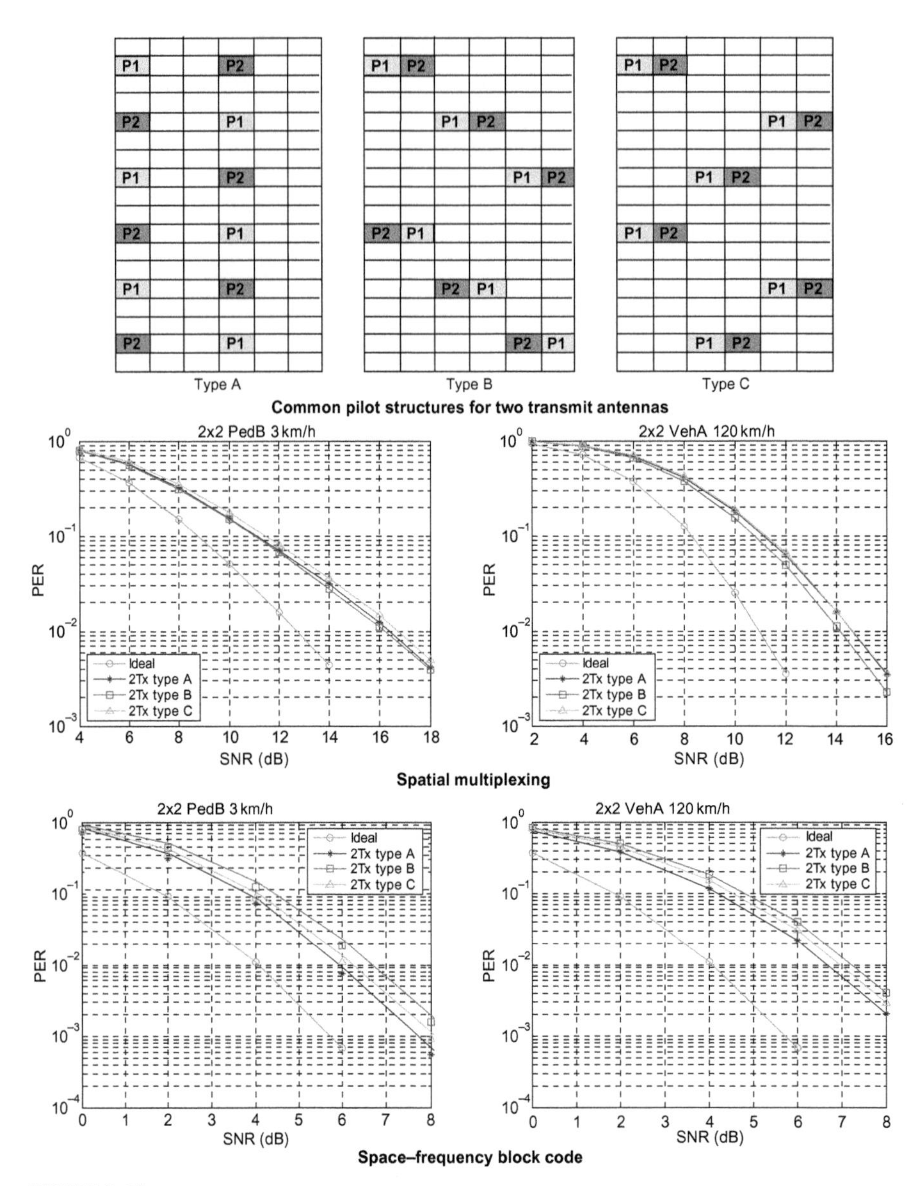

FIGURE 9.46

Channel estimation accuracy of various pilot patterns for two transmit antennas and QPSK-1/2 in different mobility and channel conditions [140].

pilot pattern has unbalanced power across antennas and symbols. The pilot tone powers have been boosted by 3 dB relative to data sub-carriers [140].

The accuracy of channel estimation using common pilots affects the performance of downlink/uplink data and control channel decoding. Figure 9.46 shows the link-level performance of spatial multiplexing with the example pilot patterns, where type B provides the best performance under low and high mobility conditions. The ideal channel estimation is used as a reference. Also shown in Figure 9.46 is the link-level performance of space-frequency block codes (SFBCs) with the example pilot patterns. In this case, type A outperforms the other structures. It can be concluded that type A is the best among three example patterns. Note that the example pilot patterns satisfy the sampling theorem constraint of Eq. (9.30). This example underlines the criteria for the design of robust and high performance pilot structures for OFDM-based cellular system operation in a wide range of mobility and channel conditions. The time/frequency characteristics of a number of typical radio channel models, mobility classes, and the associated time–frequency sampling rate requirements are summarized in Table 9.9.

Table 9.9 Characteristics of Typical Radio Channel Models ($T_u = 66.67$ μs and $T_{symbol} = 71.36$ μs)

Channel Model	RMS Delay Spread (ns) ν_{RMS}	Maximum Delay Spread (μs) τ_{max}	Maximum Sampling Rate in Frequency (Number of Sub-carriers) $\frac{T_u}{2\tau_{max}}$
ITU-T Pedestrian A (PedA)	45	0.41	81
ITU-T Pedestrian B (PedB)	750	3.7	9
ITU-T Vehicular A (VehA)	370.4	2.51	13
ITU-T Vehicular B (VehB)	4000	20	1
Typical Urban 6-Ray Model (TU)	1000	5	6
Vehicular Speed Range (km/h)	**Maximum Doppler Spread (Hz)** ν_{max}	**Coherence Time (ms)** T_c	**Maximum Sampling Rate in Time (Number of OFDM Symbols)** $\frac{1}{2T_{symbol}\nu_{max}}$
0–15	~27.78	15.23	252
15–120	~222.22	1.90	31
120–350	~648.15	0.65	10

9.10.2 Pilot-based channel estimation

While perfect knowledge of the radio channel can be used to find an upper bound for system performance, such knowledge is not available in practice and the channel needs to be frequently estimated. Channel estimation can be performed in various ways, including the use of frequency and/or time correlation properties of the wireless channel, blind or pilot (reference signal) based channel estimation, and adaptive or non-adaptive channel estimation. Non-parametric methods attempt to estimate the frequency response without relying on a specific channel model. In contrast, the parametric estimation methods assume a certain channel model and determine the parameters of this model. Spaced-time and spaced-frequency correlation functions, discussed earlier, are specific properties of a channel that can be incorporated in the estimation method, improving the quality of estimations. Pilot-based estimation methods are the most commonly used methods in OFDM systems which are applicable in systems where the sender transmits some known signals to the receiver. Blind estimation, on the other hand, relies on some properties of the signal and is rarely used in practical OFDM systems. Adaptive channel estimation methods are typically used for a rapid time-varying channel [69–71].

In this section, we describe a generic pilot-based non-adaptive channel estimation method. In a linear time-invariant AWGN channel, the relationship between the transmitted signal $X_k|_{k=0,1,\ldots,N_{\mathrm{FFT}}-1}$ and received signal Y_k can be expressed as $Y_k = H_k X_k + Z_k$, where Z_k is the frequency-domain noise sampled at the kth sub-carrier frequency, and H_k is the channel transfer-function sampled at the kth sub-carrier frequency. To estimate the channel, pilot symbols are used. Let us assume that every pth sub-carrier contains known pilot symbols $X_{pk}|_{kp=0,p,2p,\ldots,N_p-1}$, where N_p is the total number of pilot sub-carriers across frequency. Using the known pilot symbols X_{pk} and the received symbols Y_{pk} sampled at pilot sub-carriers, the channel transfer function $\hat{H}_k$ sampled at pilot sub-carriers can be written as follows:

$$\hat{H}_{pk} = \frac{Y_{pk}}{X_{pk}} + \frac{Z_{pk}}{X_{pk}} = H_{pk} + W_{pk} \tag{9.31}$$

where W_{pk} is a scaled-noise component at the pkth sub-carrier. Different methods such as one- or two-dimensional linear interpolation can be used to estimate the channel samples at other sub-carrier frequencies.

Channel estimation in OFDM systems is a two-dimensional problem, i.e., the channel transfer function or channel impulse response is a function of time and frequency variables. Due to the computational complexity of two-dimensional estimators, one-dimensional channel estimation methods are more practical. The idea behind one-dimensional estimators is to estimate the channel in one dimension (e.g., frequency), and then estimate the channel in another dimension (e.g., time), thus obtaining a two-dimensional channel estimate. Linear interpolation is a simple method to estimate the channel from channel samples estimated at pilot sub-carriers. This is done by linearly interpolating the channel samples at the two nearest pilot sub-carriers. Although linear interpolation provides some limited noise reduction of the channel estimates at data locations (due to averaging function), it is the simplicity of the solution that is more attractive. It must be noted that averaging window length (i.e., the number of pilots contained in the averaging window) is inversely proportional to the coherence bandwidth of the channel. The one-dimensional linear interpolation method estimates the channel by interpolating the channel transfer function between $\hat{H}_{m,k}$ and $\hat{H}_{m,l}$ (interpolation across frequency), or between $\hat{H}_{m,k}$ and $\hat{H}_{n,k}$ (interpolation across time). The algorithm shown in Figure 9.47 illustrates the procedure for channel estimation across the time–frequency grid based on the known values of the channel at the reference signal locations. Depending on the relative location of the sub-carrier in the time–frequency grid, the channel is estimated via one-dimensional linear interpolation of channel values at neighboring pilots or merely copying or extrapolation of values at the neighboring pilot, if the previous or the next pilot does not exist [160].

The channel transfer function $H_{m,k}$ at time–frequency index (m, k) can be modeled as a linear weighed sum of two-dimensional basis functions evaluated at the kth sub-carrier (frequency index) and at the mth OFDM symbol (time index):

$$H_{m,k} = \sum_{n=0}^{N-1} \alpha_n \phi_n(m, k) \tag{9.32}$$

where $\phi_n(m, k)$ denotes the nth basis function sampled at the mth OFDM symbol and at the kth sub-carrier, α_n is the coefficient of the nth basis function, and N is number of basis functions used in the linear model. By taking a one-dimensional approach, the above two-dimensional problem is reduced to one-dimensional, if one of the time or frequency indices is kept unchanged:

$$H_k = \sum_{n=0}^{N-1} \alpha_n \phi_n(k) \tag{9.33}$$

The channel samples at pilot sub-carriers shown in Eq. (9.33) can be written as:

$$\hat{H}_{pk} = \sum_{n=0}^{N-1} \alpha_n \phi_n(pk) + W_{pk} \tag{9.34}$$

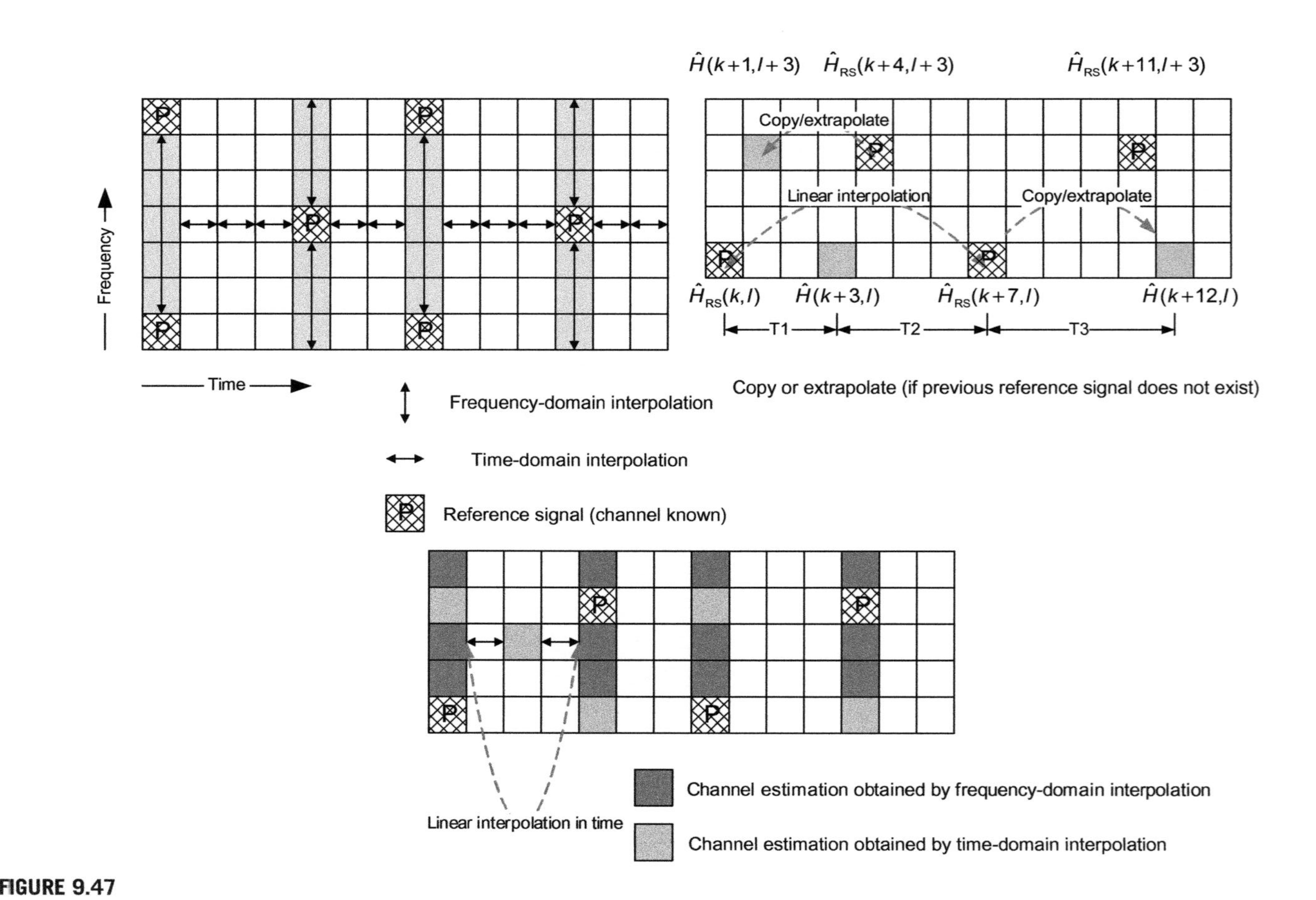

FIGURE 9.47

Two-dimensional channel estimation across frequency and time [176].

The above equation can be represented in matrix form as follows:

$$\begin{bmatrix} \hat{H}_0 \\ \hat{H}_p \\ \hat{H}_{2p} \\ \vdots \\ \hat{H}_{N_p-1} \end{bmatrix} = \begin{bmatrix} \phi_0(0) & \phi_1(0) & \phi_2(0) & \cdots & \phi_{N-1}(0) \\ \phi_0(p) & \phi_1(p) & \phi_2(p) & \cdots & \phi_{N-1}(p) \\ \phi_0(2p) & \phi_1(2p) & \phi_2(2p) & \cdots & \phi_{N-1}(2p) \\ \vdots & \vdots & \vdots & \ddots & \vdots \\ \phi_0(N_p-1) & \phi_1(N_p-1) & \phi_2(N_{p-1}) & \cdots & \phi_{N-1}(N_{p-1}) \end{bmatrix} \begin{bmatrix} \alpha_0 \\ \alpha_1 \\ \alpha_2 \\ \vdots \\ \alpha_{N-1} \end{bmatrix} + \begin{bmatrix} Z_0 \\ Z_p \\ Z_{2p} \\ \vdots \\ Z_{N_p-1} \end{bmatrix} \quad (9.35)$$

Alternatively, by using matrix notations, the above equation can be rewritten as follows:

$$\hat{\mathbf{H}} = \mathbf{\Phi}\boldsymbol{\alpha} + \mathbf{Z} \quad (9.36)$$

The least square (LS) estimate of the coefficients is calculated by minimizing the squared distance between the actual channel vector $\mathbf{H}$ and the estimated channel vector $\hat{\mathbf{H}}$ as follows:

$$\hat{\alpha}_{\mathrm{LS}} = (\mathbf{\Phi}^{\mathbf{H}}\mathbf{\Phi})^{-1}\mathbf{\Phi}^{\mathbf{H}}\hat{\mathbf{H}} = \mathbf{\Phi}^{-1}\hat{\mathbf{H}} \quad (9.37)$$

where the superscript H denotes conjugate transpose operation, hence:

$$\hat{H}_k = \sum_{n=0}^{N-1} \hat{\alpha}_n \phi_n(k) \quad (9.38)$$

A number of different options exist while selecting the basis functions such as orthogonal polynomials, Fourier series, discrete cosine/sine transform function, etc. There are a number of reasons for using orthogonal polynomials including ease of calculation of $\mathbf{\Phi}^{\mathbf{H}}\mathbf{\Phi}$ since it becomes a diagonal matrix, if matrix $\mathbf{\Phi}$ is orthogonal, resulting in ease of calculation of $\hat{\alpha}_n's$ and the inverse matrix $(\mathbf{\Phi}^{\mathbf{H}}\mathbf{\Phi})^{-1}$ with minimal computational complexity. The degree of orthogonal polynomials can be increased without changes in recursive calculation of $\hat{\alpha}_n's$. When applying the polynomial-based channel estimation, a sliding window is typically used to estimate the channel. This is done to avoid use of high-order polynomials to estimate highly frequency-selective channels. The sliding window approach allows for local approximations to the channel transfer function using a window length larger than the polynomial order, providing a better estimate.

In a special case where the OFDM channel estimation symbols are transmitted periodically and all sub-carriers over an OFDM symbol are used as pilots, the channel estimation algorithm is more simplified. In this case, the receiver uses the estimated channel transfer function to decode the received data within the block until the next pilot symbol arrives. If $Y_k = H_k X_k + Z_k|_{k=0,1,2,\ldots,N_{\mathrm{FFT}}-1}$ denotes the received signal at kth sub-carrier (pilot sub-carrier), the LS estimate of the channel transfer function in this case is given as:

$$\hat{\mathbf{H}}_{\mathbf{LS}} = (Y_1/X_1, Y_2/X_2, \ldots, Y_{N_{\mathrm{FFT}}-1}/X_{N_{\mathrm{FFT}}-1})^H = (Y_1, Y_2, \ldots, Y_{N_{\mathrm{FFT}}-1})^H; X_k = 1, \forall k \tag{9.39}$$

The channel transfer function at non-pilot sub-carriers can be estimated using interpolation in time. For convenience, we assume that the pilot sub-carriers have unity magnitude.

An alternative approach to pilot-based channel estimation is the minimum mean squared error (MMSE) method. The MMSE channel estimator exploits the second-order statistics of the channel transfer function to minimize the mean-squared error. Let $\mathbf{R_{HH}} = E(\mathbf{HH^H})$ and σ_Z^2 denote the autocorrelation matrix of the channel vector $\mathbf{H}$ and the additive white Gaussian noise variance, respectively. The MMSE estimate of the channel transfer function is given as:

$$\hat{\mathbf{H}}_{\mathrm{MMSE}} = \mathbf{R_{HH}}[\mathbf{R_{HH}} + \sigma_Z^2(\mathbf{XX^H})^{-1}]^{-1}\hat{\mathbf{H}}_{\mathbf{LS}} = \mathbf{R_{HH}}[\mathbf{R_{HH}} + \sigma_Z^2\mathbf{I}]^{-1}\hat{\mathbf{H}}_{\mathbf{LS}}; \quad \mathbf{XX^H} = \mathbf{I} \tag{9.40}$$

The MMSE estimator performs much better than LS estimators, especially under the low SNR conditions. A major drawback of the MMSE estimator is its high computational complexity, particularly if matrix inversion is necessary every time the input data changes. Figure 9.48 shows an example where the

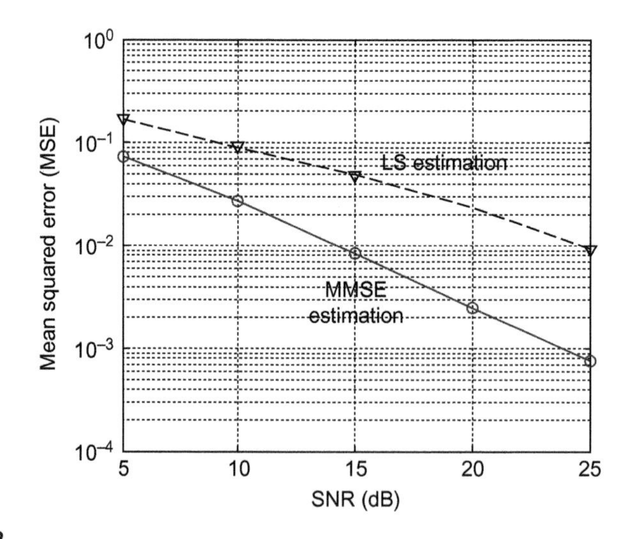

FIGURE 9.48

Example comparison of the LS and MMSE channel estimation algorithms [141].

performance of the LS and MMSE channel estimation algorithms has been compared. In this example $N_{FFT} = 64$ and the channel is modeled as a two-tap delay line filter. The superiority of MMSE over LS estimation in low SNR conditions can be seen in Figure 9.48.

9.10.3 Cell-specific reference signals

The cell-specific reference signals are used during initial cell selection for channel estimation and coherent detection of downlink physical channels. The measurements pertaining to scheduling and handover functions are based on the received signal strength of the CRS transmitted from antenna port zero (and antenna port one). The CRS sequence is a complex pseudo-random sequence, which is a length-31 Gold sequence.[11] The initialization seed for the second M-sequence used for Gold sequence generation is derived from the OFDM symbol number, slot number, and the physical cell identity. The seed is reinitialized at the beginning of each OFDM symbol within a slot. Pseudo-random sequences which are cell-specific are used to differentiate the reference signals transmitted from different cells. There is also a frequency shift that is applied to mapping of the CRS to resource elements to prevent the CRS from adjacent cells overlapping over the time–frequency grid. The generated sequence is transmitted in every slot and is spread throughout the channel bandwidth. The frequency of CRS transmission per slot and actual mapping to resource elements depend on the antenna configuration (Figure 9.49). The CRS symbols between antennas from the same cell are time-division multiplexed; the resource elements used for CRS symbols on one antenna are reserved and are not used on other antenna ports.

The CRSs are transmitted in all downlink subframes in a cell supporting PDSCH transmission. The CRSs are transmitted on one or several of antenna ports 0–3, and are only defined for sub-carrier spacing of 15 kHz. The reference-signal sequence $r_l^{n_s}(m)$ is defined as $r_l^{n_s}(m) = (1/\sqrt{2})(1 - 2c(2m)) + j(1/\sqrt{2})(1 - 2c(2m + 1))$, $\forall m = 0, 1, \ldots, 2N_{RB}^{DL} - 1$, where index n_s denotes the slot number within a radio frame, $l = 0, 1, \ldots, 6$ (or 5 in case of an extended cyclic prefix) represents the OFDM symbol number within slot n_s, and $c(i)$ is a pseudo-random sequence, which is initialized with $c_{init} = 2^{10}[7(n_s + 1) + l + 1](2N_{ID}^{cell} + 1) + 2N_{ID}^{cell} + N_{CP}$ at the start

[11]A Gold sequence is a type of binary sequence that is often used in telecommunication and satellite navigation. Gold sequences have bounded small cross-correlations within a set. A set of Gold sequences consists of $2^n - 1$ sequences each with a period of $2^n - 1$. A set of Gold sequences can be generated by taking two maximum-length sequences of the same length $2^n - 1$ such that their absolute cross-correlation is less than or equal to $2^{(n+2)/2}$, where n is the size of the linear feedback shift register used to generate the maximum length sequence [130]. The set of $2^n - 1$ (logical) exclusive OR of the two sequences in their various phases is a set of Gold codes. The highest absolute cross-correlation in this set of codes is $2^{(n+2)/2} + 1$ for even n, and $2^{(n+1)/2} + 1$ for odd n. The exclusive OR of two Gold codes from the same set is another Gold code with arbitrary phase [130].

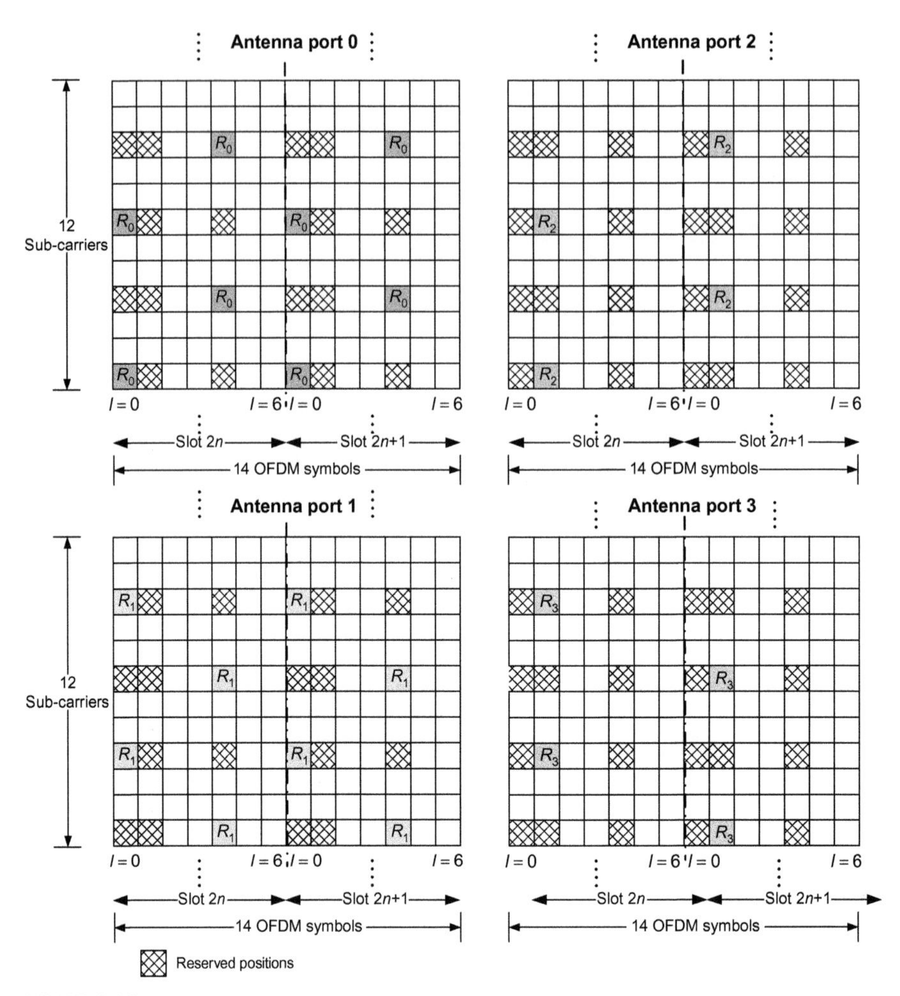

FIGURE 9.49

Mapping of cell-specific reference signals to antenna ports 0, 1, 2, and 3 for four transmit antennas over a resource block using a normal cyclic prefix [3].

of each OFDM symbol.[12] Note that $N_{CP} = 1$ for a normal cyclic prefix and $N_{CP} = 0$ for an extended cyclic prefix. The reference signal sequence $r_l^{n_s}(m)$ is mapped to complex-valued modulation symbols a_{kl}^p that are designated as reference symbols for

[12]Pseudo-random sequences are defined by a length-31 Gold sequence. The output sequence $c(n), \forall n = 0, 1, \ldots, M-1$ of length M is defined by $c(n) = (x_1(n+N_C) + x_2(n+N_C))\text{mod } 2$ in which $x_1(n+31) = (x_1(n+3) + x_1(n))\text{mod}2, x_2(n+31) = (x_2(n+3) + x_2(n+2) + x_2(n+1) + x_2(n))\text{mod } 2$, and $N_C = 1600$. The first m-sequence is initialized with $x_1(0) = 1, x_1(n) = 0, n = 1, 2, \ldots, 30$. The second M-sequence is initialized by $c_{\text{init}} = \sum_{i=0}^{30} x_2(i)2^i$ with the value depending on the application of the sequence [3].

antenna port p in slot n_s as $a_{kl}^{p} = r_l^{n_s}(m')$, where the parameters are given as follows [3]:

$$k = 6m + (v + v_{\text{shift}}) \bmod 6; \quad l = \begin{cases} 0, N_s^{\text{DL}} - 3 & \text{if } p \in \{0, 1\} \\ 1 & \text{if } p \in \{2, 3\} \end{cases}$$

$$m = 0, 1, \ldots, 2N_{\text{RB}}^{\text{DL}} - 1; \qquad m' = m + \max(N_{\text{RB}}^{\text{DL}}) - N_{\text{RB}}^{\text{DL}} \tag{9.41}$$

The variables v and v_{shift} define the position in the frequency-domain for different reference signals, where v is given by the following expression [3]:

$$v = \begin{cases} 0 & \text{if } p = 0 \text{ and } l = 0 \\ 3 & \text{if } p = 0 \text{ and } l \neq 0 \\ 3 & \text{if } p = 1 \text{ and } l = 0 \\ 0 & \text{if } p = 1 \text{ and } l \neq 0 \\ 3(n_s \bmod 2) & \text{if } p = 2 \\ 3 + 3(n_s \bmod 2) & \text{if } p = 3 \end{cases} \tag{9.42}$$

A cell-specific frequency shift is given by $v_{\text{shift}} = N_{\text{ID}}^{\text{cell}} \bmod 6$ (Figure 9.50). As shown in Figure 9.50, the resource element allocation for CRSs cycles once every six physical cell identity values, i.e., cell identity values that are integer multiples of 6 has the same resource element allocation as cell identity 0. This corresponds to one cycle for every two physical layer cell-identity groups. The resource elements

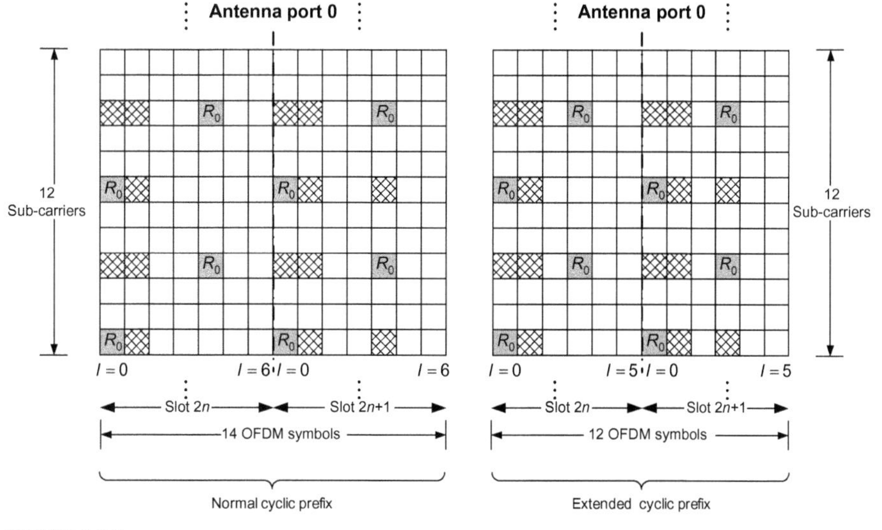

FIGURE 9.50

Cell-specific reference signal mapping for normal and extended cyclic prefix subframes [3].

(k, l) used for transmission of CRSs on an any of the antenna ports in a slot are not used for any transmission on another antenna port in the same slot and are set to zero. In an MBSFN subframe, CRSs are only transmitted in the non-MBSFN region of the MBSFN subframe. Figure 9.49 shows the CRS mapping to the resource grid. The procedure for generation of the CRSs is illustrated in Figure 9.51.

Given a different number of available OFDM symbols in a subframe with an extended cyclic prefix, the mapping of the CRSs is different than that for subframes with a normal cyclic prefix. Figure 9.50 illustrates and compares the CRS mapping in subframes with a normal and extended cyclic prefix associated with antenna port 0 and 4 transmit antennas. While the number of resource elements mapped with reference signals is the same, the CRS overhead in subframes with an extended cyclic prefix is higher due to a lower number of total sub-carriers in a resource block.

As shown in Figure 9.51, the CRSs are based on length-31 Gold pseudo-random sequences. The length-31 Gold pseudo-random sequence is generated with the seed c_{init}, based on the slot number, symbol number, cell identity, and cyclic prefix type. While the sequence itself is 230 bits long, the number of bits from the sequence selected for transmission is based on the largest channel bandwidth, which is currently 20 MHz. Therefore, $4\max(N_{\text{RB}}^{\text{DL}})$ bits are selected from

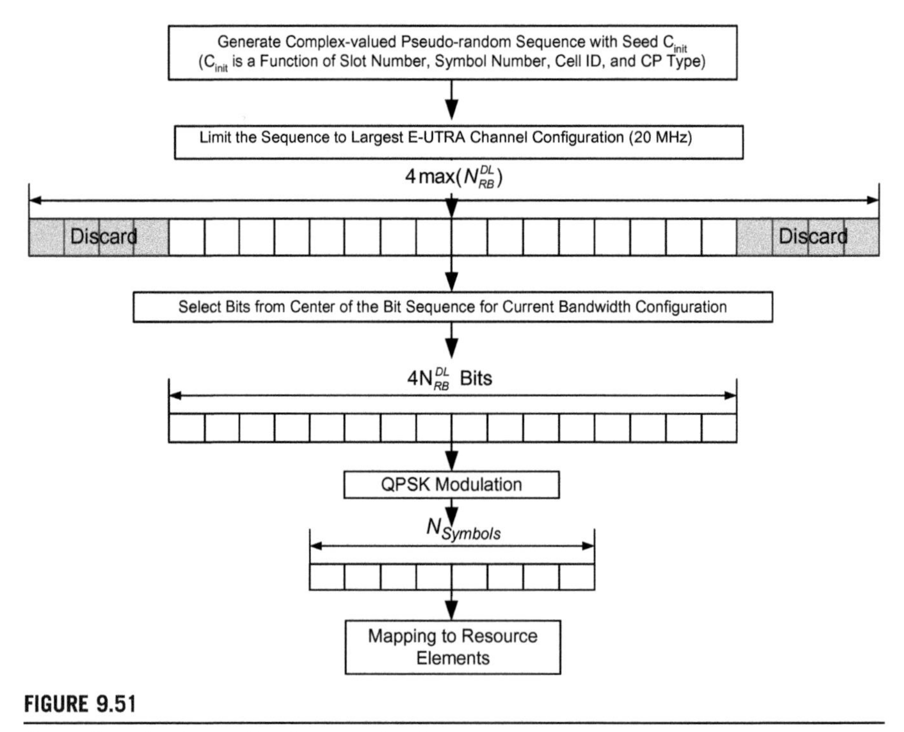

FIGURE 9.51

Generation of cell-specific reference signals.

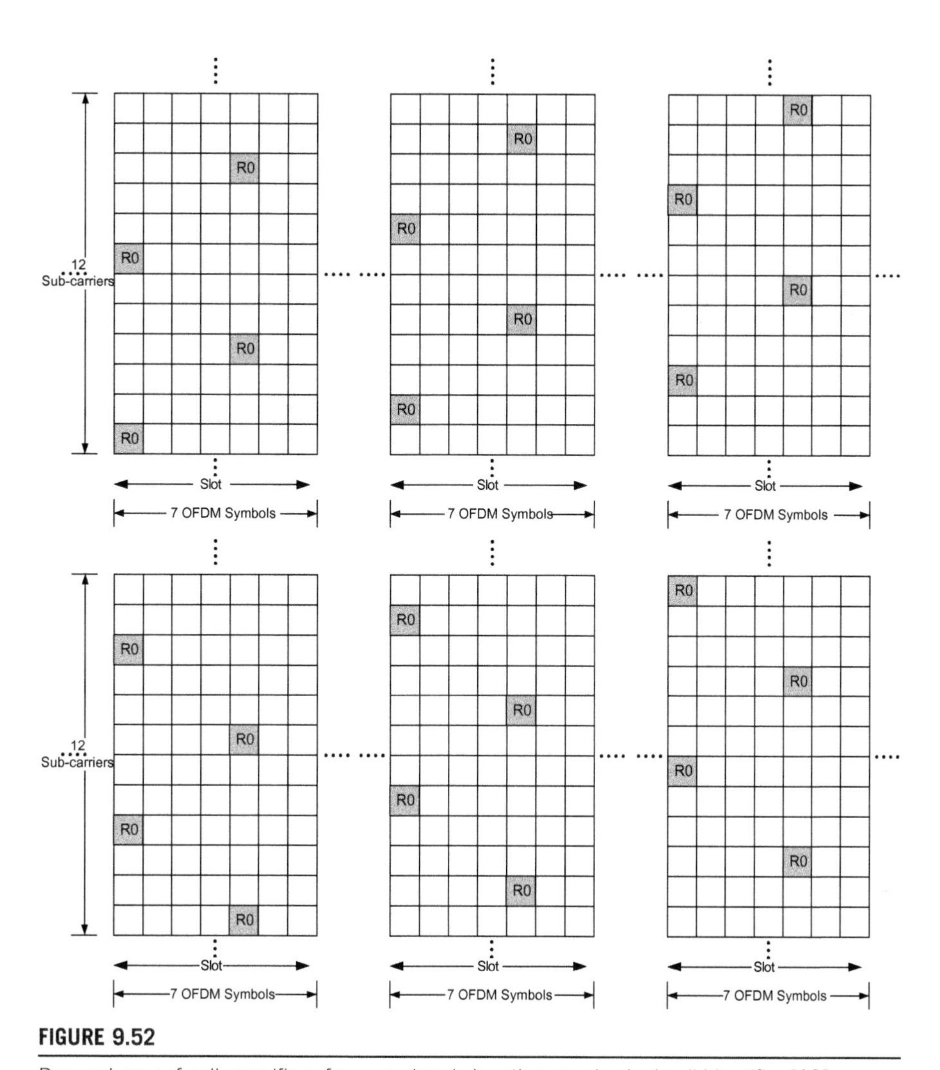

FIGURE 9.52

Dependency of cell-specific reference signals location on physical cell identifier [13].

the sequence where $\max(N_{RB}^{DL})$ is the maximum number of resource blocks in 20 MHz. From $4\max(N_{RB}^{DL})$, $4N_{RB}^{DL}$ bits are selected, where N_{RB}^{DL} is the number of resource blocks corresponding to the system bandwidth. The bits are selected from the center of the truncated sequence for the maximum channel configuration. This ensures that the same reference symbols are present around PBCH regardless of channel bandwidth. The PBCH uses the six resource blocks (72 subcarriers) at the center of the channel. The UE can use these symbols for channel estimation for PBCH detection without knowing the overall channel bandwidth. The channel bandwidth in terms of resource blocks is determined from the

PBCH. After PBCH is decoded, the UE determines the length of the reference signal used for the current channel configuration.

9.10.3.1 Received signal strength indicator measurement

Received signal strength indicator (RSSI) is the total power that UE observes across the entire system bandwidth. This includes the main signal and co-channel non-serving cell signal, adjacent channel interference, and the thermal noise within the specified frequency band. This is the power of the received signal prior to demodulation, thus the UE can measure this power without any synchronization and demodulation [6].

9.10.3.2 Reference signal received power measurement

Reference signal received power (RSRP) is defined as the linear average over the power contributions of the resource elements that carry cell-specific reference signals within the considered measurement bandwidth. For the purpose of RSRP calculation, the R_0 cell-specific reference signals are used; however, if the UE can reliably detect R_1 cell-specific reference signals, then it may use R_1 in addition to R_0 in order to calculate RSRP. The reference point for the RSRP measurement is the antenna connector of the UE. If the UE uses a receive diversity scheme, the reported RSRP value must not be lower than individual RSRP measured at any of the diversity branches. There are requirements for both absolute and relative RSRP. The absolute requirements range from ± 6 to ± 11 dB, depending on the noise level and environmental conditions. Measuring the difference in RSRP between two cells on the same frequency (intra-frequency measurement) is a more accurate operation for which the requirements vary from ± 2 to ± 3 dB. The requirements widen again to ± 6 dB when the cells are on different frequencies (inter-frequency measurement) [6].

9.10.3.3 Reference signal received quality measurement

Reference signal received quality (RSRQ) is defined as the ratio $\text{RSRQ} = N_{\text{RB}}\,\text{RSRP}/\text{RSSI}_{E\text{-UTRA}}$ (dB), where N_{RB} is the number of resource blocks over which the measurement is conducted. Note that the value of RSRQ in decibels is always negative, since the denominator is larger than the numerator. The measurements in the numerator and denominator are conducted over the same set of resource blocks. The E-UTRA carrier RSSI comprises the linear average of the total received power observed only in OFDM symbols containing the cell-specific reference symbols for antenna port zero over N_{RB} resource blocks in the measurement bandwidth by the UE from all sources, including co-channel serving and non-serving cells, adjacent channel interference, and thermal noise. If higher-layer signaling indicates certain subframes for performing RSRQ measurements, then RSSI is measured over all OFDM symbols in the designated subframes. The reference point for the RSRQ is the antenna connector of the UE. If the UE uses a receive diversity scheme, the reported RSRP value must not be lower than individual RSRQ measured at any of the diversity branches.

Table 9.10 RSRQ Mapping [6]

Reported RSRQ Value	Measured Quantity (dB)
RSRQ_00	RSRQ < −19.5
RSRQ_01	−19.5 ≤ RSRQ < −19
RSRQ_02	−19 ≤ RSRQ < −18.5
...	...
RSRQ_32	−4 ≤ RSRQ < −3.5
RSRQ_33	−3.5 ≤ RSRQ < −3
RSRQ_34	−3 ≤ RSRQ

Measuring RSRQ becomes particularly important near the cell-edge when decisions need to be made regardless of absolute RSRP to perform a handover to the target cell. Reference signal receive quality is used only during connected states. Intra- and inter-frequency absolute RSRQ accuracy varies from ±2.5 to ±4 dB, which is similar to the inter-frequency relative RSRQ accuracy of ±3 to ±4 dB.

It must be noted that RSRQ measurement provides additional information when RSRP is not sufficient to make a reliable handover or cell reselection decision. The RSRQ is the ratio of the RSRP and the RSSI, which depends on the measurement bandwidth, and thereby the number of resource blocks. The RSSI value includes the total received wideband power from all interference and thermal noise sources. Since RSRQ combines signal strength as well as interference level, its value further assists mobility management in E-UTRA. The UE typically measures RSRP or RSRQ per instructions from the network (via RRC messages) and reports the values back to the eNB. When it reports the RSRQ value, it does not use the real RSRQ value; rather it sends a non-negative value ranging from 0 to 34, and each of these values are mapped to a specific range of real RSRQ values as given in Table 9.10. The reporting range of RSRQ is defined from −19.5 dB to −3 dB with 0.5 dB resolution. As an example, assume that only reference signals are transmitted in a resource block and that data, noise, and interference are not considered. In this case, the RSRQ is equal to −3 dB. If cell-specific reference signals and the sub-carriers carrying data are equally powered, the RSRQ is equal to 1/12 or −10.79 dB.

9.10.4 UE-Specific reference signals associated with PDSCH

It was mentioned that cell-specific reference signals are transmitted on each antenna port. An antenna port is generally used as a generic term for signal transmission under identical channel conditions. The symbols that are transmitted via identical antenna ports are subject to the same channel conditions. In order to determine the characteristics of the channel from an antenna port, a UE must carry out a separate channel estimation for each antenna port. Therefore, separate reference signals that are suitable for estimating the respective channel are defined for each antenna port. There is one resource grid per antenna port. The cell-specific reference signals are used for both coherent detection and mobility-related measurements. Their structure in time and frequency ensures sufficient channel estimation accuracy. An LTE-Advanced eNB is required to support LTE Rel-8 user terminals. Since the structure of PCFICH, PHICH, PDCCH, PBCH, and PDSCH (transmit diversity only) is unchanged in Rel-10, the support of CRS is also required for LTE-Advanced UEs in order to detect these essential physical channels. The UE-specific reference signals are supported for transmission of PDSCH, and are transmitted on antenna ports $p=5$, $p=7$, $p=8$, or $p=7,8,\ldots,N_l+6$, where N_l is the number of spatial layers used for transmission of PDSCH. The UE-specific reference signals or DM-RS can be precoded and support non-codebook-based precoding. The DM-RS pattern for a higher number of layers has been extended from dual-layer format of transmission mode 8 in Rel-9 to eight layers in Rel-10. The CSI-RS is transmitted on each physical or virtual antenna port, and is only used for channel state measurement/reporting purposes. Its channel estimation accuracy can be relatively poorer compared to that of DM-RS. In Rel-8, the CRS is used for both demodulation and measurement purposes. When using precoding in conjunction with CRSs, the precoding information must be sent to the UE, and the CRS power is boosted to support various types of geometry. The Rel-8 E-UTRA downlink MIMO transmission modes are based on CRS. Figure 9.53 shows the CRS-based procedure for precoding where the cell-specific reference signals are inserted after the data is precoded. Therefore, the cell-specific reference signals do not contain information about the

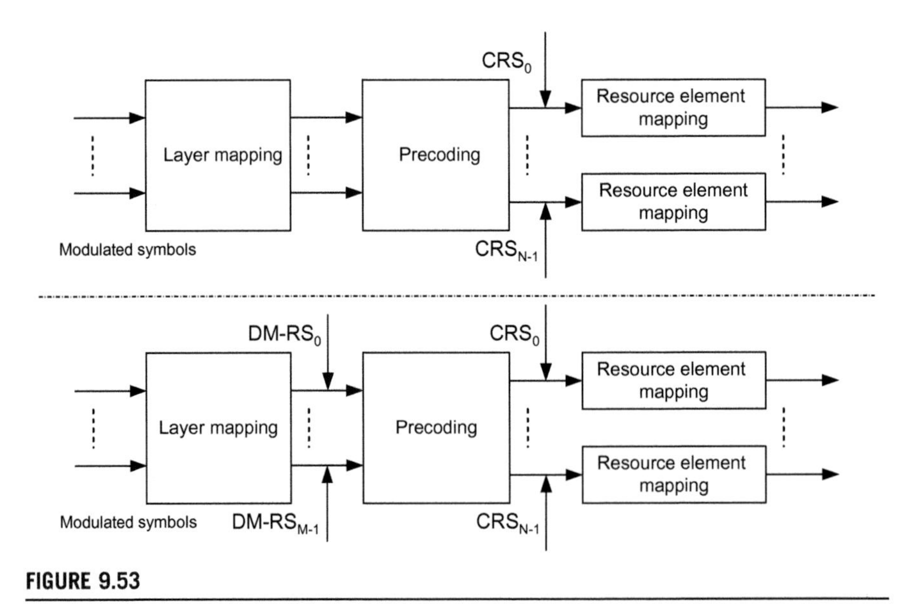

FIGURE 9.53

Comparison of the CRS and DM-RS usage in Rel-8 and Rel-10 of E-UTRA [110].

precoding scheme, and codebook information must be separately sent to the UE. In contrast, in LTE-Advanced, downlink MIMO transmission is based on DM-RS and CSI-RS. Figure 9.53 shows the DM-RS-based procedure for precoding where UE-specific reference signals are precoded in the same way as the data. Therefore, no separate precoding information is sent to the UE. Note that in LTE Rel-10, DM-RS is used for demodulation and CSI-RS is used for channel state measurement purposes. Unlike CRS-based precoding, the precoding information is not needed for demodulation.

It was stated earlier that UE-specific reference signals are used for transmission of the downlink shared channel and are transmitted on antenna port(s) $p = 5$, $p = 7$, $p = 8$ or $p = 7, 8, \ldots, N_l + 6$, where N_l is the number of layers (up to eight). It must be noted that the UE-specific reference signals on any antenna port are a valid reference for data demodulation, if the data transmission is associated with the corresponding antenna port. The UE-specific reference signals are exclusively transmitted on the resource blocks to which the corresponding user data is mapped. The UE-specific reference signal is not mapped to resource elements (k, l) that are reserved for transmission of control signaling, physical channels, or physical signals.

For antenna port 5, the UE-specific reference-signal sequence $r_{n_s}(m)$ is defined by $r_{n_s}(m) = (1/\sqrt{2})[1 - 2c(2m)] + j(1/\sqrt{2})[1 - 2c(2m + 1)], \;\forall m = 0, 1, \ldots, 12N_{\mathrm{RB}} - 1$, where N_{RB} denotes the number of resource blocks allocated to the corresponding data transmission. The pseudo-random sequence $c(i)$ was defined earlier. The pseudo-random sequence generator is initialized with $c_{\mathrm{init}} = \left(\lfloor n_s/2 \rfloor + 1\right)\left(2N_{\mathrm{ID}}^{\mathrm{cell}} + 1\right)2^{16} + n_{\mathrm{RNTI}}$ at the start of each subframe where n_{RNTI} is the 16-bit radio network temporary identifier (RNTI) of the UE. For antenna ports $p \in \{7, 8, \ldots, N_l + 6\}$, the UE-specific reference-signal sequence $r_{n_s}(m)$ is defined by $r_{n_s}(m) = (1/\sqrt{2})[1 - 2c(2m)] + j(1/\sqrt{2})[1 - 2c(2m + 1)]$, where $m = 0, 1, \ldots, 12N_{\mathrm{RB}}^{\mathrm{DL}} - 1$ (for normal cyclic prefix) or $m = 0, 1, \ldots, 16N_{\mathrm{RB}}^{\mathrm{DL}} - 1$ (for extended cyclic prefix). The pseudo-random sequence $c(i)$ is defined similar to the previous section. Prior to LTE Rel-11, the pseudo-random sequence generator was initialized with $c_{\mathrm{init}} = \left(\lfloor n_s/2 \rfloor + 1\right)\left(2N_{\mathrm{ID}}^{\mathrm{cell}} + 1\right)2^{16} + n_{\mathrm{SCID}}$ at the start of each subframe. The value of the scrambling identity n_{SCID} is zero unless specified otherwise. For a data transmission on ports 7 or 8, n_{SCID} is indicated by the DCI format 2B or 2C. In the case of DCI format 2B, $n_{\mathrm{SCID}} = 0$ if the scrambling identity field is zero. It must be noted that CQI estimation at the terminal is always based on cell-specific reference signals. In LTE Rel-11, and as a result of the introduction of the new PDCCH structure based on the UE-specific reference signals, the initialization of the DM-RS sequences associated with antenna ports $p \in \{7, 8, \ldots, N_l + 6\}$ was modified as follows: $c_{\mathrm{init}} = \left(\lfloor n_s/2 \rfloor + 1\right)\left(2N_{\mathrm{ID}}^{(n_{\mathrm{SCID}})} + 1\right)2^{16} + n_{\mathrm{SCID}}$, where $N_{\mathrm{ID}}^{(i)}|_{i=0,1}$ is defined as $N_{\mathrm{ID}}^{(i)} = N_{\mathrm{ID}}^{\mathrm{cell}}$, if no value for $N_{\mathrm{ID}}^{\mathrm{DM\text{-}RS_i}}$ is conveyed by higher layers, or if DCI format 1A is used for the DCI associated with PDSCH transmission; otherwise $N_{\mathrm{ID}}^{(i)} = N_{\mathrm{ID}}^{\mathrm{DM\text{-}RS_i}}$ [3].

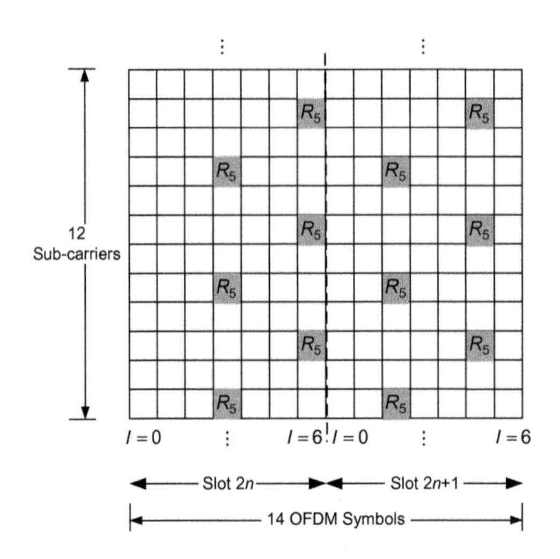

FIGURE 9.54

Mapping of UE-specific reference signals to antenna port 5 [3].

In a PRB with index n_{PRB} allocated for data transmission on antenna port 5, and assuming the normal cyclic prefix, the DM-RS $r_{n_s}(m)$ is mapped to the complex-valued modulation symbols a_{kl}^{p} over slot n_s as $a_{kl}^{5} = r_{n_s}(3l'N_{\text{RB}} + m')$, where $m' = 0, 1, \ldots, 3N_{\text{RB}} - 1$ denotes the number of UE-specific reference signal resource elements within a respective OFDM symbol allocated to user data transmission, $l' = \{0, 1\}$ for even-numbered slots, and $l' = \{2, 3\}$ for odd-numbered slots.

The UE-specific reference signal pattern for up to two layers, or alternatively rank-2 transmission, is identical to that of LTE Rel-9 to maintain backward compatibility. For up to four layers, the DM-RS pattern is obtained by extending LTE Rel-9 UE-specific reference signal pattern using a combination code-division and frequency-division multiplexing, as shown in Figure 9.55. The four precoded layers (antenna ports 7–10) are grouped into two groups of two resource elements that are spread with the same length-2 Walsh–Hadamard[13] orthogonal cover codes (OCCs) as in Rel-9. The UE-specific reference signals in different groups are frequency-multiplexed on adjacent sub-carriers. This pattern has been shown to provide reasonable performance for a variety of channel conditions and UE speeds [121].

[13]In coding theory, the Walsh–Hadamard code is a linear code over a binary alphabet that maps messages of length 2^n. The Walsh–Hadamard code is unique in that each non-zero codeword has a Hamming weight of exactly 2^{n-1}, which implies that the distance of the code is also 2^{n-1} resulting in a $[2^n, n, 2^{n-1}]_2$ code. The Walsh–Hadamard code is locally decodable, which enables recovery of parts of the original message with high probability, while only looking at a small fraction of the received codeword [27,40].

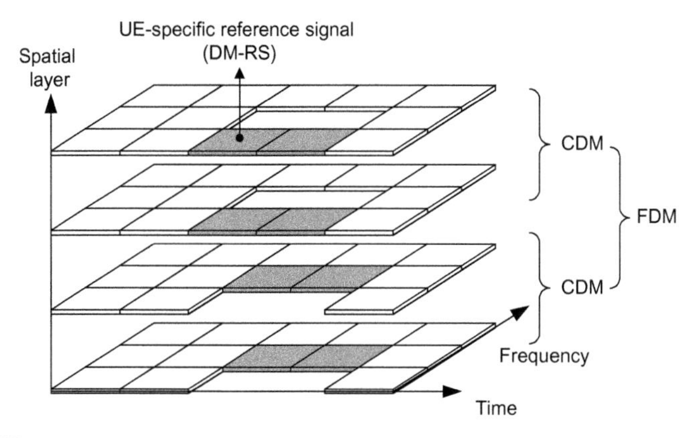

FIGURE 9.55

Code and frequency division multiplexing of the UE-specific reference signals [177].

For rank-8 transmission on antenna ports 7–14, the UE-specific reference signal pattern of Figure 9.55 is further extended using a hybrid CDM/FDM approach with two CDM groups each using a Walsh–Hadamard OCC of length 4, as shown in Figure 9.56. The UE-specific reference signals for each layer are spread across four resource elements, all having the same frequency location but different time locations within the subframe. This approach maintains the balanced power property of the rank-4 UE-specific pilot design and is optimized for pedestrian mobility class, which is the most likely use case for eight-layer transmission. The OCCs of length two are embedded in the OCC of length four to ensure backward compatibility with LTE Rel-9. The mapping of the OCCs to resource elements is shown in Figure 9.56. The mapping for odd and even resource blocks (in the frequency-domain) is different in order to provide orthogonality in both time and frequency dimensions. Such orthogonality improves the performance of channel estimation in high Doppler scenarios, and reduces the inter-carrier interference across the two CDM groups in the presence of frequency offsets. The OCC mapping also enables power balancing across OFDM symbols by peak power randomization [14]. The reference sequence for up to eight layers is the same as in Rel-9, except for antenna ports $AP_p|_{p=9,10,\ldots,14}$ where only one sequence initialization is possible.

In mathematical terms, for antenna ports $AP_p|_{p=7,8,\ldots,N_l+6}$, in a PRB with frequency-domain index n_{PRB} assigned for the corresponding user data transmission, part of the reference sequence $r(m)$ is mapped to complex-valued modulation symbols a_{kl}^p in a subframe (assuming normal cyclic prefix) as follows: $a_{kl}^p = w_p(l')$ $r[3l'\max(N_{\text{RB}}^{\text{DL}}) + 3n_{\text{PRB}} + m']$, where $w_p(i) = \overline{w}_p(i)$ if $(m' + n_{\text{PRB}}) \bmod 2 = 0$ or $w_p(i) = \overline{w}_p(3-i)$ if $(m' + n_{\text{PRB}}) \bmod 2 = 1$, where $\overline{w}_p(i)$ is given in Table 9.11. The UE-specific reference signal patterns for different transmission ranks (number of spatial layers) are shown in Figure 9.57.

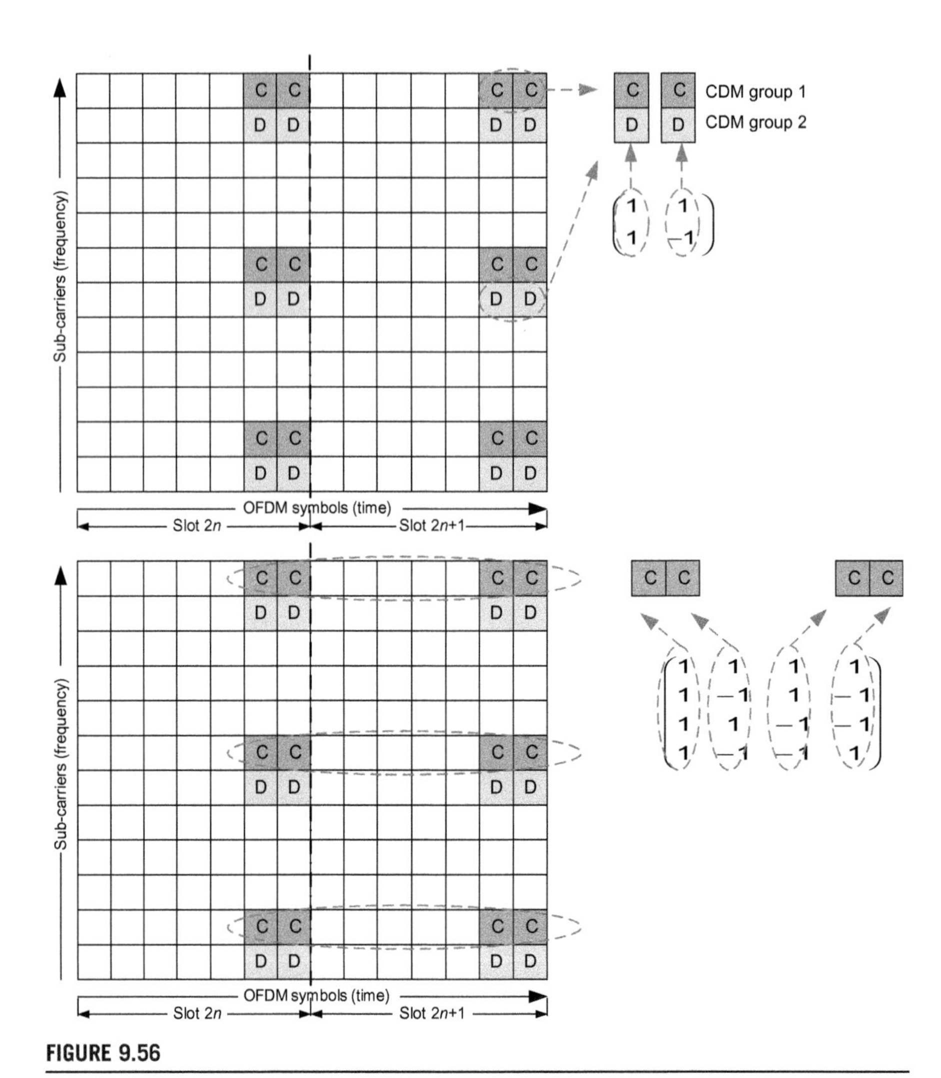

FIGURE 9.56

Length-2 and length-4 OCC mapping and DM-RS structure [121].

9.10.5 CSI reference signals

The new multi-user MIMO techniques that were proposed as part of Rel-10 and beyond require some level of CSI knowledge at the base station so that the system can dynamically adapt to the radio channel conditions experienced by the user terminals in order to optimize the performance. In TDD systems, this information is intuitively obtained from the uplink direction based on the reciprocity of the wireless channel, provided that the channel fading is sufficiently slow and transceivers at the base and mobile stations are calibrated, due to the fact that the

Table 9.11 Orthogonal Cover Code of Length 4 [3]

Antenna Port p	$[\overline{w}_p(0)\ \ \overline{w}_p(1)\ \ \overline{w}_p(2)\ \ \overline{w}_p(3)]$
7	$[+1\ \ +1\ \ +1\ \ +1]$
8	$[+1\ \ -1\ \ +1\ \ -1]$
9	$[+1\ \ +1\ \ +1\ \ +1]$
10	$[+1\ \ -1\ \ +1\ \ -1]$
11	$[+1\ \ +1\ \ -1\ \ -1]$
12	$[-1\ \ -1\ \ +1\ \ +1]$
13	$[+1\ \ -1\ \ -1\ \ +1]$
14	$[-1\ \ +1\ \ +1\ \ -1]$

same carrier frequency is used for transmission and reception. On the other hand, due to the asymmetry of FDD systems, feedback information over the uplink is required. Full CSI could cause an excessive overhead in some scenarios, thus quantized or statistical CSI are used in practice. In addition, terminal mobility can cause further degradation of the system performance, if the CSI arriving at the eNB is already outdated. Multi-antenna techniques in a multi-user scenario have the role of delivering streams of data in a spatially multiplexed manner to different users in such a way that all the degrees of freedom of a MIMO system are utilized. The idea is to perform intelligent space-division multiple access or multi-user beamforming so that the transmission of the base station is adapted to each user's instantaneous channel conditions in order to achieve the maximum gain in the direction of that user. The base station collects information on the CSI of each UE and subsequently decides on the optimal resource allocation strategy for each user.

Channel state information reference signals were introduced as part of Rel-10 to enable channel state measurement and reporting by UEs for up to eight downlink transmit-antennas to facilitate eNB precoding functions in transmission mode 9. LTE Rel-10 supports transmission of CSI-RS for 1, 2, 4, and 8 transmit antennas. The CSI-RSs also enable the UE to estimate the CSI for multiple neighbor cells in addition to its serving cell, to support multi-cell cooperative transmission schemes. The following principles have been incorporated in the design of the CSI-RS [14]:

- It is desirable to place CSI-RS at locations with uniform spacing in the frequency-domain.
- It is desirable to minimize the number of subframes containing CSI-RS in the time-domain so that a UE can estimate the CSI for different antenna ports and different cells with a minimal wake-up (active state) duty cycle when the UE is in DRX mode to reduce UE power consumption.
- The overall CSI-RS overhead involves a trade-off between accurate CSI estimation for efficient operation and minimizing the impact on legacy UEs

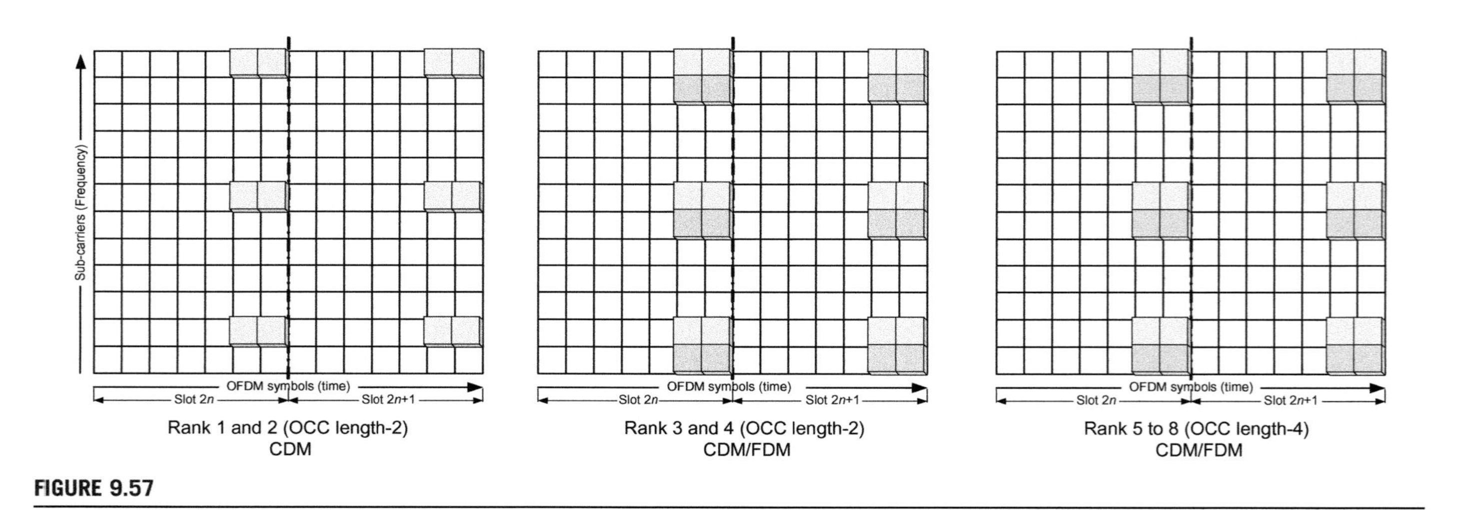

FIGURE 9.57

Mapping of UE-specific reference signals to resource blocks for different numbers of layers [177].

which are unaware of the presence of CSI-RS. Furthermore, the data sub-carriers of the legacy UE are punctured by the CSI-RS transmissions. Figure 9.58 shows that a CSI-RS density of one resource element per resource block per antenna port is a reasonable choice, as the throughput degradation compared to ideal CSI estimation is negligible.

- The CSI-RSs of different antenna ports within a cell, and ideally from other cells, should be orthogonally multiplexed to enable accurate CSI estimation.
- To ensure backward compatibility, CSI-RSs should not be mapped to the resource elements used for CRSs and control channels, as well as avoiding resource elements used for the Rel-10 UE-specific reference signals.

A set of CSI-RSs were designed in Rel-10, taking the above design considerations into account. An example CSI-RS pattern for 1, 2, 4, and 8 antenna ports is shown in Figure 9.58. The OCC codes of length 2 are used, so that CSI-RSs on two antenna ports share two resource elements on a given sub-carrier. The pattern shown in Figure 9.58 is applicable to frame structure type 1 and frame structure type 2 [121].

The CSI-RS patterns follow a nested structure, meaning that the resource elements used in the case of two CSI-RS antenna ports are a subset of those used for four and eight antenna ports; this helps to simplify the implementation. The total number of supported CSI-RS patterns is 40, which can be used to give a frequency-reuse factor of 5 between cells with eight antenna ports per cell, or a factor of 20 in the case of two antenna ports. It can be seen that collisions may occur with resource elements used for the DM-RS antenna ports defined in Rel-8 for PDSCH transmission mode 7. Therefore, it may be desirable to avoid scheduling UEs in transmission mode 7 in subframes containing CSI-RS. This pattern is designed to avoid collisions with the Rel-8 UE-specific reference signal antenna port, as this port is more suited to TDD operation where channel reciprocity can be more effectively exploited to support beamforming. However, this pattern only offers frequency-reuse factors between 3 and 12 depending on the number of antenna ports per cell [14].

The CSI-RS configuration is UE-specific. When the CSI-RS is configured, it is only present in certain subframes conforming to a given duty cycle and subframe offset. The duty cycle and offset of the subframes containing CSI-RSs and the CSI-RS pattern used in those subframes are provided to a Rel-10 UE through RRC signaling. The duty cycle and subframe offset are jointly coded, while the CSI-RS pattern is configured independently of these parameters. It should be noted that in subframes containing CSI-RSs, the rate-matching for PDSCH transmissions to Rel-10 UEs is considerate of the resource elements used for CSI-RS and the encoded PDSCH data is only mapped to the neighboring sub-carriers, whereas PDSCH transmissions to the legacy UEs are punctured by CSI-RS transmissions. Therefore, the density and periodicity of the CSI-RSs should be chosen

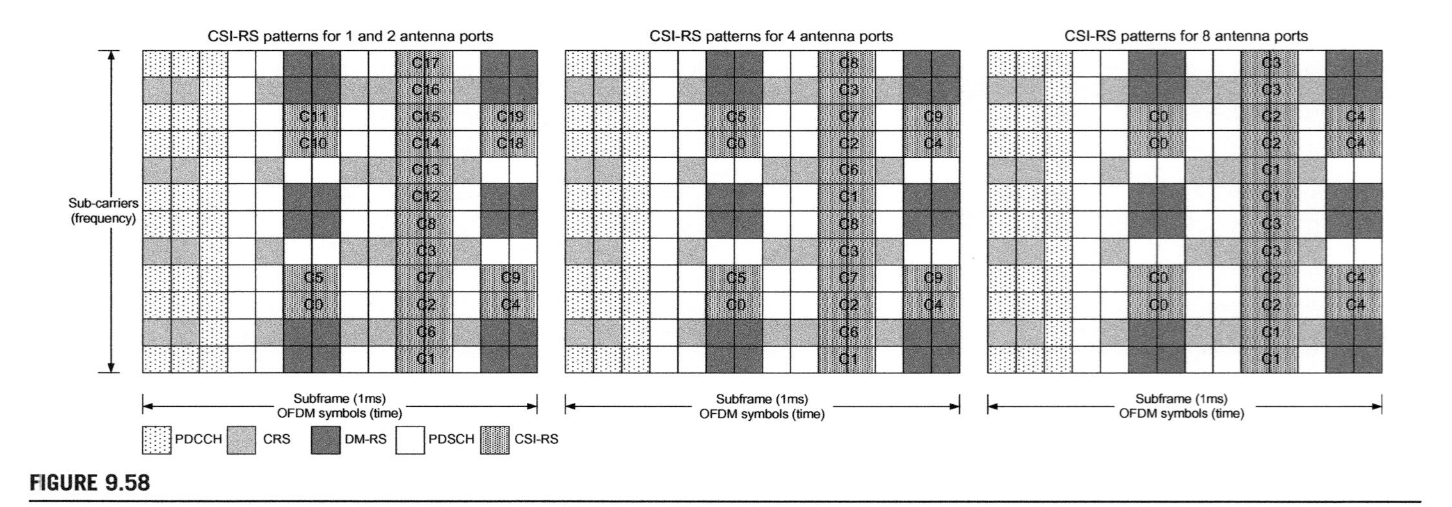

FIGURE 9.58

CSI-RS patterns for 1, 2, 4, and 8 transmit antennas [3,121].

such that the impact of puncturing[14] on the performance of these users is negligible.

In the context of cooperative MIMO, it may be possible to improve the performance of channel estimation, and especially interference estimation, by coordinating CSI-RS transmissions across multiple cells. In Rel-10 it is possible to mute a specific set of sub-carriers in data transmissions from a cell. The locations of these resource elements can be chosen to avoid colliding with CSI-RS transmissions from other cells, thus improving the inter-cell interference measurement quality. The muting pattern is signaled to the UEs using a 16-bit bitmap, where each bit corresponds to a 4-port CSI-RS configuration for frame structure type 1 or type 2. As a result, all the resource elements in the selected 4-port CSI-RS configuration are muted and the UE can assume zero transmission power on those resource elements, unless they are used to transmit CSI-RSs. In subframes in which CSI-RS muting is configured, data transmissions for Rel-10 UEs are rate-matched around the muted resource elements in the same way as for CSI-RSs. Data transmitted to Rel-8/9 UEs is punctured at the muted resource elements [15].

The CSI-RSs are transmitted on one, two, four, or eight antenna ports $AP_p = \{p = 15\}, \{p = 15, 16\}, \{p = 15, \ldots, 18\}, \{p = 15, \ldots, 22\}$ and are only defined for sub-carrier spacing of 15 kHz. The reference signal sequence is defined as $r_l^{n_s}(m) = (1/\sqrt{2})[1 - 2c(2m)] + j(1/\sqrt{2})[1 - 2c(2m+1)], \quad \forall m = 0, 1, \ldots, \max(N_{RB}^{DL}) - 1$, where n_s is the slot number within a radio frame and $l = 0, 1, \ldots, 6$ (or 5 for an extended cyclic prefix) is the OFDM symbol number within the slot. The pseudo-random sequence $c(i)$ was defined in the previous sections. The pseudo-random sequence generator is initialized with $c_{init} = 2^{10}(7(n_s + 1) + l + 1)(2N_{ID}^{CSI} + 1) + 2N_{ID}^{CSI} + N_{CP}$ at the start of each OFDM symbol where $N_{CP} = 1$ for a normal cyclic prefix and $N_{CP} = 0$, otherwise. In subframes configured for CSI-RS transmission, the reference signal sequence $r_l^{n_s}(m)$ is mapped to complex-valued modulation symbols $a_{kl}^{p} = w_\alpha r_l^{n_s}(m')$ which have been designated as reference symbols on AP_p where $m' = m + \lfloor [\max(N_{RB}^{DL}) - N_{RB}^{DL}]/2 \rfloor; m = 0, 1, \ldots, N_{RB}^{DL} - 1$ and $w_\alpha = 1$ for $p \in \{15, 17, 19, 21\}$ and $w_\alpha = (-1)^\alpha$ for $p \in \{16, 18, 20, 22\}$. The virtual CSI identifier in the above equation is equal to the cell identity unless configured by higher layers.

[14] In coding theory, puncturing is the process of removing some of the parity bits after encoding with an error-correction code. This has the same effect as encoding with an error-correction code with a higher rate, or less redundancy. However, with puncturing the same decoder can be used regardless of how many bits have been punctured, thus puncturing considerably increases the flexibility of the system without significantly increasing its complexity. In some cases, a predefined pattern of puncturing is used in an encoder. Then, the inverse operation, known as de-puncturing, is implemented by the decoder. Puncturing is used in UMTS and LTE during the rate-matching process. Puncturing is often used with the soft decoding and Viterbi algorithm [27,40].

The following parameters for CSI-RS are configured via higher layer signaling [5]:

- Number of CSI-RS ports
- CSI-RS configuration
- CSI-RS subframe configuration $I_{\mathrm{CSI-RS}}$
- Subframe configuration period $T_{\mathrm{CSI-RS}}$
- Subframe offset $T_{\mathrm{Offset-CSI-RS}}$
- UE assumption on reference PDSCH transmitted power for CSI feedback P_c. The parameter P_c is the assumed ratio of PDSCH energy per resource element (EPRE)[15] to CSI-RS EPRE when UE derives CSI feedback and takes values in the range of [−8 dB to 15 dB] in 1 dB step sizes.

Multiple CSI-RS configurations can be used in a given cell, which include zero or one configuration for which the UE can assume non-zero transmission power, and zero or more configurations for which the UE can assume zero transmission power. The UE assumes that CSI-RSs are not transmitted in the special subframe(s) in the case of frame structure type 2, in subframes where transmission of a CSI-RS would collide with transmission of synchronization signals, PBCH, or system information block type 1 messages, in subframes configured for transmission of paging messages for any UE with the cell-specific paging configuration. The resource elements (k, l) that are used for transmission of CSI-RS on any of the antenna ports $AP_p = \{15\}, \{15, 16\}, \{17, 18\}, \{19, 20\}, \{21, 22\}$ are not used for transmission of PDSCH on any antenna port in the same slot, and are not used for CSI-RS on any antenna port other than those in the above set of antennas in the same slot. The mapping of CSI-RS for 1, 2, 4, and 8 antenna ports are shown in Figure 9.58 [3].

Depending on the number of CSI-RS antenna ports, there are multiple reuse patterns on different locations which share a common base pattern. In the case of 1, 2, 4, and 8 antenna ports, there are 40, 20, 10, and 5 reuse patterns, respectively, allowing different cells to utilize different reuse patterns to avoid inter-cell CSI-RS collision. In addition, the base patterns for different numbers of CSI-RS antenna ports have a nested structure, allowing for simpler CSI-RS transmitter and receiver implementation. Each antenna port's CSI-RS is spread over two adjacent resource elements across time such that two different CSI-RSs are code-

[15]Downlink power control determines the energy per resource element (EPRE). The term resource element energy denotes the energy prior to cyclic prefix insertion. The EPRE is calculated by averaging the energy over all constellation points based on the modulation scheme applied. A UE may assume downlink cell-specific reference signal EPRE is constant across the downlink system bandwidth and constant across all subframes until different cell-specific RS power information is received. The downlink cell-specific reference-signal EPRE can be derived from the downlink reference-signal transmit-power information provided by higher layers. The downlink reference-signal transmit power is defined as the linear average over the power contributions in Watts of all resource elements that carry cell-specific reference signals within the operating system bandwidth [5].

division multiplexed over two adjacent resource elements. One of the advantages of code-division-multiplexing in this case, relative to other multiplexing schemes such as FDM, is that CDM can balance transmission power across antenna ports in the frequency-domain. The CSI-RS density is related to the channel estimation accuracy. In general, larger CSI-RS density may provide better CSI measurement accuracy while reducing downlink resource utilization. Therefore, adequate CSI-RS density, where the CSI-RS occupies minimum downlink resources while providing reasonable CSI measurement accuracy, is desirable. Since CSI-RS is only used for downlink CSI measurement, its granularity is coarse due to feedback overhead; the system performance is relatively insensitive to the channel measurement accuracy as compared to that for DM-RS. Therefore, a reduced-overhead CSI-RS transmitted only once every 5, 10, 20, 40, or 80 ms is utilized in the LTE-Advanced. A typical transmission of CSI-RS, which may consist of four antenna ports with periodicity of 10 ms, would only require an overhead equivalent to 0.2% of the entire time and frequency resources. The low overhead is achieved by allocating a single resource-element-pair per CSI-RS antenna port, except when the CSI-RS has only one antenna port, where two resource elements are allocated per resource block pair (Table 9.12). The configuration period T_{CSI-RS} and the subframe offset $T_{Offset-CSI-RS}$ for CSI-RSs are given in Table 9.13. The parameter I_{CSI-RS} can be configured separately for CSI-RSs for which the UE assumes non-zero and zero transmission power. Those subframes containing CSI-RSs must satisfy the following equation $(10n_{subframe} + \lfloor n_s/2 \rfloor - T_{Offset-\ CSI-RS}) \bmod T_{CSI-RS} = 0$ [5,14].

Although different CSI-RS patterns are likely to be used in neighboring cells to minimize CSI-RS collision, neighbor cells' downlink data signals may still significantly interfere with CSI-RS received at a cell-edge UE in a heavily loaded network in interference-limited scenarios. In addition, accurate CSI measurement for neighbor cells is also required for support of coordinated multipoint transmissions. To mitigate the interference effects and to facilitate CSI measurement of neighbor cells, LTE-Advanced supports data reference element muting, whereby data reference elements colliding with CSI-RS in a neighbor cell are transmitted with zero transmission power (i.e., muted resource elements).

Table 9.12 Relative Overhead of CSI-RS

Number of Antenna Ports	1	2	4	8
Cell-specific reference signal	4.76	9.52	14.28	–
UE-specific reference signal	7.14	7.14	14.28	14.28
CSI reference signal	$1.19/T_{CSI-RS}$	$1.19/T_{CSI-RS}$	$2.38/T_{CSI-RS}$	$4.76/T_{CSI-RS}$

Note that CSI-RS overhead is a function of CSI-RS periodicity T_{CSI-RS} in number of subframes.

Table 9.13 CSI Reference Signal Subframe Configuration [5]

CSI-RS Subframe Configuration Parameter $I_{CSI\text{-}RS}$	CSI-RS Periodicity $T_{CSI\text{-}RS}$ (Subframes)	CSI-RS Subframe Offset $T_{Offset\text{-}CSI\text{-}RS}$
0–4	5	$I_{CSI\text{-}RS}$
5–14	10	$I_{CSI\text{-}RS} - 5$
15–34	20	$I_{CSI\text{-}RS} - 15$
35–74	40	$I_{CSI\text{-}RS} - 35$
75–154	80	$I_{CSI\text{-}RS} - 75$

For an LTE-Advanced UE, the eNB applies code rate-matching around the muted resource elements so that the LTE-Advanced UE is able to decode the allocated data sub-carriers.

As mentioned earlier, a cell can be configured with one, two, four, or eight antenna-port CSI-RS. The exact CSI-RS structure, including the exact set of resource elements used for CSI-RS in a resource block, depends on the number of antenna ports for which the CSI-RS is configured within the cell and may vary for different cells. There are 40 possible positions within a resource-block pair that the CSI-RS can be mapped, and in a given cell, a subset of the corresponding resource elements is used for CSI-RS transmission. In the case of two antenna-port CSI-RS within a cell, the CSI-RS consists of two adjacent reference symbols per resource-block pair. The CSI-RS pair is separated by applying OCCs to the reference-symbol pair. Thus, there can be 20 different CSI-RS configurations based on the 40 possible reference-symbol positions.

In the case of four or eight antenna-port CSI-RS, the CSI-RS are pairwise frequency multiplexed. Thus, there is a possibility for ten or five different antenna ports for which the CSI-RS is configurations, each supporting four or eight CSI-RS, respectively. In the case of a single antenna-port CSI-RS, the same structure as for two antenna-port CSI-RS is used, although with only one of the OCCs. The overhead of a single antenna-port CSI-RS, in terms of occupied resource elements, is the same as two antenna-port CSI-RSs. In the time-domain, the CSI-RS can be transmitted with different periods, ranging from 5 to 80 ms. With a 5 ms period, the overhead per CSI-RS is roughly 0.12% per CSI-RS (one resource element per CSI-RS per resource-block pair, in every fifth subframe), thus longer periods result in lower overhead. The exact subframe in which CSI-RS is transmitted in a cell can also be configured, allowing for separation of CSI-RS transmissions among cells in the time-domain, in addition to using different sets of resource elements within the same subframe. In subframes in which CSI-RS is transmitted, the CSI-RS is transmitted in every resource block across the frequency. In other words, the CSI-RS transmission covers the entire cell bandwidth and by this means it is considered a wideband reference signal.

As mentioned above, when DM-RSs are transmitted within a resource block, the corresponding resource elements on which the reference signals are transmitted are explicitly avoided when mapping PDSCH symbols to the resource block. This restricted resource mapping, which is obviously not understood by legacy UEs, is possible as DM-RSs can be assumed to be transmitted only in resource blocks in which terminals supporting such reference signals are scheduled. In general, a legacy terminal is assumed to be never scheduled in a resource block where DM-RSs are transmitted. The situation is somewhat different for CSI-RS. As CSI-RS is transmitted within all resource blocks in the frequency-domain, it would not be practically possible to assume that legacy terminals are never scheduled in a resource block in which CSI-RS is transmitted. If the PDSCH mapping were modified to explicitly avoid the resource elements in which CSI-RS is transmitted, the mapping would not be recognized by legacy terminals. In order to alleviate this problem, the PDSCH is mapped consistent with LTE Rel-8, i.e., the mapping is not modified to avoid the resource elements on which CSI-RS is transmitted, and CSI-RS is transmitted on top of the corresponding PDSCH symbols. This will obviously impact the PDSCH demodulation performance, as some PDSCH symbols will be interfered by CSI-RS transmission. However, the remaining PDSCH symbols will not be impacted and the PDSCH will still be decodable with some acceptable performance degradation. On the other hand, if a Rel-10 terminal is scheduled in a resource block in which CSI-RS is transmitted, the PDSCH mapping is modified to explicitly avoid the resource elements on which the CSI-RS is transmitted. The remaining resource elements are used for PDSCH transmission within the cell. However, it is also possible to additionally configure one or several subsets of the resource elements as muted CSI-RS.

A muted CSI-RS has the same structure as an unmuted CSI-RS, except that nothing is actually transmitted on the corresponding resource elements. A muted CSI-RS can thus be seen as a normal CSI-RS with zero power. The motivation for inclusion of the muted CSI-RS is to enable creation of transmission holes corresponding to actual CSI-RS transmissions in neighboring cells in order to: (1) make it possible for a terminal to receive CSI-RS of neighboring cells, without being severely interfered by transmissions in its serving cell. The CSI estimation on neighboring cells would be necessary for multi-cell-transmission techniques and (2) to reduce the interference to CSI-RS transmissions in other cells. This is especially applicable to heterogeneous networks where overlapping layers with cells of substantially different transmit power are expected to coexist in the same frequency band. There can be one or multiple sets of muted CSI-RS in a cell. In the case of interference avoidance to neighboring cells, it is typically sufficient to use a single set of muted CSI-RS, as one should typically avoid interference to a single lower-power cell. When muted CSI-RS are used to alleviate intra-cell interference, in order to improve reception of CSI-RS from neighboring cells, multiple sets of muted CSI-RS are needed since neighboring cells may use different CSI-RS configurations.

9.10.6 Enhanced PDCCH demodulation reference signals

The LTE Rel-8 PDCCH was designed with the underlying assumption that each eNB with collocated antennas serves each cell in the network and different cells are distinguished through different physical cell identifiers. Furthermore, the UE relies on CRSs to detect and decode PDCCH content. However, the increasing mobile data traffic necessitates the use of dense and heterogeneous networks wherein multiple low-power remote radio heads (RRHs) with geographically displaced antennas share the same cell identity with the overlaying macro eNB. The cell capacity may be dramatically increased due to spatially split gain of RRHs. In such cases, if one common PDCCH region is shared among the macro eNB and all RRHs, the time-division-multiplexed PDCCH suffers from limited capacity when confined to three OFDM symbols. Moreover, since the legacy interleaved PDCCH spans across the entire system bandwidth, frequency-domain inter-cell interference coordination schemes for control channels transmitted by different RRHs cannot be utilized by PDCCH.

A Rel-11 work item initiated the design of a new downlink control channel structure referred to as ePDCCH which relies on DM-RSs for coherent detection and is frequency-division-multiplexed with data region. The use of DM-RSs would allow precoding and beamforming of the downlink physical control channels in the future. Figure 9.59 illustrates the structure of ePDCCH in a subframe and how it can coexist with the legacy PDCCH without affecting the interoperability with the legacy LTE systems [178]. As shown in the figure, the ePDCCH regions are time-division-multiplexed with the legacy PDCCH region, are frequency-division-multiplexed with PDSCH, and avoid other essential physical channels and signals. Since ePDCCH only relies on UE-specific reference signals for demodulation, the use of ePDCCH in MBSFN subframes would eliminate the reliance on CRS for control channel demodulation in the future releases of LTE.

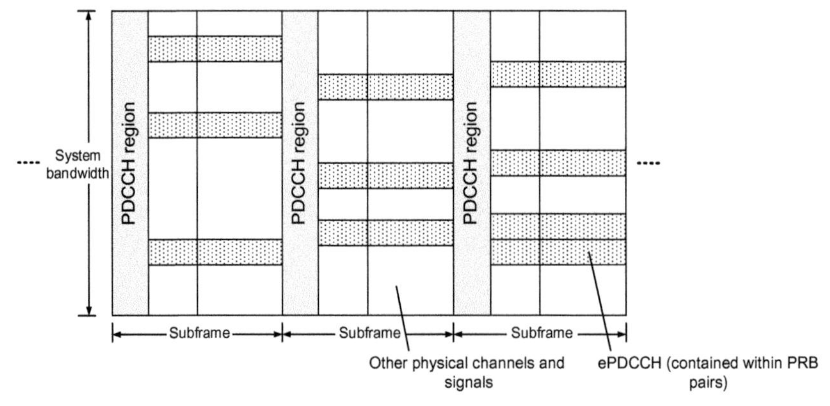

FIGURE 9.59

Enhanced PDCCH structure [149–155].

Each ePDCCH may contain multiple enhanced control channel elements (eCCEs), and multiple eCCEs can be aggregated to achieve different coding rates which depend on the channel conditions of the UE to guarantee a block error rate (BLER) of less than 1%. The existing structure of the PDCCH relies on CRS for channel estimation and coherent detection. It is shown that CRS has more overhead and is less effective when using closed-loop precoding techniques, beamforming, and MU-MIMO. The beamforming of the control channels can significantly improve the performance of the cell-edge users. The use of DM-RS or UE-specific reference signals instead of CRS would facilitate the use of beamforming for the control channels. Figure 9.60 illustrates a hypothetical beam-formed ePDCCH and depicts the localized and distributed ePDCCH mapping to the resource elements and antenna ports [149–155].

The ePDCCH DM-RSs are transmitted on antenna ports $p \in \{107, 108, 109, 110\}$ and are used as a reference for ePDCCH coherent demodulation. An ePDCCH DM-RS is not transmitted over resource element (k, l), where k denotes the frequency index and l denotes the OFDM symbol index within a slot, when one of the physical channels or physical signals are transmitted using the same resource element regardless of their antenna port. For the antenna port $p \in \{107, 108, 109, 110\}$, the reference-signal sequence $r(m)$ is defined by the equation $r(m) = (1/\sqrt{2})(1 - 2c(2m)) + j(1/\sqrt{2})(1 - 2c(2m + 1))$, where $m = 0, 1, \ldots,$

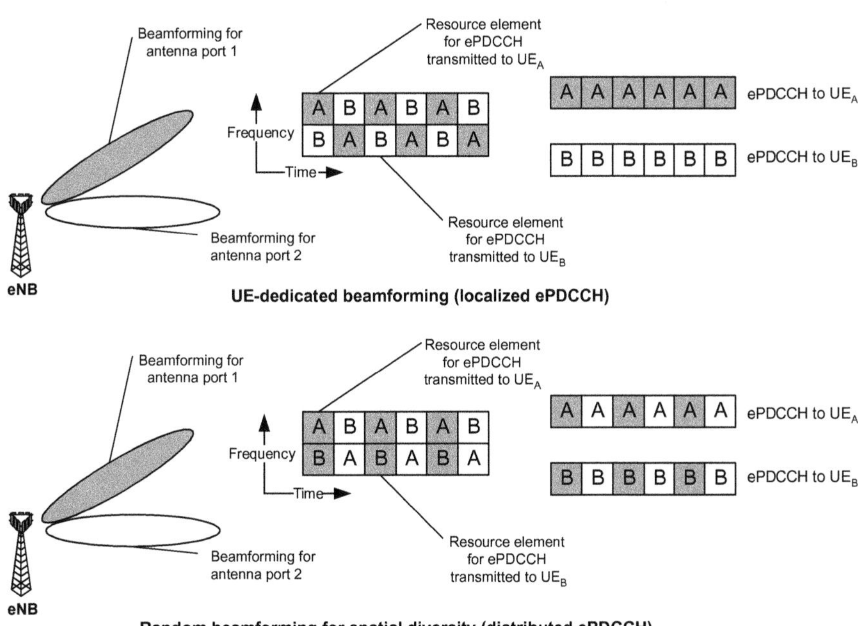

FIGURE 9.60

Possible use of beamforming for transmission of ePDCCH [149–155].

$12\max(N_{RB}^{DL}) - 1$ for a normal cyclic prefix, and $m = 0, 1, \ldots, 16\max(N_{RB}^{DL}) - 1$ for an extended cyclic prefix. The pseudo-random sequence $c(i)$ was defined in the previous sections and is initialized with the initial value $c_{init} = (\lfloor n_s/2 \rfloor + 1)\left(2N_{ID}^{ePDCCH(i)} + 1\right)2^{16} + N_{SCID}^{ePDCCH}$ at the start of each subframe where $N_{SCID}^{ePDCCH} = 2$ and $N_{ID}^{ePDCCH(i)}|_{i=0,1}$ is configured by higher layers, and index i denotes the set of ePDCCH associated with the DM-RS. For the antenna port $p \in \{107, 108, 109, 110\}$ in a PRB n_{PRB} assigned to the associated ePDCCH, the reference signal sequence $r(m)$ is mapped to complex-valued modulation symbols a_{kl}^{p} in a subframe according to the following mapping rule $a_{kl}^{p} = w_p(l')r(3l'\max(N_{RB}^{DL}) + 3n_{PRB} + m')$ for a normal cyclic prefix, where $w_p(i) = \overline{w}_p(i)$, if $(m' + n_{PRB}) \bmod 2 = 0$, and $w_p(i) = \overline{w}_p(3 - i)$ if $(m' + n_{PRB}) \bmod 2 = 1$ and the frequency index is given by $k = 5m' + 12n_{PRB} + k'$. The length-4 orthogonal cover sequence $\overline{w}_p(i)$ is defined in Table 9.14.

For an extended cyclic prefix, ePDCCH DM-RSs are not supported on antenna ports $p = 109$ and $p = 110$. The resource elements (k, l) used for transmission of ePDCCH DM-RSs to a single UE on any antenna ports in the set $S = \{107, 108\}$ or $S = \{109, 110\}$ is not used for transmission of ePDCCH on any other antenna port in the same slot, and further it is not used for DM-RSs to the same UE on any antenna port other than those in set S in the same slot. The mapping of the ePDCCH DM-RSs to resource elements within a UE-allocated resource block is identical to that corresponding to UE-specific reference signals on antenna port numbers 7–10 when replacing the antenna port numbers with 107–110 in Figure 9.57 for a normal cyclic prefix [3].

9.10.7 Positioning reference signals

Positioning is defined as the process of determining the location and/or velocity of a device using radio signals. The positioning reference signals (PRS) were introduced in LTE Rel-9 on antenna port 6, after the cell-specific reference signals were not found suitable for positioning due to the fact that the required probability of detection could not be guaranteed. A neighbor cell with its synchronization signals (primary/secondary synchronization signals) and reference

Table 9.14 The Sequence $\overline{w}_p(i)$ for Normal Cyclic Prefix [3]

Antenna Port p	$[\overline{w}_p(0)\ \overline{w}_p(1)\ \overline{w}_p(2)\ \overline{w}_p(3)]$
107	$[+1\ +1\ +1\ +1]$
108	$[+1\ -1\ +1\ -1]$
109	$[+1\ +1\ +1\ +1]$
110	$[+1\ -1\ +1\ -1]$

signals is seen as detectable, when the SINR is at least $-6\,\text{dB}$ [121,179]. Simulations during development of Rel-9 have shown that this can be only guaranteed for 70% of all cases for the third best-detected cell, which means the second best neighboring cell by assuming an interference-free environment, which cannot be ensured in a realistic scenario. The PRS design has some similarities with cell-specific reference signals as defined in LTE Rel-8. The PRS is a pseudo-random QPSK sequence that is mapped to resource elements in diagonal patterns with shifts in frequency and time to avoid collision with CRSs, and to avoid overlapping with the control channels.

The PRS is defined by its bandwidth $N_{\text{RB}}^{\text{PRS}}$ in terms of number of resource blocks used for PRS, time offset ($T_{\text{offset-PRS}}$), duration $N_{\text{subframe}}^{\text{PRS}}$ in terms of number of consecutive subframes, and its periodicity T_{PRS}. It must be noted that the PRS bandwidth is always smaller than the actual system bandwidth ($N_{\text{RB}}^{\text{PRS}} < \max(N_{\text{RB}}^{\text{DL}})$). The PRS is always mapped around the carrier center frequency and the unused DC sub-carrier in the downlink. In a subframe where PRS is configured, typically no PDSCH is transmitted. The positioning reference symbols of a certain cell can be configured to correspond to empty resource elements of the neighboring cells, thus enabling detection of the PRSs with satisfactory SIR values when receiving neighbor-cell PRSs [121].

Positioning reference signals are only transmitted in resource blocks over downlink subframes that are configured for PRS transmission. If both normal and MBSFN subframes are configured as positioning subframes within a cell, the OFDM symbols in the MBSFN subframe use the same cyclic prefix type as the first subframe in the frame. However, if MBSFN subframes are only configured to contain PRSs in a cell, the OFDM symbols configured for PRSs in the MBSFN region of these subframes use an extended cyclic prefix type. In a subframe configured for PRS transmission, the starting positions of the OFDM symbols configured for PRS transmission is identical to those in a subframe in which all OFDM symbols have the same cyclic prefix length as the OFDM symbols configured for PRS transmission. Positioning reference signals are transmitted on antenna port 6, and are not mapped to resource elements (k,l) allocated to BCH and primary and secondary synchronization channels regardless of their associated antenna ports. PRSs are only defined for sub-carrier spacing of 15 kHz.

The reference-signal sequence $r_l^{n_s}(m)$ is defined by $r_l^{n_s}(m) = (1/\sqrt{2})[1-2c(2m)] + j(1/\sqrt{2})[1-2c(2m+1)]$, $\forall m = 0,1,\ldots,2\max(N_{\text{RB}}^{\text{DL}})-1$, where $n_s = 0,1,\ldots,19$ is the slot number within a radio frame, $l = 0,1,\ldots,6$ is the OFDM symbol number within the slot. The pseudo-random sequence $c(i)$ is defined similar to the previous sections. The pseudo-random sequence generator in this case is initialized by $c_{\text{init}} = 2^{10}[7(n_s+1)+l+1](2N_{\text{ID}}^{\text{cell}}+1) + 2N_{\text{ID}}^{\text{cell}} + N_{\text{CP}}$ at the start of each OFDM symbol where $N_{\text{CP}} = 1$ for a normal cyclic prefix and $N_{\text{CP}} = 0$ for an extended cyclic prefix. The reference signal sequence $r_l^{n_s}(m)$ is mapped to complex-valued modulation symbols a_{kl}^{p} designated to the PRS on antenna port $p = 6$ such that $a_{kl}^{p} = r_l^{n_s}(m')$ where in the case of a normal cyclic prefix

Table 9.15 Positioning Reference Signal Subframe Configuration Parameters [3]

PRS Configuration Index I_{PRS}	PRS Periodicity T_{PRS} (Subframes)	PRS Subframe Offset $T_{offset-PRS}$ (Subframes)
0–159	160	I_{PRS}
160–479	320	$I_{PRS} - 160$
480–1119	640	$I_{PRS} - 480$
1120–2399	1280	$I_{PRS} - 1120$
2400–4095		Reserved

$m' = m + \max(N_{RB}^{DL}) - N_{RB}^{PRS}$. The bandwidth for PRSs and N_{RB}^{PRS} is configured by higher layers and the cell-specific frequency shift is given by $\nu_{shift} = N_{ID}^{cell} \bmod 6$.

The cell-specific subframe configuration period T_{PRS} and the cell-specific subframe offset $T_{offset-PRS}$ for transmission of PRSs are given in Table 9.15 (see Figure 9.61). The PRS configuration index I_{PRS} is configured by higher layers. Positioning reference signals are transmitted only in configured downlink subframes and, in the case of TDD duplex scheme, are not transmitted in special subframes. The PRS is transmitted in N_{PRS} consecutive subframes that are configured by higher layers. The PRS location in time is given by $(10n_{subframe} + \lfloor n_s/2 \rfloor - T_{offset-PRS}) \bmod T_{PRS} = 0$. A UE may assume that downlink PRS EPRE is constant across the PRS bandwidth and across all OFDM symbols that contain positioning reference signals in a given PRS occurrence. The structure of the positioning reference signals in the time- and frequency-domain is illustrated in Figure 9.61.

9.10.8 MBSFN reference signals

Support of MBMS in LTE has certain impacts on the physical layer design. It starts with the use of the cyclic prefix, since it is essential for the OFDM system that multipath versions of the signal arrive at the receiver within the cyclic prefix to avoid ISI. As we have multiple, but synchronized transmission from different broadcast/multicast sources, the expected delay spread is much larger than that for unicast operation. Therefore, in order to support multicast and broadcast services, MBSFN uses an extended cyclic prefix while the majority of today's LTE networks are based on a normal cyclic prefix. In the case of the mixed (or time-division-multiplexed) MBSFN and unicast operation, the subframes that are supposed to carry MBMS content are divided into a non-MBSFN and an MBSFN region. The non-MBSFN region can occupy one or two OFDM symbols at the beginning of the subframe. In this region, PCFICH, PDCCH, and PHICH will be mapped to ensure interoperability with user terminals that do not support MBSFN. For proper operation of the network, it is required to schedule terminals to receive or transmit data, page them, or provide feedback on their recent uplink transmission. In this region, CRSs and a normal cyclic prefix will be used. The MBSFN-region will carry PMCH. The extended cyclic prefix and sub-carrier spacing of

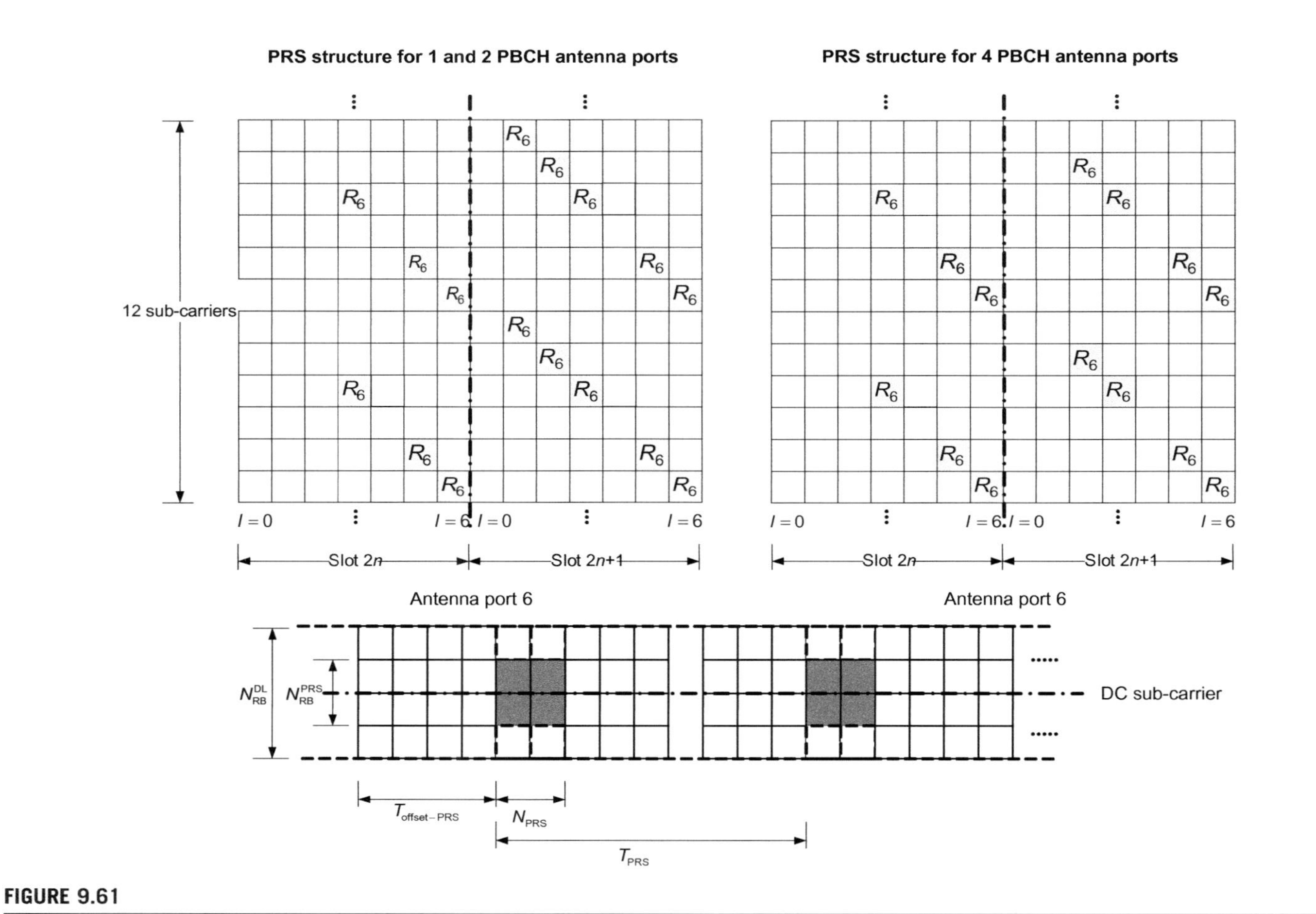

FIGURE 9.61

Structure of PRS in time- and frequency-domain [3,179].

15 kHz (or 7.5 kHz) are used in this region in order to contain the time difference of the cells that belong to an MBSFN area, while transmitting the same content to the terminals. In order to enable coherent demodulation at the terminal, as well as proper channel estimation, the use of CRSs is not sufficient. Thus new reference signals for MBMS transmission have been designed as part of Rel-9. Each cell belonging to the MBSFN area will transmit the same MBSFN reference signal pattern at the exact same time–frequency position. The MBSFN reference signals have a tighter spacing in the frequency-domain due to the frequency-selective nature of the radio channel. In contrast to the CRSs, the initialization sequence for the MBSFN reference signal does not depend on the physical cell identity, rather on the MBSFN identity provided by system information block type 13 [179].

The MBSFN reference signals are transmitted on antenna port 4 in the MBSFN region of MBSFN subframes, only if PMCH is transmitted. These reference signals are only defined for an extended cyclic prefix. The MBSFN reference-signal sequence $r_l^{n_s}(m)$ is obtained as $r_l^{n_s}(m) = (1/\sqrt{2})[1 - 2c(2m)] + j(1/\sqrt{2})\ [1 - 2c(2m+1)],\ \forall m = 0, 1, \ldots, 6\max(N_{RB}^{DL}) - 1$, where $n_s = 0, 1, \ldots, 19$ is the slot number within a radio frame, $l = 0, 1, \ldots, 5$ is the OFDM symbol number within the slot, and $c(i)$ is the pseudo-random sequence defined similar to the previous sections, which is initialized by $c_{init} = 2^9[7(n_s + 1) + l + 1](2N_{ID}^{MBSFN} + 1) + N_{ID}^{MBSFN}$ at the start of each OFDM symbol. The reference-signal sequence $r_l^{n_s}(m')$ in OFDM symbol l is mapped to complex-valued modulation symbols a_{kl}^p on antenna port $p = 4$ according to $a_{kl}^p = r_l^{n_s}(m')$, where $m' = m + 3(\max(N_{RB}^{DL}) - N_{RB}^{DL})$. Figure 9.62 illustrates the resource elements used for MBSFN reference signal transmission.

9.10.9 Reference signal overhead calculation

The reference signals regardless of their types add to the overhead and reduce the number of available resource elements in a subframe for user data allocation. The amount of overhead typically increases when multiple transmit antennas and/or an extended cyclic prefix is used. Among various types of reference signals used in LTE/LTE-Advanced, the cell-specific reference signals correspond to the highest amount of overhead, as they are always on and span the entire system bandwidth. The overhead concerning other types of reference signals varies depending on the number of antenna ports, periodicity of the configured reference signals, number of UE-allocated resource blocks, etc. The overhead is calculated as the number of resource elements in each resource block pair that is designated to the reference signals relative to the total number of resource elements in each resource block pair. The total number of resource elements is also a function of the cyclic prefix type. Table 9.16 provides the overhead figures associated with each reference signal type for FDD systems. The overhead of the reference signals in TDD systems is also a function of the downlink/uplink subframe configuration and the number of OFDM symbols in the DwPTS portion of the special subframes [13].

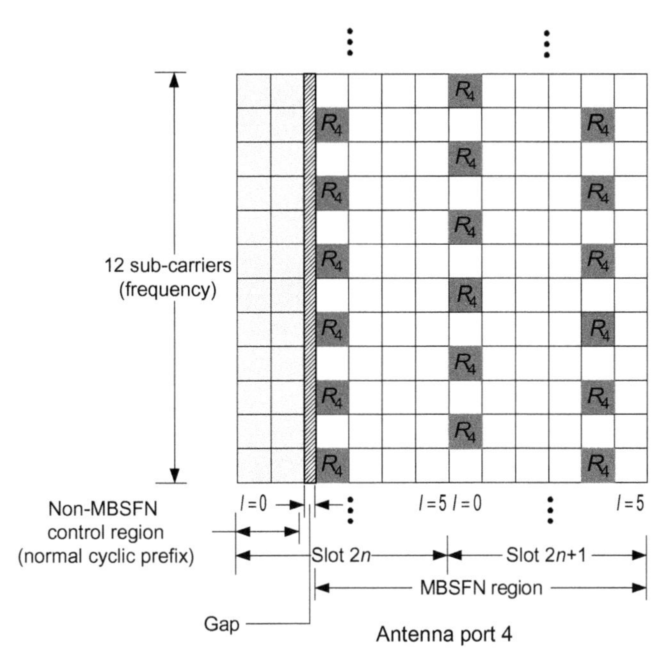

FIGURE 9.62

MBSFN reference signal structure [3,13].

9.11 Channel coding and modulation

Forward error correction (FEC) channel coding schemes are commonly used to improve the efficiency and robustness of wireless digital communication systems. On the transmitter side, an FEC encoder adds redundancy to the input data in the form of parity bits, while at the receiver, the redundancy is utilized by the FEC decoder to correct a number of channel errors. The use of FEC schemes would allow tolerance of more channel errors than otherwise can be tolerated without FEC. The use of FEC schemes enables the wireless communication systems to operate with a lower transmit power, to transmit over longer distances, to tolerate more interference, and to transmit at a higher data rate. A binary FEC encoder receives k bits at a time and produces a codeword of n bits, where $n > k$. While there are 2^n possible sequences of n bits, only 2^k codewords are used. The ratio k/n is the code rate and is denoted by R. Lower code rates corresponding to small values of R can generally correct more channel errors than higher code rates, and therefore are more robust. However, higher code rates are more bandwidth efficient because of lower overhead due to parity or redundancy bits. Thus, the selection of the code rate is a trade-off between energy efficiency and bandwidth efficiency. For every combination of code rate R, code word length n, baseband modulation scheme, channel type, and receiver noise power, there is a theoretical

Table 9.16 Reference Signal Overhead in FDD Mode [13]

Release	Type of Reference Signal	Number of Antenna Ports	Antenna Ports	Number of Reference Signals per Resource Block Pair	Overhead (%) Normal Cyclic Prefix	Overhead (%) Extended Cyclic Prefix
8	Cell-specific reference signals	1	0	8	4.8	5.6
		2	0,1	16	9.5	11.1
		4	0,1,2,3	24	14.3	16.7
	UE-specific (demodulation) reference signals	1	5	12	7.1	8.3
	MBSFN reference signals	1	4	18	–	12.5
9	Positioning reference signals	1	6	16 (14)	9.5	9.7
	UE-specific (demodulation) Reference Signals	2	7,8	12	7.1	8.3
10	UE-specific (demodulation) reference signals	1	7	12	7.1	8.3
		2	7,8	12	7.1	8.3
		4	7–10	24	14.3	16.6
		8	7–14	24	14.3	16.6
	CSI reference signals	1	15	2	$1.19/T_{CSI\text{-}RS}$	$1.39/T_{CSI\text{-}RS}$
		2	15,16	2	$1.19/T_{CSI\text{-}RS}$	$1.39/T_{CSI\text{-}RS}$
		4	15–18	4	$2.38/T_{CSI\text{-}RS}$	$2.78/T_{CSI\text{-}RS}$
		8	15–22	4	$2.38/T_{CSI\text{-}RS}$	$2.78/T_{CSI\text{-}RS}$
11	ePDCCH demodulation reference signals	4	107–110	12(24)	7.1–14.3	8.3–16.6

lower limit on the amount of energy that must be used to convey one bit of information. This limit is called the channel capacity of the communication channel [33]. Since the dawn of information theory, engineers and mathematicians have tried to construct codes that achieve performance close to Shannon capacity [27,33,39]. The introduction of convolutional turbo codes (CTCs) in 1993 followed by the invention of block turbo codes (BTCs) in 1994 closed much of the performance gap with the maximum channel capacity. Turbo codes are essentially parallel concatenated convolutional codes (PCCCs) with an internal interleaving mechanism combined with an iterative soft-decision decoding algorithm. In this section, we first review the principles of convolutional and turbo coding, and then we describe the channel coding and modulation schemes that are used in LTE/LTE-Advanced systems.

Before we proceed to the next sections, we need to clarify that detection, decision-making, hypothesis testing, and decoding are synonymous. The word detection refers to the effort to decide whether some phenomenon is present or not on the basis of some observations. For example, a radar system uses the data to detect whether a target is present. The meaning has been extended in the communication field to detect which one, among a set of mutually exclusive alternatives, is correct. Decision-making is the process of deciding between a number of mutually exclusive alternatives. Hypothesis testing is the same, and here the mutually exclusive alternatives are called hypotheses. Decoding is the process of mapping the received signal into one of the possible sets of codewords or transmitted symbols. We use the word hypotheses for the possible choices in what follows, since the word conjures up the appropriate intuitive image [100–108].

9.11.1 Principles of convolutional coding

Convolutional codes have been extensively used in telecommunication systems such as digital video, radio, mobile, and satellite communication in order to achieve reliable data communication [27]. Prior to the invention of the turbo codes, convolutional code was the most efficient coding scheme that provided the closest performance to the Shannon limit. A generic convolutional encoder (k, n, K) is characterized by a shift register with K stages, k input bits, and n output bits. The rate of the convolutional code is defined as $R = k/n$. A convolutional encoder is a discrete linear time-invariant system and can be viewed as a finite-state machine with 2^{K-1} states. Therefore, convolutional codes are linear codes which do not divide the source stream into blocks, but instead read and transmit bits continuously. The transmitted bits are a linear function of the past source bits. Usually the rule for generating the transmitted bits involves feeding the present source bit into a linear-feedback shift-register of length K, and transmitting one or more linear functions of the state of the shift register in each iteration. The resulting transmitted bit stream is the convolution of the source stream with a linear filter (Figure 9.63). The impulse–response function of this filter may have finite or infinite duration depending on the choice of feedback shift-register,

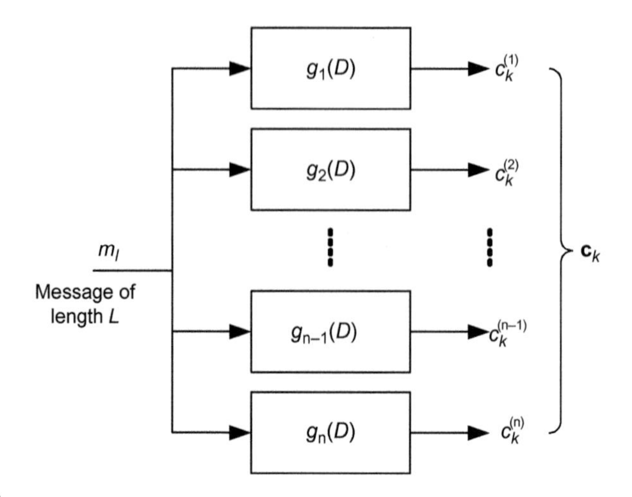

FIGURE 9.63

General form of (1/n) convolutional coder [27,40].

i.e., recursive or non-recursive convolutional codes. As an example, consider the LTE convolutional coder with $k = 1, n = 3, K = 6$, and $R = 1/3$. The linear combinations are defined via $n = 3$ generator sequences where $\mathbf{G} = \{\mathbf{g}_0, \mathbf{g}_1, \ldots, \mathbf{g}_{n-1}\}$ and $\mathbf{g}_i = \{g_{i0}, g_{i1}, \ldots, g_{i(K-1)}\}$. The generator sequences used in Rel-8 of LTE are given by $\mathbf{g}_0 = \{1\ \ 0\ \ 1\ \ 1\ \ 0\ \ 1\ \ 1\}$, $\mathbf{g}_1 = \{1\ \ 1\ \ 1\ \ 1\ \ 0\ \ 0\ \ 1\}$, and $\mathbf{g}_2 = \{1\ \ 1\ \ 1\ \ 0\ \ 1\ \ 0\ \ 1\}$ as shown in Figure 9.63. Puncturing is used to make an arbitrary k/n rate code from a basic rate 1/2 or 1/3 code by deleting some bits in the encoder output. For example, if we want to make a code with rate 2/3, we should take the basic 1/2-encoder output and transmit every second bit from the first branch and every bit from the second branch.

Traditional channel coding techniques achieve coding gain at the expense of bandwidth expansion, where coding gain is defined as the reduction in required E_b/N_0 for a given error probability. In the case of band-limited channels, bandwidth expansion is usually not possible. Trellis-coded modulation schemes use redundant non-binary modulation in conjunction with a finite state encoder. For each symbol interval, the encoder selects one of a set of M-ary waveforms, thereby generating a sequence of coded waveforms. The noisy signals at the receiver are detected and decoded by a soft-decision maximum-likelihood (ML) detector/decoder. Coding gain can be achieved without sacrificing data rate or expanding bandwidth, but at the expense of increasing decoder complexity.

A convolutional encoder, which is an example of trellis coding, is a finite-state machine, i.e., the output is a function of both the current input and the state of the encoder. A convenient way of characterizing such a state-oriented process is through a trellis diagram. Figure 9.64 illustrates a trellis diagram for a

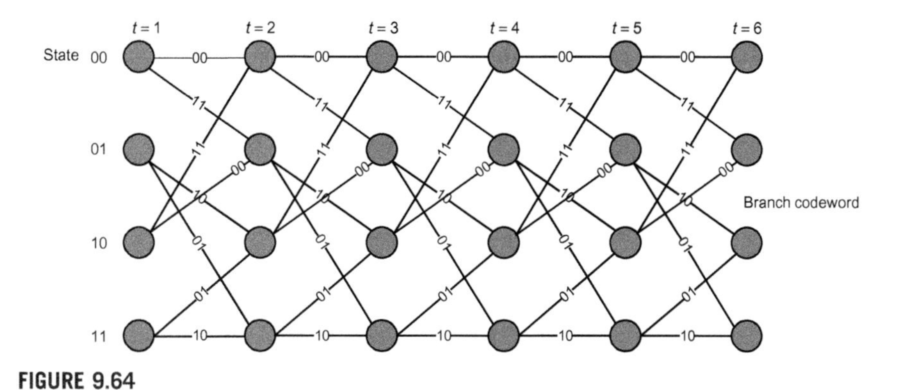

FIGURE 9.64

Four-state Trellis diagram [27,40].

hypothetical four-state convolutional encoder. The number of stages in the encoder shift register is referred to as its constraint length K, and the contents of the rightmost $K-1$ stages define the state of the encoder. Each node of the trellis represents a distinct state at a given time. A branch between any two states represents the transition between those states, and is labeled with the output code sequence associated with that transition. The most important property of the trellis diagram is that every possible sequence of message bits into the encoder corresponds to a unique path through the trellis and a unique sequence of coded bits, and vice versa. The task of the decoder is to estimate the path that the encoded message traverses through the trellis. If all input message sequences are equally likely, a decoder that will achieve the minimum probability of error is one that compares the conditional probabilities $p(\mathbf{z}_j|\mathbf{u}_m)$, where $\mathbf{z}_j$ is the received sequence and $\mathbf{u}_m$ is one of the possible transmitted sequences, and chooses the sequence corresponding to the ML. The decoder chooses $\mathbf{u}_{m'}$ if $p(\mathbf{z}_j|\mathbf{u}_{m'}) = \max_{\forall \mathbf{u}_m} p(\mathbf{z}_j|\mathbf{u}_m)$. Finding the sequence which maximizes $p(\mathbf{z}_j|\mathbf{u}_m)$ is equivalent to finding the sequence $\mathbf{u}_{m'}$ that has the greatest similarity to $\mathbf{z}_j$. Over a hard-decision channel, the similarity is often estimated by the Hamming distance between received and candidate sequences, and over a soft-decision channel it is estimated by the Euclidean distance. Since an ML decoder will choose the trellis path whose corresponding sequence $\mathbf{u}_{m'}$ is at the minimum distance to the received sequence $\mathbf{z}_j$, the ML problem is identical to the problem of finding the shortest distance through the trellis diagram [19]. A comparison of the performance of a rate 1/2 convolutional code with hard- and soft-decision decoding ($K=7$ and BPSK modulation) over an AWGN channel is shown in Figure 9.66. The advantage of a soft-decision decoding over a hard-decision decoding convolutional coder (approximately $+2$ dB coding gain) can be seen in this example. The performance of the same encoder with hard-decision decoding under Rayleigh multipath fading and AWGN channels for different number paths has been calculated and compared in Figure 9.66.

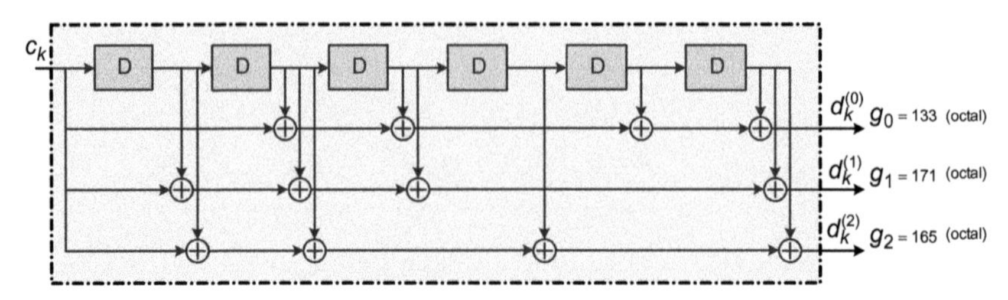

FIGURE 9.65

LTE rate 1/3 tail-biting convolutional encoder [4].

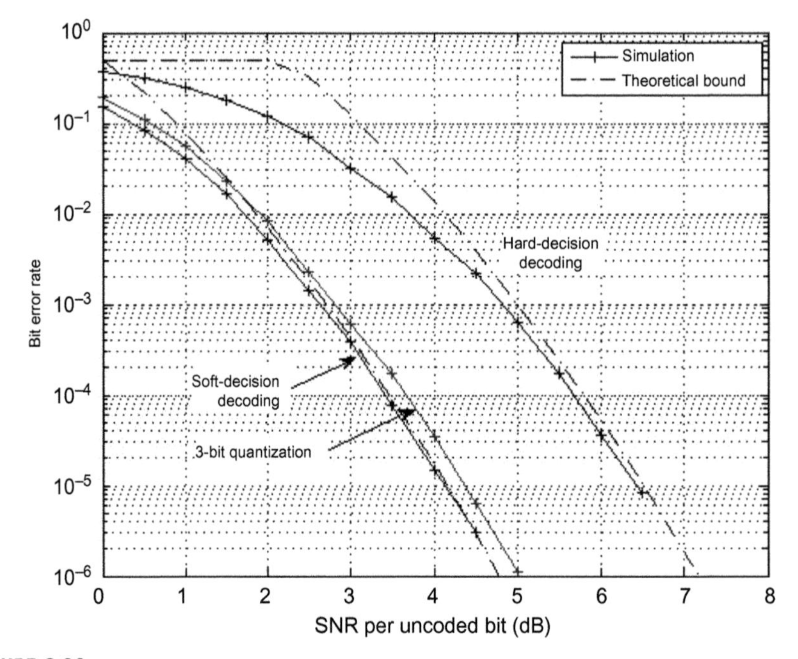

FIGURE 9.66

Performance of rate 1/2 convolutional code ($K=7$ and BPSK modulation) over AWGN channel [141].

In order to terminate a convolutional code, some bits are appended to the information sequence such that the shift-register's content returns to the initial state. Each of the k input sequences of length L bits is padded with K zeros, and these kL input sequences jointly generate $n(K+L)$ output bits. The effective rate of the terminated convolutional code is now $R_{\text{effective}} = kL/n(K+L) < k/n$, which is less than the nominal rate due to the addition of redundancy. Thus, terminating the trellis of a convolutional code is essential in the code's performance for packet-based communications.

Tail-biting convolutional coding is a technique of trellis termination which avoids the rate loss incurred by zero-tail termination at the expense of a more complex decoder. Tail-biting encoding ensures that the starting state of the encoder is the same as its ending state, and that this state value does not necessarily have to be the all-zero state. For a rate $1/n$ feed-forward encoder, this is achieved by initializing the K memory elements of the encoder with the last K information bits of a block of data of length L and ignoring the output. All of the L bits are then input to the encoder and the resulting nL output bits are used as the codeword. In comparison, the zero-tail termination method appends K zeros to a block of data to ensure the feed-forward encoder starts from and ends in the all-zero state for each block. This incurs a rate loss due to the extra tail bits (i.e., non-informational bits) that are transmitted.

The ML tail-biting decoding involves determining the best path in the trellis under the constraint that it starts and ends in the same state. One way to implement this method is to run M parallel Viterbi algorithms, where M is the number of states in the trellis, and select the decoded bits based on the Viterbi algorithm that gives the best metric. However, this makes the decoding M times more complex relative to the zero-tailed encoding.

The structure of the LTE rate 1/3 tail-biting convolutional encoder with constraint length 7 shown in Figure 9.67. The bit sequence input for a given code block to channel coding is denoted by $(c_0, c_1, \ldots, c_{K-1})$, where K is the number of bits to be encoded. The initial value of the shift register (the initial state) of the encoder is set to the values corresponding to the last 6 information bits in the input stream, so that the initial and final states of the shift register are the same. Therefore, if the content of the shift register is denoted by $(s_0, s_1, s_2, s_3, s_4, s_5)$, then the initial value of the shift register is set to $s_i = c_{(K-1-i)}$. The encoder output streams $d_k^{(0)}$, $d_k^{(1)}$, and $d_k^{(2)}$ correspond to the first, second, and third parity streams [4].

The performance of an example rate 1/2 tail-biting convolution coder in an AWGN channel with different modulation orders, coding rates, and data frame sizes using a Viterbi decoder has been evaluated and shown in Figure 9.68 where the wrap depth is a predetermined number of trellis stages used as a parameter in the figure. One way to estimate the required wrap depth for MAP decoding[16] of a

[16]There are a number of possible criteria to use in making decisions, and initially, we assume that the criterion is to maximize the probability of choosing correctly. That is, when the experiment is performed, the resulting sample point maps into a sample value h for H and a sample value y for Y. The decision-maker observes y (but not h) and maps y into a decision $\hat{H}(y)$. The decision is correct if $\hat{H}(y) = h$. In principal, maximizing the probability of choosing correctly is almost trivially simple. Given y, we calculate $p_{H|Y}(h|y)$ for each $h, 0 \leq h \leq M-1$. This is the probability that h is the correct hypothesis conditional on y. Thus the rule for maximizing the probability of being correct is to choose $\hat{H}(y)$ to be that h for which $p_{H|Y}(h|y)$ is maximized. This is denoted as $\hat{H}(y) = \arg\max_h [p_{H|Y}(h|y)]$. If the maximum is not unique, it makes no difference to the probability of being correct which maximizing h is chosen. To be explicit, we will choose the smallest maximizing h. The conditional probability $p_{H|Y}(h|y)$ is called an *a posteriori* probability, and thus the above decision rule is called the maximum a posteriori probability (MAP) rule [45].

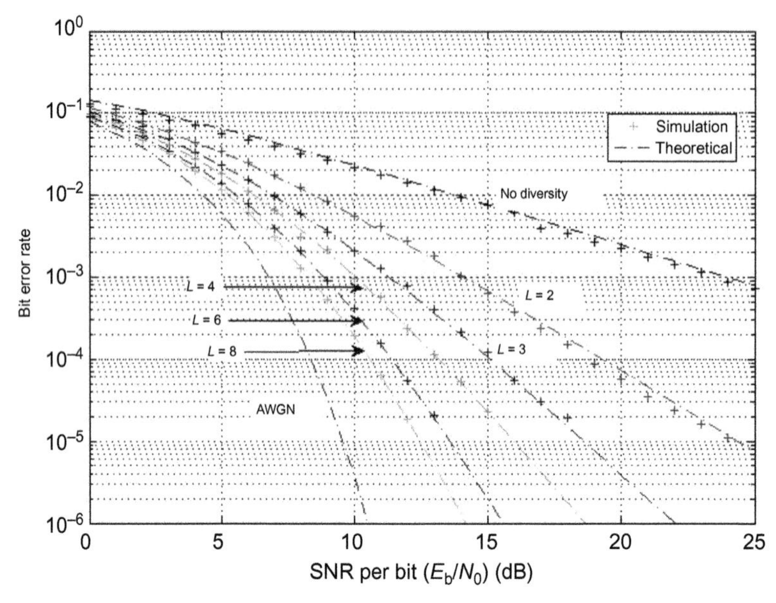

FIGURE 9.67

Performance of rate 1/2 convolutional code (K = 7 and BPSK modulation) in multi-path Rayleigh fading channel [141].

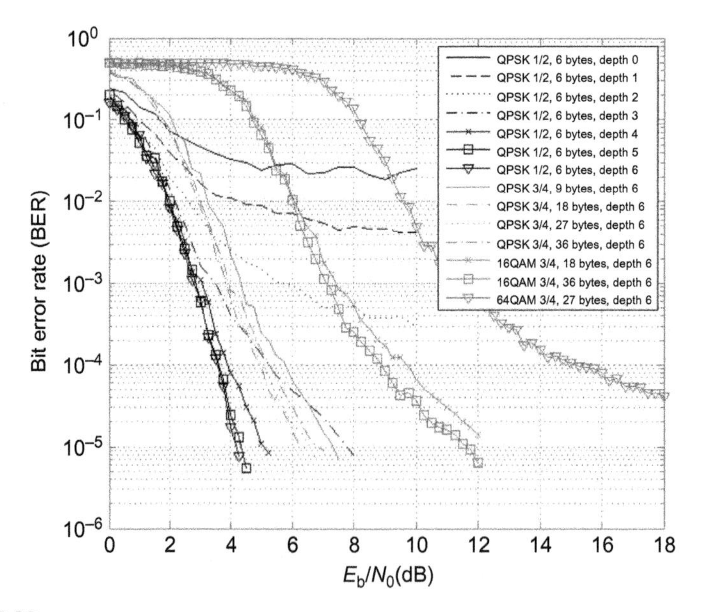

FIGURE 9.68

Performance of an example rate 1/2 tail-biting convolutional coder in an AWGN channel [144].

tail-biting convolutional code is to conduct hardware or software experiments, requiring a circular MAP decoder with a variable wrap depth to be implemented in order to measure the decoded bit error rate versus E_b/N_0 for successively increasing wrap depths. The minimum decoder wrap depth that provides the minimum probability of decoded bit error for a specified E_b/N_0 is found when further increases in wrap depth do not decrease the error probability, as shown in Figure 9.68. It can be seen that the link-level BER performance improves when increasing the depth and the performance does not significantly change after a certain limit. The effect of different data frame sizes (6, 9, 18, 27, and 36 bytes) on the BER performance has been investigated and is shown in the figure.

9.11.2 Principles of turbo coding

In information theory, one can ideally approach the Shannon limit as closely as desired using soft-decision decoding of a long block code, or a convolutional code with a large constraint length. In practice, however, decoding of these codes becomes very computationally intensive. Similar to any conventional error correcting codes, the turbo codes work by imposing a structure on the transmitted bit sequence. If the received bit sequence does not match this known structure, the receiver declares an error has occurred. If the number of errors is sufficiently small and the structure is robust, the receiver can detect the erroneous bits and reconstruct the correct bit sequence. A strong error correcting code has two key characteristics: (1) the encoder imposes a structure that maximizes the difference (or the distance) between any two valid bit sequences; and (2) the decoder utilizes all the information available at the receiver's end including the redundancy and previously unsuccessful transmissions. Thus, the decoder can determine which valid bit sequence is most likely the one that has been transmitted. The first significant difference between a turbo code and conventional code is the use of a recursive systematic encoder. A typical convolutional coder uses a non-recursive structure. By feeding one of the outputs back to the input, a recursive encoding structure is obtained. The recursive structure is systematic, i.e., the input bits appear directly as part of the encoded bit stream. Therefore, a systematic structure enables the combination of two codes in order to construct a stronger composite code. The input bit stream only needs to be transmitted once. The computed parity bits from each of the two constituent encoders are transmitted separately. The recursive structure interacts with the interleaver to give the composite code some unique performance characteristics. The optimum decoder for any code is an ML decoder. In the ML decoding, the receiver produces a probability that each received bit is either 0 or 1. The ML decoder then uses these received bit probabilities, along with its knowledge of the structure imposed by the encoder, to compute the probability of every possible transmitted bit sequence. The most probable transmitted bit sequence is then chosen as the decoded bit sequence. The ML decoding is theoretically optimal and can achieve the lowest probability of decoder failure or probability of error. However, there is no efficient way to practically compute these probabilities when the block size is large. This leads to the use of suboptimal, but realizable, decoding methods.

The second unique feature of turbo codes is their decoder, which is based on iterative application of ML decoding. The ML decoding of a convolutional code would be easy, if the constraint length (register size) is short; however, short codes are relatively weak. Combining two such codes produces a much stronger code, but now ML decoding becomes intractable. Each shorter code can be ML decoded if we assume the other code has already been decoded. This leads to an iterative decoding strategy, which is to decode the first code and then use the resulting updated probabilities to decode the second code. Once this is done, the updated results can more accurately decode the first code. This decoding strategy shares the limitations of all iterative solution methods, including the possibility of entrapment in a local minimum. The decoder output is not necessarily the true global ML solution. But the study of the corresponding link-level BER curves suggests that it tends to be a very good solution.

In a convolutional code, the input bit sequence usually has tail bits added. This initializes the encoder state to all zeros at the end of the bit sequence. In a turbo coder, this implies that a valid bit sequence cannot contain a single one. It must either be all zeros, or it must contain two or more ones. Because encoding is a linear process, this implies that any valid input sequence must differ from any other valid input sequence in at least two bits. In this case, if we consider only the first decoder's output, it differs from the correct output in two positions. However, the turbo encoder contains an interleaver. It must be noted that the two constituent encoders are operating on the same set of bits, but in a different order. Therefore, although the error bits are close together for the first decoder, they are widely separated in the second decoder's input. As a result, the second decoder's output differs from the correct output in many bit positions. It will see the incorrect bit sequence as highly implausible. This is the basic principle of operation of the turbo codes, and is the key to the high performance of turbo codes at low SNRs. Erroneous information sequences that look reasonable to one decoder are likely to be rejected by the other decoder. This is also the rationale for the error floor at higher SNRs. Since the interleaver is random, there are a few error patterns that will look plausible to both decoders. As the SNR increases, these weak patterns come to dominate the BER curve.

One of the most interesting characteristics of a turbo code is that it is not just a single code. It is, in fact, a combination of two codes that work together to achieve a synergy that would not be possible by merely using one code by itself. As shown in Figure 9.69, a turbo code is formed from the parallel concatenation of two constituent encoders separated by an interleaver. Each constituent encoder can be any form of FEC coder used for conventional data communications. However, in practice, the constituent encoders are identical convolutional

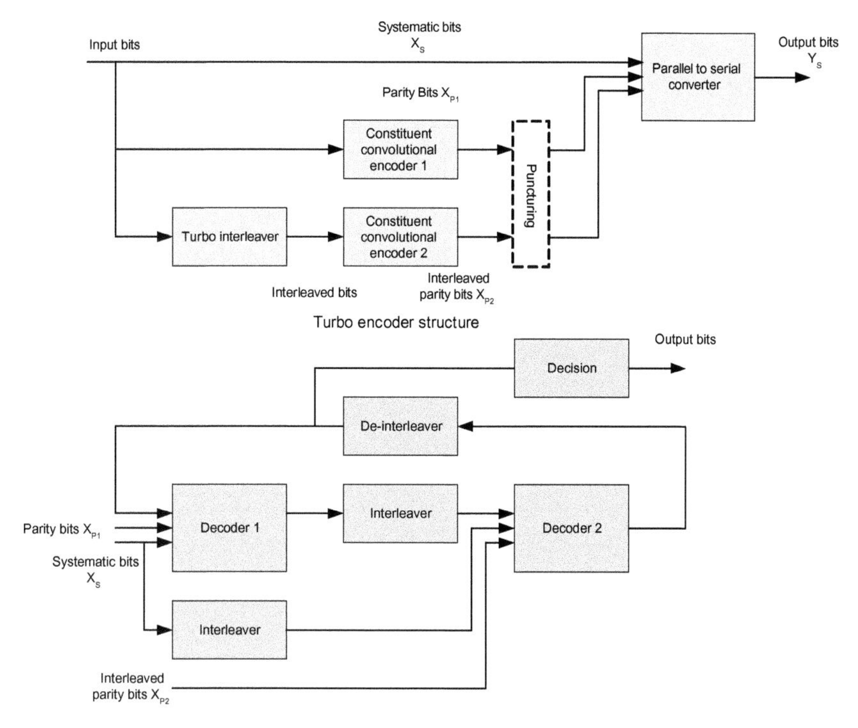

FIGURE 9.69

Generic structure of turbo encoder and iterative decoder [27,40].

encoders. As can be seen in Figure 9.69, the turbo coder consists of two identical constituent encoders. The input data stream and the parity outputs of the two parallel encoders are then serialized into a single turbo codeword. The interleaver is a critical component of the turbo code. It is a simple functional block that rearranges the order of the data bits in a prescribed, but irregular, manner. Although the same set of data bits is present at the output of the interleaver, the order of these bits has been changed. It must be noted that the interleaver used by a turbo coder is quite different than the rectangular interleavers that are commonly used in wireless systems to help mitigate the effect of deep fading. While a rectangular channel interleaver tries to space the data out according to a regular pattern, a turbo interleaver randomizes the ordering of the data in an irregular manner.

The operation of a turbo encoder and decoder can be described as follows. Let the binary input bits $\{0, 1\}$ be represented by bipolar levels $\{+1, -1\}$ and assigned to the variable d, which may take on the values $d=+1$ and $d=-1$. For an AWGN channel, the conditional probability density functions $f(x|d=-1)$

and $f(x|d=+1)$ are referred to as likelihood functions. The common hard-decision criterion, known as ML, selects the symbol $d_k=+1$ or $d_k=-1$ depending on the intercept point of received signal value x_k and the above conditional probability density functions using a fixed threshold of λ (decision point); thus, $d_k=+1$ if $x_k>\lambda$, otherwise, $d_k=-1$. Another decision rule, known as a maximum *a posteriori* probability (MAP) criterion takes into account the *a posteriori* probabilities $f(d=+1|x)$ and $f(d=-1|x)$ to construct the hypotheses H_1 and H_2 as follows:

$$f(d=+1|x) \underset{H_2}{\overset{H_1}{\gtrless}} f(d=-1|x) \tag{9.43}$$

That is, one selects hypothesis $H_1(d=+1)$, if the $f(d=+1|x)>f(d=-1|x)$; otherwise, $H_2(d=-1)$ is selected. Using the Bayes' theorem [30], the above *a posteriori* probabilities can be replaced by their equivalent expressions, yielding:

$$f(x|d=+1)p(d=+1) \underset{H_2}{\overset{H_1}{\gtrless}} f(x|d=-1)p(d=-1) \tag{9.44}$$

The likelihood ratio test is constructed as follows:

$$\frac{f(x|d=+1)p(d=+1)}{f(x|d=-1)p(d=-1)} \underset{H_2}{\overset{H_1}{\gtrless}} 1 \tag{9.45}$$

If the input bits are independent and identically distributed random variables, the above equation is simplified as follows:

$$\frac{f(x|d=+1)}{f(x|d=-1)} \underset{H_2}{\overset{H_1}{\gtrless}} 1 \tag{9.46}$$

By taking the logarithm of the above likelihood ratio, a useful metric known as log-likelihood ratio (LLR) is obtained, which is a real-valued number representing the soft-decision output of the detector:

$$L(d|x)=\log\left(\frac{f(x|d=+1)}{f(x|d=-1)}\right)+\log\left(\frac{p(d=+1)}{p(d=-1)}\right)=L(x|d)+L(d) \tag{9.47}$$

where $L(x|d)$ is the LLR of test statistics x obtained through measuring the channel output x under the condition that either $d=+1$ or $d=-1$ may have been transmitted, and $L(d)$ denotes the *a priori* LLR of the data bit d as shown in Figure 9.70. The introduction of a decoder would improve the reliability of the above decision-making process. For a systematic code, it can be shown that the LLR (soft output) expression at the output of the decoder can be written as $L_{\text{output}}(\hat{d})=L_{\text{input}}(\hat{d})+L_e(\hat{d})$, where $L_{\text{input}}(\hat{d})$ is the LLR of a data bit at the output of the detector (or the input to the decoder), and $L_e(\hat{d})$ represents additional

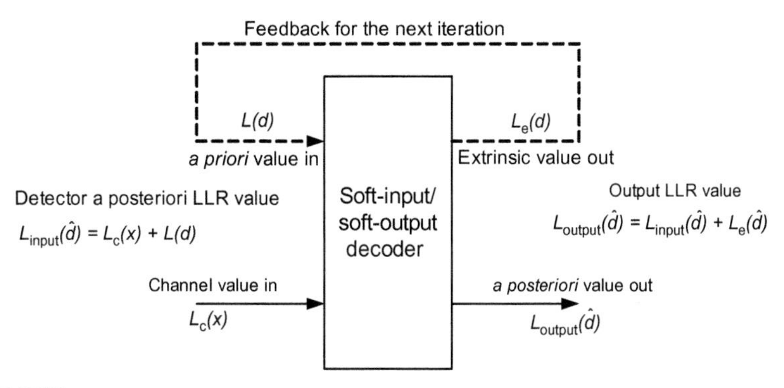

FIGURE 9.70

Illustration of soft-input/soft-output decoder for turbo codes [40,102].

information that is obtained from the decoding process (see Figure 9.70). The output sequence of a systematic decoder consists of values representing data and parity bits. Thus, the decoder output LLR can be decomposed into a data component that is associated with the detector measurement and the extrinsic component that is represented by the decoder contribution due to parity. The soft-decision component $L_{\text{output}}(\hat{d})$ is a real number that provides a hard-decision as well as the reliability of that decision. The sign of the parameter $L_{\text{output}}(\hat{d})$ denotes the hard-decision, i.e., for positive values of $L_{\text{output}}(\hat{d})$, decide $\hat{d} = +1$ and for negative values of $L_{\text{output}}(\hat{d})$, decide $\hat{d} = -1$. The magnitude of $L_{\text{output}}(\hat{d})$ denotes the reliability of the decision [98,102].

Turbo decoding relies on the exchange of probabilistic information between the two soft-input soft-output decoders shown in Figure 9.69. The extrinsic information is the result of the decoder's estimation of bit d, but not taking its own input into account. It is precisely this extrinsic information that is exchanged iteratively between the two soft-decision decoders during the decoding process. Subtracting the decoder's input from its output prevents the decoder from acting as a positive feedback amplifier and introduces stability in the feedback process. Usually, after a given number of iterations, one observes that the two decoders converge to a stable final decision for d. In practice, depending on the nature of the soft-decision decoder, scaling or clipping operations may be applied to the extrinsic information in order to ensure convergence within a small number of iterations. It is shown in the literature that the Max-Log-MAP decoding algorithm can make a good trade-off between performance and complexity, with the added advantage of not requiring any knowledge of the noise level (Figure 9.71). A stop criterion facilitates the convergence of the iterative decoding process and helps reduce the average power consumption of the decoder by reducing the average number of iterations required to decode a block without compromising performance [100–108].

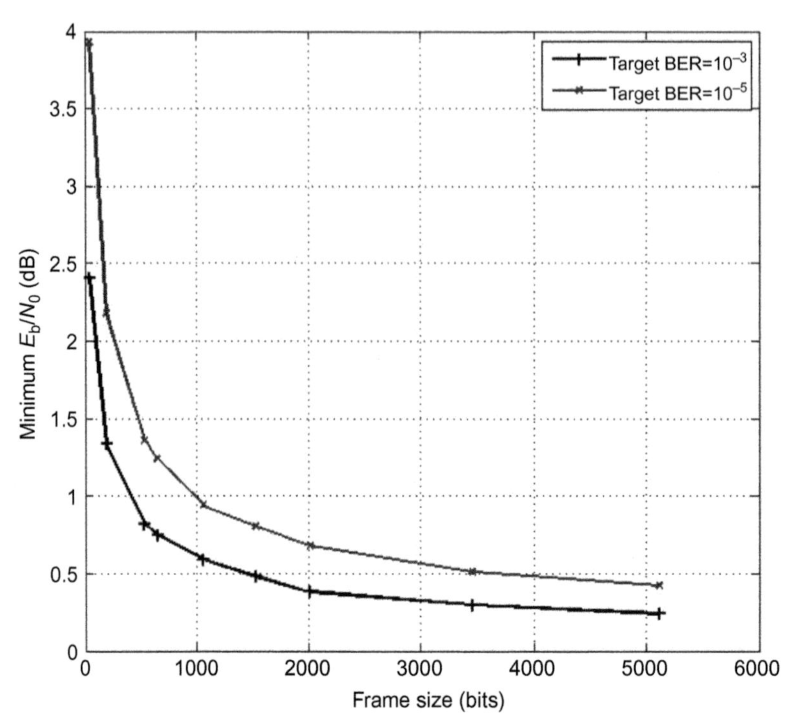

FIGURE 9.71

Minimum $E_b = N_0$ to achieve a target BER using the UMTS/LTE turbo coder ($R = 1/3$) with BPSK modulation over an AWGN channel [144].

For practical applications, it is often necessary to adapt the turbo code parameters to those of the application. The systematic bits are always present at the output of the turbo encoder; however, depending on the desired code rate, the parity bits from the two constituent encoders may be punctured prior to transmission (see Figure 9.69). The puncturing mechanism involves deleting some of the parity bits prior to transmission. As an example, if the minimum code rate of a turbo coder is 1/5, a code rate of 1/3 can be generated by deleting the second parity output of each encoder. Note that the puncturing pattern for each code rate must be known by the decoder for correct interpretation of the received bits [40].

It is shown that for smaller input frame sizes, less iteration is required; however, the performance of the turbo coder is improved by increasing the frame size. This is due to an increase in interleaver gain as the input frame size becomes larger. Figure 9.71 shows the minimum E_b/N_0 required to achieve a BER of 10^{-3} and 10^{-5} for various frame sizes using a UMTS/LTE turbo encoder with a minimum code rate of $R = 1/3$. In general, turbo codes are more suited for applications with large packet sizes and their performance degrades for small packet sizes.

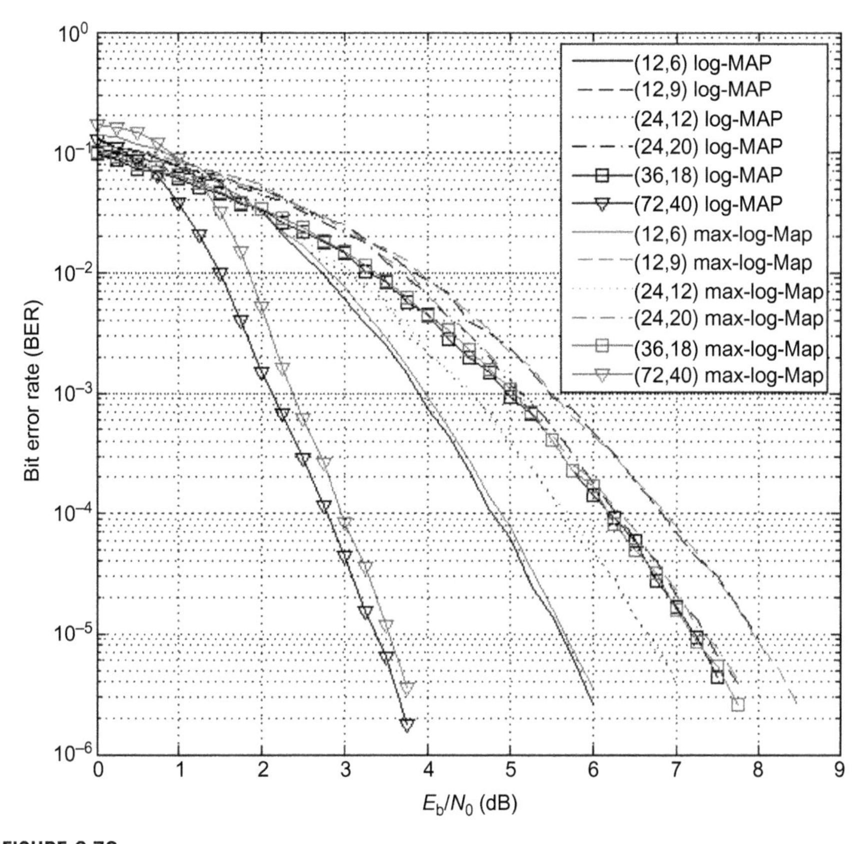

FIGURE 9.72

Comparison of the BER performance of a BTC turbo coder in AWGN channel with different code rates, frame sizes, and detector types [144].

Furthermore, the delay due to turbo interleaver might become prohibitive for delay-sensitive applications. Figure 9.72 shows the BER performance of a BTC turbo coder as a function of E_b/N_0 for various code rates, frame sizes, and detector types. In Figure 9.72, notation (n, k) denotes the number of input k and output n bytes [144].

9.11.3 Baseband digital modulation schemes

Modulation is a process by which a carrier signal is altered according to information in a message signal. For the typical baseband modulation types (PSK, DPSK, FSK, QAM, or PAM), gray-coded constellations are obtained by selecting the gray parameter in the corresponding baseband modulation function. The mapping or assignment of k information bits to $M = 2^k$ possible signal amplitudes may be done in a number of ways. The preferred assignment is the one in which the adjacent signal amplitudes differ by one binary digit since the most likely errors caused by noise involve the erroneous selection of an adjacent amplitude to the transmitted signal amplitude. In that case, only a single bit error occurs in the k-bit sequence. This mapping is referred to as gray coding. The criteria that must be taken into account when choosing a digital modulation method include the following:

- Power efficiency, i.e., the E_b/N_0 ratio for a specific error probability
- Bandwidth efficiency, i.e., the data rate per unit bandwidth
- Performance in multipath fading channels and under non-linear distortion
- Implementation cost and complexity

However, the above requirements cannot be satisfied at the same time, and some trade-off is necessary. In order to explain the latter statement, let us define a few basic terms in this context. A constellation diagram is a graphical representation of the complex envelope of each possible symbol state. The power efficiency is related to the minimum distance between the points in the constellation. The bandwidth efficiency is related to the number of points in the constellation. The gray coding is used to assign groups of bits to each constellation point. In gray coding, adjacent constellation points differ by a single bit.

The two most common digital modulation schemes used in cellular systems are M-ary PSK and M-ary QAM modulation schemes. In M-ary PSK, a group of n bits are transmitted using $M = 2^n$ different phases, whereas M-ary QAM uses a combination of amplitude and phase modulation to convey the information. Study of the literature suggests that the minimum code distance $d_{\min}$ is smaller for M-ary ASK than for M-ary PSK or M-ary QAM for the same average transmitted power, which translates into poorer error performance because the signal envelope is not constant, it results in poor performance under non-linear distortion condition. The minimum distance $d_{\min}$ is smaller for M-ary PSK compared to that for M-ary QAM for the same average transmitted power, resulting in poorer error performance. However, a constant envelope results in good performance under nonlinear distortion. The minimum distance $d_{\min}$ is larger for M-ary QAM relative to that for M-ary ASK or M-ary PSK for the same average transmitted power, which means better error performance. Nonetheless, since the signal envelope is not constant, poor error performance is expected under non-linear distortion. Pulse shaping is then used to reduce the bandwidth requirements of the modulated signal. The Nyquist technique can be used to reduce the bandwidth requirement and to eliminate ISI. The non-Nyquist techniques reduce the bandwidth requirement, but do not eliminate ISI. Note that in all M-ary modulation schemes, a group of n bits is transmitted in each signaling interval $T = T_b \log_2 M$ and the modulated signal bandwidth is inversely proportional to the signaling interval. It can be shown that the probability of error (i.e., error per bit) equations as a function of E_b/N_0 for M-ary

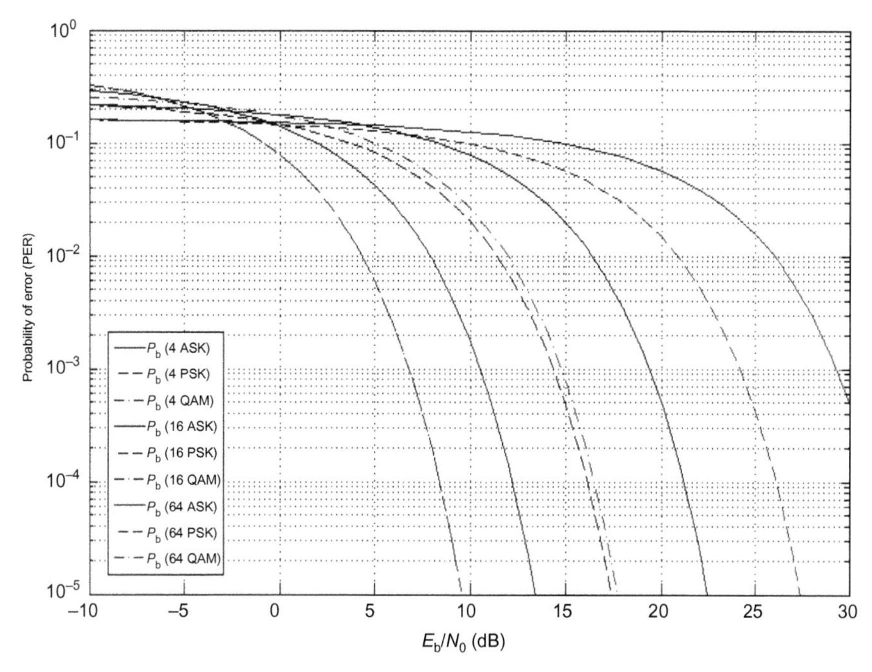

FIGURE 9.73

Probability of error as a function of energy per bit for different digital modulation schemes [159].

ASK, *M*-ary PSK, and *M*-ary QAM can be written as follows (see Figure 9.73 for graphical presentation):

$$p_b = \frac{1}{\log_2 M}\left(1 - \frac{1}{M}\right) Q\left(\sqrt{\frac{3 \log_2 M}{M^2 - 1}\frac{E_b}{N_0}}\right) \qquad M\text{-ary ASK}$$

$$p_b = \frac{1}{\log_2 M} Q\left(\sqrt{\log_2 M \frac{E_b}{N_0} \sin^2\left(\frac{\pi}{M}\right)}\right) \qquad M\text{-ary PSK} \qquad (9.48)$$

$$p_b = \frac{2}{\log_2 M}\left(1 - \frac{1}{\sqrt{M}}\right) Q\left(\sqrt{\frac{3 \log_2 M}{2(M-1)}\frac{E_b}{N_0}}\right) \qquad M\text{-ary QAM}$$

In the above equations, E_b/N_0 denotes the energy per bit, M is the modulation order, and $Q(\cdot)$ denotes the error function.[17]

[17] In mathematics, the error function is defined as $Q(x) = (2/\sqrt{\pi}) \int_0^x e^{-\tau^2} \, d\tau$. An alternative definition is sometimes referred to as the complementary error function in the literature is given as $Q(x) = (2/\sqrt{\pi}) \int_x^\infty e^{-\tau^2} \, d\tau$ [30].

The modulation mapping function in the transmission chain takes binary input $b_0b_1...b_{N_{\text{modulation order}}-1}$ with $b_i =$ "0" or "1" and produces the complex-valued modulation symbols $x = \gamma(I + jQ)$ in the output with the value of γ chosen to achieve equal average power. For each modulation, symbol, b_0 denotes the most significant bit. In the case of BPSK modulation, $N_{\text{modulation order}} = 1$; thus a single bit is mapped to the modulation symbol x with $\gamma = 1/\sqrt{2}$. In the case of QPSK modulation, $N_{\text{modulation order}} = 2$ and a pair of bits are mapped to complex-valued modulation symbol x with $\gamma = 1/\sqrt{2}$. In the case of 16QAM modulation, $N_{\text{modulation order}} = 4$ and quadruplets of bits $b_0b_1b_2b_3$ are mapped to the complex-valued modulation symbol x and $\gamma = 1/\sqrt{10}$. In the case of 64QAM modulation, $N_{\text{modulation order}} = 6$ and six information bits $b_0b_1b_2b_3b_4b_5$ are mapped to the modulation symbol x with $\gamma = 1/\sqrt{42}$. The constellation format for various modulation schemes is illustrated in Figure 9.74.

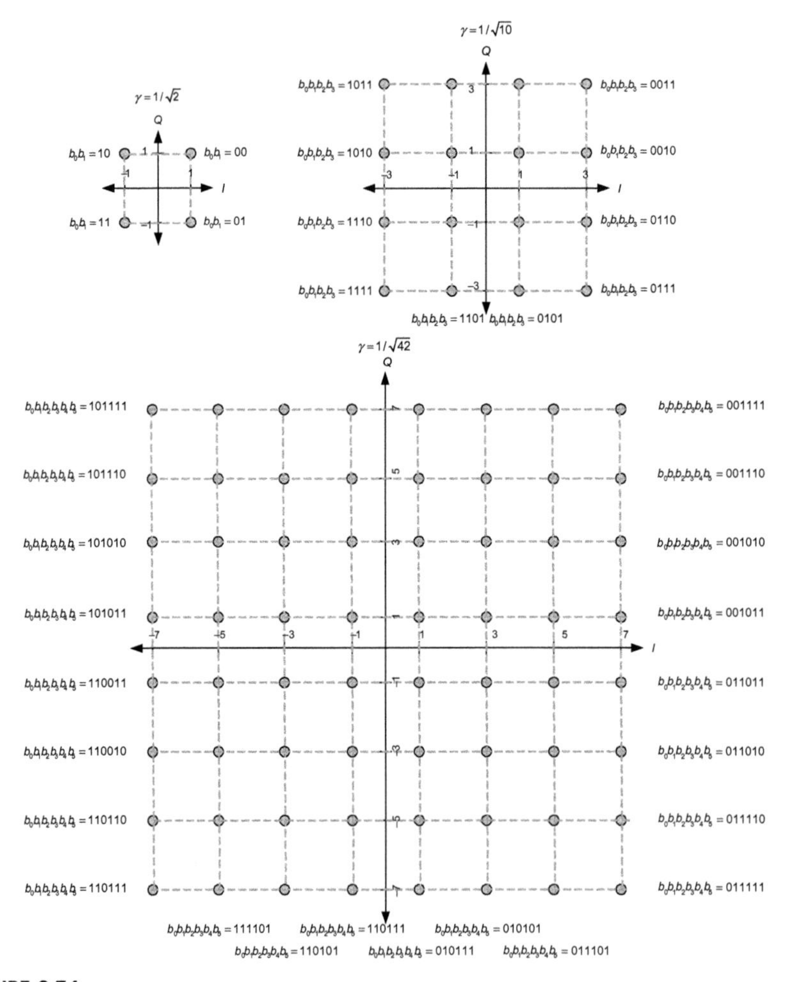

FIGURE 9.74

QPSK, 16QAM, and 64QAM constellations [3].

Figures 9.75 and 9.76 show the performance of the M-ary QAM and M-ary PSK signals in an AWGN channel in terms of probability of error as a function of SNR per bit for different values of modulation order M. It can be seen that for a given target error rate, the required SNR per bit increases with increasing modulation order, i.e., higher modulation orders can only be used in good channel conditions. The latter suggests that in poor channel conditions, more robust modulation schemes must be used.

9.11.4 Channel coding, multiplexing, and interleaving

The baseband modulation schemes supported in the downlink and uplink of LTE are QPSK, 16QAM, and 64QAM. While all three modulation schemes are supported in the downlink in all UE categories, the use of 64QAM in the uplink is only supported in the higher UE categories, and lower UE categories only support QPSK and 16QAM. The channel coding scheme for transport blocks in LTE is turbo coding similar to that of UMTS, with a basic code rate of $R = 1/3$, two eight-state constituent encoders and a contention-free quadratic permutation polynomial (QPP)[18] turbo interleaver [4]. Trellis termination is performed by taking the tail bits from the shift register feedback after all information bits are encoded. The tail bits are padded after the encoding of information bits. Before the turbo coding, transport blocks are segmented into octet-aligned segments with a maximum information block size of 6144 bits. Error detection is performed using 24-bit CRC. Link adaptation (adaptive modulation and coding) with various modulation schemes and channel coding rates is applied to the shared data channel. The same coding and modulation is applied to all groups of resource blocks

[18] In mathematics, a permutation polynomial for a given finite ring is a polynomial that performs a permutation of the elements of the ring, i.e., the mapping Z/nZ is a one-to-one mapping. In the case of finite rings Z/nZ, such polynomials have been studied and applied in the interleaver component of error detection and correction algorithms. For the finite ring Z/nZ, one can construct quadratic permutation polynomials. This is possible, if and only if, n is divisible by p^2 for some prime number p. The construction is simple, but it can produce permutations with desirable properties. As an example, consider $f(x) = 2x^2 + x$ for the ring $Z/4Z$. One can observe that $f(0) = 0$; $f(1) = 3; f(2) = 2; f(3) = 1$, hence the polynomial defines the permutation $\begin{bmatrix} 0 & 1 & 2 & 3 \\ 0 & 3 & 2 & 1 \end{bmatrix}$. Note that a ring is an algebraic structure which generalizes the main properties of the addition and the multiplication of integers, real numbers, complex numbers, and square matrices. It consists of a set together with two binary operations usually called addition and multiplication, which satisfy the following set of axioms: (1) the addition is associative and commutative, it has an identity and each element in the set has an additive inverse, thus the set is an Abelian group (or commutative group) under addition; (2) the multiplication is associative, but is not necessarily commutative, it has an identity and distributes over addition [31,32].
In general, let us assume $n = \alpha_1^{k_1}\alpha_2^{k_2}\ldots\alpha_m^{k_m}$, in which α_i is a prime number. It can be proved that any polynomial $f(x) = \beta_0 + \sum_{0<i\leq M}\beta_i x^i$ defines a permutation for the ring Z/nZ if and only if all the polynomials $f_{\alpha_k}(x) = \beta_{0_{\alpha_k}} + \sum_{0<i\leq M}\beta_{i_{\alpha_k}} x^i$ define the permutations for all rings $Z/\alpha_m^{k_m}Z$, in which $\beta_{i_{\alpha_k}}$ is the remainder of β_i modulo $\alpha_i^{k_i}$ [31,32].

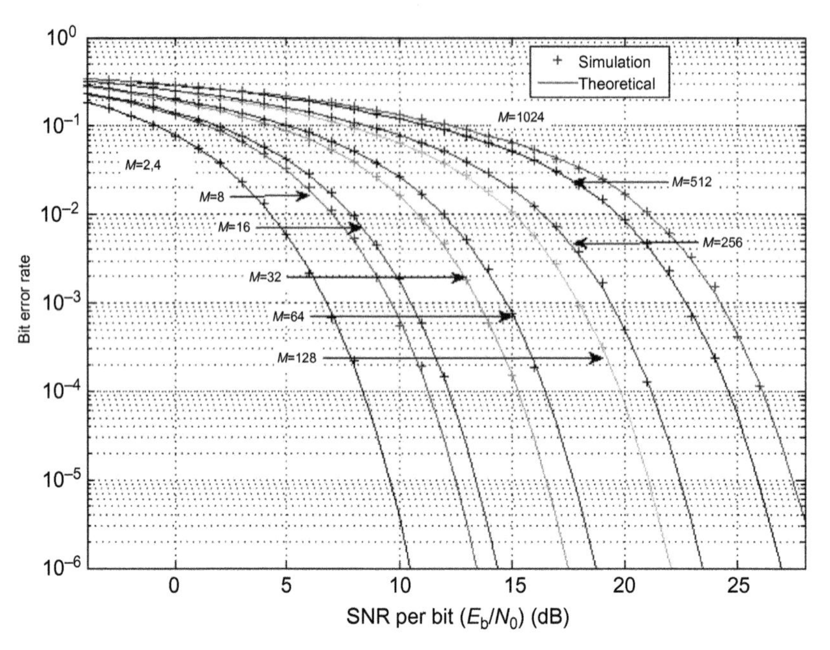

FIGURE 9.75

Performance of *M*-ary QAM in an AWGN channel [141].

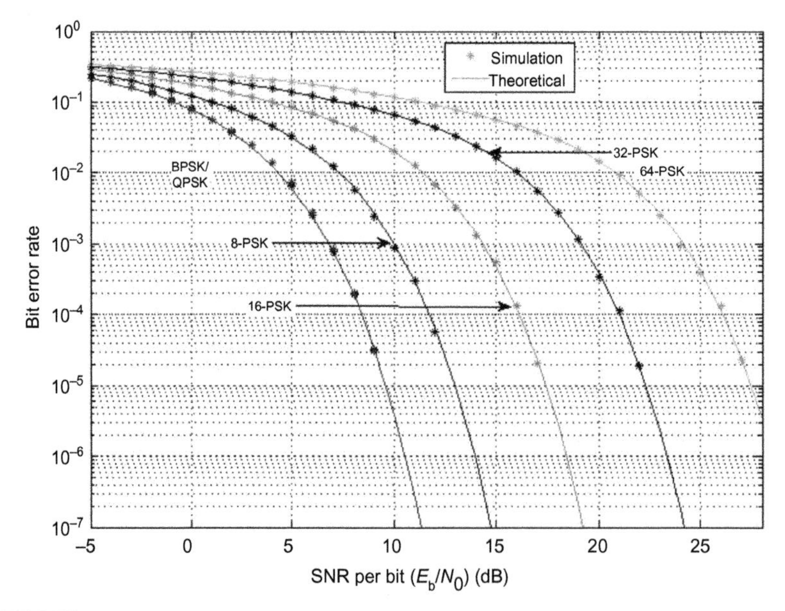

FIGURE 9.76

Performance of coherent *M*-ary PSK over an AWGN channel [141].

belonging to the same MAC PDU scheduled for one user within one TTI and within a single stream.

The user data and control information from the MAC sub-layer are encoded prior to transmission of the transport blocks over the radio link. The channel coding in E-UTRA is a combination of error detection, error correction, rate-matching, interleaving and transport channel, or control information mapping to the appropriate physical channels. As shown in Figure 9.77, the channel coding procedure for BCH transport blocks is different from that of the DL-SCH, PCH, and MCH. Depending on the type of the transport channel, a different coding scheme and coding rate is applied as given in Table 9.17. The first stage of the channel coding process is to append 16/24 bit CRC to the input transport block.

In the case of BCH, the information (i.e., MIB) is received by the channel coding unit in the form of one transport block every 40 ms. A 16-bit CRC is calculated over the entire transport block and appended to the input transport block. Let $(a_0, a_1, \ldots, a_{N_A-1})$ denote the bits in a BCH transport block delivered to the physical layer. The parity bits are denoted by $(p_0, p_1, \ldots, p_{N_P-1})$ where $N_P = 16$.

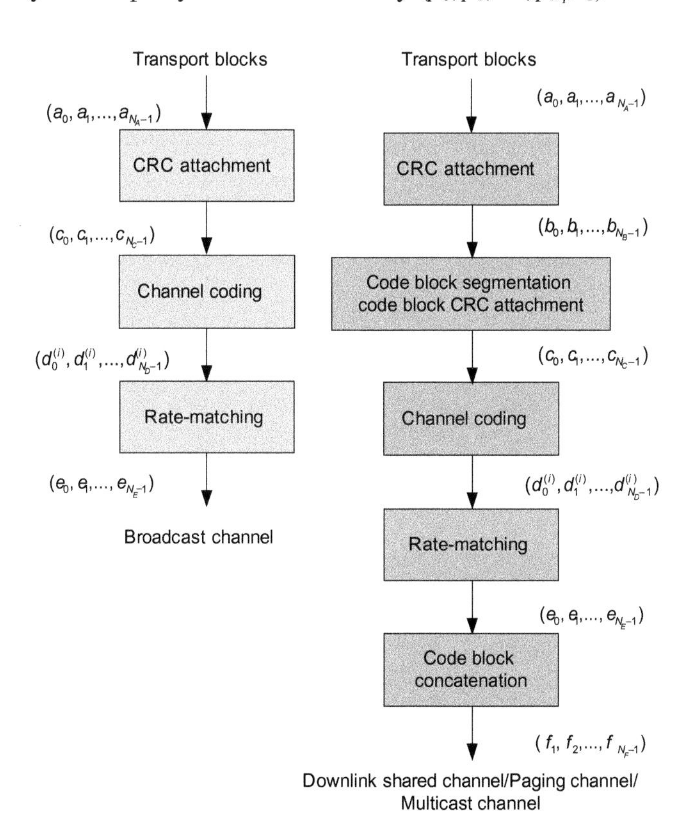

FIGURE 9.77

Channel coding procedure for downlink transport blocks [4].

Table 9.17 Channel Coding Scheme for Various Transport Channels/Control Information in E-UTRA [4]

Channel/ Information	Channel/ Information Types	Coding Scheme	Coding Rate	CRC Size (Bits)
Transport channels	Uplink shared channel (UL-SCH)	Turbo coding	1/3	24
	Downlink shared channel (DL-SCH)			24
	Paging channel (PCH)			24
	Multicast channel (MCH)			24
	Broadcast channel (BCH)	Tail-biting convolutional coding	1/3	16
Control Information	Downlink control information (DCI)	Tail-biting convolutional coding	1/3	16
	Control format indicator (CFI)	Block code	1/16	–
	HARQ indicator (HI)	Repetition code	1/3	–
	Uplink control information (UCI)	Block code	Variable	16
		Tail-biting convolutional coding	1/3	

Following the calculation of the CRC bits, they are scrambled according to the eNB multi-antenna configuration with binary sequence $(x_0, x_1, \ldots, x_{15})$ to form the output sequence $(c_0, c_1, \ldots, c_{N_C-1})$, where the first A bits of the latter sequence are formed by the input transport block and the last 16 bits are the masked CRC bits. The masking sequence is given as follows:

$$(x_0, x_1, \ldots, x_{15}) = \begin{cases} (0,0,0,0,0,0,0,0,0,0,0,0,0,0,0,0) & N_{TX} = 1 \\ (1,1,1,1,1,1,1,1,1,1,1,1,1,1,1,1) & N_{TX} = 2 \\ (0,1,0,1,0,1,0,1,0,1,0,1,0,1,0,1) & N_{TX} = 4 \end{cases} \quad (9.49)$$

where N_{TX} denotes the number of antenna ports in the eNB that are used for transmission of the PBCH. The CRC-attached information bits denoted by $(c_0, c_1, \ldots, c_{N_C-1})$ are processed through a tail-biting convolutional coder. The channel-coded bits, denoted by $(d_0^{(i)}, d_1^{(i)}, \ldots, d_{N_D-1}^{(i)})$ where $i = 0, 1, 2$ is the coded stream number to sub-block interleaver and $N_D = N_C$ is the number of bits on the ith coded stream, are fed to the rate-matching block. The use of tail-biting convolutional coding necessitates that the initial value of the shift register of the encoder be set to the values corresponding to the last 6 information bits in the input stream so that the initial and final states of the shift register are the same.

The rate-matching function for channel-coded transport channels is defined per coded block, and consists of interleaving of the three coded streams

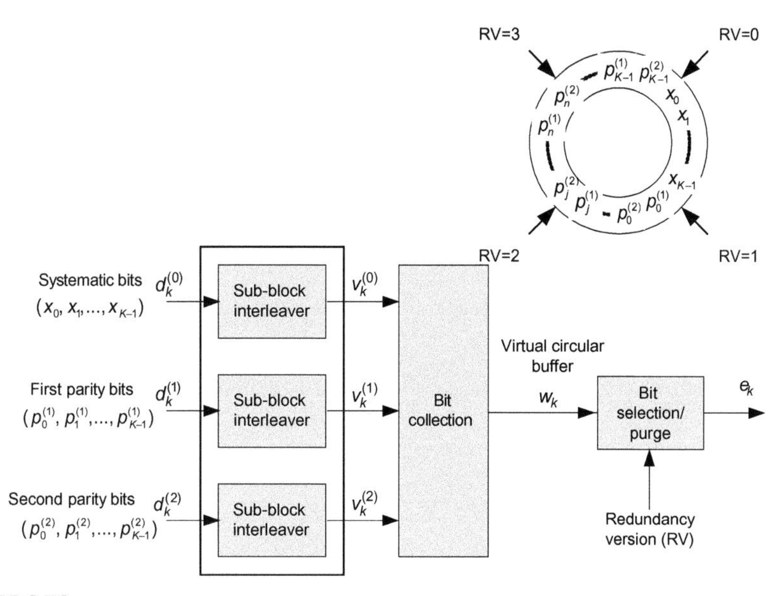

FIGURE 9.78

Rate-matching process of the coded information [4,15].

$d_k^{(0)}, d_k^{(1)}$, and $d_k^{(2)}$ followed by the collection of bits and generation of a virtual circular buffer as depicted in Figure 9.78.

The rate-matching block generates an output bit stream with the desired code rate. As the number of bits available for transmission depends on the available resources, the rate-matching algorithm is capable of producing any arbitrary code rate. The three bit streams from the channel coding module are interleaved followed by bit collection to generate a circular buffer. Bits are selected and pruned from the circular buffer to create an output bit stream with the desired code rate, as shown in Figure 9.78. The three sub-block interleavers used in the rate-matching block are identical. Interleaving is a technique to reduce the impact of burst errors on a signal so that consecutive bits of data will not be corrupted. The sub-block interleaver reshapes the coded bit sequence, row-by-row, to form a matrix with $C_{\text{sub-block}} = 32$ columns and $R_{\text{sub-block}}$ rows, where $R_{\text{sub-block}}$ is determined by finding the minimum integer that satisfies the following equation: $R_{\text{sub-block}} C_{\text{sub-block}} \leq N_D$. If $R_{\text{sub-block}} C_{\text{sub-block}} > N_D$, then N_{NULL} null bits are appended to the front of the encoded input sequence such that $R_{\text{sub-block}} C_{\text{sub-block}} = N_D + N_{\text{NULL}}$. For $d_k^{(0)}$ and $d_k^{(1)}$ bit streams, an inter-column permutation is performed on the matrix to reorder the columns as follows: 0, 16, 8, 24, 4, 20, 12, 28, 2, 18, 10, 26, 6, 22, 14, 30, 1, 17, 9, 25, 5, 21, 13, 29, 3, 19, 11, 27, 7, 23, 15, and 31 for turbo-coded transport blocks, and 1, 17, 9, 25, 5, 21, 13, 29, 3, 19, 11, 27, 7, 23, 15, 31, 0, 16, 8, 24, 4, 20, 12, 28, 2, 18, 10, 26, 6, 22, 14, and 30 for transport blocks coded by convolutional coder. For the $d_k^{(0)}$ and $d_k^{(1)}$ bit streams, the output of the block interleaver is the bit sequence readout column-by-column from the inter-column permutated matrix to create a stream $K' = R_{\text{sub-block}} C_{\text{sub-block}}$ bits long. For the $d_k^{(2)}$ bit stream, the elements in the matrix are permutated separately based on the permutation pattern shown above, but modified to create a permutation which is a function of $(R_{\text{sub-block}}, C_{\text{sub-block}}, k, K')$. This creates three interleaved bit-stream mapping $d_k^{(0)} \rightarrow v_k^{(0)}; d_k^{(1)} \rightarrow v_k^{(1)}; d_k^{(2)} \rightarrow v_k^{(2)}$. The bit collection stage creates a virtual circular buffer by combining the three interleaved encoded bit streams. The block interleaver output $v_k^{(1)}$ and $v_k^{(2)}$ are combined by successive interlacing of their values. This combination is then appended to the end of $v_k^{(0)}$ to create the circular buffer w_k as illustrated in Figure 9.79. Interlacing allows equal levels of protection for each parity sequence [4].

Following the bit collection procedure, the bits are selected and trimmed from the virtual circular buffer to create an output sequence length which meets the desired code rate. The HARQ error correction scheme is incorporated into the rate-matching algorithm of LTE/LTE-Advanced. For any desired code rate, the coded bits are sent out serially from the circular buffer from a starting location, given by the HARQ redundancy version (see Figure 9.78), wrapping around to the beginning of the buffer, if the end of the buffer is reached. The null bits are discarded. Different redundancy versions, and hence starting points, allow for the retransmission of selected data. Being able to select different starting points enables the two main methods of recovering data at the receiver in the HARQ process: (1) chase combining where retransmissions contain the same data and parity bits; (2) incremental redundancy where retransmissions contain different information so the receiver gains more knowledge upon each retransmission. The rate-matched code blocks are now concatenated back together. This is done by sequentially concatenating the rate-matched blocks together to create the output of the channel coding process $(f_1, f_2, \ldots, f_{F-1})$, as shown in Figure 9.77.

In the case of DL-SCH, PCH, and MCH transport channels, the CRC calculation and addition to the information block is similar to BCH, except that 24-bit CRC is used and no CRC masking is done since in this case, the multi-antenna configuration is part of DCI. As shown in Figure 9.80, the input bits to the code block segmentation are denoted by $(b_0, b_1, \ldots, b_{N_B-1})$, where $N_B = N_A + 24$. In LTE/LTE-Advanced, a minimum code block size of 40 bits and maximum

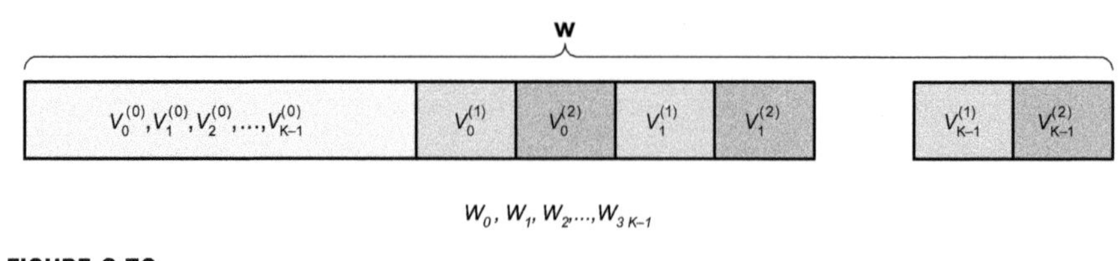

FIGURE 9.79

Structure and content of the virtual circular buffer [160].

code block size of 6144 bits are specified so that the block sizes are compatible with the block sizes supported by the turbo interleaver. If the length of the input block N_B is greater than the maximum code block size, the input block is segmented. When the input block is segmented, it is segmented into $N_C = \lceil N_B/6120 \rceil$ code blocks. Each code block is appended with a 24-bit CRC to the end. Padding bits are added to the front of each block so that code block sizes match a set of permissible turbo interleaver block sizes. If no segmentation is necessary, only one code block is generated. If N_B is less than 40 bits, then padding bits (zeros) are added to the beginning of the code block to attain a total of 40 bits. The code blocks then undergo turbo coding.

A PCCC turbo coder with two recursive convolutional encoders and a contention-free QPP interleaver is used as a channel coder in LTE/LTE/Advanced for traffic channels as shown in Figure 9.81.

The outputs of the encoder are three streams $d_k^{(0)}, d_k^{(1)}$, and $d_k^{(2)}$ in order to achieve a code rate of 1/3. As shown in Figure 9.81, the input to the first constituent encoder is the input bit stream to the turbo coding block. The input to the second constituent encoder is the output of the QPP interleaver, which is a permutated version of the input sequence. There are two output sequences from each encoder which are the systematic bits (x_k, x'_k) and parity bits (z_k, z'_k). Only one of the systematic bit sequences x_k is used as the output, as the other one x'_k is simply a permutated version of the former systematic sequence. The transfer function for each constituent encoder is given by $G(D) = [1, g_1(D)/g_0(D)]$. The first

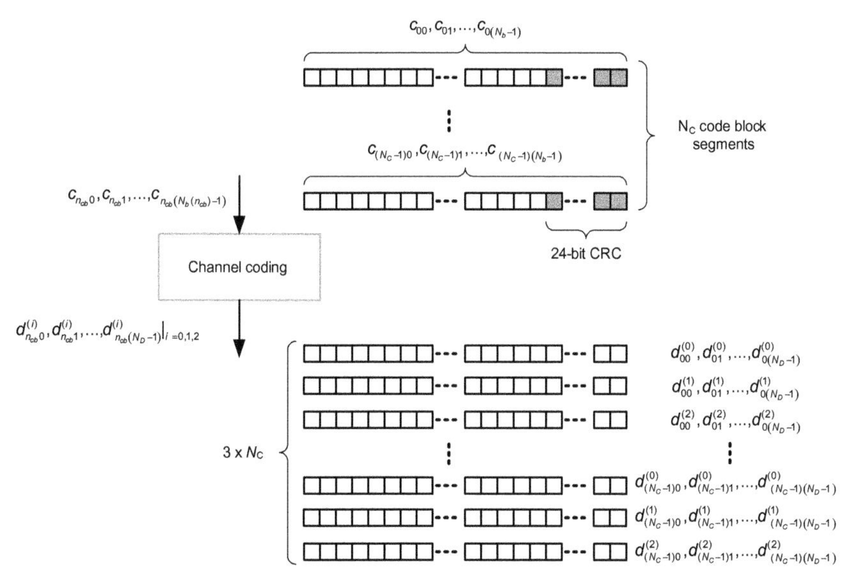

FIGURE 9.80

Code block segmentation and CRC attachment for DL-SCH [4,158,160].

element in the latter equation represents the systematic output transfer function, and the second element denotes the recursive convolutional output transfer function, where $g_0(D) = 1 + D^2 + D^3; g_1(D) = 1 + D + D^3$. The output for each sequence can be calculated using the transfer function.

The encoder is initialized with all zeros. If the code block to be encoded is the zeroth code block and the number of padding bits $N_{\text{Padding}} > 0$, then the encoder input is set to zero for $k = 0, 1, \ldots, N_{\text{Padding}} - 1$, and $d_k^{(0)}, d_k^{(1)}, k = 0, 1, \ldots, N_{\text{Padding}} - 1$ are set to null at the output. In a conventional convolutional coder, the coder is driven to an all zero state upon termination by appending zeros to the end of the input data stream. Since the decoder is aware of the start and end state of the encoder, it can decode the data. Driving a recursive coder to an all zeros state using this method is not possible. To overcome this problem trellis termination is used where, upon termination, the tail bits are fed back to the input of each encoder using a switch (see Figure 9.81). The first three tail bits are used to terminate each encoder.

The role of the interleaver is to spread the information bits such that in the event of an error, the two codes are affected differently allowing data to still be recoverable. As shown in Figure 9.81, the input bits to the turbo code QPP interleaver are denoted by $c_0, c_1, \ldots, c_{N_C-1}$, where N_C is the number of input bits. The output bits from the turbo code QPP interleaver are denoted by $c'_0, c'_1, \ldots, c'_{K-1}$. The output of the interleaver is a permutated version of the input data $c'_i = c_{\Phi(i)}, i = 0, 1, \ldots, K - 1$ where $\Phi(i) = (f_1 i + f_2 i^2) \bmod K$. The coefficients f_1

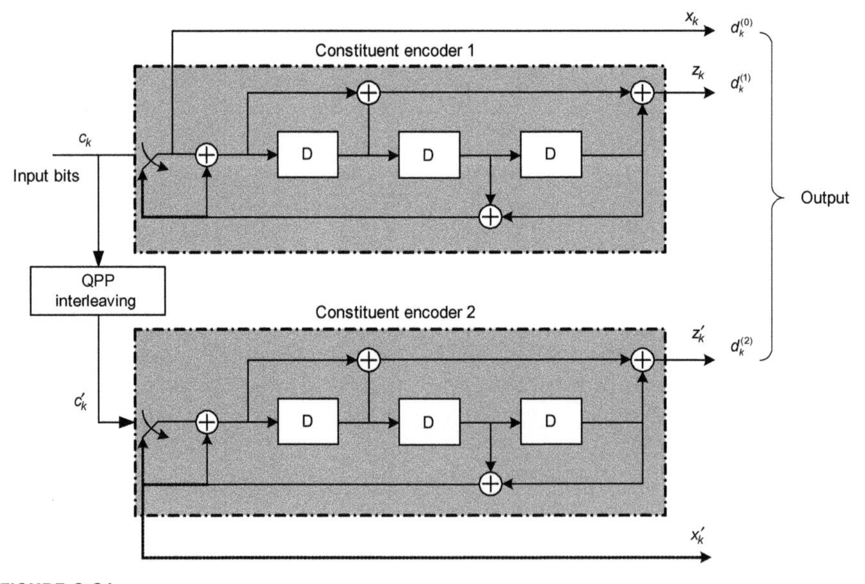

FIGURE 9.81

Structure of an LTE turbo coder [4].

and f_2 are chosen depending on the length of the input data. As an example, if $K=40$ then $f_1=3$, $f_2=10$, consequently one can find the permutation pattern as $\Phi(i)=\{0,13,6,19,12,25,18,31,24,37,30,3,36,9,2,15,8,21,14,27,20,33,26,39,32,5,38,\ldots\}$. Following processing at the rate-matching block, as explained earlier and shown in Figure 9.82, the rate-matched code blocks are sequentially concatenated to create the output of the channel coding process as $f_k, k=0,1,\ldots,N_F-1$ (Figure 9.83). The sequence of coded bits corresponding to one transport block following code block concatenation is referred to as one codeword. In the case of multiple transport blocks per TTI, the transport block to codeword mapping is performed depending on the DCI Format [3,4].

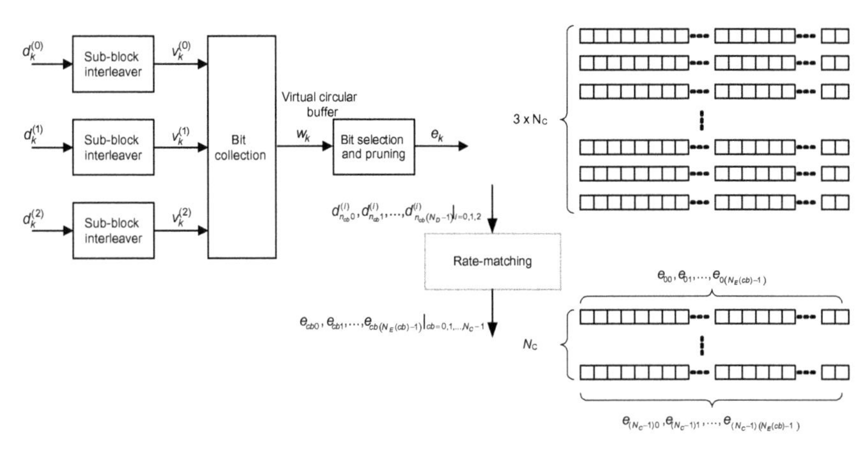

FIGURE 9.82

Illustration of the rate-matching process [158].

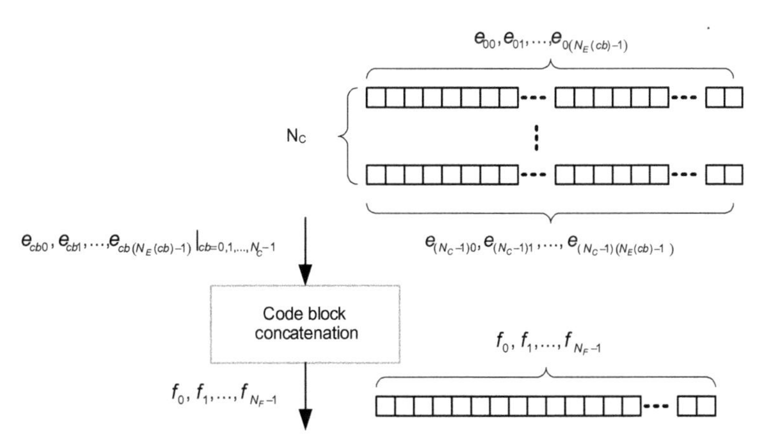

FIGURE 9.83

Illustration of the block concatenation process [158].

9.12 Downlink physical channel processing

This section describes the general physical layer processing structure through which physical channels are processed. As shown in Figure 9.84, data arrives at the coding unit in the form of a maximum of two transport blocks in each TTI. The transport blocks are appended with a CRC which is further scrambled with user identity. If the size of the transport blocks is larger than the maximum coding block size permitted by the physical layer, the code blocks are segmented, and a new CRC is calculated and affixed to each code block. Channel coding and rate-matching are later performed, and the code blocks are concatenated to create the codewords that are the input to the physical channel processing units. In order to generate the baseband signal corresponding to a downlink physical channel, the following processing steps are taken in sequence: (1) scrambling of coded bits in each of the codewords to be transmitted on a physical channel; (2) modulation of scrambled bits to generate complex-valued modulation symbols; (3) mapping of the complex-valued modulation symbols to one or several transmission layers; (4) precoding of the complex-valued modulation symbols on each layer for transmission on the antenna ports; (5) mapping of complex-valued modulation symbols for each antenna port to resource elements; and (6) generation of a complex-valued time-domain OFDM signal for each antenna port. Figure 9.84 depicts the entire physical channel processing from the point that one transport block is delivered to the physical layer by the MAC sub-layer to the point where the $I-Q$ components of the baseband signal to be transmitted on each antenna port are generated.

One or two coded transport blocks (or codewords) can be transmitted simultaneously on PDSCH depending on the precoding scheme used. The codewords undergo scrambling, modulation, layer mapping, precoding and resource element mapping procedures as shown in Figure 9.84. In the following, the function of each block in Figure 9.84 will be described. The specific details about physical layer processing pertaining to physical channels will be described in the next sections.

The block of coded bits $(b_0^l, b_1^l, \ldots, b_{(N_c^l-1)}^l)$ corresponding to codeword $l = 0, 1$, where N_c^l is the number of bits in codeword l transmitted on a physical channel in one subframe are scrambled prior to modulation, resulting in a block of scrambled

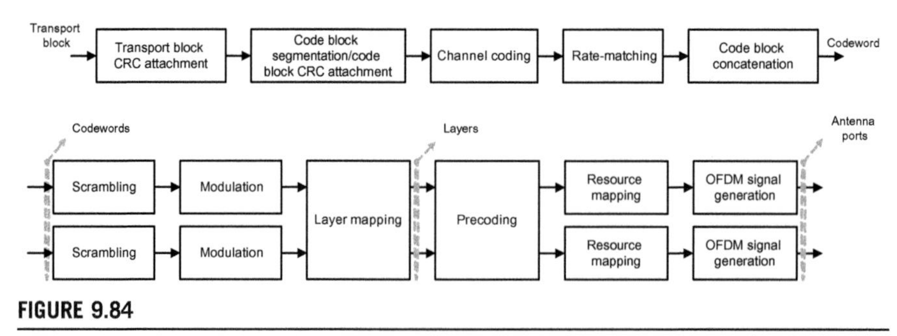

FIGURE 9.84

Overview of downlink physical channel processing in LTE/LTE Advanced [3,4].

bits $(\hat{b}_0^l, \hat{b}_1^l, \ldots, \hat{b}_{(N_c^l-1)}^l)$ in which $\hat{b}_i^l = (b_i^l + c^l(i))\bmod 2$. The scrambling sequence $c^l(i)$ was defined in previous sections. The scrambling sequence generator is initialized at the start of each subframe with the value of c_{init} whose definition depends on the transport channel type as follows: $c_{\text{init}} = n_{\text{RNTI}}2^{14} + l2^{13} + \lfloor n_s/2 \rfloor 2^9 + N_{\text{ID}}^{\text{cell}}$ for PDSCH, and $c_{\text{init}} = \lfloor n_s/2 \rfloor 2^9 + N_{\text{ID}}^{\text{MBSFN}}$ for PMCH. In the expression for PDSCH scrambling sequence initialization, n_{RNTI} corresponds to the RNTI associated with the PDSCH transmission to a specific user. The maximum number of codewords that can be transmitted in one downlink subframe is two.

For each codeword, the block of scrambled bits $(\hat{b}_0^l, \hat{b}_1^l, \ldots, \hat{b}_{(N_c^l-1)}^l)$ are modulated using one of the modulation schemes supported in the downlink, i.e., QPSK, 16QAM, 64QAM, resulting in a block of complex-valued modulation symbols $(d_0^l, d_1^l, \ldots, d_{(N_s^l-1)}^l)$. The complex-valued modulation symbols for each of the codewords are mapped to one or several layers. The complex-valued modulation symbols $(d_0^l, d_1^l, \ldots, d_{(N_s^l-1)}^l)$ for each codeword are mapped to the layers $\mathbf{x}_i = (x_i^0, x_i^1, \ldots, x_i^{N_l-1})^{\text{T}}$, $i = 0, 1, \ldots, N_s^l - 1$ where N_l is the number of layers and N_s^l is the number of modulation symbols per layer.

It will be discussed later in this chapter that LTE Rel-8/9 supports multi-antenna schemes in both downlink and uplink directions. In the downlink direction, up to four transmit antennas may be used with the maximum number of codewords of two, irrespective of the number of antennas. Advanced antenna techniques such as spatial multiplexing and beamforming were supported in the downlink to increase peak data rates and capacity. In downlink SU-MIMO, the eNB may transmit one or more layers to the same UE to take advantage of capacity gains associated with using multiple transmit/receive antennas in rich scattering environments. The UE provides the channel quality feedback and the preferred number of layers as well as the precoding matrix index to the eNB. In downlink MU-MIMO, the eNB attempts to increase capacity by transmitting to multiple users using its multiple transmit antennas at the same time and also using the same time-frequency resources. The UEs must feedback channel state information so that the eNB can schedule the UEs with sufficient channel separation together. The transmitted data is precoded to maximize throughput for users and minimize inter-user interference. In uplink direction, only MU-MIMO was supported, i.e., there is only one modulated symbol stream per UE to be received by the eNB; however, multiple UEs may transmit on the same time−frequency resources. Considering the specified UE categories for LTE Rel-8/9, one can expect two transmit antenna in the downlink and one transmit antenna in the uplink to be the baseline configuration for early LTE deployments. LTE-Advanced extended the MIMO capabilities of LTE Rel-8/9 to eight transmit antennas in the downlink and four transmit antennas in the uplink as well as new SU-MIMO and MU-MIMO modes in the downlink and uplink.

When using transmit diversity, the layer mapping is done with only one codeword, and the number of layers N_l is equal to the number of antenna ports used for transmission of the corresponding physical channel. In the case of transmit diversity with two layers, even-indexed modulation symbols are mapped to layer 0 and odd-indexed modulation symbols are mapped to layer 1. In the case of transmit diversity with four layers, the input symbols are mapped to layers sequentially as shown in Figure 9.85. If the total number of input symbols is not divisible by four ($N_s^0 \bmod 4 \neq 0$), then two null symbols are appended to the end of the symbol sequence. This creates a total number of symbols which is an integer multiple of four, as the original number of symbols will always be an integer multiple of two [3,160].

For spatial multiplexing, the mapping of complex-valued modulation symbols to spatial layers depends on the number of codewords and the number of layers, as illustrated in Figures 9.85 and 9.86 for single codeword and dual codeword layer mapping, respectively. In this case, the number of layers N_l is less than or equal to the number of antenna ports used for transmission of the corresponding physical channel. The case of a single codeword mapped to multiple layers is only applicable when the number of CRSs is four, or when the number of UE-specific reference signals is two or larger. If only one layer is used, then $x_i^0 = d_i^0$ for $i = 0, 1, \ldots, N_s^l - 1$. When one codeword is mapped to two layers, even-indexed modulation symbols are mapped to layer 0 and odd-indexed modulation symbols are mapped to layer 1. If two codewords are mapped to two layers, then codeword 0 is mapped to layer 0 and codeword 1 is mapped to layer 1. In the case of three layers, codeword 0 is mapped to layer 0, even symbols within codeword 1 are mapped to layer 1, and odd symbols within codeword 1 are mapped to layer 2. In the case of four layers, two codewords must be used. In this case, the even symbols of codeword 0 are mapped to layer 0, and the odd symbols are mapped to layer 1. The even symbols of codeword 1 are mapped to layer 2, and the odd symbols are mapped to layer 3. The above layer mapping schemes for spatial multiplexing are illustrated in Figures 9.85 and 9.86. Note that LTE Rel-10 supports downlink spatial multiplexing up to eight layers. The mapping of modulation symbols to 5, 6, 7, or 8 layers with two codewords is simply the extension of the above symbol mapping method [3].

The precoding block in Figure 9.84 receives input complex-valued vectors $\mathbf{x}_i = (x_i^0, x_i^1, \ldots, x_i^{N_l-1})^T$, with $i = 0, 1, \ldots, N_s^l - 1$ and N_s^l denoting the number of modulation symbols per layer, from the layer-mapping block and generates a block of complex-valued vectors $\mathbf{y}_i = (y_i^0, y_i^1, \ldots, y_i^{N_P-1})^T$ to be mapped to resource elements on each antenna port, where $i = 0, 1, \ldots, N_s^p - 1$, N_s^p denotes the number of symbols per antenna port, N_p is the number of antenna ports used for transmission of the corresponding physical channel, and y_i^p represents the ith symbol for antenna port p. For transmission on a single antenna port, the precoding function is defined as $y_i^p = x_i^0$ where $p \in \{0, 4, 5, 7, 8\}$ is the number of the single-antenna ports used for transmission of the physical channel, $i = 0, 1, \ldots, N_s^p - 1$, and $N_s^p = N_s^l$ [3].

The precoding function for spatial multiplexing using antenna ports associated with CRSs is only used in conjunction with layer mapping and supports two or four antenna ports. The set of antenna ports is given by $p \in \{0, 1\}$ or $p \in \{0, 1, 2, 3\}$.

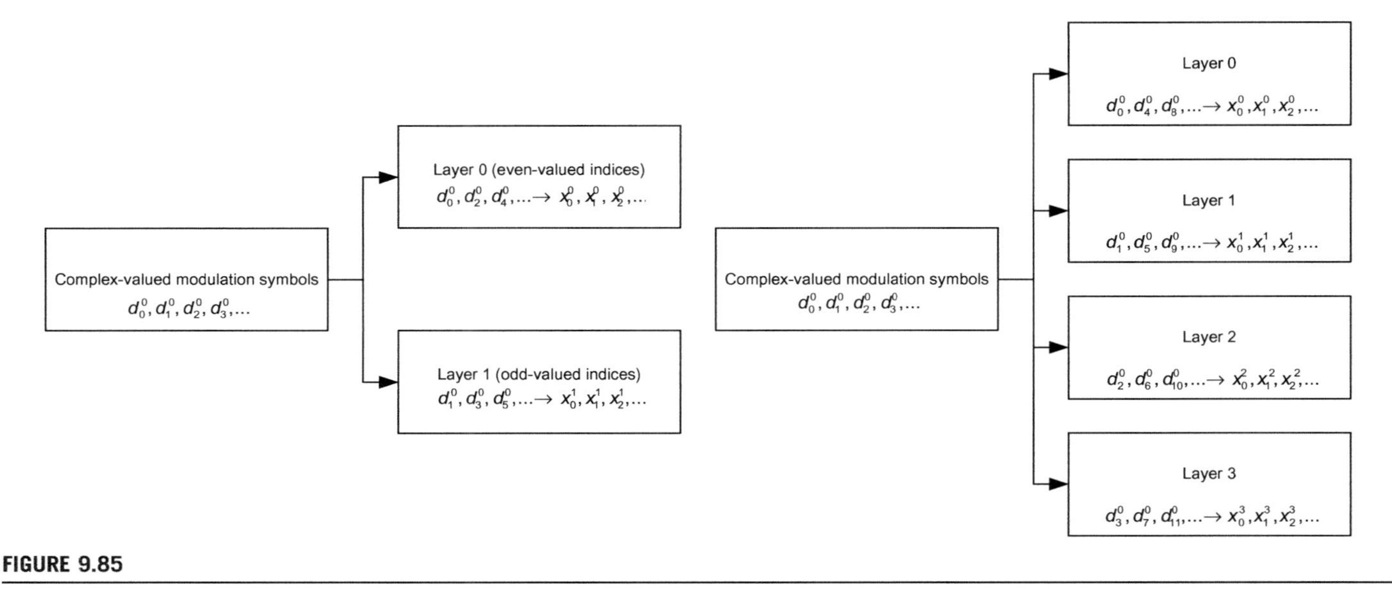

FIGURE 9.85

Layer mapping for 2 and 4 layer transmit diversity (single codeword) [3,160].

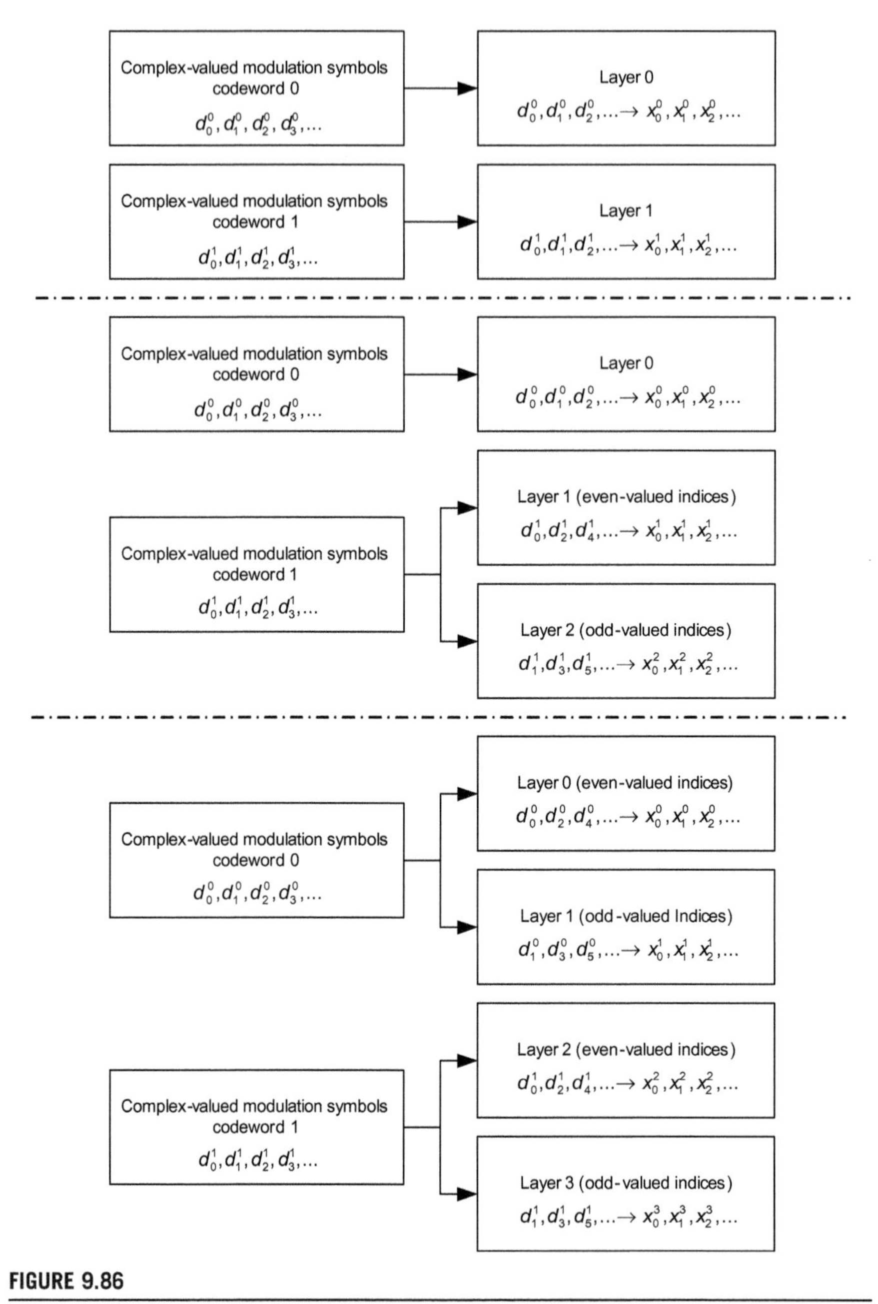

FIGURE 9.86

Layer mapping for spatial multiplexing with 2, 3, and 4 layers (dual codeword) [3,160].

The precoding function for spatial multiplexing without cyclic delay diversity (CDD) is defined as $(y_i^0, y_i^1, \ldots, y_i^{N_p-1})^{\mathrm{T}} = \mathbf{W}_i(x_i^0, x_i^1, \ldots, x_i^{N_l-1})^{\mathrm{T}}$ where the precoding matrix $\mathbf{W}_i$ is of size $N_p \times N_l$ and $i = 0, 1, \ldots, N_s^p - 1$, and $N_s^p = N_s^l$. The values of $\mathbf{W}_i$ are selected among the precoder elements in the codebook configured and stored in the eNB and the UE. The eNB can further restrict the precoder selection in the UE to a subset of the elements in the codebook.

In the case of precoding using large-delay CDD, the precoder for spatial multiplexing is defined as $(y_i^0, y_i^1, \ldots, y_i^{N_p-1})^{\mathrm{T}} = \mathbf{W}_i\mathbf{D}_i\mathbf{U}(x_i^0, x_i^1, \ldots, x_i^{N_l-1})^{\mathrm{T}}$, where the precoding matrix $\mathbf{W}_i$ is of size $N_p \times N_l$ and $i = 0, 1, \ldots, N_s^p - 1$ and $N_s^p = N_s^l$. The diagonal matrix $\mathbf{D}_i$ that is an $N_l \times N_l$ matrix which defines CDD, and the square matrix $\mathbf{U}$ is given in Table 9.18 for different numbers of transmission layers N_l. Similar to the previous scenario, the values of $\mathbf{W}_i$ are selected among the precoder elements in the codebook configured and stored in the eNB and the UE. The eNB can limit the precoder selection in the UE to a subset of the elements in the codebook.

Precoding for transmit diversity is only used in combination with layer mapping. The precoding operation for transmit diversity is defined for two and four antenna ports. For transmission on two antenna ports $p \in \{0, 1\}$, the output vector $\mathbf{y}_i = (y_i^0, y_i^1)^{\mathrm{T}}$ with $i = 0, 1, \ldots, N_s^p - 1$ is defined as follows:

$$\begin{bmatrix} y_{2i}^0 \\ y_{2i}^1 \\ y_{2i+1}^0 \\ y_{2i+1}^1 \end{bmatrix} = \frac{1}{\sqrt{2}} \begin{bmatrix} 1 & 0 & \mathrm{j} & 0 \\ 0 & -1 & 0 & \mathrm{j} \\ 0 & 1 & 0 & \mathrm{j} \\ 1 & 0 & -\mathrm{j} & 0 \end{bmatrix} \begin{bmatrix} \mathrm{Re}(x_i^0) \\ \mathrm{Re}(x_i^1) \\ \mathrm{Im}(x_i^0) \\ \mathrm{Im}(x_i^1) \end{bmatrix} \tag{9.50}$$

where $i = 0, 1, \ldots, N_s^l - 1$ and $N_s^p = 2N_s^l$.

For transmission on four antenna ports $p \in \{0, 1, 2, 3\}$ and for the purpose of CSI reporting based on antenna ports $p \in \{0, 1, 2, 3\}$ or $p \in \{15, 16, 17, 18\}$, the precoding matrix $\mathbf{W}_i$ is defined based on Householder codebook[19] $\mathbf{W}_n = \mathbf{I}_4 - 2\mathbf{u}_n\mathbf{u}_n^{\mathrm{H}}/\mathbf{u}_n^{\mathrm{H}}\mathbf{u}_n$ where $\mathbf{I}_4$ is a 4×4 identity matrix and the vector $\mathbf{u}_n$ is defined in Table 9.18.

[19]In linear algebra, a Householder transformation is defined as a linear transformation that performs a reflection about a plane or hyper-plane containing the origin. The reflection about the hyper-plane can be defined by a unit vector $\mathbf{u}_n$ which is orthogonal to the hyper-plane. The reflection of a vector $\mathbf{x}$ about this hyper-plane is defined as $\mathbf{x} - 2\langle\mathbf{u}_n, \mathbf{x}\rangle\mathbf{u}_n = \mathbf{x} - 2\mathbf{u}_n(\mathbf{u}_n{}^{\mathrm{H}}\mathbf{x})$ where $\mathbf{u}_n$ is given as a column unit vector. This is a linear transformation given by the Householder matrix $\mathbf{\Lambda} = \mathbf{I} - 2\mathbf{u}_n\mathbf{u}_n{}^{\mathrm{H}}$ in which $\mathbf{I}$ denotes the identity matrix. The Householder matrix has the following properties [31,32]:

1. It is a Hermitian matrix $\mathbf{\Lambda} = \mathbf{\Lambda}^{\mathrm{H}}$
2. It is a unitary matrix $\mathbf{\Lambda}^{-1} = \mathbf{\Lambda}^{\mathrm{H}}$ hence it is involutary $\mathbf{\Lambda}^2 = \mathbf{I}$. An involutary function is a function $f(x)$ that is its own inverse function, i.e., $f(f(x)) = x, \forall x$.
3. The eigenvalues of the Householder matrix are ± 1.
4. The determinant of a Householder reflector is $\det(\mathbf{\Lambda}) = -1$ given that the determinant of a matrix is the product of its eigenvalues.

Table 9.18 Codebook for Spatial Multiplexing with and without CDD [3]

N_l	$\mathbf{U}$	$\mathbf{D}_i$	$\mathbf{W}_i$
2	$\frac{1}{\sqrt{2}}\begin{bmatrix}1 & 1\\ 1 & e^{-j2\pi/2}\end{bmatrix}$	$\begin{bmatrix}1 & 0\\ 0 & e^{-j2\pi i/2}\end{bmatrix}$	$N_{TX}=2$: $\mathbf{W}_i=\frac{1}{\sqrt{2}}\begin{bmatrix}1 & 0\\ 0 & 1\end{bmatrix}$ $N_{TX}=4$: $\mathbf{W}_n=\mathbf{I}_4-2\mathbf{u}_n\mathbf{u}_n^H/\mathbf{u}_n^H\mathbf{u}_n$ $n=12,13,14,15$
3	$\frac{1}{\sqrt{3}}\begin{bmatrix}1 & 1 & 1\\ 1 & e^{-j2\pi/3} & e^{-j4\pi/3}\\ 1 & e^{-j4\pi/3} & e^{-j8\pi/3}\end{bmatrix}$	$\begin{bmatrix}1 & 0 & 0\\ 0 & e^{-j2\pi i/3} & 0\\ 0 & 0 & e^{-j4\pi i/3}\end{bmatrix}$	$\mathbf{W}_n=\mathbf{I}_4-2\mathbf{u}_n\mathbf{u}_n^H/\mathbf{u}_n^H\mathbf{u}_n$ $n=12,13,14,15$
4	$\frac{1}{2}\begin{bmatrix}1 & 1 & 1 & 1\\ 1 & e^{-j2\pi/4} & e^{-j4\pi/4} & e^{-j6\pi/4}\\ 1 & e^{-j4\pi/4} & e^{-j8\pi/4} & e^{-j12\pi/4}\\ 1 & e^{-j6\pi/4} & e^{-j12\pi/4} & e^{-j18\pi/4}\end{bmatrix}$	$\begin{bmatrix}1 & 0 & 0 & 0\\ 0 & e^{-j2\pi i/4} & 0 & 0\\ 0 & 0 & e^{-j4\pi i/4} & 0\\ 0 & 0 & 0 & e^{-j6\pi i/4}\end{bmatrix}$	$\mathbf{W}_n=\mathbf{I}_4-2\mathbf{u}_n\mathbf{u}_n^H/\mathbf{u}_n^H\mathbf{u}_n$ $n=12,13,14,15$

Precoding for spatial multiplexing using antenna ports with UE-specific reference signals is only applicable in conjunction with layer mapping, and supports up to eight antenna ports where the set of antenna ports is given by $p = 7, 8, \ldots, N_l + 6$. For transmission on $N_p = N_l$ antenna ports, the precoding operation is defined by $(y_i^7, y_i^8, \ldots, y_i^{N_l+6})^{\mathrm{T}} = (x_i^0, x_i^1, \ldots, x_i^{N_l-1})^{\mathrm{T}}$, where $i = 0, 1, \ldots, N_s^p - 1$ and $N_s^p = N_s^l$.

For each antenna port used for transmission of the physical channels, the block of complex-valued symbols $y_0^p, \ldots, y_{N_s^p-1}^p$ must be sequentially mapped (starting with y_0^p) to resource elements (k, l) whose coordinates satisfy the following criteria: (1) these resource elements are in the PRBs corresponding to the designated VRBs; (2) these resource elements are not used for transmission of PBCH, synchronization signals, CRSs, MBSFN reference signals, or UE-specific reference signals; (3) the UE assumes that these resource elements are not used for transmission of CSI-RSs and the DCI associated with the downlink transmission using the C-RNTI or semi-persistent C-RNTI; and (4) index l in the first slot in a subframe satisfies $l \geq l_{\mathrm{DataStart}}$, where $l_{\mathrm{DataStart}}$ is the starting OFDM symbol index for corresponding downlink transmission. Therefore, for each antenna port used for transmission of PDSCH, the block of complex-valued symbols y_i^p are mapped in sequential order to resource elements that are not occupied by the PCFICH, PHICH, PDCCH, PBCH or synchronization, and reference signals. The number of resource elements involved in this mapping is controlled by the number of resource blocks allocated to PDSCH. The symbols are mapped by increasing the sub-carrier index and mapping all available resource elements within allocated resource blocks for each OFDM symbol, as shown in Figure 9.87.

9.12.1 Physical control format indicator channel

The PCFICH carries information about the number of OFDM symbols used for transmission of PDCCHs in a subframe. In other words, the PCFICH carries a control format indicator (CFI) which indicates the number of OFDM symbols, which is typically 1, 2, or 3, except for system bandwidth less than 5 MHz, used for transmission of control channel information in each subframe. In principle, the UE can intuitively deduce the value of the CFI without using PCFICH, e.g., by multiple attempts to blindly decode the control channels assuming each possible number of symbols, but this would result in significant additional processing complexity. Three different CFI values are used in LTE/LTE-Advanced. In order to make the CFI sufficiently robust, each codeword is 32 bits long and is mapped to 16 resource elements using QPSK modulation. These 16 resource elements are organized in groups of four resource elements, known as REGs. The resource elements occupied by reference signals are not included within the REGs, which means that the total number of REGs in a given OFDM symbol depends on whether cell-specific reference signals are present. The concept of REGs (i.e., mapping in groups of four resource elements) is also used for PHICH and

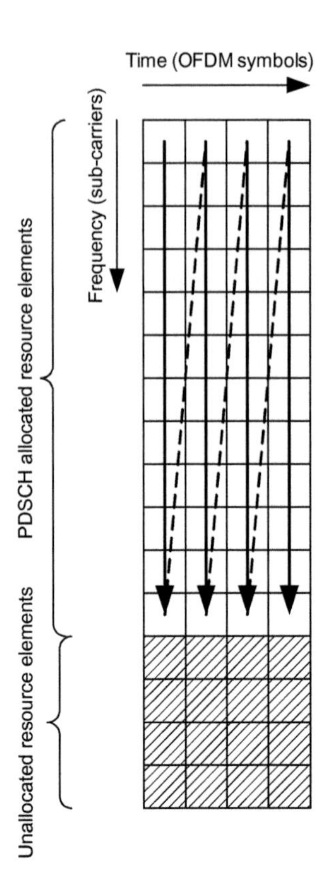

FIGURE 9.87

Resource element mapping [160].

PDCCH control channels. The PCFICH is transmitted on the same set of antenna ports as PBCH. The transmit diversity is applied if more than one antenna port is used. In order to achieve frequency diversity, the four REGs carrying the PCFICH are distributed across the frequency-domain. This is done according to a predefined pattern in the first OFDM symbol of each downlink subframe (Figure 9.88), so that the UEs can always locate the PCFICH information, which is a prerequisite for detection and decoding of the other control channels. To minimize the inter-cell interference, as well as the possibility of confusion with PCFICH information transmitted from neighboring cells, a cell-specific frequency offset is applied to the positions of the PCFICH resource elements. The offset value depends on the physical cell identity, which is derived from the primary and secondary synchronization signals. In addition, a cell-specific scrambling sequence is applied to the CFI codewords, so that the UE can preferentially receive PCFICH from the desired cell site.

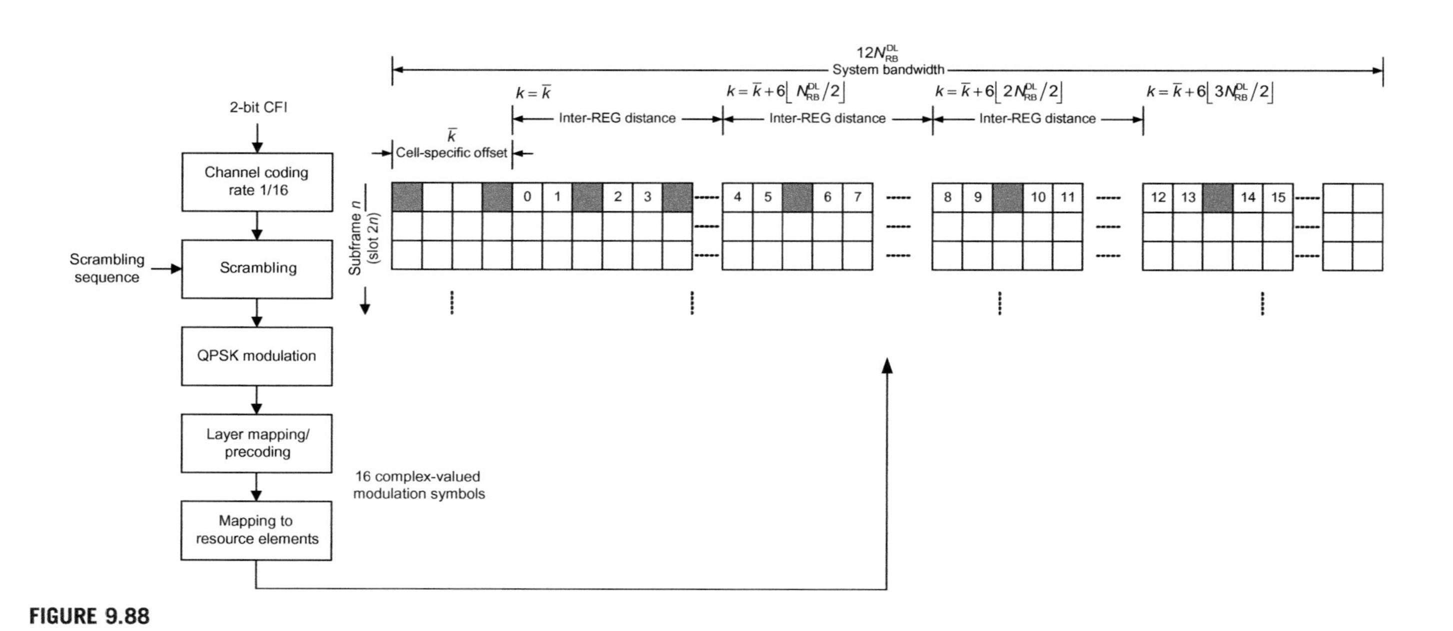

FIGURE 9.88

PCFICH physical layer processing [15,162].

As shown in Figure 9.88, the block of coded bits $(b_0, \ldots, b_{31})$ containing CFI information (i.e., 2 bits of information encoded at the rate 1/16) transmitted in one subframe are scrambled with a cell-specific sequence prior to modulation, resulting in a block of scrambled bits $(\tilde{b}_0, \ldots, \tilde{b}_{31})$, where $\tilde{b}_i = (b_i + c(i)) \bmod 2$. The scrambling sequence $c(i)$ defined in previous sections is initialized with $c_{\text{init}} = \left(\lfloor n_s/2 \rfloor + 1\right)\left(2N_{\text{ID}}^{\text{cell}} + 1\right)2^9 + N_{\text{ID}}^{\text{cell}}$ at the start of each subframe. The block of scrambled bits $(\tilde{b}_0, \ldots, \tilde{b}_{31})$ is modulated using QPSK modulation, resulting in a block of complex-valued modulation symbols $(d_0, \ldots, d_{15})$ which are mapped to spatial layers using a transmit diversity transmission mode with $N_s^0 = 16$, i.e., 16 complex-valued modulation symbols per layer and are precoded, resulting in a block of vectors $\mathbf{y}_i = (y_i^0, y_i^1, \ldots, y_i^{N_p-1})^{\text{T}}$, $i = 0, \ldots, 15$, where y_i^p represents the signal for antenna port p and $p = 0, \ldots, N_p - 1$. The number of antenna ports for cell-specific reference signals $N_p \in \{1, 2, 4\}$. The PCFICH is transmitted on the same set of antenna ports that are used for transmission of PBCH. The mapping to resource elements is defined in terms of quadruplets of complex-valued symbols. Let $Z_i^P = (y_{4i}^P, y_{4i+1}^P, y_{4i+2}^P, y_{4i+3}^P)$ denote symbol quadruplet i for antenna port p. For each of the antenna ports, symbol quadruplets are mapped in increasing order of index i to four resource-element groups in the first OFDM symbol in a downlink subframe (assuming 12 sub-carriers per resource block) in the following manner:

$$
\begin{array}{ll}
Z_0^P \text{ is mapped to the resource-element represented by} & k = \bar{k} \\
Z_1^P \text{ is mapped to the resource-element represented by} & k = \bar{k} + 6\lfloor N_{\text{RB}}^{\text{DL}}/2 \rfloor \\
Z_2^P \text{ is mapped to the resource-element represented by} & k = \bar{k} + 6\lfloor 2N_{\text{RB}}^{\text{DL}}/2 \rfloor \\
Z_3^P \text{ is mapped to the resource-element represented by} & k = \bar{k} + 6\lfloor 3N_{\text{RB}}^{\text{DL}}/2 \rfloor
\end{array}
\tag{9.51}
$$

where the additions are modulo $12N_{\text{RB}}^{\text{DL}}$, $\bar{k} = 6(N_{\text{ID}}^{\text{cell}} \bmod 2N_{\text{RB}}^{\text{DL}})$ and $N_{\text{ID}}^{\text{cell}}$ is the physical-layer cell identity. The above procedure is illustrated in Figure 9.88.

9.12.2 Physical broadcast channel

The essential system information which allows the other channels in the cell to be configured and operated is carried by PBCH. Therefore, the reliability and coverage of the PBCH is crucial for the successful operation of LTE/LTE-Advanced systems. The system information is divided into two categories: (1) the master information block as given in Table 9.19, which consists of a limited number of the most frequently transmitted parameters essential for initial access to the cell which is carried in PBCH; and (2) the other system information blocks (SIBs) which are multiplexed with unicast data and are transmitted in PDSCH. The design requirements for PBCH are as follows:

- The UE must be able to detect the PBCH without prior knowledge of the system bandwidth.
- The PBCH must have negligible impact on the total physical layer overhead.
- The PBCH must be reliably received by the cell-edge UEs.
- The PBCH must be decodable with low latency and complexity to reduce the impact on the UE battery life.

Table 9.19 Contents of the Master Information Block [9]

Parameter		Size (Bits)	Description
Downlink channel bandwidth		3	Selects one of the configured downlink system bandwidths in terms of the number of resource blocks (i.e., $N_{RB}^{DL} = 6, 15, 25, 50, 75, 100$)
PHICH configuration	PHICH Duration	1	Selects between two options "normal" and "extended" where normal refers to one OFDM symbol and extended refers to three OFDM symbols
	PHICH Resource	2	Selects one of the values of PHICH group scaling factor $N_g \in \{1/6, 1/2, 1, 2\}$
System frame number (SFN)		8	Most significant bits of the 10-bit system frame number. The two least significant bits of the SFN are deducted by the UE from the position of the PBCH (i.e., within 40 ms PBCH TTI, the first radio frame: 00, the second radio frame: 01, the third radio frame: 10, the last radio frame: 11)
Unused bits		10	Null bits for future extensions

The overall PBCH structure is shown in Figure 9.89. The detectability of the PBCH without the UE's prior knowledge of the system bandwidth is achieved by mapping the PBCH to the center of the band using 72 sub-carriers, which corresponds to the minimum system bandwidth of six resource blocks, regardless of the actual system bandwidth. The UE will have to identify the system center frequency from the synchronization signals as described in earlier sections. Minimizing the impact on total overhead is achieved by deliberately keeping the amount of information carried by the PBCH to a minimum, since achieving stringent coverage requirements for a large quantity of data would result in a high system overhead. The size of MIB it is 24 bits and it is transmitted every 40 ms. In order to achieve reliable reception, time diversity, FEC, and antenna diversity mechanisms are used to transmit PBCH. Time diversity is introduced by spreading the transmission of the PBCH over a period of 40 ms with three retransmissions at 10 ms intervals. This significantly reduces the likelihood of missing/mis-detection of MIB due to multipath fading, interference, and noise in the radio channel, even when the mobile terminal is moving at high speeds.

The FEC coding for PBCH utilizes a tail-biting convolutional coder, considering that the number of information bits to be encoded is small and turbo codes are typically not suitable for small payloads. The basic code rate is 1/3, after which a high repetition factor is applied to the systematic and parity bits such that each MIB is coded at a very low code rate (1/48 over a 40 ms period) to provide strong error protection. Antenna diversity may be utilized at the eNB and the UE. The UE performance requirements specified for LTE/LTE-Advanced assume that all UEs can achieve a level of decoding performance proportional to dual-antenna

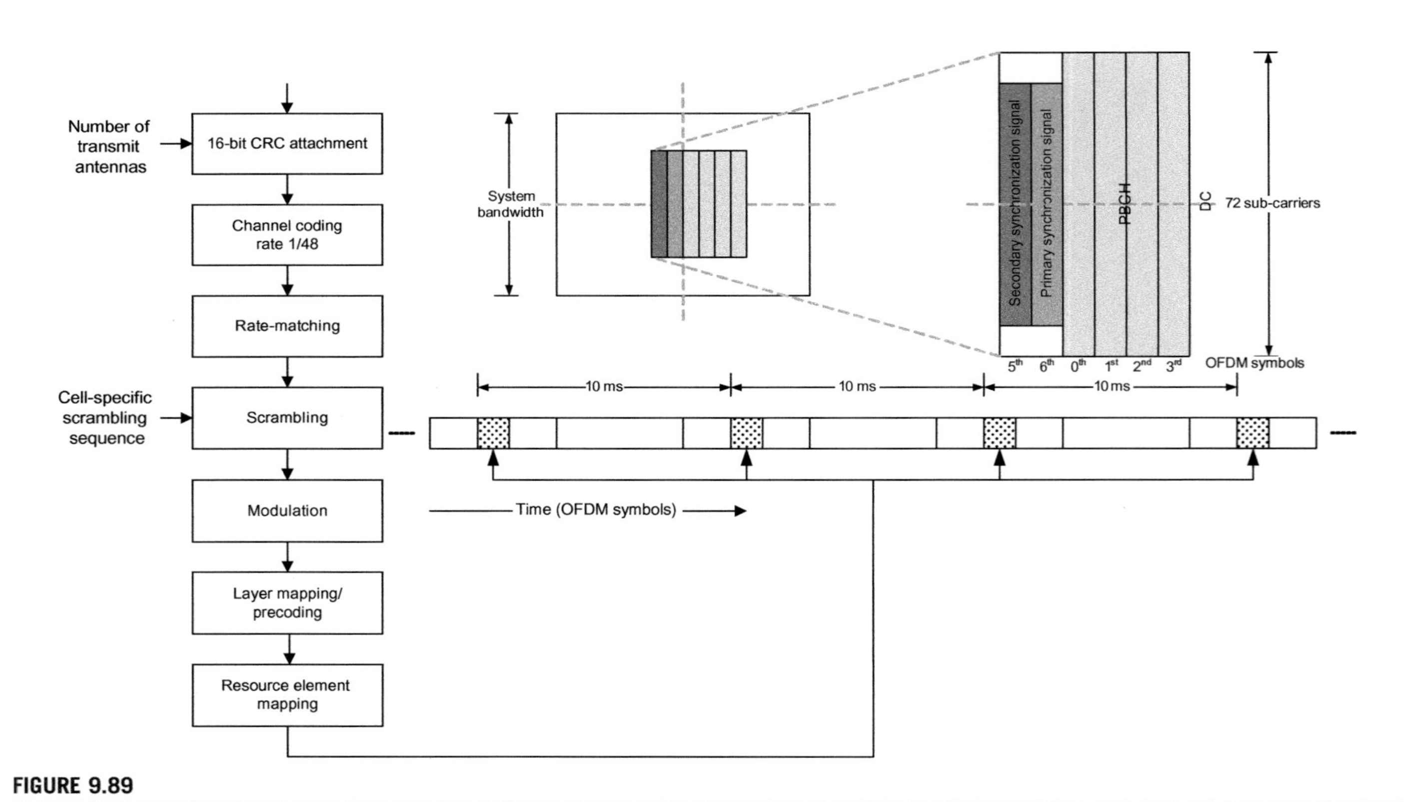

FIGURE 9.89

PBCH transmission and physical layer processing [162].

receive diversity, although it is understood that in low-frequency deployments, e.g., below 1 GHz, the advantage obtained from receive diversity is reduced due to the correspondingly higher correlation between the antennas. This enables wider or more reliable cell coverage to be achieved with fewer cell sites.

Transmit antenna diversity may be employed at the eNB to further improve coverage, depending on eNB implementation, eNB may transmit PBCH with two or four transmit antennas using an SFBC scheme. The set of resource elements designated to PBCH transmission is independent of the number of transmit antennas at the eNB, thereby any resource elements which may be used for cell-specific reference signals transmission are avoided by the PBCH, irrespective of the actual number of transmit antenna ports used by the eNB. The number of transmit antennas used by the eNB is blindly detected by the UE by decoding PBCH transmission corresponding to a different number of transmit antennas (i.e., one, two, or four). The detection of the number of transmit antennas is further facilitated by masking the CRC of each MIB with a unique masking sequence representing the number of transmit antenna ports. The 16-bit PBCH CRC masking sequence is $\langle 0,0,0,0,0,0,0,0,0,0,0,0,0,0,0,0 \rangle$,$\langle 1,1,1,1,1,1,1,1,1,1,1,1,1,1,1,1 \rangle$, or $\langle 0,1,0,1,0,1,0,1,0,1,0,1,0,1,0,1 \rangle$ for one, two, or four transmit antennas, respectively.

Achieving low latency and minimal impact on UE battery life are facilitated through use of a low code rate with repetition wherein the set of coded bits are divided into four subsets, each of which is self-decodable. Each subset of the coded bits is transmitted in a different radio frame during the 40 ms transmission period, as shown in Figure 9.89. This means that if the SINR is sufficiently good to allow the UE to successfully decode the MIB in less than four radio frames, then the UE does not need to receive the other parts of the PBCH transmission in the remainder of the 40 ms period. On the other hand, if the SINR is low, the UE can take advantage of soft-combing of other parts of the MIB transmission until MIB is successfully decoded. The UE does not know in advance the timing of the 40 ms transmission interval for each MIB on the PBCH; therefore, this information is determined implicitly from the scrambling and bit positions, which are reinitialized every 40 ms. The UE can initially determine the 40 ms timing by performing four separate decodings of the PBCH using each of the four possible phases of the PBCH scrambling code and checking the CRC for each decoding.

When a UE initially attempts to select a cell by acquiring the PBCH, different approaches may be taken to perform the necessary blind decoding. A simple approach is to perform the decoding using soft combing of the PBCH instances over four radio frames, advancing a 40 ms sliding window one radio frame at a time until the sliding window aligns with the 40 ms period of the PBCH and the decoding succeeds. However, this would result in a 40–70 ms delay before the PBCH can be decoded. A faster approach would be to attempt to decode the PBCH from the first radio frame, which should be possible provided that the SINR is sufficiently high. If the decoding fails for all four possible scrambling sequences, the PBCH from the first frame could be soft-combined with the next

PBCH instances. It is evident that the latter approach may be much faster at the expense of slightly more complexity.

More specifically, the block of coded bits $(b_0, b_1 \ldots, b_{N_{\text{bit}}-1})$ are scrambled with a cell-specific scrambling sequence prior to modulation, resulting in a block of scrambled bits $(\tilde{b}_0, \tilde{b}_1, \ldots, \tilde{b}_{N_{\text{bit}}-1})$ in which $\tilde{b}_i = (b_i + c(i)) \bmod 2$ where the cell-specific scrambling sequence $c(i)$ was defined in the previous sections. The number of bits transmitted on PBCH $N_{\text{bit}} = 1920$ for a normal cyclic prefix and $N_{\text{bit}} = 1728$ for an extended cyclic prefix. The scrambling sequence is initialized with $c_{\text{init}} = N_{\text{ID}}^{\text{cell}}$ in each radio frame, provided that $n_f \bmod 4 = 0$.

The block of scrambled bits $(\tilde{b}_0, \tilde{b}_1, \ldots, \tilde{b}_{N_{\text{bit}}-1})$ is modulated using QPSK modulation, resulting in a block of complex-valued modulation symbols $(d_0, d_1, \ldots, d_{N_s-1})$. The block of modulation symbols $(d_0, d_1, \ldots, d_{N_s-1})$ is mapped to spatial layers with $N_s^0 = N_s$ and precoded, resulting in a block of vectors $\mathbf{y}_i = (y_i^0, y_i^1, \ldots, y_i^{N_p-1})^T$ where $i = 0, 1, \ldots, N_s - 1$ and y_i^p is the signal for antenna port $p = 0, 1, \ldots, N_p - 1$ and $N_p \in \{1, 2, 4\}$.

The block of complex-valued symbols $(y_0^p, y_1^p \ldots, y_{N_s-1}^p)$ for each antenna port is transmitted during four consecutive radio frames starting in the frame whose number satisfies $n_f \bmod 4 = 0$. The symbols are mapped sequentially starting with y_0^p to resource elements (k, l) that are not reserved for transmission of reference signals. The resource mapping is done in increasing order of sub-carrier index k (i.e., frequency first) and then index l (OFDM symbol number) in slot 1 of subframe 0 in each radio frame. The resource-element indices are given by $k = 6N_{\text{RB}}^{\text{DL}} - 36 + k'; k' = 0, 1, \ldots, 71$; and $l = 0, 1, \ldots, 3$ where resource elements reserved for reference signals are excluded. The mapping procedure assumes that cell-specific reference signals for antenna ports 0–3 are used irrespective of the actual antenna configuration [3].

9.12.3 Physical hybrid-ARQ indicator channel

The PHICH carries the HARQ ACK/NACK, which indicates whether the eNB has correctly received a transmission on PUSCH. The HARQ indicator is set to "0" for a positive acknowledgment and is set to "1" for a negative acknowledgment. This information is repeated in each of three BPSK symbols. Multiple PHICHs are mapped to the same set of resource elements. These constitute a PHICH group, where different PHICHs within the same PHICH group are separated through different complex-valued orthogonal Walsh sequences. Each PHICH is uniquely identified by a PHICH index, which indicates both the group and the sequence. The sequence length is four for the normal cyclic prefix (or two for the extended cyclic prefix). As the sequences are complex-valued, the number of PHICHs in a group, i.e., the number of UEs receiving their acknowledgments on the same set of downlink resource elements, can be up to twice the sequence length. A cell-specific scrambling sequence is applied. A $3\times$ code repetition factor is applied for improved robustness, resulting in three instances of the orthogonal Walsh code being transmitted for each ACK or NACK. The error rate on the PHICH is intended to be in the order of 0.01 for ACKs, and as low as 0.0001 for NACKs. The resulting PHICH construction, including repetition and orthogonal spreading, is shown in Figure 9.90.

The PHICH duration, in terms of the number of OFDM symbols used in the time-domain, is typically one or three OFDM symbols, which is configurable and the eNB broadcasts the configuration information via the PBCH (see Table 9.19). Since the PHICH cannot extend into the PDSCH transmission region, the duration configured for the PHICH puts a lower limit on the size of the control channel region at the start of each subframe. Each of the three instances of the orthogonal code of a PHICH transmission is mapped to a REG on one of the first three OFDM symbols of each subframe, so that each PHICH is partly transmitted on each of the available OFDM symbols. This mapping is illustrated in Figure 9.90 for an example PHICH duration. The PBCH also provides the number of PHICH groups that are configured in the cell, which enables the UEs to derive the resource elements in the control region where the PDCCHs are mapped.

In order to obviate the need for additional signaling to indicate which PHICH carries the ACK/NACK response for each PUSCH transmission, the PHICH index is implicitly associated with the index of the lowest uplink resource block used for the corresponding PUSCH transmission. This relationship is such that adjacent PUSCH resource blocks are associated with PHICHs in different PHICH groups, to enable some degree of load balancing. However, this mechanism alone is not sufficient to enable multiple UEs to be allocated in the same resource blocks for a PUSCH transmission, as occurs in the case of uplink MU-MIMO, in this case, different cyclic shifts of the uplink DM-RSs are configured for the different UEs which are allocated the same time–frequency PUSCH resources, and the same cyclic shift index is then used to shift the PHICH allocations in the downlink so that each UE receives its ACK/NACK on a different PHICH. The above procedures are illustrated in Figure 9.90 [13,14].

Each PHICH group consists of 12 complex-valued symbols which require three REGs and can carry up to four (extended cyclic prefix) or eight (normal cyclic prefix) ACK/NACKs, which are identified within a PHICH group by an orthogonal code. Each PHICH group is identified by a PHICH group number and one of the eight or four orthogonal spreading codes. The starting location of the first REG of the PHICH group is derived from the physical cell identity. The other two groups are equally spaced from the first group.

The PHICH may be configured with a normal or extended duration. In a normal duration, all three REGs of a PHICH group are mapped to the first OFDM symbol. In an extended duration, the three REGs are mapped to three different OFDM symbols, as shown in Figure 9.91. The extended duration PHICH mapping provides time diversity. The duration of the PHICH is signaled via RRC and carried in the PHICH configuration information element.

As mentioned earlier, multiple PHICHs are mapped to the same set of resource elements to create a PHICH group, where PHICHs within the same

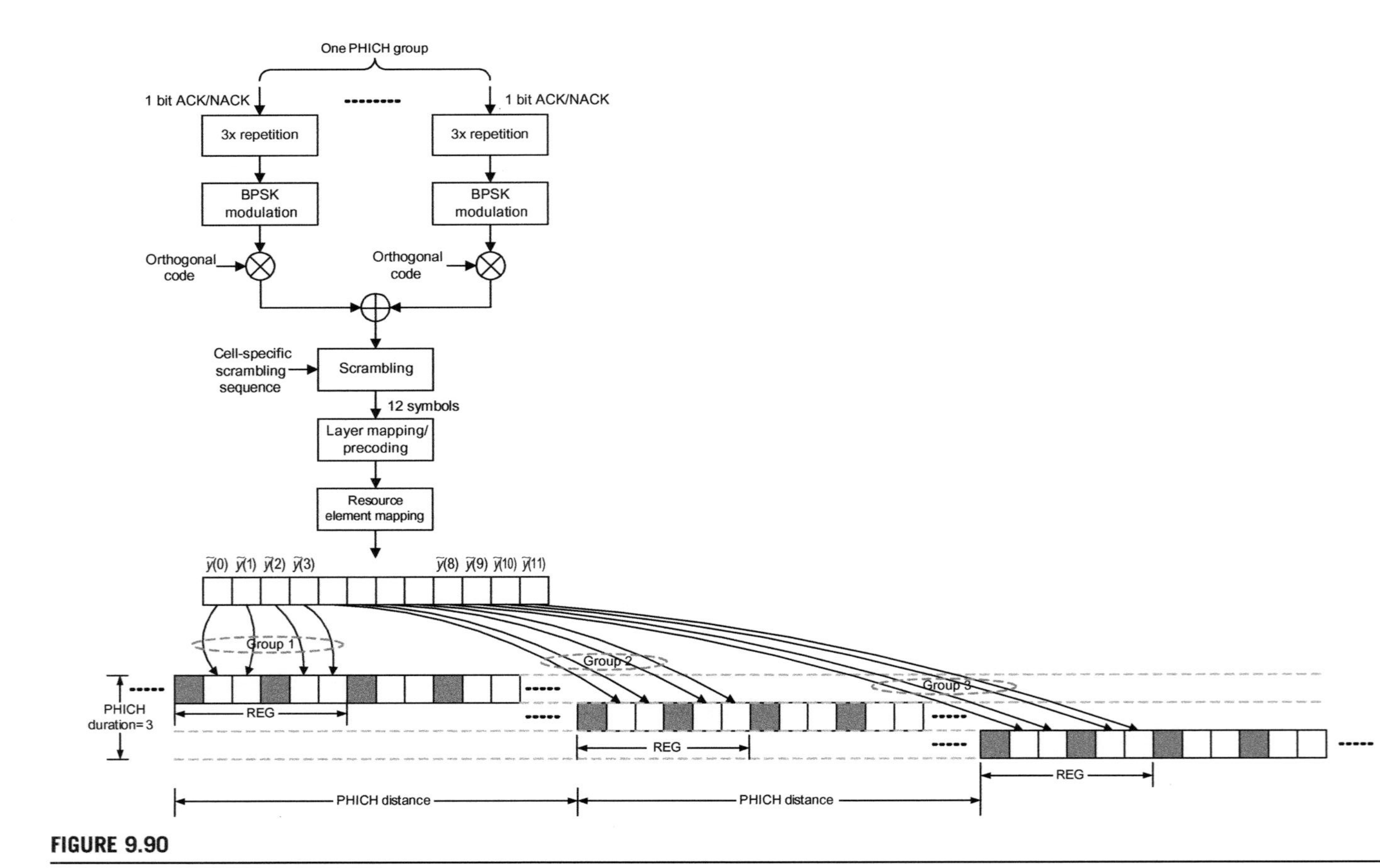

FIGURE 9.90

PHICH physical layer processing and mapping in time and frequency for an FDD system with normal cyclic prefix [15,162].

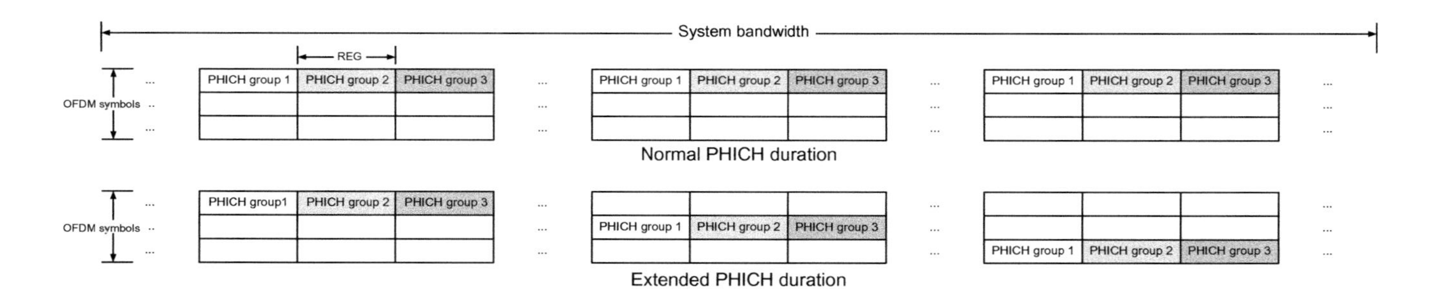

FIGURE 9.91

Illustration of normal and extended PHICH mapping.

PHICH group are separated by different orthogonal sequences. A PHICH resource is identified by the index pair $(n_{\text{PHICH}}^{\text{group}}, n_{\text{PHICH}}^{\text{sequence}})$, where $n_{\text{PHICH}}^{\text{group}}$ denotes the PHICH group number and $n_{\text{PHICH}}^{\text{sequence}}$ is the orthogonal sequence index within the group. For frame structure type 1, the number of PHICH groups $n_{\text{PHICH}}^{\text{group}}$ is constant in all subframes and is given by $N_{\text{PHICH}}^{\text{group}} = \lceil N_g(N_{\text{RB}}^{\text{DL}}/8) \rceil$ for a normal cyclic prefix and $N_{\text{PHICH}}^{\text{group}} = 2\lceil N_g(N_{\text{RB}}^{\text{DL}}/8) \rceil$ for an extended cyclic prefix, in which $N_g \in \{1/6, 1/2, 1, 2\}$ is a parameter configured by higher layers and transmitted in the form of two bits in the PBCH (see Table 9.19). The group index $n_{\text{PHICH}}^{\text{group}}$ assumes integer values between 0 and $N_{\text{PHICH}}^{\text{group}} - 1$. For frame structure type 2, the number of PHICH groups may vary from one downlink subframe to another and is given by $m_i N_{\text{PHICH}}^{\text{group}}$, where the value of parameter $m_i = 0, 1, 2$ depends on the TDD subframe configuration, e.g., $m_i = 1$ for the second subframe in TDD frame configuration 1. The group index $n_{\text{PHICH}}^{\text{group}}$ in a downlink subframe with non-zero PHICH resources assumes integer values between 0 and $m_i N_{\text{PHICH}}^{\text{group}} - 1$.

One may ask at this point how the UEs can find the PHICH corresponding to their earlier PUSCH transmissions. The UE searches for the PHICH in the fourth subframe $(n+4)$ after an uplink transport block is transmitted in uplink subframe n. The PHICH group and sequence indices are calculated based on the lowest PRB index allocated for PUSCH transmission and the cyclic shift for the PUSCH DM-RS, both of which are associated with the uplink transmission and are provided to the UE in DCI format 0 or DCI format 4. The PHICH resource (N_g) is received in the MIB or in the PHICH configuration information element via dedicated RRC signaling. The ith UE selects its PHICH group and orthogonal sequence index $(n_{\text{PHICH}}^{\text{group}}(i), n_{\text{PHICH}}^{\text{sequence}}(i))$ according to the following rules in an FDD system [13]:

$$\begin{aligned} 1 - n_{\text{PHICH}}^{\text{group}}(i) &= (I_{\text{lowest}} + n_{\text{DM-RS}}) \bmod N_{\text{PHICH}}^{\text{group}} \\ 2 - n_{\text{PHICH}}^{\text{sequence}}(i) &= \left(\lfloor I_{\text{lowest}} / N_{\text{PHICH}}^{\text{group}} \rfloor + n_{\text{DM-RS}} \right) \bmod (2N_{\text{SF}}^{\text{PHICH}}) \end{aligned} \tag{9.52}$$

where $n_{\text{DM-RS}}$ denotes the PUSCH DM-RS cyclic shift signaled in DCI format 0 or 4, and I_{lowest} is the lowest PRB index allocated to the first slot of the PUSCH transmission. In the case of a TDD system with subframe configuration 0 and PUSCH transmission in subframes 4 or 9, the PHICH group index calculated in Eq. (9.51) must be incremented by $N_{\text{PHICH}}^{\text{group}}$ [13]. For a Rel-10 UE using two transport blocks within a subframe, the parameter I_{lowest} is incremented by one when calculating the PHICH group and sequence number for the second transport block.

As shown in Figure 9.90, the block of bits $(b_0, b_1, \ldots, b_{N_{\text{bit}}-1})$ transmitted on one PHICH in one subframe are modulated using BPSK modulation, resulting in a block of complex-valued modulation symbols $(z_0, z_1, \ldots, z_{N_s-1})$ where $N_s = N_{\text{bit}}$ given the type of modulation. The block of modulation symbols $(z_0, z_1, \ldots, z_{N_s-1})$ are symbol-by-symbol multiplied by an orthogonal scrambling sequence, resulting in a sequence of scrambled modulation symbols $(d_0, d_1 \ldots, d_{N_{\mathrm{SF}}^{\mathrm{PHICH}} N_s - 1})$ where:

$$d_i = \underbrace{w(i \bmod N_{\mathrm{SF}}^{\mathrm{PHICH}})}_{\substack{\text{Orthogonal sequence symbol} \\ \text{with index } n_{\mathrm{PHICH}}^{\mathrm{sequence}}}} \quad \underbrace{(1 - 2c(i))}_{\substack{\text{Cell-specific scrambling} \\ \text{sequence symbol}}} \quad \underbrace{z\left(\lfloor i / N_{\mathrm{SF}}^{\mathrm{PHICH}} \rfloor\right)}_{\substack{\text{Modulated HARQ} \\ \text{indicator symbol}}} \quad i = 0, 1, \ldots, N_{\mathrm{SF}}^{\mathrm{PHICH}} N_s - 1 \tag{9.53}$$

The value of the spreading factor is $N_{\mathrm{SF}}^{\mathrm{PHICH}} = 4$ or 2 for a normal or extended cyclic prefix type, respectively [3]. The scrambling sequence $c(i)$ is a cell-specific scrambling sequence which was defined in the previous sections and is initialized by $c_{\mathrm{init}} = \left(\lfloor n_s/2 \rfloor + 1\right)\left(2N_{\mathrm{ID}}^{\mathrm{cell}} + 1\right)2^9 + N_{\mathrm{ID}}^{\mathrm{cell}}$ at the beginning of each subframe. The orthogonal spreading sequence $(w_0, w_1, \ldots, w_{N_{\mathrm{SF}}^{\mathrm{PHICH}} - 1})$ is defined in Table 9.20 where the sequence index $n_{\mathrm{PHICH}}^{\mathrm{sequence}}$ corresponds to the PHICH number within the PHICH group.

Since REGs contain four resource elements, each able to carry one modulation symbol, the blocks of scrambled symbols are aligned to create blocks of four symbols (Figure 9.92). In the case of a normal cyclic prefix, each of the original complex-valued modulated symbol is represented by four scrambled symbols, thus no alignment is required. In the case of an extended cyclic prefix, each of the original complex-valued modulated symbols is represented by two scrambled symbols. To create blocks of four symbols, zeros are added before or after blocks of two scrambled symbols, depending on whether the PHICH index is odd or even. This allows two groups to be combined during the resource mapping stage and mapped to one REG. Groups of four symbols are formed as illustrated in Figure 9.92.

The block of symbols $(d_0, d_1 \ldots, d_{N_{\mathrm{SF}}^{\mathrm{PHICH}} N_s - 1})$ are first aligned with REG size, and are then mapped to spatial layers and precoded, yielding a block of vectors $\mathbf{y}_i = (y_i^0, y_i^1, \ldots, y_i^{N_p - 1})^{\mathrm{T}}$ where $i = 0, 1, \ldots, N_{\mathrm{SF}}^{\mathrm{PHICH}} N_s - 1$, and y_i^p represents the signal of antenna port $p = 0, 1, \ldots, N_p - 1$. The number of antenna ports with CRSs is chosen as $N_p \in \{1, 2, 4\}$. The layer mapping and precoding operation depend on the cyclic prefix type and the number of antenna ports used for transmission of the PHICH. The PHICH is transmitted on the same set of antenna ports that are used for transmission of the PBCH.

The number of OFDM symbols used to carry PHICH is configurable by PHICH duration. The PHICH duration is either normal or extended. A normal PHICH duration allows PHICH to present only in the first OFDM symbol of the subframe. In general, an extended PHICH duration allows PHICH to be located in the first three OFDM symbols of the subframe, with some exceptions. Under those exceptions, PHICH may present in the first two OFDM symbols of the subframe. Those exceptions are as follows: (1) within subframe 1 and 6 when frame structure type 2 (TDD) is used; and (2) within MBSFN subframes.

Table 9.20 Orthogonal Spreading Sequence for PHICH Processing [3]

Sequence Index	$(w_0, w_1, \ldots, w_{N_{SF}^{PHICH}-1})$	
$n_{PHICH}^{sequence}$	Normal Cyclic Prefix $N_{SF}^{PHICH} = 4$	Extended Cyclic Prefix $N_{SF}^{PHICH} = 2$
0	[+1 +1 +1 +1]	[+1 +1]
1	[+1 −1 +1 −1]	[+1 −1]
2	[+1 +1 −1 −1]	[+j +j]
3	[+1 −1 −1 +1]	[+j −j]
4	[+j +j +j +j]	–
5	[+j −j +j −j]	–
6	[+j +j −j −j]	–
7	[+j −j −j +j]	–

Given that CFI carried by PCFICH configures the number of OFDM symbols that are used for the control region, or alternatively the size of the PDSCH region, the extended PHICH duration must not exceed the size of the control region. For example, when using an extended PHICH in subframe zero of frame structure type 1 (FDD) with 10 MHz bandwidth, the first three OFDM symbols will contain the PHICH. Therefore, the CFI must be set to 3 so that the PDSCH is not mapped to OFDM symbols 0, 1, or 2, and thus does not collide with the PHICH. As shown in Figure 9.92, the corresponding elements of each PHICH sequence are summed to create the sequence for each PHICH group where the sum is over all PHICHs in the PHICH group, and $y_i^{(p)}(n)$ represents the symbol sequence from the ith PHICH in the PHICH group. As shown in Figure 9.92, the PHICH groups are mapped to PHICH mapping units, i.e., $\bar{y}_n^p = \sum_i y_i^p(n)$ where $y_i^p(n)$ is the nth element within the ith PHICH group on antenna port p. For a normal cyclic prefix, the mapping of PHICH group m to PHICH mapping unit m' is defined as $\tilde{y}_{m'}^p(n) = \bar{y}_m^p(n)$, where $m' = m = 0, 1, ..., m_i N_{\text{PHICH}}^{\text{group}} - 1$ for frame structure type 1, and $m' = m = 0, 1, \ldots, m_i N_{\text{PHICH}}^{\text{group}} - 1$ for frame structure type 2. In other words, the PHICHs are mapped to REGs using PHICH mapping units $\tilde{y}_{m'}^p(n)$, where m' is the index of the mapping unit. For a normal cyclic prefix, each PHICH group is mapped to a PHICH mapping unit. In the case of an extended cyclic prefix, two PHICH groups are mapped to one PHICH mapping unit. Due to the location of the padding zeros added during resource group alignment, when two consecutive groups are added, the zeros of one group overlap with the data of the other group.

Each mapping unit contains 12 symbols. In order to map these 12 symbols to REGs, the mapping units are split into three groups of four symbols, or

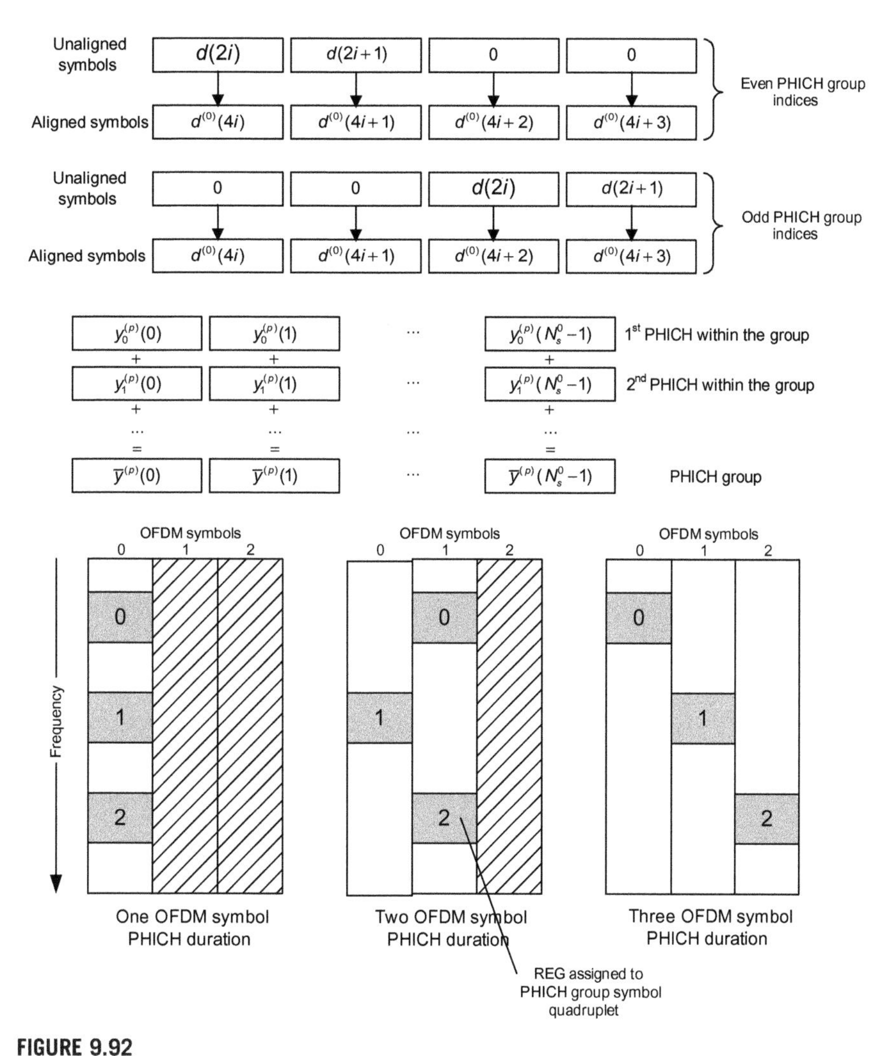

FIGURE 9.92

Illustration of PHICH symbol alignment, sequence combining, and mapping to REGs for normal and extended PHICH [160].

quadruplets. Each of the three symbol quadruplets $z^p(i) = (\tilde{y}^p(4i), \tilde{y}^p(4i+1),$ $\tilde{y}^p(4i+2)$, and $\tilde{y}^p(4i+3))$ is mapped to REG (k'_i, l'_i) such that PHICH is spread over all available OFDM symbols and resource blocks. The OFDM symbol index l'_i is set so that adjacent quadruplets are spread among the available OFDM symbols as illustrated in Figure 9.90. The sub-carrier index k'_i of the REG is based on the cell identity $N_{\text{ID}}^{\text{cell}}$ and is chosen to spread the three symbol quadruplets over the entire system bandwidth.

9.12.4 Physical downlink control channel

The physical downlink control channel is used to carry the DCI, which contains downlink/uplink scheduling decisions and power-control commands. More specifically, the DCI includes downlink scheduling assignments, PDSCH resource allocation, transport format, HARQ related information, and control information related to spatial multiplexing. The downlink scheduling assignment further includes a command for power control of the PUCCH that is used for transmission of HARQ ACK/NACK in response to downlink scheduling assignments. The DCI further includes uplink scheduling grants, PUSCH resource indication, transport format, and other HARQ related information. The uplink scheduling grant also consists of a command for power control of PUSCH. Furthermore, it carries power-control commands for a set of terminals as complementary to the commands included in the scheduling assignments/grants. It must be noted that different types of control information correspond to different DCI message sizes. For example, supporting spatial multiplexing with non-contiguous allocation of resource blocks in the frequency-domain requires larger scheduling message in comparison to an uplink grant, only allowing frequency-contiguous allocations. The DCI is subsequently categorized to different formats where a format corresponds to a certain message size, content, and usage. The LTE/LTE-Advanced DCI formats are summarized in Table 9.21. The actual message size depends on, among other factors, the cell bandwidth, thus for larger bandwidths, a larger number of bits is required to indicate the resource-block allocation. The number of CRSs in the cell and whether cross-carrier scheduling is configured will also affect the size of most DCI formats. Hence, a given DCI format may have different sizes depending on the overall configuration of the cell.

9.12.4.1 PDCCH physical layer processing

As mentioned earlier, each PDCCH carries a message known as the DCI, which includes resource assignments and other control signaling information for a UE or a group of UEs. Each PDCCH is transmitted using one or more control channel elements (CCEs), where each CCE corresponds to nine REGs (Figure 9.93). The number of REGs that are not assigned to a PCFICH or PHICH is denoted by N_{REG}. The CCEs available in the system are numbered from 0 to $N_{CCE}-1$, where $N_{CCE}=\lfloor N_{REG}/9 \rfloor$. The PDCCH supports multiple formats as listed in Table 9.21. A

Table 9.21 PDCCH Formats [3]

PDCCH Format	Number of CCEs (Aggregation Level)	Number of Resource Element Groups (REGs)	Number Of PDCCH Bits
0	1	9	72
1	2	18	144
2	4	36	288
3	8	72	576

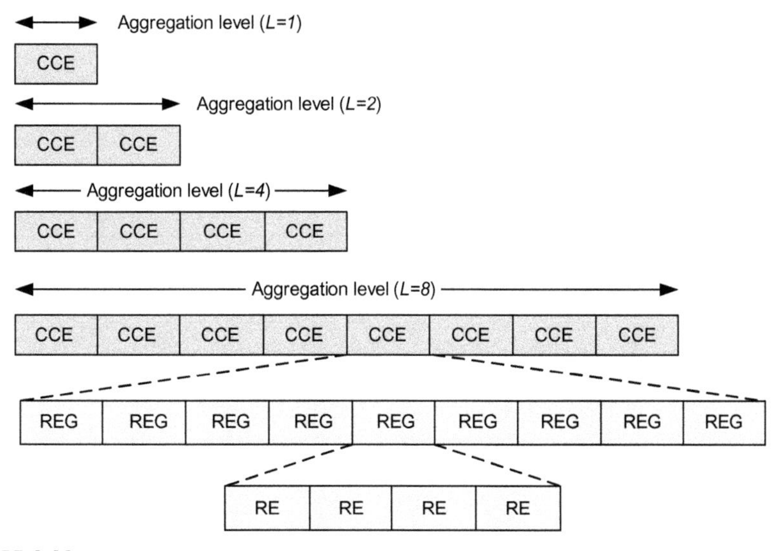

FIGURE 9.93

Illustration of various aggregation levels and CCE structure [162].

PDCCH consisting of L consecutive CCEs (aggregation level) may only start on a CCE that satisfies ($n_{\mathrm{CCE}} \bmod L = 0$), where the parameter n_{CCE} denotes the CCE index. Multiple PDCCHs can be transmitted in a subframe. Four QPSK symbols are mapped to each REG. Four PDCCH formats are supported, as listed in Table 9.21.

A cyclic redundancy check is used for error detection in the DCI messages. The entire PDCCH payload is used to calculate a set of CRC parity bits. The PDCCH payload is divided by a cyclic generator polynomial to generate 16 parity bits. These parity bits are then appended to the end of the PDCCH payload. As multiple PDCCHs relevant to different UEs can be present in one subframe, the CRC is also used to specify to which UE a PDCCH is addressed. This is done by scrambling the CRC parity bits with the corresponding RNTI of the UE. The scrambled CRC is obtained by performing a bit-wise XOR operation between the 16-bit calculated PDCCH CRC and the 16-bit RNTI. Different RNTI can be used to scramble the CRC. Some examples include:

- A UE unique identifier in the RRC_CONNECTED mode (C-RNTI).
- A paging indication identifier (P-RNTI), if the PDSCH contains paging information.
- A system information identifier (SI-RNTI), if PDSCH contains system information.

When encoding a DCI format 0 payload, containing the UE UL-SCH resource allocation, and if the UE transmit-antenna selection is configured, the RNTI-scrambled CRC undergoes a bit-wise XOR operation with an antenna selection mask. This informs the UE which transmit antenna must be used for uplink transmission. The antenna selection mask for UE antenna port 0 is $(0,0,0,0,0,0,0,0,0,0,0,0,0,0,0,0)$ and for UE antenna port 1 is $(0,0,0,0,0,0,0,0,0,0,0,0,0,0,0,1)$.

As shown in Figure 9.94, the DCI message with the attached CRC goes through tail-biting convolutional coding. LTE/LTE-Advanced uses a rate 1/3 tail-biting convolutional encoder with a constraint length $K = 7$. This means that one

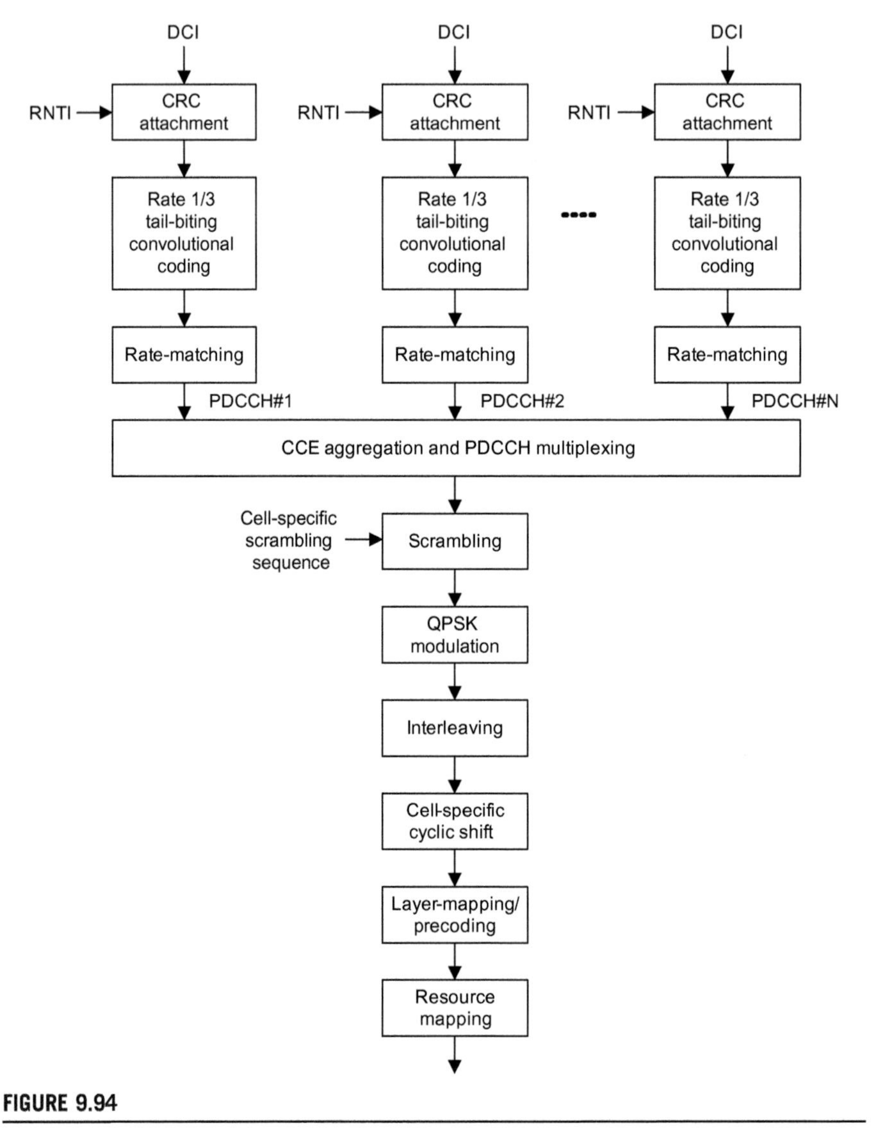

FIGURE 9.94

PDCCH physical layer processing stages [15].

in three bits of the output contain useful information while the other two add parity bits. Each output stream of the encoder is obtained by convolving the input with the impulse response of the encoder. The impulse responses are called the generator sequences of the encoder. Note that a conventional convolutional encoder initializes its internal shift register to the all-zero state and also ensures that the encoder finishes in the all-zero state, by padding the input sequence with K zeros at the end. Knowing the start and end states (i.e., all zeros) simplifies the design of the decoder (typically an implementation of the Viterbi algorithm). A tail-biting convolutional coder initializes its internal shift register to the last K bits of the current input block, rather than to the all-zero state. This means the starting and the ending states are the same, without the need to zero pad the input block. As the overhead of terminating the encoder has been eliminated, the output block contains fewer bits than a conventional convolutional coder. The drawback is that the decoder becomes more complicated, since the initial state is unknown; however, it does know the start and end states are the same. The rate-matching block creates an output bitstream with the desired code rate. The three bit streams from the convolutional encoder are interleaved, followed by bit collection to create a virtual circular buffer.

The three sub-block interleavers used in the rate-matching block are identical. Interleaving is a technique to reduce the impact of burst errors on a signal to ensure that consecutive data bits will not be corrupted. The bit collection stage creates a virtual circular buffer by concatenating the three interleaved encoded bit streams. The bits are then selected and pruned from the circular buffer to create an output sequence length which meets the desired code rate. The coded DCI messages for each control channel are multiplexed and scrambled before undergoing QPSK modulation, layer mapping, and precoding. The blocks of coded bits for each control channel are multiplexed in order to create a block of data as shown in Figure 9.96, where N_{bit}^{i} is the number of bits in the ith control channel and N_{PDCCH} is the number of control channels.

As shown in Figure 9.94, the block of coded and rate-matched bits $(b_0^i, b_1^i, \ldots, b_{N_{\text{bit}}^i-1}^i)$ on each of the control channels awaiting transmission in a subframe are multiplexed, resulting in a block of multiplexed bits $(b_0^0, b_1^0, \ldots, b_{N_{\text{bit}}^0-1}^0, b_0^1, b_1^1, \ldots, b_{N_{\text{bit}}^1-1}^1, \ldots, b_0^{N_{\text{PDCCH}}-1}, b_1^{N_{\text{PDCCH}}-1}, \ldots, b_{N_{\text{bit}}^{N_{\text{PDCCH}}-1}-1}^{N_{\text{PDCCH}}-1})$, where N_{bit}^{i} is the number of bits in the ith DCI, and N_{PDCCH} denotes the number of PDCCHs transmitted in one subframe. The block of multiplexed bits $(b_0^0, b_1^0, \ldots, b_{N_{\text{bit}}^0-1}^0, b_0^1, b_1^1, \ldots, b_{N_{\text{bit}}^1-1}^1, \ldots, b_0^{N_{\text{PDCCH}}-1}, b_1^{N_{\text{PDCCH}}-1}, \ldots, b_{N_{\text{bit}}^{N_{\text{PDCCH}}-1}-1}^{N_{\text{PDCCH}}-1})$ is scrambled with a cell-specific sequence prior to modulation (i.e., the multiplexed block of bits undergoes a bit-wise XOR operation with a cell-specific scrambling sequence), resulting in a block of scrambled bits $(\tilde{b}_0, \tilde{b}_1, \ldots, \tilde{b}_{N_{\text{total}}-1})$, in which $\tilde{b}_i = b_i \oplus c(i)$. The scrambling function provides more robustness through inter-cell interference randomization. When a UE descrambles the received bit stream with a known cell-specific scrambling sequence, interference from other cells will be descrambled

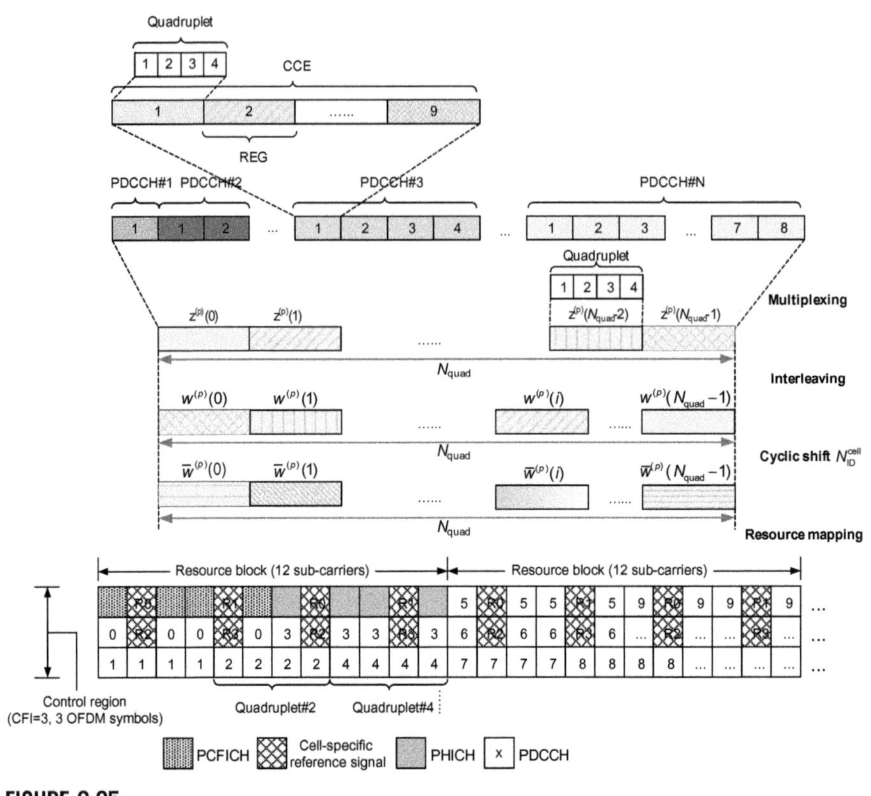

FIGURE 9.95

Example breakdown of CCE processing [162].

incorrectly and appear as uncorrelated noise. The cell-specific scrambling sequence $c(i)$ was defined in the previous sections and is initialized with $c_{\text{init}} = \lfloor n_s/2 \rfloor 2^9 + N_{\text{ID}}^{\text{cell}}$ at the start of each subframe. It must be noted that the CCE number n corresponds to bit array $(b_{72n}, b_{72n+1}, \ldots, b_{72n+71})$. If necessary, null elements are inserted in the block of bits prior to scrambling to ensure that the PDCCHs start at the permissible CCE positions, and to ensure that the length $N_{\text{total}} = 8N_{\text{REG}} \geq \sum_{i=0}^{N_{\text{PDCCH}}-1} N_{\text{bit}}^{i}$ of the scrambled block of bits matches the amount of REGs not assigned to the PCFICH or PHICH.

The block of scrambled bits $(\tilde{b}_0, \tilde{b}_1, \ldots, \tilde{b}_{N_{\text{total}}-1})$ is modulated using QPSK modulation, resulting in a block of complex-valued modulation symbols $(d_0, d_1, \ldots, d_{N_s-1})$. The block of modulated symbols $(d_0, d_1, \ldots, d_{N_s-1})$ are further mapped to $N_s^0 = N_s$ spatial layers and precoded, yielding the block of precoded vectors $\mathbf{y}_i = (y_i^0, y_i^1, \ldots, y_i^{N_p-1})^{\text{T}}$ with $i = 0, \ldots, N_s - 1$. The latter block of symbols are mapped to resources on the antenna ports used for transmission of PDCCH, where y_i^p represents the signal for antenna port p. Note that the PDCCH is transmitted on the same set of antenna ports that were used for PBCH transmission.

The mapping to resource elements are defined by operations on quadruplets of complex-valued symbols, as shown in Figure 9.95. Let $z_i^p = (y_{4i}^p, y_{4i+1}^p, y_{4i+2}^p, y_{4i+3}^p)$ denote the ith symbol quadruplet for antenna port p. The block of quadruplets $(z_0^p, z_1^p, \ldots, z_{N_{\text{quad}}-1}^p)$ where $N_{\text{quad}} = N_s/4$ are permuted (interleaved), resulting in a permuted quadruplet vector $(w_0^p, w_1^p, \ldots, w_{N_{\text{quad}}-1}^p)$. The permutation is performed similar to sub-block interleaving with the following exceptions: (1) the input and output to the interleaver are defined by symbol quadruplets instead of bits; and (2) interleaving is performed on symbol quadruplets instead of bits by substituting the bits with symbol quadruplets. The null elements at the output of the interleaver must be removed when constructing the permuted vector $(w_0^p, w_1^p, \ldots, w_{N_{\text{quad}}-1}^p)$. The block of interleaved quadruplets $(w_0^p, w_1^p, \ldots, w_{N_{quad}-1}^p)$ is then cyclically shifted, resulting in $(\overline{w}_0^p, \overline{w}_1^p, \ldots, \overline{w}_{N_{\text{quad}}-1}^p)$ where $\overline{w}_i^p = w_{(i+N_{\text{ID}}^{\text{cell}}) \bmod N_{\text{quad}}}^p$. The mapping of the block of quadruplets $(\overline{w}_0^p, \overline{w}_1^p, \ldots, \overline{w}_{N_{\text{quad}}-1}^p)$ is defined in terms of resource-element groups. Each symbol-quadruplet is sequentially mapped to an unallocated REG starting with REG $(k' = 0, l' = 0)$ such that symbol quadruplet $\overline{w}_{m'}^p$ is mapped to REG m'. The REG symbol index l' is then incremented until all the REGs at sub-carrier index $k' = 0$ have been allocated, provided that $l' < N_{\text{CFI}}$. The REG sub-carrier index k' is then incremented and the process is repeated until $k' < 12N_{\text{RB}}^{\text{DL}}$. This mapping continues until all symbol quadruplets have been allocated to REGs. The mapping is illustrated in Figure 9.95 for an example resource grid. Four transmit antenna ports and a control region size of three OFDM symbols are used to create the grid. The mapping is performed along symbols first, and then along frequency until the entire symbol quadruplets are mapped, in order to cause no significant power imbalance among the PDCCH symbols. If symbol-quadruplets were only mapped to the first OFDM symbol among OFDM symbols dedicated to the control region on a particular subframe, it would negatively impact the PA operation and performance.

Figure 9.97 shows an example mapping of CCEs in a subframe. In this example, REGs $(k' = 0, l' = 0)$ and $(k' = 6, l' = 0)$ were already allocated to PCFICH and PHICH, respectively, thus no symbol quadruplets are assigned to them. The symbol-quadruplets are first mapped to REG $(k' = 0, l' = 1)$, followed by REG $(k' = 0, l' = 2)$ and so on. The symbol-quadruplets are mapped in ascending order of index l'. As there are no REGs with $k' = 0$, the next REG allocated is REG $(k' = 4, l' = 2)$, as this REG has the lowest value of index l' that has not already been allocated. This process is repeated until all symbol quadruplets are assigned

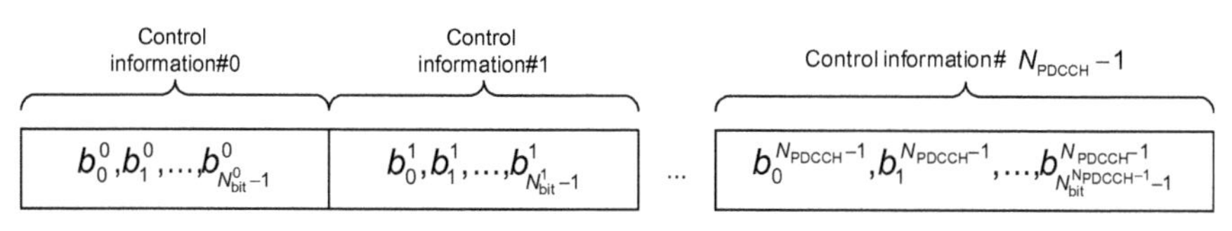

FIGURE 9.96

Illustration of PDCCH multiplexing function [160].

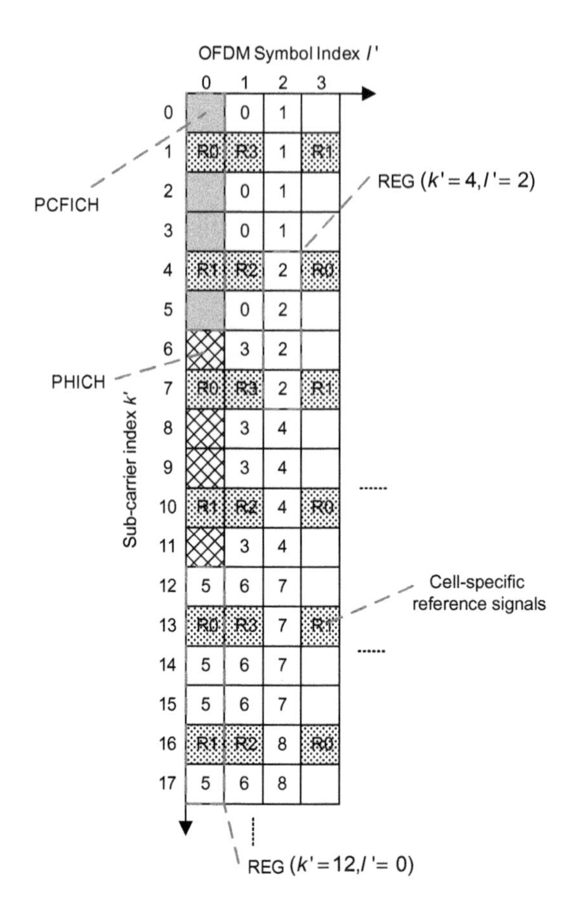

FIGURE 9.97

Example mapping of CCEs in the control region of a subframe [160].

to the available REGs [160]. Therefore, the CCE mapping process can be summarized as follows. Each symbol-quadruplet is sequentially mapped to an unallocated REG starting with REG ($k' = 0, l' = 0$). This is described as symbol quadruplet $\overline{w}^{p}_{m'}$ being mapped to REG m'. The REG symbol index k' is then incremented until all the REGs at sub-carrier index $l' = 0$ have been allocated. The REG sub-carrier index k' is then incremented and the process is repeated. This mapping continues until all symbol quadruplets have been allocated REGs. The mapping is shown in Figure 9.97 for an example resource grid. Four transmit antenna ports and control region size of three OFDM symbols are used to create the grid.

9.12.4.2 DCI formats

One PDCCH carries one DCI message with one of the formats given in Table 9.21. Given that multiple terminals can be scheduled simultaneously on both downlink and uplink, there must be a possibility to transmit multiple scheduling messages within each subframe. Each scheduling message is transmitted on a separate PDCCH, and consequently there are typically multiple simultaneous PDCCH transmissions within each cell. Furthermore, to support different radio channel conditions, link adaptation can be used, where the transmission format of the PDCCH (i.e., coding and modulation scheme, multi-antenna configuration) is adapted to cope with the severity of the radio channel conditions. For carrier aggregation, scheduling assignments/grants are transmitted individually per component carrier. Depending on the purpose of DCI message, different DCI formats are defined as follows [4]:

- Format 0: Transmission of uplink grants for PUSCH.
- Format 1: Transmission of PDSCH assignments with a single codeword.
- Format 1A: DCI Format 1A is used for compact signaling of resource assignments in single-codeword PDSCH transmissions using any PDSCH transmission mode. It is also used to allocate a dedicated preamble signature to a UE for the purpose of contention-free random access.
- Format 1B: DCI Format 1B is used for compact signaling of resource assignments in PDSCH transmissions using rank-1 closed-loop precoding. The information transmitted is the same as DCI Format 1A with the addition of an indicator for the precoding vector applied to PDSCH transmission.
- Format 1C: Transmission of PDSCH assignment using a very compact format. When format 1C is used, the PDSCH transmission is constrained to use QPSK modulation.
- Format 1D: Same as Format 1B with additional information for power offset (i.e., instead of one of the precoding vector indicator bits, there is a single bit to indicate whether a power offset is applied to the data symbols to indicate whether the transmission power is shared between two UEs) and PDSCH assignments for multi-user MIMO.
- Format 2: PDSCH assignments for closed-loop MIMO operation.
- Format 2A: PDSCH assignments for open-loop MIMO operation.
- Format 2B: PDSCH assignments for dual-layer beamforming.
- Format 2C: DCI Format 2C was introduced in Rel-10 and is used for signaling resource assignments for PDSCH transmissions using closed-loop single-user or multi-user MIMO with up to eight layers.
- Format 2D: DCI Format 2D was introduced in Rel-11 and is used for CoMP operations in conjunction with transmission mode 10.
- Format 3: Transmission of transmit power control (TPC) commands for multiple users and for PUCCH and PUSCH with 2-bit power adjustments.
- Format 3A: Transmission of TPC commands for multiple users and for PUCCH and PUSCH with 1-bit power adjustments.
- Format 4: DCI Format 4 was introduced in Rel-10 and is used for signaling resource grants for PUSCH when the UE is configured in PUSCH transmission mode 2 for uplink single-user MIMO. The information transmitted is similar to DCI Format 0, with the addition of MCS and new data indicator for a second transport block (TB) and precoding data.

Multiple UEs can be scheduled in one subframe; therefore, multiple DCI messages can be sent using multiple PDCCHs. The description, usage, and detailed information fields of various DCI formats for LTE/LTE-Advanced are given in Table 9.22.

As shown in Table 9.22, the number of bits required for resource allocation/assignment depends on the system bandwidth; therefore, the size of that information field also varies with the system bandwidth. The payload size for each DCI format including information bits and CRC are summarized in Table 9.23 for the E-UTRA supported system bandwidths. In addition, padding bits are added (if necessary) to ensure that DCI Formats 0 and 1A are of the same size, even in the case of different uplink and downlink bandwidths, in order to avoid additional complexity at the UE receiver. Furthermore, the padding bits may be added to ensure that DCI Formats 3 and 3A have the same size as DCI Formats 0 and 1A, to circumvent additional complexity at the UE receiver; to avoid potential ambiguity in identifying the correct PDCCH location; to ensure that DCI Format 1 has a different size than that of DCI Formats 0/1A, so that these formats can be easily distinguished at the UE receiver; to make sure that Format 4 has a different size from DCI Formats 1/2/2A/2B/2C, so that this format can be easily distinguished. As given in Table 9.22, in LTE Rel-10, some optional features were added that may be configured in some of the DCI formats; those bit counts are not included in Table 9.23. The DCI formats 1B/1D/2/2A/4 payload size calculations are based on two antenna ports [13]. The new DCI format 2D which has been added in Rel-11 includes the following information fields: Carrier indicator (0 or 3 bits); Resource allocation header for resource allocation type 0 or type 1 (1 bit), note that if downlink bandwidth is less than or equal to 10 resource blocks, there is no resource allocation header and resource allocation type 0 is assumed; Resource block assignment type 0/type 1 (variable number of bits depending on the system bandwidth); TPC command for PUCCH (2 bits); Downlink Assignment Index (TDD only) for uplink–downlink subframe configurations 1–6 (2 bits); HARQ process number (3 bits for FDD), (4 bits for TDD); Antenna port(s), scrambling identity and number of layers (3 bits); SRS request (0 or 1 bit) for TDD only; Modulation and coding scheme (5 bits), New data indicator (1 bit); Redundancy version (2 bits) for transport blocks 1 and 2; PDSCH resource element Mapping and Quasi Collocation Indicator (2 bits); HARQ-ACK resource offset (note that this field is present when this DCI format is carried by ePDCCH and is not present when this format is carried by PDCCH) (2 bits). The latter 2 bits are set to 0 when this format is carried by ePDCCH on a secondary cell. If both transport blocks are enabled; transport block 1 is mapped to codeword 0 and transport block 2 is mapped to codeword 1. In case one of the transport blocks is disabled; the transport block to codeword mapping is further specified [4].

9.12.4.3 Common and UE-specific search spaces and blind decoding

The PDCCH region consists of a number of CCEs which can be allocated to a PDCCH. Thc PDCCHs can be flexibly mapped to CCEs based on certain criteria.

Table 9.22 Descriptions, Usage, and Information Fields of Various DCI Formats [4,13]

DCI Format	Usage	Release	Information Fields	Size (Bits)	Description
Format 0	Scheduling of PUSCH in one uplink cell	10	Carrier indicator	0 or 3	Identifies the RF carrier to be used as the serving cell. A value of "0" implies the primary cell is used as the serving cell. This field only applies when carrier aggregation is enabled.
		10	Format 0/format 1A flag	1	Flag to differentiate between Format 0 and Format 1A.
		8	Frequency hopping flag	1	PUSCH frequency hopping flag for resource allocation type 0. This bit is used as the MSB of the resource block allocation in resource allocation type 1.
		8	Resource block allocation	7–13 variable	Resource block assignment/allocation. (In resource allocation type 0: 5, 7, 9, 11, 12, or 13 bits corresponding to 1.4, 3, 5, 10, 15, or 20 MHz system bandwidth). (In resource allocation type 1: 6, 8, 10, 12, 13, or 14 bits corresponding to 1.4, 3, 5, 10, 15, or 20 MHz system bandwidth).
		8	MCS and redundancy version	5	Modulation, coding scheme and redundancy version.
		8	New data indicator (NDI)	1	New data indicator.
		8	Transmit power control (TPC)	2	PUSCH TPC command.
		8	Cyclic shift for uplink DM-RS		
		10	Cyclic shift for uplink DM-RS and OCC index	3	Cyclic shift for uplink DM-RS (different look up tables are used for LTE Rel-8/9 and LTE Rel-10 UEs based on this value.

(Continued)

Table 9.22 (Continued)

DCI Format	Usage	Release	Information Fields	Size (Bits)	Description
		8	Aperiodic CQI request	1	Instructs the UE to provide a periodic CQI report.
		10	CSI request	1 or 2	Instructs the UE to send a CSI report, where one-bit case refers to a single cell and two-bit case applies to a UE which is configured with more than one downlink cell.
		10	SRS request	0 or 1	Instructs the UE to transmit an SRS for uplink channel measurements.
		10	Resource allocation type	1	Signals the type of uplink resource allocation.
		8	Uplink index (TDD only)	2	For TDD configuration 0, this field is the uplink index.
		8	Downlink assignment index (DAI) (TDD only)		For TDD configuration 1–6, this field is the downlink assignment index.
Format 1	Transmission of PDSCH assignments with a single codeword	10	Carrier indicator	0 or 3	Identifies the RF carrier to be used as the serving cell. A value of "0" implies the primary cell is used as the serving cell. This field only applies when carrier aggregation is enabled.
		8	Resource allocation header	0 or 1	Resource allocation header selects between resource allocation type 0 or type 1 (only exists if downlink bandwidth is greater than 10 RBs).
		8	Resource block allocation	6–25 variable	Resource block assignment/allocation (6, 8, 13, 17, 19, or 25 bits corresponding to 1.4, 3, 5, 10, 15, or 20 MHz system bandwidth).
		8	MCS	5	Modulation and coding scheme.

		8	HARQ process number	3 (FDD) 4 (TDD)	HARQ process number.
		8	New data indicator	1	HARQ new data indicator.
		8	Redundancy version	2	HARQ redundancy version.
		8	TPC command for PUCCH	2	PUCCH TPC command.
		8	Downlink assignment index (TDD only)	2	For TDD configuration 0, this field is not used. For TDD configurations 1–6, this field is interpreted as the DAI.
Format 1A	Compact signaling of resource assignments for single-codeword PDSCH transmissions for any PDSCH transmission mode	10	Carrier indicator	0 or 3	Identifies the RF carrier to be used as the serving cell. A value of "0" implies the primary cell is used as the serving cell. This field only applies when carrier aggregation is enabled.
		8	Format 0/format 1A flag	1	Flag to differentiate between Format 0 and Format 1A.
		8	Localized/ distributed VRB assignment flag	1	Selects between 0 (localized) or 1 (distributed) resource blocks.
		8	Resource block allocation	5–13 variable	Resource block assignment/allocation (5, 7, 9, 11, 12, or 13 bits corresponding to 1.4, 3, 5, 10, 15, or 20 MHz system bandwidth).
		8	MCS	5	Modulation and coding scheme.
		8	HARQ process number	3 (FDD) 4 (TDD)	HARQ process number.
		8	New data indicator	1	HARQ new data indicator.

(*Continued*)

Table 9.22 (Continued)

DCI Format	Usage	Release	Information Fields	Size (Bits)	Description
		8	Redundancy version	2	HARQ redundancy version.
		8	TPC command for PUCCH	2	PUCCH TPC command.
		8	Downlink assignment index (TDD only)	2	For TDD configuration 0, this field is not used. For TDD configurations 1–6, this field is interpreted as DAI.
		10	SRS request	0 or 1	Instructs the UE to transmit SRS for uplink channel measurements.
Format 1B	Compact signaling of resource assignments for PDSCH transmissions using closed-loop precoding with rank-1 transmission	10	Carrier indicator	0 or 3	Identifies the RF carrier to be used as the serving cell. A value of "0" implies the primary cell is used as the serving cell. This field only applies when carrier aggregation is enabled.
		8	Localized/ distributed VRB assignment flag	1	Selects between 0 (localized) or 1 (distributed) resource blocks.
		8	Resource block allocation	5–13 variable	Resource block assignment/allocation (5, 7, 9, 11, 12, or 13 bits corresponding to 1.4, 3, 5, 10, 15, or 20 MHz system bandwidth).
		8	MCS	5	Modulation and coding scheme.
		8	HARQ process number	3 (FDD) 4 (TDD)	HARQ process number.
		8	New data indicator	1	HARQ new data indicator.
		8	Redundancy version	2	HARQ redundancy version.

		8	TPC command for PUCCH	2	PUCCH TPC command.
		8	Transmitted PMI (TPMI)	2 (2 antenna ports) 4 (4 antenna ports)	Downlink PMI information, i.e., information about the codebook used for precoding PDSCH (codebook-based closed-loop MIMO).
		8	PMI confirmation for precoding	1	Indicates whether PDSCH has been precoded using the codebook indicated by TPMI or by the PMI sent by the UE.
		8	Downlink assignment index (TDD only)	2	For TDD configuration 0, this field is not used. For TDD configuration 1–6, this field is the downlink assignment index.
Format 1C	Transmission of PDSCH assignment using a very compact format	8	Gap value	0 or 1	DCI format 1C always signals resource allocation type 2 using a distributed mapping. Smaller bandwidths always use "gap1" and larger bandwidths greater than 10 MHz can select between "gap1" and "gap2."
		8	Resource block allocation	3–9 variable	Resource block assignment/allocation (3, 5, 7, 7, 8, or 9 bits corresponding to 1.4, 3, 5, 10, 15, or 20 MHz system bandwidth).
		8	Transport block size	5	DCI format 1C does not use the general TBS table, rather it uses a smaller TBS table.
Format 1D	Same as format 1B with additional information for power offset (i.e., instead of one of the precoding vector indicator	10	Carrier indicator	0 or 3	Identifies the RF carrier to be used as the serving cell. A value of "0" implies the primary cell is used as the serving cell. This field only applies when carrier aggregation is enabled.
		8	Localized/ distributed VRB assignment flag	1	Selects between 0 (localized) or 1 (distributed) resource blocks.
		8		5–13 variable	

(Continued)

Table 9.22 (Continued)

DCI Format	Usage	Release	Information Fields	Size (Bits)	Description
	bits, there is a single bit to indicate whether a power offset is applied to the data symbols to indicate whether the transmission power is shared between two UEs) and PDSCH assignments for multi-user MIMO		Resource block allocation		Resource block assignment/allocation (5, 7, 9, 11, 12, or 13 bits corresponding to 1.4, 3, 5, 10, 15, or 20 MHz system bandwidth).
		8	MCS	5	Modulation and coding scheme.
		8	HARQ process number	3 (FDD) 4 (TDD)	HARQ process number.
		8	New data indicator	1	HARQ new data indicator.
		8	Redundancy version	2	HARQ redundancy version.
		8	TPC command for PUCCH	2	PUCCH TPC command.
		8	Transmitted PMI (TPMI)	2 (2 antenna ports) 4 (4 antenna ports)	Downlink PMI information, i.e., information about the codebook used for precoding PDSCH (codebook-based closed-loop MIMO).
		8	Downlink power offset	1	Downlink power offset is used in MU-MIMO to inform the UE of its downlink transmit power relative to single-user transmit power.
		8	Downlink assignment index (TDD only)	2	For TDD configuration 0, this field is not used. For TDD configuration 1–6, this field is the downlink assignment index.
Format 2	PDSCH assignments for closed-loop	10	Carrier indicator	0 or 3	Identifies the RF carrier to be used as the serving cell. A value of "0" implies the primary cell is used as the serving cell. This field only applies when carrier aggregation is enabled.

MIMO operation	8	Resource allocation header	0 or 1	Resource allocation header selects between resource allocation type 0 or type 1 (only exists if downlink bandwidth is greater than 10 RBs).
	8	Resource block allocation	6–25 variable	Resource block assignment/allocation (6, 8, 13, 17, 19, or 25 bits corresponding to 1.4, 3, 5, 10, 15, or 20 MHz system bandwidth).
	8	TPC command for PUCCH	2	PUCCH TPC command.
	8	HARQ process number	3 (FDD) 4 (TDD)	HARQ process number.
	8	Transport block to codeword swap flag	1	Defines the relationship between transport blocks 1 and 2 relative to codewords 0 and 1. If set to "0," codeword 0 is generated from transport block 1 and codeword 1 is generated from transport block 2.
	8	MCS1	5	Modulation and coding scheme for transport block 1.
	8	New data indicator 1	1	HARQ new data indicator for transport block 1.
	8	Redundancy version 1	2	HARQ redundancy version for transport block 1.
	8	MCS2	5	Modulation and coding scheme for transport block 2.
	8	New data indicator 2	1	HARQ new data indicator for transport block 2.
	8	Redundancy version 2	2	HARQ redundancy version for transport block 2.
	8	Precoding information	3 (2-antenna ports) 6 (4-antenna ports)	Precoding information.
	8	Downlink assignment index (TDD only)	2	For TDD configuration 0, this field is not used. For TDD configuration 1–6, this field is the downlink assignment index.

(Continued)

Table 9.22 (Continued)

DCI Format	Usage	Release	Information Fields	Size (Bits)	Description
Format 2A	PDSCH assignments for open-loop MIMO operation	10	Carrier indicator	0 or 3	Identifies the RF carrier to be used as the serving cell. A value of "0" implies the primary cell is used as the serving cell. This field only applies when carrier aggregation is enabled.
		8	Resource allocation header	0 or 1	Resource allocation header selects between resource allocation type 0 or type 1 (only exists if downlink bandwidth is greater than 10 RBs).
		8	Resource block allocation	6–25 variable	Resource block assignment/allocation (6, 8, 13, 17, 19, or 25 bits corresponding to 1.4, 3, 5, 10, 15, or 20 MHz system bandwidth).
		8	TPC command for PUCCH	2	PUCCH TPC command.
		8	HARQ process number	3 (FDD) 4 (TDD)	HARQ process number.
		8	Transport block to codeword swap flag	1	Defines the relationship between transport blocks 1 and 2 relative to codewords 0 and 1. If set to "0," codeword 0 is generated from transport block 1 and codeword 1 is generated from transport block 2.
		8	MCS1	5	Modulation and coding scheme for transport block 1.
		8	New data indicator 1	1	HARQ new data indicator for transport block 1.
		8	Redundancy version 1	2	HARQ redundancy version for transport block 1.
		8	MCS2	5	Modulation and coding scheme for transport block 2.
		8	New data indicator 2	1	HARQ new data indicator for transport block 2.

		8	Redundancy version 2	2	HARQ redundancy version for transport block 2.
		8	Precoding information	0 (2 antenna ports) 2 (4 antenna ports)	Precoding information.
		8	Downlink assignment index (TDD only)	2	For TDD configuration 0, this field is not used. For TDD configuration 1–6, this field is the downlink assignment index.
Format 2B	PDSCH assignments for dual-layer beamforming	10	Carrier indicator	0 or 3	Identifies the RF carrier to be used as the serving cell. A value of "0" implies the primary cell is used as the serving cell. This field only applies when carrier aggregation is enabled.
		10	SRS request (TDD only)	0 or 1	Instructs the UE to transmit SRS for uplink channel measurements.
		9	Resource allocation header	0 or 1	Resource allocation header selects between resource allocation type 0 or type 1 (only exists if downlink bandwidth is greater than 10 RBs).
		9	Resource block allocation	6–25 variable	Resource block assignment/allocation (6, 8, 13, 17, 19, or 25 bits corresponding to 1.4, 3, 5, 10, 15, or 20 MHz system bandwidth).
		9	TPC command for PUCCH	2	PUCCH TPC command.
		9	HARQ process number	3 (FDD) 4 (TDD)	HARQ process number.
		9	Scrambling ID	1	Scrambling identity provides support for MU-MIMO by allowing separate users to reuse the same antenna port with different SCIDs.
		9	MCS1	5	Modulation and coding scheme for transport block 1.

(*Continued*)

Table 9.22 (Continued)

DCI Format	Usage	Release	Information Fields	Size (Bits)	Description
		9	New data indicator 1	1	HARQ new data indicator for transport block 1.
		9	Redundancy version 1	2	HARQ redundancy version for transport block 1.
		9	MCS2	5	Modulation and coding scheme for transport block 2.
		9	New data indicator 2	1	HARQ new data indicator for transport block 2.
		9	Redundancy version 2	2	HARQ redundancy version for transport block 2.
		9	Downlink assignment index (TDD only)	2	For TDD configuration 0, this field is not used. For TDD configuration 1–6, this field is the downlink assignment index.
Format 2C	Signaling resource assignments for PDSCH for closed-loop single-user or multi-user MIMO with up to 8 layers	10	Carrier indicator	0 or 3	Identifies the RF carrier to be used as the serving cell. A value of "0" implies the primary cell is used as the serving cell. This field only applies when carrier aggregation is enabled.
		10	SRS request (TDD only)	0 or 1	Instructs the UE to transmit SRS for uplink channel measurements.
		10	Resource allocation header	0 or 1	Resource allocation header selects between resource allocation type 0 or type 1 (only exists if downlink bandwidth is greater than 10 RBs).
		10	Resource block allocation	6–25 variable	Resource block assignment/allocation (6, 8, 13, 17, 19, or 25 bits corresponding to 1.4, 3, 5, 10, 15, or 20 MHz system bandwidth).
		10	TPC command for PUCCH	2	PUCCH TPC command.

		10	HARQ process number	3 (FDD) 4 (TDD)	HARQ process number.
		10	Antenna ports, scrambling ID and layers	3	Provides a pointer to a look-up table where the UE can find the number of layers, antenna ports, and the SCID. Scrambling identity provides support for MU-MIMO by allowing separate users to reuse the same antenna port with different SCIDs.
		10	MCS1	5	Modulation and coding scheme for transport block 1.
		10	New data indicator 1	1	HARQ new data indicator for transport block 1.
		10	Redundancy version 1	2	HARQ redundancy version for transport block 1.
		10	MCS2	5	Modulation and coding scheme for transport block 2.
		10	New data indicator 2	1	HARQ new data indicator for transport block 2.
		10	Redundancy version 2	2	HARQ redundancy version for transport block 2.
		10	Downlink assignment index (TDD only)	2	For TDD configuration 0, this field is not used. For TDD configuration 1–6, this field is the downlink assignment index.
Format 3	Transmit power control (TPC) commands for multiple users and for PUCCH and PUSCH with 2-bit power adjustment	8	TPC commands	20–28 variable	TPC commands for PUCCH and PUSCH are distinguished based the RNTI that is used to scramble CRC (20, 22, 24, 26, 26, or 28 bits corresponding to 1.4, 3, 5, 10, 15, or 20 MHz system bandwidth).

(*Continued*)

Table 9.22 (Continued)

DCI Format	Usage	Release	Information Fields	Size (Bits)	Description
Format 3A	Transmit power control (TPC) commands for multiple users and for PUCCH and PUSCH with one-bit power adjustment	8	TPC commands	21–28 variable	TPC commands for PUCCH and PUSCH are distinguished based the RNTI that is used to scramble CRC (21, 22, 25, 27, 27, or 28 bits corresponding to 1.4, 3, 5, 10, 15, or 20 MHz system bandwidth).
Format 4	Signaling resource grants for PUSCH when the UE is configured in PUSCH transmission mode 2 for uplink single-user MIMO	10	Carrier indicator	0 or 3	Identifies the RF carrier to be used as the serving cell. A value of "0" implies the primary cell is used as the serving cell. This field only applies when carrier aggregation is enabled.
		10	Resource block allocation	6–25 variable	Resource block assignment/allocation (6, 7, 10, 12, 13, or 14 bits corresponding to 1.4, 3, 5, 10, 15, or 20 MHz system bandwidth).
		10	TPC Command for Scheduled PUSCH	2	Interpretation of this command depends on whether absolute or accumulated power control is used.
		10	Cyclic shift for DM-RS and OCC		Provides a pointer to a look-up table at the UE for generation of DM-RS and OCC.
		10	Downlink Assignment Index (TDD Only)	2	For TDD configuration 0, this field is not used. For TDD configuration 1–6, this field is the downlink assignment index.

10	CSI request	1 or 2	Instructs the UE to send a CSI report. 2 bits are used when the UE is configured with more than one downlink cell; otherwise, 1 bit is used.
10	SRS request	2	Instructs the UE to transmit SRS for uplink channel measurement and selects between SRS configurations
10	Resource allocation type	1	Selects the type of uplink resource allocation (type 0 or type 1).
10	Precoding information and number of layers	3 (2 Antenna Ports) 6 (4 Antenna Ports)	Provides a pointer to a look-up table at the UE to obtain the number of layers and transmitted PMI (TPMI).
10	MCS1	5	Modulation and coding scheme for transport block 1.
10	New data indicator 1	1	HARQ new data indicator for transport block 1.
10	MCS2	5	Modulation and coding scheme for transport block 2.
10	New data indicator 2	1	HARQ new data indicator for transport block 2.

Table 9.23 DCI Formats' Payload Sizes Including 16-bit CRC and Padding Bits [13,14]

DCI Format	FDD System Payload Size (Bits)						TDD System Payload Size (Bits)					
	1.4 MHz	3 MHz	5 MHz	10 MHz	15 MHz	20 MHz	1.4 MHz	3 MHz	5 MHz	10 MHz	15 MHz	20 MHz
Format 0	37	38	41	43	43	44	39	41	43	45	46	47
Format 1	35	39	43	47	49	55	38	43	46	50	52	58
Format 1A	37	38	41	43	43	44	39	41	43	45	46	47
Format 1B	38	41	43	44	45	46	41	43	45	47	49	49
Format 1C	24	26	28	29	30	31	24	26	28	29	30	31
Format 1D	38	41	43	44	45	46	41	43	45	47	49	49
Format 2	47	50	55	59	61	67	50	53	58	62	64	70
Format 2A	44	47	52	57	58	64	47	50	55	59	61	67
Format 2B	44	47	52	57	58	64	49	51	57	61	62	68
Format 2C	46	49	54	58	61	66	50	53	58	62	64	70
Format 3	37	38	41	43	43	44	37	38	41	43	43	44
Format 3A	37	38	41	43	43	44	37	38	41	43	43	44
Format 4	46	48	50	52	53	54	48	49	52	54	55	56

The common and UE-specific PDCCHs are mapped to CCEs differently where each type has a specific set of search spaces associated with it. Each search space consists of a group of consecutive CCEs which can be allocated to a PDCCH, known as PDCCH candidate. The CCE aggregation level is given by the PDCCH format and determines the number of PDCCH candidates in a search space. The number of candidates and size of the search space for each aggregation level is given in Table 9.24.

The UE is only informed of the number of OFDM symbols which construct the control region of a subframe, and is not provided with the location of its corresponding PDCCH(s). The UE finds its PDCCH by monitoring a set of PDCCH candidates in every subframe. This process is referred to as blind decoding. The UE unmasks each control candidate's CRC using its C-RNTI. If no CRC error is detected, the UE considers that it as a successful decoding attempt and reads the control information within the candidate PDCCH. The serving eNB determines a PDCCH format to be transmitted to the UE, creates an appropriate DCI and attaches a CRC. The CRC is then masked with an RNTI corresponding to a specific user or users. If the PDCCH is meant for a specific UE, the CRC will be masked with a unique UE identifier, e.g., a cell-RNTI (C-RNTI). If the PDCCH contains paging information, the CRC will be masked with a paging indication identifier, i.e., paging-RNTI (P-RNTI). If the PDCCH contains system information, a system information identifier, i.e., a system information-RNTI (SI-RNTI), will be used to mask the CRC.

With the possibilities of different RNTIs, PDCCH candidates, DCI and PDCCH formats, a significant number of attempts may be required to successfully decode the PDCCH. To overcome this complexity, the UE first tries to blindly decode the first CCE in the control channel candidate set of a subframe. If the blind decoding fails, the UE tries to blindly decode the first 2, 4, then 8 CCEs sequentially, where the starting location is fixed for a common search space and is given by a hash function for a UE-specific search space. The UE first tries to decode PDCCHs in the common search space before trying in the UE-specific search space. When searching the common search space, it iterates for only two

Table 9.24 PDCCH Candidates Monitored by a UE [5]

Search Space $S_k^{(L)}$			Number of PDCCH Candidates N_L	DCI Format
Type	**Aggregation Level L**	**Size (Number of CCEs)**		
UE-specific	1	6	6	0, 1/1A/1B/1D, 2/2A/2B/2C
	2	12	6	
	4	8	2	
	8	16	2	
Common	4	16	4	1, 1A, 1C, 3, 3A
	8	16	2	

aggregation levels, i.e., 4 and 8, and tries to decode all PDCCH candidates for all possible common search space DCI formats (see Table 9.24). The UE-specific search is carried out on four aggregation levels, i.e., 1, 2, 4, and 8. If no CRC error is detected during a decoding attempt, the UE considers it a successful decoding and reads the decoded DCI message.

A perfect PDCCH scheduler would be able to continuously allocate resources to each UE that is chosen by the downlink and uplink schedulers up to the point where all resources for the PDCCH, PDSCH, and PUSCH simultaneously become exhausted. However, because of the mechanism used for allocating PDCCH resources, it is possible that we can no longer allocate PDCCH resources to a UE even if such resources are available. In such cases, the corresponding UE is said to be blocked, while the eNB is still able to serve subsequent UEs. As more UEs are blocked, the performance of the scheduler decreases because such UEs cannot be served. Hence, the blocking probability, together with the utilization efficiency of bandwidth and power resources, are suitable metrics to measure the performance of the system. We assume that the number of PDCCH resources (i.e., CCEs) and total PDCCH power are known for each subframe. Multiple CCEs (1, 2, 4, or 8) can be aggregated to achieve different coding rates. The aggregation level required for a particular UE depends on the channel conditions of the UE, and the size of the scheduling grant that must be used for the UE. Each UE is allocated a candidate set of PDCCH channels for each possible aggregation level. The UE must then conduct blind decoding of each of these possibilities for each of the feasible message formats to determine if it was allocated PDSCH or PUSCH resources. The starting point for the candidate set for each aggregation level varies for each UE in each subframe. However, the new starting point is always known by the eNB and the UE. For each scheduled UE, the eNB must determine a suitable aggregation level, and then must use an unallocated channel from the corresponding candidate set to signal the location and format of the transmitted data.

Let us summarize some underlying definitions concerning search spaces before proceeding to more detailed descriptions.

- *Common search space*: A common search space consists of the common control information and is monitored by all UEs in a cell. The number of CCE aggregation levels supported by the common search space is limited to four and eight, compared to the UE-specific search space where four CCE aggregation levels are possible. This reduces the burden on the UE for decoding common control information compared to decoding UE-specific control information. The common search space is used to carry important initial information including paging information, system information, and random access procedures. When searching the common search space, the decoder always starts decoding from the first CCE. This restriction further simplifies the common search space design. The decoding is done on every possible PDCCH candidate set for a given PDCCH format until it successfully decodes the PDCCH which can be present in the common search space. All

UEs monitor the common search space, which may carry paging, RACH response, system information, and uplink TPC commands. The common search space always corresponds to CCEs 0–15, i.e., four decoding candidates on aggregation level 4 (CCEs 0–3, 4–7, 8–11, 12–15), and two decoding candidates on aggregation level 8 (CCEs 0–7, 8–15). The common search space can be used for any PDCCH signaling not restricted to common PDCCH, it can be used to resolve blocking, as well. The UE may attempt a maximum of 12 blind decodings on the common search space.

- *UE-specific search space*: The UE-specific search space carries control information specific to a particular UE and is monitored by at least one UE in a cell. Unlike the common search space, the starting location of the UE-specific search space may vary for each subframe or UE. The starting location of the UE-specific search space is determined in every subframe using a hash function. In the UE-specific search space, the UE finds its PDCCH by monitoring a set of PDCCH candidates (i.e., a set of consecutive CCEs on which the PDCCH could be mapped) in every subframe. If no CRC error is detected when the UE uses its RNTI (i.e., a 16-bit value also referred to as C-RNTI) to unmask the CRC on a PDCCH, the UE determines that the PDCCH carries its own control information. The PDCCH candidate sets correspond to different PDCCH formats. There are four PDCCH formats, i.e., 0, 1, 2, or 3. If the UE fails to decode any PDCCH candidates for a given PDCCH format, then it attempts to decode candidates for other PDCCH formats. This process is repeated for possible PDCCH formats until all designated PDCCHs that can be present in the UE-specific search space are successfully decoded. A PDCCH can be transmitted on an aggregation of one or several CCEs depending on the target reliability and channel conditions. Each PDCCH carries one DCI message. The UE performs blind decoding on the received CCEs, assuming all four different aggregations of CCEs (Figure 9.98). The UE-specific search space indicates the starting offset for blind decoding; different UEs may have different offsets.

In LTE/LTE-Advanced, each scheduling grant is defined based on fixed size CCEs. Four different CCE aggregation levels are defined for the transmission of a control channel (see Table 9.21). The eNB scheduler schedules the PDCCH that should be mapped to the CCE size by applying different coding rates. For example, if the LTE scheduler uses DCI Format 1A which has a length of 28 bits, then a 16-bit CRC is added to the 28 bit information element, resulting in 44 bits. The 44 bits can be mapped to different CCE sizes depending on the desired coding rate, as follows (see Table 9.21):

- PDCCH Format 0: (CCE size = 72 bits) 44 bits are converted to 72 bits using coding rate 44/72.
- PDCCH Format 1: (CCE size = 144 bits) 44 bits are converted to 144 bits using coding rate 44/144.

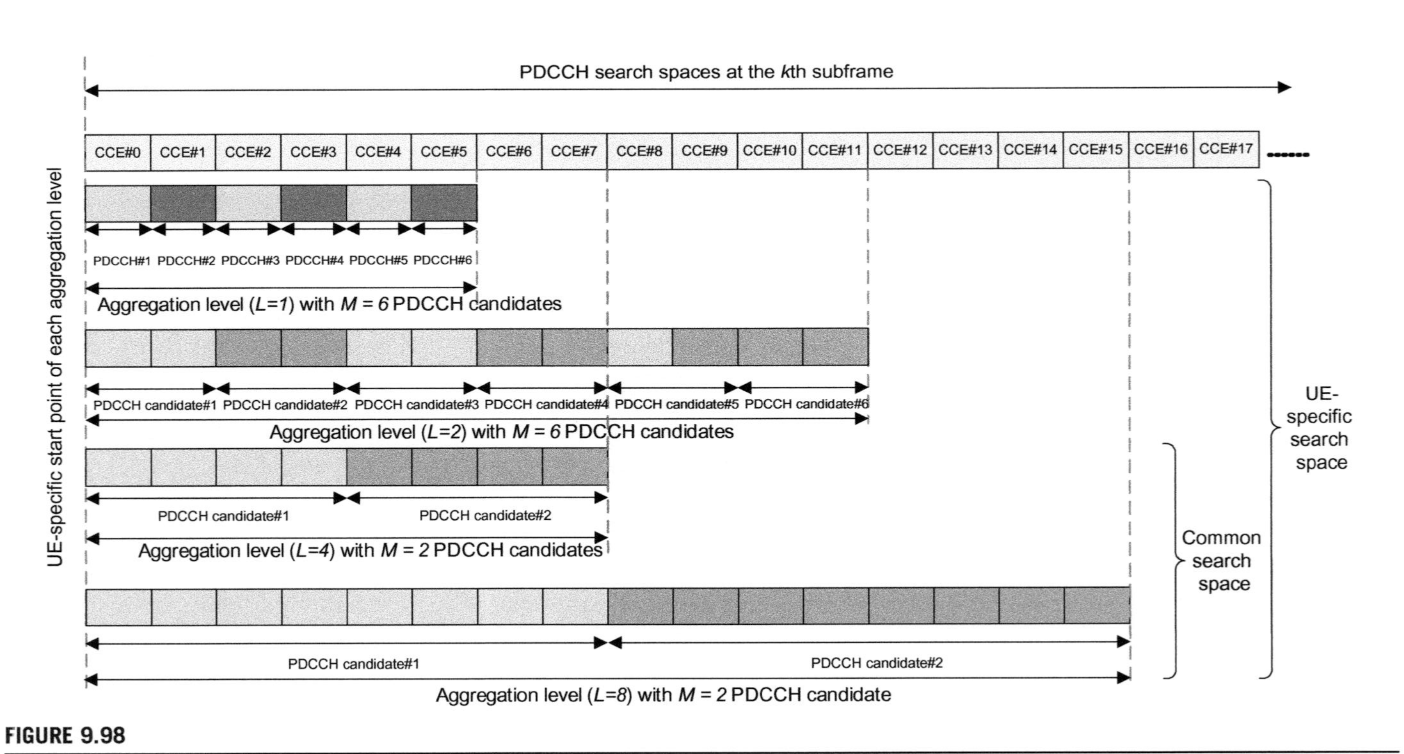

FIGURE 9.98

UE-specific/common search space (UE-specific search space starting offset is calculated based on RNTI) [15,162].

- PDCCH Format 2: (CCE size = 288 bits) 44 bits are converted to 288 bits using coding rate 44/288.
- PDCCH Format 3: (CCE size = 576 bits) 44 bits are converted to 576 bits using coding rate 44/576.

Note that the combination of PDCCH formats and CCE aggregation levels with coding rates greater than 3/4 is not supported. There are a maximum of 32 blind decoding attempts in the UE-specific search space. Assuming two payload sizes per aggregation level, the decoding attempts per payload size include six decoding attempts of 1-CCE aggregation, six decoding attempts of 2-CCE aggregation, two decoding attempts of 4-CCE aggregation, and two decoding attempts of 8-CCE aggregation. If the bandwidth is limited, then only some candidates may be available as the PDCCH region will be truncated. A PDCCH can be mapped to any candidate within its suitable search space as long as the allocated CCEs within the candidate do not overlap with a PDCCH already allocated.

The control region of each serving cell consists of a set of CCEs indexed from 0 to $N_{\text{CCE}}^{k} - 1$, where N_{CCE}^{k} denotes the total number of CCEs in the control region of the kth subframe. The UE monitors a set of PDCCH candidates on one or more activated serving cells configured for the UE for control information in every non-DRX subframe, where monitoring implies attempting to decode each of the PDCCHs in the set according to all the monitored DCI formats. The set of PDCCH candidates that a UE monitors are defined in terms of search spaces, where a search space S_k^L with aggregation level $L \in \{1, 2, 4, 8\}$ is defined by a set of PDCCH candidates. The starting point of the UE-specific search space to be monitored by the UE is determined through a hashing function.[20] For each serving cell on which the PDCCH is monitored, the starting point of the CCEs corresponding to the mth PDCCH candidate in search space S_k^L is given by $k' = [(Y_k + m') \bmod \lfloor N_{\text{CCE}}^{k}/L \rfloor]L + i$, where Y_k is defined as $Y_k = (AY_{k-1}) \bmod D$ (i.e., the integer Y_k changes in each subframe, but it is deterministically determined from the value in the previous subframe) for the UE-specific search space S_k^L with aggregation level L or $Y_k = 0$ for common search space, and aggregation levels $L = 4$ or 8. In the latter equation $Y_{-1} = n_{\text{RNTI}} \neq 0, A = 39827, D = 65537$, $k = \lfloor n_s/2 \rfloor$, and n_s is the slot number within a radio frame, and $i = 0, 1, \cdots, L-1$. For the UE-specific search space and the serving cell on which the PDCCH is monitored, if the monitoring UE is configured with a carrier indicator field (CIF), then $m' = m + M_L N_{\text{CIF}}$, where N_{CIF} is the CIF value; otherwise $m' = m$, where $m = 0, 1, \ldots, M_L - 1$, and M_L is the number of PDCCH candidates to monitor in the given search space. In the case of common search space, it is assumed that

[20] A hash function is an algorithm that maps large data sets of variable length to smaller data sets of a fixed length. Hash functions are typically used to accelerate table lookup or data comparison tasks such as finding items in a database, detecting duplicates or similar records in a large file, etc. [31].

$m' = m$. The following rules are applied to the UEs that monitor PDCCH candidates while supporting carrier aggregation [5]:

- The UE monitors one common search space at each of the aggregation levels 4 and 8 on the primary cell.
- A UE, which is not configured with a CIF, monitors one UE-specific search space at each of the aggregation levels 1, 2, 4, and 8 on each activated serving cell.
- A UE configured with a CIF monitors one or more UE-specific search spaces at each of the aggregation levels 1, 2, 4, and 8 on one or more activated serving cells configured through higher layer signaling.
- The common and UE-specific search spaces on the primary cell may overlap.
- A UE that is configured with the CIF associated with monitoring a PDCCH on serving cell monitors the PDCCH whose CRC may have been scrambled with C-RNTI in the UE-specific search space of the serving cell.
- A UE configured with the CIF associated with monitoring the PDCCH on the primary cell monitors the PDCCH whose CRC may have been scrambled with SPS C-RNTI in the UE-specific search space of the primary cell.
- The UE monitors the common search space for the PDCCH without CIF.
- For the serving cell on which the PDCCH is monitored, if the UE is not configured with a carrier aggregation, then it is required to monitor the UE-specific search space for the PDCCH without CIF; otherwise, the UE is required to monitor the UE-specific search space for the PDCCH with CIF.
- A UE is not expected to monitor the PDCCH of a secondary cell, if it is configured to monitor the PDCCH with CIF corresponding to that secondary cell in another serving cell. For the serving cell on which the PDCCH is monitored, the UE monitors PDCCH candidates at least for the same serving cell.
- A UE configured to monitor PDCCH candidates whose CRCs are scrambled with C-RNTI or SPS C-RNTI, with a common payload size and with the same first CCE index n_{CCE}, but with different sets of DCI information fields in the common and UE-specific search spaces on the primary cell, is required to assume that only the PDCCH in the common search space is transmitted by the primary cell.
- A UE configured to monitor PDCCH candidates in a given serving cell with a given DCI format size with CIF, and CRC scrambled with C- RNTI, where the PDCCH candidates may have one or more possible values of CIF for the given DCI format size, can assume that a PDCCH candidate with the given DCI format size may be transmitted in the given serving cell in any UE-specific search space corresponding to any of the possible values of CIF for the given DCI format size.

The aggregation levels defining the search spaces are listed in Table 9.24. The DCI formats that the UE is expected to monitor depend on the configured transmission mode in each serving cell. For the common search spaces, Y_k is set to 0 for the two aggregation levels $L = 4$ and $L = 8$.

9.12.5 Enhanced PDCCH

In LTE Rel-8 downlink control channel design, the PDCCH capacity is constrained by the number of available CCEs and the aggregation level associated with each scheduled PDCCH. Since the size of the time-division-multiplexed control region in each subframe is limited to three OFDM symbols, there might be some scenarios where the eNB exhausts the resources for PDCCH transmission. With the increased use of DM-RS, DCI format 2C will be more often used. DCI format 2C should use at least two CCEs for most UEs, since using only one CCE results in high coding rates (e.g., 0.81 for a 10 MHz FDD system). Moreover, for the UEs at the cell-edge configured for the CoMP mode, a large number of CCEs is needed for the PDCCH to maintain coverage, because the PDCCH cannot use CoMP to improve cell-edge performance. Furthermore, LTE Rel-11 downlink MIMO enhancement for heterogeneous network deployments requires increased PDCCH capacity and the use of virtual cell identities to overcome the limitation of the existing LTE technology in new deployment scenarios related to CoMP, HetNet, etc., where several low-power nodes or subsets of a macro-cell may share the same physical cell identity with the macro-cell.

Sharing the same cell identity by the neighboring nodes would increase intracell interference and degrade the channel estimation performance and coherent detection of data and control channels, as well as limit the use of advanced closed-loop SU-MIMO/MU-MIMO techniques. Also, the use of the same identifier by subordinate nodes would prevent transmission/reception node identification and switching. An example case is a non-uniform network deployment comprising a macro-node with several low-power RRHs using the same physical cell identifier as the macro-node. In such a case, there is one common PDCCH for the macro eNB and all RRHs, thus the capacity of the current control region might be too small for sending all PDCCHs without collision.

The use of different scrambling sequences initialized with unique cell identifiers would achieve interference randomization and help the UE to separate data and control channels transmitted by different transmission nodes. Since the reference signals (e.g., CRS, DM-RS, CSI-RS) are also scrambled with cell-specific sequences to allow the UE to conduct various channel estimations and mobility measurements on different transmission nodes, the use of unique identifiers would facilitate these functions, as well as node selection and switching in homogeneous and heterogeneous networks.

Several justifications for the design of a new downlink control structure were discussed in LTE Rel-11 contributions that are summarized below [149–155]:

- Limited PDCCH capacity: The existing PDCCH structure was designed to provide control signaling and resource assignments for UEs in a single transmission point per macro-cell deployment. Under CoMP scenario 4 (see Chapter 12), several RRHs having the same cell identity are deployed in a single macro-cell. In such a scenario, the assignments corresponding to all RRHs may have to be simultaneously transmitted. This can cause a significant increase in the number of resource assignments and shortage of CCEs per subframe. In CoMP scenario 3, each RRH has a different cell identity relative to the associated macro-cell; however, all nodes share the same time/frequency resources, which may result in significant intra-cell interference. Therefore, the detection performance of the PDCCH in subframes with strong interference may be unsatisfactory. Although the almost-blank subframe concept is supported in LTE Rel-10, it requires the network node to reduce the activity (e.g., transmission power) in the almost-blank subframe, leading to inefficient spectral utilization. Hence, the PDCCH enhancements were necessary for this scenario, as well.

- Increased resource assignment payload size: Since the initial Rel-8 PDCCH design, several improvements have been made to the standard. In particular, new transmission modes, based on UE-specific reference signals, with support for MU-MIMO were designed. It is anticipated that transmission mode 9 will be widely used in future deployments. One of the DCI formats used with transmission mode 9 is DCI format 2C. With its large payload size, the use of DCI format 2C may result in fewer resource assignments (i.e., fewer CCEs per subframe) based on the existing PDCCH structure.
- Interference avoidance/coordination on the control channels: The enhancement of MIMO performance through improved CSI feedback for certain deployment scenarios, especially the case of four transmit antennas in a cross-polarized downlink configuration, in both homogeneous and heterogeneous cases, was of special interest in LTE Rel-11. In such environments, neighboring transmission nodes may interfere with each other. The existing mechanisms in earlier releases of LTE may not be sufficient for robust transmission of control channels in dense and diverse deployments. In that case, it might be beneficial to be able to perform interference avoidance/coordination by orthogonalizing the users in neighboring cells.
- Higher user throughput: Due to capacity limits of the PDCCH, some data resources may not be allocated in a timely manner. One reason for the limitation is the use of the hashing function to map CCEs in the control region. The hashing function (in some cases) may cause two candidate sets to collide, especially when an aggregation level higher than one is chosen for each UE. As a result, the number of assignments that can be transmitted on the PDCCH will be limited, reducing the overall user throughput and increasing transmission latency.
- Use of soft frequency partitioning and fractional frequency reuse with PDCCH: The time-division-multiplexed structure of PDCCH was designed based on a single-frequency partition in each slot/subframe and assuming a frequency reuse factor of one. While hard frequency partitioning is very undesirable from a network deployment and frequency planning perspective, the soft frequency partitioning combined with fractional-frequency reuse techniques has been shown to provide more robustness and reliability for transmission of control signaling and data, due to improved interference mitigation. The use of frequency-division-multiplexing of data and control regions would allow separate power control for each channel type.

- MU-MIMO and beamforming of control channels: The existing structure of the PDCCH relies on CRS for channel estimation and coherent detection. It is shown that CRS has more overhead and is less effective when using closed-loop precoding techniques, beamforming, and MU-MIMO. The beamforming of the control channels can significantly improve the performance of the cell-edge users. The use of DM-RS instead of CRS would facilitate the use of beamforming for the control channels.
- Control overhead reduction: The resource allocation granularity is one OFDM symbol for the PDCCH, thus it amounts to approximately 7% overhead per OFDM symbol. In the frequency-domain, the control channel overhead can be significantly lower if the granularity is defined as one PRB. For instance, in a 10 MHz system, the resource allocation granularity (L1/L2 overhead per PRB) is 2% and 1% in a 20 MHz system. Therefore, the design of the ePDCCH can result in more efficient resource utilization for the control channel.

To overcome the above limitations, an enhanced physical downlink control channel (ePDCCH) was introduced in 3GPP LTE Rel-11. In LTE Rel-11, the ePDCCH can be used to convey scheduling assignments. The ePDCCH is transmitted by aggregating one or several eCCEs. Each eCCE consists of multiple enhanced REGs (eREGs). In an FDD system, the number of eREGs per eCCE $N_{\mathrm{eCCE}}^{\mathrm{eREG}}$ is four for subframes with normal cyclic prefix, and eight for subframes with an extended cyclic prefix. In a TDD system, there are four eREGs per eCCE for subframes with a normal cyclic prefix and special subframe configurations 3, 4, and 5. In addition, there are eight eREGs per eCCE in TDD systems with an extended cyclic prefix and special subframe configurations 1, 2, 3, 5, and 6 [3]. The eCCEs that construct ePDCCHs are numbered from 0 to $N_{\mathrm{ePDCCH}}^{\mathrm{eCCE}} - 1$, where the number of eCCEs per ePDCCH $N_{\mathrm{ePDCCH}}^{\mathrm{eCCE}}$ depends on the ePDCCH format as given in Table 9.25.

The eREGs are used to define the mapping of enhanced control channels to resource elements. There are 16 eREGs in each PRB pair. The eREG indexing is obtained by numbering all resource elements within a PRB pair cyclically from 0 to 15 in an increasing order of first frequency then time, skipping the resource elements designated to DM-RS for antenna ports $p = \{107, 108, 109, 110\}$ for a normal cyclic prefix or $p = \{107, 108\}$ for an extended cyclic prefix. All resource elements with number i in that PRB pair constitute eREG_i (see Figure 9.99).

As shown in Figure 9.100, the mapping of the ePDCCH to resource elements within resource block pairs can be performed in a localized or distributed manner, where the mapping defines how the eCCEs are associated with eREGs and PRB pairs. It must be noted that multiplexing of ePDCCH and PDSCH within a PRB pair is not permitted. If a UE is configured to monitor multiple ePDCCHs, then there are one or two sets of PRB pairs, which the UE is required to monitor. All ePDCCH candidates in ePDCCH set S_{m} exclusively use localized or distributed mapping which is configured by higher layers. Within ePDCCH set S_{m} in subframe i, the eCCEs available for transmission of ePDCCHs are numbered from 0 to $N_{\mathrm{eCCE}}(m, i) - 1$, where eCCE index n corresponds to eREGs numbered $(n \bmod N_{\mathrm{RB}}^{\mathrm{eCCE}}) + \mathrm{j}N_{\mathrm{RB}}^{\mathrm{eCCE}}$ in PRB index $\lfloor n/N_{\mathrm{RB}}^{\mathrm{eCCE}} \rfloor$ for localized mapping, or

Table 9.25 ePDCCH Formats [3]

ePDCCH Format	Number of eCCEs per ePDCCH ($N_{\text{ePDCCH}}^{\text{eCCE}}$)			
	Regular TDD/FDD Subframes and TDD Special Subframe Configurations 3, 4, 8 with Normal Cyclic Prefix (DCI Formats 2/2A/2B/2C/2D and $N_{\text{RB}}^{\text{DL}} \geq 25$ or Any DCI Format and $N_{\text{ePDCCH}} < 104$)		All Other Cases	
	Localized Transmission	Distributed Transmission	Localized Transmission	Distributed Transmission
0	2	2	1	1
1	4	4	2	2
2	8	8	4	4
3	16	16	8	8
4	–	32	–	16

it corresponds to eREGs numbered $\lfloor n/N_{\text{RB}}^{S_{\text{m}}} \rfloor + \text{j}N_{\text{RB}}^{\text{eCCE}}$ in PRB indices $(n + \text{j}\max[1, N_{\text{RB}}^{S_{\text{m}}}/N_{\text{eCCE}}^{\text{eREG}}]) \bmod N_{\text{RB}}^{S_{\text{m}}}$ for distributed mapping. The parameter $j = 0, 1, \ldots, N_{\text{eCCE}}^{\text{eREG}} - 1$, $N_{\text{eCCE}}^{\text{eREG}}$ is the number of eREGs per eCCE, and $N_{\text{RB}}^{\text{eCCE}} = 16/N_{\text{eCCE}}^{\text{eREG}}$ is the number of eCCEs per resource block pair (see Figure 9.99). The PRB pairs that form ePDCCH set S_{m} are numbered in ascending order from 0 to $N_{\text{RB}}^{S_{\text{m}}} - 1$ [3].

The physical layer processing of the ePDCCH can be described as follows. The block of bits $(b_0, b_1, \ldots, b_{N_{\text{bit}}-1})$ that is going to be transmitted on an ePDCCH in a subframe is scrambled, resulting in a block of scrambled bits $(\tilde{b}_0, \tilde{b}_1, \ldots, \tilde{b}_{N_{\text{bit}}-1})$ in which $\tilde{b}_i = [b_i + c(i)] \bmod 2$. The UE-specific scrambling sequence $c(i)$ defined in previous sections is initialized with $c_{\text{init}} = \lfloor n_{\text{s}}/2 \rfloor 2^9 + N_{\text{ID}}^{\text{ePDCCH}(i)}$, where index i is the ePDCCH set number, and $N_{\text{ID}}^{\text{ePDCCH}(i)}$ is provided by higher layers. The block of scrambled bits $(\tilde{b}_0, \tilde{b}_1, \ldots, \tilde{b}_{N_{\text{bit}}-1})$ is modulated using QPSK modulation, resulting in a block of complex-valued modulation symbols $(d_0, d_1, \ldots, d_{N_s-1})$. The block of complex-valued modulated symbols is mapped to a single layer and precoded according to $y_i = d_i|_{i=0,1,\ldots,N_s-1}$. The precoded block of symbols can be mapped to physical resource elements using localized or distributed mapping.

The block of complex-valued symbols $(y_0, y_1, \ldots, y_{N_s-1})$ is sequentially mapped to resource elements (k, l) starting with y_0 to the associated antenna port, provided that those physical resources are part of the eREGs assigned for transmission and are not part of a physical resource-block pair used for transmission of PBCH or synchronization signals, and are not used for CRSs, CSI-RSs for the

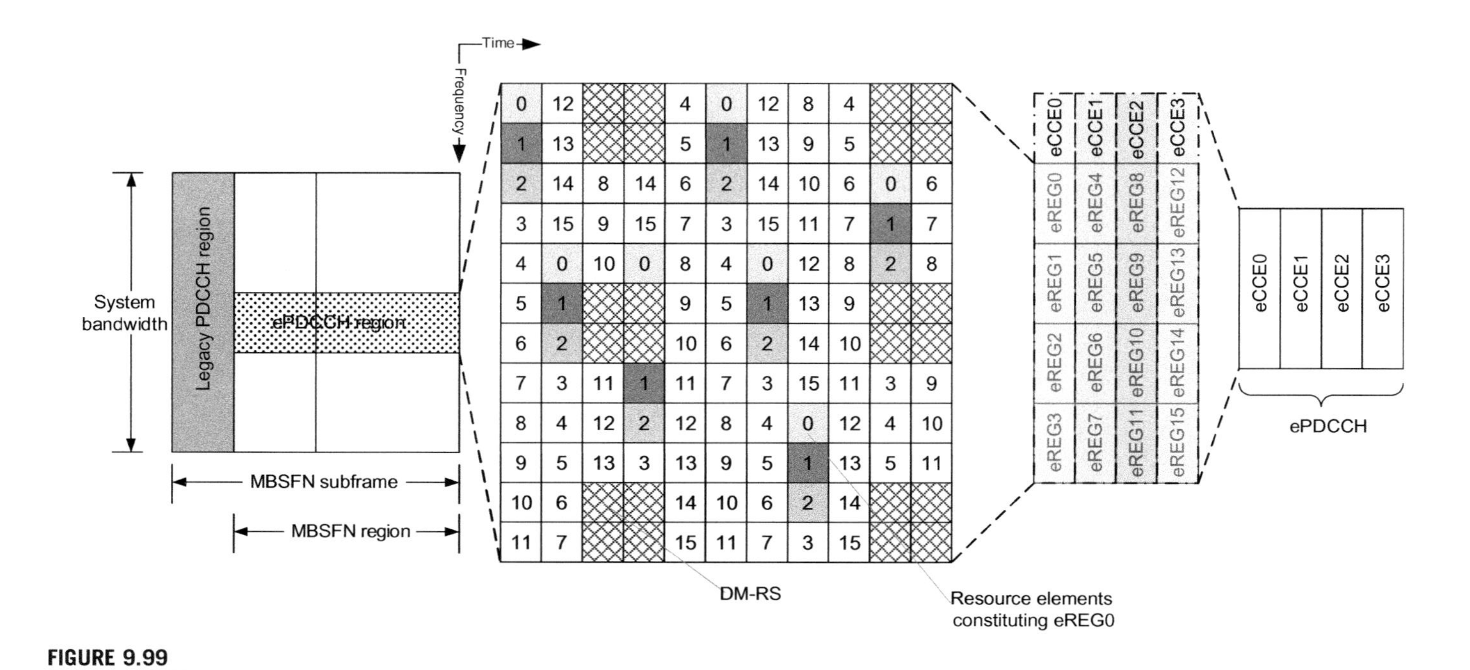

FIGURE 9.99

Structure and formation of an example ePDCCH, eCCE, and eREG [3].

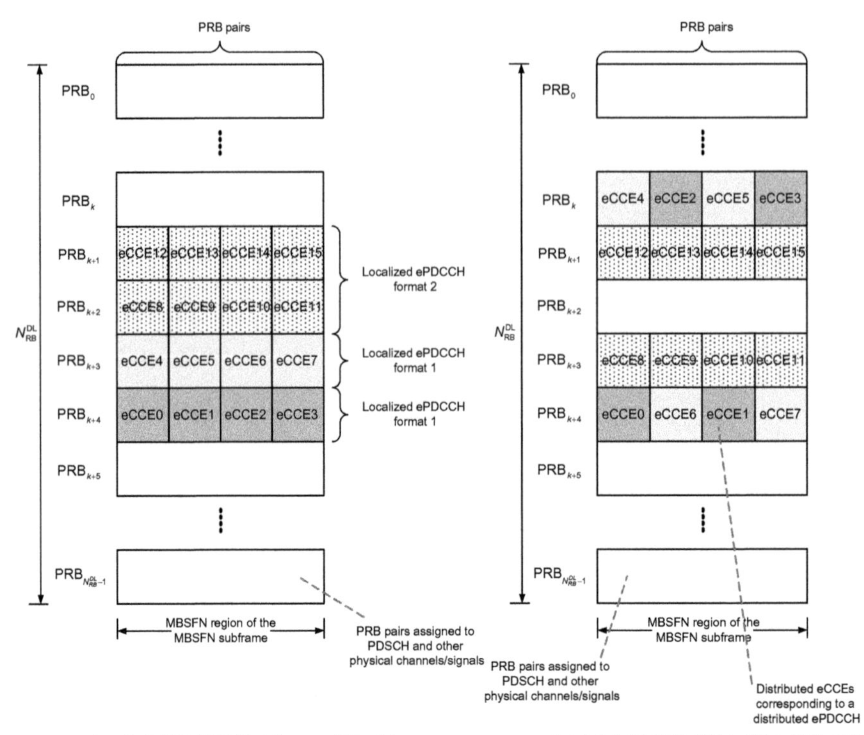

FIGURE 9.100

Structure and formation of localized and distributed ePDCCH and eCCEs [3].

specific UE; and further the index l in the first slot in a subframe stratifies the condition $l \geq l_{\text{ePDCCH}_{\text{start}}}$. The mapping to resource elements (k, l) on antenna port p is performed in increasing order of first the index k and then the index l, starting with the first slot and ending with the second slot in a subframe. For localized transmission, the single antenna port p to use for ePDCCH transmission is parameterized with parameter $n_{\text{ap}} = n_{\text{eCCE}_{\text{L}}} \bmod N_{\text{RB}}^{\text{eCCE}} + n_{\text{RNTI}} \bmod \min(N_{\text{ePDCCH}}^{\text{eCCE}}, N_{\text{RB}}^{\text{eCCE}})$, where $n_{\text{eCCE}_{\text{L}}}$ denotes the lowest eCCE index used by the referenced ePDCCH transmission in the ePDCCH set, n_{RNTI} corresponds to the RNTI associated with the ePDCCH transmission, and $N_{\text{ePDCCH}}^{\text{eCCE}}$ is the number of eCCEs per referenced ePDCCH. For FDD/TDD subframes with a normal cyclic prefix and TDD special subframe configurations 3, 4, and 8, depending on the value of parameter n_{ap} the following antenna port assignments are made [3]:

$$\begin{aligned} n_{\text{ap}} = 0 &\rightarrow p = 107 \\ n_{\text{ap}} = 1 &\rightarrow p = 108 \\ n_{\text{ap}} = 2 &\rightarrow p = 109 \\ n_{\text{ap}} = 3 &\rightarrow p = 110 \end{aligned}$$

For distributed transmission, each resource element in an eREG is associated with one out of two antenna ports in an alternating manner, where $p \in \{107, 109\}$ for a normal cyclic prefix and $p \in \{107, 108\}$ for an extended cyclic prefix.

For each serving cell, higher layer signaling can configure a UE with one or multiple ePDCCH-PRB sets for ePDCCH monitoring. Each ePDCCH-PRB set consists of a set of eCCEs indexed from 0 to $N_{\text{eCCE}}(p,k) - 1$, where $N_{\text{eCCE}}(p,k)$ denotes the number of eCCEs in ePDCCH-PRB set p in subframe k. Each ePDCCH-PRB set can be configured for either localized or distributed ePDCCH transmission. The UE is required to monitor a set of ePDCCH candidates on one or more activated serving cells configured via higher-layer signaling for control information, where monitoring implies attempting to blindly decode each of the ePDCCHs in the set according to the monitored DCI formats.

The set of ePDCCH candidates that the UE is expected to monitor are defined in terms of ePDCCH UE-specific search spaces. For each serving cell, the subframes in which the UE monitors ePDCCH UE-specific search spaces are configured via higher-layer signaling. In TDD systems using a normal cyclic prefix in downlink subframes, the UE is not expected to monitor the ePDCCH in special subframe configurations 0 and 5. A UE is not expected to monitor an ePDCCH candidate whose eCCE(s) is mapped to a PRB pair that overlaps in frequency with a transmission of either PBCH or primary/secondary synchronization signals in the same subframe.

The UE is required to monitor one common search space in every non-DRX subframe at aggregation levels 4 and 8 on the primary cell. If a UE is not configured for ePDCCH monitoring, and if it is not configured with a CIF, then it is required to monitor one PDCCH UE-specific search space at aggregation levels 1, 2, 4, 8 on each activated serving cell in every non-DRX subframe. In another case where a UE is not configured to monitor ePDCCH, but it is configured with a CIF, then it is required to monitor one or more UE-specific search spaces at aggregation levels 1, 2, 4, 8 on one or more activated serving cells that are configured through higher-layer signaling in every non-DRX subframe. Furthermore, if a UE is configured for ePDCCH monitoring on an activated serving cell, but it is not configured with CIF, then it is required to monitor one PDCCH UE-specific search space at aggregation levels 1, 2, 4, 8 on the serving cell in all non-DRX subframes where ePDCCH is not monitored on that serving cell. If a UE is configured for ePDCCH monitoring on a serving cell, and if that serving cell is activated, and if the UE is configured with a CIF, then the UE is required to monitor one or more PDCCH UE-specific search spaces at each of the aggregation levels 1, 2, 4, and 8 on that serving cell as configured by higher layer signaling in all non-DRX subframes where ePDCCH is not monitored on that serving cell. The common and PDCCH UE-specific search spaces on the primary cell may overlap. It can be concluded that in each subframe on an activated serving cell, the UE either monitors ePDCCH or PDCCH and not both of them, in order to simplify the UE search mechanism.

9.12.6 Physical downlink shared channel

The physical downlink shared channel is a downlink channel that is used for transmission of all user data, paging messages, and part of the system information which is not carried in the PBCH. The data is transmitted on the PDSCH in units of transport blocks that correspond to the MAC sub-layer PDUs. The transport blocks are received from the MAC sub-layer in each TTI where the default TTI value in LTE/LTE-Advanced is 1 ms, corresponding to one subframe duration. The maximum number of transport blocks that can be transmitted per UE per subframe is limited to two, depending on the transmission mode selected for the PDSCH for each UE. The transmission mode defines the multi-antenna configuration that is used for data transmission over the PDSCH. There are 10 transmission modes specified for LTE Rel-11 that are summarized in Table 9.26. Note that transmission mode 10 was included as part of Rel-11 enhancements to support coordinated multi-point transmission/reception; especially with multiple CSI processes configured, TM10 can be seen as enhanced TM9 with well-defined UE behavior in interference measurement via CSI interference measurement configuration [3,5].

With the exception of transmission modes 7, 8, 9, and 10, the phase reference for coherent demodulation of PDSCH is obtained from CRSs and the number of eNB antenna ports used for transmission of PDSCH is the same as the number of antenna ports used in the cell for PBCH transmission. In transmission modes 7, 8, 9, and 10, the UE-specific reference signals provide the phase reference for coherent demodulation of the PDSCH. The configured transmission mode further controls the formats of the associated downlink control signaling messages as given in Table 9.26, and the type of channel quality feedback from the UE.

The coded transport block is scrambled with a binary scrambling sequence. The sequence is based on cell identity and RNTI to make it cell- and UE-specific. The binary scrambled bits are modulated using a suitable modulation scheme considering the channel conditions and required data rate. The permissible modulation schemes for PDSCH are QPSK (2 bits/symbol), 16QAM (4 bits/symbol), or 64QAM (6 bits/symbol). Support of 64QAM is mandatory for all LTE/LTE-Advanced UE categories. The selection of modulation scheme is based on CQI feedback from the UE. The resulting codewords after modulation are mapped to spatial layers. Spatial multiplexing uses one to eight layers less than or equal to the number of antenna ports. For transmit diversity, the number of layers is equal to the number of antenna ports. The precoding matrix applies antenna weighting (phase and amplitude) for each data stream, and maps the stream to different antenna ports. In resource blocks in which UE-specific reference signals are not transmitted, PDSCH is transmitted on the same set of antenna ports used for PBCH.

All downlink resource elements are available for the PDSCH except those reserved for special purposes, such cell-specific, positioning, MBSFN, CSI, and UE-specific reference signals, synchronization signals, and control signaling. Therefore, when the UE is informed through control signaling of a resource allocation within a pair of resource blocks in a subframe, it is only the available resource elements within those resource block pairs which actually carry PDSCH

Table 9.26 Summary of PDSCH Transmission Modes in LTE/LTE-Advanced when using C-RNTI [5,13]

Release	Mode	PDSCH Transmission Scheme	DCI Format	Search Space	Channel State Information Feedback
8	1	Single-antenna transmission on antenna port 0	1	UE-specific	CQI
			1A	Common and UE-specific	
	2	Transmit diversity	1	UE-specific	CQI
			1A	Common and UE-specific	
	3	Transmit diversity	1A	Common and UE-specific	CQI and RI
		Open-loop spatial multiplexing or transmit diversity	2A	UE-specific	
	4	Transmit diversity	1A	Common and UE-specific	CQI, RI, and PMI
		Closed-loop spatial multiplexing or transmit diversity	2	UE-specific	
	5	Transmit diversity	1A	Common and UE-specific	CQI and PMI
		MU-MIMO	1D	UE-specific	
	6	Transmit diversity	1A	Common and UE-specific	CQI and PMI
		Closed-loop spatial multiplexing using single-layer transmission (closed-loop rank-1 precoding)	1B	UE-specific	
	7	Single-antenna transmission on antenna port 5 (transmission using UE-specific reference signals with a single layer)	1	UE-specific	CQI
		Single-antenna transmission using antenna port 0 or transmit diversity	1A	Common and UE-specific	

(*Continued*)

Table 9.26 (Continued)

Release	Mode	PDSCH Transmission Scheme	DCI Format	Search Space	Channel State Information Feedback
9	8	Single-antenna transmission using antenna port 0 or transmit diversity	1A	Common and UE-specific	CQI (PMI/RI if configured by eNB)
		Dual-layer transmission on antenna ports 7 and 8 or single-antenna transmission on antenna port 7 or 8 (transmission using UE-specific reference signals with up to two layers)	1B	UE-specific	
10	9	Single-antenna transmission on antenna port 0 or transmit diversity	1A	Common and UE-specific	CQI (PMI/RI if configured by eNB)
		Spatial multiplexing up to eight layers on antenna ports 7–14 (transmission using UE-specific reference signals with up to eight layers)	2C	UE-specific	
11	10	Coordinated multi-point transmission and reception (support CoMP operation and configure Rel-11 non-zero-power CSI-RS resources and interference measurement resources)	1A/2D	UE-specific	CQI (PMI/RI if configured by eNB)

data. The allocation of resource block pairs to PDSCH transmission for a particular UE is signaled to the UE by means of dynamic control signaling transmitted at the start of the relevant subframe using ePDCCH/PDCCH. The mapping of PDSCH resource elements in a subframe is illustrated in Figure 9.101. The modulation symbols are mapped to resource elements (k, l) in the allocated resource blocks. The mapping is performed first along frequency and then along symbols in time. The modulation symbols are mapped by incrementing the sub-carrier index k first along the allocated resource blocks. When the final resource block is reached, the symbol index l is incremented. After mapping, the OFDM modulation is performed across all sub-carriers on a symbol-by-symbol basis.

The paging control channel (PCCH) is a logical channel that is mapped to the paging channel (PCH) which is a transport channel. The PCH is transmitted via PDSCH similar to other data transmissions. When PCH is transmitted, a unique 16-bit identifier known as P-RNTI with a value of 0xFFFE is used to signal paging messages in the PDCCH. Multiple UEs may be paged simultaneously using a single paging allocation with the UE identity indicated within the paging message itself (Figure 9.102).

A UE is required to decode the corresponding PDSCH allocation in the same subframe (considering the restriction on the number of transport blocks signaled through RRC messages) upon detection of an ePDCCH/PDCCH of a serving cell with DCI format 1/1A/1B/1C/1D or 2/2A/2B/2C/2D intended for the UE in that subframe [5]. The UE may assume that PRSs are not present in resource blocks in which it is required to decode PDSCH when a PDCCH whose CRC is scrambled with SI-RNTI or P-RNTI with DCI format 1A or 1C is successfully detected. If the UE is configured with the CIF for a given serving cell, it must assume that the CIF is not present in any PDCCH of the serving cell in the common search space; otherwise, the UE must assume that for the given serving cell, the CIF is present in the ePDCCH/PDCCH located in the UE-specific search space, when the ePDCCH/PDCCH CRC is scrambled with C-RNTI or SPS C-RNTI.

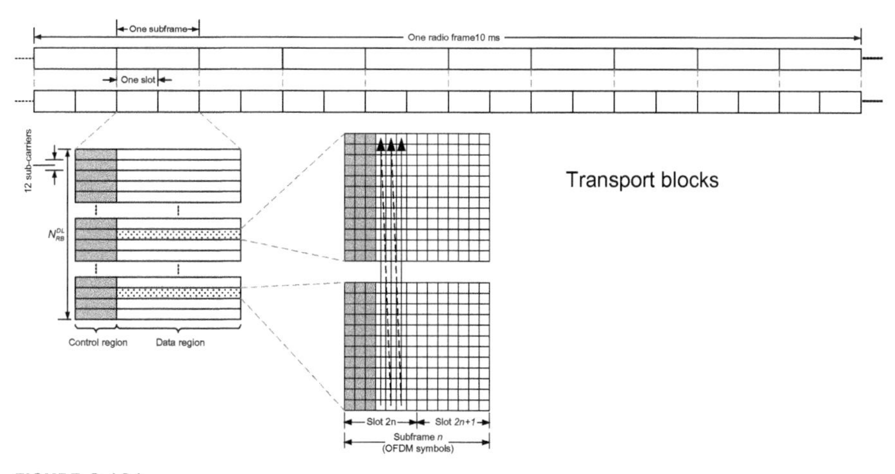

FIGURE 9.101

Illustration of PDSCH resource element mapping.

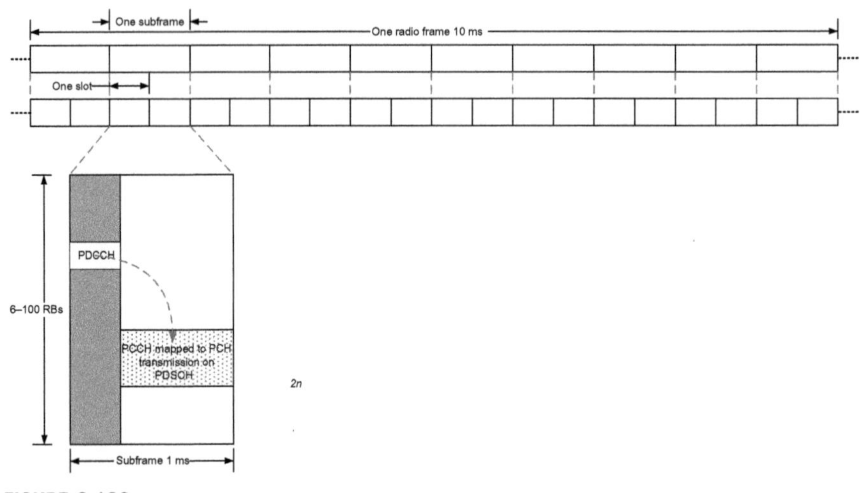

FIGURE 9.102

PCH transmission on PDSCH.

If a UE is configured by higher layers to decode the PDCCH with its CRC scrambled with SI-RNTI, the UE is required to decode PDCCH and the corresponding PDSCH according to any of the permissible combinations of DCI formats and multi-antenna schemes. The PDSCH corresponding to these PDCCHs is also scrambled with SI-RNTI. Similarly, if a UE is configured to monitor PDCCHs whose CRCs are scrambled with P-RNTI, C-RNTI, SPS C-RNTI, or RA-RNTI, the UE is required to decode the PDCCH and the corresponding PDSCH according to any of the permissible combinations of DCI formats and multi-antenna schemes. An example is given in Table 9.26 for C-RNTI [5]. Note that if the CRC of PDCCH is scrambled with broadcast identifiers such as SI-RNTI, P-RNTI, and RA-RNTI, the UE only monitors the common search space and decodes the PDSCH accordingly. For multi-user MIMO transmission of PDSCH, the UE may assume that the eNB transmission on PDSCH is on one layer. An additional power offset of $\delta_{\text{power-offset}} = -3$ dB may be applied if the downlink power offset field signaled on PDCCH with DCI format 1D is zero.

9.13 Principles of multi-antenna transmission

Design of mobile broadband wireless communication systems capable of data transmission rates in excess of 1 Gbps has been of practical interest in the past decade to enable the new generations of cellular systems and IMT-Advanced. The use of multiple antennas at the transmitter and/or receiver commonly known as the MIMO system has become a pragmatic and cost-effective approach that offers substantial gain in making 1 Gbps wireless links a reality. This section provides an overview of the MIMO wireless technology including channel models, performance limits, coding, and transceiver design. MIMO techniques can increase system throughput and transmission reliability without increasing the required bandwidth. While most communication systems suffer in multi-path channels, a MIMO system benefits from propagation over different paths through which the signals arrive at the receiver. A MIMO system typically performs better in an indoor environment while it may not properly perform in LoS environments.

The performance improvements resulting from the use of multi-antenna systems are mainly due to array gain, diversity gain, spatial multiplexing gain, and interference reduction that may be achieved through appropriate MIMO configurations. The main advantages of the multi-antenna techniques can be summarized as follows [20]:

- Array gain can be achieved through signal processing at the transmitter and/or the receiver, and results in an increase in average SINR at the receiver due to coherent combining effect. The increased SINR further results in improved coverage and user throughput. Transmit or receive array gain requires channel knowledge in the transmitter and receiver, respectively, and further depends on the number of transmit and receive antennas. The channel knowledge in the receiver is typically available (through channel estimation based on downlink reference signals), whereas CSI in the transmitter is in general more difficult to attain.
- The signal power is faded randomly in a wireless communication channel. Diversity is an effective technique to mitigate fading in a wireless channel. The diversity schemes rely on transmitting the signal over multiple and ideally independent fading channels in time, frequency, and/or space. Spatial diversity is preferred over time or frequency diversity as it does not consume additional transmission time or bandwidth. If the $N_{RX} \times N_{TX}$ links forming the MIMO channel fade independently and the transmitted signal is properly constructed, the receiver can combine the arriving signals such that the resulting signal exhibits considerably reduced amplitude variability relative to a single-input single-output (SISO) link, and a diversity gain on the order of $N_{RX} \times N_{TX}$ can be achieved. Spatial diversity gain in the absence of channel knowledge at the transmitter can be achieved using suitably designed transmit signal which is known as space–time coding.
- MIMO channels provide a linear increase in capacity as a function of $\min(N_{RX}, N_{TX})$ without requiring additional power or transmission bandwidth. This gain, referred to as spatial multiplexing gain, is realized by transmitting independent data streams from individual transmit antennas. Note that the number of independent data streams is limited to $\min(N_{RX}, N_{TX})$. Under good channel conditions and sufficiently high SINR values, the receiver can detect different data streams, yielding a linear increase in capacity.
- Co-channel interference is generated due to frequency reuse in wireless channels. When multiple antennas are used, the differentiation between the spatial signatures of the desired signal and co-channel signals can be utilized to reduce interference. While interference cancelation requires proper knowledge of the desired signal's channel, exact knowledge of the interferers'

channels may not be necessary. The interference avoidance schemes can also be implemented at the transmitter where the goal is to minimize the interference energy sent toward the co-channel users while delivering the signal to the desired user. Interference reduction allows aggressive frequency reuse, and thereby increases multi-cell capacity. It must be noted that not all advantages of MIMO schemes can be simultaneously achieved due to conflicting demands on the spatial degrees of freedom or the number of transmit and/or receive antennas. The degree to which these conflicts can be resolved depends on the signaling scheme and transceiver design.

9.13.1 Capacity of MIMO channels

As shown in Figure 9.103, a generic MIMO system consists of a MIMO transmitter with N_{TX} transmit antennas, a MIMO receiver with N_{RX} receive antennas, and $N_{RX} \times N_{TX}$ paths or channels between transmit and receive antennas. Let $x_k(t)$ denote the transmitted signal from the kth transmit antenna at time t, then the received signal at the lth antenna can be expressed as follows:

$$y_l(t) = \sum_{k=1}^{N_{TX}} h_{lk}(t)^* x_k(t) + n_l(t) \tag{9.54}$$

where $h_{lk}(t)$ and $n_l(t)$ are the channel impulse response between the kth transmit antenna, the lth receive antenna, and the additive noise at the lth receive antenna port, respectively. The above equation can be written in the frequency-domain as:

$$Y_l(\omega) = H_{lk}(\omega)X_k(\omega) + N_l(\omega) \tag{9.55}$$

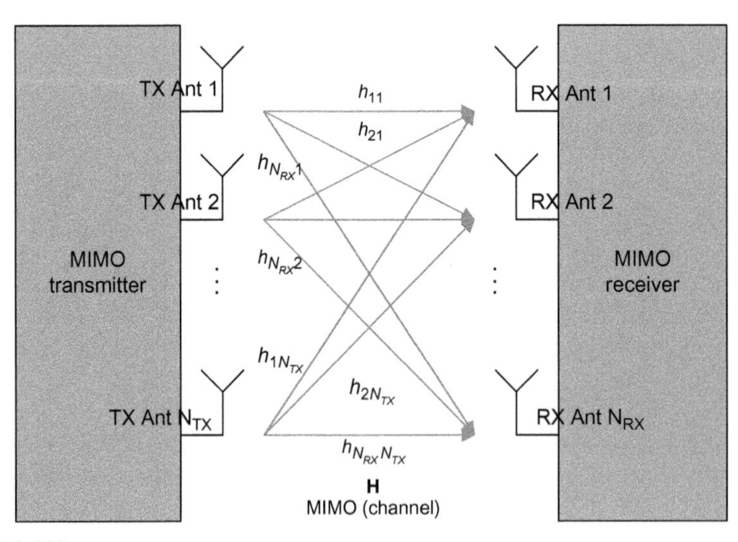

FIGURE 9.103

Illustration of the general principle of a MIMO system.

If $\mathbf{x}(\omega) = (X_1(\omega), X_2(\omega), \ldots, X_{N_{TX}}(\omega))^T$, $\mathbf{y}(\omega) = (Y_1(\omega), Y_2(\omega), \ldots, Y_{N_{RX}}(\omega))^T$, and $\mathbf{n}(\omega) = (N_1(\omega), N_2(\omega), \ldots, N_{N_{RX}}(\omega))^T$ denote the Fourier transform vectors of $\mathbf{x}_k(t)$, $\mathbf{y}_l(t)$, and $\mathbf{n}_l(t)$, respectively, then $\mathbf{y}(\omega) = \mathbf{H}(\omega)\,\mathbf{x}(\omega) + \mathbf{n}(\omega)$, where $\mathbf{H}(\omega)$ is an $N_{RX} \times N_{TX}$ channel matrix with $H_{lk}(\omega)|_{k=1,2,\ldots,N_{TX};l=1,2,\ldots,N_{RX}}$ entries.

Assuming a linear time-invariant MIMO channel, the channel input–output relationship can be further described in the discrete time-domain as follows:

$$y_l(nT) = \sum_{k=1}^{N_{TX}} \sum_{m=0}^{M-1} h_{lk}(mT)x_k[(n-m)T] + n_l(nT) \quad 0 \le n \le M-1;\ 1 \le l \le N_{RX} \tag{9.56}$$

where $x_k(n)|_{k=1,2,\ldots,N_{TX}}$ and $y_l(n)|_{l=1,2,\ldots,N_{RX}}$ represent the channel input and output time-domain signals, respectively. In the case of time-varying channels, Eq. (9.54) can be written as $y_l(t) = \sum_{k=0}^{N_{TX}-1} h_{lk}(t,\tau)x_k(\tau) + n_l(t)$, where $h_{lk}(t,\tau)$ denotes the time-varying impulse response of the lkth channel. The matrix form of Eq. (9.55) in the frequency-domain sampled at a single frequency ω_m can be written as follows:

$$\mathbf{y}(\omega_m) = \mathbf{H}(\omega_m)\mathbf{x}(\omega_m) + \mathbf{n}(\omega_m) \tag{9.57}$$

In an OFDM system, the signal processing is inherently performed in the frequency-domain. Furthermore, OFDM transforms a frequency-selective fading channel to a flat-fading channel when considering narrowband orthogonal sub-carriers. In such a system, the MIMO signal processing can be performed at each sub-carrier. This is the main reason for the suitability of MIMO extension to an OFDM system. When MIMO processing is done at each sub-carrier, the MIMO channel input–output relationship can be demonstrated as $\mathbf{y} = \mathbf{H}\mathbf{x} + \mathbf{n}$, where the channel between the transmitter and the receiver is typically modeled as a finite impulse response (FIR) filter. In this case, each tap is typically a complex-valued Gaussian random variable with exponentially decaying magnitudes. The tap delays correspond to the RMS delay spread and the channel type (e.g., low delay spread or flat fading, high delay spread or frequency selective fading). There is a new realization of the channel at every transmitted packet, if the channel remains invariant for the duration of the packet; otherwise, the variation of the channel is explicitly modeled in the signal detection. As mentioned earlier, there are $N_{RX} \times N_{TX}$ paths between the transmitter and the receiver where each channel is the sum of several FIR filters with different delay spreads. The channels may or may not be correlated. The MIMO schemes can be used with non-OFDM systems when the channel is modeled as flat fading such that:

$$y_l(nT) = \sum_{k=1}^{N_{TX}} h_{lk}x_k(nT) + n(nT) \tag{9.58}$$

An important question is to what extent MIMO techniques can increase the throughput and improve the reliability of the wireless communication systems.

This question can be answered by calculating the information theoretic capacity of a SISO channel, and comparing it with that of single-input multiple-output (SIMO), multiple-input single-output (MISO), and MIMO channels. For a memory-less SISO channel (i.e., one transmit and one receive antenna), the channel capacity is given by:

$$C_{SISO} = \log_2(1 + \gamma |h|^2) \tag{9.59}$$

where h is the normalized complex-valued gain/attenuation of a fixed wireless channel or that of a particular realization of a random channel, and $\gamma = E_s/N_0$ denotes the SNR at the receive antenna port. As the number of receive antennas increases, the statistics of channel capacity improve. Using N_{RX} receive antennas and one transmit antenna, a SIMO system is formed with a capacity given by (when the channel is unknown to the transmitter):

$$C_{SIMO} = \log_2(1 + \gamma \sum_{i=1}^{N_{RX}} |h_i|^2) \tag{9.60}$$

where h_i is the gain of the ith channel corresponding to the ith receive antenna. Note that increasing the value of N_{RX} results in a logarithmic increase in average channel capacity. It can be shown that knowledge of the channel at the transmitter in SIMO cases does not provide any capacity benefit. In the case of MISO or transmit diversity, where the transmitter does not typically have knowledge of the channel, the capacity is given by:

$$C_{MISO} = \log_2(1 + \frac{\gamma}{N_{TX}} \sum_{i=1}^{N_{TX}} |h_i|^2) \tag{9.61}$$

It is noted from the above equation that when the channel information is not available at the transmitter, $C_{MISO} < C_{SIMO}$. The power normalization factor N_{TX} ensures that the total transmit power is uniformly distributed among the transmit antennas. Furthermore, one notes the absence of an array gain in this case (MISO) compared to that of the receive diversity scenario where the energy of the multipath channels can be coherently combined. In addition, the MISO capacity has a logarithmic relationship with N_{TX} similar to that of a SIMO scheme. When the channel information is available at the transmitter, the capacity of a MISO system can approach that of a SIMO system [20].

The use of diversity at both transmitter and receiver sides gives rise to a MIMO system. The capacity of a MIMO system with N_{TX} transmit antennas and N_{RX} receive antennas is expressed as follows:

$$C_{MIMO} = \log_2\left(\det\left[\mathbf{I} + \frac{\gamma}{N_{TX}} \mathbf{H}\mathbf{H}^{\mathrm{H}}\right]\right) \tag{9.62}$$

where $\mathbf{I}$ is an $N_{RX} \times N_{RX}$ identity matrix and $\mathbf{H}$ is the $N_{RX} \times N_{TX}$ channel matrix. Note that both MISO and MIMO channel capacities are based on equal power and uncorrelated sources. It is demonstrated in the literature that the capacity of the MIMO channel increases linearly with $\min(N_{RX}, N_{TX})$ rather than logarithmically as in the case of MISO or SIMO channel capacity, since the determinant operator yields the product of non-zero eigenvalues of its channel-dependent matrix argument, each eigenvalue characterizing the SNR over a SISO eigen-channel. It will be shown later that the overall MIMO channel capacity is the sum of capacities of each of these SISO eigen-channels. The increase in capacity is dependent on properties of the channel eigenvalues. If the channel eigenvalues decay rapidly, then linear growth in capacity will not occur. However, the eigenvalues have a known limiting distribution and tend to be spaced out along the range of this distribution. Hence, it is unlikely that most eigenvalues are very small and the linear growth is indeed achieved [20].

Figures 9.104 and 9.105 show the impulse responses and the frequency responses of a large delay-spread MIMO channel (two transmit and two receive antennas) measured over the extent of a subframe. It is noted that channel characteristics vary from one path (between one transmit and one receive antenna) to another, and across time from one symbol to another. The information theoretic capacities of some MIMO channels with a different number of transmit and receive antennas are compared in Figure 9.106.

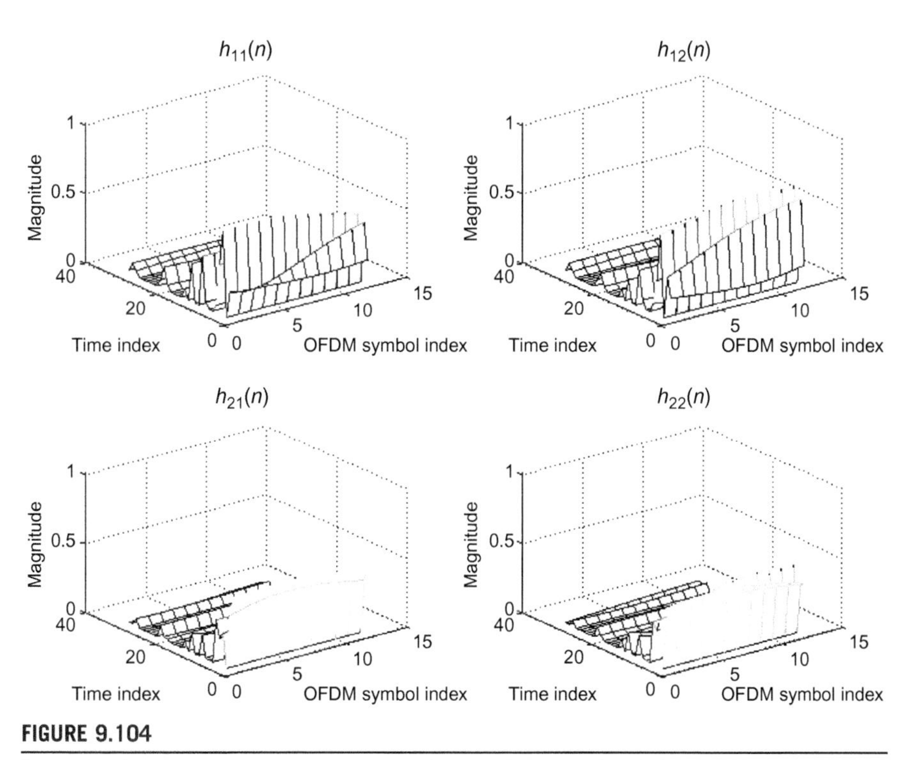

FIGURE 9.104

Example of multipath impulse responses of $N_{TX} = 2$, $N_{RX} = 2$ MIMO channel [141].

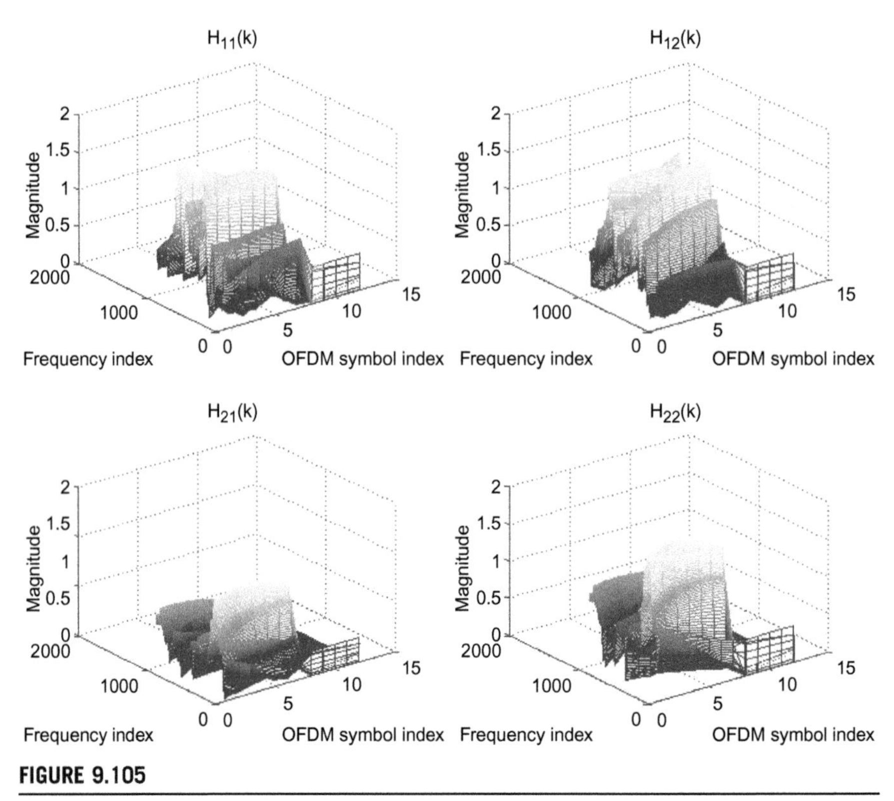

FIGURE 9.105

Example of multipath frequency responses of $N_{TX} = 2$, $N_{RX} = 2$ MIMO channel [141].

The capacity of the MIMO channel can be calculated under various conditions and different assumptions. Depending on whether the receiver has perfect channel knowledge, or whether the channel is flat fading or frequency-selective fading, different expressions for the channel capacity can be obtained. The MIMO channel capacity in Eq. (9.62) can be analyzed under two different assumptions: (1) the transmitter has no channel knowledge; and (2) the transmitter has perfect channel knowledge through feedback from the receiver or reciprocity of the downlink and uplink channels. Let us denote by Ξ the $N_{TX} \times N_{TX}$ covariance matrix of the channel input vector $\mathbf{x}$, and further assume that the channel is unknown to the transmitter, then it can be shown that the MIMO channel capacity can be written as [20]:

$$C_{MIMO} = \log_2(\det[\mathbf{I} + \mathbf{H}\Xi\mathbf{H}^{H}]) \tag{9.63}$$

where $\text{trace}(\Xi) \leq \gamma$ ensures that the total signal power does not exceed a certain limit. It can be shown that for equal transmit power and uncorrelated sources $\Xi = (\gamma/N_{TX})\mathbf{I}$, Eq. (9.63) becomes identical to Eq. (9.62). This is true when the channel matrix is unknown to the transmitter and the input signal is Gaussian-distributed, maximizing the mutual information. If the receiver measures and

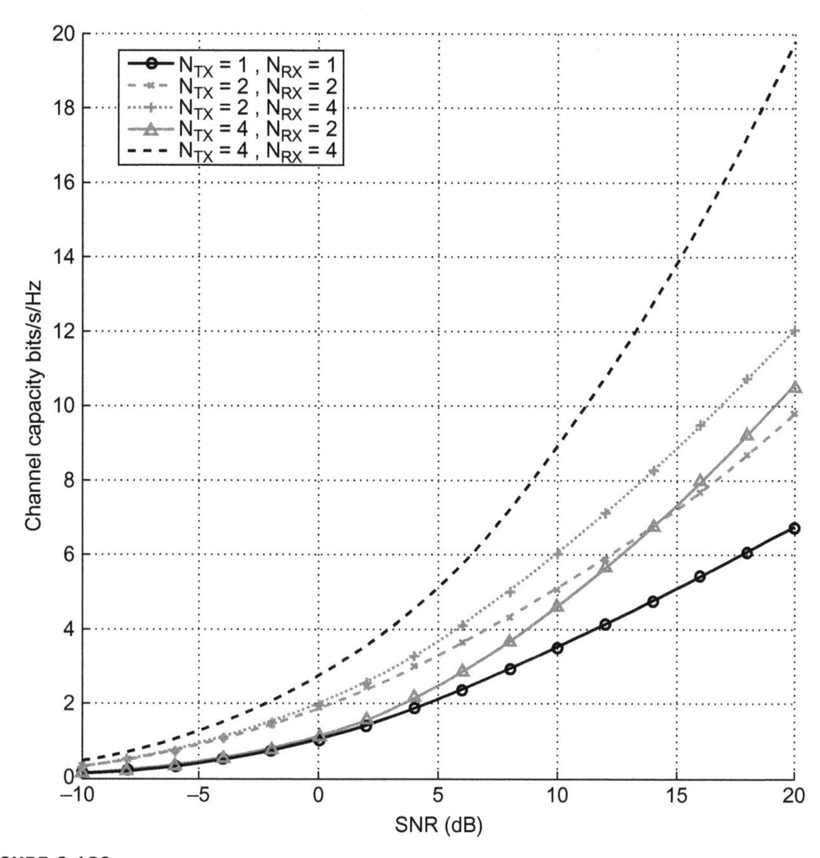

FIGURE 9.106

Comparison of the information theoretic capacity of some MIMO channels [141].

sends channel quality feedback or CSI to the transmitter, the covariance matrix Ξ is not proportional to the identity matrix, rather it is constructed from a water-filling algorithm. If one compares the capacity achieved assuming equal transmit-power and unknown channel with that of perfect channel estimation through feedback, then the capacity gain due to use of feedback is obtained. For the independent identically distributed Rayleigh fading scenario, the linear capacity growth discussed earlier will be observed. It is shown that Eq. (9.62) can be written as follows:

$$C_{\mathrm{MIMO}} = \sum_{i=1}^{\min(N_{\mathrm{TX}},N_{\mathrm{RX}})} \log_2\left(1+\frac{\gamma\lambda_i^2}{N_{\mathrm{TX}}}\right) \tag{9.64}$$

where $\lambda_i, i=1,2,\ldots,$ and $\min(N_{\mathrm{TX}},N_{\mathrm{RX}})$ are the non-zero eigenvalues of $\mathbf{HH}^{\mathrm{H}}$. We can decompose the MIMO channel into $K \le \min(N_{\mathrm{TX}},N_{\mathrm{RX}})$ equivalent parallel SISO channels using the singular value decomposition (SVD) theorem[21] [28,29]. Let $\mathbf{y} = \mathbf{Hx} + \mathbf{n}$ describe the input–output relationship of the MIMO channel where $\mathbf{y}$ is the output vector with N_{RX} components, $\mathbf{x}$ is the input vector with N_{TX} components, $\mathbf{n}$ is the additive noise vector with N_{RX} components, and $\mathbf{H}$ is the $N_{RX} \times N_{TX}$ channel matrix. Using the SVD theorem, we can show $\mathbf{H} = \mathbf{U\Sigma V}^{\mathrm{H}}$. Let $\widehat{\mathbf{x}} = \mathbf{V}^{\mathrm{H}}\mathbf{x}, \widehat{\mathbf{y}} = \mathbf{U}^{\mathrm{H}}\mathbf{y}$, and $\widehat{\mathbf{n}} = \mathbf{U}^{\mathrm{H}}\mathbf{n}$ denote the unitary transformation of the channel input and output and noise vectors, then it can be shown that $\widehat{\mathbf{y}} = \mathbf{\Sigma}\widehat{\mathbf{x}} + \widehat{\mathbf{n}}$. Since $\mathbf{U}$ and $\mathbf{V}$ are unitary matrices and $\mathbf{\Sigma} = \mathbf{diag}(\lambda_1, \lambda_2, \ldots, \lambda_{\min(N_{RX},N_{TX})}, 0, 0, \ldots, 0)$, it is clear that the capacity of this model is the same as the capacity of the model $\mathbf{y} = \mathbf{Hx} + \mathbf{n}$. However, $\mathbf{\Sigma}$ is a diagonal matrix with K non-zero elements on the main diagonal, thus $\widehat{y}_1 = \lambda_1\widehat{x}_1 + \widehat{n}_1, \ldots, \widehat{y}_K = \lambda_K\widehat{x}_K + \widehat{n}_K, \widehat{y}_{K+1} = \widehat{n}_{K+1}$. The latter equations are conceptually equivalent to K parallel SISO eigen-channels, each with signal power of $\lambda_i^2, i = 1, 2, \ldots, \min(N_{TX}, N_{RX})$. Hence, the MIMO channel capacity can be rewritten in terms of the eigenvalues of the input signal covariance matrix Ξ.

When the channel knowledge is available at the transmitter and receiver, then $\mathbf{H}$ is known and we can optimize the capacity over Ξ subject to the power constraint trace$(\Xi) \leq \gamma$. It is shown in the literature that the optimal Ξ in this case exists and is known as the water-filling solution. The channel capacity in this case is given by [20,26]:

$$C = \sum_{k=1}^{K} \log_2(\eta\lambda_i^2)^+$$
$$\gamma = \sum_{k=1}^{K} (\eta - \lambda_i^{-2})^+ \tag{9.65}$$

where $(x)^+ = x\forall x \geq 0, (x)^+ = 0\forall x < 0$ and η is a non-linear function of eigenvalues of the channel input covariance matrix. The effect of various channel conditions on the channel capacity has been extensively studied in the literature. For example, increasing the LoS signal strength at fixed SNR reduces capacity in Rician channels [20]. This can be explained in terms of the channel matrix rank

[21] The concept of decomposition of an $N \times N$ Hermitian matrix in terms of quadratic product of $N \times N$ unitary matrix composed of eigenvectors and $N \times N$ diagonal matrix of eigenvalues can be generalized to $M \times N$ complex-valued matrices of rank K. If $\mathbf{A}$ is an $M \times N$ where $(M > N)$ complex-valued matrix of rank K, then $\mathbf{A} = \mathbf{U\Sigma V}^{\mathrm{H}}$ denotes the singular value decomposition of $\mathbf{A}$. The $M \times M$ unitary matrix $\mathbf{U}$ is composed of the eigenvectors of $\mathbf{AA}^{\mathrm{H}}$, i.e., $\mathbf{U} = (\mathbf{u}_1, \mathbf{u}_2, \ldots, \mathbf{u}_m)$, where $\mathbf{AA}^{\mathrm{H}}\mathbf{u}_i = \sigma_i^2\mathbf{u}_i$. The $N \times N$ unitary matrix $\mathbf{V}$ is composed of eigenvectors of $\mathbf{A}^{\mathrm{H}}\mathbf{A}$, i.e., $\mathbf{V} = (\mathbf{v}_1, \mathbf{v}_2, \ldots, \mathbf{v}_n)$, where $\mathbf{A}^{\mathrm{H}}\mathbf{A}\mathbf{v}_i = \sigma_i^2\mathbf{v}_i$. The elements of the $M \times N$ matrix $\mathbf{\Sigma} = \begin{pmatrix} \Lambda & \mathbf{0} \\ \mathbf{0} & \mathbf{0} \end{pmatrix}$, $\Lambda = \mathbf{diag}(\sigma_1, \sigma_2, \ldots, \sigma_K)$ are the square roots of the eigenvalues of matrix $\mathbf{A}^{\mathrm{H}}\mathbf{A}$ which are referred to as the singular values of matrix $\mathbf{A}$. Therefore, matrix $\mathbf{A}$ may be written as $\mathbf{A} = \sum_{i=1}^{K} \sigma_i\mathbf{u}_i\mathbf{v}_i^{\mathrm{H}}$. The number of non-zero singular values of matrix $\mathbf{A}$ is the rank of $\mathbf{A}$. The singular values of matrix $\mathbf{A}$ are positive real numbers which satisfy $\sigma_1 \geq \sigma_2 \geq \ldots \geq \sigma_K > 0$ [29].

or through various eigenvalue properties. The issue of correlated fading is of considerable importance for implementations where the antennas are required to be closely spaced. The optimal water-filling allocation strategy is obtained when the power allocated to each spatial sub-channel is non-negative. Figure 9.107 illustrates an example where the water-filling algorithm is applied to an OFDM system assuming the number of sub-channels = 16, total power = -20 dBm, $N_0 = -80$ dBm, and bandwidth = 1 MHz. It can be observed that depending on the inverse C/N of each sub-channel, the amount of power allocated to each sub-channel differs, provided that the sum of sub-channel powers does not exceed the total available power. The power allocated to each sub-channel corresponds to parameters λ_i^2.

9.13.2 Instantaneous capacity, ergodic capacity, and outage capacity of MIMO systems

In the design of wireless communication systems, the main objective is to exploit the transmission schemes whose performance can approach the channel capacity

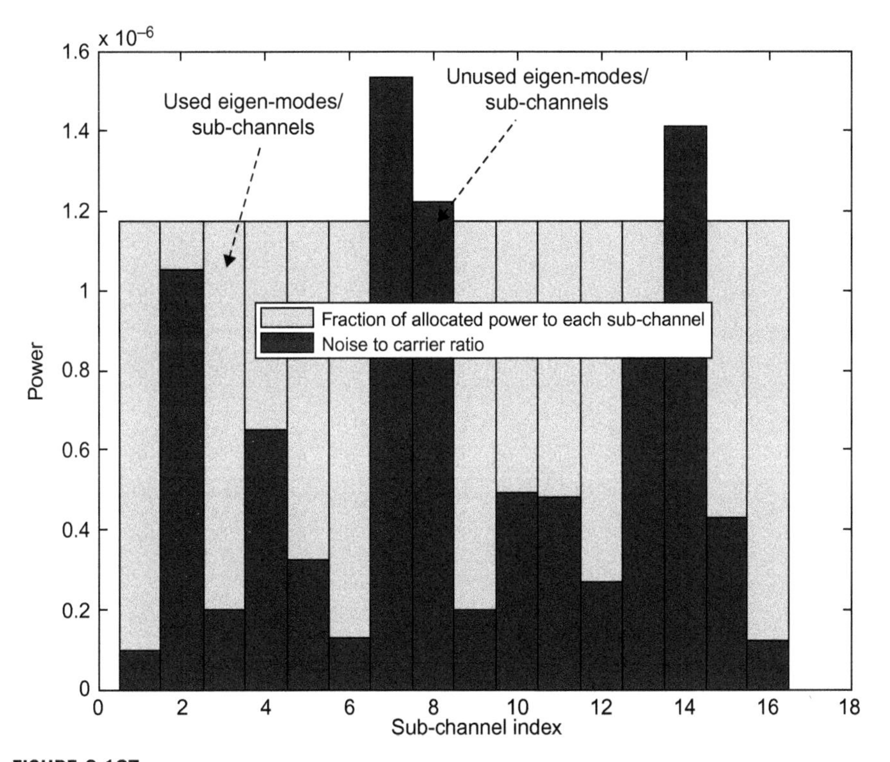

FIGURE 9.107

Example illustration of a water-filling algorithm in an OFDM system [141].

as much as possible. Therefore, it is important to understand the underlying concepts and various information theoretical definitions of channel capacity and what can be pragmatically achieved under realistic channel conditions and transceiver implementations.

Let us begin our concise study with the most generic definition of channel capacity. We denote the input and output of a memory-less SISO wireless channel with the random variables X and Y, respectively, the channel capacity is defined as [39]:

$$C = \max_{p(x)} I(X;Y) \tag{9.66}$$

where $I(X;Y)$ represents the mutual information between X and Y. Shannon's theorem [33] gives an operational meaning to the definition of the instantaneous capacity as the maximum number of bits we can transmit reliably over the channel with vanishing probability of error. The mutual information in Eq. (9.66) is maximized with respect to all possible transmit signal statistical distributions $p(x)$. Mutual information is a measure of the amount of information that one random variable contains about another variable. The mutual information between X and Y can also be written as $I(X;Y) = H(Y) - H(Y|X)$, where $H(Y|X)$ represents the conditional entropy between the random variables X and Y. The entropy of a random variable can be described as the measure of uncertainty in the random variable or the amount of information required on the average to describe the random variable. Thus, the mutual information representation of channel capacity can be described as the reduction in the uncertainty of one random variable due to the knowledge of the other. Note that the mutual information between X and Y depends on the properties of the channel through a channel matrix $\mathbf{H}$, and the properties of X through the probability distribution of X [166].

Throughout this section, it is assumed that the channel matrix $\mathbf{H}$ is random and that the receiver has perfect channel knowledge. It is also assumed that the channel is memory-less, i.e., for each use of the channel an independent realization of $\mathbf{H}$ is drawn. This means that the capacity can be computed as the maximum of the mutual information as defined earlier. The results are also valid when $\mathbf{H}$ is generated by an ergodic process, because as long as the receiver observes the $\mathbf{H}$ process, only the first order statistics are needed to determine the channel capacity.

The ergodic (mean) capacity of a random channel with $N_{RX} = N_{TX} = 1$ and an average transmit power constraint P_T can be expressed as:

$$C_{ergodic} = \mathscr{E}_{\mathbf{H}}\{\max_{p(x):P \le P_T} I(X;Y)\} \tag{9.67}$$

where P is the average power of a single codeword, transmitted over the channel, and $\mathscr{E}_{\mathbf{H}}\{.\}$ denotes the expectation over all channel realizations. Compared to the generic definition, the capacity of the channel is now defined as the maximum of the mutual information between the input and the output over all statistical distributions on the input that satisfy the power constraint. In general, the capacity of a random MIMO channel with power constraint P_T can be expressed as [20]:

$$C_{ergodic} = \mathscr{E}_{\mathbf{H}}\{\max_{p(\mathbf{x}):\mathrm{trace}(\mathbf{\Phi})\leq P_T} I(\mathbf{x};\mathbf{y})\} \tag{9.68}$$

where $\mathbf{\Phi} = \mathscr{E}[\mathbf{x}\mathbf{x}^H]$ is the covariance matrix of the transmit signal vector $\mathbf{x}$. The total transmit power is limited to P_T irrespective of the number of transmit antennas. For a fading channel, the channel matrix $\mathbf{H}$ is a stochastic process, and hence the associated channel capacity $C(\mathbf{H})$ is also a random variable. In this case, the ergodic channel capacity is defined as the average of instantaneous channel capacity over the distribution of $\mathbf{H}$. The ergodic channel capacity of the MIMO transmission scheme is given by:

$$C_{ergodic} = \mathscr{E}\left\{\max_{\mathrm{trace}(\mathbf{\Phi})=N_{TX}} \log_2\left(\det\left[\mathbf{I} + \frac{\gamma}{N_{TX}}\mathbf{H}\mathbf{\Phi}\mathbf{H}^H\right]\right)\right\} \tag{9.69}$$

where $\mathbf{\Phi}$ denotes the $N_{TX} \times N_{TX}$ covariance matrix of the channel input vector $\mathbf{x}$. According to information theoretic concepts, this capacity cannot be achieved unless channel coding is employed across an infinite number of independently fading blocks. Let us focus on the case of perfect CSI at the receiver side and no CSI at the transmitter side. Obviously, this implies that the maximization of the latter equation is now more restricted than in the previous case. Nevertheless, it has been shown in the literature that the optimal signal covariance matrix has to be chosen according to $\mathbf{\Phi} = \mathbf{I}$. This means that the antennas should transmit uncorrelated streams with the same average power. With this result, the ergodic MIMO channel capacity reduces to [20,24]:

$$C_{ergodic} = \mathscr{E}\left\{\log_2\left(\det\left[\mathbf{I} + \frac{\gamma}{N_{TX}}\mathbf{H}\mathbf{H}^H\right]\right)\right\} \tag{9.70}$$

Clearly, this is not the Shannon capacity in a true sense, since as mentioned before and with perfect channel knowledge at the transmitter one can choose a signal covariance matrix that outperforms $\mathbf{\Phi} = \mathbf{I}$ case. Nevertheless, we refer to the expression in Eq. (9.70) as the ergodic channel capacity with CSI at the receiver and no CSI at the transmitter.

The capacity under channel ergodicity was defined as the average of the maximal value of the mutual information between the transmitted and the received signals, where the maximization was carried out with respect to all possible transmit signal statistical distributions. Another measure of channel capacity that is frequently used is outage capacity. With outage capacity, the channel capacity is associated to an outage probability. Capacity is treated as a random variable which depends on the channel instantaneous response and remains constant during the transmission of a finite-length coded block of information. If the channel capacity falls below the outage capacity, there is no possibility that the transmitted block of information can be decoded with no errors, no matter which coding scheme is employed. The probability that the capacity is less than the outage capacity denoted by C_{outage} is ρ. This can be expressed in mathematical terms by $p(C < C_{outage}) = \rho$. In this case, the latter expression represents an upper bound since there is a finite probability ρ that the channel capacity is less than the outage capacity. It can also be written as a lower bound, representing the case where there is a finite probability $(1-\rho)$ that the channel capacity is higher than C_{outage}, which means $p(C > C_{outage}) = 1-\rho$. In other words, since the MIMO instantaneous channel capacity is a random variable, it is meaningful to consider its statistical distribution. A useful measure of its statistical behavior is the outage capacity. Outage analysis quantifies the level of performance (in this case capacity) that is guaranteed with a certain level of reliability. The ρ% outage capacity $C_{outage}(\rho)$ is defined as the information rate that is guaranteed for $(100-\rho)$% of the channel realizations ($p(C < C_{outage}) = \rho$). The outage capacity is often a more relevant measure than the ergodic channel capacity, because it describes in some way the quality of the channel. This is due to the fact that the outage capacity measures how the instantaneous rate supported by the channel is distributed, in terms of probability. Thus, if the rate supported by the channel is spread over a wide range, the outage capacity for a fixed probability level may become small, whereas the ergodic channel capacity may be high [166].

9.13.3 Spatial multiplexing and diversity

The main objective of space–time coding schemes is to exploit the inherent spatial (multipath) diversity in MIMO channels through properly designed space–time codes. In this section, two transmit diversity schemes, namely the Alamouti scheme and CDD, are described where both realize full spatial diversity of the channel without requiring channel knowledge at the transmitter. In the Alamouti scheme, we consider a MIMO channel with two transmit antennas and any number of receive antennas. The Alamouti scheme can be described as follows. Two different complex-valued data symbols s_1 and s_2 are transmitted simultaneously from antenna ports 1 and 2, respectively, during the first symbol period. In the second symbol period, symbols $-s_2^*$ and s_1^* are sent from antenna ports 1 and 2, respectively. It must be noted that the rate equals one in the Alamouti scheme since two independent data symbols are transmitted over two symbol periods. Let us consider the following case with two transmit antennas and one receive antenna, thereby we have two channels H_1 and H_2. The received signal at the first and second symbol period can be expressed as follows [58]:

$$\begin{aligned} y_1 &= H_1 s_1 + H_2 s_2 + n_1 \\ y_2 &= -H_1 s_2^* + H_2 s_1^* + n_2 \end{aligned} \tag{9.71}$$

Let $\mathbf{r} = (y_1, y_2^*)^H$, $\mathbf{s} = (s_1, s_2)^H$, $\mathbf{n} = (n_1, n_2)^H$, and $\mathbf{H} = \begin{bmatrix} H_1 & H_2 \\ H_2^* & -H_1^* \end{bmatrix}$, the relationship between the output and the input of the channel can be written as $\mathbf{r} = \mathbf{Hs} + \mathbf{n}$ where multiplying both sides of the latter equation by $\mathbf{H}^H$ yields

$\tilde{\mathbf{r}} = \mathbf{H}^{\mathrm{H}}\mathbf{H}\mathbf{s} + \tilde{\mathbf{n}}$. Since $\mathbf{H}$ is orthogonal, i.e., $\mathbf{H}^{\mathrm{H}}\mathbf{H} = \alpha\mathbf{I}$ where $\alpha = |H_1|^2 + |H_2|^2$; dividing both sides of the equation by α yields [58]:

$$\begin{aligned} \frac{\tilde{r}_1}{|H_1|^2 + |H_2|^2} &= s_1 + \frac{H_1^* n_1}{|H_1|^2 + |H_2|^2} + \frac{H_2 n_2}{|H_1|^2 + |H_2|^2} \\ \frac{\tilde{r}_2}{|H_1|^2 + |H_2|^2} &= s_2 + \frac{H_2^* n_1}{|H_1|^2 + |H_2|^2} + \frac{-H_1 n_2}{|H_1|^2 + |H_2|^2} \end{aligned} \tag{9.72}$$

It is assumed that the channels are independent and identically distributed with frequency-flat fading and remain constant over (at least) two consecutive symbol periods. Appropriate processing at the receiver, as shown in Eqs. (9.71) and (9.72), collapses the vector channel into a scalar channel for either of the transmitted data symbols, where $\tilde{\mathbf{r}}$ is the processed received signal corresponding to transmitted symbol $\mathbf{s}$, and $\tilde{\mathbf{n}}$ is the processed noise. Even though the CSI knowledge is not available at the transmitter, the Alamouti scheme achieves $2N_{\mathrm{RX}}$ order diversity. We note that array gain is realized only at the receiver since the transmitter does not have the CSI. The Alamouti scheme may be extended to channels with more than two transmit antennas through orthogonal space–time block coding. The link-level performance of the Alamouti scheme is compared to the SISO and SIMO cases in Figure 9.108 assuming BPSK-modulated input symbols and Rayleigh fading channel.

Delay diversity is an alternative method in which the same symbol is transmitted at different time instances such that each symbol experiences different fading. The delay diversity converts spatial diversity to frequency diversity by transmitting the data signal from the first antenna port and a delayed replica of the signal from the second antenna port. Assuming two transmit antennas and one receive antenna and that the delay induced by the second antenna equals one symbol period, the effective channel seen by the data signal is a frequency-selective fading SISO channel with impulse response $h(k) = h_1\delta(k) + h_2\delta(k-1)$, where h_1 and h_2 are complex-valued flat-fading channels. We note that the effective channel resembles a two-path SISO channel with independent fading paths and equal average path energy. An ML detector will realize full second-order diversity at the receiver. If the same signal is transmitted from two or more transmit antennas, a beamforming effect is created where in some directions the signal is attenuated, while in other directions it may be amplified. The CDD scheme mitigates this problem by transmitting cyclically shifted signals from the transmit antennas. Let us consider an OFDM symbol with constellation d_k at the kth frequency index, the time-domain signal corresponding to this symbol which is transmitted from antenna port 1 can be expressed as follows:

$$s_1(t) = \mathrm{Re}\left\{ e^{j\omega_c t} \sum_{k=-(N_{\mathrm{FFT}}-1)/2}^{(N_{\mathrm{FFT}}-1)/2} d_k\, e^{j2\pi k(t-t_g)/T_u} \right\} \quad mT_u \le t \le (m+1)T_u \tag{9.73}$$

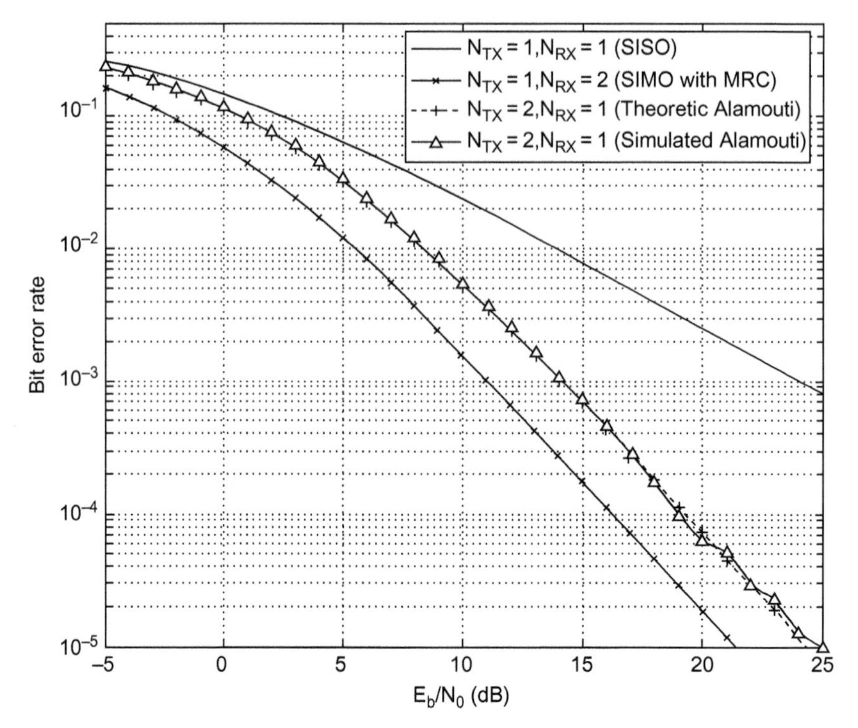

FIGURE 9.108

Performance of the Alamouti scheme [141].

Using the CDD scheme, the transmitted signal from the second antenna port is given by:

$$s_2(t) = \mathrm{Re}\left\{ e^{j\omega_c t} \sum_{k=-(N_{\mathrm{FFT}}-1)/2}^{(N_{\mathrm{FFT}}-1)/2} d_k \, e^{j2\pi k(t-t_g-T_{CDD})/T_u} \right\} \quad mT_u \le t \le (m+1)T_u \quad (9.74)$$

This is equivalent to a cyclic rotation of the signal in the time-domain. The implementation of the CDD scheme is simplified, if the cyclic shift T_{CDD} is an integer multiple of the sampling time. At the receiver, the effect of CDD at the kth index is of receiving through the channel $H_1(k) + \exp(j2\pi k T_{\mathrm{CDD}}/T_u)H_2(k)$. The advantage over the case of $H_1(k) + H_2(k)$, which is the result of transmission of the same signal over the two channels, is the prevention of the beamforming effect, i.e., directed transmission in one direction and undesired emission in the other directions.

The main objective of spatial multiplexing, as opposed to space–time diversity coding, is to maximize transmission data rate. Therefore, $\min(N_{\mathrm{TX}}, N_{\mathrm{RX}})$ independent data symbols are transmitted per symbol period such that the spatial rate is $\min(N_{\mathrm{TX}}, N_{\mathrm{RX}})$. There are several encoding options that can be used in conjunction with spatial multiplexing to achieve the increased data rate. Single codeword and multi-codeword transmission are the prominent encoding options that are typically used in wireless communication systems. In a multi-codeword scheme, the bit stream to be transmitted is de-multiplexed into separate data streams where each stream is processed through the transmission chain, i.e., independent temporal encoding, symbol mapping, and interleaving, and is then transmitted from the corresponding antennas. The antenna to stream mapping remains static over time. The spatial rate is $\min(N_{TX}, N_{RX})$ and the overall signaling rate is $n_{\text{modulation-order}} r_b \min(N_{TX}, N_{RX})$, where r_b denotes the channel coder input bit rate and $n_{\text{modulation-order}}$ denotes the QAM modulation order. The multi-codeword scheme can at most achieve N_{RX} order diversity, since any given information bit is transmitted from only one transmit antenna and received by N_{RX} receive antennas. While this is a source of sub-optimality of the multi-codeword architecture, it does simplify receiver design. The coding gain achieved by the multi-codeword scheme depends on the coding gain of the temporal code. Furthermore, a maximum array gain of N_{RX} can be realized using the multi-codeword scheme.

In the single-codeword architecture, the bit stream undergoes temporal encoding, symbol mapping, and interleaving after which it is de-multiplexed into $\min(N_{TX}, N_{RX})$ streams transmitted from the individual antennas. This form of encoding can achieve full $N_{TX} \times N_{RX}$ order diversity, provided that the temporal code is designed properly, since each information bit can be spread across all the transmit antennas. Nevertheless, the single-codeword scheme requires joint decoding of the sub-streams, which increases receiver implementation complexity compared to the multi-codeword scheme where the individual data streams can be decoded separately. The spatial rate of the single-codeword is $\min(N_{TX}, N_{RX})$, and the overall signaling rate is $n_{\text{modulation-order}} r_b \min(N_{TX}, N_{RX})$. The coding gain achieved by the single-codeword scheme will depend on the temporal channel coding characteristic and a maximum array gain of N_{RX} can be achieved [24].

Space-frequency block coding is the frequency-domain version of the space–time block codes (STBC). This family of codes is designed to achieve diversity gain by transmitting orthogonal streams which yields optimal SNR with a linear receiver. Since, in the OFDM systems, the number of OFDM symbols in a subframe may often be an odd number, while STBC (e.g., Alamouti code) operates on pairs of adjacent symbols in the time-domain, the application of STBC would not be straightforward. For SFBC transmission, the adjacent sub-carriers on each OFDM symbol are paired, encoded, and transmitted from two base station antenna ports where the encoding is defined as follows:

$$\mathbf{W} = \begin{bmatrix} x_k & x_{k+1} \\ -x_{k+1}^* & x_k^* \end{bmatrix} \tag{9.75}$$

where $(x_k, -x_{k+1}^*)$ and (x_{k+1}, x_k^*) are transmitted from antenna ports 1 and 2, respectively. Note that the spatial rate of the SFBC codes in this case is one.

9.13.4 MIMO receivers

An orthogonal space–time block coding scheme transforms a vector-detection problem into a number of scalar-detection problems. Similar concepts can be applied to frequency-selective fading MIMO channels, inspiring receiver design techniques such as zero-forcing (ZF), MMSE estimation, and (optimal) ML sequence detection [20]. The use of transmit diversity techniques such as CDD or frequency-offset diversity transforms a MISO channel into a number of SISO channels, and allows the application of SISO receiver architectures. For a general space–time trellis code, a vector Viterbi decoder can be employed. The space–time trellis coding in general provides improved performance over orthogonal space–time block coding at the expense of receiver complexity. The problem encountered by a receiver for spatial multiplexing is the presence of multi-stream interference, since the signals sent from different transmit antennas interfere with each other. Note that in spatial multiplexing, different data streams are transmitted in the same channel and occupy the same resources in time and frequency. For the sake of simplicity, we restrict our attention to the case where $N_{RX} \geq N_{TX}$.

9.13.4.1 Maximum likelihood receiver

The optimal detection for spatially multiplexed signals is to use maximum likelihood criterion. The ML receiver performs vector decoding and is optimal in the sense of minimizing the error probability. Assuming equally likely, temporally uncoded vector symbols, the ML receiver generates an estimate of the transmitted signal vector as follows:

$$\hat{\mathbf{s}} = \arg\min_{\mathbf{s}} \left\| \mathbf{y} - \sqrt{\frac{E_s}{N_{TX}}}\mathbf{Hs} \right\|^2 \tag{9.76}$$

where the minimization is performed over all possible transmit vector symbols $\mathbf{s}$. Let Z denote the alphabet size of the scalar constellation transmitted from each antenna, a complete implementation requires an exhaustive search over a total of $Z^{N_{TX}}$ vector symbols or hypotheses, making the decoding complexity of this receiver exponentially grow with the number of transmit antennas. Recent development of fast ML detection algorithms reduces the computational complexity of the ML decoder. The ML receiver realizes N_{RX} diversity order for the multi-codeword scheme and $N_{RX} \times N_{TX}$ diversity order for the single-codeword scheme [20,25].

9.13.4.2 Successive interference cancelation receiver

An alternative non-linear approach to detection of spatially multiplexed signals is successive interference cancelation (SIC) assuming that the spatially multiplexed signals are separately coded (i.e., multi-codeword transmission). As shown in Figure 9.109, in an SIC receiver, one of the spatially multiplexed signals is initially demodulated and decoded. Upon successful decoding, the data is

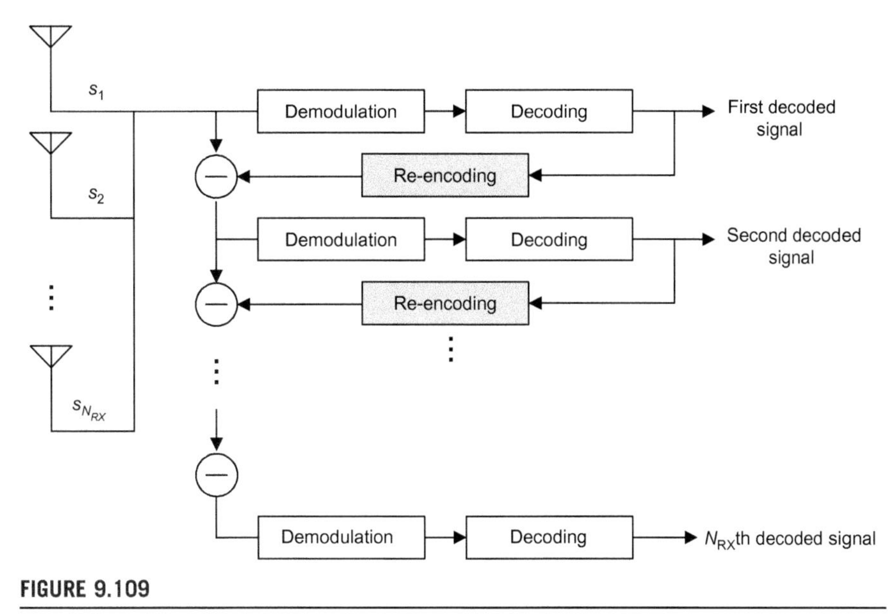

FIGURE 9.109

Structure of a successive interference cancelation MIMO receiver.

re-encoded and subtracted from the received signals. The second spatially multiplexed signal can then be demodulated and decoded in the absence of interference from the first detected signal. The subtraction would ideally improve the signal-to-interference ratio. The procedure is iterated until all spatially multiplexed signals are demodulated and decoded. It is noted that the first signals to be decoded using this approach are subject to higher interference levels compared to those that are decoded later in the process. Therefore, the ith signal to be decoded must be more robustly coded than the $(i+1)$th signal. This can be achieved in multi-codeword transmission by applying different modulation and coding schemes to different spatially multiplexed signals. The latter scheme is often referred to as per-antenna rate control (PARC) [20,25,42].

9.13.4.3 Zero-forcing receiver

The decoding complexity of the ML receiver can be reduced using a linear filter to separate the transmitted data streams and then independently decode each stream. The ZF and MMSE linear detectors are examples of such receivers. The ZF weighting matrix (filter) that decomposes the received signal into its component streams is given by the following equation [20,42]:

$$\mathbf{G}_{ZF} = \sqrt{\frac{N_{TX}}{E_s}}\mathbf{H}^{\dagger} = \sqrt{\frac{N_{TX}}{E_s}}(\mathbf{H}^{H}\mathbf{H})^{-1}\mathbf{H}^{H} \tag{9.77}$$

where $\mathbf{G}_{ZF}$ is an $N_{TX} \times N_{RX}$ matrix that inverts the channel matrix and $\mathbf{H}^{\dagger}$ denotes the Moore–Penrose inverse[22] of the channel matrix $\mathbf{H}$. The output of the ZF receiver is obtained as follows:

$$\mathbf{y} = \mathbf{s} + \sqrt{\frac{N_{TX}}{E_s}}\mathbf{H}^{\dagger}\mathbf{n} \tag{9.78}$$

which shows that the ZF receiver decomposes the channel matrix into parallel scalar channels with additive spatially colored noise. Each scalar channel is then decoded independently irrespective of noise correlation across the processed streams. The ZF receiver transforms the joint decoding problem into single stream decoding problems, thereby significantly reducing receiver complexity. This complexity reduction is achieved at the expense of noise enhancement, which in general results in a significant performance degradation (compared to the ML decoder). The diversity order achieved by each of the individual data streams equals $N_{RX} - N_{TX} + 1$. Figure 9.110 illustrates the link-level performance of a 2×2 MIMO system with a ZF receiver. It is shown that the performance is close to that of a SISO system, since the diversity order is one.

9.13.4.4 Minimum mean square error receiver

The MMSE receiver balances multi-stream interference mitigation and noise enhancement effect, and minimizes the total error. The MMSE weighting matrix (filter) is given by the following expression [20,42]:

$$\mathbf{G}_{MMSE} = \arg\min_{\mathbf{G}} \mathscr{E}\left\{ ||\mathbf{G}\mathbf{y} - \mathbf{s}||^2 \right\} = \sqrt{\frac{N_{TX}}{E_s}}\left(\mathbf{H}^{H}\mathbf{H} + \frac{N_{TX}N_0}{E_s}\mathbf{I}\right)^{-1}\mathbf{H}^{H} \tag{9.79}$$

where $\mathbf{I}$ is an $N_{RX} \times N_{RX}$ identity matrix. In low signal-to-interference plus noise ratios (i.e., SINR < 0 dB), the performance of the MMSE receiver approaches that of a matched-filter receiver, outperforming the ZF detector which tends to enhance noise. In high signal-to-interference plus noise ratios, however, the performance of the MMSE receiver approaches that of a ZF receiver and therefore realizes $N_{RX} - N_{TX} + 1$ order diversity for each data stream.

---

[22] Given a matrix $\mathbf{A}$ and vector $\mathbf{b}$ and equation $\mathbf{Ax} = \mathbf{b}$, we find the vector $\mathbf{x}$ that minimizes $\|\mathbf{Ax} - \mathbf{b}\|$. Using the orthogonality principle, the error is taken perpendicular to estimation vector such that $\mathbf{x}^H\mathbf{A}^H(\mathbf{Ax} - \mathbf{b}) = 0$ which implies that $\mathbf{A}^H\mathbf{Ax} = \mathbf{A}^H\mathbf{b}$ (since $\mathbf{x} \neq 0$), or $\mathbf{x} = (\mathbf{A}^H\mathbf{A})^{-1}\mathbf{A}^H\mathbf{b}$. One should note that $\mathbf{Ax} = \mathbf{A}(\mathbf{A}^H\mathbf{A})^{-1}\mathbf{A}^H\mathbf{b}$. The matrix $\mathbf{A}^{\dagger} = (\mathbf{A}^H\mathbf{A})^{-1}\mathbf{A}^H$ is defined as the Moore–Penrose inverse of the matrix $\mathbf{A}$. If $\mathbf{A}^H\mathbf{A}$ is not invertible, the SVD algorithm is used to calculate the matrix pseudo-inverse. In that case, we may write $\mathbf{A} = \mathbf{V}_K\mathbf{\Sigma}_K\mathbf{U}_K^H$ and seek $\mathbf{x}$ such that $||\mathbf{V}_K\mathbf{\Sigma}_K\mathbf{U}_K^H\mathbf{x} - \mathbf{b}||$ is minimized. Using the orthogonality principle, we have $\mathbf{x}^H\mathbf{U}_K\mathbf{\Sigma}_K\mathbf{V}_K^H(\mathbf{V}_K\mathbf{\Sigma}_K\mathbf{U}_K^H\mathbf{x} - \mathbf{b})$ or (assuming $\mathbf{x} \neq 0$) $\mathbf{U}_K\mathbf{\Sigma}_K^2\mathbf{U}_K^H\mathbf{x} - \mathbf{U}_K\mathbf{\Sigma}_K\mathbf{V}_K^H\mathbf{b} = 0$ such that $\mathbf{\Sigma}_K^2\mathbf{U}_K^H\mathbf{x} - \mathbf{\Sigma}_K\mathbf{V}_K^H\mathbf{b} = 0$, multiplying both sides by $\mathbf{\Sigma}_K^{-2}$, yields $\mathbf{U}_K^H\mathbf{x} - \mathbf{\Sigma}_K^{-1}\mathbf{V}_K^H\mathbf{b} = 0$. The latter is true when $\mathbf{x} = \mathbf{\Sigma}_{i=1}^{K}\left(\frac{1}{\sigma_i}\mathbf{v}_i^H\mathbf{b}\right)\mathbf{u}_i$; therefore $\mathbf{x} = \mathbf{U}_K\mathbf{\Sigma}_K^{-1}\mathbf{V}_K^H\mathbf{b}$ [29].

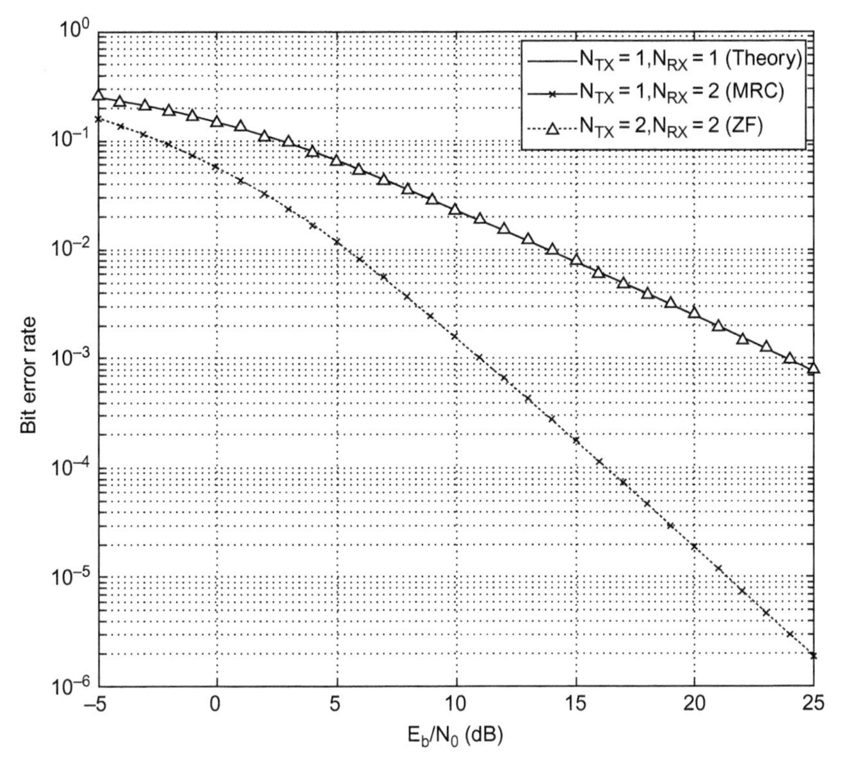

FIGURE 9.110

Link-level performance of 2 × 2 MIMO system with ZF equalizer (Rayleigh channel and BPSK modulation).

9.13.4.5 Maximal ratio combining receiver

In maximal ratio combining (MRC), the output is a weighted sum of all receiver branches, thus the coefficients $\alpha_i, i = 1, 2, \ldots,$ and N_{RX} are all non-zero. The signals are co-phased and $\alpha_i = \beta_i\, e^{-j\theta_i}, i = 1, 2, \ldots, N_{RX}$ where θ_i is the phase of the received signal on the ith branch (Figure 9.111). The magnitude of the combined output can be written as $y = \sum_{i=1}^{N_{RX}} \beta_i r_i$. Let us assume that the noise power spectral density is $N_0/2$ in each branch, which yields the combined noise at the output $n = \sum_{i=1}^{N_{RX}} \beta_i^2 N_0/2$. The SNR at the output of the combiner is given as:

$$\gamma = \frac{1}{N_0} \frac{\left(\sum_{i=1}^{N_{RX}} \beta_i r_i\right)^2}{\sum_{i=1}^{N_{RX}} \beta_i^2} \tag{9.80}$$

The objective is to find $\alpha_i, i = 1, 2, \ldots, N_{RX}$ which maximize the output SNR. It can be understood that branches with higher SNR are weighted more than those

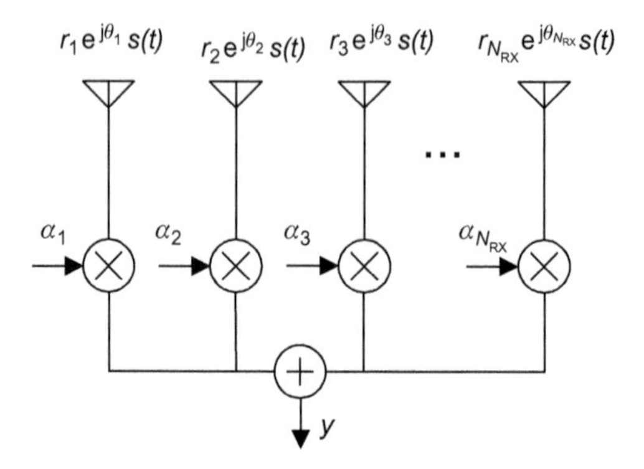

FIGURE 9.111

A generic linear combining receiver.

with lower SNR; therefore, the coefficients $\alpha_i, i = 1, 2, \ldots,$ and N_{RX} are proportional to the branch SNRs. The coefficients are found by taking partial derivatives of the output SNR expression and solving the resulting equations for the optimal weights which yields $\alpha_i = r_i^2/N_0, i = 1, 2, \ldots, N_{RX}$ and the output SNR will be the sum of the SNRs of the receiving branches. Therefore, the output SNR and the corresponding array gain increase linearly with the number of the diversity branches (i.e., the number of receive antennas).

9.13.4.6 MMSE-interference rejection combining receiver

The interference rejection combining (IRC) receiver is an effective method for improving the cell-edge user throughput, since it attempts to suppress the inter-cell interference. The IRC receiver is typically based on the minimum mean square error criteria, which requires channel estimation and covariance matrix estimation including the inter-cell interference with high accuracy. The gain from the IRC receiver is tied to the estimation of the interference signal, i.e., the covariance matrix, in terms of the downlink user throughput performance in a multi-cell environment. We define the $N_{RX} \times 1$ received signal vector $\mathbf{y}(k, l)$ corresponding to the kth sub-carrier and the lth OFDM symbol as follows: $\mathbf{y}(k, l) = \sum_{i=0}^{N_{cell}-1} \mathbf{H}_i(k, l)\mathbf{W}_{TX}^i(k, l)\mathbf{s}_i(k, l) + \mathbf{n}(k, l)$, where $\mathbf{H}_i(k, l)$ represents the $N_{RX} \times N_{TX}$ channel matrix between the ith cell and the UE, $\mathbf{W}_{TX}^i(k, l)$ represents the $N_{TX} \times N_l$ precoding matrix of the ith cell, $\mathbf{s}_i(k, l)$ denotes the $N_l \times 1$ signal vector of the ith cell, and $\mathbf{n}(k, l)$ is the $N_{RX} \times 1$ noise vector with variance σ^2. In this context, N_{TX}, N_{RX}, N_l, and N_{cell} are the number of transmit antennas of each cell, the number of receive antennas at the UE, the number of transmission layers for the UE, i.e., transmission ranks and total number of cells in the model, respectively. The zeroth cell is assumed to be the serving cell for the UE.

The received signal vector at the UE denoted by $\hat{\mathbf{s}}_0(k,l)$ is detected using a $N_l \times N_{RX}$ weighting matrix $\mathbf{W}^0_{RX}(k,l)$ expressed as $\hat{\mathbf{s}}_0(k,l) = \mathbf{W}^0_{RX}(k,l)\mathbf{y}(k,l)$. Based on MMSE criterion, the weighting matrix $\mathbf{W}^0_{RX}(k,l)$ is obtained by $\mathbf{W}^0_{RX}(k,l) = P_0\overline{\mathbf{G}}^H_0(k,l)\overline{\mathbf{R}}^{-1}_{yy}$ and $\overline{\mathbf{G}}_0(k,l) = \overline{\mathbf{H}}_0(k,l)\mathbf{W}^0_{TX}$ where P_0, $\overline{\mathbf{G}}_0$, $\overline{\mathbf{H}}_0$, and $\overline{\mathbf{R}}_{yy}$ denote the transmission power of the serving cell, the composite channel from the serving cell, the $N_{RX} \times N_{TX}$ channel matrix of the serving cell, and the covariance matrix of the received signal, respectively [180].

The baseline MMSE receiver is defined as a receiver that only suppresses the inter-layer interference, i.e., separates multiple data streams that achieve the maximum SINR for each data stream within a cell. The inter-cell interference is assumed to be AWGN at each receive antenna, making the covariance matrix a diagonal matrix. In the special case of rank-1 transmission, the MMSE receiver is equivalent to the MRC receiver. When a composite channel from the serving cell is obtained, the covariance matrix without the inter-cell interference can be represented by the composite channel from the serving cell, and the interference power from other cells and noise power. Therefore, the respective MMSE weighting matrices with ideal and realistic channel estimation are obtained as follows [180]:

$$\begin{aligned}
\mathbf{W}^{\text{Ideal}}_{\text{MMSE}}(k,l) &= P_0\mathbf{G}^H_0(k,l)[P_0\mathbf{G}_0(k,l)\mathbf{G}^H_0(k,l) + \boldsymbol{\Omega} + \sigma^2_N\mathbf{I}]^{-1} \\
\boldsymbol{\Omega} &= diag(\sigma^2_{I_0}, \sigma^2_{I_1}, \ldots, \sigma^2_{I_{N_{RX}}}) \\
\mathbf{W}^{\text{Realistic}}_{\text{MMSE}}(k,l) &= P_0\hat{\mathbf{G}}^H_0(k,l)(P_0\hat{\mathbf{G}}_0(k,l)\hat{\mathbf{G}}^H_0(k,l) + \boldsymbol{\Omega} + \sigma^2_N\mathbf{I})^{-1}
\end{aligned} \tag{9.81}$$

where $\mathbf{G}_0(k,l)$, $\hat{\mathbf{G}}_0(k,l)$, $\sigma^2_{I_j}$, and σ^2_{N} denote the ideal composite channel matrix, the estimated composite channel matrix based on the DM-RSs, which are precoded in the same manner as the data signals, the interference power from other cells at the jth receive antenna, and the noise power, respectively [180].

The MMSE-IRC receiver can suppress not only the inter-layer interference, but also the inter-cell interference when the degrees of freedom at the receiver are high, i.e., the number of receive antennas is higher than that of the desired transmission rank. If the composite channels from N_{cell} cells are perfectly known at the receiver, the ideal IRC weighting matrix is obtained as follows:

$$\mathbf{W}^{\text{Ideal}}_{\text{IRC}}(k,l) = P_0\mathbf{G}^H_0(k,l)\left(\sum_{i=0}^{N_{\text{cell}}-1} P_i\mathbf{G}_i(k,l)\mathbf{G}^H_i(k,l) + \sigma^2_{\text{N}}\mathbf{I}\right)^{-1} \tag{9.82}$$

However, it is assumed that the composite channel matrices from other $N_{\text{cell}} - 1$ cells cannot be estimated at the receiver; therefore, the covariance matrix including the inter-cell interference is obtained by statistical averaging of the receiver signal, yielding:

$$\mathbf{R}_{yy} = \mathscr{E}[\mathbf{y}(k,l)\mathbf{y}^{\text{H}}(k,l)] \tag{9.83}$$

The resulting IRC weighting matrices in the case of ideal and realistic channel estimations can be obtained as follows:

$$\begin{aligned}\mathbf{W}_{\mathrm{IRC}}^{\mathrm{Ideal}}(k,l) &= P_0\mathbf{G}_0^H(k,l)(\mathscr{E}[\mathbf{y}(k,l)\mathbf{y}^{\mathrm{H}}(k,l)])^{-1}\\ \mathbf{W}_{\mathrm{IRC}}^{\mathrm{Realistic}}(k,l) &= P_0\hat{\mathbf{G}}_0^H(k,l)(\mathscr{E}[\mathbf{y}(k,l)\mathbf{y}^{\mathrm{H}}(k,l)])^{-1}\end{aligned} \tag{9.84}$$

The accuracy of the IRC weighting depends on the accuracy of the channel estimation and covariance matrix. Therefore, in order to obtain an accurate covariance matrix, the covariance matrix must be averaged using the received samples which have gone through the same precoding matrix at the transmitter side and the same channel matrix with high correlation. More specifically, in order to ensure the use of the same precoding matrix, one can assume that the covariance matrix is averaged over one resource block using all samples except for the control signaling, the CRSs, and the CSI-RSs. Note that LTE Rel-10 allows the use of different precoding matrices for different resource blocks. Furthermore, it must be noted that some signals, such as control signaling, CRSs, and CSI-RSs, are transmitted without precoding [180].

The time-domain averaging is given more priority over frequency-domain averaging within a resource block because the channel variation within a sub-frame is smaller than that within a resource block in the frequency-domain. Two averaging methods, i.e., time-domain only, and time- and frequency-domain averaging are performed where the resulting covariance matrices are calculated as follows [180]:

$$\begin{aligned}\mathbf{R}_{yy}^{\mathrm{time}} &= \frac{1}{N_{\mathrm{time}}}\sum_{l}\mathbf{y}(k,l)\mathbf{y}^{\mathrm{H}}(k,l)\\ \mathbf{R}_{yy}^{\mathrm{time-frequency}} &= \frac{1}{N_{\mathrm{time-frequency}}}\sum_{k}\sum_{l}\mathbf{y}(k,l)\mathbf{y}^{\mathrm{H}}(k,l)\end{aligned} \tag{9.85}$$

where N_{time} and $N_{\mathrm{time-frequency}}$ denote the average number of samples for the time-domain and time- and frequency-domain averaging, respectively [180].

9.13.4.7 Sphere decoding

In its search for the ML solution, the sphere decoder (SD) evaluates all transmit signal vectors $\mathbf{x}$ fulfilling the following criterion: $||\mathbf{y}-\mathbf{H}\mathbf{x}||^2 < R^2$, where R is the search radius of a hyper-sphere. It is obvious that the selection of R is a critical issue largely influencing the complexity of any SD algorithm. Choosing R too large would lead to a sphere containing a large number of hypotheses (also referred to as candidates), leading to high detection complexity. On the other hand, choosing R too small would result in an empty sphere and the search having to be restarted with an increased radius, eventually leading to similar problems. The search for candidates fulfilling the above inequality is done by back-substitution algorithms.

Toward this goal, the Cholesky[23] or QR[24] decompositions of the channel matrix **H** may be equivalently used. Note that using a Cholesky decomposition may be advantageous in systems with more transmit than receive antennas, since the size of the upper triangular matrix **R** is limited by $\min(N_{RX}, N_{TX})$, which for $N_{RX} < N_{TX}$ leads to an under-determined problem for the QR implementation of the SD [181].

In the following, we focus on the symmetric case, i.e., $N_{RX} = N_{TX}$, and use a QR decomposition of **H**, for practicality of implementation. With $\mathbf{H} = \mathbf{QR}$, where **R** is upper-triangular and **Q** is unitary, the above inequality may be rewritten as follows:

$$||\mathbf{y}-\mathbf{Hx}||^2 < R^2 \Rightarrow ||\mathbf{Q}^H\mathbf{y}-\mathbf{Rx}||^2 < R^2 \Rightarrow ||\mathbf{y}'-\mathbf{Rx}||^2 < R^2 \qquad (9.86)$$

The above expression can be written as follows:

$$\sum_{i=l}^{N_{TX}} \left| y_i' - \sum_{j=i}^{N_{TX}} r_{ij}x_j \right|^2 < R^2 \quad l = 1, 2, \ldots, N_{TX} \qquad (9.87)$$

The detection process starts with the last layer $l = N_{TX}$, for which the latter expression reduces to $|y'_{N_{TX}} - r_{N_{TX}N_{TX}}x_{N_{TX}}|^2 < R^2$ and continues until the first layer is detected. This process is quite similar to SIC techniques, i.e., the signals from previously detected layers are subtracted from the received signal before detection within the current layer is performed. However, in SD, detection at the ith layer essentially takes on the following form:

$$|c_{ki} - x_i|^2 < \frac{R_{ki}^2}{r_{ii}^2} \qquad (9.88)$$

where c_{ki} is the search center and R_{ki} is the remainder search radius for the currently considered (incomplete) candidate c. Both SIC and SD methods depend on the estimates for previously detected layers, and hence vary from one candidate to another. The above inequality implies that not only a single, but several constellation points may be selected. The SD receiver hence performs its search in a tree like structure, which motivates the search for appropriate enumeration

[23]Given a symmetric positive-definite matrix **H**, the Cholesky decomposition is an upper triangular matrix **U** with strictly positive diagonal entries such that $\mathbf{H} = \mathbf{U}^H\mathbf{U}$ [28,29]. Alternatively, if **H** has real entries and is symmetric (or more generally, has complex-valued entries and is Hermitian) and positive-definite, then **H** can be decomposed as $\mathbf{H} = \mathbf{LL}^H$, where **L** is a lower triangular matrix with strictly positive diagonal entries and $\mathbf{L}^H$denotes the conjugate transpose of **L**. The Cholesky decomposition is unique, i.e., given a Hermitian, positive-definite matrix **H**, there is only one lower triangular matrix **L** with strictly positive diagonal entries such that $\mathbf{H} = \mathbf{LL}^H$. In the special case that **H** is a symmetric positive-definite matrix with real entries, then **L** has real entries as well [28,29].

[24]Given a matrix **H**, its QR decomposition is a matrix decomposition of the form $\mathbf{H} = \mathbf{QR}$ where **R** is an upper triangular matrix and **Q** is a unitary matrix, i.e., one satisfying $\mathbf{Q}^T\mathbf{Q} = \mathbf{I}$ where $\mathbf{Q}^T$ is the transpose of **Q** and **I** is the identity matrix [28,29].

strategies. Since we eventually consider coded transmission where extracting good quality soft outputs require finding a predefined minimum number of candidates $N_{c_{\min}}$, we take a different approach toward this problem. Our interest lies in ensuring a certain minimum number of entries N_c at the bottom of the search tree. With the entries of $\mathbf{H}$ and the noise $\mathbf{n}$ at the receiver being random variables, N_c is obviously a random variable, as well. If we allow for all hypotheses fulfilling the original inequality to be tested, we obtain an upper bound on the expected complexity. Note that whenever an empty sphere is declared, the search must be restarted with an increased radius. Since the overall detection complexity is the sum of all attempts to find the ML point, it is unfavorable to choose a small initial search radius and then increase it stepwise until $N_{c_{\min}}$ candidates are found.

For our symmetric $N_{RX} = N_{TX}$ MIMO system, the initial value for R reduces to $R^2 = 2\sigma^2 K N_{RX}$. Since in our system we normalize $2\sigma^2 = 1$ and then scale the transmitter signal constellation, this is further reduced to $R^2 = K N_{RX}$ where the critical point lies in the selection of K. It was found empirically that it would be beneficial to use a small initial search radius at low SNR that is subsequently increased with increasing SNR. The results of evaluations provided in the literature suggest that using $K = \frac{1}{60} L R_c \sqrt{\frac{E_b}{N_0}}$, where L is the number of bits per symbol and R_c is the code rate, provides a good choice that minimizes detection complexity in a wide SNR range. The middle factor ensures that the scaling of the transmitter signal set is taken into account and the last term derives from the notion that in the high SNR region, finding too many candidates becomes less probable and that the number of events with an empty sphere should be minimized. The radius can be increased by a factor of 1.5, whenever an empty sphere is declared [181].

9.13.4.8 CSI feedback

The reciprocity principle in wireless communication indicates that the channel from an antenna A to another antenna B is identical to the transpose of the channel from antenna B to antenna A, provided that the two channels are measured at the same time, the same frequency, and the same antenna locations. This principle further suggests that the transmitter can obtain information of the downlink (A to B) channel from the uplink (B to A) channel measurements, which the receiver at A can conduct. This information can involve the instantaneous channel or other channel parameters, including the channel statistics. In real full-duplex communications, however, the downlink and uplink channels cannot use identical frequency, time, and spatial instances. The reciprocity principle may still hold approximately, if the difference in any of these dimensions is relatively small compared to the channel variation across the referenced dimension.

Let us consider a scenario in which the base station measures the uplink channel during reception and uses this measurement for the CSIT of the next transmission in the downlink. In voice communications, the downlink and uplink channels to and from all the users operate in consecutive time slots. Therefore, uplink channel measurements can be made regularly using reference signals. These measurements periodically refresh the CSIT. In data communications, the downlink and uplink channels may not operate continuously; hence, specially scheduled uplink transmissions for channel measurements, known as channel sounding, are used. A subset of the users, for whom CSIT is required, is scheduled to send sounding signals. The sounding signals are orthogonal among simultaneously scheduled users, using orthogonal sub-carriers in OFDM or orthogonal codes in CDMA systems. Sounding is an effective channel measurement technique for systems with multiple transmit antennas at the base station.

One complication in using reciprocity methods is that the principle only applies to the radio channel between the antennas, while in practice the channel is measured and used at the baseband processor. Different transmit and receive RF hardware chains; therefore, become part of the downlink and uplink channels. Since these chains have different frequency transfer characteristics, reciprocity requires transmit–receive chain calibration to equalize the effects of the two chains. Calibration is expensive and has made open-loop methods less attractive in practice.

Another method of obtaining CSIT is using feedback from the UE. The channel information is measured at the receiver at B during the downlink (A to B) transmission and then sent to the transmitter at A on the uplink. In practice, the downlink transmission from a base station includes reference signals, received by all active users. These users, when instructed by the serving base station, can measure their respective channels. The users then periodically or aperiodically send their measured channel information in the uplink channel to the base station to be used in calculation of their CSIT.

The feedback transmission can either be scheduled separately or piggybacked on ongoing transmissions. In data communications, CSIT may be needed for only a subset of users, who are then scheduled to transmit their channel information. Feedback is not limited by the reciprocity requirements. However, it imposes additional system overhead by consuming radio resources. Techniques to reduce the amount of feedback have been a subject of intense study, e.g., designing vector codebooks, quantizing channel information, or selecting only the important information. Furthermore, feedback information is susceptible to channel variation due to the delay in the feedback loop. The effectiveness of feedback depends on this delay and the channel Doppler spread. For a fast time-varying channel in mobile communications, feedback techniques are usually effective up to a certain mobile speed, depending on the carrier frequency, the transmission frame length, and the round trip time. The effects of feedback delay and error have been analyzed for various precoding techniques in LTE/LTE-Advanced, revealing potentially severe performance degradation. Therefore, the optimal use of feedback must account for the information quality.

The trade-off among transmission rate, error rate, and SINR for the scenarios where the transmitter has no channel knowledge and the receiver has perfect knowledge of CSI has been studied in the literature. If the MIMO channel is assumed to be block fading, the length of the transmitted codewords is less than or equal to the channel block length, and if the channel was perfectly known to the transmitter, we could choose a signaling rate equal to or less than channel capacity and achieve error-free transmission. The coding scheme to achieve capacity consists of performing modal decomposition which decouples the MIMO channel into parallel SISO channels and then using ideal SISO channel coding. In practice, turbo codes can achieve near MIMO channel capacity performance. If the channel is unknown to the transmitter, modal decomposition is not possible. Since the channel is selected randomly according to a given fading distribution, there will always be a non-zero probability that a given transmission rate is not supported by the channel. We assume that the transmitted codeword is decoded successfully if the rate is at or below the mutual information, assuming a spatially white transmit covariance matrix, associated with the given channel realization. A decoding error is declared if the rate exceeds the mutual information. Therefore, if the transmitter does not have any knowledge of the channel, the PER would be equal to the outage probability associated with the transmission rate. We define the diversity order for a given transmission rate as [78]:

$$d(R) = -\lim_{\gamma \to \infty} \frac{\log[P_e(R, \gamma)]}{\log(\gamma)} \tag{9.89}$$

where $P_e(R, \gamma)$ is the PER corresponding to transmission rate R and SNR γ. Thus, the diversity order is the magnitude of the slope of the PER plotted as a function of the SNR on the logarithmic scale [78].

The differential feedback is a different approach to the enhancement of feedback accuracy. The differential feedback exploits the correlation between precoding matrices adjacent in time or frequencies. It only feeds back the difference between the current and the previous beamforming matrices. If the channel variation between two feedbacks is small, quantization codewords are concentrated in a small region on Grassmannian manifold[25] and not uniformly distributed over the entire beamforming space. For low-mobility users, the differential feedback may improve the feedback accuracy without increasing the number of feedback bits. The feedback starts initially and restarts periodically by sending a single feedback that fully describes the precoder. Differential feedbacks follow the start and restart feedbacks. The start and restart feedbacks employ a codebook that is

[25]A Grassmann manifold is a certain collection of vector subspaces of a vector space. In particular, $g_{n,k}$ is the Grassmann manifold of k-dimensional subspaces of the vector space $\mathbb{R}^n$. It has a natural manifold structure as an orbit space of the Stiefel manifold $v_{n,k}$ of orthonormal k-frames in G^n. One of the main features of Grassmann manifolds is that they are classifying spaces for vector bundles. A manifold is a topological space that is locally Euclidean (i.e., around every point, there is a neighborhood that is topologically the same as the open unit ball in $\mathbb{R}^n$). The Grassmannian $Gr(n, k)$ is the set of k-dimensional subspaces in an n-dimensional vector space. For example, the set of lines $Gr(n, k)$ is projective space. The real Grassmannian (as well as the complex Grassmannian) are examples of manifolds [31,32].

self-contained, e.g., the base codebook. Figure 9.112 illustrates the operation of differential feedback across time [83].

An example of differential feedback is a transformation-based differential codebook. Let t, $\mathbf{D}(t)$, and $\mathbf{V}(t)$ denote the feedback index, the corresponding feedback matrix, and the corresponding precoder, respectively. The sequential index is reset to 0 at $T_{\max}+1$. The index for the start and restart feedbacks are 0. Let $\mathbf{U}$ be a vector or a matrix and $\mathbf{Q_U}$ be a rotation matrix determined by $\mathbf{U}$. The precoders of the subsequent differential feedbacks are denoted as $\mathbf{D}(t)$ for $t=1,2,\ldots,T_{\max}$ and the corresponding precoders are given as follows:

$$\begin{aligned}\mathbf{V}(t) &= \mathbf{Q}_{\mathbf{V}(t-1)}\mathbf{D}(t)\\ \mathbf{D}(t) &= \mathbf{Q}^{H}_{\mathbf{V}(t-1)}\mathbf{V}(t)\end{aligned} \quad t=0,1,2,\ldots,T_{\max} \tag{9.90}$$

where the rotation matrix $\mathbf{Q}_{\mathbf{V}(t-1)}$ is an $N_{\mathrm{TX}}\times N_{\mathrm{TX}}$ unitary matrix derived from the previous precoder matrix $\mathbf{V}(t-1)$, and N_{TX} denotes the number of transmit antennas. The dimension of the feedback matrix $\mathbf{D}(t)$ is $N_{\mathrm{TX}}\times N_l$, where N_l denotes the number of spatial streams. The feedback matrix $\mathbf{D}(t)$ can be viewed as a description of time/frequency correlation properties which does not change over the entire frequency band and a long period of time, and $\mathbf{Q}_{\mathbf{V}(t-1)}$ can be viewed as a description of narrowband and short-term channel properties which correspond to a specific sub-band and subframe. Due to correlation between beamforming matrices across time and frequency, feedback overhead can be reduced, if the correlation property can be efficiently exploited. Let us take the time-domain correlation as an example to explain the differential feedback scheme. The hypothetical operations at the UE and at the eNB can be described as follows [18]:

1. Derivation of the differential codeword matrix at the UE, $\mathbf{D}=\hat{\mathbf{Q}}^{\mathrm{H}}_{\mathbf{V}(t-1)}\mathbf{V}(t)=[\hat{\mathbf{V}}(t-1),\hat{\mathbf{V}}^{\perp}(t-1)]^{\mathrm{H}}\mathbf{V}(t)$.
2. Quantization of the differential codeword matrix at the UE, $\hat{\mathbf{D}}=\arg\max_{\mathbf{D}_i\in\Omega}\|\mathbf{D}^{\mathrm{H}}\mathbf{D}_i\|_F$ where $\|\mathbf{D}^{\mathrm{H}}\mathbf{D}_i\|_F$ denotes the Chordal distance between

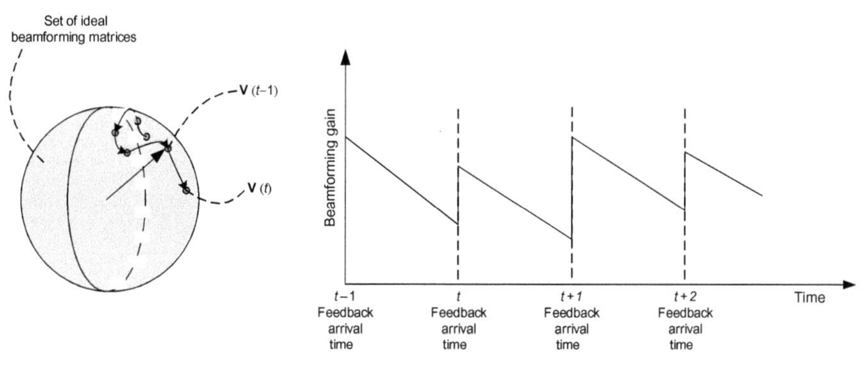

FIGURE 9.112

Example illustration of differential feedback and effect on beamforming gain.

the two codeword matrices $\mathbf{D}$ and $\mathbf{D}_i$. Therefore, the optimal codebook is selected based on the maximum distance between $\mathbf{D}$ and the codeword matrices contained in the codebook $\boldsymbol{\Omega}$.

3. Beamforming matrix reconstruction at the eNB, $\hat{\mathbf{V}}(t) = \hat{\mathbf{Q}}_{\mathbf{V}}(t-1)\hat{\mathbf{D}} = [\hat{\mathbf{V}}(t-1), \hat{\mathbf{V}}^{\perp}(t-1)]\hat{\mathbf{D}}$.
4. Beamforming at the eNB, $\mathbf{y} = \mathbf{H}\hat{\mathbf{V}}(t)\mathbf{s} + \mathbf{n}$.

In the above equations, $\mathbf{V}(t)$ and $\hat{\mathbf{V}}(t)$ denote the ideal and quantized $N_{TX} \times N_l$ beamforming matrices at time instant t, respectively, the columns of $\hat{\mathbf{V}}^{\perp}(t-1)$ have unit norm and are orthogonal to $\hat{\mathbf{V}}(t-1)$ where $\hat{\mathbf{V}}^{\perp}(t-1)$ can be calculated from $\hat{\mathbf{V}}(t-1)$ by a Householder transformation,[26] $\mathbf{D}_i$ is the codeword matrix of the differential codebook denoted by $\boldsymbol{\Omega}$. The quantization criterion maximizes the received signal power. Other criteria such as channel capacity or mean squared error can also be used. If $N_l = 1$, $\mathbf{V}(t-1)$ would be a vector (with unit norm) and the rotation matrix is given by $\mathbf{Q}_{\mathbf{V}(t-1)} = \mathbf{I} - 2\varphi\varphi^{H}; ||\varphi|| > 0$ wherein $\varphi = \exp(-j\theta)\mathbf{V}(t-1) - \varepsilon$ and θ is the phase of the first entry of $\mathbf{V}(t-1)$ and $\varepsilon = (1, 0, 0, \ldots, 0)^{T}$.

9.13.5 Precoding and beamforming

Multiple antennas at the transmitter and receiver can be used to achieve array and diversity gain instead of capacity gain. In this case, the same symbol weighted by a complex-valued scale factor is sent from each transmit antenna so that the input covariance matrix has unit rank. This scheme is referred to as beamforming. It must be noted that there are two conceptually and practically different classes of beamforming: (1) direction-of-arrival beamforming (i.e., adjustment of transmit or receive antenna directivity); and (2) eigen-beamforming (i.e., a mathematical approach to maximize signal power at the receive antenna based on certain criterion). In this section, we only consider eigen-beamforming schemes.

A classic eigen-beamforming scheme usually performs linear, single-layer, complex-valued weighting on the transmitted symbols such that the same signal is transmitted from each transmit antenna using appropriate weighting factors. In this scheme, the objective is to maximize the signal power at the receiver output. When the receiver has multiple antennas, the single-layer beamforming cannot

[26]In linear algebra, Householder transformation is defined as a linear transformation that describes a reflection about a plane or hyper-plane containing the origin. The Householder transformations are widely used in linear algebra to perform QR decompositions. This is a linear transformation given by the Householder matrix $\mathbf{Q} = \mathbf{I} - 2\mathbf{u}\mathbf{u}^{H}$ where $\mathbf{I}$ denotes the identity matrix and $\mathbf{u}$ is given as a column unit vector. The Householder matrix has the following properties: (1) it is hermitian, i.e., $\mathbf{Q} = \mathbf{Q}^{H}$; and (2) it is unitary, i.e., $\mathbf{Q}^{-1} = \mathbf{Q}^{H}$ [31,32].

simultaneously maximize the signal power at every receive antennas, hence, precoding is used for multi-layer beamforming in order to maximize the throughput of a multi-antenna system. Precoding is a generalized beamforming scheme to support multi-layer transmission in a MIMO system. Using precoding, multiple streams are transmitted from the transmit antennas with independent and appropriate weighting per each antenna such that the throughput is maximized at the receiver output.

Let us begin our concise study of eigen-beamforming and MIMO precoding using a simplified model. In a single-user MIMO (SU-MIMO) system, identity matrix precoding (for open-loop) and SVD precoding (for closed-loop) can be used to achieve link-level MIMO channel capacity. In addition, random unitary precoding can achieve the open-loop MIMO channel capacity with no signaling overhead in the uplink. The SVD precoding, on the other hand, has been shown to achieve the MIMO channel capacity when CSI is signaled to the transmitter. In a precoded SU-MIMO system with N_{TX} transmit antennas and N_{RX} receive antennas, the input–output relationship can be described as $\mathbf{y} = \mathbf{HWs} + \mathbf{n}$ where $\mathbf{s} = (s_1, s_2 \ldots, s_M)^T$ is an $M \times 1$ vector of normalized complex-valued modulated symbols, $\mathbf{y} = (y_1, y_2, \ldots, y_{N_{RX}})^T$ and $\mathbf{n} = (n_1, n_2, \ldots, n_{N_{RX}})^T$ are the $N_{RX} \times 1$ vectors of received signal and noise, respectively, $\mathbf{H}$ is the $N_{RX} \times N_{TX}$ complex-valued channel matrix, and $\mathbf{W}$ is the $N_{TX} \times M$ linear precoding matrix. The superscript "T" denotes the transpose operator.

In the receiver, a hard-decoded symbol vector $\hat{\mathbf{s}}$ is obtained by decoding the received vector $\mathbf{y}$ using a vector decoder, assuming perfect knowledge of the channel and the precoding matrices. We assume that the entries of $\mathbf{H}$ are independent and identically distributed according to $\mathbb{Z}(0, 1)$, and the entries of noise vector $\mathbf{n}$ are independent and identically distributed according to $\mathbb{Z}(0, N_0)$. The input vector $\mathbf{s}$ is assumed to be normalized, thus $\mathrm{E}[\mathbf{ss}^H] = \mathbf{I}$ where $\mathbf{I}$ is an identity matrix. Let us further assume that precoding matrix $\mathbf{W}$ is unitary, thus $\mathbf{WW}^H = \mathbf{I}$. The receiver selects a precoding matrix $\mathbf{W}_i, i = 1, 2, \ldots, N_{\text{codebook}}$ from a finite set of quantized precoding matrices $\Omega = \{\mathbf{W}_1, \mathbf{W}_2, \ldots, \mathbf{W}_{N_{\text{codebook}}}\}$ and sends the index of the chosen precoding matrix back to the transmitter over a low-delay feedback channel. There are two important issues concerning the above precoding scheme: (1) optimal selection criterion for choosing a precoding matrix from set $\boldsymbol{\Omega}$; and (2) design of codebook $\boldsymbol{\Omega}$. The matrix $\mathbf{W}_i, i = 1, 2, \ldots, N_{\text{codebook}}$ can be selected from $\boldsymbol{\Omega}$ by using either of the following optimization criteria [82]: (1) minimizing the trace of the mean squared error (MMSE-trace selection); (2) minimizing the determinant of the mean squared error (MMSE-determinant selection); (3) maximizing the minimum singular value of $\mathbf{HW}$ (singular value selection); (4) maximizing the instantaneous capacity (capacity selection); or (5) maximizing the minimum received symbol vector distance (minimum distance selection). The above selection criteria may be evaluated at the receiver using a full search over all matrices in $\boldsymbol{\Omega}$. Using distortion functions based on the selection criteria, it can be shown that the codebook $\boldsymbol{\Omega}$ is designed using Grassmannian subspace packing [82]. If MMSE-trace, singular value, or minimum distance selection is used, the codebook is designed such that $\varepsilon = \min_{\mathbf{W}_i \neq \mathbf{W}_j} ||\mathbf{W}_i\mathbf{W}_i^{\mathrm{H}} - \mathbf{W}_j\mathbf{W}_j^{\mathrm{H}}||_2$[27] is maximized. If the MMSE-determinant or capacity selection optimization method is used, the codebook is designed such that $\varepsilon = \min_{\mathbf{W}_i \neq \mathbf{W}_j} \arccos|\det(\mathbf{W}_i^{\mathrm{H}}\mathbf{W}_j)|$ is maximized [82].

The MIMO codebooks are designed based on a trade-off between performance and complexity. The following are some desirable properties of the codebooks:

1. Low-complexity codebooks can be designed by choosing the elements of each constituent matrix or vector from a small binary set, e.g., a four alphabet $(\pm 1, \pm j)$ set, which eliminates the need for matrix or vector multiplication. In addition, the nested property of the codebooks can further reduce the complexity of CQI calculation when rank adaptation is performed.
2. The base station may perform rank overriding which results in significant CQI mismatch, if the codebook structure cannot adapt to it. A nested property with respect to rank overriding can be exploited to mitigate the mismatch effects.
3. Power amplifier balance is taken into consideration when designing codebooks with constant modulus property, which may eliminate unnecessary increase in PAPR.
4. Good performance for a wide range of propagation scenarios, e.g., uncorrelated, correlated, and dual-polarized antennas, is expected from the codebook design algorithms. A DFT-based codebook is optimal for linear array with small antenna spacing since the vectors match with the structure of the transmit array response. Additionally, with an optimal selection of the matrices and the entries of the codebook (rotated block diagonal structure), significant gains can be obtained in dual-polarized scenarios.
5. Low feedback and signaling overhead are desirable from an operation and performance perspective.
6. Low memory requirement is another design consideration for the MIMO codebooks.

An alternative approach to codebook-based precoding is to use differential codebook precoding, in which the transmitter computes the precoding matrix by using a differential codebook precoding matrix specified by the receiver and the previous precoding matrix used in the transmitter, thus $\mathbf{W}_k = \mathbf{C}_{m_{\mathrm{opt}}}\mathbf{W}_{k-1}$ where each differential codebook precoding matrix $\mathbf{C}_m, m = 1, 2, \ldots, N'_{\mathrm{codebook}}$ is an $N_{\mathrm{TX}} \times N_{\mathrm{TX}}$ matrix. The receiver selects the optimum differential precoding matrix which maximizes the instantaneous capacity as follows [83]:

$$\text{Optimum} - \text{codebook} - \text{index}: m_{\mathrm{opt}} = \arg\max_{\mathrm{m}} \log_2 \det\left(\mathbf{I} + \frac{\gamma}{N_{\mathrm{TX}}}\mathbf{H}_k\mathbf{W}\mathbf{C}'_m\mathbf{C}'^{\mathrm{H}}_m\mathbf{H}_k^{\mathrm{H}}\right) \tag{9.91}$$

[27]The Euclidean norm of square matrix $\mathbf{A}$ is defined as $||\mathbf{A}||_2 = \sup_{||\mathbf{x}||=1}||\mathbf{A}\mathbf{x}||_2$. The spectral norm of matrix $\mathbf{A}$ is the largest singular value of $\mathbf{A}$ or the square root of the largest eigenvalue of the positive-semi-definite matrix $\mathbf{A}^{\mathrm{H}}\mathbf{A}$; i.e., $||\mathbf{A}||_2 = \sqrt{\lambda_{\max}(\mathbf{A}^{\mathrm{H}}\mathbf{A})}$ where $\mathbf{A}^{\mathrm{H}}$ denotes the conjugate transpose of $\mathbf{A}$ [44].

where $\{\mathbf{C}'_1,\mathbf{C}'_2,\ldots,\mathbf{C}'_{N'_{\text{codebook}}}\} = \{\mathbf{C}_1\mathbf{W}_{k-1},\mathbf{C}_2\mathbf{W}_{k-1},\ldots,\mathbf{C}_{N'_{\text{codebook}}}\mathbf{W}_{k-1}\}$ and γ is the signal-to-noise ratio. The initial precoding matrix can be set as the first M columns of the $N_{\text{TX}} \times N_{\text{TX}}$ identity matrix. One advantage of the differential precoding scheme is that the candidate precoding matrices $\{\mathbf{C}'_1,\mathbf{C}'_2,\ldots,\mathbf{C}'_{N'_{\text{codebook}}}\}$ are refined in each precoding update instance, which virtually increases the codebook size to realize finer quantization of the channel [83]. Since the differential precoding schemes compute the precoding matrix by using the previous precoding matrix, different criteria should be used to generate precoding matrices. In a slow fading environment where precoding is more effective, the channel variation during the precoding matrix update interval is relatively small. In this case, the ideal codebook matrix which is multiplied by the previous precoding matrix should be close to the identity matrix. Thus, the codebook design strategy is to optimize the codebook such that the codebook includes quasi-diagonal matrices. More specifically, codebook matrices are computed from two unitary matrices as $\mathbf{C}_{\text{m}} = \mathbf{G}^{\text{H}}(t)\mathbf{G}(t + \Delta t)$, where $\mathbf{G}(t) \in \mathbb{C}^{N_{\text{TX}} \times N_{\text{TX}}}$ is a random unitary matrix. Note that since both $\mathbf{G}(t)$ and $\mathbf{G}(t + \Delta t)$ are unitary matrices, each codebook matrix $\mathbf{C}_{\text{m}} \in \mathbb{C}^{N_{\text{TX}} \times N_{\text{TX}}}$ also has unitary properties, and small perturbation Δt produces quasi-diagonal matrices whose diagonal entries are more dominant than other entries. A set of codebooks $\{\mathbf{C}_1,\mathbf{C}_2,\ldots,\mathbf{C}_{N'_{\text{codebook}}}\}$ are generated by using various Δt values where the codebook is optimized by capacity maximization criterion.

In SVD precoding, assuming the perfect knowledge of the CSI, the channel matrix is decomposed as $\mathbf{H} = \mathbf{U}\boldsymbol{\Sigma}\mathbf{V}^{\text{H}}$, where $\mathbf{U}$ and $\mathbf{V}$ are unitary matrices (i.e., $\mathbf{U}^{\text{H}}\mathbf{U} = \mathbf{I}$, $\mathbf{V}^{\text{H}}\mathbf{V} = \mathbf{I}$) and $\boldsymbol{\Sigma}$ is a diagonal matrix containing eigenvalues of the channel matrix. If the relationship between the output and input of the channel is described as $\mathbf{y} = \mathbf{Hs} + \mathbf{n}$, then by replacing the channel matrix with its SVD form and multiplying both sides by $\mathbf{U}^{\text{H}}$, one can write $\mathbf{U}^{\text{H}}\mathbf{y} = \mathbf{U}^{\text{H}}\mathbf{U}\boldsymbol{\Sigma}\mathbf{V}^{\text{H}}\mathbf{s} + \mathbf{U}^{\text{H}}\mathbf{n}$. Further simplification of the latter equation would yield $\tilde{\mathbf{y}} = \boldsymbol{\Sigma}\mathbf{x} + \tilde{\mathbf{n}}$, if the input signal is precoded with $\mathbf{x} = \mathbf{V}^{\text{H}}\mathbf{s}$. Using SVD precoding, the channel matrix is diagonalized and the spatial interference is removed without any matrix inversion or non-linear processing. Since $\mathbf{U}$ is unitary, $\mathbf{U}^{\text{H}}\mathbf{n}$ still has the same variance as $\mathbf{n}$, thus, the SVD precoding does not result in noise enhancement. Due to substantial feedback and complexity of singular value decomposition of the channel matrix, the SVD precoding is not considered a viable precoding scheme. This procedure is illustrated in Figure 9.113.

In general, the precoding algorithms for multi-user MIMO (MU-MIMO) can be classified into linear and non-linear classes. Linear precoding schemes can achieve reasonable performance with lower complexity compared to non-linear precoding techniques. Linear precoding includes unitary and zero-forcing precoding schemes. Non-linear precoding can achieve near optimal capacity at the expense of increased complexity. Non-linear precoding is designed based on the concept of dirty paper coding (DPC), where any known interference at the transmitter can be canceled if the optimal precoding scheme is applied to the transmit signal.

Unitary precoding includes unitary and semi-unitary precoding, both of which are a simple extension of SVD precoding in SU-MIMO with the addition of the SDMA-based user scheduling technique. The SDMA-based opportunistic user

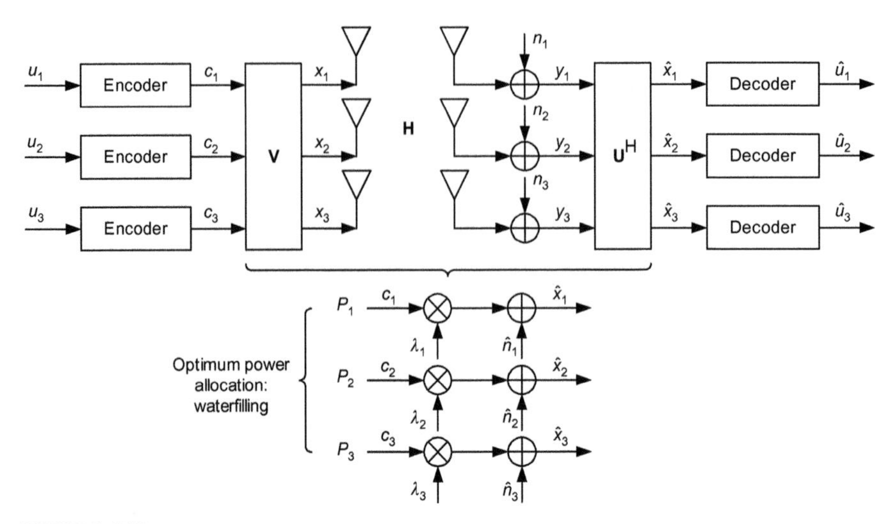

FIGURE 9.113

Illustration of the SVD concept [46].

scheduling technique groups near orthogonal users to avoid intra-group interferences at the cost of additional signaling overhead, which results in higher performance relative to the SU-MIMO. For example, it can increase diversity order to approximately the number of transmit antennas even with simple linear decoding at the receiver.

The zero-forcing precoding consists of zero-forcing and regularized zero-forcing precoding. If the transmitter has perfect knowledge of the downlink CSI, the ZF-based precoding can achieve near channel capacity performance when the number of users is large. With limited CSI at the transmitter, ZF-precoding requires increased feedback to achieve the full multiplexing gain. Hence, inaccurate CSI at the transmitter may result in significant loss of performance due to residual interference among transmit streams.

DPC is a coding technique that cancels known interference without power penalty provided that the transmitter has perfect knowledge of the interfering signals regardless of whether the receiver has knowledge of CSI. This category includes Costa precoding [89], Tomlinson-Harashima precoding [80], and the vector perturbation technique [80]. Vector perturbation uses modulo operation at the transmitter to perturb the transmitted signal vector to avoid the transmit power enhancement incurred by zero-forcing methods. The optimal perturbation method is found by solving a minimum distance problem, and thus can be implemented using sphere decoding or full search-based algorithms [73–80].

Now let us consider MU-MIMO systems and briefly study how precoding is applied in those scenarios. In the downlink direction of a precoded MU-MIMO system (alternatively known as broadcast channel in the literature) with N_{TX} transmit antennas at the base station and one receive antenna at the kth mobile station, the input–output relationship can be written as $y_k = \mathbf{h}_k^H \mathbf{x} + n_k, k = 1, 2, \ldots, N_{user}$, where

$\mathbf{x} = \sum_{i=1}^{N_{\text{user}}} s_i \mathbf{w}_i$ is the $N_{\text{TX}} \times 1$ vector of weighted transmitted symbols s_i, y_k and n_k are the received signal and noise, respectively, $\mathbf{h}_k$ is the kth $N_{\text{TX}} \times 1$ channel vector, where matrix $\mathbf{H} = (\mathbf{h}_1, \mathbf{h}_2, \ldots, \mathbf{h}_{N_{\text{user}}})^{\text{T}}$ is the $N_{\text{user}} \times N_{\text{TX}}$ complex-valued downlink channel matrix, and $\mathbf{w}_k$ is the kth $N_{\text{TX}} \times 1$ normalized linear precoding vector.

The mathematical relationship for the input and output of a precoded MU-MIMO system in the uplink (alternatively known as multiple access channel in the literature) with N_{RX} receive antennas at the base station and one transmit antenna at each user's terminal can be written as $\mathbf{y} = \sum_{k=1}^{N_{\text{user}}} s_k v_k \mathbf{h}_k + \mathbf{n}$ where $s_k v_k$ is the weighted complex-valued modulated symbol from the kth user, $\mathbf{y} = (y_1, y_2, \ldots, y_{N_{\text{RX}}})^{\text{T}}$ and $\mathbf{n} = (n_1, n_2, \ldots, n_{N_{\text{RX}}})^{\text{T}}$ are the $N_{\text{RX}} \times 1$ vectors of received signal and noise, respectively, $\mathbf{h}_k$ is the kth $N_{\text{RX}} \times 1$ channel vector, where matrix $\mathbf{H} = (\mathbf{h}_1, \mathbf{h}_2, \ldots, \mathbf{h}_{N_{\text{user}}})^{\text{T}}$ is the $N_{\text{RX}} \times N_{\text{user}}$ complex-valued uplink channel matrix. As mentioned earlier, perfect knowledge of the CSI is necessary at the transmitter in order to achieve the capacity of a MU-MIMO channel. However, in practical systems, the receiver only provides partial CSI through uplink feedback channels to the transmitter, i.e., the MU-MIMO precoding with limited feedback.

The received signal in the downlink of a MU-MIMO system with limited feedback precoding is mathematically expressed as: $y_k = \mathbf{h}_k^{\text{H}} \sum_{i=1}^{N_{\text{user}}} s_i \hat{\mathbf{w}}_i + n_k, k = 1, 2, \ldots, N_{\text{user}}$. The transmit vector for limited feedback precoding is modeled as $\hat{\mathbf{w}}_i = \mathbf{w}_i + \varepsilon_i$ where ε_i is the error vector generated as a result of the limited feedback and vector quantization, and the expression for the received signal can be rewritten as:

$$y_k = \mathbf{h}_k^{\text{H}} \sum_{i=1}^{N_{\text{user}}} s_i \mathbf{w}_i + \mathbf{h}_k^{H} \sum_{i=1}^{N_{\text{user}}} s_i \varepsilon_i + n_k, \quad k = 1, 2, \ldots, N_{\text{user}} \tag{9.92}$$

where $\mathbf{h}_k^{\text{H}} \sum_{i=1}^{N_{\text{user}}} s_i \varepsilon_i$ is the residual interference due to the limited-feedback precoding. To reduce the residual interference term, one should use more accurate CSI feedback which results in the use of more uplink resources for the feedback. It is shown in the literature that the number of feedback bits per user N_{feedback} must be increased linearly with the SNR γ_{dB} (in decibels) at the rate of $N_{\text{feedback}} = (N_{\text{TX}} - 1)\log_2 \gamma = \gamma_{\text{dB}}(N_{\text{TX}} - 1)/3$ in order to achieve the full multiplexing gain of N_{TX}. In addition, the scaling of N_{feedback} guarantees that the throughput loss relative to zero-forcing precoding with perfect CSI knowledge at the transmitter is upper-bounded by N_{TX} bps/Hz, which corresponds to approximately 3 dB power offset [78]. The throughput of a feedback-based zero-forcing system is bounded, if the SNR approaches infinity and the number of feedback bits per user is fixed. Reducing the number of feedback bits according to $N_{\text{feedback}} = \alpha \log_2 \gamma$ for any $\alpha < N_{\text{TX}} - 1$ results in a strictly inferior multiplexing gain of $N_{\text{TX}}[\alpha/(N_{\text{TX}} - 1)]$ where N_{TX} is the number of transmit antennas and γ is the SNR of the downlink channel. In order to calculate the amount of feedback required to maintain certain throughput, the difference between the feedback rates of zero-forcing precoding with perfect feedback $R_{PF-ZF}(\gamma)$ and with limited feedback $R_{LF-ZF}(\gamma)$ is required to satisfy the following constraint $\Delta R(\gamma) = R_{PF-ZF}(\gamma) - R_{LF-ZF}(\gamma) \leq \log_2 b$. In order to maintain a rate offset less than $\log_2 b$ (per user) between zero-forcing with perfect CSI and with finite-rate feedback (i.e., $\Delta R(\gamma) \leq \log_2 b, \forall \gamma$), it is sufficient to scale the number of feedback bits per user according to $N_{feedback} = \gamma_{dB}(N_{TX} - 1)/3 - (N_{TX} - 1)\log_2(b-1)$. The rate offset of $\log_2 b$ (per user) is translated into a power offset, which is a more useful metric from the design perspective. Since a multiplexing gain of N_{TX} is achieved with zero-forcing, the zero-forcing curve has a slope of N_{TX} bps/Hz/3 dB at asymptotically high SNR. Therefore, a rate offset of $\log_2 b$ bps/Hz per user corresponds to a power offset of $3\log_2 b$ decibels. To feedback $N_{feedback}$ bits through the uplink channel, the throughput of the uplink feedback channel should be larger than or equal to $N_{feedback}$, i.e., $w_{FB} \log_2(1 + \gamma_{FB}) \geq N_{feedback}$ where γ_{FB} denotes the SNR of the feedback channel. Thus, the required feedback resource to satisfy the constraint $\Delta R(\gamma) \leq \log_2 b$ can be shown to be given as follows $w_{FB} \geq [\gamma_{dB}(N_{TX} - 1)/3 - (N_{TX} - 1)\log_2(b-1)]/\log_2(1 + \gamma_{FB})$, i.e., the required feedback resource is a function of both downlink and uplink channel conditions [78].

We defined precoding as adaptive or non-adaptive weighting of the spatial streams prior to transmission from each antenna port (in a multi-antenna configuration) using a precoding matrix for the purpose of improving the reception or separation of the spatial streams at the receiver. Both feed-back and feed-forward precoder matrix selection schemes can be used in order to select the optimal weights. Feed-back precoding matrix selection techniques do not rely on channel reciprocity, rather they use feedback channels provided that the feedback latency is less than the channel coherence time. In feed-forward approaches, the necessary CSI can be theoretically obtained through direct feedback, where the CSI is explicitly signaled to the transmitter by the receiver, or sounding reference signals. The direct channel feedback methods preclude the channel reciprocity requirement, whereas channel sounding methods rely on channel reciprocity. Therefore, explicit control signaling is required for the PMI-based (feedback) schemes. However, in reciprocity-based schemes, the sounding signals in the uplink and precoded pilots in the downlink are used to assist the transmitter and receiver to appropriately select the precoding matrix. The reciprocity-based schemes have the additional advantage of not being constrained to a finite set of codebooks. Beamforming relies on long-term statistics of the radio channel and unlike reciprocity-based techniques, does not require short-term correlation between the uplink and downlink in order to properly function.

9.13.6 SU-MIMO and MU-MIMO

Single-user MIMO techniques are point-to-point schemes that improve channel capacity and reliability through the use of space–time/space–frequency codes (transmit/receive diversity) in conjunction with spatial multiplexing schemes. In a

SU-MIMO transmission, the advantage of MIMO processing is obtained from the coordination of processing among all the transmitters or receivers. In the multi-user channel, on the other hand, it is usually assumed that there is no coordination among the users. As a result of the lack of coordination among users, uplink and downlink MU-MIMO channels are different. In the uplink scenario, users transmit to the base station over the same channel. The challenge for the base station is to separate the signals transmitted by the users, using array processing or multi-user detection methods. Since the users are not able to coordinate with each other, there is not much that can be done to optimize the transmitted signals with respect to each other. If some channel feedback is allowed from the transmitter back to the users, some coordination may be possible, but it may require that each user knows all the other users' channels rather than only its own. Otherwise, the challenge in the uplink is mainly in the processing done by the base station to separate the users. In the downlink channel, where the base station simultaneously transmits to a group of users over the same channel, there is some inter-user interference for each user which is generated by the signals transmitted to other users. Using multi-user detection techniques, it may be possible for a given user to overcome the multiple access interference, but such techniques are often extremely complicated for use at the receivers. Ideally, one would like to mitigate the interference at the transmitter by carefully designing the transmitted signal. If CSI is available at the transmitter, it is aware of what interference is being created for user i by the signal it is transmitting to user j, and vice versa. The inter-user interference can be mitigated by beamforming or the use of dirty paper codes.

In general, SU-MIMO and MU-MIMO schemes are compared as follows [60–64]:

- SU-MIMO is a point-to-point link with predictable link capacity, whereas a MU-MIMO channel is a broadcast channel in the downlink direction and a multiple access channel in the uplink direction whose link-level data rates are characterized in terms of capacity regions.
- Multi-layer SU-MIMO schemes offer layer/stream diversity in the sense that if one stream has a poor SNR, the system will not necessarily experience an outage, whereas in the same situation, a MU-MIMO system will be in outage. This is due to the fact that in MU-MIMO schemes, users have typically an equal target data rate and symbol error rate on their respective links, while in SU-MIMO systems, only the sum rate of the overall link is considered, since all streams are delivered to same user.
- The MU-MIMO schemes suffer from a near-far problem due to significant differences between the path losses experienced by each user, resulting in a large deviation in the SINR of the corresponding user links. This would benefit the users with better channel conditions, while there is no near-far problem in SU-MIMO systems. The near-far problem in MU-MIMO systems may be alleviated via appropriate grouping of the users with similar channel conditions.

- The use of cooperative collocated transmit antennas in SU-MIMO schemes can facilitate the encoding at the transmitter and decoding at the receiver. In contrast, the users in a MU-MIMO scheme can cooperate in encoding at the base station in the downlink and decoding in the uplink; however, the users cannot cooperate in decoding in the downlink or encoding in the uplink directions.
- The capacity of the downlink and uplink are theoretically identical in the downlink and uplink for the SU-MIMO systems (given the same transmit power and perfect channel knowledge in the transmitter and the receiver); however, the capacity of the MU-MIMO broadcast channel and multiple access channel are not identical.
- The capacity of the SU-MIMO schemes is less impacted by lack of CSI at the transmitter, whereas the capacity of the MU-MIMO broadcast channel significantly suffers from lack of CSI at the transmitter.
- SU-MIMO suffers from limited exploitation of multi-user diversity. The number of spatial dimensions is limited by the number of antennas at the UE. There is a potential that spatial dimensions are wasted, if the UEs have fewer antennas compared to the base station.
- MU-MIMO more efficiently exploits the multi-user diversity since all spatial dimensions which are supported by the base station can be exploited. It will achieve capacity gain, if UEs have fewer antennas relative to the base station. Stronger spatial dimensions are exploited, particularly in the case of a low rank channel. The utilized spatial dimensions may be weak in the case of a low rank channel due to spatial correlation.

The advantages/disadvantages of SU-MIMO and MU-MIMO schemes are summarized in Table 9.27.

An important metric for measuring the performance of any communication channel is the information theoretic capacity. In a SU-MIMO channel, the capacity is the maximum amount of information that can be transmitted as a function of available bandwidth given a constraint on transmitted power. In SU-MIMO channels, it is common to assume that the total power distributed among all

Table 9.27 Comparison of SU-MIMO and MU-MIMO Schemes [46]

	SU-MIMO	MU-MIMO
Advantages	High user throughput High peak data rates	High system capacity Full exploitation of multi-user diversity
Disadvantages	Multiple transmit antennas at the base station are not fully exploited Multi-user diversity is not fully exploited	Degradation of peak data rates due to inter-user interference (ZF may not work perfectly due to imperfect CSIT)

transmit antennas is limited. For the MU-MIMO channel, the problem is somewhat more complex. Given the constraint on the total transmit power, it is possible to allocate varying fractions of that power to different users in the network, thus for any value of total power, different information rates are obtained. The result is a capacity region shown in Figure 9.114 for a two-user MU-MIMO channel. The maximum capacity for user 1 is achieved when 100% of the power is allocated to user 1, and for user 2, the maximum capacity is also obtained when it is allocated the full power. For every possible power distribution, there is an achievable information rate, which results in the capacity regions depicted in the figure. Two regions are shown in Figure 9.114, the larger one for the case where both users have roughly the same maximum capacity (similar channel conditions), and the other region for a case where one of the users has a much better channel condition than the other user. For N_{user} users, the capacity region is characterized by an N_{user}-dimensional hyper-region.

Let us use a simple MU-MIMO system model to demonstrate how the sum rate of the system is calculated. The simplified model of MU-MIMO broadcast channel is shown in Figure 9.115. Based on this model, the transmit vector $\mathbf{x}$ can be expressed as the weighted sum (precoded) of the input data symbols $s_k|_{k=1,2,\ldots,N_{\text{user}}}$ as follows: $\mathbf{x} = \sum_k \mathbf{p}_k s_k = \mathbf{Ps}$ where $N_{\text{user}} \times 1$ dimensional vector $\mathbf{s} = (s_1, s_2, \ldots, s_{N_{\text{user}}})^{\text{T}}$ denotes the data symbols from N_{user} users, and $\mathbf{P} = (\mathbf{p}_1, \mathbf{p}_2, \ldots, \mathbf{p}_{N_{\text{user}}})$ is the precoding matrix comprising N_{user} precoding vectors. It is assumed that the finite transmit power at the transmitter can be calculated as $P_{\text{TX}} = E\{\mathbf{x}^{\text{H}}\mathbf{x}\}$. The kth complex-valued output of the system can be written as

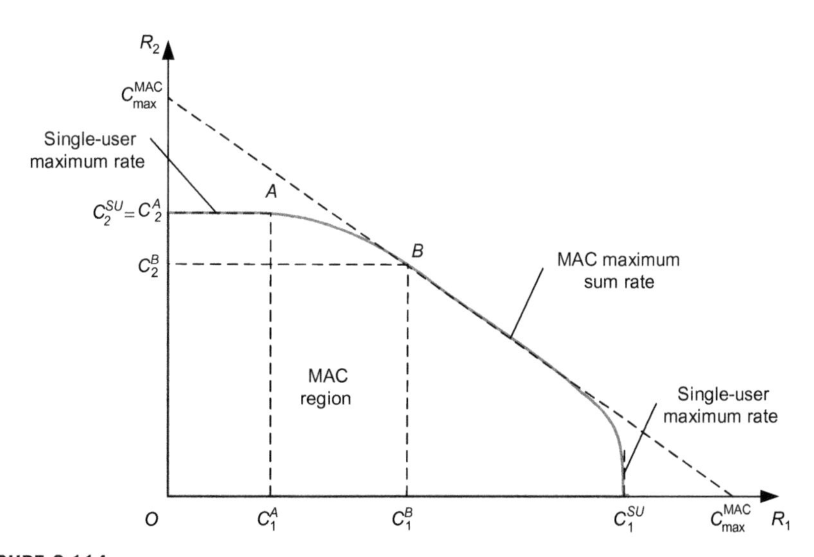

FIGURE 9.114

Capacity region of MU-MIMO multiple access channel (MAC) with two users compared to SU-MIMO [46].

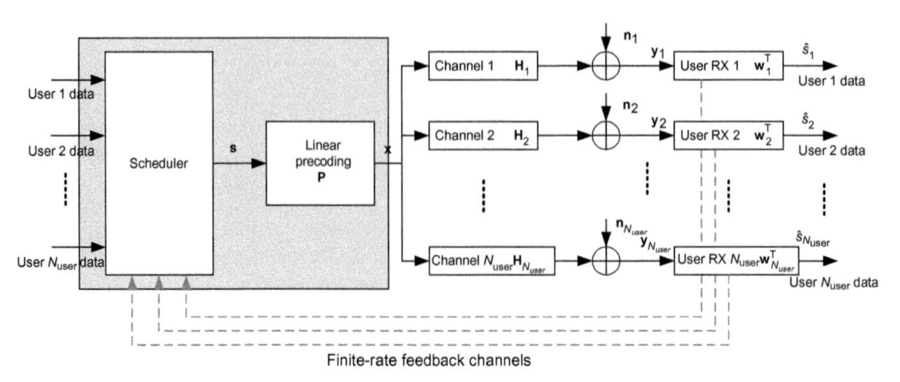

FIGURE 9.115

MU-MIMO broadcast channel model [46]

$\mathbf{y}_k = \mathbf{H}_k\mathbf{x}_k + \mathbf{n}_k \in \mathbb{C}^N$ where N denotes the dimension of vector $\mathbf{y}$. The kth branch user data can be detected using a linear MMSE receiver as $\hat{s}_k = \mathbf{w}_k^T\mathbf{x}_k \in \mathbb{C}$ in which the MMSE weighting matrix is given by [46]:

$$\mathbf{w}_k = \left(\mathbf{H}_k^*\mathbf{P}^*\mathbf{P}^T\mathbf{H}_k^T + \frac{N_{user}}{P_{TX}}\mathbf{I}_N\right)^{-1}\mathbf{H}_k^*\mathbf{p}_k^* \tag{9.93}$$

It can be further shown that the SINR at the kth output is given by:

$$\gamma_k = \frac{|\mathbf{w}_k^T\mathbf{H}_k\mathbf{p}_k|^2}{||\mathbf{w}_k||_2^2\frac{N_{user}}{P_{TX}} + \sum_{i(i\neq k)}|\mathbf{w}_k^T\mathbf{H}_k\mathbf{p}_i|^2} \tag{9.94}$$

The sum rate of the system is given as $R_{sum} = \sum_k \log_2(1+\gamma_k)$ [46].

Figure 9.116 illustrates a generic MU-MIMO scenario where N_{user} users, each having N_{TX_i} transmit antennas, transmit simultaneously to the eNB in the uplink. The maximum achievable throughput of the system is determined by the point on the N_{user}-dimensional hyper-region that maximizes the sum of all of the users' information rates, and is referred to as the sum capacity of the channel (Figure 9.117). In the case of a near-far problem where one user has a more strongly attenuated channel than other users, or some users are closer to the eNB than others, obtaining the sum capacity would cause the user with the worst channel to achieve a lower rate.

As shown in Figure 9.116, in the uplink of a MU-MIMO system, the received signal at the eNB can be written as $\mathbf{y} = \sum_{k=1}^{N_{user}}\mathbf{H}_k^H\mathbf{x}_k + \mathbf{n}$ where $\mathbf{x}_k$ is the $N_{TX_k} \times 1$ transmitted signal vector of the kth UE with N_{TX_k} transmit antennas, $\mathbf{H}_k \in \mathbb{C}^{N_{TX_k}\times N_{RX}}$ denotes the flat-fading channel matrix from the kth user to the eNB, and $\mathbf{n} = (n_1, n_2, \ldots, n_{N_{RX}}), n_k \sim \mathrm{N}(0,1)$ is an independent and identically

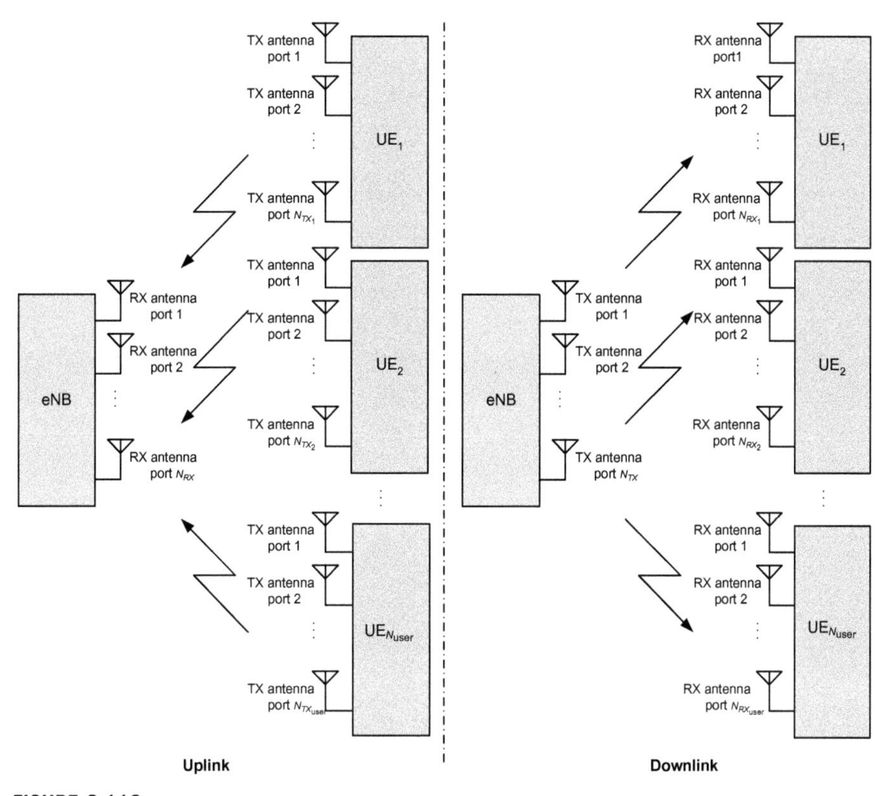

FIGURE 9.116

A generic multi-user MIMO transmission model [18].

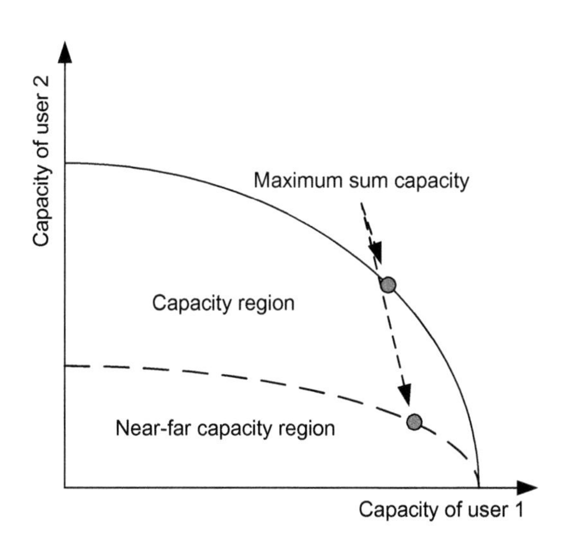

FIGURE 9.117

An example illustration of capacity region.

distributed additive white Gaussian noise vector at the eNB. We assume that the receiver k has perfect and instantaneous knowledge of the channel matrix $\mathbf{H}_k$. Note that the eNB is equipped with N_{TX} transmit and N_{RX} receive antennas.

In the downlink as illustrated in Figure 9.116, the received signal at the kth receiver can be written as $\mathbf{y}_k = \mathbf{H}_k\mathbf{x} + \mathbf{n}_k \forall k = 1, 2, \ldots, N_{user}$ where $\mathbf{H}_k \in \mathbb{C}^{N_{RX_k} \times N_{TX}}$ is the downlink channel and $\mathbf{n}_k \in \mathbb{C}^{N_{RX_k} \times 1}$ is the complex-valued additive Gaussian noise vector at the kth receiver. We assume that each receiver also has perfect and instantaneous knowledge of its own channel matrix $\mathbf{H}_k$. The transmitted signal $\mathbf{x}$ is a function of multiple users' information data, i.e., $\mathbf{x} = \sum_{k=1}^{N_{user}} \mathbf{x}_k$ where $\mathbf{x}_k$ is the signal carrying the kth user's message with covariance matrix $\mathbf{\Omega}_k = E\{\mathbf{x}_k\mathbf{x}_k^H\}$. The power allocated to the kth user is given by $\rho_k = \text{trace}(\mathbf{\Omega}_k)$. Under a sum power constraint at the eNB, the power allocation needs to maintain $\sum_{k=1}^{N_{user}} \rho_k \leq P_{total}$. Assuming a unit variance for the noise, it is now known that the capacity region for a given matrix channel realization can be written as [80]:

$$C_{DL} = \bigcup_{(\rho_1, \rho_2, \ldots, \rho_{N_{user}} | \sum \rho_k \leq P_{total})} (R_1, R_2, \ldots, R_{N_{user}}) \in \mathbb{R}^{+N_{user}}, R_l \leq \log_2 \frac{\det\left[\mathbf{I} + \mathbf{H}_i\left(\sum_{j \geq i} \mathbf{\Omega}_j\right)\mathbf{H}_i^H\right]}{\det\left[\mathbf{I} + \mathbf{H}_i\left(\sum_{j > i} \mathbf{\Omega}_j\right)\mathbf{H}_i^H\right]} \Bigg\} \tag{9.95}$$

where $\mathbb{R}^{+N_{user}}$ is the N_{user}-dimensional set of positive real numbers. The above equation may be optimized over each possible user ordering. Although difficult to realize in practice, the computation of the capacity region can be simplified using the assumption that the downlink capacity region can be calculated through the union of regions of the dual multiple access channel with all uplink power allocation vectors meeting the sum power constraint. The fundamental effect of the use of multiple antennas at either the eNB or the user terminals in increasing the channel capacity is best comprehended by examining how the sum capacity, i.e., the point obtained by the maximum $\sum_{k=1}^{N_{user}} R_k$ in the capacity region, scales with the number of active users.

The capacity region in the uplink of a MU-MIMO system can be calculated for two different cases: (1) joint decoding, i.e., the signals from a group of users are decoded in a cooperative manner; and (2) independent decoding where the signals from users are independently decoded in parallel.

The capacity of the MU-MIMO channels can be calculated using DPC techniques. The method was inspired by interference cancelation schemes. It is shown that the capacity of a channel where the transmitter has knowledge of interference is the same as if there were no interference. The DPC concept can be better comprehended by comparing the scheme to writing on a dirty paper with a properly selected ink color so that the scripts have high contrast and are clearly readable. It can be shown that the capacity of the MU-MIMO multiple-access channel in the above cases is given as follows [75]:

$$R_i \le \log_2\left[\frac{\det(\mathbf{R_y})}{\det(\mathbf{R_y}-E_{s_i}\mathbf{h}_i\mathbf{h}_i^{\mathrm{H}})}\right],\quad i=1,2,\ldots,N_{\mathrm{user}}\quad \text{(independent decoding)}$$
$$\sum_{k\in\Re} R_k \le \log_2 \det\left(\mathbf{I}_{N_{\mathrm{RX}}}+\frac{1}{N_0}\mathbf{H}_{\Re}\mathbf{R}_{\mathbf{s}_{\Re}}\mathbf{H}_{\Re}^{\mathrm{H}}\right)\quad \text{(joint decoding)} \tag{9.96}$$

In the above equations N_{RX} denotes the number of receive antennas at the base station, N_{user} is the number of users each with only one transmit antenna, $\Re$ is a subset of users whose signals are jointly decoded, $\mathbf{H}_{\Re}$ is the $N_{\mathrm{RX}} \times N_{\Re}$ complex-valued channel matrix where $N_{\Re}$ is the cardinality of the set $\Re$, $\mathbf{R}_{\mathbf{s}_{\Re}}$ denotes the covariance matrix of the signals transmitted by the users contained in the set $\Re$, $\mathbf{I}_{N_{\mathrm{RX}}}$ is an $N_{\mathrm{RX}} \times N_{\mathrm{RX}}$ identity matrix, R_k is the rate that can be reliably achieved by user k, s_i is the complex-valued symbol transmitted by user $i, i=1,2,\ldots,N_{\mathrm{user}}$, E_{s_i} is the average energy of the signal from the ith user which differs from one user to another, $\mathbf{R_y}$ is the covariance matrix of the received signal at the base station, and $\mathbf{H}=[\mathbf{h}_1,\mathbf{h}_2,\ldots,\mathbf{h}_{N_{\mathrm{user}}}]$ is the normalized $N_{\mathrm{RX}} \times N_{\mathrm{user}}$ complex-valued channel matrix composed of individual user complex-valued channel vectors $\mathbf{h}_i$.

An efficient UE pairing scheme is required at the eNB to choose the correct pair of UEs for MU-MIMO transmission in MU-MIMO systems. This pairing scheme is required to maintain minimal interference among scheduled UEs in MU-MIMO transmission. A proper pairing scheme can be designed by maximizing the Chordal distance[28] between the feedback precoding matrices of the UEs. The Chordal distance between two matrices is given in [31,32] and represented by $d_c(\mathbf{p}_i,\mathbf{p}_j)=(1/\sqrt{2})\|\mathbf{p}_i\mathbf{p}_i^{\mathrm{H}}-\mathbf{p}_j\mathbf{p}_j^{\mathrm{H}}\|_F$ where $\|.\|_{\mathrm{F}}$ denotes the Frobenius norm of the matrix. The Chordal distance generalizes the distance between two points on the unit sphere through an isometric embedding from complex Grassmann manifold $Gr(N_{\mathrm{TX}},N_l)$ to the unit sphere. Assuming an infinite number of UEs served by the current eNB, the kth UE with reported precoding matrix $\mathbf{p}_k$ will be paired with the mth UE, where the mth UE reports precoding matrix $\mathbf{p}_m$ and the Chordal distance between precoding matrices is maximized. With the maximized Chordal

[28]The asymptotic performance of a coding scheme is dominated by the shortest distance between any pair of codewords. The relevant distance measure between two codewords (vectors) $\mathbf{x}_1$ and $\mathbf{x}_2$ of an orthogonal code (of size M) for a non-coherent MIMO system is the Chordal distance defined as $d^2(\mathbf{x}_1,\mathbf{x}_2)=M-\|\mathbf{x}_1\mathbf{x}_2^{\mathrm{H}}\|_{\mathrm{F}}^2$ [31,32].

distance criterion, $\mathbf{p}_m$ stays in the anti-polar position of $\mathbf{p}_k$, hence $||\mathbf{H}_k\mathbf{p}_m||^2$ is minimized, yielding the minimized inter-UE interference. Consequently, the UE pairing scheme for MU-MIMO transmission in practical systems is designed to find the best match between two UEs (e.g., the mth UE and the kth UE) based on the reported precoding matrices and the following criterion $\mathbf{p}_m = \arg \max_{\forall \mathbf{p}_i \in \mathbf{P}_{UE}} d_c(\mathbf{p}_i, \mathbf{p}_k)$ with $\mathbf{P}_{UE}$ representing the set containing all reported precoding matrices at a certain eNB.

If the conventional AWGN channel were modified to include an additive interference term that is known at the transmitter, and the received signal were expressed as the sum of the transmitted signal + interference + noise, one could set the transmitted signal equal to the desired data minus the interference term; however, such an approach requires increased power and perfect knowledge of interference. It is shown in the literature that the capacity in such a scenario is the same as if the interference were not present; no additional power is required to cancel the interference than is used in a nominal additive white Gaussian noise channel. Therefore, writing on dirty paper, from an information theoretic perspective, is equivalent to writing on clean paper when one knows in advance where the dirt is located. This concept has been used to characterize the sum-capacity and capacity region of the multi-antenna multi-user channels. A DPC technique for the downlink of a MU-MIMO system uses the QR decomposition[29] of the channel matrix as the product of a lower triangular matrix $\mathbf{R}$ with a unitary matrix $\mathbf{Q}$ such that $\mathbf{H}=\mathbf{QR}$. The signal to be transmitted is precoded with the $\mathbf{Q}^H$ matrix, resulting in the effective channel $\mathbf{R}$. The first user of this equivalent system sees no interference from other users, i.e., the signal may be chosen irrespective of other users. The second user sees interference only from the first user; this interference is known and thus may be canceled using DPC. The other users are dealt with in a similar manner. Alternative approaches apply dirty paper techniques directly, rather than for individual users. An important difference between the MU-MIMO channel and the interference channels for which dirty paper techniques are applied is that the interference depends on the signal being designed [80].

Block diagonalization (BD) is a linear precoding technique for the downlink of MU-MIMO systems. It decomposes a MU-MIMO downlink channel into multiple parallel orthogonal SU-MIMO channels. The signal of each user is pre-processed at the transmitter using a modulation matrix that lies in the null space of all other users' channel matrices; thereby the multi-user interference in the system is effectively zero. The block diagonalization scheme is restricted

[29]In linear algebra, the QR decomposition of a matrix is defined as decomposition of a matrix into an orthogonal matrix and an upper triangular matrix. The QR decomposition is often used to solve the linear least squares problems. The QR decomposition of a real square matrix $\mathbf{A}$ is $\mathbf{A}=\mathbf{QR}$ where $\mathbf{Q}$ is an orthogonal matrix, i.e., $\mathbf{Q}^H\mathbf{Q}=\mathbf{I}$ and $\mathbf{R}$ is an upper triangular matrix. This can be generalized to a complex-valued square matrix $\mathbf{A}$ and a unitary matrix $\mathbf{Q}$. If $\mathbf{A}$ is non-singular, then this factorization is unique if one requires that the diagonal elements of $\mathbf{R}$ are positive-valued [29].

to channels where the number of transmit antennas N_{TX} is greater or equal to the total number of receive antennas in the network N_{RX}. Let us define the precoder matrices as $\mathbf{F} = [\mathbf{F}_1, \mathbf{F}_2, \ldots, \mathbf{F}_{N_{user}}] \in \mathbb{C}^{N_{RX} \times N_l}$ where $\mathbf{F}_i \in \mathbb{C}^{N_{RX} \times N_{l_i}}$ is the ith user precoder matrix. We assume that $N_l \leq N_{RX}$ is the total number of the transmitted data streams, while $N_{l_i} \leq N_{RX_i}$ is the number of data streams transmitted to the ith user. We can find the optimal precoding matrix $\mathbf{F}$ such that multi-user interference is zero by choosing a precoding matrix $\mathbf{F}_i$ that lies in the null space of the other users' channel matrices. Thereby, an MU-MIMO downlink channel is decomposed into multiple parallel independent SU-MIMO channels [80].

Minimum mean-squared-error beamforming improves the system performance by allowing a certain amount of interference especially for users equipped with a single antenna. However, it suffers from performance loss when it attempts to mitigate the interference between two closely spaced antennas, a situation that usually occurs when the user terminal is equipped with more than one receive antenna. Successive MMSE (SMMSE), which is an improved version of the MMSE precoding algorithm, was developed to mitigate this problem by successively calculating the columns of the precoding matrix for each of the receive antennas separately. Linear equalization suffers from noise enhancement, and hence has poor power efficiency in some cases. The same drawback is experienced by linear precoding, which alleviates noise by boosting the transmit power [76].

The Tomlinson–Harashima precoding (THP) algorithm is a non-linear precoding technique originally developed for multipath SISO channels. The THP algorithm is performed by moving the feedback part of the decision-feedback equalizer to the transmitter. It has also been applied for the pre-equalization of multi-user interference in MU-MIMO systems; hence, no error propagation occurs and the precoding can be performed for the interference-free channel. The MMSE precoding in combination with THP is another approach to balance the multi-user interference in order to reduce the performance loss that occurs with zero interference techniques while THP is used to reduce the multi-user interference and to improve the diversity [46].

Space-division multiple access (SDMA) is a special case of MU-MIMO where several users share the same time–frequency physical resource block due to their spatial separation. SDMA uses antenna beam direction or angle (using directional antennas) as another dimension in signal space which can be channelized and assigned to different users (Figure 9.118). Orthogonal channels can be assigned only if the angular separation between users exceeds the angular resolution of the directional antenna. In practice, SDMA is often implemented using sectorized antenna arrays. The 360° angular space is divided into a number of sectors and in each sector a highly directional antenna array is used to minimize the inter-user interference. The users within each sector can be served using multiple access schemes such as OFDMA and CDMA. The SDMA system adapts to the mobility of the users by beam-steering or assigning a user to a new channel when the user changes its geographical location [25].

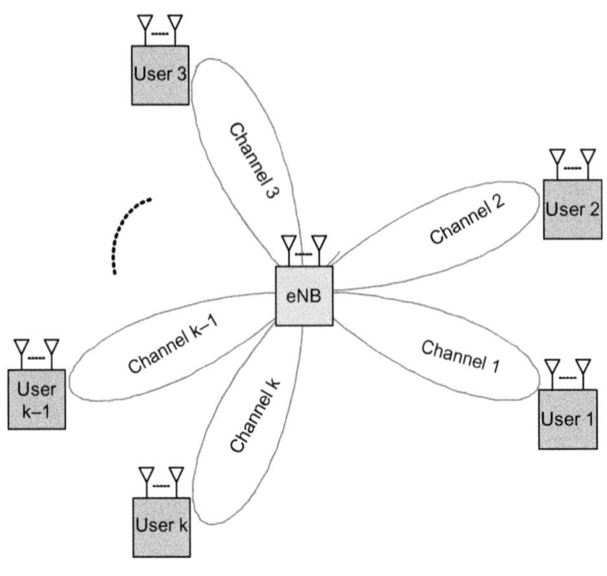

FIGURE 9.118

Illustration of the SDMA concept.

9.13.7 Collaborative MIMO and collaborative spatial multiplexing

Collaborative spatial multiplexing is a virtual MIMO technique where users transmit over the same time–frequency resource unit in the uplink. This type of spatial multiplexing improves the sector throughput without requiring multiple transmit antennas at the mobile station. The received signals from the simultaneous uplink transmission of the users are processed in the base station using maximum likelihood detection techniques. The base station's scheduler groups the users with similar channel conditions and assigns their uplink transmissions to the same time–frequency region. The operation of a collaborative spatial multiplexing scheme is conceptually similar to the regular spatial multiplexing scheme using transmit antennas that are located at different locations (user terminals are geographically apart). The throughput of collaborative spatial multiplexing theoretically increases by the number of users which are scheduled to simultaneously transmit over the same physical resource units.

The collaborative MIMO (Co-MIMO) is an extension of conventional single eNB-based MIMO techniques. It allows multiple base stations to serve one or multiple mobile stations simultaneously over the same radio resource through coordination among the participating base stations. The Co-MIMO is characterized by the following three features:

1. In Co-MIMO, each UE is jointly served by multiple base stations, which is different from conventional MIMO techniques where each UE is only served by a single eNB. Through inter-eNB coordination, the inter-cell interference among these coordinated base stations can be significantly reduced when using low-frequency reuse factors.

2. In Co-MIMO, each eNB can also simultaneously serve multiple mobile stations using the same radio resource in order to increase the system throughput. The co-channel interference among these mobile stations can be minimized through use of SDMA techniques such as beamforming or MU-MIMO schemes.
3. In Co-MIMO, SDMA processing is implemented independently within individual coordinated base stations based on the CSI between the eNB and the mobile stations that are served by the eNB. Therefore, the Co-MIMO limits the amount of information exchange between base stations, which is different from other eNB coordination approaches requiring considerable information exchange.

A central scheduler, that is a logical functional module in the network, is responsible for coordinating a group of base stations, collecting information from the participating base stations, and determining the serving relationship between base stations and mobile stations. The overall complexity of the network operation increases with increasing number of coordinated base stations in Co-MIMO scheme. The Co-MIMO scheme has the advantage of inter-cell interference reduction and spectral efficiency improvement. The inter-cell interference mitigation is realized by turning interference from neighboring cells into useful signals and separating signals for different users via MU-MIMO or SDMA techniques. The spectral efficiency improvement is achieved via interference mitigation and the number of users supported over the same time–frequency resource region. Moreover, the Co-MIMO provides a simple approach to implementation of eNB coordination and a reasonable trade-off between the performance gain and the stringent requirement on the core network traffic and implementation complexity. In practice, the user data delivery to multiple coordinated base stations in a Co-MIMO scheme can be realized in the same manner as the multicast over backhaul is done in the multicast and broadcast services. There is no additional computational complexity in base stations supporting Co-MIMO compared to the base stations which support MU-MIMO techniques [138].

In order to demonstrate the advantages of Co-MIMO, an example system-level simulation was conducted where the collaborative zone is defined over three neighboring sectors belonging to different eNBs as shown in Figure 9.119. Given that the inter-cell interference is more severe in the cell-edge, the evaluation of Co-MIMO performance is performed in the cell-edge area. The cell-edge area is characterized by the cell-edge length in the simulation, as shown in Figure 9.119. We assume a 19-cell collaborative-zone configuration in the simulation as depicted in Figure 9.119. The users are randomly dropped with a uniform distribution in the cell-edge area, one user per sector. The system performance with different sizes of cell-edge area is evaluated by setting the cell-edge length to 150, 100, and 50 m. In each collaborative region, three coordinated base stations communicate with three mobile stations located in their cell-edge areas allocating the same time and frequency resource. Each UE in the cell-edge area will be served by two coordinated base stations, and each eNB serves two mobile stations

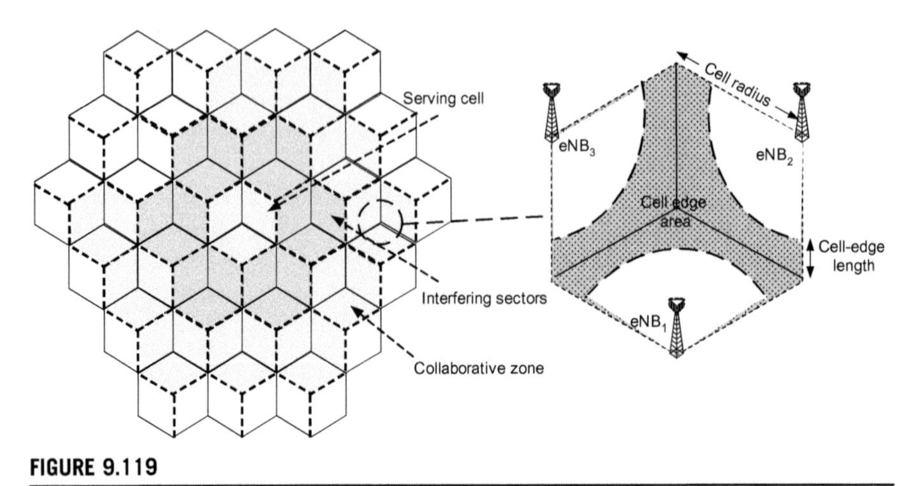

FIGURE 9.119

Example of cell configuration and cell-edge area in a collaborative zone [138].

simultaneously. The precoding matrices for the two served mobile stations are calculated independently within each single sector of a coordinated eNB. Each coordinated eNB transmits with full power and equal power allocation to the mobile stations. A single narrowband sub-carrier is assumed in the system without loss of generality. A conventional SU-MIMO scheme is used as the benchmark for comparison where three mobile stations in the cell-edge area are served independently by different base stations without any coordination in frequency reuse one deployment. As shown in Figure 9.120, even though there remains a certain amount of residual inter-cell interference, the interference power from neighboring cells within the collaborative zone is considerably reduced. If the interference from cells outside of the collaborative zone is taken into consideration, the reduction in overall interference level is less noticeable, and can be further improved by extending the collaborative zone to contain more coordinated base stations. The cumulative distribution function of the cell-edge users within the cell-edge area as a function of cell-edge length for SU-MIMO and Co-MIMO are compared and shown in Figure 9.120. It is shown that the interference power increases with increasing cell-edge length (see Figure 9.119). There is a considerable reduction in the interference power from cells within the Co-MIMO zone when using Co-MIMO techniques compared to the conventional SU-MIMO schemes. The performance gain decreases when comparing the overall interference [138].

9.13.8 Non-linear precoding and dirty-paper coding

In this section, we discuss some advanced non-linear multi-antenna transmission techniques that are not supported in current releases of LTE/LTE-Advanced, but may be considered as viable improvements in future releases of LTE standards. While the following techniques are theoretically proven to achieve higher

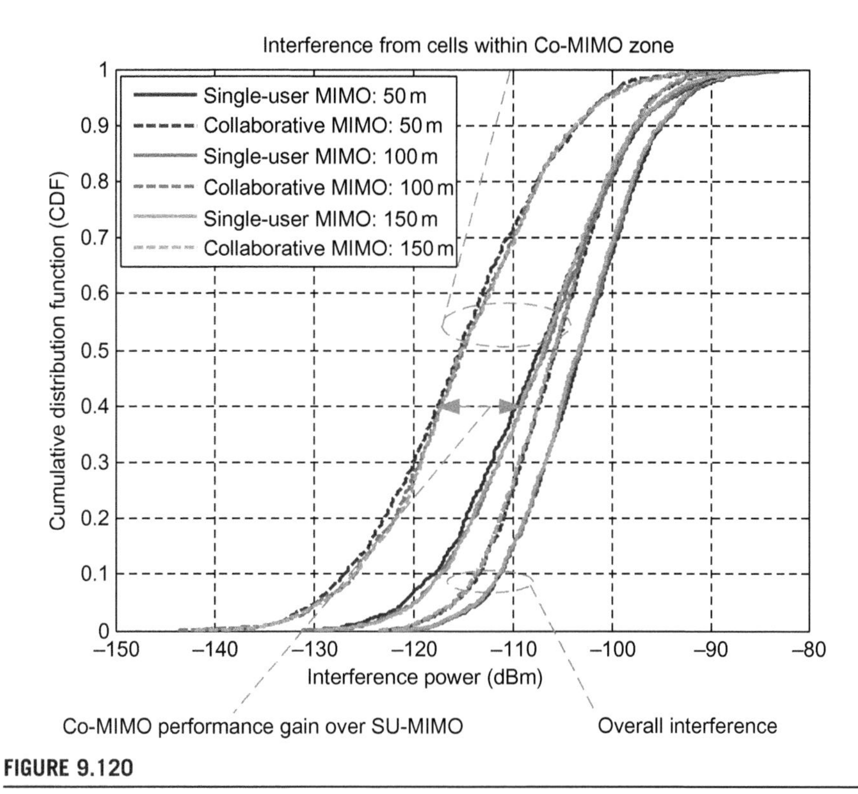

FIGURE 9.120

CDF of interference power of the cell-edge users as a function of cell-edge length [138].

capacities and user throughputs, their implementation complexities have prevented their application in real-life scenarios.

When the channel is frequency-selective, the precoder can also exploit the frequency selectivity and become frequency dependent. For single-carrier systems, non-linear precoding techniques using spatial extensions of the Tomlinson–Harashima precoder can be employed [46]. For multi-carrier systems such as OFDM, frequency-flat precoding techniques can be applied on a per sub-carrier basis. To reduce feedback overhead in OFDM, the CSIT feedback is sampled and interpolated in the frequency-domain. Exploiting the OFDM structure and sub-carrier correlation results in precoders with frequency-dependent eigen-beam directions and frequency-beam-dependent power allocation. In multi-user scenarios, partial CSIT is also highly relevant, since the channel time variation makes it impractical to attain perfect CSIT for all users. The studies have shown that the loss of degrees of freedom due to no CSIT reduces the capacity region of an isotropic vector broadcast channel to that of a scalar one. Imperfect CSIT also severely reduces the growth of the sum-rate broadcast capacity at high SNRs. Some techniques such as opportunistic scheduling, which require only an SNR feedback, can achieve an optimal throughput growth rate in broadcast channels with a large number of users. With finite-rate feedback, however, the feedback rate needs to be increased with the SNR to achieve the full multiplexing gain. Figure 9.121 illustrates two hypothetical linear and non-linear precoding schemes in a MU-MIMO scenario where $\mathbf{x} = (x_1, x_2, \ldots, x_{N_{\text{user}}})^{\mathrm{T}}$ and the precoding matrix $\mathbf{V}$ is given as follows [46]:

$$\mathbf{V} = (\mathbf{v}_1, \mathbf{v}_2, \ldots, \mathbf{v}_{N_{\text{user}}})^{\mathrm{T}} = \begin{bmatrix} v_{11} & v_{12} & \cdots & v_{1N_{\text{user}}} \\ v_{21} & v_{22} & \cdots & v_{2N_{\text{user}}} \\ \vdots & \vdots & \ddots & \vdots \\ v_{N_{\text{TX}}1} & v_{N_{\text{TX}}2} & \cdots & v_{N_{\text{TX}}N_{\text{user}}} \end{bmatrix} \tag{9.97}$$

In the non-linear scheme, it is assumed that the interference is known to the transmitter and is mitigated via DPC techniques, as discussed later in this section.

Let us consider a scenario where we have a downlink channel with one base station equipped with N_{TX} transmit antennas and N_{user} users each with one antenna as the simplest form of MU-MIMO Gaussian broadcast channel. A time sample of this channel is characterized by $\mathbf{y} = \mathbf{Hx} + \mathbf{n}$ where $\mathbf{x}$ is the signal vector transmitted in parallel by N_{TX} transmit antennas, $\mathbf{y} = (y_0, y_1, \ldots, y_{N_{\text{user}}-1})^{\mathrm{T}}$ is the vector of signals individually received by N_{user} users, and $\mathbf{n} \sim \mathbb{Z}(0,\mathbf{I})$ is an independent and identically distributed Gaussian noise vector. The channel matrix $\mathbf{H} = (\mathbf{h}_1, \mathbf{h}_2, \ldots, \mathbf{h}_{N_{\text{user}}-1})^{\mathrm{H}}$ contains the channel coefficients from the N_{TX} transmit antennas to the N_{user} users, where the row vector $\mathbf{h}_i$ is the channel as seen by the ith user. The input signal power is constrained such that trace$[\mathscr{E}(\mathbf{xx}^{\mathrm{H}})] \leq P$ where P denotes the total downlink transmitted power or the energy per channel use. Let us further assume that perfect CSI is available at the transmitter and ideal coherent receivers are utilized at the user terminals.

Under the above assumptions, a linear precoding scheme can be defined as a scheme where the transmit signal $\mathbf{x}$ is given by $\mathbf{x} = \mathbf{\Theta u}$ where $\mathbf{\Theta} \in \mathbb{C}^{N_{\text{TX}} \times N_{\text{user}}}$ and the vector $\mathbf{u} = (u_1, u_2, \ldots, u_{N_{\text{user}}-1})^{\mathrm{T}}$ contains the user coded symbols. The symbols u_i are independently generated by channel encoders for the users. Without loss of generality, we assume that $E(\mathbf{uu}^{\mathrm{H}}) = \mathbf{I}$ (i.e., normalized signal power), the power constraint implies that trace$(\mathbf{\Theta\Theta}^{\mathrm{H}}) \leq P$. The kth user SINR under linear precoding is given as follows:

$$\text{SINR}_k = \frac{\varphi_{kk} p_k}{1 + \sum_{i \neq k} \varphi_{ki} p_i} \tag{9.98}$$

where $\mathbf{\Phi} = [\varphi_{ij}]|_{i=0,1,\ldots,N_{\text{user}}-1; j=0,1,\ldots,N_{\text{TX}}-1}$ denotes a matrix whose entries are defined as $\varphi_{ij} = |\mathbf{h}_i \mathbf{w}_j^{\mathrm{H}}|^2$, $\mathbf{w}_j = \mathbf{\theta}_j / \sqrt{p_j}$ is the precoding vector, $\mathbf{\theta}_j$ denotes the jth column of $\mathbf{\Theta}$, and $p_j = |\mathbf{\theta}_j|^2$. It can be shown that assuming Gaussian codes and minimum distance decoding at the receivers, the user rates $R_k = \log_2(1 + \text{SINR}_k)$ are achievable. The transmit matrix $\mathbf{\Theta}$ can be designed in several ways. The

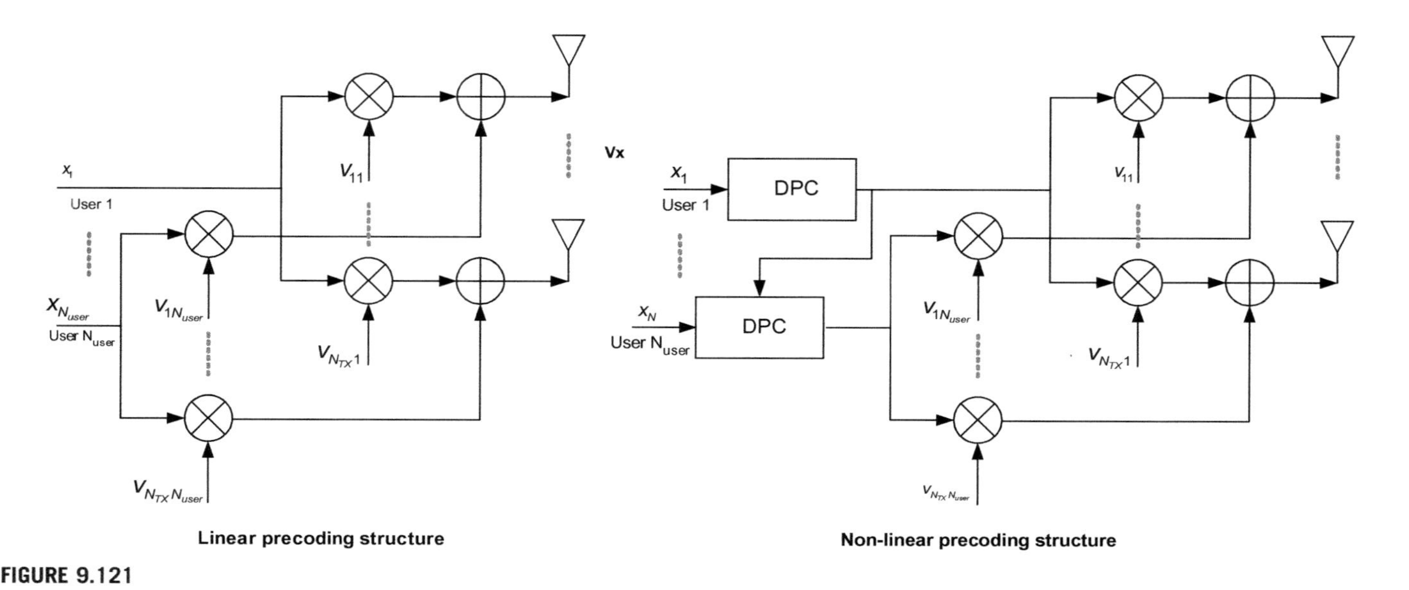

FIGURE 9.121

Illustration of linear and non-linear precoding concept [46].

throughput optimization criterion $\max_{\Theta} \sum_k R_k$ is a non-convex optimization problem that yields a maximal sum rate in a MU-MIMO scenario. The constrained maximization with respect to $\boldsymbol{\Theta}$ can be replaced by an unconstrained maximization with respect to $\hat{\boldsymbol{\Theta}}$, where the transmit signal is given by the following expression:

$$\mathbf{x} = \sqrt{\frac{P}{\text{trace}(\hat{\boldsymbol{\Theta}}\hat{\boldsymbol{\Theta}}^H)}}\hat{\boldsymbol{\Theta}}\mathbf{u} \tag{9.99}$$

The input constraint is automatically satisfied with equality for any arbitrary $\hat{\boldsymbol{\Theta}}$ [182].

A non-linear precoder, on the other hand, produces the transmitted signal $\mathbf{x}$ in the form of $\mathbf{x} = \gamma\boldsymbol{\Theta}[\mathbf{u} + \boldsymbol{\delta} - \boldsymbol{\lambda}(\mathbf{u}, \boldsymbol{\delta})]$ where $\boldsymbol{\Theta}$ is a transmit matrix such that $\text{trace}(\boldsymbol{\Theta}\boldsymbol{\Theta}^{\mathrm{H}}) \leq P$, $\boldsymbol{\delta}$ is a random dithering signal assumed to be known at the transmitter side, and $\boldsymbol{\lambda}$ is a data-dependent lattice point, i.e., $\boldsymbol{\lambda}$ is a function of $\mathbf{u}$ and $\boldsymbol{\delta}$. The normalization coefficient γ is defined as follows:

$$\gamma^2 = \frac{P}{\text{trace}(\boldsymbol{\Theta}^{\mathrm{H}}\boldsymbol{\Theta}\mathcal{E}\{(\mathbf{u} + \boldsymbol{\delta} - \boldsymbol{\lambda})(\mathbf{u} + \boldsymbol{\delta} - \boldsymbol{\lambda})^{\mathrm{H}}\})} \tag{9.100}$$

where the expectation operator corresponds to the data vectors $\mathbf{u}$ and $\boldsymbol{\delta}$ [182].

The capacity of the additive white Gaussian noise channel with interference known at the transmitter was determined by Costa [89]. In his model, as shown in Figure 9.122, the channel input vector $\mathbf{x} = (x_1, x_2, \ldots, x_K)^{\mathrm{T}}$ consisting of K real-valued components is modified by two additive memory-less Gaussian noise and interference sources. The interference source produces the vector $\mathbf{s} = (s_1, s_2, \ldots, s_K)^{\mathrm{T}}$ with normally distributed components $s_i \sim \mathbb{N}(0, \sigma_I^2)$. The interference vector is observed by the encoder before the transmission process starts. The encoder produces the code vector $\mathbf{x} = f(w, \mathbf{s})$. Here $w \in \{1, 2, \ldots, M\}$ is the index of the message that must be conveyed. All messages are equally likely, which means that $\Pr(W = w) = 1/M \quad \forall w$. All code vectors $\mathbf{x}$ should satisfy a power constraint, i.e., $P_{\mathbf{x}} = \mathcal{E}[||\mathbf{x}||^2] \leq P_{\mathrm{T}}$ where P_{T} denotes the total transmit power. The channel noise vector $\mathbf{z} = (z_1, z_2, \ldots, z_K)^{\mathrm{T}}$ has normally distributed components

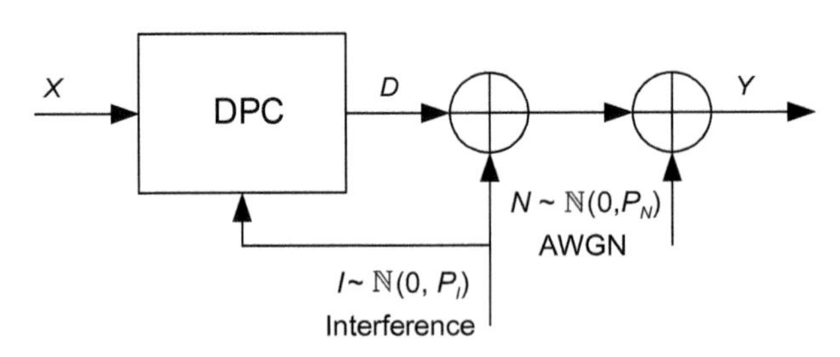

FIGURE 9.122

Costa's dirty paper coding model [89].

$z_i \sim \mathbb{N}(0, \sigma_z^2)$. The channel output vector $\mathbf{y} = (y_1, y_2, \ldots, y_K)^{\mathrm{T}} = \mathbf{x} + \mathbf{s} + \mathbf{z}$ is processed by the decoder and results in the message estimate $\hat{w} = D(\mathbf{y})$. The system has two parameters that determine its performance. For a given total transmit power $\mathrm{P_T}$, and interference and noise variances σ_I^2 and σ_z^2, the information rate $R \sim \log_2 M$ should be as large as possible, while on the other hand the expected error probability $P_e = Pr(\hat{w} \neq w)$ should be as small as possible. The largest rate that is achievable with arbitrarily small expected error probability is called the channel capacity. Costa showed that this capacity is determined only by the transmitter power P_{T} and the variance of the channel noise σ_z^2, i.e., the noise that is unknown to the transmitter. It appears that the channel capacity in this case is determined as $C = 0.5 \log_2(1 + P_{\mathrm{T}}/\sigma_z^2)$ bits/transmission. Since the interference is known to the transmitter a priori, this signaling method is known as writing on dirty paper [89]. Interference which is known to the transmitter does not cause any loss in terms of capacity.

The above concept can be generalized to the case of MU-MIMO when the interference vector is known to the transmitter a priori. Without loss of generality, let us consider the transmission model shown in Figure 9.123, in which a hypothetical base station is precoding and transmitting signals to three mobile stations each equipped with one receive antenna. Let $\mathbf{h}_k = (h_k^1, h_k^2, \ldots, h_k^{N_{\mathrm{TX}}})^{\mathrm{T}}$ denote the channel vector as seen by the kth mobile station. The corresponding precoding vector at the transmitter is represented by $\mathbf{v}_k = (v_1^k, v_2^k, \ldots, v_{N_{\mathrm{TX}}}^k)^{\mathrm{T}}$. The precoding vector $\mathbf{v}_k$ is designed such that it is orthogonal to all $\mathbf{v}_j{:}j<k$ by subtracting a linear combination of all previous $\mathbf{v}_j$ from $\mathbf{h}_k^*$ (Gram–Schmidt orthogonalization procedure[30]), i.e., $\mathbf{v}_k = \frac{\mathbf{P}_k \mathbf{h}_k^*}{\|\mathbf{P}_k \mathbf{h}_k^*\|}$ where $\mathbf{P}_1 = \mathbf{I}_{N_{TX}}$ and $\mathbf{P}_{k+1} = \mathbf{P}_k(\mathbf{I}_{N_{\mathrm{TX}}} - \mathbf{v}_k \mathbf{v}_k^{\mathrm{H}})$. The objective is to achieve the following $\mathbf{v}_l^{\mathrm{H}} \mathbf{v}_k = 0 \forall l<k, \mathbf{v}_k^{\mathrm{H}} \mathbf{v}_k = 1$, and $\mathbf{h}_k^H \mathbf{v}_j = 0 \ \forall k<j$. The latter criteria are meant to minimize the interference toward a particular mobile station from downlink transmissions to other mobile stations by appropriately selecting the precoding vectors at the transmitter. As a result, the received signal at the kth mobile station is given as follows [46]:

$$y_k = \underbrace{\sum_{i=1}^{k-1} \mathbf{h}_k^{\mathrm{T}} \mathbf{v}_i \mathbf{x}_i}_{\text{eliminated by DPC}} + \underbrace{\mathbf{h}_k^{\mathrm{T}} \mathbf{v}_k \mathbf{x}_k}_{\text{desired signal}} + \underbrace{\sum_{i=k+1}^{N_{user}} \mathbf{h}_k^{\mathrm{T}} \mathbf{v}_i \mathbf{x}_i}_{\text{eliminated by ZF}} + n_k \tag{9.101}$$

[30]In linear algebra, the Gram–Schmidt orthogonalization procedure is defined as a method for orthonormalizing a set of vectors in an inner product space (e.g., the Euclidean space $\mathbb{R}^n$). The Gram–Schmidt procedure takes a finite set of linearly-independent vectors $S = \{\mathbf{v}_1, \mathbf{v}_2, \ldots, \mathbf{v}_k\} \forall k \leq n$ and generates an orthonormal set of vectors. $S' = \{\mathbf{u}_1, \mathbf{u}_2, \ldots, \mathbf{u}_k\}$ that spans the same k-dimensional subspace of $\mathbb{R}^n$ as the original set. We define the projection operator by $\mathrm{proj}_{\mathbf{u}}(\mathbf{v}) = \frac{\langle \mathbf{u}, \mathbf{v} \rangle}{\langle \mathbf{u}, \mathbf{u} \rangle} \mathbf{u}$ where $\langle \mathbf{u}, \mathbf{v} \rangle$ denotes the inner product of the vectors $\mathbf{u}$ and $\mathbf{v}$. This operator projects the vector $\mathbf{v}$ orthogonally onto the line spanned by vector $\mathbf{u}$. The Gram–Schmidt procedure then continues as follows to generate the remaining vectors in the orthonormal set S' [44]: $\mathbf{u}_1 = \mathbf{v}_1$ and $\mathbf{u}_k = \mathbf{v}_k - \sum_{i=1}^{k-1} \mathrm{proj}_{\mathbf{u}_i}(\mathbf{v}_k)$

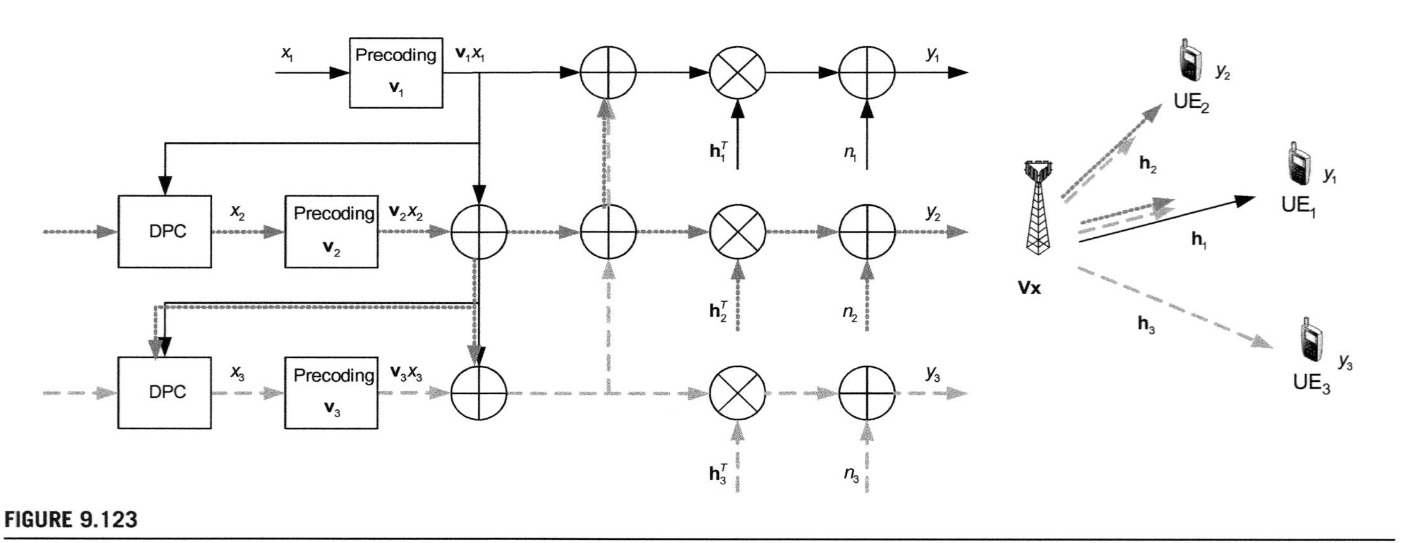

FIGURE 9.123

Sequential encoding with dirty paper coding (DPC) and ZF [46].

The sequential generation of the precoding vectors $\mathbf{v}_k$ through the Gram–Schmidt orthogonalization procedure allows optimized ordering of the users and scheduling of the users even when $N_{\text{user}} > N_{\text{TX}}$ [46]. Note that when $N_{\text{user}} > N_{\text{TX}}$, only N_{TX} users can be scheduled assuming each user has only one receive antenna. The scheduling rule in this case can be summarized as follows:

At each step $k = 1, 2, \ldots, N_{\text{TX}}$:

1. Compute precoding vector $\mathbf{v}_l$ for all users that have not been served yet.
2. For each user l, determine the effective channel $\mathbf{h}_l^{\text{T}} \mathbf{v}_l$ and the resulting sum capacity increment which would be achieved, if user l were scheduled at step k.
3. Allocate the respective resource units to the user who achieves the highest sum capacity increment, considering $C_{\text{sum-capacity}} = \sum_{k=1}^{N_{\text{user}}} C_k$, where C_k is the capacity/rate of the kth user.

9.14 LTE/LTE-Advanced downlink multi-antenna transmission schemes

Having discussed the underlying theory of multi-antenna operation, we now focus on the multi-antenna transmission techniques that have been used in LTE/LTE-Advanced. There are some practical constraints which impact the real-life performance of the theoretical MIMO systems that include limitations in practical cell deployment and antenna configurations, propagation conditions, channel characterization, measurement inaccuracies and latencies, hardware imperfections and implementation margins, and inherent implementation complexity of the multi-antenna schemes in practice. These aspects are often crucial for assessing the performance of a particular multi-antenna transmission scheme in a given propagation channel and configuration, and have been of great importance in the selection of MIMO schemes for LTE/LTE-Advanced.

The advantages of MIMO, i.e., array gain, diversity gain, and multiplexing gain, observed through link-level and system-level simulations assume ideally uncorrelated antennas and full-rank MIMO channel matrices. In this regard, the propagation environment and the antenna design (e.g., the spacing and polarization) play a significant role. If the antenna separation at the eNB and the UE is small relative to the operating wavelength, strong correlation will be observed between the spatial signatures especially in an LoS scenario, limiting the usefulness of spatial multiplexing schemes. However, this can to some extent be circumvented by means of dual-polarized antennas, which provide orthogonality even in LoS environments. Aside from use of such antennas, the condition of spatial signature independence with SU-MIMO can only be satisfied using a rich random multipath propagation environment. In diversity-based schemes, the invertibility of the channel matrix is not required, yet the entries of the channel matrix should be statistically uncorrelated. Although a greater LoS to non-LoS power ratio will tend to correlate the fading coefficients on the various antennas, this effect will be counteracted by the reduction in fading measured through the LoS component.

Another source of inconsistency between theoretical MIMO gains and practically achieved performance lies in the inability of the receiver and/or the transmitter to acquire reliable knowledge of the propagation channel. At the receiver, channel estimation is typically performed over finite samples of reference signals. In the case of beamforming and MIMO precoding, the transmitter must acquire the CSI from the receiver usually through a limited feedback channel, which causes further degradation to the available CSIT. The potential advantages of MU-MIMO over SU-MIMO include robustness with respect to the propagation environment and the preservation of spatial multiplexing gain, even in the case of UEs with a small number of antennas. However, these advantages rely on the ability of the base station to compute the required beamforming coefficients, which requires accurate CSIT; if no CSIT is available and the fading statistics are identical for all the UEs, the MU-MIMO gains disappear and the SU-MIMO becomes more advantageous. Consequently, one of the challenges in making MU-MIMO practical for cellular applications, and particularly for an FDD system, is developing efficient feedback mechanisms that allow accurate CSI to be estimated at the base station. This requires the use of appropriate codebooks for quantization.

Another issue which arises in the practical implementation of MIMO schemes is the interaction between the physical layer and the scheduling protocol. As noted earlier, in both uplink and downlink cases, the number of UEs which can be scheduled using MU-MIMO is typically limited to the number of the transmit antennas at the base station. Often one may even decide to limit the number of scheduled users to a value strictly less than the number of transmit antennas at the eNB to preserve some degrees of freedom for per-user diversity. As the number of active users typically exceeds the number of scheduled users, a sophisticated selection algorithm must be implemented to identify which set of users will be scheduled for simultaneous transmission over a particular time–frequency resource. This algorithm is not included in the specification, and various approaches are possible where a combination of rate maximization and QoS constraints will typically be considered. It is important to note that the choice of UEs that will maximize the sum rate (the sum over the individual rates for a given subframe) is one that favors UEs exhibiting not only good instantaneous SINR, but also spatial separability among their spatial signatures [14].

A 2×2 antenna configuration for MIMO is assumed as the baseline in the LTE downlink, i.e., two transmit antennas at the base station and two receive antennas at the terminal side [1,2]. Configurations with up to eight transmit or receive antennas are also supported in the specification. Different downlink MIMO modes are supported in LTE/LTE-Advanced, which can be adapted based on channel and traffic conditions, QoS requirements, and UE capability. These include transmit diversity, open-loop spatial multiplexing, codebook-based and non-codebook-based closed-loop spatial multiplexing, SU-MIMO and MU-MIMO, and precoding. The maximum number of codewords is two irrespective of the number of antennas with fixed mapping between codewords to layers. There is a semi-static switching between SU-MIMO and MU-MIMO per UE.

The objective of dynamic SU-MIMO/MU-MIMO switching is to improve system performance by allowing the network to serve a UE using either SU-MIMO or MU-MIMO. The SU/MU-MIMO switching aspect is important because, depending on the channel and traffic conditions, some UEs may be best served with SU-MIMO, while others benefit from MU-MIMO operation. As channel and traffic conditions may vary from one subframe to another, the dynamic switching aspect is important to optimize system performance. Important factors that influence the suitability of SU-MIMO versus MU-MIMO include the spatial characteristics of the MIMO channel (e.g., potential rank deficiency that could be relaxed by performing MU-MIMO), as well as the UE's typical signal-to-interference-plus-noise ratio conditions. In addition, channel and traffic variations are important factors (e.g., UE mobility or the characteristics of user traffic). MU-MIMO is more suitable when UE mobility is low and traffic conditions are stable as this improves the ability to find a suitable grouping of UEs for MU-MIMO operation.

In spatial multiplexing, up to two codewords can be mapped to different layers (i.e., up to eight layers). One codeword represents the output from one channel coder. The number of spatial layers available for transmission is equal to the rank of the channel matrix.[31] Precoding in the transmitter side is used to support spatial multiplexing. This is achieved by multiplying the signal with a precoding matrix prior to transmission. The optimum precoding matrix is selected from a predefined codebook which is known to both the eNB and UE. The optimum precoding matrix is the one which maximizes the capacity. The UE estimates the channel and selects the optimum precoding matrix. This feedback is provided to the eNB. Depending on the available bandwidth, this information is made available per resource block or group of resource blocks, since the optimum precoding matrix may vary between resource blocks. The network may configure a subset of the codebooks that the UE is able to select from. In the case of UEs with high velocity, the quality of the feedback may deteriorate. Thus, an open-loop spatial

[31]The column rank of matrix $\mathbf{P}$ is the maximum number of linearly independent column vectors of $\mathbf{P}$. The row rank of matrix $\mathbf{P}$ is the maximum number of linearly independent row vectors of $\mathbf{P}$. Equivalently, the column rank of $\mathbf{P}$ is the dimension of the column space of $\mathbf{P}$, whereas the row rank of $\mathbf{P}$ is the dimension of the row space of $\mathbf{P}$. An important property in linear algebra is that for any matrix, the column rank and the row rank are always equal. Therefore, the number of linearly independent rows or columns is simply referred to as the rank of matrix $\mathbf{P}$. It can be easily verified that the rank of $\mathbf{P}$ and $\mathbf{P}^T$ are equal. The rank is also defined as the dimension of the image of the linear transformation $\mathbf{y} = \mathbf{P}\mathbf{x}$. If a linear operator on a vector space with infinite-dimension has finite-dimensional range (i.e., a finite-rank operator), then the rank of the operator is defined as the dimension of the range. The rank of an $M \times N$ matrix is always less than or equal to min (M, N). A matrix is said to be full rank, if its rank is equal to min (M, N); otherwise, the matrix is rank deficient [29].

multiplexing mode is also supported which is based on predefined settings for spatial multiplexing and precoding. In the case of four antenna ports, different precoders are assigned cyclically to the resource elements. The eNB will select the optimum MIMO mode and the precoding configuration. The information is conveyed to the UE as part of the downlink control information on the PDCCH.

LTE Rel-8 supports spatial multiplexing with a maximum of four layers in SU-MIMO. In LTE-Advanced, the maximum number of layers for SU-MIMO was extended to eight in order to increase the peak spectral efficiency. Aside from the increased number of spatial layers, the SU-MIMO in Rel-8 and Rel-10 differ in the use of beamforming or closed-loop precoding. The SU-MIMO in Rel-8 was developed under the codebook-based precoding framework where transmit precoding vectors/matrices at the eNB are restricted within a finite set. This is because the demodulation is based on the cell-specific reference signals that are not precoded. Therefore, the precoding matrix at the eNB has to be explicitly signaled to the UE in the physical downlink control channel to enable MIMO decoding. It is obvious that codebook-based precoding limits the eNB's precoding flexibility, which is particularly important for MU-MIMO that relies on beamforming to mitigate the inter-user interference. The latter inflexibility motivated design of non-codebook-based precoding schemes in LTE-Advanced through the introduction of multi-layer demodulation reference signals. Because the demodulation reference signals are precoded with the same precoding vector/matrix as the user data/control information, the effective composite channel after precoding can be readily measured with demodulation reference signals, obviating the need to explicitly signal the transmit precoder to the UE. Furthermore, the eNB may choose arbitrary precoding vectors/matrices and achieve greatly improved precoding flexibility for both SU-MIMO and MU-MIMO schemes. Although non-codebook-based precoding has a small performance difference relative to codebook-based precoding for SU-MIMO, it is proven to be the key enabling feature for MU-MIMO in LTE-Advanced. The CSI feedback in LTE is based on implicit feedback where the UE reports a set of recommended MIMO transmission metrics (RI/PMI/CQI). The same feedback scheme is applied in LTE-Advanced for both SU-MIMO and MU-MIMO.

The CSI feedback allows downlink transmission to be adapted based on the instantaneous channel conditions. In LTE Rel-8, the CSI measurement is based on cell-specific reference signals, which is also used for data demodulation. In contrast, the CSI measurement in LTE-Advanced is based on a set of newly designed CSI-RS signals, which are low duty-cycle and low-density, and further allow a higher reuse factor compared to cell-specific reference signals. The feedback mechanisms of LTE Rel-8 and LTE Rel-10 are both based on the implicit feedback framework that has been established and tested since early 3GPP releases. The UE measures the downlink channel quality through reference signals and feeds back the CSI in the form of recommended transmission formats, which includes RI, PMI, and CQI metrics. In LTE, the CQI is defined as a set of transport block sizes, each of which translates to a maximum code rate and modulation order that can be received by the UE at a certain block error rate. As a criterion for testing the CQI report accuracy, when the reported code rate and modulation order is used for actual data transmission, the UE must be able to decode the data with a BLER of less than 0.01.

In order to support MIMO operation, the UE can be configured to report PMI and RI, in addition to CQI reports. In the case of codebook-based spatial multiplexing, precoding at the eNB is applied relative to the phase of the cell-specific reference signals for each antenna port. As a result, if the UE knows the precoding matrices that could be applicable (as defined in the configured codebook), and if it knows the transfer function of the channels from the transmit antenna ports, it can determine which codebook is the most suitable under the current channel conditions and signals this to the eNB. The preferred codebook, whose index constitutes the PMI report, is the precoder that maximizes the aggregate number of data bits which could be received across all layers. The number of resource blocks used to calculate each PMI depends on the feedback mode configured by the eNB as follows (Figure 9.124):

- Wideband PMI: The UE reports a single PMI corresponding to the preferred precoder assuming transmission over the entire system bandwidth. This is applicable for PMI feedback that is configured to be sent periodically on PUCCH, and also for PMI feedback that is sent aperiodically in conjunction with UE-selected sub-band CQI reports on PUSCH.
- Sub-band PMI: The UE reports one PMI for each sub-band over the system bandwidth, where the sub-band size is between 4 and 6 resource blocks depending on the system bandwidth (Table 9.28). This is used in scenarios where PMI feedback is sent aperiodically on PUSCH in addition to a wideband CQI report. If the eNB intends to schedule a wideband data

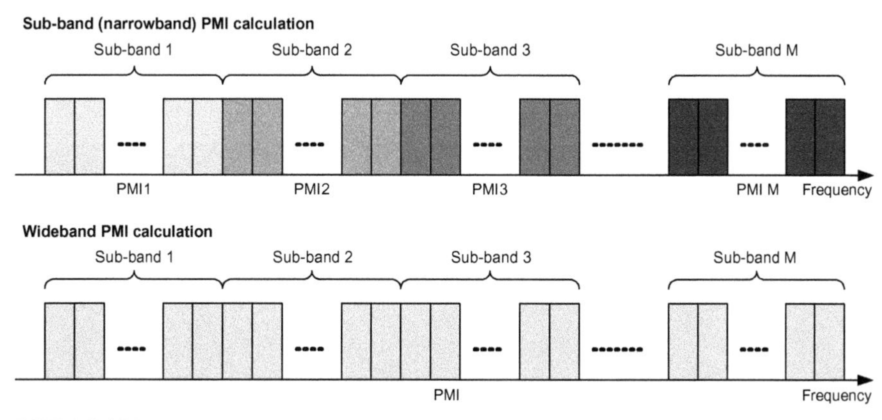

FIGURE 9.124

Illustration of wideband and sub-band PMI concept.

Table 9.28 Sub-band Sizes as a Function of System Bandwidth Applicable to the Best-M Method [5]

System Bandwidth (Number of Resource Blocks)	Sub-band Size k (Number of Resource Blocks)	Number of Preferred Sub-bands M
6–7	N/A	N/A
8–10	2	1
11–26	2	3
27–63	3	5
64–110	4	6

transmission to a UE in a frequency-selective channel, the precoder used by the eNB can change from one sub-band to another within one transport block.

- UE-selected sub-band PMI (Best-M): The UE selects a set of M preferred sub-bands, each consists of k resource blocks, and reports a single PMI corresponding to the preferred precoder assuming transmission over all of the M selected sub-bands. This is applicable in conjunction with UE-selected sub-band CQI reports on PUSCH. The eNB is not obligated to use the precoder requested by the UE; however, if the eNB selects another precoder, then it has to adjust the transmission format according to the inconsistency with the reported CQI. The eNB may also restrict the set of precoders which the UE may evaluate and report. This is known as codebook subset restriction. It enables the eNB to prevent the UE from reporting precoders which may not be useable, e.g., in some eNB antenna configurations or in correlated fading scenarios. In the case of open-loop spatial multiplexing, codebook subset restriction is equivalent to a restriction on the rank which the UE may report.

9.14.1 Transmit diversity

In LTE/LTE-Advanced, transmit diversity is only defined for two and four transmit antennas and one codeword. In order to maximize diversity gain, the antennas typically need to be uncorrelated, so they need to be sufficiently separated relative to the operating wavelength or have different polarizations. Transmit diversity is useful in a number of scenarios where the SNR is low, or for applications with low delay tolerance. Diversity schemes are also suitable for channels for which no uplink feedback is available (e.g., multimedia broadcast multicast services, physical broadcast channel, and synchronization signals), or when the feedback is not sufficiently accurate such as high-speed scenarios. Transmit diversity is also used as a fallback scheme in LTE/LTE-Advanced when other multi-antenna schemes fail. In LTE/LTE-Advanced, the multi-antenna schemes are independently configured for the physical control channels and the traffic channels, and in the case of traffic channels, it is configured independently per UE. In this section, we discuss space–frequency block codes and frequency-switched transmit diversity (FSTD) diversity schemes. Transmit diversity may be utilized for transmission of PBCH, PDCCH, and for PDSCH, if the transmit diversity mode is configured for a UE. Another transmit diversity technique which is commonly associated with OFDM is CDD. The CDD is not used in LTE as a diversity scheme, but rather as a precoding scheme for spatial multiplexing on the PDSCH.

LTE Rel-8 supports downlink transmissions on one, two, or four cell-specific antenna ports, each corresponding to one, two, or four cell-specific reference signals, where each reference signal corresponds to one antenna port. An additional antenna port associated with one UE-specific reference signal is available, as well. This antenna port can be used for conventional beamforming, especially in the case of TDD operation. The number of downlink spatial layers was increased to eight through addition of demodulation reference signals and CSI reference signals in LTE-Advanced. Transmit diversity with rank-1 is supported through Alamouti-based space–time codes in the frequency-domain. The encoding operation is performed over space and frequency dimensions so the output block $\mathbf{X}_n$ from the precoder is limited to consecutive data resource elements over a single OFDM symbol. If single-codeword transmission is assumed, the modulated symbols of a single codeword are mapped to all layers. The transmit-diversity schemes for two and four cell-specific antenna ports are also supported. Note that these transmit-diversity schemes can be used for PBCH, PDCCH, and PCFICH. Furthermore, the number of cell-specific antenna ports used to encode the PBCH is the same as the total number of configured cell-specific antenna ports, and those antenna ports are also used for other control channels. Thus, all UEs must support up to four cell-specific antenna ports and the corresponding transmit-diversity schemes. For the case of two antenna ports, the output from the precoder is given as follows:

$$\begin{matrix}\text{Antenna Port } 0\rightarrow \\ \text{Antenna Port } 1\rightarrow\end{matrix}\ \mathbf{X}_n = \begin{bmatrix} s_{n1} & s_{n2} \\ -s_{n2}^* & s_{n1}^* \end{bmatrix} \tag{9.102}$$

where the rows correspond to the antenna ports and the columns to consecutive data resource elements over the same OFDM symbol. When four antenna ports are utilized, a combination of SFBC and FSTD is used to provide robustness against the correlation between channels from different transmit antennas and for less-complex UE receiver implementation. The output from the precoder is given as follows:

$$\begin{matrix}\text{Antenna Port } 0\rightarrow \\ \text{Antenna Port } 1\rightarrow \\ \text{Antenna Port } 2\rightarrow \\ \text{Antenna Port } 3\rightarrow\end{matrix}\ \mathbf{X}_n = \begin{bmatrix} s_{n0} & s_{n1} & 0 & 0 \\ 0 & 0 & s_{n2} & s_{n3} \\ -s_{n1}^* & s_{n0}^* & 0 & 0 \\ 0 & 0 & -s_{n3}^* & s_{n2}^* \end{bmatrix} \tag{9.103}$$

The precoder in Eq. (9.103) consists of two SFBC codes which are transmitted on antenna ports 0, 2, and 1, 3, respectively. The reason for distributing a single SFBC code in such an interlaced manner on every other antenna port instead of consecutive antenna ports is related to the fact that the first two cell-specific antenna ports have a higher reference signal density than the other reference signals; therefore, the channel estimation accuracy may be lower on the third and fourth antenna ports. The above transmit-diversity scheme can be used for all downlink channels other than PHICH. For the latter case, four different ACK/NACK bits are multiplexed using orthogonal codes with a spreading factor of four over a group of four sub-carriers, and the resulting group is repeated three times in the frequency-domain to achieve frequency-diversity gain. To maintain the orthogonality between different codes in each repetition of four sub-carriers, antenna switching is not applied within each repetition. Instead, the set of antennas changes across different repetitions.

9.14.2 Spatial multiplexing

Let us review a few terms before turning our focus on spatial multiplexing schemes in LTE/LTE-Advanced. As mentioned earlier, a spatial layer refers to one of the streams generated by spatial multiplexing, or alternatively a mapping of symbols to the transmit antenna ports. Each layer is identified by a precoding vector of size equal to the number of transmit antenna ports. The rank is defined as the number of transmission layers. A codeword represents an independently encoded data block corresponding to a single transport block delivered to the physical layer by the MAC sub-layer in the transmitter and protected with a CRC. For ranks greater than one, two codewords may be transmitted (as shown in Figure 9.125). Note that the number of codewords is always less than or equal to the number of layers, which in turn is always less than or equal to the number of antenna ports.

In principle, an SU-MIMO spatial multiplexing scheme can either use a single codeword mapped to all the available layers or multiple codewords mapped to one or more spatial layers. The advantage of using one codeword is the obvious reduction in the amount of control signaling required for CQI reporting where only a single value would be needed for all layers, and for HARQ ACK/NACK

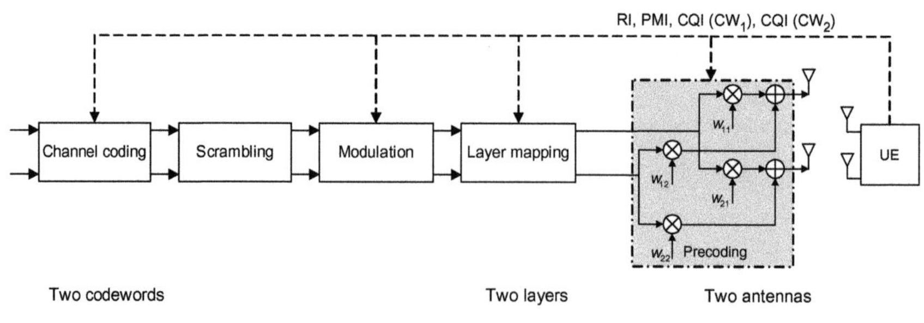

FIGURE 9.125

Example of closed-loop spatial multiplexing with precoding rank = 2 [46].

where only one ACK or NACK would be signaled per subframe per UE. In that case, the MLD receiver is optimal in terms of minimizing the BLER. At the opposite extreme, a separate codeword could be mapped to each of the layers. The advantage of such mapping is that significant gains can be achieved using an SIC receiver at the expense of additional control signaling. An MMSE-SIC receiver is shown to approach the Shannon capacity. Note that an MMSE receiver is more practical for both transmitter structures (i.e., single-codeword and multi-codeword). For LTE/LTE-Advanced, at most two codewords are used even though four layers may be transmitted. The codeword-to-layer mapping is static, since only minimal gains were shown for a dynamic mapping method. Note that in LTE/LTE-Advanced, all resource blocks associated with the same codeword use the same modulation and coding scheme (MCS), even if a codeword is mapped to multiple layers.

The PDSCH transmission modes for open-loop and closed-loop spatial multiplexing use precoding vectors derived from a predefined codebook to form the transmit layers. Each codebook consists of a set of predefined precoding matrices. In the case of closed-loop spatial multiplexing, the UE sends the index of the most desirable entry from a predefined codebook as feedback to the eNB. The preferred precoder is the weighting matrix which would maximize the capacity based on the receiver capabilities. In a single-cell and interference-free environment, the UE will typically indicate the precoder that would result in a transmission with an effective SNR that most closely follow the largest singular value of its estimated channel matrix. An example algorithm by which a UE selects the codebook which is the best fit to the propagation channel at a specific time is as follows [158]:

1. Calculate metric $\zeta(i) = \mathbf{W}_i(\mathbf{H}^{\mathrm{H}}\mathbf{H})\mathbf{W}_i^{\mathrm{H}}$ for every element in the codebook, where $\mathbf{H}$ is the channel matrix and $\mathbf{W}_i$ is the ith precoding matrix/vector.
2. Select the codebook element which minimizes the above metric $\mathrm{PMI} = \arg\min_i(\zeta(i))$.
3. Report the index of the selected codebook element (PMI) to the eNB.

Some important properties of the codebooks utilized in the precoding are as follows [14,15]:

- The LTE/LTE-Advanced precoders only perform phase corrections and no amplitude changes. This is to ensure that the power amplifiers connected to each antenna are uniformly loaded. One exception is the use of the identity matrix as the precoder. Although the identity matrix precoder may completely switch off one antenna on one layer, since each layer is still connected to one antenna at constant power, the effect across the layers is still constant modulus toward the power amplifier.
- The LTE/LTE-Advanced precoders were designed to exploit the nested property. The nested property is a method of organizing the codebooks of different ranks so that the lower rank codebook is formed by a subset of the

higher rank codebook vectors. This property simplifies the CQI calculation across different ranks. It ensures that the precoded transmission for a lower rank is a subset of the precoded transmission for a higher rank, thereby reducing the number of calculations required for the UE to generate the feedback. For example, if a specific index in the codebook corresponds to columns 1, 2, and 3 from the precoder **W** in the case of a rank-3 transmission, then the same index in the case of rank-2 transmission consists of either columns 1 and 2, or columns 1 and 3 from **W**.

- The LTE/LTE-Advanced precoders were designed to minimize complex computations. The codebook associated with dual antenna transmission consists of ± 1 and $\pm j$, which eliminates the need for any complex multiplications. The four transmit antenna codebook contains ± 1 and $\pm j$ along with magnitude scaling to enhance the performance.

Figures 9.126 and 9.127 illustrate effect of precoding on the antenna radiation pattern using precoding matrices $\mathbf{W}_0$ and $\mathbf{W}_1$ (closed-loop spatial multiplexing

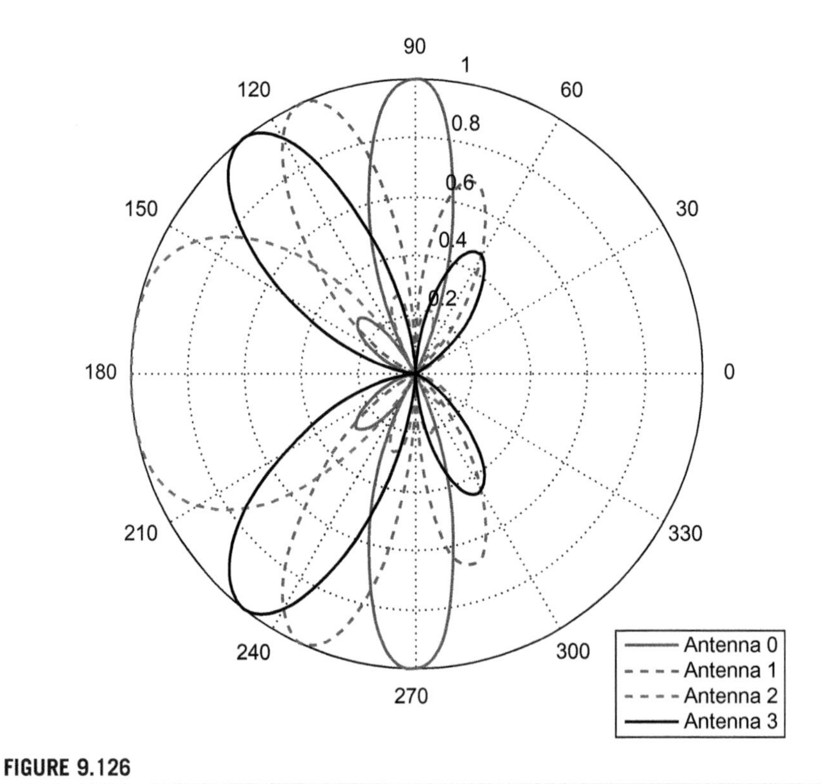

FIGURE 9.126

Effect of precoding on the antenna radiation pattern using precoding matrix $\mathbf{W}_0$ (closed-loop spatial multiplexing with four transmit antennas and rank-4 transmission).

with four transmit antennas and rank-4 transmission), respectively. The precoding matrices are defined as follows:

$$\mathbf{W}_0 = \begin{pmatrix} 0.5 & 0.5 & 0.5 & 0.5 \\ 0.5 & 0.5 & -0.5 & -0.5 \\ 0.5 & -0.5 & 0.5 & -0.5 \\ 0.5 & -0.5 & -0.5 & 0.5 \end{pmatrix}$$

and

$$\mathbf{W}_1 = \begin{pmatrix} 0.5 & -0.5j & -0.5 & 0.5j \\ 0.5j & 0.5 & 0.5j & 0.5 \\ -0.5 & -0.5j & 0.5 & 0.5j \\ -0.5j & 0.5 & -0.5j & 0.5 \end{pmatrix}$$

It can be seen that the eigen-beamforming changes the radiation pattern of the transmit antennas to direct the transmit power in a certain direction consistent with instantaneous propagation channel conditions in order to maximize the signal-to-noise ratio at the UE receiver.

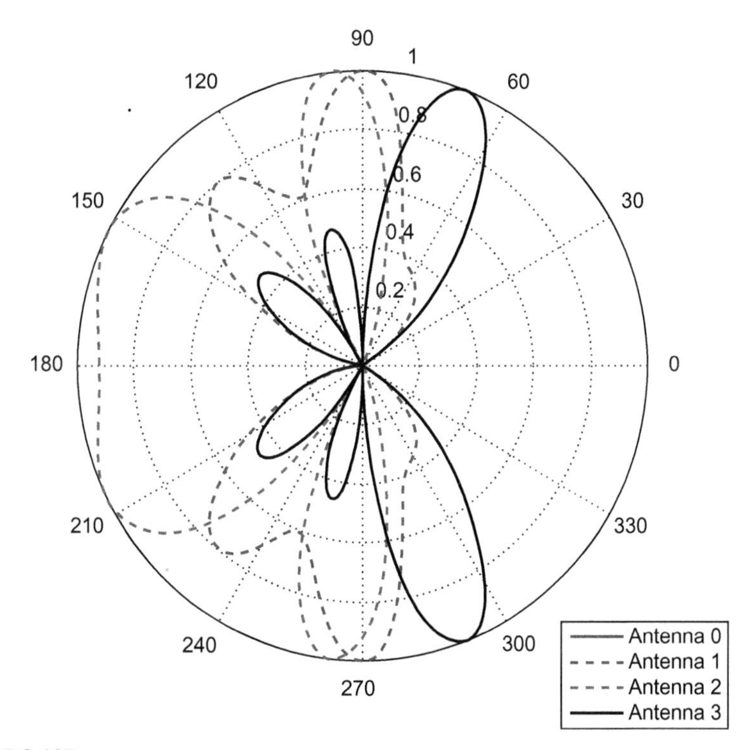

FIGURE 9.127

Effect of precoding on the antenna radiation pattern using precoding matrix $\mathbf{W}_1$ (closed-loop spatial multiplexing with four transmit antennas and rank-4 transmission).

9.14.3 Cyclic delay diversity

In the case of open-loop spatial multiplexing, the UE does not feedback PMI; rather, it sends the rank of the channel. In this mode, if the rank used for PDSCH transmission is greater than one, LTE uses a CDD scheme where the same set of OFDM symbols on the same set of sub-carriers are transmitted from multiple transmit antennas with different delays applied to each antenna. The delay is applied before the cyclic prefix is added, thereby guaranteeing that the delay is cyclic over the FFT size. Adding a time delay is equivalent to applying a phase shift in the frequency-domain. As the same time delay is applied to all sub-carriers, it corresponds to a linear phase shift across the sub-carriers. Each sub-carrier experiences a different beamforming pattern as the non-delayed sub-carrier from one antenna interferes constructively or destructively with the delayed version from another antenna. The diversity effect of CDD originates from the fact that linearly phase-shifted sub-carriers are affected differently in a multipath channel, thus artificially increasing the frequency-selectivity of the channel. Although this approach does not optimally exploit the channel in the way that ideal precoding would be matching the precoding matrix to the eigenvectors of the channel, it does help to ensure that any destructive fading is constrained to individual sub-carriers rather than affecting the entire transport block. This can be particularly beneficial if the channel information at the transmitter is unreliable, e.g., due to the limited feedback or high UE mobility. The fact that the delay is added before the cyclic prefix insertion suggests that any delay value can be used without increasing the overall delay spread of the channel. In contrast, if the delay were added after the addition of the cyclic prefix, then the range of applicable delays would be limited in order to ensure that the delay spread of the delayed symbol is not more than the maximum delay spread of the channel. The equivalence of time-delay and phase-shift implies that the CDD operation can be implemented as a frequency-domain precoder for the affected antennas, where the precoder phase is changed per sub-carrier basis according to a fixed linear function.

In LTE Rel-8, precoding for spatial multiplexing using antenna ports with cell-specific reference signals is only used in combination with layer mapping for spatial multiplexing. Spatial multiplexing supports two or four antenna ports and the set of antenna ports used is $p = 0, 1$ or $p = 0, 1, 2, 3$, respectively with p denoting the antenna port number. The precoding for spatial multiplexing without CDD is defined as $(y_i^0, y_i^1, \ldots, y_i^{N_p-1})^{\mathrm{T}} = \mathbf{W}_i (x_i^0, x_i^1, \ldots, x_i^{N_l-1})^{\mathrm{T}}$ where $\mathbf{W}_i$ is the $N_p \times N_l$ precoding matrix, N_p is the number of antenna ports, N_l denotes the number of layers, $i = 0, 1, \ldots, N_s^p - 1$, and $N_s^p = N_s^l$. For spatial multiplexing, the value of $\mathbf{W}_i$ is selected among the precoder elements in the codebook configured in the eNB and the UE. The eNB can further confine the precoder selection in the UE to a subset of the elements in the codebook using codebook subset restrictions [3].

For large-delay CDD, precoding for spatial multiplexing is defined as $(y_i^0, y_i^1, \ldots, y_i^{N_p-1})^{\mathrm{T}} = \mathbf{W}_i \mathbf{D}_i \mathbf{U} (x_i^0, x_i^1, \ldots, x_i^{N_l-1})^{\mathrm{T}}$ (Figure 9.128), where $\mathbf{D}_\mathrm{i}$ and $\mathbf{U}$ are $N_l \times N_l$ matrices supporting CDD as given in Table 9.29 for different numbers

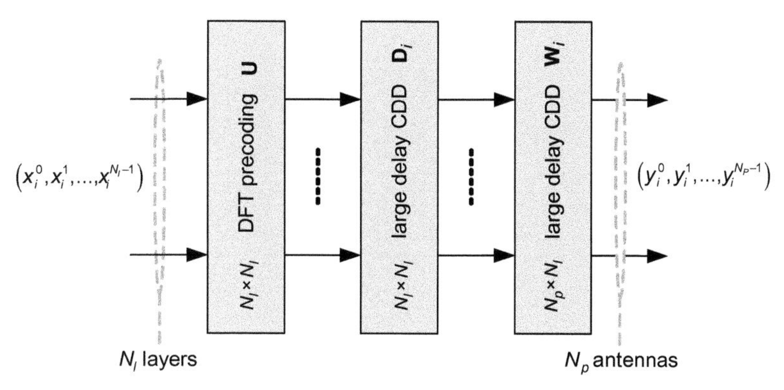

FIGURE 9.128

Open-loop spatial multiplexing with N_p antenna ports and N_l transmission layers [3].

Table 9.29 Large-Delay Cyclic Delay Diversity [3]

Number of Layers N_l	U	D_i
2	$\frac{1}{\sqrt{2}}\begin{bmatrix} 1 & 1 \\ 1 & e^{-j2\pi/2} \end{bmatrix}$	$\begin{bmatrix} 1 & 0 \\ 0 & e^{-j2\pi i/2} \end{bmatrix}$
3	$\frac{1}{\sqrt{3}}\begin{bmatrix} 1 & 1 & 1 \\ 1 & e^{-j2\pi/3} & e^{-j4\pi/3} \\ 1 & e^{-j4\pi/3} & e^{-j8\pi/3} \end{bmatrix}$	$\begin{bmatrix} 1 & 0 & 0 \\ 0 & e^{-j2\pi i/3} & 0 \\ 0 & 0 & e^{-j4\pi i/3} \end{bmatrix}$
4	$\frac{1}{2}\begin{bmatrix} 1 & 1 & 1 & 1 \\ 1 & e^{-j2\pi/4} & e^{-j4\pi/4} & e^{-j6\pi/4} \\ 1 & e^{-j4\pi/4} & e^{-j8\pi/4} & e^{-j12\pi/4} \\ 1 & e^{-j6\pi/4} & e^{-j12\pi/4} & e^{-j18\pi/4} \end{bmatrix}$	$\begin{bmatrix} 1 & 0 & 0 & 0 \\ 0 & e^{-j2\pi i/4} & 0 & 0 \\ 0 & 0 & e^{-j4\pi i/4} & 0 \\ 0 & 0 & 0 & e^{-j6\pi i/4} \end{bmatrix}$

of layers. For two antenna ports, the precoder is selected according to $\mathbf{W}_i = \mathbf{C}_1$ where $\mathbf{C}_1$ denotes the precoding matrix corresponding to precoder index zero in Table 9.30. For four antenna ports, the UE may assume that the eNB cyclically assigns different precoders to different vectors $(x_i^0, x_i^1, \ldots, x_i^{N_l-1})^T$ on the physical downlink shared channel. A different precoder is used every N_l vectors, where N_l denotes the number of transmission layers in the case of spatial multiplexing. In particular, the precoder is selected according to $\mathbf{W}_i = \mathbf{C}_k$, where k is the precoder index given by $k = \left(\lfloor i/N_l \rfloor \bmod 4\right) + 1 \in \{1, 2, 3, 4\}$, and $\mathbf{C_1}$, $\mathbf{C_2}$, $\mathbf{C_3}$, $\mathbf{C_4}$ denote precoder matrices. When two antenna ports are configured, the number of codewords is equal to the transmission rank, and codeword n is mapped to layer n.

Table 9.30 Codebook for Transmission on Antenna Ports 0 and 1 [3]

Codebook Index	Number of Layers N_l	
	1	2
0	$\frac{1}{\sqrt{2}}\begin{bmatrix}1\\1\end{bmatrix}$	$\frac{1}{\sqrt{2}}\begin{bmatrix}1 & 0\\0 & 1\end{bmatrix}$
1	$\frac{1}{\sqrt{2}}\begin{bmatrix}1\\-1\end{bmatrix}$	$\frac{1}{2}\begin{bmatrix}1 & 1\\1 & -1\end{bmatrix}$
2	$\frac{1}{\sqrt{2}}\begin{bmatrix}1\\j\end{bmatrix}$	$\frac{1}{2}\begin{bmatrix}1 & 1\\j & -j\end{bmatrix}$
3	$\frac{1}{\sqrt{2}}\begin{bmatrix}1\\-j\end{bmatrix}$	–

For transmission on two antenna ports ($p = 0,1$), the precoding matrix $\mathbf{W}_i$ is selected from Table 9.30 or a subset of it. The open-loop spatial multiplexing may be used when reliable PMI feedback is not available at the eNB, e.g., when the UE moves fast or when the feedback overhead on the uplink is extremely high. The open-loop spatial multiplexing with N_l layers and N_p antenna ports ($N_p \geq N_l$) is illustrated in Figure 9.128. The feedback consists of the RI and the CQI in open-loop spatial multiplexing. In contrast to the closed-loop spatial multiplexing, the eNB only determines the transmission rank, and a fixed set of precoding matrices are applied cyclically across all the scheduled sub-carriers in the frequency-domain. It is noted that in the open-loop spatial multiplexing mode, the transmit diversity scheme is applied when the transmission rank is set to one [3].

9.14.4 Beamforming and MU-MIMO

In LTE/LTE-Advanced, two beamforming techniques are supported for the PDSCH transmission: (1) closed-loop rank-1 precoding; and (2) beamforming with UE-specific reference signals. In closed-loop rank-1 precoding, only a single layer is transmitted, and it can also be seen as a special case of the SU-MIMO spatial multiplexing using codebook-based precoding. In this mode, the UE sends CSI feedback (PMI) to the eNB to indicate suitable precoding to apply for the beamforming. In a beamforming scheme using UE-specific reference signals, the precoding is not restricted to a predefined codebook, thus the UE will rely on DM-RS as opposed to cell-specific reference signals for PDSCH demodulation. Rel-8 of LTE only supported single-layer beamforming and the scheme was extended to support dual-layer beamforming in Rel-9. These transmission modes are used to extend cell coverage by concentrating the eNB power in the directions in which the radio channels provide the strongest paths to the UE. They are typically implemented with an array of closely spaced antenna elements (or pairs of cross-polarized elements) for creating directional transmissions (Figures 9.129 and 9.130).

The geometric design of the antenna array significantly affects the radiation characteristics. This is discussed here using the example of conventional base

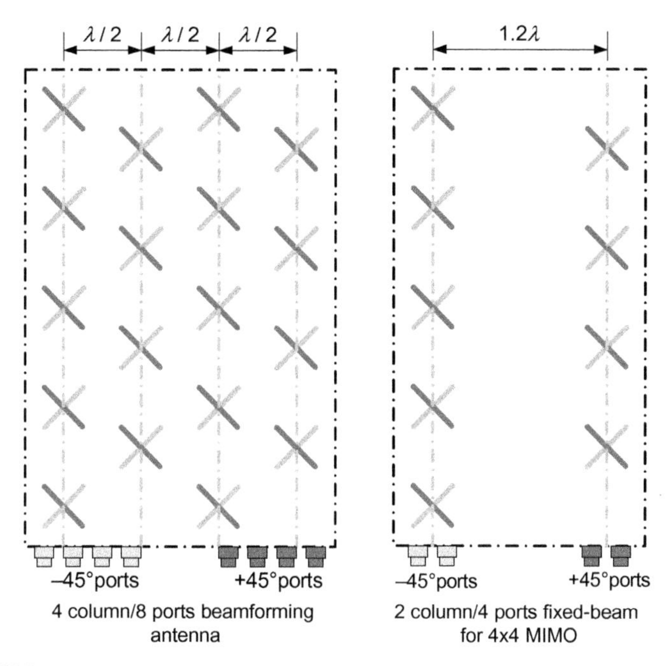

FIGURE 9.129

Practical configuration of beamforming and fixed-beam quad-port cross-polarized physical antenna elements [146].

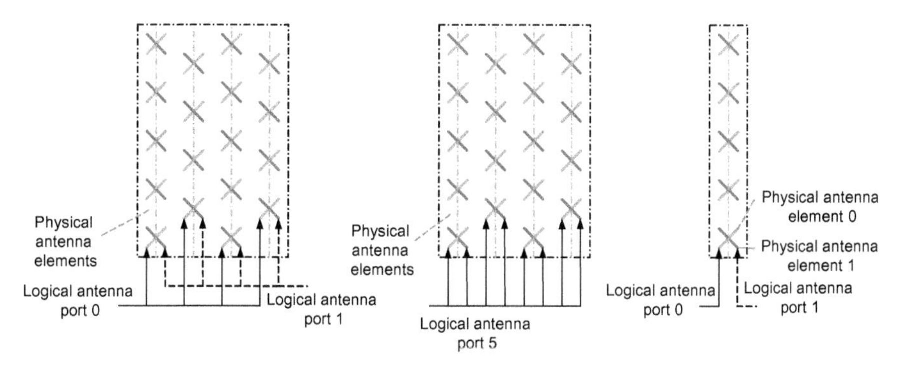

FIGURE 9.130

Example mapping of one/two antenna ports to multiple cross-polarized physical antenna elements [13].

station antennas. At present, conventional passive base station antennas are typically made up of multiple cross-polarized elements. In the y-axis, multiple elements are combined in order to satisfy the cell coverage requirements. All elements that have the same polarity radiate the same signal (see Figures 9.127 and 9.128). Particularly relevant for MIMO and beamforming is the arrangement of the cross-polarized elements and the columns in the x-axis. In Figure 9.129, the antenna at the left consists of two elements arranged at 90° relative to each other (cross-polarized). Each polarization column represents an antenna element that can transmit a different signal. This makes it possible to transmit four signals with a compact antenna arrangement, such as in 4×4 MIMO or transmit diversity. Analogously, the antenna shown in the figure can radiate four independent signals ($4 \times N$ MIMO), while the antenna at the right can radiate eight independent signals ($2 \times N$ MIMO). The antennas shown in Figure 9.129 can be used for beamforming. Beamforming requires correlated channels, i.e., elements with the same polarization (+45° or −45°) must be used. Also, the distance between the columns should be small relative to the wavelength. Beamforming can be carried out with two antenna elements (columns with the same polarization) in the antenna array shown in the left side of Figure 9.129, or with four antenna elements in the layout on the right side. Base station antenna designs are constantly evolving. Active antennas are an important trend that allow seamless integration of beamforming concepts, e.g., by implementing dedicated transceivers for the required number of antenna elements [164].

The mapping of the (logical) antenna ports to physical antenna elements of the eNB may be realized in different ways depending on the multi-antenna scheme that is used. Figure 9.130 shows three example cases where one or two antenna ports are mapped to an array of cross-polarized physical antenna elements. As shown in the figure, the array of physical antennas radiates one signal corresponding to antenna port 5 or two signals corresponding to antenna ports 0 and 1. In the case of dual antenna port transmission, the UE sees two downlink transmissions, one associated with antenna port 0 and another one associated with antenna port 1. The physical antenna mapping and arrangement are transparent to the UE. The composite radiation pattern of the eNB transmit antennas are determined based on the combination of individual antenna radiation pattern, array antenna radiation pattern, and the precoding weights (vectors) that are applied for eigen-beamforming. Two example composite radiation patterns are shown in Figures 9.126 and 9.127.

In the case where UE-specific reference signals are used, due to the fact that no precoding codebook is used (non-codebook-based transmission), an arbitrary number of transmit antennas is theoretically supported; in practical deployments up to eight antennas are typically used (e.g., an array of four cross-polarized pairs). The signals from the correlated antenna elements are phased appropriately so that they add up constructively at the location of the UE. The UE is not aware that it is receiving a precoded directional transmission as opposed to a cell-wide transmission, except that the UE is instructed to use UE-specific reference signals as the phase reference for demodulation. To the UE, the phased array of antennas appears as one antenna port. In LTE Rel-8, only a single UE-specific antenna port and associated reference signal is defined. As a consequence the beamformed PDSCH can only be transmitted with a single spatial layer. The support of two UE-specific antenna ports was introduced in Rel-9, thus enabling beamforming with two spatial layers with direct one-to-one mapping between the layers and antenna ports.

The UE-specific reference signals in LTE Rel-9 were designed to be orthogonal in order to facilitate separation of the spatial layers at the receiver. The two associated PDSCH layers can then be transmitted to a single user (SU-MIMO) in good channel conditions to increase the individual data rate, or to multiple users (MU-MIMO) to increase the system capacity. The LTE Rel-10 further extended the support for beamforming with an increased number of UE-specific reference signals and spatial layers. When two layers are transmitted over multiple spatial streams, they share the available transmit power of the eNB, thus the increase in data rate comes at the expense of a reduction of coverage. This makes rank adaptation, which is used in the codebook-based spatial multiplexing transmission mode, an important mechanism for the dual-layer beamforming. In this case, the UE is configured to feedback RI in conjunction with the PMI. It should be noted that beamforming in LTE can only be applied to PDSCH and not to the downlink control channels, although the coverage of a traffic channel can be extended through beamforming, the overall cell coverage may still be limited by the control channels. Similar to any other precoding technique, the beamforming schemes in LTE require reliable CSIT, which can ideally be obtained either via UE feedback or based on estimation of the uplink channel condition.

In the single-layer beamforming mode of Rel-8, the UE does not feed back any precoding-related information, and the eNB extracts this information from the uplink, e.g., using DoA estimations.[32] It is worth noting that in this case

---

[32]If the position of the UE is known, the beamforming weights can be adapted accordingly to optimize transmission for the UE. Therefore, signal classification algorithms such as multiple signal classification (MUSIC) can be used in the base station to determine the DoA for the UE signal and thus determine its location. A uniform linear array (ULA) antenna array is typically used, where the distance d between the individual antennas is the same and $d \leq \lambda/2$. This type of array can be seen as a spatial filtering and sampling in the signal space. Just as the Nyquist criterion applies to sampling a signal over time, the distance here must be $d \leq \lambda/2$ in order to determine the DoA [164].
The MUSIC algorithm estimates the frequency content of a signal or autocorrelation matrix using an eigen-space method. This method assumes that a signal $x(n)$ consists of L complex exponentials in the presence of additive white Gaussian noise. Given an $M \times M$ autocorrelation matrix $\mathbf{R_x}$, if the eigenvalues are sorted in decreasing order, the eigenvectors corresponding to the L largest eigenvalues span the signal subspace. Note that for $M = L + 1$, the result from MUSIC algorithm is identical to that of Pisarenko's method. The general idea is to use averaging to improve the performance of the Pisarenko's estimator. The frequency estimation function for MUSIC is given by $\hat{P}_{\mathrm{MU}}(e^{j\omega}) = \left(\sum_{i=L+1}^{M} \left|\mathbf{e}^{\mathrm{H}}\mathbf{v}_i\right|^2\right)^{-1}$ where $\mathbf{v}_i$'s are the noise eigenvectors and vector $\mathbf{e} = (1, e^{j\omega}, e^{j2\omega}, \ldots, e^{j(M-1)\omega})^{\mathrm{T}}$. The locations of the L largest peaks of the estimation function give the frequency estimates for the L signal components [28].

calibration of the eNB RF paths may be necessary. The dual-layer beamforming mode of LTE Rel-9 provides the possibility for the UE to help the eNB to derive the beamforming precoding weights by sending PMI feedback in the same way as for codebook-based closed-loop spatial multiplexing. For beamforming, this is particularly suitable for FDD deployments where channel reciprocity cannot be exploited as effectively as in the case of TDD. Nevertheless, this PMI feedback can only provide partial CSIT, since it is based on measurements of the cell-specific reference signals which, in beamforming deployments, are usually transmitted from multiple antenna elements with broad radiation patterns; furthermore, the PMI feedback uses the same codebook as for closed-loop spatial multiplexing; therefore it is constrained by quantization granularity. For CQI feedback, the eNB must take into account the fact that in the beamforming modes the channel quality experienced by the UE on the beam-formed PDSCH resources will typically be different than on the cell-specific reference signals based on which the CQI is calculated. The UE-specific reference signals cannot be used for CQI estimation because they are only in existence in the resource blocks assigned to a specific user. It must be noted that beamforming also reduces the interference to neighboring cells. In addition, the beamforming may be coordinated among neighboring eNBs to take advantage of the spatial interference created by beamforming in multiple cells [14,15].

Other algorithms may be used in which the optimum beamforming precoder is determined by channel estimation. In a TDD system, uplink and downlink are on the same frequency, thus the channel characteristics are the same based on the reciprocity principle. That is why a feedback is not needed from the UE when a suitable uplink signal is present that the base station can use to estimate the channel. In the case of LTE TDD, the uplink sounding reference signal can be used.

In LTE Rel-8, a UE configured in MU-MIMO mode assumes that an eNB transmission on the PDSCH would be performed on one layer using one of the rank-1 codebook entries defined for two or four antenna ports. Therefore, the MU-MIMO transmission is limited to a single layer and single codeword per scheduled UE, and the RI and PMI feedback are similar to a SU-MIMO rank-1 transmission. The eNB can simultaneously schedule multiple UEs on the same resource blocks, typically by pairing UEs that report orthogonal PMIs. The eNB can therefore signal to those UEs (using one bit in the DCI format 1D message on PDCCH) that the PDSCH power is reduced by 3 dB relative to the cell-specific reference signals. This assumes that the transmission power of the eNB is equally divided between the two codewords and implies that in practice a maximum of two UEs may be concurrently scheduled in Rel-8 MU-MIMO mode. The CQI reporting for MU-MIMO in Rel-8 does not take into account any interference from transmissions to jointly scheduled users. The CQI values reported are equivalent to those that would be reported for a rank-1 SU-MIMO transmission. In LTE Rel-9, the support for MU-MIMO is enhanced by the introduction of dual-layer beamforming in conjunction with precoded UE-specific reference signals. This eliminates the constraint on the use of the feedback for precoding and provides more flexibility to utilize other approaches to MU-MIMO user separation, e.g., according to a zero-forcing beamforming criterion. Note that with an increasing number of UEs, zero-forcing beamforming converges to unitary precoding. The use of precoded DM-RSs eliminates the need for explicitly signaling the precoder to the UEs. Instead, the UEs estimate the effective channel, i.e., the combination of the precoding matrix and physical channel. The DM-RS design in Rel-9 allows for up to four UEs to be scheduled for MU-MIMO transmissions [14].

9.15 PDSCH transmission modes

Each PDSCH transmission mode defines a specific set of physical layer procedures and configurations based on which the transport blocks are processed and the complex-valued baseband signals for transmission on the antenna ports are generated. There are 10 transmission modes that have been defined for LTE/LTE-Advanced of which transmission modes 1–7 apply to LTE Rel-8, transmission mode 8 was defined for LTE Rel-9, transmission mode 9 was developed in Rel-10, and transmission mode 10 was introduced in LTE Rel-11. The transmission modes in LTE/LTE-Advanced can be explained as follows [5,14,15]:

- *Transmission mode 1* uses only one transmit antenna and is a SISO or SIMO mode.
- *Transmission mode 2* utilizes transmit diversity and is the default MIMO mode in LTE. It sends the same information via various antennas, whereby each antenna stream uses different space–time coding and different frequency resources. This improves the signal-to-noise ratio and makes transmission more robust. In LTE, transmit diversity is used as a fallback option for some transmission modes, such as when spatial multiplexing cannot be used. Control channels, such as PBCH and PDCCH, are also transmitted using transmit diversity. For two antennas, a frequency-based version of the Alamouti code [58] or SFBC is used, while for four antennas, a combination of SFBC and FSTD is used.
- *Transmission mode 3* supports spatial multiplexing of two to four layers that are multiplexed to two to four antennas, respectively, in order to achieve higher data rates. It requires less UE feedback on the channel condition, i.e., no precoding matrix indicator is included and is used when channel information is missing or when the channel rapidly changes, e.g., for UEs moving at high velocity. In addition to the precoding, the signal is fed to each antenna with a specific delay (CDD), thus artificially creating frequency diversity. For two transmit antennas, a fixed precoding (codebook index 0 is used), while for four antennas, the precoders are cyclically switched.
- *Transmission mode 4* supports spatial multiplexing with up to four layers that are multiplexed to up to four antennas, respectively, in order to achieve higher data rates. To allow channel estimation at the receiver, the base station

transmits cell-specific reference signals, distributed over various resource elements and over various time slots. The UE sends a response regarding the channel condition, which includes information about which precoding is preferred from a predefined codebook. This is accomplished using an index (PMI) to the codebook, which is a table with possible precoding matrices that is known to both sides.

- *Transmission mode 5* is similar to mode 4. It uses codebook-based closed-loop spatial multiplexing; however, one layer is dedicated for each UE.
- *Transmission mode 6* is a special type of closed-loop spatial multiplexing in mode 4. In contrast to mode 4, only one layer is used (rank-1). The UE estimates the channel and sends the index of the most suitable precoding matrix back to the base station. The base station sends the precoded signal via all antenna ports. The precoding in the baseband changes the magnitude and phase of the signals to different antennas and results in beamforming effect. With four transmit antennas, there are 16 different beamforming diagrams. This implicit beamforming effect is distinguished from classical beamforming used in transmission modes 7 and 8, that are aiming at achieving a direct impact on the antenna pattern, e.g., for covering particular areas of a cell.
- *Transmission mode 7* uses UE-specific reference signals. Both data and reference signals are transmitted using the same antenna precoding. Because the UE requires only the UE-specific reference signal for demodulation of the PDSCH, the data transmission for the UE appears to have been received from only one transmit antenna, and the UE does not see the actual number of transmit antennas. Therefore, this transmission mode is also called single antenna port (port 5). The transmission appears to be originated from a single virtual antenna port 5. The LTE standard does not specify any methods for determining the beamforming parameters. Other methods such as beam-switching are also possible. The number of antennas and the antenna architecture are not specified and are implementation specific.
- *Transmission mode 8* specifies dual-layer beamforming which allows the base station to weight two layers individually at the antennas so that beamforming can be combined with spatial multiplexing for one or more UEs. As in mode 7, UE-specific reference signals are used in this mode. Since the same resource elements are used, the reference signals must be coded differently so that the UE can distinguish between them. Because two layers are used, both layers can be assigned to one UE (SU-MIMO), or the two layers can be assigned to two separate UEs (MU-MIMO).
- *Transmission mode 9* was introduced in Rel-10 and uses non-codebook-based precoding which supports up to eight layers using UE-specific reference signals.
- *Transmission mode 10* was introduced in Rel-11 to support CoMP operation and to configure Rel-11 non-zero-power CSI-RS resources and interference measurement resources. This new transmission mode was defined to provide Rel-11 UEs capable of operating on new carrier types, which do not carry CRS, to conduct the CSI measurement and feedback necessary for CoMP

essential functions. More specifically, transmission mode 10 is used for all channel measurements on Rel-11 CSI-RS resources, all interference measurements on Rel-11 interference measurement resources, and PDSCH demodulation based on UE-specific reference signals.

The performance of LTE Rel-8 transmission modes in terms of BLER and throughput as a function of SNR for a number of antenna configurations have been simulated and are shown in Figures 9.131 and 9.132. In these figures, OLSM and CLSM refer to open-loop and closed-loop spatial multiplexing,

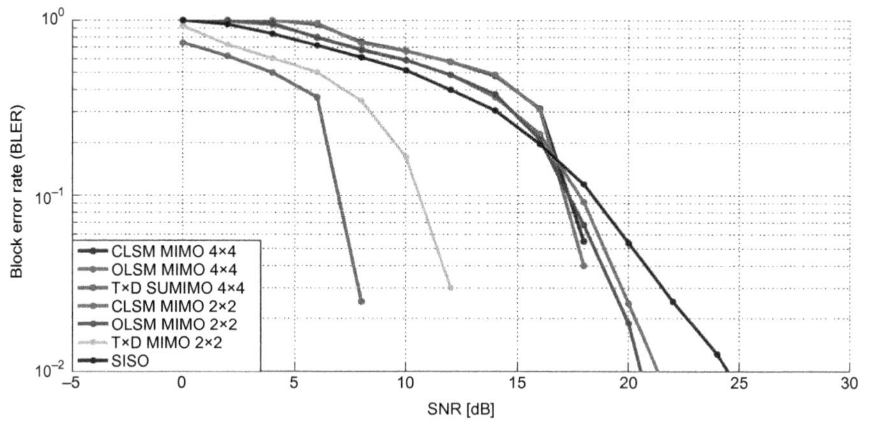

FIGURE 9.131

BLER of LTE transmission modes (modified ITU-VehA 120 km/h,CQI9,800 Subframes) [141,142].

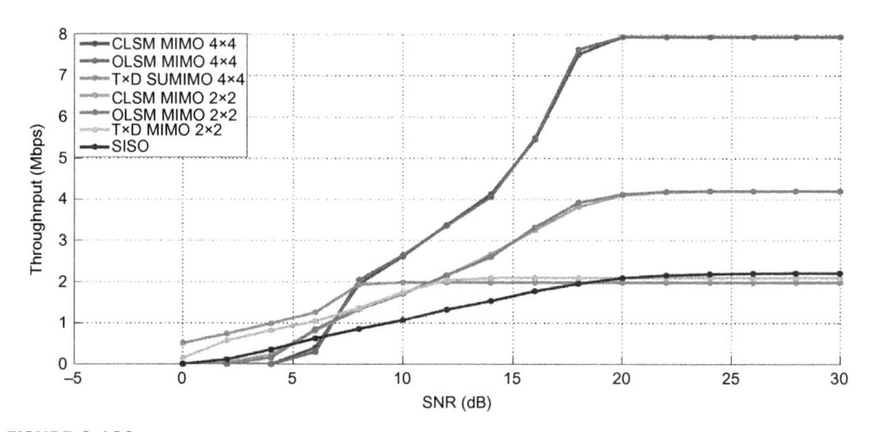

FIGURE 9.132

Throughput of LTE transmission modes (modified ITU-VehA 120 km/h) [141,142].

respectively (transmission modes 3 and 4), and TxD denotes transmit diversity (transmission mode 2) [141,142].

When a UE is configured to decode PDCCH with CRC that has been scrambled by C-RNTI, the UE is required to decode PDCCH and the corresponding PDSCH according to combinations defined in Table 9.31. Note that PDSCH transmission corresponding to these PDCCHs is scrambled by C-RNTI. Similar PDSCH transmission specifications when configured with other UE identifiers such as SPS C-RNTI are given in 3GPP TS 36.213 [5].

The UE can be semi-statically configured via higher layer signaling to receive PDSCH data transmissions signaled via ePDCCH according to one of the transmission modes 1−10. In transmission mode 10, a UE can be configured with scrambling identities $n_{\mathrm{ID}}^{\mathrm{DM\text{-}RS}_i}|_{i=0,1}$ for UE-specific reference signal generation to decode PDSCH according to a detected PDCCH/ePDCCH with CRC scrambled by the C-RNTI with DCI format 2D intended for the UE [5].

A UE in transmission mode 10 can be configured with one or more CSI processes per serving cell by higher layers. Each CSI process is associated with a CSI-RS resource and a CSI-interference measurement (CSI-IM) resource. In transmission mode 10, PUSCH PMI feedback/CQI feedback modes 1-2, 2-2, and 3-1 are supported, if the UE is configured by the eNB to report PMI/RI and the number of CSI-RS ports is greater than one; otherwise, PUSCH PMI feedback/CQI feedback modes $2-0$, and $3-0$ are supported, if the UE is configured without PMI/RI reporting or the number of CSI-RS ports is equal to one. A UE in transmission mode 10 can be configured by higher layers for multiple periodic CSI reports corresponding to one or more CSI processes per serving cell on PUCCH.

In LTE Rel-11 the concept of quasi-collocated antenna ports was introduced in order to support distributed antenna systems. Two antenna ports are quasi-collocated, if the large-scale properties of the channel over which a symbol on one antenna port is conveyed can be inferred from the channel over which a symbol on the other antenna port is conveyed. The large-scale properties include one or more of delay spread, Doppler spread, Doppler shift, average gain, and average delay. The UE does not assume that two antenna ports are quasi-collocated unless specified otherwise.

As a result, when a Rel-11 UE is configured in transmission mode 10 on a given serving cell, it can be configured with up to 4 parameter sets by higher layer signaling in order to decode PDSCH according to a detected PDCCH/ ePDCCH with DCI format 2D which is intended for the UE and the given serving cell. The UE then uses the parameter set according to the value of the PDSCH Resource Element Mapping and Quasi-Collocation indicator field in the detected PDCCH/ePDCCH with DCI format 2D for determining the PDSCH resource element mapping and antenna port quasi-collocation, provided that the UE is configured with Type B quasi-collocation. For PDSCH without a corresponding PDCCH/ePDCCH, the UE uses the parameter set indicated in the PDCCH/ ePDCCH with DCI format 2D corresponding to the associated semi-persistent scheduling activation for determining the PDSCH resource element mapping and antenna port quasi-collocation [5].

Table 9.31 Characteristics of PDSCH Modes when Configured with C-RNTI [5]

Transmission Mode	Release	DCI Format	PDCCH Search Space	Transmission Format of PDSCH	Number of Codewords	CSI Feedback from UE	Number of Layers	Number of Antenna Ports	Type of Reference Signals
1	8	1A	Common and UE-specific	Single-antenna port	1	CQI	1	1 (AP0)	Cell-specific
		1	UE-specific	Single-antenna port			1	1 (AP0)	Cell-specific
2	8	1A	Common and UE-specific	Transmit diversity	1	CQI	2,4	2,4	Cell-specific
		1	UE-specific	Transmit diversity			2,4	2,4	Cell-specific
3	8	1A	Common and UE-specific	Transmit diversity	1	CQI,RI	2,4	2,4	Cell-specific
		2A	UE-specific	Large delay CDD or transmit diversity	2		2,4	2,4	Cell-specific
4	8	1A	Common and UE-specific	Transmit diversity	1	CQI,RI,PMI	2,4	2,4	Cell-specific
		2	UE-specific	Closed-loop spatial multiplexing or transmit diversity	2		2,4	2,4	Cell-specific
5	8	1A	Common and UE-specific	Transmit diversity	1	CQI,PMI	2	2	Cell-specific
		1D	UE-specific	Multi-user MIMO	1		2	2	Cell-specific
6	8	1A	Common and UE-specific	Transmit diversity	1	CQI,PMI	2	2	Cell-specific
		1B	UE-specific	Closed-loop spatial multiplexing using a single transmission layer	1		1	1	UE-specific

(*Continued*)

Table 9.31 (Continued)

Transmission Mode	Release	DCI Format	PDCCH Search Space	Transmission Format of PDSCH	Number of Codewords	CSI Feedback from UE	Number of Layers	Number of Antenna Ports	Type of Reference Signals
7	8	1A	Common and UE-specific	If the number of PBCH antenna ports is one, single-antenna on AP0; otherwise, transmit diversity	1	CQI	1,2,4	1 (AP0) or 2,4	Cell-specific
		1	UE-specific	Single-antenna	1		1	1 (AP5)	UE-specific
8	9	1A	Common and UE-specific	If the number of PBCH antenna ports is one, single-antenna on AP0; otherwise, transmit diversity	1,2	CQI	1,2,4	1 (AP0) or 2,4	UE-specific
		2B	UE-specific	Dual layer transmission on AP7 and AP8 or single-antenna on AP7 or AP8	1,2		1,2	1 (AP7 or AP8) 2 (AP7 and AP8)	UE-specific
9	10	1A	Common and UE-specific	Non-MBSFN subframe: if the number of PBCH antenna ports is one, single-antenna on AP0; otherwise, transmit diversity MBSFN subframe: single-antenna on AP7	1,2	CQI (PTI, PMI, and RI if configured by the eNB)	Non-MBSFN: 1,2,4 MBSFN: 1	Non-MBSFN: 1 (AP0),2,4 MBSFN: 1 (AP7)	UE-specific

		2C	UE-specific	Up to eight layer transmission on AP7-14 or single-antenna on AP7 or AP8	1,2		Up to 8	Up to 8 (AP7-AP14) 1 (AP7 or AP8)	UE-specific
10	11	1A	Common and UE-specific	Non-MBSFN subframe: if the number of PBCH antenna ports is one, single-antenna on AP0; otherwise, transmit diversity MBSFN subframe: single-antenna on AP7	1,2	CQI (PTI, PMI, and RI if configured by the eNB)	Non-MBSFN: 1,2,4 MBSFN: 1	Non-MBSFN: 1 (AP0),2,4 MBSFN: 1 (AP7)	UE-specific
		2D	UE-specific	Up to eight layer transmission on AP7-14 or single-antenna on AP7 or AP8	1,2		Up to 8	Up to 8 (AP7-AP14) 1 (AP7 or AP8)	UE-specific

9.16 OFDM signal generation

The continuous time-domain signal $s_l^p(t)$ on antenna port p in OFDM symbol l in a downlink slot (assuming 12 sub-carriers per resource block) is defined as follows [3]:

$$s_l^p(t) = \sum_{k=-\lfloor 6N_{RB}^{DL} \rfloor}^{-1} a_{k^-}^p(l) e^{j2\pi k\Delta f(t-N_{CP}(l)T_s)} + \sum_{k=1}^{\lceil 6N_{RB}^{DL} \rceil} a_{k^+}^p(l) e^{j2\pi k\Delta f(t-N_{CP}(l)T)} \quad (9.104)$$

where $0 \le t < (N_{CP} + N_{FFT})T_s$, $k^- = k + \lfloor 12N_{RB}^{DL} \rfloor$, $k^+ = k + \lfloor 6N_{RB}^{DL} \rfloor - 1$, and the parameter $N_{FFT} = 2048$ for $\Delta f = 15$ kHz sub-carrier spacing. The OFDM symbols in a slot are transmitted in increasing order of index l starting at $l = 0$, where OFDM symbol $l > 0$ starts at time $\sum_{l'=0}^{l-1}[N_{CP}(l') + N_{FFT}]T_s$ within the slot. In the case where the first OFDM symbol(s) in a slot use a normal cyclic prefix and the remaining OFDM symbols use an extended cyclic prefix (e.g., MBSFN subframes), the starting position of the OFDM symbols with an extended cyclic prefix are identical to those in a slot where all OFDM symbols use an extended cyclic prefix. Therefore, there will be a part of the time slot between the two cyclic prefix regions where the transmitted signal is not specified.

9.17 Timing advance

In LTE/LTE-Advanced, when a UE attempts to establish a RRC connection with the eNB, it transmits a random access preamble from which the eNB estimates the transmission time (Figure 9.133). Then the eNB transmits a random access response which includes a timing advance command, based on which the UE adjusts the uplink frame transmit timing. Transmission of the uplink radio frame number k from the UE starts $(N_{TA} + N_{TA}^{offset})T_s$ seconds before the start of the corresponding downlink radio frame at the UE where $0 \le N_{TA} \le 20,512$ (samples) and $N_{TA}^{offset} = 0$ for frame structure type 1, and $N_{TA}^{offset} = 624$ (samples) for frame structure type 2. Note that there may be no transmission in some slots in a radio frame. Upon reception of a timing advance command, the UE is required to adjust its uplink transmission timing for the PUCCH, PUSCH, and SRS of the primary cell. The timing advance command indicates the change of the uplink timing

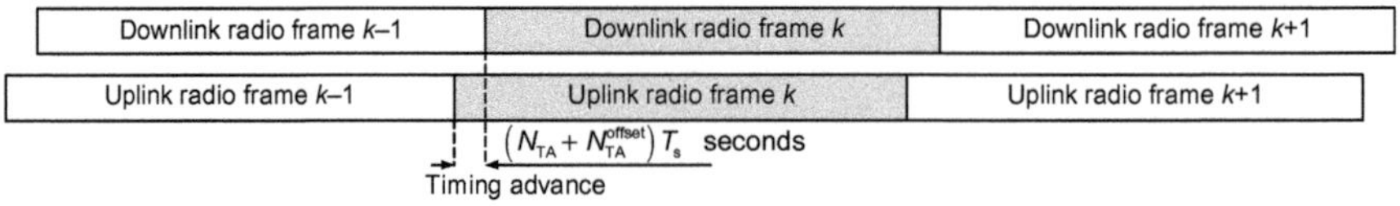

FIGURE 9.133

Relative timing of uplink and downlink radio frames [3,5].

relative to the current uplink timing in multiples of $16T_s$ (samples). The start timing of the random access preamble is discussed in Chapter 10. In LTE Rel-10, the uplink transmission timing for the PUSCH and SRS of a secondary cell is the same as the primary cell.

In the case of random access response, an 11-bit timing advance command parameter $T_A = 0, 1, 2, ..., 1282$ indicates the N_{TA} values where the amount of the timing alignment is given by $N_{TA} = 16T_A$. In other cases, a 6-bit timing advance command parameter $T_A = 0, 1, 2, ..., 63$ indicates the adjustment of the current N_{TA} value where the updated N_{TA} value is given by $N_{TA}^{New} = N_{TA}^{Old} + 16(T_A - 31)$. Note that the adjustment of N_{TA} value by a positive or a negative offset indicates advancing or deferring the uplink transmission time by a given amount, respectively. For a timing advance command received in subframe n, the timing adjustment is applied from subframe $n + 6$ onward. When the UE's uplink PUCCH, PUSCH, and SRS transmissions in subframe n and subframe $n + 1$ are overlapping due to the timing adjustment, the UE transmits the entire subframe n and does not transmit the overlapped portion of subframe $n + 1$. If the received downlink timing changes and the uplink–downlink offset is not compensated or is only partially compensated by the uplink timing adjustment without timing advance command, the UE is expected to change N_{TA} accordingly [3]

After initial uplink synchronization via the random access channel, the UE maintains its uplink frame timing alignment using the timing advance control element sent as part of a MAC PDU on PDSCH that allows the eNB to periodically send timing commands to the UE. The network may also send timing commands as part of a random access response message. The timing advance is a 6-bit number that adjusts timing (advances or delays) relative to the current UE timing, when sent as a MAC control element. The timing advance for random access response is an 11-bit quantity that specifies absolute uplink frame timing relative to the downlink frame. The eNB calculates the desired timing advance based on uplink transmissions from the UE. The eNB also configures the UE's timing alignment timer. If the UE does not receive a timing advance MAC control element before the timing alignment timer expires, it is no longer considered to be uplink-synchronized with the serving cell. When this timer expires, the UE must flush HARQ buffers and reinitialize the HARQ processes. The UE must further release all PUCCH and SRS resources. If any uplink transmission is required, the UE must perform a contention-based random access procedure in order to obtain timing alignment and an uplink grant before it can transmit in the uplink [5].

After the UE has been synchronized to the downlink transmissions from the eNB, the initial timing advance is set by means of the random access procedure in order to establish uplink synchronization. This requires the UE to transmit a random access preamble from which the eNB estimates the uplink timing and responds with an 11-bit initial timing advance command contained in the random access response message. This allows the timing advance to be configured by the eNB with a granularity of 0.52 μs from 0 up to a maximum of 0.67 ms corresponding to a cell radius of 100 km. A cell range of 100 km is sufficient for most practical rural deployment scenarios. The granularity of 0.52 μs enables the uplink transmission timing to be set with an accuracy within the length of the uplink SC-FDMA symbol cyclic prefix. This granularity is also significantly finer than the length of a cyclic shift of the uplink reference signals. Studies have shown that timing misalignment of up to 1 μs does not cause any significant degradation in system performance due to increased intra-cell interference. Thus the granularity of 0.52 μs is sufficiently fine to allow for additional timing errors arising from the uplink timing estimation in the eNB and the accuracy with which the UE sets its initial transmission timing.

Following the initial timing advance establishment for each UE in the cell, it will then need to periodically update the uplink timing in order to counter changes in the arrival time of the uplink signals at the eNB. Such changes may arise from different causes including the movement of a UE (causing the propagation delay to change at a rate dependent on the relative velocity of the UE and the eNB), sudden changes in propagation delay due to changing multipath profile (such changes typically occur most frequently in dense urban environments as the UEs move around the corners of buildings), oscillator frequency drifts in the UE, and Doppler frequency shift arising from the movement of the UE especially in LoS propagation, resulting in an additional frequency offset of the uplink signals received at the eNB [14].

The update of the timing advance to mitigate the aforementioned effects is performed by a closed-loop mechanism whereby the eNB measures the received uplink timing and issues timing advance update commands to instruct the UE to adjust its transmission timing accordingly relative to its current transmission timing. In deriving the timing advance update commands, the eNB may measure any uplink signal including sounding reference signals, demodulation reference signals, channel quality indicators, HARQ feedback, or uplink data transmissions. In general, wideband uplink signals enable more accurate timing estimation. The advantage of accurate timing estimation has to be traded off with the uplink overhead caused by use of wideband reference signals. In addition, cell-edge UEs are typically power-limited, and thus are bandwidth-limited for a given uplink SINR. The timing estimation accuracy of narrowband uplink signals can be improved through averaging multiple measurements over time and interpolating the resulting power delay profile.

The eNB must balance the overhead of sending regular timing advance update commands to all the UEs in the cell against the UE's ability to transmit as soon as data becomes available in its transmit buffer. The eNB thus configures a timer for each UE, which the UE restarts each time a timing advance update command is received; if the UE does not receive another timing advance update command before the timer expires, it must then consider its uplink to have lost synchronization. In that case, in order to avoid the risk of generating interference to uplink transmissions from other UEs, the UE is not permitted to make another uplink transmission without first transmitting a random access preamble to reinitialize the uplink timing adjustment. One further use of timing advance is to create a

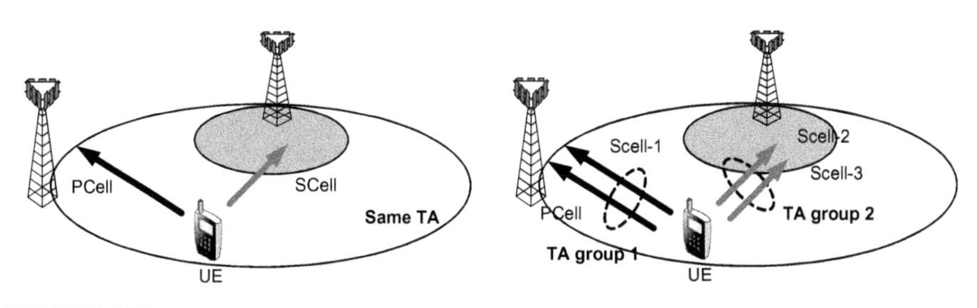

FIGURE 9.134

Illustration of single and multiple timing advance concepts [165].

switching time between uplink reception at the eNB and downlink transmission for TDD and H-FDD operation. This switching time can be generated by applying an additional timing advance offset to the uplink transmissions, to increase the amount of timing advance beyond what is required to compensate for the round-trip propagation delay [14].

Heterogeneous networks (HetNet) comprising macro-cells and small-cell base stations are considered as a key enabler for performance improvement in future cellular networks. LTE Rel-11 has specified carrier aggregation with multiple uplink timing advances and other enhancements to support non-collocated cells, e.g., multiple uplink power control instances or improved sounding reference signals. One of the scenarios of interest in HetNet is the use of remote radio heads (or geographically separated RF transceivers) connected via fiber-optic to a central baseband processing unit. A variation of this scenario, shown in Figure 9.134, would involve macro-cells which overlay small cells where both simultaneously use the same component carriers, requiring inter-cell interference coordination for proper operation (e.g., using an LTE Rel-10 feature known as enhanced inter-cell interference coordination (eICIC) which utilizes almost blank subframes (ABS)). In these scenarios, multiple timing advances are required to support non-collocated cells with carrier aggregation. Assuming synchronization to the macro-cell's PCell is already obtained, the UE has to subsequently synchronize to the SCell of the other site. Therefore, the PCell eNB will request a RACH on the SCell immediately after SCell activation (see Figure 9.134). The RACH request is sent via PDCCH signaling from the PCell. Enhanced MAC control elements are used to signal multiple timing advance values using previously reserved fields in the header.

In the case where several RF carriers require the same timing advance, these carriers will be grouped in what is known as a timing advance group (TAG).[33] The TAGs will be configured by eNB RRC signaling. This will reduce the overall

[33]A timing advance group is a group of serving cells that is configured by RRC signaling and uses the same timing reference cell and the same timing advance value. Primary timing advance group (pTAG) is the timing advance group containing the PCell. Secondary timing advance group (sTAG) is the timing advance group not containing the PCell [7].

complexity and signaling overhead for multiple TAs in case carrier aggregation is being used at each transmission point. For each TAG, a reference carrier for synchronization will be configured as a timing reference.

In carrier aggregation scenarios where two or more component carriers are aggregated to provide wider transmission bandwidths, a UE may simultaneously receive or transmit on one or multiple cells depending on its capabilities. A UE with single timing advance capability for carrier aggregation can simultaneously receive and/or transmit on multiple RF carriers corresponding to multiple serving cells sharing the same timing advance (multiple serving cells grouped in one TAG). A UE with multiple timing advance capability for carrier aggregation can simultaneously receive and/or transmit on multiple RF carriers corresponding to multiple serving cells with different timing advances (multiple serving cells grouped in multiple TAGs). It must be noted that each TAG contains at least one serving cell with configured uplink where the mapping of each serving cell to a TAG is configured by RRC signaling. A UE which does not support carrier aggregation can transmit/receive on a single RF carrier corresponding to one serving cell (one serving cell in one TAG).

LTE-Advanced supports contiguous and non-contiguous aggregation of component carriers where each RF carrier is limited to a maximum of 110 resource blocks in the frequency-domain using the LTE Rel-8/9 OFDM parameters. It is possible to configure a UE to aggregate a different number of RF carriers originating from the same eNB and of possibly different bandwidths in the uplink and the downlink. The number of TAGs that can be configured depends on the TAG capability of the UEs.

The UE is required to adjust uplink transmission timing for PUCCH/PUSCH/SRS of the primary cell based on the received timing advance command for a TAG containing the primary cell. The uplink transmission timing for PUSCH/SRS of a secondary cell is the same as the primary cell, if the secondary cell and the primary cell belong to the same TAG. In the case where a carrier-aggregation-enabled UE receives a timing advance command for a TAG that does not contain the primary cell, then it is required to adjust its uplink transmission timing for PUSCH/SRS of all the secondary cells in the TAG based on the received timing advance command.

9.18 Power control and link adaptation

Power control is a mechanism where the transmit power of the downlink or uplink control or traffic channels are adjusted at the base station or at mobile stations, based on instructions from the serving base station such that with minimal impact on the reliability of the downlink/uplink transmissions and throughput, the inter-user/inter-cell interference among users and base stations are reduced. Therefore, power control can be considered as a link adaptation mechanism that is utilized for interference mitigation in cellular systems. While increasing the transmit power over a communication link has certain advantages such as higher SNR at the receiver, which reduces the bit error rate and allows higher data rate and results in greater spectral efficiency as well as more protection against signal attenuation over fading channels, a higher transmit power; however, it has several drawbacks including increased power consumption of the transmitting device, reducing the UE battery life, and increased interference to other users in the same or adjacent frequency bands. The following sections describe the power control algorithms that are incorporated in LTE/LTE-Advanced.

9.18.1 Basic concepts

LTE/LTE-Advanced provide uplink power control mechanisms to compensate the effects of path loss, shadowing, fast fading, and implementation loss. The uplink power control is further used to mitigate inter-cell and intra-cell interference, thereby enhancing the overall throughput and reducing the effective UE power consumption. The uplink power control includes open-loop and closed-loop power control. The base station transmits necessary power control information through transmission of power control messages. The parameters of the power control algorithm are optimized on a system-wide basis by the eNB, and are broadcast periodically or trigged by events. The UE provides the necessary information through higher-layer control messages to the serving eNB in order to enable uplink power control. The eNB can exchange necessary information with neighboring base stations through backhaul to support uplink power control to facilitate the handover process.

The power control scheme may not be effective in high mobility scenarios for compensating the effects of a fast fading channel due to variation of the channel impulse response. As a result, the power control is used to mitigate the distance-dependent path loss, shadowing, and implementation loss. The uplink power control takes into consideration the MIMO transmission mode and whether a single user or multiple users are supported on the same resource at the same time. The open-loop power control compensates the channel variations and implementation loss without requiring frequent interactions with the serving eNB. The UE can determine the transmit power based on the transmission parameters sent by the eNB, uplink channel quality, downlink CSI, or the interference knowledge obtained from downlink transmissions. The open-loop power control provides a coarse initial transmit power setting for the mobile station before establishing connection with the base station.

It is perceived that rate control is more efficient than power control under certain conditions [15]. Rate control in principle implies that the power amplifier is always transmitting at full power and therefore is efficiently utilized. On the other hand, power control often results in inefficient utilization of the power amplifier because the transmission power is often less than its maximum. In practice, the radio-link data rate is controlled by adjusting the modulation scheme and/or the channel coding rate. In good channel conditions, the value of E_b/N_0 at the receiver is high, and the main limitation of the data rate is the bandwidth of the radio link. In such conditions, use of higher-order modulation, e.g., 16QAM or 64QAM, together with a high coding rate, is more appropriate for link adaptation. Similarly, in the case of poor channel conditions, the use of QPSK and low-rate coding is preferred. Link adaptation by means of rate control is referred to as adaptive modulation and coding (AMC).

A power control mechanism takes into consideration the serving eNB target link SINR and/or interference level to other cells/sectors for mitigating inter-cell interference. In order to achieve the target SINR, the serving eNB path loss can be fully or partially compensated based on a trade-off between overall system throughput and cell-edge performance. The mobile station's transmit power is adjusted in order to ensure the level of interference is less than the permissible interference level. The closed-loop power control, on the other hand, compensates channel variations through periodic power-control commands from the serving eNB. The base station measures the uplink CSI and interference level using uplink data and/or control channel transmissions, and sends power control commands to the mobile stations. Upon receiving the power control command from the eNB, the UE adjusts its uplink transmit power. The closed-loop power control is active during data and control channel transmissions.

A UE is expected to maintain the transmit power density (i.e., total transmit power normalized by transmission bandwidth) for each data and control channel below a certain level that is determined by the maximum permissible power level for the UE, emission mask, and other regulatory constraints. In other words, when the number of active logical resource units assigned to a particular user is reduced, the total transmitted power must be reduced proportionally by the UE in the absence of any additional change of power control parameters. When the number of resource blocks is increased, the total transmitted power must be proportionally increased such that the transmitted power level does not exceed the permissible power levels specified by 3GPP and the regulatory specifications [1]. For interference level control, the information about the current interface level of each eNB may be shared among the base stations via backhaul.

The (uplink) transmit power control (TPC) in mobile communication systems is meant to balance the transmitted energy per bit in order to maintain the link quality corresponding to the minimum QoS requirements, to minimize interference to other users in the system, and to minimize the power consumption of the mobile terminal. In achieving these goals, the power control has to adapt to the characteristics of the propagation channel, including path loss, shadowing, and fast fading, as well as overcoming interference from other users both within the same cell and in neighboring cells.

The requirements for uplink interference management in LTE are different from the operational requirements in the uplink of the CDMA systems. In CDMA systems, the uplink is not orthogonal and the primary source of interference which has to be managed is intra-cell interference among different intra-cell users. The uplink users in a CDMA system share the same time–frequency resources and they all contribute to a rise in interference above thermal noise level at the base station receiver; a phenomenon typically known as rise over thermal (RoT),[34] that ought to be carefully controlled and shared among users. The primary mechanism for increasing the uplink data rate for a given user is to reduce the spreading factor and increase the transmission power accordingly, consuming a larger proportion of the total available RoT in the cell. In contrast, in LTE, the uplink is theoretically orthogonal and intra-cell interference management is consequently less critical than that of a CDMA system.

The main mechanisms for adapting the uplink data rate in LTE are varying the transmit bandwidth and varying the MCS, while the transmit power spectral density (PSD) could typically remain approximately constant for a given MCS. Moreover, in a CDMA system, the power control was primarily designed with continuous transmission in mind for circuit-switched services, while in LTE, different UEs are scheduled at 1 ms transmission time intervals. This is reflected in the fact that power control in CDMA is periodic, while LTE allows for larger power steps which do not have to be periodic, with a minimum loop delay of about 5 ms. With the above considerations in mind, the power control scheme provided in LTE employs a combination of open-loop and closed-loop control, which theoretically requires less feedback than a closed-loop scheme, because the closed-loop feedback is only needed to compensate for cases when the UE's own estimate of the required power setting is not adequate.

A typical mode of operation for power control in LTE involves setting a coarse operating point for the transmit PSD based on path loss estimation. This would provide a suitable PSD setting for reference MCS in the prevailing path loss and shadowing conditions. Faster adaptation can then be applied around the open-loop operating point by closed-loop power control. This can control the interference level and adapt the transmit power setting according to the channel conditions including fast fading. However, due to the orthogonal nature of the LTE uplink, the LTE closed-loop power control does not need to be as fast as in CDMA systems. In fact, the fastest and most frequent adaptation of the uplink transmissions is by means of the uplink scheduling grants, which vary the transmission bandwidth and accordingly the total transmit power, together with setting the MCS, in order to attain the desired uplink data rate. With this combination,

---

[34]In CDMA systems, the rise over thermal (RoT) indicates the ratio between the total power received from wireless sources at a base station and the thermal noise. In order to decode the data at the base station, a minimum SINR at the receiver must be guaranteed. RoT metric is a measure of the uplink loading. By increasing the number of transmitting UEs and their transmit power, the level of interference in the uplink increases. This interference is perceived by the base station as noise adversely affecting the SINR. The base station controls the interference level by adjusting the UE uplink assignments and transmit-power [40].

the power control scheme in LTE in practice provides support for more than one mode of operation [14].

The initial open-loop operating point for the transmit power per resource block depends on a number of factors including the inter-cell interference and cell loading. It can be further divided into two components: (1) a semi-static base-level value comprising a nominal power level that is common for all UEs in the cell (measured in dBm per resource block) and a UE-specific offset; and (2) an open-loop path loss compensation component. Different base levels can be configured for PUSCH data transmissions depending on the scheduling mode (i.e., dynamic or semi-persistent scheduling), which in principle allows different BLER operating points to be used for each scheduling mode. One possible use for different BLER operating points is to achieve a lower probability of retransmission when using semi-persistent scheduling. The UE-specific offset component of the base level enables the eNB to correct the systematic offsets of the UE's transmission power setting. The path loss compensation component is based on the UE's estimate of the downlink path loss, which can be derived from the UE's measurement of RSRP and the known transmit power of the downlink reference signals, which is broadcast by the eNB.

In order to obtain a reasonable indication of the uplink path loss, the UE should filter the downlink path loss estimate with a suitable time window to remove the effect of fast fading but not shadowing. Typical filter lengths are between 100 and 500 ms [14]. For PUSCH and SRS, the degree to which the uplink PSD is adapted to compensate for the path loss can be set by the eNB, on a scale from "no compensation" to "full compensation." This is an important feature of power control in LTE and is known as fractional power control; it is configured by means of a fractional path loss compensation factor, referred to as α. In principle, the combination of the base level and the path loss compensation component allow the eNB to configure the degree to which the UE responds to the path loss. At one extreme, the eNB could configure the base level to the lowest level (−126 dBm) and rely entirely on the UE's path loss measurement to raise the power toward the cell-edge; alternatively, the eNB can set the base level to a higher value, possibly in conjunction with only partial path loss compensation. Disregarding the UE-specific offset, the range of the base level for PUSCH (−126 to +23 dBm per RB) is intended to cover the full range of target SINR values for different degrees of path loss compensation, transmission bandwidths, and interference levels. For example, the highest value of the base level corresponds to the maximum permissible transmission power of an LTE UE and would typically only be used, if path loss compensation was not used. The lowest value of base level for PUSCH −126 dBm is relevant to a case when full path loss compensation is used and the uplink transmission and reception conditions are optimal. In general, the maximum path loss that can be compensated depends on the required SINR and the transmission bandwidth. Note that this assumes full path loss compensation and ignores the dynamic offset [14]. In other words, LTE defines the setting of the UE transmit power for PUSCH by the following equation [5]:

$$P_{\text{PUSCH}} = \min[P_{T_{\max}}, 10\log_{10} M + P_0 + \alpha P_{\text{L}} + \delta_{\text{MCS}} + f(\Delta_i)]\ [\text{dBm}] \qquad (9.105)$$

where

- $P_{T_{\max}}$ is the maximum allowed transmit power that depends on the UE power class (i.e., UE maximum output power which is typically $+23$ dBm for a power class 3 UE).
- M denotes the number of physical resource blocks assigned for uplink transmission.
- P_0 is a cell/UE-specific base level parameter that is signaled through higher-layer signaling.
- α represents the path loss compensation factor, which is a 3-bit cell-specific parameter in the range of [0 1] which is signaled through higher-layer signaling, where $\alpha = 0$ corresponds to "no compensation" and $\alpha = 1$ corresponds to "full compensation."
- P_{L} is the downlink path loss estimate, which is calculated by the UE based on RSRP measurement.
- δ_{MCS} is cell/UE-specific MCS defined in 3GPP specifications for LTE.
- $f(\Delta_i)$ is a UE-specific function which indicates the use of accumulated or absolute correction value and Δ_i is the closed-loop correction value. The UE-specific TPC commands can operate in two different modes: (1) accumulative TPC commands that are available for PUSCH, PUCCH, and SRS; and (2) absolute TPC commands that are available for PUSCH only. In the case of PUSCH, the switch between these two modes is configured semi-statically for each UE by RRC signaling, i.e., the mode cannot be changed dynamically. With the accumulative TPC commands, each TPC command signals a power step relative to the previous level. This is the default mode and is particularly suitable for fine-tuning of the transmission power, and for situations where a UE receives power control commands in groups of successive subframes. This mode is similar to the closed-loop power control in CDMA systems.

In the above equation $P_0 = \alpha(\text{SNR}_0 + P_{\text{N}}) + (1 - \alpha)(P_{T_{\max}} - 10\log_{10} M_0)$ [dBm] where:

- SNR_0 is the open-loop SNR target.
- P_{N} is the noise power per resource block.
- M_0 defines the number of resource blocks for which the SNR target is set with full-power.

Some of the parameters in the above equations are broadcast by the eNB, i.e., the same value for all the users in the cell. The UE sets its initial transmission power based on received parameters from the eNB and path loss calculated by the UE per se. It is worthwhile to note that Δ_i is signaled by the eNB to any UE after it sets its initial transmit power, i.e., Δ_i has no contribution in the initial setting of the UE transmit power. The expression, based on which a UE sets its initial power, can be obtained from the above equations by ignoring δ_{MCS} and the closed-loop correction factor, thus $P_{PUSCH} = 10 \log_{10} M + P_0 + \alpha P_L$ [dBm] where the parameter M denotes the total number of resource blocks scheduled by the eNB. The power assignment for transmission at the UE is performed on the basis of per resource block, and each resource block contains an equal amount of power. Therefore, by neglecting M, the expression used by the UE to assign power to each resource block is given by $\text{PSD}_{TX} = P_0 + \alpha P_L$ [dBm/RB].

The above equation represents the transmit SPD of a resource block expressed in dBm/RB. Note that each resource block corresponds to the 180 kHz subchannel. The parameter PSD_{TX} is a helpful means to explain the basic difference between conventional and fractional power control schemes. Setting the path loss compensation factor α to one (full compensation of path loss) leads to conventional power control, whereas the use of values $0 < \alpha < 1$ (fractional compensation of path loss) leads to fractional power control. Note that $\alpha = 0$ (no compensation of path loss) leads to no power control, i.e., all users will use the maximum allowed transmission power.

When $\alpha = 1$, it can be shown that $P_0 = \text{SNR}_0 + P_N$ [dBm], thus $\text{PSD}_{TX} = P_0 + P_L$ [dBm/RB], taking into account the path loss, the received PSD at the eNB is given by $\text{PSD}_{RX} = P_0$ [dBm/RB]. It is important to note that the received PSD at the eNB is equal to P_0; therefore, a conventional power control scheme leads all users to equal SPD. This scheme is widely used in cellular systems which are not using an orthogonal transmission scheme in the uplink, such as CDMA-based systems. One of the advantages of this power control scheme is that it removes the near-far problem which is typical of CDMA systems, since it equalizes the power level of the UEs before receiving at the base station. It can be shown that under these conditions, the received PSD is the same for all users independent of their path loss for a given target SNR.

The fractional power control scheme allows the user terminals' transmissions to be received with variable PSD depending on their path loss, i.e., the user with good channel conditions will be received with high PSD. Using $0 < \alpha < 1$, the PSD is given by $\text{PSD}_{TX} = P_0 + \alpha P_L = \alpha(\text{SNR}_0 + P_N) + (1 - \alpha)P_{max} + \alpha P_L$ [dBm/RB]. In contrast to conventional power control, which allows full compensation of path loss, fractional power control compensates for a fraction of the path loss and this is the reason for the term "fractional power control." The received PSD can be obtained by taking path loss into account and is given by $\text{PSD}_{RX} = P_0 + (\alpha - 1)P_L$ [dBm/RB]. The received PSD in the case of a conventional power control scheme results in P_0, whereas the case of a fractional power control scheme has an additional term, where both P_0 and $(\alpha - 1)P_L$ are cell-specific and broadcast to the UEs by the eNB, which means that they are the same for all the UEs. Thus, P_L is the key factor that allows users to be received with different power spectral densities [15].

The fractional path loss compensation factor α can be seen as a tool to trade off the fairness of the uplink scheduling against the total cell capacity. Full path loss compensation maximizes fairness for cell-edge UEs. However, when considering multiple cells together as a system, the use of partial path loss compensation can increase the total system capacity in the uplink, as fewer resources are consumed, ensuring the success of transmissions from cell-edge UEs and less inter-cell interference is caused to neighboring cells. Path loss compensation factors around 0.7–0.8 typically give a close-to-maximal uplink system capacity (typically around 15–25% greater than can be achieved with full path loss compensation) without causing significant degradation to the cell-edge data rate. The target received PSD for a given MCS is reduced as the path loss increases, so that cell-edge UEs cause less inter-cell interference. Inter-cell interference is of particular concern for UEs located near the edge of a cell, as they may disrupt the uplink transmissions in the neighboring cells [14].

Open-loop power control is the ability of the UE transmitter to set its uplink transmit power to a specified value suitable for the eNB receiver. The uplink transmit power is given as $P_{OL} = \min[P_{max}, 10\log_{10} M + P_0 + \alpha P_L]$ [dBm], which is set by open-loop power control. The choice of α depends on whether a conventional or fractional power control scheme is used. Using $\alpha = 1$ leads to conventional open-loop power control, while $0 < \alpha < 1$ leads to fractional open-loop power control. The estimate of the path loss is obtained after measuring the RSRP and then the calculation for transmission power is performed based on the above equation.

Closed-loop power control is the ability of the UE to adapt the uplink transmit-power according to the closed-loop correction values received from the eNB, also known as TPC commands. The TPC commands are transmitted by the eNB to the UE, based on the closed-loop target signal-to-interference and noise ratio and the measured received SINR. In a closed-loop power control mechanism, the uplink receiver at the eNB estimates the SINR of the received signal and compares it with the target SINR value. When the received SINR value is below the target SINR, a TPC command is transmitted to the UE to instruct the UE to increase the transmitter power. Otherwise, the TPC command will cause the UE transmitter power to decrease. The LTE closed-loop power control mechanism operates around the open-loop operation point. The UE adjusts its uplink transmission power based on the TPC commands it receives from the eNB when the uplink power setting is performed at the UE using open-loop power control.

For the low-rate PUCCH (carrying HARQ ACK/NACK and CQI reports), full path loss compensation is always applied, because PUCCH transmissions from different UEs are code-division-multiplexed. Full path loss compensation facilitates good control of the interference among different users, and thereby helps to maximize the number of users which can be accommodated simultaneously on PUCCH. A different base level value P_0 is also provided for PUCCH compared to that used for PUSCH, assuming a reference PUCCH format. For SRS, an additional semi-static offset relative to the PUSCH power operating point may be configured through RRC signaling.

Downlink power control determines the energy per resource element (EPRE) which implies the energy prior to cyclic prefix insertion. The term resource element energy also implies the average energy taken over all constellation points for the modulation scheme in use. Uplink power control determines the average power over a SC-FDMA symbol, on which the physical channel is transmitted.

9.18.2 Downlink power allocation

The cell-specific reference signals are embedded into the overall system bandwidth at certain resource elements. In the frequency-domain, every sixth subcarrier carries a reference signal. The reference signal pattern is a pseudo-random sequence, whose generation depends on the cell identity and the cyclic prefix. Furthermore, a frequency shift is applied that is based on a modulo-6 operation. The mapping of the pattern is therefore cell-specific, but the spacing is always six sub-carriers. In the time-domain, every fourth OFDM symbol carries reference symbols. Due to their importance, reference signals are the highest powered components within the downlink signal. The power level for the reference signals is signaled within system information to the device. The reference signal power level is cell-specific and is in the range of -60 to $+50$ dBm per 15 kHz. It is a requirement that the LTE base station transmits all reference signals with constant power over the entire bandwidth. The power of all other signal components (synchronization signals, PBCH, PCFICH, PDCCH, PDSCH, and PHICH) is set relative to this value. As mentioned above, there are OFDM symbols that contain reference signals and there are OFDM symbols that do not contain reference signals.

The relative PDSCH power for these symbols is given by two different parameters ρ_A and ρ_B (see Figure 9.135). For the majority of cases, ρ_A corresponds to the parameter P_A that is signaled via higher layers. Only for some special cases such as transmit diversity with four antennas or MU-MIMO, ρ_A is computed differently. Parameter P_A is device specific, is obtained as part of the *RRCConnectionSetup* message, and can take one out of eight different values. In Figure 9.135, it is assumed that P_A is -4.77 dB. Parameter P_B is related to the cell-specific reference signal power and cannot be changed dynamically. It can take one out of four integer values. Depending on the number of transmit antennas (1, 2, or 4), each value corresponds to a certain ratio and thus a power offset. LTE Rel-8/9 networks that are currently deployed worldwide are supporting a 2×2 MIMO baseline configuration. Let us assume $P_B = 3$, in that case, the resource element carrying data in that OFDM symbol where reference signals are present is transmitted with an additional offset of 3 dB compared to

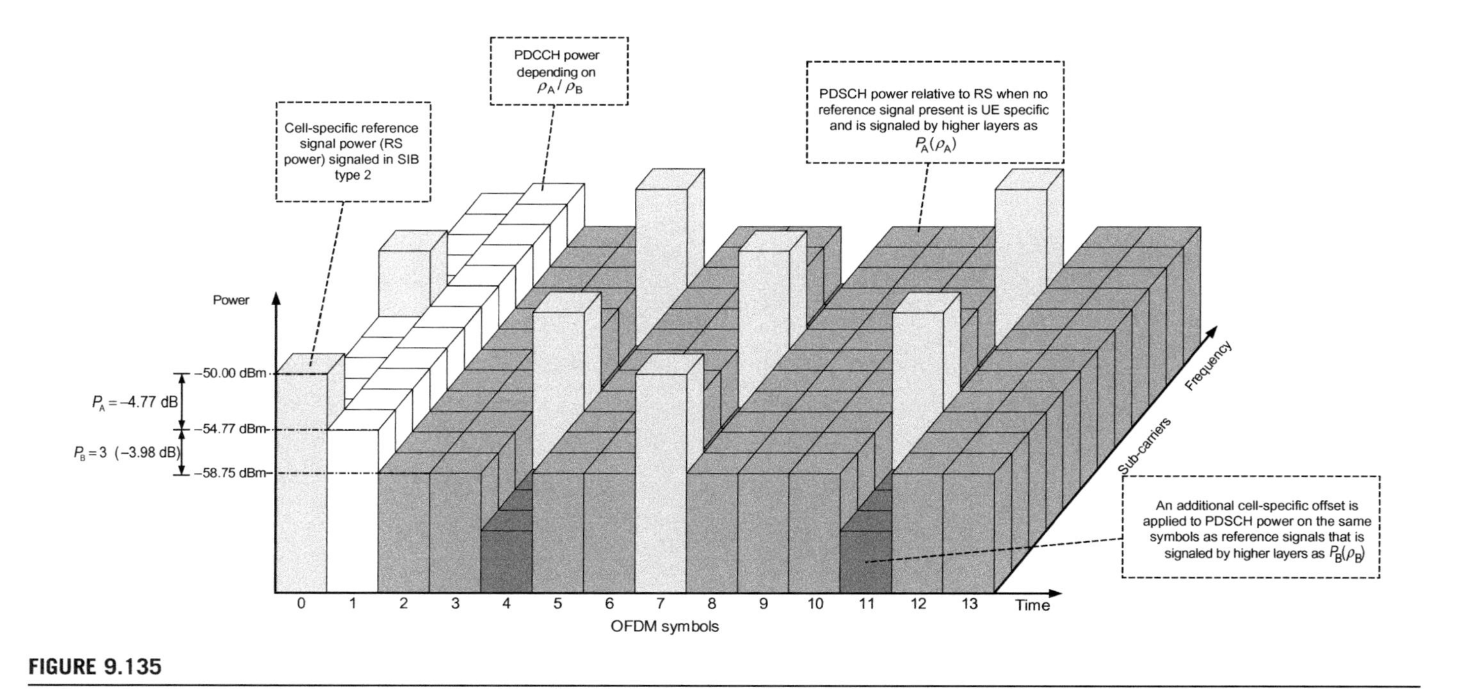

FIGURE 9.135

Example of LTE Rel-8 downlink power allocation [134].

symbols without reference signals. For only one transmit antenna (SISO), $P_B = 3$ translates to -3.98 dB as shown in Figure 9.135. The overall goal is to have a constant power for all OFDM symbols to avoid power variations at the UE receiver. With less PDSCH power given by P_B, the boost of reference signals is compensated, compared to OFDM symbols that do not contain reference signals. The PDSCH power always depends on the resource allocation, i.e., the number of allocated resource blocks. Resource allocation can change from one subframe to another subframe, thus P_A can also change on a subframe-by-subframe basis. While incorporating P_A and P_B, it is ensured that the overall OFDM symbol power remains constant, even when the PDSCH allocation is changed [5].

9.18.3 PUSCH power control

Starting with LTE Rel-10 and in the case of carrier aggregation, it is possible to transmit multiple PUSCH on different component carriers simultaneously, or it is possible to simultaneously transmit PUSCH and PUCCH on the same or different component carriers (this was not allowed in previous releases of LTE). In principle, each physical channel is separately and independently power controlled. However, in the case of concurrent transmission of multiple physical channels from the same terminal, the total transmit power for all physical channels may occasionally exceed the maximum terminal output power setting according to the terminal power class. The basic strategy is to ensure reliable transmission of all control signaling by allocating an appropriate amount of power to the respective physical channels. The remaining power is then assigned to the rest of the physical channels. For each uplink component carrier configured for a terminal, there is also an associated and explicitly configured maximum per-carrier transmit power, which may be different for different component carriers. This may be explained as follows. In many practical scenarios, the terminal will not be scheduled for uplink transmission on all of its configured component carriers, so the terminal should be able to transmit with its maximum output power. As will be seen in this section, the power control of each physical channel explicitly ensures that the total transmit power for a given component carrier does not exceed maximum per-carrier transmit-power; however, separate power-control algorithms do not ensure that the total transmit power for all component carriers does not exceed the maximum terminal output power. The latter limitation will be enforced by a subsequent power scaling applied to the physical channels that are concurrently transmitted. This power scaling is carried out in a way that control signaling channels have higher priority relative to traffic channels during concurrent transmission in the uplink.

If the UE does not simultaneously transmit PUSCH and PUCCH on the serving cell k, i.e., the kth component carrier because component carriers are referred to as cells in LTE-Advanced, then the UE transmit power $P^k_{\text{PUSCH}}(i)$ for PUSCH transmission in the ith subframe is given as follows:

$$P^k_{\text{PUSCH}}(i) = \min \left[\underbrace{P^k_{\max}(i)}_{\substack{\text{Maximum Power per} \\ \text{Component Carrier}}}, \underbrace{10\log_{10}(M^k_{\text{PUSCH}}(i))}_{\text{Bandwidth Factor}} + \underbrace{P^k_0(j) + \alpha_k(j)P^k_L}_{\substack{\text{Basic Open-loop} \\ \text{Operating Point}}} + \underbrace{\Delta^k_{\text{TF}}(i) + f_k(\Delta_{\text{TPC}}(i))}_{\text{Dynamic Offset}} \right] \text{[dBm]} \tag{9.106}$$

If the UE transmits PUSCH and PUCCH simultaneously on the serving cell k, then the UE transmit power $P^k_{\text{PUSCH}}(i)$ when transmitting PUSCH in the ith subframe is given as follows:

$$P^k_{\text{PUSCH}}(i) = \min \left[\underbrace{10\log_{10}(\hat{P}^k_{\max}(i) - \hat{P}_{\text{PUCCH}}(i))}_{\substack{\text{Maximum available power per} \\ \text{component carrier}}}, \underbrace{10\log_{10}(M^k_{\text{PUSCH}}(i))}_{\text{Bandwidth Factor}} + \underbrace{P^k_0(j) + \alpha_k(j)P^k_{\text{L}}}_{\substack{\text{Basic open-loop} \\ \text{operating point}}} + \underbrace{\Delta^k_{\text{TF}}(i) + f_k(\Delta_{\text{TPC}}(i))}_{\text{Dynamic offset}} \right] \text{[dBm]} \tag{9.107}$$

In the above equations:

- $P^k_{\max}(i)$ is the configured UE transmit power in the ith subframe for serving cell k, implying that this channel power is calculated and set for every subframe.
- $\hat{P}_{\text{PUCCH}}(i)$ is the linear value of $P_{\text{PUCCH}}(i)$.
- $M^k_{\text{PUSCH}}(i)$ is the number of resource blocks allocated for the UE in the ith subframe and on serving cell k.
- $P^k_0(j)$ is a parameter composed of the sum of a component $P^k_{0\ NOMINAL\_PUSCH}(j)$ provided through higher-layer signaling for $j = 0,1$ and another component $P^k_{0\ UE\_PUSCH}(j)$ that is provided by higher layers for $j = 0,1$ in the kth serving cell.
- P^k_L is the downlink path loss estimated via RSRP measurements by the UE on serving cell k in dB.
- $\Delta^k_{\text{TF}}(i)$ is the MCS-dependent component, referred to as transport format in the LTE specifications, which allows the transmitted power per resource block to be adapted according to the information data rate.
- i denotes the subframe number.
- $j = 0$ or 1.
- $f_k(\Delta_{\text{TPC}}(i))$ denotes a TPC command and $f_k(.)$ represents an accumulation function in the case of accumulative TPC commands.

The above power control equation allows the UE's transmit power to be controlled with a granularity of 1 dB within the range of -40 dBm to $+23$ dBm. The maximum transmission power of a UE may be subject to additional restrictions, such as the maximum transmission power in a cell may be restricted to a lower level by RRC signaling, or in some configurations, reductions in maximum output power may be applied in order to satisfy in-band and out-of-band emission requirements. For LTE Rel-10 UEs which support carrier aggregation in the uplink, the transmission power of an individual carrier or uplink physical channel may have to be scaled down according to certain rules in order to satisfy the total output power constraints. The UE accuracy in setting total transmission power depends on the length of time since the last uplink transmission has taken place, and the size of the required change in transmission power [15].

It was mentioned above that if the total transmit power of the UE exceeds $\hat{P}_{T_{\max}}(i)$, the UE scales down the power of PUSCH for the serving cell k in the ith subframe $\hat{P}^{k}_{\mathrm{PUSCH}}(i)$ such that the condition $\sum_k a_i \hat{P}^{k}_{\mathrm{PUSCH}}(i) \le (\hat{P}_{T_{\max}}(i) - \hat{P}_{\mathrm{PUCCH}}(i))$ is satisfied where $\hat{P}_{\mathrm{PUCCH}}(i)$ is the linear value of $P_{\mathrm{PUCCH}}(i)$, $\hat{P}^{k}_{\mathrm{PUSCH}}(i)$ is the linear value of $P^{k}_{\mathrm{PUSCH}}(i)$, $\hat{P}_{T_{\max}}(i)$ is the linear value of the UE total configured maximum output power $P_{T_{\max}}$ in the ith subframe, and $0 \le a_i \le 1$ is the scaling factor of $\hat{P}^{k}_{\mathrm{PUSCH}}(i)$ for serving cell k. In a hypothetical scenario, when the UE is scheduled to transmit PUSCH containing UCI on the jth serving cell and transmit PUSCH without UCI in the remaining serving cells, and if the total transmit power of the UE is going to exceed $\hat{P}_{T_{\max}}(i)$, the UE is required to scale $\hat{P}^{k}_{\mathrm{PUSCH}}(i)$ for the serving cells without UCI in the ith subframe such that the following condition is satisfied [5]:

$$\sum_{k \neq j} a_i \hat{P}^{k}_{\mathrm{PUSCH}}(i) \le (\hat{P}_{T\max}(i) - \hat{P}^{j}_{\mathrm{PUSCH}}(i)) \tag{9.108}$$

where $\hat{P}^{j}_{\mathrm{PUSCH}}(i)$ is the PUSCH transmit power for the cell with UCI and a_i is the scaling factor of $\hat{P}^{k}_{\mathrm{PUSCH}}(i)$ for serving cell k containing no UCI. In this case, the power scaling is not applied to $\hat{P}^{j}_{\mathrm{PUSCH}}(i)$ unless $\sum_{k \neq j} a_i \hat{P}^{k}_{\mathrm{PUSCH}}(i) = 0$ and the total transmit power of the UE would still exceed $\hat{P}_{T_{\max}}(i)$. Note that values of the scaling coefficients a_i are the same across serving cells when $a_i > 0$, but for certain serving cells a_i may be set to zero. In another hypothetical scenario where the UE has simultaneous PUCCH and PUSCH transmission with UCI on the jth serving cell and PUSCH transmission without UCI in the remaining serving cells, if the total transmit power of the UE would exceed $\hat{P}_{T_{\max}}(i)$, the UE sets $\hat{P}^{k}_{\mathrm{PUSCH}}(i)$ according to the following criteria [5]:

$$\hat{P}^{j}_{\mathrm{PUSCH}}(i) = \min[\hat{P}^{j}_{\mathrm{PUSCH}}(i), \hat{P}_{T_{\max}}(i) - \hat{P}_{\mathrm{PUCCH}}(i)] \tag{9.109}$$

and

$$\sum_{k \neq j} a_i \hat{P}^{k}_{\mathrm{PUSCH}}(i) \le (\hat{P}_{T\max}(i) - \hat{P}_{\mathrm{PUCCH}}(i) - \hat{P}^{j}_{\mathrm{PUSCH}}(i)) \tag{9.110}$$

9.18.4 PUCCH power control

The appropriate level of the received signal power for PUCCH transmission is simply determined by the minimum power necessary to achieve a satisfactory block error rate in decoding of the uplink control signaling information transmitted on PUCCH. However, it is important to take the following considerations into account: (1) In general, decoding performance is not determined by the received signal strength, but rather by the received signal-to-interference-plus-noise ratio, i.e., an appropriate received signal power depends on the interference level at the receiver, where the interference level depends on network deployment and is time-variant since loading of the network varies across time. (2) There are different PUCCH formats which are used to carry a different types of uplink control information. The different PUCCH formats thus carry a different number of information bits per subframe and the information they carry may also have different error-rate requirements. The required received SINR may differ in different PUCCH formats, something that needs to be taken into account when setting the PUCCH transmit power in a given subframe. The power control for PUCCH can be described by the following expression [5]:

$$P_{\mathrm{PUCCH}}(i) = \min\left[\underbrace{P^k_{\max}(i)}_{\substack{\text{Maximum available}\\ \text{power for PUCCH}}},\ \underbrace{P_0 + P^k_{\mathrm{L}}}_{\text{Basic operating point}} + \underbrace{\Delta_{\mathrm{PUCCH\text{-}FORMAT}}}_{\substack{\text{Format-dependent}\\ \text{power adjustment}\\ \text{factor}}} + \underbrace{\delta_{\mathrm{PUCCH}}(i)}_{\substack{\text{Dynamic adjustment}\\ \text{based on TPC}\\ \text{commands}}}\right] \tag{9.111}$$

where:

- $P^k_{T_{\max}}(i)$ is the configured UE transmit power in the ith subframe on serving cell k.
- $P_{\mathrm{PUCCH}}(i)$ denotes the PUCCH transmit power in a given subframe.
- P^k_{L} represents the downlink path loss as estimated by the terminal based on downlink RSRP.
- P_0 is a cell-specific parameter that is broadcast as part of the system information and can be seen as the desired or target received signal power. As discussed earlier, the required received power depends on the receiver noise and uplink interference level.
- $\delta_{\mathrm{PUCCH}}(i)$ is a UE-specific dynamic correction value, also referred to as a TPC command, included in the PDCCH with DCI format 1/1A/1B/1D/2/2A/2B/2C/2D for the primary cell or sent jointly coded with other UE-specific PUCCH correction values on a PDCCH with DCI format 3/3A whose CRC is scrambled with TPC-PUCCH-RNTI.

In practice, it is not possible to allow P_0 to vary with interference level for the following reasons: (1) the terminal does not read the system information continuously and thus it does not have access to the updated P_0 value; and (2) the uplink path loss estimate is derived from downlink measurements and is not fully accurate due to differences between the instantaneous downlink and uplink propagation conditions, as well as measurement inaccuracies. Thus, in practice, P_0 may reflect the average interference level or the relatively constant noise level.

In order to reflect different SINR requirements for various PUCCH formats, the PUCCH power control expression includes the term $\Delta_{\text{PUCCH-FORMAT}}$ which adds a format-dependent power offset to the transmit power. The power offsets are defined such that a baseline PUCCH format, i.e., the format corresponding to the transmission of a single HARQ ACK/NACK, has an offset equal to 0 dB, while the offsets for the remaining formats can be explicitly configured by the network. For example, PUCCH format 1 with QPSK modulation, carrying two simultaneous acknowledgment and used in the case of downlink spatial multiplexing, should have a power offset of 3 dB, reflecting the fact that twice as much power is needed to communicate two acknowledgment bits as opposed to one. It is possible for the network to directly adjust the PUCCH transmit power by providing the terminal with explicit power control commands that adjust the term $\delta_{\text{PUCCH}}(i)$ in the above expression. The power control commands are accumulative, i.e., each received TPC command increases or decreases the term $\delta_{\text{PUCCH}}(i)$ by a certain amount. The power control commands for PUCCH can be provided to the terminal via two different means: (1) by including a power control command in each downlink assignment (i.e., the terminal receives a TPC command every time it is scheduled on the downlink); and (2) the power control commands can also be sent using DCI format 3/3A that simultaneously provides power control commands to multiple terminals. In practice, such TPC commands are typically transmitted on a regular basis and can be used to adjust the PUCCH transmit power prior to periodic uplink channel-state reports. The power control command carried within the uplink scheduling grant consists of 2 bits, corresponding to four different update steps $(-1, 0, +1, +3\ \text{dB})$. The same is true for the power control command carried on DCI format 3/3A designated for power control. On the other hand, when the PDCCH is configured to use DCI format 3, each TPC command consists of a single bit, corresponding to the update steps -1 and $+1$ dB. In the latter case, twice as many terminals can be power controlled by a single PDCCH. The reason for including a 0 dB value (no change of PUCCH transmit power) as a power-control step is that the power control command is included in every downlink scheduling assignment and sometimes it is desirable not to update the PUCCH transmit power for each assignment.

9.18.5 SRS power control

The UE sets the transmit power of the sounding reference signal in the ith subframe and the kth component carrier as follows [5]:

$$P^k_{\text{SRS}}(i) = \min[P^k_{\max}(i), P^k_{\text{SRS\_OFFSET}}(m) + 10\log_{10}(M^k_{\text{SRS}}) + P^k_0(j) + \alpha_k(j)P^k_L + f_k(i)]\ [\text{dBm}] \tag{9.112}$$

where:

- $P^k_{\max}(i)$ is the configured UE transmit power in the ith subframe on serving cell k.
- $P^k_{\text{SRS\_OFFSET}}(m), m = 0, 1$ is a 4-bit parameter semi-statically configured by higher layers for the serving cell.
- M^k_{SRS} denotes the bandwidth of the SRS transmission in subframe i for serving cell k in number of resource blocks.
- $f_k(i)$ is the current PUSCH power control adjustment state for serving cell k.
- $P^k_0(j)$ and $\alpha_c(j)$ terms are the same as PUSCH power control with $j = 1$.

If the total transmit power of the UE for the sounding reference signal is going to exceed $\hat{P}^k_{\max}(i)$, i.e., the maximum allotted power for the kth component carrier, the UE scales down $\hat{P}^k_{\text{SRS}}(i)$ in subframe i such that $\sum_k a_i \hat{P}^k_{\text{SRS}}(i) \le \hat{P}^k_{\max}(i)$ is satisfied where $\hat{P}^k_{\text{SRS}}(i)$ denotes the linear value of $P^k_{\text{SRS}}(i)$ and $\hat{P}^k_{\max}(i)$ is the linear value of $P^k_{\max}$ defined in 3GPP TS 36.211 [3] in the ith subframe, and $0 < a_i \le 1$ are the scaling factor of $\hat{P}^k_{\text{SRS}}(i)$. Note that α_i values are the same across serving cells.

9.18.6 PRACH power control

The UE must be synchronized in the uplink before it can send signaling or data. Given that the UEs within a cell's coverage area are geographically distributed at different distances from the eNB location, each will exhibit a different propagation delay when accessing the network. Without uplink synchronization, subframes transmitted from different UEs would likely overlap and interfere with each other.

To overcome this problem, the UE performs uplink synchronization procedures on the RACH. The process is initiated by sending RACH preambles using the physical random access channel (PRACH). The random access procedure may be contention-based or contention-free. In the contention-based procedure, the UE selects a preamble from an available set of two groups. If two UEs select the same preamble in the same RACH resource, a collision can result that must be resolved before either UE can access the system. Contention-based RACH is generally used during initial access or if the UE has lost uplink synchronization. With contention-free synchronization, the UE uses a dedicated preamble and it is possible that it had already connected to the network and must resynchronize following a handover to a new cell. During the random access procedure, the preamble index, the target preamble received power (denoted by *Preamble_Received_Target_Power* parameter), the corresponding RA-RNTI, and the PRACH resource are signaled by the higher layers. The preamble transmission power P_{PRACH} is determined by $P_{\text{PRACH}} = \min[P^k_{\max}(i), \textit{Preamble\_Received\_Target\_Power} + P^k_L]$ [dBm] where $P^k_{\max}(i)$ is the configured UE transmit power for the ith subframe of the primary cell, and P^k_{L} is the downlink path loss estimate calculated by the UE for the primary cell.

During the RACH procedure, the UE transmits a preamble randomly selected from 64 possible preamble sequences divided into two groups. The group from which the UE selects the preamble depends on the size of the message that it would like to transmit and on the calculated path loss to the eNB. Once the preamble is transmitted, the UE monitors the downlink after three subframes following the end of the preamble transmission for a duration controlled by a configurable window. If no downlink response is detected, then the preamble is retransmitted with a power that is increased by a configurable parameter. This process continues until the maximum number of preamble transmissions is reached. If the preamble is successfully detected by the eNB, then the downlink random access response (RAR) message is sent allocating resources for the UE to transmit its message or enforcing a back-off. In the final step, contention resolution must be performed for a contention-based access attempt. The UE sets its initial power level to the following value: *Preamble_Initial_Received_Target_Power* + *Delta_Preamble* (see Table 9.32 for the permissible values). The parameter *Delta_Preamble* depends on the type of preamble format. The UE selects the first available RACH resource for the transmission. If there is no response within (three subframes + *RA_Response_Window* subframes), the UE assumes that the transmission has been unsuccessful. It selects another RACH preamble and performs the back-off procedure. The power level is increased by the amount *Power_Ramp_Step*. This process can be repeated until the *Preamble_Transmission_Counter* timer expires, in that case, an access failure is declared to the upper layers.

9.18.7 Power headroom

In order to assist the eNB to schedule/allocate the uplink resources to different UEs, it is important that the UE can report its available power headroom to the eNB. The eNB can use the power headroom reports (PHRs) to determine how much uplink bandwidth per subframe the UE can use. This can help to avoid allocating uplink transmission resources to UEs which are unable to properly use them.

The UE PHR MAC control element indicates the power headroom available in the UE. The PHR is encoded in 6 bits with a reporting range from +40 to −23 dB in steps of 1 dB. Positive values indicate the difference between the maximum UE transmit power and current UE transmit power. Negative values

Table 9.32 *Delta_Preamble* Values [8]

Preamble Format	Delta_Preamble Value
0	0 dB
1	0 dB
2	−3 dB
3	−3 dB
4	8 dB

indicate the difference between the maximum UE transmit power and the calculated UE transmit power (Figure 9.136). The UE transmit power is calculated based on whether the UE were to transmit according to the current grant with allocated HARQ and RV configuration. The PHR can be configured to be sent periodically or when the downlink path loss changes by a specified amount. For periodic reporting, a report is trigged by the expiration of the *Periodic_PHR_Timer*, which can be configured with values between 10 ms and infinity. For threshold reporting, a PHR is triggered when the path loss changes by the value of *DL_PathlossChange* parameter (1, 3, 6, or infinity dB), provided that the configurable timer *Prohibit_PHR_Timer* has expired. The *Prohibit_PHR_Timer* starts when a PHR has been transmitted and can have values between 0 and 1000 ms.

A PHR can only be sent in subframes in which a UE has an uplink grant, i.e., the report is related to the subframe in which it is transmitted. Therefore, the PHR is a prediction rather than a direct measurement. The UE cannot directly measure its actual transmission power headroom for the subframe in which the report is transmitted and relies on reasonably accurate calibration of the UE's power amplifier output, especially at high output powers when information about power headroom is critical to the system performance.

There are two different types of PHRs. A Type 1 PHR report indicates the power headroom assuming PUSCH-only transmission on the component carrier, while a Type-2 PHR report assumes joint PUSCH and PUCCH transmission, which is a feature added in LTE Rel-10.

9.18.8 Link adaptation

In cellular systems, the quality of the signal received by a UE depends on the channel quality from the serving cell, the level of interference from other cells, and the noise level. To optimize system capacity and coverage for a given transmission power, the transmitter attempts to adapt (and optimize) the information rate and the transmission format, as well as the multi-antenna mode for each user

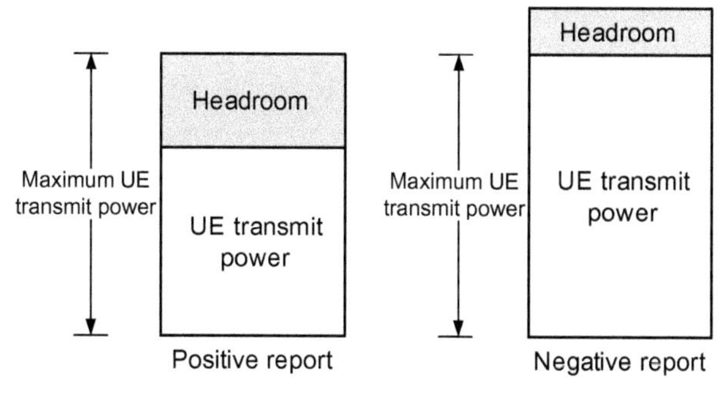

FIGURE 9.136

Power headroom interpretation.

according to the variations in received signal quality at the UE. This is commonly referred to as link adaptation and is typically based on AMC, power control, and adaptation of multi-antenna modes. Low-order modulation such as BPSK and QPSK are more robust and can tolerate higher levels of interference, but provide lower bit rates. High-order modulation such as 16QAM and 64QAM offer higher bit rates, but nonetheless are more prone to errors due to their higher sensitivity to interference, noise, and channel estimation errors; therefore, they are more useful, when the SINR is sufficiently high.

For a given modulation scheme, the code rate can be chosen depending on the channel conditions. A lower code rate can be used in poor channel conditions and a higher code rate can be used when SINR is sufficiently high. The adaptation of the code rate is achieved by applying puncturing to increase the code rate or code repetition to reduce the code rate. In LTE, the modulation and channel coding rates are constant over the allocated frequency resources for a given user, and time-domain channel-dependent scheduling and AMC are supported instead. In addition, when multiple transport blocks are transmitted to one user in a given subframe using multi-layer MIMO, each transport block can use an independent MCS. Furthermore, the UE can be configured to report CQIs to assist the eNB in selecting an appropriate MCS to use for the downlink transmissions. The CQI reports are derived from the downlink received signal quality, typically based on measurements of the downlink reference signals. It is important to note that the reported CQI is not a direct indication of SINR. Instead, the UE reports the highest MCS that it can decode with a BLER not exceeding 10%. Thus the information received by the eNB takes into account the characteristics of the UE's receiver and not just the prevailing channel quality. A UE that is designed with advanced receivers such as MLD or SIC may report a higher channel quality and depending on the characteristics of the eNB's scheduler, can receive a higher data rate [15].

A simple method by which a UE can choose an appropriate CQI value could be based on a set of BLER thresholds, as illustrated in Figure 9.137. The UE will periodically or aperiodically report the CQI value corresponding to the MCS that ensures a BLER less than 10% based on the measured received signal quality.

The CQI metric is generally used in selecting the best MCS under current channel conditions. The CQI measurement typically consists of four basic steps: measuring SINR over the bandwidth of interest (sub-band or wideband), introducing measurement error to the measured SINR, converting the SINR value to discrete CQI values, and finally reporting the CQI using PUCCH or PUSCH. The linear SINR is calculated for each set of resource blocks using the received cell-specific reference signal power and the total interference in every measurement period. The measured linear SINR value for the nth resource block is converted to logarithmic scale as follows $\mathrm{SINR}_{\mathrm{dB}}(n) = 10\log_{10}[\mathrm{SINR}_{\mathrm{linear}}(n)] + \varepsilon_{\mathrm{dB}}$ where $\varepsilon_{\mathrm{dB}}$ is a Gaussian distributed error with zero-mean and parameterized variance, which is added to the measured (ideal) SINR. The SINR values are quantized and converted to discrete CQI values $\mathrm{CQI}_{\mathrm{dB}}(n) = \Delta_Q \lfloor \mathrm{SINR}_{\mathrm{dB}}(n)/\Delta_Q + 0.5 \rfloor$. The CQI is measured at integer multiples of transmission time interval. The

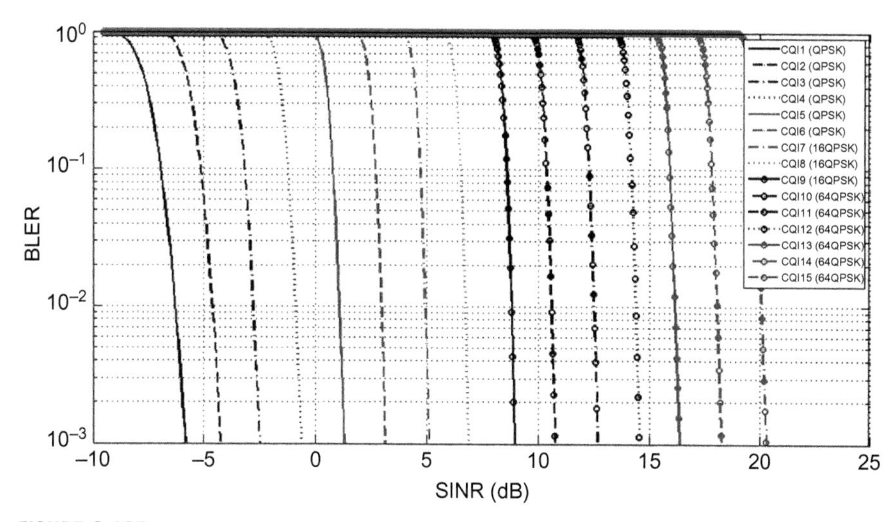

FIGURE 9.137

Link-level BLER as a function of SINR with CQI (MCS) as a parameter [142].

measured CQI values are reported with a certain delay using a specific CQI reporting scheme. It is possible to change the granularity of the basic reporting scheme by changing the number of CQI reports per TTI. Full feedback reporting is done by measuring and reporting individual CQI values for all resource blocks. The minimum granularity is achieved by wideband CQI, which is an average value calculated over all resource blocks.

The granularity of the CQI reports is determined by defining a number of sub-bands $N_{\text{sub-band}}$, each consists of $k_{\text{sub-band}}$ contiguous physical resource blocks. The value of $k_{\text{sub-band}}$ depends on the type of CQI report and is a function of the system bandwidth (see Table 9.28). In each case, the number of sub-bands spans the entire system bandwidth and is given by $N_{\text{sub-band}} = \lceil N_{\text{RB}}^{\text{DL}}/k_{\text{sub-band}} \rceil$ where $N_{\text{RB}}^{\text{DL}}$ is the number of resource blocks over the entire system bandwidth. The CQI reports can be wideband, eNB-configured sub-band, or UE-selected sub-band. In the case of multiple transmit antennas at the eNB, the CQI values may be reported for both codewords.

The eNB scheduler needs to know the number of resource blocks and MCS corresponding to each CQI value in order to properly allocate the resources for each UE. With the modulation scheme obtained from Table 9.33, one would derive a certain range of MCS that can be used for each CQI index. However, in order to obtain a specific MCS and the number of resource blocks, the code rate, given in Table 9.33, is required. Nevertheless, there is not a single formula that would give you a unique value for MCS and the number of resource blocks. Therefore, there is a set of MCS and number of resource blocks that meet the modulation scheme and code rate requirement in Table 9.33. An example is provided in Table 9.34, where the CIF is assumed to be two OFDM symbols for the

Table 9.33 CQI Values and the Corresponding Code Rates [5]

CQI Index	Modulation Scheme	Coding Rate × 1024	Information Bits per Symbol
0	Out of Range		
1	QPSK	78	0.1523
2	QPSK	120	0.2344
3	QPSK	193	0.3770
4	QPSK	308	0.6016
5	QPSK	449	0.8770
6	QPSK	602	1.1758
7	16QAM	378	1.4766
8	16QAM	490	1.9141
9	16QAM	616	2.4063
10	64QAM	466	2.7305
11	64QAM	567	3.3223
12	64QAM	666	3.9023
13	64QAM	772	4.5234
14	64QAM	873	5.1152
15	64QAM	948	5.5547

Table 9.34 Example Mapping of CQI Values to Number of Resource Blocks [158]

CQI Index	Modulation Scheme	Bits/ Symbol	REs/ PRB	N_{RB}	I_{MCS}	Transport Block Size	Code Rate
0	Out of Range						
1	QPSK	2	138	20	0	536	0.101449
2	QPSK	2	138	20	0	536	0.101449
3	QPSK	2	138	20	2	872	0.162319
4	QPSK	2	138	20	5	1736	0.318841
5	QPSK	2	138	20	7	2417	0.442210
6	QPSK	2	138	20	9	3112	0.568116
7	16QAM	4	138	20	12	4008	0.365217
8	16QAM	4	138	20	14	5160	0.469565
9	16QAM	4	138	20	16	6200	0.563768
10	64QAM	6	138	20	20	7992	0.484058
11	64QAM	6	138	20	23	9912	0.600000
12	64QAM	6	138	20	25	11448	0.692754
13	64QAM	6	138	20	27	12576	0.760870
14	64QAM	6	138	20	28	14688	0.888406
15	64QAM	6	138	20	28	14688	0.888406

control region, i.e., 138 resource elements per resource block pair in each subframe, and further it is assumed that there are 20 resource blocks assigned to the UE. The transport block size (excluding the 24-bit CRC) is then determined for various values of the CQI and modulation schemes.

As we mentioned earlier, power control dynamically adjusts the transmit power to compensate for variations in the instantaneous channel condition. The objective of these adjustments is to maintain an approximately constant E_b/N_0 at the receiver in order to successfully transmit data with an acceptable block error rate. In principle, TPC increases the power at the transmitter when the radio link experiences poor radio conditions, and vice versa. Therefore, the transmit power is in essence inversely proportional to the channel quality. This results in an approximately constant data rate, regardless of the channel variations. For services such as circuit-switched voice, this is a desirable property. Transmit-power control can be seen as one type of link adaptation, i.e., the adjustment of transmission parameters, in this case the transmit power, is adapted to differences and variations in the instantaneous channel conditions to maintain the received E_b/N_0 at a desired level. However, in mobile communication scenarios and in particular packet-data transmission, there is no desire to provide a constant data rate over a radio link at all times. In such cases, an alternative to transmit-power control is link adaptation by means of dynamic rate control. Rate control does not aim at keeping the instantaneous radio-link data rate constant, regardless of the instantaneous channel conditions. Instead, with rate control, the data rate is dynamically adjusted to compensate for the varying channel conditions, i.e., the data rate is increased when good channel conditions prevail, and vice versa. Thus, rate control maintains $E_b/N_0 \sim (P_{TX}, R)$ at the desired level, not by adjusting the transmission power P_{TX}, but rather by adjusting the data rate R. It can be shown that rate control is more efficient than power control. Rate control in principle implies that the power amplifier is always transmitting at full power and therefore is efficiently utilized. On the other hand, power control results in inefficient utilization of the power amplifier, as the transmission power is typically less than its maximum value. In practice, the link-level data rate is controlled by adjusting the modulation scheme and/or the channel coding rate. In the case of good channel conditions, the energy per bit E_b at the receiver is high and the main limitation of the data rate is going to be the bandwidth of the radio link [15].

The list of modulation schemes and code rates which can be signaled by means of a CQI value is given in Table 9.33. The AMC can exploit the UE feedback by assuming that the channel fading is sufficiently slow. This requires the channel coherence time to be greater than the CQI measurement delay (i.e., the time interval between the UE's measurement of the downlink reference signals and the subframe containing the correspondingly adapted downlink transmission on PDSCH), which is typically 7–8 ms (depending on the UE speed and the operating frequency). However, a trade-off needs to be made between the amount of CQI information reported by the UEs and the accuracy with which the link adaptation can track the prevailing channel conditions. More frequent reporting of the

CQI in the time-domain allows better tracking of the channel and interference variations, while finer resolution in the frequency-domain enables better use of frequency selective scheduling gain. However, both lead to increased feedback overhead in the uplink. Therefore, the eNB can configure both the time-domain update rate and the frequency-domain resolution of the CQI to optimize the overall performance of the system.

References

3GPP Specifications[35]

[1] 3GPP TS 36.101, Evolved Universal Terrestrial Radio Access (E-UTRA); User Equipment (UE) radio transmission and reception (Release 11), December 2012.

[2] 3GPP TS 36.104, Evolved Universal Terrestrial Radio Access (E-UTRA); Base Station (BS) radio transmission and reception (Release 11), December 2012.

[3] 3GPP TS 36.211, Evolved Universal Terrestrial Radio Access (E-UTRA), Physical Channels and Modulation (Release 11), December 2012.

[4] 3GPP TS 36.212, Evolved Universal Terrestrial Radio Access (E-UTRA) Multiplexing and Channel Coding (Release 11), December 2012.

[5] 3GPP TS 36.213, Evolved Universal Terrestrial Radio Access (E-UTRA); Physical layer procedures (Release 11), December 2012.

[6] 3GPP TS 36.214, Evolved Universal Terrestrial Radio Access (E-UTRA); Physical layer – Measurements (Release 11), December 2012.

[7] 3GPP TS 36.300, Evolved Universal Terrestrial Radio Access (E-UTRA) and Evolved Universal Terrestrial Radio Access Network (E-UTRAN) Overall description, Stage 2 (Release 11), December 2012.

[8] 3GPP TS 36.321, Evolved Universal Terrestrial Radio Access (E-UTRA); Medium Access Control (MAC) Protocol specification (Release 11), December 2012.

[9] 3GPP TS 36.331 Evolved Universal Terrestrial Radio Access (E-UTRA); Radio Resource Control (RRC); Protocol specification (Release 11), December 2012.

[10] 3GPP TR 36.912 Feasibility Study for Further Advancements for E-UTRA (LTE-Advanced) (Release 11), September 2012.

[11] 3GPP TR 36.814 Feasibility Study for Further Advancements for E-UTRA (LTE-Advanced) (Release 9), March 2010.

[12] 3GPP TR 36.819 Coordinated Multi-Point Operation for LTE Physical Layer Aspects (Release 11), December 2011.

Books

[13] C. Johnson, Long Term Evolution in Bullets, second ed., Create Space Independent Publishing Platform, 2012.

[14] S. Sesia, I. Toufik, M. Baker, LTE, the UMTS Long Term Evolution: From Theory to Practice, second ed., John Wiley & Sons, 2011.

[35]3GPP Specifications can be found at the following URL: http://www.3gpp.org/ftp/Specs/archive/36_series/

[15] E. Dahlman, S. Parkvall, J. Skold, 4G: LTE/LTE-Advanced for Mobile Broadband, Academic Press, 2011.

[16] H. Holma, A. Toskala, LTE Advanced: 3GPP Solution for IMT-Advanced, John Wiley & Sons, 2012.

[17] H. Holma, A. Toskala, LTE for UMTS: Evolution to LTE-Advanced, John Wiley & Sons, 2011.

[18] S. Ahmadi, Mobile WiMAX: A Systems Approach to Understanding IEEE 802.16m Radio Access Technology, Academic Press, 2010.

[19] B. Sklar, Digital Communications: Fundamentals and Applications, second ed., Prentice Hall, 2001.

[20] A. Paulraj, et al., Introduction to Space–Time Wireless Communications, Cambridge University Press, 2008.

[21] W.C. Jakes, Microwave Mobile Communications, second ed., Wiley-IEEE Press, 1994.

[22] G. Savo, Glisic, Advanced Wireless Communications: 4G Cognitive and Cooperative Broadband Technology, second ed., Wiley Inter-science, 2007.

[23] D.E. Dudgeon, R.M. Mersereau, Multidimensional Digital Signal Processing, Prentice Hall, 1984.

[24] D. Tse, P. Viswanath, Fundamentals of Wireless Communication, Cambridge University Press, 2005.

[25] E. Biglieri, et al., MIMO Wireless Communications, Cambridge University Press, 2010.

[26] A. Goldsmith, Wireless Communications, Cambridge University Press, 2005.

[27] S. Lin, D.J. Costello, Error Control Coding, second ed., Prentice Hall, 2004.

[28] S.L. Marple, Digital Spectral Analysis with Applications, Prentice Hall, 1987.

[29] G.H. Golub, C.F. Van Loan, Matrix Computations, third ed., Johns Hopkins University Press, 1996.

[30] A. Papoulis, Probability, Random Variables and Stochastic Processes, fourth ed., McGraw Hill Higher Education, 2002.

[31] L. Rade, Mathematics Handbook for Science and Engineering, Springer, 2010.

[32] G.A. Korn, T.M. Korn, Mathematical Handbook for Scientists and Engineers: Definitions, Theorems, and Formulas for Reference and Review, Revised Edition, Dover Publications, 2000.

[33] C.E. Shannon, W. Weaver, The Mathematical Theory of Communication, University of Illinois Press, 1998.

[34] F. Dowla, Handbook of RF and Wireless Technologies, Newnes, 2003.

[35] G. Arfken, *Bessel functions*, Mathematical Methods for Physicists, third ed., Academic Press, 1985.

[36] W.G. Bickley, Bessel Functions and Formulae, Cambridge University Press, Cambridge, England, 1957.

[37] F. Bowman, Introduction to Bessel Functions, Dover, 2010.

[38] H.G. Myung, D. Goodman, Single Carrier FDMA: A New Air Interface for Long Term Evolution, Wiley & Sons, 2008.

[39] T.M. Cover, J.A. Thomas, Elements of Information Theory, John Wiley & Sons, 1991.

[40] J.G. Proakis, Digital Communications, fifth ed., McGraw-Hill, November 2007.

[41] N. Costa, S. Haykin, Multiple-Input Multiple-Output Channel Models: Theory and Practice, John Wiley & Sons, 2010.

[42] C. Oestges, B. Clerckx, MIMO Wireless Communications: From Real-World Propagation to Space–Time Code Design, Academic Press, 2007.

[43] A. Sibille, C. Oestges, A. Zanella, MIMO: From Theory to Implementation, Academic Press, 2010.
[44] I. Stakgold, M.J. Holst, Green's Functions and Boundary Value Problems, third ed., John Wiley & Sons, 2011.
[45] R.G. Gallager, Principles of Digital Communication, Cambridge University Press, 2008.

Articles

[46] G. Bauch, G. Dietl, MIMO for 3GPP LTE-Advanced and Beyond, Global Telecommunications Conference (GLOBECOM), 2010.
[47] C. Mehlfuhrer, et al., The vienna LTE simulators—enabling reproducibility in wireless communication research, EURASIP J. Adv. Signal Process. 2011 (1) (2011).
[48] B. Sklar, Rayleigh fading channels in mobile digital communication systems. I. Characterization, IEEE Commun. Mag. 35 (7) (1997).
[49] B. Sklar, Rayleigh fading channels in mobile digital communication systems. II. Mitigation, IEEE Commun. Mag. 35 (7) (July 1997).
[50] Y. Okumura, E. Ohmori, K. Fukuda, Field strength and its variability in VHF and UHF land mobile radio service, Rev. Electr. Commun. Lab. 16 (1968).
[51] M. Hata, Empirical formulae for propagation loss in land mobile radio services, IEEE Trans. Veh. Technol. VT-29 (3) (1980).
[52] Y. Wu, W.Y. Zou, Orthogonal frequency division multiplexing: a multi-carrier modulation scheme, IEEE Trans. Consum. Electron. 41 (3) (1995).
[53] H. Su, et. al., Analysis of tone reservation method for WiMAX system, International Symposium on Communications and Information Technologies, ISCIT'06, 2006.
[54] A. Osseiran, J.C. Guey, Hopping pilot pattern for interference mitigation in OFDM, IEEE 19th International Symposium on Personal, Indoor and Mobile Radio Communications, 2008.
[55] R. Nilsson, O. Edforst, M. Sandellt, P. Ola Borjesson, An analysis of two-dimensional pilot-symbol assisted modulation for OFDM, 1997 IEEE International Conference on Personal Wireless Communications Publication, 1997.
[56] H. Sampath, P. Stoica, A. Paulraj, Generalized linear precoder and decoder design for MIMO channels using the weighted MMSE criterion, IEEE Trans. Commun. 49 (12) (2001).
[57] Y.S. Choi, P.J. Voltz, F.A. Cassara, On channel estimation and detection for multicarrier signals in fast and selective rayleigh fading channels, IEEE Trans. Commun. 49 (8) (2001).
[58] S.M. Alamouti, A simple transmit diversity technique for wireless communications, IEEE J. Sel. Areas Commun. 16 (8) (1998).
[59] A. Nosratinia, T.E. Hunter, A. Hedayat, Cooperative communication in wireless networks, IEEE Commun. Mag. (2004).
[60] J.G. Andrews, Interference cancellation for cellular systems: a contemporary overview, IEEE Wirel. Commun. (2005).
[61] D. Gesbert, et al., From theory to practice: an overview of MIMO space–time coded wireless systems, IEEE J. Sel. Areas Commun. 21 (3) (2003).
[62] R.D. Murch, K. Ben Letaief, Antenna systems for broadband wireless access, IEEE Commun. Mag. (2002).
[63] S. Catreux, et al., Adaptive modulation and MIMO coding for broadband wireless data networks, IEEE Commun. Mag. (2002).

[64] A.J. Paulraj, D.A. Gore, R.U. Nabar, H. Bölcskei, An *overview of MIMO communications—a key to gigabit wireless*, Proc. IEEE 92 (2) (2004).
[65] M.G. Parker, K.G. Patersonyand, C. Tellambura, Golay complementary sequences, in: J.G. Proakis (Ed.), Wiley Encyclopedia of Telecommunications, John Wiley & Sons, 2003.
[66] D.J. Love, R.W. Heath Jr., Grassmannian beamforming for multiple-input multiple-output wireless systems, IEEE Trans. Inf. Theory 49 (10) (2003).
[67] V. Tarokh, H. Jafarkhani, On the computation and reduction of the peak-to-average power ratio in multicarrier communications, IEEE Trans. Commun. 48 (1) (2000).
[68] T.M. Schmid, D.C. Cox, Robust frequency and timing synchronization for OFDM, IEEE Trans. Commun. 45 (12) (1997).
[69] M. Pukkila, Channel Estimation Modeling, Postgraduate Course in Radio-communications, Fall 2000.
[70] M.S. Akram, Pilot-based channel estimation in OFDM systems, Master Thesis, 2007.
[71] Freescale Semiconductor Application Note, Channel Estimation in OFDM Systems, AN3059 Rev. 0, 2006.
[72] V. Srivastava, et al., Robust MMSE channel estimation in OFDM systems with practical timing synchronization, IEEE Wireless Communications and Networking Conference 2004.
[73] L. Zheng, D.N.C. Tse, Diversity and multiplexing: a fundamental tradeoff in multiple-antenna channels, IEEE Trans. Inf. Theory 49 (5) (2003).
[74] H. Weingarten, Y. Steinberg, S. Shamai, The capacity region of the gaussian multiple-input multiple-output broadcast channel, IEEE Trans. Inf. Theory 52 (9) (2006).
[75] F. Kaltenberger, et al., Capacity of linear multi-user MIMO precoding schemes with measured channel data, EURECOM, Sophia-Antipolis, France.
[76] V. Stankovic, M. Haardt, Multi-user MIMO downlink precoding for users with multiple antennas, in: Proceedings of the 12th Meeting of the Wireless World Research Forum (WWRF), Toronto, Canada, 2004.
[77] Q.H. Spencer, A.L. Swindlehurst, M. Haardt, Zero-Forcing methods for downlink spatial multiplexing in multiuser MIMO channels, IEEE Trans. Signal Process. 52 (2) (2004).
[78] N. Jindal, MIMO broadcast channels with finite-rate feedback, IEEE Trans. Inf. Theory 52 (11) (2006).
[79] Q.H. Spencer, C.B. Peel, A. Lee Swindlehurst, M. Haardt, An introduction to the multi-user MIMO downlink, IEEE Commun. Mag. (2004).
[80] D. Gesbert, et al., Shifting the MIMO paradigm, IEEE Signal Process. Mag. (2007).
[81] L. Wei, Capacity of Hybrid Open-Loop and Closed-Loop MIMO with Channel Uncertainty at Transmitter, Helsinki University of Technology, Espoo, 2008.
[82] D.J. Love, R.W. Heath Jr., Grassmannian precoding for spatial multiplexing systems, in: Proceedings of the Allerton Conference on Communication Control and Computing, Monticello, 2003.
[83] T. Abe, G. Bauch, Differential codebook MIMO precoding technique, 2007 IEEE Global Telecommunications Conference, 2007.
[84] B. Bandemer, M. Haardt, S. Visuri, Linear MMSE multi-user MIMO downlink precoding for users with multiple antennas, 17th Annual IEEE International Symposium on Personal, Indoor and Mobile Radio Communications (PIMRC), 2006.

[85] G. Mitchell, F.R. Kschischang, An augmented orthogonal code design for the non-coherent MIMO channel, 2008 24th Biennial Symposium on Communications, 2008.
[86] Y. Fan, P. Lundén, M. Kuusela, M. Valkama, Efficient semi-persistent scheduling for VoIP on EUTRA downlink, IEEE 68th Vehicular Technology Conference (VTC), Fall 2008.
[87] Goldsmith, et al., Capacity limits of MIMO channels, IEEE J. Sel. Areas Commun. 21 (2003).
[88] H. Weingarten, Y. Steinberg, S. Shamai, The capacity region of the gaussian MIMO broadcast channel, in: Proceedings of the Conference Information Sciences and Systems (CISS), Princeton, NJ, 2004.
[89] M. Costa, Writing on dirty paper, IEEE Trans. Inf. Theory 29 (1983).
[90] T. Svantesson, A.L. Swindlehurst, A performance bound for prediction of a multi-path MIMO channel, in: Proceedings of the 37th Asilomar Conference on Signals, Systems, and Computers, Session: Array Processing for Wireless Communications, Pacific Grove, CA, 2003.
[91] C.B. Peel, B.M. Hochwald, A.L. Swindlehurst, A vector-perturbation technique for near-capacity multi-antenna multi-user communication, IEEE Trans. Commun. (2003).
[92] M. Bengtsson, B. Ottersten, Optimal and suboptimal beamforming, in: L.C. Godara (Ed.), Handbook of Antennas in Wireless Communications, CRC Press, 2001.
[93] M. Schubert, H. Boche, Solution of the multiuser downlink beamforming problem with individual SINR constraints, IEEE Trans. Veh. Technol. 53 (2004).
[94] Q.H. Spencer, A.L. Swindlehurst, M. Haardt, Zero-forcing methods for downlink spatial multiplexing in multi-user MIMO channels, IEEE Trans. Signal Process. 52 (2004).
[95] G. Caire, S. Shamai, On the achievable throughput of a multi-antenna gaussian broadcast channel, IEEE Trans. Inf. Theory 49 (2003).
[96] U. Erez, S. Shamai, R. Zamir, Capacity and lattice strategies for cancelling known interference, in: Proceedings International Symposium Information Theory and its Applications, 2000.
[97] C. Windpassinger, R.F.H. Fischer, J.B. Huber, Lattice-reduction-aided broadcast precoding, in: Proceedings of Fifth ITG Conference Source and Channel Coding, 2004.
[98] Berrou, Glavieux, Thitimajshima, Near shannon limit error-correcting coding and decoding: turbo-codes, in: Proceedings of 1993 International Communication Conference, 1993.
[99] Q.H. Spencer, A.L. Swindlehurst, Channel allocation in multi-user MIMO wireless communications systems, in: Proceedings of 2004 International Communication Conference, 2004.
[100] Perez, Seghers, Costello, A distance spectrum interpretation of turbo codes, IEEE Trans. Inf. Theory 42 (6) (1996).
[101] J. Rothweller, Turbo codes, IEEE Potentials 18 (1) (1999).
[102] C. Berrou, A. Glavieux, P. Thitimajshima, Near Shannon limit error-correcting coding and decoding: turbo codes, in: Proceedings of 1993 International Communication Conference, 1993.
[103] C. Berrou, A. Glavieux, Near optimum error correcting coding and decoding: turbo-codes, IEEE Trans. Commun. 44 (10) (1996).

[104] C. Heegard, Stephen B. Wicker, Turbo Coding, Springer, 2010.

[105] C. Berrou, R. Pyndiah, P. Adde, C. Douillard, R. Le Bidan, An overview of turbo codes and their applications, The European Conference on Digital Wireless Technology, 2005.

[106] H. Ma, J. Wolf, On tail biting convolutional codes, IEEE Trans. Commun. COM-34 (2) (1986).

[107] C. Weiss, C. Bettstetter, S. Riedel, Code construction and decoding of parallel concatenated tail-biting codes, IEEE Trans. Inf. Theory 47 (1) (2001).

[108] Y.E. Wang, R. Ramesh, To bite or not to bite—a study of tail bits versus tail-biting, Seventh IEEE International Symposium Personal, Indoor and Mobile Radio Communications, PIMRC'96, vol. 2, 1996.

[109] B.E. Priyanto, H. Codina, Initial performance evaluation of DFT-Spread OFDM Based SC-FDMA for UTRA LTE Uplink, IEEE 65th Vehicular Technology Conference (VTC), 2007.

[110] J. Lee, J.-K. Han, J. Zhang, MIMO Technologies in 3GPP LTE and LTE-Advanced, EURASIP J. Wireless Commun. Networking (2009).

[111] S. Catreux, P.F. Driessen, L.J Greenstein, Attainable throughput of an interference-limited multiple-input multiple-output (MIMO) cellular system, IEEE Trans. Commun. 49 (8) (2001).

[112] I. Telatar, Capacity of multi-antenna Gaussian channels, AT&T Technical Memorandum, 1995.

[113] G.J. Foschini, M.J. Gans, On limits of wireless communications in a fading environment when using multiple antennas, Wireless Pers. Commun 1998.

[114] M.-S. Alouini, A. Goldsmith, Capacity of Nakagami multipath fading channels, Proceedings Vehicular Technology Conference, 1997.

[115] D. Chizhik, G.J. Foschini, R.A. Valenzuela, Capacities of multi-element transmit and receive antennas: correlations and keyholes, Electron. Lett. 36 (2000).

[116] D. Gesbert, H.B.̈olcskei, D. Gore, A. Paulraj, MIMO wireless channels: capacity and performance prediction, Global Telecommunications Conference (GLOBECOM), 2000.

[117] D.A. Gore, R.U. Nabar, A. Paulraj, Selecting an optimal set of transmit antennas for a low rank matrix channel, Proc. ICASSP 2000 49 (8) (2000).

[118] S. Sandhu, R.U. Nabar, D.A. Gore, A. Paulraj, Near-optimal selection of transmit antennas for a MIMO channel based on Shannon capacity, in: Proceedings of the 34th Asilomar Conference on Signals, Systems and Computers, 2000.

[119] A.F. Molisch, M.Z. Win, J.H. Winters, A. Paulraj, Capacity of MIMO systems with antenna selection, in: Proceedings of the IEEE International Conference on Communications (ICC), 2001.

[120] Z. Shen, et al., Overview of 3GPP LTE-advanced carrier aggregation for 4G wireless communications, IEEE Commun. Mag. 50 (2) (2012).

[121] Y.-H. Nam, et al., Evolution of reference signals for LTE-advanced systems, IEEE Commun. Mag. 50 (2) (2012).

[122] L. Liu, et al., Downlink MIMO in LTE-advanced: SU-MIMO vs. MU-MIMO, IEEE Commun. Mag. 50 (2) (2012).

[123] D. Lee, et al., Coordinated multipoint transmission and reception in LTE-advanced: deployment scenarios and operational challenges, IEEE Commun. Mag. 50 (2) (2012).

[124] D. Bai, et al., LTE-Advanced modem design: challenges and perspectives, IEEE Commun. Mag. 50 (2) (2012).

[125] J. Lee, Y. Kim, et al., Coordinated multipoint transmission and reception in LTE-Advanced systems, IEEE Commun. Mag. 50 (2) (2012).

[126] Z. Shen, et al., Dynamic uplink-downlink configuration and interference management in TD-LTE, IEEE Commun. Mag. 50 (2) (2012).

[127] C.S. Park, et al., Evolution of uplink MIMO for LTE-advanced, IEEE Commun. Mag. 49 (2) (2011).

[128] R. Irmer, et al., Coordinated multipoint: concepts, performance, and field trial results, IEEE Commun. Mag. 49 (2) (2011).

[129] M. Vu, A. Paulraj, MIMO Wireless Linear Precoding, IEEE Signal Process. Mag. 24 (5) (2007).

[130] R. Gold, Optimal binary sequences for spread spectrum multiplexing, IEEE Trans. Inf. Theory 13 (4) (1967).

[131] X. Cai, G.B. Giannakis, Bounding performance and suppressing inter-carrier interference in wireless mobile OFDM, IEEE Trans. Commun. 51 (12) (2003).

[132] J. Lee, H.-L. Lou, D. Toumpakaris, J.M. Cioffi, Effect of carrier frequency offset on OFDM systems for multipath fading channels, Global Telecommunications Conference (GLOBECOM), 2004.

Miscellaneous

[133] 3G Americas, MIMO and smart antennas for 3G and 4G wireless systems. <http://www.3gamericas.org>, 2010.

[134] Rohde, Schwarz Application Notes, LTE-Advanced technology introduction, 2010.

[135] 3G Americas, 3GPP mobile broadband innovation path to 4G: Release 9, Release 10 and beyond: HSPA + , SAE/LTE and LTE-Advanced, 2010.

[136] 3G Americas, HSPA to LTE-Advanced: 3GPP broadband evolution to IMT-Advanced (4G), 2009.

[137] 3G Americas, The mobile broadband evolution: 3GPP release 8 and beyond HSPA + , SAE/LTE and LTE-Advanced, 2009.

[138] IEEE 802.16m-07/244r1, Y. Song, L. Cai, K. Wu, H. Yang, Collaborative MIMO, 2007.

[139] Agilent Technologies, 3GPP long term evolution: system overview, product development, and test challenges. <http://www.agilent.com>, 2009.

[140] IEEE C802.16m-08/153, B.-C. Ihm, J. Choi, W. Lee, Pilot related to DL-MIMO, 2008.

[141] MATLAB CENTRAL, An open exchange for the MATLAB and SIMULINK user community. <http://www.mathworks.com/matlabcentral>.

[142] Vienna University of Technology, Institute of Telecommunications, LTE link-level and system-level simulators. <http://www.nt.tuwien.ac.at/about-us/staff/josep-colom-ikuno/lte-simulators>, 2012.

[143] 3GPP R1-060385 Motorola, Cubic metric in 3GPP-LTE, 2006.

[144] The Coded Modulation Library (The Iterative Solutions Coded Modulation Library (ISCML) is an open source toolbox for simulating capacity approaching codes in Matlab). <http://www.iterativesolutions.com>.

[145] 4G Americas, 4G mobile broadband evolution: release 10, release 11 and beyond—HSPA, SAE/LTE and LTE-Advanced, 2012.

[146] 4G Americas, MIMO and smart antennas for mobile broadband networks, 2012.

[147] 4G Americas, Developing and integrating a high performance HET-NET, 2012.

[148] 4G Americas, 4G mobile broadband evolution: 3GPP release 10 and beyond—HSPA + , SAE/LTE and LTE-Advanced, 2011.
[149] 3GPP R1-113297, NTT DOCOMO, Enhanced PDCCH for DL MIMO in Rel-11, 2011.
[150] 3GPP R1-112928, Ericsson, ST-Ericsson, On enhanced PDCCH design, 2011.
[151] 3GPP R1-113175, Renesas Mobile, Link-level evaluation of E-PDCCH design aspects, 2011.
[152] 3GPP R1-113194, LG Electronics, Consideration on transmit diversity scheme for enhanced PDCCH transmission, 2011.
[153] 3GPP R1-113202, Intel Corporation, Performance analysis of enhanced downlink control signaling, 2011.
[154] 3GPP R1-113238, Research in Motion, E-PDCCH transmission with transmit diversity, 2011.
[155] 3GPP R1-114302, NTT DOCOMO, DM-RS design for E-PDCCH in Rel-11, 2011.
[156] Anritsu, Future technologies and testing for fixed mobile convergence, SAE and LTE in cellular mobile communications. <http://www.anritsu.com/search/en-US/downloadssearch.aspx>, 2008.
[157] Wireless World Initiative New Radio—WINNER + , Enabling Techniques for LTE-A and beyond, Deliverable D2.2. <http://projects.celtic-initiative.org/winner + /deliverables_winnerplus.html>, 2010.
[158] LTE System Description. <http://www.sharetechnote.com>.
[159] DSPLOG Signal Processing for Communication <http://www.dsplog.com>.
[160] 4G Evolution Lab—LTE and LTE-Advanced Toolbox and Block-set: a simulation library for the 3GPP LTE and LTE Advanced specification for the Mathworks MATLAB and Simulink. <http://www.steepestascent.com>.
[161] R. Duncan, Supporting Operators with LTE Roaming, CDG International Roaming Team, August 2012.
[162] Samsung Electronics, LTE PHY specification. <http://www.slideshare.net/allabout4g/lte-rel-8-physical-layer>, 2008.
[163] Motorola White Paper, TD-LTE, Exciting alternative global Momentum. <www.motorola.com>, 2010.
[164] B. Schulz, LTE Transmission modes and beamforming, Rohde & Schwarz White Paper, 2011.
[165] E. Seidel, LTE-A Carrier Aggregation Enhancements in Release 11, NOMOR Research GmbH, Munich, Germany, 2012.
[166] B. Holter, On the Capacity of the MIMO Channel- A Tutorial Introduction, Norwegian University of Science and Technology, 2001.
[167] S. Giraud, et al., Bulk acoustic wave filters synthesis and optimization for multi-standard communication terminals, IEEE Trans Ultrason Ferroelectr Freq Control 57 (1) (2010).
[168] L.A. Coldren, R. Rosenberg, Surface-acoustic-wave resonator filters, Proc. IEEE 67 (1) (1979).
[169] Return loss, from Wikipedia, < http://en.wikipedia.org/wiki/Return_loss >.
[170] Masaaki Koiwa, et al., Multiband Mobile Terminals, NTT DoCoMo Tech. J. 8 (2) (2006).
[171] Siemens White Paper, TD-SCDMA: the Solution for TDD bands, March 2004 < http://www.tdscdma-forum.org/ >.

[172] D.C. Chu, Polyphase Codes With Good Periodic Correlation Properties, IEEE Trans Inf Theory (1972).

[173] Maximum length sequence, from Wikipedia, <http://en.wikipedia.org/wiki/Maximum_length_sequence>.

[174] Jung-In Kim, et al., SSS Detection Method for Initial Cell Searchin 3GPP LTE FDD/TDD Dual Mode Receiver, Ninenth International Symposium on Communication and Information Technologies, 2009.

[175] J. van de Beek, et al., Low-complex frame synchronization in OFDM systems, Fourth IEEE International Conference on Universal Personal Communications. 1995.

[176] ARTIST 4G Project, D2.2 Advanced receiver signal processing techniques – evaluation and characterization, 2011.

[177] Hidekazu Taoka, et al., MIMO and CoMP in LTE-Advanced, DoCoMo Techn. J. 12 (2) (2010).

[178] Sigen Ye, et al., Enhanced Physical Downlink Control Channel in LTE Advanced Release 11, IEEE Communications Magazine, 2013.

[180] Yusuke Ohwatari, et al., Performance of Advanced Receiver Employing Interference Rejection Combining to Suppress Inter-cell Interference in LTE-Advanced Downlink, IEEE Vehicular Technology Conference (VTC), 2011.

[181] Ernesto Zimmermann, et al., On the Complexity of Sphere Decoding, in Proceedings International Symposium on Wireless Personal Multimedia Communication, 2004.

[182] Federico Boccardi, et al., Precoding Schemes for the MIMO-GBC, Int. Zurich Seminar on Communications (IZS), 2006.

CHAPTER

Uplink Physical Layer Functions

10

CHAPTER CONTENTS

This chapter describes the physical layer protocols and functional processing in the LTE/LTE-Advanced uplink direction. As shown in Figure 10.1, the physical layer is the lowest protocol layer in baseband signal processing that interfaces with the physical media (in this case air-interface) through which the signal is transmitted and received. The physical layer receives MAC PDUs (transport blocks) and processes them through channel coding, interleaving, baseband modulation, layer mapping for multi-antenna transmission, eigen-precoding, resource element, and antenna mapping. The choice of the appropriate modulation and coding scheme and multi-antenna transmission mode are critical to achieve the

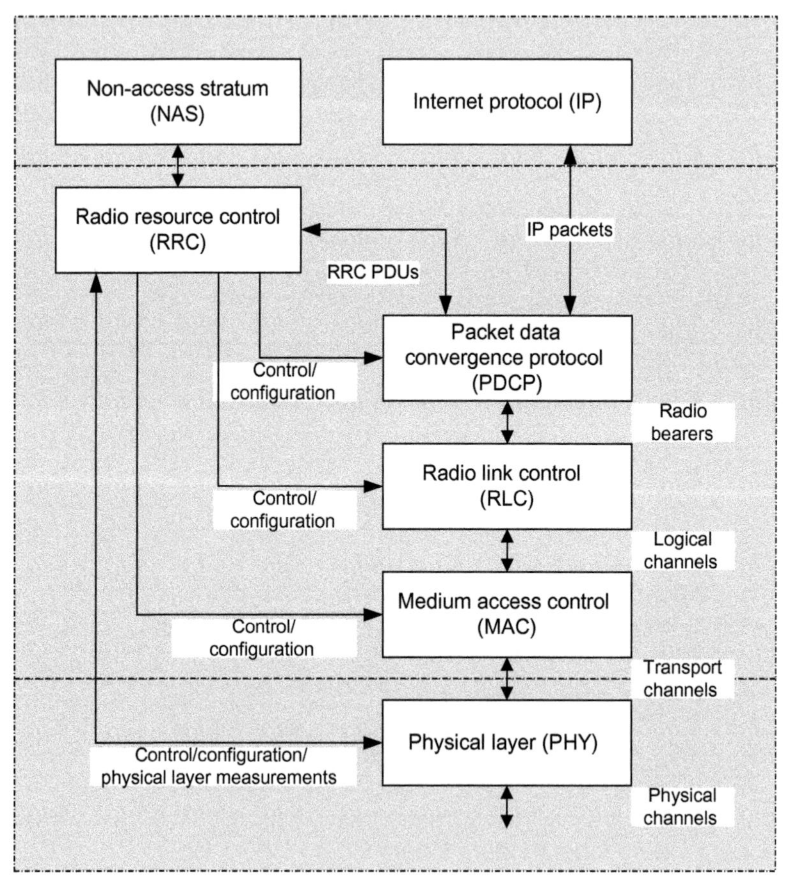

FIGURE 10.1

Overall view of LTE/LTE-Advanced physical layer.

desired reliability/robustness and system/user throughput in mobile wireless data communications. Typical mobile radio channels tend to be dispersive and time variant and exhibit severe Doppler effects, multipath delay variation, intra-cell and inter-cell interference, and fading.

A good and robust design of the physical layer ensures that the system can normally operate and overcome the above deleterious effects, and can provide the maximum throughput and lowest latency under various operating conditions. The chapters on physical layer in this book are dedicated to the systematic design of physical layer protocols and functional blocks of Rel-10/11 of LTE systems, the theoretical background on physical layer procedures, and performance evaluation of physical layer components. The theoretical background is provided to ensure that the reader understands the underlying theory governing the operation of various functional blocks and procedures. While the focus is mainly on the techniques that were incorporated in the design of the LTE/LTE-Advanced physical layer, the author has attempted to take a more generic and systematic approach to the design of physical layer for the systems beyond IMT-Advanced so that the reader can understand and apply the learnings to the design and implementation of any MIMO-OFDM-based physical layer irrespective of the radio access technology. Physical layer processes both control-plane and user-plane signals; however, due to different design requirements and reliability and performance criteria, the procedures tend to be different.

While the requirements for the design of the LTE/LTE-Advanced uplink physical layer are mainly similar to those of the downlink, there are some design attributes that are specific to the uplink physical layer including the desire for orthogonal uplink transmission by different UEs to minimize intra-cell interference and maximize the capacity, flexible data rate adaptation as a function of SINR, sufficiently low peak-to-average power ratio (PAPR) of the transmitted signal to avoid excessive cost, size, and power consumption of the UE transmit circuitry, ability to exploit the frequency diversity attainable by the wider bandwidths even when transmitting at low data rates, support for frequency-selective scheduling, and support for advanced multiple-antenna techniques in order to exploit spatial diversity and to enhance uplink capacity. The multiple-access scheme selected for the LTE uplink to satisfy the above design requirements is single-carrier frequency division multiple access (SC-FDMA). A major advantage of SC-FDMA over the direct-sequence code-division multiple access (DS-CDMA) scheme used in UMTS is that it achieves intra-cell orthogonality even in frequency-selective channels. The SC-FDMA scheme avoids strong intra-cell interference linked to DS-CDMA, which significantly reduces system capacity and limits the use of adaptive modulation. A code-division multiplexed uplink further suffers from an increased PAPR, if multi-code transmission is used from a single UE.

The use of OFDM for the LTE uplink would have been attractive given the maximal uplink/downlink functional commonality and ease of scheduling/implementation. In principle, an OFDM scheme similar to the LTE downlink could satisfy all the uplink design criteria listed above, except the low PAPR requirement. The SC-FDMA combines the desirable characteristics of OFDM with the low PAPR of single-carrier transmission schemes. Similar to OFDM, the SC-FDMA divides the transmission bandwidth into multiple parallel sub-carriers, with the orthogonality between the sub-carriers being maintained in frequency-selective channels by the use of a cyclic prefix or guard period. The use of a cyclic prefix prevents inter-symbol interference (ISI) between SC-FDMA symbols. It transforms the linear convolution of the multipath channel into a circular convolution, enabling the receiver to equalize the channel simply by scaling each sub-carrier by a complex gain factor. However, unlike OFDM, where the data symbols directly modulate each sub-carrier independently such that the amplitude of each sub-carrier at a given time instant is set by the constellation points of the baseband modulation scheme, in SC-FDMA the signal modulated on a given sub-carrier is a linear combination of the data symbols transmitted at the same time instant. Therefore, in each symbol period, the transmitted sub-carriers of an SC-FDMA signal carry a component of each modulated data symbol. This gives SC-FDMA its crucial single-carrier property, which results in the PAPR being significantly lower than pure multi-carrier transmission schemes such as OFDM [39].

10.1 Overview of uplink physical layer processing

This section provides an overview of physical layer processing in the LTE/LTE-Advanced uplink. The processing steps and the associated control signaling are depicted in Figure 10.2. The main criteria in the design of the LTE/LTE-Advanced physical layer were to increase the application throughput and capacity, reduce access latency, support higher user mobility, minimize intra-cell and inter-cell interference, improve reliability of control and data channel coverage especially at the cell-edge, and reduce the complexity and signaling overhead. The overall performance evaluations provided in Chapter 11 suggest that the above requirements have been successfully met. In the next sections, we describe in detail the functional blocks and protocols of the uplink physical layer and their interactions based on the order the information is processed, as shown in Figure 10.2.

10.2 Principles of SC-FDMA

The LTE uplink uses SC-FDMA with a cyclic prefix in order to achieve inter-user orthogonality and to enable efficient frequency-domain equalization at the receiver. The discrete Fourier transform (DFT)-spread OFDM (DFT-S OFDM) is a form of the single-carrier transmission technique, where the signal is generated in the frequency-domain similar to OFDMA, and is illustrated in Figure 10.3.

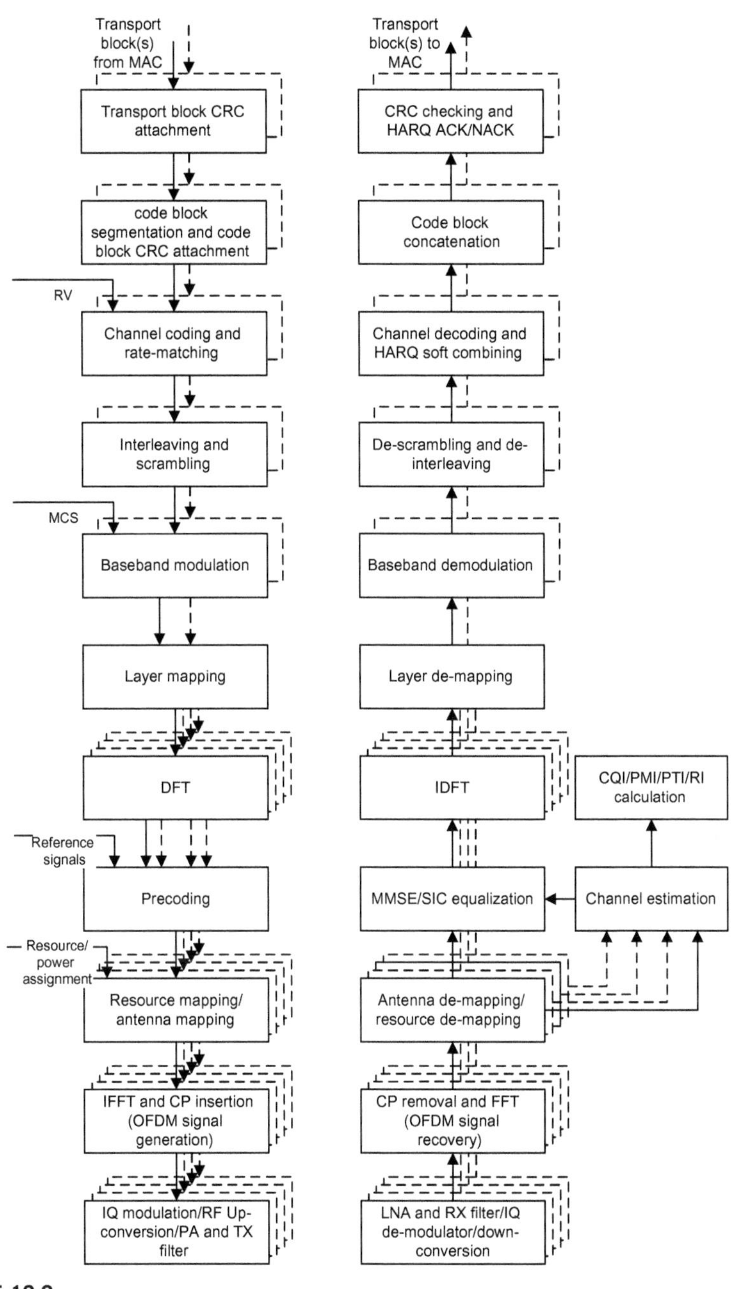

FIGURE 10.2

The LTE/LTE-Advanced uplink physical layer processing blocks [3,14,15,16].

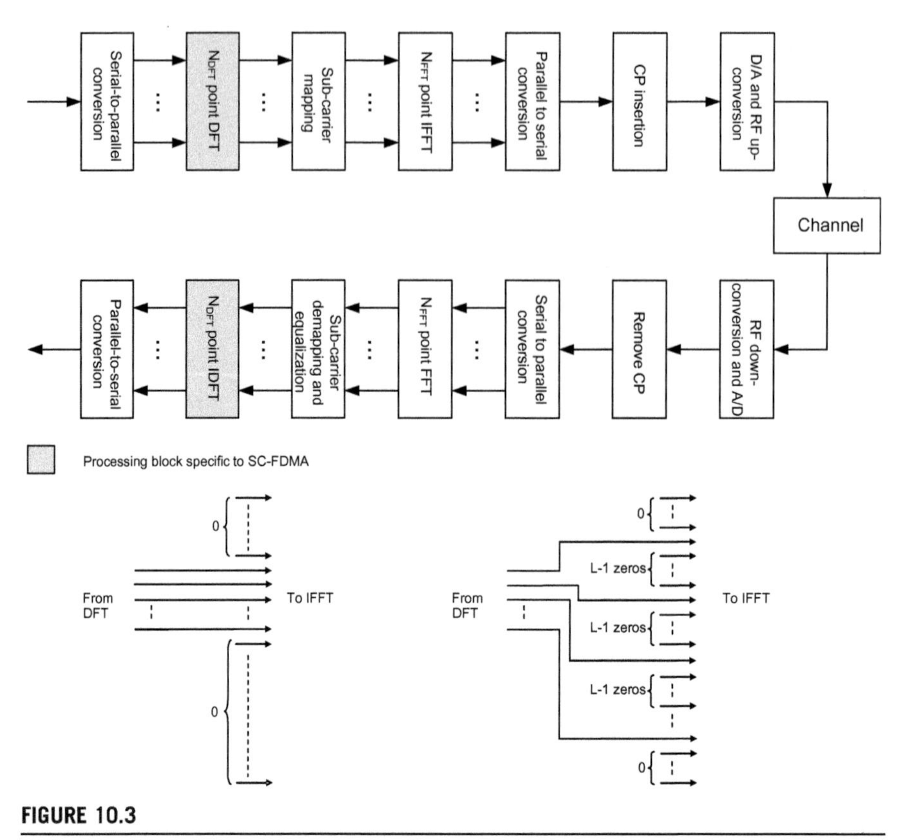

FIGURE 10.3

Transmitter/receiver structure for SC-FDMA with localized and distributed sub-carrier mapping schemes (note that $N_{DFT} < N_{FFT}$) [39].

In Figure 10.3, the common processing blocks in OFDMA and SC-FDMA are distinguished from those that are specific to SC-FDMA. This allows for a relatively high degree of commonality with the downlink OFDMA baseband processing using the same parameters, e.g., clock frequency, sub-carrier spacing, and FFT/IFFT size. The use of SC-FDMA in the uplink is mainly due to relatively inferior PAPR properties of OFDMA that result in worse uplink coverage compared to SC-FDMA. The PAPR characteristics are important for cost-effective design of a mobile station's power amplifiers (PAs).

The principles of SC-FDMA signal processing can be explained as follows. The ith transmitted symbol in an SC-FDMA system without cyclic prefix in single transmit/receive antenna case can be expressed as a vector of length N_{FFT} samples defined by $\mathbf{y} = \mathbf{F\Theta Dx}$, where $\mathbf{x} = (x_1, x_2, \ldots, x_{N_{DFT}})^T$ is an $N_{DFT} \times 1$ vector with N_{DFT} QAM-modulated symbols (the superscript "T" denotes matrix transpose operation), $\mathbf{D}$ is an $N_{DFT} \times N_{DFT}$ matrix which performs N_{DFT}-point DFT operation, $\mathbf{\Theta}$ is the $N_{FFT} \times N_{DFT}$ mapping matrix for sub-carrier assignment, and

$\mathbf{F}$ performs N_{FFT}-point IFFT operation. After the propagation through the multipath fading channel and addition of the additive white Gaussian noise and removing the cyclic prefix and going through the N_{FFT}-point FFT module, the received signal vector in the frequency-domain can be expressed as $\mathbf{z} = \mathbf{H\Theta Dx} + \mathbf{w}$, where $\mathbf{H}$ is the diagonal channel matrix and $\mathbf{w}$ is the receiver noise vector. Note that the maximum excess delay of the channel is assumed to be shorter than the cyclic prefix; therefore, the ISI can be mitigated by the cyclic prefix. The amplitude and phase distortion in the received signal due to the multipath channel is compensated by a frequency-domain equalizer (FDE) and the signal at the FDE output can be described as $\mathbf{v} = \mathbf{Cz}$, where $\mathbf{C} = \mathrm{diag}(c_1, c_2, \ldots, c_{N_{\mathrm{FFT}}})$ is the diagonal matrix of FDE coefficients. The FDE complex coefficients can be derived using minimum mean-square error (MMSE) criterion as follows:

$$c_k = \frac{H_k^*}{|H_k|^2 + \sigma_{\mathrm{n}}^2/\sigma_{\mathrm{s}}^2} \tag{10.1}$$

where k denotes the sub-carrier index, σ_{n}^2 denotes the variance of the additive noise, and σ_{s}^2 is the variance of the transmitted pilot symbol. Following the sub-carrier de-mapping function and IDFT de-spreading, an $N_{\mathrm{DFT}} \times 1$ vector $\hat{\mathbf{x}}$ containing N_{DFT} QAM-modulated symbols as an estimate to the input vector $\mathbf{x}$ is obtained at the receiver. The IDFT de-spreading block in the receiver averages the noise over each sub-carrier. A particular sub-carrier may experience deep fading in a frequency-selective fading channel. The IDFT de-spreading averages and spreads the fading effect, which results in a noise enhancement to all the QAM symbols. Therefore, the IDFT de-spreading makes DFT-S OFDM more sensitive to the noise.

As shown in Figure 10.3, the modulation symbols in blocks of N_{DFT} symbols are processed through an N_{DFT}-point DFT processor, where N_{DFT} denotes the number of sub-carriers assigned to the transmission of the data/control block. The rationale for the use of DFT precoding is to reduce the cubic metric (CM) of the transmitted signal. From an implementation point of view, the DFT size should ideally be a power of 2. However, such a constraint would limit the scheduler flexibility in terms of the amount of resources that can be assigned for an uplink transmission. In LTE, the DFT size and the size of the resource allocation is limited to products of the integers 2, 3, and 5. For example, the DFT sizes of 60, 72, and 96 are allowed, but a DFT size of 42 is not allowed [16]. Therefore, the DFT can be implemented as a combination of relatively low-complex radix-2, radix-3, and radix-5 FFT processing blocks.

The sub-carrier mapping in SC-FDMA determines which part of the spectrum is used for transmission by inserting a number of zeros (i.e., null sub-carriers inserted between the data sub-carriers) in the upper and/or lower end of the frequency region, as shown in Figure 10.3. Distributed mapping can be obtained by inserting $L-1$ zeros between each DFT output sample. A mapping with $L=1$ (as shown in Figure 10.3) corresponds to localized transmissions, i.e., transmissions where the DFT outputs are mapped to consecutive sub-carriers. In localized sub-carrier mapping, a group of N_{DFT} adjacent sub-carriers are allocated to a user, where $N_{DFT} < N_{FFT}$ results in zeros being appended to the output of the DFT spreading which results in an interpolated version of the original N_{DFT} QAM data symbols at the IFFT output of the OFDM modulator. The transmitted signal is similar to a narrowband single carrier with a cyclic prefix, which is equivalent to time-domain generation with a repetition factor of one ($L = 1$) and sinc($\cdot$) function pulse shaping. Distributed transmission, on the other hand, allocates N_{DFT} equally spaced sub-carriers (i.e., every Lth sub-carrier). Therefore, $L-1$ zeros are inserted between the N_{DFT} DFT outputs and additional zeros are appended to either side of the DFT output prior to the IFFT operation. Similar to the localized case, the zeros appended on either side of the DFT output provide sinc($\cdot$) function interpolation effect, while the zeros inserted between the DFT outputs produce waveform repetition in the time-domain. This results in a transmitted signal similar to time-domain interleaved FDMA (IFDMA) with repetition factor L and pulse shaping. Unlike the standard OFDM where the each data symbol is carried by the individual sub-carriers, the SC-FDMA transmitter carries data symbols over a group of sub-carriers transmitted simultaneously. In other words, the group of sub-carriers that carry each data symbol can be viewed as one frequency band carrying data sequentially in a standard FDMA. It can be mathematically shown that the SC-FDMA baseband time-domain samples after IDFT are the original data symbol set repeated in time-domain over the symbol period (Figure 10.4).

The performance of SC-FDMA with localized and distributed sub-carrier mapping was simulated to demonstrate the potential gain that can be obtained via sub-carrier distribution across frequency, particularly in higher mobility conditions. Figure 10.5 shows the block error rate (BLER) as a function of signal-to-noise

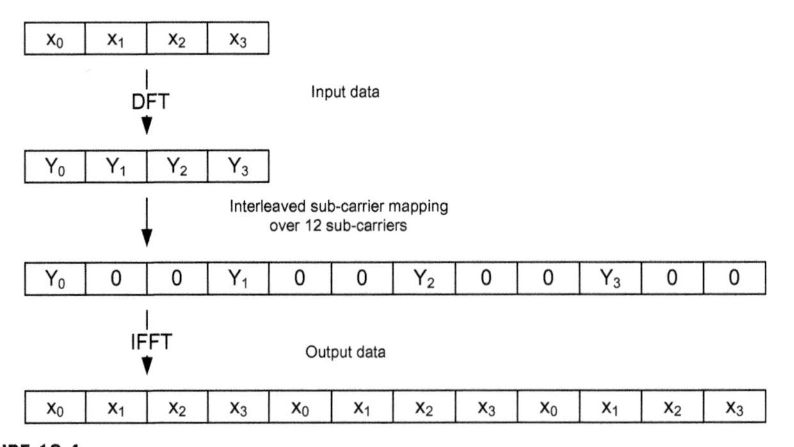

FIGURE 10.4

Time-domain representation of interleaved SC-FDMA.

ratio (SNR) for localized and distributed sub-carrier mapping. It is assumed that $N_{DFT} = 60, N_{FFT} = 512$, and $N_{CP} = 20$ (number of samples constituting the cyclic prefix). Furthermore, the evaluation is conducted with ITU-PedA and ITU-VehA channel models, MMSE equalization at the receiver, QPSK modulation, and raised-cosine pulse shaping at the transmitter. It can be observed that the distributed sub-carrier mapping is more suited for high-mobility conditions and the gain is increased with increasing mobility.

The goal of equalization is to compensate the effects of channel distortion due to frequency-selectivity and to restore the original signal. One approach to signal equalization is in the time-domain using a linear equalizer, which consists of a linear filter with an impulse response $w(t)$ operating on the received signal. By selecting different filter impulse responses, different receiver/equalizer schemes can be implemented. For example, in DS-CDMA systems, a RAKE receiver is typically used. The RAKE receiver is an equalization method in which the filter impulse response is selected to match the channel impulse response $w(t) = h^*(-t)$. Therefore, in a RAKE receiver, the filter impulse response is selected as the complex conjugate of the time-reversed channel impulse response. Selecting the receiver filter according to the maximal ratio combining (MRC) criterion as above, i.e., a channel-matched filter, the filter maximizes the post-processing SNR. However, MRC-based filtering does not provide any compensation for any radio channel frequency selectivity and no equalization. Thus, MRC-based receiver filtering is appropriate when the received signal is mainly impaired by noise or interference from other transmissions. Another alternative is to select the receiver filter to compensate the radio channel frequency selectivity. This can be achieved by selecting the receiver filter impulse response to satisfy $w(t) * h(t) = 1$, where the

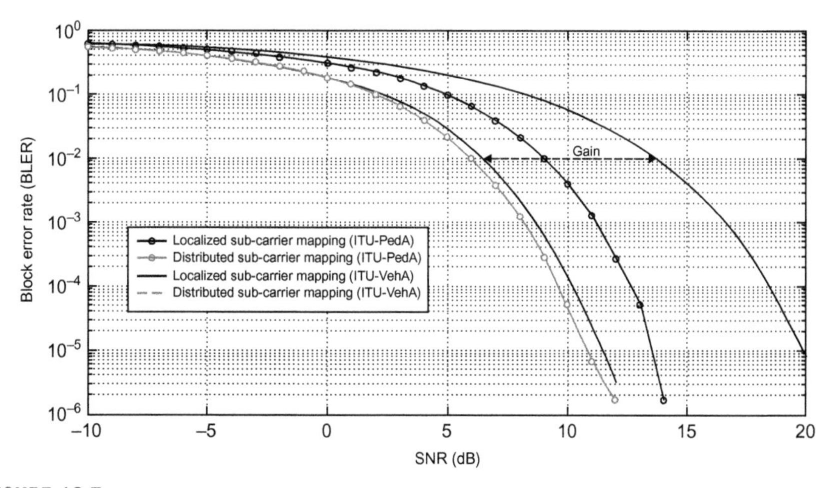

FIGURE 10.5

Comparison of localized and distributed sub-carrier mapping in SC-FDMA [136].

operator "*" denotes linear convolution. This method of filtering is known as zero-forcing (ZF) equalization, which compensates any channel frequency selectivity. However, the ZF equalization may lead to a significant increase in the noise level after equalization, degrading the overall link performance. This will be the case especially when the channel has large variations in its frequency response. Another alternative is to select a filter which provides a trade-off between signal distortion due to channel frequency selectivity and the corruption due to noise/interference, resulting in a filter impulse response that minimizes the mean-square error between the equalizer output and the transmitted signal. The linear equalizers are typically implemented as a time-discrete finite impulse response (FIR) filter with a certain number of taps. In general, the complexity of such a discrete-time equalizer increases with increasing bandwidth of the signal [16].

An alternative to time-domain equalization is frequency-domain equalization which can significantly reduce the complexity of linear equalization. In this method, the equalization is performed on a block of data. The received signal is transformed to the frequency-domain using a DFT operation. The equalization is done as a frequency-domain filtering operation, where the frequency-domain filter $W(k)$ is the DFT of the corresponding time-domain impulse response $w(n)$. The equalized frequency-domain signal is then transformed to the time-domain using an IDFT operator. For processing of each signal block of size $N = 2^m$, the frequency-domain equalization would include two N-point DFT/IDFT operations and N complex multiplications.

With the introduction of a cyclic prefix, the channel would appear as a circular convolution over a receiver processing block of size N. Therefore, there would be no need for overlap-and-discard in the receiver processing. Furthermore, the frequency-domain filter taps can now be calculated directly from an estimate of the sampled channel frequency response. Similar to the OFDM case, the drawback of using a cyclic prefix in conjunction with single-carrier transmission is the overhead in terms of extra power and bandwidth consumption. One method to reduce the relative cyclic-prefix overhead is to increase the block size N of the FDE. However, the accuracy of block equalization requires the channel to be approximately constant over a time period corresponding to the size of the processing block.

The detection procedure for an SC-FDMA signal is illustrated in Figure 10.6, and is compared with that of an OFDM waveform. The transmission through a time-dispersive or equivalently a frequency-selective channel will distort the SC-FDMA signal and an equalizer is needed to compensate for the effects of channel frequency-selectivity. However, as shown in Figure 10.6, a simple one-tap equalizer can be applied to each sub-carrier in OFDM, whereas in the case of SC-FDMA, the frequency-domain equalization function is applied to the complex-valued symbols at the output of the sub-carrier de-mapping and prior to the IDFT operation (Figure 10.7).

It must be noted that in OFDM downlink parameterization, the DC sub-carrier is unused in order to support direct conversion receiver architectures. In contrast, nulling DC sub-carrier is not possible in SC-FDMA since it affects the low

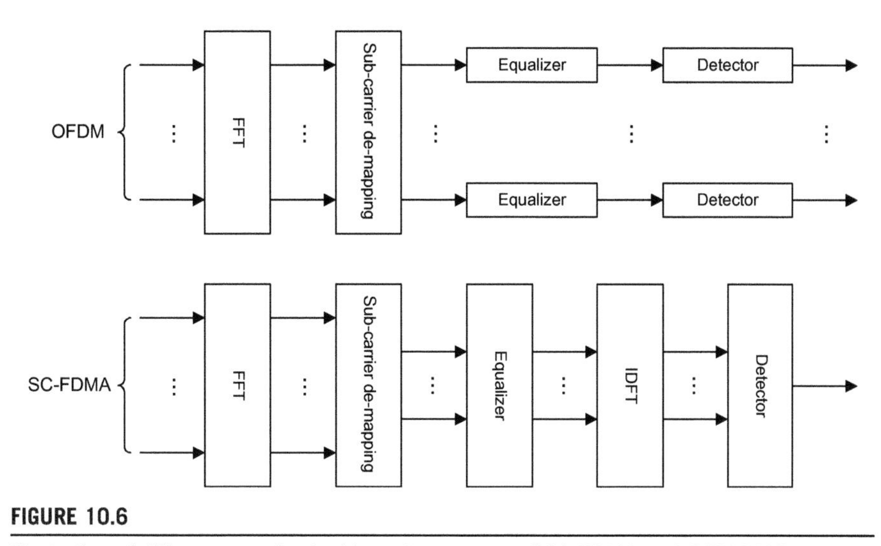

FIGURE 10.6

Illustration of different equalization/detection aspects of SC-FDMA and OFDM [39].

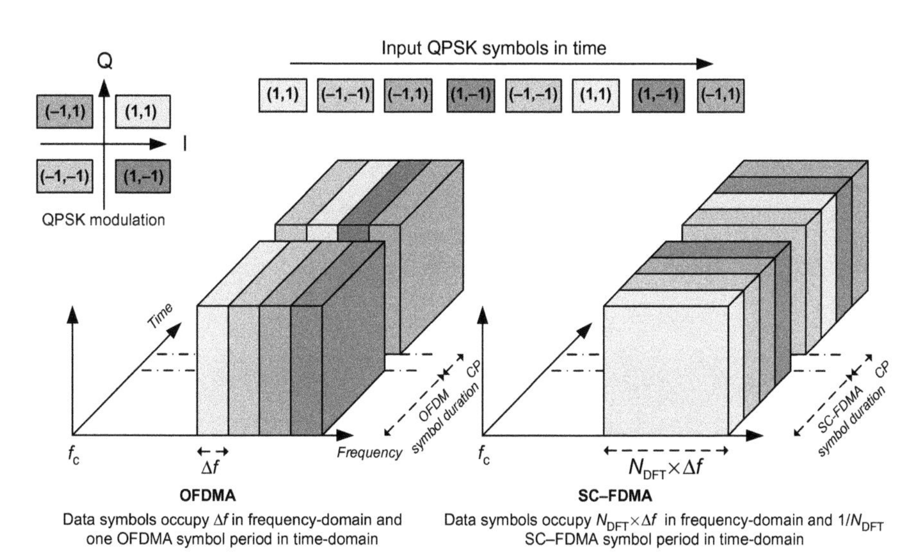

FIGURE 10.7

Comparison of OFDMA and SC-FDMA operation using QPSK modulation with $N_{DFT} = 4$ and $L = 1$. The input QPSK symbols, shown on top, are fed into the modulator from left to right [134].

CM/PAPR property of the transmit signal. Direct conversion transmitters and receivers can introduce distortion at the carrier frequency (zero frequency or DC sub-carrier in baseband) due to local oscillator leakage. In the LTE/LTE-Advanced downlink, this issue is avoided by inclusion of an unused DC sub-carrier. However, for the uplink, the same solution may adversely impact the low CM property of the transmitted signal. In order to minimize the impact of such distortion on the packet error rate and the CM/PAPR, the DC sub-carrier of the SC-FDMA signal is modulated in the same way as all the other sub-carriers but the sub-carriers are all frequency-shifted by half a sub-carrier spacing ($\Delta f/2$), resulting in an offset of 7.5 kHz relative to the DC sub-carrier. Therefore, two sub-carriers straddle the DC location; hence, the amount of distortion affecting any individual resource block is reduced by half.

In order to demonstrate the similarities and differences between OFDMA and SC-FDMA operation, let us assume that one wishes to transmit a sequence of eight QPSK symbols as shown in Figure 10.7 [134]. In the OFDMA case, assuming $N_{DFT} = 4$, four QPSK symbols would be processed in parallel, each of them modulating its own sub-carrier at the appropriate QPSK phase. After one OFDM symbol period, a guard period or cyclic prefix is inserted to mitigate the multipath effects. For SC-FDMA, each symbol is transmitted sequentially. With $N_{DFT} = 4$, there are four data symbols transmitted in one SC-FDMA symbol period. The higher-rate data symbols require four times the bandwidth, and so each data symbol occupies $N_{DFT} \times \Delta f$ Hz of the spectrum assuming a sub-carrier spacing of Δf Hz. After four data symbols, the cyclic prefix is inserted. Note the OFDMA and SC-FDMA symbol periods are the same [134].

As mentioned earlier, the PAPR of OFDM is intrinsically inferior to SC-FDMA. Figure 10.8 shows the comparison of complementary cumulative distribution functions (CCDF)[1] of OFDM and SC-FDMA PAPRs. It can be seen that the PAPR of SC-FDMA is approximately 3 dB better than that of OFDM with a probability of 0.99. In the case of 16QAM modulation, the PAPR of SC-FDMA increases relative to that of SC-FDMA with QPSK modulation, whereas in the case of OFDM, the PAPR distribution is more or less independent of the modulation scheme because the OFDM signal is the sum of a large number of independently modulated sub-carriers, thus the instantaneous power has an approximately exponential distribution, regardless of the modulation scheme applied to the different sub-carriers [16].

A new study suggests that the PAPR of an OFDM signal does not precisely predict the PA power de-rating or power capability as accurately as the CM criterion. The CM criterion has been adopted by 3GPP as a method to determine PA power de-rating because of its accuracy over a wide range of devices and signals.

---

[1]The CDF of the real-valued random variable X is defined as $x \rightarrow F_X(x) = P(X \leq x), \forall x \in \mathbb{R}$, where the right-hand side represents the probability that the random variable X takes on a real value less than or equal to x. The CDF of X can be defined in terms of the probability density function $f(x)$ as $F(x) = \int_{-\infty}^{x} f(x)\mathrm{d}x$. The CCDF, on the other hand, is defined as $P(X > x) = 1 - F_X(x)$ [31–33].

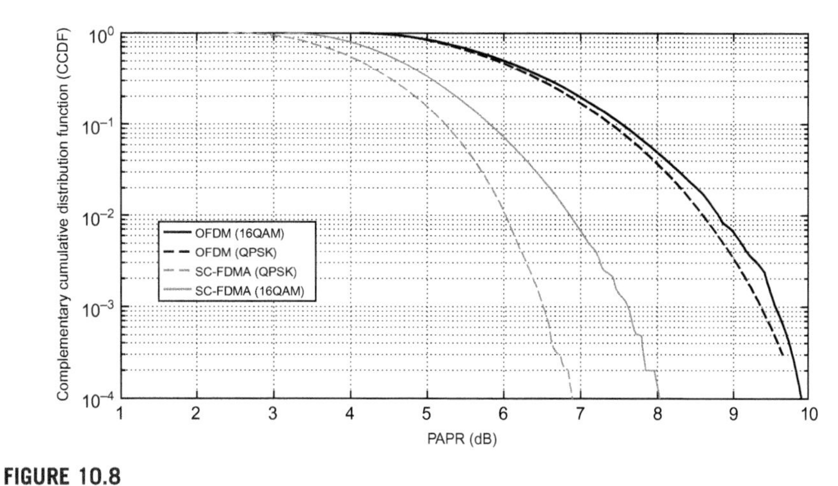

FIGURE 10.8

Comparison of OFDM and SC-FDMA PAPRs (5 MHz bandwidth) [136].

This method has proven to be more accurate for WCDMA signals compared to methods that use the statistical PAPR to predict power de-rating based on the information that has been collected from several devices for a variety of signals. The CM can be calculated using the following equation [138]:

$$\mathrm{CM} = \frac{20\log(s^3_{\mathrm{norm}_{\mathrm{RMS}}}) - 20\log(s^3_{\mathrm{norm\text{-}reference}_{\mathrm{RMS}}})}{K} \tag{10.2}$$

where $s_{\mathrm{norm}_{\mathrm{RMS}}}$ denotes the root mean-square (RMS) value of the normalized voltage waveform of the input signal, and $s_{\mathrm{norm\text{-}reference}_{\mathrm{RMS}}}$ is the RMS value of the normalized voltage waveform of the reference signal (i.e., signal generated using 12.2kbps adaptive multi-rate (AMR) speech codec [ref]). The CM is used to model the impact of nonlinearity in the PA on the adjacent-channel leakage ratio. A small value of the CM represents a small PA backoff. The CM is insensitive to the highest power values of the distribution that only have very low probabilities. Thus, methods to reduce the PAPR may not have any major impact on the CM. In practice, the value of parameter K was empirically determined to be 1.85 for a set of WCDMA signals. In other words, this equation computes the cubic power of a signal $s(t)$ and, compares it to a reference signal $s_{\mathrm{reference}}(t)$, and uses the empirical slope factor K to estimate the CM value. Although the signal vectors are shown as a function of time, the timescale is effectively removed in the RMS operation and is not a part of the computation. Thus, the raw CM result is not a function of symbol rate, or alternatively not a function of signal bandwidth. Similarly, a change in chip rate has no effect on PAPR. The bandwidth effects will be significant in this study since it includes signals with 3 dB bandwidths larger than the 3.84 MHz used for WCDMA. Table 10.1 gives the CM values corresponding to some example input signals such as OFDM, DFT-S OFDM, and

Table 10.1 Example CM Values Corresponding to Slope Factor *K* [138]

Signal					
Type	Modulation	RAW CM (dB)	*K*	Bandwidth Offset (dB)	CM (dB)
WCDMA	Voice	1.52	–	0.00	0
OFDM	16QAM	7.75	1.56	0.77	4.76
DFT-S OFDM	QPSK	3.44	1.56	0.77	2.00
DFT-S OFDM	16QAM	4.85	1.56	0.77	2.90
DFT-S OFDM	64QAM	5.18	1.56	0.77	3.11
IFDMA	QPSK	2.40	1.56	0.00	0.56
IFDMA	16QAM	4.36	1.56	0.00	1.82
IFDMA	64QAM	4.64	1.56	0.00	2.00

IFDMA which is a variant of SC-FDMA. It can be seen that DFT-S OFDM has an advantage relative to other signals shown in the table. In Table 10.1, the bandwidth offset in decibels denotes a de-rating increase of 0.77 dB from the CM regression line that can be used to estimate the actual de-rating of higher bandwidth signals. The *K* factor is empirically derived based on the measurements on various PAs.

10.3 SC-FDMA and clustered SC-FDMA

The LTE Rel-8 uplink is based on single-carrier frequency division multiple access scheme, which allocates carriers across a contiguous block of spectrum, thus limiting scheduling flexibility. LTE-Advanced introduced clustered SC-FDMA in the uplink, allowing frequency-selective scheduling of component carriers for better link-level performance. The physical uplink shared channel (PUSCH) and the physical uplink control channel (PUCCH) can be scheduled together to reduce latency. However, clustered SC-FDMA increases PAPR, leading to transmitter non-linearity issues. Furthermore, the presence of multi-carrier signals increases opportunities for co-channel and adjacent-channel interference.

LTE-Advanced enhances the uplink multiple access scheme by adopting clustered SC-FDMA. This scheme is similar to SC-FDMA but has the advantage that it allows non-contiguous (clustered) groups of sub-carriers to be allocated for transmission by a single UE, thus enabling uplink frequency-selective scheduling and improved link performance. Clustered SC-FDMA was chosen with preference to pure OFDM to avoid a significant increase in PAPR. It helps satisfy the requirement for increased uplink spectral efficiency while maintaining backward

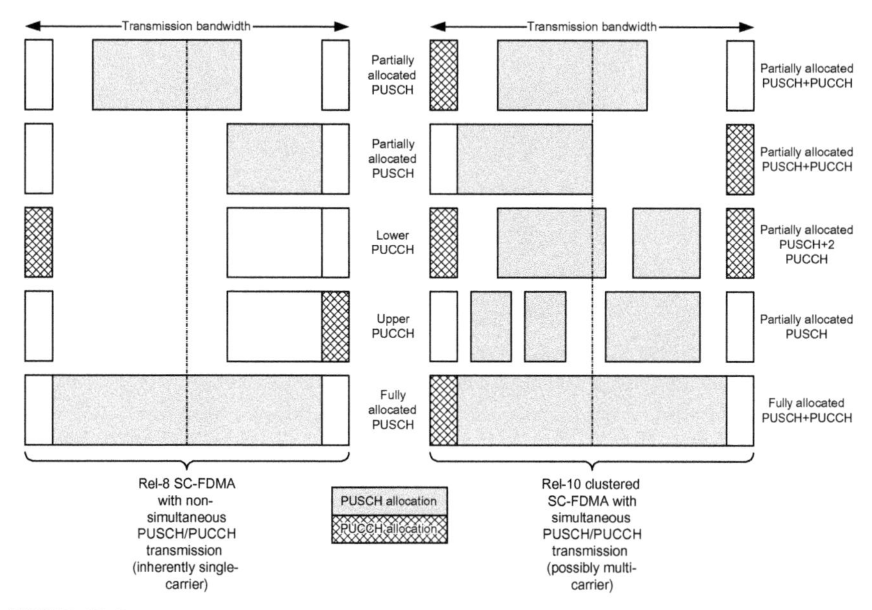

FIGURE 10.9

Comparison of LTE and LTE-Advanced uplink transmission [161].

compatibility with LTE Rel-8. There is only one transport block and one HARQ entity per scheduled component carrier. Each transport block is mapped to a single component carrier and a UE may be scheduled over multiple component carriers simultaneously using carrier aggregation.

Examples of different LTE Rel-8 and Rel-10 uplink configurations are shown in Figure 10.9. The main observation is that all LTE Rel-8 configurations are single carrier, which means that the PAPR is no greater than the underlying QPSK or 16QAM modulation format, whereas in LTE Rel-10, it is possible to transmit more than one carrier, which makes the PAPR higher than the Rel-8 cases. Note that the carriers mentioned above in the context of clustered SC-FDMA and simultaneous PUCCH/PUSCH transmission are contained within one component carrier and should not be confused with multiple component carriers and carrier aggregation. The initial specifications limit the number of SC-FDMA clusters to two, which will provide some improved spectral efficiency over a single cluster when transmitting through a frequency-selective channel with more than one distinct peak.

The clustered SC-FDMA must be distinguished from multiple SC-FDMA in the sense that clustered SC-FDMA retains a single DFT operation, but modifies the resource element mapping at the output of the DFT operation from a single cluster as used for SC-FDMA to multiple clusters which are multiplexed with $N_{\mathrm{FFT}} - N_{\mathrm{DFT}}$ zeros to form the input of the IFFT operation over N_{FFT} virtual subcarriers. The resulting waveform is no longer single-carrier but still has a low CM value. On the other hand, multiple SC-FDMA consists of a number of DFT operations, where the kth DFT block is of size N_{DFT}^{k}; the output symbols are then multiplexed with $N_{\mathrm{FFT}} - \sum_{k=1}^{N_{\mathrm{cluster}}} N_{\mathrm{DFT}}^{k}$ zeros to fit the N_{FFT}-point IFFT operation. This waveform is no longer single carrier and experiences worse CM compared to that of clustered SC-FDMA for the same number of clusters [16].

10.4 Uplink frame structure

The LTE frame structure for the FDD and TDD schemes was discussed in Chapter 9. The downlink and uplink frame structure parameters are the same, except that the time-domain underlying symbols in the downlink are OFDM symbols, whereas in the uplink the underlying symbols are SC-FDMA symbols due to the use of different multiple access methods in the uplink and downlink. In this chapter, some of the uplink-specific aspects of frame structure are repeated to set the ground for further uplink physical layer discussions as well as physical channel mappings. As mentioned in Chapter 9, LTE/LTE-Advanced downlink and uplink transmissions are organized into radio frames of duration 10 ms consisting of 10 equally-sized consecutive subframes of 1 ms each, consisting of a number of consecutive OFDM/SC-FDMA symbols. Each subframe is further divided into two equally sized slots of 0.5 ms. The number of OFDM/SC-FDMA symbols within one subframe is either 14 for a normal cyclic prefix or 12 for an extended cyclic prefix (Figure 10.10).

In an FDD scheme, all ten subframes within a radio frame contain either downlink or uplink subframes depending on the link direction. The uplink and downlink transmitters use separate frequency bands for transmissions; therefore, both downlink and uplink transmitters can transmit at all times and at the same time [3]. In a TDD scheme, the bandwidth is shared between uplink and downlink, with the sharing being performed by allotting different periods of time to uplink and downlink (i.e., time multiplexing). In LTE/LTE-Advanced, there are seven different patterns of uplink/downlink switching known as uplink–downlink configurations 0 through 6 that are shown in Table 10.2 [3].

As shown in Figure 10.10 and Table 10.2, the special subframe (subframe 1 in every uplink–downlink configuration and subframe 6 in uplink–downlink configurations 0, 1, 2, and 6) contains a portion of downlink transmission at the start of the subframe (downlink pilot time slot or DwPTS), a portion of unused symbols in the middle of the subframe (guard period), and a portion of uplink transmission at the end of the subframe (uplink pilot time slot or UpPTS). The lengths of DwPTS, GP, and UpPTS can take one of the ten combinations of values given in Table 10.3. The LTE standard specifies the lengths in terms of the sampling time T_s, but the lengths can be interpreted in terms of the number of OFDM/SC-FDMA symbols, as shown in Table 10.3.

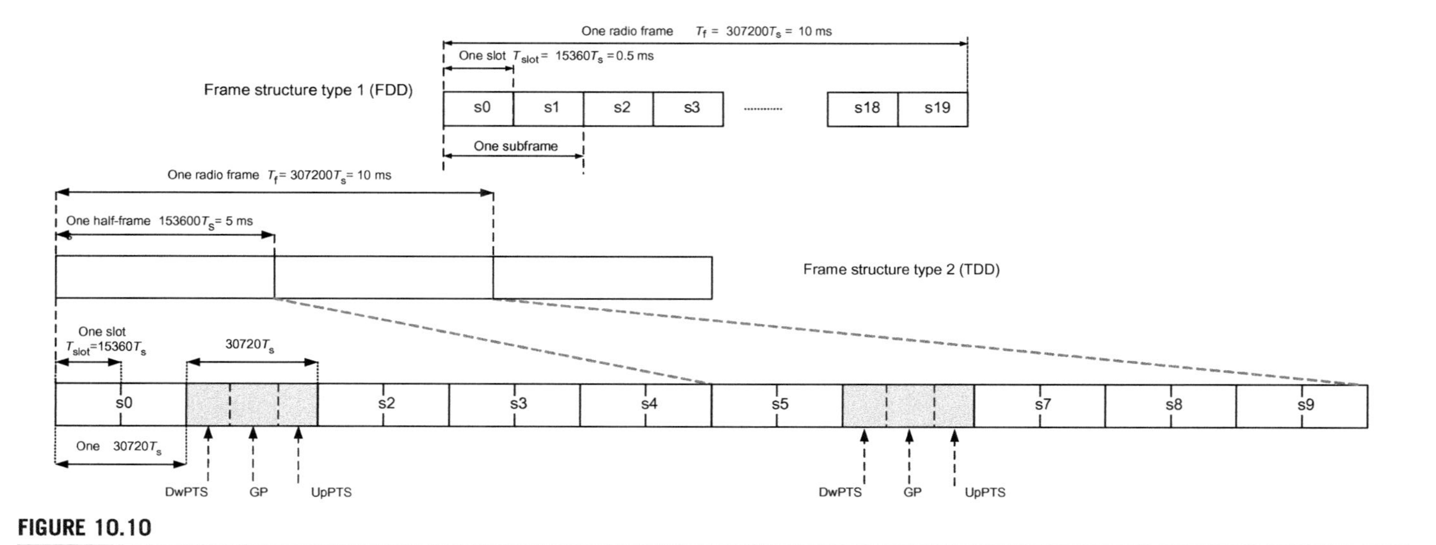

FIGURE 10.10

LTE/LTE-Advanced FDD/TDD frame structures [3].

Table 10.2 LTE/LTE-Advanced TDD Frame Configurations [3]

Uplink–Downlink Configuration	Multiplexing of Downlink, Uplink, and Special Subframes in One Radio Frame									
0	DL	SP	UL	UL	UL	DL	SP	UL	UL	UL
1	DL	SP	UL	UL	DL	DL	SP	UL	UL	DL
2	DL	SP	UL	DL	DL	DL	SP	UL	DL	DL
3	DL	SP	UL	UL	UL	DL	DL	DL	DL	DL
4	DL	SP	UL	UL	DL	DL	DL	DL	DL	DL
5	DL	SP	UL	DL	DL	DL	DL	DL	DL	DL
6	DL	SP	UL	UL	UL	DL	SP	UL	UL	DL

10.5 Physical resource structure and channel mapping

The smallest resource unit for uplink transmissions is a resource element (one sub-carrier). An uplink physical channel corresponds to a set of resource elements carrying information originating from the higher layers. The following uplink physical channels are defined:

- The physical uplink shared channel (PUSCH) carries user data and upper layer signaling as well as control information such as HARQ ACK/NACK, channel

Table 10.3 Special Subframe Configurations in TDD [3]

Special Subframe Configuration	Normal Cyclic Prefix in Downlink			Extended Cyclic Prefix in Downlink		
	DwPTS	UpPTS		DwPTS	UpPTS	
		Normal Cyclic Prefix in Uplink	Extended Cyclic Prefix in Uplink		Normal Cyclic Prefix in Uplink	Extended Cyclic Prefix in Uplink
0	3			3		
1	9			8	1	1
2	10	1	1	9		
3	11			10		
4	12			3		
5	3			8	2	2
6	9			9		
7	10	2	2	–	–	–
8	11			–	–	–
9	6			–	–	–

quality indicator (CQI), precoding matrix index (PMI), and rank indication (RI).

- The physical uplink control channel (PUCCH) carries control information such as HARQ ACK/NACK, CQI, and RI. In LTE Rel-8, the UE transmits control information on the PUCCH, when no PUSCH allocations are available to the UE. Starting from LTE Rel-10, PUCCH and PUSCH can be simultaneously transmitted by the UE. The PUCCH carries uplink control information (UCI). Each PUCCH uses one resource block in each of the two slots of the subframe. As shown in Figure 10.11, the PUCCH resource blocks are always allocated at the edge of the channel. This helps reduce out-of-band (OOB) emissions from data transmissions on inner resource blocks. Allocating the PUCCH in the center of the band would further create non-contiguous PUSCH resources, which limits contiguous resources that can be allocated to a single UE, and thus would limit the achievable peak throughput. The PUCCH resource blocks are transmitted with a hopping pattern from one edge to the other edge in each slot of the subframe. Frequency hopping provides frequency diversity for PUCCH. A maximum of four resource blocks are reserved for PUCCH in this example. The physical resources used for PUCCH depend on parameters that are provided by the higher layers. The PUCCH allocations also contain demodulation reference signals (DM-RSs) for coherent demodulation.

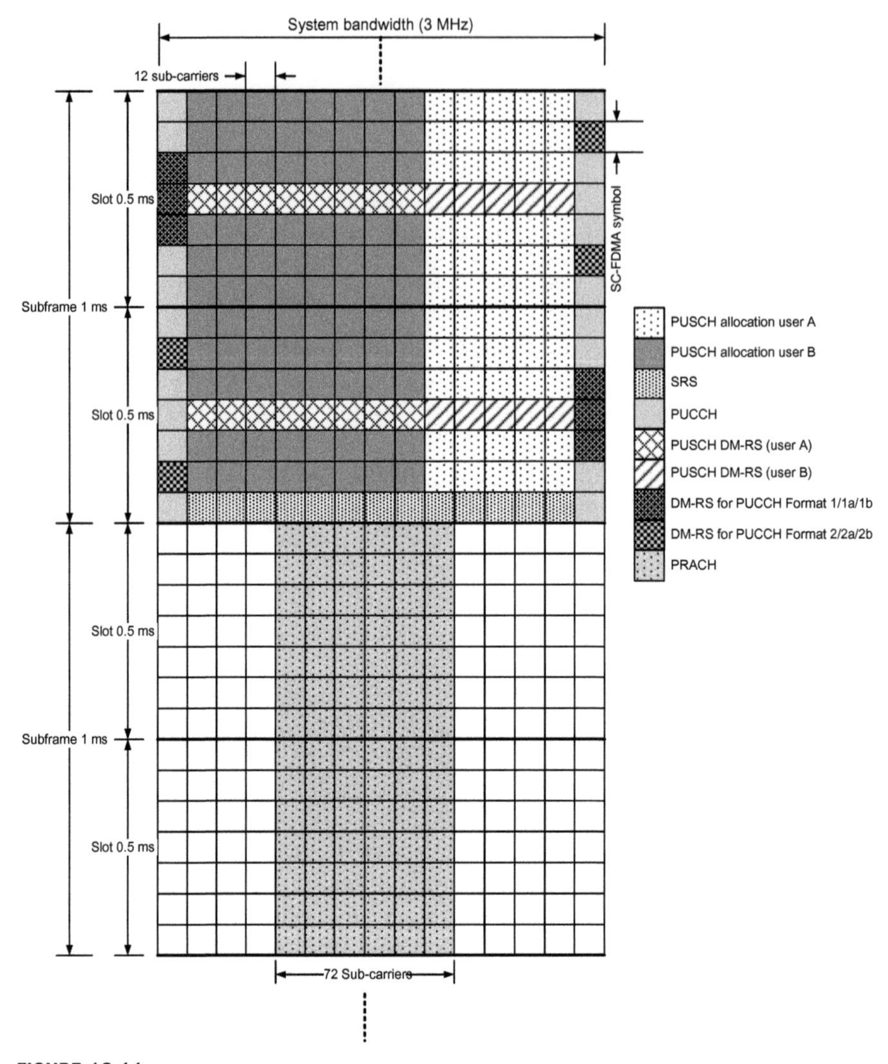

FIGURE 10.11

Example of uplink frame structure and location of the physical signals/channels.

The PUCCH allocation may contain one, two, or three DM-RS symbols. The DM-RS for each UE is generated using a cyclic shift of base Zadoff–Chu sequences. The RRC sub-layer assigns cyclic shift parameters (i.e., cyclic shift and base sequence group) to the UE during the RRC connection setup. The location of DM-RS differs based on formats and the length of the cyclic prefix.

- The physical random-access channel (PRACH) is used by non-synchronized UEs to access the network. It is further used during initial system access, handovers, and radio link failure recovery scenarios. A preamble sequence, which may be randomly selected from a group of sequences or assigned (dedicated) by the eNB, is transmitted over PRACH. The sequences are Zadoff–Chu sequences of length 839. There are 64 different cyclic shifts generated from each root Zadoff–Chu sequence with zero correlation between them. Therefore, different cyclic shifts are essentially orthogonal to each other. The PRACH resource consists of six consecutive resource blocks. The location of PRACH is indicated by an offset parameter signaled by higher-layer signaling. There is at most one PRACH resource per subframe in FDD mode. Multiple UEs can access the PRACH resource by using different preambles. The UEs may select preambles randomly, which may result in collision. Therefore, contention resolution mechanisms are necessary. The PRACH may not be present in every subframe and every radio frame. The *PRACH configuration index* parameter indicates frame number and subframe numbers in which the PRACH resource is available. The sub-carrier spacing for PRACH transmission is different from that of other channels in the uplink. The number of sub-carriers is equal to the length of preamble, which is 839 (due to longer OFDM symbol length for PRACH). The sub-carrier spacing is 1.25 kHz, resulting in a longer symbol duration of 800 μs. The choice of sub-carrier spacing also results in additional guard bands on both sides because the entire 1.08 MHz is not used for the PRACH transmission.

In addition, there are two types of reference signals in the uplink as follows:

- The demodulation reference signal is a known signal transmitted in PUSCH and/or PUCCH whenever they are transmitted and provides a coherent demodulation reference for uplink transmissions. The DM-RS provides the reference for channel estimation and coherent demodulation of uplink transmissions from the UE on PUSCH or PUCCH. The DM-RS is transmitted only in the resource blocks transmitted from the UE for PUSCH or PUCCH. It is transmitted during a specific symbol and time-division multiplexed with the resource blocks allocated to the UE for data or control information transmission. The DM-RS signal is based on Zadoff–Chu sequences with a length of up to the maximum number of resource blocks. Thirty groups of base sequences are defined where each group contains three base sequences. Group hopping patterns and sequence hopping patterns are defined, resulting in different base sequences in different slots. The base sequences are generated differently based on whether they are more or less than three resource blocks. From these base sequences, different cyclic shifts are defined. The cyclic shifts are derived from UE-specific parameters assigned in DCI format 0 and cell identity. This creates the UE-specific cyclic shifts necessary to achieve orthogonality for MU-MIMO, where the same resource block is allocated to multiple users. The cyclic shift is also derived from the slot number, resulting in different cyclic shifts in different slots for the same UE.

- The sounding reference signal (SRS) is a wideband signal transmitted by the UE for several purposes. The eNB can use SRS transmission to select the best resource blocks to allocate to the UE (i.e., frequency-selective scheduling). It may also be used for uplink power control. If uplink MIMO is supported, it may be further used for adaptive antenna switching for uplink transmit diversity. The SRS sequences are selected from the same Zadoff–Chu root sequences used for DM-RS. The cyclic shifts are configured for each UE when SRS transmission is configured. If configured, the SRS transmission always occurs in the last SC-FDMA symbol of a subframe. The SRS transmission may not be allowed in every subframe. The periodicity of subframes containing SRS may be limited to 2, 5, 10, 20, 40, 80, 160, or 320 ms [5]. The bandwidth for SRS transmission is specified in terms of number of resource blocks and may extend from 4 to 36 resource blocks, if the maximum number of resource blocks in the uplink is in the range of 6-40. The SRS bandwidth may extend from 4 to 96 resource blocks, if the maximum number of resource blocks in the uplink is in the range of 80–110 [3]. The SRS parameters (periodicity, bandwidth, etc.) can be cell-specific and broadcast from each cell. Additionally, UE-specific parameters that override cell-specific parameters may be configured through higher-layer signaling. Frequency hopping may be defined for SRS transmission, if the SRS bandwidth is smaller than the system bandwidth.

The uplink unit of resource allocation is a resource block. A resource block is a two-dimensional array of all SC-FDMA symbols in a slot and 12 sub-carriers in the frequency-domain. There are 72 (6 symbols by 12 sub-carriers) or 84 (7 symbols by 12 sub-carriers) resource elements in a resource block corresponding to extended cyclic prefix or normal cyclic prefix, respectively. The number of resource blocks depends on the channel bandwidth. There is a minimum of six resource blocks for the 1.4 MHz channel and 100 resource blocks for the 20 MHz channel. A UE is allocated a number of contiguous or non-contiguous resource blocks. Figure 10.11 illustrates an example of mapping of uplink physical channels and signals to the physical resources within a subframe for an uplink bandwidth of 3 MHz (15 resource blocks).

10.5.1 Physical resource blocks

The signal transmitted in each slot is described by one or several resource grids of $12 \times N_{RB}^{UL}$ sub-carriers and N_s^{UL} SC-FDMA symbols. The uplink resource grid is illustrated in Figure 10.12. The parameter N_{RB}^{UL} denotes the number of uplink resource blocks, which depends on the uplink transmission bandwidth, configured in the cell whose value falls in the range of $\min(N_{RB}^{UL}) \le N_{RB}^{UL} \le \max(N_{RB}^{UL})$, where $\min(N_{RB}^{UL}) = 6$ and $\max(N_{RB}^{UL}) = 110$ represent the minimum and maximum uplink bandwidths, respectively. The number of SC-FDMA symbols in a slot N_s^{UL} depends on the cyclic prefix length which is configured by the higher layer signaling. Similar to the downlink, there are seven SC-FDMA symbols for the normal cyclic prefix and six SC-FDMA symbols for extended cyclic prefix in each slot.

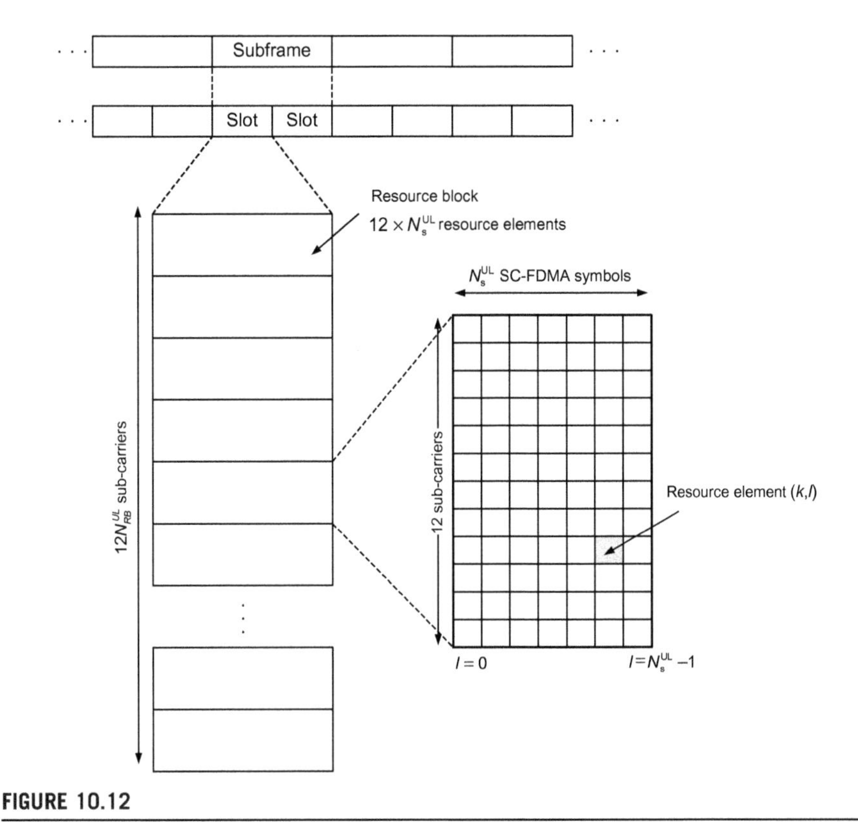

FIGURE 10.12

Uplink physical resource structure [3].

Each resource element in the resource grid within a slot is uniquely defined by the index pair (k, l), where $k = 0, 1, \ldots, 12 \times N_{RB}^{UL} - 1$ and $l = 0, 1, \ldots, N_s^{UL} - 1$ are the frequency and time-domain indices, respectively. The resource element (k, l) on antenna port p corresponds to the complex-valued symbol a_{kl}^{p}. A physical resource block (PRB) is defined as N_s^{UL} consecutive SC-FDMA symbols in the time-domain and 12 consecutive sub-carriers in the frequency-domain. Note that the number of sub-carriers in a resource block corresponding to sub-carrier spacing 7.5 kHz is 24; however, throughout this book, we assume a 15 kHz sub-carrier spacing. A PRB in the uplink thus consists of $12 \times N_s^{UL}$ resource elements, corresponding to one slot in the time-domain and 180 kHz in the frequency-domain. The PRB number in the frequency-domain n_{PRB} and the resource element (k, l) in a slot are related as $n_{PRB} = \lfloor k/12 \rfloor$.

10.5.2 Uplink resource allocation

The concept of virtual resource blocks (VRBs) or distributed resource allocation and the mapping of VRBs to PRB were discussed in Chapter 9 for downlink resource allocations. The distributed resource allocation in the downlink consists of two separate steps that include mapping from VRB pairs to PRB pairs such that frequency-consecutive VRB pairs are not mapped to frequency-consecutive PRB pairs, as well as splitting of each resource-block pair such that the two resource blocks of the resource-block pair are transmitted with a certain frequency gap over the slots of each subframe. The latter step can be seen as frequency hopping on a slot basis. In the absence of the frequency hopping, each VRB is directly mapped to a PRB. Note that localized transmissions without use of frequency hopping can achieve some degree of frequency-selective-scheduling gain, whereas localized transmission using frequency hopping yields frequency-diversity gain as well as inter-cell interference randomization [16].

The VRB concept can also be used in the uplink, allowing use of frequency-diversity transmission in the uplink. However, in the uplink, where transmission from a terminal should always be over a set of consecutive sub-carriers in the absence of clustered SC-FDMA, distribution of resource-block pairs in the frequency-domain is not possible. Therefore, uplink distributed resource allocation is only similar to the second step of downlink distributed transmission, i.e., inserting a frequency gap between the PRBs in the first and second slots of a subframe. This is referred to as PUSCH frequency hopping in the literature. There are two types of uplink frequency hopping defined for PUSCH as follows: (1) sub-band hopping according to cell-specific hopping/mirroring patterns (as shown in Figure 10.13); and (2) hopping based on explicit hopping information contained in the scheduling grant. The uplink frequency hopping is not supported for clustered SC-FDMA transmission, since sufficient frequency diversity gain can be achieved by properly assigning the location of the clusters.

In order to support sub-band frequency hopping according to cell-specific hopping/mirroring patterns, a set of consecutive sub-bands with predetermined size are defined as illustrated in Figure 10.13. It should be noted that the sub-bands do not cover the total uplink frequency band due to the fact that a number of resource blocks at the edges of the uplink frequency band are designated for transmission of PUCCH. As an example, if the overall uplink bandwidth contains 25 resource blocks and there are a total of four sub-bands, each consisting of five resource blocks, then five resource blocks at both edges of the band are not included in the hopping bandwidth and can be used for PUCCH transmission (see Figure 10.13). In the case of sub-band hopping, the set of VRBs provided in the scheduling grant are mapped to a corresponding set of PRBs according to a cell-specific hopping pattern. The physical resources to be used for transmission are obtained by shifting the VRBs provided in the scheduling grant by a number of sub-bands according to the hopping pattern, where the hopping pattern can provide different shifts for each slot. The hopping pattern is cell specific, which implies that it is the same for all terminals within a cell; however, different terminals will transmit on non a_{kl}^{p} overlapping physical resources as long as they are assigned non a_{kl}^{p} overlapping virtual resources [16].

In addition to the hopping pattern, there is also a cell-specific mirroring pattern defined in a cell. The mirroring pattern controls, on a slot basis, if mirroring

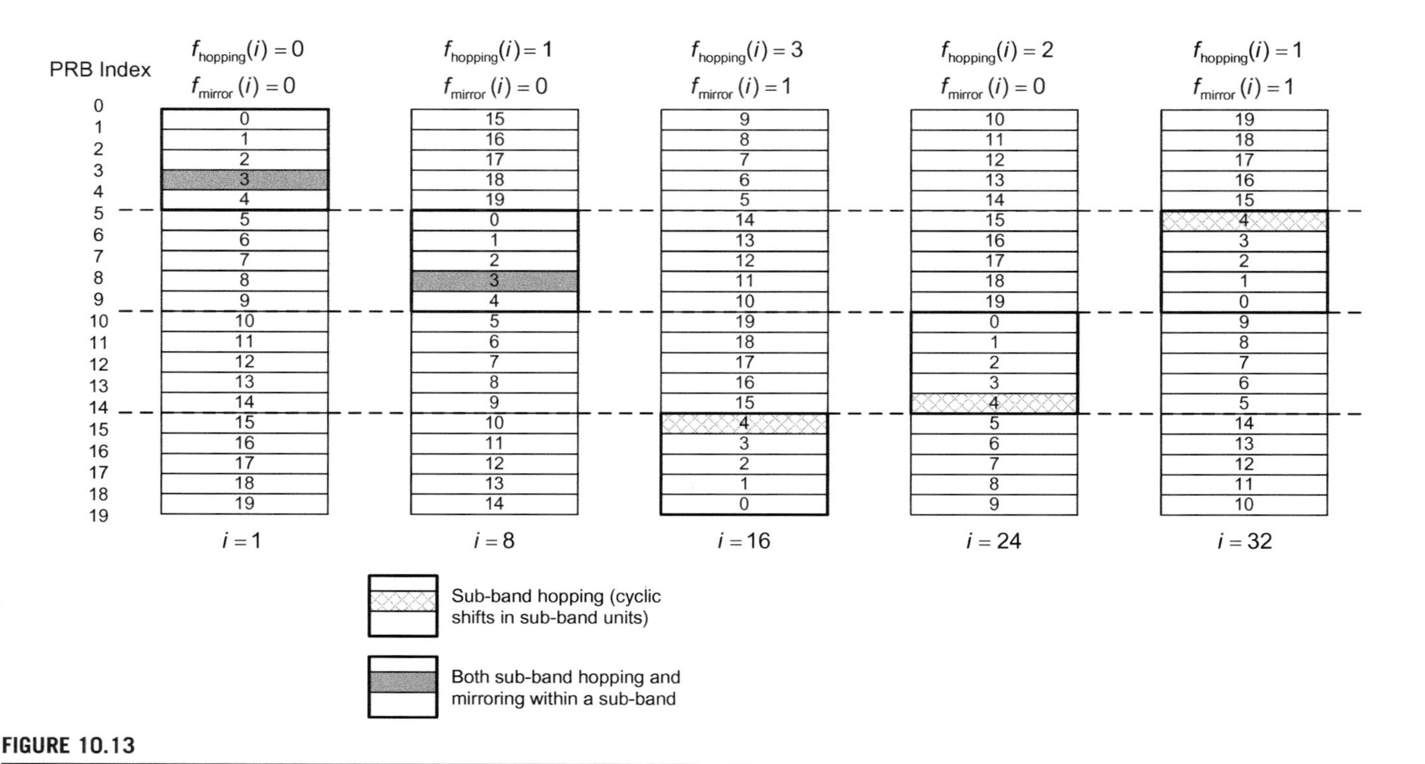

FIGURE 10.13

Illustration of an example PUSCH frequency hopping (20 resource blocks, $L_{sb} = 4$) [156].

within each sub-band should be applied to the assigned resource. In principle, mirroring suggests that the resource blocks within each sub-band are numbered right to left instead of left to right. Figure 10.13 illustrates an example of mirroring in conjunction with hopping within a sub-band. The mirroring pattern is not applied to the first slot, while mirroring is applied to the second slot. Both hopping and mirroring patterns depend on the physical layer cell identity and are typically different in neighboring cells. Furthermore, the periodicity of the hopping/mirroring patterns is one radio frame.

An alternative approach to uplink cell-specific hopping/mirroring is frequency hopping according to explicit information provided in the scheduling grant. In such cases, the scheduling grant includes information about the resource allocation for uplink transmission in the first slot and information related to the amount of offset applicable to the resource to be used for uplink transmission in the second slot relative to the resource location in the first slot. Selection between hopping according to cell-specific hopping/mirroring patterns as described above, or hopping according to explicit information in the scheduling grant, can be made dynamically. More specifically, for cell bandwidths less than 50 resource blocks, there is a flag in the scheduling grant indicating whether frequency hopping is based on cell-specific hopping/mirroring patterns or according to information in the scheduling grant. In the latter case, the hopping is always half of the hopping bandwidth. In the case of larger bandwidths in excess of 50 resource blocks, there are 2 bits in the scheduling grant. One of the combinations indicates that hopping should be according to the cell-specific hopping/mirroring patterns, while the three remaining combinations indicate hopping of 1/2, +1/4, and −1/4 of the hopping bandwidth [16].

The uplink also supports transmission time interval (TTI) bundling where four consecutive subframes are bundled together following a downlink grant. Four HARQ processes are necessary, the downlink PHICH transmissions correspond to the last subframe of the bundle, and the retransmissions are non-dynamic. The purpose of TTI bundling is to enable successful reception of the uplink transmissions in coverage-limited scenarios. The TTI bundling is configured by the RRC sub-layer.

The uplink resource allocation scheme in LTE Rel-8 is single-cluster allocation, which is contiguous resource allocation in the frequency-domain. LTE-Advanced has enabled multi-cluster uplink transmissions. The single-cluster allocations use uplink resource allocation type 0, which is identical to downlink resource allocation type 2, except that the single-bit flag indicating localized/distributed transmission is replaced by a single-bit hopping flag. The resource allocation field in the DCI provides the set of VRBs to be used for uplink transmission. The set of PRBs to use in the two slots of a subframe is controlled by the hopping flag.

The multi-cluster allocations with up to two clusters were introduced in LTE Rel-10 and use uplink resource allocation type 1. In resource allocation type 1, the starting and ending positions of two clusters of frequency-contiguous resource blocks are encoded and indicated with an index. The uplink resource allocation type 1 does not support frequency hopping since the frequency diversity is achieved through the use of the clustered allocation. It is evident that indication of two clusters of resources requires additional signaling compared to the single-cluster case. However, the total number of bits used for resource allocation type 1 is identical to that of type 0 (see Figure 10.13). This is similar to the case of downlink allocation types 0 and 1. Since frequency hopping is not supported for allocation type 1, the bit, otherwise designated to the hopping flag, can be used for extending the resource allocation field. However, despite the extension of the resource allocation field by 1 bit, the number of bits is not sufficient to provide a single-resource-block resolution in the two clusters for all bandwidths. Instead, similar to downlink resource allocation type 0, groups of resource blocks are used and the starting and ending positions of the two clusters are given in terms of group numbers. The size of the group is proportional to the uplink bandwidth. The size of the group is one for small bandwidths and four for large system bandwidths [16]. Both DCI formats 0 and 4 support uplink resource allocation types 0 and 1. The number of allocated resource blocks is always equal to one or a multiple of 2, 3, or 5. It means that 1, 2, 3, 4, 5, 6, 8, 9, 10, 12, 15, etc., are permissible resource allocation sizes in the uplink (Figure 10.14).

10.5.2.1 Uplink resource allocation type 0

The resource allocation information contained in uplink resource allocation type 0 for a scheduled UE contains a set of contiguously allocated VRB indices denoted by n_{VRB}. A resource allocation field in the scheduling grant consists of a resource indication value (RIV) parameter corresponding to the starting resource block $\mathrm{RB_{START}}$ and the length of the allocation in terms of contiguously allocated resource blocks $L \geq 1$. The RIV is defined as $\mathrm{RIV} = N_{\mathrm{RB}}^{\mathrm{UL}}(L-1) + \mathrm{RB_{START}}$ if $(L-1) \leq \lfloor N_{\mathrm{RB}}^{\mathrm{UL}}/2 \rfloor$; otherwise, the RIV is given as $\mathrm{RIV} = N_{\mathrm{RB}}^{\mathrm{UL}}(N_{\mathrm{RB}}^{\mathrm{UL}} - L + 1) + (N_{\mathrm{RB}}^{\mathrm{UL}} - 1 - \mathrm{RB_{START}})$. This is similar to resource allocation type 2 in the downlink. An example of uplink resource allocation using type 0 in 1.4 MHz (6 resource blocks) is shown in Figure 10.15.

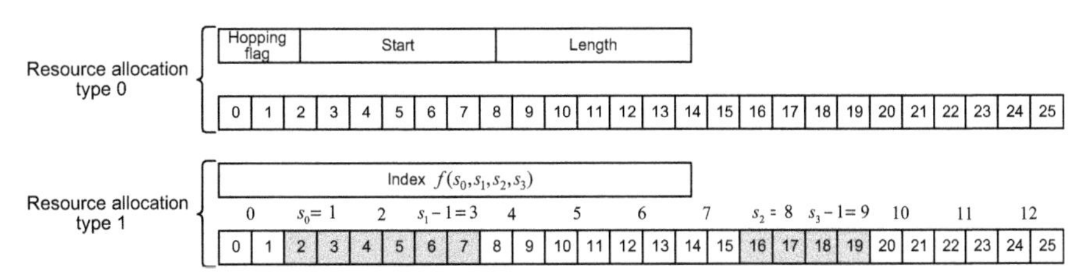

FIGURE 10.14

Illustration of uplink resource allocation types [16].

10.5.2.2 Uplink resource allocation type 1

The uplink resource allocation type 1 was introduced in LTE Rel-10 to allow allocation of non-contiguous resource-block groups (RBGs). In uplink resource allocation type 1, an indication is provided to a scheduled UE of two sets of resource blocks where each set includes one or more consecutive RBGs of size N_{RBG}, assuming a system bandwidth of $N_{\text{RB}}^{\text{UL}}$. This type of resource allocation is defined by four indices that are combined into a single combinatorial index l before being sent to the UE. A complete representation of combinatorial index l requires $\left\lceil \log_2 \left[\binom{\lceil N_{\text{RB}}^{\text{UL}}/N_{\text{RBG}} + 1 \rceil}{4} \right] \right\rceil$ bits. The bits from the resource allocation field in the scheduling grant represent index l, unless the number of bits in the resource allocation field in the scheduling grant is either smaller or larger than that required number to represent index l. If the number of bits is smaller than that required to represent l, the bits in the resource allocation field in the scheduling grant would represent the least-significant bits of l and the value of the remaining bits of l are set to zero. If the number of bits is larger than that required for representation of l, index l would be mapped to the least-significant bits of the resource allocation field in the scheduling grant. Note that resource-block allocation in this case is defined by four indices (s_0, s_1, s_2, s_3). The parameters s_0 and s_1 define the first and the last RBG within the first set of RBGs, and the parameters s_2 and s_3 define the first and the last RBG within the second set of RBGs. Therefore, the combinatorial index l corresponds to the starting and the ending index of resource-block-set 1 (s_0 and s_1), and resource-block-set 2 (s_2 and s_3), respectively, where $l = \sum_{i=0}^{M-1} \binom{N - s_i}{M - i}$, $M = 4$, and $N = \lceil N_{\text{RB}}^{\text{UL}}/N_{\text{RBG}} \rceil + 1$. If the corresponding ending RBG index is equal to the starting index, only one RBG is allocated.[2]

---

[2] The combinatorial index l is calculated based on the sorted set of M sub-band indices $\{s_i\}_{i=0,1,\ldots,M-1}$ given $1 \le s_k \le N, s_k < s_{k+1}$ as $l = \sum_{i=0}^{M-1} \binom{N - s_i}{M - i}$ where $\binom{x}{y}: x \ge y$ is the extended binomial coefficient and $l \in \left\{0, 1, \ldots, \binom{N}{M} - 1\right\}$. The uniqueness and the range of the combinatorial index are proven in the literature [162,163]. In the receiving end, the de-mapper regenerates the set of indices $\{s_i\}_{i=0,1,\ldots,M-1}$ from the received label l using the following algorithm:

$x_{\min} = 1;$
for $i = 0$ to $M - 1$
$x = x_{\min};$
$p = \binom{N - x}{M - i};$
while $p > l$
    $x = x + 1;$
    $p = \binom{N - x}{M - i};$
end
$s_i = x;$
$x_{\min} = s_i + 1;$
$l = l - p;$
end

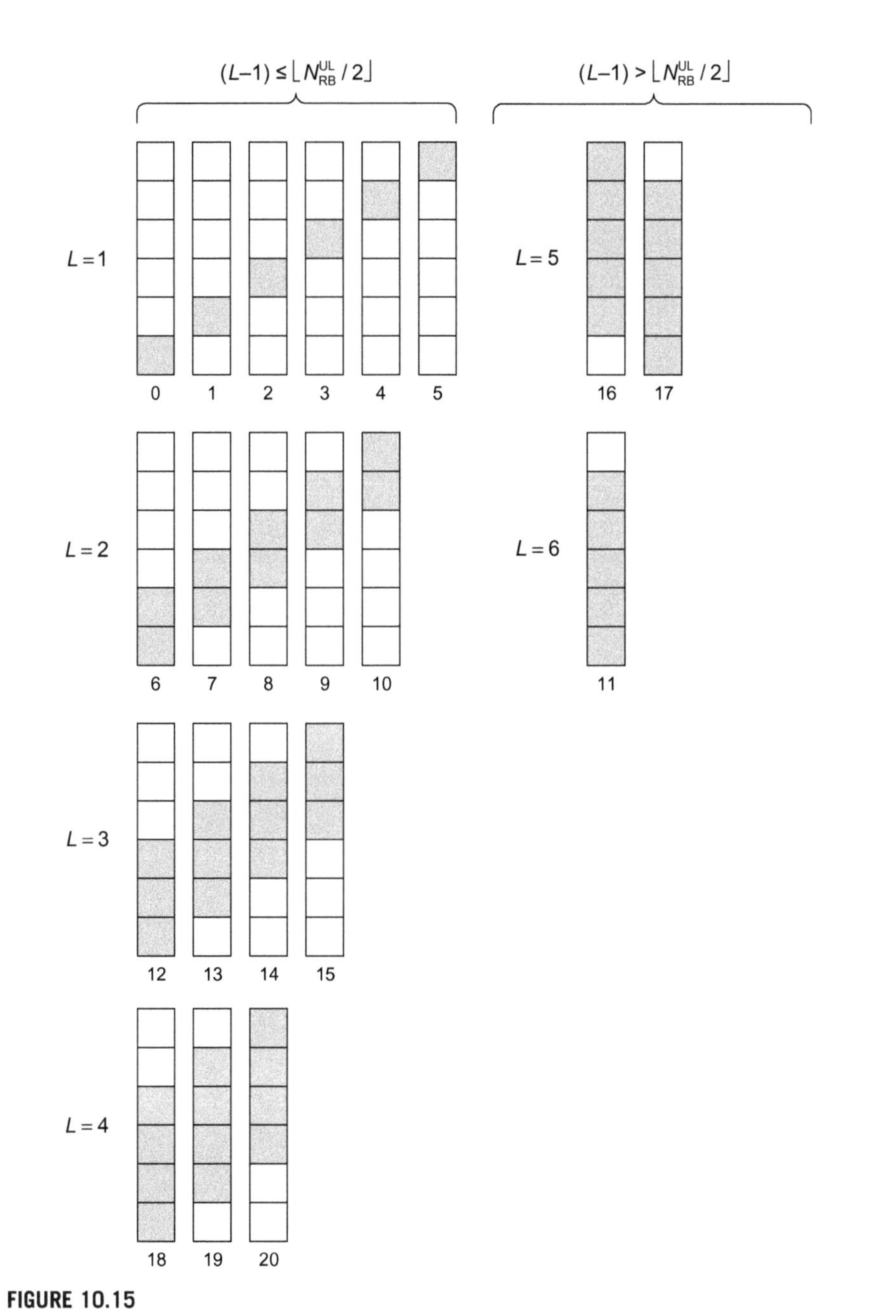

FIGURE 10.15

Illustration of an example uplink resource allocation type 0 in 1.4 MHz [14].

Table 10.4 Subframe Delay k Between PDCCH Assignment and PUSCH Transmission for TDD Mode [5]

TDD Uplink/Downlink Configuration (DL:UL)	PDCCH/PHICH Subframe Number n									
	0	1	2	3	4	5	6	7	8	9
0 (2:3)	4	6	–	–	–	4	6	–	–	–
1 (3:2)	–	6	–	–	4	–	6	–	–	4
2 (4:1)	–	–	–	4	–	–	–	–	4	–
3 (7:3)	4	–	–	–	–	–	–	–	4	4
4 (8:2)	–	–	–	–	–	–	–	–	4	4
5 (9:1)	–	–	–	–	–	–	–	–	4	–
6 (5:5)	7	7	–	–	–	7	7	–	–	5

An important consideration concerning the uplink resource allocation is the timing relevance of PDCCH uplink assignment (or PHICH transmission) and the uplink subframe in which physical resources are reserved for the UE PUSCH transmission or retransmission. In FDD mode, the uplink resource allocation using PDCCH (or PHICH transmission) for the UE in subframe n is related to PUSCH transmission/retransmission in subframe $n+4$, i.e., there is a delay of four subframes between the uplink resource assignment (or HARQ ACK/NACK) in the downlink and the corresponding uplink transmission. For TDD mode, the timing relevance of PDCCH assignment (or PHICH transmission) and PUSCH transmission/retransmission is more complicated and depends on the downlink/uplink subframe configuration. The delay (in number of subframes) between receiving an uplink resource assignment in PDCCH or HARQ ACK/NACK in PHICH in subframe n and PUSCH transmission in subframe $n+k$ (where $k \geq 4$ and is selected such that $n+k$ is an uplink subframe when the ACK/NACK can be transmitted from the UE on PUCCH or PUSCH) for TDD mode for various downlink/uplink configurations is given in Table 10.4 [5]. It can be noted that, due to certain limitations imposed by downlink/uplink subframe configuration, PUSCH transmission may be delayed by up to seven subframes.

The uplink/downlink subframe configuration 0 is a special case because the number of uplink subframes is greater than the number of downlink subframes. Due to an insufficient number of downlink subframes, the downlink portion of the special subframe is also used for downlink transmission.

Multiplexing is a method for grouping and sending several independent HARQ ACK/NACKs in the TDD mode in a single uplink subframe for the previously received transport blocks. This method allows independent retransmission of erroneous transport blocks. However, it also implies that multiple ACK/NACK bits need to be transmitted from the UE, which can potentially limit the uplink coverage. Alternatively, bundling of HARQ ACK/NACKs implies that the outcome of the decoding of downlink transport blocks from multiple downlink subframes can

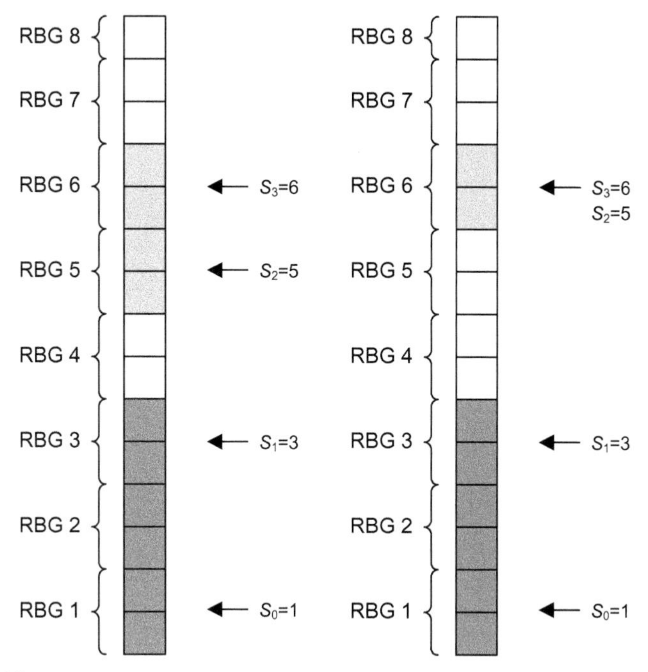

FIGURE 10.16

Example of uplink resource allocation type 1 in 3 MHz [14].

be combined into a single acknowledgement in the uplink. Combining acknowledgments related to multiple downlink transmissions into a single uplink message assumes that the terminal has not missed any of the scheduling assignments on which the ACK is based. As an example, assume that the eNB schedules the UE in two (subsequent) subframes, but the terminal misses the PDCCH transmission in the first subframe and successfully decodes the control information in the second subframe. The terminal would transmit one ACK as if it was scheduled in the second subframe only, while the eNB would interpret the ACK as if both transmissions were successfully received by the UE. To avoid such misinterpretations, the downlink assignment index (DAI) field in the scheduling assignment on the PDCCH is used. The DAI informs the terminal about the number of transmissions it should expect before generating a combined acknowledgment. If there is a mismatch between the DAI and the number of transmissions the terminal actually received, the terminal concludes that some assignments are missing. As a result, the UE does not transmit any HARQ acknowledgment [15] (Figure 10.16).

Let us consider a TDD example, downlink/uplink subframe configuration 2 includes six downlink subframes, two special subframes, and two uplink subframes. Downlink data can be transmitted within the DwPTS portion of the special subframes, which effectively results in eight downlink transmission opportunities.

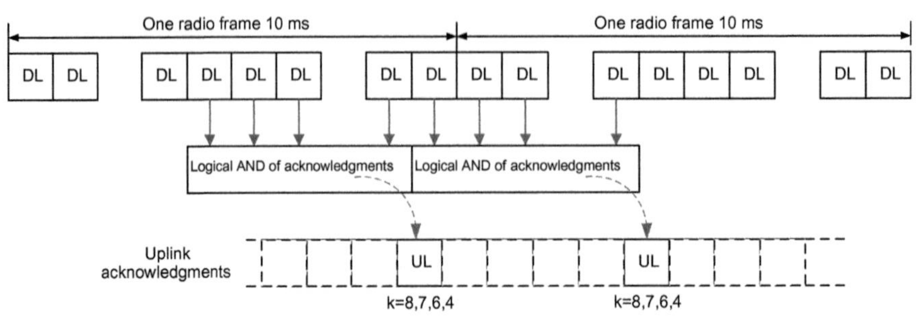

FIGURE 10.17

Illustration of HARQ acknowledgment bundling for TDD frame configuration 2 [14].

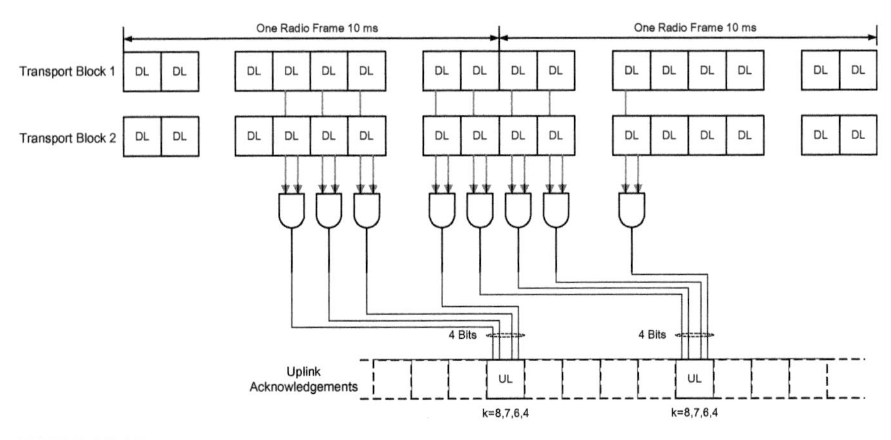

FIGURE 10.18

Illustration of HARQ ACK multiplexing for TDD frame configuration 2 [14].

If the downlink transmissions are divided into two groups and a single acknowledgment is generated from each group, each uplink subframe is required to carry a single HARQ ACK/NACK, which is called ACK/NACK bundling as depicted in Figure 10.17. On the other hand, for the same subframe configuration, 16 acknowledgments per radio frame can be reduced to eight acknowledgments using two uplink subframes. PUCCH format 1b can be used to carry four acknowledgments when channel selection is used, i.e., the PUCCH payload carries two ACKs and the selection of PUCCH resources carriers the other two ACKs, as illustrated in Figure 10.18 [14].

The use of bundled ACKs in a single uplink subframe is not limited to TDD mode. The carrier aggregation in FDD with a different number of uplink and downlink component carriers encounters the same problem. In LTE Rel-10, there is only one PUCCH that is always transmitted on the primary component carrier.

Hence, even in asymmetric carrier aggregation scenarios, the possibility to transmit more than two HARQ ACK bits in the uplink must be supported. This is performed via resource selection or by using PUCCH format 3. The ACKs in response to PUSCH transmissions are transmitted using multiple PHICHs, where the PHICH is transmitted on the same downlink component carrier as the uplink scheduling grant initiating the uplink transmission.

10.6 Uplink reference signals and channel estimation

There are two types of reference signals specified for the LTE/LTE-Advanced uplink. The uplink demodulation reference signals are used by the eNB for channel estimation and coherent demodulation of uplink physical channels (PUSCH and PUCCH). The demodulation reference signals are only transmitted in conjunction with PUSCH or PUCCH within the same bandwidth as the corresponding physical channel. Uplink demodulation reference signals are used by the eNB for channel state information (CSI) estimation, and to support uplink channel-dependent scheduling and link adaptation. The demodulation reference signals can also be used in cases when uplink transmission is needed, although there is no data to transmit. An example would be to enable the network to adjust uplink frame timing via an uplink timing alignment procedure. The uplink demodulation reference signals can also be used to estimate the downlink CSI from uplink transmission, when sufficient uplink/downlink reciprocity exists.

Since maintaining low CM value and PA efficiency for uplink transmissions is very important, the requirements for design of uplink reference signals are different compared to those in the downlink. In principle, transmitting reference signals together with other uplink transmissions from the same terminal is not efficient for the uplink. Instead, certain SC-FDMA symbols are exclusively designated for DM-RS transmission. Therefore, the reference signals are time-multiplexed with other uplink transmissions from the same terminal. The structure of the reference signal per se ensures a low CM value over the reference symbols. In the case of PUSCH transmission, a DM-RS is transmitted over the fourth SC-FDMA symbol of each uplink slot. In the case of PUCCH transmission, the number of SC-FDMA symbols used for a reference signal in a slot, as well as the exact position of these symbols, differs among different PUCCH formats [16].

In general, there is no advantage in estimating the channel outside the frequency band of the corresponding PUSCH or PUCCH transmission that is going to be coherently demodulated. The bandwidth of the reference signal corresponding to the length of the reference signal sequence is equal to the bandwidth of the corresponding PUSCH or PUCCH transmission measured in the number of subcarriers. The latter requirement suggests that, for PUSCH transmission, it should be possible to generate reference signal sequences of different length corresponding to the possible bandwidths of PUSCH transmission. However, it must be noted that the length of the reference signal sequence must be a multiple of 12 (i.e., the size of a basic PRB), since uplink resource allocations for PUSCH transmissions are always measured in terms of the number of resource blocks. The PUCCH transmission from a terminal is always carried out over a single resource block in the frequency-domain. In the case of PUCCH transmission, the length of the reference signal sequence is equal to 12. The uplink reference signals should preferably have the following properties [121]:

- Limited power variation in the frequency-domain to allow similar channel estimation quality for all frequencies, or constant amplitude in the frequency-domain for equal excitation of all the allocated sub-carriers for unbiased channel estimates.
- Low CM value, no worse than data transmissions, in the time-domain. Note that a lower CM value for reference signals relative to the data would allow power boosting of the reference signals at the cell-edge.
- Good autocorrelation property for accurate channel estimation.
- Good cross-correlation property to reduce interference among different reference signals transmitted on the same resources by neighboring cells.

We will discuss how the above requirements have been addressed in the design of the LTE/LTE-Advanced uplink reference signals.

10.6.1 Uplink demodulation reference signals

The uplink DM-RS design in LTE Rel-8 is based on extended Zadoff–Chu sequences. This choice is because of the low PAPR property of these sequences, as well as the possibility of using mechanisms for orthogonalizing multiple DM-RS sequences which are utilized in spatial multiplexing schemes. The use of different cyclic shifts of a root Zadoff–Chu sequence in the time-domain allows multiple users to transmit their (orthogonally code-division multiplexed) DM-RS to the same eNB provided that the reference signals are transmitted in the same set of resource blocks [121].

In LTE-Advanced, SU-MIMO with up to four-layer spatial multiplexing is supported in the uplink. While the uplink DM-RS corresponding to different layers can be theoretically orthogonalized via assigning different cyclic shifts to each layer, the orthogonality property of the cyclically shifted sequences may not hold in high-delay-spread scenarios. Furthermore, the orthogonality of the cyclically shifted sequences is guaranteed only when the root Zadoff–Chu sequences have identical waveform and length, which implies that the time–frequency resource allocation of MU-MIMO UEs should be identical. The uplink DM-RS in LTE-Advanced are designed to overcome these drawbacks, while maintaining backward compatibility.

In LTE/LTE-Advanced, two SC-FDMA symbols (one DM-RS per slot) are configured for DM-RS in a subframe within the allocated time–frequency resources as shown in Figure 10.19 for PUSCH transmission. Each uplink demodulation reference signal sequence $r_{uv}(k,\tau)$ associated with PUSCH transmission is defined by the time-domain cyclic shift $\tau(\lambda)$ of the base sequence $r_{uv}(k)$ according to

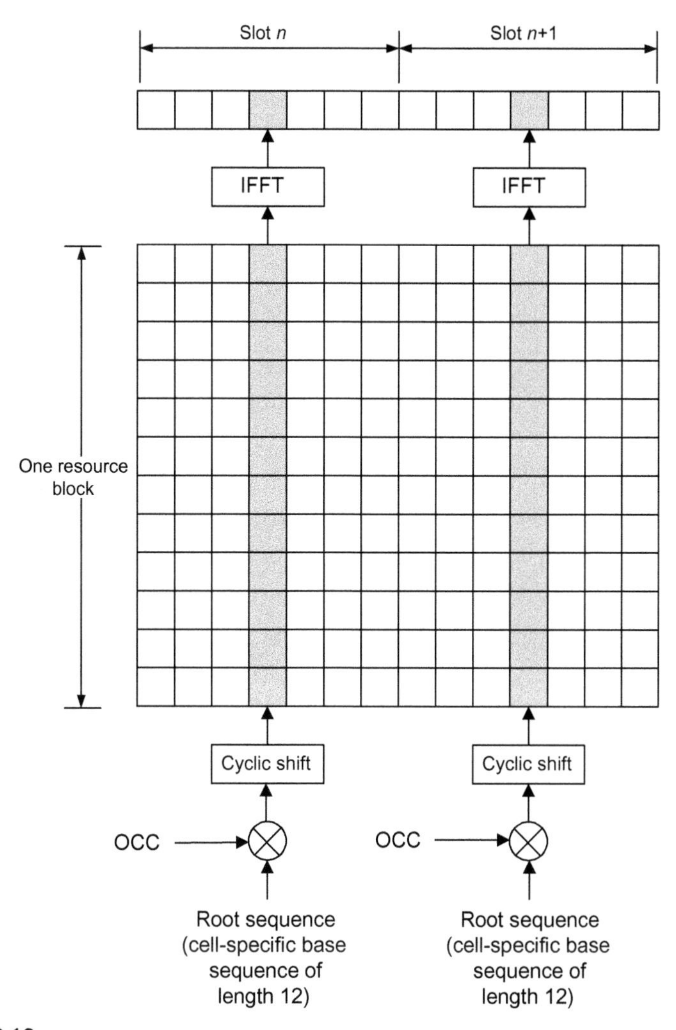

FIGURE 10.19

Uplink DM-RS generation and structure in time and frequency [121,14].

$r_{uv}(mL_{RS} + k, \tau) = w_m(\lambda)e^{j\tau(\lambda)k}\, r_{uv}(k)|0 \leq k < L_{RS} - 1, m = 0, 1,$ where $L_{RS} = 12m'$ (i.e., the length of the reference signal is an integer multiple of the basic resource-block size in frequency) is the length of the reference signal sequence, $1 \leq m' \leq \max(N_{RB}^{UL})$, and $\lambda \in \{0, 1, \ldots, N_l - 1\}$ denotes the spatial layer index. Using the latter expression, multiple reference signal sequences can be generated from a single base sequence through different values of cyclic shift $\tau(\lambda)$ [3].

The base sequences $r_{uv}(k)$ are divided into groups where the index $u \in \{0, 1, \ldots, 29\}$ represents the group number and v is the base sequence number within the group, such that each group contains one base sequence $v = 0$ for each

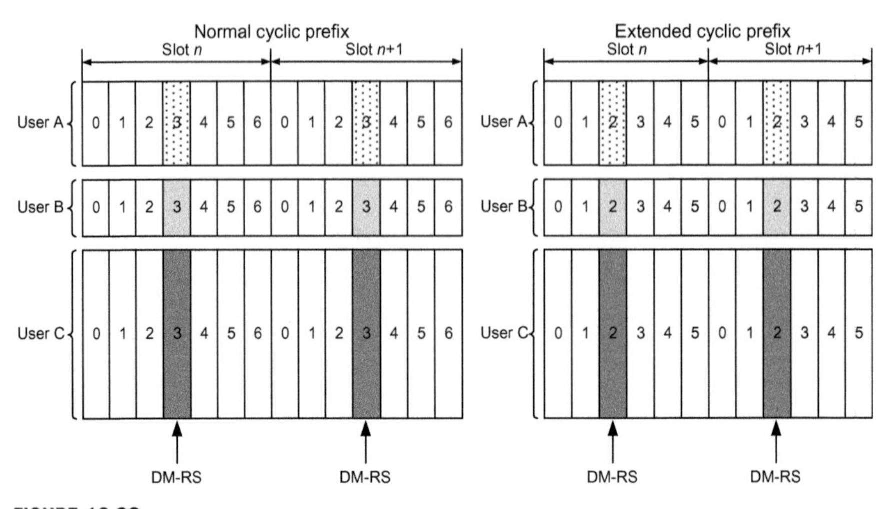

FIGURE 10.20

PUSCH DM-RS location in time and frequency [155].

length $L_{RS} = 12m'|1 \leq m' \leq 5$ and two base sequences denoted by $v \in \{0, 1\}$ of each length $L_{RS} = 12m'|6 \leq m' \leq \max(N_{RB}^{UL})$. The sequence-group number u and the index v within the group may vary across time. The definition of the base sequence $\{r_{uv}(0), r_{uv}(1), \ldots, r_{uv}(L_{RS} - 1)\}$ depends on the sequence length L_{RS} (Figure 10.21).

The time-domain cyclic shift (post IFFT in the OFDM modulation) can be interpreted as a phase rotation in the frequency-domain (pre-IFFT in the OFDM modulation). For frequency non-selective channels over 12 sub-carriers constituting a resource block, it is possible to achieve orthogonality among DM-RSs generated from the same base sequence, if $\tau(\lambda) = \frac{k\pi}{6}, k = 0, 1, \ldots, 11$, assuming the DM-RSs are synchronized in time. The orthogonality can be exploited to transmit DM-RS at the same time, using the same frequency resources without interference. In general, DM-RS generated from different base sequences will not be orthogonal; however, they will present low cross-correlation properties. To maximize the number of available Zadoff–Chu sequences, a prime length sequence is needed. The minimum sequence length in the uplink is 12 (i.e., the number of sub-carriers in a resource block), which is not prime number. Therefore, Zadoff–Chu sequences are not suitable by themselves; thus, there are effectively two types of base reference sequences: (1) those with a sequence length larger than 36 sub-carriers (spanning three or more resource blocks), which use a cyclic extension of Zadoff–Chu sequences; and (2) those with a sequence length less than 36 sub-carriers (spanning one or two resource blocks where the latter use a special QPSK sequence ($r_{uv}(k) = e^{j\phi(k)\pi/4}, \quad 0 \leq k \leq L_{RS} - 1$) [3].

For sequences of length three resource blocks and larger, the base sequence is a repetition, with a cyclic offset, of a Zadoff–Chu sequence of length N_{ZC}^{RS}, where

N_{ZC}^{RS} is the largest prime number that satisfies the condition $N_{ZC}^{RS} < L_{RS}$. Therefore, the base sequence will contain one complete length N_{ZC}^{RS} Zadoff–Chu sequence plus a fractional repetition appended on the end. At the receiver, the appropriate de-repetition can be done and the zero autocorrelation property will hold across the length N_{ZC}^{RS} vector. For sequences shorter than three resource blocks, the sequences are a composition of unity modulus complex numbers derived from computer simulation, and are defined in the LTE specifications [3].

There are a total of 30 sequence groups, each containing one sequence with length shorter than 60 sub-carriers. This corresponds to transmission bandwidths of 1, 2, 3, 4, and 5 resource blocks. Additionally, there are two sequences (one for $v = 0$ or 1) with length greater than 72 sub-carriers corresponding to transmission bandwidths of six resource blocks or more. Note that not all values of m' are allowed, where m' is the number of resource blocks used for transmission of the PUSCH DM-RS. Only values which are the product of powers of 2, 3, and 5 are valid, i.e., $m' = 2^{\alpha_1}3^{\alpha_2}5^{\alpha_3}$ and $\alpha_1, \alpha_2,$ and α_3 are non-negative integers. This is due to the fact that the DFT sizes of the SC-FDMA precoding operation are limited to values which are the product of powers of 2, 3, and 5. The DFT operation can span more than one resource block. Since each resource block has 12 sub-carriers, the total number of sub-carriers fed to the DFT will be equal to $12m'$. Given that the value of $12m'$ has to be a product of powers of 2, 3, and 5, this implies that the number of resource blocks must themselves be the product of powers of 2, 3, and 5. Therefore, values of m' such as 7, 11, 14, or 19 are not valid [16].

For a given time slot, the uplink reference signal sequences to use within a cell are taken from a specific sequence group. If the same group is used for all slots, then this is known as fixed assignment. On the other hand, if the group number u varies for all slots within a cell, this is known as group hopping. The cell system information determines which method to use, and the group assignment depends on the type of physical channel (PUSCH or PUCCH). The PUSCH demodulation reference signal is mapped to the fourth SC-FDMA symbol of the slot for a normal cyclic prefix and to every third SC-FDMA slot for an extended cyclic prefix, as shown in Figure 10.20.

The LTE Rel-10 uses codebook-based precoding for uplink SU-MIMO data transmissions, and obviously the same precoding mechanism is applied to the uplink DM-RS. As such, a receiver can directly estimate the spatially precoded channel. Note that the uplink DM-RSs are distributed in the same manner as data when clustered SC-FDMA is used. An eNB can choose root Zadoff–Chu sequences for the reference signal by considering the autocorrelations and cross-correlations of the sequences that are in use among neighboring eNBs. However, such coordination is not always practically feasible for cellular operators. Hence, a root Zadoff–Chu sequence randomization mechanism, referred to as sequence hopping and sequence-group hopping, is available depending on each operator's choice, where the root Zadoff–Chu sequence varies in every slot, if sequence-group hopping is enabled. The spreading factor $w(\lambda_i) = \pm 1 | i = 0, 1, 2, 3$ is an

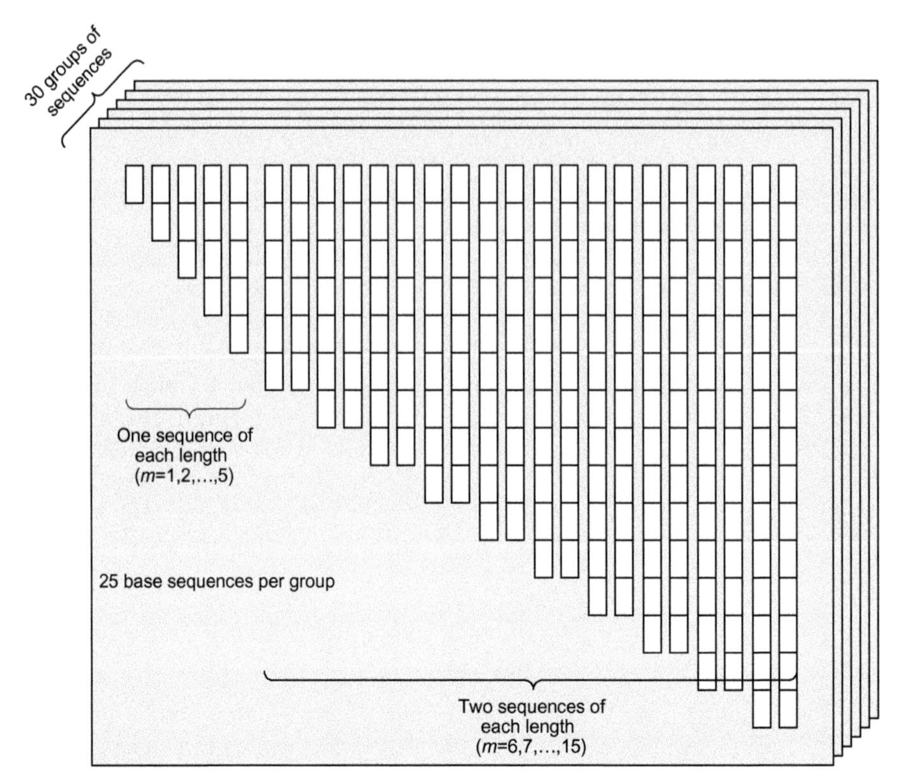

FIGURE 10.21

Illustration of base sequence grouping used for PUCCH/PUSCH DM-RS in 3 MHz [14].

orthogonal cover code (OCC), which is used to spread the reference signal sequence over two SC-FDMA symbols. The introduction of OCC was motivated in order to efficiently support SU-MIMO and MU-MIMO operation in the uplink (see Figure 10.19).

For $L_{RS} \geq 36$, the base sequence $\{r_{uv}(0), r_{uv}(1), \ldots, r_{uv}(L_{RS}-1)\}$ is given by $r_{uv}(k) = x_l(k \bmod N_{ZC}^{RS})| \quad 0 \leq k < L_{RS}-1$, where the lth-root Zadoff–Chu sequence is defined by the following expression [3]:

$$x_l(m) = -\exp\left(j\frac{\pi l m(m+1)}{N_{ZC}^{RS}}\right), \quad 0 \leq m \leq N_{ZC}^{RS}-1, \quad l = \lfloor l' + 0.5 \rfloor + (-1)^{\lfloor 2l' \rfloor} v, \quad l' = N_{ZC}^{RS}(u+1)/31 \tag{10.3}$$

The length N_{ZC}^{RS} of the Zadoff–Chu sequence is given by the largest prime number such that $N_{ZC}^{RS} < L_{RS}$. The sequence-group number u in slot n_s is defined by a group hopping pattern $f_{\text{group-hopping}}(n_s)$ and a sequence-shift pattern $f_{\text{shift-pattern}}$ according to $u = [f_{\text{group-hopping}}(n_s) + f_{\text{shift-pattern}}] \bmod 30$. There are 17 different hopping patterns and 30 different sequence-shift patterns. Sequence-group hopping can be enabled or disabled through a cell-specific parameter signaled by higher layers (see Figure 10.20) [3].

The DM-RS group hopping pattern is defined as $f_{\text{group-hopping}}(n_s) = \left[\sum_{i=0}^{7} c(8n_s + i)2^i\right] \bmod 30$ when group hopping is enabled and is set to zero, otherwise, which is the same for PUSCH and PUCCH. The pseudo-random sequence $c(i)$ was defined in Chapter 9 and is initialized with $c_{\text{init}} = \lfloor N_{\text{ID}}^{\text{RS}}/30 \rfloor$ at the beginning of each radio frame.[3] Note that the definition of the sequence-shift pattern $f_{\text{shift-pattern}}$ is different for PUCCH and PUSCH. For PUCCH, the sequence-shift pattern is given by $f_{\text{shift-pattern}}^{\text{PUCCH}} = N_{\text{ID}}^{\text{RS}} \bmod 30$. For PUSCH, the sequence-shift pattern is given by $f_{\text{shift-pattern}}^{\text{PUSCH}} = (N_{\text{ID}}^{\text{cell}} + \Delta_{\text{ss}}) \bmod 30 \quad | \Delta_{\text{ss}} \in \{0, 1, \ldots, 29\}$ and is configured via higher-layer signaling [3], if no value for $n_{\text{ID}}^{\text{PUSCH}}$ is provided by the higher layers, or if the PUSCH transmission corresponds to a *Random Access Response* or a retransmission of the same transport block as part of the contention-based random access procedure; otherwise, it is defined as $f_{\text{ss}}^{\text{PUSCH}} = N_{\text{ID}}^{\text{RS}} \bmod 30$. The parameter Δ_{ss} is cell-specific and is broadcast in SIB2 which defines an offset between 0 and 29 that can be used to select different sequence groups for cells with the same physical cell identifier $N_{\text{ID}}^{\text{cell}}$. The concept of virtual cell identity was introduced in LTE Rel-11 to overcome the limitation of earlier releases of LTE technology in new deployment scenarios related to CoMP, HetNet, etc., where several low-power nodes or subsets of a macro-cell may share the same physical layer cell identity with the macro-cell. Sharing of the same cell identifier by the neighboring nodes would increase intra-cell interference and degrade the channel estimation performance and coherent detection of data and control channels, as well as limiting the use of advanced closed-loop SU-MIMO/MU-MIMO beamforming techniques. Also, the use of the same identifier by subordinate nodes would prevent transmission/reception node identification and switching. The use of different scrambling sequences initialized with unique cell identifiers would achieve interference randomization and help the UE to separate data and control channels transmitted by different transmission nodes. Since the reference signals (e.g., CRS, DM-RS, CSI-RS) are also scrambled with cell-specific sequences to allow the UE to conduct various channel estimations and mobility measurements on different transmission nodes, the use of unique scrambling sequences would facilitate these functions as well as node selection and switching in homogeneous and heterogeneous networks.

[3]The definition of virtual cell identity $N_{\text{ID}}^{\text{RS}}$ (a new concept in LTE Rel-11) depends on the type of uplink physical channel [3]. For transmissions on PUSCH, $N_{\text{ID}}^{\text{RS}} = N_{\text{ID}}^{\text{cell}}$, if no value for $N_{\text{ID}}^{\text{PUSCH}}$ is configured by the higher layers or if the temporary C-RNTI was used to transmit the most recent uplink-related DCI for the transport block associated with the corresponding PUSCH transmission; otherwise $N_{\text{ID}}^{\text{RS}} = N_{\text{ID}}^{\text{PUSCH}}$. For transmissions on PUCCH, the virtual cell identity $N_{\text{ID}}^{\text{RS}} = N_{\text{ID}}^{\text{cell}}$, if no value for $N_{\text{ID}}^{\text{PUCCH}}$ is configured by the higher layers; otherwise $N_{\text{ID}}^{\text{RS}} = N_{\text{ID}}^{\text{PUCCH}}$. For sounding reference signal transmission, the virtual cell identity is defined as $N_{\text{ID}}^{\text{RS}} = N_{\text{ID}}^{\text{cell}}$ [3,5].

The sequence hopping is only applied to reference signals whose length $N_{RS} \geq 72$ and the base sequence number defined as $v = c(n_s)$ within the base sequence group in slot n_s, if group hopping is disabled but sequence hopping is enabled. The pseudo-random sequence $c(i)$ was defined earlier and is initialized with $c_{init} = \lfloor N_{ID}^{RS}/30 \rfloor 2^5 + f_{shift\text{-}pattern}^{PUSCH}$ at the beginning of each radio frame. For reference signals of length $N_{RS} < 72$, the base sequence number v within the base sequence group is given by $v = 0$ [3].

The cyclic shift $\tau(\lambda)$ in slot n_s is defined as $\tau(\lambda) = \pi n_{CS}(\lambda)/6$, where $n_{CS}(\lambda) = [n_{DM\text{-}RS}^{(1)} + n_{DM\text{-}RS}^{(2)}(\lambda) + n_{PN}(n_s)] \bmod 12$. The values of parameters $n_{DM\text{-}RS}^{(1)}$ and $n_{DM\text{-}RS}^{(2)}(\lambda)$ are determined based on the value of the cyclic shift for the DM-RS field in the latest uplink grant within DCI format 0 or 4, and are given in Table 10.5. The first row of Table 10.5 can be used to obtain $n_{DM\text{-}RS}^{(2)}(0)$ and $w_m(\lambda)$, if there is no uplink-related DCI for the same transport block associated with the corresponding PUSCH transmission, and if the initial PUSCH for the same transport block is semi-persistently scheduled, or if the initial PUSCH for the same transport block is scheduled by the random-access response grant. The parameter $n_{PN}(n_s)$ is given by $n_{PN}(n_s) = \sum_{i=0}^{7} c(8N_s^{UL} n_s + i)2^i$, where the cell-specific pseudo-random sequence $c(i)$ was defined in Chapter 9. The pseudo-random sequence generator in this case is initialized with $c_{init} = \lfloor N_{ID}^{cell}/30 \rfloor 2^5 + f_{shift\text{-}pattern}^{PUSCH}$ at the beginning of each radio frame, if no value for $N_{ID}^{PUSCH\text{-}DM\text{-}RS}$ is configured by the higher layers, or if the temporary C-RNTI was used to transmit the most recent uplink-related DCI for the transport block associated with PUSCH transmission; otherwise, it is defined as $c_{init} = \left\lfloor \frac{N_{ID}^{PUSCH\text{-}DM\text{-}RS}}{30} \right\rfloor 2^5 + (N_{ID}^{PUSCH\text{-}DM\text{-}RS} \bmod 30)$ [3].

The vector of the UE-specific DM-RSs is precoded according to the following equation:

$$[\tilde{r}_{PUSCH}^{(0)}, \tilde{r}_{PUSCH}^{(1)}, \ldots, \tilde{r}_{PUSCH}^{(N_P-1)}]^T = \mathbf{W}[r_{PUSCH}^{(0)}, r_{PUSCH}^{(1)}, \ldots, r_{PUSCH}^{(N_l-1)}]^T \tag{10.4}$$

where N_P denotes the number of antenna ports used for PUSCH transmission and $\mathbf{W}$ is the precoding matrix. For PUSCH transmission using a single antenna port $N_P = 1, W = 1$, and $N_l = 1$. For spatial multiplexing $N_P = 2$ or $N_P = 4$ and the precoding matrix $\mathbf{W}$ is identical to the precoding matrix used for precoding of the PUSCH in the same subframe [3].

In uplink SU-MIMO where a UE transmits multiple spatial layers simultaneously, multiple DM-RS can be orthogonally multiplexed by assigning a different cyclic shift value for each layer. However, as mentioned earlier, the orthogonality among the sequences may not be preserved when the delay spread is high due to increased correlation among the spatial layers. To overcome this limitation, the OCCs were introduced in LTE-Advanced. As shown in Figure 10.19 and further emphasized in Table 10.5, different OCCs are applied for higher-numbered layers (i.e., $\lambda \geq 3$) and lower-numbered layers (i.e., $\lambda \geq 2$), so reference signals associated with the higher rank transmissions (i.e., $N_l \geq 3$)

Table 10.5 OCC and Cyclic Shift Parameters for Uplink DM-RS [3]

Cyclic Shift Value Carried in Uplink Grant	**Layer-Dependent Cyclic Shift**					**Layer-Dependent Orthogonal Cover Code $(w_0(\lambda), w_1(\lambda))$**			
	$n^{(1)}_{DM\text{-}RS}$	$n^{(2)}_{DM\text{-}RS}(\lambda)$							
		$\lambda=0$	$\lambda=1$	$\lambda=2$	$\lambda=3$	$\lambda=0$	$\lambda=1$	$\lambda=2$	$\lambda=3$
000	0	0	6	3	9	$[1\ \ 1]$	$[1\ \ 1]$	$[1\ \ -1]$	$[1\ \ -1]$
001	2	6	0	9	3	$[1\ \ -1]$	$[1\ \ -1]$	$[1\ \ 1]$	$[1\ \ 1]$
010	3	3	9	6	0	$[1\ \ -1]$	$[1\ \ -1]$	$[1\ \ 1]$	$[1\ \ 1]$
011	4	4	10	7	1	$[1\ \ 1]$	$[1\ \ 1]$	$[1\ \ 1]$	$[1\ \ 1]$
100	6	2	8	5	11	$[1\ \ 1]$	$[1\ \ 1]$	$[1\ \ 1]$	$[1\ \ 1]$
101	8	8	2	11	5	$[1\ \ -1]$	$[1\ \ -1]$	$[1\ \ -1]$	$[1\ \ -1]$
110	9	10	4	1	7	$[1\ \ -1]$	$[1\ \ -1]$	$[1\ \ -1]$	$[1\ \ -1]$
111	10	9	3	0	6	$[1\ \ 1]$	$[1\ \ 1]$	$[1\ \ -1]$	$[1\ \ -1]$

are still sufficiently orthogonal, even though cyclic shift separation is small. Although the orthogonality provided by the OCCs may also fall apart due to high Doppler shift, the effect is minimal because higher rank transmission is mainly used for stationary or nomadic UEs [121].

The generation of LTE Rel-8/9 DM-RS sequences (no OCC is applied) for PUSCH in each uplink slot can be summarized as follows:

1. Determine whether group hopping and sequence hopping are enabled from higher layer signaling.
2. Obtain Δ_{ss} and the cyclic shift from higher layer signaling.
3. Calculate $c_{\text{init}} = \lfloor N_{\text{ID}}^{\text{cell}}/30 \rfloor$ thus find the scrambling sequence $c(i)$.
4. Obtain $n_{\text{DM-RS}}^{(1)}, n_{\text{DM-RS}}^{(2)}(\lambda)$ from Table 10.5 given the cyclic shift value is known.
5. Calculate $f_{\text{group-hopping}}(n_s) = \left[\sum_{i=0}^{7} c(8n_s + i)2^i\right] \text{mod } 30$ and $v = c(n_s)$ or 0 given n_s is known.
6. Calculate $f_{\text{shift-pattern}}^{\text{PUSCH}} = (N_{\text{ID}}^{\text{cell}} + \Delta_{ss}) \text{mod } 30$.
7. Calculate $u = \left[f_{\text{group-hopping}}(n_s) + f_{\text{shift-pattern}}\right] \text{mod } 30$.
8. Calculate $n_{\text{PN}}(n_s) = \sum_{i=0}^{7} c(8N_s^{\text{UL}} n_s + i)2^i$.
9. Calculate $n_{\text{CS}}(\lambda) = \left[n_{\text{DM-RS}}^{(1)} + n_{\text{DM-RS}}^{(2)}(\lambda) + n_{\text{PN}}(n_s)\right] \text{mod } 12$.
10. Calculate $\tau(\lambda) = \pi n_{\text{CS}}(\lambda)/6$.
11. Calculate lth-root Zadoff–Chu sequence $x_l(m) = -\exp\left(j\frac{\pi l m(m+1)}{N_{\text{ZC}}^{\text{RS}}}\right)$ given that $0 \le m \le N_{\text{ZC}}^{\text{RS}} - 1$, $l = \lfloor l' + 0.5 \rfloor + (-1)^{\lfloor 2l' \rfloor} v, l' = N_{\text{ZC}}^{\text{RS}}(u+1)/31$ *(Note that $N_{\text{ZC}}^{\text{RS}}$ is the largest prime number less than the size of PUSCH allocation or L_{RS}).*
12. Calculate $r_{uv}(mL_{\text{RS}} + k, \tau) = e^{j\tau(\lambda)k} r_{uv}(k) | 0 \le k < L_{\text{RS}} - 1, m = 0, 1$.

The MU-MIMO schemes can be categorized into two classes from a DM-RS orthogonality perspective, i.e., identical-band MU-MIMO and non-identical-band MU-MIMO [121]. Identical-band MU-MIMO is a scheme in which the frequency band allocations for MU-MIMO-capable UEs are identical and the DM-RS orthogonality can be achieved by assigning different cyclic shifts. Although this scheme is LTE Rel-8/9 compatible and can be used for multiplexing LTE Rel-8/9 and Rel-10 UEs, the scheduling restriction may adversely impact the uplink system throughput. Non-identical-band MU-MIMO is a scheme in which the frequency band allocations for MU-MIMO UEs are not identical. Although this improves frequency scheduling flexibility and may improve the uplink system throughput, it implies that the DM-RS root Zadoff–Chu sequences for the two slots of a subframe should be the same in order to utilize the OCCs. Therefore, the OCCs can be used to realize non-identical-band MU-MIMO [121].

10.6.2 Sounding reference signals

The SRS are physical signals transmitted in the uplink to enable the eNB to estimate the CSI over a range of frequencies. The estimation of the CSI assists the

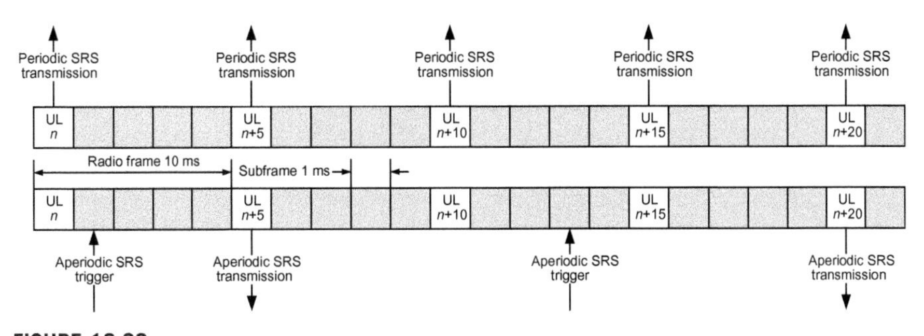

FIGURE 10.22

Illustration of periodic and aperiodic SRS transmission [121].

eNB scheduler to properly allocate radio resources to the UE, and to select different transmission parameters such as the instantaneous data rate and uplink multi-antenna transmission parameters for optimized uplink transmission (i.e., uplink channel-dependent scheduling). The SRS transmission can be further used for uplink timing estimation as well as to estimate downlink channel conditions assuming downlink/uplink channel reciprocity. When uplink/downlink channel reciprocity is assumed, SRS measurements can also be used to support downlink transmissions such as angle-of-arrival (AoA) measurements on SRS to support downlink beamforming [14]. Therefore, SRS is not necessarily transmitted together with an uplink physical channel and when transmitted in conjunction with PUSCH, the SRS may cover a different and typically larger frequency region. There are two types of SRS specified for uplink transmission, i.e., periodic and aperiodic SRS (Figure 10.22).

The subframe in which SRS is transmitted by any UE within the cell is signaled via cell-specific broadcast signaling. A 4-bit cell-specific parameter indicates 15 possible sets of subframes in which SRS may be transmitted within each radio frame. This configurability provides sufficient flexibility in adjusting the SRS overhead depending on the deployment scenario. The 16th configuration switches the SRS off in the cell, which may be appropriate for a cell serving primarily high-speed UEs. The SRS transmissions are always in the last SC-FDMA symbol in the configured subframes. The SRS and DM-RS are located in different SC-FDMA symbols. The PUSCH data transmission is not permitted on the SC-FDMA symbol designated for SRS, resulting in a worst-case sounding signal overhead of around 7% with the SRS symbol in every subframe.

The eNB may either request an individual SRS transmission from a UE, or configure a UE to transmit SRS periodically until it is no longer necessary. A single-bit UE-specific signaling parameter indicates whether the requested SRS transmission is periodic or aperiodic. If periodic SRS transmissions are configured for a UE, the periodicity may be 2, 5, 10, 20, 40, 80, 160, or 320 ms. The SRS periodicity and SRS subframe offset within the period in which the UE should transmit its SRS are configured via a 10-bit UE-specific dedicated signaling

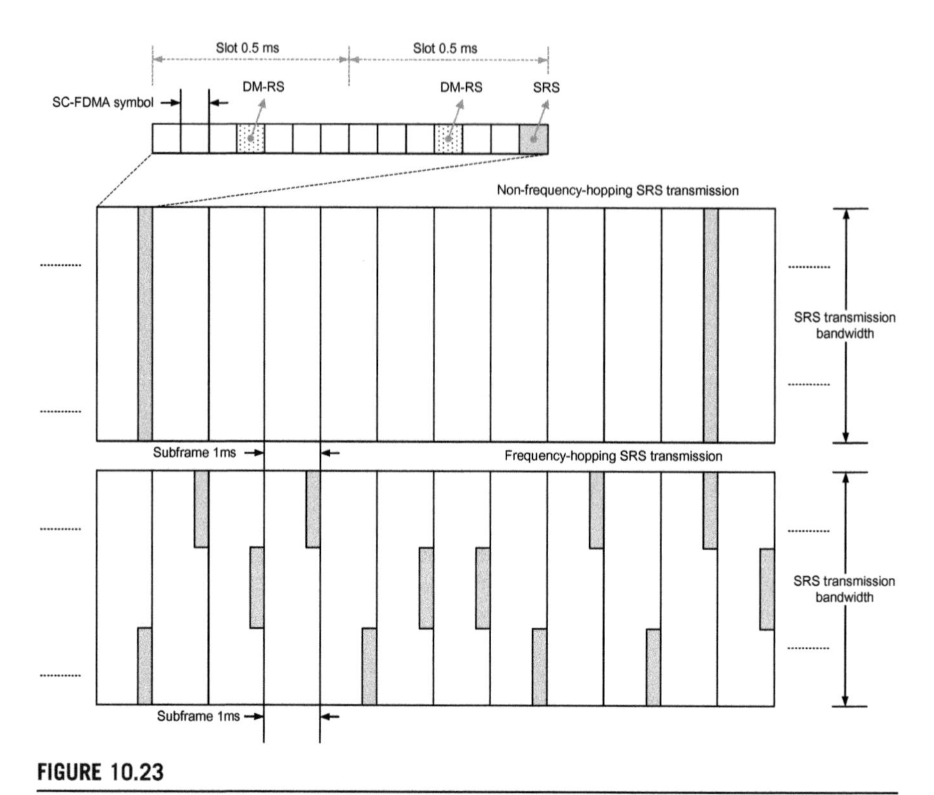

FIGURE 10.23

Non-frequency-hopping (wideband) versus frequency-hopping (sub-band) SRS [16].

parameter. There are differences between cell-specific SRS (signaled via cell-specific broadcast signaling) and UE-specific SRS (signaled via UE-specific dedicated signaling) configuration parameters, and their method of signaling to the UE noted in the forthcoming discussions.

The periodic SRS is based on the extended Zadoff–Chu sequence and is periodically transmitted in the last SC-FDMA symbol of an uplink subframe, as shown in Figure 10.23. As the example in Figure 10.23 shows, the SRS transmitted by the UEs are multiplexed in the time- and frequency-domain through configuring SRS periodicity T_{SRS}, frequency comb pattern, and SRS bandwidth. Similar to DM-RS, up to eight orthogonal SRS transmissions can be code-division multiplexed over a frequency region using different cyclic shifts of the root Zadoff–Chu sequence. The periodic nature of the LTE Rel-8 SRS transmission was not sufficiently flexible to support sounding from an increased number of UE antennas and/or an increased number of UEs, due to use of semi-static RRC signaling for UE configuration. As a result, aperiodic SRS transmission was introduced in LTE Rel-10 to complement periodic SRS transmission, where the eNB dynamically schedules a UE for SRS transmission on demand (see Figure 10.22). Therefore, the SRS resources are not tied to a single UE until RRC (re)configuration by the eNB. This mechanism allows efficient management of a fixed set of time/frequency/code SRS resources for a larger number of UEs.

Periodic and aperiodic SRS transmission share a common set of cell-specific SRS resources (subframe configuration period, subframe offset, and maximum SRS bandwidth). Different sets of UE-specific sounding parameters are independently allocated for periodic and aperiodic SRS transmission including transmission bandwidth, periodicity, frequency comb pattern, and cyclic shift. The UE is configured with a fixed set of subframes in which it may be scheduled for aperiodic SRS transmission. This reduces the eNB scheduling complexity, because in any subframe only a subset of UEs are scheduled for aperiodic SRS transmission. An example of periodic and aperiodic SRS transmission timing is shown in Figure 10.22 for comparison, where periodic and aperiodic SRS have the same subframe offset and $T_{SRS_{Periodic}} = 5$ ms. An aperiodic SRS transmission occurs in subframe $n+k$ corresponding to a trigger in subframe n, where $k \geq 4$, corresponding to a 4 ms delay constraint which is meant to maintain the minimum timing response similar to a normal downlink assignment or uplink grant.

To minimize the downlink signaling overhead, a semi-static configuration of aperiodic SRS parameters similar to that of periodic SRS was adopted in LTE Rel-8. Furthermore, the aperiodic SRS-related signaling bits are piggybacked in either a downlink assignment (scheduling downlink data transmission) or an uplink grant (scheduling uplink data transmission), in order to reduce signaling overhead. Since the aperiodic SRS is primarily used to support multi-antenna sounding, further configuration flexibility can be achieved by forming three independent sets of parameters that can be selected via a 2-bit SRS field in the uplink grant, scheduling uplink SU-MIMO data transmission. For an uplink single-antenna transmission, a single-bit aperiodic SRS field in the uplink grant is used [14,16].

There are only a few subframes in which the eNB can send an aperiodic SRS trigger, in order to satisfy the timing constraint ($k \geq 4$) for a specific aperiodic SRS transmission opportunity. The latter scheduling constraint has some impacts on some TDD frame configurations, where only a few uplink subframes are available in a radio frame. Therefore, to increase the scheduling opportunities for aperiodic SRS, LTE Rel-10 specifies a single-bit aperiodic SRS control field in some downlink assignments. Note that the uplink sounding can be used to support downlink beamforming by exploiting channel reciprocity in TDD systems.

To ensure that there is no timing mismatch between CSI estimates from multiple antennas, the UE is configured to simultaneously transmit SRS from all configured antenna ports. The orthogonality of the cyclic shifts falls apart as the frequency selectivity of the channel increases. This degradation is more pronounced as the number of antenna ports increases. Therefore, for both periodic and aperiodic SRS, a combination of the frequency comb (i.e., a multiplexing/interlacing pattern in frequency-domain) and cyclic shift can be used to multiplex the SRS sequences corresponding to four antenna ports in large delay-spread channels, while for low delay-spread channels or when the UE is equipped with

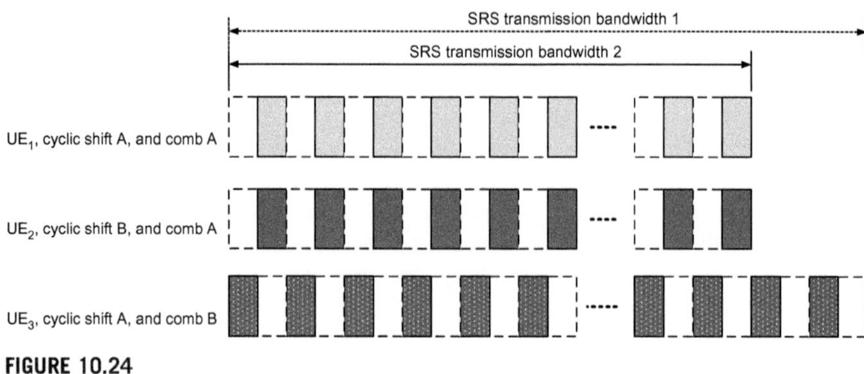

FIGURE 10.24

Multiplexing SRS transmission from different UEs [16].

two antenna ports, only different cyclic shifts are exploited to multiplex the SRS (as shown in Figure 10.24). The cyclic shifts and comb values for all antenna ports are implicitly derived from the values in each of the semi-statically configured parameter sets.

A UE transmits SRS in the serving-cell SRS resources using one of the following (UE-specific) trigger types [5]:

- Trigger type 0 via higher layer signaling.
- Trigger type 1 via DCI format 0/4/1A for FDD and TDD, and DCI format 2B/2C for TDD.

A UE may be configured with SRS parameters for trigger types 0 and 1 in each serving cell. The following SRS parameters are serving-cell specific and semi-statically configurable by higher layers for trigger type 0 and for trigger type 1 [5]:

- Transmission comb for trigger type 0 and each configuration of trigger type 1.
- Starting PRB assignment for trigger type 0 and each configuration of trigger type 1.
- Duration for trigger type 0, i.e., single or indefinite (until disabled).
- SRS configuration index I_{SRS}, SRS periodicity T_{SRS}, and SRS subframe offset $T_{\text{SRS-offset}}$ for trigger type 0 and trigger type 1.
- SRS bandwidth for trigger type 0 and each configuration of trigger type 1.
- Frequency hopping bandwidth for trigger type 0.
- Cyclic shift for trigger type 0 and each configuration of trigger type 1.
- Number of antenna ports for trigger type 0 and each configuration of trigger type 1.

A UE may be configured to transmit SRS on N_p antenna ports of a serving cell where N_p may be configured via higher layer signaling. For PUSCH transmission mode 1, $N_p \in \{0, 1, 2, 4\}$ and for PUSCH transmission mode 2, $N_p \in \{0, 1, 2\}$

when two antenna ports are configured for PUSCH and $N_p \in \{0, 1, 4\}$ when four antenna ports configured for PUSCH.

A UE may be configured for SRS transmission on multiple antenna ports of a serving cell within one SC-FDMA symbol of the same subframe. The SRS transmission bandwidth and starting PRB assignment are the same for all the configured antenna ports of a given serving cell [5]. The UE-specific parameters for SRS trigger type 0 in a serving cell, i.e., periodicity T_{SRS} and subframe offset $T_{\text{SRS-offset}}$ are defined in Table 10.6 for FDD and TDD modes. An example of SRS parameter configuration for an FDD system is shown in Figure 10.25 for better understanding of the effect of each SRS parameter on the SRS transmission. The periodicity of SRS transmission, in number of subframes, is serving-cell specific and is selected from the set $T_{\text{SRS}} \in \{2, 5, 10, 20, 40, 80, 160, 320\}$. For the SRS periodicity value of $T_{\text{SRS}} = 2$ ms in TDD, two SRS resources are configured in half-frame containing uplink subframes. The UE-specific parameters for SRS trigger type 1 in a serving cell, i.e., periodicity T_{SRS} and subframe offset $T_{\text{SRS-offset}}$, are given in Table 10.6 for FDD and TDD modes. The periodicity of the SRS transmission, in number of subframes, is serving-cell specific and is selected from the set $T_{\text{SRS}} \in \{2, 5, 10\}$. For the SRS periodicity value of $T_{\text{SRS}} = 2$ ms in TDD, two SRS resources are configured in a half-frame containing UL subframes in a serving cell [5].

The cell-specific SRS subframe configuration period T_{SFC} and the cell-specific subframe offset Δ_{SFC} for the transmission of SRS are listed in Table 10.7, for frame structure types 1 and 2, where the SRS subframe configuration parameter is provided by the higher layers. The sounding reference signal is transmitted in subframes whose numbers satisfy $\lfloor n_s/2 \rfloor \bmod T_{\text{SFC}} \in \{\Delta_{\text{SFC}}\}$. For frame structure type 2, SRS is transmitted only in configured uplink subframes or UpPTS portion of the special subframe. Note that the cell-specific SRS parameters $(T_{\text{SFC}}, \Delta_{\text{SFC}})$ and UE-specific SRS parameters $(T_{\text{SRS}}, T_{\text{SRS-offset}})$ are not identical.

The SRS sequence is defined by $r_{\text{SRS}}(n, \overline{p}) = r_{\text{uv}}(n, \tau_{\overline{p}})$, where $r_{\text{uv}}(n, \tau_{\overline{p}})$ is a cyclically shifted Zadoff–Chu sequence defined in the previous section, u is the PUCCH sequence-group number, and v is the base sequence number. The antenna-dependent cyclic shift $\tau_{\overline{p}}$ of the SRS is given as $\tau_{\overline{p}} = \frac{\pi}{4}\left(n_{\text{SRS}} + \frac{8\overline{p}}{N_p}\right) \bmod 8 |_{\overline{p} \in \{0, 1, \ldots, N_p - 1\}}$, where $n_{\text{SRS}} = \{0, 1, 2, 3, 4, 5, 6, 7\}$, is configured separately for periodic and each configuration of aperiodic sounding by the higher-layer signaling for each UE, $\overline{p}$ is sequential numbering of the antenna ports, and N_p is the number of antenna ports used for SRS transmission.

The SRS sequence is scaled with scaling factor β_{SRS} in order to conform to the transmit power P_{SRS} and is then mapped sequentially starting with $r_{\text{SRS}}(0, \overline{p})$ to resource elements (k, l) on antenna port $\overline{p}$ such that $a_{[2k' + k_0(\overline{p})]l} = \frac{1}{\sqrt{N_p}} \beta_{\text{SRS}} r_{\text{SRS}}(k', \overline{p})|_{k'=0,1,\ldots,N_{\text{SRS}}-1}$, where N_p is the number of antenna ports used for SRS transmission and the relation between the index $\overline{p}$ and the antenna port p is described in 3GPP TS 36.211 [3]. The set of antenna ports used for SRS transmission is configured independently for periodic and each configuration of aperiodic sounding. The quantity $k_0(p)$ is the frequency-domain starting position

Table 10.6 UE-Specific SRS Periodicity and Subframe Offset for SRS Trigger Types 0 and 1 [5]

SRS Trigger Type	SRS Configuration Index I_{SRS}	SRS Periodicity T_{SRS} (ms)	SRS Subframe Offset $T_{SRS\text{-}offset}$
		FDD	
0	0–1	2	I_{SRS}
	2–6	5	$I_{SRS}-2$
	7–16	10	$I_{SRS}-7$
	17–36	20	$I_{SRS}-17$
	37–76	40	$I_{SRS}-37$
	77–156	80	$I_{SRS}-77$
	157–316	160	$I_{SRS}-157$
	317–636	320	$I_{SRS}-317$
	637–1023	Reserved	Reserved
		TDD	
	0	2	0, 1
	1	2	0, 2
	2	2	1, 2
	3	2	0, 3
	4	2	1, 3
	5	2	0, 4
	6	2	1, 4
	7	2	2, 3
	8	2	2, 4
	9	2	3, 4
	10–14	5	$I_{SRS}-10$
	15–24	10	$I_{SRS}-15$
	25–44	20	$I_{SRS}-25$
	45–84	40	$I_{SRS}-45$
	85–164	80	$I_{SRS}-85$
	165–324	160	$I_{SRS}-165$
	325–644	320	$I_{SRS}-325$
	645–1023	Reserved	Reserved
1		FDD	
	0–1	2	I_{SRS}
	2–6	5	$I_{SRS}-2$
	7–16	10	$I_{SRS}-7$
	17–31	Reserved	Reserved
		TDD	
	0	2	0, 1
	1	2	0, 2
	2	2	1, 2

(Continued)

Table 10.6 (Continued)

SRS Trigger Type	SRS Configuration Index I_{SRS}	SRS Periodicity T_{SRS} (ms)	SRS Subframe Offset $T_{SRS\text{-}offset}$
		FDD	
	3	2	0, 3
	4	2	1, 3
	5	2	0, 4
	6	2	1, 4
	7	2	2, 3
	8	2	2, 4
	9	2	3, 4
	10–14	5	$I_{SRS} - 10$
	15–24	10	$I_{SRS} - 15$
	25–31	Reserved	Reserved

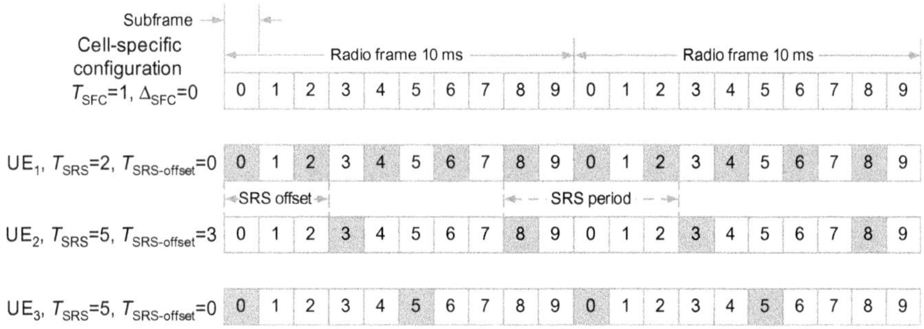

FIGURE 10.25

SRS parameter configuration example (frame structure type 1) [14].

of the SRS, and $N_{SRS}(B_{SRS}) = 6m_{SRS}(B_{SRS})$ is the bandwidth-dependent length of the SRS sequence (see Table 10.8 for the permissible values). The cell-specific SRS bandwidth configuration $C_{SRS} \in \{0, 1, 2, 3, 4, 5, 6, 7\}$ and the UE-specific SRS bandwidth $B_{SRS} \in \{0, 1, 2, 3\}$ are signaled to the UE via higher layers.

10.7 Uplink physical channel processing

The uplink multiple access scheme (SC-FDMA) in LTE employs separate physical channels for transmission of data and control signaling. The structure of the uplink physical channels is designed to make efficient use of the available time–frequency resources, and to effectively support multiplexing between data and control signaling. LTE Rel-8/9 utilizes few uplink multiple antenna techniques,

Table 10.7 Cell-Specific SRS Parameters for Frame Structure Type 1 and Type 2 [3]

SRS Subframe Configuration	Frame Structure Type 1		Frame Structure Type 2	
	Configuration Period T_{SFC}**(ms)**	**Transmission Offset** Δ_{SFC}**(ms)**	**Configuration Period** T_{SFC}**(Subframes)**	**Transmission Offset** Δ_{SFC}**(ms)**
0	1	{0}	5	{1}
1	2	{0}	5	{1,2}
2	2	{1}	5	{1,3}
3	5	{0}	5	{1,4}
4	5	{1}	5	{1,2,3}
5	5	{2}	5	{1,2,4}
6	5	{3}	5	{1,3,4}
7	5	{0,1}	5	{1,2,3,4}
8	5	{2,3}	10	{1,2,6}
9	10	{0}	10	{1,3,6}
10	10	{1}	10	{1,6,7}
11	10	{2}	10	{1,2,6,8}
12	10	{3}	10	{1,3,6,9}
13	10	{0,1,2,3,4,6,8}	10	{1,4,6,7}
14	10	{0,1,2,3,4,5,6,8}	Reserved	Reserved
15	Reserved	Reserved	Reserved	Reserved

Table 10.8 SRS Bandwidth Configuration Parameters [3]

SRS Bandwidth Configuration C_{SRS}	Uplink Bandwidth N_{RB}^{UL}	SRS-Bandwidth $B_{SRS}=0$		SRS-Bandwidth $B_{SRS}=1$		SRS-Bandwidth $B_{SRS}=2$		SRS-Bandwidth $B_{SRS}=3$	
		m_0	N_0	m_1	N_1	m_2	N_2	m_3	N_3
0	$6 \le N_{RB}^{UL} \le 40$	36	1	12	3	4	3	4	1
1		32	1	16	2	8	2	4	2
2		24	1	4	6	4	1	4	1
3		20	1	4	5	4	1	4	1
4		16	1	4	4	4	1	4	1
5		12	1	4	3	4	1	4	1
6		8	1	4	2	4	1	4	1
7		4	1	4	1	4	1	4	1
0	$40 < N_{RB}^{UL} \le 60$	48	1	24	2	12	2	4	3
1		48	1	16	3	8	2	4	2
2		40	1	20	2	4	5	4	1
3		36	1	12	3	4	3	4	1
4		32	1	16	2	8	2	4	2
5		24	1	4	6	4	1	4	1
6		20	1	4	5	4	1	4	1
7		16	1	4	4	4	1	4	1
0	$60 < N_{RB}^{UL} \le 80$	72	1	24	3	12	2	4	3
1		64	1	32	2	16	2	4	4
2		60	1	20	3	4	5	4	1
3		48	1	24	2	12	2	4	3
4		48	1	16	3	8	2	4	2
5		40	1	20	2	4	5	4	1
6		36	1	12	3	4	3	4	1
7		32	1	16	2	8	2	4	2
0	$80 < N_{RB}^{UL} \le 110$	96	1	48	2	24	2	4	6
1		96	1	32	3	16	2	4	4
2		80	1	40	2	20	2	4	5
3		72	1	24	3	12	2	4	3
4		64	1	32	2	16	2	4	4
5		60	1	20	3	4	5	4	1
6		48	1	24	2	12	2	4	3
7		48	1	16	3	8	2	4	2

including closed-loop antenna selection and a simple form of MU-MIMO. More advanced multi-antenna techniques including closed-loop spatial multiplexing for SU-MIMO and transmit diversity for control signaling were introduced in LTE Rel-10. This section describes the theoretical and implementation aspects of the procedures that are used for processing the uplink physical channels.

Figure 10.26 illustrates different phases of the physical channel processing. For uplink carrier aggregation, different component carriers are associated with separate transport channels with separate physical layer processing. Most of the uplink physical processing steps outlined in Figure 10.26 are similar to the corresponding steps in the downlink physical channel processing. The physical processing of the transport channels can be summarized as follows:

- CRC calculation and insertion per transport block: The CRC is calculated for and appended to each uplink transport block.
- Code block segmentation and code block CRC insertion: Similar to downlink, code block segmentation, including code block CRC insertion, is applied to transport blocks larger than 6144 bits.
- Channel coding: A rate 1/3 turbo code with quadratic permutation polynomial (QPP)-based interleaving is used for the uplink.
- Rate matching and HARQ functions: The physical layer part of the uplink rate-matching and HARQ functionality are essentially the same as downlink, with sub-block interleaving and insertion into a circular buffer, followed by bit selection with four redundancy versions (RVs). Note that there are some important differences between downlink and uplink HARQ protocols, such as asynchronous versus synchronous and adaptive versus non-adaptive HARQ schemes.
- Bit-level scrambling: The goal of the scrambling function in the uplink is to randomize the interference and to ensure that the processing gain provided by the channel coding can be fully utilized.
- Data modulation: QPSK, 16QAM, and 64QAM modulation schemes can be used for the uplink transport channel.
- DFT precoding: The modulation symbols form blocks of N_{DFT} symbols and are fed through a DFT module, where N_{DFT} corresponds to the number of subcarriers assigned to transmission. The DFT precoding is done to reduce the CM of the transmitted signal. From an implementation point of view, the DFT size should be a power of 2. However, such a constraint would limit the scheduler flexibility in terms of the amount of resources that could be assigned to an uplink transmission. On the other hand, all possible DFT sizes should ideally be allowed for more flexibility. The LTE/LTE-Advanced have chosen to limit the DFT size and the size of the resource allocation to products of the integers 2, 3, and 5.
- Resource element mapping: Mapping of precoded complex-valued symbols to resource elements.
- Antenna mapping and generation of a complex-valued time-domain SC-FDMA signal for each antenna port: In the first release of LTE, only

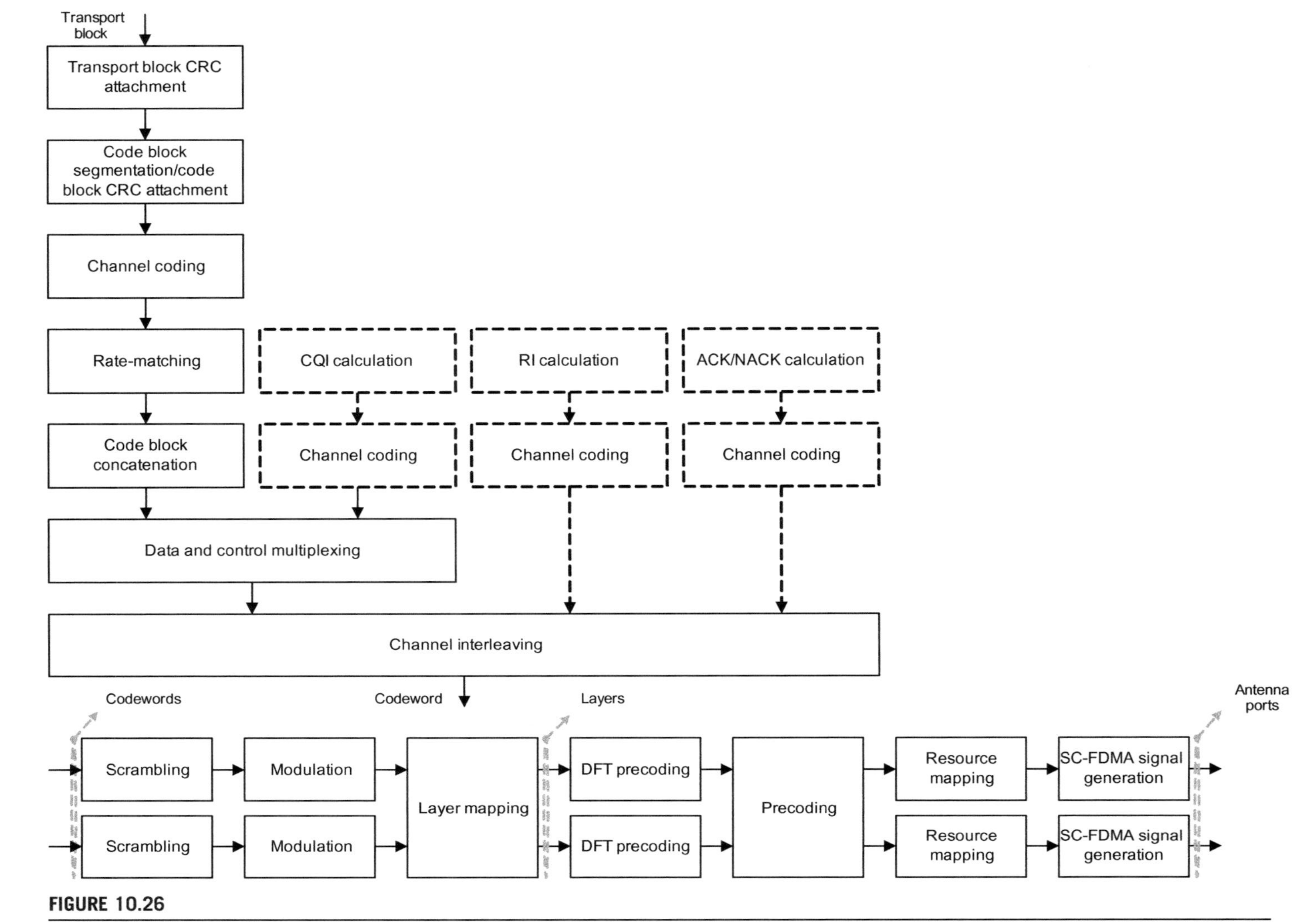

FIGURE 10.26

Uplink physical layer procedures for transport block processing [3].

single-antenna transmission was supported in the uplink. However, as part of LTE Rel-10 improvement of multi-antenna transmission, antenna precoding with up to four antennas ports is now supported in the uplink.

10.7.1 Physical uplink control channel

Uplink control signaling is typically used for the following purposes:

- Control signaling associated with user data, which is always transmitted together with uplink data and is used in the processing of that data. Examples include transport format indications, new data indicators (NDI), and MIMO parameters.
- Control signaling not associated with data is transmitted independent of any uplink data packets. Examples include HARQ ACK/NACK for downlink data packets, CQIs to support link adaptation, and MIMO feedback such as rank indicators and precoding matrix indicators for downlink transmissions, as well as scheduling requests (SRs) for uplink transmissions.

In LTE, the use of an orthogonal uplink multiple access scheme implies that the eNB controls resource allocation and determination of the uplink transmission parameters for the UEs. When simultaneous uplink PUSCH data and control signaling are scheduled, the control signaling is normally multiplexed with data prior to the DFT spreading, in order to preserve the single-carrier low CM property of the uplink transmission (see Figure 10.26). The PUCCH is used by a UE to transmit any necessary control signaling only in subframes in which the UE does not have any allocated resources for PUSCH transmission. In the design of PUCCH, the low PAPR and low CM property has been taken into consideration. The control signaling information on PUCCH is transmitted in a frequency region that is configured at the edges of the uplink system bandwidth. In order to minimize the resources needed for transmission of control signaling, PUCCH in LTE is designed to exploit frequency diversity, i.e., each PUCCH transmission in one subframe comprises a single resource block at or near one edge of the system bandwidth, followed by a second resource block in the second slot of the subframe at or near the opposite edge of the system bandwidth, where the two resource blocks are referred to as a PUCCH region. This design can achieve frequency diversity gain of approximately 2 dB compared to transmission in the same resource block throughout the subframe [16] (Figure 10.27). A snapshot of an uplink slot carrying a single PUCCH is shown in Figure 10.28 [153].

One important aspect of LTE transmitter design is the necessity to minimize unwanted emissions. Because LTE is deployed in the same frequency bands as UMTS and other legacy 3G cellular technologies, the 3GPP specifications [1] regulate the UE emissions to minimize interference and to ensure compatibility between different radio systems. The primary concern is the control of spurious emissions, which can occur at any frequency band. However, new challenges arise at the band edges, where the transmitted signal must comply with rigorous

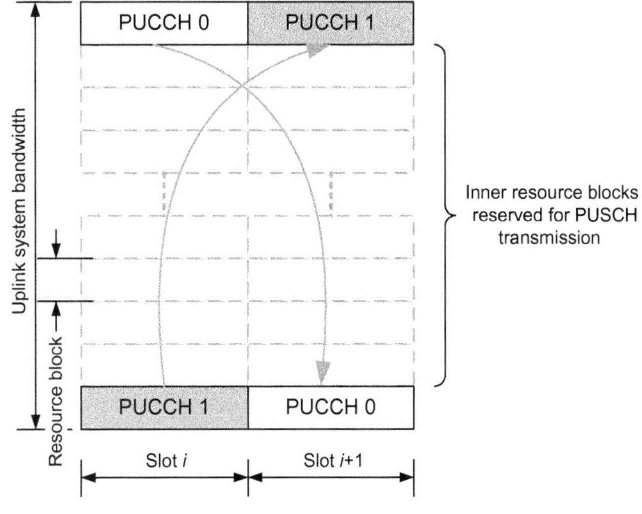

FIGURE 10.27

Example of PUCCH resource-block allocation [3,16].

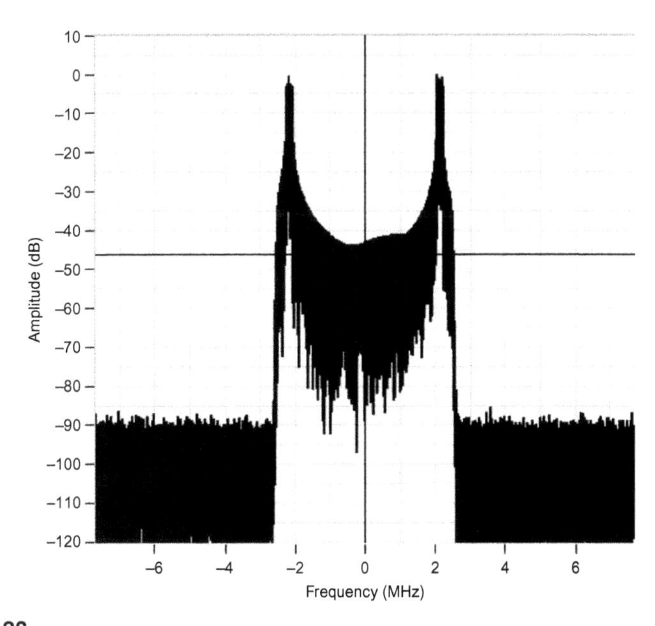

FIGURE 10.28

Snapshot of an uplink slot carrying a single PUCCH [153].

power leakage requirements. With LTE supporting channel bandwidths up to 20 MHz, and with many bands too narrow to support more than a few channels, a large proportion of the LTE channels will be at the edge of the band. Controlling transmitter performance at the edge of the band requires a design with filtering to attenuate OOB emissions without affecting in-channel performance. Factors such as cost, power efficiencies, physical size, and location in the transmitter block diagram are also important considerations. Ultimately the LTE transmitter must meet all specified limits for unwanted emissions, including limits on the amount of power that leaks into adjacent channels, as defined by the adjacent-channel leakage power ratio (ACLR).

The uplink control signaling must be transmitted regardless of whether the UE has data to send or whether any uplink resources are allocated for the terminal. As a result, there are two methods for transmission of the uplink control signaling as follows:

- Non-simultaneous transmission of PUCCH and PUSCH: When the terminal does not have any scheduling grant for PUSCH transmission in the current subframe, the uplink control signaling is transmitted on PUCCH. In uplink carrier aggregation scenarios, PUCCH is only transmitted on the primary component carrier. Consequently, data transmissions on multiple component carriers in the downlink can be acknowledged on a single uplink component carrier.
- Simultaneous transmission of PUCCH and PUSCH: When the terminal has a scheduling grant for PUSCH transmission in the current subframe, the uplink control signaling is time-multiplexed with the coded data on the PUSCH prior to DFT precoding and SC-FDMA signal generation (see Figure 10.26). Since the terminal has already been assigned radio resources, there is no need to support transmission of the SR in this case. Instead, scheduling information can be included in the MAC headers. In carrier aggregation scenarios, all control signaling is transmitted on the primary component carrier, i.e., uplink control signaling cannot be distributed across multiple component carriers.

One must differentiate between the above cases in consideration of efficient operation of the uplink PA and maximization of the uplink coverage. However, when there is sufficient power available in the terminal, simultaneous transmission of PUSCH and PUCCH can be used with minimal impact on the uplink coverage. The possibility of simultaneous PUSCH and PUCCH transmission was introduced in LTE Rel-10 which added more flexibility at the expense of slight increase in the CM value. Simultaneous transmission of PUSCH and PUCCH can always be avoided by using the basic mechanism introduced in LTE Rel-8. It should be noted that while PUCCH is only transmitted on the primary component carrier, as the primary component carrier is specified on a per-terminal basis, from a network perspective, PUCCH resources are used on multiple component carriers [16].

Since the size of a resource block in any subframe might be too large for control signaling payload transmitted by a single terminal, and in order to efficiently utilize the resources designated for control signaling transmission, multiple terminals can share the same resource block. This can be done by assigning different orthogonal phase rotations (cyclic shifts) of a cell-specific length-12 frequency-domain sequence to the UEs simultaneously using the resource block. Note that a linear phase rotation in the frequency-domain is equivalent to applying a cyclic shift in the time-domain. The resource used by PUCCH is not only specified in the time–frequency-domain by the resource-block pair, but also by the phase rotation applied. There are up to 12 different phase rotations specified, providing up to 12 different orthogonal sequences from each cell-specific sequence. However, in the frequency-selective channels, only some of the 12 phase rotations can be used in order to maintain orthogonality. Typically, up to six rotations are considered usable in a cell from a radio propagation perspective, although inter-cell interference may result in a smaller number of useful phase rotations from an overall system perspective. Higher-layer signaling is used to configure the number of phase rotations used in a cell [16]. There are three formats defined for PUCCH, formats 1, 2, and 3, each capable of carrying a different payload size.

In summary, placing the control regions at the edges of the system bandwidth has a number of advantages, including the following [15]:

- The frequency diversity achieved through frequency hopping is maximized by allowing hopping from one edge of the band to the other.
- OOB emissions[4] are smaller, if a UE only transmits on a single resource block per slot compared to multiple resource blocks. The PUCCH regions can serve as a virtual guard band between the wider-bandwidth PUSCH transmissions of adjacent carriers, and can therefore improve coexistence.
- Using control regions on the band edges maximizes the achievable PUSCH data rate, as the entire central portion of the band can be allocated to a single UE. If the control regions were in the central portion of a carrier, a UE bandwidth allocation would be limited to one side of the control region in order to maintain the single-carrier characteristic of the signal, thus limiting the maximum achievable data rate.
- Control regions on the band edges impose fewer constraints on the uplink data scheduling, both with and without inter-subframe and/or intra-subframe frequency hopping.

[4]Unwanted emissions consist of out-of-band emissions and spurious emissions. Spurious emissions are emissions caused by unwanted transmitter effects such as harmonics emissions, parasitic emissions, intermodulation products, and frequency conversion products. Out-of-band emissions are unwanted emissions outside the channel bandwidth resulting from the modulation process and non-linearity in the transmitter. If unwanted emissions are not sufficiently suppressed, they can considerably impair other intra-band radio services.

10.7.2 PUCCH formats

The physical uplink control channel carries uplink control information. The UCI can also be transmitted via PUSCH. The UCI typically consists of (1) SR; (2) HARQ ACK/NACK in response to downlink data packets received on PDSCH (one ACK/NACK bit transmitted in the case of single-codeword downlink transmission, while two ACK/NACK bits are used in the case of two codeword downlink transmission); and (3) CSI, which includes CQI as well as the MIMO-related feedback consisting of RI and PMI (20 bits per subframe are used for the CSI). The amount of UCI that a UE can transmit in a subframe depends on the number of SC-FDMA symbols available for transmission of control signaling excluding the SC-FDMA symbols used for reference signal transmission for coherent detection of PUCCH. For frame structure type 2, PUCCH is not transmitted in the UpPTS field. It must be noted that concurrent transmission of PUCCH and PUSCH in the same subframe by a UE is not allowed in LTE Rel-8/9. Support for simultaneous transmission of PUCCH and PUSCH in the same subframe from the same UE was added in LTE Rel-10, provided that it is configured by the higher layers. The PUCCH supports the multiple formats given in Table 10.9. Formats 2a and 2b are only supported for a normal cyclic prefix. Using the PUCCH formats, the following combinations of UCI on PUCCH are supported [5]:

- Format 1 is used for carrying positive SR.
- Format 1a is used for transmitting 1-bit HARQ ACK/NACK, or in case of FDD for sending 1-bit HARQ ACK/NACK with positive SR.
- Format 1b is used for carrying 2-bit HARQ ACK/NACK, or for transmitting 2-bit HARQ ACK/NACK with positive SR. PUCCH format 1b is further used for sending up to 4-bit HARQ ACK/NACK with channel selection, when the UE is configured with more than one serving cell, or in the case of TDD when the UE is configured with a single serving cell.
- Format 2 is used for sending a CSI report when it is not multiplexed with HARQ ACK/NACK. PUCCH format 2 is also used for transmission of a CSI report multiplexed with HARQ feedback for an extended cyclic prefix.
- Format 2a is used for carrying a CSI report multiplexed with 1-bit HARQ ACK/NACK for a normal cyclic prefix.
- Format 2b is used for sending a CSI report multiplexed with 2-bit HARQ ACK/NACK for a normal cyclic prefix.
- Format 3 is used for transmitting up to 10-bit HARQ ACK/NACK for FDD, and for up to 20-bit HARQ ACK/NACK for TDD. PUCCH format 3 is also used for sending up to 11-bit corresponding to 10-bit HARQ ACK/NACK and 1-bit positive/negative SR for FDD, and for up to 21-bit corresponding to 20-bit HARQ ACK/NACK and 1-bit positive/negative SR for TDD.

All PUCCH formats use a cyclic shift $N_{CS}(n_s, l)$, which used to be cell-specific until LTE Rel-11, which is a function of the SC-FDMA symbol number l and the slot number n_s according to $N_{CS}(n_s, l) = \sum_{i=0}^{7} c(8N_s^{UL} n_s + 8l + i)2^i$, where

Table 10.9 PUCCH Formats and Their Usage [3]

PUCCH Format	Modulation Scheme	Number of Bits per Subframe	Usage	Multiplexing Capacity[a] (UE/ Resource Block)
1	N/A	N/A	SR	36, 18[b], 12
1a	BPSK	1	ACK/NACK	36, 18[b], 12
1b	QPSK	2	ACK/NACK	36, 18[b], 12
2	QPSK	20	CQI (20 coded bits)	12, 6[b], 4
2a	QPSK + BPSK	21	CQI (20 coded bits) + ACK/NACK (1 bit)	12, 6[b], 4
2b	QPSK + QPSK	22	CQI (20 coded bits) + ACK/NACK (2 bits)	12, 6[b], 4
3	QPSK	48	Multiple ACK/NACK bits for carrier aggregation (up to 10 ACK/NACK bits in FDD systems with carrier aggregation or up to 20 bits ACK/NACK bits + 1 SR bit in TDD systems with carrier aggregation)	5

[a]*The combination of time-domain cyclic shifts and time-domain orthogonal codes determines the number of UEs which can be multiplexed within a single pair of PUCCH resource blocks. The total number of multiplexing possibilities is defined by the product of the number cyclic shifts and the number of the orthogonal codes.*
[b]*Typical value with six different rotations (choosing every other cyclic shift).*

the pseudo-random sequence $c(i)$ is initialized with $c_{\text{init}} = N_{\text{ID}}^{\text{RS}}$, where $N_{\text{ID}}^{\text{RS}}$ denotes the virtual cell identifier,[5] at the beginning of each radio frame. The definition of the virtual cell identifier $N_{\text{ID}}^{\text{RS}}$ depends on the type of transmission. For transmissions associated with PUSCH $N_{\text{ID}}^{\text{RS}} = N_{\text{ID}}^{\text{cell}}$, if no value for $N_{\text{ID}}^{\text{PUSCH}}$ is configured by higher layers if PUSCH transmission corresponds to a *Random Access Response* or a retransmission of the same transport block as part of the contention-based random access procedure; otherwise $N_{\text{ID}}^{\text{RS}} = N_{\text{ID}}^{\text{PUSCH}}$. For transmissions associated with PUCCH $N_{\text{ID}}^{\text{RS}} = N_{\text{ID}}^{\text{cell}}$, if no value for $N_{\text{ID}}^{\text{PUCCH}}$ is configured by higher layers; otherwise $N_{\text{ID}}^{\text{RS}} = N_{\text{ID}}^{\text{PUCCH}}$. The virtual cell identifier for SRS is defined as $N_{\text{ID}}^{\text{RS}} = N_{\text{ID}}^{\text{cell}}$ [3].

The physical resource allocation for PUCCH depends on two parameters, $N_{\text{RB}}^{(2)}$ and $N_{\text{CS}}^{(1)}$, that are signaled by the higher layers. The variable $N_{\text{RB}}^{(2)} \geq 0$ denotes the

[5]A simple approach is to use the PUCCH resource offset parameter to facilitate frequency-domain separation between inter-cell orthogonal and non-orthogonal resources. A virtual cell identifier may be used in the newly reserved PUCCH resource pool.

bandwidth in terms of the number of resource blocks that are available to be used for the PUCCH format 2/2a/2b transmission in each slot. The variable $N_{CS}^{(1)}$ denotes the number of cyclic shifts used for PUCCH format 1/1a/1b in a resource block which is used for a combination of PUCCH formats 1/1a/1b and 2/2a/2b. The value of $N_{CS}^{(1)}$ is an integer multiple of Δ_{shift}^{PUCCH} within the range of $\{0, 1, \ldots, 7\}$, where Δ_{shift}^{PUCCH} is signaled via the higher layers. There is no mixed resource block if $N_{CS}^{(1)} = 0$. There is only one resource block in each slot that can support a mix of PUCCH formats 1/1a/1b and 2/2a/2b. The resources used for transmission of PUCCH formats 1/1a/1b, 2/2a/2b, and 3 are represented by non-negative indices $n_{PUCCH}^{(1,\bar{p})}, n_{PUCCH}^{(2,\bar{p})} < 12N_{RB}^{(2)} + \left\lceil \frac{N_{CS}^{(1)}}{8} \right\rceil (10 - N_{CS}^{(1)})$ and $n_{PUCCH}^{(3,\bar{p})}$, respectively.

The mapping of various PUCCH formats to PUCCH regions is shown in Figure 10.29. The mapping of PUCCH to PRBs follows the steps below:

1. Start from resource blocks in the outer edge of the band toward middle of the band.
2. Allocate PUCCH format 2/2a/2b resources.
3. Allocate PUCCH formats containing mixed ACK/NACK and CQI.
4. Allocate PUCCH format 1/1a/1b resources.

It can be seen in the figure that PUCCH format 2/2a/2b is mapped and transmitted on the outermost band-edge resource blocks followed by a mixed PUCCH resource block of CSI format 2/2a/2b and SR/HARQ ACK/NACK format 1/1a/1b, and then by a PUCCH SR/HARQ ACK/NACK format 1/1a/1b allocation toward the center of the band. The number of PUCCH resource blocks available for use by CSI format 2/2a/2b is signaled to the UEs in the cell through broadcast signaling.

10.7.2.1 PUCCH formats 1, 1a, 1b

The structure of PUCCH format 1 is illustrated in Figure 10.31. PUCCH format 1 is used for transmission of HARQ ACK/NACK (i.e., a single-bit HARQ feedback corresponding to one downlink component carrier is used to generate one BPSK-modulated symbol, and in the case of downlink spatial multiplexing, 2 bits of HARQ feedback are used to generate one QPSK-modulated symbol) and for sending a positive SR, where the same constellation point as for a HARQ NACK is used. The BPSK/QPSK-modulated symbol is then used to generate the signal for transmission in each of the two PUCCH slots. In each of the seven SC-FDMA symbols of a slot assuming a normal cyclic prefix, a length-12 sequence obtained via phase rotation of the cell-specific base sequence is transmitted. Three SC-FDMA symbols are used for reference signals to enable channel estimation and coherent detection by the eNB, and the remaining four SC-FDMA symbols are modulated to generate the BPSK/QPSK-modulated symbols (as shown in Figure 10.31). In principle, the BPSK/QPSK-modulated symbol could directly

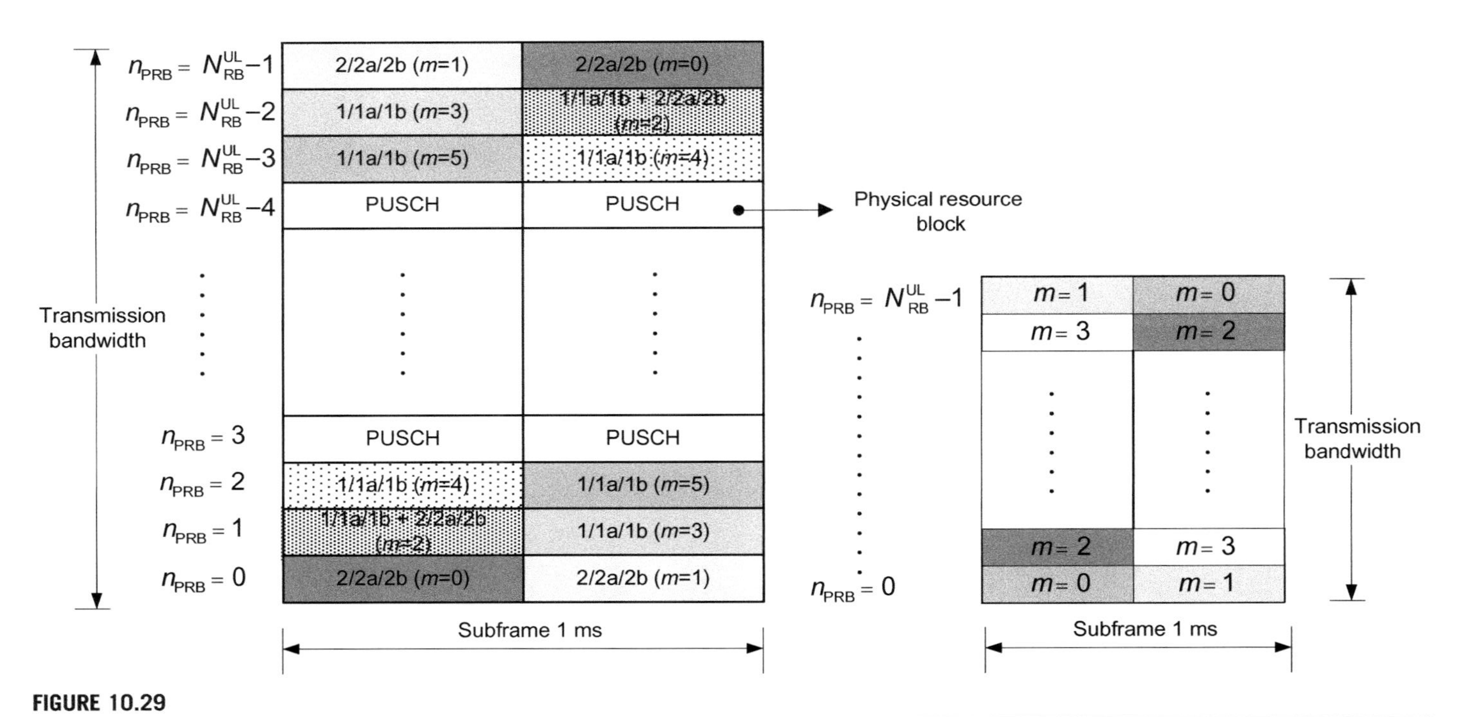

FIGURE 10.29

Example mappings of PUCCH to PRBs in a subframe [3].

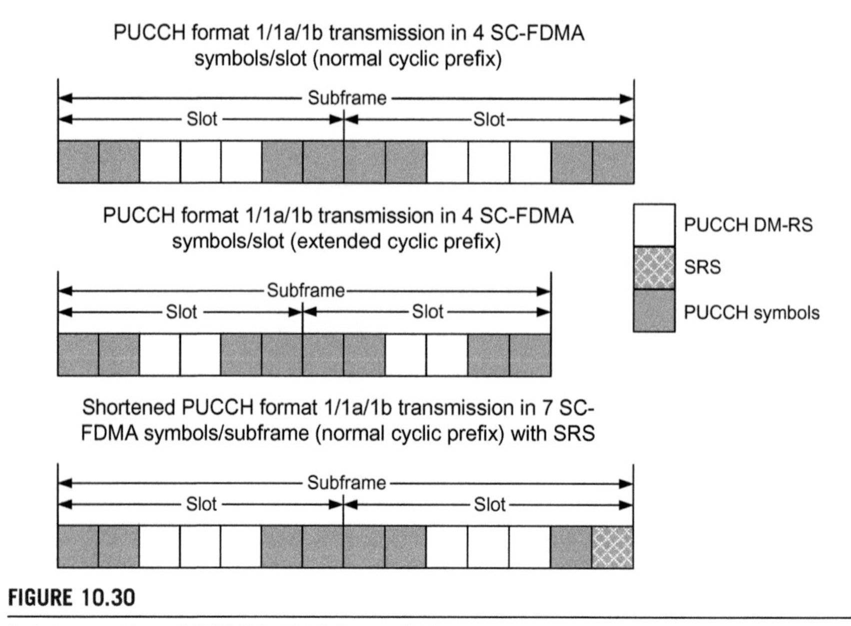

FIGURE 10.30

Relative position of PUCCH, DM-RS, and SRS symbols.

modulate the rotated length-12 sequence used to differentiate terminals transmitting on the same time–frequency resource. However, this would result in unnecessarily lower PUCCH capacity. Therefore, the BPSK/QPSK-modulated symbol is multiplied by a length-4 orthogonal cover code. Multiple terminals may transmit on the same time–frequency resource using the same phase-rotated sequence and can still be separated through different OCCs. In order to estimate the channels for the respective terminals, the reference signals are also scrambled by appropriate orthogonal cover sequences with different length, i.e., three for the case of a normal cyclic prefix and two for an extended cyclic prefix. Each cell-specific sequence can be used for up to 36 different terminals (assuming all 12 phase rotations or cyclic shifts are available, whereas typically 6 of the cyclic shifts are utilized), there is a threefold improvement in the PUCCH capacity compared to the case when no cover sequence is used. The cover sequences are three Walsh sequences of length 4 for the data part and three DFT sequences of length 3 for the reference signals [3,14,16].

A PUCCH format 1 resource, used for either a HARQ feedback or an SR, is represented by a single scalar resource index. From the index, the phase rotation and the orthogonal cover sequence are derived. The use of a phase rotation of a cell-specific sequence, together with orthogonal sequences as described earlier, provides orthogonality between different terminals in the same cell when transmitting PUCCH on the same set of resource blocks. Therefore, there is ideally no intra-cell interference, which helps improve the overall performance. However,

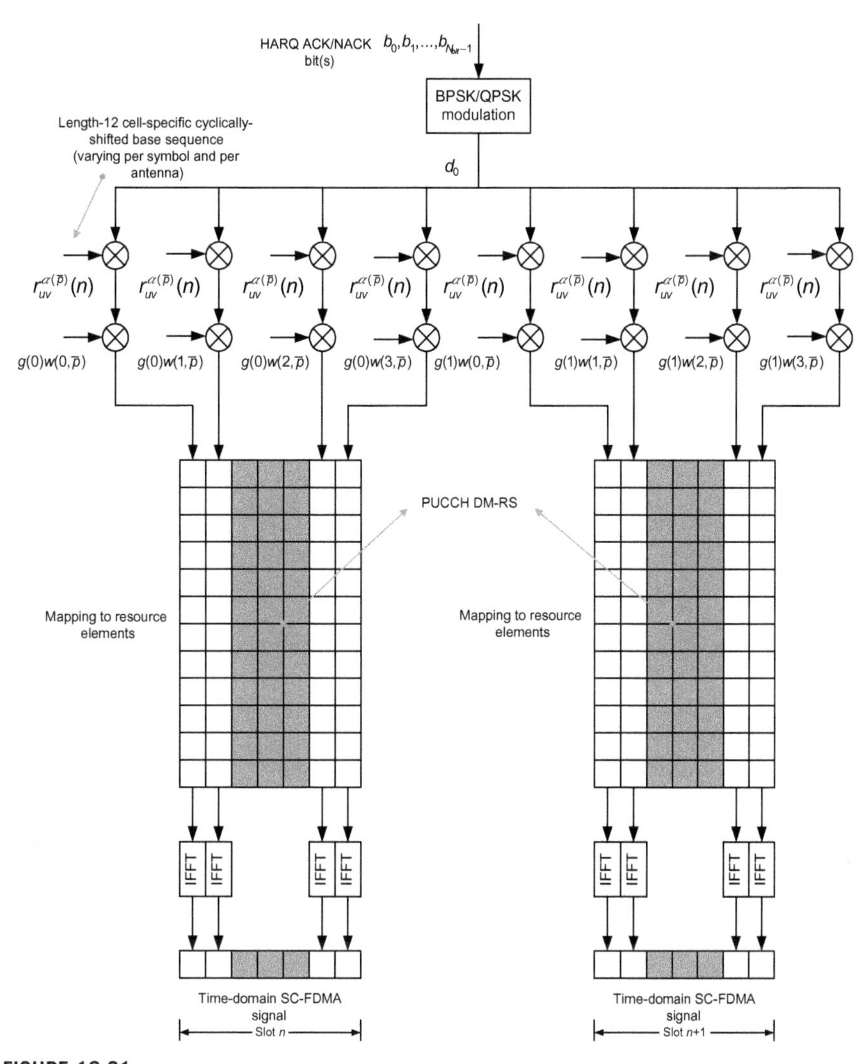

FIGURE 10.31

Physical layer processing for PUCCH format 1/1a/1b (HARQ ACK/NACK and normal cyclic prefix).

PUCCH typically suffers from inter-cell interference since different sequences used in neighboring cells are non-orthogonal. In order to randomize the inter-cell interference, the phase rotation of the sequence used in a cell is changed on a symbol-by-symbol basis in a slot according to a hopping pattern derived from the physical layer cell identity. In addition, slot-level hopping is applied to the orthogonal cover and phase rotation to further randomize the interference.

The UE scheduling request is carried by PUCCH format 1 and is signaled by the presence or absence of PUCCH format 1 transmission. For PUCCH formats 1a and 1b, 1 or 2 bits are transmitted depending on the HARQ feedback. The block of bits $b_0, b_1, \ldots, b_{N_{\text{bit}}-1}$ are modulated using BPSK (format 1a) or QPSK (format 1b) modulation, resulting in a complex-valued symbol d_0. Note that the number of bits is $N_{\text{bit}} = 1$ for format 1a and $N_{\text{bit}} = 2$ for format 1b carrying HARQ ACK/NACK. The complex-valued symbol d_0 is multiplied by a cyclically shifted sequence of length $N_{\text{s}}^{\text{PUCCH}} = 12$, $r_{\text{uv}}^{\alpha(\bar{p})}(n)$ for each of the N_p antenna ports used for PUCCH transmission according to the following equation [3]:

$$y^{\bar{p}}(n) = \frac{1}{\sqrt{N_p}} d_0 r_{\text{uv}}^{\alpha(\bar{p})}(n)\big|_{n=0,1,\ldots,N_{\text{s}}^{\text{PUCCH}}-1} \tag{10.5}$$

where $r_{\text{uv}}^{\alpha(\bar{p})}(n)$ is the cyclically shifted Zadoff–Chu base sequence. The antenna-port-specific cyclic-shift $\alpha(\bar{p})$ varies across symbols and slots. The block of complex-valued symbols $y^{\bar{p}}(0), y^{\bar{p}}(1), \ldots, y^{\bar{p}}(N_{\text{s}}^{\text{PUCCH}} - 1)$ are scrambled by scrambling sequence $g(n_{\text{s}})$ and are block-wise spread with the antenna-port-specific orthogonal sequence $w(i, \bar{p})$ according to the following equation [3]:

$$z^{\bar{p}}(m' N_{\text{SF}}^{\text{PUCCH}} N_{\text{s}}^{\text{PUCCH}} + m N_{\text{s}}^{\text{PUCCH}} + n) = g(n_{\text{s}}) w(m, \bar{p}) y^{\bar{p}}(n) \tag{10.6}$$

where $m = 0, 1, \ldots, N_{\text{SF}}^{\text{PUCCH}} - 1; n = 0, 1, \ldots, N_{\text{s}}^{\text{PUCCH}} - 1; m' = 0, 1$ and $g(n_{\text{s}}) = 1$ if $n'_{\bar{p}}(n_{\text{s}}) \bmod 2 = 0$; otherwise $g(n_{\text{s}}) = e^{j\pi/2}$, $N_{\text{SF}}^{\text{PUCCH}} = 4$ for both slots of normal PUCCH format 1/1a/1b and $N_{\text{SF}}^{\text{PUCCH}} = 4$ for the first slot and $N_{\text{SF}}^{\text{PUCCH}} = 3$ for the second slot of shortened PUCCH format 1/1a/1b (see Figure 10.30).

In addition to dynamic scheduling using PDCCH, a UE can be semi-persistently scheduled according to a specific pattern. In that case, there will be no PDCCH to assign the PUCCH resources in the uplink. The configuration of the semi-persistent scheduling includes information on the PUCCH index to be used for uplink HARQ feedback. Note that a UE can use PUCCH resources only when it has been scheduled in the downlink. The amount of PUCCH resources required for HARQ feedback does not necessarily increase if the number of terminals in the cell increases; rather it is a function of the number of control channel elements (CCEs) in the downlink control signaling.

For downlink carrier aggregation, when there are multiple simultaneous downlink allocations for a single terminal, multiple feedback bits are transmitted in the uplink (i.e., 1 or 2 bits in the case of spatial multiplexing per downlink component carrier). PUCCH format 1b can be used to support more than 2 bits in the uplink by using resource selection. Let us assume that 4 bits are going to be transmitted in the uplink. With resource selection, 2 bits would indicate which PUCCH resource should be used and the remaining 2 bits are transmitted using the normal PUCCH structure, but on the resource addressed by the first 2 bits. One resource is derived from the first CCE using the same rule used in the absence of carrier aggregation, assuming that the scheduling assignment is transmitted on the primary cell. The remaining resources are semi-statically configured by RRC signaling. For more than 4 bits, resource selection is less efficient and PUCCH format 3, which was introduced in LTE Rel-10 for carrier aggregation, is utilized.

Unlike HARQ feedback, whose occurrence is expected and predicable by the eNB, a specific terminal's need for uplink resources is unpredictable, which is why a contention-based mechanism has been provisioned in LTE to allow uplink resource scheduling request. The random-access mechanism is used for SRs. A contention-based mechanism may work well for a small number of UEs; however, by increasing the number of UEs simultaneously requesting resources, the probability of collision increases. Therefore, LTE further provides a contention-free scheduling-request mechanism on PUCCH, where each terminal in the cell is assigned a reserved resource on which it can transmit a request for uplink resources. The contention-free resource is represented by a PUCCH resource index every *n*th subframe. In other words, a request for uplink resources can be made by means of the random-access channel; however, due to the probability of collisions during congestion periods, an alternative method is provided using PUCCH format 1. Each UE in the cell is assigned a specific resource index mapping, a resource which can be used every *n*th subframe in order to transmit a scheduling request. Therefore, if PUCCH energy is detected, the eNB will consider it as an SR from the corresponding UE. Since each UE will have a specific resource allocated, there is no probability of collision. However, as a drawback, the number of available PUCCH resources is reduced.

A single SR is deemed sufficient for the case of carrier aggregation. The discussion earlier has considered transmission of either a HARQ feedback or an SR. However, there are cases where the UE needs to concurrently transmit HARQ feedback and an SR. In such cases, the HARQ ACK/NACK is transmitted on the scheduling-request resource. As we discussed earlier, this is possible because the scheduling request carries no explicit information. By comparing the amount of energy detected on the ACK resource and the scheduling-request resource for a specific terminal, the eNB can determine whether the terminal is requesting resources, as well.

10.7.2.2 PUCCH formats 2, 2a, 2b

The CSI reports are used to provide the eNB with an estimate of the downlink communication channel observed by the UE in order to assist channel-dependent scheduling (Figure 10.32). The CSI report consists of multiple bits transmitted in one subframe and is transmitted using PUCCH format 2. Channel status reports include CQI, RI, and PMI. The HARQ ACK/NACK can also be transmitted with channel status information. There are two forms of channel coding for this purpose, one for the channel quality information (i.e., CQI) only and another one for the combination of CQI and HARQ feedback. The channel quality indicator represents the recommended modulation scheme and coding rate that should be used for the downlink transmission. The RI provides information about the rank of the channel, which is used to determine the optimal number of layers that should be used for the downlink transmission which is only used for spatial multiplexing modes. The PMI provides information about which precoding matrix to use, which is only used in closed-loop spatial multiplexing modes. PUCCH format 2, as shown in Figure 10.32, is generated by phase rotation of a cell-specific base sequence (same as PUCCH format 1), followed by a block coding function using a punctured Reed–Muller code[6] and QPSK modulation of the CSI. There are ten QPSK symbols to be transmitted in a subframe where the first five complex-valued modulation symbols are transmitted in the first slot and the remaining in the next slot. In PUCCH format 2/2a/2b, two SC-FDMA symbols are used as reference signals to allow coherent detection and demodulation at the eNB. In the remaining modulation symbols, each QPSK-modulated symbol is multiplied by a phase-rotated length-12 cell-specific base sequence, inverse Fourier-transformed, and the resulting signal is transmitted over the corresponding SC-FDMA symbol. For an extended cyclic prefix, there are six SC-FDMA symbols per slot; therefore, only one reference symbol per slot is allocated. The transmission of PUCCH format 2 relies on phase rotation of a cell-specific sequence similar to that for PUCCH format 1, which allows the two formats to be transmitted in the same resource block. Note that orthogonal cover sequences are not used for PUCCH format 2. The rotation angles for different symbols carrying PUCCH format 2 are the hopping across symbol and slot boundaries for interference randomization [16].

Resources for transmission of PUCCH format 2 can be represented by an index, which can be interpreted as a channel number that is configurable via higher-layer signaling for each terminal to transmit its CSI report. The CSI reports on PUCCH are periodic. There are also aperiodic reports that are only transmitted on PUSCH. In case of carrier aggregation, multiple periodic reports, one per component carrier, are transmitted that are offset in time such that reports for two different component carriers do not collide (Table 10.10).

---

[6]Reed–Muller codes are a family of linearerror-correcting codes used in communication systems. The special cases of Reed–Muller codes include Hadamard codes, Walsh–Hadamard codes, and Reed–Solomon codes. Reed–Muller codes are denoted by an $RM(d,r)$ notation, where d is the order of the code and r determines the length of code $n=2^r$. Reed–Muller codes are related to binary functions on field $GF(2^r)$ over the elements $\{0,1\}$. It can be shown that $RM(0,r)$ codes are repetition codes of length, rate $R=1/n$ and minimum distance $d_{\min}=n$, $RM(1,r)$ codes are parity check codes of length $n=2^r$, rate $R=(r+1)/n$ and minimum distance $d_{\min}=n/2$. Reed–Muller codes have the following properties [ref]:

1. The set of all possible exterior products of up to d of v_i form a basis for $\mathbf{F}_2^n$.
2. The rank of $RM(d,r)$ code is defined as $\sum_{s=0}^{r}\binom{d}{s}$
3. $RM(d,r)=RM(d,r-1)|RM(d-1,r-1)$, where '|'. denotes the bar product of two codes.
4. $RM(d,r)$ has minimum Hamming weight2^{d-r}.

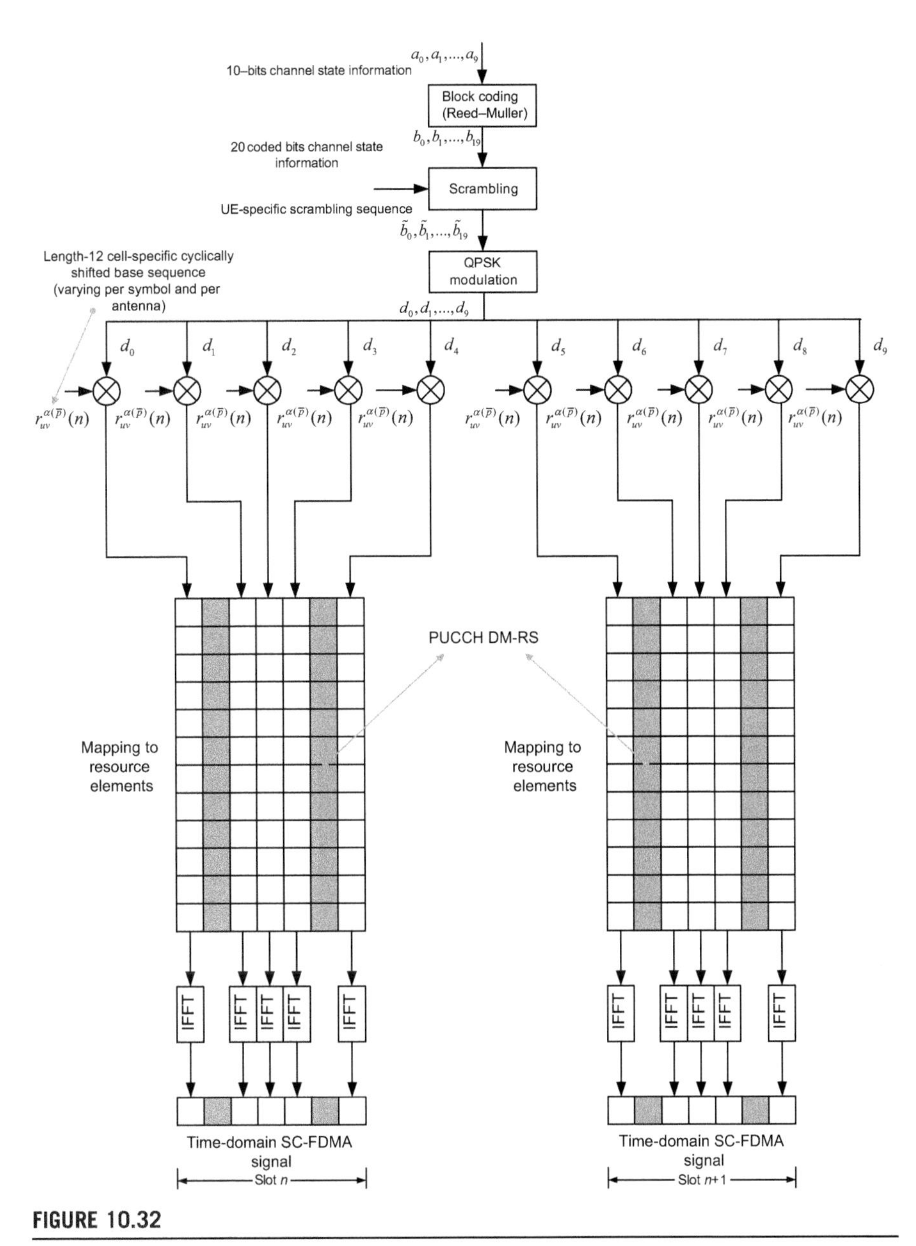

FIGURE 10.32

Physical layer processing for PUCCH format 2/2a/2b (normal cyclic prefix, CQI only) [16].

The structure of PUCCH format 2 in one subframe with normal cyclic prefix is shown in Figure 10.32. The first and the fifth SC-FDMA symbols are used for PUCCH DM-RS transmission with a normal cyclic prefix, whereas in the case of an extended cyclic prefix only one reference symbol is transmitted on the third

Table 10.10 Size of the PUCCH Region [15]

System Bandwidth (MHz)	Number of Resource Blocks per Subframe	Number of PUCCH Regions
1.4	2	1
3	4	2
5	8	4
10	16	8
20	32	16

SC-FDMA symbol. The number of reference symbols per slot results from a trade-off between channel estimation accuracy and the sustainable code rate for the UCI bits. For a small number of UCI bits at a low SINR operating point (for a typical 1% target error rate), improving the channel estimation accuracy by using more reference symbols is more desirable than using a lower channel code rate. However, with larger numbers of UCI bits, the required SINR increases and the higher code rate resulting from a larger overhead of reference symbols becomes more critical, thus fewer reference symbols are utilized. In consideration of the above facts, two reference symbols per slot were found to provide the best trade-off in terms of performance and overhead, given the payload sizes required. Therefore, ten CSI bits are channel coded with a rate 1/2 punctured $(20, k)$ Reed–Muller code to generate 20 coded bits, which are then scrambled, similar to PUSCH data, with a length-31 Gold sequence[7] prior to QPSK modulation. One QPSK-modulated symbol (according to Table 10.11) is transmitted on each of the ten SC-FDMA symbols in the subframe by multiplying with a cyclically shifted base sequence of length-12 prior to OFDM modulation. The 12 cyclic shifts allow 12 different UEs to be orthogonally multiplexed on the same PUCCH format 2 resource blocks. The processing of a PUCCH DM-RS reference sequence, i.e., transmitted on the second and sixth SC-FDMA symbols for the normal cyclic prefix and the fourth symbol for the extended cyclic prefix, is similar to the frequency-domain CSI signal sequence without the CSI data modulation. In order to provide inter-cell interference randomization, cell-specific

[7]A Gold sequence is a type of binary sequence that is often used in telecommunication and satellite navigation signal processing. Gold sequences have bounded small cross-correlations within a set. A set of Gold sequences consists of $2^n - 1$ sequences each with a period of $2^n - 1$. A set of Gold sequences can be generated by taking two maximum-length sequences of the same length $2^n - 1$ such that their absolute cross-correlation is less than or equal to $2^{(n+2)/2}$, where n is the size of the linear feedback shift register used to generate the maximum length sequence [165]. The set of $2^n - 1$ logical XOR or of the two sequences in their various phases is a set of Gold codes. The highest absolute cross-correlation in this set of codes is $2^{(n+2)/2} + 1$ for even n and $2^{(n+1)/2} + 1$ for odd n. The logical XOR of two Gold codes from the same set is another Gold code.

Table 10.11 Modulation Symbols for PUCCH Format 1a/1b and PUCCH Format 2a/2b [3]

PUCCH Format	$b_0, b_1, \ldots, b_{N_{bit}-1}$	d_0
1a	0	1
	1	−1
1b	00	1
	01	−j
	10	j
	11	−1
PUCCH Format	$b_{20}, b_{21}, \ldots, b_{N_{bit}-1}$	d_{10}
2a	0	1
	1	−1
2b	00	1
	01	−j
	10	j
	11	−1

symbol-level cyclic-shift hopping is used. Intra-cell interference randomization is achieved by cyclic-shift re-mapping in the second slot [15].

In mathematical terms, a block of coded bits $b_0, b_1, \ldots, b_{19}$ is scrambled with a UE-specific scrambling sequence, resulting in a block of scrambled coded bits $\tilde{b}_0, \tilde{b}_1, \ldots, \tilde{b}_{19}$ according to $\tilde{b}(i) = [b(i) + c(i)]\text{mod } 2$, where the scrambling sequence $c(i)$ was defined in Chapter 9 and is initialized with $c_{\text{init}} = \left(\lfloor n_s/2 \rfloor + 1\right)\left(2N_{\text{ID}}^{\text{cell}} + 1\right)2^{16} + n_{\text{RNTI}}$ at the start of each subframe, where n_{RNTI} denotes the 16-bit temporary UE identity in the connected mode.[8] The block of scrambled bits $\tilde{b}_0, \tilde{b}_1, \ldots, \tilde{b}_{19}$ are QPSK modulated, resulting in a block of complex-valued modulation symbols $d_0, d_1, \ldots, d_9$. Each complex-valued symbol $d_0, d_1, \ldots, d_9$ is multiplied by a cyclically shifted length $N_{\text{s}}^{\text{PUCCH}} = 12$ sequence $r_{\text{uv}}^{\alpha(\tilde{p})}(n)$ for each of the N_p antenna ports used for PUCCH transmission according to the following equation [3]:

$$z^{(\tilde{p})}(N_{\text{s}}^{\text{PUCCH}} n + i) = \frac{1}{\sqrt{N_p}} d(n) r_{\text{uv}}^{\alpha(\tilde{p})}(i)\Big|_{n=0,1,\ldots,9;\, i=0,1,\ldots,11} \tag{10.7}$$

[8] Pseudo-random sequences are defined by a length-31 Gold sequence. The output sequence $c(n), \forall n = 0, 1, \ldots, M-1$ of length M is defined by $c(n) = (x_1(n + N_C) + x_2(n + N_C))\text{mod } 2$ in which:
$x_1(n + 31) = (x_1(n + 3) + x_1(n))\text{mod } 2$;
$x_2(n + 31) = (x_2(n + 3) + x_2(n + 2) + x_2(n + 1) + x_2(n))\text{mod } 2$, and $N_C = 1600$. The first m-sequence is initialized with $x_1(0) = 1, x_1(n) = 0, n = 1, 2, \ldots, 30$. The second m-sequence is initialized by $c_{\text{init}} = \sum_{i=0}^{30} x_2(i)2^i$ with the value depending on the application of the sequence.

where $r_{uv}^{\alpha(\overline{p})}(n)$ is an antenna-port-dependent cyclically shifted Zadoff–Chu sequence of length 12. The resources used for transmission of PUCCH format 2/2a/2b are identified by resource index $n_{PUCCH}^{(2,\overline{p})}$ in which the cyclic shift $\alpha_{\overline{p}}(n_s, l)$ is determined according to $\alpha_{\overline{p}}(n_s, l) = \frac{\pi n_{cs}^{\overline{p}}(n_s, l)}{6}$, where $n_{cs}^{\overline{p}}(n_s, l) = [n_{cs}^{cell}(n_s, l) + n'_{\overline{p}}(n_s)] \bmod 12$ [3]. A UE is semi-statically configured via higher-layer signaling to periodically report different CQI, PMI, and RI types in PUCCH format 2 using a resource index $n_{PUCCH}^{(2)}$, which indicates both the PUCCH region and the cyclic shift to be used. The allocation index for PUCCH format 2/2a/2b transmission regions is given by $m = \left\lfloor n_{PUCCH}^{(2)}/12 \right\rfloor$ and the assigned cyclic shift is given by $n_{RS}^{PUCCH} = n_{PUCCH}^{(2)} \bmod 12$. The simultaneous transmission of uplink HARQ feedback and CSI on PUCCH can be enabled by UE-specific higher-layer signaling. However, if simultaneous transmission is not enabled or supported and the UE needs to transmit HARQ feedback on PUCCH in the same subframe in which a CSI report has been configured, the CSI is dropped and only HARQ feedback is transmitted. In subframes where the eNB scheduler allows for simultaneous transmission of CSI and HARQ feedback on PUCCH from a UE, the CSI and the 1- or 2-bit HARQ feedback are multiplexed in the same PUCCH resource block without compromising the low cubic-metric property of single-carrier signal.

The time and frequency resources that can be used by the UE to report CSI which consists of CQI, PMI, precoding type indicator (PTI),[9] and/or RI are controlled by the eNB. For spatial multiplexing, the UE determines an RI corresponding to the number of useful transmission layers. A UE is semi-statically configured via higher-layer signaling to periodically send feedback on different CSI metrics (CQI, PMI, PTI, and/or RI) on the PUCCH using the reporting modes given in Table 10.12 [5].

Two periodic CSI reporting modes are supported in LTE/LTE-Advanced as follows:

1. *Wideband reports*, where the average channel quality across the entire cell bandwidth is calculated and represented with a single CQI value. In this case, a single PMI value for the entire bandwidth is further reported, if PMI reporting is enabled.

[9]PTI was introduced in LTE Rel-10 and allows a UE to dynamically switch between different types of PMI reports to the serving eNB. It is applicable to transmission mode 9 which uses a two-stage approach to PMI reporting. The first stage corresponds to reporting wideband and long-term channel characteristics (PMI_1), and the second stage concerns frequency-selective short-term channel properties (PMI_2). A PTI value of "0" indicates that the UE has detected PMI_1 is changing, thus it needs to send an update to the eNB. A PTI value of "1," on the other hand, indicates that PMI_1 is not changing and the report should be on PMI_2, which means that a number of sub-band PMI_2 values as opposed to a single wideband value should be sent to the eNB [14].

Table 10.12 CQI and PMI Feedback Types and Corresponding Transmission Modes for PUCCH CSI Reporting [5]

		PMI Feedback Type	
		No PMI	Single PMI
PUCCH CQI Feedback Type	Wideband (Wideband CQI)	Mode 1-0 TM1 TM2 TM3 TM7 TM8[a] TM9[b] TM10[c]	Mode 1-1 TM4 TM5 TM6 TM8[a] TM9[b] TM10[c]
	UE selected (sub-band CQI)	Mode 2-0 TM1 TM2 TM3 TM7 TM8[a] TM9[b] TM10[c]	Mode 2-1 TM4 TM5 TM6 TM8[a] TM9[b] TM10[c]

[a]*TM 8: Mode 1-1 or 2-1, if the UE is configured with PMI/RI reporting; otherwise, modes 1-0 or 2-0, if the UE is configured without PMI/RI reporting.*
[b]*TM 9: Mode 1-1 or 2-1, if the UE is configured with PMI/RI reporting and the number of CSI-RS ports is greater than one; otherwise, mode 1-0 or 2-0, if the UE is configured without PMI/RI reporting or the number of CSI-RS ports is equal to one.*
[c]*TM10: Mode 1-1 or 2-1, if the UE is configured with PMI/RI reporting and the number of CSI-RS ports is greater than one; otherwise modes 1-0 or 2-0, if the UE is configured without PMI/RI reporting or the number of CSI-RS ports is equal to one.*

2. *UE-selected reports*, where the total bandwidth of a cell is divided into one to four bandwidth parts, with the number of bandwidth parts derived from the cell bandwidth as given in Table 10.13. For each bandwidth part, the terminal selects the best sub-band within that part. The sub-band size ranges from four to eight resource blocks. Since the supported PUCCH payload size is limited, the reporting cycles through the bandwidth parts and in one subframe reports the wideband CQI and PMI (if enabled) for that bandwidth part, as well as the best sub-band and the CQI for that sub-band. The RI (if enabled) is reported in a separate subframe.

The periodic CSI reporting mode for each serving cell is configured via higher-layer signaling. For wideband periodic CQI reporting, the reporting period can be configured as $\{2, 5, 10, 16, 20, 32, 40, 64, 80, 128, 160\}$ subframes, or it can

Table 10.13 PUCCH Reporting Type Payload Size per PUCCH Reporting Mode and Mode State [5]

PUCCH Reporting Type	Report Contents	Mode State	PUCCH Reporting Modes			
			Mode 1-1 (Bits/ Bandwidth Part)	Mode 2-1 (Bits/ Bandwidth Part)	Mode 1-0 (Bits/ Bandwidth Part)	Mode 2-0 (Bits/ Bandwidth Part)
1	Sub-band CQI	RI = 1	N/A	4 + L	N/A	4 + L
		RI > 1	N/A	7 + L	N/A	4 + L
1a	Sub-band CQI/ PMI_2	$N_p = 8$ RI = 1	N/A	8 + L	N/A	N/A
		$N_p = 8$ 1 < RI < 5	N/A	9 + L	N/A	N/A
		$N_p = 8$ RI > 4	N/A	7 + L	N/A	N/A
2	Wideband CQI/ PMI	$N_p = 2$ RI = 1	6	6	N/A	N/A
		$N_p = 4$ RI = 1	8	8	N/A	N/A
		$N_p = 2$ RI > 1	8	8	N/A	N/A
		$N_p = 4$ RI > 1	11	11	N/A	N/A
2a	Wideband PMI_1	$N_p = 8$ RI < 3	N/A	4	N/A	N/A
		$N_p = 8$ 2 < RI < 8	N/A	2	N/A	N/A
		$N_p = 8$ RI = 8	N/A	0	N/A	N/A
2b	Wideband CQI/ PMI_2	$N_p = 8$ RI = 1	8	8	N/A	N/A
		$N_p = 8$ 1 < RI < 4	11	11	N/A	N/A

		$N_p = 8$ RI = 4	10	10	N/A	N/A
		$N_p = 8$ RI > 4	7	7	N/A	N/A
2c	Wideband CQI/ PMI_1/PMI_2	$N_p = 8$ $RI = 1$	8	N/A	N/A	N/A
		$N_p = 8$ $1 < \mathrm{RI} \leq 4$	11	N/A	N/A	N/A
		$N_p = 8$ $4 < \mathrm{RI} \leq 7$	9	N/A	N/A	N/A
		$N_p = 8$ RI = 8	7	N/A	N/A	N/A
3	RI	$N_p = 2, 4$ two-layer spatial multiplexing	1	1	1	1
		$N_p = 8$ two-layer spatial multiplexing	1	N/A	N/A	N/A
		$N_p = 4$ four-layer spatial multiplexing	2	2	2	2
		$N_p = 8$ four-layer spatial multiplexing	2	N/A	N/A	N/A
		Eight-layer spatial multiplexing	3	N/A	N/A	N/A
4	Wideband CQI	$RI = 1, \mathrm{RI} > 1$	N/A	N/A	4	4
5	RI/PMI_1	$N_p = 8$ two-layer spatial multiplexing	4	N/A	N/A	N/A
		$N_p = 4$ and eight-layer spatial multiplexing	5			
6	RI/PTI	$N_p = 8$ two-layer spatial multiplexing	N/A	2	N/A	N/A
		$N_p = 8$ four-layer spatial multiplexing	N/A	3	N/A	N/A
		$N_p = 8$ eight-layer spatial multiplexing	N/A	4	N/A	N/A

Table 10.14 Relationship of Sub-band Size and Bandwidth Parts with System Bandwidth [5]

System Bandwidth N_{RB}^{DL} (Resource Blocks)	Sub-band Size L_{SB} (Resource Blocks)	Bandwidth Parts N_{BP}
6–7	N/A	N/A
8–10	4	1
11–26	4	2
27–63	6	3
64–110	8	4

be set to off. For the UE-selected sub-band CQI, a CQI report in a certain subframe of the serving cell describes the channel quality in a particular part or parts of the bandwidth of that serving cell. The bandwidth parts are indexed in order of increasing frequency and decreasing sizes starting at the lowest frequency. The CQI and PMI payload sizes of each PUCCH CSI reporting mode are given in Table 10.14. While the wideband feedback mode is similar to that sent via PUSCH, the UE-selected sub-band CQI transmitted via PUCCH is different. In this case, the total number of sub-bands N_{SB} is divided into fractions called bandwidth parts. The value of the parameters (N_{SB}, L_{SB}, N_{BP}) depends on the system bandwidth as summarized in Table 10.14. In this case, one CQI value is computed and reported for a single selected sub-band from each bandwidth part, along with the corresponding sub-band index. The concept of periodic CQI reporting modes (wideband and sub-band) in time- and frequency-domains is illustrated in Figure 10.33.

For each serving cell, there are a total of N_{SB} equally sized sub-bands corresponding to system bandwidth N_{RB}^{DL} where $N_{SB} = \lfloor N_{RB}^{DL}/L_{SB} \rfloor$ sub-bands of size L_{SB}. A bandwidth part bp is frequency-consecutive and consists of N_{bp} sub-bands where N_{BP} bandwidth parts cover the system bandwidth. If the total number of bandwidth parts $N_{BP} = 1$, i.e., the wideband measurement, then $N_{bp} = \lceil N_{RB}^{DL}/L_{SB} \rceil$; otherwise, if $N_{BP} > 1$, then $N_{bp} = \lceil N_{RB}^{DL}/L_{SB}/N_{BP} \rceil$ depending on the values of these parameters, each bandwidth part is scanned in sequential order according to increasing frequency. For UE-selected sub-band feedback, a single sub-band out of N_{bp} sub-bands of a bandwidth part is selected along with the corresponding L bit label indexed in order of increasing frequency, where $L = \lceil \log_2 \lceil N_{RB}^{DL}/L_{SB}/N_{BP} \rceil \rceil$.

For each serving cell, the CQI/PMI reporting period P_{CQI} in number of subframes and relative reporting offset N_{offset}^{CQI} in number of subframes are configured by higher layers. The RI reporting period P_{RI} and relative reporting offset N_{offset}^{RI} are determined based on a parameter provided by higher layers. The relative RI

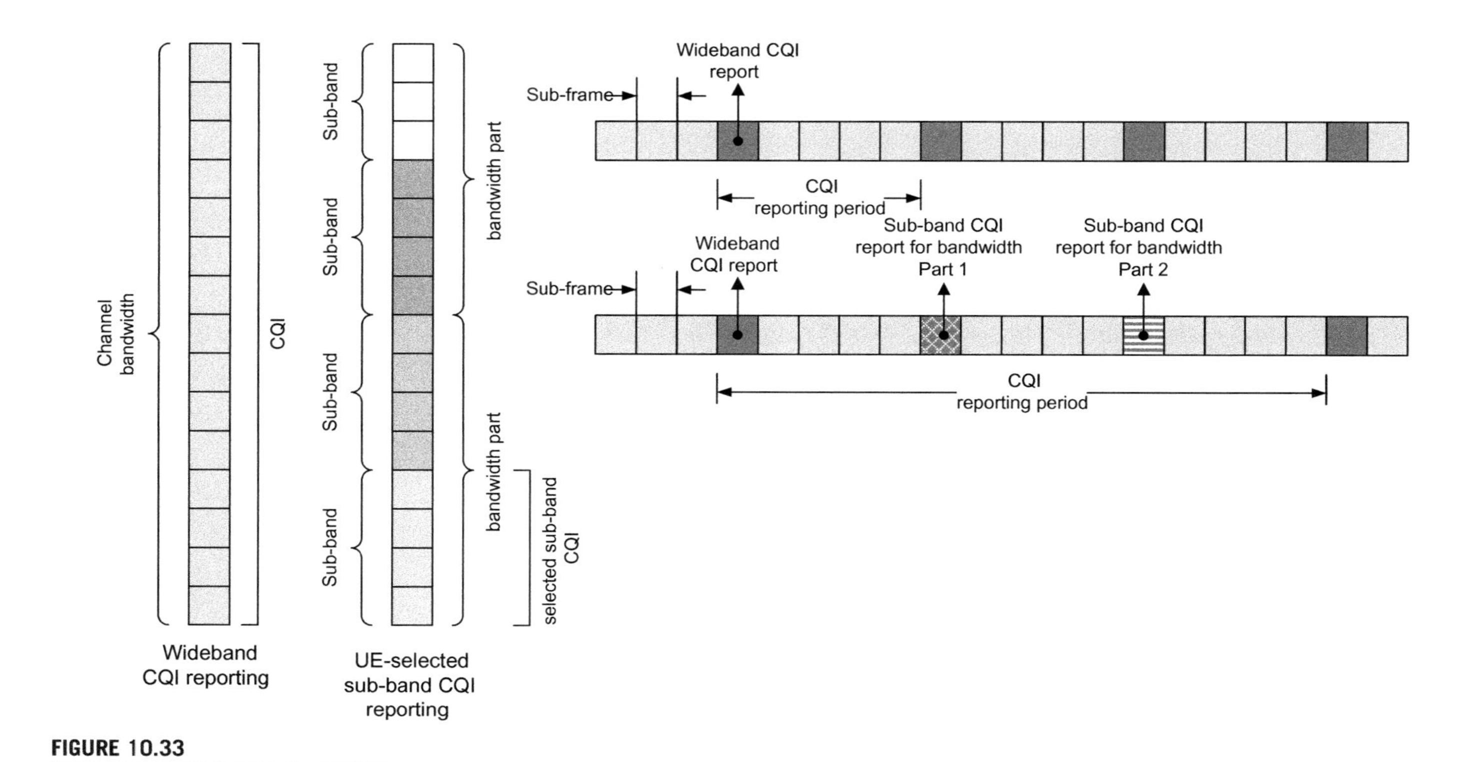

FIGURE 10.33

Illustration of the concept of periodic CQI reporting modes in time and frequency [14].

Table 10.15 Reporting Period, Reporting Offset, and Configuration Index of CQI/PMI and RI [5]

CQI/PMI Configuration Index	CQI/PMI Reporting Period P_{CQI}	CQI/PMI Reporting Offset N_{offset}^{CQI}
$0 \leq \alpha \leq 1$	2	α
$2 \leq \alpha \leq 6$	5	$\alpha-2$
$7 \leq \alpha \leq 16$	10	$\alpha-7$
$17 \leq \alpha \leq 36$	20	$\alpha-17$
$37 \leq \alpha \leq 76$	40	$\alpha-37$
$77 \leq \alpha \leq 156$	80	$\alpha-77$
$157 \leq \alpha \leq 316$	160	$\alpha-157$
$\alpha = 317$	Reserved	
$318 \leq \alpha \leq 349$	32	$\alpha-318$
$350 \leq \alpha \leq 413$	64	$\alpha-350$
$414 \leq \alpha \leq 541$	128	$\alpha-414$
$542 \leq \alpha \leq 1023$	Reserved	
RI configuration index	RI reporting period P_{RI}	RI reporting offset N_{offset}^{RI}
$0 \leq \beta \leq 160$	1	$-\beta$
$161 \leq \beta \leq 321$	2	$-(\beta-161)$
$322 \leq \beta \leq 482$	4	$-(\beta-322)$
$483 \leq \beta \leq 643$	8	$-(\beta-483)$
$644 \leq \beta \leq 804$	16	$-(\beta-644)$
$805 \leq \beta \leq 965$	32	$-(\beta-805)$
$966 \leq \beta \leq 1023$	Reserved	

reporting offset values are chosen in the range of $N_{offset}^{RI} \in \{0, -1, \ldots, -(P_{CQI}-1)\}$. If a UE is configured to report more than one CSI subframe set, then different sets of parameters corresponding to different subframe sets will be provided to the UE. When wideband CQI/PMI reporting is configured, the reporting instances satisfy the expression $(10n_f + \lfloor n_s/2 \rfloor - N_{offset}^{CQI}) \bmod P_{CQI} = 0$. When RI reporting is configured, the reporting interval is an integer multiple P_{RI} of period P_{CQI} subframes. The reporting instances for RI are subframes that satisfy the equation $(10n_f + \lfloor n_s/2 \rfloor - N_{offset}^{CQI} - N_{offset}^{RI}) \bmod (P_{CQI}P_{RI}) = 0$. When both wideband CQI/PMI and sub-band CQI reporting are configured, the reporting instances satisfy the equation $(10n_f + \lfloor n_s/2 \rfloor - N_{offset}^{CQI}) \bmod P_{CQI} = 0$. Figure 10.33 illustrates an example of wideband and sub-band CQI reporting. Reporting period, reporting offset, and configuration index of CQI/PMI and RI reports are listed in Table 10.15 [5].

In the case of collision of a CSI report with PUCCH reporting type 3, 5, or 6 of one serving cell with a CSI report with PUCCH reporting type 1, 1a, 2, 2a, 2b, 2c, or 4 of the same serving cell, the CSI report with PUCCH reporting type

(1, 1a, 2, 2a, 2b, 2c, or 4) has lower priority and will be dropped. If the UE is configured with more than one serving cell, it transmits a CSI report corresponding to only one serving cell in any given subframe.

10.7.2.3 PUCCH format 3

For downlink carrier aggregation and when there are simultaneous transmissions on multiple downlink component carriers, multiple HARQ feedback bits are generated and sent by the UE. As discussed earlier, PUCCH format 1b in conjunction with resource selection can be used in this case; however, if the number of HARQ feedback bits is greater than four, this method is not going to be an efficient solution from a performance perspective. In order to support efficient transmission of a larger number of HARQ feedback bits, PUCCH format 3, which was introduced in LTE Rel-10, can be used. The main distinction of PUCCH format 3, as illustrated in Figure 10.34, is DFT-precoded OFDM, i.e., the same transmission scheme used for the uplink transmission. The HARQ feedback bits, one or two per downlink component carrier, depending on the transmission mode configured for that particular component carrier, are concatenated with an SR bit into a sequence of bits where the bits corresponding to unscheduled transport blocks are set to zero. The concatenated bits are block coded and scrambled using a cell-specific scrambling sequence to randomize inter-cell interference.

The resulting 48 bits are QPSK-modulated and divided into two groups, one per slot, i.e., 12 QPSK symbols per slot. PUCCH format 3 employs two SC-FDMA symbols in each slot (with normal cyclic prefix) for reference signal transmission. In each slot, the block of 12 DFT-precoded QPSK symbols is transmitted in the remaining SC-FDMA symbols. To further randomize the inter-cell interference, a cyclic shift is applied to the block of 12 QPSK symbols prior to DFT precoding. The cyclic shifts vary from one symbol to another. To increase the multiplexing capacity, an orthogonal sequence of length 5 is multiplied by each of the five SC-FDMA symbols carrying data in a slot. As a result, up to five terminals may share the same resource-block pair for PUCCH format 3. Different orthogonal cover sequences are used in the two slots to improve the performance in high-Doppler scenarios. To enable channel estimation for different transmissions sharing the same resource block, different reference signal sequences are used.

The orthogonal cover sequences are obtained as five DFT sequences. A Walsh sequence of length 4 may be used in the second slot of a subframe in order to make the last SC-FDMA symbol available when SRS is configured in that subframe. Similar to other PUCCH formats, a PUCCH format 3 resource can be represented by a single index from which the orthogonal sequence and the resource-block number can be derived. A terminal can be configured with four different resources for PUCCH format 3. In the scheduling grant for a secondary component carrier, 2 bits are used to inform the terminal of the four resources. In that case, the scheduler can avoid PUCCH collisions between different terminals by assigning different resources to different UEs. Due to significant differences in

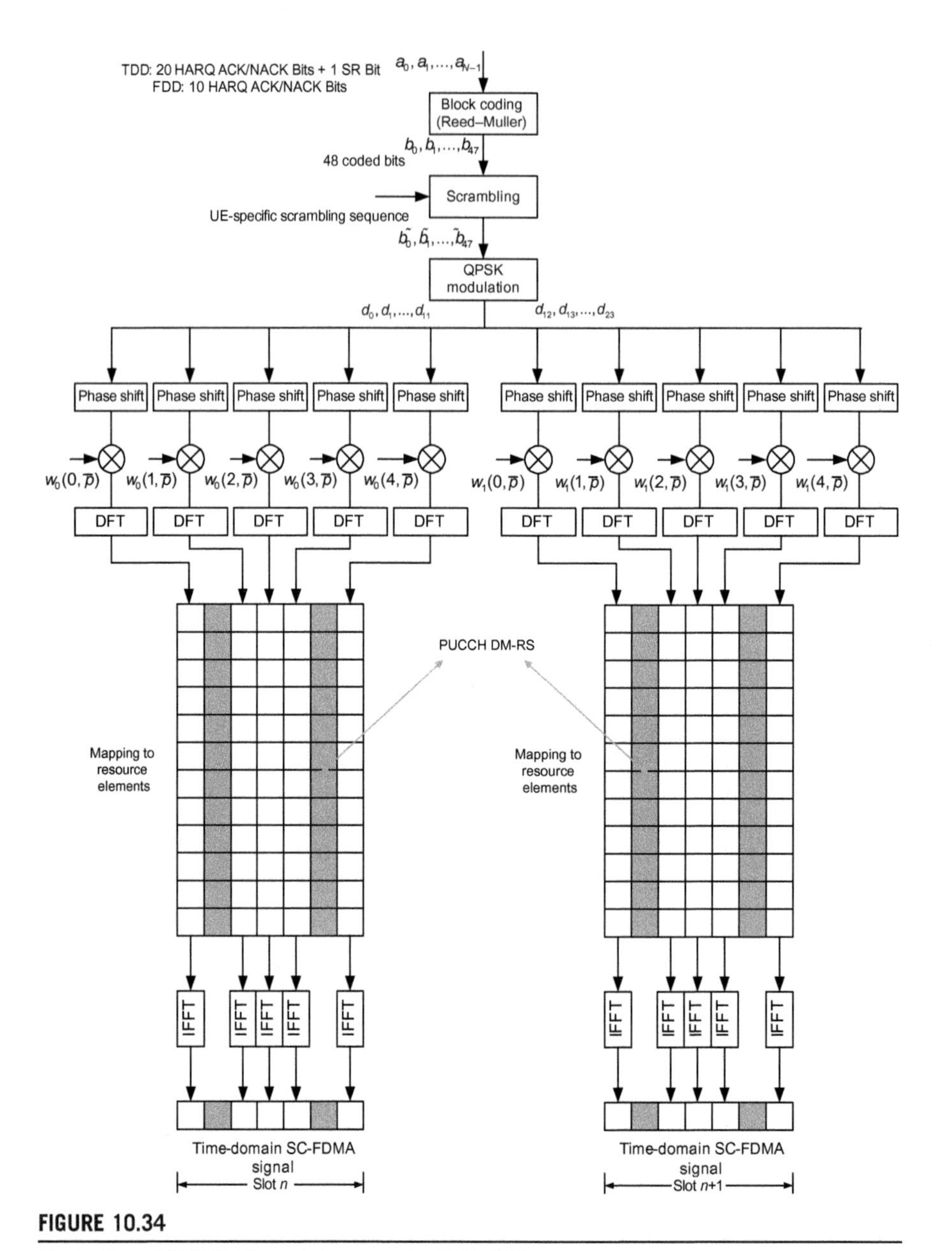

FIGURE 10.34

Illustration of PUCCH format 3 processing [14–16].

the structure of PUCCH format 3 compared to other PUCCH formats, resource blocks cannot be shared between them.

The physical layer processing of PUCCH format 3 can be described as follows. The block of coded bits $b_0, b_1, \ldots, b_{N_{\text{bit}}-1}$ are scrambled with a UE-specific

Table 10.16 Values of PUCCH Format 3 Orthogonal-Spreading Sequences [3]

Sequence Index n	Orthogonal Sequence $[w_0(0), w_1(0), \ldots, w_n(N_{SF}^{PUCCH}-1)]$	
	$N_{SF}^{PUCCH}=5$	$N_{SF}^{PUCCH}=4$
0	$[1 \quad 1 \quad 1 \quad 1 \quad 1]$	$[+1 \quad +1 \quad +1 \quad +1]$
1	$[1 \quad e^{j2\pi/5} \quad e^{j4\pi/5} \quad e^{j6\pi/5} \quad e^{j8\pi/5}]$	$[+1 \quad -1 \quad +1 \quad -1]$
2	$[1 \quad e^{j4\pi/5} \quad e^{j8\pi/5} \quad e^{j2\pi/5} \quad e^{j6\pi/5}]$	$[+1 \quad +1 \quad -1 \quad -1]$
3	$[1 \quad e^{j6\pi/5} \quad e^{j2\pi/5} \quad e^{j8\pi/5} \quad e^{j4\pi/5}]$	$[+1 \quad -1 \quad -1 \quad +1]$
4	$[1 \quad e^{j8\pi/5} \quad e^{j6\pi/5} \quad e^{j4\pi/5} \quad e^{j2\pi/5}]$	–

scrambling sequence, resulting in a block of scrambled coded bits $\bar{b}_0, \bar{b}_1, \ldots, \bar{b}_{N_{bit}-1}$ such that $\bar{b}_i = (b_i + c(i)) \bmod 2$, where the scrambling sequence $c(i)$ was defined in Chapter 9 and is initialized with $c_{init} = (\lfloor n_s/2 \rfloor + 1)(2N_{ID}^{cell}+1)2^{16} + n_{RNTI}$ at the start of each subframe where n_{RNTI} denotes the UE temporary identifier C-RNTI. The block of scrambled bits is then QPSK-modulated, resulting in a block of complex-valued modulation symbols $d_0, d_1, \ldots, d_{N_s-1}$ in which $N_s = N_{bit}/2 = 12$. The complex-valued symbols $d_0, d_1, \ldots, d_{N_s-1}$ is block-wise spread with the orthogonal sequences $w_0(i,\bar{p})$ and $w_1(i,\bar{p})$, yielding $N_{SF_0}^{PUCCH} + N_{SF_1}^{PUCCH}$ sets containing 12 values where:

$$y_n^{\bar{p}}(i) = \begin{cases} w_0(\bar{n},\bar{p})\exp\left(j\pi\left\lfloor n_{cs}^{cell}(n_s,l)/64\right\rfloor/2\right)d_i & n < N_{SF_0}^{PUCCH} \\ w_1(\bar{n},\bar{p})\exp\left(j\pi\left\lfloor n_{cs}^{cell}(n_s,l)/64\right\rfloor/2\right)d_{i+12} & n \geq N_{SF_0}^{PUCCH} \end{cases} \tag{10.8}$$

where $\bar{n} = n \bmod N_{SF_0}^{PUCCH}$; $n = 0, 1, \ldots, N_{SF_0}^{PUCCH} + N_{SF_1}^{PUCCH} - 1$; $N_{SF_0}^{PUCCH} = N_{SF_1}^{PUCCH} = 5$; and $i = 0, 1, \ldots, 11$ in both slots in a subframe using normal PUCCH format 3 and $N_{SF_0}^{PUCCH} = 5$; $N_{SF_1}^{PUCCH} = 4$ in a subframe using shortened PUCCH format 3 to accommodate the SRS placement. The orthogonal sequences $w_0(i,\bar{p})$ and $w_1(i,\bar{p})$ are Walsh sequences whose values are given in Table 10.16.

Each set of the generated complex-valued symbols following the spreading function are cyclically shifted according to $\tilde{y}_n^{\bar{p}}(i) = y_n^{\bar{p}}([i + n_{cs}^{cell}(n_s,l)] \bmod 12)$, where $n_{cs}^{cell}(n_s,l)$ is a cell-specific cyclic shift, n_s is the slot number within a radio frame, and l is the SC-FDMA symbol number within a slot. The cyclically shifted sets of complex-valued symbols are DFT-precoded according to:

$$z^{\bar{p}}(12n+k) = \frac{1}{\sqrt{12N_p}}\sum_{i=0}^{11}\tilde{y}_n^{\bar{p}}(i)e^{-j\frac{\pi ik}{6}}; k=0,1,\ldots,11; n=0,1,\ldots,N_{SF_0}^{PUCCH}+N_{SF_1}^{PUCCH}-1 \tag{10.9}$$

where N_p is the number of antenna ports used for PUCCH format 3 transmission, resulting in a block of complex-valued symbols $z_0^{\bar{p}}, z_1^{\bar{p}}, \ldots, z_{11\times(N_{SF_0}^{PUCCH}+N_{SF_1}^{PUCCH})}^{\bar{p}}$. The

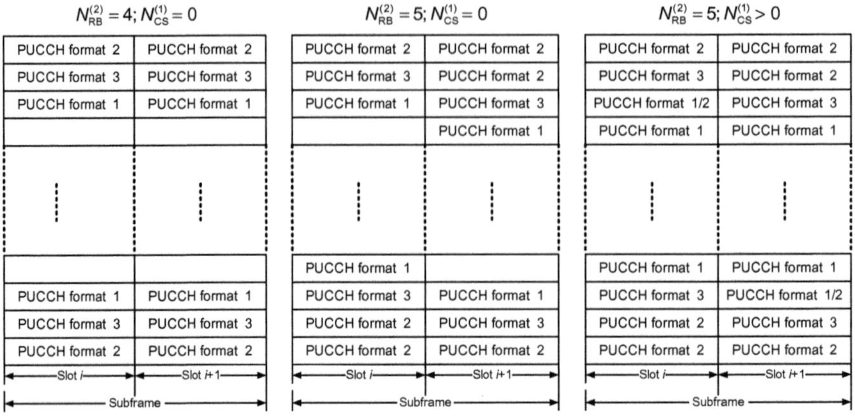

$N_{RB}^{(2)} = 4; N_{CS}^{(1)} = 0$

Slot i	Slot $i+1$
PUCCH format 2	PUCCH format 2
PUCCH format 3	PUCCH format 3
PUCCH format 1	PUCCH format 1
⋮	⋮
PUCCH format 1	PUCCH format 1
PUCCH format 3	PUCCH format 3
PUCCH format 2	PUCCH format 2

Subframe

$N_{RB}^{(2)} = 5; N_{CS}^{(1)} = 0$

Slot i	Slot $i+1$
PUCCH format 2	PUCCH format 2
PUCCH format 3	PUCCH format 2
PUCCH format 1	PUCCH format 3
	PUCCH format 1
⋮	⋮
PUCCH format 1	
PUCCH format 3	PUCCH format 1
PUCCH format 2	PUCCH format 3
PUCCH format 2	PUCCH format 2

Subframe

$N_{RB}^{(2)} = 5; N_{CS}^{(1)} > 0$

Slot i	Slot $i+1$
PUCCH format 2	PUCCH format 2
PUCCH format 3	PUCCH format 2
PUCCH format 1/2	PUCCH format 3
PUCCH format 1	PUCCH format 1
⋮	⋮
PUCCH format 1	PUCCH format 1
PUCCH format 3	PUCCH format 1/2
PUCCH format 2	PUCCH format 3
PUCCH format 2	PUCCH format 2

Subframe

FIGURE 10.35

Example of resource-block allocation for various PUCCH formats [14].

amplitudes of the resulting complex-valued symbols are scaled in order to conform to the PUCCH transmit power, and are mapped in sequence starting with $z_0^{(\tilde{p})}$ to resource elements. PUCCH uses one resource block in each of the two slots in a subframe. Within the PRB used for PUCCH format 3 transmission, the mapping of complex symbols $z_i^{(\tilde{p})}|_{0,1,\ldots,11\times(N_{SF_0}^{PUCCH}+N_{SF_1}^{PUCCH})}$ to resource elements (k, l) on antenna port $\tilde{p}$ that are not reserved for transmission of reference signals is in increasing order of the first index and then the second index, i.e., frequency first, starting with the first slot in the subframe. The relation between the antenna index $\tilde{p}$ and the antenna port number p is given in 3GPP TS 36.211 [3]. In case of simultaneous transmission of SRS and PUCCH format 1, 1a, 1b, or 3 when there is one configured serving cell, a shortened PUCCH format is used where the last SC-FDMA symbol in the second slot of a subframe is skipped. Figure 10.35 shows an example of resource mapping for various PUCCH formats [14].

Precoder-based multi-layer transmission is only applied to uplink data transmission on PUSCH. LTE Rel-10 allows a terminal with multiple transmit antennas to use multi-antenna transmission for uplink control signaling on PUCCH, in order to achieve diversity gain. In that case, two-antenna transmit diversity is used for PUCCH transmission. The scheme is referred to as spatial orthogonal resource transmit diversity (SORTD). The main concept of SORTD is to transmit the uplink control signaling using different resources (time, frequency, and/or code) on different antennas. In principle, the PUCCH transmissions from the two antennas will be identical to PUCCH transmissions from two different terminals using different resources. Thus, SORTD provides additional diversity at the expense of using twice as many PUCCH resources compared to non-SORTD transmission. For four physical antennas at the terminal, implementation-specific antenna virtualization is used, where a transparent scheme is used to map the two-antenna-port signal to four physical antennas [16]. Figure 10.36 illustrates

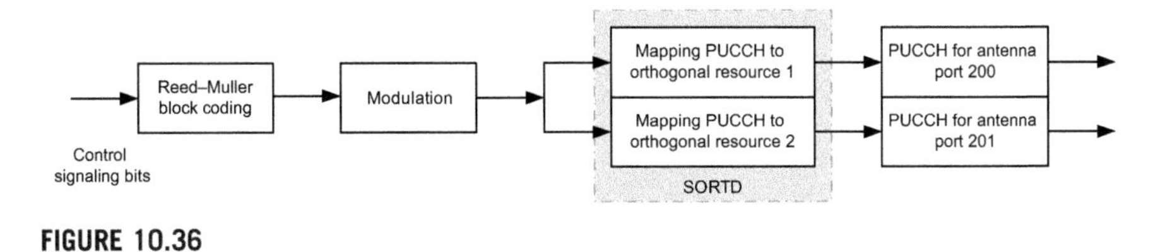

FIGURE 10.36

Illustration of SORTD concept [14,166].

PUCCH transmission using the SORTD concept [14]. SORTD is applicable to PUCCH formats 2/2a/2b and 3. The transmit diversity is not supported in LTE Rel-10 for PUCCH format 1b with channel selection. If the transmit antennas are uncorrelated, the SORTD scheme provides the optimal diversity performance at the expense of consuming extra resources which may reduce the PUCCH multiplexing capacity [15].

As mentioned earlier, PUCCH format 3 supports transmission of 48 coded bits. The actual number of HARQ feedback bits depends on the number of configured cells (active component carriers), the configured transmission modes of each carrier, and the HARQ ACK/NACK bundling window size (i.e., the number of downlink subframes associated with a single uplink subframe applicable to TDD systems). In FDD mode, the maximum payload size of ten HARQ ACK/NACK bits is supported corresponding to five component carriers each configured for MIMO transmission. In TDD mode, PUCCH format 3 supports HARQ feedback payload size of up to 20 bits. If the number of HARQ ACK/NACK bits is greater than 20, spatial bundling (i.e., a logical AND) of the ACK/NACK bits corresponding to the two codewords within a downlink subframe is performed for each of the serving cells. The maximum payload size carried by PUCCH format 3 in LTE Rel-10 is 21 bits, in which 20 bits are associated with HARQ ACK/NACK and 1 bit corresponds to SR. The HARQ feedback bits are concatenated in ascending order of the downlink component-carrier index. For payload sizes less than or equal to 11 bits, Reed–Muller channel codes from LTE Rel-8 are used in conjunction with circular-buffer rate-matching. When the payload size is larger than 11 bits, alternate ACK/NACK bits are input to two separate Reed–Muller encoders [15].

In order to alleviate inter-cell interference effects, cell-specific scrambling per SC-FDMA symbol was introduced. The PUCCH resource to be used for format 3 is signaled explicitly to the UE. A set of four resources is configured by RRC signaling, of which one resource is assigned dynamically for each ACK/NACK occurrence using an indicator transmitted in the transmit power control (TPC) field of the PDCCH corresponding to PDSCH on the secondary component carrier(s) [15]. The PDCCH assignments on the secondary component carriers in a given subframe indicate the same value. If transmit diversity is used for PUCCH format 3, higher-layer signaling is used to configure four pairs of PUCCH resources. The PDCCH assigning resources for the secondary cell PDSCH indicates one of these pairs to be used by the two antenna ports. If no PDCCH corresponding to PDSCH on a secondary cell is received in a given subframe and a single PDSCH is received on the primary cell, a UE that is configured for PUCCH format 3 would instead use the LTE Rel-8 PUCCH format 1a or 1b.

10.7.3 Physical uplink shared channel

The PUSCH is used to transmit the uplink shared channel (UL-SCH) and L1/L2 control information. The UL-SCH is the transport channel used for transmitting uplink data (a transport block). The L1/L2 control signaling carries HARQ acknowledgments for received downlink shared channel (DL-SCH) transport blocks, channel quality reports, and SRs. The physical layer processing stages are shown in Figure 10.37. The information bits are channel coded by a turbo coder with a basic code rate of 1/3, which is adapted to a suitable code rate through a rate-matching process. This is followed by symbol-level channel interleaving which follows a simple time-first mapping where adjacent data symbols are mapped first to the adjacent SC-FDMA symbols in the time-domain and then across the sub-carriers. The coded and interleaved bits are then scrambled by a length-31 Gold code prior to modulation mapping, DFT-spreading, sub-carrier mapping, and OFDM modulation. The signal is frequency-shifted by half a sub-carrier prior to transmission, to avoid the distortion caused by the DC sub-carrier being concentrated in one resource block. The modulation schemes supported in the uplink are QPSK, 16QAM, and 64QAM (optional for lower UE categories). To generate the PUSCH payload, each transport block undergoes transport block coding. The encoding process includes 24-bit CRC calculation, code block segmentation and 24-bit CRC attachment to code segments (if any), turbo encoding, and rate-matching with redundancy and code block concatenation. Furthermore, coding of control signaling information is carried out with the resulting codewords time-multiplexed and interleaved with the UL-SCH data. The control information includes CQI, PMI, RI, and HARQ feedback.

The transport block size for PUSCH data transmission is signaled in the corresponding resource grant in PDCCH, i.e., DCI format 0 or LTE Rel-10 DCI format 4 for SU-MIMO. Combined with the MCS, the transport block size reflects the effective code rate. The permissible transport-block-size indices, MCSs, and RVs for PUSCH are given in Table 10.17. In most cases, a generally linear range of code rates are available for each resource allocation size. However, one exception is an index which allows a transport block size of 328 bits in a single resource-block allocation with QPSK modulation, which corresponds to a code rate greater than unity. This is primarily designed to support cell-edge VoIP transmissions using only one resource block per subframe, allowing the UE's power spectral density to be maximized for improved coverage, as well as using the TTI bundling where the transmission is repeated in four consecutive subframes, together with typically three retransmissions at 16 ms intervals [16].

Following the above summary, let us discuss the physical layer processing of the UL-SCH in more detail. When a transport block $a_0, a_1, \ldots, a_{N_A-1}$ of length N_A

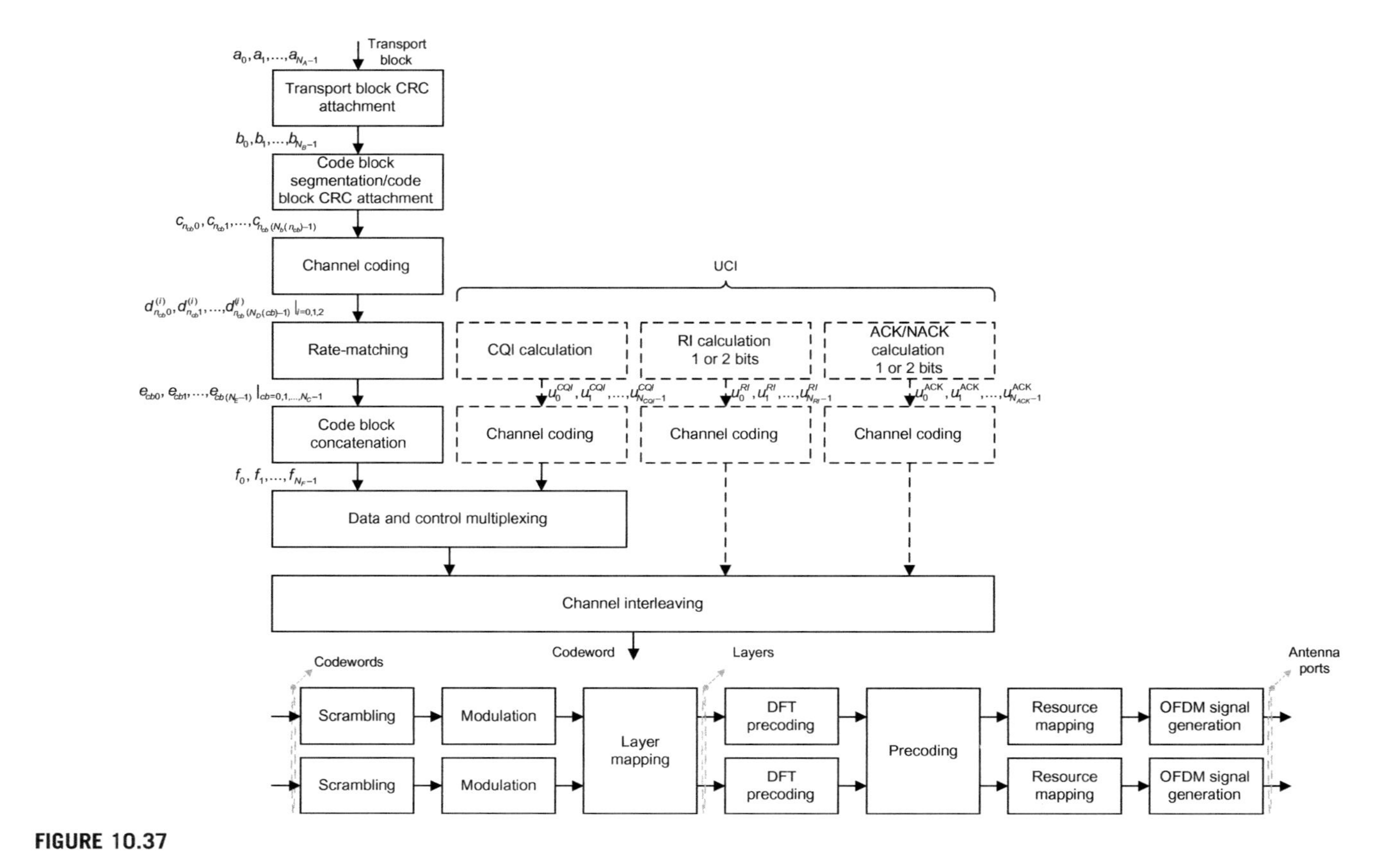

FIGURE 10.37

Physical layer processing for uplink shared channel [15,153].

Table 10.17 Modulation, Transport Block Size Index, and Redundancy Version for PUSCH [5]

MCS Index	Modulation Order	Transport Block Size Index	Redundancy Version
0	2	0	0
1	2	1	0
2	2	2	0
3	2	3	0
4	2	4	0
5	2	5	0
6	2	6	0
7	2	7	0
8	2	8	0
9	2	9	0
10	2	10	0
11	4	10	0
12	4	11	0
13	4	12	0
14	4	13	0
15	4	14	0
16	4	15	0
17	4	16	0
18	4	17	0
19	4	18	0
20	4	19	0
21	6	19	0
22	6	20	0
23	6	21	0
24	6	22	0
25	6	23	0
26	6	24	0
27	6	25	0
28	6	26	0
29	Reserved		1
30			2
31			3

is delivered to the physical layer, it is appended by 24 parity bits $p_0, p_1, \ldots, p_{23}$ where the entire transport block is used to calculate the CRC parity bits. The lowest order information bit a_0 is mapped to the most significant bit of the transport block. The 24 parity bits are calculated using the generator polynomial

$G_{\text{CRC-24A}}(D)$.[10] The CRC-attached bits $b_0, b_1, \ldots, b_{N_B-1}$ are fed to the code block segmentation unit, where $N_B = N_A + 24$ is the number of bits in the transport block including CRC (see Figure 10.37).

In LTE/LTE-Advanced, a minimum and maximum code block size is specified so that the block sizes are compatible with the block sizes supported by the turbo interleaver, where the minimum code block size is 40 bits and the maximum code block size is 6144 bits. If the length of the CRC-added input block is greater than the maximum code block size, the input block is segmented. When the input block is segmented, it is segmented into $N_C = \left\lceil N_B/(\underbrace{6144}_{\text{Maximum TB Size}} - \underbrace{24}_{\text{CRC Length}}) \right\rceil$ code blocks. Each code block is appended with 24-bit CRC. Padding bits are appended to the beginning of the segment (if required) so that the code block sizes match a set of valid turbo interleaver block sizes. If no segmentation is needed, then only one code block is produced. If N_B is less than the allowed minimum code block size, then zeros are added to the beginning of the code block in order to attain a total of 40 bits. Figure 10.38 illustrates the code segmentation and second CRC attachment process.

The code block segments, following the code block segmentation function, are denoted by $c_{n_{\text{cb}}0}, c_{n_{\text{cb}}1}, \ldots, c_{n_{\text{cb}}(N_b(n_{\text{cb}})-1)}$, where n_{cb} denotes the code block number and $N_b(n_{\text{cb}})$ is the number of bits in the code block number n_{cb}. The code blocks are appended with 24-bit CRC and are fed to the channel coding block. The total number of code blocks is denoted by N_C and each code block is individually turbo-coded.

The code block segments then go through turbo coding. Turbo coding is a form of forward error correction which improves the transmission reliability and robustness by adding redundancy to the code. The turbo encoder scheme used is a parallel concatenated convolutional code (PCCC)[11] with two recursive convolutional coders and a contention-free QPP interleaver, as shown in Figure 10.39. The encoder generates three output streams $d_k^{(0)}, d_k^{(1)}, d_k^{(2)}$, achieving a code rate of 1/3. Turbo-coded blocks are delivered to the rate-matching block. As shown in Figure 10.39, the input to the first constituent encoder is the input bit stream to the turbo coding block. The input to the second constituent encoder is the output of the QPP interleaver, i.e., a permutated version of the input sequence. The role of the interleaver is to spread the information bits such that in the event of a burst error, the two code streams are affected differently allowing data to be recoverable. There are two output sequences from each encoder: a systematic (x_k, x'_k) output and a parity (z_k, z'_k) output. Only one of the systematic sequences, x_k is used

[10] $G_{\text{CRC-24A}}(D) = D^{24} + D^{23} + D^{18} + D^{17} + D^{14} + D^{11} + D^{10} + D^7 + D^6 + D^5 + D^4 + D^3 + D + 1$.

[11] The first class of turbo code was the PCCC. Since the introduction of the original parallel turbo codes in 1993, many other classes of turbo code have been discovered, including serial versions and repeat-accumulate codes. Iterative turbo decoding methods have also been applied to more conventional FEC systems, including Reed-Solomon corrected convolutional codes [28,54].

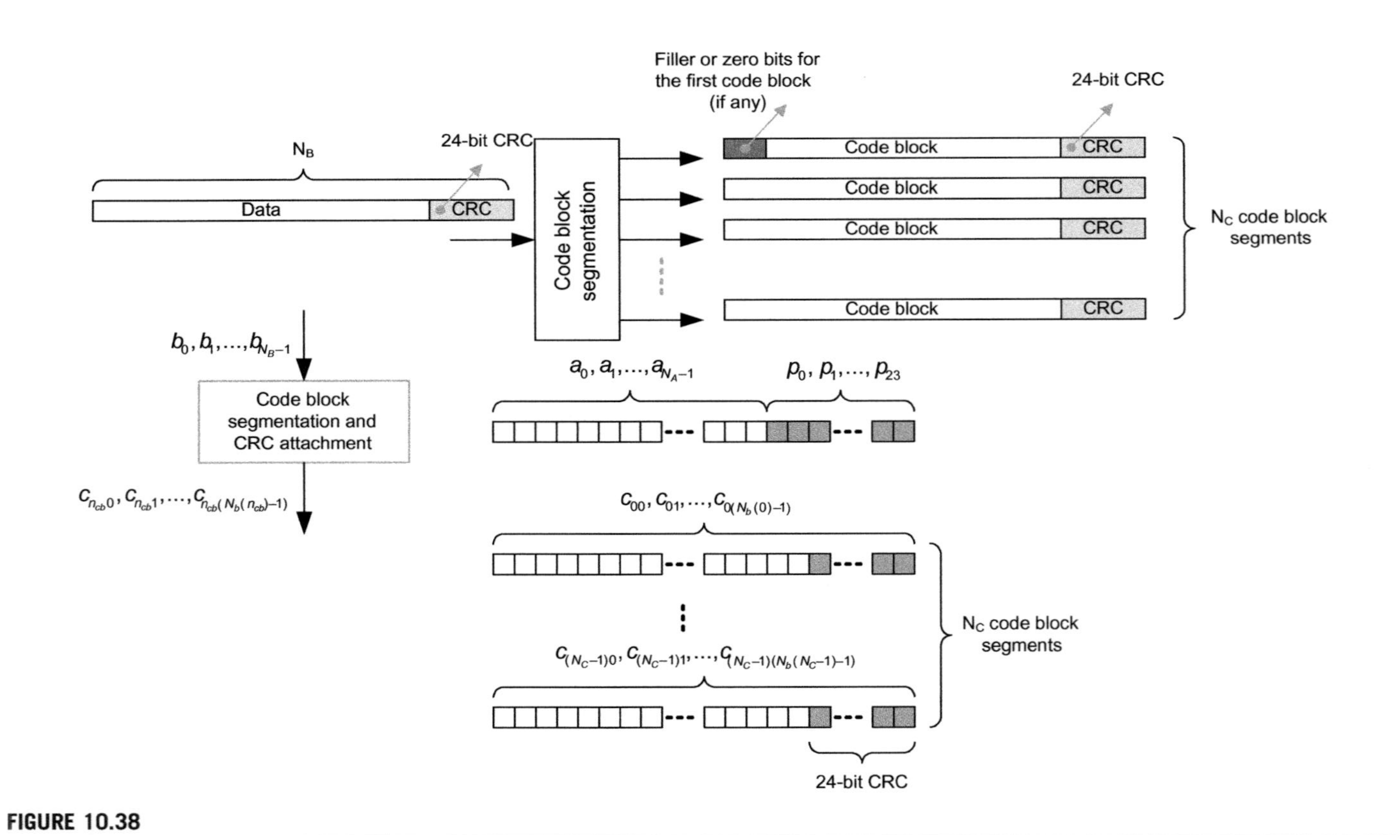

FIGURE 10.38

Code block segmentation and CRC attachment [4,153,155].

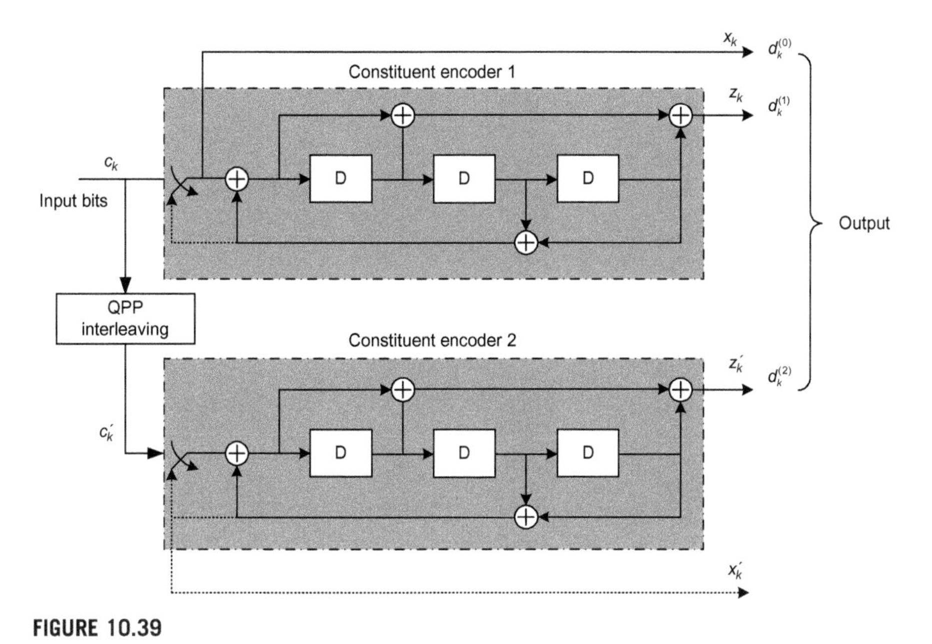

FIGURE 10.39

LTE/LTE-Advanced turbo encoder structure [4].

as an output, as the other x'_k is simply a permutated version of the chosen systematic sequence. After the turbo coding procedure, the output bits are denoted by $d^{(i)}_{n_{\text{cb}}0}, d^{(i)}_{n_{\text{cb}}1}, \ldots, d^{(i)}_{n_{\text{cb}}(N_D-1)}|_{i=0,1,2}$, where $N_D = N_{\text{cb}} + 4$ denotes the number of bits in the ith coded stream for code block number n_{cb}.

Note that a standard convolutional coder initializes its internal registers to an all-zeros state and ensures that the coder finishes in an all-zeros state by padding the end of the input sequence with $L_{\text{constraint-length}}$ zeros. Since the decoder has prior knowledge of the start and end state of the encoder, it can decode the data. Driving a recursive coder to an all-zeros state using this method is not possible. To overcome this problem trellis termination is used. A tail-biting convolutional coder initializes its internal shift registers to the last $L_{\text{constraint-length}}$ bits of the current input block, rather than an all-zeros state. This means that the start and end states are the same without the need to append zeros to the input block, eliminating the unnecessary overhead to terminate the coder by padding the input with zeros. The drawback to this method is that the decoder becomes more complicated since the initial state is unknown. However, it is known that the start and end sequences are the same [28]. Note that the number of output bits N_D following tail-biting convolutional coding with rate 1/3 is $N_D = N_{\text{cb}}$. The tail-biting convolutional coding with different rates is used for coding of control information in the downlink and uplink [4] (Figure 10.40).

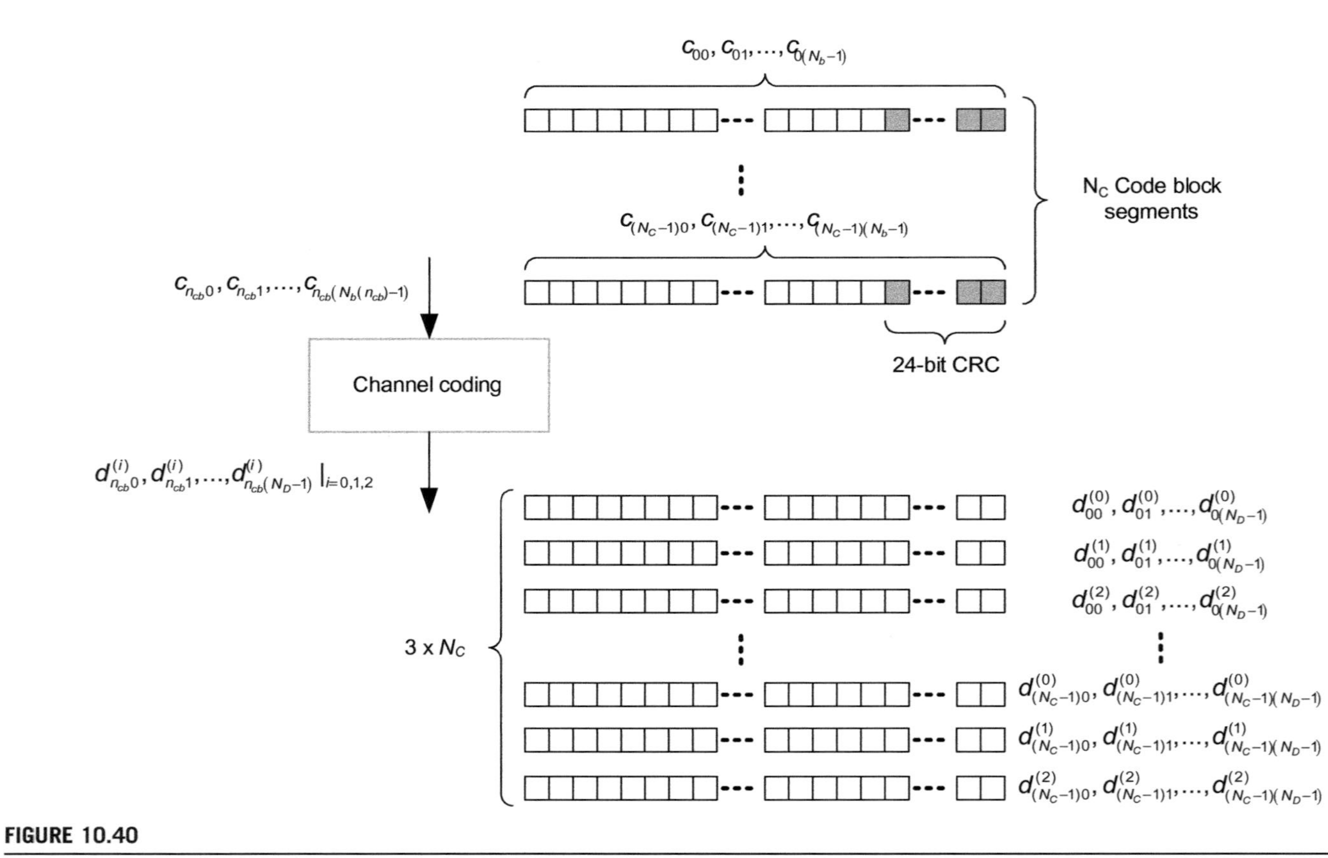

FIGURE 10.40

Demonstration of the channel coding process in LTE/LTE-Advanced [153].

The rate-matching block creates an output bit stream with a desired code rate. As the number of bits available for transmission depends on the available resources, the rate-matching algorithm is capable of producing any arbitrary rate. The 3-bit streams from the turbo encoder are interleaved, followed by bit collection to create a circular buffer. Bits are selected and pruned from the buffer to create an output bit stream with the desired code rate. The process is illustrated in Figure 10.41. The three sub-block interleavers used in the rate-matching block are identical. Interleaving is a technique to reduce the impact of burst errors on a signal, as consecutive bits of data will not be corrupted. The sub-block interleaver reshapes the encoded bit sequence in a row-by-row fashion to form a matrix. The bit collection stage creates a virtual circular buffer by combining the three interleaved encoded bit streams. Bits are then selected and pruned from the circular buffer to create an output sequence length which meets the desired code rate.

The HARQ error correction scheme is incorporated into the rate-matching algorithm of LTE/LTE-Advanced. For any desired code rate, the coded bits are output serially from the circular buffer from a starting location, given by the RV, wrapping around to the beginning of the buffer, if the end of the buffer is reached and null bits are discarded. Different redundancy versions and hence starting points allow retransmission of selected data. Being able to select different starting points enables the two main methods of reassembling data at the receiver in the HARQ process. In chase combining, retransmissions contain the same data and parity bits, whereas in incremental redundancy, the retransmissions contain different redundancy information so that the receiver gains knowledge upon each retransmission. The rate-matched code blocks are now concatenated back together. This is done by sequentially concatenating the rate-matched blocks together to create the output of the channel coding.

The output of the rate-matching function $e_{\text{cb}0}, e_{\text{cb}1}, \ldots, e_{\text{cb}(N_E(cb)-1)}$, where cb denotes the code block index and $N_E(\text{cb})$ is the number of rate-matched bits for code block number cb, serves as input to the code block concatenation module, which sequentially concatenates the rate-matched blocks together to create the output of the channel coding $f_0, f_1, \ldots, f_{N_F-1}$, where N_F represents the total number of coded bits for transmission of the given transport block over N_l transmission layers. The latter excludes the bits used for control information when it is multiplexed with UL-SCH data. The control information comprises CQI (and/or PMI), HARQ ACK/NACK, and RI. Different coding rates for the control information are achieved by allocating a different number of coded symbols for its transmission (Figure 10.42).

As shown in Figure 10.37, when control data is transmitted in PUSCH, the channel coding procedures for HARQ feedback $u_0^{\text{ACK}}, u_1^{\text{ACK}}, \ldots, u_{N_{\text{ACK}}-1}^{\text{ACK}}$, RI information $u_0^{\text{RI}}, u_1^{\text{RI}}, \ldots, u_{N_{\text{RI}}-1}^{\text{RI}}$, and CQI information $u_0^{\text{CQI}}, u_1^{\text{CQI}}, \ldots, u_{N_{\text{CQI}}-1}^{\text{CQI}}$ are performed independently.

An important issue related to transmission of control signaling on PUSCH is how to maintain the performance/reliability of control signaling at the desired level. It is noted that power control will set the SINR target of PUSCH according

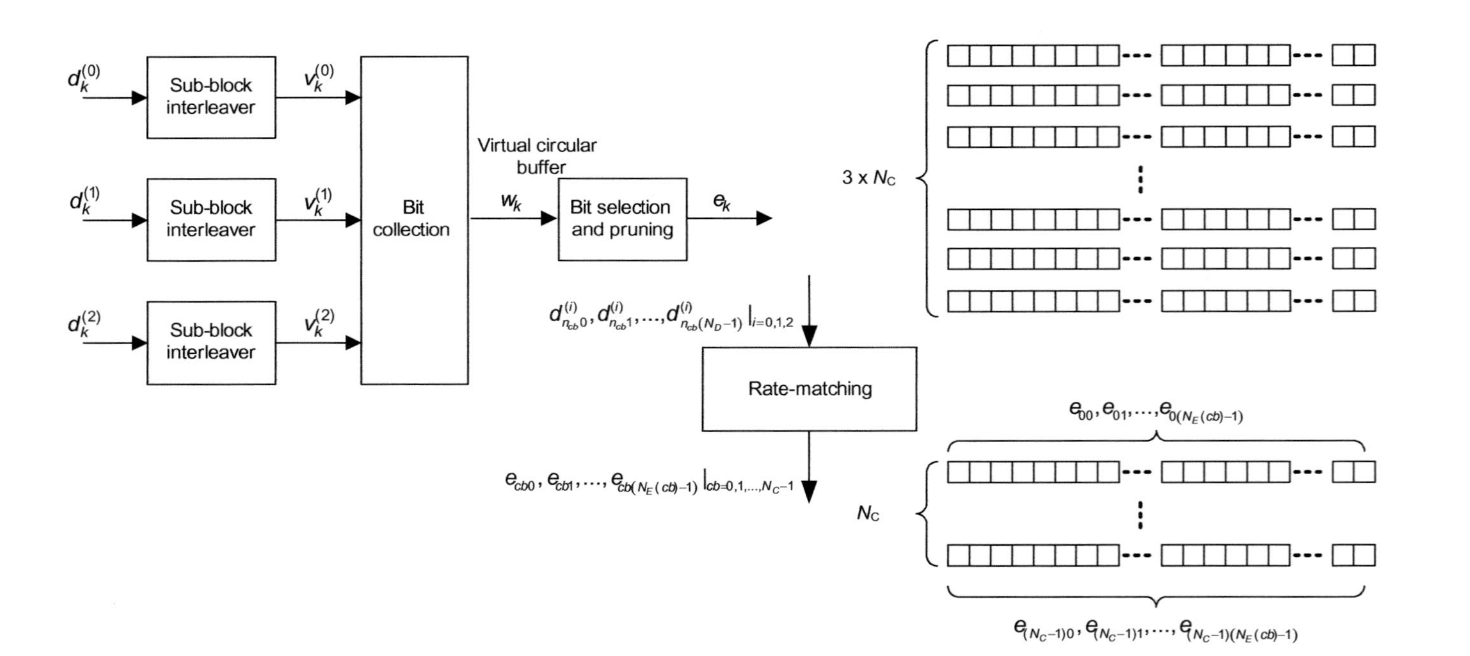

FIGURE 10.41

LTE/LTE-Advanced rate-matching procedures [4].

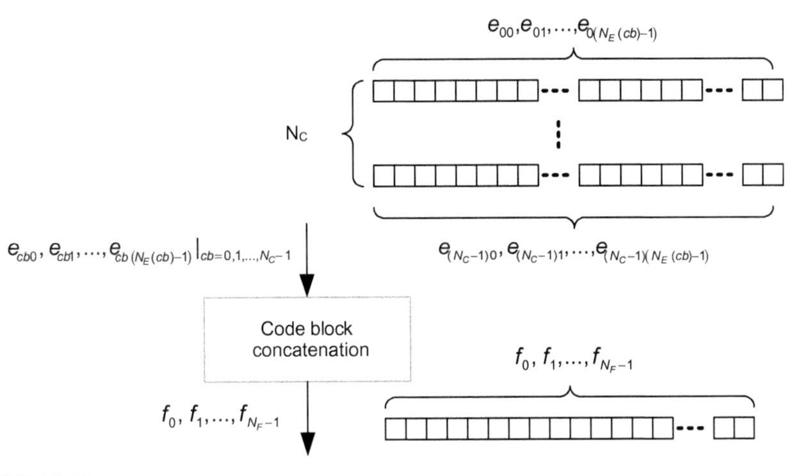

FIGURE 10.42

Illustration of code block concatenation.

to the data channel. Therefore, the control channel has to adapt to the SINR operating point set for data. One way to adjust the available resources would be to apply different power offset values for data and control parts. The problem with the latter power adjustment scheme is that the single-carrier property of the uplink signal is compromised. Therefore, this scheme is not used in the LTE uplink control and data multiplexing. Instead, a scheme based on a variable coding rate for the control information is used. This is achieved by varying the number of coded symbols for control channel transmission. To minimize the overall overhead related to inclusion of the control signaling information, the size of physical resources allocated to control transmission is scaled according to PUSCH quality. Note that the coding rate of control signaling, in this case, is given implicitly by the MCS of PUSCH data.

When the UE transmits HARQ feedback or RI bits, it is required to determine the number of coded modulation symbols per layer N_Q for HARQ ACK/NACK or RI. When only a single transport block is transmitted in PUSCH containing HARQ feedback or RI bits, the number of modulation symbols per layer is determined using the following equation [4]:

$$N_Q = \min\left(\left\lceil \frac{N_x N_{\mathrm{sc}}^{\mathrm{PUSCH\text{-}initial}} N_s^{\mathrm{PUSCH\text{-}initial}} \beta_{\mathrm{offset}}^{\mathrm{PUSCH}}}{\sum_{\mathrm{cb}=0}^{N_C-1} N_{\mathrm{cb}}} \right\rceil,\ 4\,N_{\mathrm{sc}}^{\mathrm{PUSCH}}\right) \tag{10.10}$$

where $N_x = N_{\mathrm{ACK}}$ or N_{RI} is the number of HARQ ACK/NACK bits or RI bits, $N_{\mathrm{sc}}^{\mathrm{PUSCH}}$ is the number of sub-carriers associated with scheduled PUSCH transmission over one SC-FDMA symbol, and $N_s^{\mathrm{PUSCH\text{-}initial}}$ is the number of SC-FDMA symbols per subframe for initial PUSCH transmission for the same transport block, given by $N_{s}^{\text{PUSCH-initial}} = \left[2(N_{s}^{\text{UL}} - 1) - N_{\text{SRS}}\right]$, in which $N_{\text{SRS}} = 1$, if the UE transmits PUSCH and SRS in the same subframe for initial transmission, or if the PUSCH resource allocation for initial transmission even partially overlaps with the cell-specific SRS subframe and bandwidth configuration, or if the subframe for initial transmission is a UE-specific type-1 SRS subframe; otherwise $N_{\text{SRS}} = 0$. The parameters $N_{\text{sc}}^{\text{PUSCH-initial}}$, N_C, and N_{cb} are obtained from the initial PDCCH/ePDCCH for the same transport block. For the case where two transport blocks are transmitted in PUSCH containing HARQ feedback or RI bits, the number of modulation symbols per layer is determined by $N_Q = \max[\min(N'_Q, 4N_{\text{sc}}^{\text{PUSCH}}), \min(N_Q)]$ [5].

The offset parameter $\beta_{\text{offset}}^{\text{PUSCH}}$ is used to adjust the quality of control signals for the PUSCH data channel. It is a UE-specific parameter configured by higher layer signaling. Different control channels need their own offset-parameter setting. There are some issues that need to be taken into consideration when configuring the offset parameter as follows [17]:

- BLER operation point for the PUSCH data.
- BLER operation point for the L1/L2 control signaling.
- Difference in coding gain between control and data parts, due to different coding schemes and different coding block sizes (no coding gain with 1-bit ACK/NACK).

Different BLER operating points for data and control parts are due to the fact that HARQ is used for the data channels, whereas the control channels do not benefit from HARQ. The higher the difference in the BLER operating point between data and control channels, the larger is the offset parameter (and vice versa). Similar behavior also relates to the packet size. The highest offset values are needed with ACK/NACK signals due to the lack of coding gain.

The number of coded symbols N_Q used by the UE to transmit the HARQ feedback is determined using the number of HARQ bits (1 or 2 depending on the number of codewords in use), the scheduled PUSCH bandwidth expressed as a number of sub-carriers, the number of SC-FDMA symbols per subframe for the initial PUSCH transmission, and information obtained from the initial PDCCH/ePDCCH for the same transport block. Each positive ACK is encoded as "1", and NACK is encoded as "0." If HARQ feedback consists of a single bit of information u_0^{ACK}, corresponding to a single codeword, then it is encoded depending on the modulation order used for the transport block, e.g., if the transport block is modulated using 16QAM, then encoded HARQ feedback is represented as $\left[u_0^{\text{ACK}} \; * \; * \; *\right]$. If the HARQ feedback consists of 2 bits of information $\left[u_0^{\text{ACK}} \; u_1^{\text{ACK}}\right]$, where u_0^{ACK} and u_1^{ACK} correspond to the first and second codewords, respectively, and $u_2^{\text{ACK}} = (u_0^{\text{ACK}} + u_1^{\text{ACK}}) \bmod 2$, then they are encoded depending on the number of coded symbols, e.g., if $N_Q = 2$ then the encoded HARQ feedback is represented as $\left[u_0^{\text{ACK}} \; u_1^{\text{ACK}} \; u_2^{\text{ACK}} \; u_0^{\text{ACK}} \; u_1^{\text{ACK}} \; u_2^{\text{ACK}}\right]$. For some values of N_Q, placeholders "*" are used to scramble the HARQ ACK/NACK

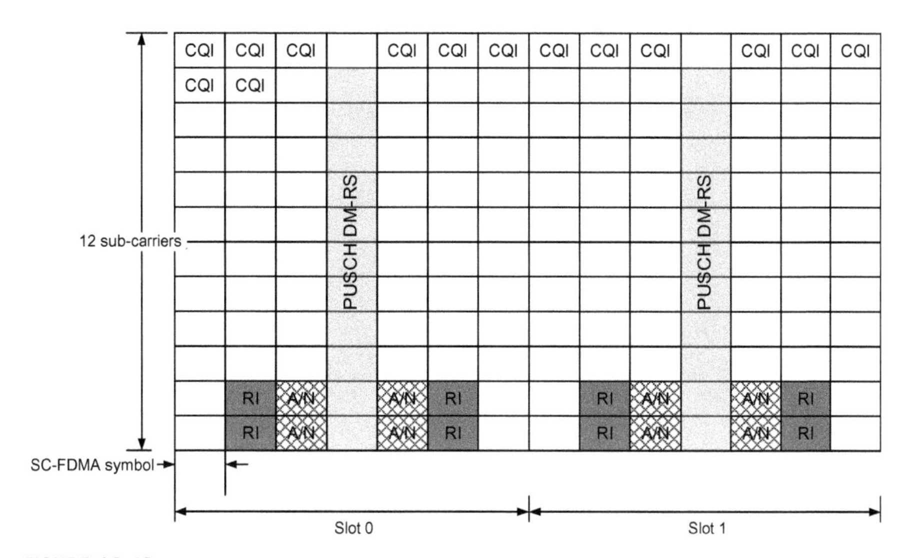

FIGURE 10.43

Multiplexing of control and data bits in PUSCH [15,16].

bits in such a way as to maximize the Euclidean distance of the modulation symbols carrying the HARQ information [4].

Let us further elaborate on the control and data multiplexing scheme on PUSCH. When a UCI is transmitted in a subframe in which the UE has been allocated resources on PUSCH, the UCI is multiplexed with the UL-SCH data prior to DFT operation in order to limit the uplink transmit power. In LTE Rel-8/9, UE is never allocated PUCCH and PUSCH in the same subframe. The number of resource elements reserved for transmission of the control bits in PUSCH is based on the MCS assigned for PUSCH and offset parameters $\beta_{\text{CQI-offset}}$, $\beta_{\text{HARQ-ACK-offset}}$, or $\beta_{\text{RI-offset}}$, which are semi-statically configured by higher layers. This would allow the use of different code rates for different types of UCI. The PUSCH data and UCI are never mapped to the same resource element. The UCI is mapped in such a way that it is present in both slots of the subframe. Since the eNB has prior knowledge of a UCI transmission, it can easily demultiplex the UCI and data. An example of multiplexing of CQI/PMI, HARQ ACK/NACK, and RI with the PUSCH data symbols to uplink resource elements is shown in Figure 10.43. As shown in the figure, the CQI/PMI symbols are mapped first followed by the data symbols. In this example, the RI bits occupy four symbols around DM-RS. The RI symbols are multiplexed with data symbols. The HARQ indicator (HI) ACK/NACK symbols are punctured into data symbols around DM-RS.

A modulation order similar to that used for UL-SCH data is applied to CQI/PMI. For small CQI and/or PMI report sizes up to 11 bits, a $(32, k)$ simplex code,[12] similar to the one used for PUCCH, is used, with optional circular repetition of encoded data and no CRC attached. For large CSI reports with greater than 11 bits, an 8-bit CRC is calculated and attached, and channel coding and rate-matching is performed using the tail-biting convolutional code.

The modulated HARQ ACK/NACK symbols are mapped immediately around the DM-RS position and modulated CQI/PMI symbols are time-first mapped across the subframe. The HARQ feedback resources are punctured into data starting from the lowest sub-carrier in the resource block. The maximum number of resources for HARQ ACK/NACK is four SC-FDMA symbols. In the case of 2-bit HARQ ACK/NACK, a $(3, 2)$ simplex coding is performed (see Table 10.18 for the mapping). The CQI resources are placed at the beginning of the data resources. For a CQI payload less than or equal to 11 bits, $(32, k)$ simplex code is used. For a CQI payload larger than 11 bits, tail-biting convolutional code is utilized. The RI bits are positioned next to the HARQ feedback bits in the resource block, irrespective of the presence or absence of HARQ ACK/NACK bits. In case of a 2-bit RI, a $(3, 2)$ simplex coding is employed.

The HARQ ACK/NACK resources are mapped to SC-FDMA symbols by puncturing the UL-SCH data. The positions next to the DM-RS are used in order to improve channel estimation. The maximum amount of resource for HARQ ACK/NACK is four SC-FDMA symbols. The coded RI symbols are placed next to the HARQ ACK/NACK symbol positions regardless of HARQ ACK/NACK presence in a given subframe. The modulation of 1- or 2-bit

Table 10.18 (3,2) Simplex Code for 2-Bit HARQ ACK/NACK and RI [15]

Input Bits	Output Bits
00	000
01	011
10	101
11	110

[12]Simplex codes belong to the family of linear error-correcting or error-detecting block codes that are simply implemented as polynomial codes by means of shift registers. Considered as (n, k) codes, they have a codeword length of $2^k - 1$. The minimum Hamming distance of binary simplex codes is equal to 2^{k-1}. They can be regarded as Reed–Muller codes shortened by one digit and are identical to the m-sequences of length 2^{k-1}, together with the zero word. They are called simplex because their codewords form a simplex, i.e., a finite graph of k points (the vertices) or a geometric figure, in which every vertex is connected to every other vertex, e.g., a triangle, in Hamming space (a Hamming space is the set of all 2^n binary strings of length n) [28].

HARQ ACK/NACK or RI is such that the Euclidean distance of the modulation symbols carrying HARQ ACK/NACK and RI is maximized; therefore, the outermost constellation points of the higher-order modulations are used, resulting in increased transmit power for ACK/NACK/ and RI transmission relative to the average PUSCH data power. The coding of the RI and CQI/PMI are performed separately, with the UL-SCH data being rate-matched around the RI resource elements similar to the case of CQI/PMI. The reference CQI/PMI MCS is computed from the CSI payload size and resource allocation. The channel coding and rate-matching of the control signaling without UL-SCH data is the same as that of multiplexing control with UL-SCH data. It is possible for the network to configure a UE to repeat each HARQ ACK/NACK transmission in multiple consecutive subframes in order to improve the reliability of the ACK/NACK transmission from power-limited UEs at the cell-edge, particularly in large cell deployments. The number of subframes over which each ACK/NACK transmission is repeated is signaled to the UE by a UE-specific parameter. Therefore, four different channel coding methods may be applied to control signals that are transmitted on PUSCH as follows:

- Repetition coding when 1-bit ACK/NACK is transmitted.
- $(3, 2)$ simplex coding when 2-bit ACK/NACK or RI are transmitted.
- $(32, k)$ Reed–Muller block codes are used when CQI/PMI information is less than 11 bits.
- Tail-biting convolutional coding with basic code rate of 1/3 is used when CQI/PMI information is greater than 11 bits.

As shown in Figure 10.37, the block of bits $b_0^{\mathrm{cw}}, b_1^{\mathrm{cw}}, \ldots, b_{N_{\mathrm{bit}}(\mathrm{cw})}^{\mathrm{cw}}$, in which $N_{\mathrm{bit}}(\mathrm{cw})$ denotes the number of bits transmitted in codeword cw on PUSCH in one subframe, are scrambled with a UE-specific scrambling sequence prior to modulation, resulting in a block of scrambled bits $\overline{b}_0^{\mathrm{cw}}, \overline{b}_1^{\mathrm{cw}}, \ldots, \overline{b}_{N_{\mathrm{bit}}(\mathrm{cw})}^{\mathrm{cw}}$. Note that each codeword is bit-wise multiplied with an orthogonal sequence and a UE-specific scrambling sequence to create a sequence of symbols for each codeword. The parameter $N_{\mathrm{bit}}(\mathrm{cw})$ is the number of bits transmitted using codeword cw. The scrambling sequence is a pseudo-random sequence based on length-31 Gold sequence which is initialized using the radio network temporary identifier associated with PUSCH transmission n_{RNTI}, the cell identifier $N_{\mathrm{ID}}^{\mathrm{cell}}$, the slot number within the radio frame n_s, and the codeword index at the start of each subframe as $c_{\mathrm{init}} = n_{\mathrm{RNTI}} 2^{14} + cw 2^{13} + \lfloor n_s/2 \rfloor 2^9 + N_{\mathrm{ID}}^{\mathrm{cell}}|_{\mathrm{cw}=0,1}$. Scrambling with a cell-specific sequence serves the purpose of inter-cell interference mitigation. When a base station descrambles a received bit stream with a known cell-specific scrambling sequence, interference from other cells will be descrambled incorrectly and therefore only appear as uncorrelated noise. The scrambled codeword undergoes QPSK, 16QAM, or 64QAM modulation to generate complex-valued symbols, which provides the flexibility to optimize the data transmission depending on the channel conditions.

For each codeword cw, the block of scrambled bits $\overline{b}_0^{\mathrm{cw}}, \overline{b}_1^{\mathrm{cw}}, \ldots, \overline{b}_{N_{\mathrm{bit}}(\mathrm{cw})}^{\mathrm{cw}}$ is modulated as described in 3GPP TS 36.211 [3], resulting in a block of complex-valued symbols $d_0^{\mathrm{cw}}, d_1^{\mathrm{cw}}, \ldots, d_{N_s(\mathrm{cw})}^{\mathrm{cw}}$. The complex-valued modulation symbols for each of the codewords to be transmitted are mapped to one or two layers. Complex-valued modulation symbols $d_0^{\mathrm{cw}}, d_1^{\mathrm{cw}}, \ldots, d_{N_s(\mathrm{cw})}^{\mathrm{cw}}$ associated with codeword cw are mapped to spatial layers $\mathbf{x}(i) = [x_0(i), x_1(i), \ldots, x_{N_l-1}(i)]^T |_{i=0,1,\ldots,N_s^l-1}$, where N_l is the number of layers and N_s^l is the number of modulation symbols per layer.

The precoding function in the uplink is not the same as that in the downlink (multi-antenna) precoding in LTE Rel-8/9. The block of complex-valued symbols $d_0^{\mathrm{cw}}, d_1^{\mathrm{cw}}, \ldots, d_{N_s(\mathrm{cw})}^{\mathrm{cw}}$ is divided into $N_s(cw)/N_{\mathrm{sc}}^{\mathrm{PUSCH}}$ sets. Each set of size $N_{\mathrm{sc}}^{\mathrm{PUSCH}} = N_{\mathrm{DFT}}$ corresponds to one SC-FDMA symbol. A DFT is then applied to each set, which is essentially the precoding part of the SC-FDMA modulation. The size of the DFT, i.e., the value of $N_{\mathrm{sc}}^{\mathrm{PUSCH}} = N_{\mathrm{DFT}}$, must be a prime number that is a product of 2, 3, and/or 5 and satisfies the following equation $N_{\mathrm{RB}}^{\mathrm{PUSCH}} = 2^{\alpha_2} 3^{\alpha_3} 5^{\alpha_5} \leq N_{\mathrm{RB}}^{\mathrm{UL}}$, where α_2, α_3, and α_5 are a set of non-negative integers. The final stage in PUSCH physical layer processing is to map the symbols to the allocated physical resource elements. The allocation sizes are limited to values which are multiples of 2, 3, and 5 as described earlier. The latter limitation is imposed by the precoding stage. The symbols are mapped in increasing order beginning with sub-carriers, then SC-FDMA symbols. The SC-FDMA symbols carrying DM-RS or SRS are avoided during the mapping process. Figure 10.44 illustrates the order of mapping of the output of the precoder to the allocated resource blocks.

The codeword processing structure shown in Figure 10.37 includes enhancements that were added as part of LTE Rel-10, where the complex-valued symbols are mapped to one or more spatial layers following the baseband modulation. The layer mapping function maps the codeword(s) to spatial layer (s). For transmission on a single antenna port, a single layer is used and the antenna mapping is defined by $x_0(i) = d_0(i)|_{i=0,1,\ldots,N_s^l-1}$. For uplink spatial multiplexing, the number of layers N_l is less than or equal to the number of antenna ports N_p used for transmission of the PUSCH. The case of a single codeword mapped to multiple layers is only applicable when there are four antenna ports for PUSCH transmission [3].

As illustrated in Figure 10.37, the modulation symbols in units of $N_{\mathrm{sc}}^{\mathrm{PUSCH}} = 12 N_{\mathrm{RB}}^{\mathrm{PUSCH}}$ symbols are fed to $N_{\mathrm{sc}}^{\mathrm{PUSCH}}$ – point DFT, where $N_{\mathrm{sc}}^{\mathrm{PUSCH}}$ denotes the number of sub-carriers assigned to the transmission of PUSCH. The reason for the precoding is to reduce the cubic metric of the transmitted signal. For LTE, the DFT size and thus the size of resource allocation is limited to products of integer powers of 2, 3, and 5 ($N_{\mathrm{RB}}^{\mathrm{PUSCH}} = 2^{\alpha_2} 3^{\alpha_3} 5^{\alpha_5} \leq N_{\mathrm{RB}}^{\mathrm{UL}}$, where α_2, α_3, and α_5 are non-negative integers). For each layer, the block of complex-valued symbols $x^{(l')}(0), x^{(l')}(1), \ldots, x^{(l')}(N_s^l - 1)|_{l'=0,1,\ldots,N_l-1}$ is divided into $N_s^l/N_{\mathrm{sc}}^{\mathrm{PUSCH}}$ sets, each corresponding to one SC-FDMA symbol. The transform

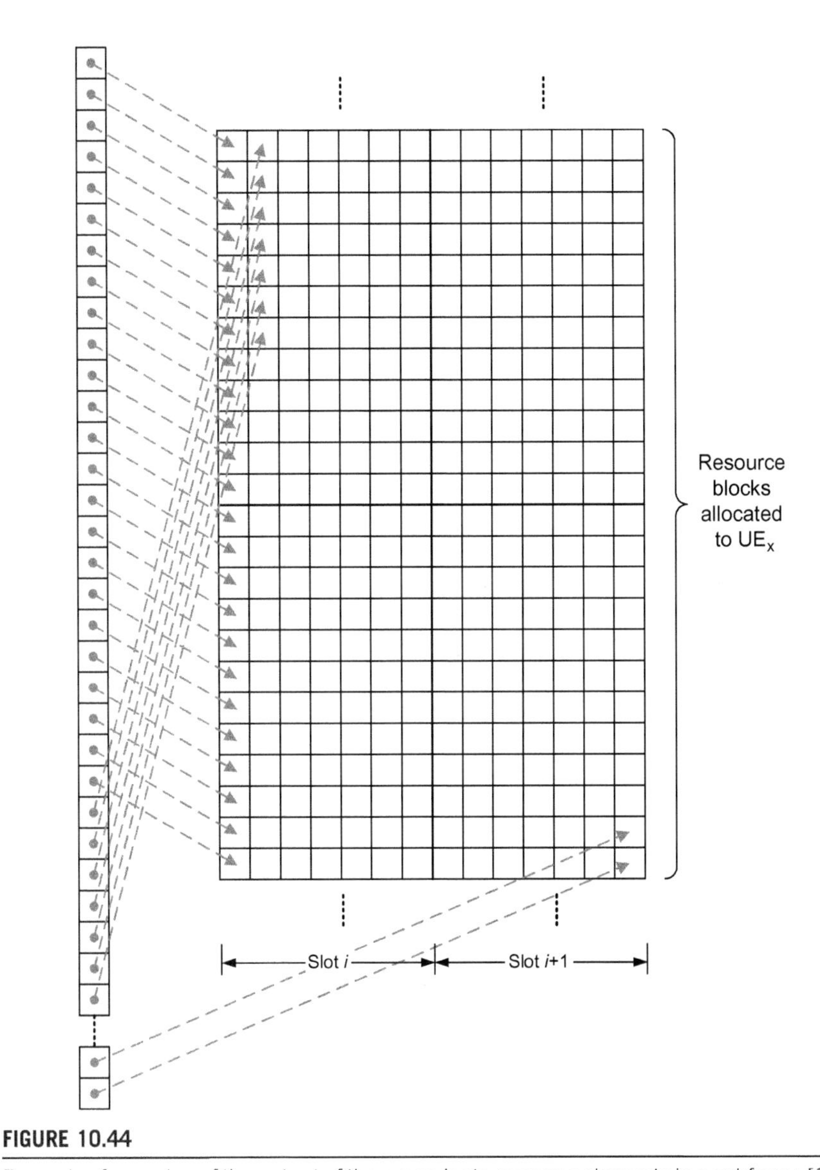

FIGURE 10.44

Example of mapping of the output of the precoder to resource elements in a subframe [155].

precoding is applied to each set, resulting in a block of complex-valued symbols $y^{(l')}(0), y^{(l')}(1), \ldots, y^{(l')}(N_s^l - 1)$ as follows (Table 10.19):

$$y^{(l')}(lN_{sc}^{PUSCH} + k) = \frac{1}{\sqrt{N_{sc}^{PUSCH}}} \sum_{i=0}^{M_{sc}^{PUSCH}-1} x^{(l')}(lN_{sc}^{PUSCH} + i)$$

$$\exp\left(-j\frac{2\pi ik}{N_{sc}^{PUSCH}}\right)\Big|_{k=0,1,\ldots,N_{sc}^{PUSCH}-1;\,l=0,1,\ldots,N_s^l/N_{sc}^{PUSCH}-1}$$

Table 10.19 Codeword-to-Layer Mapping for Uplink Spatial Multiplexing [3]

Number of Layers N_l	Number of Codewords N_{CW}	Layer Mapping $i = 0, 1, \ldots, N_s^l - 1$
1	1	$x_0(i) = d_0(i)$
2	1	$x_0(i) = d_0(2i)$ $x_1(i) = d_0(2i+1)$
2	2	$x_0(i) = d_0(i)$ $x_1(i) = d_1(i)$
3	2	$x_0(i) = d_0(i)$ $x_1(i) = d_1(2i)$ $x_2(i) = d_1(2i+1)$
4	2	$x_0(i) = d_0(2i)$ $x_1(i) = d_0(2i+1)$ $x_2(i) = d_1(2i)$ $x_3(i) = d_1(2i+1)$

The precoder module takes as input a block of vectors $\mathbf{y}(i) = (y^{(0)}(i), y^{(1)}(i), \ldots, y^{(N_l-1)}(i))^{\mathrm{T}}|_{i=0,1,\ldots,N_s^l-1}$ from the transform precoder, and generates a block of vectors $\mathbf{z}(i) = \mathbf{W}\mathbf{y}(i) = (z^{(0)}(i), \ z^{(1)}(i), \ldots, z^{(N_p-1)}(i))^{\mathrm{T}}|_{i=0,1,\ldots,N_s^p-1}$ that are mapped to resource elements. The precoding matrix $\mathbf{W}$ (codebook) is an $N_p \times N_l$ matrix and is given for $N_p = 2$ in Table 10.20 [3].

For each antenna port $\overline{p}$ that is used for transmission of PUSCH in a subframe, the block of complex-valued symbols $z^{(\overline{p})}(0), z^{(\overline{p})}(1), \ldots, z^{(\overline{p})}(N_s^p - 1)$ are multiplied with an amplitude scaling factor β_{PUSCH} in order to adjust the transmit power P_{PUSCH} and are then mapped sequentially, starting with $z^{(\overline{p})}(0)$, to PRBs on antenna port $\overline{p}$ assigned for transmission of PUSCH. The relation between the index $\overline{p}$ and the antenna port number p is given in 3GPP TS 36.211 [3]. The resource element (k, l) that is assigned to PUSCH information is required to meet certain criteria including the following:

- It is not used for transmission of reference signals.
- It is not part of the last SC-FDMA symbol in a subframe, if the UE transmits SRS in the same subframe.
- It is not part of the last SC-FDMA symbol in a subframe configured with cell-specific SRS, if the PUSCH transmission partly or fully overlaps with the cell-specific SRS bandwidth.
- It is not part of an SC-FDMA symbol reserved for possible SRS transmission in a UE-specific aperiodic SRS subframe.

The mapping to physical resource elements, as shown in Figure 10.44, is in increasing order of first, the sub-carrier (frequency) index k, and then the symbol

Table 10.20 Precoding Codebook for Transmission with Two Antenna Ports [3]

Codebook Index	Number of Layers	
	Single Layer	Dual Layer
0	$\frac{1}{\sqrt{2}}\begin{bmatrix}1\\1\end{bmatrix}$	$\frac{1}{\sqrt{2}}\begin{bmatrix}1 & 0\\0 & 1\end{bmatrix}$
1	$\frac{1}{\sqrt{2}}\begin{bmatrix}1\\-1\end{bmatrix}$	–
2	$\frac{1}{\sqrt{2}}\begin{bmatrix}1\\j\end{bmatrix}$	–
3	$\frac{1}{\sqrt{2}}\begin{bmatrix}1\\-j\end{bmatrix}$	–
4	$\frac{1}{\sqrt{2}}\begin{bmatrix}1\\0\end{bmatrix}$	–
5	$\frac{1}{\sqrt{2}}\begin{bmatrix}0\\1\end{bmatrix}$	–

(time) index l, starting with the first slot in the subframe. If uplink frequency hopping is disabled or the resource blocks allocated to PUSCH transmission are not contiguous in frequency, the set of PRBs to be used for transmission are identified by $n_{\text{PRB}} = n_{\text{VRB}}$, where n_{VRB} is obtained from the uplink scheduling grant. If uplink frequency hopping with a predefined hopping pattern is enabled, the set of PRBs used for PUSCH transmission in slot n_s is identified by the scheduling grant along with a predefined pattern according to the following equation [3]:

$$\underbrace{n'_{\text{PRB}}(n_s)}_{\substack{\text{Allocated PRB}\\ \text{with adjusted}\\ \text{indexing starting at}\\ \text{0 following}\\ \text{PUCCH allocation}}} = \left(\underbrace{n'_{\text{VRB}}}_{\substack{\text{Allocated}\\ \text{VRB with adjusted indexing}\\ \text{starting at 0 following}\\ \text{PUCCH allocation}}} + \underbrace{f_{\text{hopping}}(i)L_{\text{sub-band}}}_{\text{Hopping function}} + \underbrace{\left((L_{\text{sub-band}}-1)-2(n'_{\text{VRB}} \bmod L_{\text{sub-band}})\right)f_{\text{mirror}}(i)}_{\text{Mirroring function}} \right) \bmod (L_{\text{sub-band}} N_{\text{sub-band}}) \quad (10.11)$$

In the above equation $i = \lfloor n_s/2 \rfloor$ for the case of inter-subframe hopping and $i = n_s$ for the case of intra- and inter-subframe hopping, $n_{\text{PRB}}(n_s) = n'_{\text{PRB}}(n_s) + \lceil N_{\text{RB}}^{\text{offset}}/2 \rceil | N_{\text{sub-band}} > 1$, and $n'_{\text{VRB}} = n_{\text{VRB}} - \lceil N_{\text{RB}}^{\text{offset}}/2 \rceil | N_{\text{sub-band}} > 1$. The parameter n_{VRB} is obtained from the scheduling grant and PUSCH hopping offset $N_{\text{RB}}^{\text{offset}}$ is provided by the higher layers. The size of each sub-band is given as $L_{\text{sub-band}} = \lfloor (N_{\text{RB}}^{\text{UL}} - N_{\text{RB}}^{\text{offset}} - N_{\text{RB}}^{\text{offset}} \bmod 2)/N_{\text{sub-band}} \rfloor | N_{\text{sub-band}} > 1$. The number

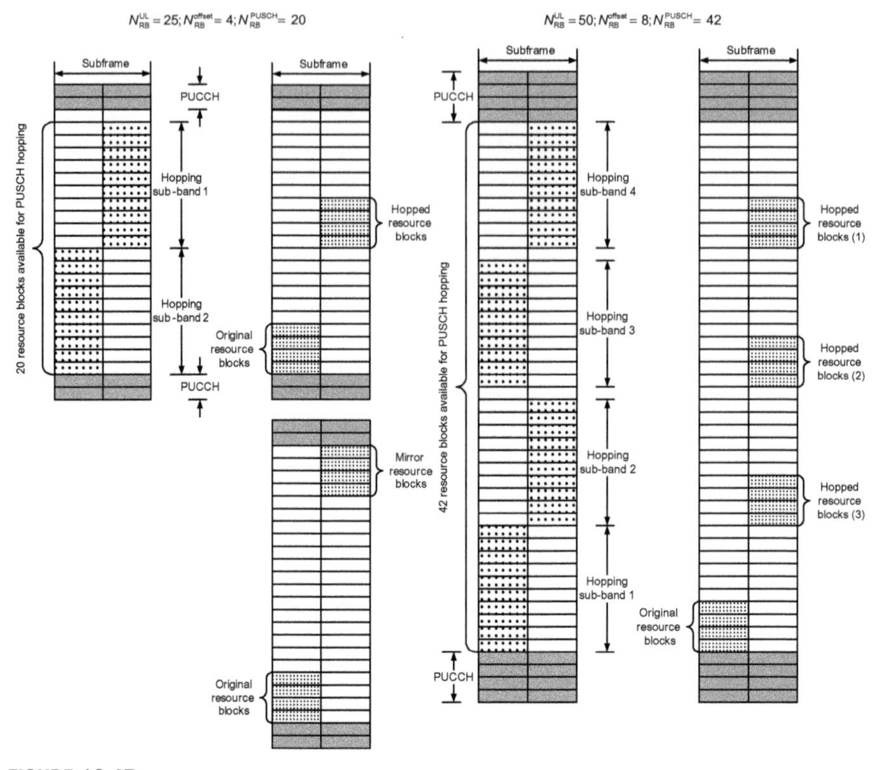

FIGURE 10.45

Example of PUSCH resource hopping/mirroring in 5 and 10 MHz channels [14].

of sub-bands $N_{\text{sub-band}}$ is signaled by the higher layers. The binary function $f_{\text{mirror}}(i) \in \{0, 1\}$ determines if the mirroring function is used.

In the above equation, the component identified as hopping function shifts the VRB allocation by an integer number of sub-bands. The shift is defined using a pseudo-random sequence which can be calculated by the UE and the eNB. In type 1 frame structure, the pseudo-random sequence is initialized using the physical layer cell identity $N_{\text{ID}}^{\text{cell}}$, whereas in type 2 frame structure, the pseudo-random sequence is initialized using a combination of the physical layer cell identity and the radio frame number $(2^9(n_f \bmod 4) + N_{\text{ID}}^{\text{cell}})$. The mirroring component in the above equation mirrors the VRB allocation within the sub-band. Mirroring is applied when the binary pseudo-random function $f_{\text{mirror}}(i) = 1$, which results in reflecting a VRB allocation starting "x" resource blocks from the bottom of the sub-band around the center of the sub-band such that the mirrored VRB allocation starts "x" resource blocks from the top of the sub-band (Figure 10.45). The pseudo-random mirroring component is initialized similar to the hopping component. Modular arithmetic is applied to the result of the hopping pattern function to provide wrap around, which ensures that the resource blocks after hopping remain within the boundaries of the subframe.

Note that in addition to the hopping pattern, there is also a cell-specific mirroring pattern defined in a cell. The mirroring pattern controls, on a slot-by-slot basis, whether mirroring within each sub-band should be applied to the assigned resource blocks. In principle, mirroring implies that the resource blocks within each sub-band are numbered in reversed order. Figure 10.45 illustrates an example of PUSCH resource allocations assuming 5 and 10 MHz channel bandwidth. In this example, PUCCH occupies four or eight resource blocks (depending on the bandwidth) in each slot, and the UE is informed through higher-layer signaling that those resource blocks are not available for hopping. The PUSCH hopping offset N_{RB}^{offset} is used for this purpose. The value of the hopping offset is rounded up to an even number, if the received value is not even.

As shown in Figure 10.45, there are two hopping modes specified in LTE/LTE-Advanced: (1) inter-subframe hopping where the resource blocks allocated by DCI format 0 are used for even-numbered transmission of uplink transport blocks, while hopped resource blocks are used for odd-numbered transmission of uplink resource blocks; and (2) intra- and inter-subframe hopping where the resource blocks allocated by DCI format 0 are used in the first time slot within the subframe while the hopped resource blocks are used in the second time slot [14].

10.7.4 Physical random-access channel

The physical random access channel is used by non-synchronized UEs during initial system access, handover, and prior to downlink data transmission. The UE transmits a preamble sequence over PRACH, which may be randomly selected from a group of sequences assigned by the eNB. The structure of the sequence is shown in Figure 10.46. The sequences are Zadoff–Chu sequences of length 839 for the FDD mode (139 for TDD mode). There are 64 different cyclic shifts that are generated from each root Zadoff–Chu sequence with zero cross-correlation. Therefore, different cyclic shifts of the base sequence are mutually orthogonal. The PRACH resource consists of six consecutive resource blocks. The location of PRACH is indicated by an offset parameter signaled by higher layers. There is one PRACH region per subframe in the FDD mode. Multiple UEs can access the PRACH region using different preambles. The UEs may select preambles randomly, which may result in a collision. Therefore, contention resolution mechanisms are necessary. The PRACH may not always present in every subframe and every radio frame. The PRACH configuration parameter indicates frame number and subframe numbers in which the PRACH resource is available.

The sub-carrier spacing for PRACH transmission is different from that of other channels in the uplink. The number of sub-carriers is equal to the length of preamble (839 for the FDD mode and 139 for TDD). The sub-carrier spacing is 1.25 kHz, resulting in longer symbol duration of 800 μs for the FDD mode (133 μs for TDD mode). The sub-carrier spacing also results in an effective guard band because the entire 1.08 MHz (six resource blocks) is not used for PRACH transmission.

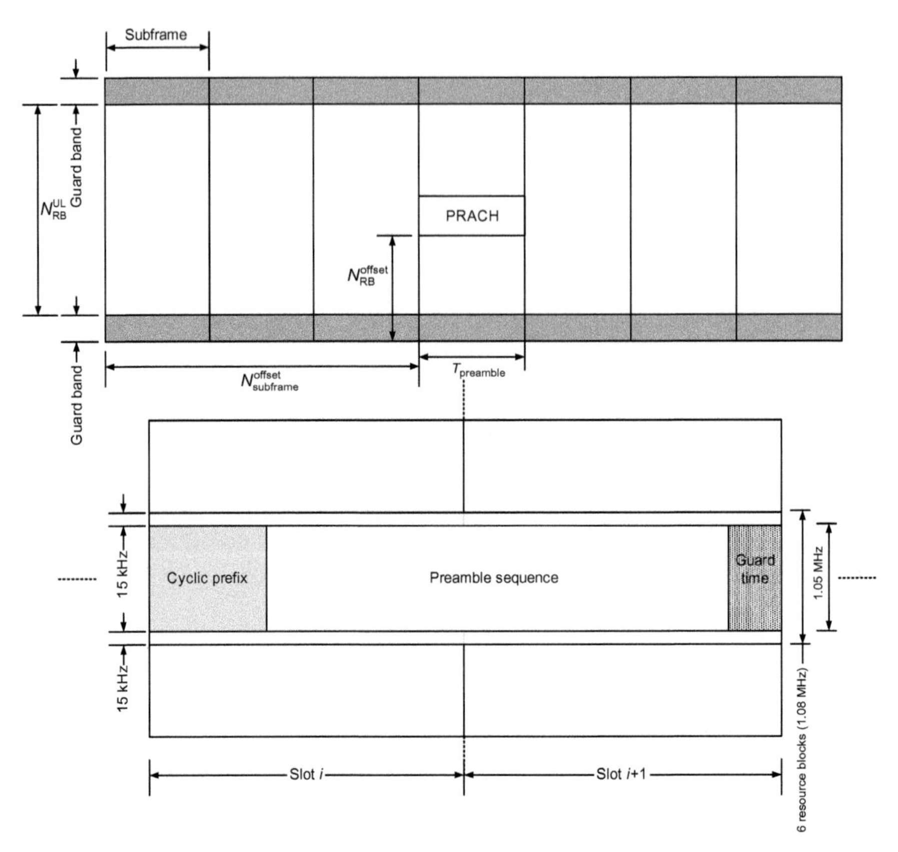

FIGURE 10.46

Structure of PRACH in time- and frequency-domain [14,153].

The PRACH preamble is an OFDM-based signal, but it is generated using a different structure from other uplink transmissions; most notably it uses narrower sub-carrier spacing which is not orthogonal to the PUSCH, PUCCH, and SRS; therefore, those channels will suffer from some interference from the PRACH. However, the sub-carrier spacing used by PRACH is an integer sub-multiple of the sub-carrier spacing used for the other channels, thereby PUSCH, PUCCH, and SRS do not interfere on the PRACH. Prior to initiation of the non-synchronized physical random-access procedure, the physical layer receives random-access channel parameters (i.e., PRACH configuration and time/frequency position), and parameters for determining the root sequences and their cyclic shifts in the preamble sequence set for the primary cell (i.e., index to logical root sequence table, cyclic shift N_{CS}, and whether set type is unrestricted or restricted) from the higher layers.

The PRACH preamble consists of a cyclic prefix, useful part of the sequence, and then a guard period which is simply an unused portion of time up to the end of the last subframe occupied by PRACH. This guard period allows for timing

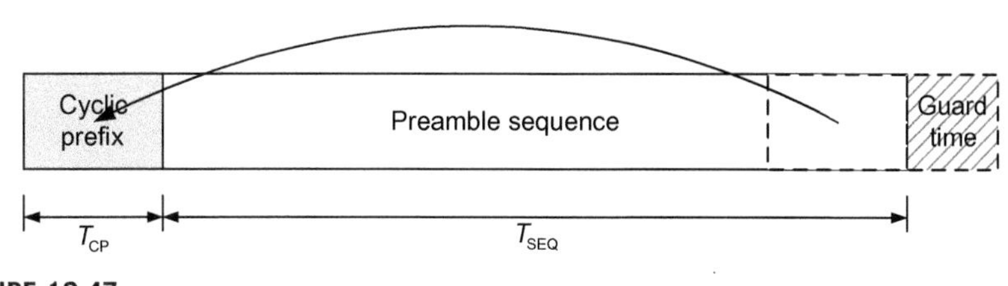

FIGURE 10.47

PRACH preamble structure in time.

uncertainty due to variability of the UE to eNB distance. Therefore, the size of the guard period determines the cell radius, as any propagation delay exceeding the guard time (GT) would cause the random-access preamble to overlap with the following subframe at the eNB receiver. The use of OFDM transmission with a cyclic prefix allows for implementation of an efficient frequency-domain-based receiver in the eNB to perform PRACH detection.

The UE aligns the start of the random-access preamble with the start of the corresponding uplink subframe at the UE, assuming a timing advance of zero. The preamble length is shorter than the PRACH slot in order to accommodate a gap for GT in order to absorb the variable propagation delay. Figure 10.48 shows two preambles received at the eNB with different timings depending on the propagation delay. The PRACH preamble sequence is optimized relative to its periodic autocorrelation property [15]. The random-access preamble, illustrated in Figure 10.47, consists of a cyclic prefix of length T_{CP} and a sequence of length T_{SEQ}. The parameter values are given in Table 10.21 and are configured by higher layers.

The transmission of the random-access preamble is triggered by the MAC sublayer and is restricted to certain time and frequency resources. These resources are identified in increasing order of the subframe number within the radio frame and the PRBs in the frequency-domain such that index 0 corresponds to the lowest numbered PRB and subframe within the radio frame. The PRACH resources within the radio frame are identified by a PRACH resource index. For frame structure type 1 with preamble formats 0–3, there is a maximum of one random-access resource per subframe. The start of the random-access preamble is aligned with the start of the corresponding uplink subframe at the UE assuming $N_{TA}=0$, where N_{TA} is the length of the timing advance. The first PRB N_{RA} allocated to the PRACH transmission corresponding to preamble formats 0, 1, 2, and 3 is defined as $N_{RA}=N_{RA}^{offset}$, where the PRACH frequency offset parameter $0 \le N_{RA}^{offset} \le N_{RB}^{UL}-6$ is defined as a PRB number configured by the higher layers (see Figure 10.46). For frame structure type 2 supporting PRACH preamble formats 0–4, there is a possibility for multiple random-access resources in an uplink subframe (or UpPTS for preamble format 4) depending on the uplink/downlink ratio. The PRACH sequence duration T_{SEQ} is chosen taking the following considerations into account (Table 10.22):

- There is a trade-off between sequence length and overhead, where on the one hand, the sequence length must be as long as possible to maximize the number

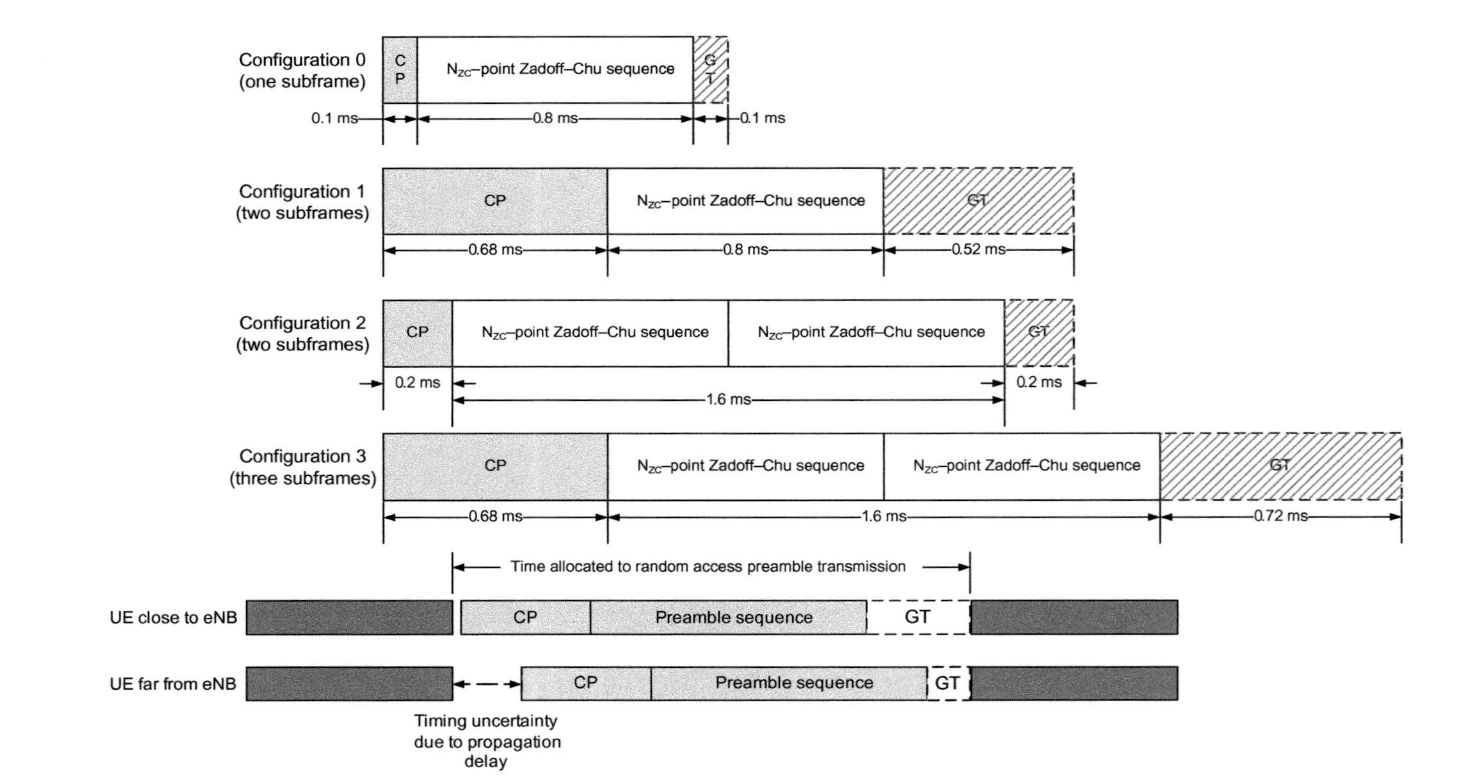

FIGURE 10.48

Illustration of PRACH preamble formats and mitigation of the impact of variable propagation delay on PRACH preamble [16].

Table 10.21 PRACH Preamble Parameters [3,14]

PRACH Preamble Format	Cyclic Prefix T_{CP}(μs)	Sequence Duration T_{SEQ}(μs)	Guard Time (μs)	Preamble Length (ms)	Maximum Supported Cell Range (km)
0	103.13	800	96.88	1	14.5
1	684.38	800	515.53	2	77.3
2	203.13	1600	196.88	2	29.5
3	684.38	1600	715.63	3	100.2
4†	14.58	133	9.38	0.16	1.4

†*Applicable to frame structure type 2 and special subframe configurations with UpPTS lengths of* 142.71 μs *(2 symbols with normal cyclic prefix) and* 166.67 μs *(2 symbols with extended cyclic prefix).*

of available orthogonal preambles, while on the other hand, the sequence must fit within a subframe in order to minimize the PRACH overhead. The lower bound for T_{SEQ} should support unambiguous round-trip time estimation for a UE located at the edge of the largest expected cell size including the maximum delay spread expected in such large cells.

- The length of the cyclic prefix must encompass the maximum round-trip delay and the maximum delay spread of the channel. Furthermore, for a cell-edge UE, the delay-spread energy at the end of the preamble is replicated at the end of the cyclic prefix, and is therefore within the preamble interval. Consequently, there is no need to include the maximum delay spread in the GT. Therefore, instead of locating the sequence in the center of the PRACH slot, it is shifted later by half of the maximum delay spread, allowing the maximum round-trip delay to be increased by the same amount. Note that, similar to a regular OFDM symbol, the residual delay spread at the end of the preamble from a cell-edge UE leaks into the next subframe, but this is taken care of by the cyclic prefix at the start of the next subframe to avoid any ISI.
- The sub-carrier spacing of PRACH and PUSCH must be numerically proportional. An additional benefit of this property is the possibility to reuse the FFT module from the SC-FDMA signal processing for PUSCH to implement the large DFT block involved in the PRACH transmitter and receiver. It is further desirable to minimize the orthogonality loss in the frequency-domain between the preamble sub-carriers and the sub-carriers of the adjacent uplink data transmissions. This is achieved if the PUSCH data symbol sub-carrier spacing Δf_{PUSCH} is an integer multiple of the PRACH sub-carrier spacing Δf_{RA}.
- Under a noise-limited scenario, which is typical in low density and medium to large suburban or rural cells, coverage performance can be estimated from a link budget calculation. Assuming an Okumura–Hata empirical model of

Table 10.22 Frame Structure Type 1 Random-Access Configuration [3]

Configuration Index	Preamble Format	System Frame Number (SFN)	Subframe Number	Configuration Index	Preamble Format	System Frame Number (SFN)	Subframe Number
0	0	Even	1	32	2	Even	1
1	0	Even	4	33	2	Even	4
2	0	Even	7	34	2	Even	7
3	0	Any	1	35	2	Any	1
4	0	Any	4	36	2	Any	4
5	0	Any	7	37	2	Any	7
6	0	Any	1, 6	38	2	Any	1, 6
7	0	Any	2,7	39	2	Any	2,7
8	0	Any	3, 8	40	2	Any	3, 8
9	0	Any	1, 4, 7	41	2	Any	1, 4, 7
10	0	Any	2, 5, 8	42	2	Any	2, 5, 8
11	0	Any	3, 6, 9	43	2	Any	3, 6, 9
12	0	Any	0, 2, 4, 6, 8	44	2	Any	0, 2, 4, 6, 8
13	0	Any	1, 3, 5, 7, 9	45	2	Any	1, 3, 5, 7, 9
14	0	Any	0, 1, 2, 3, 4, 5, 6, 7, 8, 9	46	N/A	N/A	N/A

15	0	Even	9	47	2	Even	9
16	1	Even	1	48	3	Even	1
17	1	Even	4	49	3	Even	4
18	1	Even	7	50	3	Even	7
19	1	Any	1	51	3	Any	1
20	1	Any	4	52	3	Any	4
21	1	Any	7	53	3	Any	7
22	1	Any	1, 6	54	3	Any	1, 6
23	1	Any	2,7	55	3	Any	2,7
24	1	Any	3, 8	56	3	Any	3, 8
25	1	Any	1, 4, 7	57	3	Any	1, 4, 7
26	1	Any	2, 5, 8	58	3	Any	2, 5, 8
27	1	Any	3, 6, 9	59	3	Any	3, 6, 9
28	1	Any	0, 2, 4, 6, 8	60	N/A	N/A	N/A
29	1	Any	1, 3, 5, 7, 9	61	N/A	N/A	N/A
30	N/A	N/A	N/A	62	N/A	N/A	N/A
31	1	Even	9	63	3	Even	9

distance-dependent path loss, the PRACH signal power received at the eNB can be written as follows [15]:[13]

$$\underbrace{P_{\mathrm{RA}}(R)}_{\substack{\text{PRACH signal power}\\ \text{at the eNB}}} = \underbrace{P_{\max}}_{\substack{\text{UE transmitter}\\ \text{EIRP}}} + \underbrace{G_{\mathrm{RX}}}_{\substack{\text{eNB receiver}\\ \text{antenna gain}\\ \text{(including cable loss)}}} - \underbrace{P_{\mathrm{L}}(R)}_{\substack{\text{Distance-dependent}\\ \text{path loss}}} - \underbrace{P_{\mathrm{PL}}}_{\substack{\text{Penetration}\\ \text{loss}}} - \underbrace{P_{\mathrm{FM}}}_{\substack{\text{Log-normal}\\ \text{fade margin}}} \tag{10.12}$$

In the above equation, variable R is the cell radius in kilometers. From the above path-loss estimate, the initial PRACH transmission power is obtained by adding a configurable offset. The random-access mechanism allows power ramping where the actual PRACH transmission power is increased following each unsuccessful random-access attempt. For the first attempt, the PRACH transmission power is set to the initial PRACH power. In most cases, this is sufficient for the random-access attempt to be successful. However, if the random-access attempt fails, the PRACH transmission power for the next attempt is increased by a configurable step size to increase the likelihood of the next attempt being successful. Since the random-access preamble is orthogonal to the user data, the need for power ramping to control intra-cell interference is more relaxed than in other systems with non-orthogonal random-access, and in many cases the transmission power is set such that the first random-access attempt with a high likelihood is successful, which results in lower access latencies.

As mentioned earlier, PRACH uses a narrower sub-carrier spacing than normal uplink transmission, specifically 1250 Hz for random-access preamble formats 0–3, and 7500 Hz for random-access preamble format 4. The ratio of the normal uplink sub-carrier spacing to PRACH sub-carrier spacing is 12 for random-access preamble formats 0–3 and 2 for random-access preamble format 4. The PRACH is designed to fit in the same bandwidth as six resource blocks of normal uplink transmission, i.e., 72 sub-carriers at 15,000 Hz sub-carrier spacing (1.08 MHz). This simplifies scheduling gaps in normal uplink transmission to allow for PRACH transmission opportunities. Therefore, there are 864 sub-carriers for random-access preamble formats 0–3, and 144 sub-carriers for random-access preamble format 4. However, in practice, random-access preamble formats 0–3 use 839 active sub-carriers, and random-access preamble format 4 uses 139 active sub-carriers.

---

[13]Effective isotropic radiated power (EIRP), also known as equivalent isotropic radiated power, is the measured radiated power in a single direction (i.e., for fixed azimuth and elevation angles). Typically, for antenna radiation pattern measurement, if a single value of EIRP is given, this will be the maximum value of the EIRP over all measured angles. EIRP can also be interpreted as the amount of power that a perfectly isotropic antenna would need to radiate to achieve the same measured value.

Similar to normal uplink SC-FDMA transmission, there is a shift equal to half sub-carrier spacing (7500 Hz), which for PRACH is equivalent to six sub-carrier shift. A further sub-carrier offset (seven sub-carriers for random-access preamble formats 0–3 and two sub-carriers for random-access preamble format 4) centers the PRACH transmission within the 1.08 MHz transmission bandwidth. In practice, PRACH sequence is an OFDM-based reconstruction of a Zadoff–Chu sequence in the time-domain. The OFDM modulator is used to position the Zadoff–Chu sequence in the frequency-domain, i.e., to place six resource blocks of PRACH in six consecutive resource blocks starting from a particular PRB denoted as N_{RA}. The output of the OFDM modulator in the time-domain is a Zadoff–Chu sequence, if the input to the OFDM modulator is a Zadoff–Chu sequence in the frequency-domain. Therefore, N_{ZC} active sub-carriers are set to the values of an N_{ZC}-point DFT of an N_{ZC}-sample Zadoff–Chu sequence.

The cyclically shifted Zadoff–Chu sequences have several properties including the following: the amplitude of the sequences is constant, which ensures efficient PA utilization and maintains the low PAPR properties of the single-carrier transmitter; the sequences further have ideal cyclic autocorrelation, which is important for obtaining an accurate timing estimation at the eNB. Finally, the cross-correlation of different preambles based on cyclic shifts of the same Zadoff–Chu root sequence is zero at the receiver as long as the cyclic shift N_{CS} used when generating the preambles is larger than the maximum round-trip propagation time in the cell plus the maximum delay spread of the channel. Therefore, due to the ideal cross-correlation property, there is no intra-cell interference from multiple random-access attempts using preambles derived from the same Zadoff–Chu root sequence. To support different cell sizes, the cyclic shift N_{CS} is broadcast as part of the system information. Therefore, in smaller cells, a small cyclic shift can be configured, resulting in a larger number of cyclically shifted sequences from each root sequence. For cell sizes under 1.5km, all 64 preambles can be generated from a single root sequence. In larger cells, a larger cyclic shift is configured in order to generate 64 preamble sequences, since a greater number of root Zadoff–Chu sequences are in use in the cell. Although the larger number of root sequences is not an issue, the zero cross-correlation property only holds between shifts of the same root sequence and from an interference perspective, it is advantageous to use as few root sequences as possible.

The detection of the random-access preamble, in principle, is based on the cross-correlation of the received signal with the root Zadoff–Chu sequences. One disadvantage of Zadoff–Chu sequences is the difficulty in separating a frequency offset from the distance-dependent delay. A frequency offset results in an additional correlation peak in the time-domain, a correlation peak that corresponds to a spurious terminal-to-base-station distance. In addition, the actual correlation peak is attenuated. At low-frequency offsets, this effect is negligible and the detection performance is hardly affected. However, at high Doppler frequencies, the spurious correlation peak can be larger than the true peak. This results in erroneous detection, i.e., the correct preamble may not be detected or the delay estimate may be incorrect. To avoid the ambiguities resulting from spurious correlation peaks, the set of preamble sequences generated from each root sequence can be restricted. Restrictions imply that only some of the sequences that can be generated from a root sequence are used to define random-access preambles. The location of the spurious correlation peak relative to the actual peak depends on the root sequence; thus different restrictions should be applied for different root sequences. The restrictions are broadcast by the eNB as part of the system information [16].

The transmission opportunities for PRACH are allocated in time first and then in frequency, if time multiplexing is not sufficient to hold all opportunities of a PRACH configuration needed for a certain density value D_{RA} without overlapping in time, for random-access preamble formats 0–3, frequency multiplexing is performed according to the following equation: $N_{RA} = N_{RA}^{offset} + 6\lfloor \frac{f_{RA}}{2} \rfloor$ if $f_{RA} \bmod 2 = 0$; otherwise $N_{RB}^{UL} - 6 - N_{RA}^{offset} - 6\lfloor \frac{f_{RA}}{2} \rfloor$, where N_{RB}^{UL} is the total number of uplink resource blocks (uplink bandwidth), f_{RA} is a frequency resource index within the considered time instance, N_{RA} is the first PRB allocated to the PRACH transmission, and PRACH frequency offset parameter $0 \leq N_{RA}^{offset} \leq N_{RB}^{UL} - 6$ is the first PRB available for PRACH expressed as a PRB number configured by higher layers (see Figure 10.46). Each random-access preamble occupies a bandwidth corresponding to six consecutive resource blocks for both frame structure types.

The random-access preambles are generated from Zadoff–Chu sequences with a zero-correlation zone. This means that the PRACH sequences are generated from one or several root Zadoff–Chu sequences. The network configures the set of preamble sequences which the UE is allowed to use. There are 64 preambles available in each cell. The set of 64 preamble sequences in a cell is obtained by including, in the order of increasing cyclic shift, all the available cyclic shifts of a root Zadoff–Chu sequence with the logical index *RACH_ROOT_SEQUENCE*, where the parameter *RACH_ROOT_SEQUENCE* is broadcast as part of the system information. Additional preamble sequences, in case 64 preambles cannot be generated from a single root Zadoff–Chu sequence, are obtained from the root sequences with the consecutive logical indices until all 64 sequences are found. The logical root sequence order is cyclic, i.e., the logical index 0 is consecutive to 837. The uth root Zadoff–Chu sequence is defined by:

$$x_u(n) = \exp\left(-j\frac{\pi u n(n+1)}{N_{ZC}}\right); 0 \leq n \leq N_{ZC} - 1 \tag{10.13}$$

where the length N_{ZC} of the Zadoff–Chu sequence is either 839 for random-access preamble formats 0–3, or 139 for random-access preamble format 4. From the uth root Zadoff–Chu sequence, random-access preambles with zero correlation zones (ZCZs)[14] of length $N_{CS}-1$ are defined by cyclic shifts according to $x_{uv}(n)=x_u[(n+C_v)\text{mod } N_{ZC}]$, in which the value of the cyclic shift depends of the type of the sequence set. The unrestricted set or restricted sets are distinguished based on higher layer signaling.

The parameter N_{CS} is given in 3GPP TS 36.211 [3] for preamble formats 0–3 and 4. The parameter *ZCZ configuration* is provided by the higher layers (Table 10.23). The parameter *high-speed flag* is also provided by higher layers and determines whether unrestricted set or restricted set is used (Table 10.24).

The PRACH preamble sequence duration $T_{SEQ}=1/\Delta f_{RA}$ is derived from the required PRACH preamble sequence energy to thermal noise power ratio E_s/N_0 in order to meet the target miss detection and false alarm probabilities, as follows: $T_{SEQ}=\frac{N_0 N_F}{P_{RA}(R)}\left(\frac{E_s}{N_0}\right)$, where N_0 is the thermal noise power density (in mW/Hz) and N_F is the noise figure of the eNB receiver. Typical target values for miss detection and false alarm probabilities are 10^{-2} and 10^{-3}, respectively. The 1600 μs preamble sequence of formats 2 and 3 is implemented by repeating the baseline 800 μs preamble sequence. These formats can provide up to 3dB link budget improvement, which is useful in large cells and/or to balance PUSCH/PUCCH and PRACH coverage at low data rates (Figure 10.49).

The PRACH slots shown in Figure 10.50 can be configured to occur in up to 16 different patterns or resource configurations. Depending on the number of UEs attempting to access the system in the same time interval, one or more PRACH resources may be allocated per PRACH slot period. The eNB is required to process the PRACH signal as soon as it is detected so that Msg2 of the RACH procedure can be sent within the required time window. If there is more than one PRACH resource per PRACH period, it is desirable to multiplex the PRACH resources in time rather than in frequency. This helps to avoid excessive processing at the eNB [15]. The available slot configurations are designed to simplify

---

[14] Definition of ZCZ sequences: let S denote a sequence set with M sequences of period L. Generally, the set of ZCZ sequences are characterized by the period of sequences L, the number of sequences in the set M, the length of the ZCZ L_{ZCZ}, and the number of phases P. It can be shown that the following mathematical upper bound concerning L_{ZCZ} exist $L_{ZCZ}\leq L/M-1$. The set S can be represented as $S=\{S_0,S_1,\ldots,S_p,\ldots,S_{M-1}\}$ and $S_p=\{s_0^p,s_1^p,\ldots,s_q^p,\ldots,s_{L-1}^p\}$, where $0\leq p\leq M-1$ and $0\leq q\leq L-1$, S_p and s_q^p denote a sequence and a sequence element, respectively. If all of the sequences in the set S satisfy the following autocorrelation and cross-correlation properties, then S is referred to as a set of ZCZ sequences or a ZCZ sequence set $\rho_{S_p}(l)=\sum_{q=0}^{L-1}s_q^p s_{(q+l)\text{mod } L}^{p^*}=\begin{cases}0 & -L_{ZCZ}\leq l\leq -1;1\leq l\leq L_{ZCZ}\\ \varepsilon_p & l=0\end{cases}$ and $\rho_{S_pS_{p'}}(l)=\sum_{q=0}^{L-1}s_q^p s_{(q+l)\text{mod } L}^{p'^*}=0$. In the above equation, symbol "*" denotes complex conjugate, $\rho_{S_p}(l)$ represents the periodic autocorrelation function of the sequence S_p, and $\rho_{S_pS_{p'}}(l)$ denotes the periodic cross-correlation function between the sequences S_p and $S_{p'}$, ε_p is the energy of the sequence S_p. If all of the elements of the sequence S_p are normalized complex-valued numbers, then the sequence energy ε_p is equal to L. In general, M, L, and L_{ZCZ} are referred to as the family size of the ZCZ sequence set, the period of the sequences, and the length of the ZCZ, respectively [167].

Table 10.23 PRACH Parameters [3,10]

Parameter	Size (bits)	Signaling Method	Description
PRACH time-domain configuration index	6	Broadcast (SIB2)	Provides information about the PRACH time-domain (subframe) configuration
PRACH frequency offset	7	Broadcast (SIB2)	Provides information about the PRACH frequency-domain configuration. Only applicable to TDD mode
Preamble format	2	Broadcast (SIB2)	For TDD there is also a short preamble, i.e., a total of five preamble formats
Root-sequence index	10	Broadcast (SIB2)	Used to generate cyclically shifted Zadoff–Chu sequences
Zero-correlation-zone length	4	Broadcast (SIB2)	Different interpretation depending on high-speed flag
High-speed flag	1	Broadcast (SIB2)	Selects between restricted and unrestricted preamble sets
PRACH transmit power setting	power Ramping Step 2 preamble Initial Received Target Power 4	Broadcast (SIB2)	*powerRampingStep* (0, 2, 4, or 6 dB), *preambleInitialReceivedTargetPower* (−120, −118, −116, −114, −112, −110, −108, −106, −104, −102, −100, −98, −96, −94, −92, or −90 dBm)

Table 10.24 Classification of PRACH Preambles to Restricted and Unrestricted Sets [3]

ZCZ Configuration Parameter	Format 4	Formats 0–3	
		N_{CS}	
	N_{CS}	Unrestricted Set	Restricted Set
0	2	0	15
1	4	13	18
2	6	15	22
3	8	18	26
4	10	22	32
5	12	26	38
6	15	32	46
7	N/A	38	55
8	N/A	46	68
9	N/A	59	82
10	N/A	76	100
11	N/A	93	128
12	N/A	119	158
13	N/A	167	202
14	N/A	279	237
15	N/A	419	–

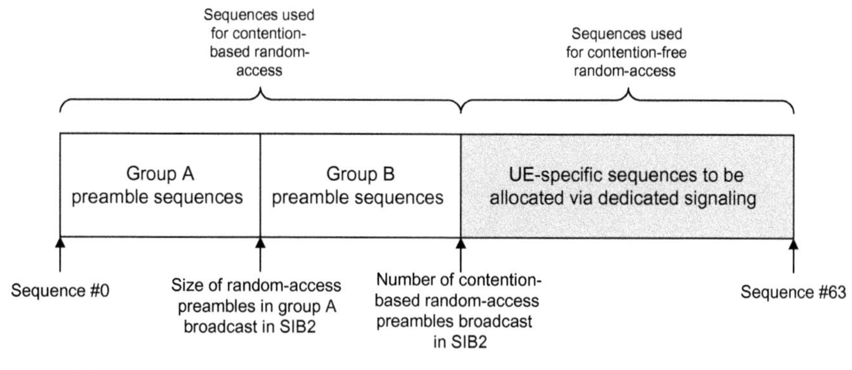

FIGURE 10.49

Classification of random-access preamble sequences [14].

PRACH receiver which may be used for multiple cells of an eNB, assuming a periodic pattern with period 10 or 20 ms.

The random-access preamble sequence can be generated at the system sampling rate by means of a large IDFT unit. The DFT block in Figure 10.51 is optional because the sequence can be directly mapped to the IDFT input in the frequency-domain. The cyclic shift can be implemented either in the time-domain after the IDFT or in the frequency-domain before the IDFT through a phase shift. For all possible system sampling rates, both cyclic prefix and sequence duration correspond to an integer number of samples. The method of Figure 10.51 does not require any time-domain filtering in the baseband, but requires large IDFT sizes (up to 24,576 for a 20 MHz spectrum allocation), which are practically cumbersome. Therefore, another option for generating the preamble consists of using small-size IDFT and shifting the preamble to the required frequency location through time-domain up-sampling and filtering (hybrid frequency/time-domain sequence generation method). Given that the preamble sequence length is 839, the smallest IFFT size that can be used is 1024. The size of the random-access cyclic prefix and preamble sequence duration has been chosen to provide an integer number of samples at the system sampling rate. The cyclic prefix can be inserted before the up-sampling and time-domain frequency shift, in order to minimize the intermediate storage requirements.

Given that the sampling rate for a 20 MHz LTE system is 30.72 MHz, and considering that the random-access preamble spans 0.8 ms, it can be concluded that the number of samples in time is equal to 24,576. Furthermore, it was mentioned that the PRACH sub-carrier spacing is 1.25 kHz, while the sub-carrier spacing for PUSCH and PUCCH is 15 kHz. As shown in Figure 10.51, in order to maintain the same sampling rate, a 12×2048 point DFT operation is needed for the PRACH signal generation at the transmitter side, if the entire processing is done in the frequency-domain. An alternative approach is to use time-domain signal generation and extraction which involves up-sampling and filtering operations at the transmitter. The drawback of

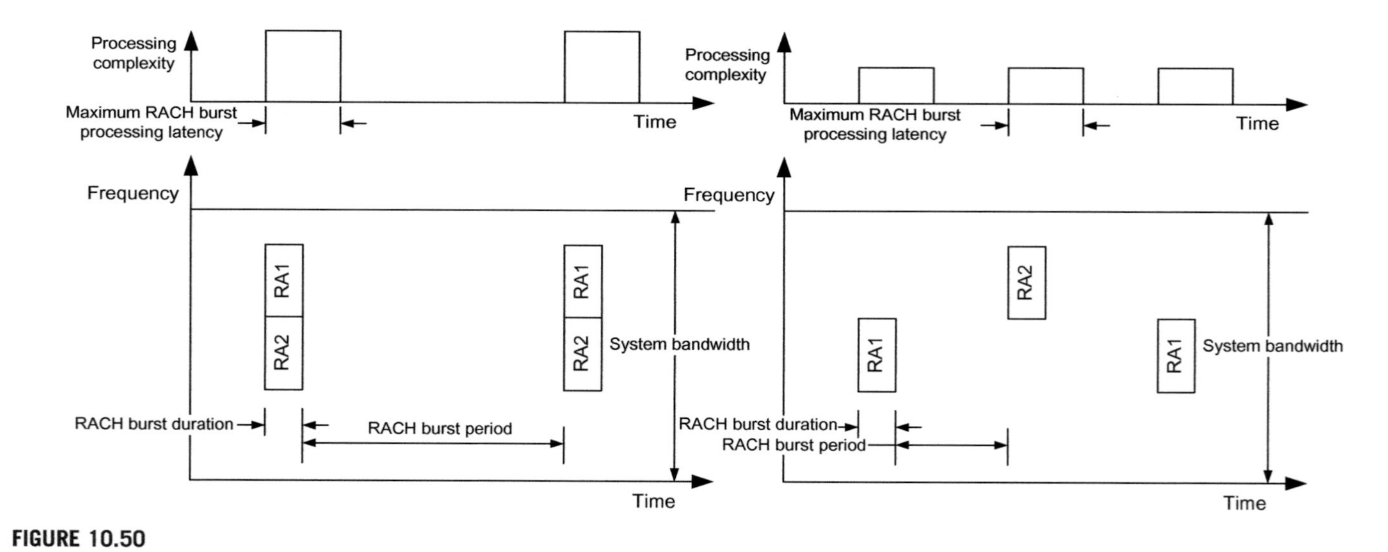

FIGURE 10.50

Example of PRACH resource configuration and processing [15].

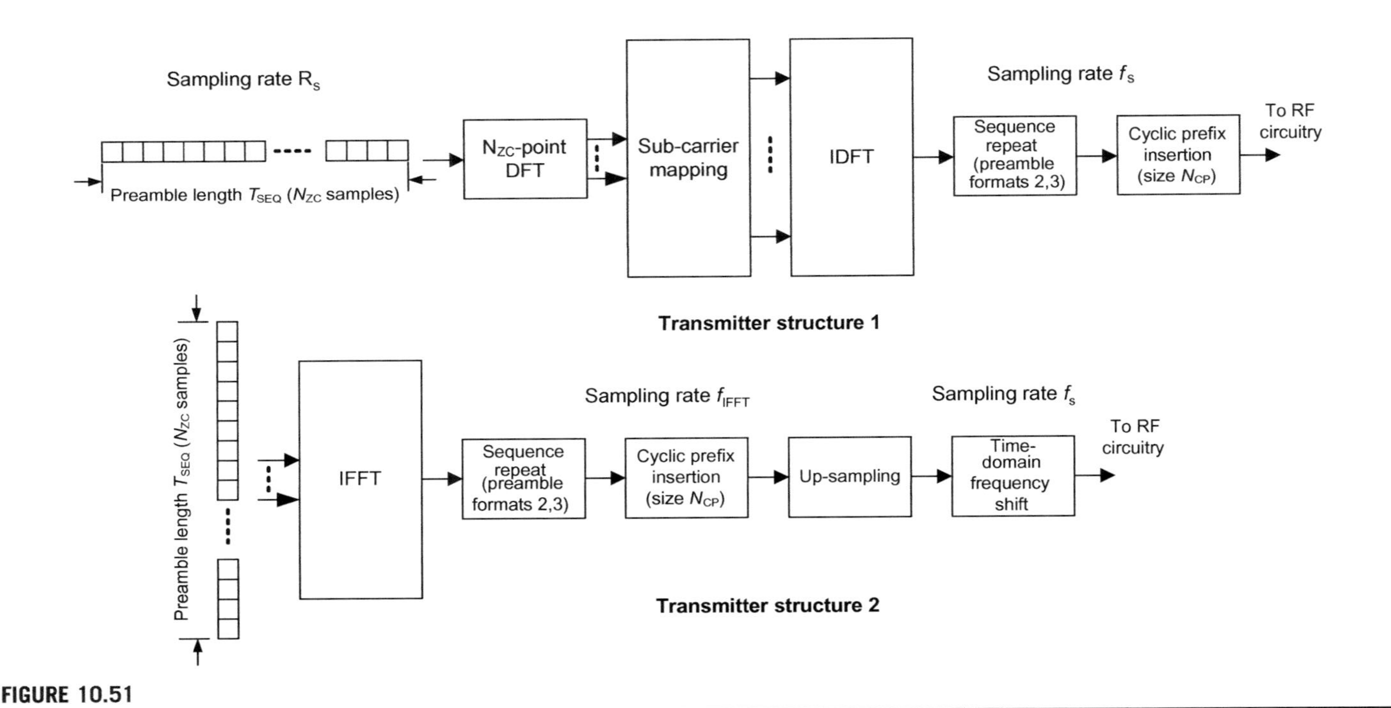

FIGURE 10.51

PRACH transmitter structure options [15,168].

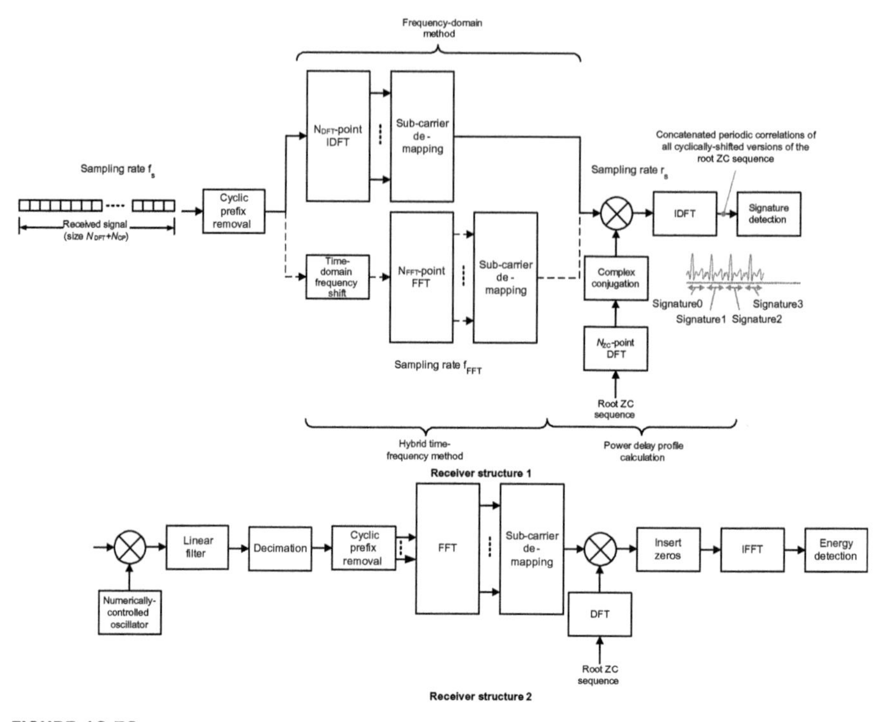

FIGURE 10.52

PRACH receiver structure options [15,168].

time-domain implementation is that the up-sampling from 1.08 MHz to the system sampling rate of 30.72 MHz is difficult to implement.

The implementation of the PRACH signal at the eNB receiver can take a frequency-domain or a hybrid time/frequency-domain approach. As illustrated in Figure 10.52 as an example, the common parts to both approaches are the cyclic prefix removal, which always occurs at the front end at the system sampling rate, the power delay profile calculation, and signature detection. The two approaches differ only in the computation of the sub-carriers carrying the PRACH signal(s). The frequency-domain method computes the full range of sub-carriers used for uplink transmission over the system bandwidth from 800-μs-long received input samples. As a result, the PRACH sub-carriers are directly extracted from the set of uplink sub-carriers, which does not require any frequency shift or time-domain filtering but involves an extremely large DFT computation. Note that even though DFT size $N_{DFT} = n2^m$, thus allowing fast and efficient DFT computation algorithms, the DFT computation cannot start until the complete sequence is stored in memory, which increases the processing delay.

In order to reduce the implementation complexity, particularly for the number of multiplications in the PRACH detector at the eNB, frequency-domain processing may be adopted for PRACH preamble detection. As shown in Figure 10.52, the received signal is first processed in the time-domain, is then transformed to the frequency-domain by FFT, and is multiplied by the DFT-transformed RACH sequences (i.e., cross-correlation calculation). The cross-correlation is obtained by transforming back to the time-domain by IFFT, followed by zero insertion. Figure 10.52 further illustrates the main components of another RACH preamble receiver structure using the DFT-based (frequency-domain) method. In the receiver, the down-conversion block uses a numerically controlled oscillator, which shifts the frequency band of PRACH to the lowest frequency band. The down-converted signal is filtered with a linear filter to prevent aliasing effect after decimation.

The hybrid time/frequency-domain method extracts the desired PRACH signal using a time-domain frequency shift along with down-sampling and anti-aliasing filtering. The down-sampling ratio and the corresponding anti-aliasing filter characteristics are chosen to provide a number of PRACH time samples suitable for an FFT or simple DFT computation at a sampling rate which is an integer fraction of the system sampling rate. Unlike the frequency-domain approach, the hybrid time/frequency-domain computation can start as soon as the first samples have been received, reducing the processing latency.

The analog random-access signal $s(t)$ is defined as follows:

$$s(t)=\beta_{\mathrm{PRACH}}\sum_{k=0}^{N_{\mathrm{ZC}}-1}\sum_{n=0}^{N_{\mathrm{ZC}}-1} x_{\mathrm{uv}}(n)e^{-\mathrm{j}(2\pi nk/N_{\mathrm{ZC}})}\exp[\mathrm{j}2\pi(k+\varphi+\gamma(k_0+1/2))\Delta f_{\mathrm{RA}}(t-T_{\mathrm{CP}})] \tag{10.14}$$

where $0\leq t<T_{\mathrm{SEQ}}+T_{\mathrm{CP}}$, β_{PRACH} is an amplitude scaling factor in order to ensure conformance to the transmit power P_{PRACH}, and $k_0=12N_{\mathrm{RA}}-6N_{\mathrm{RB}}^{\mathrm{UL}}$. The location of PRACH in the frequency-domain is controlled by the parameter N_{RA}. The factor $\gamma=\Delta f/\Delta f_{\mathrm{RA}}$ accounts for the difference in sub-carrier spacing between the random-access preamble and uplink data transmission. The sub-carrier spacing for the random-access preamble $\Delta f_{\mathrm{RA}}=1250$ Hz for random-access preamble formats 0–3, and $\Delta f_{\mathrm{RA}}=7500$ Hz for random-access preamble format 4. The fixed offset determining the frequency-domain location of the random-access preamble within the PRBs $\phi=7$ for random-access preamble formats 0–3, and $\phi=2$ for random-access preamble format 4 [3].

10.8 SC-FDMA signal generation

The time-domain analog signal $s_l^p(t)$ for antenna port p over SC-FDMA symbol l in an uplink slot is defined as follows (applicable to all uplink physical signals and physical channels except PRACH):

$$s_l^p(t)=\sum_{k=-\lfloor 6N_{\mathrm{RB}}^{\mathrm{UL}}\rfloor}^{\lceil 6N_{\mathrm{RB}}^{\mathrm{UL}}\rceil-1} a_{k'}^{p}(l)\exp\{\mathrm{j}\pi(2k+1)\Delta f[t-N_{\mathrm{CP}}(l)T_{\mathrm{s}}]\},0\leq t<(N_{\mathrm{CP}}(l)+N_{\mathrm{FFT}})T_{\mathrm{s}} \tag{10.15}$$

where $k' = k + \lfloor 6N_{\mathrm{RB}}^{\mathrm{UL}} \rfloor$, $N_{\mathrm{FFT}} = 2048$, $\Delta f = 15$ kHz, and α_{kl}^{p} is the complex-valued content of resource element (k, l) on antenna port p. The SC-FDMA symbols in a slot are transmitted in increasing order of symbol index l, starting with $l = 0$, where SC-FDMA symbol $l > 0$ starts at time $\sum_{l'=0}^{l-1}[N_{\mathrm{CP}}(l') + N]T_{\mathrm{s}}$ within the slot [3].

10.9 LTE/LTE-Advanced uplink multi-antenna transmission schemes

Simultaneous transmission from multiple transmit antennas from a single UE was not supported in LTE Rel-8/9. In order to simplify the UE implementation and to reduce power consumption, a single-power amplifier was assumed to be available at the UE. However, LTE Rel-8/9 provided support for open/closed-loop antenna-selection transmit diversity in the uplink from UEs which are equipped with multiple transmit antennas. LTE Rel-8/9 also supported virtual MIMO in the uplink. More advanced multi-antenna techniques, including closed-loop spatial multiplexing for SU-MIMO and transmit diversity for control signaling, were introduced in LTE Rel-10. Uplink closed-loop antenna selection for up to two transmit antennas is supported as an optional UE capability in LTE and is configurable by higher layers. If a UE indicates support for uplink antenna selection, the eNB may configure and schedule the UE based on that capability.

LTE/LTE-Advanced use the concept of antenna ports to differentiate between logical and physical antennas. Logical antennas or antenna ports are mapped to physical antennas based on certain deployment considerations. Prior to LTE Rel-10, there was only one transmit antenna on the UE, thus the notion of antenna ports did not apply. However, LTE Rel-10 extended the multi-antenna capabilities of the uplink and provided support for spatial multiplexing up to four transmit antennas for PUSCH and dual-antenna transmit diversity for PUCCH. Therefore, a new set of logical antennas or antenna ports have been defined that are listed in Table 10.25. As an example, PUSCH transmission, including its DM-RS as well as SRS transmission, is associated with antenna ports $\{10\}, \{20, 21\}$, and $\{40, 41, 42, 43\}$, for one, two, and four transmit antennas, respectively. The virtual antenna port $\bar{p} \in \{0, 1, 2, 3\}$ is used in this book and in 3GPP specifications to refer to the actual uplink antenna ports. Note that each antenna port has its own reference signals, enabling the eNB to estimate the channel from each UE transmit antenna.

Before proceeding to the next sections, recall that a spatial layer is the term used in LTE for one of the different streams generated by spatial multiplexing. A layer can be described as a mapping of complex-valued symbols onto the transmit antenna ports. Each layer is identified by a precoding vector/matrix of size equal to the number of transmit antenna ports, and can be associated with a radiation pattern. The rank of the transmission is the number of layers transmitted. A codeword is an independently encoded data block, corresponding to a single transport block delivered from the MAC sub-layer in the transmitter to the physical layer and protected with a CRC.

Table 10.25 Virtual and Physical Antennas in the Uplink [3]

Physical Channel/Signal	Virtual Antenna Port $\bar{p}$	Uplink Antenna Port p — Number of Physical Antennas: 1	2	4
PUSCH and DM-RS for PUSCH	0	10	20	40
	1	–	21	41
	2	–	–	42
	3	–	–	43
SRS	0	10	20	40
	1	–	21	41
	2	–	–	42
	3	–	–	43
PUCCH and DM-RS for PUCCH	0	100	200	–
	1	–	201	–

10.9.1 Transmit antenna selection

When closed-loop antenna selection is enabled, the eNB indicates which antenna should be used for PUSCH transmission, by implicitly coding this information in the uplink scheduling grant (i.e., DCI format 0 or 4). The 16-bit CRC is scrambled (modulo-2 addition) by one of two antenna selection masks $\langle 0,0,0,0,0,0,0,0,0,0,0,0,0,0,0,0\rangle$ for the UE's first transmit antenna, and/or $\langle 0,0,0,0,0,0,0,0,0,0,0,0,0,0,0,1\rangle$ for the second antenna [4]. The antenna selection mask is applied in addition to the UE identity (RNTI) masking, identifying the UE for which the scheduling grant is intended. This implicit encoding avoids the use of explicit antenna selection bits which would result in an increased overhead for UEs not supporting (or not configured for) transmit antenna selection. The UE identity can be detected directly from the 15 least-significant bits of the decoded mask with no need to use the transmitted antenna selection mask (the 16th bit).

When using the adaptive HARQ, the antenna indicator, using CRC masking, is always sent in the uplink grant to indicate which antenna to use. For example, for a high-Doppler UE with adaptive HARQ, the eNB might instruct the UE to alternate between the transmit antennas or, alternatively, to select the primary antenna. In typical UE implementations, a transmit antenna gain imbalance of 3–6 dB between the secondary and primary antennas is not uncommon. Alternatively, when using non-adaptive HARQ, the UE may choose either antenna; therefore, in low-Doppler scenarios, the UE may use the same antenna

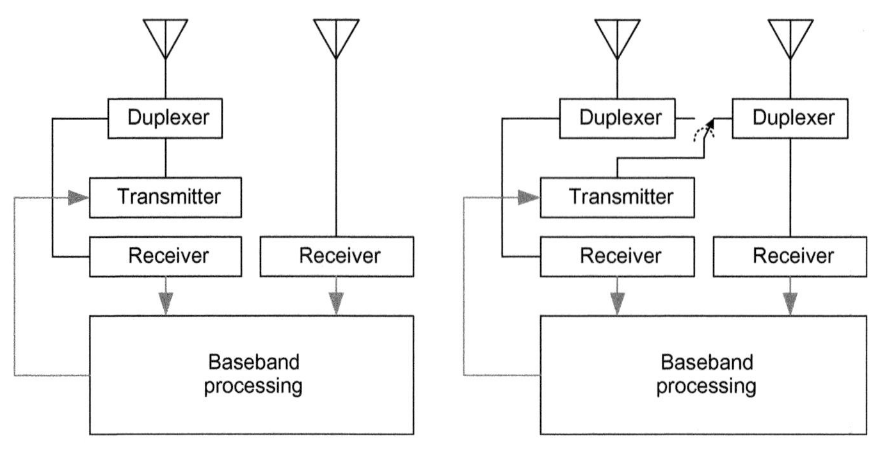

FIGURE 10.53

Illustration of a UE front end with/without transmit antenna selection [14].

signaled in the uplink grant, while in high-Doppler conditions the UE may switch between antennas or select the primary antenna. For a large number of retransmissions with non-adaptive HARQ, the antenna indicated in the uplink grant may not be the best and it is better to let the UE select the antenna. If the eNB instructs the UE to use a specific antenna for the retransmissions, it can use adaptive HARQ. If the eNB enables a UE's closed-loop antenna selection capability, the SRS transmissions alternate between the transmit antennas in successive SRS transmission subframes, irrespective of enablement of the frequency hopping feature [15].

It must be noted that receive diversity is mandatory for all UE categories in LTE, and as such there is more than one antenna on the UE that may be used for transmit antenna selection, as shown in Figure 10.53, with minimal changes to implementation. The transmit antenna selection scheme can be useful, if the UE is held in such a way that one transmit antenna is covered by hand and the other antenna is not covered. Although the inclusion of an RF switch may introduce some insertion loss, this drawback may be overcome by the benefits of the diversity scheme in some practical scenarios. It is not mandatory for the UEs to support transmit antenna selection. The UE capability information message can be used to inform the network of whether the UE supports this feature. The eNB can instruct the UEs supporting this feature to use either open-loop or closed-loop transmit antenna selection.

10.9.2 Virtual MU-MIMO (cooperative MIMO)

Another uplink MIMO mode that was introduced in LTE Rel-8/9 was MU-MIMO. The legacy uplink MU-MIMO is based on transmission from multiple UEs on the same set of resource blocks, each using a single transmit antenna. From the point of view of an individual UE, this is a single-antenna transmission scheme; however, from an eNB perspective, this is a MU-MIMO scheme that requires more sophisticated scheduling and uplink reception. In order to support uplink MU-MIMO, LTE Rel-8/9 provides orthogonal DM-RS using different cyclic shifts to allow the eNB to obtain independent channel estimations for the uplink from each UE. A cell can assign up to eight different cyclic shifts using a 3-bit PUSCH cyclic shift offset in the uplink scheduling grant; therefore, uplink MU-MIMO of up to eight UEs can be theoretically supported in a cell. The inter-cell or cooperative MU-MIMO is supported in LTE by assigning the same base sequence groups and/or reference signal hopping patterns to the different cells.

The main objective of MU-MIMO is to increase the overall system throughput. The uplink MU-MIMO is transparent to the UE and has no effect on the UE implementation. The cooperative or virtual MIMO in the uplink requires the eNB to select and group the spatially uncorrelated UEs that are going to share the same resource blocks, to schedule the UEs within the group for uplink transmission over PDCCH, and to separate and decode uplink transmissions from each UE within the group. Each UE within the group will be assigned a different cyclic shift for the corresponding DM-RS to allow the eNB to estimate the channel from each UE and to decode the corresponding payload. The use of different cyclic shifts provides orthogonality among the UEs within the group, and enables the eNB to separate them as long as the same number of resource blocks is used for the UEs within group. The orthogonality will be lost if each UE is allocated a different number of resource blocks. While LTE Rel-8/9 defined only cyclic shifts to differentiate the UEs in the uplink, LTE Rel-10 introduced an additional level of differentiation using OCCs, as discussed earlier. Therefore, OCC will allow the UEs within the same MU-MIMO group to be allocated a different number of resource blocks.

10.9.3 Closed-loop spatial multiplexing

Closed-loop spatial multiplexing with up to four antennas was introduced as part of the LTE Rel-10 improvements in the uplink. The physical antenna elements required for uplink spatial multiplexing might be already available on the UEs that support spatial multiplexing in the downlink, which can be shared in both directions via an RF switch, e.g., a UE that supports 2×2 spatial multiplexing in the downlink already has two antenna elements. Nevertheless, the uplink spatial multiplexing further requires additional transmission chains and PAs in the uplink that operate simultaneously. The eNB provides the UE with instructions on the number of spatial layers and the codebooks or precoding matrices that must be used. The criterion for selecting the precoding matrix is to maximize the SNR at the eNB receiver [14].

The design principles of spatial multiplexing with precoding for the uplink data transmission are similar to those for the downlink. The precoding operation can mathematically be expressed as multiplication of an $N_{RX} \times 1$ DFT-spread signal vector by an $N_{TX} \times N_{RX}$ precoding matrix, which is chosen from a predefined set of matrices, also known as a codebook. Note that the ith column vector of the precoding matrix represents the antenna-spreading weight of the ith layer. The main purpose of precoding is to match the precoding matrix with the channel characteristics in order to increase the received signal power and reduce inter-layer interference, thereby improving the SINR of each layer.

Let us assume that the eNB has N_{RX} receive antennas and the eigen-beamforming module precodes each layer with one of $\min(N_{TX}, N_{RX})$ eigenvectors of the channel correlation matrix in order to improve the capacity. In practice, the choice of the precoder must be signaled from the eNB to the UE; therefore, the number of precoding matrices (i.e., codebook size) is limited by the size of the downlink control signaling. The use of a finite codebook results in a reduced precoding gain. If the alphabet for the non-zero elements of the precoding matrix is confined to $(\pm 1, \pm j)$, the matrix calculations involved in the selection of the codebook are significantly simplified. Furthermore, the constant-modulus property of the QPSK alphabet enables all PAs to transmit with equal power level. This balanced power distribution would result in maximization of the PA efficiency, regardless of which precoding matrix is used. A drawback of this alphabet constraint is further reduction of the precoding gain.

The design of a precoding codebook used for downlink MIMO is different from that in the uplink in the sense that the uplink codebook structure should not increase the CM of the transmitted signal relative to single-antenna transmission (measured at each individual antenna port). The Householder codebook of downlink MIMO linearly combines multiple layers and inevitably increases the CM. A design constraint on uplink codebooks is that each row should have at most one non-zero element. In other words, each transmit antenna is not allowed to convey more than one layer; therefore, no linear combination is allowed. This constraint, while it preserves the CM, gives rise to further reduction of the precoding gain. One of the objectives of the uplink precoder design is to optimize the precoding gain under the CM preserving constraint [122].

It is recognized from an information theory perspective that the codebooks with maximum chordal distance are optimal. However, the UE antennas are likely to be practically correlated due to proximity; therefore, the UE antenna configuration and indexing, antenna spacing, and polarization should be carefully considered in selecting the codebook vectors. Since each layer is assigned to a separate group of UE antennas, antenna grouping with lower inter-layer correlation is preferred, leading to lower inter-layer interference. This is also consistent with the eigen-beamforming principle in that eigenvectors of channel correlation tend to have elements with larger amplitude in one antenna group than in the other antenna groups. Under the above constraints, it can be concluded that the full-rank codebooks consist of the identity matrices. For non-full-rank codebooks, it is desirable to maximize the precoding gain while preserving the low CM property.

Since PUSCH DM-RS is precoded with uplink data at the same time, it is impossible to select the precoder (i.e., the rank and precoding matrix) based on DM-RS, since only a subset of the spatial channel dimensions is observed.

Table 10.26 Codebook for Uplink Transmission on Two Antenna Ports {20, 21}[3]

Codebook Index	Number of Layers	
	$N_l = 1$	$N_l = 2$
0	$\frac{1}{\sqrt{2}}\begin{bmatrix}1\\1\end{bmatrix}$	$\frac{1}{\sqrt{2}}\begin{bmatrix}1 & 0\\0 & 1\end{bmatrix}$
1	$\frac{1}{\sqrt{2}}\begin{bmatrix}1\\-1\end{bmatrix}$	–
2	$\frac{1}{\sqrt{2}}\begin{bmatrix}1\\j\end{bmatrix}$	–
3	$\frac{1}{\sqrt{2}}\begin{bmatrix}1\\-j\end{bmatrix}$	–
4	$\frac{1}{\sqrt{2}}\begin{bmatrix}1\\0\end{bmatrix}$	–
5	$\frac{1}{\sqrt{2}}\begin{bmatrix}0\\1\end{bmatrix}$	–

Instead, non-precoded SRS transmission is the conventional method for the eNB to efficiently select an appropriate precoder for the uplink data. The SRS is antenna-specific, as opposed to layer-specific in the case of DM-RS; therefore, the required number of orthogonal resources is the same as the number of transmit antennas. There exists a great amount of similarity between DM-RS and SRS. The SRS is derived from the same cell-specific base sequence as DM-RS, and the SRS orthogonality is further provided through different cyclic shifts [122].

In the case of two transmit antennas and rank 1 transmission, the codebook consists of four constant-modulus vectors and two antenna selection vectors (Table 10.26). The inclusion of constant-modulus vectors enables the signals from multiple UE antennas to be combined constructively at the receiving eNB by introducing a relative phase shift of 0°, 90°, 180°, or 270° between UE antennas. In the example shown in Figure 10.54, a phase shift of 270° maximizes the SNR of received signal due to constructive combining, whereas a phase shift of 90° minimizes the SNR due to destructive combining. On the other hand, the inclusion of antenna selection vectors reduces transmit power and is intended for UE power saving. For example, if UE antennas experience significant gain imbalance, one of the antenna selection vectors can be used to switch off the low-gain antenna [122].

The case of four transmit antennas and rank-1 codebook involves 16 constant-modulus vectors for constructive combining and eight antenna-selection vectors for UE power saving. For constant-modulus vectors, four relative phase shifts are considered between the first two antennas. Additional relative phase shifts are applied between the first and third antennas, and between the second and fourth antennas, accounting for imperfect antenna polarization. The rank-2 codebook (given in Table 10.27) covers every possible antenna grouping. For the rank-3

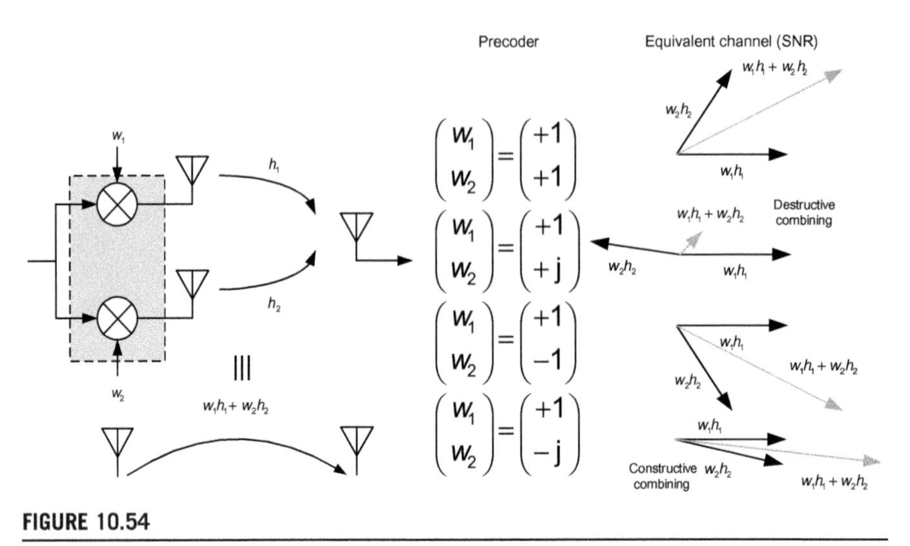

FIGURE 10.54

Illustration of the effect of precoding on SNR (rank 1 transmission) [122].

codebook, the first layer is assigned to two transmit antennas, whereas layers 2 and 3 are each assigned one antenna. Every antenna grouping (six groupings in total) for the first layer is covered, along with an intra-group phase shift of 0° or 180°. It should be noted that no inter-group phase shift is considered in the rank-2 and rank-3 codebooks, since it does not contribute to layer separation. The performance of the system with these codebooks is similar to the Householder codebook used in the downlink MIMO, unless the UE antennas are highly correlated. Note that the Householder codebook has a similar codebook size, while its alphabet is 8PSK and the CM is not concerned. Therefore, it can be concluded that the loss of precoding gain (compared to eigen-beamforming) largely originates from the codebook size constraint, rather than the alphabet restriction and the limitation concerning the CM of the transmitted signal. The precoder is defined by rank and PMI, which are selected by the eNB based on uplink channel measurements and signaled to the UE through the rank indicator and PMI fields in an uplink grant. The rank and PMI may be selected to maximize the throughput.

10.9.4 Transmit diversity

Multi-antenna transmit-diversity techniques are used to provide reliability and to improve coverage for uplink control channels. The transmit diversity scheme introduced for PUCCH in LTE Rel-10 was designed to ensure backward compatibility with the LTE Rel-8 PUCCH design. A dual-antenna transmit diversity scheme for HARQ and SR (PUCCH format 1/1a/1b) is utilized by transmitting the same control information from different UE antennas by using different orthogonal resources, thus providing no multiplexing gain. This is referred to as

Table 10.27 Codebook for Transmission on Four Antenna Ports {40, 41, 42, 43} [3]

Codebook Index	Number of Layers $N_l = 2$				Codebook Index	Number of Layers $N_l = 2$			
0–3	$\frac{1}{2}\begin{bmatrix}1 & 0\\1 & 0\\0 & 1\\0 & -j\end{bmatrix}$	$\frac{1}{2}\begin{bmatrix}1 & 0\\1 & 0\\0 & 1\\0 & j\end{bmatrix}$	$\frac{1}{2}\begin{bmatrix}1 & 0\\-j & 0\\0 & 1\\0 & 1\end{bmatrix}$	$\frac{1}{2}\begin{bmatrix}1 & 0\\-j & 0\\0 & 1\\0 & -1\end{bmatrix}$	8–11	$\frac{1}{2}\begin{bmatrix}1 & 0\\0 & 1\\1 & 0\\0 & 1\end{bmatrix}$	$\frac{1}{2}\begin{bmatrix}1 & 0\\0 & 1\\1 & 0\\0 & -1\end{bmatrix}$	$\frac{1}{2}\begin{bmatrix}1 & 0\\0 & 1\\-1 & 0\\0 & 1\end{bmatrix}$	$\frac{1}{2}\begin{bmatrix}1 & 0\\0 & 1\\-1 & 0\\0 & -1\end{bmatrix}$
4–7	$\frac{1}{2}\begin{bmatrix}1 & 0\\-1 & 0\\0 & 1\\0 & -j\end{bmatrix}$	$\frac{1}{2}\begin{bmatrix}1 & 0\\-1 & 0\\0 & 1\\0 & j\end{bmatrix}$	$\frac{1}{2}\begin{bmatrix}1 & 0\\j & 0\\0 & 1\\0 & 1\end{bmatrix}$	$\frac{1}{2}\begin{bmatrix}1 & 0\\j & 0\\0 & 1\\0 & -1\end{bmatrix}$	12–15	$\frac{1}{2}\begin{bmatrix}1 & 0\\0 & 1\\0 & 1\\1 & 0\end{bmatrix}$	$\frac{1}{2}\begin{bmatrix}1 & 0\\0 & 1\\0 & -1\\1 & 0\end{bmatrix}$	$\frac{1}{2}\begin{bmatrix}1 & 0\\0 & 1\\0 & 1\\-1 & 0\end{bmatrix}$	$\frac{1}{2}\begin{bmatrix}1 & 0\\0 & 1\\0 & -1\\-1 & 0\end{bmatrix}$

spatial orthogonal resource transmit diversity and is illustrated in Figure 10.36. In LTE Rel-8/9, the orthogonal resources are available through either various cyclic shifts of length-12 base sequences or OCCs of length-4 for transmission of control information of uplink control channels. The DM-RS for the uplink control channel also uses the orthogonal resources through the cyclic shifts of length-12 base sequences or OCCs of length-3.

In general, there are theoretically 36 orthogonal resources available for the uplink control channel in each subframe. Assuming perfect orthogonality, the signals from multiple transmit antennas are received separately and combined constructively. In the case of four transmit antennas and from the eNB perspective, the UE transmissions use two orthogonal resources, i.e., the cyclic shift and the OCCs, and it is up to the UE implementation if and how the transmission is distributed over the four transmit antennas. One possibility is to allow the third and fourth antennas to use the same orthogonal resources allocated to the first and second antennas, respectively, and transmit the same information. Another possibility is to transmit only on two of the four UE antennas. The same transmit diversity scheme is also applied for the channel quality information report (PUCCH format 2/2a/2b) [122]. If the UE transmit antennas are uncorrelated, SORTD provides the best diversity performance at the expense of consuming multiple orthogonal resources, reducing PUCCH multiplexing capability in general.

10.9.5 Uplink MIMO receivers

The practical constraints against implementing spatial multiplexing particularly in the uplink are the cost of multiple transmit and receive radio chains and the form-factor limitation of multi-antennas for handheld devices. By making use of advanced signal-processing techniques in the baseband processing stage, high-performance MIMO transmission can be achieved. A number of MIMO receiver algorithms can be used, depending on the requirements pertaining to receiver complexity.

In principle, the downlink MIMO channel is a broadcast channel. In this case, the individual data streams transmitted from each antenna are mixed with other transmit data streams at each receive antenna. To separate the mixed data streams at the receiver, designers can multiply the (mixed-stream) received signals with a matrix that is the inverse of the MIMO channel matrix. This is called a zero-forcing receiver. It is a simple MIMO receiver that often suffers performance loss at higher noise and interference levels. To minimize performance loss, the inversion of the MIMO channel matrix operation is adjusted according to the interference or noise level. This is called an MMSE receiver. But both the ZF and MMSE receivers typically have up to 2 dB performance loss compared with maximum likelihood detection (MLD) receivers.

Improved reception of spatial multiplexing MIMO transmission requires an exhaustive search of MIMO signal constellation combinations. The MIMO transmission process is modeled at the receiver in such a way that a specific complex modulation constellation is generated for each transmit antenna stream. The constellations are then applied to the MIMO channel input. At the MIMO channel output, the corresponding received signal is generated for each receive antenna. At this point, the Euclidean distance is computed for the modeled MIMO channel output signal relative to the received signal. In this way, the different modulation constellations at the MIMO channel input constitute different hypothesis tests. If all possible combinations of constellations at the MIMO channel input can be emulated, their associated MIMO channel outputs can be generated, and the Euclidean distance can be computed against the received signal. The minimum Euclidian distance associated with the constellation combination hypothesis provides the most likely decoding. Such an exhaustive constellation search at the MIMO channel input is the optimal MIMO receiver.

Optimal receiver performance, however, comes at the cost of receiver complexity. Performing all the hypothesis tests requires $2^{M \times N_{\mathrm{TX}}}$ complex operations, where M is the modulation order and N_{TX} is the number of transmit antennas. As an example, for 64QAM with two transmit antennas, the MLD decoding complexity is $2^{12} = 4096$ complex operations. For 64QAM with four transmit antennas, the MLD decoding complexity is $2^{24} = 16,777,216$ complex operations. On the other hand, the exhaustive constellation-search-based MLD receiver provides a good modular structure. The MLD receiver is suitable for application-specific integrated-circuit implementation. Several fast constellation search algorithms are available such as the sphere decoding algorithm [169]. The MLD-based MIMO receivers have many advantages compared with ZF and MMSE receivers. The MLD receiver performance advantages are significant when the MIMO channels are correlated, i.e., small-form-factor user devices require a compact packaging of multiple antennas, resulting in a high level of antenna correlation as well as a correlated MIMO channel. The ZF and MMSE receivers, conversely, can both suffer sharp performance losses when MIMO channels are correlated.

In addition, because of the limitations of small-form-factor user devices, it is difficult to achieve the same antenna gain for the two receive antennas embedded in the device. In the case of 4 dB antenna gain imbalance, the performance loss for the MLD receiver is negligible. Another advantage is diversity gain. The MLD receiver can not only separate the transmit data streams, but also can achieve the receive diversity gain for multiple receive antennas. Furthermore, the receiver can support a very high-mobility environment including reception at speeds exceeding 120km/h. Reception of the spatial multiplexing MIMO transmissions can be optimized by selecting the best receiver operation, based on the MIMO channel condition and the modulation constellation, to reduce the average processing power required. This reduces the power consumption of the user device and extends its battery life.

Since an MMSE receiver requires fewer computations than the MLD receiver, under certain transmission conditions, the performancc difference between the MMSE receiver and the MLD is negligible. We can measure and test the correlation condition of the MIMO channel before we decide which receiver should be used. If the MIMO channel correlation is very low or is orthogonal, the MMSE receiver can be used with no performance loss, while reducing the computational complexity by avoiding the use of the MLD receiver. Similarly, we can use the MMSE receiver when the SNR is high.

The advanced receiver can be further extended by successive interference cancellation (SIC) receiver architecture. The rationale behind the SIC receiver is that the signal quality of the multiple transmit data streams are not the same, because of the fading variations of the MIMO channel. We can successfully demodulate the first data stream, re-modulate it, and then subtract the first data stream from the received input combination in the absence of interference from the first data stream. We can then demodulate the second data stream successfully with a simple maximum ratio combining receiver.

Due to the nature of the MIMO channel, the eigenvalue of the MIMO channel matrix represents the signal strength for the data stream. For the two-transmit and two-receive MIMO configuration, different modulations for the different data streams can be scheduled for transmission at the base station. The user device measures the signal quality (typically one strong data stream and one weak data stream) and uses the feedback channel to request the coding and modulation for each data stream transmitted from the base station. If the user device implements an SIC receiver, the signal-quality measurement can be based on the receiver operation. In this case, the weak data stream's signal quality is measured in the absence of interference from the strong data stream. Therefore, the user device can request the increased coding and modulation transmission level for the second data stream. This can improve the throughput of the second data stream, increasing the overall throughput of the user device.

In the LTE/LTE-Advanced uplink, the use of SC-FDMA for multiple-access introduces ISI in dispersive channels. As a result, the ideal MIMO receiver should address both ISI and spatial multiplexing interference. There are a number of receiver implementation options including linear minimum mean-square error (LMMSE) equalization and SIC that can be used for MIMO detection in the LTE uplink.

In general, the downlink $N_{RX} \times 1$ received-signal vector $\mathbf{y}(k,l)$ at the kth subcarrier and the lth OFDM symbol can be expressed as the sum of the desired signal $\mathbf{H}_1(k,l)\mathbf{s}_1(k,l)$ from the serving eNB and the interference signals $\mathbf{H}_j(k,l)\mathbf{s}_j(k,l)|_{\forall j \neq 1}$ from other base stations and the noise vector $\mathbf{n}(k,l)$ as follows:

$$\mathbf{y}(k,l) = \mathbf{H}_1(k,l)\mathbf{s}_1(k,l) + \sum_{j=2}^{N_{eNB}} \mathbf{H}_j(k,l)\mathbf{s}_j(k,l) + \mathbf{n}(k,l) \tag{10.16}$$

where $\mathbf{s}_j(k,l)|_{j=1,2,\ldots,N_{eNB}}$ and $\mathbf{H}_j(k,l)|_{j=1,2,\ldots,N_{eNB}}$ represent the $N_{TX} \times 1$ transmitted signal vector and the $N_{RX} \times N_{TX}$ channel matrix between the jth cell and the UE including the contribution from both receiver branches. In the above equation $\mathbf{H}_j(k,l) = \begin{pmatrix} \mathbf{H}_j^1(k,l) & \mathbf{H}_j^2(k,l) \end{pmatrix}^T$ in which $\mathbf{H}_j^i(k,l)$ denotes the $N_{TX} \times 1$ channel matrix as seen at the ith receive antenna. The $N_l \times 1$ recovered signal vector at the UE denoted by $\hat{\mathbf{s}}_1(k,l)$ is detected using the $N_l \times N_{RX}$ receiver weighting matrix $\mathbf{W}_1(k,l)$ as follows:

$$\hat{\mathbf{s}}_1(k,l) = \mathbf{W}_1(k,l)\mathbf{y}_1(k,l) \tag{10.17}$$

The above generic model will be exploited in the receiver algorithms described in the following sections.

10.9.5.1 LMMSE receiver

An LMMSE receiver is a widely used solution which generates an MMSE estimate for each time-domain symbol. The MMSE has been used as a baseline receiver in LTE/LTE-Advanced evaluations due to its simplicity. The detection problem in an LMMSE receiver reduces to finding a weighting matrix such that $\hat{\mathbf{s}}_1(k,l) = \mathbf{W}_1(k,l)\mathbf{y}_1(k,l)$, where the weighting matrix is given as follows [25–27,43,44]:

$$\mathbf{W}_1(k,l) = \hat{\mathbf{H}}_1^H(k,l)\mathbf{R}^{-1}; \quad \mathbf{R} = P_1\hat{\mathbf{H}}_1(k,l)\mathbf{H}_1^H(k,l) + \sigma^2\mathbf{I} \tag{10.18}$$

where $\hat{\mathbf{H}}_j(k,l)$ and σ^2 denote the estimated channel matrix and noise power, respectively, P_1 is the transmission power of the serving cell and is given by $P_1 = E[|\mathbf{s}_1(k,l)|^2]$.

Due to the linearity and unitary properties of DFT, an equivalent equalization method would be to obtain a frequency-domain MMSE estimate for each symbol in the frequency-domain, and then take the IDFT to obtain the time-domain MMSE symbol estimation. The MMSE estimation for each frequency-domain symbol is obtained by linearly combining received signals collected from multiple receive antennas in the frequency-domain, using MMSE combining weights $\mathbf{W}(k) = [\mathbf{H}(k)\mathbf{H}^H(k) + \mathbf{R}(k)]^{-1}\mathbf{H}(k)$, where k is the sub-carrier index, $\mathbf{H}(k)$ is a vector representing the frequency response for the layer signal of interest (one element per receive antenna), and $\mathbf{R}(k)$ is the impairment covariance matrix encompassing the spatial correlation between the impairment components. The performance of this simple MMSE equalizer asymptotically approaches the theoretical capacity in a SIMO channel. However, in a MIMO channel, LMMSE equalizer performance is far from the SIMO capacity due to the presence of spatial multiplexing interference [127].

10.9.5.2 MRC receiver

The weighting matrix in an MRC receiver for rank-1 transmission can be represented as follows [25–27,43,44]:

$$\mathbf{W}_1(k,l) = \hat{\mathbf{H}}_1^H(k,l) \tag{10.19}$$

where $\hat{\mathbf{H}}_1(k,l)$ is the estimated channel matrix.

10.9.5.3 MMSE-IRC receiver

As an enhanced receiver, the MMSE-IRC receiver can suppress not only the inter-stream interference, but also the inter-cell interference when the degrees of freedom at the receiver are sufficient, i.e., the number of receiver antennas is greater than the number of desired data streams. The MMSE-IRC receiver weighting matrix can be expressed as follows:

$$\mathbf{W}_1(k,l) = \hat{\mathbf{H}}_1^H(k,l)\mathbf{R}^{-1} \tag{10.20}$$

where $\hat{\mathbf{H}}_1(k,l)$ and $\mathbf{R}$ denote the estimated channel matrix and channel covariance matrix, respectively. To obtain the MMSE-IRC receiver weighting matrix, the covariance matrix including the sources of inter-cell interference must be estimated. Various schemes can be considered for this purpose, as described below.

In a covariance matrix estimation scheme based on cell-specific reference signals, the covariance matrix is estimated at the location of the CRSs using the following equations:

$$\begin{aligned} \mathbf{R} &= P_1\hat{\mathbf{H}}_1(k,l)\hat{\mathbf{H}}_1^H(k,l) + \frac{1}{N_{\mathrm{sp}}}\sum_{(k,l)\in\{(k_{\mathrm{CRS}},l_{\mathrm{CRS}})\}} \tilde{\mathbf{r}}(k,l)\tilde{\mathbf{r}}^H(k,l) \\ \tilde{\mathbf{r}}(k,l) &= \mathbf{r}(k,l) - \hat{\mathbf{H}}_1(k,l)\mathbf{s}_1(k,l) \end{aligned} \tag{10.21}$$

where P_1 is the transmission power of the serving cell which is given by $P_1 = E[|\mathbf{s}_1(k,l)|^2]$ and N_{sp} is the number of sampling resource elements.

In a DM-RS-based covariance matrix estimation scheme, which is applicable when DM-RS are configured, the covariance matrix is estimated at DM-RS resource elements using the following equations:

$$\begin{aligned} \mathbf{R} &= P_1\hat{\mathbf{H}}_1(k,l)\hat{\mathbf{H}}_1^H(k,l) + \frac{1}{N_{\mathrm{sp}}}\sum_{(k,l)\in\{(k_{\mathrm{DM\text{-}RS}},l_{\mathrm{DM\text{-}RS}})\}} \tilde{\mathbf{r}}(k,l)\tilde{\mathbf{r}}^H(k,l) \\ \tilde{\mathbf{r}}(k,l) &= \mathbf{r}(k,l) - \hat{\mathbf{H}}_1(k,l)\mathbf{s}_1(k,l) \end{aligned} \tag{10.22}$$

In a data-based covariance matrix estimation scheme, the covariance matrix is estimated at PDSCH resource elements using the following equations:

$$\mathbf{R} = \frac{1}{N_{\mathrm{sp}}}\sum_{(k,l)\in\{(k_{\mathrm{PDSCH/DM\text{-}RS}},l_{\mathrm{PDSCH/DM\text{-}RS}})\}} \mathbf{r}(k,l)\,\mathbf{r}^H(k,l) \tag{10.23}$$

where $\{(k_{\mathrm{PDSCH/DM\text{-}RS}}, l_{\mathrm{PDSCH/DM\text{-}RS}})\}$ denotes the set of resource elements assigned to DM-RS or PDSCH data sub-carriers.

10.9.5.4 SIC receiver

An SIC receiver is a good candidate for MIMO reception. It can be shown that SIC receivers with per-layer rate control can achieve MIMO capacity in non-dispersive channels [127]. The operation of an SIC receiver can be described as follows. The receiver first detects the signal sent by the first layer and after successful detection, it cancels the interference contributed by the detected signal before it detects the other layers. The process is repeated until all the layers are detected. Typically, each layer is canceled only when the CRC check is successful. With perfect per-layer rate control, the transmission rate for each layer is chosen to satisfy sufficiently low BLER; consequently, the SIC receiver can cancel interference contributed by previously detected layers. An SIC receiver typically consists of a number of LMMSE equalizers. The role of these MMSE equalizers is to suppress the residual interference. The cancelation of other layers is reflected in the impairment correlation, and thus in the MMSE combining weights. The cancelation of the detected layers allows the MMSE equalizer to use its available degree of freedom to better suppress other dominant interfering signals, resulting in a reduced interference level and a higher SINR.

In practice, there are some limitations pertaining to an SIC receiver operation e.g., an SIC receiver does not address the ISI problem. Although frequency-domain MMSE equalization is effective in suppressing ISI, there is an opportunity to achieve small performance improvement by properly canceling ISI. Moreover, turbo encoding and CRC checking is not necessarily performed on a per-layer basis, rather per-codeword rate control and CRC checking is used for uplink MIMO. For example, for the rank-4 transmission, the first and second layers are mapped to the same codeword and share the same CRC bits. Hence, these two layers need to be decoded jointly rather than individually. As a result, prior to the decoding stage, the second layer receiver cannot cancel the interference contributed by the first layer, resulting in high intra-codeword interference [127].

10.10 Uplink transmission modes

In order to support uplink SU-MIMO, the concept of transmission mode for PUSCH was introduced in the LTE Rel-10 to differentiate between single-antenna and spatial multiplexing transmission. The uplink transmission modes are as follows:

- PUSCH transmission mode 1, in which uplink transmission is from a single antenna port.
- PUSCH transmission mode 2, where uplink transmission is from multiple antenna ports. In this mode, the UE can be configured to transmit from either two or four antenna ports. If a UE is configured in PUSCH transmission mode 2, it can transmit up to two transport blocks per subframe.

The eNB signals the uplink transmission mode to the UE via an RRC connection setup, RRC connection reconfiguration, or RRC connection reestablishment message. Transmission mode 2 allows dynamic switching between single-antenna and spatial multiplexing transmission [14]. The transport blocks are mapped to one or more layers according to the same rules discussed for the downlink in LTE Rel-8. Each transport block is independently acknowledged through PHICH. The PHICH index for the first codeword is the same as that used in LTE Rel-8 for single-codeword PUSCH transmission, i.e., associated with the lowest PRB index of the PUSCH resource allocation. The PHICH index for the second codeword is related to the first by an offset. The PDCCH resource grant for uplink MIMO transmissions includes two independent NDIs to indicate whether a retransmission is anticipated. The closed-loop codebook-based precoding is used for PUSCH in a similar way to the LTE Rel-8 PDSCH in transmission mode 4. The UE is instructed by the eNB to use a specific rank and precoding matrix signaled within the uplink grant [15]. The uplink transmission modes are applicable when UE is addressed by its C-RNTI. The closed-loop spatial multiplexing is not supported when the UE is addressed by an SPS-RNTI [14]. Transmission mode 1 is the default uplink transmission mode for a UE unless the UE is configured for another uplink transmission mode via higher-layer signaling. When a UE configured in transmission mode 2 receives a DCI format 0 (uplink scheduling grant), it assumes that PUSCH transmission is associated with transport block 1 and that transport block 2 is disabled.

10.11 Uplink feedback in LTE/LTE-Advanced

In wireless communications, CSI refers to channel properties of a wireless communication link. This information describes how a signal propagates from the transmitter to the receiver and represents the combined effect of scattering, multipath fading, signal power attenuation with distance, etc. The knowledge of CSI at the transmitter and/or the receiver makes it possible to adapt data transmission to current channel conditions, which is crucial for achieving reliable and robust communication with high data rates in multi-antenna systems. The CSI is often required to be estimated at the receiver, and usually quantized and fed back to the transmitter. The downlink channel can be estimated from uplink reference signals in TDD systems under certain conditions due to reciprocity. In general, the transmitter and receiver can observe different CSI.

There are two types of CSI, i.e., instantaneous CSI and statistical CSI. In instantaneous CSI or short-term CSI, the current channel conditions are known, which can be interpreted as knowing the impulse response of a digital filter. This provides an opportunity to adapt the transmitted signal to the impulse response and thereby to optimize the received signal for spatial multiplexing or to achieve low bit error rates. In statistical CSI or long-term CSI, the statistical characteristics or statistics of the channel are known. The latter information may include the type of fading distribution, the average channel gain, the line-of-sight component, and the spatial correlation. Similar to the instantaneous CSI, this information can be used for optimization of transmission parameters.

The CSI estimation accuracy is practically limited by how fast the channel conditions are varying. In fast-fading channels where the channel conditions may vary rapidly during transmission of a single information symbol, only statistical CSI is reasonable. On the other hand, in slow-fading scenarios, the instantaneous CSI can be estimated with reasonable precision and used for transmission adaptation for a period of time before becoming obsolete. In practical scenarios, the available CSI is often manifested as instantaneous CSI with some estimation/quantization error combined with some statistical information. Recall that in a narrowband flat-fading channel with multiple transmit and receive antennas, the system is modeled as $\mathbf{y} = \mathbf{Hx} + \mathbf{n}$, where $\mathbf{y}$ and $\mathbf{x}$ are the received and transmitted signal vectors, respectively, $\mathbf{H}$ and $\mathbf{n}$ are the channel matrix and the receiver noise vector, respectively. The noise is often modeled as circular symmetric complex normal random variable with statistical distribution $\mathbf{n} \sim Z(\mathbf{0}, \mathbf{R_n})$ where the mean value is zero and the noise covariance matrix $\mathbf{R_n}$ can be estimated. In the case of instantaneous CSI, the channel matrix $\mathbf{H}$ is known; however, due to channel estimation errors, the channel information can be represented as $\hat{\mathbf{H}} = \mathbf{H} + \mathbf{e}$ in which $\text{vec}(\hat{\mathbf{H}}) \sim Z(\text{vec}(\mathbf{H}), \mathbf{R_e})$, where $\hat{\mathbf{H}}$ denotes the estimated channel matrix and $\mathbf{R_e}$ is the estimation error covariance matrix. The vectorization operator[15] $\text{vec}(\cdot)$ is used to obtain the column representation of matrix $\mathbf{H}$. In the case of statistical CSI, the statistics of $\mathbf{H}$ are known. For instance, in a Rayleigh fading channel, this corresponds to $\text{vec}(\mathbf{H}) \sim Z(\mathbf{0}, \mathbf{R_H})$ for a known channel covariance matrix $\mathbf{R_H}$ [170].

The uplink feedback in LTE/LTE-Advanced for support of downlink data transmission consists of RI, PMI, PTI, and CQI. The UE uses the UCI to provide the eNB with feedback that includes CQI, PMI, PTI, RI, scheduling requests, and HARQ acknowledgments. The UE sends the uplink feedback to the eNB using PUCCH or PUSCH. As mentioned earlier, LTE Rel-8/9 does not support simultaneous transmission of PUCCH and PUSCH in the uplink.

Figure 10.55 illustrates the various forms of CSI reporting. The CSI reporting is divided into periodic and aperiodic categories. The periodic CSI reporting uses PUCCH unless data transmission is already scheduled on PUSCH. The aperiodic CSI reporting is always transmitted via PUSCH. The aperiodic reporting can be triggered using a flag within the DCI that is used to allocate PUSCH resources. Since the capacity of PUSCH is greater than that of PUCCH, larger control information payload can be included in aperiodic CSI feedback.

The scheduling request is not explicitly sent on PUSCH, rather the UE includes a buffer status report (BSR) MAC control element in the uplink MAC PDUs. The BSR MAC control element provides the eNB with information regarding the amount of data in the UE's local buffer awaiting transmission. The LTE Rel-10 allows simultaneous transmission of PUCCH and PUSCH in the same uplink subframe. In this case, the UCI and feedback can be sent on PUCCH, whereas aperiodic CSI reports and user data can be transmitted on PUSCH.

---

[15]In linear algebra and matrix theory, the vectorization of a matrix is a linear transformation which converts the matrix into a column vector. More specifically, the vectorization of an $M \times N$ matrix $\mathbf{H}$, denoted by $\text{vec}(\mathbf{H})$, in which $\text{vec}(\mathbf{H})$ is an $MN \times 1$ column vector is obtained by stacking the columns of the matrix $\mathbf{H}$ on top of each other, as follows: $\text{vec}(\mathbf{H}) = (h_{11}, h_{21}, \ldots, h_{M1}; h_{12}, h_{22}, \ldots, h_{M2}; \ldots; h_{1N}, h_{2N}, \ldots, h_{MN})^{\mathrm{T}}$ where h_{ij} represents the (i,j) element of matrix $\mathbf{H}$ and the superscript T denotes the transpose operation. The vectorization operation can be interpreted as the isomorphism $\mathbf{H}^{M \times N} := \mathbf{H}^{M} \otimes \mathbf{H}^{N} \cong \mathbf{H}^{MN}$ between these vector spaces (of matrices and vectors) in coordinates [30–33].

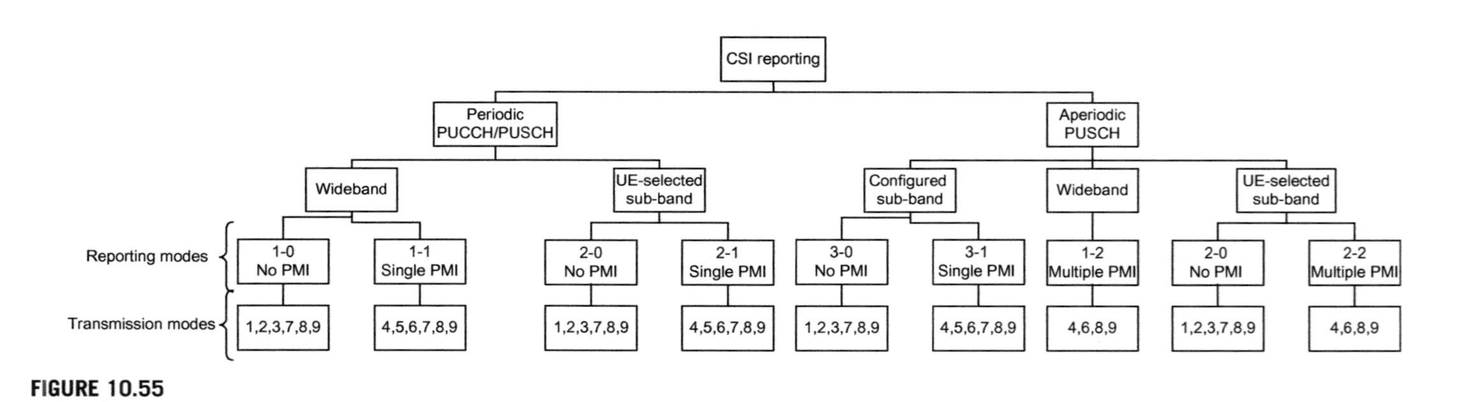

FIGURE 10.55

Classification of CSI reporting modes [14].

The RI indicates the number of spatial layers which can be supported by the UE under current channel conditions. It has been observed in the course of LTE evaluation that the frequency-selective RI reporting does not provide significant performance benefit; therefore, only one wideband RI is reported for the entire bandwidth [122]. However, the reporting of PMI and CQI can be in the form of wideband or sub-band (frequency-selective). For RI values equal to one, only one CQI is reported for each reporting unit in frequency, which is either wideband or sub-band in the case of frequency-selective reporting. For RI values greater than one, the CQI for each codeword is reported for the closed-loop spatial multiplexing, as different codewords may be assigned to different layers, while only one CQI is reported for the open-loop spatial multiplexing, as one codeword interacts with all layers. The PMI indicates the preferred precoding candidate for the corresponding frequency unit, e.g., a particular sub-band or the entire system bandwidth, and is selected from the possible precoding candidates depending on the number of transmit antennas and according to the RI value.

The CQI metric indicates a combination of the maximum data rate and the modulation scheme, which can achieve BLER less than 10% at the UE after detection, assuming that the reported rank and PMI are applied to the time–frequency resource (see Figure 10.57). Given this definition of CQI, PMI, and RI, the UE can report the maximum data rate that it can receive and demodulate, taking into account its receiver capability and channel conditions. In the case of the frequency-selective PMI/CQI reporting, the UE reports a PMI/CQI for each sub-band (Figure 10.56). For the non-frequency-selective wideband PMI/CQI reporting, the UE reports a single wideband PMI/CQI corresponding to the entire bandwidth. In the frequency-selective reporting mode, the sub-band CQI is reported as a differential value with respect to the wideband CQI in order to reduce the signaling overhead. The sub-band CQI reports are encoded differentially with respect to the wideband CQI using 2 bits as follows: $\Delta CQI_{\text{sub-band}} = CQI_{\text{sub-band(index)}} - CQI_{\text{wideband}}(\text{index})$, where the permissible sub-band differential CQI offset values are $\{\leq -1, 0, 1, \geq 2\}$. In best-$M$ average reporting, the UE first estimates the channel quality for each sub-band. Then it selects the M best values and reports to the eNB a single average CQI corresponding to the MCS/TBS that the UE can receive correctly, assuming that the eNB schedules the UE on those M sub-bands. The parameter M depends on the system bandwidth and corresponds to approximately 20% of the system bandwidth. Figure 10.56 shows the principle of the best-M average CQI measurement and reporting scheme. The CQI value for the M selected sub-bands for each codeword is encoded differentially using 2 bits relative to the corresponding wideband CQI as follows: $\Delta CQI = CQI_{\text{average-bestM-sub-band}}(\text{index}) - CQI_{\text{wideband}}(\text{index})$, where the possible differential CQI values are $\{\leq 1, 2, 3, \geq 4\}$. It must be noted that the subtraction of the wideband CQI value prior to quantization of sub-band CQI values reduces the dynamic range and thus the quantization error.

When the frequency-selective CQI reporting is configured, sub-band CQIs as well as wideband CQI are reported, and the wideband CQI serves as the baseline

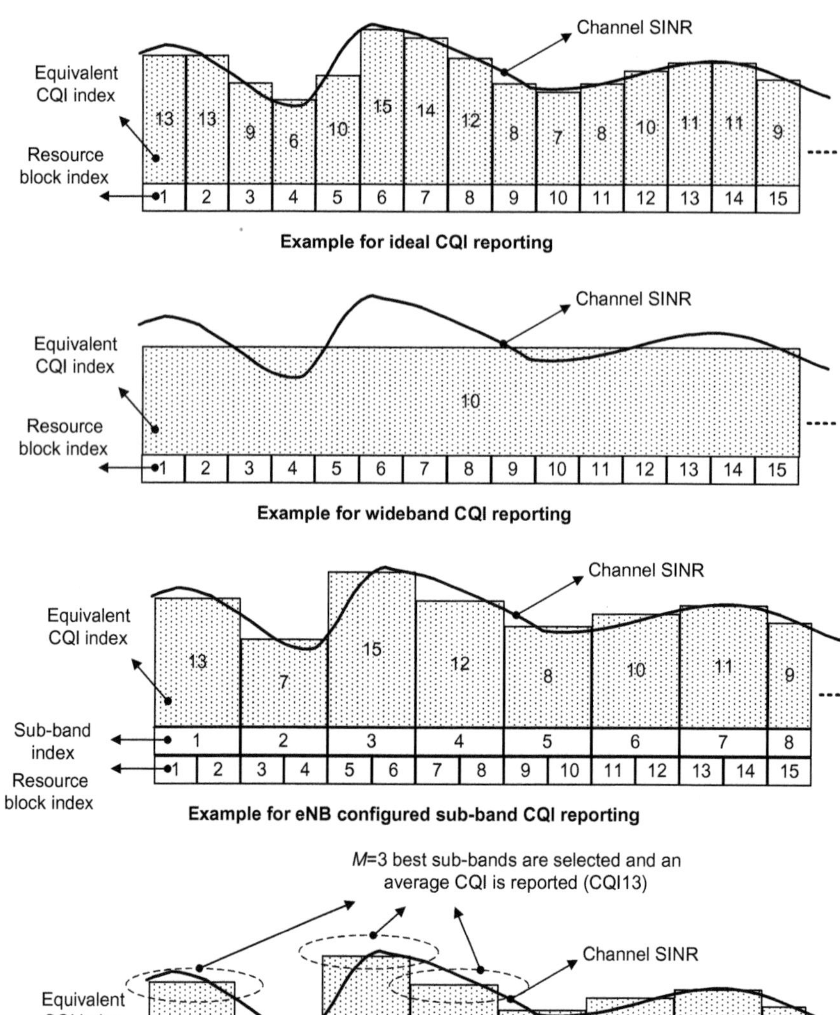

FIGURE 10.56

Graphic comparison of various CQI reporting methods [170].

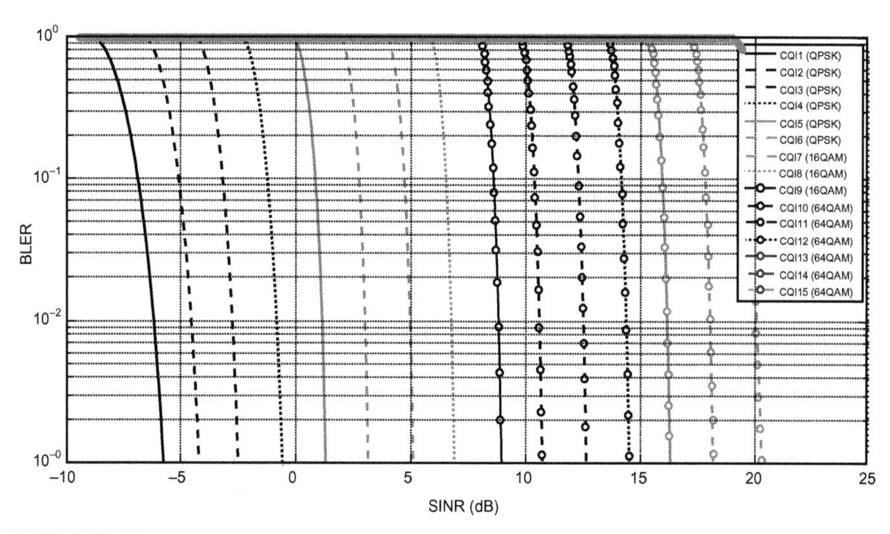

FIGURE 10.57

Link-level variation of BLER versus SINR with CQI as a parameter [137].

for recovering the downlink channel condition over the entire frequency band. The frequency-selective reporting results in a large signaling overhead. In the cases where the uplink overhead is the limiting factor, the eNB can also configure non-frequency-selective CQI/PMI reports, to mitigate various channel conditions and different antenna configurations, while keeping the signaling overhead at an appropriate level. Thus, various feedback modes are specified depending on the frequency selectivity of the CQI and the PMI reports.

The physical channels that can be used for uplink feedback signaling are PUCCH and PUSCH. The uplink feedback via PUSCH is used to accommodate a large amount of feedback information in a single reporting instance. For example, the reporting on PUSCH can include RI, wideband CQI per codeword, and the PMI for each sub-band. Note that the rank indicator is separately encoded from the other fields using 1 bit for the case of two transmit antennas and 2 bits for the case of four transmit antennas. The number of bits for the other information fields is determined according to the rank indicator. Another example is the reporting of the RI, wideband PMI, wideband CQI per codeword, and the sub-band differential CQI for each sub-band per codeword. It is noted that simultaneous reporting of sub-band differential CQI and sub-band PMI is not supported due to excessive signaling overhead.

The physical uplink control channel, on the other hand, is designed for transmission of small amounts of control signaling information. Therefore, for PUCCH reporting modes, separate time instances are used for reporting RI, wideband CQI/PMI, and sub-band CQI of the sub-band selected by the UE. The spatial differential CQI represents the difference between the CQIs of two codewords and is defined for wideband CQI and sub-band CQI. In the case of sub-band CQI feedback, each reporting instance corresponds to a group of sub-bands, from which the UE selects the sub-band with the best CQI. Due to these characteristics, the eNB may configure periodic reporting of a limited amount of feedback information on PUCCH, while triggering the UE to report a large amount of detailed feedback information on PUSCH, if accurate channel information is necessary for transmission of large amounts of data in the downlink.

The set of sub-bands S that the UE is required to conduct measurements for CQI reporting spans the entire downlink system bandwidth. A sub-band is a set of L_{sb} contiguous PRBs, where L_{sb} is dependent on the system bandwidth. Note that the last sub-band in set S may have fewer than k contiguous PRBs depending on N_{RB}^{DL}. The number of sub-bands for system bandwidth N_{RB}^{DL} is defined by $N_{sb} = \lceil N_{RB}^{DL}/L_{sb} \rceil$. The sub-bands are indexed in the order of increasing frequency and non-increasing size starting at the lowest frequency while taking the following rules into consideration [5]:

- For transmission modes 1, 2, 3, and 5 as well as transmission modes 8, 9, and 10 without PMI/RI reporting, transmission mode 4 with RI equal to one, and transmission modes 8, 9, and 10 with PMI/RI reporting and RI equal to one, a single 4-bit wideband CQI is reported.
- For transmission modes 3 and 4 as well as transmission modes 8, 9, and 10 with PMI/RI reporting, CQI is calculated assuming transmission of one codeword for RI equal to one and two codewords for RI greater than one.
- For transmission mode 4 with RI greater than one as well as transmission modes 8, 9, and 10 with PMI/RI reporting, PUSCH-based triggered reporting includes reporting of a wideband CQI which comprises a 4-bit wideband CQI for codeword 0 and a 4-bit wideband CQI for codeword 1.
- For transmission mode 4 with RI greater than one 1 as well as transmission modes 8, 9, and 10 with PMI/RI reporting, PUCCH-based reporting includes reporting a 4-bit wideband CQI for codeword 0 and a wideband spatial differential CQI. The wideband spatial differential CQI value includes a 3-bit wideband spatial differential CQI value for codeword 1 offset level where codeword 1 offset level is the difference between the wideband CQI index for codeword 0 and wideband CQI index for codeword 1.

A UE is semi-statically configured by higher layers to feed back CQI, PMI, and corresponding RI on the same PUSCH using one of the CSI reporting modes provided in Table 10.28. Note that CQI/PMI feedback modes via PUSCH are not supported for a serving cell with bandwidth $N_{RB}^{DL} \leq 7$ [5].

As given in Table 10.28 and shown Figure 10.55, the aperiodic CQI/PMI feedback types reported via PUSCH can be categorized and described as follows [5]:

- *Wideband feedback*
  - Mode 1-2: For each sub-band, a preferred precoding matrix is selected from the codebook subset assuming transmission is scheduled only in that

Table 10.28 CQI and PMI Feedback Types for PUSCH/PUCCH CSI Reporting Modes [5]

	PMI Feedback Types via PUSCH			PMI Feedback Types via PUCCH	
	No PMI	**Single PMI**	**Multiple PMI**	**No PMI**	**Single PMI**
Wideband (wideband CQI)	–	–	Mode 1-2	Mode 1-0	Mode 1-1
UE Selected (sub-band CQI)	Mode 2-0	–	Mode 2-2	Mode 2-0	Mode 2-1
Higher-layer configured sub-band CQI)	Mode 3-0	Mode 3-1	–	–	–

sub-band. The UE reports one wideband CQI value per codeword, which is calculated assuming the use of the corresponding selected precoding matrix in each sub-band and transmission on a set of sub-bands. The UE further reports the selected PMI for a set of sub-bands, except for transmission modes 9 and 10 configured with eight CSI-RS ports where the first PMI is reported for the set of sub-bands, and the second PMI is reported for each sub-band in the set.

- *Higher-layer configured sub-band feedback*
  - Mode 3-0: The UE reports a wideband CQI value which is calculated assuming transmission on a set of sub-bands. The UE also reports one sub-band CQI value for each set of sub-bands. The sub-band CQI value is calculated assuming transmission is scheduled only in that sub-band. Both wideband and sub-band CQI represent channel quality for the first codeword, even if the rank indicator is greater than one.
  - Mode 3-1: A single precoding matrix is selected from the codebook subset assuming transmission on a set of sub-bands. The UE reports one sub-band CQI value per codeword for each set of sub-bands, which are calculated assuming the use of a single precoding matrix in all sub-bands and assuming transmission is scheduled in the corresponding sub-band. The UE reports a wideband CQI value per codeword, which is calculated assuming the use of a single precoding matrix in all sub-bands, and transmission is scheduled on a set of sub-bands. The UE reports the selected single precoding matrix indicator, except for transmission modes 9 and 10 configured with eight CSI-RS ports where two precoding matrix indicators are reported.
- *UE-selected sub-band feedback*
  - Mode 2-0: The UE selects a set of M preferred sub-bands of size L_{sb} (where L_{sb} and M are given in Table 10.29 for each system bandwidth) within the set of sub-bands S. The UE also reports one CQI value

Table 10.29 Sub-band Size L_{sb} and Number of Sub-bands M in S as a Function of System Bandwidth [5]

Number of Downlink Resource Blocks N_{RB}^{DL}	Sub-band Size L_{sb} (Resource Blocks)	Number of Sub-bands M
6–7	N/A	N/A
8–10	2	1
11–26	2	3
27–63	3	5
64–110	4	6

reflecting transmission only over the M selected sub-bands determined in the previous step. The CQI represents channel quality for the first codeword, even when RI > 1. Additionally, the UE reports one wideband CQI value which is calculated assuming transmission on a set of sub-bands S. The wideband CQI represents channel quality for the first codeword, even when RI > 1. For transmission mode 3, the reported CQI values are calculated conditioned on the reported RI. For other transmission modes, they are reported conditioned on rank 1.

- Mode 2-2: The UE performs joint selection of the set of M preferred sub-bands of size L_{sb} within the set of sub-bands S and a preferred single precoding matrix selected from the codebook subset that is preferred to be used for transmission over the M selected sub-bands. The UE reports one CQI value per codeword reflecting transmission only over the selected M preferred sub-bands and using the same selected single precoding matrix in each of the M sub-bands. Except for transmission modes 9 and 10 with eight CSI-RS ports configured, the UE also reports the selected single precoding matrix indicator preferred for the M selected sub-bands. A UE further reports the selected single precoding matrix indicator for all sub-bands in the set S. For transmission modes 9 and 10 with eight CSI-RS ports configured, the UE reports the first precoding matrix indicator for all sub-bands in the set S. A UE further reports a second precoding matrix indicator for all sub-bands in the set S and a second precoding matrix indicator for the M selected sub-bands A single precoding matrix is selected from the codebook subset assuming transmission on a set of sub-bands S. A UE provides the wideband CQI value per codeword which is calculated assuming the use of a single precoding matrix in all sub-bands and transmission on a set of S sub-bands S. For transmission modes 4, 8, 9, and 10, the reported PMI and CQI values are calculated conditioned on the reported RI. For other transmission modes, they are reported conditioned on rank 1. For all UE-selected sub-band feedback modes, the UE further provides the positions of the M selected sub-bands using a combinatorial index [5].

A UE can be semi-statically configured by higher layers to periodically send different CSI (CQI, PMI, PTI, and/or RI) on PUCCH using the reporting modes given in Table 10.28. The periodic CSI reporting mode for each serving cell is configured by higher-layer signaling. For the UE-selected sub-band CQI, a CQI report in a certain subframe of a certain serving cell describes the channel quality in a particular part or in particular parts of the bandwidth of that serving cell, described subsequently as bandwidth part or parts. The bandwidth parts are indexed in the order of increasing frequency and decreasing sizes starting at the lowest frequency. The following CQI/PMI and RI reporting types with distinct periods and offsets are supported for the PUCCH CSI reporting modes given in Table 10.28 [5]:

- Type 1 report supports CQI feedback for the UE selected sub-bands.
- Type 1a report supports sub-band CQI and second PMI feedback.
- Type 2, 2b, and 2c report supports wideband CQI and PMI feedback.
- Type 2a report supports wideband PMI feedback.
- Type 3 report supports RI feedback.
- Type 4 report supports wideband CQI.
- Type 5 report supports RI and wideband PMI feedback.
- Type 6 report supports RI and PTI feedback.

10.12 Uplink transmitter and receiver chain

An example of a transmitter/receiver chain of an LTE modem is depicted in Figures 10.58 and 10.59 [171–173]. Note that the modem implementation presented in this section only includes the physical layer processing functions assuming higher layer protocols are processed by another processor. The digital front-end unit is the interface between the analog front-end and digital baseband modules in the wireless systems. The function of the digital front end (block of digital filters) is typically to perform gain control, sampling rate conversion, pulse shaping, matched filtering, and phase correction. The digital front end is functionally simple relative to other baseband blocks; however, it consumes a significant portion of the chip area and power. As an OFDM system, LTE is sensitive to carrier frequency offset (CFO) which causes inter-carrier interference. The timing synchronization and fractional CFO estimation/correction can mitigate the effects of CFO and the oscillator phase noise. Integer frequency offset estimation is then applied to remove misalignment, which is an integer multiple of the sub-carrier spacing. Although a large portion of CFO can be compensated, there is still a residual frequency offset due to estimation errors. The residual frequency offset is estimated based on the frequency-domain symbols after the FFT operation in the receiver.

Similar to other OFDM systems, reference signals are inserted during sub-carrier mapping in both time and frequency dimensions. The reference signals

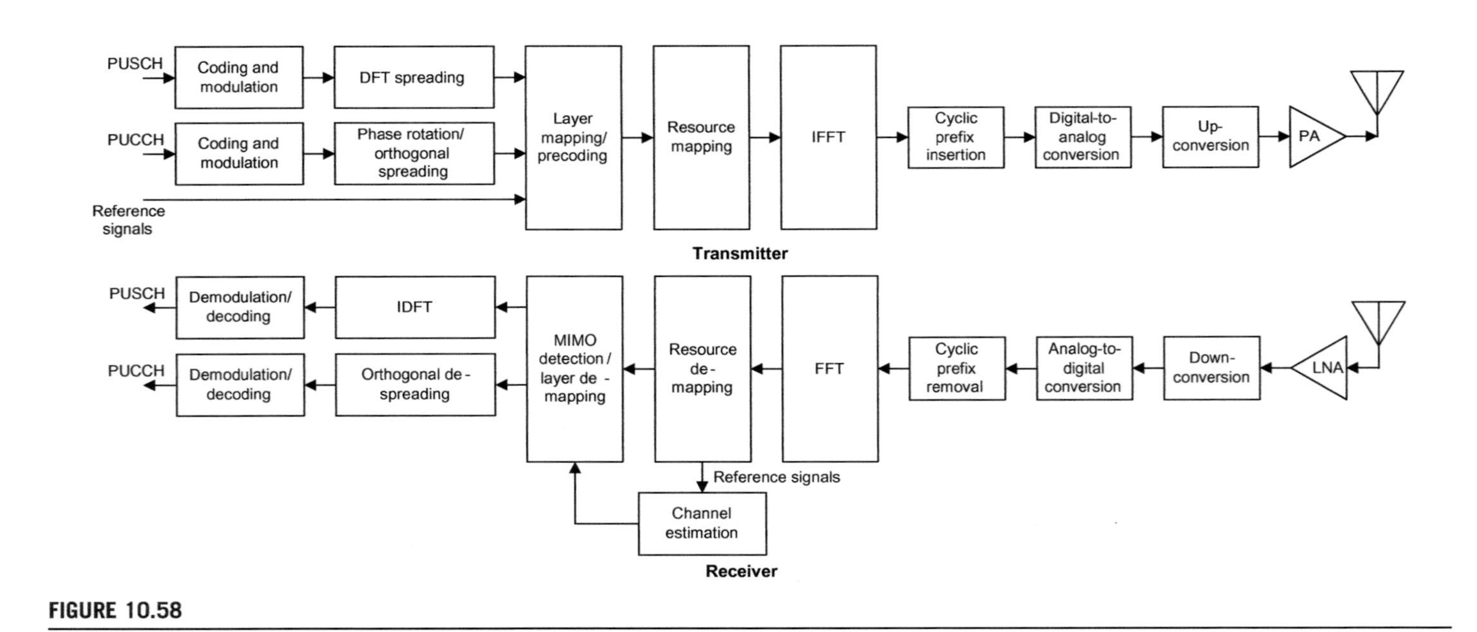

FIGURE 10.58

Example of uplink transmission and reception chains [14].

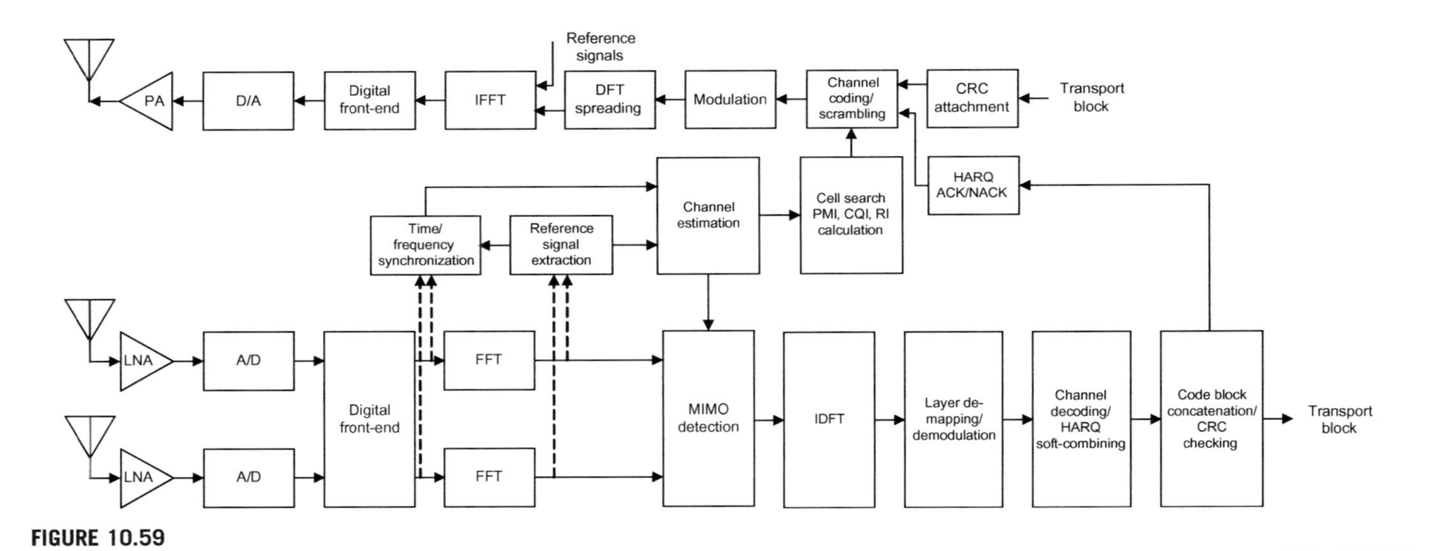

FIGURE 10.59

Block diagram of an example LTE UE modem (physical layer) [173].

transmitted from multiple antennas are mutually orthogonal (provided that the antennas are uncorrelated), which means the channel impulse response between different transmit-receive antenna pairs can be separately estimated. This assumption would reduce high complexity of MIMO channel estimation at the cost of lower spectral efficiency. For MIMO systems, a major challenge is the separation and detection of the transmitted symbols at the receiver given the amount of operations involved (e.g., matrix inversion and multiplication/division). Among different detection algorithms, maximum likelihood detection is an optimum detector that computes the following likelihood function:

$$L(b_i|\mathbf{r}) = \log\left(\frac{\sum_{\mathbf{s}:b_i(\mathbf{s})=1}\exp(-\sigma_0^{-2}||\mathbf{r}-\hat{\mathbf{H}}\mathbf{s}||^2)}{\sum_{\mathbf{s}:b_i(\mathbf{s})=0}\exp(-\sigma_0^{-2}||\mathbf{r}-\hat{\mathbf{H}}\mathbf{s}||^2)}\right) \tag{10.24}$$

In the above equation, $b_i(\mathbf{s})$ denotes the ith bit of the input sequence $\mathbf{s}$, σ_0^2 is the noise variance, and $\mathbf{r} = \mathbf{Hs} + \mathbf{n}$ is the received signal. The functional block diagram in Figure 10.59 shows an example of the internal data flows through an LTE UE with two receive antennas and one transmit antenna. The RF signal is received by the receiver antennas, down-converted, scaled by an automatic gain control, and digitized by an analog-to-digital converter (A/D) within the analog and digital front-end module including a sample rate converter and a numerically controlled oscillator.

The baseband processor receives the digitized signal in the form of complex-valued samples from the A/D and delivers the decoded bit stream to a higher layer protocol and application processor. It first removes the cyclic prefix, transforms the signal into frequency-domain by an FFT operation, performs channel estimation, equalization, log-likelihood ratio (LLR) calculation, HARQ combining, channel decoding, and CRC check for different physical channels. In addition, control information for closed-loop MIMO operation is generated, i.e., CQI, PMI, and RI. In the uplink direction, PUSCH data is channel coded, DFT spread, and mapped to resource elements together with modulated PUCCH and modulated reference signals. Following IFFT and cyclic prefix insertion procedures, the data is sent to transmitter front end and RF circuitry. For random access, the corresponding PRACH is processed. In a typical receiver, synchronization is usually initially conducted to acquire information such as the beginning of frame and the CFO calculation and correction [173].

The solution to Eq. (10.24) requires calculation of the likelihood function over the entire set of possible transmitted vectors, resulting in a significant amount of complexity in practice. However, since maximum likelihood provides the best theoretical performance, it is commonly used as a benchmark when comparing other algorithms. Linear detection algorithms such as MMSE have very low complexity. The MMSE equalizer is defined as follows: $\hat{s}_{\text{MMSE}} = \mathbf{W}_{\text{MMSE}}\mathbf{r} = (\hat{\mathbf{H}}^H\hat{\mathbf{H}}+\sigma_0^2\mathbf{I})^{-1}\hat{\mathbf{H}}^H\mathbf{r}$. The MMSE detection is the most commonly used detection scheme and involves matrix inversion which can be efficiently performed through direct inversion with sufficient numerical stability. Despite the low complexity,

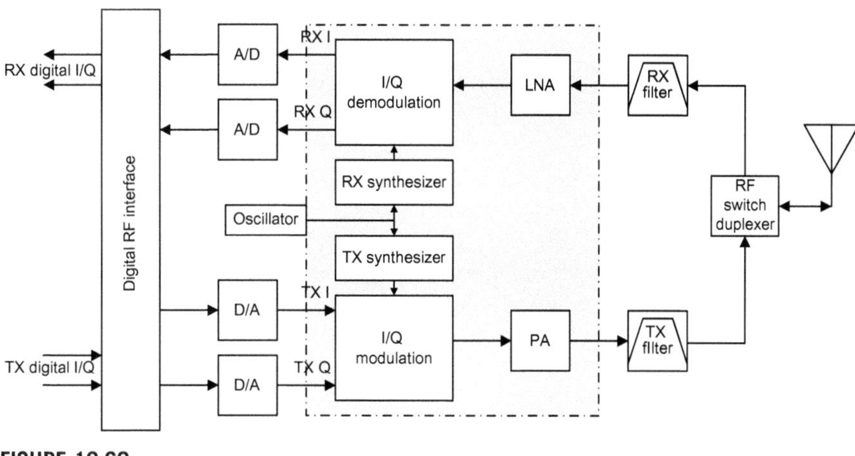

FIGURE 10.60

Simplified block diagram of RF transmitter and receiver in an LTE device [159].

MMSE detection has relatively poor performance, especially in slow fading channels. As a trade-off between detection performance and implementation complexity, soft-output MIMO detectors and sphere decoders can be used [169].

As shown in Figures 10.59 and 10.60, the radio signals received by the RF front end are down-converted to analog baseband signals (using a zero-IF converter), then they are converted to digital baseband signals using A/D converters. The digital front end applies filtering to the baseband signals and FFT is applied to transform the time-domain signal into frequency-domain where channel estimation and symbol detection are performed. In LTE, the channel estimator uses the reference signals to estimate the channel matrix $\mathbf{H}$ for all data sub-carriers. The estimated channel matrix $\hat{\mathbf{H}}$ and the received symbol vector $\mathbf{r}$ belonging to the current user will be extracted for MIMO detection. The above functions are performed by the baseband processor. The estimated channel matrix is conveyed to the MIMO detector together with the received symbols extracted from data sub-carriers. The coefficients are fed to the detector to compute the LLR soft-output $L(b_i^k)$ (i.e., ith bit of the input vector on the kth sub-carrier). For example, in case of MMSE detection, the detector fetches the MMSE weighting matrix $\mathbf{W}_{\mathrm{MMSE}}$ from the local memory and multiplies the MMSE weighting matrix $\mathbf{W}_{\mathrm{MMSE}}$ with the received symbol vectors $\mathbf{r}$ in order to compute $\hat{s}$ and to map $\hat{s}$ to LLR values. The LLR values are then passed to the FEC decoder where it will be deinterleaved and processed by the channel decoder to generate the output bits. The FEC module contains the soft buffer for chase combining in HARQ and a parallel turbo decoder. Although many baseband processing functions can be assigned to a programmable processor, there are still some tasks which need to be implemented in dedicated hardware to meet the performance and power constraints. In that case, hardware accelerators are added to the on-chip system. They are typically controlled by a number of control registers which can be set by the baseband processor.

The RFICs in the UE may include different portions of the schematic block diagram shown in Figure 10.60, with some containing the transmitter or receiver only, and others combining transmitter and receiver in one component. Traditionally RFICs have used analog baseband I/Q inputs and outputs. However, digital I/Q interfaces are becoming more common. The mobile industry processor interface (MIPI) alliance has developed the DigRF standard [174], which describes the high-speed digital baseband to RFIC interface for new generation of cellular radios, which includes embedded control protocols for the RFIC.

References

3GPP Specifications[16]

[1] 3GPP TS 36.101, Evolved Universal Terrestrial Radio Access (E-UTRA); User Equipment (UE) radio transmission and reception (Release 11), December 2012.

[2] 3GPP TS 36.104, Evolved Universal Terrestrial Radio Access (E-UTRA); Base Station (BS) radio transmission and reception (Release 11), December 2012.

[3] 3GPP TS 36.211, Evolved Universal Terrestrial Radio Access (E-UTRA), Physical Channels and Modulation (Release 11), December 2012.

[4] 3GPP TS 36.212, Evolved Universal Terrestrial Radio Access (E-UTRA) Multiplexing and Channel Coding (Release 11), December 2012.

[5] 3GPP TS 36.213, Evolved Universal Terrestrial Radio Access (E-UTRA); Physical layer procedures (Release 11), December 2012.

[6] 3GPP TS 36.214, Evolved Universal Terrestrial Radio Access (E-UTRA); Physical layer – Measurements (Release 11), December 2012.

[7] 3GPP TS 36.300, Evolved Universal Terrestrial Radio Access (E-UTRA) and Evolved Universal Terrestrial Radio Access Network (E-UTRAN) Overall description, Stage 2 (Release 11), December 2012.

[8] 3GPP TS 36.306, Evolved Universal Terrestrial Radio Access (E-UTRA); User Equipment (UE) Radio Access Capabilities (Release 11), December 2012.

[9] 3GPP TS 36.321, Evolved Universal Terrestrial Radio Access (E-UTRA); Medium Access Control (MAC) Protocol specification (Release 11), December 2012.

[10] 3GPP TS 36.331 Evolved Universal Terrestrial Radio Access (E-UTRA); Radio Resource Control (RRC); Protocol specification (Release 11), December 2012.

[11] 3GPP TR 36.912 Feasibility Study for Further Advancements for E-UTRA (LTE-Advanced) (Release 11), September 2012.

[12] 3GPP TR 36.814 Feasibility Study for Further Advancements for E-UTRA (LTE-Advanced) (Release 9), March 2010.

[13] 3GPP TR 36.819 Coordinated Multi-Point Operation for LTE Physical Layer Aspects (Release 11), December 2011.

[16] 3GPP specifications can be found at the following URL: http://www.3gpp.org/ftp/Specs/archive/36_series/.

Books

[14] C. Johnson, Long Term Evolution in Bullets, second ed., Create Space Independent Publishing Platform, 2012.

[15] S. Sesia, I. Toufik, M. Baker, LTE, the UMTS Long Term Evolution, From Theory to Practice, second ed., John Wiley & Sons, 2011.

[16] E. Dahlman, S. Parkvall, J. Skold, 4G, LTE/LTE-Advanced for Mobile Broadband, Academic Press, 2011.

[17] H. Holma, A. Toskala, LTE Advanced, 3GPP Solution for IMT-Advanced, John Wiley & Sons, 2012.

[18] H. Holma, A. Toskala, LTE for UMTS, Evolution to LTE-Advanced, John Wiley & Sons, 2011.

[19] S. Ahmadi, Mobile WiMAX, A Systems Approach to Understanding IEEE 802.16m Radio Access Technology, Academic Press, 2010.

[20] B. Sklar, Digital Communications, Fundamentals and Applications, second ed., Prentice Hall, 2001.

[21] A. Paulraj, et al., Introduction to Space-Time Wireless Communications, Cambridge University Press, 2008.

[22] W.C. Jakes, Microwave Mobile Communications, second ed., Wiley-IEEE Press, 1994.

[23] S.G. Glisic, Advanced Wireless Communications, 4G Cognitive and Cooperative Broadband Technology, second ed., Wiley Inter-science, 2007.

[24] D.E. Dudgeon, R.M. Mersereau, Multidimensional Digital Signal Processing, Prentice Hall, 1984.

[25] D. Tse, P. Viswanath, Fundamentals of Wireless Communication, Cambridge University Press, 2005.

[26] E. Biglieri, et al., MIMO Wireless Communications, Cambridge University Press, 2010.

[27] A. Goldsmith, Wireless Communications, Cambridge University Press, 2005.

[28] S. Lin, D.J. Costello, Error Control Coding, second ed., Prentice Hall, 2004.

[29] S.L. Marple, Digital Spectral Analysis with Applications, Prentice Hall, 1987.

[30] G.H. Golub, C.F. Van Loan, Matrix Computations, third ed., Johns Hopkins University Press, 1996.

[31] A. Papoulis, Probability, Random Variables and Stochastic Processes, fourth ed., McGraw Hill Higher Education, 2002.

[32] L. Rade, Mathematics Handbook for Science and Engineering, Springer, 2010.

[33] G.A. Korn, T.M. Korn, Mathematical Handbook for Scientists and Engineers, Definitions, Theorems, and Formulas for Reference and Review (Revised Edition), Dover Publications, 2000.

[34] C.E. Shannon, W. Weaver, The Mathematical Theory of Communication, University of Illinois Press, 1998.

[35] F. Dowla, Handbook of RF and Wireless Technologies, Newnes, 2003.

[36] G. Arfken, *Bessel Functions*, Mathematical Methods for Physicists, third ed., Academic Press, 1985.

[37] W.G. Bickley, Bessel Functions and Formulae, Cambridge University Press, Cambridge, England, 1957.

[38] F. Bowman, Introduction to Bessel Functions, Dover, 2010.

[39] H.G. Myung, D. Goodman, Single Carrier FDMA, A New Air Interface for Long Term Evolution, John Wiley & Sons, 2008.
[40] T.M. Cover, J.A. Thomas, Elements of Information Theory, John Wiley & Sons, 1991.
[41] J.G. Proakis, Digital Communications, fifth ed., McGraw-Hill, 2007.
[42] N. Costa, S. Haykin, Multiple-Input Multiple-Output Channel Models, Theory and Practice, John Wiley & Sons, 2010.
[43] C. Oestges, B. Clerckx, MIMO Wireless Communications, from Real-World Propagation to Space-Time Code Design, Academic Press, 2007.
[44] A. Sibille, C. Oestges, A. Zanella, MIMO, from Theory to Implementation, Academic Press, 2010.
[45] I. Stakgold, M.J. Holst, Green's Functions and Boundary Value Problems, third ed., John Wiley & Sons, 2011.
[46] R.G. Gallager, Principles of Digital Communication, Cambridge University Press, 2008.
[47] K. Fazel, G.P. Fettweis, Multi-Carrier Spread-Spectrum, Kluwer Academic Publishers, 1997.

Articles

[48] G. Bauch, G. Dietl, MIMO for 3GPP LTE-Advanced and beyond, Global Telecommunications Conference (GLOBECOM), 2010.
[49] C. Mehlfuhrer, et al., The Vienna LTE simulators—enabling reproducibility in wireless communication research, EURASIP J. Adv. Signal Process, (1) (2011).
[50] H.G. Myung, J. Lim, D.J. Goodman, Single carrier FDMA for uplink wireless transmission, IEEE Veh. Technol. Mag. 1 (3) (2006).
[51] H. Ekström, et al., Technical solutions for the 3G long-term evolution, IEEE Commun. Mag. 44 (3) (2006).
[52] D. Falconer, et al., Frequency domain equalization for single-carrier broadband wireless systems, IEEE Commun. Mag. 40 (4) (2002).
[53] H. Sari, et al., Transmission techniques for digital terrestrial TV broadcasting, IEEE Commun. Mag. 33 (2) (1995).
[54] C. Berrou, A. Glavieux, P. Thitimajshima, Near Shannon limit error correcting coding and decoding, "turbo codes", in: Proceedings of the IEEE International Conference on Communications, Geneva, Switzerland, May 1993.
[55] 3GPP R1-050397, E-UTRA uplink numerology and frame structure 3GPP TSG RAN WG1, meeting 41, Athens, Greece, 2005.
[56] 3GPP R1-050584, E-UTRA uplink numerology and design, 3GPP TSG RAN WG1, meeting 41bis, Sophia Antipolis, France, 2005.
[57] 3GPP R1-050971, Single carrier uplink options for E-UTRA, 3GPP TSG RAN WG1, meeting 42, London, UK, 2005.
[58] U. Sorger, I. De Broeck M. Schnell, Interleaved FDMA—a new spread-spectrum multiple-access scheme, In: Proceedings of the IEEE International Conference on Communications (ICC), 1998.
[59] K. Bruninghaus, H. Rohling, Multi-carrier Spread Spectrum and its Relationship to Single-Carrier Transmission, In: Proceedings of the IEEE Vehicular Technology Conference (VTC), May 1998.

[60] D. Galda, H. Rohling, A Low Complexity Transmitter Structure for OFDM-FDMA Uplink Systems, in: Proceedings of the IEEE Vehicular Technology Conference (VTC), May 2002.
[61] R. Dinis, D. Falconer, C.T. Lam, M. Sabbaghian, A Multiple Access Scheme for the Uplink of Broadband Wireless Systems, in: Proceedings of the IEEE Global Telecommunications Conference, November 2004.
[62] H. Sari, G. Karam I. Jeanclaude, Frequency-Domain Equalization of Mobile Radio and Terrestrial Broadcast Channels, in: Proceedings of the IEEE Global Telecommunications Conference, November 1994.
[63] D. Falconer, S.L. Ariyavisitakul, A. Benyamin-Seeyar, B. Eidson, Frequency domain equalization for single-carrier broadband wireless systems, IEEE Commun. Mag. 40 (2002).
[64] V. Nangia, K.L. Baum, Experimental broadband OFDM system—field results for OFDM and OFDM with frequency domain spreading, in: Proceedings of the IEEE International Conference on Vehicular Technology, Vancouver, BC, Canada, 2002.
[65] 3GPP R1-062061, Uplink DC subcarrier distortion considerations in LTE, 3GPP TSG RAN WG1, meeting 46, Tallinn, Estonia, 2006.
[66] 3GPP R1-050702, DFT-spread OFDM with pulse shaping filter in frequency domain in evolved UTRA uplink, 3GPP TSG RAN WG1, meeting 42, London, UK, 2005.
[67] 3GPP R1-051434, Optimum family of spectrum-shaping functions for PAPR reduction in SC-FDMA, 3GPP TSG RAN WG1, meeting 43, Seoul, Korea, 2005.
[68] D.C. Chu, Poly-phase codes with good periodic correlation properties, IEEE Trans. Inform. Theory (1972).
[69] B.M Popovic, Generalized chirp-like poly-phase sequences with optimal correlation properties, IEEE Trans. Inform. Theory 38 (1992) 1406–1409.
[70] 3GPP R1-060878, E-UTRA SC-FDMA uplink pilot/reference signal design & TP, 3GPP TSG RAN WG1, meeting 44bis, Athens, Greece, 2006.
[71] 3GPP R1-080576, Way forward on the sequence grouping for UL DM-RS, 3GPP TSG RAN WG1, meeting 51bis, Sevilla, Spain, 2008.
[72] 3GPP R1-072584, Way Forward for PUSCH RS, 3GPP TSG RAN WG1, meeting 49, Kobe, Japan, 2007.
[73] 3GPP R1-072585, Way forward for PUCCH RS, 3GPP TSG RAN WG1, meeting 49, Kobe, Japan, 2007.
[74] 3GPP R1-074278, Hopping and planning of sequence groups for uplink RS, 3GPP TSG RAN WG1, meeting 50bis, Shanghai, China, 2007.
[75] 3GPP R1-080719, Hopping patterns for UL RS, 3GPP TSG RAN WG1, meeting 52, Sorrento, Italy, 2008.
[76] 3GPP R1-081133, WF on UL DM-RS hopping pattern generation, 3GPP TSG RAN WG1, meeting52, Sorrento, Italy, 2008.
[77] 3GPP R1-071341, Uplink reference signal planning aspects, 3GPP TSG RAN WG1, meeting 48bis, St. Julian's, Malta, 2007.
[78] 3GPP R1-080983, Way forward on the cyclic shift hopping for PUCCH, 3GPP TSG RAN WG1, meeting 52, Sorrento, Italy, 2008.
[79] 3GPP R1-073756, Benefit of non-persistent UL sounding for frequency hopping PUSCH, 3GPP TSG RAN WG1, meeting 50, Athens, Greece, 2007.
[80] 3GPP R1-082199, Physical-layer parameters to be configured by RRC, 3GPP TSG RAN WG1, meeting 53, Kansas City, MI, 2008.

[81] 3GPP R1-072671, Uplink channel interleaving, 3GPP TSG RAN WG1, meeting 49bis, Orlando, FL, 2007.
[82] B. Classon et al., Multi-dimensional adaptation and multi-user scheduling techniques for wireless OFDM systems, in: Proceedings of the IEEE International Conference on Communications, Anchorage, Alaska, 2003.
[83] 3GPP R1-071340, Considerations and recommendations for UL sounding RS, 3GPP TSG RAN WG1, meeting 48bis, St Julian's, Malta, March 2007.
[84] 3GPP R1-073756, Benefit of non-persistent UL sounding for frequency hopping PUSCH, 3GPP TSG RAN WG1, meeting 50, Athens, Greece, 2007.
[85] A. Ghosh et al., Uplink control channel design for 3GPP LTE, in: Proceedings of the IEEE International Symposium on Personal, Indoor and Mobile Radio Communications, Athens, Greece, 2007.
[86] 3GPP R1-070084, Coexistance simulation results for 5 MHz E-UTRA/UTRA FDD uplink with revised simulation assumptions, 3GPP TSG RAN WG4, meeting 42, St. Louis,MI, 2007.
[87] S. Zhou, G.B. Giannakis, C.L. Martret, Chip-interleaved block-spread code division multiple access, IEEE Trans. Commun. 50 (2002).
[88] 3GPP R1-080190, Embedding ACK/NACK in CQI reference signals and receiver structures, 3GPP TSG RAN WG1, meeting 51bis, Sevilla, Spain, 2008.
[89] 3GPP R1-073564, Selection of orthogonal cover and cyclic shift for high speed UL ACK channels, 3GPP TSG RAN WG1, meeting 50, Athens, Greece, 2007.
[90] 3GPP R1-080035, Joint proposal on uplink ACK/NACK channelization, 3GPP TSG RAN WG1, meeting 51bis, Sevilla, Spain, 2008.
[91] 3GPP R1-062072, Uplink reference signal multiplexing structures for E-UTRA, 3GPP TSG RAN WG1, meeting 46, Tallinn, Estonia, 2006.
[92] 3GPP R1-080983, Way forward on the cyclic shift hopping for PUCCH, 3GPP TSG RAN WG1, meeting 52, Sorrento, Italy, 2008.
[93] 3GPP R1-080931, ACK/NACK channelization for PRBs containing both ACK/NACK and CQI, 3GPP TSG RAN WG1, meeting 52, Sorrento, Italy, 2008.
[94] 3GPP R1-104135, Confirmation of UE behavior in case of simultaneously-triggered SR, SRS and CQI, 3GPP TSG RAN WG1, meeting 61bis, Dresden, Germany, 2010.
[95] 3GPP R1-081928, Way forward on indication of UE antenna selection for PUSCH, 3GPP TSG RAN WG1, meeting 53, Kansas City, MI, May 2008.
[96] 3GPP R1-082109, UE transmit antenna selection, 3GPP TSG RAN WG1, meeting 53, Kansas City, MI, May 2008.
[97] 3GPP R1-062175, Random access burst design for E-UTRA, 3GPP TSG RAN WG1, meeting 46, Tallinn, Estonia, 2006.
[98] R.L. Frank, S.A. Zadoff, R. Heimiller, Phase shift pulse codes with good periodic correlation properties, IEEE Trans. Inform. Theory 7 (October 1961).
[99] D.C. Chu, Poly-phase codes with good periodic correlation properties, IEEE Trans. Inform. Theory 18 (1972).
[100] 3GPP R1-072325, Multiple values of cyclic shift increment NCS, 3GPP TSG RAN WG1, meeting 49, Kobe, Japan, 2007.
[101] 3GPP R1-073624, Limitation of RACH sequence allocation for high mobility cell, 3GPP TSG RAN WG1, meeting 50, Athens, Greece, 2007.

[102] 3GPP R1-040642, Comparison of PAR and cubic metric for power de-rating, 3GPP TSG RAN WG1, meeting 37, Montreal, Canada, 2004.
[103] 3GPP R1-060023, Cubic metric in 3GPP-LTE, 3GPP TSG RAN WG1, LTE ad-hoc meeting, Helsinki, Finland, 2006.
[104] 3GPP R1-071517, RACH sequence allocation for efficient matched filter implementation, 3GPP TSG RAN WG1, meeting 48bis, St Julian's, Malta, 2007.
[105] 3GPP R1-071409, Efficient matched filters for paired root Zadoff–Chu sequences, 3GPP TSG RAN WG1, meeting 48bis, St Julian's, Malta, 2007.
[106] 3GPP R1-063377, UL timing control accuracy and update rate, 3GPP TSG RAN WG1, meeting 47, Riga, Latvia, 2006.
[107] 3GPP R1-072841, Simulation of uplink timing error impact on PUSCH, 3GPP TSG RAN WGI, meeting 49bis, Orlando, FL, 2007.
[108] M.P.J. Baker, T.J. Moulsley, Power control in UMTS release 99, in: Proceedings of the First IEE International Conference on 3G Communications, 2000.
[109] 3GPP R1-080881, Range and representation of Delta_MCS, 3GPP TSG RAN WG1, meeting 52, Sorrento, Italy, 2008.
[110] J. Lee, J.-K. Han, J. Zhang, MIMO technologies in 3GPP LTE and LTE-advanced, EURASIP J. Wireless Commun. Netw. (2009).
[111] S. Catreux, P.F. Driessen, L.J Greenstein, Attainable throughput of an interference-limited multiple-input multiple-output (MIMO) cellular system, IEEE Trans Commun. 49 (8) (2001).
[112] I. Telatar, Capacity of multi-antenna Gaussian channels, AT&T Tech. Memo. (1995).
[113] G.J. Foschini, M.J. Gans, On limits of wireless communications in a fading environment when using multiple antennas, Wireless Personal Commun. (1998).
[114] M.-S. Alouini, A. Goldsmith, Capacity of Nakagami multipath fading channels, in: Proceedings of the Vehicular Technology Conference, 1997.
[115] D. Chizhik, G.J. Foschini, R.A. Valenzuela, Capacities of multi-element transmit and receive antennas: correlations and keyholes, Electron. Lett. 36 (2000).
[116] D. Gesbert, H. Bölcskei, D. Gore, A. Paulraj, MIMO wireless channels: capacity and performance prediction, Global Telecommunications Conference (GLOBECOM), 2000.
[117] D.A. Gore, R.U. Nabar, A. Paulraj, Selecting an optimal set of transmit antennas for a low rank matrix channel, Proc. ICASSP 2000 49 (8) (2000).
[118] S. Sandhu, R.U. Nabar, D.A. Gore, A. Paulraj, Near-optimal selection of transmit antennas for a MIMO channel based on Shannon capacity, in: Proceedings of the 34th Asilomar Conference on Signals, Systems and Computers, 2000.
[119] A.F. Molisch, M.Z. Win, J.H. Winters, A. Paulraj, Capacity of MIMO systems with antenna selection, in: Proceedings of the IEEE International Conference on Communications (ICC), 2001.
[120] Z. Shen, et al., Overview of 3GPP LTE-advanced carrier aggregation for 4G wireless communications, IEEE Commun. Mag. 50 (2) (2012).
[121] Y.-H. Nam, et al., Evolution of reference signals for LTE-advanced systems, IEEE Commun. Mag. 50 (2) (2012).
[122] L. Liu, et al., Downlink MIMO in LTE-advanced: SU-MIMO vs. MU-MIMO, IEEE Commun. Mag. 50 (2) (2012).

[123] D. Lee, et al., Coordinated multipoint transmission and reception in LTE-advanced: deployment scenarios and operational challenges, IEEE Commun. Mag. 50 (2) (2012).
[124] D. Bai, et al., LTE-advanced modem design: challenges and perspectives, IEEE Commun. Mag. 50 (2) (2012).
[125] J. Lee, Kim, et al., Coordinated multipoint transmission and reception in LTE-advanced systems, IEEE Commun. Mag. 50 (2) (2012).
[126] Z. Shen, et al., Dynamic uplink–downlink configuration and interference management in TD-LTE, IEEE Commun. Mag. 50 (2) (2012).
[127] C.S. Park, et al., Evolution of uplink MIMO for LTE-Advanced, IEEE Commun. Mag. 49 (2) (2011).

Miscellaneous

[128] 3G Americas, MIMO and smart antennas for 3G and 4G wireless systems. <http://www.3gamericas.org>, May 2010.
[129] Rohde & Schwarz Application Notes, LTE-Advanced Technology Introduction, March 2010.
[130] 3G Americas, 3GPP Mobile Broadband Innovation Path to 4G: Release 9, Release 10 and Beyond: HSPA + , SAE/LTE and LTE-Advanced, February 2010.
[131] 3G Americas, HSPA to LTE-Advanced: 3GPP Broadband Evolution to IMT-Advanced (4G), September 2009.
[132] 3G Americas, The Mobile Broadband Evolution: 3GPP Release 8 and Beyond HSPA + , SAE/LTE and LTE-Advanced, February 2009.
[133] IEEE 802.16m-07/244r1, Yang Song, Liyu Cai, Keying Wu, and Hongwei Yang, Collaborative MIMO, November 2007.
[134] Agilent Technologies, 3GPP long term evolution: system overview, product development, and test challenges. <http://www.agilent.com>, 2009.
[135] IEEE C802.16m-08/153, Bin-Chul Ihm, Jinsoo Choi and Wookbong Lee, "Pilot Related to DL-MIMO", March 2008.
[136] MATLAB CENTRAL, An open exchange for the MATLAB and SIMULINK user community. <http://www.mathworks.com/matlabcentral>.
[137] Vienna University of Technology, Institute of Telecommunications, LTE link-level and system-level simulators. <http://www.nt.tuwien.ac.at/about-us/staff/josep-colom-ikuno/lte-simulators>, 2012.
[138] 3GPP R1-060385 Motorola, Cubic Metric in 3GPP-LTE, February 2006.
[139] The Coded Modulation Library (The Iterative Solutions Coded Modulation Library (ISCML) is an open source toolbox for simulating capacity approaching codes in Matlab). <http://www.iterativesolutions.com>.
[140] 4G Americas, 4G Mobile Broadband Evolution: Release 10, Release 11 and Beyond - HSPA, SAE/LTE and LTE-Advanced, October 2012.
[141] 4G Americas, MIMO and Smart Antennas for Mobile Broadband Networks, October 2012.
[142] 4G Americas, Developing & Integrating a High Performance HET-NET, October 2012.
[143] 4G Americas, 4G Mobile Broadband Evolution: 3GPP Release 10 and Beyond - HSPA + , SAE/LTE and LTE-Advanced, February 2011.

[144] 3GPP R1-113297, NTT DOCOMO, Enhanced PDCCH for DL MIMO in Rel-11, October 2011.

[145] 3GPP R1-112928, Ericsson, ST-Ericsson, On enhanced PDCCH design, October 2011.

[146] 3GPP R1-113175, Renesas Mobile, Link-level evaluation of E-PDCCH design aspects, October 2011.

[147] 3GPP R1-113194, LG Electronics, Consideration on transmit diversity scheme for enhanced PDCCH transmission, October 2011.

[148] 3GPP R1-113202, Intel Corporation, Performance analysis of enhanced downlink control signaling, October 2011.

[149] 3GPP R1-113238, Research in Motion, E-PDCCH transmission with transmit diversity, October 2011.

[150] 3GPP R1-114302, NTT DOCOMO, DM-RS design for E-PDCCH in Rel-11, November 2011.

[151] Anritsu, Future technologies and testing for fixed mobile convergence, SAE and LTE in Cellular mobile communications. <http://www.anritsu.com/search/en-US/downloadssearch.aspx>, 2008.

[152] Wireless World Initiative New Radio—WINNER + , Enabling techniques for LTE-A and beyond, deliverable D2.2. <http://projects.celtic-initiative.org/winner + /deliverables_winnerplus.html>, 2010.

[153] LTE system description. <http://www.sharetechnote.com>.

[154] DSPLOG signal processing for communication. <http://www.dsplog.com>.

[155] 4G Evolution Lab—LTE and LTE-Advanced Toolbox and Block-set: a simulation library for the 3GPP LTE and LTE Advanced specification for the Mathworks MATLAB and Simulink. <http://www.steepestascent.com>.

[156] Samsung electronics, LTE PHY specification. <http://www.slideshare.net/allabout4g/lte-rel-8-physical-layer>, 2008.

[157] MOTOROLA WHITE PAPER, TD-LTE, exciting alternative global momentum. <www.motorola.com>, 2010.

[158] B. Schulz, LTE Transmission modes and beamforming, Rohde & Schwarz White Paper, 2011.

[159] M. Lew, Real-world stimulus-response testing for LTE transmitter RF components, agilent technologies. <http://www.microwavejournal.com/articles/print/9158-real-world-stimulus-response-testing-for-lte-transmitter-rf-components>, 2010.

[160] Adaptive Multi-Rate (AMR) Feature Descriptions, Agilent Technologies <http://wireless.agilent.com/rfcomms/refdocs/gsm/gprsla_amr_bse_config.html>

[161] Jinbiao Xu, LTE-Advanced Signal Generation and Measurement Using SystemVue, Agilent Technologies, Inc., December 2010.

[162] 3GPP R1-080182, Labelling of UE-selected subbands on PUSCH, Huawei, January 2008.

[163] 3GPP R1-080555, Labelling complexity of UE selected subbands on PUSCH, Huawei, January 2008.

[164] Ben Cooke, Reed-Muller Error Correcting Codes, MIT Undergraduate Journal of Mathematics, Number 1, June 1999 <http://www-math.mit.edu/phase2/UJM/vol1/>.

[165] R. Gold, "Optimal binary sequences for spread spectrum multiplexing", IEEE Transactions on Information Theory, Vol. 13 , Issue 4, February 1967.

[166] Hidekazu Taoka et al., “MIMO and CoMP in LTE-Advanced”, NTT DoCoMo Tech Journal, September 2010.

[167] Hideyuki Torii et al., “A New Class of Zero-Correlation Zone Sequences”, IEEE Transactions on Information Theory, Vol. 50, No. 3, March 2004.

[168] Technical Note, “An Overview of LTE PRACH Detection for the PC20x”, picoChip felexible wireless, January 2008.

[169] Ernesto Zimmermann et al., “On the Complexity of Sphere Decoding”, in Proceedings International Symposium on Wireless Personal Multimedia Communication, 2004.

[170] Mathew William Churchman Barata, LTE Performance Evaluation with Realistic Channel Quality Indicator Feedback, Universitat Politècnica de Catalunya, April 2011.

[171] J.P.T. Dobbelsteen, Mapping an LTE Baseband Receiver on a Multi-Core Architecture, Eindhoven University of Technology, november 2009.

[172] Di Wu et al., “System Architecture for 3GPP LTE Modem using a Programmable Baseband Processor”, International Symposium on System-on-Chip, 2009.

[173] Jens Berkmann et al., “On 3G LTE Terminal Implementation - Standard, Algorithms, Complexities and Challenges”, International Wireless Communications and Mobile Computing Conference (IWCMC ’08), 2008.

[174] DigRF(SM) Specifications, MIPI Alliance, <http://www.mipi.org/specifications/digrfsm-specifications>.

CHAPTER

Link-Level and System-Level Performance of LTE-Advanced

11

CHAPTER CONTENTS

International telecommunication union-radio telecommunications (ITU-R) was responsible for defining of a global standard for the fourth generation of mobile communication systems known as international mobile telecommunications (IMT)-Advanced. One of the technologies included in this global standard was the LTE-Advanced proposed by 3GPP. The IMT-Advanced systems were required to fulfill a set of requirements specified by ITU-R working party 5D [12–16]. The first set of requirements was verified through inspection of the proposed technologies. Those requirements included the support of scalable bandwidths in the IMT spectrum, a wide range of services and inter-system handover capability with at least with one of systems in the IMT-2000 family. The satisfaction of the second set of requirements was confirmed analytically by conducting numerical calculations of peak spectral efficiency, user-plane and control-plane latency, and handover interruption times. The third set of requirements was required to be evaluated via link-level and system-level simulations initially by the proponent and later by designated external evaluation groups. These requirements were cell spectral efficiency, cell-edge user spectral efficiency, mobility, and VoIP capacity.

In order to reduce the complexity of the process, simulations are often divided into two stages or levels of abstraction known as link-level and system-level. The link-level simulations are used to assess the performance of the physical layer and those higher layer aspects directly related to the radio interface. In link-level simulation, a single-cell radio link is modeled, including some specific features such as synchronization, modulation, channel coding, channel fading, channel estimation, demodulation, multi-antenna processing, etc. On the other hand, a system-level simulator allows evaluation of the performance of a network comprising multiple cells and moving mobile stations. At this level, system modeling encompasses a set of base stations and all their associated mobile terminals. Both the signal level received by each user and other users' interferences are modeled, taking into account the propagation losses and channel fading effects. Signal to interference plus noise ratio (SINR) is calculated for each active user considering the current network configuration and relative random placement of the base stations and mobile stations. These SINR values can be then translated to block error rate (BLER) or effective throughput values using models whose development is based on the results obtained in the link-level simulations. This interaction between link- and system-level simulators is usually referred to as link-to-system mapping or link-level abstraction models.

After developing the simulation platform, a calibration process is performed. The calibration is of paramount importance to ensure the validity of the results obtained with the simulation platform. Calibration is performed through comparison of the outcomes of the developed simulator with the outcomes of the simulation platforms of other sources. The calibration process involves comparison and adjustment of downlink wideband SINR, also known as geometry, and coupling loss distributions under common assumptions. Following the first step, the calibration further includes evaluation and comparison of downlink and uplink spectral efficiencies, user throughput distributions, and SINR distributions for a basic LTE system configuration.

The link-level and system-level simulations are used to evaluate the performance of mobile radio access technologies under various operating conditions and deployment scenarios. While the simulations do not model the entire deployment parameters and propagation conditions that may be involved in a practical scenario due to increased computational complexity of the model, the statistical modeling of the parameters and estimation/measurement errors should be sufficiently accurate such that the simulation results are a faithful representation of the performance in an actual deployment [51]. More specifically:

- *Link-level simulations* would facilitate the study of channel estimation, tracking, and prediction algorithms, as well as synchronization algorithms, MIMO performance, adaptive modulation and coding, and feedback techniques. Furthermore, receiver structures (e.g., MMSE, ZF, MLD, SIC), modeling of channel encoding and decoding, and physical-layer modeling which is crucial for system-level simulations are typically analyzed at the link-level. Although MIMO broadcast channels have been investigated quite extensively over the past decade, there are still many open theoretical and practical questions that need to be resolved. For example, LTE offers the flexibility to adjust many transmission parameters, but it is not clear how to exploit the available degrees of freedom to achieve the optimum performance under different deployment scenarios.
- *System-level simulations* focus more on network-related issues, such as resource allocation and scheduling, multi-user handling, mobility management, admission control, interference management, and network planning optimization. Furthermore, in a multi-user system, such as LTE, it is not directly clear which figure of merits (metrics) should be used to evaluate the performance of the system. The classical measures of (un)coded bit error rate, (un)coded BLER, and throughput may not accurately reflect multi-user properties. More comprehensive measures of the LTE performance include fairness, multi-user diversity, or degree of freedom. However, these theoretical concepts have to be mapped to performance values that can be verified by means of simulations.

As shown in Figure 11.1, an example of an LTE link-level simulation platform comprises three basic building blocks; namely, transmitter, channel model, and receiver. Depending on the type of simulation, one or several instances of these basic building blocks are utilized. The transmitter and receiver blocks are linked by the channel model, which is used to transmit the downlink data, while signaling and uplink feedback are ideally assumed to be error free. Since signaling is more strongly protected than data, by means of lower coding rates and/or lower-order modulations, the assumption of error-free signaling is in fact fairly realistic. Equivalently, errors on the signaling channels will only occur when the data channels are already facing severe performance degradation. In the downlink,

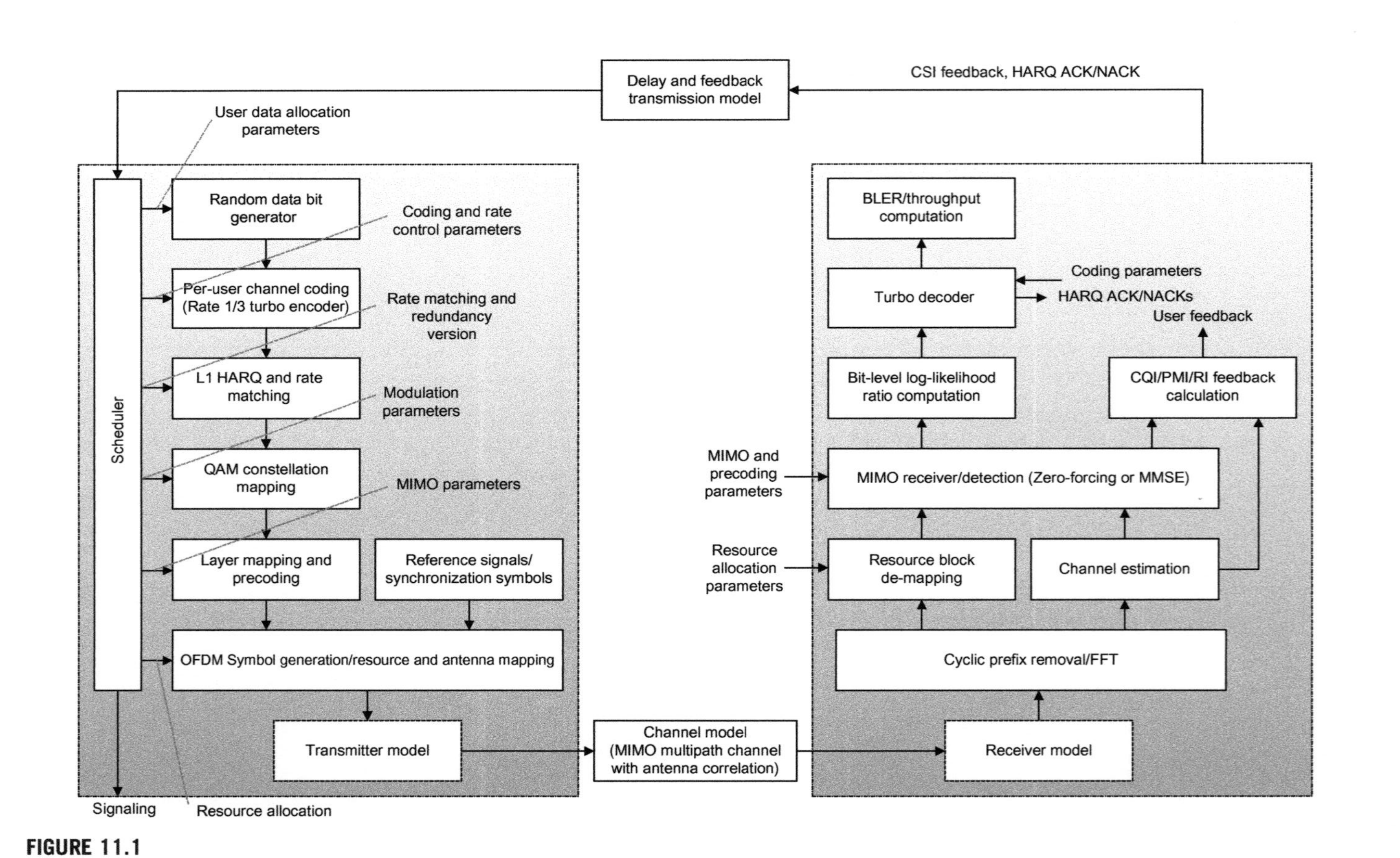

FIGURE 11.1

General structure of an LTE downlink link-level simulation platform [8,9,45].

the signaling information generated by the transmitter to the receiver contains coding, HARQ, scheduling, and precoding parameters. In the uplink, channel quality indicator (CQI), precoding matrix index (PMI), and rank indicator (RI) are signaled, which together form the channel state information (CSI) feedback [45].

In system-level simulations as illustrated in Figure 11.2, the performance of the entire network is evaluated. In an LTE system, such network consists of a number of eNBs that cover a specific geographical area in which several mobile stations are randomly dropped. The system-level simulation platform (shown in Figure 11.2) consists of a link measurement model and a link performance model. The link measurement model measures the link quality, obtained via the UE CSI measurement reports, which is required to perform link adaptation, resource allocation, and MIMO mode selection. An appropriate link quality measure is evaluated per sub-carrier. Based on the SINR metric, the UE calculates the uplink feedback (PMI, RI, and CQI), which is used as input for link adaptation at the eNB. The scheduling algorithm assigns resources to users in order to optimize the performance of the system based on CSI feedback. The scheduling is a key

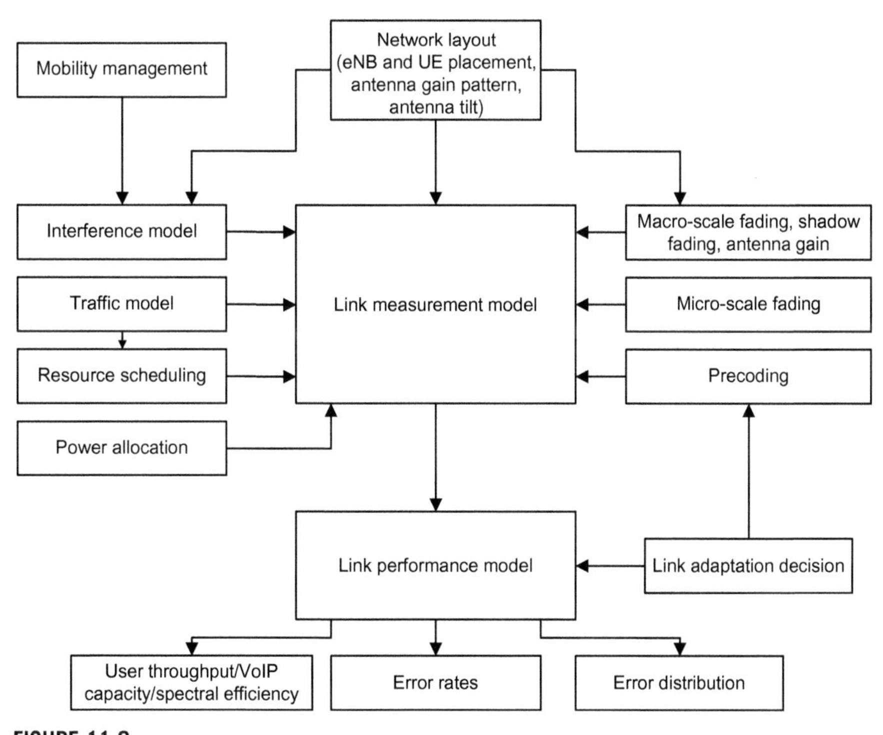

FIGURE 11.2

General structure of an LTE system-level simulation platform [45].

component in the proper functioning of the radio interface. Two types of schedulers can be identified: the opportunistic, where the scheduler considers the state of the radio channel to make the best allocation possible; and the non-opportunistic, where the allocation has no knowledge of the radio channel's state.

Based on the link measurement model outputs, the link performance model predicts the BLER of the link according to the receiver SINR and the transmission parameters (e.g., modulation and coding). In order to generate the network topology, transmission sites are generated. The scheduler assigns physical resources, precoding matrices, and selects a suitable modulation and coding scheme for each active UE attached to an eNB. The actual assignment depends on the scheduling algorithm and the UE feedback. At the UE side, the received post-processing SINR is calculated in the link measurement model. The SINR is determined by the signal, interference, and noise power levels, which are dependent on the cell layout that is defined by the eNB relative positions, large-scale (macroscopic, macro-scale) path loss, shadow fading, and the time-variant small-scale (microscopic, micro-scale) fading parameters. The CQI feedback is calculated based on the sub-carrier SINRs and the target transport BLER. The CQI reports are generated by SINR-to-CQI mapping and made available to the eNB via a feedback channel with adjustable delay where the feedback imperfections are modeled. At the transmitter, the appropriate MCS is selected by the CQI value to achieve the targeted BLER during the transmission. In particular, in high mobility scenarios, the feedback delay caused by computation and signaling timings can lead to performance degradation, if the channel state changes rapidly during the delay period. In the link performance model, an additive white Gaussian noise (AWGN)-equivalent SINR γ_{AWGN} is obtained via mutual information effective signal to interference and noise ratio mapping (MIESM). In a second step, γ_{AWGN} is mapped to the BLER via AWGN link performance curves. The BLER value acts as a probability for computing HARQ ACK/NACKs, which are combined with the transport-block size to compute the link throughput. The simulation output consists of traces, containing link throughput and error rates for each user, as well as cell aggregate throughput, from which statistical distributions of throughputs and errors can be extracted [45].

The large-scale path loss and the shadow fading are modeled as position-dependent maps. The large-scale path loss is calculated according to the available models and combined with the antenna gain pattern of the corresponding eNB. An inter-site map correlation for shadow fading is similarly obtained. While the large-scale path loss and the shadow fading are modeled as a position-dependent trace, the small-scale fading is modeled as a time-dependent trace. The calculation of the latter trace is based on the transmitter precoding, the small-scale fading MIMO channel matrix, and the receive filter. The small-scale fading trace consists of the signal power and the interference power after the receive filter.

The link-level and system-level evaluation of the IMT-Advanced candidates was comprehensively performed in strict compliance with the technical parameters and the methodology specified by ITU-R [12–16]. Each requirement was independently evaluated, except for the cell and cell-edge user spectral efficiencies criteria that were jointly assessed using the same system-level simulation; consequently, the candidates were required to simultaneously satisfy the corresponding minimum requirements. Furthermore, the system-level simulation setup used in the assessment of the mobility requirement was the same as that used for the evaluation of cell spectral efficiency and cell-edge user spectral efficiency. In general, the evaluation of the IMT-Advanced candidate technologies followed the following principles [14]:

1. Use of reproducible methods including computer simulation, analytical/theoretical approach, and inspection.
2. Technical evaluation against performance targets that were set for each test environment.
3. Consistency between self-evaluations and technical descriptions provided in technology description templates.
4. Use of unified methodology, software, and common parameter sets by the evaluation groups including channel models, link-level parameters, and link-to-system mapping.
5. Evaluation of multiple proposals using one simulation platform by each evaluation group to ensure comparability of the results.
6. Inclusion of an L1/L2 overhead in the evaluation of cell spectral efficiency, cell-edge user spectral efficiency, and VoIP capacity.
7. Independent drop of the users with uniform distribution over the coverage area where each mobile user corresponds to an active session that runs for the duration of the drop.
8. Random assignment of line-of-sight (LoS) *and* non-line-of-sight (NLoS) path loss models to the users.
9. Cell assignment to a user is based on the air-interface cell selection schemes.
10. Fading signal and fading interference are computed from each mobile station to each cell, and from each cell to each mobile station (in both directions on an aggregated basis).
11. Consideration of the interference over thermal (IoT) constraint in the uplink design such that the average IoT value remains less than or equal to 10 dB.
12. In a full-buffer traffic model, packets are not blocked upon their arrival; that is, buffer sizes are assumed to be infinite.
13. Data and VoIP packets are scheduled with appropriate packet schedulers for full-buffer and VoIP traffic models separately.
14. Modeling of channel quality feedback delay, feedback channel errors, packet errors, and real channel estimation (as opposed to ideal channel estimation) effects and retransmission of erroneous packets.
15. Realistic modeling of the overhead channels, that is, the overhead due to feedback and control channels.

For each drop, the system-level simulations were conducted and the process was repeated with the users dropped at new random geographical locations within the network. A large number of drops were simulated to ensure convergence in the system performance metrics, and the width of confidence intervals of the user and system performance metrics was calculated and reported. The confidence interval[1] and the associated confidence level indicate the reliability of the estimated parameter value [23,53]. The confidence level is the probability that the true parameter value is within the confidence interval such that the higher the confidence level, the larger the confidence interval. All cells in the network were simulated assuming dynamic channel models and use of a 19-cell wrap-around scheme to model the inter-cell interference except for the indoor environment.

This chapter describes the link-level and system-level evaluation of 3GPP LTE-Advanced against ITU-R requirements for IMT-Advanced systems using the methodology specified by ITU-R [14]. The theoretical background, including definition of performance metrics, channel models, physical-layer abstraction schemes, and the traffic models are provided to ensure in-depth understanding of the evaluation process and the results.

11.1 Definition of the performance metrics

11.1.1 Link budget

Link budget is a commonly used metric to evaluate the coverage of a cellular system in the downlink or uplink. In order to calculate the link budget, one must account for all gains and losses from the transmitter to the receiver over the air-interface. It accounts for the attenuation of the transmitted signal due to propagation as well as the antenna gains, cable and implementation loss, and other miscellaneous losses. The time-varying channel gains such as fading are taken into account by adding some margin depending on the anticipated severity of its effects. The amount of

[1]A confidence interval provides an estimated range of values which is likely to include an unknown population parameter, the estimated range being calculated from a given set of sample data. If independent samples are taken repeatedly from the same population and a confidence interval is calculated for each sample, then a certain percentage (confidence level) of the intervals will include the unknown population parameter. Confidence intervals are more informative than the simple results of hypothesis tests since they provide a range of plausible values for the unknown parameter.
In mathematical terms, let X denote a random sample from a probability distribution with statistical parameters α, which is a quantity to be estimated, and β, representing quantities that are not of immediate interest. A *confidence interval* for the parameter α, with confidence level or confidence coefficient γ, is an interval with random endpoints $(f_1(X), f_2(X))$, determined by the pair of random variables $f_1(X)$ and $f_2(X)$, with the property $\Pr_{\alpha,\beta}(f_1(X) < \alpha < f_2(X)) = \gamma \forall \alpha, \beta$. The number γ, with typical values close to but not greater than one, is sometimes given in the form $1-\gamma'$, where γ' is a small non-negative number close to zero. In the latter expression, $\Pr_{\alpha,\beta}(x)$ indicates the probability distribution of X characterized by (α, β). An important part of this definition is that the random interval $(f_1(X), f_2(X))$ covers the unknown value α with a high probability no matter what the true value of α actually is [23,53].

margin required can be reduced by the use of mitigating techniques, such as antenna diversity or frequency hopping. An abstract description of the link budget in any direction (downlink/uplink) has the following form [14,26]:

$$\text{Received power (dBm)} = \text{Transmitted power (dBm)} + \text{Gains (dB)} - \text{Losses (dB)} \quad (11.1)$$

Link budget evaluation is a well-known method for initial cell planning that needs to be carried out separately for the downlink and uplink. Although the link budget can be calculated separately for each link, it is the combination of the links that determines the performance of the system. Using the margins in the link budget, the expected signal-to-noise ratio can be evaluated at given distances. Using these results, the noise-limited range can be evaluated for the system. The typical link budget parameters are summarized in Table 11.1.

11.1.2 User and sector throughput

Although a user may be sufficiently covered within certain areas of a cell for a given service, when multiple users are in a cell, the radio resources (time, frequency, space, and power) must be shared among the active users. It can be expected that a user's average data rate may be reduced by a factor of N_{user} when

Table 11.1 Typical Link Budget Parameters [14,26]

Link Budget Parameters	Link Budget Parameters
Carrier frequency/channel bandwidth	Number of receive antennas
BS/MS antenna heights (m)	Receive antenna gain (dBi)
Area coverage (km)	Receiver cable, connector, or body losses (dB)
Type of service (VoIP, full-buffer data, etc.)	Receiver noise figure (dB)
MCS	Thermal noise density (dBm/Hz)
Multipath channel model	Receiver interference density (dBm/Hz)
Mobile speed (km/h)	Required SNR (dB)
Number of transmit antennas	Receiver implementation margin (dB)
Maximum transmitter power per antenna (dBm)	Fast fading margin including scheduler gain (dB)
Transmitter power amplifier back-off (dB)	HARQ combining gain (dB)
Transmit antenna gain (dBi)	Handover gain (dB)
Transmit array gain (dB)	BS/MS diversity gain (dB)
Control channel power boosting gain (dB)	Lognormal shadow fading standard deviation (dB)
Data carrier power loss due to pilot/control boosting (dB)	Shadow fading margin (dB)
Transmitter cable, connector, combiner, body losses (dB)	Penetration loss (dB)

there are N_{user} active users, assuming resources are equally shared among active users and there is no multi-user diversity gain, relative to that of a single user. Therefore, any coverage assessment for any particular service must be coupled with the number of active users in the cell and their associated quality of service requirements. If the users with poor channel conditions are provided with more radio resources, it would adversely impact the total cell throughput. Thus, there is a trade-off between coverage and capacity. The performance metric that is typically used is the number of admissible users, parameterized by the service minimum bit rate, maximum tolerable latency, permissible outage probability, etc. It is assumed that simulation statistics are collected from sectors belonging to the test cells of a 19-cell deployment scenario in order to model the effects of intercell interference. In the following, downlink and uplink throughputs are separately evaluated and the corresponding metrics are differentiated by superscripts "DL" and "UL," respectively.

The user throughput is defined as the ratio of the number of information bits that the user successfully received divided by the total simulation time. If user k has $p_k^{\mathrm{DL(UL)}}$ downlink (uplink) packet transmissions during the simulation period T_{sim}, if there are $q_{ik}^{\mathrm{DL(UL)}}$ packets in the ith transmission, and if b_k^{ij} denotes the number of the correctly received bits in the jth packet; then the average user throughput for user k is given as follows:

$$R_k^{\mathrm{DL(UL)}} = \frac{\sum_{i=1}^{p_k^{\mathrm{DL(UL)}}} \sum_{j=1}^{q_{ik}^{\mathrm{DL(UL)}}} b_k^{ij}}{T_{\mathrm{sim}}} \tag{11.2}$$

The average user throughput is defined as the sum of the user throughput of each user in the cell divided by the total number of users in the cell. Let N_{user} denote the number of users in the sector/cell and user $k = 1, 2, \ldots, N_{\mathrm{user}}$ has throughput $R_k^{\mathrm{DL(UL)}}$, then DL (or UL) sector throughput is defined as:

$$R^{\mathrm{DL(UL)}} = \sum_{k=1}^{N_{\mathrm{user}}} R_k^{\mathrm{DL(UL)}} \tag{11.3}$$

The user throughput per transmission is the number of correctly received bits per packet divided by the duration of the transmission. If user k has $p_k^{\mathrm{DL(UL)}}$ downlink (uplink) transmissions, if there are $q_{ik}^{\mathrm{DL(UL)}}$ packets in the ith transmission, and if there are b_k^{ij} correctly received bits in the jth packet, then the average transmission throughput can be defined as follows:

$$R_k^{\mathrm{DL(UL)}} = \frac{1}{p_k^{\mathrm{DL(UL)}}} \left(\sum_{i=1}^{p_k^{\mathrm{DL(UL)}}} \frac{\sum_{j=1}^{q_{ki}^{\mathrm{DL(UL)}}} b_k^{ij}}{\Delta T_{ki}^{\mathrm{DL(UL)}}} \right) \tag{11.4}$$

where $\Delta T_{ki}^{\mathrm{DL(UL)}}$ denotes the duration of the ith transmission to user k. The average user throughput per transmission is defined as the sum of the user throughputs per transmission divided by the number of users in the cell.

11.1.3 Outage throughput

The outage throughput $\eta_{outage}(R_{min})$ is defined as the percentage of users with data rate R_k^{DL} less than a predefined minimum data rate R_{min}. The cell-edge user throughput is defined as the fifth percentile point of the cumulative distribution function (CDF) of users' average throughput per transmission assuming that 95% of the users are expected to achieve a certain throughput per transmission regardless of their geographical location in the cell.

11.1.4 Packet delay

If the *j*th packet of the *i*th transmission belongs to user *k*, if the packet arrives at the BS (or MS) MAC at time instant $T_{arrival}^{kij}$ and is delivered to the MS (or BS) MAC sublayer at time instant $T_{departure}^{kij}$, then the packet delay can be calculated as follows:

$$D_{kij}^{DL(UL)} = T_{arrival}^{kij} - T_{departure}^{kij} \tag{11.5}$$

The downlink and uplink delays are denoted by superscript "DL" or "UL," respectively. The packets that are dropped or erased may not be included in the analysis of packet delays depending on the traffic model. For example, in the modeling of traffic for delay sensitive applications, packets may be dropped, if packet transmissions are not completed within a specified delay bound. The impact of the dropped packets is included in the packet loss rate. The CDF of the packet delay per user provides a basis in which maximum latency, second percentile, and average latency as well as jitter can be derived. The second percentile point of the CDF of packet delay denotes the packet delay value for which 98% of packets have a delay less than that value. As an example, the VoIP capacity is defined such that the percentage of users in outage is less than 2% where a user is assumed to have experienced a service outage, if less than 98% of the VoIP packets have been delivered successfully to the user within a one-way radio access delay bound of 50 ms.

The average packet delay is defined as the average interval between packets originated at the source (either MS or BS) and received at the destination (either BS or MS) in a system for a given duration of transmission. The average packet delay for user *k*, $\overline{D}_k^{DL(UL)}$ is given by:

$$\overline{D}_k^{DL(UL)} = \frac{\sum_{i=1}^{p_k} \sum_{j=1}^{q_{ki}} (T_{departure}^{kij} - T_{arrival}^{kij})}{\sum_{i=1}^{p_k} q_{ki}} \tag{11.6}$$

The CDF of users' average packet delay is the cumulative distribution of the average packet delay observed by all users in the cell. The packet loss ratio per user is defined as:

$$\text{Packet loss ratio} = 1 - \frac{\text{Total number of successfully delivered packets}}{\text{Total number of packets}} \tag{11.7}$$

where the total number of packets includes packets that were transmitted over the air-interface and packets that were dropped prior to transmission. The data throughput of a BS is defined as the number of information bits per second that a cell can successfully deliver or receive via the air-interface using appropriate scheduling algorithms.

11.1.5 Spectral efficiency

We consider both physical-layer spectral efficiency and MAC sublayer spectral efficiency as important performance indicators for a cellular system. The physical-layer spectral efficiency is the system throughput measured at the interface of the physical layer and the MAC sublayer, thus including physical-layer overhead but excluding MAC and upper layer protocol overheads. The MAC sublayer spectral efficiency represents the system throughput measured at the interface of the MAC sublayer and the upper layers, thus including both the physical layer and MAC protocol overheads. The MAC efficiency of the system is evaluated by dividing the MAC sublayer spectral efficiency by the physical-layer spectral efficiency. The average cell spectral efficiency is defined as:

$$\mathrm{SE}_{\mathrm{cell}} = \frac{R_{\mathrm{aggregate}}}{\mathrm{BW}_{\mathrm{effective}}} \tag{11.8}$$

where $R_{\mathrm{aggregate}}$ denotes the aggregate cell throughput and $\mathrm{BW}_{\mathrm{effective}}$ is the effective channel bandwidth. The effective channel bandwidth is defined as $\mathrm{BW}_{\mathrm{effective}} = \alpha_{\mathrm{DL-UL}}\mathrm{BW}_{\mathrm{transmission}}$ where $\mathrm{BW}_{\mathrm{transmission}}$ denotes the channel bandwidth, and $\alpha_{\mathrm{DL-UL}}$ is the downlink/uplink ratio in the TDD systems. Note that for FDD systems $\alpha_{\mathrm{DL-UL}} = 1$ and for TDD systems with DL:UL ratio of 2:1; $\alpha_{\mathrm{DL-UL}} = 2/3$ for the downlink and 1/3 for the uplink, respectively.

The CDF of SINR is defined as the CDF for the signal to interference plus noise ratio measured at the BS for each MS on the uplink, or at the MS for the BS in the downlink. This metric would allow comparison between different reuse scenarios, network loading conditions, multi-antenna algorithms, resource allocation, power control schemes, etc.

11.1.6 Jain fairness index

Fairness measures or metrics are used in network engineering to determine whether users or applications are receiving a fair share of system resources. There are several mathematical and conceptual definitions of fairness among which the Jain index is a commonly used metric defined as follows [44]:

$$J(x_1, x_2, \ldots, x_n) = \frac{\left(\sum_{i=1}^{n} x_i\right)^2}{n \sum_{i=0}^{n} x_i^2} \tag{11.9}$$

The Jain index rates the fairness of a set of values where there are n users and x_i is the throughput for the ith connection. The result ranges from $J = 1/n$ (worst case) to $J = 1$ (best case), and it is maximum when all users receive the same allocation. This index is $J = k/n$ when k users equally share the resources, and the other $n - k$ users receive zero allocation. This metric identifies under-utilized channels and is not excessively sensitive to atypical network flow patterns. Alternatively, the Jain index can be defined as:

$$J = \frac{\overline{R}^2_{\text{throughput}}}{\overline{R}^2_{\text{throughput}} + \sigma_{R_{\text{throughput}}}} \tag{11.10}$$

where $\overline{R}_{\text{throughput}}$ and $R_{\text{throughput}}$ denote the average throughput and throughput values, respectively.

11.1.7 Geometry (wideband SINR distribution)

In the calibration of the large-scale parameters of the channel model, all the multipath effects are not considered. Simulations are conducted to obtain for each user a path gain and a wideband SINR value. The path gain is defined as the average signal gain between an MS and its serving BS. It includes distance attenuation, shadowing, and antenna gains. The wideband SINR is the ratio of the average power received from the serving cell and the average interference power received from other cells plus noise. Sometimes this measure is called the geometry factor, since it presents a great dependency with the position of the user in relation to the position of the base stations. Then, the calibration of this measure ensures not only the calibration of the channel model, but also the calibration of the network layout. In addition, it involves the calibration of the physical layer considered in the simulation platform functional description, since the SINR is calculated in this layer.

11.1.8 Effective interference over thermal

In code division multiple access (CDMA) systems, the rise over thermal (RoT) indicates the ratio between the total power received from wireless sources at a base station and the thermal noise. In order to decode the data at the base station, a minimum SINR at the receiver must be guaranteed. The RoT metric is a measure of the uplink loading. By increasing the number of transmitting UEs and their transmit power, the level of interference in the uplink increases. This interference is perceived by the base station as noise adversely affecting the SINR. The base station controls the interference level by adjusting the UE uplink assignments and transmit power [25]. Similarly, IoT reflects the effective interference received by the base station. The definition of effective IoT measure takes the interference suppression capabilities of the base station receiver into consideration.

In order to formulate a mathematical expression for effective IoT, we consider a single-input multiple-output (SIMO) system and further assume that the base station has a linear receiver, such as maximal ratio combining (MRC) or MMSE with N_{RX} antennas. The received signal at the mth receive antenna can be expressed as $r_m = h_m x + e_m + n_m$ where h_m is the channel between the single transmit antenna and the mth receive antenna, x is the transmitted symbol, e_m is the sum of the interfering signals, and n_m is the thermal noise at the mth receive antenna. The receiver processing consists of estimating the transmitted symbol x by appropriately weighting the received signal at each antenna with a complex-valued weight w_m^H, and then summing the weighted signals. The sum-weighted output signal is given by $y = \sum_{m=1}^{N_{RX}} w_m^H r_m = \sum_{m=1}^{N_{RX}} w_m^H (h_m x + e_m + n_m)$. Alternatively, using vector notations, one can write the latter equation as $\mathbf{y} = \mathbf{w}^H(\mathbf{h}x + \mathbf{e} + \mathbf{n})$.

The effective SINR at the receiver output is given by the ratio of the power of the desired part of the filtered received signal $\mathbf{w}^H\mathbf{h}x$ to the power of the undesired part of the weighted received signal $\mathbf{w}^H(\mathbf{e} + \mathbf{n})$ which under the assumption that noise and interference are independent can be written as:

$$\text{SINR}_{\text{effective}} = \frac{\mathbf{w}^H\mathbf{h}\mathbf{h}^H\mathbf{w}}{\mathbf{w}^H(\mathbf{Q} + \mathbf{N})\mathbf{w}} \tag{11.11}$$

where $\mathbf{Q} = \varepsilon\{\mathbf{e}\mathbf{e}^H\}$ is the covariance matrix of the interference, and $\mathbf{N} = \varepsilon\{\mathbf{n}\mathbf{n}^H\}$ is the covariance matrix of the thermal noise. Note that typically $\mathbf{N} = \sigma^2\mathbf{I}$ where σ^2 is the noise variance per antenna and $\mathbf{I}$ is the $N_{RX} \times N_{RX}$ identity matrix. The denominator in Eq. (11.11) is defined as the power of the interference and thermal noise. $P_{I+N_{\text{effective}}} = \mathbf{w}^H(\mathbf{Q} + \mathbf{N})\mathbf{w}$. The effective IoT can now be defined as the ratio between the total effective noise and interference power to the effective thermal noise power:

$$IoT_{\text{effective}} = \frac{P_{I+N_{\text{effective}}}}{P_{N_{\text{effective}}}} = \frac{\mathbf{w}^H(\mathbf{Q} + \mathbf{N})\mathbf{w}}{\mathbf{w}^H\mathbf{N}\mathbf{w}} \tag{11.12}$$

Since receiver weights typically vary over frequency and time, the effective IoT should be calculated per frequency and time instant, and then averaged. The average is taken over allocated resources and unused resources are excluded.

11.1.9 Cell spectral efficiency and cell-edge user spectral efficiency

Cell spectral efficiency SE_{cell} is defined as the aggregate throughput of all users, that is, the number of correctly received bits delivered to the upper layers at the MAC service access point over a certain period of time, divided by the channel bandwidth divided by the number of cells. The channel bandwidth is defined as the effective bandwidth multiplied by the frequency reuse factor, where the effective bandwidth is the operating bandwidth that is appropriately scaled by the uplink/downlink ratio. The cell spectral efficiency is measured in bits/s/Hz/cell.

If d_i denotes the number of correctly received bits by user i (in the downlink) or from user i (in the uplink) in a system comprising N_{user} active users and N_{cell} cells, and if B denotes the channel bandwidth and T the time over which the data bits are received. The cell spectral efficiency can be expressed as follows:

$$\text{SE}_{\text{cell}} = \frac{\sum_{i=1}^{N_{\text{user}}} d_i}{TBN_{\text{cell}}} \tag{11.13}$$

The (normalized) user throughput is defined as the average user throughput, that is, the number of correctly received bits by users delivered to upper layers at the MAC service access point over a certain period of time, divided by the channel bandwidth, and is measured in bits/s/Hz. The cell-edge user spectral efficiency is defined as the 5% point of the CDF of the normalized user throughput. Let $d_i, T_i,$ and B denote the number of correctly received bits by user i, the active session duration for user i, and the channel bandwidth, respectively. The (normalized) user throughput of user i is defined as:

$$R^{(i)}_{\text{user-throughput}} = \frac{d_i}{T_i B} \tag{11.14}$$

11.2 Calculation of static and dynamic overhead

The calculation and analysis of the signaling and control channels' overhead, as well as the physical-layer overhead associated with synchronization and broadcast channels, guard times and guard bands, reference signals, and that of the cyclic prefix are important in determining the actual performance of a cellular system. In general, the overhead channels are essential for proper operation of a cellular system and statically (independent of the number of the users in the cell) or dynamically (depending on the number of the users in the cell) consume some radio resources and make them permanently or temporarily unavailable for data transmission. There are two types of overhead channels: static and dynamic. A static overhead channel requires fixed base station power, time slot, and/or bandwidth. On the other hand, a dynamic overhead channel requires base station power, time, and/or bandwidth which dynamically change over time as a function of the number of active users.

The layer 1 (L1) and layer 2 (L2) overhead are accounted for in time, frequency, or space for calculation of system performance metrics, such as spectral efficiency, user throughput, VoIP capacity, etc. Examples of L1 overhead include synchronization signals, guard sub-carriers and DC sub-carrier, guard time (in TDD systems), reference signals, and cyclic prefix. Examples of L2 overheads include common and dedicated control channels, HARQ ACK/NACK feedback, channel quality feedback, random access channel, packet headers and sub-headers, as well as cyclic redundancy check (CRC). The power allocation/boosting should

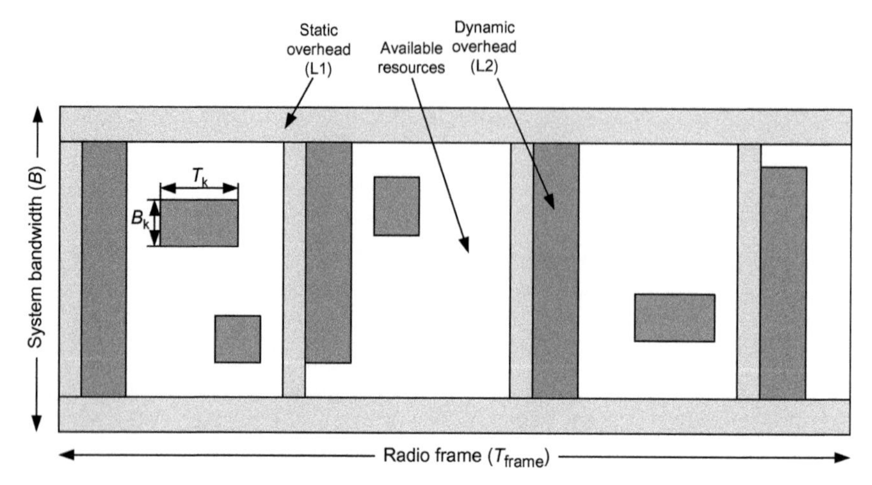

FIGURE 11.3

Illustration of the static and dynamic overhead.

also be accounted for in modeling resource allocation for control channels. Based on the example two-dimensional time-frequency model shown in Figure 11.3, the L1 and L2 overhead can be calculated as:

$$O_{L1+L2} = \lim_{T_{frame} \to \infty} \frac{\sum_{k \in S(L1) \cup S(L2)} B_k T_k}{BT_{frame}}, \quad 0 \leq O_{L1+L2} < 1 \tag{11.15}$$

where B denotes the system bandwidth and T_{frame} is the radio frame size, B_k is the resource occupied by the kth L1 (or L2) overhead channel across frequency dimension, and T_k denotes the required time resource for transmission of the kth L1 (or L2) overhead channel in time.

Using this model, the total resources consumed by the L1 overhead over a relatively long period of time is given as:

$$O_{L1} = \lim_{T_{frame} \to \infty} \frac{1}{BT_{frame}} \sum_{k \in S(L1)} B_k T_k, \quad 0 \leq O_{L1} < 1 \tag{11.16}$$

where S(L1) denotes the set of resources used for static L1 overhead channels. The total resources consumed by the L2 overhead over a relatively long period of time is given as:

$$O_{L2} = \lim_{T_{frame} \to \infty} \frac{1}{BT_{frame}(1 - O_{L1})} \sum_{k \in S(L2)} B_k T_k, \quad 0 \leq O_{L2} < 1 \tag{11.17}$$

Note that the amount of resources that have already been used by the L1 overhead must not be recounted in calculation of O_{L2} and that $O_{L1+L2} \neq O_{L1} + O_{L2}$. Furthermore, S(L2) denotes the set of resources used for L2 overhead channels

The average spectral efficiency inclusive of the effect of the L1/L2 overhead can be obtained as $SE_{average}(1 - O_{L1+L2})$ where $SE_{average}$ denotes the average spectral efficiency that has been obtained via system-level simulation excluding the L1/L2 overhead.

11.3 Traffic models

11.3.1 Statistical model for conversational speech

VoIP refers to real-time delivery of coded voice packets across a network using the Internet protocols. Several robust voice codecs for encoding conversational speech have been developed and discussed in the literature, including ITU-T G.729 (8 kbps) [17] and adaptive multi-rate (AMR) codec (4.75–12.2 kbps) [5,27]. In the following discussion, a VoIP session is defined as the entire user call time, and a typical conversation is characterized by periods of active speech or talk spurts followed by silence periods. Figure 11.4 illustrates a two-state Markov model[2] for conversational speech [23,24,54].

The steady-state condition of the model requires that $P_0 = a/(a + c)$ and $P_1 = c/(a + c)$ where P_0 and P_1 are the probabilities of being in state 0 and state 1, respectively. As shown in Figure 11.4, the probability of a transition from state 1 (the active speech state) to state 0 (the inactive state) while in state 1 is equal to a, whereas the probability of a transition from state 0 to state 1 while in state 0 is c. The model is updated at the speech encoder frame rate $R = 1/T$, where T is the

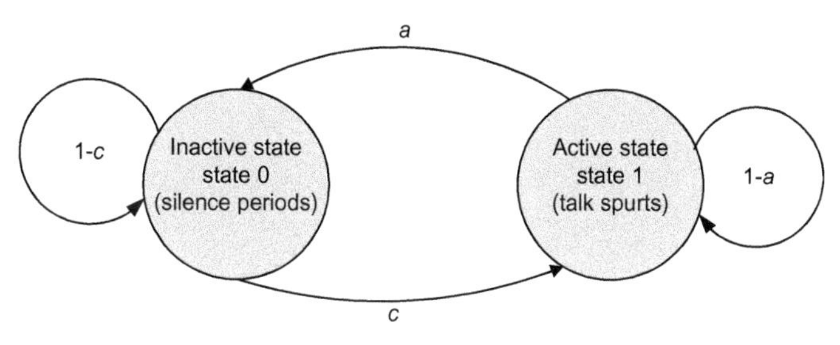

FIGURE 11.4

Two-state Markov model for speech [23,24].

[2]A sequence $X_1, X_2, \ldots, X_{n-1}, X_n$ of random variables is called a Markov sequence, if for any n, the conditional probability of X_n satisfies the following condition $F(X_n|X_{n-1}, X_{n-2}, \ldots, X_1) = F(X_n|X_{n-1})$, that is, the conditional distribution of X_n given $X_{n-1}, X_{n-2}, \ldots, X_1$ have been observed is equal to the conditional distribution X_n assuming only X_{n-1} has been observed. The transitional densities of a Markov sequence satisfy the Chapman–Kolmogorov equation. A random process whose future probabilities are determined by its most recent values is called a Markov process. A stochastic process $x(t)$ is called Markov, if for any n and $t_1 < t_2 < \ldots < t_n$ we have $P(x(t_n) \leq x_n | x(t_{n-1}), x(t_{n-2}), \ldots, x(t_1)) = P(x(t_n) \leq x_n | x(t_{n-1}))$[23,24,54].

encoder frame duration which is typically 20 ms for most speech encoders. Packets are generated at time intervals $iT+\tau$ where τ is the packet arrival delay, and i denotes the encoder frame index. During the active state, packets of fixed sizes are generated at these time intervals, while the model is updated at regular frame intervals. The size of the packets and the rate at which the packets are sent depend on the corresponding voice codecs and compression schemes. The voice activity factor (VAF) λ is defined as $\lambda = P_1 = c/(a+c)$. A talk-spurt is defined as the time period τ_{TS} between entering the active state and leaving the active state. The probability that a talk spurt has duration nT is given by $P(\tau_{\text{TS}} = nT) = a(1-a)^{n-1}, n = 1, 2, \ldots$ whereas the probability that a silence period has duration mT is given as $P(\tau_{\text{SP}} = mT) = c(1-c)^{m-1}, m = 1, 2, \ldots$. The average talk spurt duration μ_{TS} (in number of speech frames) is defined as $\mu_{\text{TS}} = \varepsilon[\tau_{\text{TS}}] = 1/a$. The mean silence period duration μ_{SP} (in number of speech frames) is given by $\mu_{\text{SP}} = \varepsilon[\tau_{\text{SP}}] = 1/c$. The distribution of the time period τ_{AE} (in number of speech frames) between successive active state entries is defined as the convolution of the distributions of τ_{SP} and τ_{TS}:

$$P(\tau_{\text{AE}} = nT) = \frac{c}{c-a} a(1-a)^{n-1} + \frac{a}{a-c} c(1-c)^{n-1}, \quad n = 1, 2, \ldots \tag{11.18}$$

It must be noted that τ_{AE} can be further considered as the time between MAC layer resource reservations, provided that a single reservation is made per user per talk-spurt. Note that in practice, very small values of τ_{AE} may not result in separate reservation requests. Since the transitions from state 1 to state 0 and vice versa are independent, the mean time μ_{AE} between active state entries is the sum of the mean time in each state, that is, $\mu_{\text{AE}} = \mu_{\text{TS}} + \mu_{\text{SP}}$. Therefore, the mean rate of arrival $\overline{R}_{\text{AE}}$ of transitions into the active state can be expressed as $\overline{R}_{\text{AE}} = 1/\mu_{\text{AE}}$.

The voice capacity calculation assumes the use of 12.2 kbps mode of 3GPP AMR codec with a 50% VAF provided that the percentage of users in outage is less than 2%, where a user is defined to have experienced outage if more than 2% of the VoIP packets are dropped, erased, or otherwise not delivered successfully to the user within the delay bound of 50 ms. The packet delay is defined based on the 98th percentile of the CDF of all individual users' 98th percentiles of packet delay (i.e., the 98th percentile of the packet delay CDF is first determined for each user and then the 98th percentile of the CDF that describes the 98th percentiles of the individual user delay is obtained). The VoIP capacity is measured in terms of the number of active users/MHz/cell. It is the minimum of the capacity calculated for either downlink or uplink divided by the effective bandwidth in the respective direction. In other words, the effective bandwidth is the operating bandwidth normalized appropriately considering the uplink/downlink ratio (in TDD systems).

During each VoIP session, a user will be in the active or inactive state. The duration of time that the user stays in each state is exponentially distributed. In the active or inactive state, voice packets of fixed sizes will be generated at

Table 11.2 VoIP Traffic Model Parameters [4,26]

VoIP Model Attribute	Statistical Distribution	Parameters	Probabilistic Distribution
Active/inactive state duration	Exponential	$\mu = 1.25$ s	$f_x = \lambda e^{-\lambda x}, x \geq 0$ $\lambda = 1/\mu$
Probability of state transition	N/A	$c = 0.01$, $d = 0.99$	N/A
Packet arrival delay jitter (downlink only)	Laplacian	$\beta = 5.11$ ms	$f_x = \frac{1}{2\beta} e^{\frac{-\lvert\tau\rvert}{\beta}}$ $-80\text{ ms} \leq \tau \leq 80\text{ ms}$

intervals of $iT + \tau$ seconds, where T is the voice-codec frame size which is typically 20 ms, τ is the network delay jitter (delay variations) in the downlink, and i is the VoIP frame number. In the uplink direction, τ is equal to zero. Given that the range of the delay jitter is limited to 120 ms, the model may be implemented by generating packets at times $iT + \tau'$ seconds where $\tau' = \tau + 80$ (ms) denotes a positive delay value (Table 11.2). The air-interface delay is measured as the time interval between the packet arrival time $iT + \tau'$ to successful reception and decoding of the packet at the receiver. Statistical distribution and parameters associated with the VoIP traffic model are summarized in Table 11.2. The assumptions that were used for VoIP capacity calculation are shown in Table 11.3.

11.3.2 Full-buffer traffic model

In the full-buffer user traffic model, all users in the system are assumed to have data to send or receive at any given time. In other words, there is always a constant amount of data that needs to be transferred, as opposed to bursts of data with probabilistic arrival time distribution. This model allows the assessment of the spectral efficiency of the system independent of the actual user traffic distribution model. A user is in outage if the residual packet error rate after HARQ retransmissions exceeds 1%.

11.4 Link-to-system mapping (PHY abstraction)

The objective of physical-layer abstraction is to model link-level performance and to reduce the amount of computations and complexity in system-level simulations. The motivation for abstraction of the physical layer is that simulating the physical layer across multiple base stations and mobile stations in a cellular network simulation platform can be computationally cumbersome and practically impossible [46,50]. The abstraction must be reasonably accurate, computationally simple,

Table 11.3 EESM/MIESM Parameters and MSE for all CQIs in a Bandwidth of Four PRBs [50]

CQI	Code Block Size	Approximate Code Rate	EESM					MIESM		
			β	MSE	α_1	α_2	MSE	α_1	α_2	MSE
1	72	0.072	0.372	0.003	0.363	0.360	0.003	0.278	0.276	0.002
2	112	0.113	0.531	0.004	0.519	0.515	0.004	0.389	0.385	0.003
3	192	0.185	0.865	0.004	0.858	0.855	0.003	0.633	0.629	0.002
4	312	0.297	1.403	0.014	1.439	1.460	0.012	1.056	1.067	0.008
5	456	0.435	1.636	0.076	1.647	1.675	0.073	1.111	1.112	0.045
6	608	0.585	1.646	0.023	1.636	1.621	0.022	1.035	1.019	0.013
7	768	0.368	3.418	0.159	3.748	4.123	0.057	1.065	1.121	0.002
8	1008	0.477	5.042	0.182	5.160	5.491	0.121	1.034	1.065	0.008
9	1248	0.600	6.122	0.106	6.307	6.649	0.066	0.937	0.980	0.013
10	1440	0.455	1.010	0.553	9.790	12.262	0.064	0.942	1.150	0.008
11	1728	0.553	0.399	0.713	11.489	14.827	0.024	0.556	0.727	0.010
12	2048	0.650	0.621	0.658	17.669	23.625	0.051	0.700	0.935	0.017
13	2368	0.753	0.962	0.825	24.081	32.971	0.069	0.760	1.017	0.038
14	2688	0.852	7.933	0.382	22.622	28.049	0.007	0.657	0.821	0.006
15	2880	0.925	15.859	0.077	25.245	28.103	0.034	0.840	0.985	0.024

Table 11.4 Baseline Configuration Parameters for IMT-Advanced Evaluations [14]

Simulation Parameters	Indoor Hotspot (InH)	Urban Micro-Cellular (UMi)	Urban Macro-Cellular (UMa)	Rural Macro-Cellular (RMa)
Baseline Parameters				
Base station antenna height (m)	6 (on the ceiling)	10 (below rooftop)	25 (above rooftop)	35 (above rooftop)
Number of BS antennas (receive/transmit)	Up to 8/up to 8	Up to 8/up to 8	Up to 8/up to 8	Up to 8/up to 8
Base station transmit power (dBm)	24 (40 MHz) 21 (20 MHz)	41 (10 MHz) 44 (20 MHz)	46 (10 MHz) 49 (20 MHz)	46 (10 MHz) 49 (20 MHz)
Mobile station transmit power (dBm)	21	24	24	24
Number of MS antennas (receive/transmit)	Up to 2/up to 2	Up to 2/up to 2	Up to 2/up to 2	Up to 2/up to 2
Minimum distance between MS and BS (m)	≥ 3	≥ 10	≥ 25	≥ 35
RF carrier frequency (GHz)	3.4	2.5	2	0.8
Outdoor-to-indoor penetration loss	N/A	See Annex 1, Table A1-2 of [14]	N/A	N/A
Outdoor to in-car penetration loss	N/A	N/A	9 dB (lognormal, $\sigma = 5$ dB)	9 dB (lognormal, $\sigma = 5$ dB)
Parameters for Analytical Assessment of Peak Spectral Efficiency				
Number of BS antennas (receive/transmit)	Up to 4/up to 4	Up to 4/up to 4	Up to 4/up to 4	Up to 4/up to 4
Number of MS antennas (receive/transmit)	Up to 4/up to 2	Up to 4/up to 2	Up to 4/up to 2	Up to 4/up to 2

(Continued)

Table 11.4 (Continued)

Simulation Parameters	Indoor Hotspot (InH)	Urban Micro-Cellular (UMi)	Urban Macro-Cellular (UMa)	Rural Macro-Cellular (RMa)
Parameters for System-Level Simulations				
Network layout	Indoor floor	Hexagonal grid	Hexagonal grid	Hexagonal grid
Inter-site distance (m)	60	200	500	1732
Channel model	Indoor hotspot channel model	Urban micro-cellular channel model	Urban macro-cellular channel model	Rural macro-cellular channel model
User distribution	Randomly and uniformly distributed over area	Randomly and uniformly distributed over area 50% pedestrian users and 50% users indoor	Randomly and uniformly distributed over area 100% vehicular users	Randomly and uniformly distributed over area 100% high-speed vehicular users
User mobility	All mobile stations have fixed and identical speed with randomly and uniformly distributed direction	All mobile stations have fixed and identical speed with randomly and uniformly distributed direction	All mobile stations have fixed and identical speed with randomly and uniformly distributed direction	All mobile stations have fixed and identical speed with randomly and uniformly distributed direction
MS speed (km/h)	3	3	30	120
Inter-site interference	Explicitly modeled	Explicitly modeled	Explicitly modeled	Explicitly modeled
BS noise figure (dB)	5	5	5	5
MS noise figure (dB)	7	7	7	7
BS antenna gain (bore-sight) (dBi)	0	17	17	17
MS antenna gain (dBi)	0	0	0	0
Thermal noise level (dBm/Hz)	− 174	− 174	− 174	− 174

Parameters for Assessment of Cell Spectral Efficiency and Cell-Edge User Spectral Efficiency				
Traffic model	Full buffer	Full buffer	Full buffer	Full buffer
System bandwidth	2 × 20 MHz (FDD) 40 MHz (TDD)	2 × 10 MHz (FDD) 20 MHz (TDD)	2 × 10 MHz (FDD) 20 MHz (TDD)	2 × 10 MHz (FDD) 20 MHz (TDD)
Number of users/cell	10	10	10	10
Parameters for Evaluation of VoIP Capacity				
Traffic model	VoIP	VoIP	VoIP	VoIP
System bandwidth	2 × 5 MHz (FDD) 10 MHz (TDD)	2 × 5 MHz (FDD) 10 MHz (TDD)	2 × 5 MHz (FDD) 10 MHz (TDD)	2 × 5 MHz (FDD) 10 MHz (TDD)
Simulation duration for a single drop(s)	20	20	20	20
Parameters for Link-Level Simulation (Mobility Requirement)				
Traffic model	Full buffer	Full buffer	Full buffer	Full buffer
Channel model	Indoor hotspot channel model	Urban micro-cellular channel model	Urban macro-cellular channel model	Rural macro-cellular channel model
System bandwidth (MHz)	10	10	10	10
Number of users/cell	1	1	1	1

relatively independent of channel models, and extensible to interference models and multi-antenna processing methods. The system-level simulations are used to characterize the average user/system performance, which may be useful in providing insights for deployment of the system in terms of cell and frequency planning. For such simulations, the average performance of a system is quantified using the topology and macro channel characteristics to calculate a wideband SINR distribution across the cell (alternatively known as geometry). Each subscriber's SINR distribution is then mapped to the highest MCS that could be supported, based on link-level SINR tables which include fast fading characteristics of the channel. The link-level SINR-BLER look-up tables serve as the PHY abstraction for predicting average link-level performance. The instantaneous channel conditions are used to improve the performance of the cellular systems. Channel-dependent scheduling and adaptive coding and modulation are examples of channel-adaptive schemes employed to improve system performance. Therefore, properly designed system-level simulation methodologies do explicitly model the dynamic behavior of the system [28,37,50,51].

In system-level simulations, an encoded data/control packet may be transmitted over a time-frequency selective channel. In OFDM-based systems, the encoded block is transmitted over several sub-carriers, the post-processing SINR values of the pre-decoded streams are thus non-uniform, and the channel gains of sub-carriers can be time-varying. As a result, in the transmission of a large encoded packet, the encoded symbols with unequal SINR ratios at the input of the decoder are considered due to the time or frequency selectivity of the channel impulse or frequency response over the duration of packet transmission.

The goal of a PHY abstraction or a link-to-system mapping is to predict the encoded BLER for a given received channel realization across the OFDM sub-carriers that are used to transmit the channel-coded blocks. In order to predict the performance, the post-processing SINR values at the input to the forward error correction (FEC) decoder are considered as input to the PHY abstraction. As the link-level curves are generated assuming a frequency flat channel response at a given SINR, an effective SINR is defined to accurately map the system-level SINR to the link-level curves in order to determine the resulting BLER. The latter mapping is known as effective SINR mapping (ESM) in the literature. The ESM PHY abstraction compresses the vector of received SINR values to a single effective SINR value, which can then be further mapped to a BLER value. Several ESM approaches to predict the instantaneous link-level performance have been studied in the literature. Examples include mean instantaneous capacity, exponential effective SINR mapping (EESM), and mutual information (MI) ESM. Each of these PHY abstractions uses a different function to map the vector of SINR values to a single number. Given the instantaneous EESM SINR, mean capacity or mutual information effective SINR, the BLER for each MCS is calculated using a suitable mapping function. The general process of link-to-system mapping is illustrated in Figure 11.5.

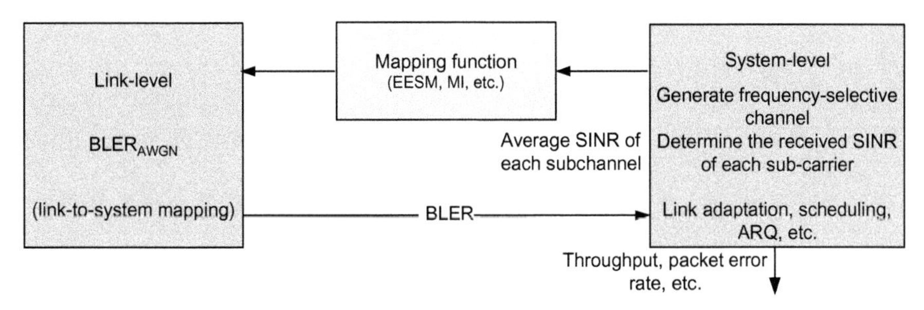

FIGURE 11.5

Link-to-system mapping process [26].

Given an experimental BLER measured in a multi-state channel with a specific MCS, the effective SINR of that channel is defined as the SINR that would produce the same BLER, with the same MCS, in AWGN conditions. In general, for a given multi-state channel with N_{FFT} different SINR measurements $\{\mathrm{SINR}_1, \mathrm{SINR}_2, \mathrm{SINR}_3, \ldots, \mathrm{SINR}_{NFFT}\}$, the effective SINR can be described as follows:

$$\mathrm{SINR}_{\mathrm{effective}} = \alpha_1 \Phi^{-1}\left(\frac{1}{N_{\mathrm{FFT}}}\sum_{n=1}^{N_{\mathrm{FFT}}} \Phi\left(\frac{\mathrm{SINR}_n}{\alpha_2}\right)\right) \tag{11.19}$$

where $\mathrm{SINR}_{\mathrm{effective}}$ denotes the effective SINR, SINR_n is the SINR measured at the nth sub-carrier, N_{FFT} is the number of symbols in a coded block, or the number of sub-carriers used in an OFDM system, and $\Phi(.)$ is an invertible function. In the case of the mutual information-based ESM the function $\Phi(.)$ is derived from the constrained capacity, whereas in the case of EESM, the function $\Phi(.)$ is derived from the Chernoff bound[3] on the probability of error [55]. The EESM model uses the mapping function $\Phi(x) = 1 - \exp(-x)$ for all modulation schemes and in many cases, the constants α_1 and α_2 are assumed to be equal $\alpha_1 = \alpha_2 = \beta$. The parameter β can be interpreted as a shift in $\Phi(.)$ function to adapt the model to different MCSs. If a MIMO scheme is utilized assuming use of linear processing (ZF or MMSE) in the receiver, one still can use the effective SINR concept

[3]In the theory of probability and stochastic processes, the Chernoff bound provides exponentially decreasing bounds (asymptotic) on tail distributions of sums of independent random variables compared to the first or second moment based tail bounds such as Markov's inequality or Chebyshev inequality, which only yield power-law bounds on tail decay. Let $\mathbf{x}_1, \mathbf{x}_2, \ldots, \mathbf{x}_n$ denote discrete and independent random variables with probability $p > 1/2$. Then, the probability of simultaneous occurrence of more than $n/2$ of the events x_k has an exact value P, where $P = \sum_{i=\lfloor n/2 \rfloor}^{n} \binom{n}{i} p^i (1-p)^{n-i}$. The Chernoff bound provides a lower bound for P given by $P \geq 1 - \exp[-2n(p-1/2)^2]$. The Chernoff bound applies to a class of random variables and provides exponential falloff of probability with distance from the mean. The critical condition that is needed for a Chernoff bound is that the random variable must be a sum of independent indicator random variables [55].

by taking the MIMO post-processing SINR as the set of measurements $\{SINR_1, SINR_2, SINR_3, \ldots, SINR_{NFFT}\}$ of the multi-state channel. Assuming that the receiver has a good estimation of the MIMO channel matrix, it can calculate the post-processing SINR based on well-known expressions. In spatial multiplexing mode, the antenna correlation is translated into degradation of the post-processing SINR, thus assuming that the channel is known, the effective SINR model may also capture the effect of antenna correlation. The parameters α_1 and α_2 depend on the modulation scheme, the code block size, and the code rate, but should be valid for SISO and MIMO with linear equalization regardless of the MIMO antenna correlation. Based on the above discussion, in the case of EESM, the effective SINR in snapshot i can be written as follows:

$$\mathrm{SINR}^i_{\mathrm{effective}}(\beta) = \beta \ln\left(\frac{1}{N_{\mathrm{FFT}}}\sum_{n=1}^{N_{\mathrm{FFT}}} \exp\left(\frac{\mathrm{SINR}^i_n}{\beta}\right)\right) \tag{11.20}$$

If M denotes the number of simulated snapshots, the value of β can be obtained as follows:

$$\hat{\beta} = \arg\min_{\beta}\left\{\sum_{i=1}^{N_{\mathrm{snapshots}}} [\log(\mathrm{BLER}_i) - \log(\mathrm{BLER}_R(\mathrm{SINR}^i_{\mathrm{effective}}(\beta)))]^2\right\} \tag{11.21}$$

where $\mathrm{BLER}_R(.)$ is the reference BLER curve in the AWGN channel for the current MCS and code block size. That is, to obtain β, we minimize the mean squared error between the estimated BLER and the simulated BLER for the set of simulated snapshots using numerical methods. By using the logarithmic BLER error in Eq. (11.21), the minimization algorithm tries to obtain minimal error at low BLER, which is the region of interest. For MIESM Eq. (11.19) is rewritten as follows:

$$\mathrm{SINR}^i_{\mathrm{effective}}(\alpha_1, \alpha_2) = \alpha_1 \ln\left(\frac{1}{N_{\mathrm{FFT}}}\sum_{n=1}^{N_{\mathrm{FFT}}} \exp\left(\frac{\mathrm{SINR}^i_n}{\alpha_2}\right)\right) \tag{11.22}$$

In which, the calculation of α_1 and α_2 involves a two-dimensional minimization of the mean square error (MSE) as follows:

$$(\hat{\alpha}_1, \hat{\alpha}_2) = \arg\min_{\alpha_1, \alpha_2}\left\{\sum_{i=1}^{N_{\mathrm{snapshots}}} [\log(\mathrm{BLER}_i) - \log(\mathrm{BLER}_R(\mathrm{SINR}^i_{\mathrm{effective}}(\alpha_1, \alpha_2)))]^2\right\} \tag{11.23}$$

In the case of MIESM, there is an additional challenge that the mutual-information function $\Phi(.)$ must be inverted numerically. Table 11.3 lists the obtained EESM/MIESM parameters as well as the MSE for each CQI. The values of the parameters in each case are in good agreement with those previously published in the literature [50,51]. The main advantage of EESM over MIESM is that the nonlinear weighting function $\Phi(.)$ and its inverse function used to calculate the effective SINR are closed-form and easy to compute. The performance of EESM to predict the BLER in the MIMO high-correlation scenarios can be tested in two ways, that is, using a single parameter (one-dimension minimization) or using two different parameters. The advantage of using a single parameter is that minimization effort is simple and there is a single minimum. This is the reason that there are two sets of parameters for EESM shown in Table 11.3. The values of the EESM parameters in Table 11.3 are in good agreement with those published in the literature for QPSK and 16QAM and deviate for 64QAM [28,34,50,51]. Comparing the MSEs corresponding to MIESM and EESM in the table suggests that MIESM outperforms EESM and provides more accurate link-abstraction relative to reference link-level curves at the expense of more complexity.

The accuracy of a mutual information-based metric depends on the equivalent channel over which this metric is defined. The capacity is defined as the mutual information based on a Gaussian channel with Gaussian distributed inputs. The modulation constrained capacity is the mutual information of a symbol channel, that is, constrained by the input symbols drawn from a complex set. The computation of the mutual information per coded bit can be derived from the received symbol-level mutual information which is regarded as received bit mutual information rate (RBIR) link-to-system mapping. An alternative approach directly arrives at the bit-level mutual information and is known as the mean mutual information per bit (MMIB) PHY abstraction method. The procedure of the MIESM approach is illustrated in Figure 11.6. Given a set of N received encoder symbol SINRs from the system-level simulation, denoted as $\{\text{SINR}_1, \text{SINR}_2, \text{SINR}_3, \ldots, \text{SINR}_N\}$; a mutual-information metric is calculated. Based on the computed MI metric, an equivalent SINR is obtained and used to look-up the BLER value [34,50,51].

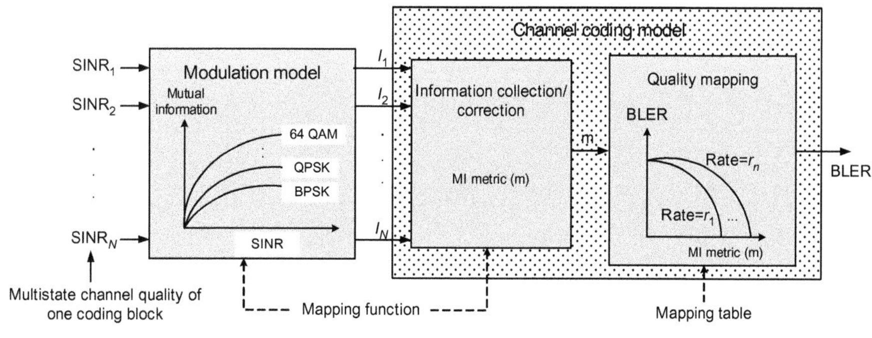

FIGURE 11.6

Procedure for MIESM link-to-system mapping method [26].

11.5 IMT-Advanced test environments

For evaluation of the IMT-Advanced candidates, the ITU-R WP 5D defined several test environments where each test environment is characterized with certain user mobility, path loss and channel models, and system configuration parameters. The evaluation of candidate radio access technologies was performed in selected scenarios of the following test environments [12–14]:

- Indoor test environment models isolated cells at offices and/or in hotspots comprising stationary and pedestrian users. In this model, the emphasis is on very small cells, very high user throughputs, and high user density inside buildings.
- Urban micro-cellular test environment with higher user density was defined to model pedestrian and slow vehicular users. The micro-cellular test environment's focus is on small cells and high user densities which demonstrates typical traffic load in city centers and dense urban areas. The key characteristic of this test environment is high traffic loads as well as outdoor and outdoor-to-indoor coverage. This scenario is interference-limited. A continuous cellular layout and the associated interference are assumed and radio access points are located below rooftop level.
- Urban macro-cellular environment (base coverage urban) was defined to model coverage for pedestrian users up to fast-moving vehicular users. The base coverage urban test environment focuses on large cells and continuous coverage. The main characteristics of this test environment are continuous and ubiquitous coverage in urban areas. This scenario is interference-limited, using macro-cells with radio access points above rooftop level. In an urban macro-cell test environment, mobile stations are located outdoor at street level and fixed base station antennas are located clearly above surrounding building heights. Therefore, NLoS or obstructed sight can be considered in this scenario.
- High-speed or rural test environment models high-speed vehicular and trains. The high-speed test environment focuses on larger cells and continuous coverage. The key characteristics of this test environment are continuous wide area coverage supporting high-speed vehicles. This scenario will therefore be noise-limited and/or interference-limited and uses macro-sized cells.

11.6 Network layout for system-level simulations

In the rural or high-speed, macro- and micro-cellular environments, no specific topographical details are taken into consideration. Base stations are placed in regular grids conforming to a hexagonal layout. A basic hexagonal layout with three cells/sectors per site is shown in Figure 11.7 where antenna bore-sight, cell range, and inter-site distance are also illustrated. The system-level simulations use a wrap-around model with 19 sites each comprising 3 cells. The users are dropped

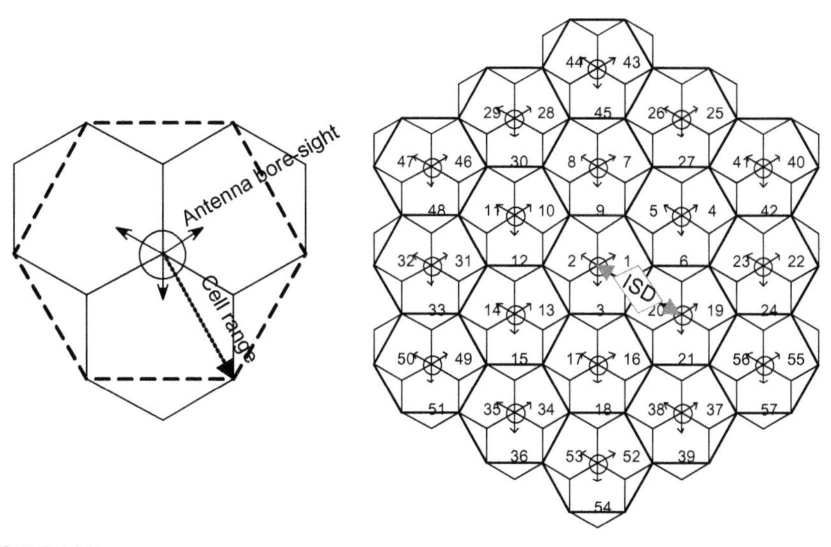

FIGURE 11.7

Illustration of a hexagonal cellular network layout [14].

uniformly over the entire coverage area (Monte Carlo method[4]) in order to model the inter-cell interference. Depending on the configuration being simulated and required output, the effect of the surrounding outer cells may be disregarded. In such cases, only 19 cells of the center cluster may be modeled. For the cases where modeling of the interfering outer cells is necessary for accuracy of the results, the wrap-around structure with the seven cluster network can be used. A cluster is defined as six displacements of the center hexagon. In the wrap-around inter-cell interference model, the network is extended to a cluster of networks consisting of seven replicas of the original hexagonal network, with the original hexagonal network in the middle while the other six copies are attached to it symmetrically on six sides. The number of mobile stations is predetermined for each sector/cell, where each MS location is randomly distributed with uniform distribution (see Figure 11.7).

The serving cell of each MS is determined in two steps due to the wrap-around nature of the cell layout. The first step is to determine the 19 shortest-distanced cells for each MS from all seven logical cell clusters, and the second step is to determine the serving cell/sector among the nearest 19 cells for each MS based on the strongest link according to the path loss and shadowing. To determine the shortest-distanced cell for each MS, the distances between the MS

[4]Monte Carlo methods are a class of computational algorithms that rely on repeated random sampling to compute their results. Monte Carlo methods are often used in simulating physical and mathematical systems. Because of their reliance on repeated computation of random or pseudorandom quantities, these methods are most suited to calculation by a computer and tend to be used when it is unfeasible or impossible to find a closed-form result with a deterministic algorithm [56].

and all logical cell clusters are calculated and the 19 cells with a shortest distance in all seven cell clusters are selected. The serving cell for each MS provides the strongest link with a strongest received long-term signal power. It should be noted that the shadowing experienced on the link between MS and cells located in different clusters is the same.

11.7 IMT-Advanced evaluation methodology and baseline configurations

The IMT-Advanced candidate technologies were required to be evaluated under at least three test environments. ITU-R WP 5D specified a common evaluation methodology based on which the candidates were required to be configured and assessed against the IMT-Advanced minimum system requirements. Table 11.4 provides the common parameters for evaluation of radio access technologies as specified by ITU-R recommendations [6,12–14].

The BS antennas in each sector are assumed to have antenna patterns in azimuth and elevation directions that are given by:

$$\begin{aligned} A_e(\phi) &= -\min\left[12\left(\frac{\phi-\phi_{\text{tilt}}}{\phi_{3\text{dB}}}\right)^2, A_m\right] \quad \text{Elevation direction} \\ A(\theta) &= -\min\left[12\left(\frac{\theta}{\theta_{3\text{dB}}}\right)^2, A_m\right] \quad \text{Azimuth direction} \end{aligned} \tag{11.24}$$

where $A(\theta)$ is the relative antenna gain (in dB) in the direction of $-180° \leq \theta \leq 180°$, $\theta_{3\text{ dB}}$ is the 3 dB beam-width ($\theta_{3\text{dB}} = 70°$),$A_m = 20$ dB is the maximum attenuation, $A_e(\phi)$ is the relative antenna gain (in dB) in the elevation direction $-90° \leq \phi \leq 90°$, $\phi_{3\text{dB}}$ is the elevation 3 dB beam-width ($\phi_{3\text{dB}} = 15°$), and ϕ_{tilt} is the tilt angle. The value of the antenna tilt angle is assumed to be $\phi_{\text{tilt}} = 0°$ for InH, $\phi_{\text{tilt}} = 12°$ for UMi, $\phi_{\text{tilt}} = 12°$ for UMa, and $\phi_{\text{tilt}} = 6°$ for RMa test environments.

The combined antenna pattern at angles off the cardinal axes is computed as $-\min[-(A(\theta)+A_e(\phi)), A_m]$. The antenna bearing is defined as the angle between the main antenna lobe center and a line directed due east given in degrees. The bearing angle increases in a clockwise direction. The center directions of the main antenna lobe in each sector point to the corresponding side of the hexagon. For an indoor test environment, the BS is assumed to have an omni-directional antenna pattern. The mobile station is assumed to have omni-directional antennas.

11.8 Link-level and system-level channel models

The channel behavior is described by its long- and short-term fading characteristics where the former often depends on the geometrical location of a user in a wireless network and the latter defines the time-variant spatial channels.

In general, there are two ways of modeling a channel, deterministic and stochastic. The deterministic category encompasses all models that describe the propagation channel for a specific transmitter location, receiver location, and environment. Deterministic channel models are site-specific, as they clearly depend on the location of transmitter, receiver, and the properties of the environment. They are therefore most suitable for network planning and deployment. In many cases, it is not possible or desirable to model the propagation channel in a specific environment. Especially for system testing and evaluation, it is more appropriate to consider channels that reflect "typical," "best case," and "worst case" propagation scenarios. A stochastic channel model thus reflects the statistics of the channel impulse responses and during the actual simulation; impulse responses are generated as realizations according to those statistics[18–22,29–33,35–43].

Essential to the evaluation of multiple-antenna techniques, which are the key enabling technology for IMT-Advanced systems, is the modeling of MIMO channels that can be represented as double-directional channels or as vector (or matrix) channels. The former representation is more related to the physical propagation effects, while the latter is more focused on the mathematical effect of the channel on the system. The double-directional model is a physical model in which the channel is constructed from summing over multiple waves or rays. Thus, it can also be referred to as a "ray-based model." The vector or matrix channel is a mathematical or analytical model in which the space-time channel as seen by the receiver is constructed mathematically, assuming certain system and antenna parameters. In this approach, the channel coefficients are correlated random processes in both space and time, where the correlation is defined mathematically.

The ITU-R WP 5D specified channel models for the evaluation of IMT-Advanced candidate technologies which consisted of primary and extension modules. The framework of the primary module is based on the WINNER II channel model,[5] which applies the same approach as the 3GPP/3GPP2 spatial channel model (SCM).[6] The extension module extends the capabilities of the IMT-Advanced channel model to cover additional deployment scenarios beyond that of IMT-Advanced candidate technology evaluations, allowing the use of modified parameters to generate large-scale parameters. The ITU-R WP 5D specified channel model is a geometric stochastic model. It does not explicitly specify the locations of the scatterers, rather the directions of the rays generated as a result of scattering objects. The geometric modeling of the radio channel enables separation of propagation parameters and antennas. The channel parameters for individual snapshots are determined stochastically based on statistical distributions extracted from channel measurements. Antenna geometries and radiation patterns

[5]Information on WINNER II channel models can be found at http://www.ist-winner.org/deliverables.html.

[6]Information on the 3GPP/3GPP2 spatial channel model can be found in 3GPP TR 25.996 *spatial channel model for multiple input multiple output* (MIMO), September 2003.

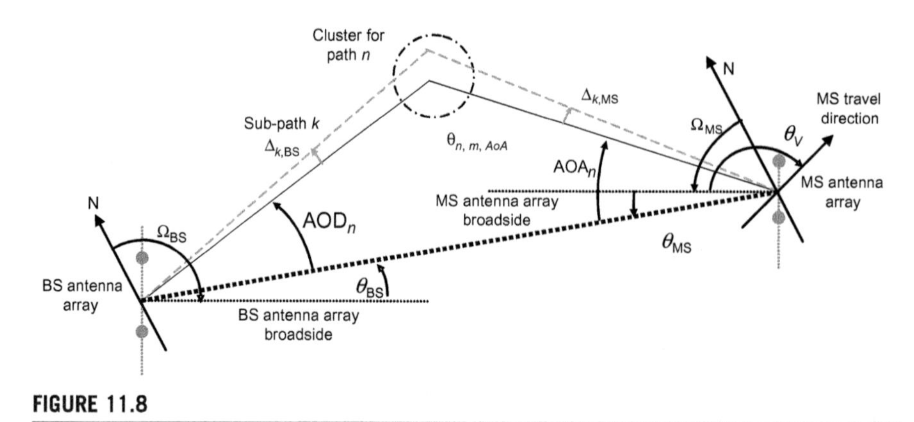

FIGURE 11.8

Illustration of angular parameters in the MIMO channels [1,14].

can be defined properly by the user of the model. As shown in Figure 11.8, channel realizations are generated through the application of the geometrical principle by summing the rays (plane waves) with specific small-scale parameters, such as delay, power, angle-of-arrival (AoA), and angle-of-departure (AoD). The superposition results in correlation between antenna elements and temporal fading with geometry-dependent Doppler spectrum. A number of rays constitute a cluster. A cluster and a propagation path diffused in space are equivalent in delay and angle domains. A generic MIMO channel model is applied to all test scenarios. The time-variant impulse response matrix of the MIMO channel can be written as (assuming N_{TX} transmit antennas at the BS and N_{RX} receive antennas at the MS) $\mathbf{H}(t;\tau) = \sum_{n=1}^{N} \mathbf{H}_n(t;\tau)$ where τ denotes the delay and $N = N_{RX} \times N_{TX}$ is the number of multipaths. The channel impulse response is composed of the antenna array response matrices $\mathbf{F}_{TX}$ and $\mathbf{F}_{RX}$ for the transmitter and the receiver, respectively, as well as the dual-polarized propagation channel response matrix $\mathbf{h}_n$ for the nth cluster, hence:

$$\mathbf{H}_n(t;\tau) = \iint \mathbf{F}_{RX}(\phi)\, \mathbf{h}_n(t;\tau,\varphi,\phi)\, \mathbf{F}_{TX}^T(\varphi)\, d\varphi\, d\phi \tag{11.25}$$

The channel from transmit antenna element s to receive antenna element u for cluster n can be expressed as follows:

$$H_{u,s,n}(t;\tau) = \sum_{m=1}^{M} \begin{array}{l} \begin{bmatrix} \mathbf{F}_{RX,u,V}(\phi_{n,m}) \\ \mathbf{F}_{RX,u,H}(\phi_{n,m}) \end{bmatrix}^T \begin{bmatrix} \alpha_{n,m,VV} & \alpha_{n,m,VH} \\ \alpha_{n,m,HV} & \alpha_{n,m,HH} \end{bmatrix} \begin{bmatrix} \mathbf{F}_{TX,s,V}(\varphi_{n,m}) \\ \mathbf{F}_{TX,s,H}(\varphi_{n,m}) \end{bmatrix} \times \exp(j2\pi\upsilon_{n,m}t)\delta(\tau-\tau_{n,m}) \\ \times \exp(j2\pi\lambda_0^{-1}(\bar{\phi}_{n,m}\cdot\bar{r}_{RX,u}))\exp(j2\pi\lambda_0^{-1}(\bar{\varphi}_{n,m}\cdot\bar{r}_{TX,s})) \end{array} \tag{11.26}$$

where $\mathbf{F}_{RX,u,V}$ and $\mathbf{F}_{RX,u,H}$ denote the antenna element u field patterns for vertical and horizontal polarizations, respectively, $\alpha_{n,m,VV}$ and $\alpha_{n,m,VH}$ are the complex-valued gains of vertical-to-vertical and horizontal-to-vertical polarizations of ray n, m, respectively, λ_0 represents the wave length of the RF carrier, $\overline{\varphi}_{n.m}$ AoD unit vector, $\overline{\phi}_{n.m}$ is the AoA unit vector, $\overline{r}_{TX,s}$ and $\overline{r}_{rx,u}$ are the location vectors of element s and u, respectively, and $\upsilon_{n,m}$ denotes the Doppler frequency component of ray n, m [1,14,26]. If the radio channel is dynamically modeled, the above-mentioned small-scale parameters are going to be time-variant.

In Figure 11.8, the placement of the MS with respect to each BS is to be determined according to the cell layout. From this placement, the distance between the MS and the BS d and the LoS directions with respect to the BS and MS, θ_{BS} and θ_{MS} can be determined. Note that θ_{BS} and θ_{MS} are defined relative to the broadside directions. The MS antenna array orientations Ω_{MS} are independent and identically distributed and are obtained from a uniform 0−360° distribution.

The generic MIMO channel model is a stochastic model with two levels of unpredictability: large-scale parameters, such as shadow fading, delay, and angular spreads that are drawn randomly from tabulated distribution functions; and small-scale parameters, such as delay, power, and direction of arrival and departure that are selected randomly according to tabulated distribution functions and large-scale parameters. An infinite number of different realizations of the channel model can be generated randomly selecting different initial phases for the scattering objects.

The general approach in the SCM is based on the geometry of a network layout. The large-scale parameters, such as path loss and shadowing factor, are generated according to the geometric positions of the BS and MS. Then the statistical channel behavior is defined using distribution functions of delay and angle, and also by the power delay and angular profiles. Typically, an exponential power delay profile and Laplacian power angular profile are assumed with the function completely defined once the root mean-square (RMS) delay spread and angular spread are specified. The RMS delay and angular spread parameters can be random variables with a mean and standard deviation. The RMS delay and angular spread can be mutually correlated, together with other large-scale parameters, such as shadowing factors. According to the exact profile and distribution functions defined by the particular RMS delay and angular spread values, a finite number of channel taps are randomly generated with a per-tap delay, mean power, mean AoA and AoD, and RMS angular spread. They are defined in a way such that the overall power profile and distribution function are satisfied. Each tap is the contribution of a number of rays (plane waves) arriving at the same time, with each ray having its own amplitude, AoA, and AoD. The number of taps and their delay and angles may be randomly defined, but a reduced-complexity model can specify the delays, mean powers, and angles of the channel taps in a predetermined manner when typical values are often chosen. Similar to the well-known tapped-delay-line version of the wide-sense stationary uncorrelated scattering model, where the power delay profile is fixed, a spatial tapped-delay-line reduced-complexity model additionally defines the spatial information, such as per-tap mean AoA, AoD, and per-tap angular spread, and the power angular profile. Spatial tapped-delay-line models are also referred to as cluster-delay-line models as each tap is modeled as the effect of a cluster of rays arriving at the same time. Each tap suffers from fading in space and over time. The spatial fading process will satisfy a predetermined power angular profile.

The realization of a time-varying spatial channel can be performed in two ways:

- Ray-based: The channel coefficient between each transmit and receive antenna pair is the summation of all rays at each tap and at each time instant, according to the antenna configuration, gain pattern, AoA, AoD of each ray. The temporal channel variation depends on the traveling speed and direction relative to the AoA/AoD of each ray.
- Correlation-based: The antenna correlation for each tap is computed first according the per-tap mean AoA/AoD, per-tap power angular profile, and antenna configuration parameters (e.g., spacing and polarization). The per-tap Doppler spectrum depends on the traveling speed and direction relative to the mean per-tap AoA/AoD, as well as the per-tap power angular profile. The MIMO channel coefficients at each tap can then be generated mathematically by transforming typically the independent and identically distributed Gaussian random variables according to the antenna correlation and the temporal correlation (correspondingly the particular Doppler spectrum).

The generic channel model is based on the drop concept. The system-level simulation is performed in a sequence of "drops," where a "drop" is defined as one simulation run over a certain time period when using the generic model. A drop (or snapshot of a channel segment) is a simulation instance where the random properties of the channel remain constant except for the fast fading caused by the changing phases of the rays. The constant properties during a single drop are the power, delay, and direction of the rays. In a simulation, the number and the length of drops are selected based on the evaluation requirements and the deployment scenario. The generic channel model allows simulation of several statistically independent drops in order to obtain a statistical representation of system performance. The channel model parameters for evaluation of IMT-Advanced technologies in various test environments are shown in Table 11.5. In Table 11.5, DS is the RMS delay spread, ASD is the RMS azimuth spread of departure angles, ASA denotes the RMS azimuth spread of arrival angles, SF represents the shadow fading, and parameter K denotes the Rician K-factor. IMT-Advanced primary module path loss model parameters for a hexagonal cell layout are summarized in Table 11.6.

The break point distance in the above table is defined as $d'_{BP} = 4h'_{BS}h'_{MS}f_c/c$ where f_c is the center frequency (Hz), $c = 3.0 \times 10^8 \ m/s$ is the propagation velocity in free space, h'_{BS} and h'_{MS} are the effective antenna heights at the BS and the MS, respectively. The effective antenna heights h'_{BS} and h'_{MS} are computed as follows: $h'_{BS} = h_{BS} - 1$ and $h'_{MS} = h_{MS} - 1$ where h_{BS} and h_{MS} are the actual antenna heights. The effective height in urban environments is assumed to be equal to 1.0m. In the above table, PL_b denotes basic path loss, PL_{B1} is loss of UMi

Table 11.5 Channel Model Parameters for IMT-Advanced Test Environments [4,14]

Channel Model Parameters		InH		UMi			UMa		RMa	
		LoS	NLoS	LoS	NLoS	Outdoor-to-Indoor	LoS	NLoS	LoS	NLoS
Delay spread (DS) $\log_{10}$(s)	μ	−7.70	−7.41	−7.19	−6.89	−6.62	−7.03	−6.44	−7.49	−7.43
	σ	0.18	0.14	0.40	0.54	0.32	0.66	0.39	0.55	0.48
AoD spread (ASD) $\log_{10}$(degrees)	μ	1.60	1.62	1.20	1.41	1.25	1.15	1.41	0.90	0.95
	σ	0.18	0.25	0.43	0.17	0.42	0.28	0.28	0.38	0.45
AoA spread (ASA) $\log_{10}$(degrees)	μ	1.62	1.77	1.75	1.84	1.76	1.81	1.87	1.52	1.52
	σ	0.22	0.16	0.19	0.15	0.16	0.20	0.11	0.24	0.13
Shadow fading (SF) (dB)	σ	3	4	3	4	7	4	6	4	8
K-factor (*K*) (dB)	μ	7	N/A	9	N/A	N/A	9	N/A	7	N/A
	σ	4	N/A	5	N/A	N/A	3.5	N/A	4	N/A
Cross-correlations	ASD vs. DS	0.6	0.4	0.5	0	0.4	0.4	0.4	0	−0.4
	ASA vs. DS	0.8	0	0.8	0.4	0.4	0.8	0.6	0	0
	ASA vs. SF	−0.5	−0.4	−0.4	−0.4	0	−0.5	0	0	0
	ASD vs. SF	−0.4	0	−0.5	0	0.2	−0.5	−0.6	0	0.6
	DS vs. SF	−0.8	−0.5	−0.4	−0.7	−0.5	−0.4	−0.4	−0.5	−0.5
	ASD vs. ASA	0.4	0	0.4	0	0	0	0.4	0	0
	ASD vs. *K*	0	N/A	−0.2	N/A	N/A	0	N/A	0	N/A
	ASA vs. *K*	0	N/A	−0.3	N/A	N/A	−0.2	N/A	0	N/A
	DS vs. *K*	−0.5	N/A	−0.7	N/A	N/A	−0.4	N/A	0	N/A
	SF vs. *K*	0.5	N/A	0.5	N/A	N/A	0	N/A	0	N/A

(*Continued*)

Table 11.5 (Continued)

Channel Model Parameters		InH		UMi			UMa		RMa	
		LoS	NLoS	LoS	NLoS	Outdoor-to-Indoor	LoS	NLoS	LoS	NLoS
Delay distribution		Exponential	Exponential	Exponential	Exponential	Exponential	Exponential	Exponential	Exponential	Exponential
AoD and AoA distribution		Laplacian		Wrapped Gaussian			Wrapped Gaussian		Wrapped Gaussian	
Delay scaling parameter r_τ		3.6	3	3.2	3	2.2	2.5	2.3	3.8	1.7
Cross-polarization ratio (dB)	μ	11	10	9	8.0	9	8	7	12	7
Number of clusters		15	19	12	19	12	12	20	11	10
Number of rays per cluster		20	20	20	20	20	20	20	20	20
Cluster ASD		5	5	3	10	5	5	2	2	2
Cluster ASA		8	11	17	22	8	11	15	3	3
Per cluster shadowing standard deviation ζ (dB)		6	3	3	3	4	3	3	3	3
Correlation distance (m)	DS	8	5	7	10	10	30	40	50	36
	ASD	7	3	8	10	11	18	50	25	30
	ASA	5	3	8	9	17	15	50	35	40
	SF	10	6	10	13	7	37	50	37	120
	K	4	N/A	15	N/A	N/A	12	N/A	40	N/A

Table 11.6 IMT-Advanced Primary Module Path Loss Models (f_c Is in GHz, All Distances Are in Meters, and a Hexagonal Cell Layout Is Assumed) [14]

Scenario		Path Loss PL (dB)	Shadow Fading Standard Deviation (dB)	Applicability Range (m)/Antenna Height (m)
InH	LoS	$16.9\ \log_{10}(d) + 32.8 + 20\ \log_{10}(f_c)$	$\sigma = 3$	$3 < d < 100$ $h_{BS} = 3 - 6$ $h_{MS} = 1 - 2.5$
	NLoS	$43.3\ \log_{10}(d) + 11.5 + 20\ \log_{10}(f_c)$	$\sigma = 4$	$10 < d < 150$ $h_{BS} = 3 - 6$ $h_{MS} = 1 - 2.5$
UMi	LoS	$22.0\ \log_{10}(d) + 28.0 + 20\ \log_{10}(f_c)$ $40\ \log_{10}(d_1) + 7.8 - 18\ \log_{10}(h'_{BS}) -$ $18\ \log_{10}(h'_{MS}) + 2\ \log_{10}(f_c)$	$\sigma = 3$	$10 < d_1 < d'_{BP}$ $d'_{BP} < d_1 < 5000$ $h_{BS} = 10; h_{MS} = 1.5$
	NLoS	$36.7\ \log_{10}(d) + 22.7 + 26\ \log_{10}(f_c)$	$\sigma = 4$	$10 < d < 2000$ $h_{BS} = 3$ $h_{MS} = 1 - 2.5$
	Outdoor-to-indoor	$PL_b + PL_{tw} + PL_{in}$ $PL_b = PL_{B1}(d_{out} + d_{in});\quad PL_{tw} = 20;\quad PL_{in} = 0.5 d_{in}$	$\sigma = 7$	$10 < d_{out} + d_{in} < 1000$ $0 < d_{in} < 25$
UMa	LoS	$22.0\ \log_{10}(d) + 28.0 + 20\ \log_{10}(f_c)$ $40.0\ \log_{10}(d_1) + 7.8 - 18.0\ \log_{10}(h'_{BS})$ $- 18.0\ \log_{10}(h'_{MS}) + 2.0\ \log_{10}(f_c)$	$\sigma = 4$	$10 < d_1 < d'_{BP}$ $d'_{BP} < d_1 < 5000$ $h_{BS} = 25; h_{MS} = 1.5$
	NLoS	$161.04 - 7.1\ \log_{10}(W) + 7.5\ \log_{10}(h)$ $- (24.37 - 3.7(h/h_{BS})^2)\log_{10}(h_{BS})$ $+ (43.42 - 3.1\ \log_{10}(h_{BS}))(\log_{10}(d) - 3)$ $+ 20\ \log_{10}(f_c) - (3.2(\log_{10}(11.75 h_{MS}))^2 - 4.97$	$\sigma = 6$	$10 < d < 5000\ h_{BS} = 25;\ h_{MS} = 1.5;$ $W = 20; h = 205 < h < 50;$ $5 < W < 50\ 10 < h_{BS} < 150;$ $1 < h_{MS} < 10$

(Continued)

Table 11.6 (Continued)

Scenario		Path Loss PL (dB)	Shadow Fading Standard Deviation (dB)	Applicability Range (m)/Antenna Height (m)
RMa	LoS	$PL_1 = 20\ \log_{10}(40\pi d f_c/3) + \min(0.03h^{1.72}, 10)$	$\sigma = 4$	$10 < d < d_{BP}$ $d_{BP} < d < 10000$
		$\log_{10}(d) - \min(0.044h^{1.72}, 14.77) + 0.002\log_{10}(h)d$	$\sigma = 6$	$h_{BS} = 35; h_{MS} = 1.5\ W = 20; h = 5$
		$PL_2 = PL_1(d_{BP}) + 40\ \log_{10}(d/d_{BP})$		
	NLoS	$161.04 - 7.1\ \log_{10}(W) + 7.5\ \log_{10}(h)$ $-(24.37 - 3.7(h/h_{BS})^2)\log_{10}(h_{BS})$ $+ (43.42 - 3.1\ \log_{10}(h_{BS}))(\log_{10}(d) - 3)$ $+ 20\ \log_{10}(f_c) - (3.2\ (\log_{10}(11.75\ h_{MS}))^2 - 4.97)$	$\sigma = 8$	$10 < d < 5000$ $h_{BS} = 35;$ $h_{MS} = 1.5W = 20; h = 5$

outdoor scenario, PL_{tw} is penetration loss through the wall, PL_{in} denotes the loss inside, d_{out} is the distance from the BS to the wall next to the MS location, d_{in} denotes the perpendicular distance from the wall to the MS which is evenly distributed between 0 and 25 m, and the break point distance is defined as $d_{BP} = 2\pi h_{BS} h_{MS} f_c / c$.

Spatial correlation is often said to degrade the performance of multi-antenna systems and impose a limit on the number of antennas that can effectively fit in a small mobile device. This seems intuitive as the spatial correlation decreases the number of independent channels that can be created by precoding. When modeling spatial correlation, it is useful to employ the Kronecker model, where the correlation between transmit and receive antennas are assumed independent and separable. This model is reasonable when the main scattering appears close to the antenna arrays and has been validated by outdoor and indoor measurements. Assuming Rayleigh fading, the Kronecker model means that the channel matrix can be represented as $\mathbf{H} = \mathbf{R}_{RX}^{1/2} \mathbf{H}_{channel} (\mathbf{R}_{TX}^{1/2})^T$ where the elements of $\mathbf{H}_{channel}$ are independent and identically distributed as circular symmetric complex Gaussian with zero-mean and unit variance. The important part of the model is that $\mathbf{H}_{channel}$ is pre-multiplied by the receive-side spatial correlation matrix $\mathbf{R}_{RX}$ and post-multiplied by the transmit-side spatial correlation matrix $\mathbf{R}_{TX}$. Equivalently, the channel matrix can be expressed as $\mathbf{H} \sim Z(0, \mathbf{R}_{TX} \otimes \mathbf{R}_{RX})$ where $\otimes$ denotes the Kronecker product.

Using the Kronecker model, the spatial correlation depends directly on the eigenvalue distributions of the correlation matrices $\mathbf{R}_{RX}$ and $\mathbf{R}_{TX}$. Each eigenvector represents a spatial direction of the channel and its corresponding eigenvalue describes the average channel/signal gain in this direction. For the transmit-side, the correlation matrix $\mathbf{R}_{TX}$ describes the average gain in a spatial transmit direction, while the receive-side correlation matrix $\mathbf{R}_{RX}$ describes a spatial receive direction. High spatial correlation is represented by a large eigenvalue spread in $\mathbf{R}_{TX}$ or $\mathbf{R}_{RX}$, implying that some spatial directions are statistically stronger than the others. On the other hand, low spatial correlation is represented by a small eigenvalue spread in $\mathbf{R}_{TX}$ or $\mathbf{R}_{RX}$, which implies that almost the same signal gain can be expected from all spatial directions [57].

Let us now focus on a specific case of downlink transmission from a base station to a mobile station. Denoting the nth snapshot of the spatial correlation matrices at the BS and the MS by $\mathbf{R}_{BS,n}$ and $\mathbf{R}_{MS,n}$, the per-tap spatial correlation is determined as the Kronecker product[7] of the BS and MS's antenna correlation matrices:

[7]Given an $m \times n$ matrix $\mathbf{A}$ and a $p \times q$ matrix $\mathbf{B}$, the Kronecker product $\mathbf{C} = \mathbf{A} \otimes \mathbf{B}$, also called the matrix direct product, is an $mp \times nq$ matrix with elements defined by $c_{\alpha\beta} = a_{ij} b_{kl}$, where $\alpha \equiv p(i-1)+k$ and $\beta \equiv q(j-1)+l$. The matrix direct product provides the matrix of the linear transformation induced by the vector space tensor product of the original vector spaces. More precisely, suppose that operators $S: V_1 \to W_1$ and $T: V_2 \to W_2$ are given by $S(x) = Ax$ and $T(y) = By$. Then $S \otimes T: V_1 \otimes V_2 \to W_1 \otimes W_2$ is determined by $S \otimes T(x \otimes y) = (Ax) \otimes (By) = (A \otimes B)(x \otimes y)$[57].

$$\mathbf{R}_n = \mathbf{R}_{\mathrm{BS},n} \otimes \mathbf{R}_{\mathrm{MS},n} \tag{11.27}$$

We denote the number of receive antennas by N_{RX} and the number of transmit antennas by the N_{TX}. If cross-polarized antennas are present at the receiver, it is assumed that $N_{\mathrm{RX}}/2$ receive antennas have the same polarization, while the remaining $N_{\mathrm{RX}}/2$ receive antennas have orthogonal polarization. Likewise, if cross-polarized antennas are present at the transmitter, it is assumed that the $N_{\mathrm{TX}}/2$ transmit antennas have the same polarization; while the remaining $N_{\mathrm{TX}}/2$ transmit antennas have orthogonal polarization. It is further assumed that the antenna arrays are composed of pairs of collocated antennas with orthogonal polarization. With these assumptions, the per-tap channel correlation is determined as follows:

$$\mathbf{R}_n = \mathbf{R}_{\mathrm{BS},n} \otimes \mathbf{\Gamma} \otimes \mathbf{R}_{\mathrm{MS},n} \tag{11.28}$$

Where $\mathbf{R}_{\mathrm{MS},n}$ is an $N_{\mathrm{RX}} \times N_{\mathrm{RX}}$ matrix if all receive antennas have the same polarization, or an $N_{\mathrm{RX}}/2 \times N_{\mathrm{RX}}/2$ matrix if the receive antennas are cross-polarized. Likewise, $\mathbf{R}_{\mathrm{BS},n}$ is an $N_{\mathrm{TX}} \times N_{\mathrm{TX}}$ matrix if all the transmit antennas have the same polarization, or an $N_{\mathrm{TX}}/2 \times N_{\mathrm{TX}}/2$ matrix if the transmit antennas are cross-polarized. Matrix $\mathbf{\Gamma}$ is a cross-polarization matrix based on the cross-polarization defined in the cluster-delay-line models. Matrix $\mathbf{\Gamma}$ is a 2×2 matrix if cross-polarized antennas are used at the transmitter or at the receiver. It is a 4×4 matrix if cross-polarized antennas are used at both the transmitter and the receiver.

Table 11.7 defines the correlation matrices for the eNB and UE in E-UTRA with a different number of transmit/receive antennas at each entity [4].

For the scenarios with more antennas at either eNB, UE, or both, the channel spatial correlation matrix can be expressed as the Kronecker product of $\mathbf{R}_{\mathrm{eNB}}$ and $\mathbf{R}_{\mathrm{UE}}$ according to $\mathbf{R}_{\mathrm{spatial}} = \mathbf{R}_{\mathrm{eNB}} \otimes \mathbf{R}_{\mathrm{UE}}$. The parameters α and β for different antenna correlation types are given in Table 11.8.

Table 11.7 eNB/UE Correlation Matrix [4,58]

Entity	Number of Antennas		
	1	2	4
eNB	$R_{\mathrm{eNB}} = 1$	$R_{\mathrm{eNB}} = \begin{pmatrix} 1 & \alpha \\ \alpha^* & 1 \end{pmatrix}$	$R_{\mathrm{eNB}} = \begin{pmatrix} 1 & \alpha^{1/9} & \alpha^{4/9} & \alpha \\ \alpha^{1/9^*} & 1 & \alpha^{1/9} & \alpha^{4/9} \\ \alpha^{4/9^*} & \alpha^{1/9^*} & 1 & \alpha^{1/9} \\ \alpha^* & \alpha^{4/9^*} & \alpha^{1/9^*} & 1 \end{pmatrix}$
UE	$R_{\mathrm{UE}} = 1$	$R_{\mathrm{UE}} = \begin{pmatrix} 1 & \beta \\ \beta^* & 1 \end{pmatrix}$	$R_{\mathrm{UE}} = \begin{pmatrix} 1 & \beta^{1/9} & \beta^{4/9} & \beta \\ \beta^{1/9^*} & 1 & \beta^{1/9} & \beta^{4/9} \\ \beta^{4/9^*} & \beta^{1/9^*} & 1 & \beta^{1/9} \\ \beta^* & \beta^{4/9^*} & \beta^{1/9^*} & 1 \end{pmatrix}$

Table 11.8 Values of α and β for Different Antenna Correlations [4]

Low Correlation		Medium Correlation		High Correlation	
α	β	α	β	α	β
0	0	0.3	0.9	0.9	0.9

Table 11.9 EPA/EVA/ETA Multipath Power Delay Profiles [4,58]

Tap Excess Delay (ns)	Relative Power (dB)	Tap Excess Delay (ns)	Relative Power (dB)	Tap Excess Delay (ns)	Relative Power (dB)
EPA		EVA		ETU	
0	0.0	0	0.0	0	−1.0
30	−1.0	30	−1.5	50	−1.0
70	−2.0	150	−1.4	120	−1.0
90	−3.0	310	−3.6	200	0.0
110	−8.0	370	−0.6	230	0.0
190	−17.2	710	−9.1	500	0.0
410	−20.8	1090	−7.0	1600	−3.0
–	–	1730	−12.0	2300	−5.0
–	–	2510	−16.9	5000	−7.0

Table 11.10 Delay Profiles of LTE Channel Models [4,58]

Model	Number of Channel Taps	RMS Delay Spread (ns)	Maximum Tap Excess Delay (ns)
EPA	7	45	410
EVA	9	357	2510
ETU	9	991	5000

The multipath fading channel model specifies three different delay profiles which are representative of low, medium, and high delay spread environments, that is, extended pedestrian A (EPA) model, extended vehicular A (EVA) model, and extended typical urban (ETU) model [52]. The multipath delay profiles for these channels are shown in Table 11.9 and the associated number of channel taps, RMS delay spread, and maximum tap excess delays are provided in Table 11.10. In addition to a multipath delay profile, a maximum Doppler frequency is specified for each multipath fading propagation condition as follows: EPA 5 Hz ($\nu_{max} = 5$ Hz), EVA 70 Hz ($\nu_{max} = 70$ Hz), and ETU 300 Hz ($\nu_{max} = 300$ Hz).

11.9 LTE-Advanced link-level and system-level performance

The 3GPP LTE-Advanced was a candidate for IMT-Advanced whose performance was fully characterized based on the ITU-R defined evaluation methodology. The following sections summarize the performance characterization of the 3GPP LTE-Advanced as part of the IMT-Advanced submission.

11.9.1 Cell spectral efficiency and cell-edge spectral efficiency

Cell spectral and cell-edge spectral efficiencies were evaluated through extensive simulations conducted by a number of 3GPP member companies. The simulation results were reported based on specific 3GPP LTE-Advanced configurations, that is, downlink and uplink MU-MIMO, downlink and uplink CoMP, and uplink SU-MIMO, for both FDD and TDD duplex schemes in various test scenarios. It must be noted that during LTE Rel-10 development and prior to IMT-Advanced submission, CoMP was an incomplete study item in 3GPP and no standard technique was specified for CoMP at that time; therefore, CoMP simulation results presented in this chapter, which were provided by a number of 3GPP member companies, did not use standard techniques and may not agree with each other in various conditions. Nevertheless, the CoMP study item was later completed and CoMP was specified as a feature in LTE Rel-11 specifications that will be discussed in the next chapter [11].

The performance differences among contributing sources can be explained by implementation-specific functionalities at the transmitter and the receiver that are not explicitly specified by the standard, such as receiver type, scheduling algorithms, and error modeling. In the results reported for the downlink, the size of the control channel (L OFDM symbols) and the number of the MBSFN subframes are the factors that affect the overhead. Note that in 3GPP submission, the control channel overhead is statically modeled, that is, the downlink control channels occupy a fixed number of OFDM symbols irrespective of the number of users, although it is expected that the size of the physical downlink control channel dynamically varies with the number of active users in the cell. In the MBSFN subframes and in the MBSFN region, there are no cell-specific reference signals (CRS), which effectively reduces the overhead.

Channel estimation and receiver types are the factors that affect the demodulation performance. Availability of CSI at the eNB is the assumption that impacts the transmit signal processing at the eNB for MU-MIMO and CoMP schemes. In the tables for the uplink, physical uplink control channel bandwidth is the factor that affects the overhead. Each value in Tables 11.11 and 11.12 is obtained as an average of all samples provided by different companies. The results suggest that the requirements are fulfilled with $L = 3$ corresponding to the largest overhead except for the system bandwidth of 1.4 MHz where the requirements cannot be met. If the control overhead

Table 11.11 LTE-Advanced Downlink Cell Spectral and Cell-Edge Spectral Efficiencies [2,6]

Scheme and Antenna Configuration	Test Environment	Duplex Scheme	ITU-R Requirement Cell Spectral Efficiency/Cell-Edge Spectral Efficiency	Cell Spectral Efficiency (bits/s/Hz/cell)			Cell-Edge Spectral Efficiency (bits/s/Hz)		
				$L=1$	$L=2$	$L=3$	$L=1$	$L=2$	$L=3$
Rel-8 SU-MIMO 4 × 2(A)	InH	FDD	3/0.1	4.8	4.5	4.1	0.23	0.21	0.19
MU-MIMO 4 × 2 (C)			3/0.1	6.6	6.1	5.5	0.26	0.24	0.22
Rel-8 SU-MIMO 4 × 2 (A)		TDD	3/0.1	4.7	4.4	4.1	0.22	0.20	0.19
MU-MIMO 4 × 2 (C)			3/0.1	6.7	6.1	5.6	0.24	0.22	0.20
MU-MIMO 4 × 2 (C)	UMi	FDD	2.6/0.075	3.5	3.2	2.9	0.10	0.096	0.087
MU-MIMO 4 × 2 (A)			2.6/0.075	3.4	3.1	2.8	0.12	0.11	0.099
CS/CB-CoMP 4 × 2 (C)			2.6/0.075	3.6	3.3	3.0	0.11	0.099	0.089
JP-CoMP 4 × 2 (C)			2.6/0.075	4.5	4.1	3.7	0.14	0.13	0.12
MU-MIMO 8 × 2 (C/E)			2.6/0.075	4.2	3.8	3.5	0.15	0.14	0.13
MU-MIMO 4 × 2 (C)		TDD	2.6/0.075	3.5	3.2	3.0	0.11	0.096	0.089
MU-MIMO 4 × 2 (A)			2.6/0.075	3.2	2.9	2.7	0.11	0.10	0.095

(*Continued*)

Table 11.11 (Continued)

Scheme and Antenna Configuration	Test Environment	Duplex Scheme	ITU-R Requirement Cell Spectral Efficiency/Cell-Edge Spectral Efficiency	Cell Spectral Efficiency (bits/s/Hz/cell)			Cell-Edge Spectral Efficiency (bits/s/Hz)		
				$L=1$	$L=2$	$L=3$	$L=1$	$L=2$	$L=3$
CS/CB-CoMP 4 × 2 (C)			2.6/0.075	3.6	3.3	3.1	0.10	0.092	0.086
JP-CoMP 4 × 2 (C)			2.6/0.075	4.6	4.2	3.9	0.10	0.092	0.085
MU-MIMO 8 × 2 (C/E)			2.6/0.075	4.2	3.9	3.6	0.12	0.11	0.099
MU-MIMO 4 × 2 (C)	UMa	FDD	2.2/0.06	2.8	2.6	2.4	0.079	0.073	0.066
CS/CB-CoMP 4 × 2 (C)			2.2/0.06	2.9	2.6	2.4	0.081	0.074	0.067
JP-CoMP 4 × 2 (A)			2.2/0.06	3.0	2.7	2.5	0.080	0.073	0.066
CS/CB-CoMP 8 × 2 (C)			2.2/0.06	3.8	3.5	3.2	0.10	0.093	0.084
MU-MIMO 4 × 2 (C)		TDD	2.2/0.06	2.9	2.6	2.4	0.079	0.071	0.067
CS/CB-CoMP 4 × 2 (C)			2.2/0.06	2.9	2.6	2.4	0.083	0.075	0.070
JP-CoMP 4 × 2 (C)			2.2/0.06	3.6	3.3	3.1	0.090	0.082	0.076
CS/CB-CoMP 8 × 2 (C/E)			2.2/0.06	3.7	3.3	3.1	0.10	0.093	0.087

Rel-8 SU-MIMO 4 × 2 (C)	RMa	FDD	1.1/0.04	2.3	2.1	1.9	0.081	0.076	0.069
Rel-8 SU-MIMO 4 × 2 (A)			1.1/0.04	2.1	2.0	1.8	0.067	0.063	0.057
MU-MIMO 4 × 2 (C)			1.1/0.04	3.9	3.5	3.2	0.11	0.099	0.090
MU-MIMO 8 × 2 (C)			1.1/0.04	4.1	3.7	3.4	0.13	0.12	0.11
Rel-8 SU-MIMO 4 × 2 (C)		TDD	1.1/0.04	2.0	1.9	1.8	0.072	0.067	0.063
Rel-8 SU-MIMO 4 × 2 (A)			1.1/0.04	1.9	1.7	1.6	0.057	0.053	0.049
MU-MIMO 4 × 2 (C)			1.1/0.04	3.5	3.2	3.0	0.098	0.089	0.083
MU-MIMO 8 × 2 (C/E)			1.1/0.04	4.0	3.6	3.4	0.12	0.11	0.10
Rel-8 Single-Layer BF 8 × 2 (E)			1.1/0.04	2.5	2.3	2.1	0.11	0.10	0.093

Table 11.12 3GPP LTE-Advanced Uplink Cell Spectral and Cell-Edge Spectral Efficiencies [2,6]

Scheme and Antenna Configuration	Test Environment	Duplex Scheme	ITU-R Requirement Cell Spectral Efficiency/Cell-Edge Spectral Efficiency	Cell Spectral Efficiency (bits/s/Hz/cell)	Cell-edge Spectral Efficiency (bits/s/Hz)
Rel-8 SIMO 1 × 4 (A)	InH	FDD	2.25/0.07	3.3	0.23
Rel-8 SIMO 1 × 4 (C)			2.25/0.07	3.3	0.24
Rel-8 MU-MIMO 1 × 4 (A)			2.25/0.07	5.8	0.42
SU-MIMO 2 × 4 (A)			2.25/0.07	4.3	0.25
Rel-8 SIMO 1 × 4 (A)		TDD	2.25/0.07	3.1	0.22
Rel-8 SIMO 1 × 4 (C)			2.25/0.07	3.1	0.23
Rel-8 MU-MIMO 1 × 4 (A)			2.25/0.07	5.5	0.39
SU-MIMO 2 × 4 (A)			2.25/0.07	3.9	0.25
Rel-8 SIMO 1 × 4 (C)	UMi	FDD	1.8/0.05	1.9	0.073
Rel-8 MU-MIMO 1 × 4 (A)			1.8/0.05	2.5	0.077
MU-MIMO 2 × 4 (A)			1.8/0.05	2.5	0.086
Rel-8 SIMO 1 × 4 (C)		TDD	1.8/0.05	1.9	0.070
Rel-8 MU-MIMO 1 × 4 (A)			1.8/0.05	2.3	0.071
MU-MIMO 2 × 4 (A)			1.8/0.05	2.8	0.068
MU-MIMO 1 × 8 (E)			1.8/0.05	3.0	0.079

Rel-8 SIMO 1 × 4 (C)	UMa	FDD	1.4/0.03	1.5	0.062
CoMP 1 × 4 (A)			1.4/0.03	1.7	0.086
CoMP 2 × 4 (C)			1.4/0.03	2.1	0.099
Rel-8 SIMO 1 × 4 (C)		TDD	1.4/0.03	1.5	0.062
CoMP 1 × 4 (C)			1.4/0.03	1.9	0.090
CoMP 2 × 4 (C)			1.4/0.03	2.0	0.097
MU-MIMO 1 × 8 (E)			1.4/0.03	2.7	0.076
Rel-8 SIMO 1 × 4 (C)	RMa	FDD	0.7/0.015	1.8	0.082
Rel-8 MU-MIMO 1 × 4 (A)			0.7/0.015	2.2	0.097
CoMP 2 × 4 (A)			0.7/0.015	2.3	0.13
Rel-8 SIMO 1 × 4 (C)		TDD	0.7/0.015	1.8	0.080
Rel-8 MU-MIMO 1 × 4 (A)			0.7/0.015	2.1	0.093
CoMP 2 × 4 (A)			0.7/0.015	2.5	0.15
MU-MIMO 1 × 8 (E)			0.7/0.015	2.6	0.10

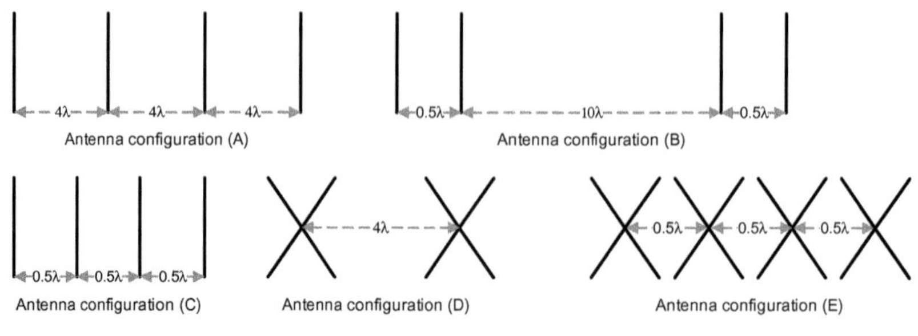

FIGURE 11.9

Illustration of antenna configurations used in LTE evaluations.

assumption is relaxed, that is the number of active users decreases, $L = 1, 2$ can be considered and thereby the performance can be further improved. The tables show that LTE Rel-8 with SU-MIMO 4×2 configuration can already fulfill the ITU-R requirements. The tables also show that further performance improvements can be achieved using additional technical features.

In the following tables, various antenna configurations have been utilized. There are four or eight transmit antennas with the following configurations: (A) uncorrelated co-polarized, that is, co-polarized antennas are four-wavelength apart; (B) grouped co-polarized, that is, two groups of co-polarized antennas where there is a 10-wavelength distance between the center of each group and 0.5-wavelength separation between antennas within each group; (C) correlated and co-polarized, that is, 0.5-wavelength between antennas; (D) uncorrelated and cross-polarized, that is, columns with $\pm 45°$ linearly polarized antennas and separated by four wavelengths; and (E) correlated and cross-polarized, that is, columns with $\pm 45°$ linearly polarized antennas and separated by 0.5-wavelengths. These configurations are illustrated in Figure 11.9.

The downlink and uplink cell spectral and cell-edge spectral efficiencies in various test environments for the TDD and FDD duplex schemes are shown in Tables 11.11 and 11.12, respectively. In these tables, the acronyms "JP-CoMP," "CS-CoMP," and "CB-CoMP" refer to joint processing, coordinated scheduling, and coordinated beamforming multi-point transmission schemes, respectively. The joint processing refers to a CoMP scheme where data is available at each of the geographically separated transmission points and physical downlink shared channel (PDSCH) transmission occurs from multiple points, whereas coordinated scheduling/beamforming is an alternative CoMP technique where data is only available at serving cell and data transmission is from that point, but user scheduling/beamforming decisions are made with coordination among different cells.

11.9.2 VoIP capacity

While most of the modern packet-switched air-interfaces do not specify voice codecs, calculation of the VoIP capacity requires explicit assumptions about a particular voice codec and its parameters, such as the frame rate, real-time transport protocol (RTP) payload size, and discontinuous transmission (DTX) capability. The VoIP capacity of LTE-Advanced was calculated assuming use of AMR 12.2 kbps codec with a 50% speech activity factor. It was further assumed that the percentage of users in outage is less than 2% where a user is defined to have experienced outage if less than 98% of the VoIP packets have been delivered successfully to the user within a one-way radio access delay bound of 50 ms. The assumptions and the configuration parameters used in the system-level simulations for VoIP are as described in Table 11.13. The number of active VoIP users was calculated through extensive system-level simulations that were conducted by a number of 3GPP member companies. Table 11.14 shows the VoIP capacity results in the indoor, micro-cellular, macro-cellular, and high-speed test environments for FDD and TDD duplex schemes. The results suggest that LTE-Advanced satisfies the ITU-R requirements in various deployment scenarios. The antenna configurations are illustrated in Figure 11.9.

11.9.3 Mobility

IMT-Advanced specified four mobility classes based on UE speeds as follows: Stationary (0 km/h); Pedestrian (0 to 10 km/h); Vehicular (10 to 120 km/h); and High-speed vehicular (120 to 350 km/h). In order to evaluate LTE-Advanced against mobility requirement, the methodology that is set forth in Report ITU-R M.2135-1 [14] is followed, where in the first step the system-level simulation is performed for each test environment in the uplink to obtain the uplink SINR distribution (see Figure 11.10). From the uplink SINR distribution the 50% point of CDF is obtained for each test environment. In the second step, link-level simulation is conducted to generate the spectral efficiency versus SINR curves for each test environment. The spectral efficiency values include the effect of control channel overhead from the uplink system-level simulations for each test scenario. The mobility requirement in each mobility class is satisfied, if the uplink traffic channel

Table 11.13 Assumptions for VoIP Capacity Calculation [14,26]

Parameter	Value
Codec	AMR12.2 kbps mode
Encoder frame length	20 ms
VAF	50%
Payload	Active: 33 bytes (octet-aligned mode)
	Inactive: 7 bytes
	SID packet is sent every 160 ms during silence intervals
Protocol overhead with compressed header	10 bits + padding (RTP-pre-header)
	4 bytes (RTP/UDP/IP) 2 bytes (radio link control (RLC)/ security) 16 bits (CRC)

Table 11.14 LTE-Advanced VoIP Capacity [6]

Antenna Configuration	Duplex Scheme	Test Environment	ITU-R Requirement	VoIP Capacity (Active Users/MHz/Cell)
Antenna configuration (A)	FDD	InH	50	140
		UMi	40	80
		UMa	40	68
		RMa	30	91
Antenna configuration (C)		InH	50	131
		UMi	40	75
		UMa	40	69
		RMa	30	94
Antenna configuration (A)	TDD	InH	50	137
		UMi	40	74
		UMa	40	65
		RMa	30	86
Antenna configuration (C)		InH	50	130
		UMi	40	74
		UMa	40	67
		RMa	30	92

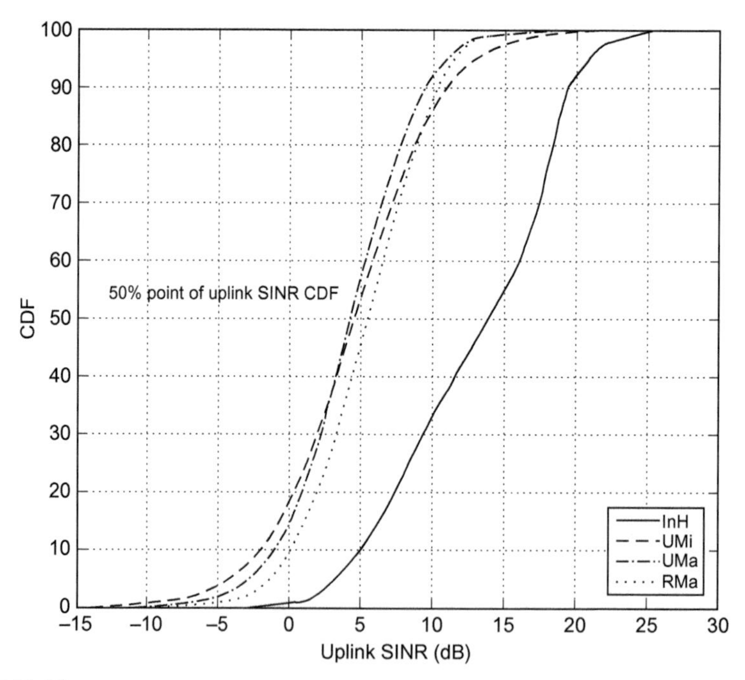

FIGURE 11.10

Statistical distribution of the uplink SINR in various test environments [6].

link-level data rate, normalized by bandwidth, is greater than the minimum requirements shown in Table 11.15, when the user is moving at the maximum speed in that mobility class in each of the test environments [14].

It can be concluded from Table 11.15 that the evaluated LTE-Advanced configurations can fulfill the ITU-R requirements concerning the mobility in all test environments.

11.9.4 Peak spectral efficiency

The peak spectral efficiency of LTE-Advanced is calculated based on the following assumptions: 20 MHz bandwidth; one OFDM symbol for downlink control signaling; CRS corresponding to one and four antenna ports; UE-specific reference signals corresponding to 24 and zero resource elements per resource-block pair. It was further assumed that physical broadcast channel and synchronization sequences occupy a total of 564 (Rel-10 MU-MIMO/CoMP with six MBSFN subframes per 10 ms) and 528 (Rel-8 downlink SU-MIMO) resource elements per radio frame for downlink eight- and four-layer spatial multiplexing, respectively. In addition, DL/UL configuration 1 (2 DL subframes, 1 special subframe, 2 uplink subframes) and special-subframe configuration 4 (12 DwPTS, 1 GP, 1 UpPTS, UpPTS is used for sounding reference signal transmission) are assumed for the TDD mode. The results shown in Table 11.16 suggest that LTE Rel-8 already fulfills the ITU-R requirements for downlink peak spectral efficiency. The table also shows the additional performance improvement that can be achieved using advanced features, for example, downlink eight-layer spatial multiplexing.

The uplink peak spectral efficiency is calculated assuming 20 MHz bandwidth, PUCCH occupying two pairs of resource blocks per subframe, and physical random access channel consuming six resource-block pairs per radio frame. The same TDDDL/UL and special-subframe configurations are assumed as for the downlink peak spectral efficiency calculation. Table 11.16 shows that the extension of LTE Rel-8 with two-layer spatial multiplexing can fulfill the ITU-R requirement in the uplink.

11.9.5 User-plane/control-plane latency and handover interruption time

The LTE user-plane one-way access latency for a scheduled UE consists of fixed node-processing delays, which include radio frame alignment and 1 ms transmission time interval (TTI) duration. Using the latency calculation model shown in Figure 11.11, and assuming that the number of HARQ processes is 8 for the FDD mode, the one-way latency is given as $T_{\text{USER-PLANE}} = 4 + 8p$ where p is the probability of HARQ retransmissions or the error probability of the first HARQ retransmission. While the minimum latency of $T_{\text{USER-PLANE}} = 4$ ms is achieved for $p = 0$, a more realistic value of $T_{\text{USER-PLANE}} = 4.8$ ms is obtained for $p = 0.1$.

Table 11.15 3GPP LTE Uplink Link-Level Simulation Results for the Mobility Requirements [2,6,14]

Path Loss Model (LoS/NLoS)	Duplex Scheme	Test Environment	ITU-R Requirements	Median SINR (dB)	Uplink Spectral Efficiency (bits/s/Hz)
Antenna configuration 1 × 4 (NLoS)	FDD	InH	1.0	13.89	2.56
		UMi	0.75	4.54	1.21
		UMa	0.55	4.30	1.08
		RMa	0.25	5.42	1.22
Antenna configuration 1 × 4 (LoS)		InH	1.0	13.89	3.15
		UMi	0.75	4.54	1.42
		UMa	0.55	4.30	1.36
		RMa	0.25	5.42	1.45
Antenna configuration 1 × 4 (NLoS)	TDD	InH	1.0	13.89	2.63
		UMi	0.75	4.54	1.14
		UMa	0.55	4.30	0.95
		RMa	0.25	5.42	1.03
Antenna configuration 1 × 4 (LoS)		InH	1.0	13.89	3.11
		UMi	0.75	4.54	1.48
		UMa	0.55	4.30	1.36
		RMa	0.25	5.42	1.38

Table 11.16 3GPP LTE-Advanced Peak Spectral Efficiency [2]

Scheme	Duplex Scheme	Direction	Spectral Efficiency (bits/s/Hz)
ITU-R requirement	FDD	Downlink	15
Rel-8 four-layer spatial multiplexing			16.3
Eight-layer spatial multiplexing			30.6
ITU-R requirement		Uplink	6.75
Two-Layer spatial multiplexing			8.4
Four-Layer spatial multiplexing			16.8
ITU-R requirement	TDD	Downlink	15
Rel-8 four-layer spatial multiplexing			16.0
Eight-Layer spatial multiplexing			30.0
ITU-R requirement		Uplink	6.75
Two-Layer spatial multiplexing			8.1
Four-Layer spatial multiplexing			16.1

The user-plane one-way latency for a scheduled UE in the TDD mode consists of fixed node-processing delays, radio frame alignment, and TTI duration. The latency component can be seen in Figure 11.11 for the downlink and uplink. Using the latency model shown in Figure 11.11, the total one-way processing time is 2.5 ms, T_{FA} is the radio frame alignment which depends on various configurations of the TDD frame structure, and the TTI duration is 1 ms, hence, the user-plane latency of the TDD mode can be written as $T_{USER\text{-}PLANE} = 3.5 + T_{FA} + pT_{RTT}$ where T_{RTT} is the average HARQ round-trip time and p is the error probability of the first HARQ transmission.

Table 11.17 shows the user-plane latency component breakdown in downlink and uplink for different TDD DL/UL configurations when $p = 0.1$ is assumed. It is shown that in all cases, LTE can meet the ITU-R requirements for user-plane latency.

The above analysis further shows that the 5 ms user-plane latency requirement can be satisfied in the TDD mode in uplink and downlink using the DL/UL configuration 6 only when $p = 0$ is assumed.

In order to calculate the control-plane latency, the model shown in Figure 11.12 is used. The transition from the RRC_IDLE to the RRC_CONNECTED state in LTE comprises several steps that are shown in sequential order in Table 11.18.

While the results of the analysis suggest that LTE Rel-8 satisfies the ITU-R requirements, using the improved features and protocols introduced in later releases would allow reduction of the control-plane latency to 50 ms. Note that since the non-access stratum (NAS) setup process can be executed in parallel with the RRC connection setup, it does not appear in the total latency calculation.

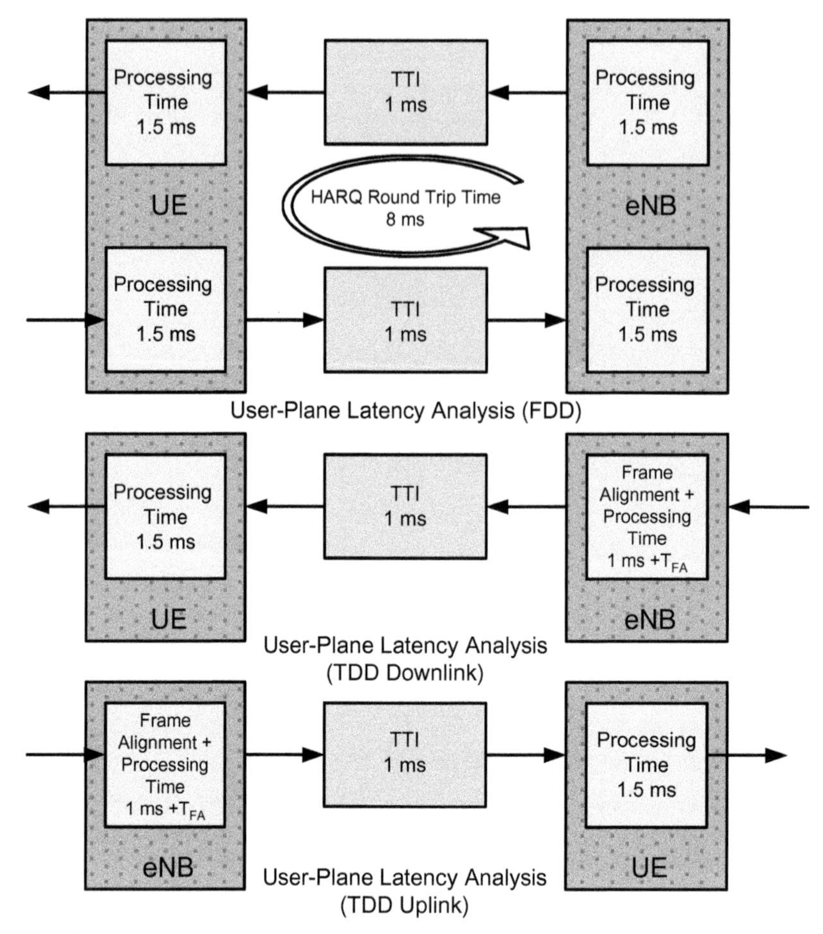

FIGURE 11.11

LTE user-plane latency calculation model [2].

The analysis shown in Table 11.18 shows that any of the LTE configurations can fulfill the ITU-R requirements related to control-plane latency.

The calculation of the user-plane latency was carried out for synchronized (or prescheduled) UEs. In the case where the UE is not prescheduled or when the UE is required to synchronize in order to be allocated uplink resources, reduced RACH scheduling period, shorter PUCCH cycle, and reduced processing delays can be applied to satisfy the control-plane latency requirements.

The LTE-Advanced general handover procedure is based on that of LTE Rel-8, which was described in Chapter 4, and is illustrated in Figure 11.13. The model shown in the figure has been used for calculation of the handover interruption time. As shown in the figure, once the handover command has been processed by

Table 11.17 TDD Mode User-Plane Latency Analysis with 10% HARQ Retransmission Probability [2]

Step	Procedure	Direction	DL/UL Configuration						
			0	1	2	3	4	5	6
1	eNB processing delay (ms)	Downlink	1	1	1	1	1	1	1
2	Frame alignment (ms)		1.7	1.1	0.7	1.1	0.8	0.6	1.4
3	TTI duration (ms)		1	1	1	1	1	1	1
4	UE processing delay (ms)		1.5	1.5	1.5	1.5	1.5	1.5	1.5
5	HARQ retransmission (ms)		0.1×10	0.1×10.2	0.1×9.8	0.1×10.5	0.1×11.6	0.1×12.4	0.1×11.2
	Total one-way delay (ms)		6.2	5.62	5.18	5.65	5.46	5.34	6.02
1	UE processing delay (ms)	Uplink	1	1	1	1	1	1	1
2	Frame alignment (ms)		1.1	1.7	2.5	3.3	4.1	5	1.4
3	TTI duration (ms)		1	1	1	1	1	1	1
4	eNB processing delay (ms)		1.5	1.5	1.5	1.5	1.5	1.5	1.5
5	HARQ retransmission (ms)		0.1×11.6	0.1×10	0.1×10	0.1×10	0.1×10	0.1×10	0.1×11.5
	Total one-way delay (ms)		5.76	6.2	7	7.8	8.6	9.5	6.05

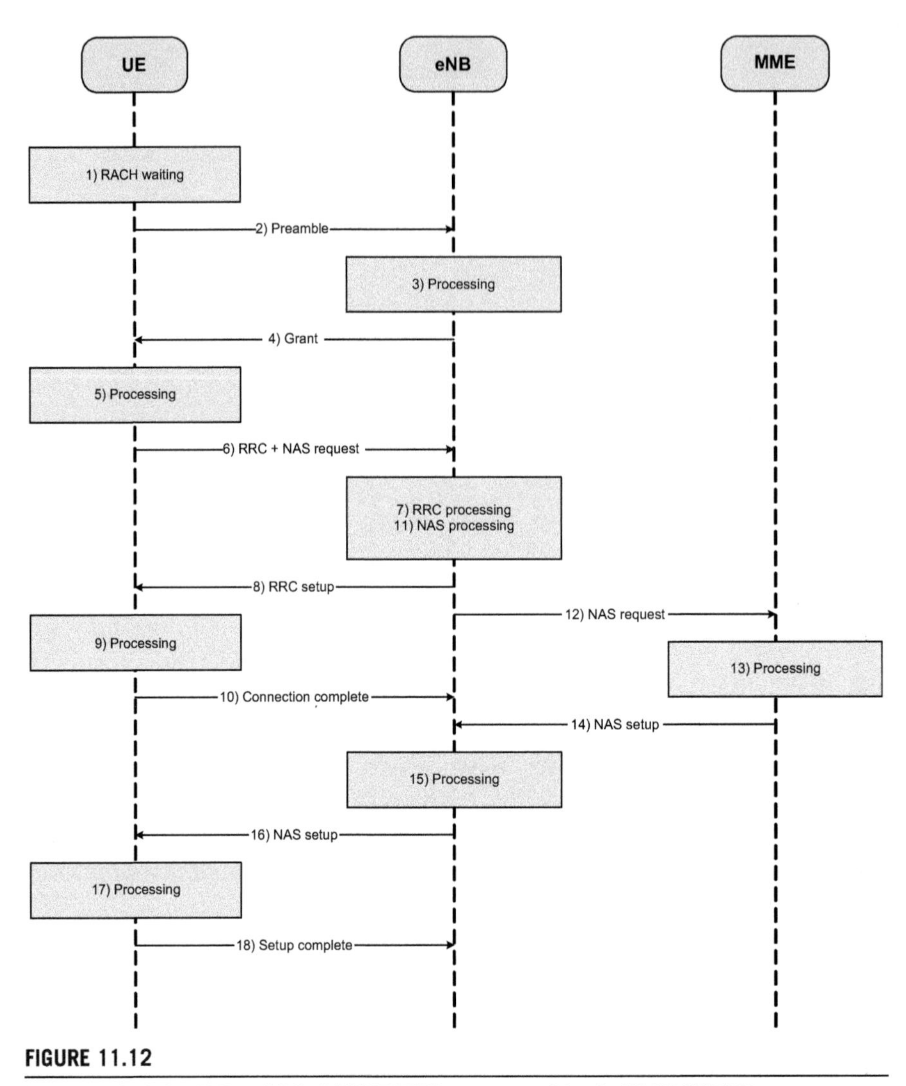

FIGURE 11.12

Analysis of RRC_IDLE to RRC_CONNECTED state transition in 3GPP LTE [2].

the UE, the UE detaches from the source eNB and stops receiving data. This marks a point in time where the user-plane connectivity is interrupted. The UE then performs frequency synchronization with the target eNB depending on whether the target cell is operating on the same carrier frequency as the current serving cell. Since the UE has already identified and measured the target cell, the corresponding delay will be negligible. The UE then performs downlink synchronization. Although baseband and RF timing alignments are part of the delay budget, since the UE has already acquired downlink synchronization with the target

Table 11.18 Analysis of Control-Plane Latency Components [2]

Step	Procedure	Processing Time (ms)
1	Average delay due to RACH scheduling period (1 ms RACH cycle)	0.5
2	RACH preamble	1
3–4	Preamble detection and transmission of random access response (time between the end RACH transmission and UE's reception of scheduling grant and timing adjustment)	3
5	UE processing delay (decoding of scheduling grant, timing alignment, and C-RNTI assignment + L1 encoding of RRC connection request)	5
6	Transmission of RRC and NAS request	1
7	Processing delay in eNB (L2 and RRC)	4
8	Transmission of RRC connection setup and UL grant	1
9	Processing delay in the UE (L2 and RRC)	12
10	Transmission of RRC connection setup complete	1
11	Processing delay in eNB (Uu→S1-C)	–
12	S1-C transfer delay	–
13	MME Processing Delay (including UE context retrieval of 10 ms)	–
14	S1-C transfer delay	–
15	Processing delay in eNB (S1-C→Uu)	4
16	Transmission of RRC security mode command and connection reconfiguration + TTI alignment	1.5
17	Processing delay in UE (L2 and RRC)	16
	Control-plane delay	50

cell in conjunction with previous measurements and can relate the target cell downlink timing to the source cell downlink timing with a time offset, the corresponding delay is going to be less than 1 ms. The data forwarding between the two eNBs is initiated before the UE moves and establishes connection to the target cell, and because the backhaul is faster than the radio interface, forwarded data is already awaiting transmission in the target cell when the user-plane with the target cell is reestablished. The latter delay component thereby does not affect the overall delay.

Based on the analysis shown in Table 11.19, the minimum handover interruption time is 10.5 ms for the FDD mode and 12.5 ms for the TDD mode. Note that this delay does not depend on the frequency of the target cell as long as the cell frequency has already been measured by the UE during the measurement gap.

As explained earlier, the minimum handover interruption time is 10.5 ms for FDD and 12.5 ms for TDD regardless of the frequency of the target cell and

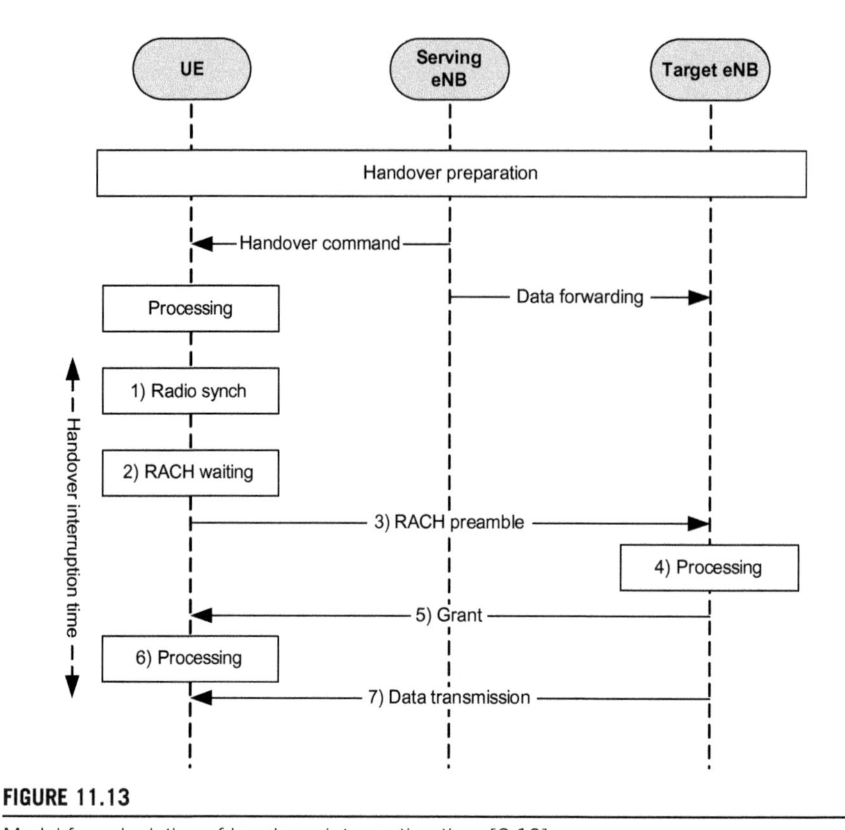

FIGURE 11.13

Model for calculation of handover interruption time [2,10].

measurements. Note that in the FDD mode, the average delay due to RACH scheduling is 0.5 ms assuming 1 ms RACH periodicity, whereas in the TDD mode, the minimum delay is obtained with configuration 0 (i.e., six normal uplink subframes); however, when taking into account the average waiting time for a downlink subframe to receive the random access response and to initiate transmission of downlink data, TDD configuration 1 (with random access preambles in special subframes) offers the shortest handover interruption time.

11.9.6 Estimation of the L1/L2 overhead

The downlink L1/L2 overhead of the LTE-Advanced includes different types of reference signals, that is, CRS transmitted within each resource block, UE-specific demodulation reference signals (DM-RS), and the reference signals that are specifically utilized in the estimation of CSI (CSI-RS). The overhead channels further include control signaling channels that are transmitted in the first L OFDM symbols ($L = 1$, 2, or 3) of each subframe ($L = 4$ in the case of 1.4 MHz bandwidth system bandwidth) and the synchronization signals, as

Table 11.19 Breakdown of Handover Interruption Time [2]

Step	Procedure	FDD Mode Processing Time (ms)	TDD Mode Processing Time (ms)
1	Radio synchronization to the target cell	1	1
2	Average delay due to RACH scheduling period assuming 1 ms periodicity	0.5	2.5
3	RACH preamble	1	1
4–5	Preamble detection and transmission of random access response (time between the end RACH transmission and UE's reception of scheduling grant and timing adjustment)	5	5
6	Decoding of scheduling grant and timing alignment	2	2
7	Transmission of downlink data	1	1
	Handover interruption time	10.5	12.5

well as physical broadcast control channel. The protocol headers corresponding to layer-2 sub-layers (MAC/RLC/PDCP) were also included in the self-evaluation results. The overhead due to CRS and L1/L2 control signaling depend on the number of antenna ports and the number of OFDM symbols used for the control region. The number of overhead resource elements per resource-block pair (168 resource elements in total) and the corresponding relative overhead is shown in Table 11.20 for non-MBSFN and MBSFN subframes. The overhead (due to L1/L2 control signaling and CRS) in Table 11.20 is shown in terms of the number of resource elements per resource-block pair and in percentages.

In the case of MBSFN subframes, the overhead is reduced. All subframes in a radio frame are assumed to be either regular or MBSFN type in overhead estimation. The total overhead depends on the fraction of MBSFN subframes. In the self-evaluation results, a fraction of 0% and 60% was assumed. The overhead due to UE-specific reference signals depends on the number of UE-specific antenna-ports and is 12 resource elements (7.1%) for one and two antenna ports, and 24 resource elements (14.3%) for three to eight antenna ports. Note that the overhead due to UE-specific reference signals is only present in resource blocks in which UE-specific reference signals are transmitted.

The relative overhead due to synchronization signals and physical broadcast channel depends on the operation bandwidth. A total of 528 resource elements per 10 ms radio frame are used for the synchronization signals and physical broadcast channel for four CRS antenna ports, corresponding to approximately 0.6% and 0.3% overhead for 10 and 20 MHz operating bandwidth, respectively.

Table 11.20 LTE-Advanced L1/L2 Overhead [2,6]

Control-Region Size	Subframe Type	FDD Duplex Scheme Number of Cell-Specific Antenna Ports			TDD Duplex Scheme Number of Cell-Specific Antenna Ports		
		1	2	4	1	2	4
$L = 1$	Regular subframes	18 (10.7%)	24 (14.3%)	32 (19.0%)	18 (10.7%)	24 (14.3%)	32 (19.0%)
$L = 2$		30 (17.9%)	36 (21.4%)	40 (23.9%)	30 (17.9%)	36 (21.4%)	40 (23.9%)
$L = 3$		42 (25%)	48 (28.6%)	52 (31.0%)	42 (25%)	48 (28.6%)	52 (31.0%)
$L = 1$	MBSFN subframes	12 (7.1%)	12 (7.1%)	12 (7.1%)	12 (7.1%)	12 (7.1%)	12 (7.1%)
$L = 2$		24 (14.2%)	24 (14.2%)	24 (14.2%)	24 (14.2%)	24 (14.2%)	24 (14.2%)
$L = 3$		36 (21.4%)	36 (21.4%)	36 (21.4%)	36 (21.4%)	36 (21.4%)	36 (21.4%)

The overhead due to protocol headers is proportional to data packet size and is approximately 2.7%, 0.51%, and 0.32% for physical-layer data rates of 1, 10, and 100 Mbps, respectively. This overhead was not included in the self-evaluation results. The relative overhead due to CSI-RS depends on the number of antennas and periodicity. In a typical case, it is about 0.12% per antenna port (0.48% for four antenna ports and 0.96% for eight antenna ports). The relative overhead due to UE-specific reference signals is estimated at approximately 7% in case of rank 1 and rank 2 transmission, and 14% for rank 3 to rank 8 transmission.

In the uplink, the L1/L2 overhead includes the demodulation reference symbols that are used in uplink channel estimation for coherent demodulation and are transmitted once every 0.5 ms. It further includes the sounding reference signal used for uplink channel state estimation at the eNB and the L1/L2 control signaling transmitted on configurable amount of resource blocks, as well as L2 control overhead due to random access, uplink time-alignment control, power headroom reports, and buffer-status reports. The L2 overhead associated with the PDU headers (MAC/RLC/PDCP) is further taken into consideration in the total overhead calculation. The amount of overhead due to DM-RS is approximately 14%, corresponding to one SC-FDMA symbol in each slot. The relative overhead is estimated independent of the rank of the transmission. The amount of SRS overhead depends on the sounding reference signal transmission interval and the bandwidth. Using a 10 ms SRS transmission interval and full-band SRS, the relative overhead is approximately 0.7%. The amount of uplink resources reserved for random access depends on the PRACH configuration, for example, a typical case with PRACH preamble format 0 is six resource blocks per radio frame, resulting in a relative overhead of 0.6%, 1.2%, and 2.4% for a channel bandwidth of 20, 10, and 5 MHz, respectively.

The relative overhead due to uplink timing-alignment control depends on the configuration and the number of active UEs within a cell. The absolute overhead is typically less than 32 bits/s per UE. The amount of overhead for buffer-status reports depends on the configuration; assuming a continuous data and reporting interval of 10–20 ms, the absolute overhead is 0.8–3.2 kbps. The amount of overhead due to protocol headers depends on the data packet size and is approximately 2.7%, 0.51%, and 0.32% for L1 data rates of 1, 10, and 100 Mbps, respectively. The above overhead calculations are based on a normal cyclic prefix length. In the case of MBSFN subframes in the TDD mode, all the subframes are assumed to be of MBSFN type in the overhead estimation. The total overhead depends on the fraction of MBSFN subframes. In self-evaluation results, a fraction of 0% and 33% has been assumed for a TDDDL:UL ratio of 3:2 (i.e., TDDDL/UL configuration 1).

In the TDD mode, the relative overhead due to synchronization signals and physical broadcast channel depends on the operation bandwidth and DL/UL configuration. A total of 528 resource elements per 10 ms radio frame are used for the synchronization signals and the broadcast channel when using four CRS antenna ports. For a DL:UL ratio of 3:2, the latter overhead translates into approximately 1.0% and 0.5%, for 10 and 20 MHz operation bandwidth, respectively. The amount of SRS overhead depends on the SRS transmission interval and bandwidth as well as the usage of UpPTS in the special subframe. For a DL:UL ratio of 3:2 with four DL/UL switching points per 10 ms radio frame, the SRS transmission interval of 5 ms and full-band SRS within UpPTS, the relative overhead is approximately 3.45%. The amount of uplink resources reserved for random access depends on the configuration of the random access channel. In a typical case with a DL:UL ratio of 3:2 and PRACH preamble format 0, i.e., six resource blocks per radio frame are used for the PRACH, resulting in a relative overhead of 1.4%, 2.8%, and 5.6% for channel bandwidth of 20, 10, and 5 MHz, respectively.

11.9.7 Link budget

The LTE-Advanced link budget under various deployment scenarios specified by ITU-R was evaluated as part of the IMT-Advanced submission (Table 11.21). Table 11.21 summarizes the link budget of the TDD mode. It is shown that LTE-Advanced control and data channel coverage meets the IMT-Advanced link budget requirements.

11.9.8 Evaluation of LTE-Advanced against LTE Rel-10 requirements

This section provides the self-evaluation results of LTE-Advanced against LTE Rel-10 requirements [3], where the target values for cell spectral efficiency and cell-edge spectral efficiency are specified for 3GPP case 1 channel model [4] with 2×2, 4×2, and 4×4 antenna configurations in the downlink, and 1×2 and 2×4 antenna configurations in the uplink. Table 11.22 shows the downlink/uplink cell and cell-edge spectral efficiency results under 3GPP Case 1 configuration for the FDD and TDD modes. The results suggest that, if LTE Rel-8 is extended with MU-MIMO 4×2 in the downlink, the LTE Rel-10 performance targets can be fulfilled. The table also includes the advanced features by which the performance can be further improved. The uplink performance results indicate that LTE Rel-8 SIMO and LTE-Advanced CoMP, SU-MIMO, and MU-MIMO schemes can fulfill the LTE Rel-10 requirements. In 3GPP case 1 scenario, the center frequency is 2.0 GHz, the transmission bandwidth is 2×10 MHz for the FDD mode and 20 MHz for the TDD mode, inter-site distance is 500 m, the penetration loss is 20 dB, and all users are assumed moving at 3 km/h.

11.10 Miscellaneous evaluations

In order to investigate the usage and advantages of various LTE downlink MIMO schemes in realistic deployment scenarios, link-level and system-level simulations were performed under certain assumptions [47–49]. In order to

Table 11.21 LTE-Advanced Link Budget in Various Deployment Scenarios for the TDD Mode [2,6,15]

Parameter	InH		UMi		UMa		RMa	
	Downlink	Uplink	Downlink	Uplink	Downlink	Uplink	Downlink	Uplink
System Configuration								
Carrier frequency (GHz)	3.4	3.4	2.5	2.5	2.0	2.0	0.8	0.8
BS antenna heights (m)	6	6	10	10	25	25	35	35
MS antenna heights (m)	1.5	1.5	1.5	1.5	1.5	1.5	1.5	1.5
Cell area reliability for control channels	95%	95%	95%	95%	95%	95%	95%	95%
Cell area reliability for control channels	90%	90%	90%	90%	90%	90%	90%	90%
Transmission bit rate for control channel (kbps)	46.93	1.60	44.80	1.60	44.80	1.60	44.80	1.60
Transmission bit rate for data channel (Mbps)	4.65	0.37	2.25	0.074	2.25	0.074	2.25	0.074
Target packet error rate for control channels	1%	1%	1%	1%	1%	1%	1%	1%
Target packet error rate for data channels	10%	10%	10%	10%	10%	10%	10%	10%
Spectral efficiency (bits/s/Hz) for data	0.23	0.52	0.22	0.21	0.22	0.21	0.22	0.21
Path loss model	NLoS	NLoS	NLoS	NLoS	NLoS	NLoS	NLoS	NLoS
Mobile speed (km/h)	3	3	3	3	30	30	120	120
Feeder loss (dB)	2	2	2	2	2	2	2	2
Transmitter								
Number of transmit antennas	2	1	2	1	2	1	2	1
Maximum transmit power per antenna (dBm)	21	21	41	24	46	24	46	24
Total transmit power (dBm)	24	21	44	24	49	24	49	24
Transmitter antenna gain (dBi)	0	0	17	0	17	0	17	0

(Continued)

Table 11.21 (Continued)

Parameter	InH		UMi		UMa		RMa	
	Downlink	Uplink	Downlink	Uplink	Downlink	Uplink	Downlink	Uplink
Transmitter array gain (dB)	0	0	0	0	0	0	0	0
Control channel power boosting gain (dB)	0	0	0	0	0	0	0	0
Data channel power loss due to pilot/control boosting (dB)	0	0	0	0	0	0	0	0
Cable, connector, combiner, body losses, etc. (dB)	3	1	3	1	3	1	3	1
Control channel EIRP (dBm)	21	20	58	23	63	23	63	23
Data channel EIRP (dBm)	21	20	58	23	63	23	63	23
Receiver								
Number of receive antennas	2	4	2	4	2	4	2	4
Receiver antenna gain (dBi)	0	0	0	17	0	17	0	17
Cable, connector, combiner, body losses, etc. (dB)	1	3	1	3	1	3	1	3
Receiver noise figure (dB)	7	5	7	5	7	5	7	5
Thermal noise density (dBm/Hz)	– 174	– 174	– 174	– 174	– 174	– 174	– 174	– 174
Receiver interference density for control channels (dBm/Hz)	– 174	– 174.9	– 169.3	– 161.7	– 169.3	– 161.7	– 169.3	– 161.7
Receiver interference density for data channels (dBm/Hz)	– 174	– 174.9	– 169.3	– 165.7	– 169.3	– 165.7	– 169.3	– 165.7
Total noise plus interference density for control channels (dBm/Hz)	– 167	– 168	– 165	– 161	– 165	– 161	– 165	– 161
Total noise plus interference density for data channels (dBm/Hz)	– 167	– 168	– 165	– 164	– 165	– 164	– 165	– 164

Occupied channel bandwidth for control channels (MHz)	36	0.18	18	0.18	18	0.18	18	0.18
Occupied channel bandwidth for data channels (MHz)	36	1.8	18	0.90	18	0.9	18	0.90
Effective noise power for control channels (dBm)	− 91	− 115	− 92	− 108	− 92	− 108	− 92	− 108
Effective noise power for control channels (dBm)	− 91	− 105	− 92	− 104	− 92	− 104	− 92	− 104
Required SNR for the control channel (dB)	− 4.2	− 10.6	− 4.2	− 10.5	− 4.2	− 10.1	− 4.2	− 9.9
Required SNR for the data channel (dB)	− 1.6	− 2.3	− 1.7	− 5.5	− 1.7	− 5.1	− 1.7	− 4.8
Receiver implementation margin (dB)	2	2	2	2	2	2	2	2
HARQ gain for control channel (dB)	0	0	0	0	0	0	0	0
HARQ gain for data channel (dB)	0.5	0.5	0.5	0.5	0.5	0.5	0.5	0.5
Receiver sensitivity for control channel (dBm)	− 94	− 124	− 95	− 117	− 95	− 117	− 95	− 116
Receiver sensitivity for data channel (dBm)	− 91	− 106	− 93	− 108	− 93	− 108	− 93	− 108
Hardware link budget for control channel (dB)	115	144	153	157	158	157	158	156
Hardware link budget for data channel (dB)	113	126	151	148	156	148	156	148
Calculation of Available Path Loss								
Lognormal shadow fading standard deviation (dB)	4	4	4	4	6	6	8	8
Shadow fading margin for control channels (dB)	2.8	2.8	3.1	3.1	5	5	5	5
Shadow fading margin for data channels (dB)	0.9	0.9	1.3	1.3	8.1	8.1	10.5	10.5
BS selection/macro-diversity gain (dB)	0	0	0	0	4.9	4.9	6.7	6.7

(*Continued*)

Table 11.21 (Continued)

Parameter	InH		UMi		UMa		RMa	
	Downlink	Uplink	Downlink	Uplink	Downlink	Uplink	Downlink	Uplink
Penetration loss (dB)	0	0	0	0	0	0	0	0
Other gains (dB)	0	0	0	0	9	9	9	9
Available path loss for control channel (dB)	111	138	149	151	0	0	0	0
Available path loss for data channel (dB)	111	122	148	144	140	136	137	134
Range/Coverage Efficiency Calculation								
Maximum range for control channel (m)	100.0	100.0	1405.2	1621.2	1175.6	978.0	3210.9	2634.3
Maximum range for data channel (m)	100.0	100.0	1383.9	1062.7	1258.4	714.4	3565.5	1975.1
Coverage area for control channel (km^2/site)	0.031	0.031	6.20	8.26	4.34	3.01	32.4	21.8
Coverage area for data channel (km^2/site)	0.031	0.031	6.02	3.55	4.97	1.60	39.9	12.3

Table 11.22 Self-Evaluation of LTE-Advanced Against LTE Rel-10 Requirements [4]

Scheme and Antenna Configuration	Direction	Duplex Scheme	Rel-10 Requirements (Cell Spectral Efficiency/Cell-Edge Spectral Efficiency)	Cell Spectral Efficiency (bits/s/Hz/cell) $L=3$	Cell-Edge Spectral Efficiency (bits/s/Hz) $L=3$
MU-MIMO 2 × 2 (C)	Downlink	FDD	2.4/0.07	2.69	0.090
JP-CoMP 2 × 2 (C)			2.4/0.07	2.70	0.104
MU-MIMO 4 × 2 (C)			2.6/0.09	3.43	0.118
CS/CB-CoMP 4 × 2 (C)			2.6/0.09	3.34	0.129
JP-CoMP 4 × 2 (C)			2.6/0.09	3.87	0.162
MU-MIMO 4 × 4 (C)			3.7/0.12	4.69	0.203
CS/CB-CoMP 4 × 4 (C)			3.7/0.12	4.66	0.205
JP-CoMP 4 × 4 (C)			3.7/0.12	5.19	0.269
MU-MIMO 2 × 2 (C)		TDD	2.4/0.07	2.88	0.113
JP-CoMP 2 × 2 (C)			2.4/0.07	3.15	0.130
MU-MIMO 4 × 2 (C)			2.6/0.09	3.76	0.151
JP-CoMP 4 × 2 (C)			2.6/0.09	4.64	0.199
MU-MIMO 4 × 4 (C)			3.7/0.12	4.97	0.209

(*Continued*)

Table 11.22 (Continued)

Scheme and Antenna Configuration	Direction	Duplex Scheme	Rel-10 Requirements (Cell Spectral Efficiency/Cell-Edge Spectral Efficiency)	Cell Spectral Efficiency (bits/s/Hz/cell)	Cell-Edge Spectral Efficiency (bits/s/Hz)
				$L=3$	$L=3$
CS/CB-CoMP 4 × 4 (C)			3.7/0.12	5.06	0.244
JP-CoMP 4 × 4 (C)			3.7/0.12	6.61	0.330
Rel-8 SIMO 1 × 2 (C)	Uplink	FDD	1.2/0.04	1.33	0.047
CoMP 1 × 2 (C)			1.2/0.04	1.40	0.051
SU-MIMO 2 × 4 (C)			2.0/0.07	2.27	0.091
Rel-8 SIMO 1 × 2 (C)		TDD	1.2/0.04	1.24	0.045
CoMP 1 × 2 (C)			1.2/0.04	1.51	0.051
SU-MIMO 2 × 4 (C)			2.0/0.07	2.15	0.090
MU-MIMO 2 × 4 (C)			2.0/0.07	2.59	0.079

Table 11.23 System-Level Simulation Parameters (Other Parameters Same as [4] Case 1 Scenario)

Parameter/Feature	Value/Configuration
Bandwidth	1.4 MHz
Frequency	2.1 GHz
Network layout	19-cell wrap-around (tri-sector with 3D antenna models)
Scheduler algorithm	Proportional fair
Channel model	WINNER +
Receiver type	Zero-forcing
Simulation length	5000 subframes
Number of UEs/Cell	10
Path loss model	$PL = 128.1 + 37.6 \log_{10}R$
Feedback delay	Modeled

reduce the simulation run time, a 1.4 MHz system bandwidth was assumed. Figures 11.14–11.19 show snapshots of the simulation results. Table 11.23 provides the assumptions that were made in addition to baseline configuration for 3GPP case 1 that is included in 3GPP TR 36.814 [4]. The UEs are randomly dropped in the cell, but all UEs are assumed to be in the outdoor area. MIESM link-to-system mapping scheme is used in the simulations.

Figure 11.14 shows the LTE-Advanced downlink throughput versus SNR, as well as downlink spectral efficiency CDF for SISO and closed-loop spatial multiplexing (CLSM) transmission mode 4 with various antenna configurations.

The CDF of wideband SINR (geometry) with macroscopic path loss with and without additional lognormal-distributed space-correlated shadow fading is illustrated in Figure 11.15. The downlink spectral efficiency (bits/s/Hz) for various MIMO configurations with lognormal-distributed space-correlated shadow fading effect is shown in Figure 11.16. This figure shows how spectral efficiency is distributed geographically across the cells and the network. In particular, the cell-edges and the effect of MIMO to improve cell-edge performance must be noted. Figure 11.17 illustrates the sector SINR calculated with distance-dependent macroscopic path loss and additional lognormal-distributed space-correlated shadow fading. In this case we assume a SISO configuration. Figure 11.18 depicts downlink spectral efficiency (bits/s/Hz) without the effect of shadow fading. The geographical distribution of sector SINR calculated with distance-dependent macro-scale path loss is shown in Figure 11.19. Figure 11.19 further shows CQI mapping across the cell. As expected, the lower modulation order and coding rates are used near cell-edges, and as we move toward the eNB, higher CQI values are reported by the UEs. The results closely correlate with the expectations of the system behavior as well as those published in the literature under given assumptions.

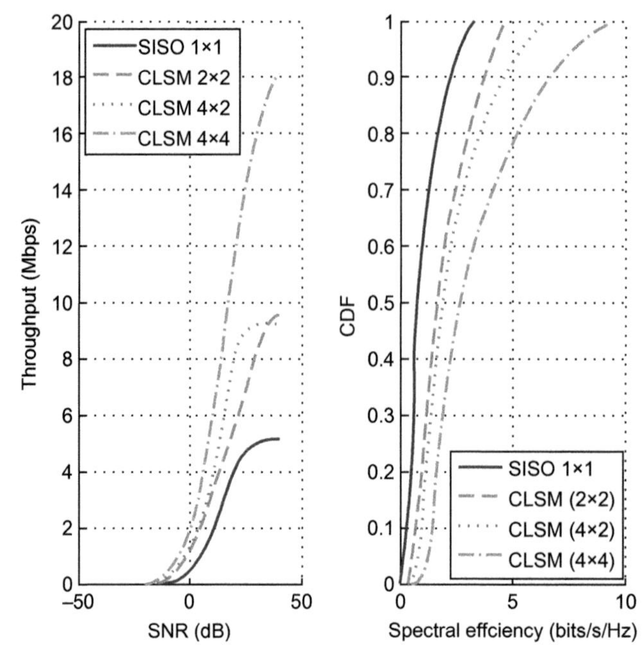

FIGURE 11.14

LTE-Advanced downlink throughput versus SNR and spectral efficiency CDF for various MIMO modes (1.4 MHz bandwidth) [45].

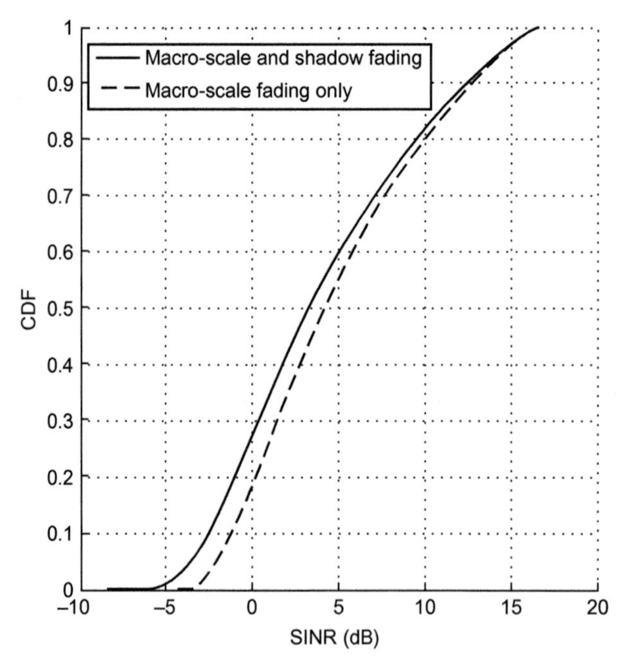

FIGURE 11.15

Distribution of downlink wideband SINR [45].

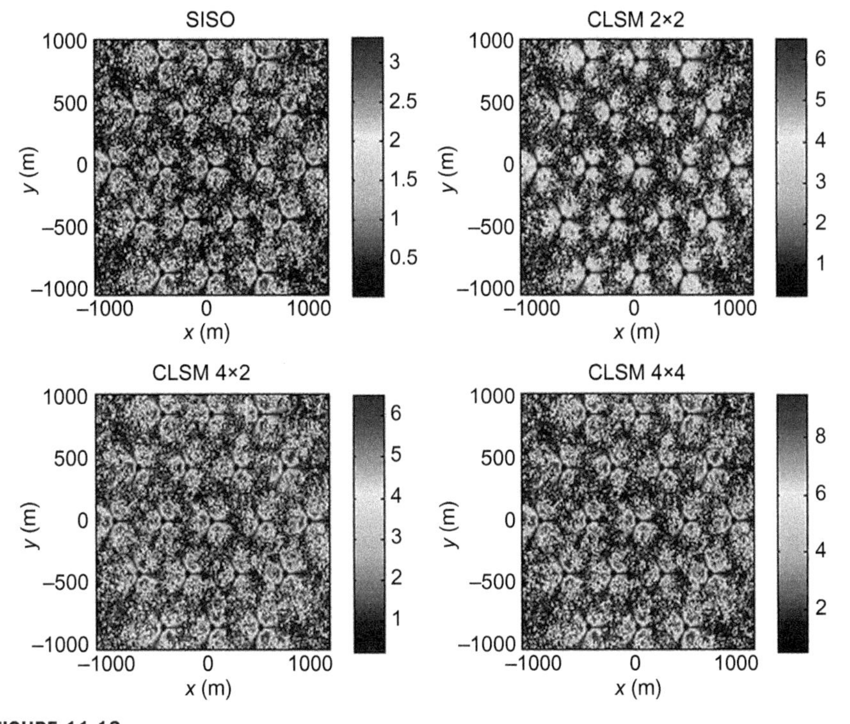

FIGURE 11.16

Spectral efficiency (bits/s/Hz) for various MIMO configurations (1.4 MHz bandwidth, with shadow fading) [45]

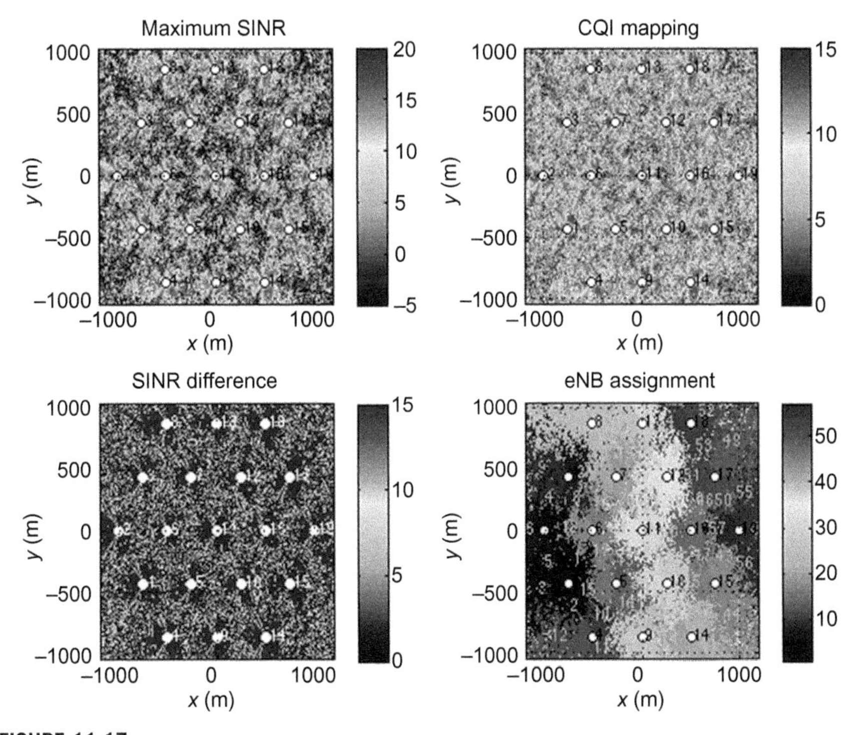

FIGURE 11.17

Sector SINR calculated with distance-dependent macroscopic path loss and additional lognormal-distributed space-correlated shadow fading (1.4 MHz bandwidth, SISO) [45]

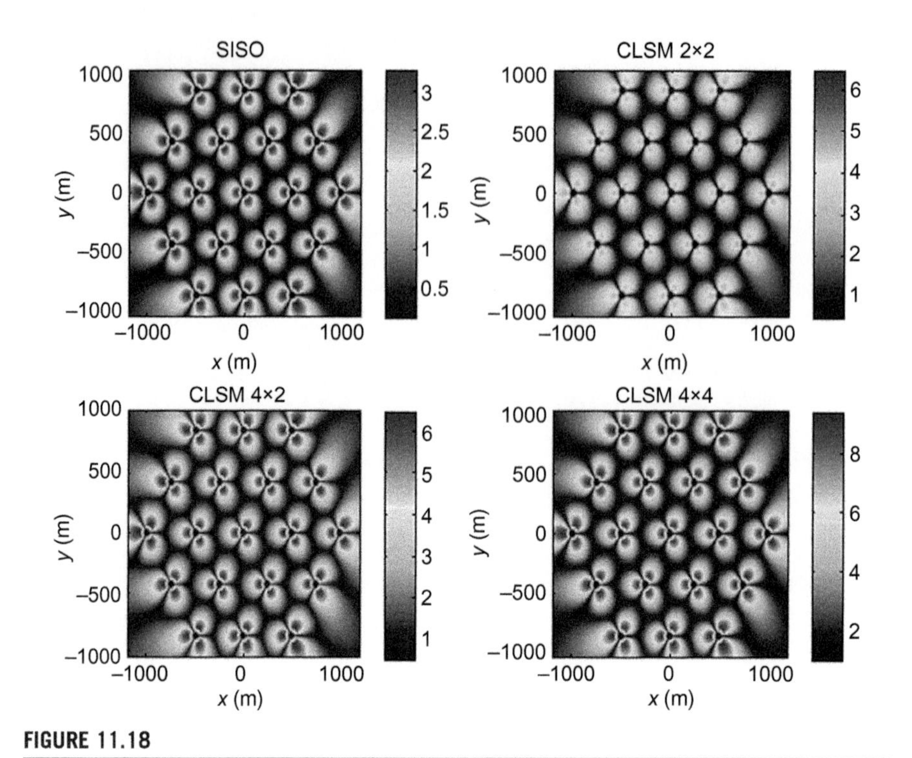

FIGURE 11.18

Spectral efficiency (bits/s/Hz) (1.4 MHz bandwidth and no shadow fading) [45]

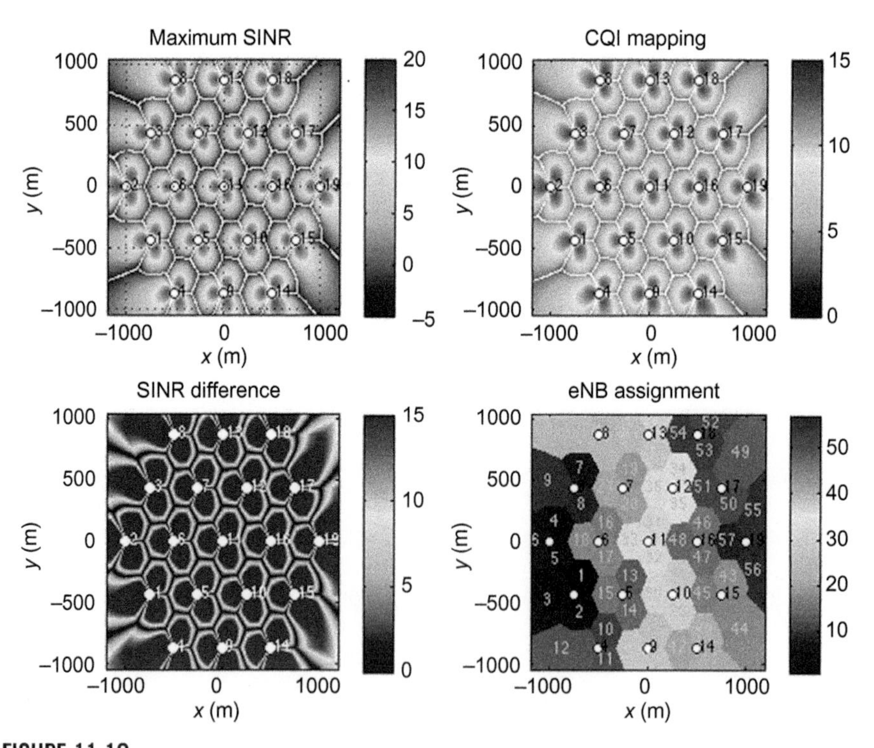

FIGURE 11.19

Sector SINR calculated with distance-dependent macro-scale path loss and CQI mapping (1.4 MHz bandwidth, SISO) [45]

References

3GPP Specifications

[1] 3GPP TR 25.996, Spatial Channel Model for Multiple Input Multiple Output (MIMO) Simulations (Release 11), September 2012.

[2] 3GPP TR 36.912, Feasibility Study for Further Advancements for E-UTRA (LTE-Advanced) (Release 11), September 2012.

[3] 3GPP TR 36.913, Requirements for further advancements for Evolved Universal Terrestrial Radio Access (E-UTRA) (LTE-Advanced) (Release 11), September 2012.

[4] 3GPP TR 36.814, Further Advancements for E-UTRA Physical Layer Aspects, March 2010.

[5] 3GPP TS 26.101, AMR Speech Codec Frame Structure (Release 11), September 2012.

[6] 3GPP RP-090738, TR 36.912 Annex A3: Self-Evaluation Results, September 2009.

[7] ETSI Technical Report 101 112 v3.2.0, Universal Mobile Telecommunications System (UMTS); Selection procedures for the choice of radio transmission technologies of the UMTS (UMTS 30.03 version 3.2.0), April 1998.

[8] 3GPP TS 36.211, Evolved Universal Terrestrial Radio Access (E-UTRA), Physical Channels and Modulation (Release 11), December 2012.

[9] 3GPP TS 36.212, Evolved Universal Terrestrial Radio Access (E-UTRA) Multiplexing and Channel Coding (Release 11), December 2012.

[10] 3GPP TS 36.331 Evolved Universal Terrestrial Radio Access (E-UTRA); Radio Resource Control (RRC); Protocol specification (Release 11), December 2012.

[11] 3GPP TR 36.819 Coordinated Multi-Point Operation for LTE Physical Layer Aspects (Release 11), December 2011.

ITU Specifications

[12] Report ITU-R M.2133, Requirements, evaluation criteria and submission templates for the development of IMT-Advanced, November 2008.

[13] Report ITU-R M.2134, Requirements related to technical performance for IMT-Advanced radio interface(s), November 2008.

[14] Report ITU-R M.2135-1, Guidelines for evaluation of radio interface technologies for IMT-Advanced, December 2009.

[15] Report ITU-R M.2012, Detailed specifications of the terrestrial radio interfaces of International Mobile Telecommunications Advanced (IMT-Advanced), January 2012.

[16] Recommendation ITU-R M.1225, Guidelines for evaluation of radio transmission technologies for IMT-2000, 1997.

[17] Recommendation ITU-T G.729: Coding of Speech at 8 kbit/s Using Conjugate-Structure Algebraic-Code-Excited Linear Prediction (CS-ACELP), January 2007.

Books

[18] L. Rade, Mathematics Handbook for Science and Engineering, Springer, Berlin Heidelberg, 2010.

[19] W.H. Tranter, et al., Principles of Communication Systems Simulation with Wireless Applications, Prentice Hall, 2004.

[20] J.D. Parsons, The Mobile Radio Propagation Channel, John Wiley and Sons, 2000.

[21] W.C. Jakes, Microwave Mobile Communications, John Wiley & Sons, New York, NY, 1974.

[22] M. Patzold, Mobile Fading Channels, John Wiley & Sons, 2002.

[23] A. Papoulis, Probability, Random Variables, and Stochastic Processes, fourth ed., McGraw Hill, 2002.

[24] A.T. Bharucha-Reid, Elements of the Theory of Markov Processes and Their Applications, McGraw-Hill, 1960.
[25] S. Sesia, I. Toufik, M. Baker, LTE, the UMTS Long Term Evolution: From Theory to Practice, second ed., John Willey & Sons, 2011.

Articles

[26] IEEE 802.16m-08/004r5, IEEE 802.16m Evaluation Methodology Document (EMD), March 2009.
[27] IETF RFC 3267, Real-Time Transport Protocol (RTP) Payload Format and File Storage Format for the Adaptive Multi-Rate (AMR) and Adaptive Multi-Rate Wideband (AMR-WB) Audio Codecs, J. Sjoberg, et al., June 2002.
[28] Rodriguez-Herrera, et al., Link-To-System mapping techniques using a spatial channel model, IEEE 62nd Vehicular Technology Conference 2005, vol. 3, 2005.
[29] 3GPP-3GPP2 Spatial Channel Ad-hoc Group, Spatial Channel Model Text Description, v7.0, August 2003.
[30] WINNER Project IST-4-027756, WINNER D1.1.2 v1.2, WINNER II Channel Models. <https://www.ist-winner.org>, September 2007.
[31] WINNER Project IST-2003-507581, WINNER D5.4 v. 1.4, Final Report on Link Level and System Level Channel Models. <https://www.ist-winner.org>, November 2005.
[32] WINNER Project, IST-2003-507581, WINNER D1.3 version 1.0, Final usage scenarios, June 2005.
[33] 3GPP2 TSG-C30-20061204-062A, cdma2000 Evaluation Methodology (v6), December 2006.
[34] S. Tsai, A. Soong, Effective-SNR Mapping for Modeling Frame Error Rates in Multiple-State Channels, 3GPP2-C30-20030429-010, April 2003.
[35] Digital mobile radio towards future generation systems, COST Action 231 Final Report, EUR 18957, 1999.
[36] Y. Oda, K. Tsunekawa, M. Hata, Advanced LoS path-loss model in microcellular mobile communications, IEEE Trans. Antennas Propag. 51 (2003).
[37] 3GPP R1-061001, Ericsson, LTE Channel Models and Link Simulations, March 2006.
[38] Source Code for a MATLAB/ANSI-C Implementation of the WINNER Phase I Channel Model. <https://www.ist-winner.org/phase_model.html>, August 2006.
[39] WINNER Project IST-WINNER II Deliverable D1.1.1 v1.0, WINNER II Interim Channel Models, December 2006.
[40] M. Steinbauer, A.F. Molisch, E. Bonek, The double-directional radio channel, IEEE Antennas Propag. Mag. (2001).
[41] G.J. Foschini, M.J. Gans, On limits of wireless communications in a fading environment when using multiple antennas, Wirel. Pers. Commun. 6 (1998).
[42] P. Almers, et al., Survey of channel and radio propagation models for wireless MIMO systems, EURASIP J. Wirel. Commun. Networking (2007).
[43] W. Dong, et al., Cluster Identification and Properties of Outdoor Wideband MIMO Channel, IEEE 66th Vehicular Technology Conference, VTC-2007, September 2007.
[44] R. Jain, D.M. Chiu, W. Hawe, A Quantitative Measure of Fairness and Discrimination for Resource Allocation in Shared Computer Systems, DEC Research Report TR-301, 1984.

[45] Technical University of Vienna LTE Simulators available at <http://www.nt.tuwien.ac.at/ltesimulator/>.

[46] J.C. Ikuno, M. Wrulich, M. Rupp, System level simulation of LTE networks, in: Proceedings of the 71st IEEE Vehicular Technology Conference, Taipei, Taiwan, May 2010.

[47] H. Claussen, Efficient modeling of channel maps with correlated shadow fading in mobile radio systems, September 2005.

[48] L. Hentila, et al., MATLAB implementation of the WINNER Phase II Channel Model ver1.1. <https://www.ist-winner.org/phase2model.html>, December 2007.

[49] L. Thiele, et al., Modeling of 3D Field Patterns of Down-tilted Antennas and Their Impact on Cellular Systems, in: International ITG Workshop on Smart Antennas, Berlin, Germany, February 2009.

[50] J. Olmos, et al., Link Abstraction Models Based on Mutual Information for LTE Downlink, Universitat Politècnica de Catalunya and Universidad Politécnica de Valencia, 2010.

[51] D. Martín-Sacristán, et al., LTE-Advanced System Level Simulation Platform for IMT-Advanced Evaluation, Universitat Politècnica de València, 2011.

[52] R. Jain, Channel models: a tutorial, WiMAX Forum AATG, February 2007.

[53] Confidence interval, Wikipedia, <http://en.wikipedia.org/wiki/Confidence_interval>.

[54] Markov Sequence, Wolfram MathWorld, <http://mathworld.wolfram.com/MarkovSequence.html>.

[55] H. Chernoff, A measure of asymptotic efficiency for tests of a hypothesis based on the sum of observations, Ann. Math. Stat. 23 (4) (1952).

[56] Monte Carlo method, Wikipedia, <http://en.wikipedia.org/wiki/Monte_Carlo_method>.

[57] Kronecker Product, Wolfram MathWorld, <http://mathworld.wolfram.com/KroneckerProduct.html>.

[58] Propagation Channel Models, Steepest Ascent Ltd, <http://www.steepestascent.com/content/mediaassets/html/LTE/Help/PropagationConditions.html>.

CHAPTER

Coordinated Multipoint Transmission and Reception (CoMP)

12

CHAPTER CONTENTS

The MIMO or network MIMO techniques are used for improving sector throughput and cell-edge throughput through multi-cell collaborative precoding, network coordinated beamforming, or inter-cell interference nulling. When a mobile station is at the cell-edge, it may be able to receive signals from multiple cell sites, and the mobile station's transmission may also be received at multiple cell sites. If the data transmission and signaling from multiple cell sites can be coordinated, the downlink performance can be significantly improved. This coordination can be similar to the interference avoidance techniques or the case where the same data is transmitted from multiple cell sites. In the uplink, since the signal can be received by multiple cell sites, the system can take advantage of coordinated multi-point reception to significantly improve the link performance. The coordinated multi-point (CoMP) transmission/reception specified in long-term evolution (LTE) Rel-11 is a method of MIMO transmission for interference reduction, which enables features such as network synchronization, cell- and user-specific reference signals, feedback of multi-cell channel state information (CSI), and synchronous data transfer between the base stations that can be used for interference mitigation and to achieve macro-diversity gain [23–26].

The LTE Rel-11 specifications provide support for coordinated transmission in the downlink and coordinated reception in the uplink. The use cases include homogeneous configurations, where the points are different cells, as well as heterogeneous configurations, where a set of low-power nodes (e.g., remote radio heads (RRH) or pico-cells) are located in the geographical area served by a macro-cell. For coordinated transmission in the downlink, the signals transmitted from multiple transmission points are coordinated to improve the received strength of the desired signal at the UE or to reduce the co-channel interference. The main advantage of coordinated reception in the uplink is to ensure that the uplink signal from the UE is reliably received by the network while limiting uplink interference, taking into account the existence of multiple reception points. Note that the mechanisms discussed in this chapter apply to both FDD and TDD modes of operation [5].

In this chapter, we will discuss the theoretical and practical aspects of coordinated multipoint transmission/reception not only from the radio access network perspective, but also from the core network and backhaul point of view and will investigate whether some of the proposed schemes would realistically offer any gains when their backhaul implications and cost of implementation and operation are considered.

12.1 Interference cancellation in cellular networks

In the downlink of an LTE system, the UE often suffers from different types of interference. In the case of single-user MIMO (SU-MIMO), the signal received by the UE includes a linear combination of multiple streams, which causes inter-stream interference. On the other hand, in a multi-user MIMO (MU-MIMO) scenario, where a single cell communicates with many UEs simultaneously using the same time–frequency resources, the signal received by each UE suffers from interference caused by the signals intended for other UEs served by the same cell. This interference is known as inter-user interference. A UE at the cell-edge suffers from inter-cell interference, i.e., interference caused by the signals sent from a neighbor cell to its UEs. The neighbor cell might belong to the same site as the serving cell (intra-site) or to a different site (inter-site). Different interference cancellation schemes are applied at the receiver to mitigate the interference in the above cases. While the inter-stream interference cancellation can be performed at the UE without requiring any additional information, inter-user interference cancellation may require knowledge of the precoding matrices of the other UEs that one wishes to cancel. Moreover, the inter-cell interference cancellation needs an estimate of the channels seen between the neighbor interfering cells and the UE. The above interference models are illustrated in Figure 12.1 [26].

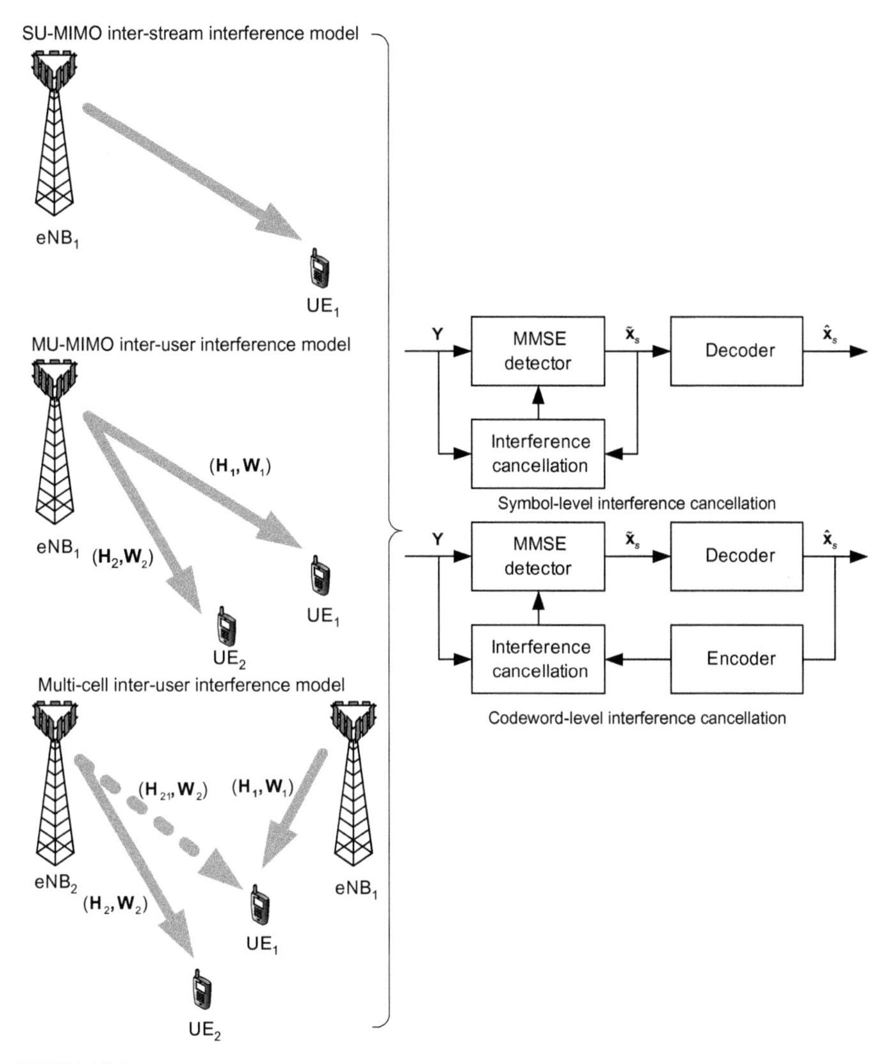

FIGURE 12.1

Illustration of interference models and cancelation at symbol and codeword level [26]. MMSE, minimum mean square error.

Let us consider a scenario where eNBs and UEs have multiple transmit/receive antennas (i.e., a MIMO channel). As shown in Figure 12.1 and in the case of a SU-MIMO inter-stream interference model, the signal received at the UE includes, in addition to noise, a linear combination of multiple streams. The beam is generated by using a non-identity precoding matrix. For simplicity, we assume that the channel is flat. In order to mitigate the inter-stream interference at the

UE, interference cancellation can be applied either at the modulated symbol level, where we cancel the estimated interfering stream before decoding, or at the codeword level where we cancel the interfering decoded codeword (see Figure 12.1).

Interference cancellation techniques have been widely studied in the literature [23–26]. While successive interference cancellation scheme processes each stream successively by canceling the previously decoded streams, parallel interference cancellation scheme processes all stream in parallel and cancel their interference after they have all been decoded independently. In order to avoid the propagation of interference cancellation error, each stream can be canceled only if it was correctly decoded by checking the corresponding decoded packet cyclic redundancy check (CRC). Successive interference cancellation can also be done iteratively, i.e., after the first iteration over all streams; we can further improve the quality of the decoded streams by canceling the residual errors estimated in the first iteration, and so on. There are methods proposed in the literature for reducing the complexity of iterative interference cancellation algorithms. In Figure 12.1, and in the case of the MU-MIMO inter-user interference model, the cell serves both UEs simultaneously using the same time–frequency resources. The first user UE_1 receives, in addition to its own data, interference due to data intended for the second user UE_2, i.e., inter-user interference. In order to cancel the inter-user interference, UE_1 needs to know the precoding matrix $\mathbf{W_2}$ that was used for UE_2. In general, $\mathbf{W_2}$ is not known by UE_1, and some additional signaling would be required in order to make $\mathbf{W_2}$ available at UE_1.

In Figure 12.1 and in the case of the multi-cell MU-MIMO interference model, UE_1 at the cell-edge is served by the eNB_1 through channel $\mathbf{H_1}$ and with the precoding matrix $\mathbf{W_1}$, and suffers from inter-cell interference, i.e., interference from the neighbor cell which serves another user UE_2 with the precoding matrix $\mathbf{W_2}$. In order to cancel the inter-cell interference, UE_1 needs an estimate of the channel $\mathbf{H}_{21}$ and the precoding matrix $\mathbf{W_2}$ intended for UE_2. The channel $\mathbf{H}_{21}$ can be estimated by UE_1 based on some reference signals sent from the neighbor cell. The precoding matrix $\mathbf{W_2}$ is in general not known by UE_1 and would require some additional signaling from eNB_2.

Let us consider the case of multi-cell inter-user interference shown in Figure 12.1. It can be shown that the signal to interference plus noise ratio (SINR) at UE_1 can be formulated as follows:

$$\mathrm{SINR}_1(\mathbf{W}_1, \mathbf{W}_2) = \frac{|\mathbf{H}_1 \mathbf{W}_1|^2}{\sigma_n^2 + \sigma_I^2 + |\mathbf{H}_{21} \mathbf{W}_2|^2} \tag{12.1}$$

In Eq. (12.1), $\mathbf{W}_1$and $\mathbf{W}_2$ represent the beamforming optimization matrices/vectors at the eNBs; $\mathbf{H}_1$ and $\mathbf{H}_{21}$ denote information that is not typically available at the CoMP scheduling unit, but measurable at the UEs; σ_n^2 denotes the receiver noise variance; and σ_I^2 indicates the interference from eNBs outside the coordinated set. Thus, the inter-cell interference propagated through channels $\mathbf{H}_{21}$ and $\mathbf{H}_{12}$ results in degradation of SINR measured at UE_1 and UE_2, respectively.

As mentioned above, the UE in the downlink of an LTE system can suffer from inter-stream, inter-user, and/or intra-site/inter-cell interference. While the UE is able to perform inter-stream interference cancellation without any additional information, it needs the precoding matrices of other UEs served by the same cell in an MU-MIMO system in order to cancel the interuser interference, and it additionally requires an estimate of the channels of the interfering neighboring cells in order to be able to cancel the inter-cell interference. An important use case for successive interference cancellation is MU-MIMO, but the principles can also be generalized to CoMP schemes. Multi-user MIMO is well known to improve the spectral efficiency by allowing a group of users to simultaneously share the same radio resources. In addition, MIMO provides antenna gain that improves the overall performance. The capacity region of the multi-user channel can be theoretically increased by MIMO through the so-called multiplexing gain. For the spatially separated (uncorrelated) users, the multiplexing gain can be achieved if CSI is available at the base station.

12.2 Distributed antenna systems

A distributed antenna system (DAS) is a network of spatially separated antennas connected to a common source via a transport medium that provides wireless service within a geographic area (cell) or a structure (indoor coverage). The basic notion of DAS was to split the transmit power among several antenna elements which are geographically separated, in order to provide coverage over the same area as a centralized antenna system, but with reduced total power and improved reliability. Thus, a set of centralized antennas radiating at high power are replaced by a group of low-power antenna elements to cover the same area. This concept was initially described in a paper by Saleh et al. [9]. Different variants of DAS have recently been deployed by selected network operators worldwide [10,16]. Examples include systems comprising RRH which consist of multiple outdoor units in which only the radio frequency (RF) front-end functionalities are implemented. The RRHs are connected to the baseband processing part of the base station through a bi-directional high-capacity/low-latency link. The DAS scheme can be implemented as a multi-airlink/multi-frequency/multi-wireless network, or a fiber-optic-based network. While the basic DAS concept is very similar to RRH, it can be differentiated with the latter features based on its greater flexibility and capacity that can be dynamically adapted to and optimized for varying traffic and user mobility conditions. The DAS enables independent operation of multiple frequency bands and multiple protocols across a single access network; consequently, multiple operator network sharing can be supported by a DAS [8,10–14,18]. Recent studies have shown that in addition to coverage improvements, a DAS can also provide other advantages such as reduced power and increased system capacity in a single-cell environment. Each distributed antenna module is connected to a home base station (or central unit) via dedicated wires, fiber optics, or an exclusive RF link. Although the connections via the same RF link used in a cell are possible, the connections with the same RF link construct information-theoretic relay channels are categorized as another research area called cooperative communications. Since the distributed antenna modules and the home base station together construct a macroscopic multiple-input single-output (MISO) vector channel, the DAS can be also interpreted as a macroscopic multiple-antenna system [19].

The advantage of the DAS scheme is that less power is consumed in overcoming penetration and shadowing losses, since a line-of-sight channel is often present, leading to reduced fading depths and reduced delay spread. A DAS can be implemented using passive splitters and feeders, and active repeater amplifiers can be included to overcome the feeder loss. If a given area is covered by N distributed antenna elements rather than a single antenna, then the total radiated power is reduced by approximately a factor of $N^{1-n/2}$, and the power per antenna is reduced by a factor of $N^{n/2}$, where a path loss model with path loss exponent n[1] is assumed. As an alternative, the total area covered could be extended for a given limit of effective radiated power, which may be important to ensure compliance with regulatory limits. In other words, for a given target radiated power, an N-element antenna system will have increased coverage area over a single-antenna system by a factor of $N^{1-2/n}$. Given the same coverage area, the maximum path loss for an N-element distributed antenna system is reduced by a factor of $N^{n/2}$. The transmit power of user devices can then be reduced by a factor of $N^{n/2}$, resulting in a more energy-efficient uplink and lower battery consumption. The use of distributed antenna systems is expected to reduce inter-cell interference and to improve SINR especially for users near cell boundaries, which are often performance limiting users, compared to conventional cellular systems in an interference-limited multi-cell environment. As a result, distributed antenna systems achieve lower symbol error probability and higher capacity than conventional cellular systems. Recent studies suggest that distributed antenna architectures could be an effective solution for reducing inter-cell interference in an interference-limited cellular environment.

Let us look at a distributed antenna system from mathematical point of view. As shown in Figure 12.2, in a distributed antenna system the main baseband processing modules such as channel cards are centralized at a location (central processing unit), and are connected with distributed antenna modules. Each distributed antenna module is physically connected with a home base station via a dedicated link. The general architecture of a DAS in a multi-cell environment is illustrated in Figure 12.2, where a cell is covered by a small base station and six

[1]In the study of wireless communications, path loss can be represented by the path loss exponent, whose value is normally in the range of 2–4, where 2 is for propagation in free space and 4 is for a relatively lossy environment and for the case of full specular reflection from the earth's surface. In some environments, such as buildings, stadiums, and other indoor environments, the path loss exponent can reach values in the range of 4–6. On the other hand, a tunnel may act as a waveguide, resulting in a path loss exponent of less than 2.

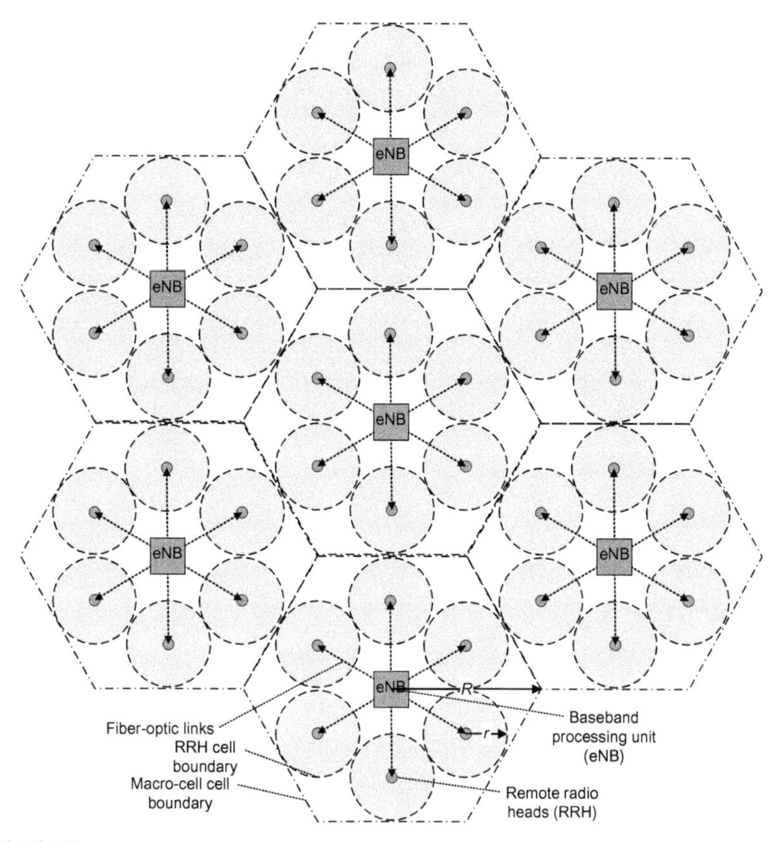

FIGURE 12.2

Example of the architecture of a distributed antenna system with one-tier interference model.

distributed antenna modules. In contrast, the same area is covered by only a single high-power base station in traditional cellular systems. The actual number of distributed antenna modules would be determined by coverage, user densities, and other deployment/economic/environmental factors. Let $P_i^{(j)}$ denote the total transmit power of the ith distributed antenna module of the jth cell, where the small base station of each cell and home cell are indexed by $i = 0$ and $j = 0$, respectively. Let us further assume that $\sum_{i=1}^{N_{\text{unit}}} P_i^{(j)} = P \; \forall j$, where P is the total transmit power of the conventional base station for a fair comparison. The radius of a cell is R, and the coverage radius of a distributed antenna module is r. We further assume a seven cell cellular structure with universal frequency reuse, where a given cell is surrounded by one continuous tier of six cells, as shown in Figure 12.2.

Let us assume (without loss of generality) that there are six RRHs in each cell, and further in each site, downlink signals are transmitted through all the distributed antenna modules and the home base station. In this case, the multiple distributed antenna modules and the home base station together construct a macroscopic MISO vector channel given by [19]:

$$\mathbf{h} = \Big(\sqrt{PL_0^{(0)}}h_0^{(0)} \quad \sqrt{\mathrm{PL}_1^{(0)}}h_1^{(0)} \quad \sqrt{\mathrm{PL}_2^{(0)}}h_2^{(0)} \quad \sqrt{\mathrm{PL}_3^{(0)}}h_3^{(0)} \quad \sqrt{\mathrm{PL}_4^{(0)}}h_4^{(0)} \quad \sqrt{\mathrm{PL}_5^{(0)}}h_5^{(0)} \quad \sqrt{\mathrm{PL}_6^{(0)}}h_6^{(0)}\Big) \tag{12.2}$$

In Eq. (12.2), $h_i^{(j)}$ denotes short-term fading channel from the ith distributed antenna module in the jth cell which is an independent and identically distributed complex Gaussian random variable $h_i^{(j)} \sim \mathbb{C}(0,1)$. The fading channel is assumed to be static during orthogonal frequency division multiplexing (OFDM) symbol duration. The coefficient $\mathrm{PL}_i^{(j)}$ denotes path loss and shadow fading from the ith distributed antenna module in the jth cell. The transmitted signal vector is represented as $\mathbf{x} = (x_0^{(0)}, x_1^{(0)}, \ldots, x_6^{(0)})^{\mathrm{T}}$, where each entry $x_i^{(j)}$ represents the transmitted signal from the ith distributed antenna module in the jth cell with $\mathscr{E}\{|x_i^{(j)}|^2\} = P_i^{(j)}$. Under these conditions, the received signal at a mobile station in a given OFDM symbol duration is given by [19]:

$$y = \underbrace{\mathbf{hx}}_{\text{Desired signal}} + \underbrace{\sum_{j=1}^{6}\sum_{i=0}^{6} h_i^{(j)}\sqrt{\mathrm{PL}_i^{(j)}}x_i^{(j)}}_{\text{Inter-cell interference}} + \underbrace{n}_{\text{Receiver noise}} \tag{12.3}$$

where n is an additive white Gaussian noise with normal distribution $n \sim \mathcal{N}(0, \sigma_n^2)$. Since the number of interfering sources is sufficiently large and the interfering sources are independent of each other, the interference plus noise can be assumed to be a complex Gaussian random variable z with variance $\sigma_z^2 = \sum_{j=1}^{6}\sum_{i=0}^{6}\mathrm{PL}_i^{(j)}P_i^{(j)} + \sigma_n^2$ according to the central limit theorem[2], whereas the interference plus noise power of the conventional cellular system is given by $\sigma_c^2 = \sum_{j=1}^{6}\mathrm{PL}_0^{(j)}P_0^{(j)} + \sigma_n^2$. When only a single distributed antenna module or the

[2]In probability theory, the central limit theorem states that, given certain conditions, the mean of a sufficiently large number of independent random variables, each with finite mean and variance, will be approximately normally distributed. Let $X_1, X_2, \ldots, X_n$ be a set of n independent random variables and let each X_i have an arbitrary probability distribution $p(x_1, x_2, \ldots, x_N)$ with mean μ_i and a finite variance σ_i^2. Then, the normal form variable $X_{\text{normal}} = \left(\sum_{i=1}^{N}(x_i - \mu_i)\right)/\sqrt{\sum_{i=1}^{N}\sigma_i^2}$ has a limiting cumulative distribution function which approaches a normal distribution. Under additional conditions on the distribution of the addend, the probability density itself is also normal with mean $\mu = 0$ and variance $\sigma^2 = 1$. If conversion to normal form is not performed, then the random variable $X = \left(\sum_{i=1}^{N}x_i\right)/N$ is normally distributed with $\mu_X = \mu_x$ and $\sigma_X = \sigma_x/\sqrt{N}$ [28].

home base station is selected for transmission (based on the criterion of minimizing propagation path loss), the received signal is given by [19]:

$$y = \underbrace{h_m^{(0)} \sqrt{\mathrm{PL}_m^{(0)}} x_m^{(0)}}_{\text{Desired signal}} + \underbrace{\sum_{j=1}^{6} h_i^{(j)} \sqrt{PL_i^{(j)}} x_i^{(j)}}_{\text{Inter-cell interference}} + \underbrace{n}_{\text{Receiver noise}} \tag{12.4}$$

where $m = \arg\min_{k=0,1\ldots 6}(\mathrm{PL}_0^{(j)}, \mathrm{PL}_1^{(j)}, \ldots, \mathrm{PL}_6^{(j)})$ and qj is randomly selected from the set $\{0, 1, \ldots, 6\}$. The interference plus noise can be shown to be a complex Gaussian random variable u with variance $\sigma_u^2 = \sum_{j=1}^{6} \mathrm{PL}_{qj}^{(j)} P_{qj}^{(j)} + \sigma_n^2$. Since other cells adopt the same transmission scheme as the home cell, only one of the interfering sources in the other cells can be randomly selected in a dynamic point switching scheme. If the CSI is known only at the receiver and the channel is ergodic, the ergodic capacity at a given location of the mobile station under study can be written as follows:

$$C_{\text{ergodic}} = \mathscr{E}_{\mathbf{h}} \left\{ \log_2 \left(1 + \frac{1}{\sigma_z^2} \mathbf{h} \boldsymbol{\Phi} \mathbf{h}^H \right) \right\} \tag{12.5}$$

where $\boldsymbol{\Phi}$ is the covariance matrix of the transmitted signal vector $\mathbf{x}$ given by $\boldsymbol{\Phi} = \mathrm{diag}(P_0^{(0)}, P_1^{(0)}, \ldots, P_6^{(0)})$. When the channel is known in the transmitter side, the ergodic capacity at a given location of the mobile station is given by:

$$C_{\text{ergodic}} = \mathscr{E}_{\mathbf{h}} \left\{ \max_{\varphi_{ii} \le P_i^{(0)} \forall i} \log_2 \left(1 + \frac{1}{\sigma_z^2} \mathbf{h} \boldsymbol{\Phi} \mathbf{h}^H \right) \right\} \tag{12.6}$$

where φ_{ii} is the ith diagonal element of matrix $\boldsymbol{\Phi}$. The maximum capacity is achieved by proper design of the transmit power spectrum matrix $\boldsymbol{\Phi}$ under individual power constraints of the distributed antenna modules and the home base station. If power cooperation among the distributed antenna modules and the home base station is allowed, this problem reduces to a standard water-filling problem in a macroscopic MISO vector channel [19].

The above study suggests that the distributed antenna systems effectively reduce inter-cell interference and improve SINR especially for the cell-edge users, which are typically performance limiting users, compared to conventional cellular systems in an interference-limited multi-cell environment. As a result, distributed antenna systems achieve lower symbol error probability and higher capacity than conventional cellular systems; thus, distributed antenna architectures appear to be an effective solution for reducing inter-cell interference in an interference-limited cellular environment.

12.3 Theory of CoMP operation

Future cellular networks are required to provide high-data-rate services to a large number of users, which necessitates high spectral efficiency over the entire cell coverage area. In order to address this problem, it is important to ensure robustness of the radio interface to interference, in particular inter-cell interference, which appears when the same radio resources are reused in different cells in an uncoordinated manner. The inter-cell interference particularly degrades the performance of users located in the cell-edge areas, which causes an inconsistent level of performance among cell-edge and inner-cell users. One possible solution to alleviate this performance discrepancy is to employ coordinated multi-point transmission and reception, which refers to a system where the transmission or reception at multiple, geographically separated antenna sites is dynamically coordinated in order to improve system performance. The coordination can be distributed either by means of direct communication between different sites, or by means of a central coordinating node [34–47].

The notion of clustering (grouping of the cooperating cells) is essential to coordinated beamforming and relaxed coherent joint processing. Indeed, the term CoMP suggests that a number of base stations share some knowledge about users. However, as the number of users and eNBs increases, the signaling overhead required for the inter-eNB information exchange and the amount of feedback needed from the users dramatically increase. Therefore, cooperation should be limited to a small number of cells/sites. To achieve this goal, the network is divided into clusters of cooperating cells. Cluster selection is important to the performance of cooperation strategies. Cluster formation may be static, if the clusters remain fixed in time, or dynamic (see Figures 12.7 and 12.8). Selection may be performed in a network-centric or in a user-centric manner. The network-centric clustering typically divides the network into a set of disjoint clusters of eNBs such that one eNB can belong only to one cluster. In contrast, in the user-centric clustering approach, one eNB may belong to more than one cluster, depending on the parameter under consideration. From the user point of view, this means that, in a given cell, each user may have a different set of cooperating eNBs. The concept of clustering is similar to cooperative sets and/or measurement sets, which will be defined later.

CoMP architectures may include centrally located or distributed processing among the collaborating nodes. The use of geographically distributed antennas in CoMP and the introduction of multiple transmission nodes is conceptually similar to the distributed antenna systems. In particular, in a CoMP scheme exploiting DAS principles, there are several possible transmission strategies including the power-weighted transmission scheme and the single transmit-selection scheme.

Assuming linear transceiver processing, a CoMP system with N antennas is ideally able to accommodate up to N streams without becoming interference limited. The inter-stream interference can be controlled or even completely eliminated by proper precoder selection. This is especially true in the coherent joint processing case, where user data is conveyed from multiple base station antenna nodes over a large virtual MIMO channel. The coherent multi-user multi-cell precoding techniques have strict requirements in terms of signaling and measurements. In addition to the complete channel knowledge of all jointly processed links, a stringent synchronization across the transmitting nodes and centralized entities performing scheduling and computation of joint precoding weights is required in order to avoid carrier phase drifting at different transmit nodes. A large amount of data needs to be exchanged between the network nodes. Therefore, broadband links, such as optical fibers or dedicated radio links, are needed.

Another form of coordinated transmission is dynamic multi-cell scheduling and interference avoidance, where the network nodes coordinate their transmissions (precoder design, scheduling) in order to minimize the inter-cell interference. Carrier phase coherence between the transmit nodes is not required, since each data stream is transmitted from a single transmission point. As a result, the non-coherent coordinated multi-cell transmission approaches have somewhat looser requirements on the coordination and the backhaul, but could potentially need centralized resource management mechanisms.

In order to understand how the above coordinated transmission schemes are mathematically formulated, consider an example of a centralized non-codebook-based coordinated beamforming scenario with static cooperating clusters. Let us assume a general downlink system model for the network consisting of N_{cell} cells and N_{user} single-antenna UEs per cell. This scenario can be considered a traditional cellular network, in which only one transmitter per cell is used, or a CoMP network, in which more than one transmitter per cell is used. We refer to this model as a CoMP system made of N_{unit} single-antenna remote units per cell. In general, the downlink system model of a CoMP cellular network can be expressed as $\mathbf{y}=\mathbf{HWx}+\mathbf{n}$, where $\mathbf{H}\in\mathbb{C}^{N_{user}\times N_{cell}N_{unit}}$, and $\mathbf{W}\in\mathbb{C}^{N_{cell}N_{unit}\times N_{user}}$ are the CoMP processing matrices, which depend on the particular CoMP configuration, $\mathbf{x}=(x_1,x_2,\ldots,x_{N_{user}})^T\in\mathbb{C}^{N_{user}\times 1}$ is a complex vector that contains the transmitted signal, $\mathbf{y}=(y_1,y_2,\ldots,y_{N_{user}})^{\mathrm{T}}\in\mathbb{C}^{N_{user}\times 1}$ is a complex vector containing N_{user} signals received by N_{user} UEs, and $\mathbf{n}\in\mathbb{C}^{N_{user}\times 1}$ is the noise vector. The ijth entry $\{h_{ij}^{(k)}\}$ of matrix $\mathbf{H}$ represents the complex-valued channel gain between the ith UE $(1\le i\le N_{user})$ and the jth remote unit $(1\le j\le N_{unit})$ belonging to the kth cell $(1\le k\le N_{cell})$. The jith element $\{w_{ji}^{(k)}\}$ of the matrix $\mathbf{W}$ is the square root of the transmission power applied at the jth remote unit $(1\le j\le N_{unit})$, transmitting toward the ith UE $(1\le i\le N_{user})$, belonging to the kth cell $(1\le k\le N_{cell})$. In this model, we assume a flat fading channel as typically observed at each subcarrier in an OFDM system. The above model can be extended to the case of remote units and mobile terminals equipped with multiple antennas. In the following, we assume that there are two cells in the network, and there are three remote units per cell, and further there are two users in each cell, thus $N_{cell}=2$, $N_{unit}=3$, and $N_{user}=2$. Using the above hypotheses, one can rewrite the received signal equation as follows [24]:

$$\begin{pmatrix} y_1 \\ y_2 \end{pmatrix} = \begin{pmatrix} \mathbf{H}^{(1)} & \mathbf{H}^{(2)} \end{pmatrix} \begin{pmatrix} \mathbf{W}^{(1)} \\ \mathbf{W}^{(2)} \end{pmatrix} \begin{pmatrix} x_1 \\ x_2 \end{pmatrix} + \begin{pmatrix} n_1 \\ n_2 \end{pmatrix} \quad (12.7)$$

which can be further expanded as follows:

$$\begin{pmatrix} y_1 \\ y_2 \end{pmatrix} = \begin{pmatrix} h_{11}^{(1)} & h_{12}^{(1)} & h_{13}^{(1)} & h_{11}^{(2)} & h_{12}^{(2)} & h_{13}^{(2)} \\ h_{21}^{(1)} & h_{22}^{(1)} & h_{23}^{(1)} & h_{21}^{(2)} & h_{22}^{(2)} & h_{23}^{(2)} \end{pmatrix} \begin{pmatrix} w_{11}^{(1)} & w_{12}^{(1)} \\ w_{21}^{(1)} & w_{22}^{(1)} \\ w_{31}^{(1)} & w_{32}^{(1)} \\ w_{11}^{(2)} & w_{12}^{(2)} \\ w_{21}^{(2)} & w_{22}^{(2)} \\ w_{31}^{(2)} & w_{32}^{(2)} \end{pmatrix} \begin{pmatrix} x_1 \\ x_2 \end{pmatrix} + \begin{pmatrix} n_1 \\ n_2 \end{pmatrix} \tag{12.8}$$

where y_i is the signal received by the ith UE and x_j denotes the signal transmitted by the jth remote unit. The particular expressions of $\mathbf{W} \in \mathbb{C}^{N_{\text{cell}} N_{\text{unit}} \times N_{\text{user}}}$ for CoMP transmit-selection and power-weighted schemes when only one user per cell is considered are as follows:

$$\mathbf{W}_{\text{transmit-selection}} = \begin{pmatrix} 0 & 0 \\ w_{\text{ji}}^{(1)} & 0 \\ 0 & 0 \\ 0 & 0 \\ 0 & w_{\text{ji}}^{(2)} \\ 0 & 0 \end{pmatrix} \quad \mathbf{W}_{\text{weighted-power}} = \begin{pmatrix} w_{11}^{(1)} & 0 \\ w_{21}^{(1)} & 0 \\ w_{31}^{(1)} & 0 \\ 0 & w_{12}^{(2)} \\ 0 & w_{22}^{(2)} \\ 0 & w_{32}^{(2)} \end{pmatrix} \tag{12.9}$$

In the above expressions, and for the CoMP transmit-selection scheme, the only non-zero term $w_{\text{ji}}^{(k)}$ for $k = 1, 2$ is selected by a certain criterion such as minimizing the path loss between any of the remote units transmitting toward the mobile user which is being served, i.e., $w_{\text{ji}}^{(k)} = \min_j \{PL(j, i)\}|_{k=1,2}$ for the ith mobile user. In the CoMP power-weighted scheme $\sum_{j=1}^{N_{\text{unit}}} w_{\text{ji}}^{(k)} = P_{\text{total}}$ for $k = 1, 2$ where P_{total} is the total transmit power of the conventional base station, for the ith mobile user.

Consider a coordination cluster with N_{TP} transmission points, each equipped with an antenna array comprising N_{TX} antenna elements, and N_{user} UEs equipped with N_{RX} receive antennas. Assuming an OFDM transmission, the baseband signal received by the ith UE at the kth subcarrier in the downlink is given by:

$$\mathbf{y}_i(k) = \mathbf{H}_i(k)\mathbf{x}(k) + \mathbf{n}_i(k) \tag{12.10}$$

where the $N_{\text{RX}} \times N_{\text{unit}} N_{\text{TX}}$ matrix $\mathbf{H}_i(k)$ is the composite channel between all transmission points and the ith UE; $\mathbf{n}_i(k)$ represents thermal noise vector including inter-cluster interference which is assumed to be white Gaussian distributed with covariance matrix $\mathscr{E}\{\mathbf{n}_i(k)\mathbf{n}_i^H(k)\} = \sigma_i^2\mathbf{I}$; $\mathbf{x}$ is an $N_{\text{unit}}N_{\text{TX}} \times 1$ vector representing the sum signal transmitted from all transmission points, given by:

$$\mathbf{x}(k) = \sum_{i=1}^{N_{\text{user}}} \mathbf{x}_i(k) = \sum_{i=1}^{N_{\text{user}}} \sqrt{p_i}\mathbf{u}_i s_i(k) \tag{12.11}$$

where $s_i(k)$ is a modulation symbol drawn from a unit-variance symbol alphabet, which is transmitted to the ith UE using the transmit power p_i and beamforming vector $\mathbf{u}_i$, respectively. The beamforming vectors are normalized to have unit power $\mathscr{E}(\mathbf{u}_i^H\mathbf{u}_i) = 1$ for $i = 1, 2, \ldots, N_{\text{user}}$. The average SINR for the ith UE, assuming an MRC receiver type, is given by [24]:

$$\text{SINR}_i(\mathbf{U},\mathbf{p}) = \frac{\mathbf{p}_i\mathbf{u}_i^H\mathbf{R}_i\mathbf{u}_i}{\sum_{j\neq i}\mathbf{p}_j\mathbf{u}_j^H\mathbf{R}_i\mathbf{u}_j + \sigma_i^2} \tag{12.12}$$

where $\mathbf{U}$ is the transmit beamforming matrix given by $\mathbf{U} = (\mathbf{u}_1 \quad \mathbf{u}_2 \quad \ldots \quad \mathbf{u}_{N_{\text{user}}})$, $\mathbf{p}_i$ is the vector of transmit powers, and $\mathbf{R}_i$ is the transmit correlation matrix for the ith UE, given by $\mathbf{R}_i = \mathscr{E}\{\mathbf{H}_i^H(k)\mathbf{H}_i(k)\} = \text{diag}(R_{i1}, R_{i2}, \ldots, R_{iN_{\text{unit}}})$. Since the transmission points are separated by large distances, their antennas are mutually uncorrelated. The considered coordinated beamforming scheme is a modified form of the multi-user beamforming algorithm, in which the beamforming vectors and power allocations are obtained by maximizing the jointly achievable SINR margin as follows [24]:

$$C = \max_{\mathbf{U},\mathbf{p}} \min_{i\in\{1,2,\ldots,N_{\text{user}}\}} \frac{\text{SINR}_i(\mathbf{U},\mathbf{p}_i)}{\gamma_i} \tag{12.13}$$

The above optimization is performed under per-transmitter power constraint:

$$\sum_{i\in S_n} p_i \leq P_{\max} \quad n\in\{1, 2, \ldots, N_{\text{unit}}\} \tag{12.14}$$

where γ_i is the target SINR for the ith UE, S_n is the set of UEs connected to the nth transmission point, and $P_{\max}$ is the maximum transmission point transmit power. The beamforming vectors and downlink power allocations are found iteratively using the uplink–downlink duality theorem, which states that the downlink broadcast channel has a virtual dual uplink multiple access channel which has the same SINR achievable regions as the downlink, and the same beamforming vectors to achieve the SINRs in both links [24].

It can be expected that the most appropriate scenarios for coordinated beamforming have low angular spreads which are found in the urban, suburban, and rural environments with above-rooftop antenna deployment. Due to low feedback requirements, coordinated beamforming is especially of interest in the FDD mode. The joint processing might be difficult to realize in a practical system due to the stringent feedback requirements. It can also be expected that coordinated beamforming is robust to user mobility, hence making it an attractive alternative

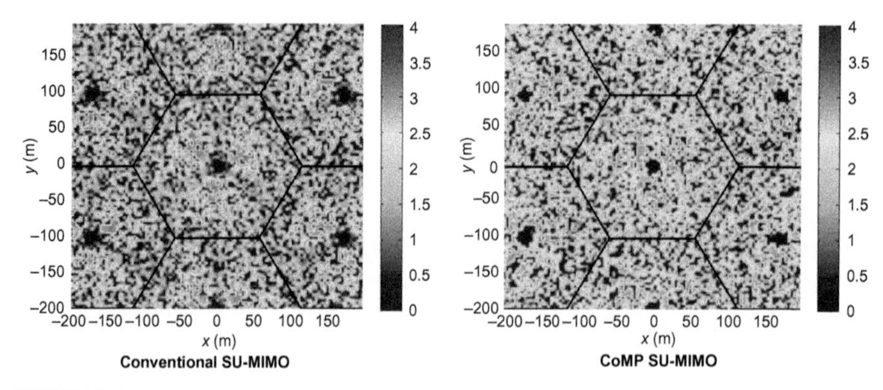

FIGURE 12.3

Performance of intra-eNB CoMP with SU-MIMO and implicit feedback relative to conventional SU-MIMO.

both in low and high mobility scenarios [24]. The performance (cell spectral efficiency) of an intra-eNB CoMP with SU-MIMO and implicit feedback relative to a conventional SU-MIMO scheme have been simulated and are shown in Figure 12.3. The improvement of cell spectral efficiency due to use of CoMP in different areas of the cell and in particular the cell-edge areas can be seen in Figure 12.3. The simulation assumptions are consistent with those recommended in 3GPP TR 36.814 [4].

12.4 CoMP architectures

In LTE Rel-8, MIMO transmission in each cell is controlled independent of that of the neighbor cells. Inter-cell interference coordination (ICIC) was supported by means of coordination message exchanges between base stations via a standardized inter-eNB interface known as X2 [6,7]. The goal of ICIC is to provide the scheduler at one cell with information about the current or prospective interference condition at its neighbors. However, in reality, the transfer latency of these coordination messages depends on the backhaul technology and is not negligible. While ICIC techniques are primarily designed for semi-static coordination, CoMP techniques target more dynamic coordination and may require much lower latency. The macro-cells constitute the basis of a mobile network's coverage; the deployment of low-power nodes (e.g., pico- or femto-cells or relay nodes) within the macro-cell is expected to be an effective solution in the future to address ever-increasing mobile traffic demands. Such networks, referred to as heterogeneous networks, are characterized by severe inter-cell interference between the macro and the low-power nodes due to their proximity and different transmit power classes. The inter-cell interference effect is even more detrimental when the effective cell size of the low-power node is expanded to balance the traffic

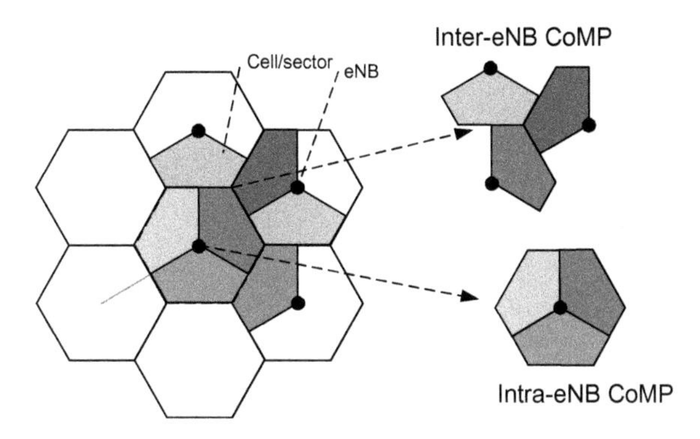

FIGURE 12.4

Illustration of inter-eNB and intra-eNB CoMP scenarios.

load between the macro-cell and the low-power node cell; thus, the UE may be associated with a cell that does not provide the strongest received signal power.

In general, and as shown in Figure 12.4, CoMP transmission/reception techniques can be classified into the following two major categories:

1. Inter-eNB CoMP, wherein coordination amongst eNBs is done via the X2 interface with typical feedback delay of 5–30 ms. This method requires standardization of the feedback channel with consideration for overhead and delay reduction.
2. Intra-eNB CoMP, wherein transmission points are usually connected via wideband/low-delay optical fibers with typical feedback delay of 5 ms and with minimal impact on the X2 interface.

LTE Rel-11 has focused on the intra-eNB CoMP scenarios in order to limit the scope of specification impacts on lower protocol layers.

One recent deployment trend that is foreseen to be widely applied to LTE networks consists of splitting the base station functionalities into a baseband unit, which performs in particular the scheduling and the baseband processing functions, and a number of RRHs responsible for all RF transmission/reception aspects. The baseband processing unit is typically located at the center of the cell and is connected via optical fiber to the RRHs. In addition to suppressing the feeder loss (since the power amplifier is immediately close to the antenna), this approach allows the baseband processing unit to manage different radio sites in a central manner. Furthermore, having geographically separated RRHs controlled from the same location enables either centralized baseband processing units jointly managing the operation of several radio sites, or the exchange of very low-latency coordination messages among individual baseband processing units. The RRH deployments facilitate the tight coordination between transmission/reception points that is required for some CoMP schemes. The applicability and benefits of CoMP techniques depend to a great extent on the backhaul characteristics (latency and capacity), which determine the type of CoMP processing and the corresponding

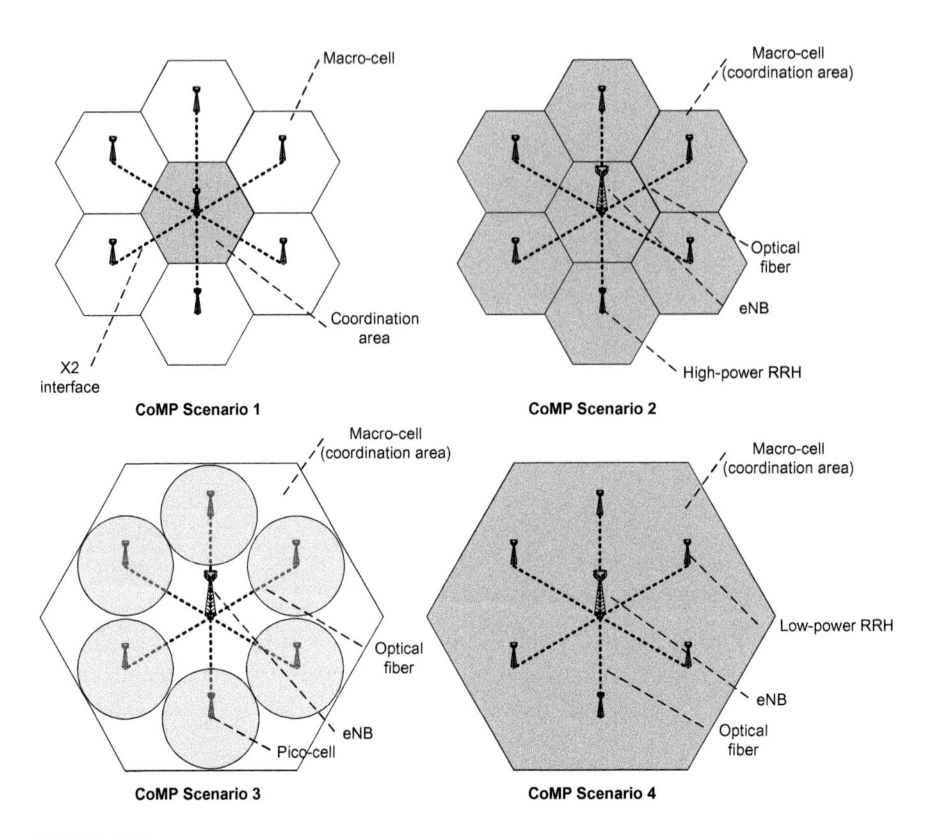

FIGURE 12.5

Illustration of CoMP scenarios [5].

performance. In order to account for the various possible network topologies and backhaul characteristics, the study of CoMP in 3GPP focused on the following scenarios (Figure 12.5) [5,27]:

- CoMP scenario 1 (homogeneous macro network with intra-site CoMP): Coordination among the cells (sectors) controlled by the same macro base station (where no backhaul connection is needed).
- CoMP scenario 2 (homogeneous macro network with inter-site CoMP): Coordination between cells belonging to different radio sites from a macro network.
- CoMP scenario 3 (heterogeneous network with low-power pico-cells within the macro-cell coverage area): Coordination between a macro-cell and low-power transmit/receive points within its coverage, each point controlling its own cell (with its own cell identity).
- CoMP scenario 4 (heterogeneous network with low-power RRH): Coordination between a macro-cell and low-power transmit/receive points

within its coverage, each point controlling its own cell, except that the low-power transmit/receive points constitute geographically distributed antennas of the macro-cell, and are thus all associated with the macro-cell identity.

CoMP scenario 1 is defined for homogeneous networks where the coordination area is restricted to the cells of a single site controlled by an eNB. Although the extent of coordination is limited to the cells of the same site, CoMP scenario 1 has the advantage of requiring no inter-eNB interface among different sites. Such an advantage would make CoMP scenario 1 attractive to the network operators.

CoMP scenario 2 is for homogeneous networks where the coordination area is expanded from the CoMP scenario 1 to include the cells of different sites. CoMP scenario 2 may be realized by having a single eNB control the RRHs at different sites, or by having multiple eNBs at different sites to coordinate with each other. CoMP scenario 2 is depicted in Figure 12.5 with high-power RRHs at different sites controlled by a single eNB. Depending on the number of cells involved and the latency of connections between the sites, which are both dependent on network implementation, different levels of performance gain over CoMP scenario 1 may be obtained.

CoMP scenario 3 relies on a heterogeneous network model where macro-cells with high transmission power and pico-cells with low transmission power are connected. In practice, this scenario may be realized by a single eNB controlling the high-power RRH of the macro-cell and the low-power RRHs of pico-cells within the macro-cell coverage area. Another possible network implementation is that the eNBs of the macro-cell and the pico-cells are different and coordinate with each other. Figure 12.5 illustrates CoMP scenario 3 with low-power RRHs at different sites controlled by a single eNB.

CoMP scenario 4 is defined for heterogeneous networks where low-power RRHs are distributed throughout the coverage area of a macro-cell. The difference between this scenario and the other intra-eNB scenarios is that all low-power RRHs in this case share the same physical cell identity with the macro-cell. Therefore, it can be construed that the low-power RRHs form a set of distributed antennas in the macro-cell area. Since the coordination is done among distributed antennas within the same cell, depending on the size of macro-cell, there may not be a need for conventional mobility support such as handover procedures among the RRHs. Furthermore, since each RRH is not an independent cell, a low-delay and high-capacity backhaul connection such as optical fiber can provide the radio connection between the macro-eNB and the RRHs. CoMP scenarios 3 and 4 are expected to find applications in metropolitan areas where network deployment is dense and RRHs with different transmission power levels can be deployed to improve control/data coverage and capacity [17].

A set of geographically collocated antennas that correspond to a particular sectorization scheme are traditionally configured as a cell. A UE attempts to connect to a single cell at any given time, if the measured received reference signal strength from that cell is higher relative to that from other cells. This cell then becomes its serving cell. Given the new definition of CoMP scenarios as introduced earlier, the antennas configured as a cell may not be geographically collocated. The term transmission point can then be used to refer to a set of collocated antennas, and a cell can correspond to one or more such transmission points. Note that a single geographical site may contain multiple transmission points in the case of sectorization, with one transmission point corresponding to one sector. CoMP techniques can also be defined in a more straightforward manner as the coordination between transmission points.

A mobile station is said to be connected to a cooperating set/cluster if its serving cell belongs to the cluster. We assume no coordination between clusters, which also means that a cell belongs to one, and only one, cluster. The serving cell selection in LTE Rel-8 is based on the long-term average reference signal received power (RSRP) measurements. The cell with the strongest RSRP acts as the serving cell. The serving cell also defines the cluster to which the mobile is connected. Since each cell uniquely belongs to one cluster, the serving cluster is determined once the serving cell is selected. It is assumed that the measurement set of a UE is a subset of the cluster.

12.4.1 Air-interface and network support for CoMP schemes

In order to enable CoMP operation, the network architecture must be able to provide sufficient coordination and synchronization among different transmission points and/or cells/sites. If inter-eNB coordination is enabled, user data and/or CSI will be exchanged and shared between eNBs; thus, low-latency and high-capacity backhaul would be required to support inter-eNB CoMP deployment scenarios. The impacts on the radio interface mainly depend on type of CoMP scheme that is utilized, which may include, but is not limited to, the following [5]:

- CSI measurement and feedback mechanisms at the UE, which may include reporting dynamic channel conditions between transmission points in the CoMP measurement set and the UE. Depending on the duplex scheme, channel reciprocity may be exploited to facilitate channel estimation at the transmission points.
- Low-latency and configurable reporting mechanisms to facilitate the decision-making process concerning the set of participating transmission points.
- Static and dynamic coordination and tight synchronization are required prior to transmission of the signal over multiple transmission points.
- Reference signals with unique signatures are required to enable inter-cell/intra-cell interference and channel measurements. Depending on the CoMP scheme, the LTE Rel-10 CSI reference signal (CSI-RS) structures were found to be suitable to provide support for certain interference measurement and characterization [33].
- Improved control signaling design may be required to support signaling and feedback for certain CoMP algorithms. LTE Rel-11 provided the enhanced physical downlink control channel (ePDCCH) and other downlink control

signaling improvements, as well as transmission mode 10 to accommodate the CoMP design and operation.

Different impacts on the network architecture are foreseen depending on the CoMP scenario being deployed, including coordination among base stations, use of radio over fiber links to connect different transmission/reception points, or coordinated relaying, affecting both data-plane and control-plane protocols. The optical fiber links have a small impact on the current network architecture since the interface between base station control unit and the associated RRH can be either proprietary or based on an open standard interface such as the Common Public Radio Interface (CPRI)[3] standard which defines the interface between entities known as RECs and local or RRH referred to as radio equipment (RE) [29]. It is necessary to further study the transfer of data and control information over optical fiber interfaces. Cooperation among distributed base stations has a higher impact on the network. A central control/scheduling/processing unit may be necessary depending on the required coordination scheme.

Other important aspects to be considered when deploying CoMP in a network include centralized/decentralized management of UEs connected to multiple transmit/receive points. Most proposals addressing multi-antenna schemes often assume power constraints on the sum of power transmitted by different antennas. This assumption may not hold when each antenna in a multi-antenna base station is powered by its own power amplifier and is limited by the linearity of that amplifier, or when the system uses distributed antennas belonging to the same or coordinated base stations, and each one is subject to its own power limits and emission constraints [23].

12.4.2 Downlink CoMP

A set of geographically separated points directly and/or indirectly involved in downlink data transmission to a UE over a time–frequency resource is defined as a CoMP cooperating set. Note that this set may be transparent to the UE. The transmission points in a cooperating set are either directly participating in transmission of data in the time–frequency resource, or indirectly participating in transmission,

[3]The CPRI is an initiative to define a publicly available specification that standardizes the protocol interface between the radio equipment control (REC) and the radio equipment (RE) in wireless base stations. This allows interoperability of equipment from different vendors and preserves the software investment made by wireless service providers. Conventional base stations are located adjacent to the antenna in a small building at the base of the antenna tower. Finding suitable sites can be a challenge because of the footprint required for the building, the need for structural reinforcement of rooftops, and the availability of both primary and backup power sources. CPRI allows the use of a distributed architecture where base stations, containing the REC, are connected to remote radio heads via low-loss fiber links that carry the CPRI data. This architecture reduces costs for service providers because only the remote radio heads containing the REs need to be situated in environmentally challenging locations. The base stations can be centrally located in less challenging locations where footprint, climate, and availability of power are more easily managed [29].

which make them candidate points for data transmission that do not transmit data but contribute in making decisions on the user scheduling/beamforming in the time–frequency resource. The CoMP transmission points (TPs) are defined as a set of points transmitting data to a UE. It is understood that CoMP TPs are a subset of the CoMP cooperating set. In general, the downlink CoMP schemes may be classified into one of the following categories (see Figure 12.6) [5]:

- Joint processing (JP) where data for a UE is available at more than one transmission point in the CoMP cooperating set for a time–frequency resource. A subcategory of JP CoMP is the joint transmission (JT) in which data is simultaneously transmitted from multiple points (part of or an entire CoMP cooperating set) to a single UE or multiple UEs in a time–frequency resource. Alternatively, the data to a UE can be simultaneously transmitted from multiple points, e.g., to (coherently or noncoherently) improve the received signal quality and/or data throughput. Another subgroup of JP CoMP is known as dynamic point selection (DPS)/muting where data is transmitted from one point (within the CoMP cooperating set) in a time–frequency resource. The transmitting/muting point may change from one subframe to another, including varying over the resource block pairs within a subframe. The data is available simultaneously at multiple points. This includes dynamic cell selection (DCS). The DPS may be combined with JT, in which case, multiple points can be selected for data transmission in the time–frequency resource.
- Coordinated scheduling/beamforming (CS/CB) in which data for a UE is only available and is transmitted from one point in the CoMP cooperating set, i.e., downlink data is transmitted from that point for a time–frequency resource but user scheduling/beamforming decisions are made with coordination among points corresponding to the CoMP cooperating set. In this case, the transmission points are selected semi-statically. Semi-static point selection (SSPS) is a method where data is transmitted to a specific UE from one point at a time. The transmitting point may only change in a semi-static manner. Muting may be applied in a dynamic and semi-static manner to the above transmission schemes (Figure 12.6).

The CS/CB scheme reduces the interference level experienced by a UE terminal by appropriately selecting the beamforming weights of interfering points to steer the interference toward the null space of the interfered UE. The JT scheme allows one or more neighboring points to transmit the desired signal rather than interfering signals from the point of view of the selected UE. The transmission point selection scheme enables the UE to be dynamically scheduled via the most appropriate transmission point by exploiting changes in the channel slow/fast fading conditions. Hybrid methods may also be implemented in a network to cope with different types of interference. A combination of JP and CS/CB schemes is possible in the downlink. Data for a UE may be available only in a subset of points in the CoMP cooperating set for a particular time–frequency resource, but user scheduling/beamforming decisions are made with coordination among points

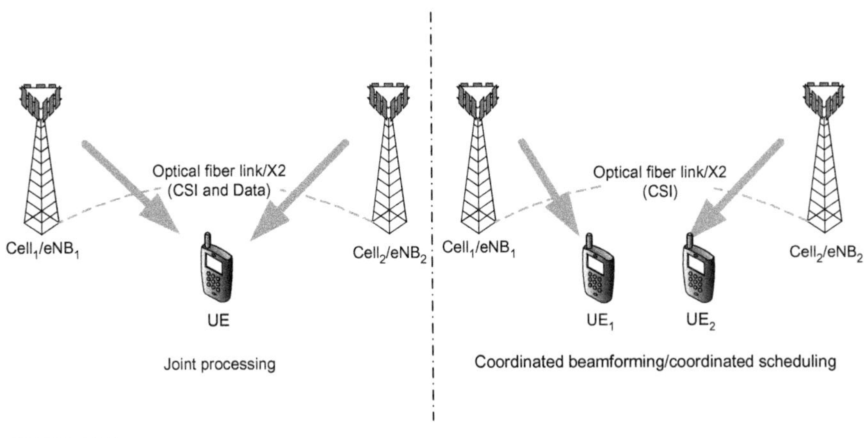

FIGURE 12.6

Illustration of downlink CB/CS and JT CoMP concepts.

corresponding to the CoMP cooperating set. For example, some points in the cooperating set may transmit data to the target UE according to a JP scheme, while other transmission points in the cooperating set may perform one form of CS/CB scheme. For the JT scheme, CoMP transmission points may include multiple points in the CoMP cooperating set at each subframe for a certain time–frequency resource, whereas in the case of CS/CB, DPS, SSPS, a single point in the CoMP cooperating set is the CoMP transmission point at each subframe for a certain time–frequency resource. For SSPS, this CoMP transmission point can change semi-statically within the CoMP cooperating set. Examples of static and dynamic formation of CoMP clusters (cooperating sets) are shown in Figures 12.7 and 12.8.

12.4.3 Uplink CoMP

The uplink CoMP schemes may be divided into one of the following categories:

- Joint reception (JR) in which physical uplink shared channel (PUSCH) transmitted by the UE is received jointly at multiple points which are part of or the entire CoMP cooperating set at a certain time, e.g., to improve the received signal quality.
- CS/CB where user scheduling and precoding decisions are made with coordination among points within the CoMP cooperating set. The user data is only intended for one recipient point in the set.

Similar to the downlink, the uplink CoMP cooperating set is a set of geographically separated points that may be designated for data reception from a UE. The uplink CoMP reception points are a set of points for receiving data from a UE. The CoMP reception points are a subset of the CoMP cooperating set in the uplink. For the JR scheme, CoMP reception points may include multiple points in

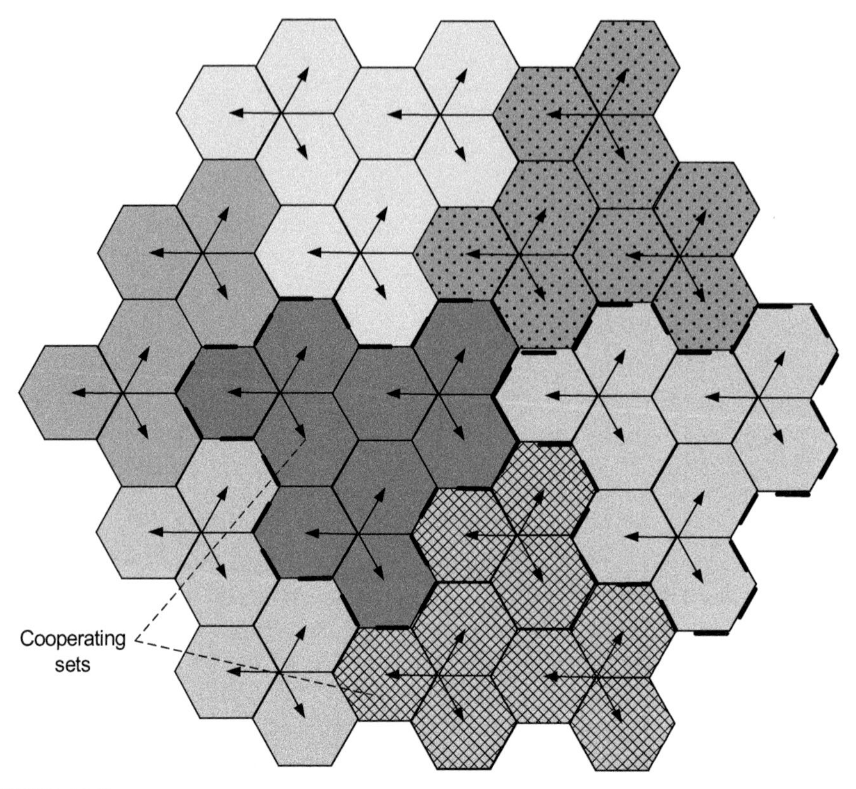

FIGURE 12.7

Illustration of static CoMP clusters (Scenario 2, 9 cells per cooperating set) [5].

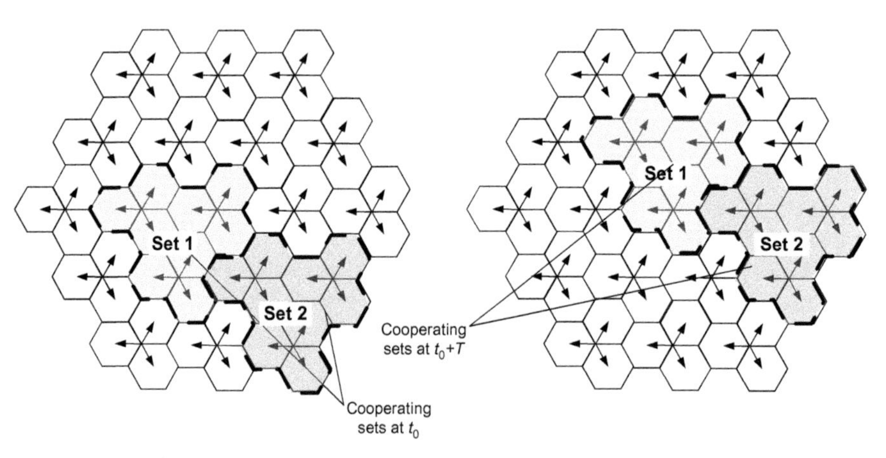

FIGURE 12.8

Dynamic clustering of CoMP (change of cooperating CoMP sets across time).

the CoMP cooperating set at each subframe for a certain time–frequency resource, whereas for uplink CS/CB, a single point in the CoMP cooperating set is the CoMP reception point at each subframe for a certain time–frequency resource [5].

In other words, uplink CoMP schemes can be divided into two major groups. The first group uses dynamic selection of the best reception point according to short-term channel quality. The second category uses joint reception, where several reception points act as a joint receiver with a larger number of antennas. Joint reception is more complex to implement, but it offers larger gains relative to DPS. One of the most important aspects of coordinated uplink reception is the decoupling of the points that transmit downlink control from those that receive uplink data. Important aspects of the implementation pertain to coordination and exchange of information among reception points in terms of selection of reception points, computation of the receive combiner, coordinated scheduling, and exchange of received signal samples (e.g., quantized received signals, locally combined signals, or log-likelihood ratios). These functions are typically centralized in the baseband processing unit connected to the coordinated eNBs via optical fiber links [17].

In uplink coordinated multi-point reception, the UE signals which were previously treated as interference at a reception points, other than the serving node, would become desired signals. Uplink reference signals that are used for channel estimation in the demodulation process of PUSCH are very sensitive to interference. In LTE Rel-8/9/10, the assignment of the pseudo-random base sequences for scrambling the reference signals of a UE was made by a single serving cell. It was possible to assign orthogonal sequences to the UE, whose reference signals were received by the same receiving point, via either different cyclic shifts of the same base sequence or orthogonal cover codes. At the same time, several interference randomization techniques improved the inter-cell interference generated by uncoordinated uplink scheduling assignments across multiple neighboring cells. These randomization techniques include sequence-group hopping, sequence hopping, and different cyclic shifts [1–3]. The opportunity offered by CoMP to receive the UE's signals at one cell when the UE's transmission is intended for another cell created new constraints for assigning orthogonal sequences to the users allocated in overlapping time–frequency resources [17].

LTE Rel-11 introduced new capabilities for configuring reference signal initialization parameters. In particular, the initialization of the pseudo-random sequences is no longer associated with the physical cell identity, but rather is explicitly signaled as a UE-specific virtual cell identity [1]. This enables the assignment of the same base sequence with different cyclic shifts to the UE which is connected to different cells, allowing virtual cells to be created within the coverage area of a macro-cell deployed with RRHs, as in CoMP scenario 4. Furthermore, it is possible to configure the initialization of the cyclic shift hopping in a dedicated manner for each UE, which enables grouping of UEs by orthogonal cover codes across different cells even when these UEs are assigned different base sequences [17].

12.4.4 CSI measurement and feedback in CoMP-enabled networks

A downlink CoMP measurement set is defined as a group of transmission points relative to which the UE measures and reports the CSI. The UE reports may select points for which actual feedback is transmitted. The measurement set typically encompasses cells for which the mobility-management-related measurements are conducted for the purpose of cell selection/reselection and handover. The proper operation of any CoMP scenario depends on the accuracy and timeliness of the CSI measurement and feedback in the network. Some CoMP schemes are more affected by CSI inaccuracy and feedback delays. LTE Rel-11 has specified a general control signaling framework that enables all CoMP schemes to properly operate by enhancing and extending the features that existed in the previous releases. The CoMP measurement set is alternatively defined as a set of CSI-RS resources on which the UE is required to measure and report the CSI that may include rank indicator, precoding matrix index, and channel quality indicator. There is typically a unique correspondence between a CSI-RS resource and a transmission point. As shown in Figure 12.9, the CoMP measurement set is a subset of the CoMP cluster/cooperating set ($S_{\text{measurement-set}} \subset S_{\text{cooperating-set}}$). The size of a CoMP measurement set that can be configured for a UE is limited to three CSI-RS resources in order to reduce the uplink signaling overhead. The UE at different locations may observe different sets of strong-signal transmission points; thus, the configuration of a CoMP measurement set is UE-specific. Due to UE mobility, CoMP measurement sets dynamically change and need to be reconfigured when necessary to ensure optimal CoMP performance. To assist the network in managing the CoMP measurement set of each UE, a CoMP resource management procedure based on CSI-RS measurement has been defined in LTE Rel-11. While the CoMP resource management concept can be applied to all CoMP scenarios, it is particularly useful for CoMP scenario 4 where the transmission points are not identifiable from their transmitted cell-specific reference signals, making it impossible to compare the received signal strengths based on cell-specific reference signals.

The CoMP resource management procedure requires the UE to perform CSI-RS-based received signal quality measurements on a set of CSI-RS resources that are candidates for inclusion in the CoMP measurement set. The set of CSI-RS resources explicitly configured by the network for CoMP resource management is called the CoMP resource management set. The latter set can be configured independently of the CoMP measurement set [17]. The measurement of CSI-RS received power is similar to that defined for RSRP measurement on cell-specific reference signals. Nevertheless, there are some differences between the two types of measurements. The UE is only required to conduct measurement on CSI-RS resources in the CoMP resource management set. Therefore, the UE does not autonomously detect CSI-RS resources, resulting in reduced UE operational complexity. The CoMP resource management measurement/reporting is event-driven and the triggers can be either absolute, i.e., a measurement report is triggered

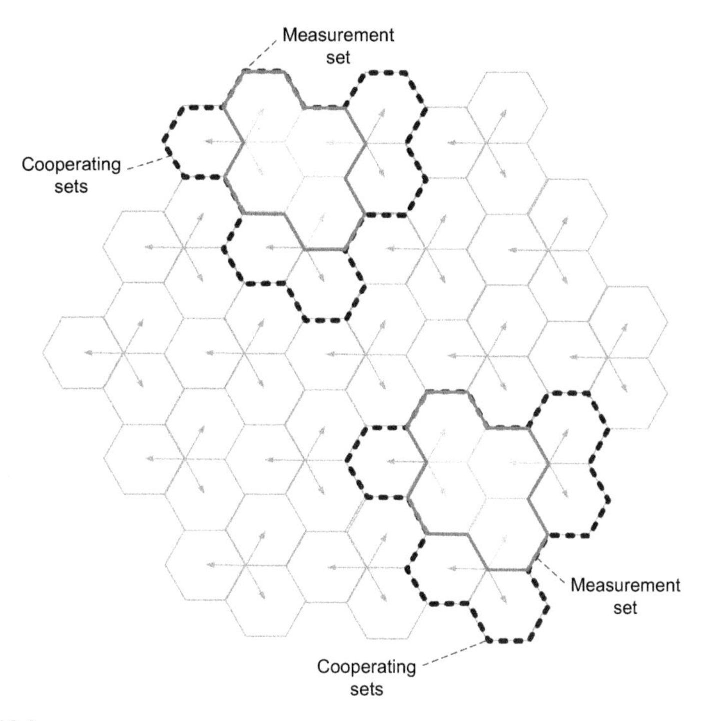

FIGURE 12.9

Example illustration of cooperating and measurement sets.

when the received signal power goes beyond an absolute threshold, or relative, i.e., the received signal power is compared to that of another CSI-RS resource.

To ensure the CSI-RS received power measurement is immune to interference and to ensure the accuracy of the measurement, the network can mute the transmissions on the resource elements from the neighboring TPs that correspond to the CSI-RS resource elements of the transmission point being measured. When performing CSI-RS received power measurements, the UE may assume that the timing of the received CSI-RS is the same as that derived from the primary and secondary synchronization signals of the serving cell. This allows the UE to use a single timing derived from the synchronization signals of the serving cell to receive CSI-RS that might be transmitted from a different TP. While this simplifies UE implementation, the impact on the accuracy of CSI-RS received power depends on the extent of the timing error. Studies suggest that there is no significant impact on CSI-RS measurement provided that a timing error is limited to ± 3 ms [17,21,27].

Since a UE needs to measure CSI-RS to generate CSI feedback and receive UE-specific reference signals to coherently demodulate PDSCH, it is imperative that the CSI-RS and DM-RS are transmitted such that the interference among the reference signals of multiple TPs is sufficiently randomized. LTE Rel-10 addresses this requirement by defining unique initialization seeds for generating the scrambling sequences of each transmission point/cell's reference signals based on the physical cell identity. This approach may not work for CoMP scenario 4 since multiple TPs share the same cell identity. In order to provide sufficient interference randomization, LTE Rel-11 replaced the use of the physical cell identifier for initialization of the scrambling sequences with a virtual cell identifier, which is signaled to the UE using higher layer signaling, allowing flexible deployment of transmission points in CoMP scenario 4, and providing sufficient interference randomization for DM-RS and CSI-RS. In the case of CSI-RS, the interference among CSI-RSs of different TPs is further reduced by allocating different time or frequency resources for CSI-RS transmissions. The use of a virtual cell identifier provides interference randomization when the same time and frequency resources are used across different cells for CSI-RS transmissions [1].

The feedback scheme for CoMP in LTE Rel-11 is defined per CSI-RS resource, where the UE reports the CSI individually for multiple CSI-RS resources. Note that each CSI-RS resource is transmitted from one transmission point. As a result, when the UE is configured with multiple CSI-RS resources and performs per CSI-RS resource feedback, it reports individual CSIs for each of the TPs transmitting the configured CSI-RS resources. There is at least one feedback process associated with each CSI-RS resource within the CoMP measurement set. Each feedback process provides the eNB with the measured CSI (measured on a particular CSI-RS resource). The CSI feedback can be configured to be periodically reported, or scheduled for aperiodic reporting. For CS/CB, separate CSI feedback for at least two TPs can be configured where one of the processes would be used for reporting the CSI for the serving TP, and the other(s) can be used for reporting the CSI associated with the interfering TP(s). Similarly, for a dynamic point switching scheme, multiple CSI feedback processes can be configured, allowing the network to dynamically switch the transmitting TPs. The above feedback mechanism can also support other CoMP schemes including non-coherent JT, where precoding is applied individually to the transmit antennas belonging to each TP, and coherent JT in a non-standardized form by aggregating multiple TPs under a single CSI-RS resource.

The reader's attention is drawn to the fact that there are two types of feedback in general to support CoMP operation, as follows:

1. Explicit feedback which is characterized by having a channel part and a noise-and-interference part. The channel part includes reports for each point in the UE's measurement set that is reported in a given subframe that contains one or several channel properties. The channel properties include instantaneous channel matrix either in the form of full matrix or main eigen component(s) of the channel matrix; and transmit channel covariance either in the form of full covariance matrix or main eigen component(s) of the channel covariance matrix. Inter-point channel properties may also be reported including a noise-and-interference part, total interference received power, total received signal covariance matrix, and covariance matrix of the noise and interference.

2. Implicit feedback is based on certain hypotheses at the UE, and the feedback is generated depending on the operation scheme, e.g., SU-MIMO/MU-MIMO, single cell/point or coordinated transmission, and single-point CS/CB or multi-point (JP) transmission. Under this category, other forms of feedback may be considered such as precoding matrix index (PMI) with fine quantization granularity, wideband/subband PMI, and channel quality indicator (CQI). Note that CQI only accounts for interference outside the CoMP measurement set or relative received power between CoMP transmission points.

LTE Rel-11 does not specify any standardized support for coherent JT. Another important feature that has been included in LTE Rel-11 to support CoMP is the interference measurement resource (IMR), which is a set of resource elements where a UE can be configured to measure interference. Prior to LTE Rel-11, interference measurement was mainly implementation-specific and based on cell-specific reference signals. Such interference measurement schemes may not be applicable to CoMP because in CoMP scenario 4, multiple TPs may share a common cell identity. In that case, TPs within a cell are not distinguishable since they are transmitting the same cell-specific reference signals. Moreover, the interference measured using cell-specific reference signals does not allow the network to adaptively control the interference measured by the UE. As an example, if the effect of a particular TP shutdown is desired, the network can configure two separate IMRs for two feedback processes. One of the IMRs would capture the interference with the TP turned off, while the other the IMR would capture the interference with the TP turned on. The configuration of the IMR is done on a per-UE basis via higher layer signaling. Depending on the measurement reports from the UE, the network can determine which transmission points should be added to, removed from, or replaced in the CoMP measurement set in order to select the set of transmission points with the strongest signal measured at the UE [3].

12.5 Effects of X2 latency on CoMP performance

The X2 latency is defined as the accumulation of the time needed to send a message across the X2 interface, as well as the required time to process the X2 message in the eNB [20,30]. The non-zero X2-latency affects HARQ ACK/NACK round trip time and the CSI feedback delay. The HARQ ACK/NACK round trip time is defined as the time interval between the first transmission of PDSCH using a HARQ process identifier and the earliest retransmission/fresh transmission using the same HARQ process identifier. The CSI feedback delay is the time interval between the CQI measurement period at the UE and the next downlink data transmission from the eNB that is based on the knowledge of this CQI. Since both CQI as well as HARQ information need to be transmitted from the serving cells to the master cell of a cluster across the X2 interface, the X2 delay impacts both HARQ delay as well as CQI delay. The ACK/NACK delay determines the number of HARQ processes that are needed to guarantee continuous transmission (efficient channel utilization), if the mobile station is scheduled all the time. Therefore, the impact of long HARQ ACK/NACK delays is important for low to medium cell loads. In highly loaded cells, the number of required HARQ processes decreases, since each mobile station attains less access to the channel. This impact is therefore dependent on the cell load. Due to the increased CQI delay, the CQI contains less information about the channel at the time when it is accessed. This means that it becomes more difficult to adapt the MCS and other transmission parameters to the instantaneous channel conditions.

In order to investigate the effect of X2 latency on the inter-eNB JT CoMP operation, let us assume that a sufficient number of HARQ processes are available such that the impact of the X2 delay on the HARQ ACK/NACK delay can be neglected. This is ensured by setting the number of available HARQ processes to infinity. To investigate the impact of X2 latency on CQI delay, X2 delays of 0, 1, 3, 5, 10, 15, and 20 ms were simulated. Figure 12.10 shows the loss in average user throughput for various X2 delays relative to an ideal interface with zero delay. The losses are given for the 5th, 50th, and 95th percentile of the cumulative distribution function (CDF) of average user throughput without X2 delay. It can be seen that the losses become larger with increasing values of X2 delay. This indicates that inter-eNB JT CoMP is very sensitive to X2 delay.

12.6 Performance of CoMP scenarios

The performance of intra-eNB CoMP techniques has been extensively studied and evaluated in 3GPP and documented in Ref. [5] as part of the CoMP feasibility studies. The detailed simulation assumptions and simulation models used in this evaluation are described in 3GPP TR 36.814 and TR 36.819 [4,5]. The evaluations were conducted by various sources in the following deployment scenarios [31,32]:

1. Scenario 1: Homogeneous network with intra-site CoMP.
2. Scenario 2: Homogeneous network with high transmit-power RRHs.
3. Scenario 3: Heterogeneous network with low-power RRHs within the macro-cell coverage, where transmission/reception points are identified by different cell IDs.
4. Scenario 4: Heterogeneous network with low-power RRHs within the macro-cell coverage where the transmission/reception points created by the RRHs have the same cell IDs as the macro-cell.

Despite the fact that there was a detailed evaluation methodology that was used by all companies participating in the evaluations, there were discrepancies in the reported results among sources because they used different assumptions concerning (unspecified parameters) channel estimation error modeling, channel reciprocity modeling, the feedback/SRS mechanisms, scheduling algorithm, and the

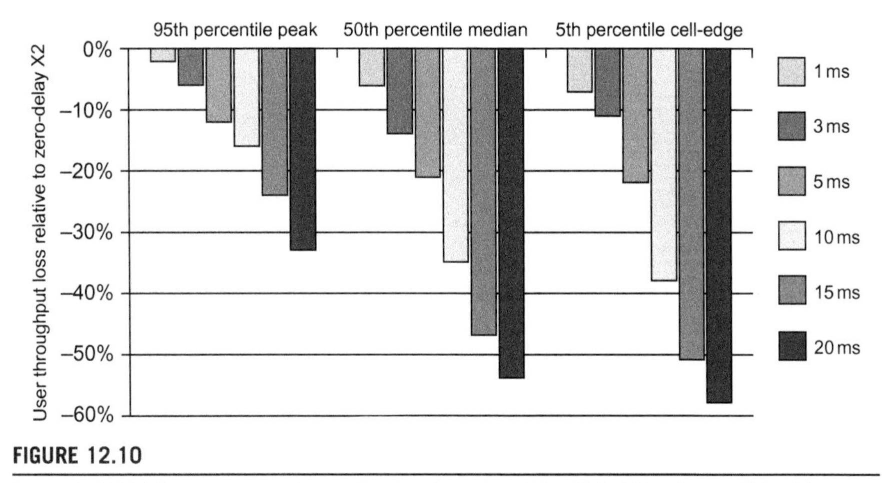

FIGURE 12.10

Impact of X2 latency on user throughput using inter-eNB JT CoMP scheme [15,22].

receiver architecture. In the full-buffer/non-full-buffer evaluations, cell average spectral efficiency in bits/s/Hz/cell and 5% user spectral efficiency in bits/s/Hz/user were evaluated and reported. The user spectral efficiency is defined as the amount of data (file size) per time needed to download data per occupied bandwidth. The time needed to download the data starts when the packet is received in the transmit buffer, and ends when the last bit of the packet is correctly delivered to the receiver. The serving cell spectral efficiency represents the total amount of data for all users during the entire observation time per the number of cells and per occupied bandwidth.

It was observed from the evaluation results for scenarios 1 and 2 that in the downlink, CoMP schemes offer performance gains in homogeneous networks. In the uplink, considerable gain is achievable with a CoMP JR scheme in scenarios 1 and 2. The results were obtained based on ideal and realistic assumptions. It was concluded from the evaluation results in scenarios 3 and 4 that CoMP can provide performance benefits in heterogeneous networks both with full-buffer and FTP traffic models. Based on the evaluation results on the impact of backhaul constraints (capacity/latency), it was concluded that the performance of CoMP schemes relying on spatial information exchange is sensitive to the delay between two transmission points; however, the level of sensitivity depends on the CoMP scheme.

The macro-cell area average spectral efficiency in Figure 12.11 is defined as the sum of cell throughputs in the macro-cell area, which also includes pico-cells/RRHs deployed in the macro-cell area. The results of evaluations show that performance gains can be achieved using CoMP compared to the traditional single-cell schemes in both homogeneous and heterogeneous networks. CoMP performance gains are noted for UEs severely affected by interference. These UEs are typically found at the cell-edges in homogeneous networks. In general, JT CoMP

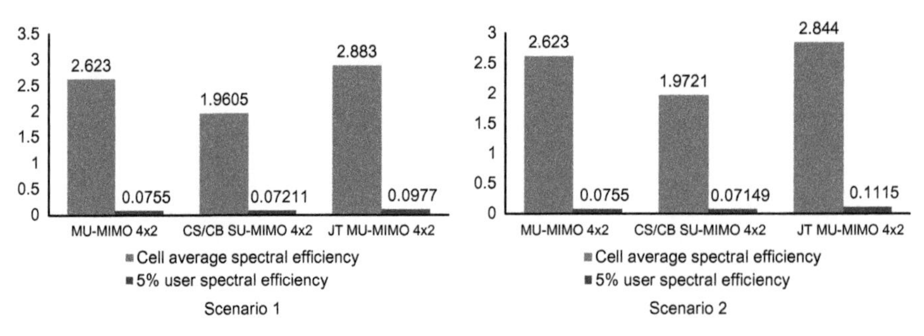

FIGURE 12.11

Comparison of the absolute performance downlink MU-MIMO, CS/CB SU-MIMO, and JT MU-MIMO (FDD, full-buffer, 3GPP spatial channel model, cross-polarized antennas) [4,5].

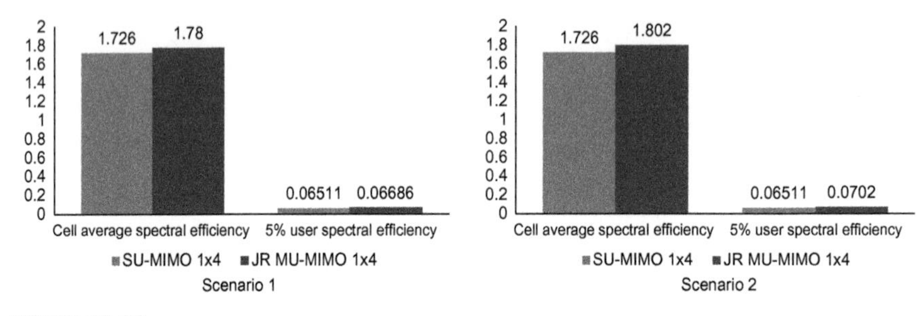

FIGURE 12.12

Comparison of the absolute performance uplink SU-MIMO and JR MU-MIMO (FDD, full-buffer, 3GPP spatial channel model, cross-polarized antennas) [4,5].

techniques provide larger performance gains relative to CS/CB CoMP schemes, but at the expense of larger backhaul overheads and more stringent requirements on backhaul capacity and latency (see Figure 12.11). These observations suggest that CoMP techniques are a promising trend for LTE-Advanced networks. CoMP will thereby contribute to providing the end users with a more consistent experience of mobile broadband across the network coverage areas (Figure 12.12).

The relative performance gains of CoMP scenarios 3 and 4 over HetNet with and without enhanced ICIC (eICIC) were evaluated and are shown in Figure 12.13.

The impact of lower capacity/higher latency links connecting the transmission points was also investigated. The evaluation results were obtained for CoMP in a homogeneous network [5]. Figure 12.14 shows the evaluation results for SU-MIMO CS/CB and MU-MIMO JT CoMP schemes with varying levels of backhaul delay. The CSI feedback delay is assumed to be 5 ms. The results suggest that the JT CoMP scheme demonstrates larger performance variation, and is more sensitive to backhaul delays compared to the CS/CB CoMP scheme.

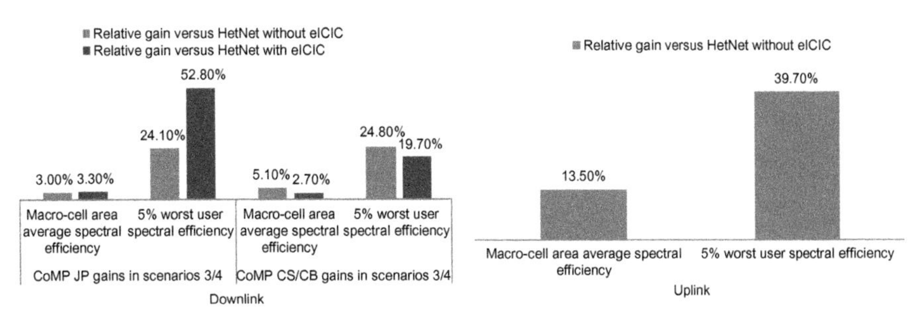

FIGURE 12.13

Relative performance gain of downlink/uplink CoMP in scenarios 3/4 (FDD, full-buffer, configuration 1) [4,5].

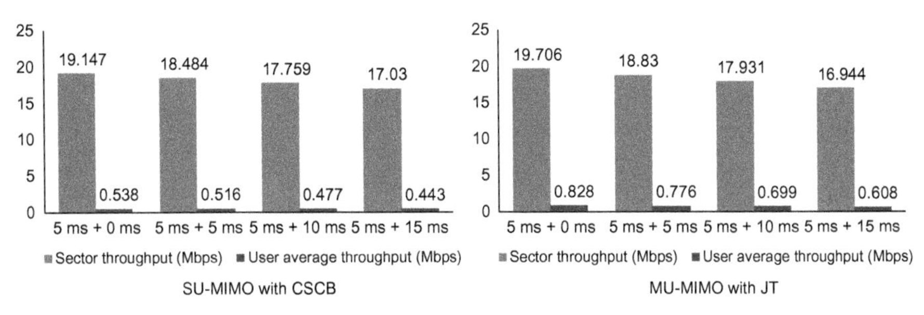

FIGURE 12.14

Effect of CSI feedback delay + backhaul delay on sector and user average throughput [4,5].

References

3GPP Specifications

[1] 3GPP TS 36.211, Evolved Universal Terrestrial Radio Access (E-UTRA); Physical Channels and Modulation (Release 11), December 2012.

[2] 3GPP TS 36.212, Evolved Universal Terrestrial Radio Access (E-UTRA); Multiplexing and Channel Coding (Release 11), December 2012.

[3] 3GPP TS 36.213, Evolved Universal Terrestrial Radio Access (E-UTRA); Physical Layer Procedures (Release 11), December 2012.

[4] 3GPP TR 36.814, Further Advancements for E-UTRA; Physical Layer Aspects (Release 9), March 2010.

[5] 3GPP TR 36.819, Coordinated Multi-point Operation for LTE Physical Layer Aspects (Release 11), December 2011.

[6] 3GPP TS 36.420, X2 General Aspects and Principles (Release 11), September 2012.

[7] 3GPP TS 36.423, X2 Application Protocol (X2AP) (Release 11), December 2012.

Books

[8] Y. Zhang, H. Hu, J. Luo, Distributed Antenna Systems: Open Architecture for Future Wireless Communications, Auerbach Publications, June 2007.

Articles

[9] A.M. Saleh, A.J. Rustako, R.S. Roman, Distributed antennas for indoor radio communications, IEEE Trans. Commun. 35 (December 1987).

[10] The distributed antenna systems forum. <http://thedasforum.org>.

[11] B. Chow, et al., Radio-over-fiber distributed antenna system for WiMAX bullet train field trial, IEEE Mobile WiMAX Symposium, California, USA, July 2009.

[12] Distributed antenna systems. <http://www.alino.com/Info/DistributedAntennaSystems/das.htm>.

[13] Jiwon Kang, Bin-Chul Ihm, Wookbong Lee, Distributed antenna system for future 802.16, LG Electronics, IEEE C802.16-10/0018, March 2010.

[14] Distributed antenna systems. <http://en.wikipedia.org/wiki/Distributed_Antenna_System>.

[15] S. Brueck, et al., Centralized scheduling for joint transmission coordinated multipoint in LTE-Advanced, International ITG Workshop on Smart Antennas (WSA 2010), 2010.

[16] O. Simeone, Distributed Antenna Systems: Open Architecture for Future Wireless Communications, New Jersey Institute of Technology, August 2006.

[17] J. Lee, et al., Coordinated multipoint transmission and reception in LTE-Advanced systems, IEEE Commun. Mag. (November 2012).

[18] S. Firouzabadi, A. Goldsmith, Downlink Performance and Capacity of Distributed Antenna Systems, Stanford University, September 2011.

[19] W. Choi, et al., Downlink performance and capacity of distributed antenna systems in a multi-cell environment, IEEE Trans. Wireless Commun. 6 (1) (January 2007).

[20] J. Robson, Guidelines for LTE backhaul traffic estimation, NGMN White Paper, NGMN Alliance, July 2011.

[21] R. Irmer, et al., Coordinated multipoint: concepts, performance, and field trial results, IEEE Commun. Mag. (February 2011).

[22] White Paper, Backhauling X2, Cambridge Broadband Networks Limited.

[23] P. Komulainen, M. Boldi, D1.4 initial report on advanced multiple antenna systems, wireless world initiative new radio—WINNER+. <http://projects.celtic-initiative.org/winner+/deliverables_winnerplus.html>, January 2009.

[24] S. Mayrargue, D1.8 intermediate report on CoMP (Coordinated Multi-Point) and relaying in the framework of CoMP, wireless world initiative new radio—WINNER+. <http://projects.celtic-initiative.org/winner+/deliverables_winnerplus.html>, January 2010.

[25] A. Osseiran, D2.2 enabling techniques for LTE-A and beyond, wireless world initiative new radio—WINNER+. <http://projects.celtic-initiative.org/winner+/deliverables_winnerplus.html>, July 2010.

[26] A. Tyrrell, H. Schöneich, Definitions and architecture requirements for supporting interference exploitation techniques, advanced radio interface technologies for 4G systems ARTIST4G. <https://ict-artist4g.eu/projet/deliverables>, August 2010.

[27] D. Lee, et al., Coordinated multipoint transmission and reception in lte-advanced: deployment scenarios and operational challenges, IEEE Commun. Mag. (February 2012).

[28] Wolfram MathWorld, Central limit theorem. <http://mathworld.wolfram.com/CentralLimitTheorem.html>.

[29] Common public radio interface (CPRI). <http://www.cpri.info/>.

[30] 3GPP R1-111174, Backhaul Modeling for CoMP, Orange, Telefónica, February 2011.

[31] 3GPP R1-111944, Phase 1 CoMP simulation evaluation results and analysis for full buffer, LG Electronics, May 2011.

[32] 3GPP R1-111944, Phase 1 CoMP results, Samsung, May 2011.

[33] Y.H. Nam, et al., Evolution of reference signals for LTE-advanced systems, IEEE Commun. Mag. 50 (2) (February 2012).

[34] S. Brueck, Backhaul Requirements for Centralized and Distributed Cooperation Techniques, ITG Heidelberg, July 2010. <http://www.ikr.uni-stuttgart.de/Content/itg/fg524/Meetings/2010-07-08-Heidelberg/index.html>.

[35] P. Marsch, S. Khattak, G. Fettweis, A framework for determining realistic capacity bounds for distributed antenna systems, Proceedings of the IEEE Information Theory Workshop, Chengdu, China, October 2006.

[36] K.M. Karakayli, G.J. Foschini, R.A. Valenzuela., Network coordination for spectrally efficient communications in cellular systems, IEEE Wireless Commun. 13 (4) (August 2006).

[37] P. Marsch, G. Fettweis, On multi-cell cooperative transmission in backhaul-constrained cellular systems, Ann. Télécomm. 63 (5–6) (May 2008).

[38] V. Jungnickel, et al., Interference aware scheduling in the multiuser MIMO-OFDM downlink, IEEE Commun. Mag. 47 (6) (June 2009).

[39] V. Jungnickel, et al., Field trials using coordinated multi-point transmission in the downlink, Third IEEE International Workshop on Wireless Distributed Networks, IEEE PIMRC, September 2010.

[40] P. Marsch, Coordinated Multi-Point Under a Constrained Backhaul and Imperfect Channel Knowledge, PhD thesis, Technische Universitat Dresden, March 2010.

[41] A. Müller, P. Frank, Cooperative interference prediction for enhanced link adaptation in the 3GPP LTE uplink, IEEE VTC, Spring 2010.

[42] A.Müller, P.Frank, Performance of the LTE uplink with intra-site joint detection and joint link adaptation, IEEE VTC, Spring 2010.

[43] M. Grieger, et al., Field trial results for a coordinated multi-point (CoMP) uplink in cellular systems, Proceedings of the ITG/IEEE Workshop on Smart Antennas, Bremen, Germany, February 2010.

[44] V. Kotzsch, G. Fettweis, Interference analysis in time and frequency asynchronous network MIMO OFDM systems, IEEE WCNC 2010, Sydney, Australia, April 2010.

[45] J. Giese, M.A. Awais, Performance upper bounds for coordinated beam selection in LTE-advanced, Proceedings of the ITG/IEEE Workshop on Smart Antennas, Bremen, Germany, February 2010.

[46] G. Fettweis et al., Field trial results for LTE-advanced concepts, Proceedings of the IEEE ICASSP, Dallas, TX, March 2010.

[47] L. Thiele, V. Jungnickel, T. Haustein, Interference management for future cellular OFDMA systems using coordinated multi-point transmission, IEICE Trans. Commun. (December 2010) Special issue on wireless distributed networks.

CHAPTER

Carrier Aggregation

13

CHAPTER CONTENTS

LTE Rel-8/9 specified system bandwidths of 1.4, 3, 5, 10, 15, and 20 MHz to meet different spectrum and deployment requirements. Support of wider bandwidths up to 100 MHz was one of the distinctive features of IMT-Advanced systems. The IMT-Advanced systems targeted peak data rates in excess of 1 Gbps for low mobility and 100 Mbps for high mobility scenarios [32,33]. In order to support wider transmission bandwidths, LTE Rel-10 introduced the carrier aggregation concept, where two or more component carriers with arbitrary bandwidths belonging to the same or different frequency bands could be aggregated. The support of system bandwidths up to 100 MHz would allow a substantial increase in link-level peak data rate, as well as in the system capacity and user throughput. Another advantage of carrier aggregation is to allow efficient use of fragmented spectrum in the form of a virtually wideband spectrum. The fragmented spectrum may belong to the same or different frequency bands and may be of different bandwidths. LTE Rel-11 further extended the advantages of carrier aggregation by configuring extended carriers and carrier segments. Furthermore, carrier aggregation can be used to mitigate inter-cell interference in heterogeneous networks [30,31].

Using the carrier aggregation scheme, it would be possible to simultaneously schedule a user on multiple component carriers for downlink or uplink data transmission, resulting in some challenges in resource scheduling and load balancing across the network. In a non-contiguous inter-band carrier aggregation scenario, where the aggregated carriers belong to different frequency bands, the fading characteristics might be different between component carriers; as a result, the coverage may vary significantly from one carrier to another. At different locations in the cell, some users may only be scheduled on fewer carriers, while other users may have access to the entire carrier set, which may have a negative impact on sustained user throughput within the coverage area of an eNB.

To allow smooth network migration and upgrades to new releases, it was essential to ensure backward compatibility of LTE-Advanced with LTE Rel-8/9, as well as to support mixed operation of the legacy LTE systems and LTE Rel-10 on the same carrier in an operator's network. LTE Rel-10 carrier aggregation satisfied this requirement by configuring each component carrier in a backward compatible manner with LTE Rel-8/9, and by further supporting at least one of the bandwidths compatible with LTE Rel-8/9. The backward compatibility implies that the complete set of LTE Rel-8/9 downlink physical channels and signals are transmitted on each component carrier using LTE Rel-8/9 signaling procedures, including broadcast channel, synchronization signals, reference signals, and control channels. This would also allow reuse of the LTE Rel-8/9 RF designs and implementations at the eNB and the UE. Although we will primarily focus on carrier aggregation for FDD systems, carrier aggregation is fully supported in the TDD systems, as well.

In this chapter, we will take a pragmatic and systematic approach to describe carrier aggregation principles and its impact on different protocol layers. Various aspects of carrier aggregation including deployment scenarios, physical and MAC layer support, as well as higher-layer signaling and RF transceiver design challenges, will be discussed and examples will be provided.

13.1 Principles of carrier aggregation

In LTE-Advanced carrier aggregation terminology, a component carrier is often referred to as a serving cell, is assigned its own cell identifier, and is managed as a serving cell by the higher layers. Each individual RF carrier is known as a component carrier. A component carrier can be downlink and uplink or downlink only, but it cannot be an uplink-only RF carrier for obvious reasons. The number of downlink component carriers may be different than that of the

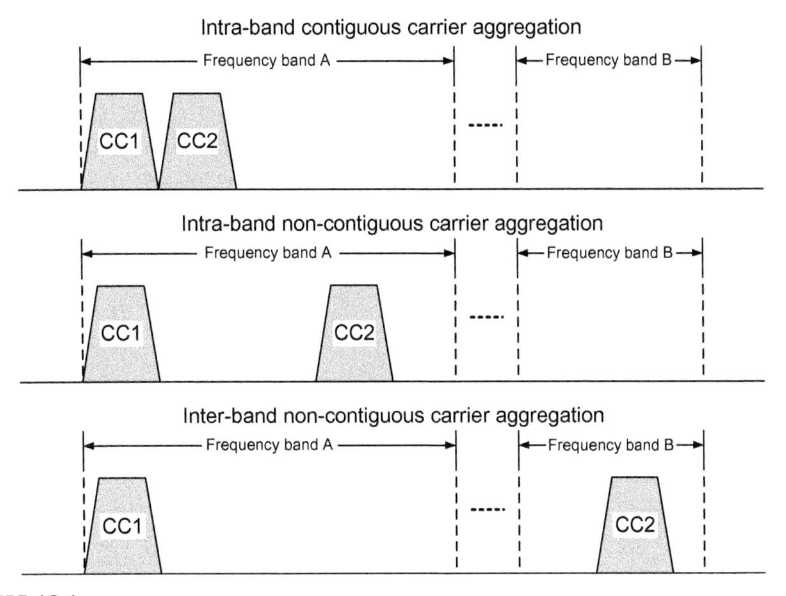

FIGURE 13.1

Carrier aggregation scenarios [16,18,20].

uplink. In general, a different number of component carriers can be aggregated for the downlink and uplink. The UE may support carrier aggregation only in the downlink or in both directions. Up to five component carriers, possibly each with different bandwidth can be aggregated, allowing transmission bandwidths of up to 100 MHz [12,14]. In FDD systems, a serving cell comprises a pair of different carrier frequencies for downlink and uplink transmissions, whereas for TDD, a serving cell is defined as a single-carrier frequency where downlink and uplink transmissions occur in different transmission time intervals. Given that component carriers do not have to be contiguous in frequency which enables utilization of fragmented spectrum, operators with a fragmented spectrum can provide high data-rate services based on the availability of a virtually wide bandwidth even though they do not own a single wideband spectrum allocation.

Each UE has a single serving cell that provides all necessary control information and functions, such as non-access stratum mobility information, security establishment, RRC connection establishment/reconfiguration, etc. This serving cell is referred to as the primary cell (PCell). Other auxiliary cells or component carriers are referred to as secondary cells (SCells).

As shown in Figure 13.1, there are three different carrier aggregation scenarios, as follows [11]:

1. Intra-band contiguous carrier aggregation: This type of carrier aggregation uses a single frequency band. It is the simplest form of carrier aggregation from an implementation point of view. In this scenario, the

carriers are contiguous and the frequency spacing between center frequencies of contiguously aggregated RF carriers is a multiple of 300 kHz in order to be compatible with the 100 kHz frequency raster of LTE Rel-8/9 and preserve orthogonality of the subcarriers with 15 kHz spacing.

2. Intra-band non-contiguous carrier aggregation: In this scenario, the aggregated carriers are not adjacent and the multi-carrier signal cannot be treated as a single signal; therefore, two transceivers may be required. This adds significant complexity particularly to the UE where form-factor, power consumption, and cost are the major concerns.
3. Inter-band non-contiguous: This type of carrier aggregation uses different frequency bands which might be appealing to network operators due to the possibility of using fragmented frequency bands. In this scenario, the UE may have to use multiple transceivers, having unavoidable impact on cost, performance, and power consumption. In addition, there are also further complexities resulting from the requirements to reduce intermodulation and cross-modulation impairments created by the transceivers. This type of carrier aggregation may improve mobility by exploiting radio propagation characteristics of different bands.

An enhanced carrier aggregation work item was approved for LTE Rel-11, where as part of that work item, new non-backward-compatible partially configured carrier types could be defined for LTE Rel-11 and beyond to address the ever-increasing demand for higher capacity and higher data rates in cellular networks [19]. The new carrier types are going to be useful in practical deployments where the size of the available frequency block does not match the provisioned bandwidths of 1.4, 3, 5, 10, 15, and 20 MHz in LTE Rel-8/9. If the size of the frequency block is smaller than 5 MHz (corresponding to 25 resource blocks), the number of resources on the carrier segment or extension carrier may not be adequate to efficiently transmit user data and synchronization signals, broadcast and control channels. Although it is desirable to reduce the overhead of the new carrier types (i.e., extension carrier and carrier segment), in practical deployment scenarios, if the frequency gap between the reference carrier and the extension/segment carrier is excessively large, the time and frequency synchronization and slot/subframe alignment of the carriers in a carrier set may not be maintained without including synchronization signals on the extension carrier/carrier segments.

Any fully configured backward compatible carrier in a carrier set may be extended by an extension carrier or appended by segment carriers. The design of the new carrier types has been required to support the operation in the following scenarios, but not necessarily equally optimized for both cases:

- Synchronized carriers: The base carrier and the new carrier types are synchronized in time and frequency to the extent that no separate synchronization is needed in the receiver.

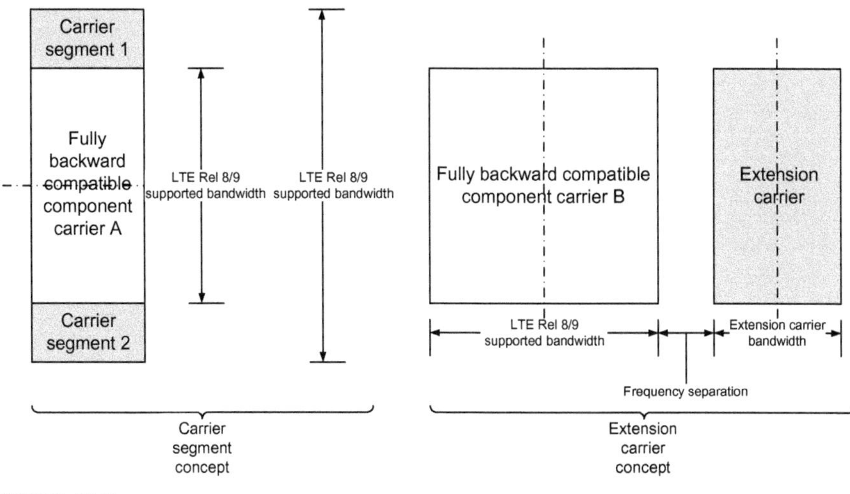

FIGURE 13.2

Illustration of the carrier segments and extension carriers [15].

- Unsynchronized carriers: The base carrier and the new carrier types are not synchronized with the same degree of accuracy as the synchronized carriers.

Note that synchronization is considered from the perspective of the UE receiver. Energy efficiency, flexible spectrum usage, heterogeneous network deployments, and machine type communications were identified as key objectives for the enhanced carrier aggregation project in 3GPP [19]. The considerations for the design of the additional carrier types that could assist in addressing the above goals mainly focused on operation with no cell-specific reference signals, enhanced control signaling with self-contained reference signals, and configurability of synchronization and other reference signal types when needed.

The new carrier types are defined as follows [15]:

- Carrier segment: The carrier segments are defined as the bandwidth extensions of an LTE Rel-8 compatible component carrier, provided that the total number of resource blocks is limited to 110. The combined RF carrier (i.e., LTE Rel-8 RF carrier + carrier segments as shown in Figure 13.2) utilizes the same backward compatible mechanisms of accessing the additional resources on the carrier segments as well as the base RF carrier. As an example, if two 2.5 MHz carrier segments are combined with a 5 MHz LTE Rel-8 RF carrier, the resulting 10 MHz RF carrier is going to be utilized as a 10 MHz LTE Rel-8 RF carrier. The use of carrier segments in the above manner would allow more efficient utilization of small fragments of spectrum without having to transmit additional overhead due to synchronization, broadcast, and control channels on the carrier segments. The notion of carrier segment allows for aggregation of additional resource blocks to a component carrier, while still retaining the backward compatibility of the original carrier, Note that carrier segments are always adjacent and linked to one component carrier and cannot be used as stand-alone. They do not provide synchronization signals, system information, or paging, and therefore cannot be used for random access or camping. They support the same HARQ process and transmission mode as the component carrier to which they are linked.

- Extension carrier: The extension carrier is a carrier that cannot be operated as a single stand-alone RF carrier and must be part of a component carrier set where at least one of the carriers in the set is a stand-alone backward compatible component carrier. Depending on whether the extension carriers are adjacent or non-adjacent to a component carrier carrying synchronization signals, the extension carrier may or may not be required to carry synchronization signals. If the extension carrier does not carry the synchronization signals, then it must be located close to a component carrier with synchronization signals for reliable synchronization. No legacy control channel PDCCH, PHICH, PCFICH, and no cell-specific reference signals is another possible configuration for the extension carriers which allows reduced overhead and more efficient use of radio resources on that carrier. The extension carrier supports a separate HARQ process, PDCCH, and transmission mode, which is configured from its linked component carrier.

13.2 Deployment scenarios

Carrier aggregation enables various network deployments. In general, carrier aggregation is used to improve data rates for users within overlapped areas of the cells (cell boundaries). However, carrier aggregation can also be used to mitigate inter-cell interference in heterogeneous networks (see Chapter 14). The following network deployment scenarios were considered during the development of LTE-Advanced. Although exemplified with two component carriers at frequencies f_{c_1} and f_{c_2} as shown in Figure 13.3, the concept can be generalized to any number of component carriers [7,22,23,29]:

- Deployment Scenario 1: In this case, cells with carrier frequencies f_{c_1} and f_{c_2} are (geographically) collocated, and their coverage is overlaid with f_{c_1} and f_{c_2} in the same frequency band. They provide approximately the same coverage due to similar path loss characteristics within the same band. This carrier aggregation scenario achieves higher data rates throughout the cell where both layers provide sufficient coverage and mobility. An example scenario is the case where $f_{c_1} = 2000$ MHz and $f_{c_2} = 800$ MHz are of the same band where aggregation is possible between the overlaid cells.

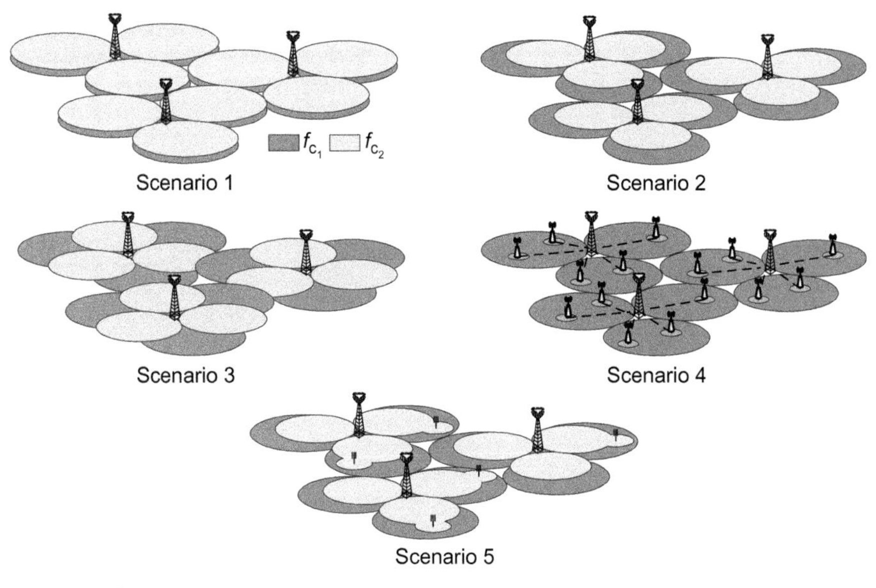

FIGURE 13.3

Carrier aggregation deployment scenarios [7].

- Deployment Scenario 2: In this case, the cells with carrier frequencies f_{c_1} and f_{c_2} are collocated and overlaid with f_{c_1} and f_{c_2} in different frequency bands. Different coverage is provided on different carriers due to the larger path loss in the higher frequency band. Mobility is typically supported on the carrier in the lower frequency band which further provides sufficient coverage. The carrier in the higher frequency band is used to improve data rates and throughput. The cells f_{c_1} and f_{c_2} are collocated and overlaid, but f_{c_2} has smaller coverage due to larger path loss. In other words, only f_{c_1} provides sufficient coverage and f_{c_2} is used to improve throughput. An example scenario would be the case where $f_{c_1} = \{800\text{ MHz}, 2000\text{ MHz}\}$ and $f_{c_2} = \{3500\text{ MHz}\}$ where aggregation is possible between the overlaid cells.
- Deployment Scenario 3: In this case, the cells comprising carrier frequencies f_{c_1} and f_{c_2} are collocated with f_{c_1} and f_{c_2} in different frequency bands. The antennas for cells of f_{c_2} are directed to the cell boundaries of f_{c_1} to improve the cell-edge data rates and user throughput. Due to the larger path loss, coverage holes exist for cells in the higher frequency band on which mobility management is typically not performed. The carrier aggregation is supported in areas with overlapping coverage and the mobility is based on f_{c_1} coverage. An example would be the case where $f_{c_1} = \{800\text{ MHz}, 2000\text{ MHz}\}$ and $f_{c_2} = \{3500\text{ MHz}\}$ in which aggregation is possible between the overlaid cells.

- Deployment Scenario 4: In this case the cells associated with carrier frequency f_{c_1} provide macro coverage and remote radio heads (RRHs) corresponding to carrier frequency f_{c_2} are used to improve throughput at hot spots. The mobility is performed based on the cell coverage of frequency f_{c_1}. In this deployment scenario, the carrier frequencies f_{c_1} and f_{c_2} are usually of different bands. The carrier aggregation is applicable to users within the coverage of RRHs and the underlying macro-cells. An example would be the case where $f_{c_1} = \{800\ \text{MHz}, 2000\ \text{MHz}\}$ corresponds to a larger cell and $f_{c_2} = \{3500\ \text{MHz}\}$ corresponds to smaller cell, in which aggregation is possible between the overlaid cells.
- Deployment Scenario 5: This case is similar to the second scenario where frequency-selective repeaters or distributed antenna systems are additionally deployed to extend the coverage for one of the carrier frequencies. It is expected that f_{c_1} and f_{c_2} cells of the same eNB can be aggregated where coverage overlaps.

13.3 Physical and MAC layer aspects of carrier aggregation

In this section, we discuss the physical and MAC layer aspects of carrier aggregation, and we will particularly focus on the impacts of carrier aggregation on the system operation at the lower protocol layers. Note that scheduling and use of the RF carriers are related to physical and MAC layers, and to a great extent are transparent to the upper layers.

13.3.1 Physical layer aspects

The physical layer is in charge of processing and transmission of data and control signaling between the radio access network and the user device. The LTE-Advanced carrier aggregation feature enables concurrent data transmission on multiple component carriers (up to five), where the transmission schemes are largely inherited from LTE Rel-8/9 data transmission procedures per component carrier, including the multiple access scheme; modulation and coding; and multi-antenna processing. Some additional UE functionalities beyond those of LTE Rel-8/9 are supported in LTE Rel-10, including data transmissions over two clusters of contiguous bandwidth on the same carrier where the data is precoded by a single DFT unit (see Chapter 10 for a description of clustered SC-FDMA), simultaneous transmission of a control channel and a data channel in the PCell (see Chapter 10 for simultaneous PUCCH and PUSCH transmission), and data transmissions from multiple antennas with spatial multiplexing [3–6]. Nevertheless, improving the downlink and uplink control signaling to efficiently support data transmission was the main challenge in the design of LTE-Advanced carrier aggregation [29].

The PDCCH provides information on resource allocation; modulation and coding scheme HARQ; and other parameters for downlink and uplink data transmission on the physical downlink/uplink shared channel. It is possible to configure each component carrier to transmit the PDCCH for scheduling PDSCH/PUSCH on the same carrier frequency and associated linked uplink carrier, where the linkage of the downlink and uplink carriers is conveyed in SIB2 [10]. In addition, the PDCCH on a component carrier can be configured to schedule PDSCH and PUSCH transmissions on other component carriers through a cross-carrier scheduling feature. Cross-carrier scheduling was primarily developed to support heterogeneous networks comprising a combination of macro-eNBs and low-power nodes (e.g., pico-cell, femto-cell, and RRHs) where significant intercell interference may arise when those networks are deployed on the same carrier [30,31]. Since the PDCCH is transmitted across the entire bandwidth of the respective carrier, interference coordination methods based on fractional frequency reuse may not be adequate in reducing the inter-cell interference on the PDCCH. Carrier aggregation can address this issue by configuring the UE connected to different network nodes to receive the PDCCH on different carriers, thereby reducing or even eliminating inter-cell interference on the PDCCH. With cross-carrier scheduling, only one component carrier needs to be protected for the PDCCH transmission, which can be used to allocate resources on other component carriers.

Cross-carrier scheduling enables a PDCCH on a component carrier to be configured in order to schedule PDSCH and PUSCH transmissions on another component carrier using a 3-bit carrier indicator field (CIF) which is inserted at the beginning of the downlink control information (DCI) messages. The carrier aggregation can reduce or even eliminate inter-cell interference on the PDCCH using cross-carrier scheduling. Figure 13.4 shows a hypothetical heterogeneous network scenario where a component carrier at f_1 is used by the macro-cell at full power and by the pico-cells at reduced power; whereas component carrier f_2 is used by the macro-cell at reduced power and by the pico-cells at nominally low power. In this case, cross-carrier scheduling from component carrier f_2 allows inter-cell interference mitigation among the macro- and pico-cells.

Cross-carrier scheduling is also an effective mechanism to schedule data transmissions on carriers of small bandwidth from a carrier with a larger bandwidth, considering that the PDCCH reliability targets are more challenging to achieve on carriers with small bandwidth due to limited frequency diversity gain. The PDCCH search space consists of a set of PDCCH candidates over which the UE performs blind decoding operations to determine whether there is any PDCCH addressed to it in a subframe. A common PDCCH search space is monitored by all UEs only in their respective PCells for receiving system information, paging, power control, and random access-related control information. Each UE also monitors a UE-specific PDCCH search space primarily carrying scheduling grants for downlink and uplink data transmissions. The start of the UE-specific PDCCH search space is a function of the UE's radio network temporary identifier (RNTI).

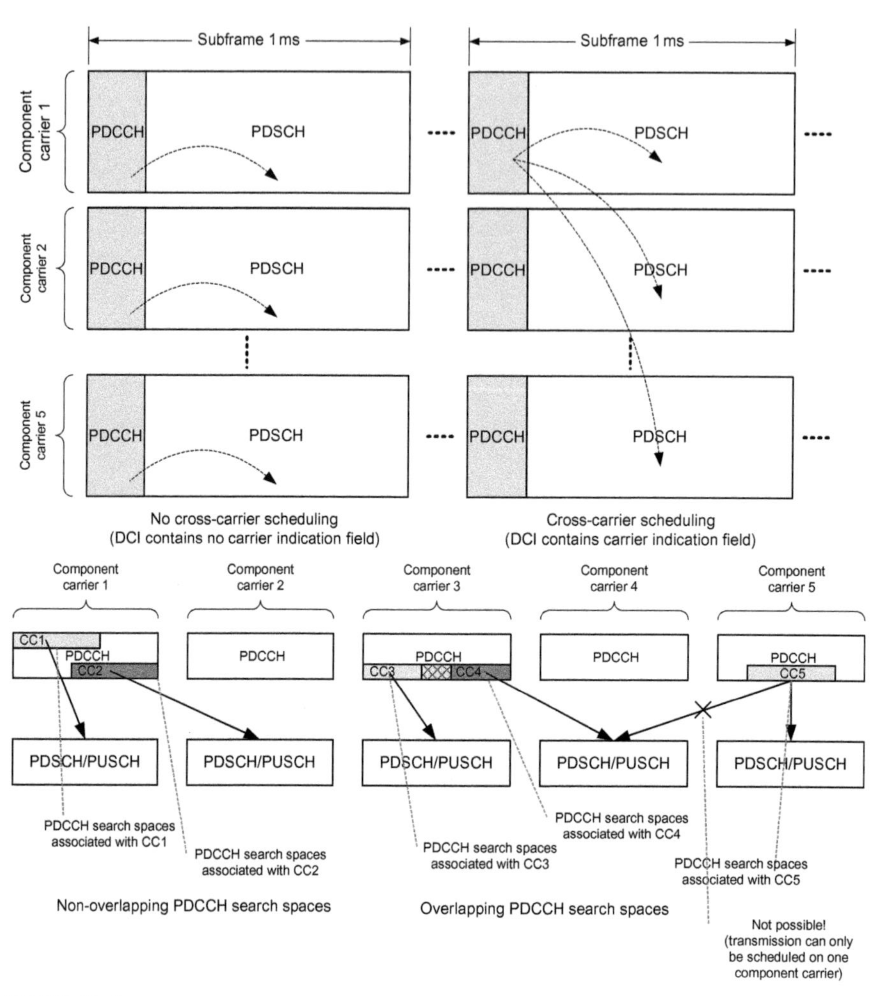

FIGURE 13.4

Illustration of the cross-carrier scheduling concept in carrier aggregation [13,14].

Since the same set of resources for PDCCH transmission is shared by all UEs, a PDCCH for a UE may not be transmitted due to partial overlapping with another PDCCH belonging to a different UE; this condition is referred to as PDCCH blocking. To maintain the PDCCH blocking probability at LTE Rel-8/9 level, an LTE Rel-10/11 UE monitors one UE-specific PDCCH search space for each activated component carrier. For a UE that is not configured with cross-carrier scheduling, the UE-specific PDCCH search space on each component carrier is the same as that in LTE Rel-8/9, whereas for a UE that is configured with cross-carrier scheduling, the UE-specific PDCCH search space is defined by a

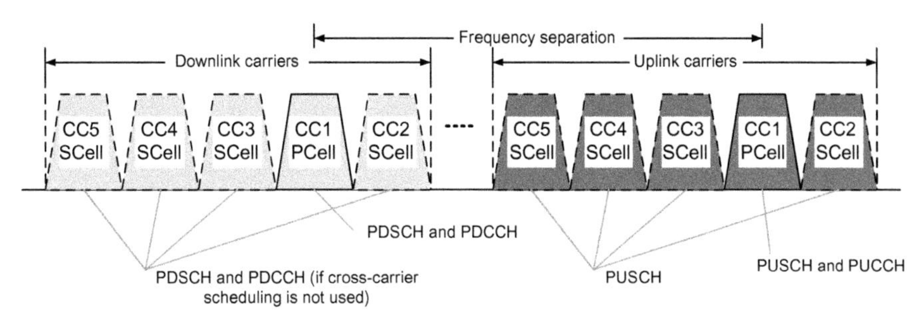

FIGURE 13.5

Symmetric carrier aggregation in the downlink and uplink (example) [16].

combination of the UE's RNTI and the serving cell index. The number of PDCCH blind decoding operations performed by a UE scales (approximately) linearly with the number of its activated component carriers [29]. If carrier aggregation is configured, the maximum number of blind decoding operations that a UE performs is 44 for the PCell plus 32 for each active downlink SCell. An additional 16 blind decodings are needed for each uplink component carrier that is configured for uplink MIMO operation. When cross-carrier scheduling is configured, CIF is only present in the PDCCH messages in the UE-specific search space and not in the common search space [13].

The PDSCH/PUSCH on a component carrier can only be scheduled from one component carrier. Therefore, for each PDSCH/PUSCH component carrier, there is an associated component carrier, configured via UE-specific RRC signaling, where the corresponding DCI can be transmitted. Figure 13.4 illustrates an example, where PDSCH/PUSCH transmissions on component carrier 1 are scheduled using PDCCHs transmitted on component carrier 1. In this case, since cross-carrier scheduling is not used, there is no carrier indicator in the corresponding DCI formats. The PDSCH/PUSCH transmissions on component carrier 2 are cross-carrier scheduled from PDCCHs transmitted on component carrier 1. Therefore, the DCI formats in the UE-specific search space for component carrier 2 include the CIF. Note that the transmissions on a component carrier can be scheduled by the PDCCHs on one component carrier; thus, component carrier 4 cannot be scheduled by PDCCHs on component carrier 5, given that there is a semi-static association between component carriers 3 and 4 for PDCCH transmission in this example [14].

As mentioned earlier, there is one UE-specific search space per aggregation level and component carrier used for PDSCH/PUSCH. This is illustrated in Figure 13.4, where PDSCH/PUSCH transmissions on component carrier 1 are scheduled using PDCCHs transmitted on component carrier 1. There is no CIF in the UE-specific search space for component carrier 1 since cross-carrier scheduling is not used. For component carrier 2, on the other hand, a carrier indicator field exists in the UE-specific search space, given that component carrier 2 is cross-carrier scheduled from the PDCCHs transmitted on component carrier 1. Search spaces for different component carriers may overlap in some subframes. In Figure 13.4, this happens for the UE-specific search spaces for component carriers 3 and 4. The terminal can identify the two search spaces independently, assuming that a CIF is used for component carrier 4, but is not used for component carrier 3. If the UE-specific and common search spaces corresponding to different component carriers happen to overlap for some aggregation levels when cross-carrier scheduling is configured, the terminal only needs to monitor the common search space. The reason for this is to avoid ambiguities arising from this condition; if the component carriers have different bandwidths, a DCI format in the common search space may have the same payload size as another DCI format in the UE-specific search space corresponding to another component carrier [14].

The PCFICH is transmitted in each subframe and conveys a control format indicator (CFI) field, which indicates the number of OFDM symbols carrying control information in each subframe (up to three OFDM symbols per subframe). The UE derives the starting OFDM symbol used for the PDSCH transmission based on the decoded CFI value. For the PDSCH scheduled by PDCCH on the same component carrier, the UE is required to decode the PCFICH in order to determine the starting OFDM symbol used for the PDSCH. When cross-carrier scheduling is utilized, the starting OFDM symbol for the PDSCH transmission is configured by the network through higher layers, and the UE is not required to decode PCFICH on that component carrier.

The physical HARQ indicator channel carries HARQ feedback (ACK/NACK) information associated with a PUSCH transmission. The PHICH structure in LTE Rel-10/11 carrier aggregation is the same as that for LTE Rel-8/9. The PHICH corresponding to a PUSCH transmission is always transmitted on the component carrier where the PDCCH schedules the PUSCH transmission.

The uplink control information (UCI) includes HARQ feedback corresponding to one or more PDSCHs, channel state information, and scheduling request. Similar to LTE Rel-8/9, the UCI can be transmitted on the PUCCH, if there is no transmission on PUSCH in a subframe, or transmitted on PUSCH, if there is a scheduled transmission on PUSCH. LTE-Advanced further supports simultaneous PUCCH and PUSCH transmissions in a subframe (see Figure 13.6). This allows the eNB to flexibly control the performance of the PUCCH and PUSCH independently and to avoid the UCI overhead on PUSCH by utilizing existing PUCCH resources. The PUCCH can only be transmitted on the PCell, since it typically has more reliable link quality and coverage relative to the SCells (see Figure 13.5). When applicable, UCI is always transmitted on a single PUSCH.

HARQ feedback indicates whether a transport block is correctly received and decoded. LTE Rel-8/9 supports transmitting a maximum of two and four ACK/NACKs in an uplink subframe for FDD and TDD, respectively. Given independent HARQ processes per component carrier, the maximum number of HARQ feedback bits in a subframe in response to PDSCH transmissions can be 10 bits

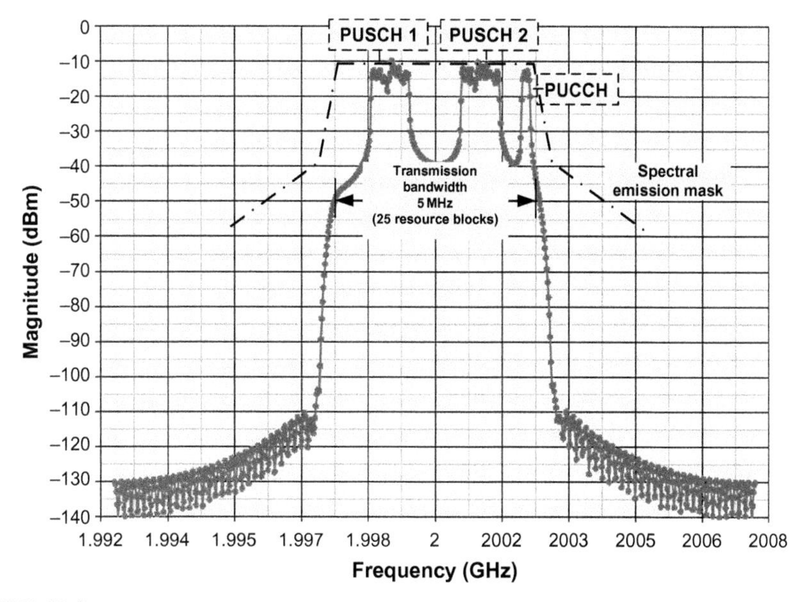

FIGURE 13.6

Snapshot of clustered SC-FDMA (simultaneous PUSCH and PUCCH transmission) in the uplink [17,18].

for FDD and up to 90 bits for TDD (assuming five component carriers with MIMO transmission modes per component carrier and TDD subframe configuration 5). Reliable transmission of large number of HARQ ACK/NACK bits in an uplink subframe requires a substantial amount of resources and sufficiently high signal to interference plus noise ratio. Therefore, HARQ feedback bundling/multiplexing schemes are designed for TDD systems to limit the maximum number of ACK/NACK bits in an uplink subframe to 20. Multiple HARQ feedback bits are bundled by a logical AND operation to produce one HARQ feedback bit per component carrier and per subframe for certain TDD subframe configurations. The PUCCH format 3 was introduced in LTE Rel-10 to support HARQ feedback for a UE configured with downlink carrier aggregation. The PUCCH format 3 uses a SC-FDMA waveform with five SC-FDMA symbols per slot carrying HARQ feedback information and one or two SC-FDMA symbols per slot (depending on the size of the cyclic prefix) for transmission of demodulation reference signals, to enable coherent demodulation of HARQ feedback (see Chapter 10 for more information on PUCCH format 3). An orthogonal cover code of length 5 is applied to the encoded HARQ feedback bits in the corresponding SC-FDMA symbols, allowing multiplexing of up to five UEs in one resource block. Assuming QPSK modulation, k HARQ feedback bits are encoded using the $(32, k)$ Reed–Muller code with circular rate matching into 48 encoded HARQ acknowledgment bits, which are transmitted in a subframe comprising two slots.

LTE-Advanced further supports the HARQ feedback multiplexing scheme for TDD, where two HARQ feedback bits are generated per component carrier, representing the number of consecutively correctly received PDSCHs on the component carrier. PUCCH format 1b can be used to carry four acknowledgments when channel selection is used, i.e., the PUCCH payload carries two acknowledgments and the selection of PUCCH resources carries the other two acknowledgments. The use of bundled acknowledgments in a single uplink subframe is not limited to the TDD duplex mode. The carrier aggregation in FDD with a different number of uplink and downlink component carriers encounters the same problem. In LTE Rel-10, there is only one PUCCH that is always transmitted on the primary component carrier. Hence, even in asymmetric carrier aggregation scenarios, the possibility to transmit more than two HARQ acknowledgment bits in the uplink must be supported. This is performed via resource selection or by using PUCCH format 3. The acknowledgments in response to PUSCH transmissions are transmitted using multiple PHICHs, where the PHICH is transmitted on the same downlink component carrier as the uplink scheduling grant which initiated the uplink transmission.

The channel-state information feedback enables the eNB to perform PDSCH scheduling including resource allocation, MCS selection, transmission rank adaptation, and the choice of precoding matrix. To accommodate different channel conditions and interference levels among different component carriers, CSI is fed back for each component carrier. Both periodic and aperiodic CSI feedbacks are supported in LTE Rel-10/11, where periodic CSI is transmitted on a set of semi-statically configured subframes, and aperiodic CSI is dynamically scheduled by the eNB via the PDCCH. Periodic CSI feedback is independently configured for each component carrier and transmitted on the PUCCH using PUCCH format 2. Since the payload size of PUCCH format 2 is limited, the periodic CSI for only one component carrier can be unconditionally transmitted in a subframe. A priority based on the periodic CSI reporting type determines the component carrier for which periodic CSI is reported in the event that periodic CSI feedback of multiple carriers collide in a subframe. If the collisions involve the same periodic CSI reporting type, the priority is according to ascending cell index. Aperiodic CSI feedback is transmitted on PUSCH and typically carries more CSI bits than periodic CSI feedback. LTE Rel-10/11 supports aperiodic CSI feedback for a single or multiple component carriers in a subframe so that the eNB can balance the CSI accuracy and feedback overhead [29].

Uplink power control is a mechanism to instruct the UEs to determine the PUSCH/PUCCH transmission power with the objective of ensuring that the target reception reliability of that channel is met. The uplink power control supported in LTE Rel-8/9/10 consists of an open-loop component adjusting for the path loss between the UE and its serving eNB, and of a closed-loop component based on transmission power control commands in the PDCCH. The uplink power control processes are independent among the respective activated cells and each process is the same as LTE Rel-8/9. The resulting transmission power for PUSCH or

PUCCH is referred to as nominal transmission power. The key differentiation with respect to the LTE Rel-8/9 uplink power control operation occurs when the sum of the nominal transmission powers from the UE exceeds the total maximum output power for which the UE is configured in a specific subframe. Depending on whether the UE is configured for simultaneous PUCCH and PUSCH transmissions, on the UCI types for transmission, and on the existence of any PUSCH transmission, the UE may transmit in a subframe one PUCCH, one PUSCH with UCI, one or more PUSCH transmissions without UCI, or any other combination. As control information is more important than data information and does not benefit from the use of HARQ, the UE transmission power is distributed by descending priority to the PUCCH, PUSCH with UCI, and PUSCH without UCI, where the nominal transmission power for a channel with higher priority is first guaranteed until the UE total maximum output power is reached. For one or more PUSCH transmissions without UCI, the UE scales each respective nominal PUSCH transmission power by the same factor, considering that the same quality of service for data transmission is required on all component carriers [29].

The eNB may not know the exact UE transmission power due to the possibility of missing power control commands, UE-specific implementation of maximum power reduction (MPR),[1] and additional maximum power reduction (A-MPR)[2] in order to meet regulatory requirements. To provide the eNB with more accurate information on the UEs transmission power, power headroom reporting was supported in LTE Rel-8/9. The UE may report type 1 and/or type 2 power headroom for an activated cell. The type 1 and type 2 power headroom reports indicate the amount of remaining the UEs transmission power beyond those already used for PUSCH, or PUSCH and PUCCH transmission in a subframe, respectively [29].

13.3.2 MAC layer aspects

From the MAC perspective, the carrier aggregation simply provides additional conduits, thus the MAC sublayer plays the role of a multiplexing entity for the aggregated component carriers. Figure 13.7 shows an overview of the downlink

---

[1]The purpose of MPR is to allow the UE, in some configurations, to lower its maximum output power in order to meet the general requirements on signal quality and out-of-band (OOB) emissions. Note that MPR is an allowance and the UE is not required to use it [13].

[2]The eNB may inform the UE of the possibility of further lowering its maximum power by signaling an additional MPR (A-MPR). The need for an A-MPR occurs with certain combinations of E-UTRA bands, channel bandwidths, and transmission bandwidths for which the UE must meet additional (more stringent) requirements for spectrum emission mask and spurious emissions. As with MPR, the A-MPR is an allowance, not a requirement, and it applies in addition to MPR. Regardless of whether the UE makes use of the allowed MPR and A-MPR, the additional requirements for spectrum emission mask and spurious emissions that are signaled by the network always apply. The reason for the complex set of conditions for the relaxation is the expected intermodulation products which may fall into adjacent bands which have different levels of sensitivity (e.g., public safety bands) [13].

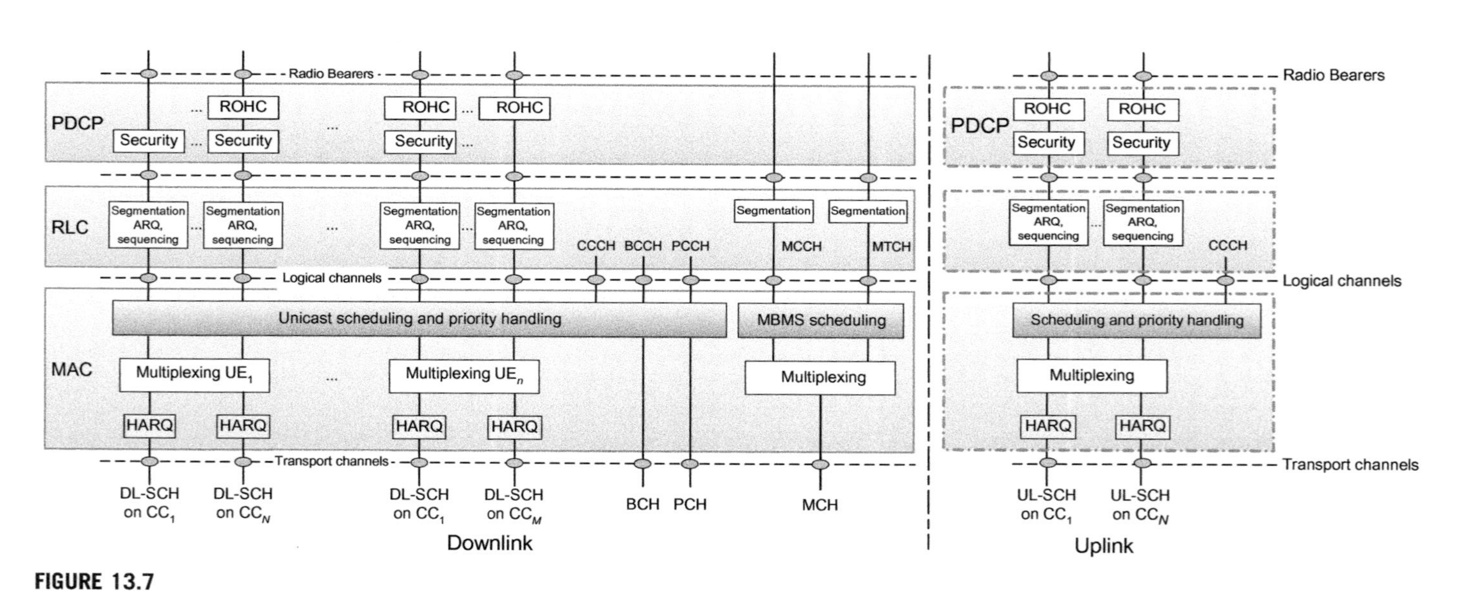

FIGURE 13.7

LTE-Advanced downlink/uplink layer 2 structure when carrier aggregation is configured [7].

user-plane protocol stack at the base station, as well as the corresponding mapping of the most essential radio resource management functionalities for carrier aggregation. Each user has at least one radio bearer which is referred to as the default radio bearer. The exact mapping of data to the default bearer is up to the operator policy and is configured via the traffic flow template. In addition to the default radio bearer, users may have additional bearers configured. There is one PDCP and RLC entity per radio bearer, including functionalities such as robust header compression, security, segmentation, and ARQ. As mentioned in the previous chapters, the interface between the RLC sublayer and MAC sublayer is referred to as the logical channel.

There is one MAC entity per user, which controls the multiplexing of data from all logical channels to the user, and further controls how this data is transmitted on the available component carriers. As illustrated in Figure 13.7, there is a separate HARQ entity per component carrier, which essentially means that if data transmission is on the *i*th component carrier, the HARQ retransmissions in case of erroneous reception must be performed on the *i*th component carrier. The interface between the MAC sublayer and the physical layer, which are referred to as transport channels, is also separate for each component carrier. The transport blocks sent on different component carriers can be transmitted with independent MCSs as well as different MIMO schemes. The latter allows data on one component carrier to be transmitted with open-loop transmit diversity (transmission mode 2), whereas data on another component carrier sent with dual-layer beamforming (transmission mode 8). There is independent link adaptation per component carrier in order to take advantage of more accurate link adaptation on different component carriers according to their respective propagation and radio channel conditions. Semi-persistent scheduling can only be configured for the PCell, and PDCCH allocations for the PCell can override the corresponding semi-persistent scheduling resource allocation.

The LTE Rel-8/9 control-plane protocol stack also applies to LTE-Advanced with multiple component carriers, meaning that there is one RRC per user, independent of the number of component carriers. Similarly, idle mode mobility procedures of LTE Rel-8/9 also apply in a network supporting carrier aggregation. It is also possible for a network to configure only a subset of component carriers for idle mode camping [26].

E-UTRA in general supports a synchronized uplink by means of un uplink timing advance (TA) adjustment procedure. The carrier aggregation in LTE Rel-10 only supports a single timing advance value which is applied to all component carriers. This means that the base station transceivers for different carriers should be at the same location to avoid different propagation delays. The use of RRHs, distributed antennas, and repeaters was limited. The signal should be received within the cyclic prefix length for a correct reception by a regular UE. The timing advance group (TAG) was introduced in LTE Rel-11 for supporting multiple timing advances typically encountered in inter-band carrier aggregation scenarios. A TAG includes one or more serving cells with the same uplink timing advance and the same downlink timing reference cell. If a TAG contains the PCell, it is referred to as the primary timing advance group (pTAG). If a TAG contains only SCell(s), it is denoted as the secondary timing advance group (sTAG). There is one timing reference cell and one time alignment timer (TAT) per TAG, and each TAT may be configured with a different value. For pTAG, the PCell is used as the timing reference cell, whereas for sTAG, the UE may use any activated SCell from the same sTAG as the timing reference cell. From an RF requirement point of view, the number of component carriers is limited to two for LTE Rel-11; thus, if the sTAG is configured, there is only one SCell in the sTAG [7].

The initial uplink timing alignment of the sTAG is obtained by an eNB initiated random access procedure similar to the pTAG. The SCell in a sTAG can be configured with random access channel resources, and the eNB may instruct the UE to perform random access on the SCell. The Msg2 (or random access response) in response to the SCell preamble is transmitted on the PCell using an RA-RNTI that conforms to the LTE Rel-8 RACH procedure. The grant in Msg2 is valid for the SCell in which the preamble was transmitted. The UE stops transmission of the random access preamble on the SCell when reaching the maximum number of transmissions. However, the UE will not indicate a random access problem to upper layers, if the maximum number of preamble transmissions is reached for the random access procedure on the SCell. The UE tracks the downlink frame timing change of the SCell and adjusts the uplink transmission timing following the timing advance commands from the eNB [7].

13.4 Protocol and signaling aspects of carrier aggregation

In order to enable mobility when supporting carrier aggregation, a UE handles a component carrier in the same way as it would deal with another carrier frequency in LTE Rel-8/9. The UE performs various measurements as instructed by the eNB including intra- and inter-frequency measurements which are used for selecting LTE carriers, as well as inter-RAT measurements which are meant for non-LTE RAT selection. As of LTE Rel-10/11, the following event-triggered reporting criteria are defined [10]:

- Event A1: Serving cell becomes better than absolute threshold.
- Event A2: Serving cell becomes worse than absolute threshold.
- Event A3: Neighbor cell becomes better than an offset relative to the serving cell.
- Event A4: Neighbor cell becomes better than absolute threshold.
- Event A5: Serving cell becomes worse than one absolute threshold and neighbor cell becomes better than another absolute threshold.
- Event A6: Intra-frequency neighbor becomes better than an offset relative to an SCell (LTE Rel-10/11 only).

For inter-RAT mobility, the following criteria for measurement reporting are defined [10]:

- Event B1. Neighbor cell becomes better than absolute threshold.
- Event B2. Serving cell becomes worse than one absolute threshold and neighbor cell becomes better than another absolute threshold.

The new measurement event A6 was introduced for LTE-Advanced carrier aggregation to compare the neighbor cells' component carriers to the current SCell, thus allowing reconfiguration of SCells.

The discontinuous reception (DRX) mode was defined in LTE Rel-8/9. If one or more SCells are configured for a UE in addition to the PCell, the same DRX mode is applied to all serving cells. According to network configuration and ongoing HARQ processes, the UE determines the DRX activity cycle which is common to all serving cells. This means that the active times for PDCCH monitoring are identical across all downlink component carriers (see Figure 13.8).

The SCell activation/deactivation is an efficient mechanism to reduce the UE power consumption in LTE Rel-10/11 carrier aggregation in addition to DRX. On a deactivated SCell, the UE neither receives downlink signals nor transmits any uplink signal. The UE is also not required to perform measurements on a deactivated SCell. Deactivated SCells can be used as a path loss reference for measurements in uplink power control. It is assumed that these measurements would be less frequent while the SCell is deactivated in order to conserve the UE power. On the other hand, for an activated SCell, the UE performs normal activities for downlink reception and uplink transmission. Activation and deactivation of SCells is controlled by the eNB. As shown in Figure 13.8, the SCell activation/deactivation is performed when the eNB sends an activation/deactivation command in the form of a MAC control element. A timer may also be used for automatic deactivation, if no data or PDCCH messages are received on an SCell for a certain period of time. This is the only case in which deactivation can be executed autonomously by the UE. Serving cell activation/deactivation is performed independently for each SCell, allowing the UE to be activated only on a particular set of SCells. Activation/deactivation is not applicable to the PCell because it is required to always remain activated when the UE has an RRC connection to the network [29].

As already mentioned, if the UE is configured with one or more SCells, the network may activate and deactivate the configured SCells. The PCell is always activated. The network activates and deactivates the SCell(s) by sending an activation/deactivation MAC control element as described in Chapter 8. Furthermore, the UE maintains an *sCellDeactivationTimer* per configured SCell and deactivates the associated SCell upon its expiration. The same initial timer value is applied to each instance of the *sCellDeactivationTimer* and it is configured by RRC signaling. The configured SCells are initially deactivated upon addition and after a handover [9].

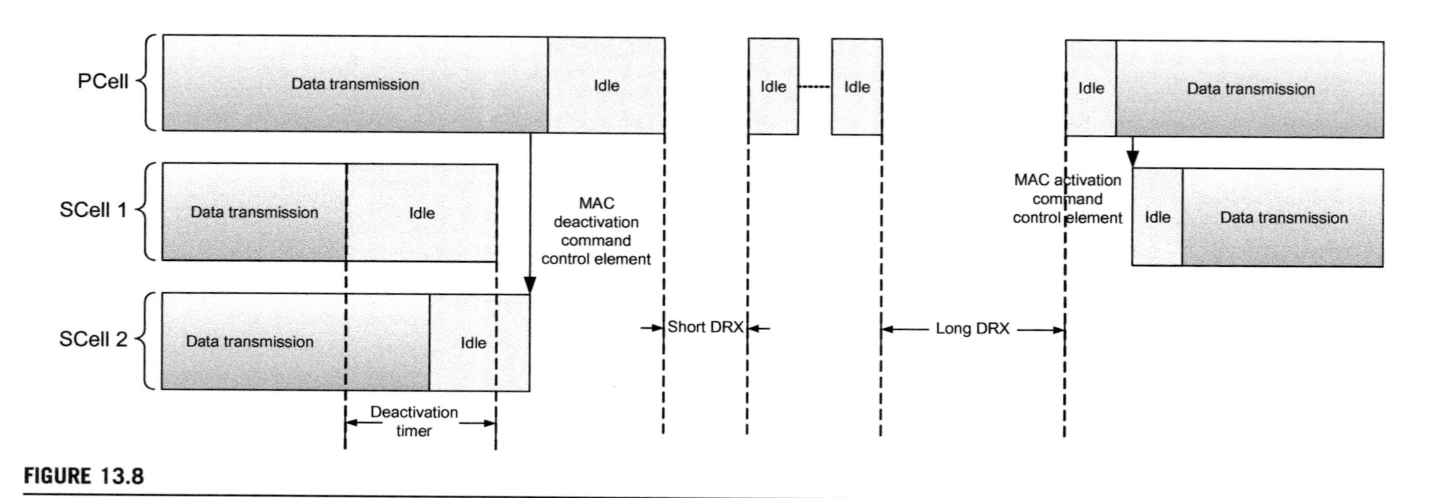

FIGURE 13.8

Illustration of SCell activation/deactivation procedure [20].

As mentioned earlier, the activation or deactivation of downlink component carriers are performed through a MAC control element. A command to activate a component carrier takes effect eight subframes after the receipt of the activation/deactivation command, i.e., if the MAC control element is received in subframe n, then the SCells are activated/deactivated starting from subframe $n + 8$. There is also a timer-based mechanism for deactivation such that a terminal may deactivate SCells after a configurable time with no activity. Note that the primary component carrier is always active. In the uplink, there is no explicit means of activation for uplink component carriers. However, whenever a downlink component carrier is activated or deactivated, the corresponding uplink component carrier is also activated or deactivated [14].

13.5 Radio resource management aspects of carrier aggregation

Cell management is the control procedure enabling the network to add, remove, or change SCells, or to switch the PCell of a UE. A UE in the RRC_IDLE state establishes RRC connection on a serving cell, which automatically becomes its PCell. Depending on the carrier where initial access is performed, different UEs in a network with carrier aggregation capability may have different PCells. With the RRC connection established on the PCell, the network can further configure one or more SCells for a carrier-aggregation-capable UE, considering the UEs capability [8] to meet the UE's data service requirements. The necessary information, including system information of an SCell is conveyed to the UE via dedicated RRC signaling. Addition, removal, or reconfiguration of SCells for a UE is performed via dedicated RRC signaling (see Figure 13.9). The network can further change the PCell of a UE to improve the link quality of the PCell on which critical control information is sent, or to provide load balancing among different SCells. The change of PCell in LTE Rel-10 carrier aggregation can only be performed via a handover procedure, and it does not necessarily require the UE to switch to single-carrier operation. Intra-LTE handover in LTE-Advanced allows the target PCell to configure one or more SCells for the UE to use immediately after handover [29].

The carrier aggregation additional SCells cannot be activated immediately at the time of RRC connection establishment, since there is no provision in the RRC connection setup procedure for SCells. SCells are added or removed from the set of serving cells through the RRC connection reconfiguration procedure. Note that, since intra-LTE handover is treated as an RRC connection reconfiguration, SCell handover is supported. The carrier aggregation-related information sent by the eNB pursuant to the RRC connection reconfiguration procedure can be summarized as follows [20]:

- Cross-carrier scheduling configuration: This indicates if scheduling for the referenced SCell is managed by that SCell or by another cell.

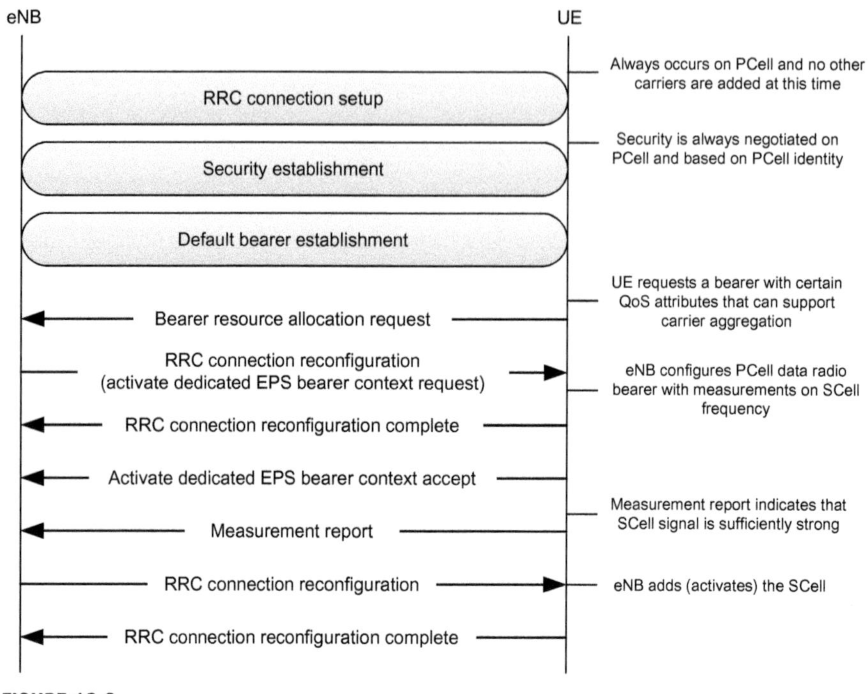
eNB
UE
Always occurs on PCell and no other carriers are added at this time
RRC connection setup
Security is always negotiated on PCell and based on PCell identity
Security establishment
Default bearer establishment
UE requests a bearer with certain QoS attributes that can support carrier aggregation
Bearer resource allocation request
RRC connection reconfiguration (activate dedicated EPS bearer context request)
eNB configures PCell data radio bearer with measurements on SCell frequency
RRC connection reconfiguration complete
Activate dedicated EPS bearer context accept
Measurement report indicates that SCell signal is sufficiently strong
Measurement report
RRC connection reconfiguration
eNB adds (activates) the SCell
RRC connection reconfiguration complete

FIGURE 13.9

Summary of message exchange for addition of SCells [20].

- SCell PUSCH configuration: This indicates whether resource block group hopping is utilized on the SCell.
- SCell uplink power control configuration: This carries a number of primitives related to the SCell uplink transmit power command, including the path loss reference-linking parameter.
- SCell CQI reporting configuration: This carries a number of primitives related to CQI measurement reporting for SCells.

13.6 RF and implementation aspects of carrier aggregation

Although not considered a problem for the base station, carrier aggregation will undoubtedly pose major difficulties for the UE, which must handle multiple (simultaneously operating) transceivers. The addition of simultaneous non-contiguous transmitters creates a highly challenging radio environment in terms of spurious signal emissions and self-blocking. Carrier aggregation brings new technical challenges, especially for the LTE Rel-10/11 UE implementation. The complexity of the UE's RF front-end module can vary greatly, depending on which type(s) of carrier aggregation are supported, with contiguous carrier aggregation being the least complex scenario. In LTE Rel-10 devices, contiguous carrier aggregation will not exceed two component carriers or 40 MHz maximum bandwidth. It may be possible to support this configuration with a single wideband transceiver in the UE. For non-contiguous component carrier allocations, the UE will have to use multiple transceivers or a single, extremely wide wideband transceiver. Using multiple transceivers may be more realistic in the sense that such configuration requires only the addition of parallel paths to process each frequency band, as in existing multiband devices. However, using multiple transceivers also increases the size, cost, and power consumption of the mobile device [12–14].

In a wideband transceiver, a single transceiver must process the multi-band non-contiguous carrier aggregation using wideband RF components. There are two main issues with this approach. First, as the bandwidth increases, the effective noise bandwidth increases, thus resulting in increased noise power. Second, with a wider bandwidth, more unwanted signals are likely to be received from other sources. Therefore, for non-contiguous carrier aggregation, most proposals today are leaning toward the use of multiple transceivers instead of a single wideband approach. In addition to increasing the RF front-end complexity, using simultaneous non-contiguous transmitters creates a highly challenging radio environment in terms of spurious signal management and self-blocking. Because these challenges particularly impact UE design, more work needs to be done before interband carrier aggregation can be successfully introduced in the uplink [14].

Simultaneous transmit or receive with mandatory MIMO support would aggravate the challenge of antenna design. The frequency bands designated for the deployment of LTE are mostly the existing IMT-2000 bands, or the bands where the legacy systems, including UMTS and GSM, were deployed. The frequency bands in some regions are defined in a technology-neutral manner, which means that coexistence among different technologies is a necessity. In general, the requirements for LTE operation in different frequency bands do not impose any specific requirements on the radio-interface design. However, there are implications for the RF requirements and the way they are defined in order to support coexistence between operators in the same geographical area, collocation of base station equipment of different operators, and coexistence with services in adjacent frequency bands and across country borders [14].

Operators may deploy LTE or other IMT-2000 technologies such as HSPA or EDGE in their respective bands. Such coexistence requirements to a large extent are developed within 3GPP, but there may also be regional requirements defined by regulatory bodies in some frequency bands. There are in many cases limitations to where base station equipment can be deployed. The base station sites are often shared between operators, or an operator will deploy multiple technologies in one site. This imposes additional requirements on base station transceivers. The use of the RF spectrum is regulated through international agreements governed by ITU-R. As a result, there are requirements for coordination among operators in different countries, and for coexistence with services in adjacent frequency bands. Coexistence between operators of TDD systems in the same band is provided by inter-operator synchronization in order to avoid interference between downlink and uplink transmissions of different operators. Through the release-independent principle, it is possible to design terminals based on an early release of 3GPP specifications that support a frequency band added in a later release [14].

The requirements for carrier aggregation are defined for carrier aggregation configurations with associated bandwidth combination sets. For inter-band carrier aggregation, a *carrier aggregation configuration* is a combination of operating bands, each supporting a carrier aggregation bandwidth class. For intra-band contiguous carrier aggregation, a carrier aggregation configuration is a single operating band supporting a carrier aggregation bandwidth class. For each carrier aggregation configuration, requirements are specified for all bandwidth combinations contained in a *bandwidth combination set*, which is indicated per supported band combination in the UE radio access capability [8]. A UE can indicate support of several bandwidth combination sets per band combination [1].

13.6.1 Temporal/spectral masks and unwanted emissions

E-UTRA needs to ensure user orthogonality in the time-domain similar to UMTS systems. In UMTS, the requirements for user orthogonality were based on ON/OFF masks, which defined the allowable transmit power during the OFF and ON periods of transmission. In E-UTRA, similar requirements exist for the eNB for FDD and TDD operation modes. However, for the LTE UE, the requirements are considerably more complex than those for UMTS, due to the characteristics of the SC-FDMA scheme used for the uplink transmissions. Figure 13.10 shows the general ON/OFF mask in time which applies whenever the UE is required to switch on or off. Note that although the requirement is given in terms of a mask, it actually applies to the average power during the ON and OFF periods. In UMTS the transient period 20 μs is centered on the slot boundary, whereas in LTE, the transient period is in general shifted into the next subframe. Therefore, the first few SC-FDMA samples are more susceptible to corruption by insufficient transmit power or inter-UE interference, whereas the samples at the end of the subframe are protected. For PRACH, the transient periods are located outside the preamble in order to protect the entire random access preamble. The same principle applies to sounding reference signals to allow reliable uplink channel sounding without impairments arising from transient periods. This is also applicable to discontinuous transmission measurement gaps [1,13].

For TDD, where two sounding reference signals can be transmitted on adjacent symbols in the UpPTS field, the transient periods are located between the two sounding reference signals. The general temporal masks apply in the case of transitions into and out of the OFF power state. A general output power dynamic requirement also exists for continuous transmission at slot boundaries where frequency hopping occurs, and at subframe boundaries for either frequency hopping or power changes. In all these cases, the transient periods are symmetric around

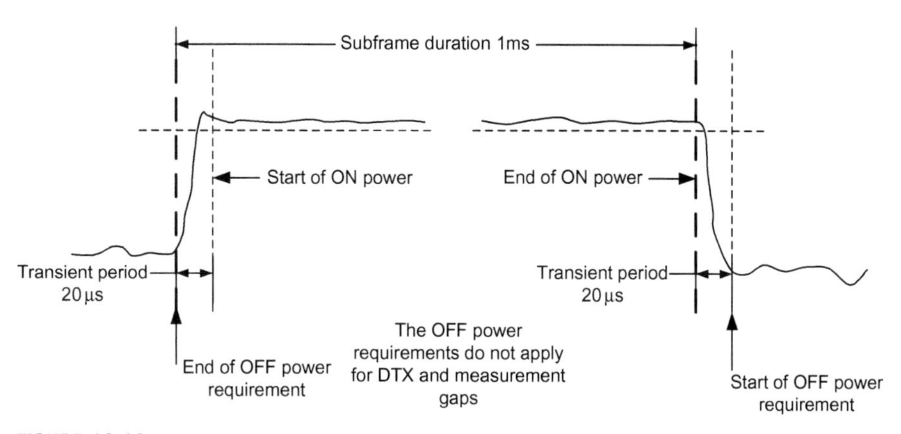

FIGURE 13.10

General ON/OFF temporal mask [1].

the slot or subframe boundaries. The transient power masks are applied in transition from the PUCCH/PUSCH to sounding reference signal with DTX after sounding reference signal, from the PUCCH/PUSCH to sounding reference signal to PUCCH/PUSCH, from DTX to sounding reference signal to PUSCH/PUCCH, or from PUCCH/PUSCH to DTX in a sounding reference signal symbol to PUCCH/PUSCH [1,13].

A radio transmitter is required not to transmit any signal outside its designated transmission band. However, in practice, all radio transmitters do transmit unwanted signals outside their designated transmission bands. The LTE specifications define two separate types of unwanted emissions: OOB emissions and spurious emissions. OOB emissions are unwanted emissions immediately outside the channel bandwidth resulting from the modulation process and non-linearity in the transmitter, but excluding spurious emissions. Spurious emissions are emissions which are caused by unwanted transmitter effects such as harmonics emission, parasitic emission, intermodulation products,[3] and frequency conversion products, but exclude OOB emissions. The OOB emissions requirement for the eNB transmitter is specified both in terms of adjacent channel leakage power ratio (ACLR) and operating band unwanted emissions. The operating band unwanted emissions define all unwanted emissions in the downlink operating band plus the frequency ranges 10 MHz above and 10 MHz below the band. Unwanted emissions outside of this frequency range are limited by a spurious emissions requirement. For an eNB supporting multi-carrier or intra-band contiguous carrier

[3]Intermodulation or intermodulation distortion (IMD) is the amplitude modulation of signals containing two or more different frequencies in a system with non-linearities. The intermodulation between each frequency component will form additional signals at frequencies that are not just at harmonic frequencies (integer multiples) of either, but also at the sum and difference frequencies of the original frequencies, and at multiples of those sum and difference frequencies.

aggregation, the unwanted emission requirements apply to channel bandwidths of the outermost carrier larger than or equal to 5 MHz. The occupied bandwidth is the width of a frequency band such that, below the lower and above the upper frequency limits, the mean powers emitted are each equal to a specified percentage $\beta/2$ of the total mean transmitted power. The value of $\beta/2$ is set to 0.5%. This requirement also applies during the transmitter ON period [1].

ACLR is the ratio of the filtered mean power centered on the assigned channel frequency to the filtered mean power centered on an adjacent channel frequency. The requirements apply outside of the edges of the RF bandwidth regardless of the type of transmitter considered, i.e., single carrier or multi-carrier. It applies to all transmission modes provisioned by the manufacturer's specification. In addition, for an eNB operating in non-contiguous spectrum, the ACLR applies to the first adjacent channel inside any sub-block gap with a sub-block gap size $W_{gap} \geq 15$ MHz. The ACLR requirement for the second adjacent channel applies inside any sub-block gap with a sub-block gap size $W_{gap} \geq 20$ MHz. This requirement is also applied during the transmitter ON period [1].

Since OOB emissions occur close to the desired signal transmission, increasing the power level of the desired signal will usually increase the level of the unwanted emissions. On the other hand, reducing the transmit power is usually an effective method for reducing the OOB emissions, thus providing one possible solution to network-signaled power-reduction requirements. The OOB emissions are an inevitable by-product of the modulation process, and are often caused by non-linearities in power amplifiers. In fixed-bandwidth radio systems, the OOB emission requirements were defined with respect to the center frequency of the transmission. Since E-UTRA supports variable bandwidth, it is more convenient to define OOB requirements with respect to the edge of the channel bandwidth rather than the center of the channel, as shown in Figure 13.11. In LTE, the OOB emissions are defined by means of spectrum emission masks and ACLR requirements.

The spectrum emission mask is a power attenuation function defined for out-of-channel emissions relative to the in-channel power spectrum. The spectrum emission mask of the UE applies to frequencies within Δf_{OOB} of the edge of the assigned LTE channel bandwidth, as shown in Figure 13.11. For intra-band contiguous carrier aggregation the spectrum emission mask of the UE applies to frequencies Δf_{OOB} starting from the $\pm$edge of the aggregated channel bandwidth (see Figure 13.11). For intra-band contiguous carrier aggregation and bandwidth class C, the power of any UE emissions are required not to exceed the levels specified in Table 13.1 within the specified channel bandwidth [1].

13.6.2 Operating bands for carrier aggregation

E-UTRA was designed to operate in the frequency bands defined in [1,2]. The requirements were defined for 1.4, 3, 5, 10, 15, and 20 MHz bandwidths with a specific configuration in terms of number of physical resource blocks. Using carrier

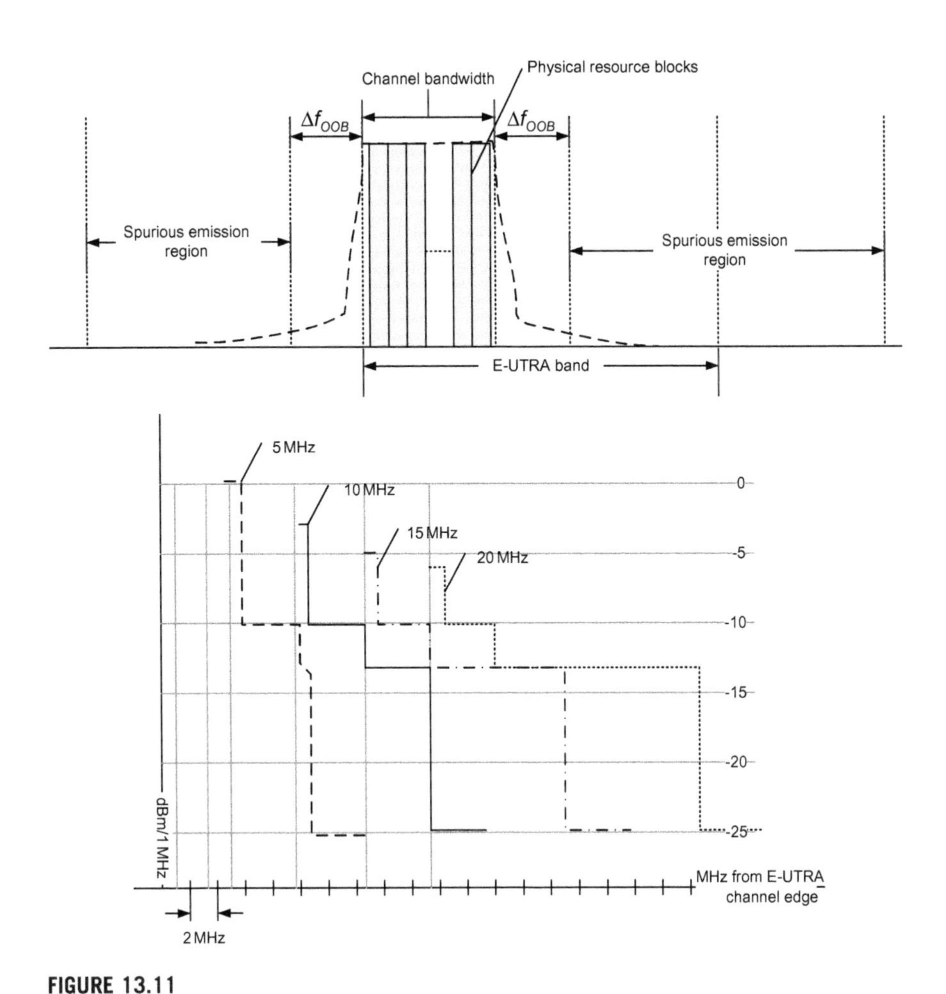

FIGURE 13.11

Transmitter RF spectrum and UE spectrum emission mask [1,13].

aggregation, a number of contiguous and/or non-contiguous frequency bands can be aggregated to create a virtually larger bandwidth. The channel raster is 100 kHz, which means the center frequency must be a multiple of 100 kHz [1]. To support transmission in paired and unpaired spectrums, two duplexing schemes are supported, i.e., FDD, allowing both full and half-duplex terminal operation, as well as TDD.

Figure 13.12 illustrates the relation between the channel bandwidth ($BW_{Channel}$) and the transmission bandwidth, i.e., the number of permissible physical resource blocks (N_{RB}). The channel edges are defined as the lowest and highest frequencies of the carrier separated by the channel bandwidth, i.e., $f_C \pm \frac{1}{2}BW_{Channel}$. In the case of carrier aggregation, the aggregated channel

Table 13.1 General E-UTRA Carrier Aggregation Spectrum Emission Mask for Bandwidth Class C [1]

	Spectrum Emission Limit (dBm/$BW_{channel\text{-}CA}$)				
Δf_{OOB}(MHz)	50 RB +100 RB (29.9 MHz)	75 RB + 75 RB (30 MHz)	75 RB + 100 RB (34.85 MHz)	100 RB + 100 RB (39.8 MHz)	Measurement Bandwidth (MHz)
± 0–1	−22.5	−22.5	−23.5	−24	0.030
± 1–5	−10	−10	−10	−10	1
± 5–29.9	−13	−13	−13	−13	1
± 29.9–30	−25	−13	−13	−13	1
± 30–34.85	−25	−25	−13	−13	1
± 34.85–34.9	−25	−25	−25	−13	1
± 34.9–35		−25	−25	−13	1
± 35–39.8			−25	−13	1
± 39.8–39.85			−25	−25	1
± 39.85–44.8				−25	1

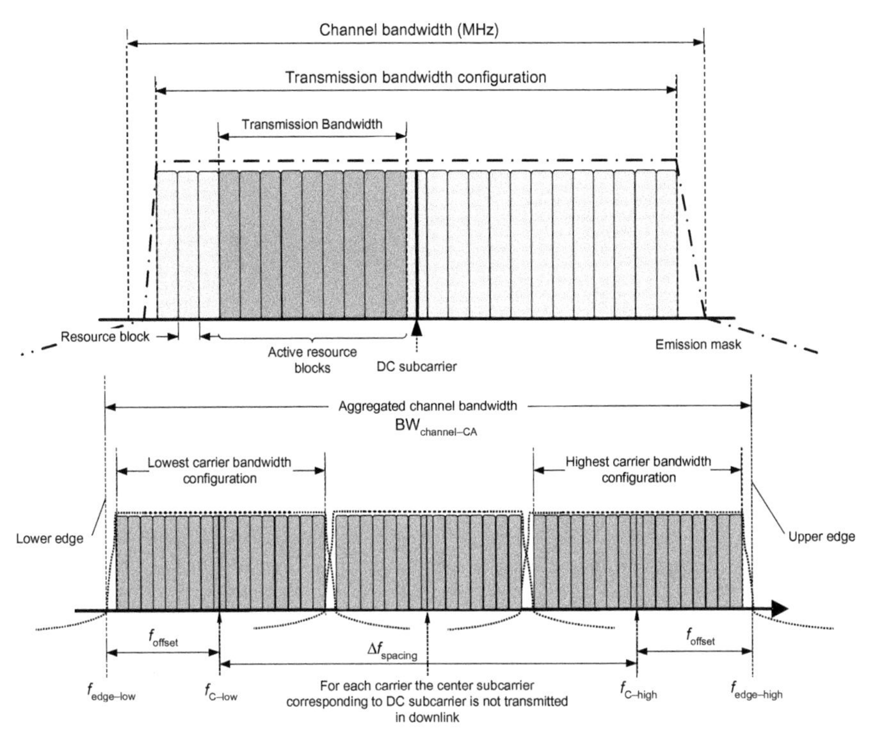

FIGURE 13.12

Relationship between channel bandwidth and transmission bandwidth in E-UTRA [2].

bandwidth $\mathrm{BW}_{\text{Channel-CA}}$ is defined as $\mathrm{BW}_{\text{Channel\_CA}} = f_{\text{edge-high}} - f_{\text{edge-low}}$ (see Figure 13.12). The lower band edge $f_{\text{edge-low}}$ and the upper band edge $f_{\text{edge-high}}$ of the aggregated channel bandwidth are used as the reference points in frequency for transmitter and receiver requirements, and are subsequently defined as follows: $f_{\text{edge-low}} = f_{C\text{-low}} - f_{\text{offset-low}}$ and $f_{\text{edge-high}} = f_{C\text{-high}} + f_{\text{offset-high}}$. The lower and upper frequency offsets depend on the transmission bandwidth configurations of the lowest and highest assigned edge component carriers, and are defined as $f_{\text{offset-low}} = 0.18N_{\text{RB-low}}/2 + BW_{\text{Guard-band}}$ and $f_{\text{offset-high}} = 0.18N_{\text{RB-high}}/2 + BW_{\text{Guard-band}}$ where $N_{\text{RB-low}}$ and $N_{\text{RB-high}}$ denote the transmission bandwidth configurations for the lowest and highest assigned component carrier, respectively, and $BW_{\text{Guard-band}}$ denotes the nominal guard band size. Note that the 0.18 multiplier in the equations is the bandwidth of a physical resource block in MHz. The aggregated transmission bandwidth is defined as the number of the aggregated resource blocks within the entire assigned aggregated channel bandwidth, and is defined per carrier aggregation band class.

As shown in Figure 13.12, the spacing between carriers depends on the deployment scenario and the size of the frequency block available to the network operator, as well as the channel bandwidth. The nominal channel spacing between two adjacent E-UTRA carriers is defined as $\Delta f_{\text{spacing}} = (BW_{\text{Channel(1)}} + BW_{\text{Channel(2)}})/2$ where $BW_{\text{Channel(1)}}$ and $BW_{\text{Channel(2)}}$ parameters are the channel bandwidths of the two respective E-UTRA carriers. The channel spacing can be adjusted to optimize performance in a particular deployment scenario. For intra-band contiguous carrier aggregation bandwidth, the nominal channel spacing between two adjacent E-UTRA component carriers is specified as follows [1,2]:

$$\Delta f_{\text{spacing-CA}}\,(\text{MHz}) = 0.3\left\lfloor \frac{BW_{\text{Channel(1)}} + BW_{\text{Channel(2)}} - 0.1\left|BW_{\text{Channel(1)}} - BW_{\text{Channel(2)}}\right|}{0.6} \right\rfloor$$

The channel spacing for intra-band contiguous carrier aggregation can be adjusted to any integer multiple of 300 kHz (less than the nominal channel spacing) to optimize performance in a particular deployment scenario. The E-UTRA channel raster (including the carrier aggregation scenarios) is 100 kHz for all bands, which means that the carrier center frequency must be an integer multiple of 100 kHz.

The carrier frequency in the downlink/uplink directions is designated by the E-UTRA absolute radio frequency channel number (EARFCN) in the range of 0–65,535. The relation between EARFCN and the downlink carrier frequency (in MHz) is given by $f_{\text{DL}} = f_{\text{DL-low}} + 0.1(N_{\text{DL}} - N_{\text{Offset-DL}})$, where $f_{\text{DL-low}}$, N_{DL}, and $N_{\text{Offset-DL}}$ parameters are given in [1,2]. Similarly, the relationship between EARFCN and the uplink carrier frequency (in MHz) for the uplink is defined as $f_{\text{UL}} = f_{\text{UL-low}} + 0.1(N_{\text{UL}} - N_{\text{Offset-UL}})$, where $f_{\text{UL-low}}$, N_{UL}, and $N_{\text{Offset-UL}}$ parameters are provided in [1,2]. Note that N_{DL} and N_{UL} in the latter equations are the downlink and uplink EARFCN, respectively. Figure 13.13 shows a snapshot of four adjacent 20 MHz component carriers chosen from the 3.5 GHz band that are aggregated with the adjacent center frequency spacing set to 20.1 MHz (a multiple of 300 kHz). This figure also shows the spectrum of two 20 MHz component carriers chosen from Band 7 in 2600 MHz that are aggregated with the center frequency spacing set to 20.1 MHz [18].

The channel numbers that designate carrier frequencies which are extremely close to the operating band edges are not used. This implies that the first 7, 15, 25, 50, 75, and 100 channel numbers at the lower operating band edge, and the last 6, 14, 24, 49, 74, and 99 channel numbers at the upper operating band edge are not utilized for channel bandwidths of 1.4, 3, 5, 10, 15, and 20 MHz, respectively.

A sub-block is defined as one contiguous allocated block of spectrum for transmission and reception by the same eNB. There may be multiple instances of sub-blocks within an RF bandwidth. Sub-block bandwidth is defined as the bandwidth of one sub-block. The frequency gap between two consecutive sub-blocks within an RF bandwidth is referred to as sub-block gap, where the RF requirements in the gap are based on coexistence for uncoordinated operation. The lower sub-block edge of the sub-block bandwidth $BW_{\text{Channel-block}}$ is defined as $f_{\text{edge\_block\_low}} = f_{C\_\text{block\_low}} - f_{\text{offset}}$. The upper sub-block edge of the sub-block

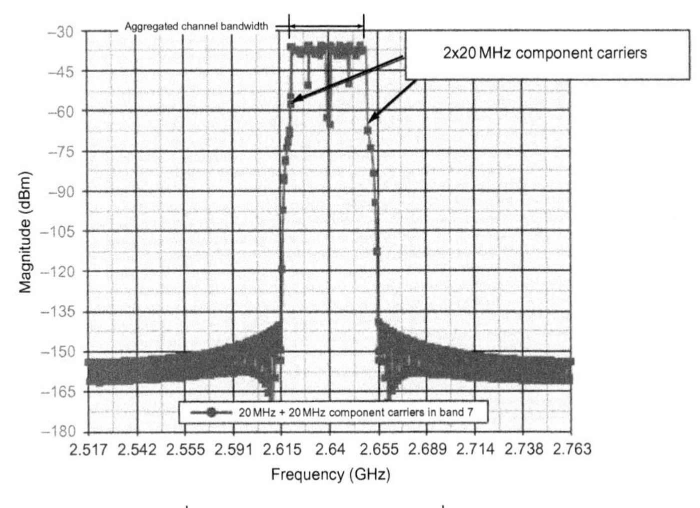

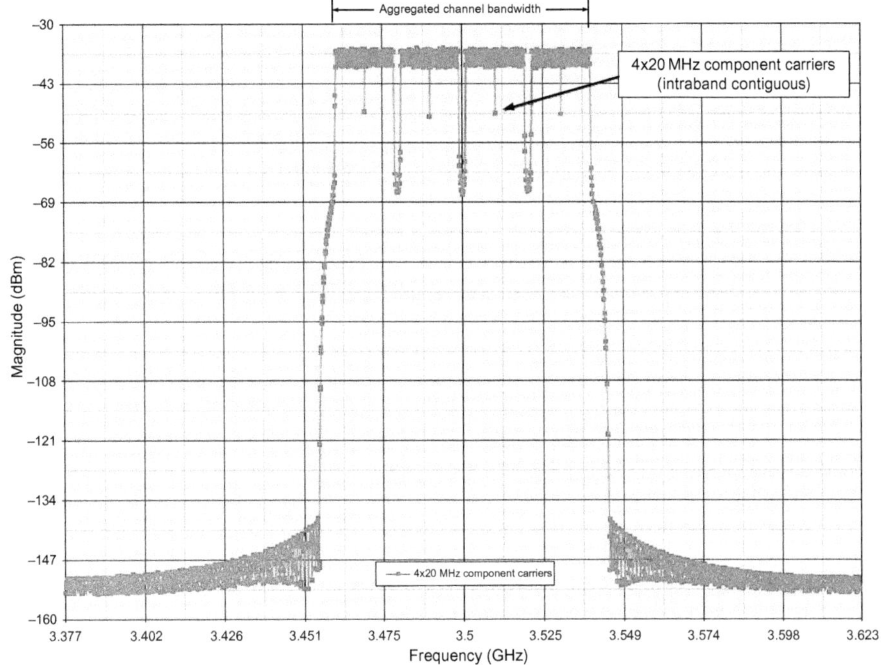

FIGURE 13.13

Snapshot of spectrum of two and four aggregated adjacent component carriers [17,18].

bandwidth is defined as $f_{edge\_block\_high} = f_{C\_block\_high} + f_{offset}$. The sub-block bandwidth $BW_{Channel\text{-}block}$ is alternatively defined as $BW_{Channel\_block} = f_{edge\_block\_high} - f_{edge\_block\_low}$ in MHz [2].

LTE Rel-10 specified the radio frequency aspects of inter-band carrier aggregation only for FDD, assuming aggregation of bands 1 and 5. However, network operators are interested in various band combinations, as shown in Table 13.2, depending on their available spectrum resources. Radio frequency properties of each band combination must be specified to comply with the regulations. In addition, inter-band carrier aggregation for TDD needs to be supported in future releases with the possibility of having different TDD uplink–downlink subframe configurations on different bands to coexist with the already deployed TDD systems in that band.

A carrier aggregation bandwidth class is defined by the aggregated transmission bandwidth configuration and the maximum number of component carriers supported by a UE. Table 13.3 shows supported bandwidth classes for LTE Rel-11 UE as defined in [1]. Figure 13.14 illustrates the standard notation used by the UE to indicate its support of carrier aggregation for a particular frequency band or band combination to the network. The example shows that in order to indicate support for intra-band, non-contiguous carrier aggregation in band 5 based on bandwidth class A, the notation would be CA_5A_5A. It informs the network that this device is able to receive (or transmit) two separate carriers in frequency band 5, each with a maximum bandwidth of 100 resource blocks, i.e., 20 MHz. Another example is CA_1C which indicates that the UE can operate with two contiguous component carriers with a maximum of 200 resource blocks which can be in the form of 15 + 15 or 20 + 20 MHz. Note that there might be different bandwidth combinations for each E-UTRA carrier aggregation configuration [1]. Table 13.4 summarizes some important parameters associated with UE categories as of LTE Rel-11.

13.6.3 Transmitter characteristics for carrier aggregation

The transmitter characteristics not only define RF requirements for the desired signal transmitted from the UE and the eNB, but also for the unavoidable unwanted emissions outside the transmission bandwidth. These requirements are fundamentally categorized as follows: (1) output power-level requirements set the limits for the maximum permissible transmitted power corresponding to dynamic variation of the power level during transmission; (2) transmitted signal quality requirements define the cleanliness of the transmitted signal and also the relationship between multiple transmitter branches; and (3) unwanted emissions requirements set the limits for all emissions outside of the transmitted carrier(s) operating band which are tightly coupled to regulatory requirements and coexistence with other systems. The following is a list of major UE and eNB transmitter attributes [14]:

- eNB: Output power level (maximum output power, output power dynamics, and on/off power for TDD), transmitted signal quality (frequency error, error vector magnitude (EVM), and time alignment between transmitter branches),

Table 13.2 Carrier Aggregation Bands and Band Aggregation as of LTE Rel-11 [1]

Band Combination	Carrier Aggregation Scenario	Uplink Frequency Range (MHz) $f_{UL\text{-}low} - f_{UL\text{-}high}$	Downlink Frequency Range (MHz) $f_{DL\text{-}low} - f_{DL\text{-}high}$	Duplex Mode
1	Intra-band contiguous	1920–1980	2110–2170	FDD
7		2500–2570	2620–2690	FDD
40		2300–2400	2300–2400	TDD
41		2496–2690	2496–2690	TDD
1 and 5	Inter-band	1920–1980	2110–2170	FDD
		824–849	869–894	
1 and 18		1920–1980	2110–2170	FDD
		815–830	860–875	
1 and 19		1920–1980	2110–2170	FDD
		830–845	875–890	
1 and 21		1920–1980	2110–2170	FDD
		1447.9–1462.9	1495.9–1510.9	
2 and 17		1850–1910	1930–1990	FDD
		704–716	734–746	
2 and 29		1850–1910	1930–1990	FDD
		N/A (downlink only SCell)	717–728	
3 and 5		1710–1785	1805–1880	FDD
		824–849	869–894	
3 and 7		1710–1785	1805–1880	FDD
		2500–2570	2620–2690	
3 and 8		1710–1785	1805–1880	FDD
		880–915	925–960	
3 and 20		1710–1785	1805–1880	FDD
		832–862	791–821	
4 and 5		1710–1755	2110–2155	FDD
		824–849	869–894	
4 and 7		1710–1755	2110–2155	FDD
		2500–2570	2620–2690	
4 and 12		1710–1755	2110–2155	FDD
		699–716	729–746	
4 and 13		1710–1755	2110–2155	FDD
		777–787	746–756	
4 and 17		1710–1755	2110–2155	FDD
		704–716	734–746	
4 and 29		1710–1755	2110–2155	FDD
			717–728	

(Continued)

Table 13.2 (Continued)

Band Combination	Carrier Aggregation Scenario	Uplink Frequency Range (MHz) $f_{UL\text{-}low} - f_{UL\text{-}high}$	Downlink Frequency Range (MHz) $f_{DL\text{-}low} - f_{DL\text{-}high}$	Duplex Mode
		N/A (downlink only SCell)		
5 and 12		824–849	869–894	FDD
		699–716	729–746	
5 and 17		824–849	869–894	FDD
		704–716	734–746	
7 and 20		2500–2570	2620–2690	FDD
		832–862	791–821	
8 and 20		880–915	925–960	FDD
		832–862	791–821	
11 and 18		1427.9–1447.9	1475.9–1495.9	FDD
		815–830	860–875	

unwanted emissions (operating band unwanted emissions, ACLR, spurious emissions, occupied bandwidth, and transmitter intermodulation).

- UE: Output power level (transmit power, output power dynamics, and power control), transmitted signal quality (frequency error and transmit modulation quality), unwanted emissions (spectrum emission mask, ACLR, spurious emissions, occupied bandwidth, and transmit intermodulation).

The ACLR is directly related to the operating point of the power amplifier. In general, leakage into adjacent channels increases abruptly as the power amplifier is driven into its non-linear operating region at the highest output power levels due to the intermodulation products. As a result, it is important that the peak output power of the UE does not drive the power amplifier too far into its non-linear region. On the other hand, most power amplifiers are designed to operate efficiently only in a small region at the top of the linear operating region. This corresponds to the rated output power of the power amplifier below which the efficiency drops abruptly. Since high efficiency is crucial to ensuring a long battery life for the UE, it is also desirable to maintain the power amplifier operating as close as possible to the top of the linear operating region. If the ACLR and spectrum mask requirements cannot be met, typically the UE output power must be reduced to bring the leakage to acceptable levels. This can be achieved without excessive loss of efficiency by reducing the peak output power of the power amplifier, a process known as derating. The amount of derating required is highly dependent on the peak to average power ratio (PAPR) and bandwidth of the transmitted signal. In general, for any given channel bandwidth and power amplifier, transmissions with a larger occupied bandwidth create more OOB emissions,

Table 13.3 Carrier Aggregation Bandwidth Classes and Corresponding Nominal Guard Bands [1]

Carrier Aggregation Bandwidth Class	Aggregated Transmission Bandwidth Configuration	Maximum Number of Component Carriers	Nominal Guard Band $BW_{\text{Guard Band}}$
A	$N_{\text{RB-Aggregated}} \leq 100$	1	$0.05BW_{\text{Channel-1}}$
B	$N_{\text{RB-Aggregated}} \leq 100$	2	–
C	$100 < N_{\text{RB-Aggregated}} \leq 200$	2	$0.05\max(BW_{\text{Channel-1}}, BW_{\text{Channel-2}})$
D	$200 < N_{\text{RB-Aggregated}} \leq 300$	–	–
E	$300 < N_{\text{RB-Aggregated}} \leq 400$	–	–
F	$400 < N_{\text{RB-Aggregated}} \leq 500$	–	–

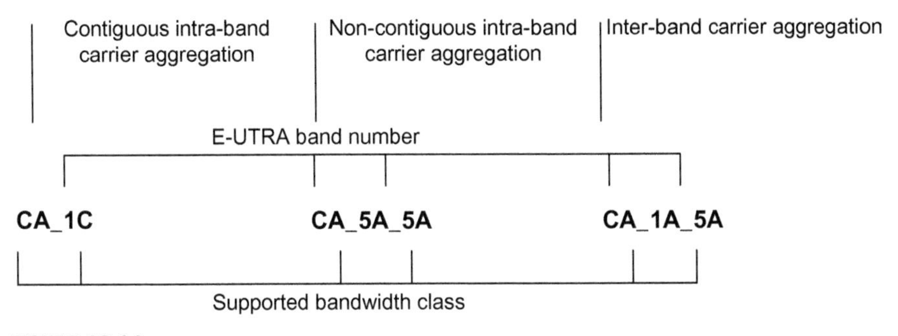

FIGURE 13.14

Standard notation for carrier aggregation support (type, frequency band, and bandwidth) [1,16].

resulting in larger adjacent channel leakage than transmissions with lower occupied bandwidth. The increase in OOB emissions from the larger occupied bandwidth of the LTE signal is mainly due to increased adjacent channel occupancy by third and fifth order intermodulation products (see Figure 13.15). For an LTE uplink transmitted signal, certain combinations of resource block allocations and modulation schemes create more OOB emissions than others [13].

For low-bandwidth applications such as VoIP, which typically use QPSK modulation, no power derating is required from the normal-rated maximum UE output power, ensuring broad coverage for such applications, since full output power of the power amplifier can be used to counter the path loss effect at the cell-edge. Moreover, the structure of the LTE uplink signal, with control information being usually positioned at the channel edge and high-bandwidth data transmissions toward the middle of the band, helps to improve OOB emissions and reduce ACLR. Taking the above considerations into account, the total power

Table 13.4 UE Categories as of LTE Rel-11 [8]

UE Category	Peak Data Rate Downlink/ Uplink (Mbps)	Maximum Number of DL-SCH Transport Block Bits Received within a TTI	Maximum Number of Bits of a DL-SCH Transport Block Received within a TTI	Total Number of Soft Channel Bits in the Downlink	Maximum Number of Supported Layers for Spatial Multiplexing in the Downlink	Maximum Number of UL-SCH Transport Block Bits Transmitted within a TTI	Maximum Number of Bits of a UL-SCH Transport Block Transmitted within a TTI	Support for 64QAM in Uplink	Total Layer 2 Buffer Size (Bytes)	Maximum Number of Bits of an MCH Transport Block Received within a TTI
Category 1	10/5	10,296	10,296	250,368	1	5160	5160	No	150,000	10,296
Category 2	50/25	51,024	51,024	1,237,248	2	25,456	25,456	No	700,000	51,024
Category 3	100/50	102,048	75,376	1,237,248	2	51,024	51,024	No	1,400,000	75,376
Category 4	150/50	150,752	75,376	1,827,072	2	51,024	51,024	No	1,900,000	75,376
Category 5	300/75	299,552	149,776	3,667,200	4	75,376	75,376	Yes	3,500,000	75,376
Category 6	300/50	301,504	149,776 (4 layers) 75,376 (2 layers)	3,654,144	2 or 4	51,024	51,024	No	3,300,000	(75,376)
Category 7	300/150	301,504	149,776 (4 layers) 75,376 (2 layers)	3,654,144	2 or 4	102,048	51,024	No	3,800,000	(75,376)
Category 8	1200/600	2,998,560	299,856	35,982,720	8	1,497,760	149,776	Yes	4,220,0000	(75,376)

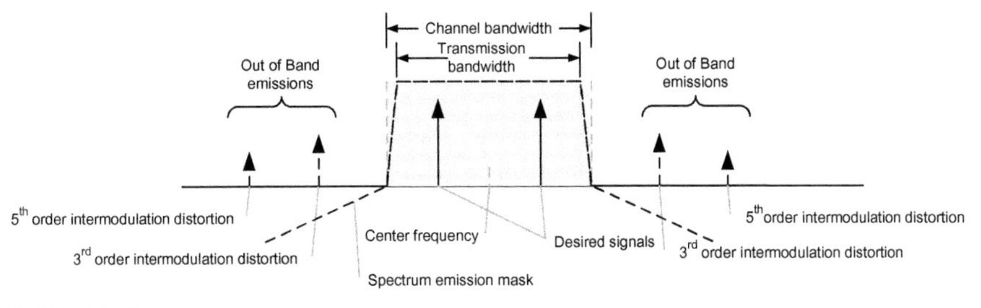

FIGURE 13.15

Illustration of the IMD components [20].

derating required to meet a target ACLR value can be decomposed into a term corresponding to the occupied bandwidth as a proportion of the channel bandwidth, and a term corresponding to the waveform of the transmitted signal. Small resource assignments at the edges of the band behave as tones and hence may produce highly concentrated IMD products [14]. Therefore, for simultaneous PUSCH and PUCCH transmission, the spectral emission mask is expected to be the limiting requirement, as it is for LTE Rel-8 with full resource allocation (see Figure 13.6).

As mentioned earlier, the ACLR metric defines the ratio of the power transmitted in the assigned channel bandwidth to the power of the unwanted emissions transmitted in an adjacent channel. There is a corresponding receiver requirement known as adjacent channel selectivity (ACS), which defines a receiver's ability to suppress a signal in an adjacent channel. Therefore, from the receiver's point of view, the unwanted emissions as a result of transmissions in the adjacent bands in the desired signal's receiver bandwidth are represented by ACLR, and the ability of the receiver to suppress the interfering signal in the adjacent channel is defined by ACS. The two parameters when combined define the total leakage between two transmissions on adjacent channels. That ratio is called the adjacent channel interference ratio (ACIR), and is defined as the ratio of the power transmitted on one channel to the total interference received by a receiver on the adjacent channel due to both transmitter RF filter imperfection (ACLR) and receiver RF filter imperfection (ACS) [14].

In LTE, the EVM is required to be less than 17.5% for QPSK, 12.5% for 16QAM and 8% for 64QAM (applicable to the downlink only). The EVM values are designed to correspond to no more than 5% loss in average and cell-edge throughputs in typical deployment scenarios. At the link level, EVM is equivalent to signal to noise ratio loss [13]. The occupied bandwidth is defined as the bandwidth containing 99% of the total integrated mean power of the transmitted spectrum on the assigned channel. The occupied bandwidth for all transmission bandwidth configurations (resource blocks) must be less than the channel bandwidth. For intra-band contiguous carrier aggregation the occupied bandwidth is a measure of the bandwidth containing 99% of the total integrated power of the transmitted spectrum. The occupied bandwidth must be less than the aggregated channel bandwidth [1].

The output power dynamics are impacted by the UE transmitter architecture, which may be based on single or multiple power amplifiers. Figure 13.16 illustrates various options for UE transmitter architecture which can be used to support different carrier aggregation scenarios. As shown in Figure 13.16 for alternative 1, if two component carriers are contiguous, then a single IFFT module can be used to generate both OFDM signals. A single RF mixer can be used to convert the baseband signal frequency to RF frequency followed by a single power amplifier to amplify the power of the transmitted signal. If two component carriers are non-contiguous but within the same frequency band, then alternatives 2 or 3 shown in Figure 13.16 may be used as a general transmitter architecture. However, for non-contiguous component carriers in different frequency bands, separate IFFT modules, mixers, and power amplifiers would be required to process the OFDM signals, unless a multi-band power amplifier is utilized.

13.6.4 Receiver characteristics for carrier aggregation

Although the requirements for LTE receivers are similar to those defined for UMTS, they are defined differently due to the flexible bandwidth properties of LTE. The receiver RF requirements are primarily divided into three categories: (1) sensitivity and dynamic range requirements for receiving the desired signal; (2) receiver susceptibility to interfering signals defining receiver's vulnerability to different types of interfering signals at different frequency offsets; and (3) unwanted emission limits. The following is a list of attributes for the UE and eNB receivers [14]:

- eNB: Sensitivity and dynamic range (reference sensitivity, dynamic range, and in-channel selectivity), receiver susceptibility to interfering signals (OOB blocking, in-band blocking, narrowband blocking, ACS, and receiver intermodulation), and unwanted emissions from the receiver (receiver spurious emissions). The spurious emissions power is the power of emissions generated or amplified in a receiver that appear at the eNB receiver antenna connector. The requirements apply to all eNBs with separate receive and transmit antenna ports. The test for the FDD eNB must be performed when transmitter and receiver are on, with the transmit port properly terminated. For the TDD eNB with a common receive and transmit antenna port, the requirement applies during the transmitter OFF period. The blocking characteristic is a measure of the receiver's ability to receive a desired signal at its assigned channel in the presence of an unwanted interferer. Narrowband blocking is similar to ACS, but with interfering signals consisting of only one resource block.
- UE: Sensitivity and dynamic range (reference sensitivity power level and maximum input level), receiver susceptibility to interfering signals (OOB blocking,

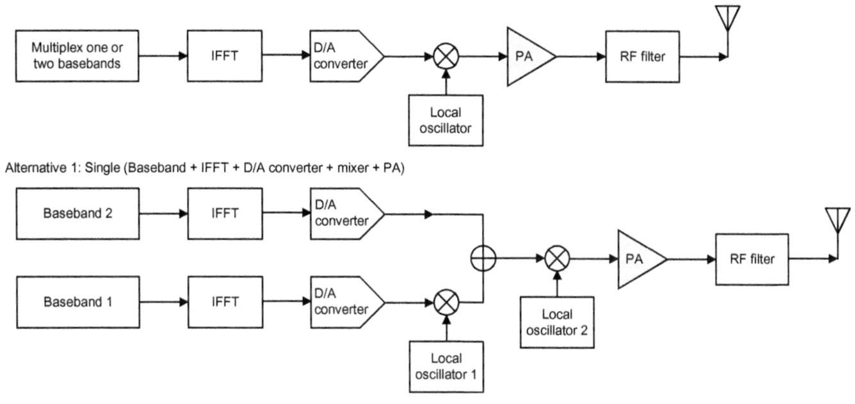

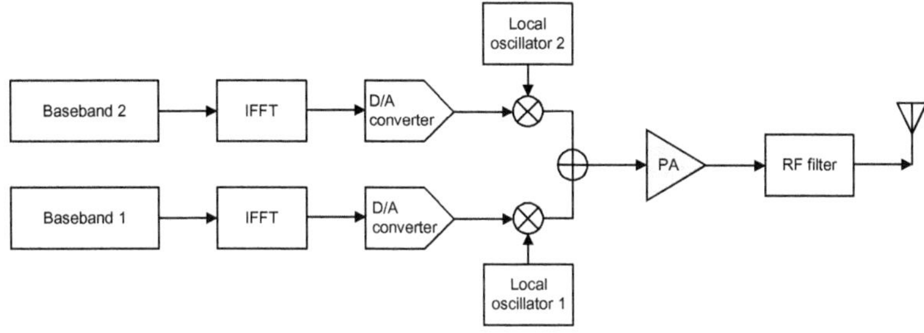

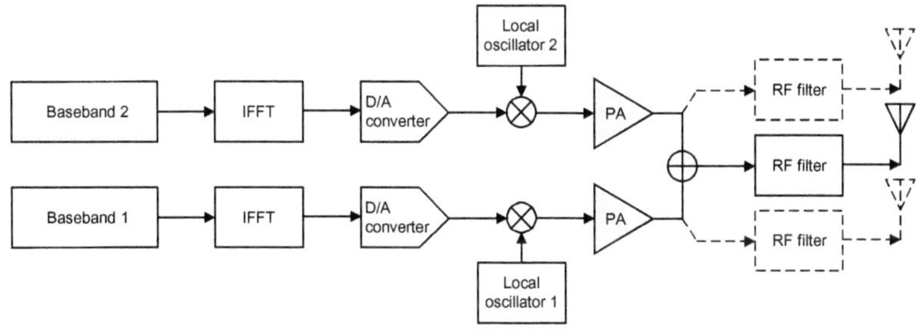

FIGURE 13.16

Example transmitter implementation for intra-band/inter-band carrier aggregation [11].

spurious response, in-band blocking, narrowband blocking, ACS, and intermodulation characteristics), and unwanted emissions from the receiver (receiver spurious emissions). Third and higher order mixing of the two interfering RF signals can produce an interfering signal in the band of the desired channel. Intermodulation response rejection is a measure of the capability of the receiver to receive a wanted signal on its assigned channel frequency in the presence of two interfering signals (a continuous wave and E-UTRA signals).

The LTE RF receiver requirements were partly derived from UMTS. This was meant to ensure that the implementation of an LTE RF receiver is not going to be more complex than that of UMTS in order to reduce redesigning efforts. The main differences between the LTE and UMTS RF receiver requirements arise from the variable channel bandwidth and the different multiple access schemes. The receiver RF requirements are based on a number of specific assumptions for testing purposes including integrated antennas with 0 dBi gain; two antenna ports; test signals of equal power level applied to each antenna port; and use of a maximal ratio combining to combine the signals. It is assumed that the signals are received from independent AWGN channels, thus signal addition would provide 3 dB diversity gain.

LTE supports different MCSs; however, the RF receiver specifications are defined for only two MCSs, referred to as reference channels, which are at the extremes of the available range, in order to reduce the number of conformance tests which have to be performed. The low-SNR reference channel uses QPSK with a code rate of 1/3, while the high SNR reference channel uses 64QAM with a code rate of 3/4. For each of the reference channels, the SINR requirements, at which 95% throughput is achieved, are specified.

When the LTE receiver is tested in full-duplex FDD mode, the transmitter must also be operating so that signal leakage from the transmitter to the receiver due to insufficient isolation of the duplexer is taken into consideration. This condition does not apply to half-duplex FDD or TDD operation. In practice, the transmitted signal leakage interferes with the receiver not only because of the power of the fundamental components, but also because of the OOB phase noise of the transmitted signal when it falls into the receiver band. In LTE, the spurious emissions from the UE transmitter in its own receive band are required to be −47 dBm or less, measured in a 1 MHz bandwidth. This corresponds to −107 dBm/Hz. The maximum transmit power for an LTE mobile device is +23 dBm for power class 3, thus the spurious emissions requirement is −130 dBc/Hz. The LTE receiver sensitivity is measured with the transmitter operating at the full power for its class. In order to allow the transmitter to degrade the receiver sensitivity by no more than 0.5 dB, the transmitter noise needs to be 9 dB below the noise floor at the antenna connector. If we assume that the spurious emissions of the transmitter in the corresponding receiver band are just at the limit of the specifications, it can be shown that the duplexer isolation needs to be at least 78 dB for the LTE FDD mode. However, making duplexers smaller and cheaper is often achieved at the expense of compromised isolation, and typical duplexers might provide just 45−55 dB isolation [14]. Therefore, the transmit signal could substantially desensitize the receiver. This is not acceptable, thus the LTE transmitter has to achieve spurious emissions about 20 dB lower than the transmitter requirements within the receiver band. The transmit power at the input to the low-noise amplifier is equal to the transmit power at the antenna plus the duplexer attenuation from transmit port to antenna (about 2 dB) minus the duplexer transmit-to-receiver isolation (~50 dB). Therefore, at the maximum LTE UE output power of +23 dBm, the mean transmitter power leakage to the low-noise amplifier is −25 dBm [13].

This value is calculated based on the average transmit power, but the transmitted signal contains amplitude modulation, and therefore the peak signal will be higher. The transmitter leakage problem presents a particularly challenging requirement, if applied to RF bands with a small frequency separation. In FDD systems, when performing a selectivity or blocking test with a single interfering signal in one of the adjacent channels, the receiver is also exposed to its own transmitted signal as a second interferer, which, in the worst case scenario, can cause intermodulation, in which the interferers are mixed together to generate in-band IMD products.

The maximum input level is the maximum mean received signal strength, measured at each antenna port, at which there is sufficient SINR for a specified modulation scheme to meet a minimum throughput requirement. In LTE, the maximum input level is only specified for the high SINR reference channel, assuming that the high SINR reference channel is about 4 dB higher due to the PAPR of the signal. The downlink requirement is for a maximum average input level of −25 dBm at any channel bandwidth with full resource block allocation. In FDD systems, the requirement must be met when the transmitter is set at 4 dB below the maximum output power. If PAPR is measured on a bitwise basis and plotted as a complementary cumulative probability distribution, then for the LTE downlink it reaches a value of around 11 dB for QPSK and 11.5 dB for 64QAM, over a window of 106 bits. In OFDM, the PAPR is dominated by the multi-carrier nature of the signal, and therefore does not vary much with modulation [13].

The primary purpose of the reference sensitivity requirement is to verify the receiver noise figure, which is a measure of how much the receiver's RF signal chain degrades the SNR of the received signal. For this reason, a low-SNR transmission scheme using QPSK is chosen as a reference channel for the reference sensitivity test. The reference sensitivity is defined at a receiver input level where the throughput is 95% of the maximum throughput for the reference channel. For the base station, reference sensitivity could potentially be defined for a single resource block up to a group covering all resource blocks. Considering the complexity, a maximum granularity of 25 resource blocks has been chosen, which means that for channel bandwidths larger than 5 MHz, sensitivity is verified over multiple adjacent 5 MHz blocks, while it is only defined over the full channel for smaller channel bandwidths. For the UE, reference sensitivity is defined for the full channel bandwidth signals and with all resource blocks allocated to the desired signal. For channel bandwidths greater than 5 MHz in some operating bands, the nominal reference sensitivity needs to be met with a minimum number of allocated resource blocks [14]. In other words, the reference sensitivity level is the minimum mean received signal strength applied to both antenna ports (assuming two receive antennas) at which there is sufficient SINR for the specified modulation scheme to meet a minimum throughput requirement of 95% of the maximum possible value. It is measured with the transmitter at full power. The reference sensitivity is a range of values that can be calculated as $P_{Reference\_Sensitivity} = 10\log(kT_0B) + NF + SINR + L_{Implementation\_Margin} - 3$ (dBm), where kT_0B is the thermal noise power in dBm, B is the noise equivalent bandwidth ($\sim$system bandwidth), NF is the maximum receiver noise figure, $SINR$ is the minimum SINR for the chosen MCS, $L_{Implementation\_Margin}$ is the implementation margin, and -3 represents the receive diversity gain [13]. Note that the noise floor for any receiver is defined as $Noise\_Floor = 10\log(kT_0B) + NF = -174 + NF + 10\log B$ (dBm), where $NF = SNR_{in} - SNR_{out}$ (in dB) is a measure of SINR degradation by the components in the RF signal path, including the RF filters and low-noise amplifier. The thermal noise power density $kT_0 = -174$ dBm/Hz is measured at typical room temperature of 290°K. The LTE specifications require $NF \leq 9$ dB for the UE and $NF \leq 5$ dB for the eNB, nevertheless, commercial RF components and UE receivers can achieve lower than these limits [13].

The UE receiver dynamic range requirement is meant to ensure that the receiver can operate at received signal levels considerably higher than the reference sensitivity. The base station dynamic range testing assumes the presence of increased interference and corresponding higher signal levels, thereby testing the effects of different receiver impairments. In order to drive the receiver to the worst condition, a higher SNR transmission scheme using 16QAM is applied for the test. In order to further drive the receiver to processing higher signal levels, an interfering AWGN signal at 20 dB above the assumed noise floor is added to the received signal. The dynamic range requirement for the UE is specified as a maximum signal level at which the throughput requirement is met.

Blocking corresponds to the scenario when strong interfering signals received outside the operating band (OOB blocking) or inside the operating band (in-band blocking), but not adjacent to the desired signal. In-band blocking includes interferers in the first 20 MHz outside the band for the base station and the first 15 MHz for the UE. The scenarios are modeled with a continuous wave signal for the OOB case and an E-UTRA signal for the in-band case. There are additional base station's blocking requirements for the scenario when the base station is co-located with another base station in a different operating band. For the UE, a fixed number of exceptions are allowed from the OOB blocking requirement, for each assigned frequency channel and at the respective spurious response frequencies. At those frequencies, the UE must comply with the more relaxed spurious response requirement. Note that these effects and satisfaction of the requirements become more challenging when carrier aggregation is supported.

ACS is the case when there is a strong signal in the channel adjacent to the desired signal, and is closely related to the corresponding ACLR requirement. For the UE, the ACS is specified for two cases with a lower and a higher signal level. In-channel selectivity is the case when there are multiple received signals of different received power levels inside the channel bandwidth, where the performance of the weaker desired signal is verified in the presence of the stronger interfering signal. In-channel selectivity is only specified for the base station. Narrowband blocking is an adjacent strong narrowband interferer, which in the requirement is modeled as a single resource block LTE signal for the base station and a continuous wave signal for the UE. Receiver intermodulation occurs when there are two interfering signals in the spectral proximity of the desired signal, where the interferers are one continuous wave and one LTE signal. The interferers are placed in frequency in such a way that the main intermodulation product falls within the desired signal's channel bandwidth. There is also a narrowband intermodulation requirement for the base station where the continuous wave signal is very close to the wanted signal, and the LTE interferer is a single resource block signal. For all requirements except in-channel selectivity, the wanted signal uses the same reference channel as in the corresponding reference sensitivity requirement. When the interference is added, the same 95% relative throughput is met as for the reference channel, but at a desensitized higher desired signal level [14].

References

3GPP Specifications[4]

[1] 3GPP TS 36.101, Evolved Universal Terrestrial Radio Access (E-UTRA); User Equipment (UE) Radio Transmission and Reception (Release 11), December 2012.
[2] 3GPP TS 36.104, Evolved Universal Terrestrial Radio Access (E-UTRA); Base Station (BS) Radio Transmission and Reception (Release 11), December 2012.
[3] 3GPP TS 36.211, Evolved Universal Terrestrial Radio Access (E-UTRA), Physical Channels and Modulation (Release 11), December 2012.
[4] 3GPP TS 36.212, Evolved Universal Terrestrial Radio Access (E-UTRA) Multiplexing and Channel Coding (Release 11), December 2012.
[5] 3GPP TS 36.213, Evolved Universal Terrestrial Radio Access (E-UTRA); Physical Layer Procedures (Release 11), December 2012.
[6] 3GPP TS 36.214, Evolved Universal Terrestrial Radio Access (E-UTRA); Physical Layer—Measurements (Release 11), December 2012.
[7] 3GPP TS 36.300, Evolved Universal Terrestrial Radio Access (E-UTRA) and Evolved Universal Terrestrial Radio Access Network (E-UTRAN) Overall Description, Stage 2 (Release 11), December 2012.
[8] 3GPP TS 36.306, Evolved Universal Terrestrial Radio Access (E-UTRA); User Equipment (UE) Radio Access Capabilities (Release 11), December 2012.
[9] 3GPP TS 36.321, Evolved Universal Terrestrial Radio Access (E-UTRA); Medium Access Control (MAC) Protocol Specification (Release 11), December 2012.
[10] 3GPP TS 36.331 Evolved Universal Terrestrial Radio Access (E-UTRA); Radio Resource Control (RRC); Protocol Specification (Release 11), December 2012.
[11] 3GPP TR 36.815, Further Advancements for E-UTRA; LTE-Advanced Feasibility Studies in RAN WG4 (Release 9), June 2010.

[4]3GPP specifications can be found at the following URL:http://www.3gpp.org/ftp/Specs/archive/

Books

[12] C. Johnson, *Long Term Evolutionin Bullets*, second ed., Create Space Independent Publishing Platform, July 2012.
[13] S. Sesia, I. Toufik, M. Baker, *LTE, the UMTSLong Term Evolution, From Theory to Practice*, second ed., John Wiley & Sons, August 2011.
[14] E. Dahlman, S. Parkvall, J. Skold, *4G, LTE/LTE-Advancedfor Mobile Broadband*, Academic Press, May 2011.
[15] X. Zhang, X. Zhou, *LTE-Advanced Air Interface Technology*, CRC Press, September 2012.

Articles

[16] A. Roessler, M. Kottkamp, M. Sandra, *Carrier Aggregation—(One) Key Enabler for LTE-Advanced*, Rohde & Schwarz, October 2012.
[17] J. Xu, *LTE-Advanced Signal Generation and Measurement Using SystemVue*, Agilent Technologies, December 2010.
[18] Application Note, Introducing LTE-Advanced, Agilent Technologies, March 2011.
[19] C. Hoymann, et al., A lean carrier for LTE, IEEE Commun. Mag. (February 2013).
[20] Application Note, Understanding Carrier Aggregation, Anritsu, February 2013.
[21] N. Miki, et al., Carrier aggregation for bandwidth extensions in LTE-advanced, NTT DoCoMo Tec. J. 12 (2) (September 2010).
[22] E. Seidel, LTE-A Carrier Aggregation Enhancements in Release 11, NOMOR Research GmbH, Munich, Germany, August, 2012.
[23] M. Iwamura, et al., Carrier aggregation framework in 3GPP LTE-advanced, IEEE Commun. Mag. (August 2010).
[24] C. M.Park, et al., System level performance evaluation of various carrier aggregation scenarios in LTE-advanced, 15th International Conference on Advanced Communications Technology (ICACT), January 2013.
[25] A. Li, et al., Search space design for cross-carrier scheduling in carrier aggregation of LTE-advanced system, Proceedings of 2011 IEEE ICC.
[26] M. Abduljawad, et al., Carrier aggregation in long term evolution-advanced, 2012 IEEE Control and System Graduate Research Colloquium (ICSGRC 2012).
[27] K.I. Pedersen, et al., Carrier aggregation for LTE-advanced: functionality and performance aspects, IEEE Commun. Mag. (June 2011).
[28] E. Seidel, 3GPP LTE—A Standardization in Release 12 and Beyond, NOMOR Research GmbH, Munich, Germany, January 2013.
[29] Z. Shen, et al., Overview of 3GPP LTE-advanced carrier aggregation for 4G wireless communications, IEEE Commun. Mag. (February 2012).
[30] A. Damnjanovic, et al., A survey on 3GPP heterogeneous networks, IEEE Wireless Commun. (June 2011).
[31] A. Ghosh, et al., Heterogeneous cellular networks: from theory to practice, IEEE Commun. Mag. (June 2012).
[32] Report ITU-R M.2133, Requirements, evaluation criteria and submission templates for the development of IMT-advanced, November 2008.
[33] Report ITU-R M.2134, Requirements related to technical performance for IMT-advanced radio interface(s), November 2008.

CHAPTER

14 Enhanced Inter-cell Interference Coordination and Multi-radio Coexistence

CHAPTER CONTENTS

In this chapter, we discuss two important features that were included in the recent releases of LTE to address inter-cell interference problems in LTE systems and in-device inter-radio interference issues, where both problems adversely affect the system capacity, user throughput, and normal operation of the LTE system.

The proliferation of Internet-connected mobile devices will continue to drive growth in data traffic in an exponential order, forcing network operators to dramatically increase the capacity of their networks. The ever-increasing demand for mobile communication services has made it difficult for the conventional homogeneous networks to provide the required capacity and user throughput as they have approached their theoretical limits. One of the most promising and cost-effective solutions to improve the system capacity is to increase the node density, i.e., to deploy low-power nodes such as relay, pico-cell, femto-cell, and remote radio heads in homogeneous networks. The new network architecture is known as heterogeneous networks (HetNets). The deployment of low-power nodes can

LTE-Advanced.

provide a more flexible and more effective way to eliminate the coverage holes in macro-cells and improve the system capacity. Therefore, it is considered as a promising approach in the next generation wireless communication systems such as systems beyond IMT-Advanced [16,17]. One of the key issues in the new network architecture is the handover performance deterioration. In co-channel HetNets, the cell-edge UEs can experience more serious interference. The basic principle of inter-cell interference coordination (ICIC) is to avoid high-power transmission on time–frequency resources on which cell-edge users are scheduled in neighboring cells, users that would otherwise experience high interference and correspondingly low data rates. In the next sections, we study the causes and remedies of inter-cell interference in the HetNets [2].

We will then discuss signaling and procedures for in-device coexistence (IDC) and for interference avoidance. An increasing number of UEs are equipped with multiple radio transceivers for LTE, Wi-Fi, Bluetooth (BT), Global Navigation Satellite System (GNSS), and so on. The studies conducted for LTE Rel-11 IDC work item have shown that existing radio resource management mechanisms in some cases are not adequate to overcome the coexistence issues, and some enhanced signaling and other procedures are necessary to avoid or mitigate the collocated multi-radio interference in the identified usage scenarios. Therefore, a new IDC indication message has been defined in LTE Rel-11. This message enables the UE to alert the network to an interference issue, and provide information regarding the direction and nature of the interference, which may be identified in either the time- or frequency-domain. Upon receipt of the IDC message, the network will take appropriate steps to alleviate the problem by reallocating radio resources or configuring discontinuous reception mode parameters [1].

14.1 Intra-cell/inter-cell interference mitigation methods

The classic design principle of cellular communications systems is to divide a geographical area into non-overlapping cells, each served by a single base station. The intercell interference is a serious problem of such systems, in particular when frequency reuse one is utilized. One way to reduce the inter-cell interference within the system is to employ efficient interference averaging mechanisms that organize transmission such that the interference is evenly distributed among all users.

There are three major frequency reuse schemes in cellular systems for mitigating inter-cell interference: hard frequency reuse; fractional frequency reuse (FFR); and soft frequency reuse. A hard frequency reuse scheme splits the system bandwidth into a number of non-overlapping sub-bands according to a predetermined reuse factor, and lets neighboring cells transmit on different sub-bands (scenario A in Figure 14.1). As shown in Figure 14.2, an FFR scheme partitions the given bandwidth into an inner and an outer partition. It then allocates the inner part to the users closer to the base station with reduced power and applies a

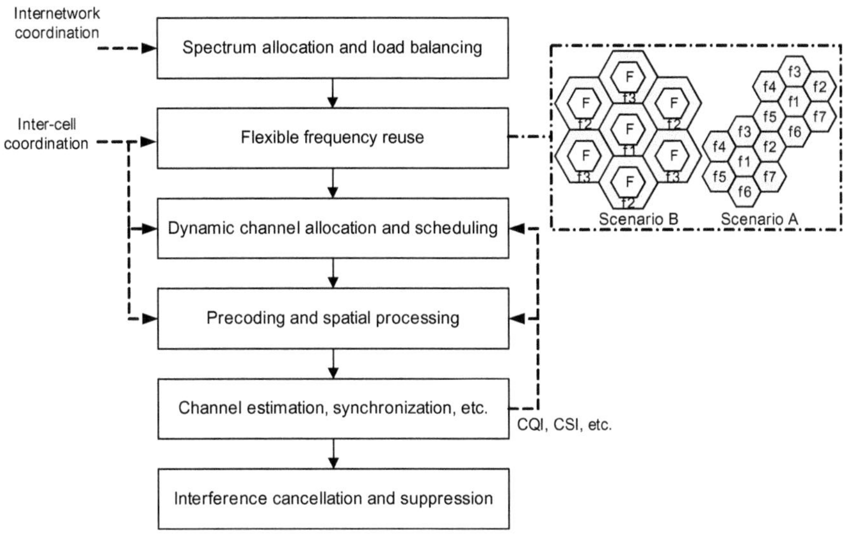

FIGURE 14.1

Co-channel interference mitigation schemes [12].

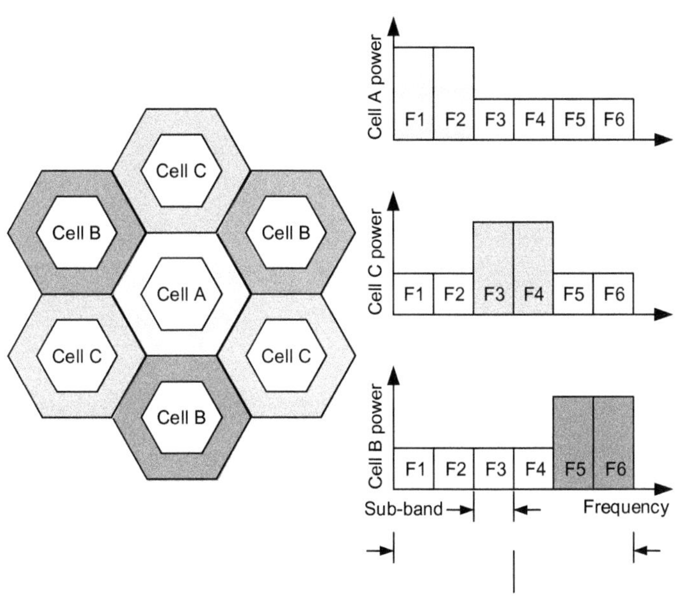

FIGURE 14.2

Illustration of fractional frequency reuse concept [17].

frequency reuse factor of 1, i.e., the inner part is completely reused by all base stations. For users closer to the cell-edge, a fraction of the outer part of the bandwidth is dedicated with a frequency reuse factor greater than 1 (see scenario B in Figure 14.1). With soft frequency reuse, the overall bandwidth is shared by all base stations, i.e., a reuse factor of 1 is applied, but for the transmission on each sub-band, the base stations are restricted to a certain power limit.

Depending on whether noise or inter-cell interference is the dominant effect, the cellular system is referred to as *noise-limited* or *interference-limited.* In an interference-limited system, the carrier to interference plus noise ratio (CINR) can approximately be replaced by the carrier to interference ratio (CIR), and the inter-cell interference plays an important role in terms of achievable system capacity and coverage. In a network deployment with a smaller frequency reuse factor, severe inter-cell interference may occur in the system. In fact, the closer the user is to the cell-edge, the more severe the inter-cell interference can be. Due to a large distance between the transmitter and the receiver, the desired signal power P_{C_j} is then relatively weak compared to the inter-cell interference signal strength. By increasing the frequency reuse factor, the inter-cell interference can be effectively limited, however, at the expense of sacrificing the available bandwidth for each cell and more complex network planning. In addition, in order to increase the data rate and spectral efficiency without requiring more spectrum resources in cellular networks, two traditional spatial reuse mechanisms, cell splitting and sectorization are often utilized. Several advanced antenna techniques such as MIMO and adaptive beamforming can also be used to enhance cellular system performance.

When properly selected transmission format and link adaptation are employed, the capacity of cellular systems depends essentially on the average quality of the received radio signals, which can be characterized by a CINR metric, defined as follows:

$$\mathrm{CINR} = \frac{P_{C_j}}{P_N + \sum_{k \neq j} P_{I_k}}$$

where P_{C_j} is the received carrier signal power from the jth (serving) cell, P_{I_k} denotes the received co-channel interference signal from the kth cell (other than the serving cell), and P_N is the noise power. The received signal power P_{C_j} is determined by the distance between the transmitter and the receiver, the transmission power, and the associated path loss caused by the environment. The most dominant noise type is thermal noise originated from hardware components in the system. The background radiation originating from cosmic or terrestrial sources picked up by the receive antennas also adds to the overall noise in the receiver. As for the perceived interference, actually, the interference P_{I_k} includes not only co-channel interference, which is also known as inter-cell interference, but also intra-cell interference. Since inter-cell interference is the dominant component in an OFDMA system (due to orthogonality of the users in time and frequency) compared to the intra-cell interference, our focus in this section will therefore be only on the inter-cell interference. The inter-cell interference depends on the number of co-channel cells, the distance between these cells, the transmission power, and the terrain characteristics, in which the first two factors depend on the choice of the frequency reuse factor.

The minimization of the inter-cell interference is desirable to fulfill the following objectives: (i) maximization of the overall network capacity; (ii) maximization of the cell-edge throughput; and (iii) enhancements in terms of both the overall network capacity and the cell-edge performance. Interference mitigation techniques can be generally classified into three major categories: interference cancellation, interference randomization, and interference coordination techniques.

Inter-cell interference cancellation techniques have been investigated and deployed with varying degrees of success in terrestrial mobile networks for more than 20 years. The basic principle of inter-cell interference cancellation techniques is to regenerate the interfering signals and subsequently subtract them from the desired signal. Various inter-cell interference cancellation techniques have been proposed in the literature, and they are mainly characterized as filter-based and multi-user detection (MUD) schemes [8–10]. Filter-based approaches try to mitigate intercell interference by means of linear filters and interference models. In contrast, MUD directly includes the interfering signals in the decoding process. This is done by jointly decoding the desired signal and the interfering signals, or by decoding and subtracting the interfering signals from the signal of interest. Inter-cell interference cancellation techniques can also be jointly used with multi-user MIMO techniques, if the UEs are equipped with multiple-receive antennas. From an implementation viewpoint, interference cancellation does not require any modification of the standards, making it an attractive technique. Although inter-cell interference cancellation promises significant gains and its algorithms are mature, it is considered mostly as a technique for the uplink due to processing complexity which scales exponentially with the number of active users. Furthermore, it requires exchange of information in real-time between the base stations about every few milliseconds to maximize the system gain. Moreover, inter-cell interference cancellation can potentially improve channel performance to that of AWGN if accurate channel estimation is available [8–10].

In contrast to inter-cell interference cancellation, which tries to eliminate the effect of interfering sources in the received signal, inter-cell interference randomization, which is also known as interference whitening or interference averaging, attempts to make the inter-cell interference appear as background noise, i.e., it averages the interference across the data symbols of a data block or the entire frequency band. The approaches include frequency hopping and interleave division multiple access. However, these methods randomize the interference into white Gaussian noise, which cannot reduce interference in nature. Therefore, these approaches can hardly achieve a substantial performance improvement [8–10].

ICIC techniques, also known as inter-cell interference avoidance, in OFDMA radio access systems have recently gained much attention from both the academia and the standardization communities. In order to reduce inter-cell interference in cellular networks, the ICIC methods apply restrictions to resource management in a coordinated manner among neighboring cells. Resources for coordination can be space (with beamforming antennas), time, frequency, code resources, or transmit power. The restrictions can be in the form of time/frequency resource availability to the resource manager, and/or restrictions on the transmit power that can be applied to certain time–frequency resources. The impact on the system architecture as a result of support of ICIC schemes can be mainly divided into four major categories depending on the degree of distribution: global; distributed; decentralized; and local ICIC schemes.

Global ICIC schemes require an omniscient central entity with full system knowledge, which is capable of acquiring the global system state in zero time, and can distribute scheduling and other decisions instantly to all network nodes. Distributed ICIC schemes rely on one or more central components, which exchange information potentially with all nodes in the network. The central components thereby collect state information and distribute information relevant for coordination to the network nodes, e.g., coordinated multi-point transmission/reception. Decentralized ICIC schemes are distinguished from distributed schemes by having no central entity involved in the coordination process. The coordination is performed exclusively by communication and information exchange among the network nodes. Local ICIC schemes are based mainly on information that is available locally in each eNB. There is no coordination-related communication and information exchange among neighboring eNBs. The local information includes feedback received from the UEs about the strongest interferers. Information exchange among eNBs or with a central coordinator could result in higher performance gains at the expense of increased system complexity [8–10].

A global ICIC scheme is practically unfeasible, although it may potentially attain the highest performance improvements among all ICIC schemes. Local ICIC schemes require no separate network equipment and provide the most flexibility with respect to the network design, although using local ICIC schemes may not achieve the highest performance gain.

An interference mitigation scheme based on beamforming transmits a narrow beam toward the desired user instead of transmitting in the conventional omnidirectional or sectorized manner [12]. This scheme is capable of reducing the interference imposed on users of the system, particularly benefiting the cell-edge users. Therefore, the classic beamforming technique is sometimes also referred to as interference alignment. Beamforming techniques may be further classified into two types, namely, non-adaptive and adaptive beamforming, where both types require partial or full channel state information feedback. Adaptive beamforming techniques adapt the antenna complex-valued weights according to diverse optimization criteria by exploiting the channel knowledge. A particularly beneficial adaptive beamforming technique is the well-known eigen beamforming, where the antenna weights are chosen to be the eigenvector of the largest eigenvalue of the spatial MIMO channel covariance matrix. On the other hand, a beam-switching technique uses the best-oriented beam among a number of preselected beams, where each array-weight vector is a function of the beam direction, as well as of the antenna element's position. In a typical 120° antenna pattern sector generated by a uniformly spaced linear array having four antenna elements, a total of eight beams may be created, where the adjacent beams are evenly spaced by an angle of 13° [12]. The index of the best beam vector resulting from the maximum mean channel gain measured at the receiver is signaled back to the transmitter, where the channel gain is averaged over a certain time period, a set of subcarriers, and over the spatial domain constituted by the receive antennas, depending on the accuracy of the CSI feedback as well as depending on the affordable feedback bit rate.

The combination of more advanced opportunistic transmission and beamforming techniques may attain further performance improvement in multi-user environments. The concept of beam hopping may be further utilized to intelligently avoid directional interference. Regardless of the use of the adaptive or non-adaptive variant, when transmitting directionally by exploiting the maximum equivalent channel gain, the received signal power and the SINR are enhanced, or alternatively, the transmitted power may be reduced despite maintaining a certain received SINR. As a result, the interference imposed on the adjacent cells is reduced, and the overall interference level in the entire system is reduced. In contrast to the directivity exploited by the above-mentioned beamforming techniques, the benefits of transmit diversity may be limited by transmitting the same signal from different antennas, provided that they experience independent fading. Since the transmit power required for maintaining a certain error probability may be reduced as a benefit of transmit diversity, the interference imposed becomes lower [10].

The concept of distributed antennas has been exploited to ensure that the antenna elements are sufficiently geographically dispersed and far apart to guarantee that the signals generated by each antenna are uncorrelated. The advantage of employing distributed antennas is the increased macro-diversity effect, which is particularly beneficial for the cell-edge users, which would otherwise suffer from severe co-channel interference. In contrast to the conventional single-cell transmission, where each user is served by its local base station only and suffers from co-channel interference caused by the adjacent cells, in the distributed antenna multi-cell transmission, each user is served by several coordinated base stations and each base station jointly processes multiple users' information, which is shared via the backhaul using a high-capacity low-latency optical fiber link. This leads to an equivalent distributed MU-MIMO system [10].

14.2 Heterogeneous networks

One of the most promising and cost-effective solutions to improve the system capacity is to deploy low-power nodes such as relays, pico-cells, femto-cells, and remote radio heads overlaid by macro-cell networks. The new network architecture is known as heterogeneous networks (as shown in Figure 14.3) can provide a flexible and effective way to eliminate the coverage holes in macro-cells and improve the system capacity. One of the key issues in the new network architecture is the handover performance deterioration in co-channel HetNets, where the cell-edge UEs can experience more serious interference [2,13–15].

The output power of the low-power nodes varies depending on the type of deployment and usage scenario. If the low-power nodes are intended for outdoor deployments, their transmit power ranges from 250 mW to approximately 2 W. They do not require an air-conditioning unit for the power amplifier, and are much lower in cost than traditional macro base stations, where their transmit power typically varies between 5 and 40 W. Femto base stations are meant for indoor use, and their transmit power is typically 100 mW or less. Unlike pico base stations, femto base stations may be configured with a restricted association, allowing access only to its closed subscriber group (CSG) members, or alternatively they can be configured as an open subscriber group (OSG), allowing full or partial access to the public. A network that consists of a mix of macro-cells and low-power nodes, where some may be configured with restricted access and some may lack wired backhaul, is referred to as a HetNet [11,16].

Cellular systems have been traditionally deployed as homogeneous networks using a macro-node-centric planning process. A homogeneous cellular system is a network of base stations in a planned layout and a collection of user terminals, in which all the base stations have similar transmit power levels, antenna patterns, receiver noise floors, and similar backhaul connectivity to the packet data network. Furthermore, all base stations offer unrestricted access to user terminals in

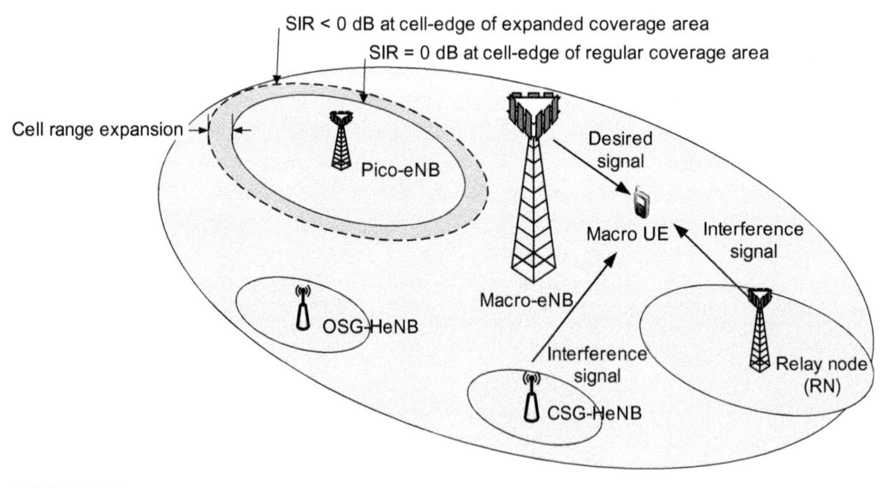

FIGURE 14.3

An example HetNet topology [22].

the network, and can serve approximately the same number of user terminals, all of which carry similar data flows with similar QoS requirements.

The locations of the macro base stations are carefully chosen through network planning, and the base station settings are properly configured to maximize the coverage and to control the interference between base stations. As the data traffic grows and the RF environment changes, the network relies on cell splitting or additional carriers to overcome capacity and link budget limitations in order to maintain uniform user experience. However, this deployment process is complex and iterative. Moreover, site acquisition for macro base stations with towers is becoming more difficult in a dense urban environment, thus a more flexible deployment model is needed for operators to improve broadband user experience in a ubiquitous and cost-effective way.

In a homogeneous network comprising macro-cells, each mobile terminal is served by the base stations with the strongest signal strength, while the unwanted signals received from other base stations are usually treated as interference. In a HetNet, such principles can lead to significantly suboptimal performance. In such systems, smarter resource coordination among base stations, better server selection strategies, and more advanced techniques for efficient interference management can provide substantial gains in throughput and user experience as compared to a conventional approach of deploying macro-cellular network infrastructure.

A pico base station is characterized by a substantially lower transmit power relative to a macro base station. The potentially large disparity between the transmit power levels of macro and pico base stations implies that in a mixed macro/pico deployment, the downlink coverage of a pico base station is much smaller than that of a macro base station. This is not the case for the uplink, where the strength of the signal received from a user terminal depends on the terminal transmit power, which is the same for uplink paths from the terminal to different base stations. Therefore, the uplink coverage of all the base stations is similar and the uplink handover boundaries are determined based on channel gains. This can potentially create a mismatch between downlink and uplink handover boundaries, and make the base station to device association or cell selection more difficult in HetNets, compared to homogenous networks, where downlink and uplink handover boundaries are closely matched (Figure 14.4).

If the cell selection is predominantly based on downlink signal strength, as in LTE Rel-8, the usefulness of pico base stations will be greatly diminished. In this scenario, the larger coverage of high-power base stations limits the benefits of cell splitting by attracting a large fraction of user terminals toward macro base stations based on signal strength without having enough macro base station resources to efficiently serve these user terminals, while lower power base stations may not be serving any user terminals. Therefore, from the point of view of network capacity, it is desirable to balance the load between macro and pico base stations by expanding the coverage of pico base stations and subsequently increasing cell splitting gains [11,16].

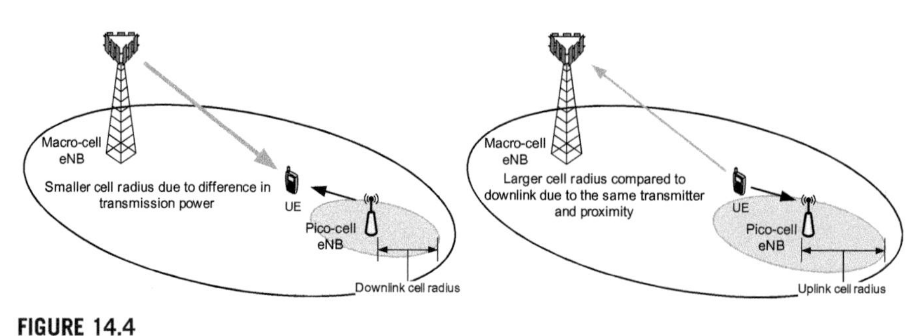

FIGURE 14.4

Illustration of uplink/downlink imbalance in HetNets [11].

Balanced load distribution between macro and pico base stations can be achieved if base station to device association is based on path loss (associating the base station with the minimum path loss rather than the base station with the maximum downlink signal strength), and a fixed partitioning of resources between the macro and pico base stations. As a result, more users are associated with the pico base stations, and the network is able to provide more equitable distribution of radio resources to each user. The effect is even more pronounced in hotspot layouts where users are clustered around pico base stations. Capacity gains can be achieved through sharing of the resources allocated for low-power base stations, while sufficient coverage is provided by high-power base stations on the resources that are allocated to them.

In a HetNet, in order for a user terminal to obtain service from a low-power base station in the presence of macro base stations with stronger downlink signal strength, the pico base station needs to perform interference coordination, for control and traffic channels, with the dominant macro interferers, and the user terminals need to support advanced receivers for interference cancellation. In the case of femto base stations, only the owner or subscribers of the femto base station may be allowed to access the femto base stations. For user terminals that are close to these femto base stations, but barred from accessing them, the interference caused by the femto base stations to the user terminals can be particularly severe, making it difficult to establish a reliable downlink communication to these user terminals. Therefore, as opposed to homogeneous networks, where resource sharing is a good transmission strategy, femto networks necessitate more coordination via resource partitioning across base stations to manage inter-cell interference.

The introduction of low-power nodes in a macro network creates imbalance between uplink and downlink coverage (see Figure 14.4). Due to the larger transmit power of the macro base station, the handover boundary is shifted closer to the low-power node, which can lead to severe uplink interference problems as the UEs served by macro base stations create strong interference to the low-power nodes. Given the relatively small footprint of low-power nodes, even in the case of the most optimized placement, low-power nodes may become underutilized due to geographic changes in data traffic. The performance of a mixed deployment of macro-, pico-, and open femto-cells was evaluated in the literature [1,11,17,22], which showed that the limited coverage of low-power nodes is the main reason for limited performance gain in HetNets. Furthermore, some of the deployed femto-cells may have enforced restricted association, which effectively creates a coverage hole and aggravates the interference problem. In order to cope with the interference, it is necessary to introduce techniques that can adequately address these issues. Due to restricted association, a CSG femto-cell creates a downlink coverage hole at the vicinity of the femto-cell for non-member UEs. In order to reduce the effect of the coverage hole, a femto-cell could adapt the transmitted power intelligently such that interference is contained within the intended coverage area such as a residential site. Since the femto-cell coverage and femtocell interference to the macro-UE are directly proportional to the femto-to-macro signal strength ratio, a femto-cell could set the proper transmit power by estimating the additional interference that could be tolerated by a non-member UE near the femto-cell. As a result, the femto-cell maintains a similar coverage area regardless of its location within the overlaid macro-cell.

Pico-cells are eNBs with lower transmit power than traditional macro-cells. They are typically equipped with omnidirectional antennas and are deployed indoors or outdoors, often in a planned manner. Their transmit power ranges from 250 mW to approximately 2 W for outdoor deployments, while it is typically 100 mW or less for indoor deployments. Since pico-cells are regular eNBs from an architecture perspective, they can benefit from X2-based ICIC.

In light of the above challenges in the operation of HetNets, an ICIC scheme is critical to HetNet deployment. A basic ICIC technique involves resource coordination among interfering base stations, where an interfering base station avoids use of some resources in order to enable control and data transmissions to the victim user terminal. More generally, interfering base stations can coordinate their transmission power and/or spatial beams with each other in order to enable control and data transmissions to their corresponding user terminals.

The resource partitioning can be performed in time-domain, frequency-domain, or spatial-domain. Time-domain partitioning can better adapt to user distribution and network load variations, and is the most attractive method for spectrum-constrained environments. For example, a macro base station can choose to reserve some of the subframes in each radio frame for use by pico base stations, based on the number of user terminals served by pico and macro base stations, and/or based on the data rate requirements of the user terminals. Frequency-domain partitioning offers less granular resource allocation and flexibility, but is a viable method especially in an asynchronous network. Spatial-domain partitioning can be supported by a CoMP transmission/reception scheme, which was discussed in Chapter 12. For time-domain resource partitioning, a macro base station can use almost blank subframes (ABS) to reserve some subframes for pico nodes. The macro base station continues transmission of legacy common control channels during almost blank subframes to maintain backward compatibility with legacy user terminals. The user terminals can cancel interference on common control channels within almost blank subframes caused either by higher power macro base stations or by nearby femto stations that they are prohibited to access. An interference cancellation receiver can handle colliding and non-colliding reference signal scenarios, and can eliminate the need for cell planning for heterogeneous deployment [22].

The studies in 3GPP concerning HetNet mobility improvement have continued in LTE Rel-11 [2]. The comprehensive simulations conducted in these studies concluded that handover performance of HetNets was not as good as homogeneous macro-cell deployments. Furthermore, it was shown that the UE speed has significant impact on the handover performance, and that careful DRX configurations are required to avoid negative impact on handover performance. The objective in LTE Rel-12 is thus to enhance handover performance in HetNet scenarios (e.g., minimization of handover failure rate, minimization of ping-pong effects, recovery from radio link failure), as well as supporting UE mobility and longer DRX cycles. The enhancement of mobility in small-cell scenarios with focus on inter-frequency schemes will also be part of the LTE Rel-12 development.

14.3 Enhanced inter-cell interference coordination (eICIC)

The basic principle of ICIC is to avoid high-power transmission on time–frequency resources on which cell-edge users are scheduled in neighboring cells, users that would otherwise experience high interference and correspondingly low data rates. In LTE Rel-8, ICIC is performed over the frequency-domain by transmission of messages across a standardized backhaul inter-eNB interface such as X2. Because the latency over X2 is typically in the order of a few milliseconds, it is expected that any updates or reconfiguration to the ICIC messages are relatively infrequent (in the order of seconds). The LTE Rel-8 frequency-domain ICIC in the downlink is based on event-triggered message exchange between eNBs approximately every 200 ms to allow adaptive FFR. Uplink ICIC is based on event-triggered *overload indicator* and *high interference indicator* messages exchanged between the eNBs approximately every 20 ms. In the context of ICIC, the following metrics are defined [11,7]:

- *High interference indicator (HII)*: The HII provides information about a set of resource blocks on which an eNB is going to schedule transmissions for cell-edge terminals, i.e., resource blocks on which a neighboring cell can expect higher inter-cell interference. Although it is not explicitly specified how an eNB should respond to an HII message (or any other ICIC-related X2 signaling) received from a neighboring eNB, a reasonable action for the receiving eNB would be to avoid scheduling its own cell-edge terminals on the same resource blocks, reducing the uplink interference to cell-edge transmissions in its own cell, as well as in the cell from which HII message was received. The HII is an indicator which is sent as a bitmap with one bit per resource block/sub-band that the serving cell intends to use for scheduling cell-edge UEs, potentially causing high inter-cell interference to the neighboring cell. An eNB can send HII with different, neighbor-cell-specific contents to different neighbor cells.

- *Relative narrowband transmit power (RNTP)*: For the downlink, the relative narrowband transmit power is defined to support ICIC operation. The RNTP is similar to the HII in the sense that it provides information for each resource block whether the relative transmit power of that resource block exceeds a threshold level. Similar to the HII, a neighboring cell can use the information provided by the received RNTP when scheduling its own terminals, especially terminals in the cell-edge that are more likely to be interfered by the neighboring cell. The RNTP message is signaled using a bitmap wherein each RNTP bit value indicates whether the corresponding resource block pair is limited by a transmit power threshold. Upon receipt of the RNTP message, the recipient eNBs can take into account this information while making their scheduling decisions for subsequent subframes. One such example may be that the recipient eNB avoids scheduling UEs in resource blocks where the source eNB (e.g., macro-cell eNB) is transmitting above a certain power limit.
- *Overload indicator (OI)*: The overload indicator is an indicator exchanged over the X2 interface, reflecting the uplink interference plus noise level (low, medium, or high) of a resource block measured by an eNB. Although the reaction of the recipient eNB upon receipt of the OI is not standardized, a possible reaction to the OI by the recipient eNB would be to limit the maximum transmit power of a UE scheduled on resource blocks indicated by the OI to have high interference.

During LTE Rel-10 standardization, 3GPP developed two approaches to coordinate inter-layer interference, primarily targeting the aforementioned deployment scenarios. Both approaches rely on a semi-static resource partitioning across layers, but take two different approaches to ICIC of control and data channels. In one approach, interference avoidance is addressed by means of ICIC in the frequency-domain, relying on carrier aggregation (CA) as introduced in LTE Rel-10. In the other approach, interference avoidance is addressed by means of ICIC in the time-domain, relying on ABS. Both approaches target scenarios with low-power nodes, assuming time-synchronization with the macro-eNB.

14.3.1 CA-based eICIC schemes

In CA, a terminal may simultaneously receive/transmit data and control on multiple components carriers (CCs), where a component carrier is fully backward compatible with LTE Rel-8. A terminal supporting carrier aggregation can be configured by higher layer signaling to enable cross-carrier scheduling on certain component carriers. This implies that a terminal receiving a downlink assignment on one component carrier may receive associated data on another component carrier. One of the main motivations for introducing cross-carrier scheduling was to enhance operations in HetNets in a multi-carrier deployment. The basic idea in CA-based ICIC is to create a protected component carrier for reliable reception of downlink physical signals, system information, and control channels at victim layers, but data can be received on any configured downlink component carrier via cross-carrier scheduling. This, in conjunction with RNTP at aggressor layers, forms the basis of CA-based eICIC. As shown in Figure 14.5, assume a macro−pico deployment scenario with two downlink component carriers, CC1 and CC2. In this particular scenario, a macro-cell creates a protected component carrier by reducing its transmit power on CC1 in order to enable larger picocell coverage on CC1. On CC2, the macro-cell transmits with its nominal output power, implying that the size of the pico-cell associated with CC2 will be reduced in comparison with the pico-cell coverage on CC1. The dashed areas in the figure represent cell range expansion zones, i.e., areas with reduced reception reliability of the downlink control channels, PDCCH, PHICH, and PCFICH [11,22].

As shown in Figure 14.5, it is obvious that without use of eICIC, a terminal configured with CC1 and CC2 can simultaneously receive from both carriers only in the following scenarios: (i) it operates within the coverage of the pico-cell associated with CC2 when connected to the pico-cell; (ii) it operates within the coverage of the macro-cell on CC1 when connected to the macro-cell. However, with cross-carrier scheduling a pico-cell-edge user can receive PDCCH on the protected CC1, whereas the associated PDSCH is received on CC2, on which the macro-cell will have reduced its transmit power on PDSCH in resource blocks known by the pico-cell via X2 signaling of RNTP. It is also shown that the operation of the CA-based eICIC, in which the macro-cell avoids transmissions or reduces its transmission power of PDSCH in resource blocks in the upper band of CC2, the users operating in the cell range expansion zones receive data via cross-carrier scheduling, whereas users closer to the eNB sites do not need to rely on data reception via cross-carrier scheduling. A UE typically associates with a base station that exhibits strong downlink SINR. If the macro-cell transmission interferes with that of a low-power node, the UE may not be able to select the low-power node. With cell range expansion, a UE can associate with a low-power node despite the fact that the downlink SINR may be much lower than 0 dB. Time-domain resource partitioning enables load balancing between high- and low-power nodes. Note that resource partitioning is dynamically adapted to network load. Capacity gains in HetNets are achieved by off-loading traffic from macro-cells to pico-cells, and the amount of off-loading is increased by cell range expansion. Cell range expansion virtually increases the footprint of the low-power node. The operating signal-to-interference ratio (SIR) in the expanded cell range may be less than 0 dB. The ABS facilitate serving UEs in the expanded cell range by the victim nodes. However, interference from cell-specific reference signals and other control signals within ABS limits the cell range expansion. LTE Rel-10 supports cell range expansion with SIR = −6 dB, which was further decreased to −9 dB in LTE Rel-11 [11,22].

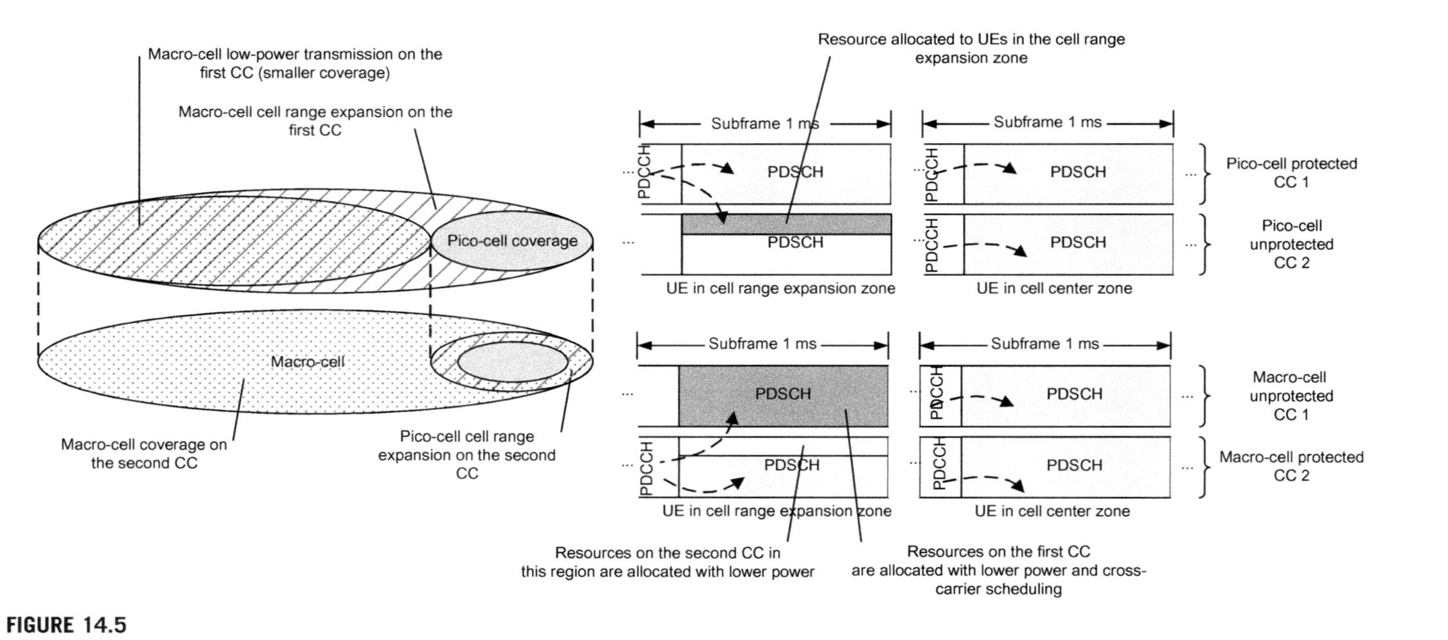

FIGURE 14.5

Illustration of the CA-based eICIC concept [11].

In CA-based eICIC, the primary cell associates with a protected component carrier. Pico-cell users would then have CC1 as their primary cell, whereas macro-cell users would have CC2 as their primary cell. The CRS transmission from an aggressor cell (necessary for backward compatibility with legacy LTE systems) will potentially create inter-cell interference toward victim users. Therefore, LTE Rel-10 introduced transmission mode 9 for which demodulation of PDSCH relies entirely on the UE-specific reference signals for data demodulation, and CSI reference signals for CSI measurement. Through RNTP signaling, DM-RS collisions across different cells can be avoided since DM-RS are only transmitted in resource blocks carrying PDSCH. Such UE-specific configuration is useful in HetNet scenarios where the CRS reception might be deteriorated. Furthermore, a terminal can be configured to conduct measurements corresponding to mobility and radio link monitoring on its primary cell only, i.e., the cell with the best CRS quality [5–7].

14.3.2 Non-CA-based eICIC

The underlying concept of time-domain eICIC is that an aggressor layer creates protected subframes for a victim layer by reducing its transmission activity in certain subframes. To do so, the aggressor eNB reduces its transmission power of some downlink signals, or alternatively mutes their transmission during a set of low-interference subframes designated as ABS whose occurrences are known in advance at the coordinated eNBs. The ABS patterns can be constructed by configuring MBSFN subframes and/or by not scheduling unicast traffic, or alternatively by reducing transmit power in certain subframes. During ABS, the aggressor eNB does not transmit PDSCH but may transmit CRS, critical control channels, and broadcast and paging information. If the victim cell schedules its UEs in subframes that overlap with aggressor cell ABS, the victim cell protects its UEs from strong inter-cell interference, thereby increasing the likelihood of successful PDCCH/PDSCH reception (Figure 14.6).

Backward compatibility is ensured by transmitting essential physical signals, system information, and paging in the same way as in LTE Rel-8. This implies that CRS transmission at the aggressor layers will create interference on either PDSCH signal or CRS signals at victim layers even during ABS, although the CRS interference on PDSCH can be reduced by configuring MBSFN subframes, since the CRS is not transmitted in the MBSFN region of MBSFN subframes. Furthermore, the primary and secondary synchronization signals, as well as system information, are transmitted in the zero and fifth subframes of a radio frame, and will thus be inter-layer interference sources that cannot be avoided by introducing ABS. In the case of a macro–femto deployment, CSG femto-cells represent the aggressor layer, whereas in the case of a macro–pico deployment, macro-cells represent the aggressor layer. With the time-domain ICIC, a macro-cell eNB may schedule users during protected subframes when they are in the proximity of CSG femto-cells, whereas a pico-cell eNB may schedule their cell-edge users only during protected subframes (Figure 14.7). The information

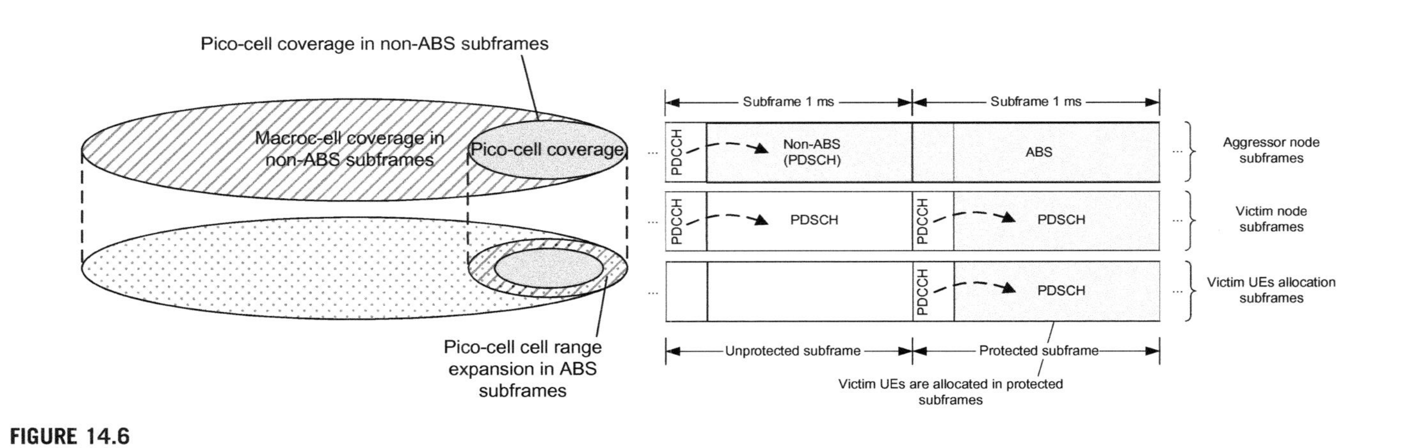

FIGURE 14.6

Illustration of the non-CA-based eICIC concept [11].

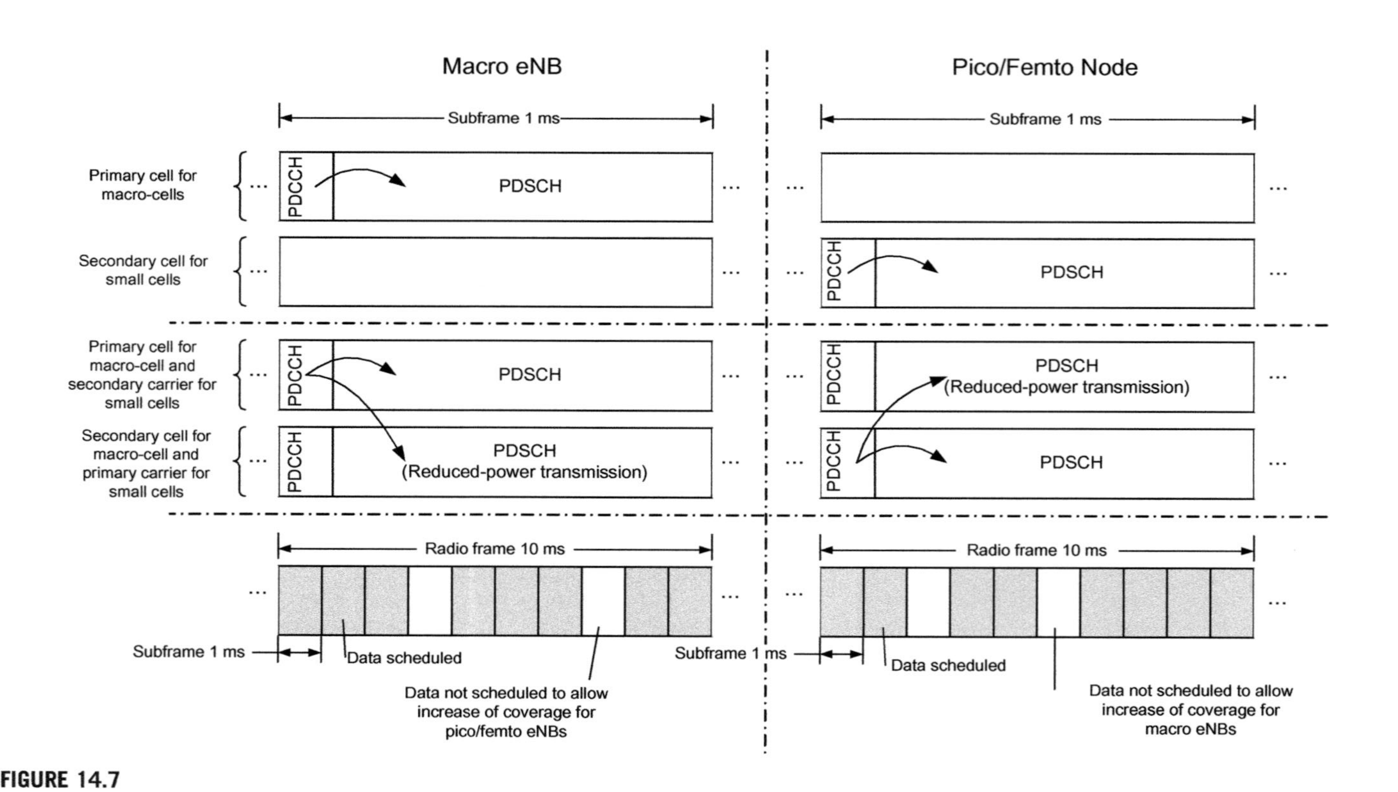

FIGURE 14.7

Illustration of CA-based and non-CA-based eICIC methods [16].

regarding the subframes on which coordinated low-interference transmission occurs is exchanged between coordinating eNBs via bitmap patterns. Upon receiving the bitmap pattern either via customized network operation administration and maintenance, or via the X2 interface, the recipient eNB can schedule data for its victim UEs on subframes that overlap with the aggressor ABS transmissions, as shown in Figure 14.7. In order to enable this, the time-domain eICIC requires time-synchronized eNB transmissions at least at the subframe boundaries. In an FDD system, for optimal operation with a synchronous uplink HARQ in a HetNet, a bitmap of length 40 bits (which is equivalent to a 40-ms periodicity) is used. Additional interference mitigation through subframe shifting is required to avoid cross-layer interference from broadcast channel and synchronization signals. It is also desirable to configure paging occasions, and scheduled system information, in non-ABS at the aggressor layer and in protected subframes at the victim layer [11].

In LTE Rel-8, a UE may average channel and interference estimates over multiple subframes to derive a CSI feedback. When time-domain eICIC is used, the interference level of these two types of subframes, i.e., protected and unprotected subframes, can be significantly different. In order to ensure that the UE does not average interference across the two different types of subframes, the eNB can configure two subframe subsets such that the UE only averages the channel and interference estimates over subframes within the subset, but not across the two subsets. The UE then reports the CSI measurements separately for the two subsets either periodically (with the respective configured reporting periods for each subset) or aperiodically (i.e., one of the CSI measurements from two subframe subsets is reported via PUSCH, when triggered by PDCCH) [11].

In the connected mode, where a UE's mobility is controlled by the network, the network may perform an inter-frequency or inter-RAT handover to a different carrier frequency when the network determines that a non-CSG UE (range-extended pico-cell UE) is in the proximity of a CSG femto-cell based on the UE's measurement reports. In the idle mode, where the mobility is controlled by the UE within the network, the UE autonomously performs an inter-frequency or inter-RAT cell reselection to a base station on a different carrier when the UE moves close to a dominant interferer. Thus, in both cases, an inter-frequency/inter-RAT handover can potentially create LTE coverage loss, possibly leading to LTE service discontinuity on a collocated carrier. To avoid occurrence of LTE coverage holes, LTE Rel-10 eICIC developed mobility support and radio link monitoring (RLM) [11].

In LTE, the UE performs RRM measurements to manage its network connection and to facilitate handover. In the connected mode, the UE reports RSRP and/or RSRQ of the serving and the neighboring cells to the serving eNB (see Chapter 9 for a description of RSRP and RSRQ). The network then performs a handover, transferring the UE to a neighbor cell when the serving cell's RSRP falls below the neighbor cell's RSRP plus an offset to ensure UE service continuity during mobility. In LTE Rel-10, a victim UE can be configured to perform resource-specific RRM measurements only on protected subframes. The resource-specific RRM measurements can improve accuracy of RSRP/RSRQ reporting by eliminating persistent interference on the CRS. Note that the collision of CRSs transmitted from different cells can be avoided by properly assigning the physical cell identities.

Radio link monitoring is a measurement function for ensuring that UEs maintain time synchronization and can reliably receive their control information. In LTE, the UE continuously evaluates the quality of the serving eNB in the downlink, and ensures that it is time synchronized to the serving eNB. If downlink synchronization is lost, the UE must suspend its uplink transmissions and should try to recover its existing link or to find a new serving cell with which a reliable downlink channel can be established. The RLM functionality comprises out-of-sync and in-sync evaluation procedures where the UE monitors CRS quality of the serving cell in the presence of co-channel interference. If the UE's serving cell is not the strongest measured cell due to severe interference, the UE can potentially declare a radio link failure. In order to ensure predictable UE behavior, LTE Rel-10 provides a mechanism so that the serving eNB can configure resource-specific RLM measurements at the UE for recording its radio link quality as experienced by the UE in protected subframes. This ensures that the link quality is tied to that of the protected subframes, and the UEs can continue to maintain their connection in HetNet scenarios [11].

In summary, time-domain eICIC may be utilized for pico-cell users which are served in the edge of the serving pico-cell, e.g., for traffic off-loading from a macro-cell to a pico-cell, allowing such UEs to receive service from the pico-cell on the same frequency layer. The inter-cell interference in this case may be mitigated by the macro-cell(s) utilizing ABS to protect the corresponding pico-cell's subframes from interference. A UE served by a pico-cell uses the protected resources for cell measurements (RRM), RLM, and CSI measurements. For a UE served by a pico-cell, the RRM/RLM/CSI measurement resource restriction may allow more accurate measurement of the pico-cell under severe interference from the macro-cell(s). The pico-cell may selectively configure the RRM/RLM/CSI measurement resource restriction only for those UEs subject to strong inter-cell interference. Furthermore, for a UE served by a macro-cell, the network may configure a RRM measurement resource restriction for neighbor cells in order to facilitate mobility from the macro-cell to a pico-cell [3].

14.3.3 Further enhanced ICIC

Due to time limitations, some enhanced ICIC techniques were deprioritized in LTE Rel-10. Particularly, the impact of legacy transmissions on control and data channel demodulation when using ABS was left for consideration in LTE Rel-11, leading to a further enhanced ICIC (FeICIC) scheme development. The work on the FeICIC scheme attempted to identify the UE performance requirements, possible air-interface changes, and eNB signaling improvements. In order to significantly improve detection of physical cell identity and system information in the presence of dominant interferers, the study focused on UE receiver implementation, UE performance requirements, and eNB signaling required for UE measurement/ reporting, as well as network assisting information for CRS interference cancellation and usage of reduced-power ABS. In LTE Rel-11, the enhanced downlink physical control channel has been developed to allow improved control channel inter-cell interference mitigation.

Co-channel deployments are of interest due to cost-effectiveness and high spectral efficiency. The resource partitioning can effectively mitigate interference from the traffic channels. However, inter-cell interference originated from the synchronization and control channels as well as CRSs will persist, since transmission of these signals is necessary for legacy support. Subframe time shifting can be utilized in FDD systems to avoid collision of the control/synchronization channels between eNBs of different power classes that require partitioning, but it cannot be applied to TDD systems. There is no network solution for CRS interference mitigation for either FDD or TDD systems. Without additional interference mitigation of the aforementioned channels and CRS, only limited SIR bias values (approximately 6–10 dB) can be supported, which limits the potential for cell range expansion. It must be ensured that UEs are able to detect and acquire a weak cell and measure and report the measurements to the network, which is a prerequisite for handover and cell range expansion. The CRS interference mitigation has an important role in system performance. Strong CRS interference, even though present only in a relatively small fraction of resource elements, can significantly degrade turbo code performance and the overall SINR; therefore, the potential gains of cell range expansion can be severely reduced. Given that the aforementioned interference sources are all broadcast at full power targeting UEs at the cell-edge, it must be ensured that a robust UE-based interference cancellation solution is feasible, which decodes the strongest signal, performs channel gain estimation toward the interfering cell, cancels the interfering signal, and continues the procedure until the desired signals of the serving cell are acquired.

As mentioned earlier, adaptive inter-cell interference management in LTE Rel-10 is enabled by coordination of resources used for scheduling data traffic through the X2 interface. The granularity of the negotiated resource is a single subframe (Figure 14.8). The resource partitioning enables cell range expansion through use of a bias for cell range determination. In a typical case, cell range expansion is enabled to improve system capacity, and a cell bias is applied to low-power nodes. The bias value refers to a threshold that triggers handover between two cells [16]. A positive bias means that the UE will be handed over to a pico-cell as soon as the difference in the signal strength from the macro-cell and pico-cell falls below a threshold. The macro base station informs the pico base station which resources will be utilized for scheduling a macro-UE, and which subframes are assigned as ABS. Therefore, the low-power nodes are informed of the interference pattern from the high-power base station, and can

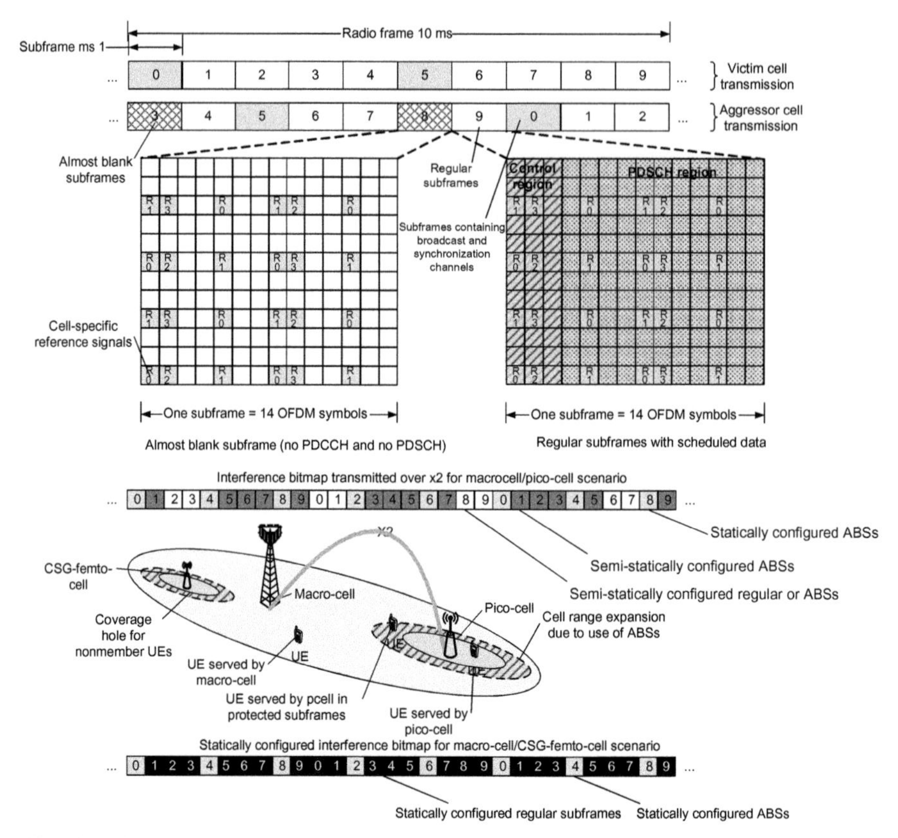

FIGURE 14.8

Illustration of the static/semi-static interference management eICIC schemes for FDD systems in macro-cell/pico-cell and macro-cell/CSG-femto-cell scenarios [16].

schedule the UE in the extended cell areas on subframes protected from high-power interference. Interference coordination is performed by means of a bitmap, where each bit in the bitmap is mapped to a single subframe (see Figure 14.8). The size of the bitmap is 40 bits, suggesting that the interference pattern repeats every 40 ms. Depending on the traffic volume, the pattern can change every 40 ms. The aggressor node effectively controls which resources can be used to serve the UE in the cell range expansion zones. An example of coordination between a macro-cell and pico-cell is shown in Figure 14.8. Note that the cell range expansion may be desirable for a macro base station as well. A typical use case for cell range expansion in the macro base station is to reduce the number of handovers. In a scenario with a large number of high-mobility users, the number of handovers in the network can become problematic, particularly in the deployment of pico base stations, the cells become small. The same resource partitioning schemes can then be utilized to allow high-mobility macro-UEs to remain attached to a macro-cell while inside the coverage area of a pico-cell [16]. In this case, the pico base station needs to restrict scheduling data traffic on some resources. In a macro–femto scenario (see Figure 14.8), due to the lack of an X2 interface between the eNB and the HeNB, only a static network-configured solution is feasible when CSG-femto nodes are deployed on the same frequency as the macro-cell. Due to the static nature of the solution, the resource partitioning pattern cannot be adapted to the instantaneous characteristics of actual data traffic [16].

14.4 Multi-radio coexistence and interference avoidance

There has been a significant increase in the use of wireless devices equipped with a number of different radio technologies collocated on a single hardware platform in recent years, where each radio technology (depending on availability and coverage) can facilitate certain user applications/network services. A typical configuration of multi-radio mobile terminals consists of one or more wide-area RATs such as HSPA +, LTE, mobile WiMAX, cdma2000, and a number of short-range wireless technologies such as Bluetooth and Wi-Fi operating in the industrial, scientific, and medical (ISM) bands, as well as a GNSS or global positioning system (GPS) receiver for positioning. When these radios operate simultaneously in adjacent or overlapping RF spectrums, it results in interference from transmission of one radio to the reception of the other radios. This condition is referred to as in-device coexistence interference. IDC interference occurs when the frequency separation of less than 50 MHz is allotted between UMTS/LTE and the ISM-band radios (Table 14.1). With a frequency separation of less than 20 MHz, typically 50 dB of isolation is required to avoid IDC effects. The small form factor of mobile terminals usually provides only 10–30 dB isolation. As a result, an effective solution is required to mitigate the IDC interference without degrading the performance of the wireless access systems collocated on the user terminal [1,18].

Table 14.1 Experimental Results Concerning Minimum Frequency Gap Requirements [1]

		Minimum Frequency Gap Requirement (MHz)		
		Antenna Isolation		
Aggressor	**Victim**	**10 dB**	**15 dB**	**20 dB**
LTE Band 40	Wi-Fi	58	52	50
Wi-Fi	LTE Band 40	56	50	46
LTE Band 7	Wi-Fi	60	52	50

The commercial applications of GNSS receivers typically use the L1 band (1575.42 MHz) for location-based services. The uplink direction of LTE Band 13 (777–787 MHz) and Band 14 (788–798 MHz) can disrupt the operation of a GNSS receiver using the L1 band. This is because the second harmonic of Band 13 (i.e., 1554–1574 MHz) and the second harmonic of Band 14 (i.e., 1576–1596 MHz) are extremely close to the L1 band [1].

14.4.1 Collocated multi-radio coexistence issues

As shown in Figure 14.9, the lower portion of the ISM band is very close to LTE Band 40 (TDD). In the case of LTE-BT simultaneous operation, the LTE transmitter causes interference to the BT receiver, and the BT transmitter causes interference to the LTE receiver by desensitizing the receiver. Similar interference issues exist for LTE-Wi-Fi concurrent operation. It is also shown in Figure 14.9 that there is less than 20 MHz frequency separation between BT and LTE Band 7 (FDD uplink). Thus, the LTE transmitter causes interference to the BT receiver. There is no impact on the LTE receiver from the BT transmitter, because the

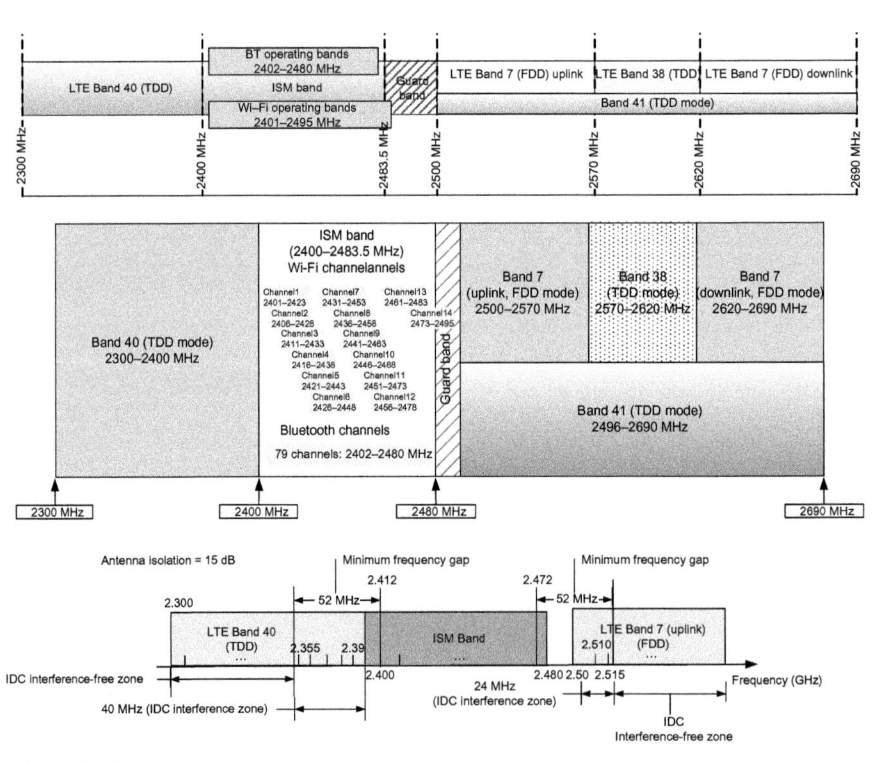

FIGURE 14.9

E-UTRA frequency bands around 2.4 GHz ISM band [1].

corresponding LTE Band 7 (FDD downlink) is far away from the ISM band. There is only 5 MHz separation between Wi-Fi and the LTE Band 7 uplink; however, since Wi-Fi operation is limited to channel 13 in most countries except Japan, there is often a frequency separation of 17 MHz. In the case of LTE-Wi-Fi coexistence, only the Wi-Fi receiver will be affected by the LTE uplink transmitter [1].

Interference between LTE Bands 7 and 40/41 with the 2.4 GHz ISM band is significant when operating in adjacent bands, where current RF filtering schemes are not adequate to mitigate the problem in a number of practical scenarios according to 3GPP RAN Working Group 4 studies [1]. The IDC interference level and its impact on the receiver performance depends on transmit power and receiver blocking characteristic of each radio, and physical characteristics of transceivers (e.g., filter frequency responses, antenna isolation). LTE Rel-8/9/10 already defined some mechanisms such as DRX mode and measurement gaps which would allow mitigation of some coexistence scenarios; however, these schemes are solely controlled by the eNB. The network-controlled UE assisted (i.e., providing triggers) and UE-based autonomous solutions were considered as part of the LTE Rel-11 IDC work item. The interference between LTE and the 2.4-GHz ISM band is significant when operating in adjacent bands, as shown in Figure 14.9. Those problems can be summarized as follows [1]:

- LTE Band 40 to ISM interference: LTE deployed in the upper 30 MHz of Band 40 can desensitize[1] the entire ISM band. LTE in any channel in Band 40 can desensitize the lower 20 MHz of the ISM band.
- ISM to LTE Band 40 interference: ISM in the lower 20 MHz can desensitize all channels in LTE Band 40. ISM in any part of its band can desensitize the upper 20 MHz of LTE Band 40.
- LTE Band 7 to ISM interference: LTE Band 7, in the uplink direction, deployed in 2510 MHz channel can desensitize the entire ISM band.
- With LTE deployed in Band 40, in some scenarios, and BT as a master can cause about 44% downlink error rates, and BT as a slave can cause about 55% downlink error rate.
- With LTE deployed in Band 7, the BT receiver error rate is 100% with high data rate applications on LTE.
- LTE Band 7/13/14 transmitter induces interference to GNSS receiver.

As shown in Figure 14.9, there are 14 channels in the ISM band designated for Wi-Fi operation. Each channel has 5 MHz separation from other channels,

---

[1]Desensitization is a common performance degradation indicator, where the desensitization is relative to an assumed reference sensitivity value. The desensitization is approximated as $10\log_{10}x$, where x is the ratio between the coexistence interference and the noise floor at the receiver sensitivity level. Using desensitization as the performance measure provides an indication of the degradation that can be expected when the victim system is in its most vulnerable state at the edge of its coverage, and thus may be descriptive of a worst case scenario. The error vector magnitude (EVM) measure can also indicate potential degradation in receiver performance because a signal with large EVM would likely be incorrectly decoded at the receiver.

with the exception of channel 14 where separation is 12 MHz. Channel 1 starts with 2401 MHz and channel 14 ends at 2495 MHz. Different countries have different policies for the number of allowed channels of Wi-Fi operation [21]. Since E-UTRA Band 7 is an FDD band, there is no impact on the LTE receiver from the Wi-Fi transmitter; however, the Wi-Fi receiver is interfered by the LTE uplink transmitter in this frequency band. Bluetooth operates in 79 channels, each having 1 MHz bandwidth in the ISM band. The first channel starts at 2402 MHz, and the last channel ends at 2480 MHz [20]. The LTE Band 40 and BT will adversely affect each other, and the transmission of LTE Band 7 in the uplink direction will affect BT reception, as well.

14.4.2 In-device coexistence scenarios

In order to find an efficient and effective solution for the IDC issues that were mentioned above, it is important to identify the use cases that are commonly encountered. This is due to the fact that different usage models result in different assumptions about the behavior of the LTE transceiver and other collocated radio technologies, which eventually impact solutions. The IDC scenarios of interest in this study are as follows [1]:

1. *LTE + BT earphone (VoIP service)*: In this scenario, LTE is used for VoIP, and the BT earphone is used to send or receive voice packets. The activities of LTE and BT transceivers are going to be very similar in this case because of the end-to-end latency requirement.
2. *LTE + BT earphone (multimedia service)*: In this scenario, multimedia content is downloaded via the LTE link, and audio is routed to a BT headset, where the activities of LTE and BT are correlated. The requirements for scheduling/unscheduled periods for a typical streaming application can be set based on the LTE and BT characteristics. Activity time for BT can vary dynamically with a streaming application. The BT audio stream typically uses the advanced audio data profile (A2DP) [20], in which transmission delays greater than 60 ms can cause audible playback artifacts; consequently, the scheduling period of LTE should not exceed this limit. The latency requirement is less stringent on the LTE side in this case, and depends on the QoS class of the content. The maximum unscheduled period for LTE can be as long as 150 ms. However, in order not to degrade the LTE performance, it is desirable to minimize the LTE unscheduled period, and the smallest unscheduled period is determined by the on-time needed by the BT transceiver to sustain the data rate, depending on the link condition. This number typically ranges from 15 to 60 ms. As a result, the LTE scheduling period should be less than 60 ms, and the unscheduled period should be between 15 and 60 ms [1].
3. *LTE + Wi-Fi portable router*: In this scenario, LTE is considered a backhaul link providing access to the Internet, and the connectivity is shared by other

local users using Wi-Fi. The Wi-Fi transceiver is operated as an access point and has full control over frequency channel and transmitting power. Given the ability of the Wi-Fi transceiver to select the frequency channel, it may be possible to avoid interference to/from Wi-Fi by moving the Wi-Fi signal away from the LTE band, as shown in Figure 14.10. In the downlink, the worst case latency will be for a packet arriving at the eNB at the beginning of the LTE unscheduled period; thus, the latency is the sum of the LTE unscheduled period (waiting time for LTE scheduling) and the LTE scheduling period (waiting time for Wi-Fi scheduling). In the uplink, the scheduling/unscheduled periods can be made as small as 1 ms to minimize latency, but this is not desirable due to the impact on retransmissions and other sensitive timings in LTE and Wi-Fi operation. In order to fulfill the latency requirements of

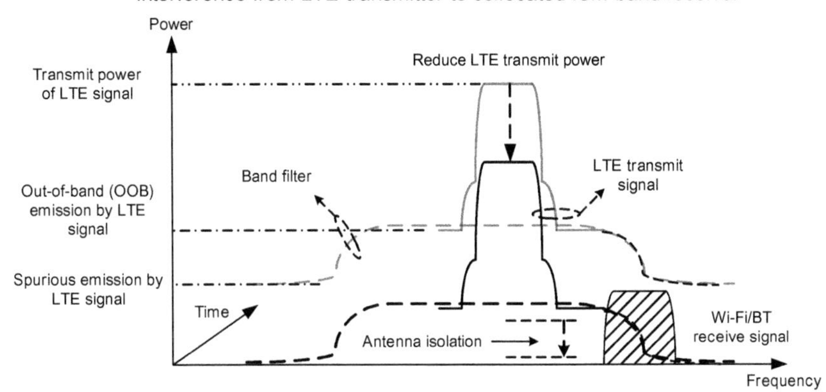

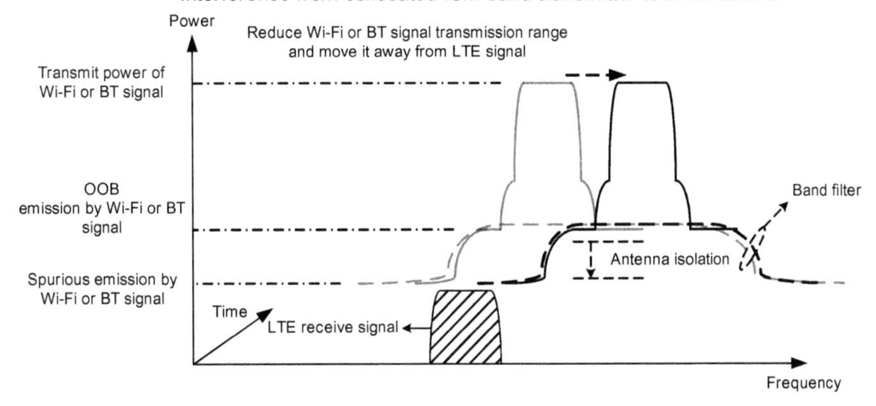

FIGURE 14.10

Illustration of IDC issues and some remedies [1].

common services under this scenario, the scheduling periods and unscheduled periods should be typically not more than 20–60 ms.

4. *LTE + Wi-Fi off-load*: In this scenario, an LTE UE can also connect to Wi-Fi to off-load traffic, and the Wi-Fi transceiver operates as a terminal in the infrastructure mode. It is difficult for the Wi-Fi transceiver to change the configured frequency channel in this case. In addition, the Wi-Fi radio is required to scan for the beacon signal transmitted from the Wi-Fi access point every 102.4 ms to maintain connection. The scheduling/unscheduled period requirements pertaining to this scenario differ from the previous scenario. One difference is about Wi-Fi beacon reception by the UE in Wi-Fi client mode, which implies alignment of the LTE unscheduled period with the Wi-Fi beacons. Furthermore, the scheduling period of LTE should less than 100 ms in order to facilitate beacon reception. In this case, the packets only go through one radio link (either Wi-Fi for off-load packets or LTE for normal packets), resulting in somewhat larger scheduling/unscheduled intervals that can meet the latency requirements. The ratio of the scheduling/unscheduled intervals should correspond to the traffic volume of the normal and offloaded traffic.
5. *LTE + GPS receiver*: In this scenario, the LTE UE is also equipped with a GPS receiver to support location-based services.

14.4.3 In-device coexistence and interference avoidance

The LTE network-controlled UE-assisted solutions which were studied during LTE Rel-10 can be categorized as follows [1]:

- Frequency division multiplexing (FDM) based solutions include moving the LTE signal away from the ISM band, or moving the ISM signal away from the LTE frequency band (see Figure 14.10).
- Time division multiplexing (TDM) based solutions include DRX mode, HARQ process reservation, and uplink scheduling restriction.
- Power control-based solutions include LTE power control or ISM-band radio power control (see Figure 14.10).
- UE autonomous solutions include time-domain remedies such as LTE denials for infrequent short-term events, LTE denials for ISM data packets, and ISM denials for LTE important signals reception.

The RF filtering in the transmit side reduces the out-of-band (OOB) spurious emissions of the aggressor leaking into the receive band of the victim. The receive filter reduces the blocking effect due to the transmissions in the band of the aggressor technology. When two radios operate in adjacent bands, there can be limited attenuation of each filter in the transition band that overlaps with the designated band of the other technology. This can lead to a large interference to the victim, due to spurious emissions and imperfect blocking performance. In some cases, such as LTE Band 40 and the ISM band starting at 2400 MHz, there is no guard band between the frequencies of the two technologies. There are further challenges in RF design due to variations in filter response across manufacturing processes, as well as a wide temperature range over which the terminal must operate [18].

As shown in Figure 14.11, in a typical UE implementation, the baseband processing modules of UMTS/LTE and ISM-band radios are separated, whether within the same chip from the same vendor or different chips from the same or different manufacturers. In either case, the two chips may exchange handshake signals, referred to as inter-chip handshaking and signaling, to convey real-time status and control information including advanced notice of future activities. This advanced notification signal can be used as a trigger to notify the serving eNB of forthcoming IDC conditions. The eNB can suspend normal UE LTE operation during the ISM-band radio activities and resume normal operation once the IDC condition is resolved. The IDC coordination module (typically proprietary) resides in the UE, as shown in Figure 14.11, and is responsible for coordination and prioritization among different collocated radios, obtaining and interpreting the coordination information from different radios, and generating control commands to different radio modules including but limited to generation of the IDC trigger messages in the uplink (to be transmitted via LTE transceiver module), interpretation of the IDC acknowledgment messages, and so on.

The IDC solutions for interference avoidance are mainly considered for the UE in RRC connected mode. The idle mode operation itself is not considered a problem since the UE can just stop ISM-band radio transmissions at important LTE events such as paging. There are three interference avoidance approaches that can be taken to address the IDC issues: (1) uncoordinated, where different radio technologies within the same UE operate independently without any internal coordination; (2) UE only, where there is an internal coordination between different radio technologies within the UE, which means that the activities of one radio are known by other radios. However, in this case the network is not

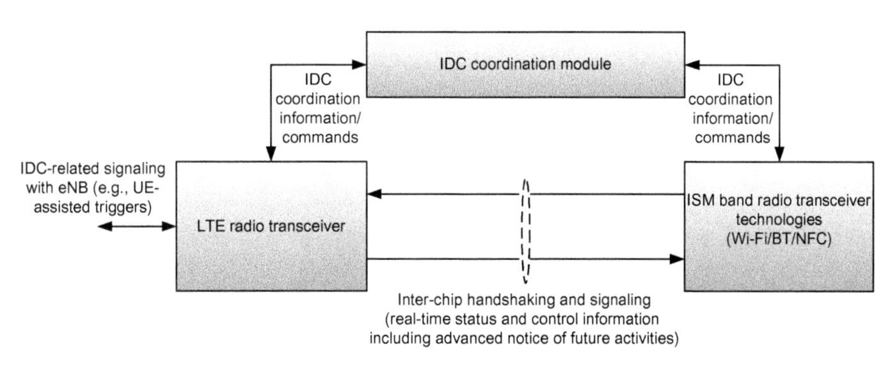

FIGURE 14.11

Example block diagram of the UE internal IDC signaling and coordination module.

aware of the coexistence issues experienced by the UE and is not involved in the coordination; and (3) network-coordinated with UE assistance, where different radio technologies within the UE are aware of possible coexistence problems and the UE can inform the network about such problems. The latter method is illustrated in Figure 14.11, and has become the basis for the solutions that have been developed in LTE Rel-11 for collocated multi-radio coexistence problems.

Based on the above assumption that different collocated radio modules can handshake and exchange (advance) information/notification about their current or upcoming activities, we now study the solutions that were developed in 3GPP to mitigate collocated multi-radio coexistence issues. It must be noted that other standardization organizations such as Wi-Fi Alliance [21] and Bluetooth Special Interest Group [20] have also taken steps on their side to find solutions that can alleviate or ideally resolve the in-device coexistence and interference problems. For example, BT devices may by default configure themselves as masters; thus, in-device BT may only behave as a BT slave under certain scenarios. On the 3GPP side, the standard solution in LTE Rel-11 enables the UE, with the assistance from the serving eNB, to resolve the IDC interference issues in situations where the actions taken by BT/Wi-Fi radios are not sufficient to solve the problem [18].

Based on RF analysis results, one straightforward solution to effectively mitigate or reduce the IDC interference is to let radio transceivers work in different frequency bands with sufficient frequency separation. The basic concept of an FDM-based solution is to move the victim (or aggressor) signal farther away from the aggressor (or victim) signal and increase the frequency separation for better filtering performance. An LTE FDM-based solution is to move the LTE signal farther away from the ISM band by performing inter-frequency handover within LTE according to the existing procedure specified in [3,4]. On the contrary, in the ISM FDM-based solution, the ISM radio signal is moved away from the LTE frequency band in the frequency-domain by switching the Wi-Fi channel in the portable router scenario, or by making use of the BT adaptive frequency hopping (AFH) feature to select a set of BT channels away from the LTE operating band (see Figure 14.9). The LTE FDM-based solution is applicable to all usage models as long as an alternative LTE frequency channel is available. In some deployments, using the FDM-based solution to resolve IDC interference problems may not be possible or desirable. As an example, the LTE FDM-based solution may not be applicable when all available LTE bands are impacted by ISM radio activities, the UE is in an area with only one LTE frequency band, or the UE does not have good channel quality in the alternative bands. It is also possible that the operator has alternative frequency bands, but those bands are congested due to a large number of active UEs. To ensure efficient network operation, it can be beneficial to balance load between RF bands and retain the UE in the current frequency, even if the LTE receiver is disturbed by the ISM signals. In that case, time multiplexing of LTE and ISM-band radios is more desirable [18].

In LTE TDM-based solutions, the basic idea is to time-multiplex the transmission or reception of different radios by either adapting the LTE transmission/reception timing to that of ISM-band radios, or vice versa. In LTE, DRX was adopted as a TDM-based solution for IDC interference avoidance. In the DRX mode, the UE is allowed to switch its LTE receiver off and stop monitoring the PDCCH. Since uplink transmissions are scheduled on the PDCCH, the gaps in PDCCH reception imply that uplink transmissions cannot be scheduled when the UE is in DRX mode. As a result, the time is divided into scheduled and unscheduled periods from the LTE device perspective. Unscheduled time can be used for ISM-band communication. To facilitate the DRX solution for coexistence of LTE and other radios, the UE should provide assisting information to the eNB.

The UE provides the eNB with a desired TDM pattern. For example, the parameters related to the TDM pattern can consist of periodicity of the TDM pattern and scheduling period (or unscheduled period), as shown in Figure 14.12. It is up to the eNB to decide and signal a suitable DRX configuration to the UE based on a UE suggested TDM pattern and other criteria. The scheduling period corresponds to the active time of DRX operation, while unscheduled period corresponds to the inactive time. The eNB would attempt to guarantee the unscheduled period using appropriate downlink/uplink scheduling, sounding reference signal transmission configuration, DRX command MAC control element usage, and so on. During the inactive time, the UE is allowed to delay the initiation of dedicated

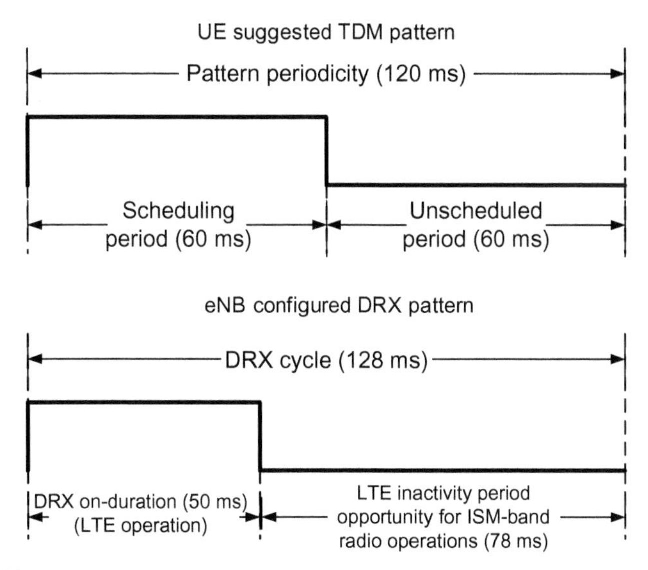

FIGURE 14.12

Example UE suggested TDM pattern and eNB configured DRX pattern [1].

scheduling request and/or random access procedure. Figure 14.12 illustrates an example where the eNB signals a DRX configuration based on a UE suggested pattern [1].

In the LTE + Wi-Fi off-load usage model introduced earlier, long scheduling and unscheduled periods are desired to enable completion of pending data transmissions. However, long periods would increase the latency of packet transmission. In addition, since in this scenario Wi-Fi acts as a terminal, it should be able to receive a beacon signal transmitted once every 102.4 ms. As a result, a suitable TDM pattern suggested by the UE should be 128 ms, where LTE occupies the first 60 ms. However, due to a limited number of DRX patterns in LTE, the eNB may configure the UE with a long DRX cycle of 128 ms and "on duration" of 40 ms. It should be noted that because of pending HARQ retransmissions in the LTE side, transition from scheduling time to unscheduled time is not unexpected. Therefore, LTE transmissions may stop before expiration of the on duration timer.

In an LTE + BT voice scenario, scheduling and unscheduled periods for each radio alternate much faster than in the Wi-Fi coexistence scenario. In enhanced synchronous connection oriented (eSCO) mode in BT [20], the transmission interval is 3.75 ms having six transmit/receive slots; thus, in order to maintain the minimum QoS requirements for voice conversation, the network should ensure at least a pair of correct BT transmit/receive instances in each BT interval of 3.75 ms. The remainder of the time can be allocated to LTE radio activities. The UE indicates to the network the desired subframes reservation bitmap (10 bits, one for each subframe). For short-term subframe reservation patterns, it is desirable that the pattern is consistent with uplink/downlink HARQ timing so that none of the HARQ processes is interrupted. Based on the assisting information, the network configures suitable DRX parameters.

In certain cases where the IDC interference condition cannot be predicted in advance, it is impossible to perform inter-frequency handover, or to provide unscheduled periods via DRX configuration. For example, one such condition may be the case where an unpredicted BT connection setup occurs and the DRX cannot be configured early enough to accommodate this critical communication. In another scenario, some critical signaling may occur irregularly where the UE can neither perform handover to an alternative band or be configured with a specific DRX configuration. Therefore, there are conditions where FDM-based or TDM-based schemes may not be applicable, and the UE has to deny LTE uplink transmissions autonomously to protect important signaling on other radios. However, the denial rate should be very low so that it does not significantly impact LTE operation. In addition to the TDM-based solution performed under the control of an LTE network, the UE may also use time-multiplex solutions through internal coordination. For example, the UE may terminate Wi-Fi or BT transmission when the LTE is receiving critical downlink signals. This would be useful to prevent disruption in the LTE signaling procedure as a result of unpredictable IDC interference from the ISM-band radios. Consequently, the

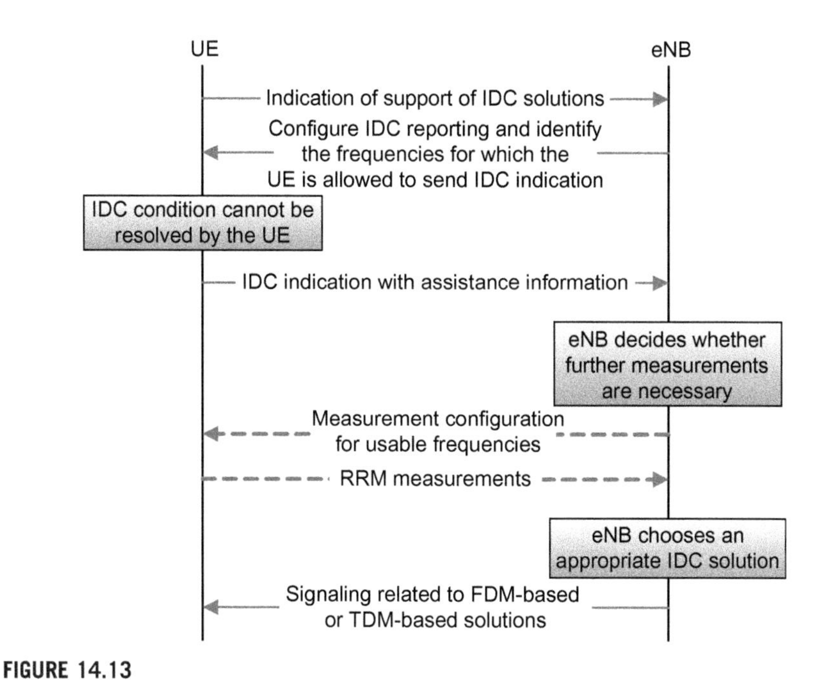

FIGURE 14.13

Signaling procedure for network-coordinated UE-assisted IDC resolution [18].

UE must carefully assess the IDC conditions before deciding on the best mitigation strategy in order not to compromise the performance and normal operation of the collocated radios.

Figure 14.13 illustrates the signaling procedure in network-coordinated UE-assisted IDC interference mitigation solution. When a UE experiences an IDC condition that cannot be locally resolved and network assistance is required, the UE sends an IDC indication via dedicated RRC signaling to report the problem [3]. The detail of the IDC indication trigger is left to UE implementation. The UE may rely on the existing LTE measurements and/or UE internal coordination. The IDC indication should be triggered based on ongoing IDC interference on the serving or non-serving frequencies, instead of assumptions or predictions of potential interference. A UE that supports IDC functionality indicates this capability to the network, and the network can then configure the UE through dedicated signaling whether the UE is allowed to send an IDC indication. The UE may only send an IDC indication for E-UTRA uplink/downlink RF carriers for which a measurement object is configured [4].

The term "ongoing IDC interference" should be treated as a general guideline by the UE. For the serving frequency, the ongoing interference may be related to the interference caused by an aggressor radio to a victim radio during either active data exchange or upcoming data activity which is expected in a few hundred milliseconds. For a non-serving frequency, the ongoing interference may be an anticipation of the LTE radio becoming an aggressor or a victim, if it makes handover to a non-serving frequency. Similarly, concerning activation of secondary cells in the case of carrier aggregation, the ongoing interference may be an anticipation of the LTE radio becoming an aggressor or a victim. Ongoing interference may affect several subframes/slots. When the eNB is notified of an IDC condition through an IDC indication from the UE, the eNB can choose to apply an FDM-based or a TDM-based solution. To assist the eNB in selecting an appropriate solution, all necessary/available assisting information for both methods is sent together as part of IDC indication message to the eNB. The IDC assisting information contains the list of E-UTRA carriers suffering from ongoing interference, the direction of the interference and depending on the scenario; it may also contain preferred time-multiplex patterns or parameters to enable appropriate DRX configuration for the UE on the serving E-UTRA carrier. The IDC indication is also used to update the IDC assistance information, including the cases when the UE no longer encounters the IDC condition. In the case of inter-eNB handover, the IDC assisting information is transferred from the source eNB to the target eNB. The UE should only indicate that the IDC condition is over when it is not experiencing an IDC condition any more. The UE should not resend the same IDC indication message to the network, but it may resend the same IDC indication after handover. The IDC condition can be divided into three phases: (i) the UE detects the start of IDC condition but does not initiate the transmission of the IDC indication to the eNB; (ii) the UE has initiated the transmission of the IDC indication to the eNB and no solution is configured by the eNB to resolve the IDC issue; (iii) the eNB has provided the UE with a solution that resolves the IDC issue (Table 14.2) [3].

In addition, once configured by the network, the UE can autonomously deny LTE uplink transmission in all phases to protect ISM-band transmission in special circumstances, if other solutions cannot be used. On the other hand, it is assumed that the UE also autonomously denies ISM-band transmission to ensure LTE connectivity to the eNB in order to perform LTE procedures, e.g., RRC connection reconfiguration and paging. The network may configure a long-term denial rate by dedicated RRC signaling to limit the amount of LTE uplink autonomous denials. Otherwise, the UE must not perform any LTE uplink autonomous denials [3].

In summary, the IDC interference level and its impact on the receiver performance depends on transmit power and receiver blocking characteristic of each radio and physical characteristics of transceivers (e.g., filter responses, antenna isolation). The following conclusions were drawn from the analyses performed in 3GPP related to this work item:

- For most cases, we observe that frequency-domain solutions, i.e., moving to other frequencies, and RF filtering can sufficiently suppress the coexistence interference.

Table 14.2 RRM, RLM, CSI Measurements in Different Phases of IDC Interference [3]

Phases of IDC Interference	RRM Measurements	RLM Measurements	CSI Measurements
Phase 1	It is up to the UE implementation and RRM measurement requirements	It is up to the UE implementation and RLM measurement requirements	It is up to the UE implementation and CSI measurement requirements
Phase 2 (If the UE determines that the network does not provide a solution that resolves its IDC problem, it performs measurements)	The UE is required to ensure that the measurements are free of IDC interference	The UE is required to ensure that the measurements are free of IDC interference effects. The UE should maintain LTE connectivity in this phase, meaning that RLM measurements are not impacted by IDC interference. If no solution is provided within a time, which is up to UE implementation, the UE may declare RLF or it may continue to deny the ISM transmission	–
Phase 3 (If the IDC indication message reports the IDC condition on a neighbor frequency, the UE performs RRM measurements for that frequency)	The UE is required to ensure that the measurements are free of IDC interference	The UE is required to ensure that the measurements are free of IDC interference effects	–

Table 14.3 LTE Blocking Levels for Bluetooth and Wi-Fi Receivers for Different LTE Transmit Powers and Operating Bands [1]

LTE Transmit Power (dBm)	Blocking with FBAR Filter (dBm)				
	2300–2370 MHz	2360–2380 MHz	2380–2400 MHz	2500–2520 MHz	2520–2570 MHz
23	−34	−28	7	−37	−40
21	−36	−30	5	−39	−42
19	−38	−32	3	−41	−44
17	−40	−34	1	−43	−46
15	−42	−36	−1	−45	−48
13	−44	−38	−3	−47	−50
11	−46	−40	−5	−49	−52
9	−48	−42	−7	−51	−54
7	−50	−44	−9	−53	−56
5	−52	−46	−11	−55	−58
3	−54	−48	−13	−57	−60
1	−56	−50	−15	−59	−62
−1	−58	−52	−17	−61	−64
−3	−60	−54	−19	−63	−66
−5	−62	−56	−21	−65	−68
−7	−64	−58	−23	−67	−70
−9	−66	−60	−25	−69	−72

- For the uppermost region of LTE Band 40 (2380–2400 MHz), LTE transmitting with the maximum power of 23 dBm can block the Wi-Fi or BT signal in the entire ISM band, thus limiting the maximum LTE transmit power below 23 dBm, moving LTE signal away from ISM, or TDM-based solution needs to be considered (Table 14.3).
- For 2300–2380 MHz in LTE Band 40, Wi-Fi or BT desense due to the LTE interference may be acceptable except for the lower 20 MHz of the ISM band, since the performance of the state-of-the-art film bulk acoustic resonator (FBAR) filtering[2],

[2]FBAR filters are a form of bulk acoustic wave (BAW) filter that have superior performance with steeper rejection curves compared to surface acoustic wave (SAW) filters. The FBAR filters also feature 0.3–0.5 dB less insertion loss, resulting in up to 50 mA less current consumption, thereby improving battery life and talk time of the mobile devices. A thin-film bulk acoustic resonator is a device consisting of a piezoelectric material contained between two electrodes and acoustically isolated from the surrounding medium. The FBAR devices using piezoelectric films with thicknesses ranging from several micrometers down to tenths of micrometers resonate in the frequency range of approximately 100 MHz to 10 GHz. Aluminum nitride and zinc oxide are two common piezoelectric materials used in FBARs [19].

device OOB/spurious emissions, sensitivity, and blocking of implementations are typically better than specification limits. Additionally, limiting the number of data/control bearing resource blocks and assigning resources away from the ISM band, which directly impact the OOB emissions, can help reduce desensitization to the lower 20 MHz of the ISM band.

Table 14.3 shows LTE blocking for Wi-Fi and BT receivers for different LTE transmit powers and operating bands in E-UTRA Band 40 and Band 7. As shown in Table 14.3, a cell-edge UE with maximum transmitting power of 23 dBm in the uppermost 20 MHz segment of Band 40 can result in a maximum of 7 dBm OOB blocking interference[3] in the Wi-Fi/BT receiver. The LTE transmit power level needs to be limited for the simultaneous operation with ISM-band reception, if the LTE transceiver is operated in the upper 20 MHz of Band 40. The maximum allowed LTE transmit power for coexistence varies depending on the blocking characteristics of the Wi-Fi/BT receivers [1].

The RF analyses conducted in 3GPP related to the IDC interference between ISM-band and LTE technologies, and the conclusions that were made and documented as a result of those analyses were mainly based on four information sources. The analyses and measurements presented in 3GPP TR 36.816 [1] indicate that for some IDC scenarios, significant degradation on both LTE and ISM-band radios can occur despite the use of state-of-the-art RF filtering technology. The precise quantitative results differed from one source to another due to different assumptions in the analyses or the measurement approaches. Nonetheless, the conclusions were consistent in that at least a significant fraction of spectrum is highly desensitized when the other technology is transmitting. The assumptions used by those sources in the simulation or measurements presented to 3GPP are summarized in Table 14.4.

[3]Radio receiver blocking or receiver desensitization is an important parameter for any radio receiver. A good radio receiver blocking or desensitization performance is particularly important in the scenarios where a number of radios of various forms are used in close proximity to each other. With wireless communications being used for everything from Wi-Fi to cellular communications and Bluetooth, as well as many more traditional applications, there are many instances where two radios operate very close to each other, and the receiver blocking performance will be very critical. Blocking rejection is the ability of a receiver to tolerate an OOB signal and avoid blocking. A good automatic gain control design is part of achieving good blocking rejection. When a very strong OOB signal appears at the input of a receiver, it is often found that the sensitivity is reduced. The effect arises because the front-end amplifiers are saturated as a result of the strong OOB signal. This often occurs when a receiver and transmitter are operated from the same site and the transmitter signal is exceedingly strong. When this occurs, it has the effect of suppressing all the other signals trying to pass through the amplifier, giving the effect of a reduction in gain. Blocking is generally specified as the level of the unwanted signal at a given offset, which will give a 3-dB reduction in gain. Depending on the type of receiver, the values for blocking will vary considerably. As a reference point, a good radio receiver should be able to withstand blocking signals of about 10 dBm before this effect becomes noticeable.

Table 14.4 Assumptions for the RF Analyses in 3GPP IDC Studies [1]

Parameter	Analysis 1	Analysis 2	Analysis 3	Analysis 4
LTE frequency band	40 and 7	40 and 7	40	40 and 7
ISM radio	BT, Wi-Fi	Wi-Fi	Wi-Fi	BT, Wi-Fi
Interference directions considered for B and 40	LTE to BT/ Wi-Fi; BT/Wi-Fi to LTE	LTE to Wi-Fi; Wi-Fi to LTE	LTE to Wi-Fi	LTE to BT/ `Wi-Fi; BT/Wi-Fi to LTE
Interference mechanisms	Spurious emission and blocking	Spurious emission and blocking	Spurious emission	Spurious emission and blocking
RF filtering method	FBAR	No filters external to test setup	Commercially available filter (typical/minimum)	FBAR
Antenna isolation (dB)	12	15, 20, 25	12	12
LTE transmitter power (dBm)	23	23	N/A	0, 15, 23
Wi-Fi transmitter power (dBm)	20	20	20	20, 14.5
BT transmitter power (dBm)	10	N/A	N/A	4, 0
LTE RSSI as victim (dBm)	−94	−70	−94	−94 (Band 40) −92 (Band 7)
Wi-Fi RSSI (dBm)	−79	−50	N/A	−89, −76
BT RSSI (dBm)	−90	N/A	N/A	−70
LTE bandwidth (MHz)	20	25−100 resource blocks (over 20)	20	20
Wi-Fi bandwidth (MHz)	22	22	22	22
BT bandwidth (MHz)	1	N/A	N/A	1
Performance measure	Desensitization (dB)	EVM	Desensitization (dB)	Desensitization (dB)

References

3GPP Specifications[4]

[1] 3GPP TR 36.816, Evolved Universal Terrestrial Radio Access (E-UTRA), Study on Signaling and Procedure for Interference Avoidance for In-Device Coexistence (Release 11), December 2011.
[2] 3GPP TR 36.839, Evolved Universal Terrestrial Radio Access (E-UTRA), Mobility Enhancements in Heterogeneous Networks (Release 11), December 2012.
[3] 3GPP TS 36.331, Evolved Universal Terrestrial Radio Access (E-UTRA), Radio Resource Control (RRC), Protocol specification (Release 11), December 2012.
[4] 3GPP TS 36.300, Evolved Universal Terrestrial Radio Access (E-UTRA) and Evolved Universal Terrestrial Radio Access Network (E-UTRAN) Overall description, Stage 2 (Release 11), December 2012.

Books

[5] C. Johnson, Long Term Evolution in Bullets, second ed., Create Space Independent Publishing Platform, July 2012.
[6] S. Sesia, I. Toufik, M. Baker, LTE, the UMTS Long Term Evolution, From Theory to Practice, second ed., John Wiley & Sons, August 2011.
[7] E. Dahlman, S. Parkvall, J. Skold, 4G, LTE/LTE-Advanced for Mobile Broadband, Academic Press, May 2011.

Articles

[8] IST-4-027756 WINNER II, D4.7.1 Interference averaging concepts, June 2007.
[9] IST-4-027756 WINNER II, D4.7.2 Interference avoidance concepts, June 2007.
[10] IST-4-027756 WINNER II, D4.7.3 Smart antenna based interference mitigation, June 2007.
[11] L. Lindbom et al., Enhanced Inter-cell Interference Coordination for Heterogeneous Networks in LTE-Advanced: A Survey, Cornell University Library (eprint), December 2011.
[12] R. Zhang, L. Hanzo, Co-channel interference mitigation:active and passive techniques, IEEE Vehicular Technology Magazine, December 2010.
[13] S. Xu1, J. Han, T. Chen, Enhanced inter-cell interference coordination in heterogeneous networks for LTE-Advanced, IEEE 75th Vehicular Technology Conference, May 2012.
[14] R. Ghaffar, R. Knopp, Fractional frequency reuse and interference suppression for OFDMA networks, Proceedings of the Eighth International Symposium on Modeling and Optimization in Mobile, Ad Hoc and Wireless Networks (WiOPT), June 2010.
[15] Y. Peng, W. Yang, Y. Zhang, Y. Zhu, Mobility performance enhancements for LTE-Advanced heterogeneous networks, IEEE 23rd International Symposium on Personal, Indoor and Mobile Radio Communications (PIMRC), 2012.
[16] A. Damnjanovic, et al., A survey on 3GPP heterogeneous networks, IEEE Wireless Communications, June 2011.
[17] A. Ghosh, et al., Heterogeneous cellular networks: from theory to practice, IEEE Communications Magazine, June 2012.

[4]3GPP specifications can be found at the following URL: http://www.3gpp.org/ftp/Specs/archive/.

[18] Z. Hu, et al., Interference avoidance for in-device coexistence in 3GPP LTE-Advanced: challenges and solutions, IEEE Communications Magazine, November 2012.

[19] Thin-film bulk acoustic resonator, Wikipedia. <http://en.wikipedia.org/wiki/Thin_film_bulk_acoustic_resonator>.

[20] Bluetooth Core Specifications, Bluetooth Special Interest Group. <https://www.bluetooth.org/Technical/Specifications/adopted.htm>.

[21] Wi-Fi Alliance Published Specifications. <https://www.wi-fi.org/knowledge-center/published-specifications>.

[22] Stefan Brück, 3G/4G mobile communications systems, Qualcomm Technologies Inc. <http://www.ant.uni-bremen.de/en/courses/mcs/>.

CHAPTER

Positioning and Multimedia Broadcast/Multicast Services 15

CHAPTER CONTENTS

In this chapter, we take a top-down systems approach to describe two important service enablers of LTE/LTE-Advanced that both started in LTE Rel-9 and have been enhanced in the later releases. The UE positioning and location-based services, as well as the evolved multimedia broadcast/multicast services (E-MBMS), are considered the key enablers of various services in modern cellular systems. There are numerous applications that rely on these functionalities to provide the users with the best quality of experience and ease of access to location-optimized and customized content.

This chapter begins with the description of the positioning standard for LTE/ LTE-Advanced systems. We review the positioning requirements, positioning architectures, and signaling. The coordinate systems and the transformations used in positioning systems are described in detail. There are many different use cases for positioning information. The positioning functions may be used internally by the EPS, value-added network services, the UE itself, or through the network by a third party. While the location-based services may be used for emergency services (which may be mandated), the feature has not been exclusively developed for emergencies.

We will then describe the E-MBMS feature in LTE/LTE-Advanced. The E-MBMS bearer service is the service provided by the packet-switched domain to E-MBMS user services in order to deliver IP multicast[1] datagrams to multiple receivers using minimal network and radio resources. An E-MBMS user service is a service provided to the end user by means of the MBMS bearer service, and possibly other EPS bearers. The E-MBMS is a point-to-multipoint service in which data is transmitted from a single source entity to multiple recipients. Transmitting the same data to multiple recipients allows the sharing of network resources. The E-MBMS for EPS bearer service supports broadcast mode over E-UTRAN and UTRAN. The E-MBMS is realized by the addition of a number of new features and new functional entities to the 3GPP network architecture.

An E-MBMS counting function was introduced that allows counting the number of connected mode users that are either receiving or are interested in receiving particular MBMS services. In LTE Rel-11, E-UTRAN MBMS is enhanced to ensure service continuity when multi-carrier functionality is deployed in an MBMS service area. The MBMS services may be deployed on different frequencies over different geographic areas. LTE Rel-11 enhancements allow the network to inform MBMS-capable devices of information related to MBMS deployment, such as carrier frequencies and service area identities (SAIs). In LTE Rel-11, an MBMS-capable device can inform the network of its interest in MBMS services by indicating the carrier frequencies associated with the MBMS services of interest. The MBMS interest indication by the device also allows the device to indicate the priority between MBMS service and unicast service. The network uses the MBMS interest indication provided by the device for mobility management decisions so that the device is always able to use its receiver at the appropriate carrier frequency to ensure continuity of MBMS services. In idle mode, an MBMS-capable device can prioritize a particular carrier frequency during cell reselections depending on the availability of the MBMS service of interest on that carrier frequency. To ensure

[1]IP multicast is a method of transmitting IP datagrams to a group of interested receivers in a single transmission. It is often employed for streaming media applications on the Internet and private networks. The method is the IP-specific version of the general concept of multicast networking. It uses especially reserved multicast address blocks in IPv4 and IPv6. In IPv6, IP multicast addressing replaces broadcast addressing as implemented in IPv4. IP multicast is described in IETF RFC 1112 [20].

MBMS service continuity in connected mode, the MBMS interest indications received from the device are signaled to the target cell as part of handover preparation procedure [4].

15.1 Positioning and location-based services

Positioning defines the process of determining the position and/or velocity of a device using radio signals. Location-based services are considered a significant feature of the modern cellular networks which allow the users to locate themselves and each other, as well as enable the mobile applications to offer services and information customized to the current users' locations. In addition to the commercial usage, there are also public safety applications for positioning. As an example, an emergency call made from a mobile device can be tracked and sufficiently accurate location of the caller can be determined, even though the caller may not know where he/she is. Many government regulatory organizations around the world have made support of accurate positioning functionality mandatory for all mobile devices. In the United States, it is mandatory for network operators to support FCC's Enhanced 911[2], making the location of a cell phone available to emergency call dispatchers. With this mandate, the FCC has defined accuracy requirements for different position estimation methods. Cellular network operators are responsible for positioning wireless terminals with an accuracy of 50 and 150 m in 67% and 95% of all positioning attempts, respectively. These figures are valid for UE-based positioning methods, i.e., where the positioning method involves functionality located in the UE such as global positioning system (GPS) receiver hardware. The requirements for network-based methods, without such UE functionality, are 100 m (67%) and 300 m (90%) [17]. These are stringent accuracy requirements that need to be fulfilled and guaranteed no matter what position estimation technology is used, or in which environment the position

[2]The FCC's wireless Enhanced 911 (E911) requirements seek to improve the effectiveness and reliability of wireless 911 services by providing the dispatchers with additional information on wireless 911 calls. The FCC's wireless E911 rules apply to all wireless licensees, broadband personal communications service (PCS) licensees, and certain specialized mobile radio (SMR) licensees. The FCC has divided its wireless E911 program into two phases: Phase I and Phase II. Under Phase I, the FCC requires carriers, within 6 months of a valid request by a local public safety answering point (PSAP) to provide the PSAP with the telephone number of the originator of a wireless 911 call and the location of the cell site or base station transmitting the call. Under Phase II, the FCC requires wireless carriers, within 6 months of a valid request by a PSAP to begin providing information that is more precise, specifically, the latitude and longitude of the caller. This information must meet FCC accuracy standards, generally within 50–300 m, depending on the type of technology used. The deployment of E911 requires the development of new technologies and upgrades to local 911 PSAPs, as well as coordination among public safety agencies, wireless carriers, technology vendors, equipment manufacturers, and local wireline carriers [17].

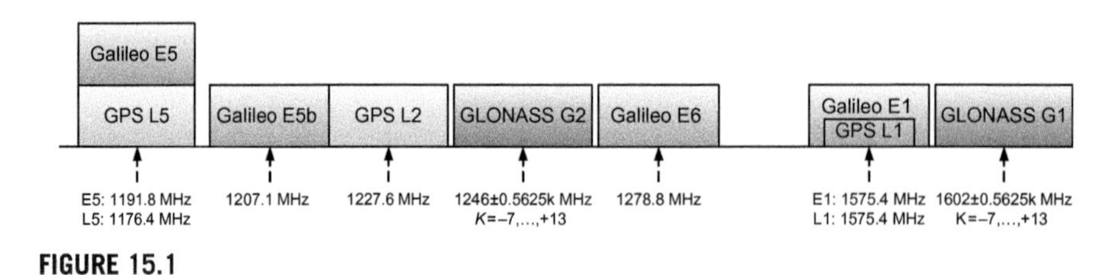

FIGURE 15.1

Frequency bands used by prominent positioning methods [15].

is estimated (i.e., indoor or outdoor). Determining a user's device position is traditionally based on satellite-based position estimation using Global Navigation Satellite Systems (GNSS) such as GPS. Figure 15.1 provides the frequency bands of the prominent global positioning systems.

The majority of modern mobile devices have an integrated GNSS receiver. To properly estimate the position, the receiver needs to have an unobstructed line of sight to at least four satellites, and this is one of the main drawbacks of GNSS-based methods. In indoor environments, line of sight reception of the low-power radio signals transmitted from the medium-earth-orbit satellites is not guaranteed, if not impossible. To overcome this limitation, especially in an environment with poor signal conditions, Assisted-GNSS (A-GNSS) methods have been developed, e.g., Assisted-GPS. In A-GPS, a cellular network uses network resources to provide assistance data that helps the device to locate and utilize the available GPS satellites faster. A-GNSS reduces start-up and acquisition times, increases sensitivity, and reduces the UE's power consumption. While there is often an acceptable coverage by a cellular network to determine the A-GPS information, this does not help if the satellites cannot be detected due to extremely weak reception of GPS signals inside a building. In that case, position estimation using radio signals is more feasible. In GSM, position estimation based on radio signals has been around for many years. As mobile broadband technology advances and the demand for more accurate positioning and location-based services increases, more sophisticated and accurate positioning methods have been developed. In this section, we will only focus on the positioning methods developed for E-UTRA.

The positioning methods supported in LTE mainly rely on the high-level network architecture as shown in Figure 15.2. The positioning network architecture has been defined in such a way that it is generally independent from the underlying network. There are three main entities in this architecture, location service (LCS) client, LCS server, and LCS target. An LCS client is the entity requesting the service and in the majority of the cases is located in the LCS target. This service obtains the location information by sending a request to the server. The location server is a physical or logical entity that collects measurements and other location information from the device and base station, and assists the device with measurement and estimation of its position. The server processes the request from the client and provides the client with the requested information [15,18].

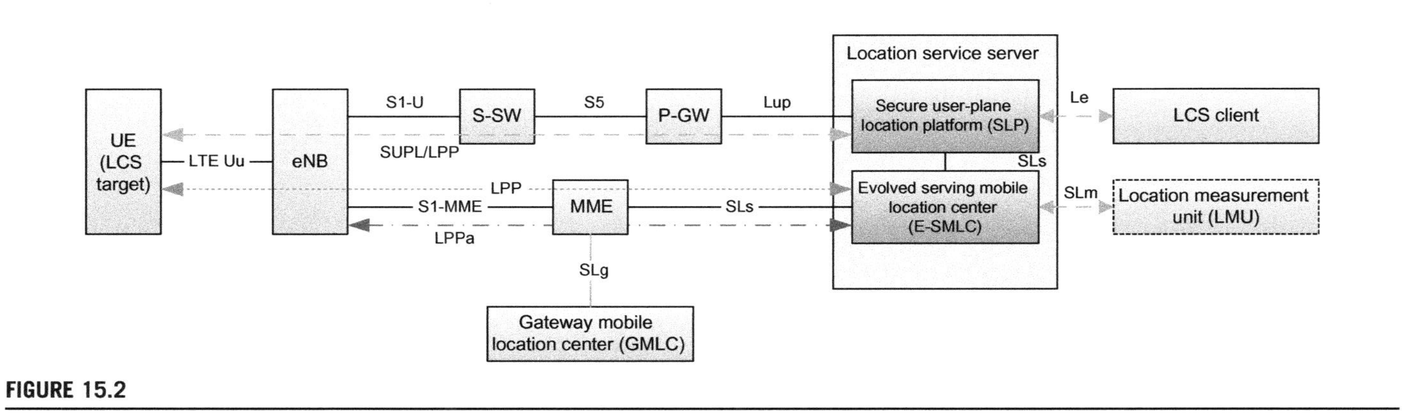

FIGURE 15.2

E-UTRA location service architecture [7,15].

15.1.1 E-UTRA positioning network architecture

Some location-based services are targeted at individual users, while others are focused on aggregated user behavior that will be of value to enterprises and application and content providers. In general, the services can be grouped into the following categories [12]:

- Regulatory-based applications such as emergency call services.
- Consumer applications, such as social networking, service finder, and navigation services.
- Business applications such as fleet management or social networking tailored to business needs.
- Third-party applications that enable merchants or application and content providers to target specific end users for their advertising or services.
- Network operator applications that help detect internal problems, or optimize networks and business capabilities.

Given that location-based services have generally diverse requirements, the communication service providers must carefully identify and prioritize a particular service's requirements before choosing methods of gathering and delivering location information. The following list summarizes some of the most important general requirements [12]:

- Providing the required degree of location accuracy.
- Promptness of response to a location request.
- Scalability to support a large number of user locations.
- Support for real-time, periodic, or event-triggered location updates.
- Ease of implementation.
- Battery consumption for mobile devices.
- Impact on network resources.
- Differentiation of the network provider service from application and content providers offerings.

The EPS provides several methods for delivering location information when it is determined. Each method has certain capabilities, strengths, and constraints, which necessitate satisfaction of the unique requirements of each specific service.

The user-plane method requires minimal interaction with the underlying wireless access technologies; thus, this method can be used ubiquitously across LTE and the legacy networks. It is also scalable, which makes it well suited to support commercial location-based services with less complexity, cost, and impact on the network than the control-plane solution. This solution relies on user-plane data bearers to transfer location-assistance and positioning-related messages between the secure user-plane location[3] (SUPL)-enabled terminal (SET) and the SUPL location platform

[3]SUPL is an IP-based protocol for assisted-GPS in order to receive positioning information of GPS satellites faster via IP connections instead of slow GPS satellite links.

(SLP). An SUPL client is required on the user device. Lack of direct connection to the access network is a disadvantage for supporting positioning methods, which require assistance data from the network. In addition, because this solution can only enable single-user location requests, it is not suitable for applications that need periodic reporting on all user locations within a cell or zone. Depending on the location technology used, the user-plane solution can support a high level of accuracy to meet the needs of the majority of location-based services targeted at the individual user, such as turn-by-turn navigation and fleet management.

In the control-plane method, which was specified in LTE Rel-9, positioning data is transferred over the control-plane, between the UE, the eNB, and the positioning elements, i.e., evolved serving mobile location center/gateway mobile location center (GMLC). This method provides convenient access to assistance data from the network, and enables more reliable service performance than the user-plane solution. Since the control-plane method does not require a client on the UE, it can be used to support emergency services on LTE networks for limited service devices. On the other hand, this solution has less scalability than the user-plane, only supports single-user location requests, and is more difficult to interwork with legacy networks.

The 3GPP legacy technologies (GSM and UMTS) have supported LCSs network architecture for the purpose of positioning mobile devices since 3GPP Rel-4. With the introduction of EPS in 3GPP Rel-8, control-plane location-service architecture for the EPS was introduced in 3GPP Rel-9. Figure 15.2 illustrates the control-plane location service architecture for the EPS, which introduced new interfaces SLg and SLs, and allowed the EPS control-plane element mobility management entity to interconnect with the location service core network elements, making location services using the positioning functionality provided by the E-UTRAN possible.

The location service architecture follows a client/server model with the GMLC acting as the location server providing location information to location service clients. The GMLC sends location requests to the evolved serving mobile location center (E-SMLC) through the MME to retrieve this location information. The E-SMLC is responsible for interaction with the UE through E-UTRAN to obtain the UE position estimate or to obtain position measurements that help the E-SMLC to estimate the UE position. Note that the GMLC interaction over the interfaces connected to it other than SLg was already available before 3GPP Rel-9 for GSM and UMTS access.

The location service clients can be part of the core network or external to the core network, and can also reside in the UE or be attached to the UE. Depending on the location of the location service client, the location request initiated by the location service client may be a mobile-originated, mobile-terminated, or network-induced location request. Emergency location service is possible even if the UE does not have a valid service subscription due to regulatory mandates. Support of location service-related functionality in the E-UTRAN, MME, and UE are optional. The location service is applicable to any target UE, no matter whether the UE supports location service. The following is a description of various LCS architectural elements and interfaces shown in Figure 15.2 [18]:

- *Gateway mobile location center*: This entity receives and processes location requests from location service clients; obtains routing information from the home subscriber server; performs registration authorization; communicates information needed for authorization, location service requests, and location information with other GMLC; checks the target UE privacy profile settings; and depending on roaming support, it may assume the role of requesting GMLC, visited GMLC, or home GMLC.
- *Evolved serving mobile location center*: This entity manages the overall coordination and scheduling of resources required for the location determination of a UE that is attached to E-UTRAN. It calculates the final location and estimates velocity, as well as the location accuracy. It interacts with the UE to exchange location information applicable to the UE-assisted and the UE-based positioning methods. It interacts with the E-UTRAN to exchange location information applicable to the network-assisted and the network-based positioning methods.
- *Mobility management entity*: This entity is responsible for UE subscription authorization. It coordinates location services and positioning requests. It handles charging and billing, and performs E-SMLC selection (e.g., based on network topology to balance the load on E-SMLC, location service client type, and requested QoS). It is further responsible for authorization and operation of the location service.
- *Home subscriber server*: This entity is used for storage of location services subscription data and routing information.
- *Location measurement unit*: The location measurement unit (LMU) conducts measurements and communicates the measurements to an E-SMLC. All positioning measurements obtained by an LMU are supplied to the E-SMLC that made the request. A UE positioning request may involve measurements by multiple LMUs. The SLm interface between the E-SMLC and the LMU is used for uplink positioning. It is used to transport SLm-application protocol (AP) messages over the E-SMLC-LMU interface. Both stand-alone LMU and LMU integrated into an eNB are supported in LTE Rel-11.

Two protocols are used for the overall information exchange concerning location services: (i) LTE positioning protocol (LPP); and (ii) LTE positioning protocol annex (LPPa). The latter protocol is used for communication between the location server and the eNB. The base station (when using the observed time difference of arrival (OTDOA) method) is in charge of properly configuring the radio signals that are used by the terminal for positioning measurements, i.e., positioning reference signals (PRSs). It further provides information to the E-SMLC; it enables the device to conduct inter-frequency measurements, if required; and based on the E-SMLC request, it conducts measurements and sends the results to the server. The LPP can be used in the user-plane and control-plane methods. The LPP is a point-to-point protocol that allows multiple connections to different devices. The exchanged LPP messages and information can be divided into four categories:

1. UE positioning capability information transfer to the E-SMLC.
2. Positioning assistance data delivery from the E-SMLC to the device.
3. Location information transfer.
4. Session management.

The SLm-AP (application protocol), terminated between the E-SMLC and the LMU, is used to support the following functions: (i) delivery of target UE configuration data from the E-SMLC to the LMU; and (ii) request of positioning measurements from the LMU and delivery of positioning measurements to the E-SMLC. The SLm-AP is directly used between the E-SMLC and the LMU.

The MME receives a request for location service associated with a particular target UE from another entity (e.g., GMLC, eNB, or UE), or the MME by itself decides to initiate some location service on behalf of a particular target UE (e.g., for an IP multimedia subsystem (IMS) emergency call from the UE). The MME then sends a location service request to the E-SMLC. The E-SMLC processes the location service request, which may include transferring assistance data to the target UE to help UE-based and/or UE-assisted positioning. The E-SMLC then returns the result of the location service to the MME, e.g., a position estimate for the UE and/or an indication of any assistance data transferred to the UE. In the case of a location service requested by an entity other than the MME (e.g., UE, eNB, or E-SMLC), the MME returns the location service result to this entity.

The UE positioning is an access network function (e.g., GERAN, UTRAN, and E-UTRAN). An access network may support one or more UE positioning method, which may be the same or different from another access network. In E-UTRAN, the following UE positioning methods are supported: Cell ID (CID)-based; enhanced Cell ID (E-CID)-based; downlink OTDOA; and network-assisted GNSS positioning methods. Determining the position of a UE involves two main steps: (i) radio signal measurements; and (ii) position and optionally velocity estimation based on the measurements. The signal measurements may be conducted by the UE or the E-UTRAN. Both TDD and FDD radio interfaces will be supported in E-UTRAN positioning methods. The basic signals measured for terrestrial position methods are typically the E-UTRA radio transmissions. Other transmissions such as general radio navigation signals, including those from global navigation satellites systems, can also be measured. The position estimation may be performed in the UE or in the E-SMLC. In UE-assisted positioning, the UE performs the downlink radio measurements and the E-SMLC estimates the UE position, while in UE-based positioning, the UE performs both downlink radio signal measurements and position estimation. The UE may require some assistance from the network in the form of assistance data in order to perform the downlink measurements, and these are provided by the network either autonomously or upon UE request. The E-UTRAN positioning capabilities are intended to be forward compatible to other access types and other positioning methods, in an effort to reduce the amount of additional positioning support needed in the future [7].

15.1.2 Network-based and UE-based positioning methods

A network service provider needs to select an appropriate method for determining the location of a user device, based on the service requirements that have been identified. There are three primary network-enabled modes of operation for determining a user's location. In the UE-based mode, the user device obtains location measurements with assistance from the network and calculates the user's position. In the device-assisted mode, the user device provides positioning measurements, which the network uses to calculate the user's location. In the network-based mode, the network calculates the user's position without involving the device. The following location technologies support one or more of the above modes.

The differentiation between supported positioning methods is based on two facts. First, what the "measurement entity" is, which could only be the device or the base station. Second, what the "position estimation entity" is. The terms "-based" and "-assisted" refer to the node that is responsible for performing the positioning calculation (which may also provide measurements), and a node that provides measurements (but does not perform the positioning calculation), respectively. Thus, an operation in which measurements are provided by the UE to the E-SMLC to be used in the calculation of a position estimation is described as "UE-assisted" (and alternatively "E-SMLC-based"), whereas one in which the UE calculates its own position is described as "UE-based." With that knowledge, the positioning methods supported in LTE can be further categorized. Table 15.1 provides an overview of supported positioning methods in LTE.

In Table 15.1, A-GPS refers to the case where the network assists GPS-enabled mobile devices to improve the performance of their GPS receivers by providing assistance information such as reference time, visible satellite list, reference position, and satellite ephemeris. In general, A-GPS provides the highest accuracy of any network-enabled technique. However, it is not reliable in indoor or in dense urban or high-rise building environments, in which case A-GPS can be supplemented with other techniques such as downlink OTDOA or E-CID. In CID-based methods, a user device is located within the coverage area of its serving eNB, typically to the specific cell/sector within the eNB. Although this method is the least accurate one, it is the easiest to implement, and highly scalable. In E-CID-based methods, the position of the UE is determined to a finer level compared to the CID-based method, using additional radio signal measurements. This method has low accuracy (50–1000 m) depending on the size of the cell, but it is easier to implement than other methods, and is generally available across diverse vendor products and networks. Downlink OTDOA uses the measured timing of downlink signals received from multiple eNBs to locate the user device in relation to neighboring eNBs. In dense urban and indoor environments, OTDOA can be used to supplement A-GPS, provided that the user device can

Table 15.1 E-UTRA UE Positioning Methods [7]

3GPP Release	Method	UE-based	UE-Assisted (E-SMLC-Based)	LMU-Assisted/ E-SMLC-Based	eNB-Assisted	Secure User-Plane Location
9	A-GNSS	Yes Measurement: UE Estimation: UE	Yes Measurement: UE Estimation: LCS server	No	No	Yes (UE-based and UE-assisted)
9	Downlink OTDOA	No	Yes Measurement: UE Estimation: LCS server	No	No	Yes (UE-assisted)
9	Enhanced Cell-ID	No	Yes Measurement: UE Estimation: LCS server	No	Yes Measurement: eNB Estimation: LCS server	Yes (UE-assisted)
11	Uplink TDOA	No	No	Yes	No	No

detect PRSs transmitted from three or more eNBs. The uplink time difference of arrival (UTDOA) has been specified as part of LTE Rel-11, and has minimal impact on air-interface resources. This uplink positioning method makes use of the measured timing at multiple LMUs of uplink signals transmitted from the UE. The LMU measures the timing of the received signals using assistance data received from the positioning server, and the resulting measurements are used to estimate the location of the UE.

15.1.3 Positioning methods in LTE/LTE-Advanced

The positioning feature provides a means to determine the geographic location and/or velocity of the UE based on measuring radio signals. The position information may be requested by and reported to a client (e.g., an application) associated with the UE, or by a client within or attached to the core network. The position information is reported in standard formats such as those for cell-based or geographical coordinates, together with the estimated errors (uncertainty) in the position and velocity estimation as well as the positioning scheme(s) used to obtain the position estimate.

The uncertainty of the position measurement is dependent on network implementation and is controlled by the network operator. The uncertainty may vary between networks, as well as from one area within a network to another. The uncertainty may be hundreds of meters in some areas and only a few meters in others. In the event that a particular position measurement is provided through a UE-assisted process, the uncertainty may also depend on the capabilities of the UE. The uncertainty of the position information is dependent on the method used, the position of the UE within the coverage area, and the activity of the UE. Several design options of the E-UTRAN system (e.g., size of cell, adaptive antenna technique, path loss estimation, timing accuracy, and eNB surveys) allow the network operator to choose a suitable and cost-effective UE positioning method. The following assumptions are made for the operation of UE positioning in E-UTRAN [7]:

- The positioning method is equally applicable to TDD and FDD systems.
- The provision of the UE positioning function in EPS is optional through support of the specified methods in the eNB and the E-SMLC.
- UE positioning is applicable to any target UE, no matter if the UE supports location service, but with restrictions on the use of certain positioning methods depending on UE capability [6].
- The positioning information may be used for internal system operations to improve system performance.
- The location service architecture and functions include the option to accommodate several techniques of measurement and processing to ensure adaptability to the future technologies and service requirements.
- The LMU aspects are implementation-specific.

The following sections describe the specific positioning methods that are specified in E-UTRAN specifications as of LTE Rel-11.

15.1.3.1 Principles of GPS

The GPS is a space-based satellite navigation system that provides location and time information in all weather conditions, anywhere on or near the earth where there is an unobstructed line of sight to four or more GPS satellites. The system provides critical capabilities to military, civil, and commercial users around the world. It is maintained by the United States government, and is freely accessible to anyone with a GPS receiver.

A GPS receiver calculates its position by precisely timing the signals sent by GPS satellites in medium earth orbits. Each satellite continuously transmits messages that include the time at which the message was transmitted, and the satellite position at the time of message transmission. The receiver uses the messages it receives to determine the transit time of each message and computes the distance to each satellite. Each of these distances and satellites' locations define a sphere. The receiver is on the surface of each of these spheres when the distances and the satellites' locations are correctly calculated. These distances and satellites' locations are used to compute the location of the receiver using the navigation equations. This location is then displayed, perhaps with a moving map display or latitude and longitude; elevation information may be included. Many GPS units show derived information such as direction and speed, calculated from position changes.

In typical GPS operation, four or more satellites must be visible to obtain an accurate result. Four sphere surfaces typically do not intersect; thus, we can confidently conclude that when we solve the navigation equations to find an intersection, this solution gives us the position of the receiver along with accurate time, thereby eliminating the need for a very large, expensive, and power-consuming clock. The very accurately computed time is mainly used only for display and is not utilized in many GPS applications, which use only the location. However, a number of applications for GPS do use this highly accurate timing.

The receiver uses the messages received from satellites to determine the satellites' positions and the transmission times. Let (x_i, y_i, z_i, t_i) denote the coordinates of the ith satellite position and the transmission time, where the subscript $i = 1, 2, 3, \ldots, N$ is the satellite number, assuming $N \geq 4$. When the transmission time of message reception is indicated by the local clock t_r, the true reception time is going to be $t_r + t_{bias}$, where t_{bias} denotes the receiver's clock bias. The message transit time is $t_r + t_{bias} - t_i$. Assuming the message is traveling at the speed of light, the distance traveled is calculated as $(t_r + t_{bias} - t_i)c$. Having the knowledge of the distance between the receiver and the satellite, and the satellite's position, implies that the receiver is on the surface of a sphere centered at the satellite's position with radius equal to this distance. Thus, the receiver is at or near the intersection of the surfaces of the spheres if it receives signals from more than one satellite. In the ideal case with no errors, the receiver is at the intersection of the surfaces of the spheres [13].

The clock error or bias t_{bias} is the amount that the receiver's clock is off. The receiver has four unknowns, the three coordinates of the GPS receiver position, and the clock bias (x,y,z,t_{bias}). The equations of the sphere surfaces are given by $(x-x_i)^2+(y-y_i)^2+(z-z_i)^2=(t_r+t_{\text{bias}}-t_i)^2c^2 \quad i=1,2,3,\ldots,N$. These equations can be solved by algebraic or numerical methods.

Each GPS satellite continuously broadcasts a navigation message on the L1 C/A and L2 P/Y frequencies (see Figure 15.1) at a rate of 50 bps. Each message takes 750 s to be completely received. The message structure has a basic format of a 1500-bit long frame comprising five subframes, each subframe having 300 bits (6 s) long. Subframes 4 and 5 are sub-commutated 25 times each, so that a complete data message requires the transmission of 25 full frames. Each subframe consists of 10 words, each 30 bits long. Thus, with 300 bits in a subframe, times 5 subframes in a frame, times 25 frames in a message, each message is 37,500 bits long. At a transmission rate of 50 bps, it would take 750 s to transmit the entire almanac message. Each 30-s frame begins precisely on the minute or half-minute as indicated by the atomic clock on each satellite [13].

The first subframe of each frame encodes the week number and the time within the week, as well as the data about the condition of the satellite. The second and the third subframes contain the ephemeris or the precise orbit for the satellite. The fourth and fifth subframes contain the almanac, which contains coarse orbit and status information for up to 32 satellites in the constellation, as well as data related to error correction. Thus, in order to obtain an accurate satellite location from the transmitted message, the receiver must demodulate the message from each satellite it includes in its solution for 18–30 s. In order to collect the transmitted almanacs, the receiver must demodulate the message for 732–750 s [13].

All satellites broadcast at the same frequencies. Signals are encoded using code division multiple access (CDMA) codes, allowing messages from individual satellites to be distinguished from each other based on unique encodings for each satellite that the receiver must be aware of. Two distinct types of CDMA encodings are used: the coarse/acquisition code, which is accessible by the general public; and the precise code, which is encrypted so that only the US military can access it. The ephemeris is updated every 2 hours and is generally valid for 4 hours, with provisions for updates every 6 hours or longer in non-nominal conditions. The almanac is updated typically every 24 hours. Figure 15.3 illustrates the overall structure of the GPS frame.

It would take 12.5 min to download the entire almanac, and it would take even longer if the signal quality becomes poor and the receiver fails to correctly decode any portions of this data. To determine the location of the GPS satellites, two types of data are required by the GPS receiver: the almanac and the ephemeris. This data is continuously transmitted by the GPS satellites, and the GPS receiver collects and stores this data. The almanac data is data that describes the

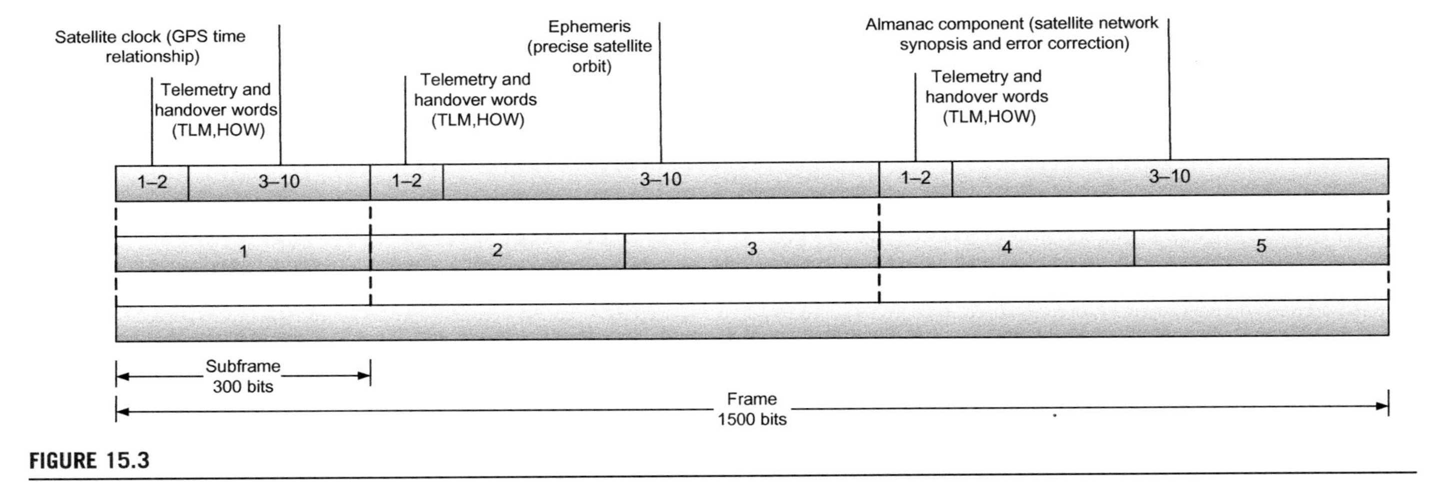

FIGURE 15.3

Overall structure of the GPS frame [16].

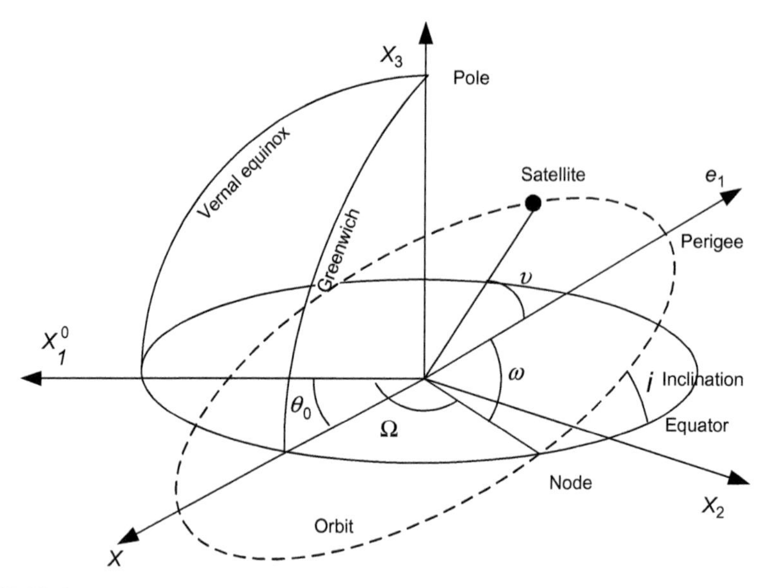

FIGURE 15.4

Illustration of almanac parameters [16].

orbital information of the satellites. Each satellite will broadcast almanac data for all satellites. The GPS receiver uses this data to determine which satellites are going to be visible in the sky. It can then determine which satellites should be tracked. With the almanac data, the receiver can focus on those satellites it can see, and exclude those that are over the horizon or out of view. Figure 15.4 illustrates the GPS almanac parameters.

The almanac parameters consist of size and shape of orbit (represented by semi-major axis a and eccentricity e), orientation of the orbital plane in the inertial system (represented by right ascension of ascending node Ω, argument of perigee ω, and inclination i), and position of the satellite in the orbital plane (represented by epoch of perigee T_0).

Ephemeris data provides the GPS receiver with the position of each GPS satellite at any time throughout the day (Figure 15.5). Each satellite will broadcast its own ephemeris data showing the orbital information for that satellite only. Given that the ephemeris data provides very precise orbital and clock correction data necessary for accurate positioning, its validity is much shorter. It is broadcast in three 6-s blocks and repeated every 30 s. The validity of data is implementation-specific and is usually up to 4 hours. The ephemeris data includes the following list [13,16]:

- $\sqrt{a}$ square root of the semi-major axis.
- e eccentricity.
- i_0 inclination angle at the reference time.

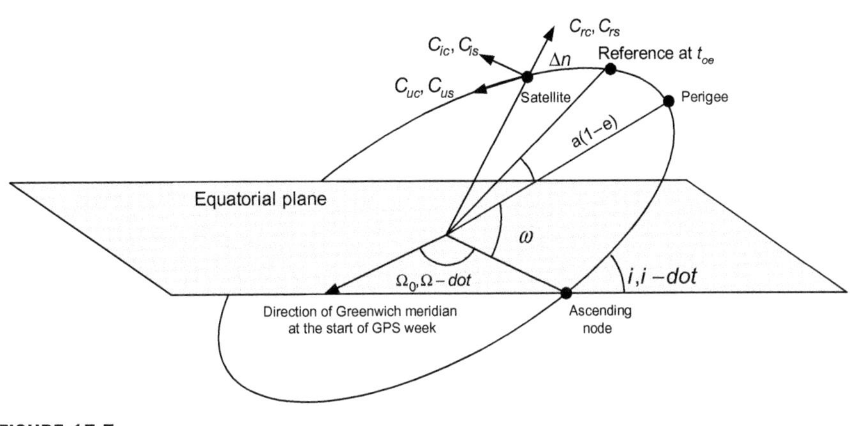

FIGURE 15.5

Illustration of ephemeris information [16].

- Ω_0 longitude of the ascending node at the beginning of the GPS week.
- ω argument of perigee.
- M_0 mean anomaly at the reference time.
- Δn correction to the mean motion computed using $\sqrt{a}$.
- *i–dot* the rate of change of inclination with time.
- *Ω–dot* the rate of change of the right ascension of the ascending node with time.
- C_{uc}, C_{us} amplitude of the correction terms for the computed argument of latitude.
- C_{rc}, C_{rs} amplitude of the correction terms for the computed orbit radius.
- C_{ic}, C_{is} amplitude of the correction terms for the computed inclination angle.
- t_{oe} ephemeris reference time.

15.1.3.2 A-GNSS positioning

GNSS refers to satellite navigation systems that provide autonomous geospatial positioning with global or regional coverage. Different GNSS methods can be used separately or in combination to determine the location of a UE. In the E-UTRAN, the GNSS is designed to work with assistance data provided by the network. A-GNSS uses signals broadcast by satellites to determine the positions of UEs equipped with GNSS receivers.

Two types of assistance data are provided to improve the positioning speed and accuracy which include data assisting the measurements (e.g., reference time, visible satellite list, satellite signal Doppler, code phase Doppler, and code phase search windows), and data assisting position calculation (e.g., reference time, reference position, satellite ephemeris, and clock corrections). A-GNSS provides improved accuracy relative to the stand-alone GNSS, and it can further reduce the UE GNSS start-up and acquisition times, it can increase the UE GNSS sensitivity, and it can allow the UE to consume less power with the GNSS-capable receiver put in idle mode when it is not needed. The A-GNSS methods can be operated in UE-assisted mode, where the UE performs GNSS measurements and sends them to the E-SMLC to calculate its position, or the UE-based mode, where the UE performs GNSS measurements and calculates its own location.

15.1.3.3 Observed time difference of arrival positioning

The OTDOA positioning method is based on measuring the time difference of arrival (TDOA) on special reference signals, embedded into the overall downlink signal, received from different eNBs. Each TDOA measurement describes a hyperbola, where the two focal points (F_1,F_2) are at the two measured eNBs. The measurement needs to be conducted at least for three pairs of base stations. The position of the device is the intersection of the three hyperbolas for the three measured base stations (A-B, A-C, and B-C), as shown in Figure 15.6.

The measurement conducted between a pair of eNBs is defined as the reference signal time difference (RSTD). The measurement is defined as the relative timing difference between a subframe received from the *j*th (neighboring) cell and the corresponding subframe from the *i*th (serving) cell. These measurements are conducted on the PRSs and the results are reported back to the location server, where the calculation of the position takes place. PRSs were introduced in

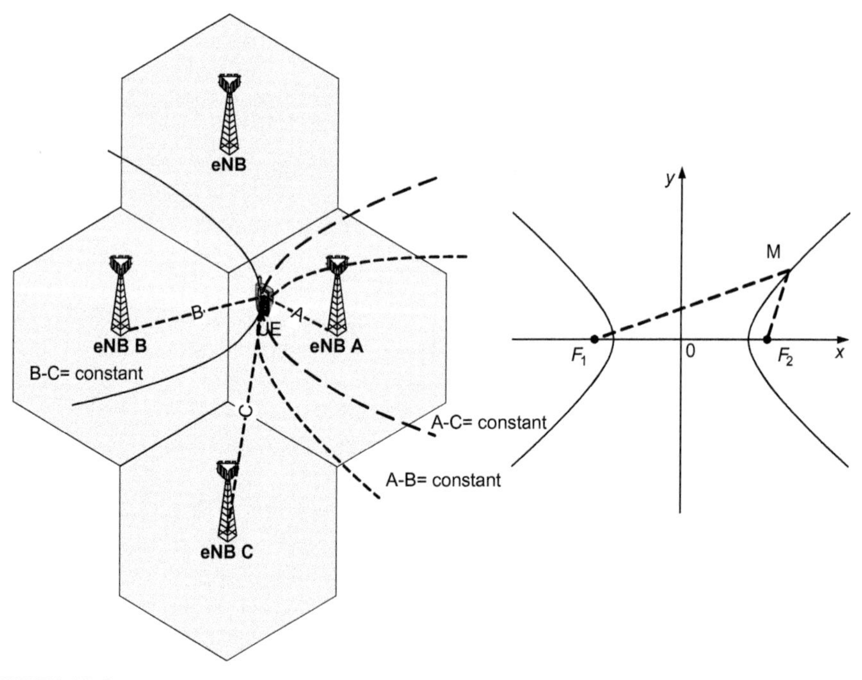

FIGURE 15.6

Illustration of downlink OTDOA positioning method [15].

LTE Rel-9 associated with antenna port 6, since the performance of cell-specific reference signals was not deemed sufficient for positioning. The reason is that the required high probability of detection could not be guaranteed. A neighbor cell with its synchronization signals (primary/secondary synchronization signals) and reference signals is seen as detectable, when the post-processing SINR is at least −6 dB. The results of the simulations performed during standardization of LTE Rel-9 suggested that this could be only guaranteed for 70% of all cases for the third best-detected cell, means second best neighboring cell. This is not enough, and further it assumed an interference-free environment, which cannot be ensured in a realistic scenario. However, PRSs still have some similarities with cell-specific reference signals as defined in LTE Rel-8. It is a pseudo-random QPSK sequence that is mapped in diagonal patterns with shifts in frequency and time to avoid collision with cell-specific reference signals and the control channels. The structure of the PRSs is described in Chapter 9. The PRSs can be muted on certain occasions to further reduce inter-cell interference. Therefore, PRS configuration and PRS muting is provided via the LPP protocol from the location server [15].

At least three timing measurements from geographically dispersed base stations with a good geometry are needed to solve for two coordinates of the UE, where the good geometry means that no two branches of the distinct hyperbolas intersect twice so that a unique solution can be found. In practice, a larger number of measurements, typically at least six to seven, are desirable. The position calculation is based on the multilateration approach by which an intersection of hyperbolas is found, where a hyperbola for a pair of cells corresponds to a set of points with the same RSTD for the two cells. The OTDOA positioning method is illustrated in Figure 15.6, where a UE is receiving downlink signals from the eNBs A, B, and C, and performing RSTD measurements τ_{AB}, τ_{AC}, and τ_{BC} which form three intersecting hyperbolic strips, the widths of which correspond to the RSTD measurement uncertainties. The distance equivalents of the RSTD measurements are d_{AB}, d_{AC}, and d_{BC}, respectively. Assuming the UE is located at (x,y) in a Cartesian coordinate system, and eNBs are located at $(x_A, y_A), (x_B, y_B)$, and (x_c, y_c), respectively, the system of equations to be solved in the least squares sense to find two-dimensional coordinates (x,y) is as follows:

$$\begin{aligned}
\sqrt{(x-x_A)^2-(y-y_A)^2}-\sqrt{(x-x_B)^2-(y-y_B)^2} &= d_{AB}-\Delta d_{AB}\\
\sqrt{(x-x_A)^2-(y-y_A)^2}-\sqrt{(x-x_C)^2-(y-y_C)^2} &= d_{AC}-\Delta d_{AC}\\
\sqrt{(x-x_B)^2-(y-y_B)^2}-\sqrt{(x-x_C)^2-(y-y_C)^2} &= d_{BC}-\Delta d_{BC}
\end{aligned}$$

Accurate knowledge of the transmitter locations and timing offsets is needed to solve the above equations for the UE's coordinates. Determining transmitter timing is nontrivial at the required accuracy level. The advantage of OTDOA is that synchronization between the eNBs and the UE is not required, unlike when the time of arrival measurement is used. Although RSTD measurements can be performed on a number of downlink signals, e.g., cell-specific reference signals or synchronization signals, it has been recognized that the existing signals suffer from poor reception, which is critical for OTDOA when multiple remote neighbor cells have to be detected. Therefore, to ensure the possibility of positioning measurements of a proper quality, and for a sufficient number of distinct locations, PRS having transmission patterns with an effective reuse of six has been introduced in LTE Rel-9. The interference to PRS can be further minimized, allowing simultaneous PRS transmissions for groups of cells with either orthogonal PRS patterns or low inter-cell interference. In OTDOA, the UE receiver has to deal with PRS from neighbor cells much weaker than those received from the serving cell. Furthermore, without the approximate knowledge of when the measured signals are expected to arrive in time and what is the exact PRS pattern, the UE would need to search the signals blindly, which would impact the time and accuracy of the measurements. To facilitate UE measurements, the network transmits assistance data to the UE, including a neighbor cell list with CIDs, the number of antenna ports, the number of consecutive downlink subframes per positioning occasion, PRS transmission bandwidth, expected RSTD, and the estimated uncertainty [11]. It should be noted that the detection of at least three eNBs is an absolute minimum. In practice, additional measurements are often necessary to be able to suppress outliers in the measurements. Such outliers may arise due to non-line-of-sight propagation. Thus, the detection capability of a TDOA method can often be the limiting factor, which makes it important to understand how the detection performance is computed.

15.1.3.4 E-CID-based positioning

The downlink OTDOA is the method of choice for dense urban and indoor environments where the A-GNSS method may not be able to provide reliable availability and satisfactory performance. Another method of positioning in LTE is the E-CID based on the cell of origin concept. With cell of origin, the position of the device is estimated using the knowledge of the geographical coordinates of its serving base station. The knowledge of the serving cell can be obtained after performing a tracking area update, or through paging. The position accuracy in this case is linked to the cell size, as the location server is only aware that the device is served by this base station. This method would of course not be able to fulfill the accuracy requirements mandated by the FCC [17]. Therefore, the E-CID method was defined in LTE, mainly for devices with no integrated GNSS receiver. In addition to the use of geographical coordinates of the serving base station, the position of the device is estimated more accurately by performing measurements on radio signals.

The E-CID method can take three approaches, based on different types of measurements: (i) E-CID with estimation of the distance from one base station; (ii) E-CID with measurement of the distance from three base stations; and (iii) E-CID by measuring the angle of arrival (AoA) from at least two base stations. In the first two cases, the possible measurement can be reference signal received power (RSRP), a standard quality measurement for LTE Rel-8 terminals, TDOA, and the measurement of the timing advance or round-trip time.

In the first case, the position accuracy would be just a circle. The second and third methods can provide position accuracy of a point, when measuring more radio sources. In the first and second cases, the measurements are conducted by the UE; thus, they are considered UE-assisted. In the third case, the measurements are performed by the base station; thus, it is eNB-assisted. Ever since LTE Rel-9 was introduced, the timing advance measurement has been enhanced; as a result, there are now two types of measurements: Type 1 and Type 2. The Type 2 measurement relies on the timing advance estimated from receiving a PRACH preamble during the random access procedure. Type 1 is defined as the sum of the receive–transmit timing difference at the eNB (positive or negative value), and the receive–transmit timing difference at the terminal, which is always a positive value. The base station first measures its own timing difference and instructs the device to adjust its uplink timing using a timing advance MAC command (see Chapter 8). The UE measures and reports the receive–transmit timing difference, as well. Both timing differences allow the calculation of the timing advance Type 1, corresponding to the round-trip time. The round-trip time is reported to the location server, where the distance to the base station is calculated using $d = cT_{\mathrm{RTT}}/2$.

The round-trip time and timing advance can be used for distance estimation; nevertheless, they do not provide any information on direction. This can only be obtained through an AoA measurement. The AoA is defined as the estimated angle of a UE with respect to a reference direction which is geographical north, positive in a counterclockwise direction, as seen from an eNB. The base station can usually estimate this angle on any uplink signal; however, typically, sounding reference signals (SRSs) are used for this purpose. The antenna array configuration has an impact on AoA measurements, the larger the array size, the higher the accuracy. With a linear array of equally spaced antenna elements, the received signal at any adjacent element is phase-rotated by a fixed angle whose value is a function of the AoA as well as the antenna element spacing and carrier frequency [15].

15.1.3.5 UTDOA positioning

In the uplink positioning methods, the UE's position is estimated based on timing measurements of uplink radio signals conducted at different LMUs, along with the knowledge of the geographical coordinates of the LMUs. The time required for a signal transmitted by a UE to reach an LMU is proportional to the length of the transmission path between the UE and the LMU. A set of LMUs are designated to sample the UE's signal at the same time [7].

In order to obtain uplink measurements, the LMUs need to know the characteristics of the SRSs transmitted by the UE for the time period required to conduct the uplink measurement. These characteristics should be static over the periodic transmission of SRS during the uplink measurements. Therefore, the E-SMLC will inform the serving eNB to configure the UE to transmit SRSs for the purpose of uplink positioning. It is up to the eNB to make the final decision on resources to be assigned, and to communicate this configuration information back to the E-SMLC so that E-SMLC can configure the LMUs. The eNB may decide (e.g., in case no resources are available) to configure no resources for the UE, and to report the empty resource configuration to the E-SMLC [7].

The conceptual difference between the UTDOA and OTDOA positioning methods is that the latter requires multiple transmit points, while the former utilizes multiple receive points at different locations, although the position calculation principle is the same. In order to perform UTDOA timing measurements on user data, one reference receiver decodes the UE signals and forwards the sequence to cooperating receivers. This procedure is relatively complex and requires a significant amount of signaling. Therefore, SRSs have been selected for UTDOA measurements. For UTDOA positioning, periodically transmitted SRSs are scheduled in a non-dynamic manner to allow a sufficiently long time for the measurements. This reduces the need for signaling of the full scheduling information to communicate the SRS configuration parameters. The SRS occupies the last SC-FDMA symbol in an uplink subframe. The bandwidth and the periodicity of SRS transmission can be configured to meet the positioning requirements. The SRS-based measurements for positioning may be performed and processed directly at a base station, or the operation may be performed by specific measurement units that may share the antennas with the base stations. The main disadvantage of the UTDOA method relative to OTDOA is a potential reception problem due to the power control of the UE transmissions. The UEs close to their serving base station transmit at a lower power level to avoid unnecessary interference to other UEs; this is known as the near–far problem. The consequence is that such signals may not be strong enough to reach the required uplink signal strength for UTDOA measurements at neighbor cells.

15.2 Evolved multimedia broadcast/multicast service

As we discussed earlier, the E-MBMS bearer service is the service provided by the packet-switched domain to E-MBMS user services in order to deliver IP multicast datagrams to multiple receivers by minimal use of network and radio resources. An E-MBMS user service is a service provided to the end user by means of the E-MBMS bearer service, and possibly other EPS bearers. The E-MBMS is a point-to-multipoint service in which data is transmitted from a single source entity to multiple recipients over a large geographical area. Transmitting the same content to multiple recipients allows the sharing of network resources. The E-MBMS for EPS bearer service supports broadcast mode over E-UTRAN and UTRAN. The E-MBMS is realized by the addition of a number of new features and functional entities to the LTE network architecture. In the next sections, we describe the E-MBMS network architecture, user-plane and control-plane protocols, and interfaces, as well as E-UTRAN functions and protocols that support MBMS services in LTE/LTE-Advanced systems.

15.2.1 E-MBMS network architecture

MBMS is a bearer service for simultaneous transmission of content over a large number of synchronized cells to all (or a group of) UEs subscribed to the service over a common bearer. In that case, the resulting signal will appear to the UE as one transmission over a time-disperse radio channel. This is concept is known as multimedia broadcast multicast service single frequency network (MBSFN). A group of cells that transmit the same content to multiple users forms an MBSFN area. Multiple cells can belong to a single MBSFN area, and each cell can be part of up to eight MBSFN areas. There can be up to 256 different MBSFN areas defined in a network, each one with a unique identity. Once defined, MBSFN areas will not change dynamically. It is not required that a terminal simultaneously receive content from multiple MBSFN areas. In other words, an MBSFN area is a geographical area comprising one or several cells that transmit the same content. For example, in Figure 15.7, which illustrates a hypothetical MBSFN network, cells 2 and 3 belong to both MBSFN areas 1 and 2. The MBSFN areas are static and do not vary over time. The usage of MBSFN transmission requires not only time synchronization among the cells participating in an MBSFN area, but also usage of the same set of radio resources in each of the cells for a particular service. This coordination is the responsibility of the multi-cell/multicast coordination entity (MCE), which is a logical node in the radio access network that controls resource allocation and transmission parameters across the cells in the MBSFN area [10].

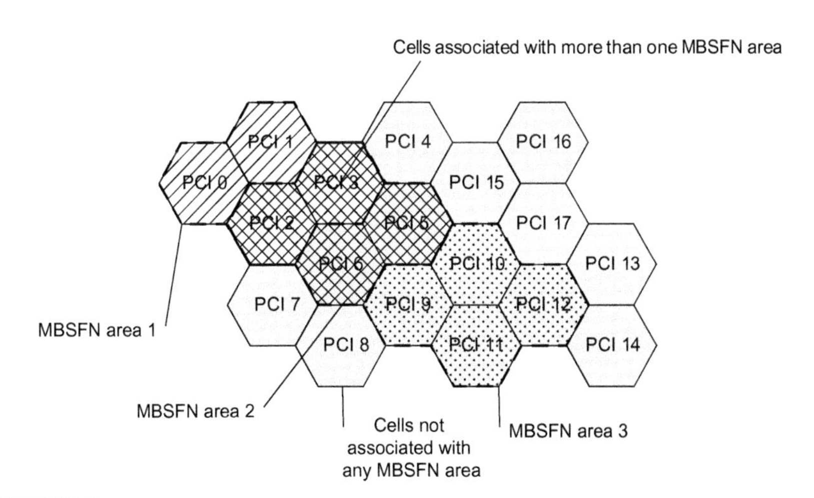

FIGURE 15.7

Illustration of MBSFN areas.

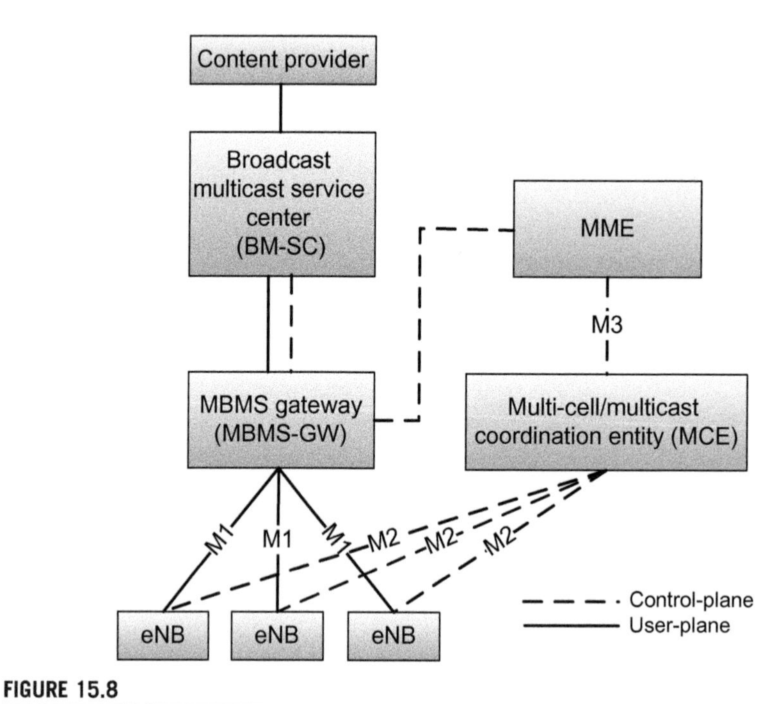

FIGURE 15.8

E-MBMS network architecture [5].

Figure 15.8 illustrates the E-MBMS network architecture. As shown in the figure, the MCE can control multiple eNBs, each controlling one or more cells. The broadcast multicast service center (BM-SC), located in the core network, is responsible for authorization and authentication of content providers, charging, and overall configuration of the data flow through the core network. The MBMS gateway (MBMS-GW) is a logical node handling multicast of IP packets from the BM-SC to all eNBs involved in transmission in an MBSFN area. It also manages session control signaling via the MME. From the BM-SC, the MBMS data is forwarded using IP-based multicast, a method of sending IP packets to multiple receiving network nodes in a single transmission, via the MBMS-GW to the cells. The MBMS is not only efficient from a radio-interface point of view, but it also saves resources in the transport network by avoiding unicast transmission of the same packet to multiple nodes unless necessary. This can lead to significant savings in the transport network.

More specifically, an MBMS-enabled network consists of the following entities and interfaces [5]:

- *Multi-cell/multicast coordination entity*: The MCE is a logical entity whose functions are admission control and allocation of the radio resources used by all eNBs in the MBSFN area for multi-cell MBMS transmissions. The MCE decides whether to establish radio bearer(s) for the new MBMS service(s), if the radio resources are not sufficient for the corresponding MBMS service(s), or may allocate radio resources from other radio bearer(s) of ongoing MBMS service(s) according to allocation and retention priority (ARP). In addition to allocation of the time/frequency radio resources, the MCE controls the radio configuration, e.g., modulation and coding scheme, counting and acquisition of counting results for MBMS service(s), resumption of MBMS session(s) within MBSFN area(s) based on the ARP and/or the counting results for the corresponding MBMS service(s), suspension of MBMS session(s) within MBSFN area(s) based on the ARP and/or on the counting results for the corresponding MBMS service(s). In the case of distributed MCE architecture, the MCE manages the above functions for a single eNB of an MBSFN area.

- *E-MBMS gateway*: The MBMS-GW is a logical entity that is located between the BM-SC and eNBs whose functions include sending/broadcasting of MBMS packets to each eNB participating in the service. The MBMS-GW uses IP multicast as the method of forwarding MBMS user data to the eNBs. The MBMS-GW performs MBMS session control signaling (i.e., session start/update/stop) toward the E-UTRAN via MME.
- *MCE–MME interface:* The application part of the MCE–MME (M3) control-plane interface allows MBMS session control signaling on E-UTRAN radio access bearer (E-RAB) level (i.e., does not convey radio configuration data). The procedures include MBMS session start and stop. The stream control transmission protocol (SCTP)[4] is used as the signaling transport method.
- *MCE–eNB interface*: The application part of the MCE–eNB (M2) control-plane interface conveys radio configuration data for the eNBs configured with multi-cell transmission mode and session control signaling. The SCTP protocol is used as signaling transport.
- *MBMS–GW–eNB interface*: The user-plane interface MBMS-GW-eNB (M1) carries IP packets between the two network entities and uses IP multicast for point-to-multipoint delivery of user packets.

There are two MBMS architectures defined in 3GPP. In the "distributed MCE architecture," the MCE is part of the eNB and the M2 interface is maintained between the MCE and the corresponding eNB, whereas in the "centralized MCE architecture" (shown in Figure 15.8), the MCE is a logical entity and can be deployed as a stand-alone physical entity or collocated in another physical entity

[4]SCTP is a transport layer protocol, performing a similar role as the commonly used protocols transmission control protocol (TCP) and user datagram protocol (UDP). It provides some of the service features of both protocols, i.e., it is message-oriented like UDP and ensures reliable and in-sequence transport of messages with congestion control like TCP. The protocol is defined by IETF RFC 4960 [19]. The SCTP protocol may be characterized as message-oriented, i.e., it transports a sequence of messages comprising a group of bytes, rather than transporting an unbroken stream of bytes as does TCP. As in UDP, a sender sends a message in SCTP in one operation, and that exact message is passed to the receiving application process in one operation. In contrast, TCP is a stream-oriented protocol, transporting streams of bytes reliably and in order.

such as eNB. In both cases, the M2 interface is maintained between the MCE and all eNB(s) belonging to the corresponding MBSFN area [5].

15.2.2 Basic definitions

In the context of MBMS, the following definitions are used [5]:

- *MBSFN synchronization area* is defined as an area of the network where all eNBs can be synchronized and perform MBSFN transmissions. MBSFN synchronization areas are capable of supporting one or more MBSFN areas. On a given frequency layer, an eNB can only belong to one MBSFN synchronization area. MBSFN synchronization areas are independent from the definition of MBMS service areas.
- *MBSFN transmission or a transmission in MBSFN mode* is a simulcast transmission technique realized by transmission of identical waveforms at the same time from multiple cells (macro diversity). An MBSFN transmission from multiple cells within the MBSFN area is considered like a single transmission from a UE point of view.
- *MBSFN area* is defined as a geographical area consisting of a group of cells within an MBSFN synchronization area of a network, which are coordinated to perform an MBSFN transmission. Except for the reserved cells, all cells within an MBSFN area contribute to the MBSFN transmission and advertise its availability. The UE may only need to consider a subset of nodes in the MBSFN areas that are configured to broadcast the desired service.
- *MBSFN area reserved cell* is a cell within an MBSFN area which does not contribute to the MBSFN transmission. The cell may be allowed to transmit for other services, but at restricted power on the resources allocated for the MBSFN transmission.
- Each *Synchronization sequence* (SYNC) protocol data unit contains a time stamp which indicates the start time of the synchronization sequence. For an MBMS service, each synchronization sequence has the same duration which is configured in the BM-SC and the MCE.
- The *synchronization period* provides the time reference for the indication of the start time of each synchronization sequence. The time stamp which is provided in each SYNC PDU is a relative value which refers to the start time of the synchronization period. The duration of the synchronization period is configurable.

15.2.3 E-MBMS user-plane and control-plane protocols

The overall user-plane protocol termination of content synchronization is shown in Figure 15.9. This structure is based on the functional allocation for unicast, and the SYNC protocol is defined additionally on the transport network layer to

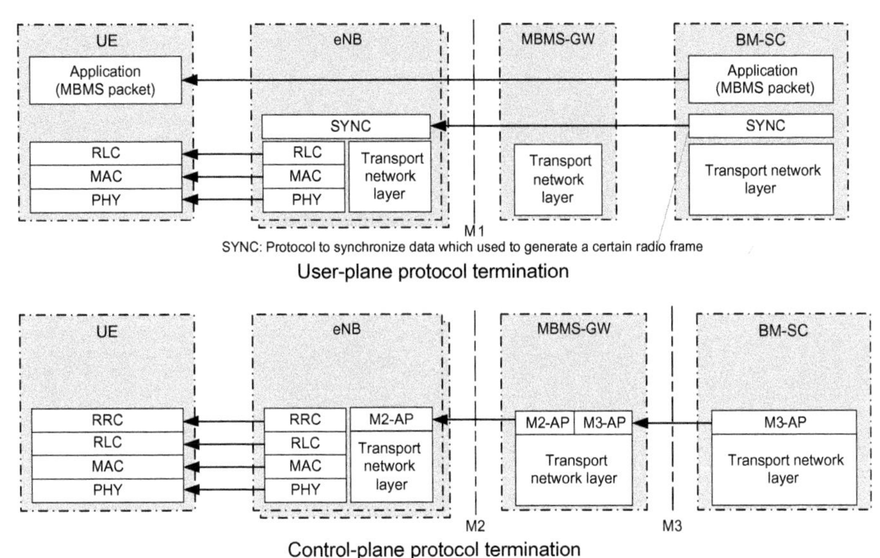

FIGURE 15.9

E-MBMS user-plane and control-plane protocol structure [5].

support the content synchronization mechanism. The SYNC is a protocol for transporting additional information that enables eNBs to identify the timing for radio frame transmission and detect packet loss. Each MBMS service uses its own SYNC entity. The SYNC protocol is applicable to downlink and is terminated in the BM-SC. The multicast control logical channel is terminated in the eNB on the network side. The control-plane protocol structure and termination is further shown in Figure 15.9.

15.2.4 Physical and MAC layer aspects of E-MBMS

The MBSFN transmission is made possible through use of a new transport channel known as a multicast channel (MCH). As shown in Figure 15.10, two types of MBMS-related logical channels are multiplexed to MCH, i.e., the multicast traffic channel (MTCH) and multicast control channel (MCCH). The MTCH is a logical channel that is used to carry MBMS data corresponding to a certain MBMS service. If the number of services provided in an MBSFN area is large, multiple MTCHs can be configured. For MBSFN transmission, no acknowledgments are transmitted by the terminals, no RLC retransmissions can be used, and consequently, the RLC unacknowledged mode is utilized. The MCCH is a logical channel that is used to transport control information necessary for reception of a certain MBMS service, including the subframe allocation and MCS for each MCH. There is one MCCH per MBSFN area. Similar to MTCH, the RLC sublayer uses unacknowledged mode [2].

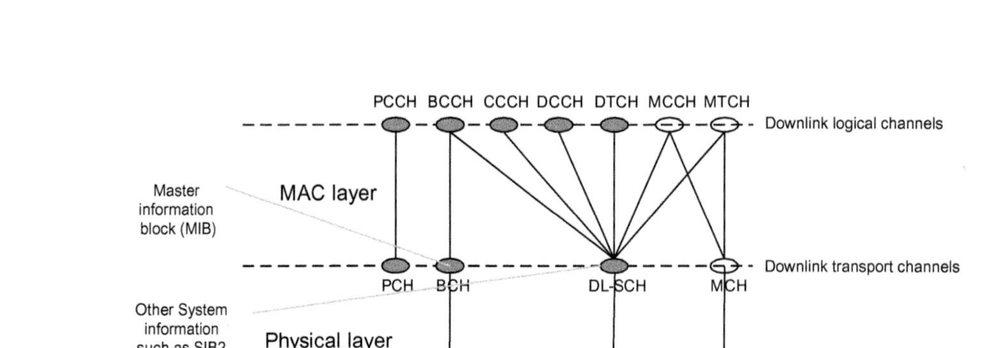

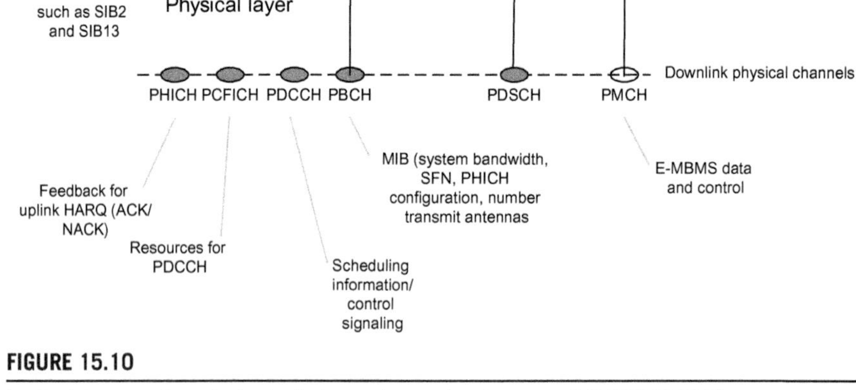

FIGURE 15.10

Logical, transport, and physical channel mapping to support E-MBMS [3,5].

Multiple MTCHs and one MCCH can be multiplexed in the MAC sublayer to form an MCH transport channel. The MAC header contains information about MTCH/MCCH multiplexing, such that the terminal can demultiplex the information upon reception. The MCH is transmitted using MBSFN in one MBSFN area. The transport channel processing for MCH is similar to the downlink shared channel with some exceptions. In the case of MBSFN transmission, the same data is transmitted with the same transport format using the same physical resource from multiple cells typically belonging to different eNBs. Thus, the MCH transport format and resource allocation cannot be dynamically adjusted by the eNB. As described above, the transport format is instead determined by the MCE and signaled to the terminals as part of the information sent on MCCH. As the MCH transmission is simultaneously targeting multiple terminals, and therefore no feedback is used, HARQ is not used for MCH transmission. As already mentioned, multi-antenna transmission (transmit diversity and spatial multiplexing) does not apply to MCH transmission. The physical multicast channel (PMCH) scrambling is unique in each MBSFN area, i.e., identical for all cells participating in MBSFN transmission. The MCH is mapped to PMCH and transmitted in MBSFN subframes (see Chapter 9 for a description of MBSFN subframes). An MBSFN subframe consists of two parts: a control region, used for transmission of non-MBSFN unicast control signaling; and an MBSFN region, used for transmission of MCH. Unicast control signaling may be needed in an MBSFN subframe, e.g., to schedule uplink transmissions in the next subframe or for MBMS-related signaling [10].

In the case of MBSFN transmission, the cyclic prefix should not only cover the main part of the actual channel time dispersion, but also the timing difference between the transmissions received from the cells involved in the MBSFN transmission. Therefore, MBSFN transmissions, which can take place in the MBSFN regions only, use an extended cyclic prefix. As there are different cyclic prefix lengths and subcarrier spacing defined for LTE, there are different modes for MBMS that take advantage of these configurations. In a mixed mode of MBMS and unicast transmission, the subcarrier spacing is 15 kHz, and resources (subframes) are shared between MBMS data and unicast traffic/control signaling. In a dedicated MBMS mode or single-cell scenario, the RF carrier is only used for MBMS transmission. In this case, a different subcarrier spacing of 7.5 kHz can be used, offering a larger cyclic prefix of 33.3 μs which leads to a further improved broadcast over large cells (that exhibit larger delay spreads).

If a normal cyclic prefix is used for normal subframes, and also in the control region of MBSFN subframes, there will be a small gap between the two parts of an MBSFN subframe. The reason is to keep the start time of the MBSFN region fixed, irrespective of the cyclic prefix length used for the control region. The MCH is transmitted by means of MBSFN from a group of cells that are part of the corresponding MBSFN area. Thus, from the UE's perspective, the radio channel through which the MBMS content has been propagated is the aggregation of the channels of each cell within the MBSFN area. In order to estimate the channel for coherent demodulation of MCH, the UE cannot rely on the cell-specific reference signals transmitted from each cell; rather, in order to coherently demodulate MCH, dedicated MBSFN reference signals are inserted within the MBSFN part of the MBSFN subframe (see Chapter 9 for a description of MBSFN reference signals). These reference signals are transmitted by means of MBSFN over a set of cells that form the MBSFN area, i.e., they are transmitted in the same time–frequency resources and with the same reference signal values from each cell. As a result, the channel estimation based on these reference symbols correctly reflects the overall aggregated channel corresponding to MCH transmission of all cells that are part of the MBSFN area. The MBSFN transmission in conjunction with specific MBSFN reference signals can be seen as transmission using antenna port 4.

The MBSFN-enabled UE can assume that all MBSFN transmissions within a given subframe correspond to the same MBSFN area. Therefore, the UE may choose to interpolate over all MBSFN reference signals within a given MBSFN subframe when estimating the aggregated MBSFN channel. In contrast, MCH transmissions in different subframes may correspond to different MBSFN areas. The frequency-domain density of MBSFN reference signals is much higher than the corresponding density of cell-specific reference signals, given that the aggregated channel of all cells involved in the MBSFN transmission (due to large coverage area) is equivalent to a highly time-dispersive or highly frequency-selective channel. There is only a single MBSFN reference signal in MBSFN subframes. Thus, multi-antenna transmission such as transmit diversity and spatial multiplexing is not supported for MCH transmission. The main argument for not supporting any standardized transmit diversity scheme for MCH transmission is that the high frequency selectivity of the aggregated MBSFN channel in itself provides substantial frequency diversity [10].

15.2.5 E-MBMS single frequency network transmission

Multiple MBMS services can be mapped to the same MCH transport channel. One MCH contains data belonging to only one MBSFN area. An MBSFN area contains one or more MCHs. All MCHs have the same coverage area. For MCCH and MTCH, the UE does not perform RLC reestablishment upon cell change between the cells belonging to the same MBSFN area. Within the MBSFN subframes, all MCHs within the same MBSFN area occupy a pattern of subframes, not necessarily adjacent in time, which is common for all of these MCHs and which is referred to as a common subframe allocation (CSA) pattern. As shown in Figure 15.11, the CSA pattern is periodically repeated with the CSA period. The MCH subframe allocation (MSA) for every MCH carrying MTCH is defined by the CSA pattern, the CSA period, and the MSA-end, which are all signaled on MCCH. The MSA-end indicates the last subframe of the MCH within the CSA period. Subsequently, the MCHs are time multiplexed within the CSA period, which ultimately defines the interleaving between the MCHs. It is further possible for MCHs not to use all MBSFN resources signaled as part of the Rel-8 MBSFN signaling. Furthermore, such MBSFN resources can be shared for more than one purpose (e.g., MBMS, positioning). During one MCH scheduling period (MSP), which is configurable per MCH, the eNB multiplexes different MTCHs and (optionally) MCCH for transmission on MCH. The MCH scheduling information (MSI) is provided per MCH to indicate which subframes are used by each MTCH during the MSP [5].

In the example provided in Figure 15.11, the scheduling period for the first MCH is 16 frames, corresponding to one CSA period, and the scheduling information for this MCH is therefore transmitted once every16 frames. The scheduling period for the second MCH is 32 frames, corresponding to two CSA periods, and the scheduling information is transmitted once every 32 frames. The MSPs can range from 80 ms to 10.24 s.

The MCCH information is transmitted periodically with a fixed repetition period, and changes to the MCCH information can only occur at specific times. Whenever MCCH information is changed, which can only be done at the beginning of a new modification period, as shown in Figure 15.12, the network notifies the UEs about the upcoming MCCH information change in the preceding MCCH modification period. The notification mechanism uses the PDCCH for this purpose. An 8-bit bitmap, where each bit represents a certain MBSFN area, is

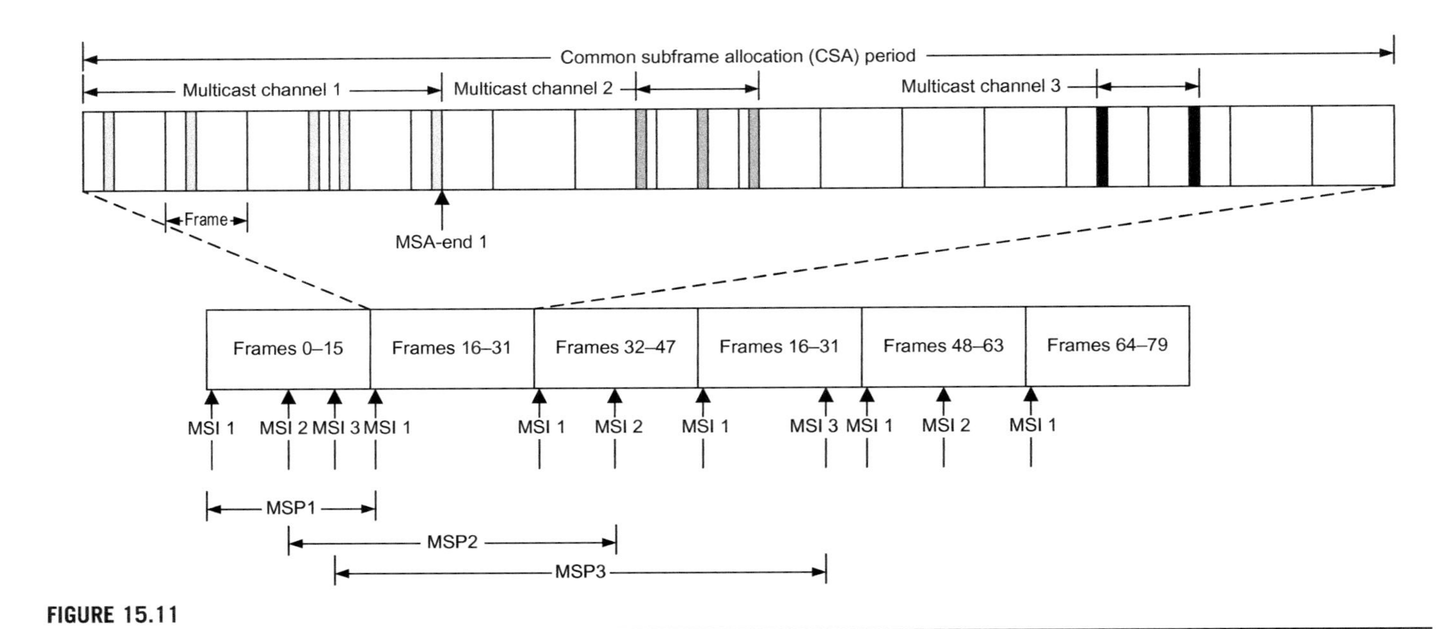

FIGURE 15.11

Example scheduling of MBMS content [10].

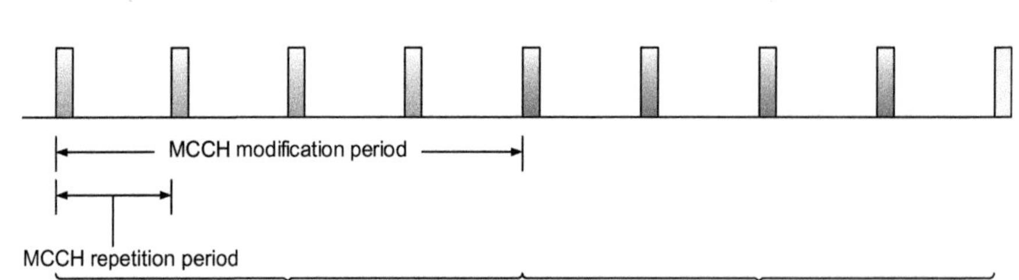

FIGURE 15.12

Example MCCH transmission schedule [1].

transmitted in the PDCCH in an MBSFN subframe using DCI format 1C and the M-RNTI (see Chapter 8 for a description of multicast RNTI). The notification bitmap is only transmitted when there are changes in the services, as well as to indicate a counting request in an MBSFN area. The purpose of notification indicators and modification periods is to maximize the amount of time the UE may sleep to reduce power consumption. In the absence of any changes to MCCH information, a terminal not subscribed to E-MBMS may enter the DRX mode, and only wake up when the notification indicator is transmitted. The periodic transmission of MCCH is useful in order to support mobility, because a terminal entering a new MBSFN area or a terminal that has missed the first MCCH transmission does not have to wait until the start of a new modification period to receive MCCH information [10].

System information block type 2 (SIB2) carries all relevant information about common and shared channels in LTE. LTE Rel-9 has extended the scope of this information to include MBMS-specific information. The new information element *MBSFN-SubframeConfig* defines which radio frames contain subframes that can be used for MBMS (see the example in Figure 15.13). The MBSFN subframes can be used by all MBSFN areas. The E-MBMS information in SIB2 includes the radio frame allocation period and a radio frame allocation offset, as well as the subframe allocation mode, which can be one or four consecutive radio frames containing MBSFN subframes. A maximum of six subframes in each radio frame can be used for MBMS transmission. This is due to the fact that zero and fifth subframes carry synchronization signals and broadcast channel, and subframes 0, 4, 5, and 9 can be used for paging depending on the paging cycle. A bitmap is used to indicate which of the six possible subframes are used for MBMS transmission.

System information block type 13 (SIB13) provides MBSFN identity (MBSFN ID), non-MBSFN region length (one or two OFDM symbols), and MCCH configuration. The MCCH configuration provides information on the repetition period for MCCH, the MCCH offset as well as the actual subframe where MCCH is transmitted, and the MCS used for MCCH. It is important to note that for MCCH, only four MCSs are allowed: MCS indices 2, 7 (both QPSK); 13 (16QAM); and 19 (64QAM). The MCCH always defines one MBSFN area and carries a single

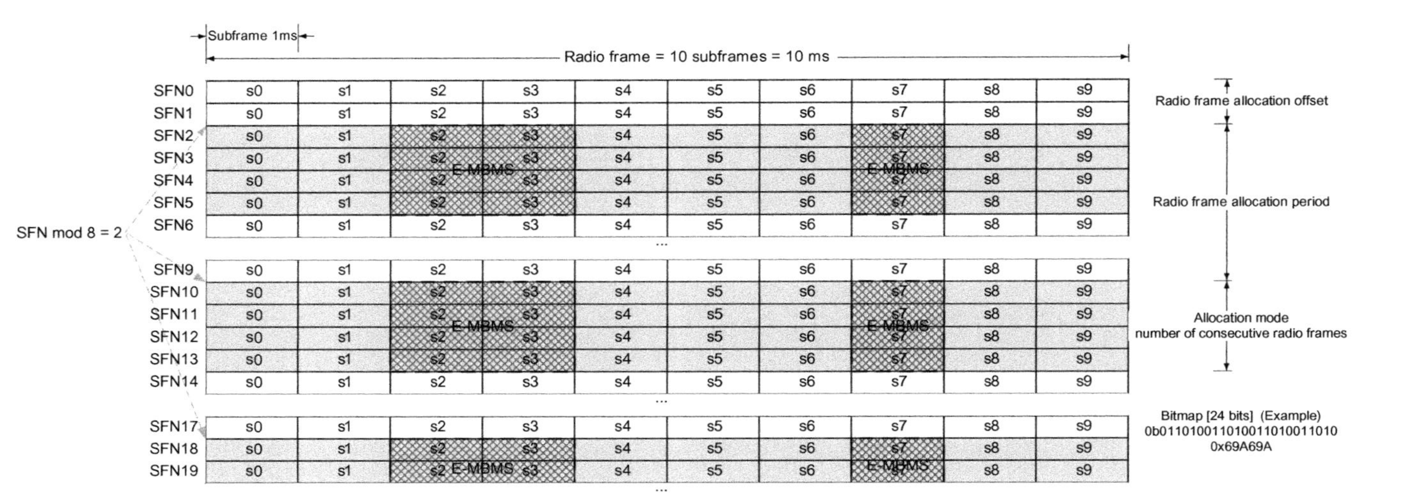

FIGURE 15.13

Example interpretation of E-MBMS information in SIB2 [15].

message, i.e., *MBSFNAreaConfiguration*. This information element provides all required information for scheduling MBMS services, including all PMCH belonging to that particular MBSFN area and the configuration parameters for the sessions that are carried by the corresponding PMCH. There are up to 15 PMCH, where each one has up to 29 MTCHs. For each PMCH, there is information about which MBSFN subframe carries that PMCH (start, end), as well as the MCS that has been utilized.

In general, the control information for the UEs supporting E-MBMS is separated as much as possible from unicast control information. Most of the E-MBMS control information is provided on MCCH. The E-UTRA employs one MCCH logical channel per MBSFN area. In the case where the network configures multiple MBSFN areas, the UE acquires the MBMS control information from the MCCHs that are configured to identify whether the services which the UE is interested to receive are ongoing. The behavior of the UE when it is unable to simultaneously receive MBMS and unicast services depends on the UE implementation. As of LTE Rel-11, an E-MBMS-capable UE is only required to support reception of a single MBMS service at a time, and reception of more than one MBMS service is implementation specific. The MCCH carries the *MBSFN Area Configuration* message, which indicates the ongoing E-MBMS sessions as well as the corresponding radio resource configuration. The MCCH may also carry the *MBMS Counting Request* message, when E-UTRAN needs to count the number of UEs in the RRC_CONNECTED state that are receiving or are interested in receiving specific MBMS services. A limited amount of E-MBMS control information is provided on the broadcast control channel. This is primarily the information needed to acquire the MCCH(s), and is carried via SIB13. An MBSFN area is identified by the *mbsfn-AreaId* in SIB13. During mobility, the UE assumes that the MBSFN area is continuous when the source cell and the target cell broadcast the same value in the *mbsfn-AreaId* [1].

15.2.6 Mobility support in E-MBMS

The E-MBMS mobility is based on the same procedures that are specified for unicast services. There are no additional mechanisms to instruct the E-MBMS-capable UEs to select a carrier frequency providing E-MBMS. Furthermore, there are no provisions to reduce the service interruption time when moving from one MBSFN area to another. Nevertheless, the E-UTRAN can assign the highest cell reselection priority to the carrier frequency that provides E-MBMS to facilitate the transfer of ongoing MBMS service during E-MBMS-subscribed UE mobility. The E-MBMS mobility procedures would allow the UE to start or continue receiving MBMS service(s) via MBSFN when moving across cell(s). The E-UTRAN procedures provide support for service continuity with respect to mobility within the same MBSFN area. Within the same geographic area, MBMS services can be provided on more than one frequency, and the frequencies used to provide MBMS services may change from one geographic area to another within a PLMN. The E-MBMS-subscribed UEs can be either in the RRC_IDLE state performing cell reselection, or in the RRC_CONNECTED state obtaining target cell MTCH information from the target cell MCCH. To avoid the need for acquiring the E-MBMS related system information and MCCH of neighbor frequencies, the UE is provided with assistance information on the frequency of MBMS services via MBSFN as follows [1,5]:

- *User service description*: The application layer provides temporary mobile group identities (TMGIs) for each service, the session start and end time, the frequencies, and the MBMS SAIs belonging to the MBMS service area.
- *System information*: MBMS and non-MBMS cells use *System Information Block Type 15* to provide the MBMS SAIs of the current frequency, and of each neighbor frequency. The E-MBMS SAIs of the neighboring cells may be provided via X2 signaling.

15.2.7 E-MBMS counting

The MBMS counting in LTE is used to determine if there is a sufficient number of UEs interested in receiving a service to enable the operator to decide whether it would be efficient to deliver the service via MBSFN. It allows the operator to choose between enabling or disabling MBSFN transmission for the service. The MBMS counting applies only to the UEs in the connected mode. Enabling and disabling MBSFN transmission is realized by the MBMS service suspension and resumption function. The counting scheme can be applied to a service already provided by MBSFN in an MBSFN area (an ongoing service), or a service that has not been provided via MBSFN in an MBSFN area (service not started yet), which may be a service provided via unicast bearers [5].

The counting procedure is initiated by the network. Initiation of the counting procedure results in a request to each eNB within the MBSFN area to send a counting request (extended MCCH message), which contains a list of TMGIs, which uniquely identifies an MBMS bearer service. A single globally unique TMGI is allocated by the BM-SC per MBMS bearer service requiring UE feedback. The connected mode UEs which are receiving or are interested in the services will respond with a *Counting Response* message, which includes short MBMS service identities (unique within the MBSFN service area), and may optionally include the information to identify the MBSFN area. The network has the means to disable UE counting per service, and the UE is able to report on multiple MBMS services via a single *Counting Response* message. It is not necessary to retransmit the *Counting Response* message when the UE moves within the same MBSFN area. The network obtains one response from the UE related to one *Counting Request* message, which is broadcast for one modification period. The UE cannot automatically indicate to a network a change of interest in MBMS service(s). The network counts the UE's interest per service [5].

References

3GPP Specifications[5]

[1] 3GPP TS 36.331, Evolved Universal Terrestrial Radio Access (E-UTRA), Radio Resource Control (RRC), Protocol Specification (Release 11), December 2012.

[2] 3GPP TS 36.322, Evolved Universal Terrestrial Radio Access (E-UTRA), Radio Link Control (RLC), Protocol Specification (Release 11), December 2012.

[3] 3GPP TS 36.321, Evolved Universal Terrestrial Radio Access (E-UTRA), Medium Access Control (MAC), Protocol Specification (Release 11), December 2012.

[4] 3GPP TS 36.304, Evolved Universal Terrestrial Radio Access (E-UTRA), UE Procedures in Idle Mode (Release 11), December 2012.

[5] 3GPP TS 36.300, Evolved Universal Terrestrial Radio Access (E-UTRA) and Evolved Universal Terrestrial Radio Access Network (E-UTRAN) Overall Description, Stage 2 (Release 11), December 2012.

[6] 3GPP TS 36.306, Evolved Universal Terrestrial Radio Access (E-UTRA), User Equipment (UE) Radio Access Capabilities (Release 11), December 2012.

[7] 3GPP TS 36.305, Evolved Universal Terrestrial Radio Access Network (E-UTRAN), Stage 2 Functional Specification of User Equipment (UE) Positioning in E-UTRAN (Release 11), December 2012.

Books

[8] C. Johnson, Long Term Evolution in Bullets, second ed., Create Space Independent Publishing Platform, July 2012.

[9] S. Sesia, I. Toufik, M. Baker, LTE, the UMTS Long Term Evolution, From Theory to Practice, second ed., John Wiley & Sons, August 2011.

[10] E. Dahlman, S. Parkvall, J. Skold, 4G, LTE/LTE-Advanced for Mobile Broadband, Academic Press, May 2011.

[11] R. Zekavat, R.M. Buehrer, Handbook of Position Location: Theory, Practice, and Advances, Wiley-IEEE Press, November 2011.

Articles

[12] S. Cherian, A. Rudrapatna, A primer on location technologies in LTE networks, Alcatel-Lucent, July 2012. <http://www2.alcatel-lucent.com/techzine/a-primer-on-location-technologies-in-lte-networks>.

[13] Global Positioning System, Wikipedia <http://en.wikipedia.org/wiki/Global_Positioning_System>.

[14] MSF Services Working Group, MSF Whitepaper on Location Services in LTE Networks, Multi Service Forum, April 2010 <http://www.msforum.org/>.

[15] M. Kottkamp, et al., LTE Release 9 Technology Introduction White paper, Rohde & Schwarz, December 2011.

[16] LTE System Description. <http://www.sharetechnote.com/>.

[17] Federal Communications Commission (FCC), Wireless E911 location accuracy requirements, June 2011. <http://www.fcc.gov>.

[5] 3GPP Specifications can be found at the following URL: http://www.3gpp.org/ftp/Specs/archive/

[18] 4G Americas White Paper, 4G Mobile Broadband Evolution: Release 10, Release 11 and Beyond, October 2012. <http://www.4gamericas.org/>.
[19] R. Stewart, Stream control transmission protocol, IETF RFC 4960, September 2007. <http://www.rfc-editor.org/rfc/rfc4960.txt>.
[20] S. Deering, Host extensions for IP multicasting, IETF RFC 1112, August 1989. <http://www.ietf.org/rfc/rfc1112.txt>.

Index

Note: Page numbers followed by "*f*", and "*t*" refers to figures and tables respectively.

D

J

K

L

M

N

O

P

Q

R

S

CPI Antony Rowe
Chippenham, UK
2018-09-04 17:57